Palaeozoic Climate Cycles:
Their Evolutionary and Sedimentological Impact

Geological Society books refereeing procedures

The Society makes every effort to ensure that the scientific and production quality of its books matches that of its journals. Since 1997, all book proposals have been refereed by specialist reviewers as well as by the Society's Books Editorial Committee. If the referees identify weaknesses in the proposal, these must be addressed before the proposal is accepted.

Once the book is accepted, the Society Book Editors ensure that the volume editors follow strict guidelines on refereeing and quality control. We insist that individual papers can only be accepted after satisfactory review by two independent referees. The questions on the review forms are similar to those for *Journal of the Geological Society*. The referees' forms and comments must be available to the Society's Book Editors on request.

Although many of the books result from meetings, the editors are expected to commission papers that were not presented at the meeting to ensure that the book provides a balanced coverage of the subject. Being accepted for presentation at the meeting does not guarantee inclusion in the book.

More information about submitting a proposal and producing a book for the Society can be found on its website: www.geolsoc.org.uk.

It is recommended that reference to all or part of this book should be made in one of the following ways:

GĄSIEWICZ, A. & SŁOWAKIEWICZ, M. (eds) 2013. *Palaeozoic Climate Cycles: Their Evolutionary and Sedimentological Impact*. Geological Society, London, Special Publications, **376**.

BOJAR, A.-V., NEUBAUER, F. & KOEBERL, C. 2013. Geochemical record of Late Devonian to Early Carboniferous events, Palaeozoic of Graz, Eastern Alps, Austria. *In:* GĄSIEWICZ, A. & SŁOWAKIEWICZ, M. (eds) *Palaeozoic Climate Cycles: Their Evolutionary and Sedimentological Impact*. Geological Society, London, Special Publications, **376**, 87–108. First published online September 19, 2013, http://dx.doi.org/10.1144/SP376.15.

GEOLOGICAL SOCIETY SPECIAL PUBLICATION NO. 376

Palaeozoic Climate Cycles: Their Evolutionary and Sedimentological Impact

EDITED BY

A. GĄSIEWICZ
Polish Geological Institute, Poland

and

M. SŁOWAKIEWICZ
University of Bristol, UK

2013
Published by
The Geological Society
London

THE GEOLOGICAL SOCIETY

The Geological Society of London (GSL) was founded in 1807. It is the oldest national geological society in the world and the largest in Europe. It was incorporated under Royal Charter in 1825 and is Registered Charity 210161.

The Society is the UK national learned and professional society for geology with a worldwide Fellowship (FGS) of over 10 000. The Society has the power to confer Chartered status on suitably qualified Fellows, and about 2000 of the Fellowship carry the title (CGeol). Chartered Geologists may also obtain the equivalent European title, European Geologist (EurGeol). One fifth of the Society's fellowship resides outside the UK. To find out more about the Society, log on to www.geolsoc.org.uk.

The Geological Society Publishing House (Bath, UK) produces the Society's international journals and books, and acts as European distributor for selected publications of the American Association of Petroleum Geologists (AAPG), the Indonesian Petroleum Association (IPA), the Geological Society of America (GSA), the Society for Sedimentary Geology (SEPM) and the Geologists' Association (GA). Joint marketing agreements ensure that GSL Fellows may purchase these societies' publications at a discount. The Society's online bookshop (accessible from www.geolsoc.org.uk) offers secure book purchasing with your credit or debit card.

To find out about joining the Society and benefiting from substantial discounts on publications of GSL and other societies worldwide, consult www.geolsoc.org.uk, or contact the Fellowship Department at: The Geological Society, Burlington House, Piccadilly, London W1J 0BG: Tel. +44 (0)20 7434 9944; Fax +44 (0)20 7439 8975; E-mail: enquiries@geolsoc.org.uk.

For information about the Society's meetings, consult *Events* on www.geolsoc.org.uk. To find out more about the Society's Corporate Affiliates Scheme, write to enquiries@geolsoc.org.uk.

Published by The Geological Society from:
The Geological Society Publishing House, Unit 7, Brassmill Enterprise Centre, Brassmill Lane, Bath BA1 3JN, UK

The Lyell Collection: www.lyellcollection.org
Online bookshop: www.geolsoc.org.uk/bookshop
Orders: Tel. +44 (0)1225 445046, Fax +44 (0)1225 442836

British Library Cataloguing in Publication Data

A catalogue record for this book is available from the British Library.
ISBN 978-1-86239-357-8
ISSN 0305-8719

Distributors
For details of international agents and distributors see:
www.geolsoc.org.uk/agentsdistributors

Typeset by Techset Composition India (P) Ltd., Bangalore and Chennai, India.
Printed by Berforts Information Press Ltd, Oxford, UK

Contents

Case studies: Middle to Late Permian palaeoclimatic impacts

Case studies: Permian-Triassic transition

Late Palaeozoic environmental changes: an introduction

ANDRZEJ GĄSIEWICZ[1]* & MIROSŁAW SŁOWAKIEWICZ[1,2]

[1]*Polish Geological Institute, Rakowiecka 4, 00-975 Warsaw, Poland*

[2]*School of Chemistry, University of Bristol, Organic Geochemistry Unit, Cantock's Close, Bristol BS8 1TS, UK*

**Corresponding author (e-mail: Andrzej.Gasiewicz@pgi.gov.pl)*

The idea of this volume arose from our proposal to the palaeoclimatic session held at the European Geosciences Union General Assembly 2011 Conference in Vienna (Austria). Although the original scope was broader, the included contributions focus on palaeoenvironmental changes during the late Palaeozoic.

Late Palaeozoic history is especially important in that it records an increase of tectonics/diastrophism, frequent sea-level fluctuations and climatic changes, all finally leading to the most geocratic time of Earth's history with the formation of the supercontinent Pangaea.

This volume includes a number of articles that discuss the fossil record around the Late Palaeozoic Ice Age, and its decline during the Carboniferous with its unusual plant proliferation soon after, when the hottest and driest period in Earth's history (in the Permian) caused the development of huge 'sand seas' on the land. Finally, a biotic event, the largest mass extinction in Earth History at the Permian–Triassic transition, ends the Palaeozoic world. This volume addresses two key issues – the palaeoclimate and depositional cycles – both of which have a strong impact on the record of palaeoenvironmental changes and the understanding of environmental perturbations.

The vigorous current debate on the future climate of the globe inevitably increases our interest in past climates. From a geological perspective, climatic perturbations occurred many times before humans were on the planet. Climate varies on several timescales, depends on many high-order factors including tectonic uplift and the orbital wandering of the Earth, and as a result it changes through time. Consequently, palaeoclimatic reconstructions are growing in their importance as they enable us to use the past as a key to the future.

The Earth's system experiences various cycles of very different origin, scale, frequency, amplitude, phase and basic level. From a geological perspective, the cyclicity appears unidirectional, like Brownian motion. However, in the past, the complex interaction of these cycles has led to catastrophic changes in Earth's systems, such as those during the late Palaeozoic, when palaeoclimatic and geotectonic cycles coincided. These cycles are also influencing our lives, and will do so in our future, so their impact needs to be understood. The geological archives record both the climatic and cyclic phenomena on local to regional scales, including in the late Palaeozoic, as documented in this volume. Understanding these records will improve our predictions about climate phenomena in the future. They also show that the present climatic perturbations are insignificant from a geological point of view.

Changes in the late Palaeozoic Earth system are illustrated in this volume by case studies of fossil records from North and South America, East and South Asia, South Africa and Australia, and as regional reviews (section one). The methods include palaeoenvironmental analysis, including palaeobiological, sedimentological, stratigraphic (sequence stratigraphy), palaeomagnetic and geochemical methods. They describe various environmental settings from palaeodeserts, through marine-associated giant salt pans, to shallow to deep marine records. The articles report new data from less known study areas, provide up-to-date data ordering, and expand our knowledge of the evolution of marine and terrestrial environments, climatic changes and biotic events in late Palaeozoic time.

This special publication is divided into five chapters, the first comprising articles focused on the broad descriptive character of Late Carboniferous to Early Triassic palaeoclimatic changes and stratigraphy (with the latest part of this period in Europe, i.e. Zechstein). Subsequent articles include more detailed studies of the geological periods grouped into the latest Devonian to Early Carboniferous (section two), Late Carboniferous to Early Permian (section three), Middle to Late Permian (section four), and finally the Permian–Triassic transition (section five). A brief synopsis of individual papers is given below.

Wopfner (2012) reviews Westphalian–Early Triassic palaeoclimatic conditions between Samfrau (comprising the southern margins of South America and South Africa, extending via Antarctica

From: GĄSIEWICZ, A. & SŁOWAKIEWICZ, M. (eds) 2013. *Palaeozoic Climate Cycles: Their Evolutionary and Sedimentological Impact*. Geological Society, London, Special Publications, **376**, 1–4.
First published online October 17, 2013, http://dx.doi.org/10.1144/SP376.23

to eastern Australia) and Tethys. He concludes that the fundamental changes in Gondwanan depositional environments resulted not from the slight drift of Gondwana to the equator or modifications of the global ocean, but from evolution of the solar system and its impact in terms of cosmic ray flux, magnetic intensity, gravitational vectors and their variation over the course of the galactic year.

Wang *et al.* (2013) discuss Upper Carboniferous to Lower Permian shallow-water carbonates and their high-frequency cyclothems. Two biotic events occurring at the Mid-Carboniferous and Sakmarian–Artinskian boundaries in South China are consistent with two major regressions, probably associated with climate change caused by the glaciations of Gondwana. Biotic crises and faunal evolution of the Early–Middle Permian of the Uralian range and Southern Primorye area are also the subject of papers by **Kossovaya *et al.* (2013)** and **Kossovaya & Kropatcheva (2013)**.

Chemostratigraphy is used as a tool to understand not only biotic events but also facies and palaeoclimatic changes across stratigraphic boundaries. This approach is applied by **Bojar *et al.* (2013)**, **Laya *et al.* (2013)**, **Nielsen *et al.* (2012)**, **Fio *et al.* (2013)** and **Hiete *et al.* (2013)** to document climatically impacted successions in Europe and South America. When there is a lack of fossils, as in the Rotliegend or Zechstein strata, magnetostratigraphy may be a reliable method for stratigraphic correlation of regionally extended depositional sequences. This tool is used by **Szurlies (2013)**, who provides a new basin-scale Late Permian magnetostratigraphic correlation of the Southern Permian Basin of Europe.

Zubin-Stathopoulos *et al.* (2012) present the latest data on Late Pennsylvanian–Early Permian carbonate environments from the NW coast of Pangaea recorded in western Canada. Their study shows that the major global climatic shift across the Asselian–Sakmarian boundary did not affect carbonate sedimentation in their study area. A warm, protected sea, which developed at *c.* 20°N palaeolatitude during the Asselian and Sakmarian, provided a place for a photozoan association when the low-latitude NW coast of Pangaea was dominated by cool-water carbonates. Similarly, **Dineen *et al.* (2012)**, **Tabor *et al.* (2013)**, **Hanken & Nielsen (2013)**, **Isbell *et al.* (2012)**, **Kiersnowski (2013)** and **Czapowski & Tomassi-Morawiec (2013)** use geological strata to understand climate evolution, salinity control in sedimentary basins, or changes in vegetation on Pangaea during the late Palaeozoic in South and North America, and Europe.

In a series of papers on cyclostratigraphy, more evidence is presented on transgressive–regressive sequences triggered by variations in the Gondwanan ice sheets (**Ueno *et al.* 2012**; **Waksmundzka 2013**) from the Carboniferous to Early Permian. **Gebhardt & Hiete (2013)** found that cyclicity in the non-marine Upper Carboniferous deposits of the Saale Basin (Germany) was driven climatically, whereas **Gąsiewicz (2013)** challenged the existing 'cyclic' model of the Zechstein Main Dolomite (Upper Permian) carbonates deposition. Instead, he suggests that it was not global eustatic influxes but a combination of local relative sea-level fluctuations induced by climate impact and variation of the palaeonvironmental factors that determined the development of platformal carbonates.

Słowakiewicz & Gąsiewicz (2013) present a different approach regarding climate control on the deposition, distribution and productivity of potential source rocks for hydrocarbons. As the source of the Zechstein Main Dolomite petroleum deposits in the Southern Permian Basin is still enigmatic (partly Carboniferous and intraformational, i.e. sourced from the Main Dolomite carbonate source rocks), the authors use existing archival database and palaeoclimate models to understand organic matter productivity and its distribution in Late Permian time.

The papers' shape owe a lot to the work of the reviewers: Anna Antoshkina, Benoit Beauchamp, David Bond, Matthew Clapman, Mark Cocker, Ivan Dragičević, Geir Elvebakk, Paweł Filipiak, Clinton Foster, Reinhard Gast, Donald Gautier, John Geissman, Mark Geluk, Philip Heckel, Jason Hilton, John Isbell, Ganqing Jiang, Nathan A. Jude, Dennis Kent, Olga Kossovaya, Berit Legler, Carlos O. Limarino, Jacek Matyszkiewicz, Tom McKie, Horng-Sheng Mii, Marek Narkiewicz, Jesper Nielsen, Alejandra Pagani, Tadeusz M. Peryt, Catherine Reid, Barry Richards, Daniela Schmidt, Michael Schramm, Guang Shi, Stanisław Skompski, Ian Somerville, Masatoshi Sone, Tetsuo Sugiyama, Tore Amund Svånå, Arturo Taboada, Neil Tabor, Stephanie Thomas, Maurice E. Tucker, Greg Wahlman, Xunlian Wang, Colin Waters, Dieter Weyer, Pat Wilde and Anna Wysocka.

The idea of this volume has also benefited from reviews of the book proposal provided by Greg Retallack (University of Oregon), Marco Vecoli (University of Lille) and John Marshall (University of Southampton).

Finally, we would like to thank the GSL Publishing House Staff for their extensive work on this volume and the Polish Geological Institute for partial sponsorship of the book (project 62.9711.1202.00.1).

References

Bojar, A.-V., Neubauer, F. & Koeberl, C. 2013. Geochemical record of Late Devonian to Early Carboniferous events Palaeozoic of Graz, Eastern Alps, Austria. *In*: Gąsiewicz, A. & Słowakiewicz, M. (eds) *Palaeozoic Climate Cycles: Their Evolutionary and Sedimentological Impact*. Geological Society, London, Special

Publications, **376**. First published online September 19, 2013, http://dx.doi.org/10.1144/SP376.15

Czapowski, G. & Tomassi-Morawiec, H. 2013. Palaeogeographic and palaeoclimate factors of salinity fluctuations in the eastern part of the Late Permian (Zechstein) European Basin: case study from the salt basin in Poland. *In*: Gąsiewicz, A. & Słowakiewicz, M. (eds) *Palaeozoic Climate Cycles: Their Evolutionary and Sedimentological Impact*. Geological Society, London, Special Publications, **376**. First published online September 19, 2013, updated October 9, 2013, http://dx.doi.org/10.1144/SP376.16

Dineen, A. A., Fraiser, M. L. & Isbell, J. L. 2012. Palaeoecology and sedimentology of Carboniferous glacial and post-glacial successions in the Paganzo and Río Blanco basins of northwestern Argentina. *In*: Gąsiewicz, A. & Słowakiewicz, M. (eds) *Palaeozoic Climate Cycles: Their Evolutionary and Sedimentological Impact*. Geological Society, London, Special Publications, **376**. First published online November 26, 2012, http://dx.doi.org/10.1144/SP376.3

Fio, K., Sremac, J., Vlahović, I., Velić, I. & Spangenberg, J. E. 2013. Permian deposits and the Permian–Triassic boundary in Croatia: palaeoclimatic implications based on palaeontological and geochemical data. *In*: Gąsiewicz, A. & Słowakiewicz, M. (eds) *Palaeozoic Climate Cycles: Their Evolutionary and Sedimentological Impact*. Geological Society, London, Special Publications, **376**. First published online February 27, 2013, http://dx.doi.org/10.1144/SP376.8

Gąsiewicz, A. 2013. Climatic control on the Late Permian Main Dolomite (Ca2) deposition in northern margin of the Southern Permian Basin and implications to its internal cyclicity. *In*: Gąsiewicz, A. & Słowakiewicz, M. (eds) *Palaeozoic Climate Cycles: Their Evolutionary and Sedimentological Impact*. Geological Society, London, Special Publications, **376**. First published online October 17, 2013, http://dx.doi.org/10.1144/SP376.22

Gebhardt, U. & Hiete, M. 2013. Continental Upper Carboniferous red beds in the Variscan intermontane Saale Basin, central Germany: orbital forcing detected by wavelet analysis. *In*: Gąsiewicz, A. & Słowakiewicz, M. (eds) *Palaeozoic Climate Cycles: Their Evolutionary and Sedimentological Impact*. Geological Society, London, Special Publications, **376**. First published online August 6, 2013, http://dx.doi.org/10.1144/SP376.12

Hanken, N.-M. & Nielsen, J. K. 2013. Upper Carboniferous–Lower Permian *Palaeoaplysina* build-ups on Svalbard: the influence of climate, salinity and sea-level. *In*: Gąsiewicz, A. & Słowakiewicz, M. (eds) *Palaeozoic Climate Cycles: Their Evolutionary and Sedimentological Impact*. Geological Society, London, Special Publications, **376**. First published online September 19, 2013, http://dx.doi.org/10.1144/SP376.17

Hiete, M., Röhling, H.-G., Heunisch, C. & Berner, U. 2013. Facies and climate changes across the Permian–Triassic boundary in the North German Basin: insights from a high-resolution organic carbon isotope record. *In*: Gąsiewicz, A. & Słowakiewicz, M. (eds) *Palaeozoic Climate Cycles: Their Evolutionary and Sedimentological Impact*. Geological Society, London, Special Publications, **376**. First published online February 27, 2013, http://dx.doi.org/10.1144/SP376.9

Isbell, J. L., Henry, L. C., Reid, C. M. & Fraiser, M. L. 2012. Sedimentology and palaeoecology of lonestone-bearing mixed clastic rocks and cold-water carbonates of the Lower Permian Basal Beds at Fossil Cliffs, Maria Island, Tasmania (Australia): Insight into the initial decline of the late Palaeozoic ice age. *In*: Gąsiewicz, A. & Słowakiewicz, M. (eds) *Palaeozoic Climate Cycles: Their Evolutionary and Sedimentological Impact*. Geological Society, London, Special Publications, **376**. First published online November 26, 2012, http://dx.doi.org/10.1144/SP376.2

Kiersnowski, H. 2013. Late Permian Eolian Sand Seas from the Polish Rotliegend Basin in the context of palaeoclimatic periodicity. *In*: Gąsiewicz, A. & Słowakiewicz, M. (eds) *Palaeozoic Climate Cycles: Their Evolutionary and Sedimentological Impact*. Geological Society, London, Special Publications, **376**. First published online September 26, 2013, updated October 9, 2013, http://dx.doi.org/10.1144/SP376.20

Kossovaya, O. L. & Kropatcheva, G. S. 2013. Extinction of Guadalupian rugose corals: an example of biotic response to the Kamura event (southern Primorye, Russia). *In*: Gąsiewicz, A. & Słowakiewicz, M. (eds) *Palaeozoic Climate Cycles: Their Evolutionary and Sedimentological Impact*. Geological Society, London, Special Publications, **376**. First published online August 6, 2013, http://dx.doi.org/10.1144/SP376.13

Kossovaya, O., Vachard, D. & Izart, A. 2013. Climatic impact on the reef biota in the Cisuralian and Guadalupian (Permian), East European Platform. *In*: Gąsiewicz, A. & Słowakiewicz, M. (eds) *Palaeozoic Climate Cycles: Their Evolutionary and Sedimentological Impact*. Geological Society, London, Special Publications, **376**. First published online September 25, 2013, http://dx.doi.org/10.1144/SP376.19

Laya, J. C., Tucker, M. E., Gröcke, D. R. & Perez-Huerta, A. 2013. Carbon, oxygen and strontium isotopic composition of low-latitude Permian carbonates (Venezuelan Andes): climate proxies of tropical Pangea. *In*: Gąsiewicz, A. & Słowakiewicz, M. (eds) *Palaeozoic Climate Cycles: Their Evolutionary and Sedimentological Impact*. Geological Society, London, Special Publications, **376**. First published online February 28, 2013, http://dx.doi.org/10.1144/SP376.10

Nielsen, J. K., Błażejowski, B., Gieszcz, P. & Nielsen, J. K. 2012. Carbon and oxygen isotope records of Permian brachiopods from relatively low and high palaeolatitudes: climatic seasonality and evaporation. *In*: Gąsiewicz, A. & Słowakiewicz, M. (eds) *Palaeozoic Climate Cycles: Their Evolutionary and Sedimentological Impact*. Geological Society, London, Special Publications, **376**. First published online November 26, 2012, http://dx.doi.org/10.1144/SP376.6

Słowakiewicz, M. & Gąsiewicz, A. 2013. Palaeoclimatic imprint on the distribution and genesis of Zechstein Main Dolomite (Upper Permian) petroleum source rocks in Poland: Sedimentological and geochemical rationales. *In*: Gąsiewicz, A. & Słowakiewicz, M. (eds) *Palaeozoic Climate Cycles:*

Their Evolutionary and Sedimentological Impact.Geological Society, London, Special Publications, **376**. First published online September 25, 2013, http://dx.doi.org/10.1144/SP376.21

Szurlies, M. 2013. Late Permian (Zechstein) magnetostratigraphy in Western and Central Europe. *In*: Gąsiewicz, A. & Słowakiewicz, M. (eds) *Palaeozoic Climate Cycles: Their Evolutionary and Sedimentological Impact*. Geological Society, London, Special Publications, **376**. First published online February 27, 2013, http://dx.doi.org/10.1144/SP376.7

Tabor, N. J., Romanchock, C. M., Looy, C. V., Hotton, C. L., DiMichele, W. A. & Chaney, D. S. 2013. Conservatism of Late Pennsylvanian vegetational patterns during short-term cyclic and long-term directional environmental change, western equatorial Pangaea. *In*: Gąsiewicz, A. & Słowakiewicz, M. (eds) *Palaeozoic Climate Cycles: Their Evolutionary and Sedimentological Impact*. Geological Society, London, Special Publications, **376**. First published online September 19, 2013, http://dx.doi.org/10.1144/SP376.14

Ueno, K., Hayakawa, N., Nakazawa, T., Wang, Y. & Wang, X. 2012. Pennsylvanian–Early Permian cyclothemic succession on the Yangtze Carbonate Platform, South China. *In*: Gąsiewicz, A. & Słowakiewicz, M. (eds) *Palaeozoic Climate Cycles: Their Evolutionary and Sedimentological Impact*. Geological Society, London, Special Publications, **376**. First published online November 29, 2012, http://dx.doi.org/10.1144/SP376.5

Waksmundzka, M. I. 2013. Carboniferous coarsening-upward and non-gradational cyclothems in the Lublin Basin (SE Poland): palaeoclimatic implications. *In*: Gąsiewicz, A. & Słowakiewicz, M. (eds) *Palaeozoic Climate Cycles: Their Evolutionary and Sedimentological Impact*. Geological Society, London, Special Publications, **376**. First published online September 19, 2013, http://dx.doi.org/10.1144/SP376.18

Wang, X., Qie, W., Sheng, Q., Qi, Y., Wang, Y., Liao, Z., Shen, S. & Ueno, K. 2013. Carboniferous and Lower Permian sedimentological cycles and biotic events of South China. *In*: Gąsiewicz, A. & Słowakiewicz, M. (eds) *Palaeozoic Climate Cycles: Their Evolutionary and Sedimentological Impact*. Geological Society, London, Special Publications, **376**. First published online August 6, 2013, http://dx.doi.org/10.1144/SP376.11

Wopfner, H. 2012. Late Palaeozoic–Early Triassic deposition and climates between Samfrau and Tethys: A review. *In*: Gąsiewicz, A. & Słowakiewicz, M. (eds) *Palaeozoic Climate Cycles: Their Evolutionary and Sedimentological Impact*. Geological Society, London, Special Publications, **376**. First published online November 26, 2012, http://dx.doi.org/10.1144/SP376.4

Zubin-Stathopoulos, K. D., Beauchamp, B., Davydov, V. I. & Henderson, C. M. 2012. Variability of Pennsylvanian–Permian Carbonate Associations and implications for NW Pangea Palaeogeography, east-central British Columbia, Canada. *In*: Gąsiewicz, A. & Słowakiewicz, M. (eds) *Palaeozoic Climate Cycles: Their Evolutionary and Sedimentological Impact*. Geological Society, London, Special Publications, **376**. First published online November 26, 2012, http://dx.doi.org/10.1144/SP376.1

Late Palaeozoic–Early Triassic deposition and climates between Samfrau and Tethys: A review

H. WOPFNER

Department of Geology, University of Cologne, Zülpicher Strasse 49a, 50674 Köln, Germany (e-mail: h.i.wopfner@t-online.de)

Abstract: In the period between the Westphalian and the Early Triassic, climatic conditions across Gondwana changed from the Gondwana-wide glaciation in the Late Carboniferous–Early Permian to warm to hot, semi-humid to arid conditions by the Late Permian–Early Triassic. In Afro-Arabia, glaciers advanced to 40° palaeolatitude creating a steep gradient between them and the temperate-humid environment in the vicinity of the equator. Synchronous deglaciation associated with a eustatic rise of sea level in the Early Sakmarian was followed by a flattening of the climate gradient and increased humidity, expressed by blackwater rivers and coal swamps between the South Pole and about 50° palaeolatitude and increasing aridity towards the equator. Brief poleward advances of hot aridity are evident around the Kungurian. Expansion of the hot climate zone to the south took place in the Middle Permian, commencing with a humid to semi-humid phase in the south and semi-arid to arid conditions towards the north. A short pluvial event is locally indicated at the beginning of the Triassic, but warm semi-arid to semi-humid conditions prevailed in the south and hot aridity near the equator. There are variations of that general theme in eastern Gondwana, but the fundamental changes in climate-controlled depositional environments occurred almost synchronously over the whole of Gondwana, including Antarctica. The slight drift of Gondwana towards the equator is therefore an unlikely cause for these changes, as are modifications of ocean–continent geometries or of atmospheric composition. As the main driver was temperature it is suggested that governing control, not only of global temperature but also of the intensity of geotectonic activities, was imposed on the solar system by forces emanating from our galaxy (cosmic ray flux, magnetic intensity, gravitational vectors) and their variation over the course of a galactic year (the time for the solar system to orbit the core of the galaxy).

During the end of the Palaeozoic/beginning of the Mesozoic, some of the more dramatic climatic events in Earth's history were experienced. These included the great Gondwana glaciation in the Late Carboniferous and the Early Permian, the formation of the large coal deposits in all parts of Gondwana, the formation of important oil and gas reservoirs in Arabia, central Australia, America and the North Sea and, finally, the creation of the enormous evaporate deposits right across Euramerica in the Late Permian. This period also experienced the great extinction event at the end of the Palaeozoic. Within Gondwana its depositional record is synonymous with the Pangea Megasequence (PMS) (Veevers 1990; Wopfner & Jin 2009).

This review deals with the climatic developments as witnessed by climate-sensitive sedimentary rocks of Pangea Megasequences in central Gondwana between its Tethyan margin and the Samfrau orogenic belt, concentrating on the Afro-Arabian–South Alpine segment of Gondwana (Fig. 1). The term 'Samfrau' was introduced by Du Toit (1927) as 'Samfrau Geosyncline', referring to the active fold belt which edged the Panthalassa margin of Gondwana in the Palaeozoic. A comprehensive description of that region was presented in Veevers & Powell (1994). This review includes terranes attached originally to the northern margin of Gondwana, but now incorporated within the Alpine System.

Hardly anyone would question the extreme climatic variations experienced during the Permian, but there is no general agreement about their governing control. Many researchers see the causes for the onset of glaciations and the climatic changes following the collapse of the glaciers as simply the migration of Gondwana across the South Pole and the subsequent northwards drift of Gondwana (e.g. Crowell & Frakes 1970; Crowell 1978, 1995; Visser 1995; López-Gamundi 1997; Ziegler *et al.* 1997).

However, scrutiny of glacial and post-glacial depositional sequences raises the question of whether the concept of climate control due to Gondwana's drift across the South Pole may require modification (Dickins 1985, 1997; Collinson 1997).

Tectonic framework

Figure 1 represents a generalized reconstruction of Gondwana at around the Carboniferous/Permian boundary, showing the major occurrences of Late Palaeozoic depocentres including those of the

From: Gąsiewicz, A. & Słowakiewicz, M. (eds) 2013. *Palaeozoic Climate Cycles: Their Evolutionary and Sedimentological Impact*. Geological Society, London, Special Publications, **376**, 5–32.
First published online November 26, 2012, http://dx.doi.org/10.1144/SP376.4

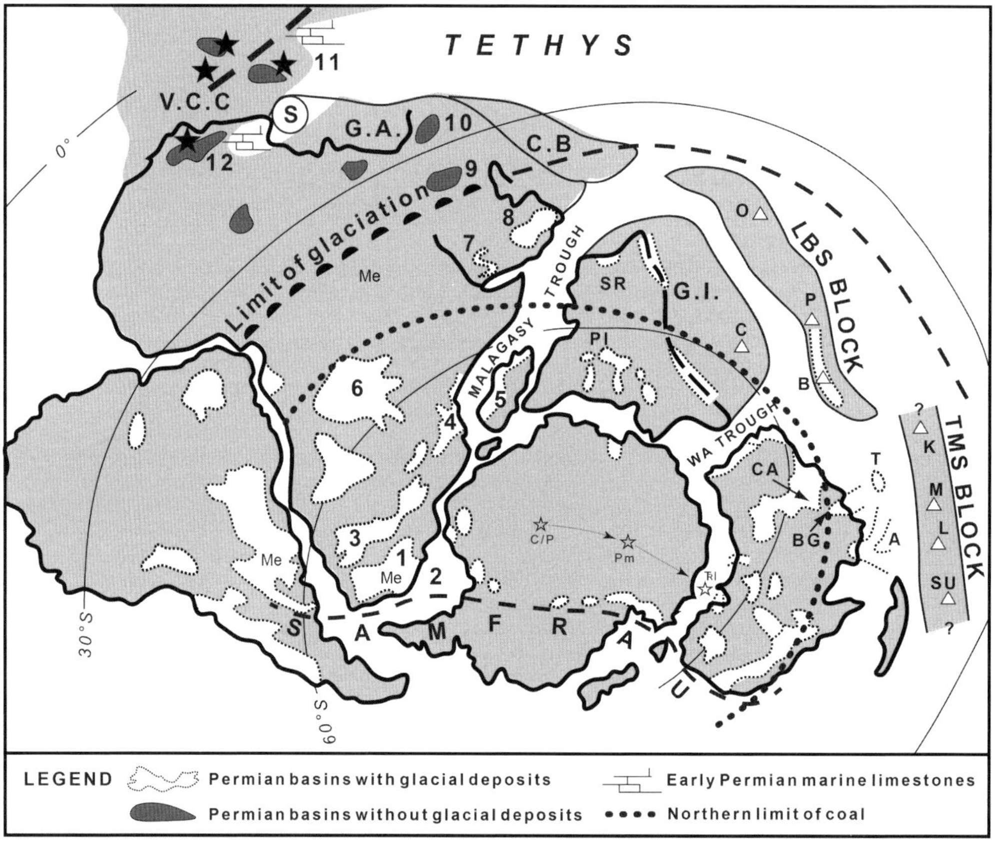

Fig. 1. Reconstruction of Gondwana at the beginning of the Permian, modified after Wopfner & Jin (2009). The distribution of basins with Late Carboniferous–Early Triassic basin fill (=Pangea Megasequence PMS) is depicted, distinguishing between basins with and without glacigene deposits. The 'northern' limit of glaciation established in Afro-Arabia is extended by a dashed line to show the approximate limit within the realm of Tethys. Numerals identify the following basins: 1: Karoo; 2: Falkland; 3: Kalahari–Luangwa; 4: Ruhuhu–Tanga; 5: Morondawa; 6: Congo; 7: Yemen; 8: South Oman; 9: Saudi Arabia; 10: Levante; 11: Collio–Etsch–Carnia (South Alpine Terrane); 12: Morroco. A: Arafura Basin; B: Baoshan Basin; BG: Bonaparte Gulf Basin; CA: Canning Basin; C.B.: Cimmerian Blocks (narrow sense); G.A.: Greater Apulia; G.I.: Greater India (after Patzelt *et al.* 1996); K: Kungshui (northern Tengchong Basin); L: Langkawi Archipelago; M: South Thai Basin (Phuket); Me: finds of Mesosaurus; O: Orba Lake (W. Tibet); P: Poindo, Lhasa Basin; Pl: Peninsular India; Q: Qomolangma (Mt Everest); S: Sicily; SR: Salt Range, Pakistan; SU: Sumatra; T: Tengchong Terrane; V.C.C.: Variscan continental crust. Black stars indicate Early Permian acid volcanism. LBS: Lhasa–Baoshan–Shan; TMS: Tengchong–Malay–Sumatra. Open stars identify position of South Pole, modified after Powell & Li (1994) at C/P: Carboniferous–Permian boundary; Pm: Middle Permian; Trl: Early Triassic.

Kimmerian and Alpine terranes. As mentioned above, the Panthalassa margin of Gondwana was governed by the compressive system of the Samfrau Geosyncline. This fold system extended from southern South America and South Africa via Antarctica to eastern Australia (Veevers & Powell 1994; Lopez-Gamundi 1997). The fold movements generally commenced in the Early Carboniferous and extended, with interruptions, into the Triassic. In South Africa, they created the Cape Fold System (see Catuneanu *et al.* 2005).

The compressive conditions along the Panthalassa edge of Gondwana contrasted with the northern margin of Gondwana where extension and transtension prevailed. It was the realm of the nascent Tethys, the cradle of the future Alpine–Himalayan mountain belt (Sengör 1998). In the course of the opening of Neo-Tethys, the Kimmerian terranes split off Gondwana and positioned themselves on its northern shore prior to the cataclysm of the Alpine–Himalayan orogeny. These terranes include some parts of Afghanistan, the

Lhasa-Baoshan-Shan terranes, the Tengchong-Thai/Malay terrane and the North Sumatra block of Wopfner & Jin (2009) or Sibumasu of Metcalfe (1988). Their Gondwana origin is identified by deposits of the Gondwana glaciation and by characteristic faunal and floral assemblages (Jin 1994; Shi *et al.* 1995; Wopfner 1996; Archbold 2001; Metcalfe 2002).

The land bridge between Africa and Europe created by the Hercynian suturing was modified in the Early Permian by the inception of extensional systems. Major volcanic rifts, pull-apart basins and other distension structures heralded the opening of the western Tethys and the development of its southern shelf which bore most of the South and East Alpine facies realm (Stampfli *et al.* 1998). Good evidence for the Early Permian rift movements associated with the opening of the Tethys is known from many parts in North Africa (Guiraud *et al.* 2005) and the adjoining Adria Plate, where the lower crust was intruded by gabbros during the Permian (Muttoni *et al.* 2003). The Khenifra and Mechraâ Ben Abbou basins and the western region of the High Atlas in Morocco (Biron 1982; El Wartiti *et al.* 1990) may serve as examples. Rift structures controlled the development of Permian basins of the Southern Alps (Wopfner 1984; Cassinis 1986; Venturini 1990; Cassinis & Ronchi 2001). This Alpine realm is viewed as part of the passive margin of greater Apulia (Brandner & Keim 2011) or it is placed on the Adria Plate which may be viewed as a terrane attached to or just off the northern margin of Afro-Arabia. The palaeo-position of the Adria Plate in relation to Afro-Arabia is still very much debated. Muttoni *et al.* (2003) placed it north of Algeria in modern coordinates, whereas Wortmann *et al.* (2001) preferred a slightly more easterly position. Considering certain similarities in facies and tectonic development, the present author selected a palaeo-position of the South Alpine Terrane NE of Morocco, just off Greater Apulia (Fig. 1). The terrane remained on the southern shores of Tethys until its closure during the Cainozoic, when it was incorporated within the Alpine fold belt.

The area discussed in this review is therefore contrasted by opposing tectonic systems: convergence dominated the Panthalassa margin and expansion controlled the edge along the Tethys (Catuneanu *et al.* 2005). Two major rift systems, the Malagasy and the West Australian Troughs, propagated from the Tethyan margin at the beginning of the Permian (Wopfner 1994). A structural section across the Malagasy Trough is shown in Figure 2.

Selecting Afro-Arabia for this review takes account of the fact that the shape of this part of Gondwana has not changed substantially since the Late Palaeozoic. Except for the opening of the Red Sea and the separation of Malagasy, the only significant modifications took place along the northern margin of Africa. The major Cretaceous–Tertiary rift structures with their important petroleum systems of central Africa (Schull 1988; Genik 1993; Guiraud *et al.* 2005) had little influence on the overall shape of the continent. Proposing a Pangea 'B' reconstruction, Muttoni *et al.* (2003) suggested a right lateral shear displacement in northwestern Morocco roughly along the South Atlas Fault Zone. This had no influence on the position of the basins discussed in this review which, except for the South Alpine Terrane, were all situated to the south of that shear zone.

The depositional sequences on which the climatic interpretation is based are described in the following sections.

Africa south of the equator

The post-Namurian succession south of the equator (i.e. Pangea Megasequence) is referred to the Karoo Supergroup and its time equivalents. In the southern part of South Africa, Karoo sedimentation took place in the foredeep of the Samfrau convergence (Catuneanu *et al.* 2005), but the northern part extended onto the craton. The structural setting changed from there to the north. A complex system of generally N–NE-trending rift basins extended from Namibia/Botswana along the eastern parts of Africa to Tanzania and western Madagascar (Wopfner 1994; Ring 1995; Piqué *et al.* 1999), whereas large sag basins developed in the Congo, in Angola and in Gabon. These two settings were separated by a NE-trending structural high, also termed 'Median Thermal Swell' (Catuneanu *et al.* 2005), which apparently was a focus of glaciation in central Africa (Rust 1975; Fig. 3).

South Africa

The Gondwana glaciation has left indubitable traces over many parts of Africa and Arabia (Figs 1 & 3). By far, the most widespread records of the glacial event are known from the Karoo Basin in South Africa. Various aspects have been discussed, among others, by Stratten (1970), Rust (1975), Visser (1986, 1989, 1995, 1997) and Smith *et al.* (1993), to mention a few.

The beginning of the glaciation in southern Africa is still under discussion. Field evidence for the lower limit is provided in the southern Karoo Basin where tillites of the Dwyka Group rest unconformably on strata of the Namurian Lake Menz Formation (Fig. 4). The formation consists of thinly bedded, tidal flat and subtidal deposits that show no evidence for any glacial activity. This conforms to the observation of Visser (1997) that the oldest

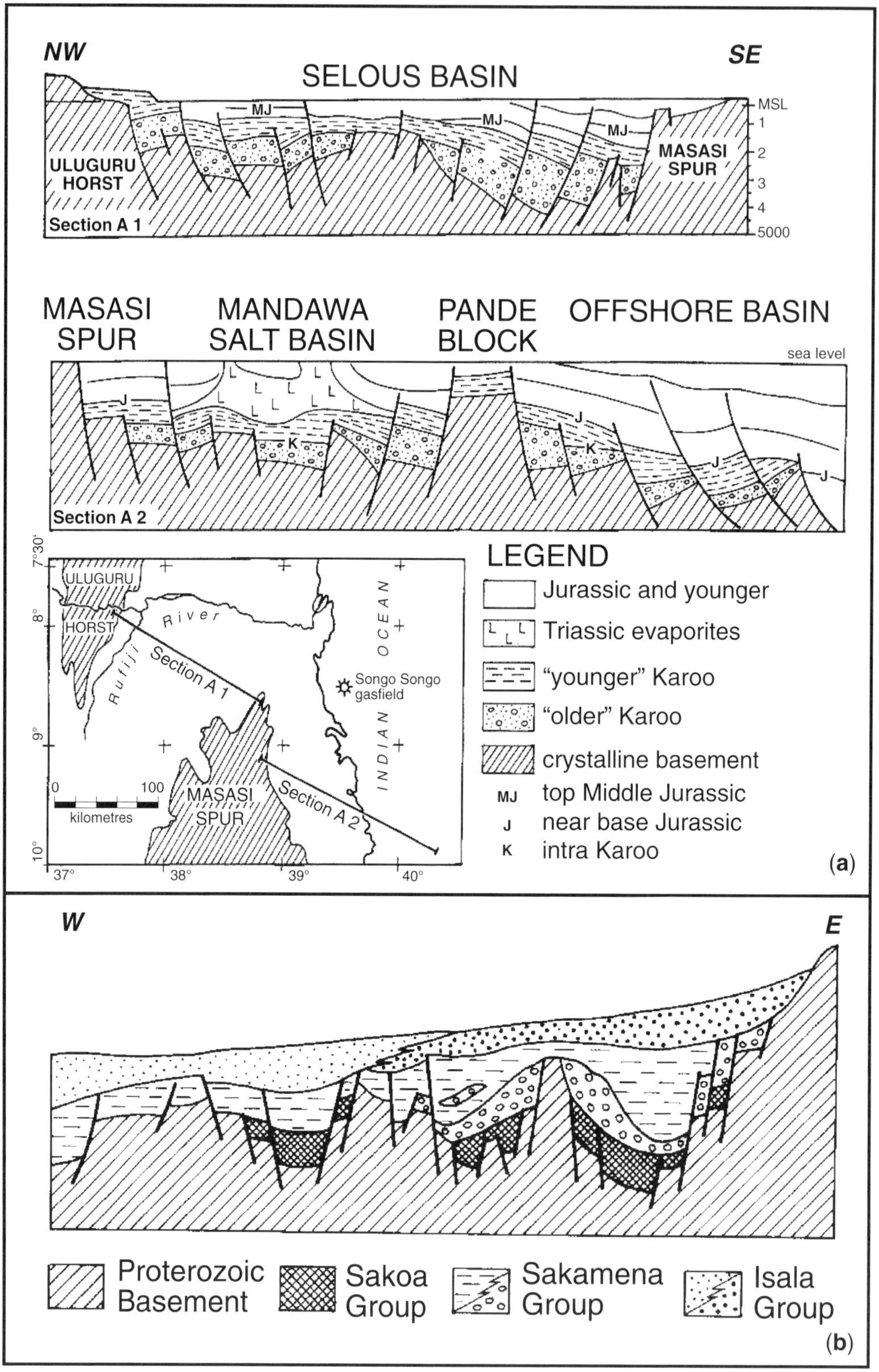
NW
SE
SELOUS BASIN
MJ
MJ
MJ
MSL
1
2
3
4
5000
ULUGURU HORST
MASASI SPUR
Section A 1
MASASI SPUR
MANDAWA SALT BASIN
PANDE BLOCK
OFFSHORE BASIN
sea level
J
K
J
K
J
J
Section A 2
7°30'
8°
9°
10°
37°
38°
39°
40°
ULUGURU
HORST
River
Rufiji
Section A 1
Section A 2
MASASI SPUR
INDIAN OCEAN
Songo Songo gasfield
0
100
kilometres
LEGEND
Jurassic and younger
Triassic evaporites
"younger" Karoo
"older" Karoo
crystalline basement
MJ top Middle Jurassic
J near base Jurassic
K intra Karoo
(a)
W
E
Proterozoic Basement
Sakoa Group
Sakamena Group
Isala Group
(b)

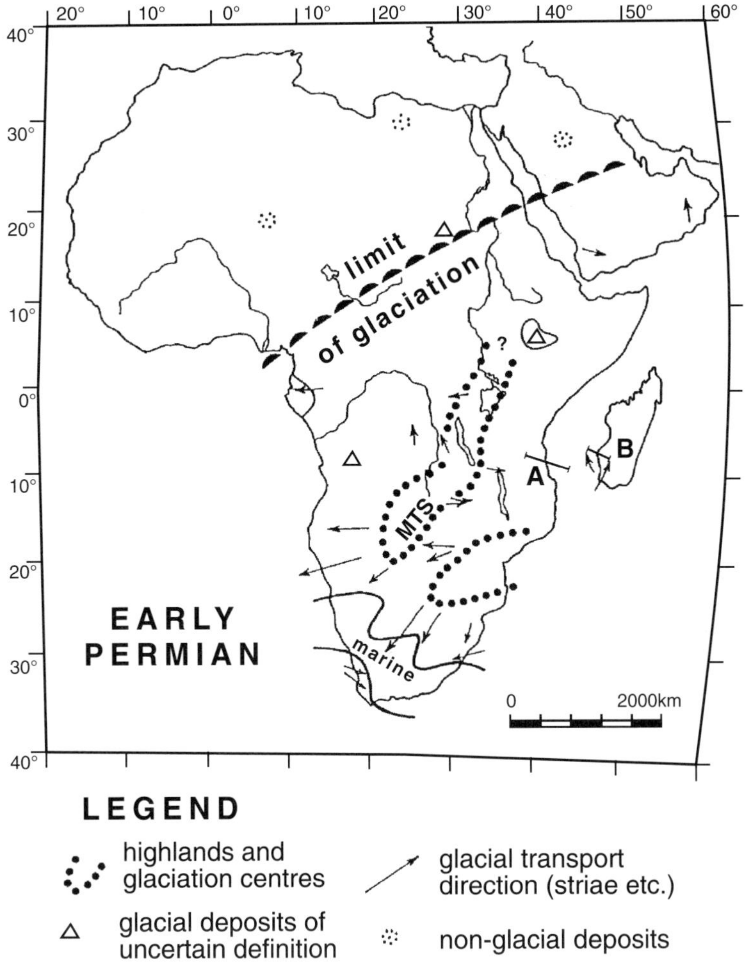

Fig. 3. Map of Afro-Arabia showing northern limit of glaciation and position of the Median Thermal Swell (MTS). The ridge separated the rift basins in the east from the sag basins in the west and supported the accumulation of regional centres of glaciation. The directions of ice movements, modified after Rust (1975), support the existence of local high lands as foci of glaciation. The lines 'A' and 'B' show the position of the sections depicted in Figure 2.

diamictites in South Africa are of Westphalian age. They are found in the western Karoo Basin and are topped by a mudstone unit which he termed 'Deglaciation Sequence One' (DS 1). To propose an earlier onset of glaciation based on indirect evidence such as 'ice push deformation' is, in the view of this author, questionable.

The second glacial maximum started in the Stephanian. It reached its greatest expansion in the Asselian and was responsible for the maximum ice cover and the maximal glacial advance within the Permo-Carboniferous ice age. Depositional environment within the foredeep was glaciomarine, but on the cratonic shelf further north striated pavements covered by lodgement and washout tills and alluvial fans indicate deposition ranging from sub-aerial conditions to sedimentation in shallow periglacial lakes (Fig. 5).

Fig. 2. Composite structural section across Malagasy Trough demonstrates complex, syndepositional rifting during the Permo-Carboniferous sedimentation cycle. (**a**) Interpreted seismic reflection survey from the Uluguru Horst onto the East African continental shelf, *c.* 300 km south of Dar es Salaam. (**b**) Structure of the Morondava Basin of western Madagascar which mirrors the Tanzanian profile. Both sections demonstrate intensive faulting during deposition of Karoo equivalent strata. The position of the sections is shown on the map in Figure 3 (after Wopfner 2002).

Fig. 4. Dwyka Tillite (in foreground) rests unconformably on thinly bedded Lake Menz Formation of Namurian age. Exposure photographed near Floriskraal, south of Laingsburg in the Great Karoo.

Bangert *et al.* (1999) identified four deglaciation sequences in Karoo successions of southern Namibia and western South Africa and obtained SHRIMP dates on zircons contained in interbedded layers of ashfall tuffs. For the Ganigobis Shale Member which appears to top DS 1 of Visser (1997), they obtained ages of 302 ± 3.0 Ma and 299.2 ± 3.2 Ma (Fig. 5).

The present author views these deglaciation sequences as products of interglacial intervals and differentiates between them and the final deglaciation event in the Sakmarian. Some of the Karoo glaciers issued into the sea and some extended SW onto South America (López-Gamundi 1997; Visser 1997). The Falkland Islands, which bear widespread sedimentary evidence of the Gondwana glaciations, are generally placed east of southernmost South Africa (López-Gamundi & Rosello 1998; see Fig. 1). The Stephano-Asselian glacial maximum was terminated in the early Sakmarian (DS 2 of Visser 1997). It is associated with a eustatic rise of sea level demonstrated by the presence of *Eurydesma* (Dickins 1985). In adjacent basins of South America and the Falkland Islands, final deglaciation is evidenced by the Piedra Azul and the Port Sussex formations, respectively (López-Gamundi & Rosello 1998).

There is no direct evidence for glacial activity after the Early Sakmarian deglaciation event, although Visser (1997) postulated a third and final glaciation in the vicinity of the palaeo-south-pole, deduced from 'large discharge of melt water'. Considering evidence from Antarctica (Miller & Isbell 2010), a post-Sakmarian glacial event in Antarctica seems very unlikely (see Climate history).

The Ecca Group contains the terminal deposits of the glacigene succession and comprises clastic sequences of the foredeep including turbidites, regularly banded or banked siliciclastics, organic-rich mudstones and cherts of a deep marine environment (in ascending order): Prince Albert, White Hill, Collingham and Laingsburg (Ripon) formations. The shallow marine Waterford and Fort Brown formations conclude the Ecca succession in the southern Karoo Basin (Fig. 5). Bangert *et al.* (1999) obtained SHRIMP ages from the basal Prince Albert Formation of 2.88 ± 3.0 Ma and 289.6 ± 3.8 Ma. Thin tuffaceous layers in the Collingham Formation gave a radiometric age of 270 ± 1 Ma (Catuneanu *et al.* 2005). These authors, using an older timescale, placed the 270 Ma time within the Sakmarian. According to the timescale accepted by the Subcommission on Permian Stratigraphy (Wardlaw 1999) however, this date falls into the latest Kungurian. Other more recent timescales (Menning *et al.* 2006) place this date still higher in the stratigraphic column. Fildani *et al.* (2009) obtained additional SHRIMP age dates from ash beds intercalated in turbidites of the upper Ecca Group east of 24° longitude. These demonstrate that 'the marine uppermost Laingsburg Formation of the Ecca Group are contemporaneous with the fluvial upper Beaufort Group of the east-central Karoo Basin', suggesting marked diachronism and fundamentally different depositional conditions between the foredeep and the cratonic shelf. A generalized space/time distribution of the formations is shown in the stratigraphic columns in Figure 5.

On the cratonic shelf situated NE and north of the foredeep, the Ecca succession is represented by Pietermaritzburg, Vreyheid and Volkrust formations and their lateral equivalents (Fig. 5; Smith *et al.* 1993; Catuneanu *et al.* 2005). The Pietermaritzburg Formation is a monotonous sequence of black to dark grey, in parts richly carbonaceous siltstones and shales, apparently of freshwater origin. It constitutes the typical deglaciation facies in the eastern Karoo Basin and may be correlated with similar deglaciation deposits in East and Central Africa and southern Arabia. It is succeeded by the Vryheid Formation which harbours the major coal deposits of southern Africa. The formation comprises a lower coal-bearing unit consisting of stacked sandstone to silt sequences, the end members of which are thick coal seams, and an upper shale-dominated coal unit. The formation was deposited in prograding deltas, associated fluvial-paludal plains and coastal environments (Tavener-Smith 1982; Cairncross & Cadle 1987). The lower coal beds are of Artinskian–Early Kungurian age, the upper unit of Rodian–Wordian age (Fig. 5). Proxy signals indicate a warm and humid climate (Cairncross 2001), thus refuting previously held believes of glacial proximity of coal swamps (Cairncross & Cadle 1987).

The succeeding Beaufort Group covers the time interval from the Late Wordian to the end of the Anisian (Fig. 5). It consists dominantly of fluvial,

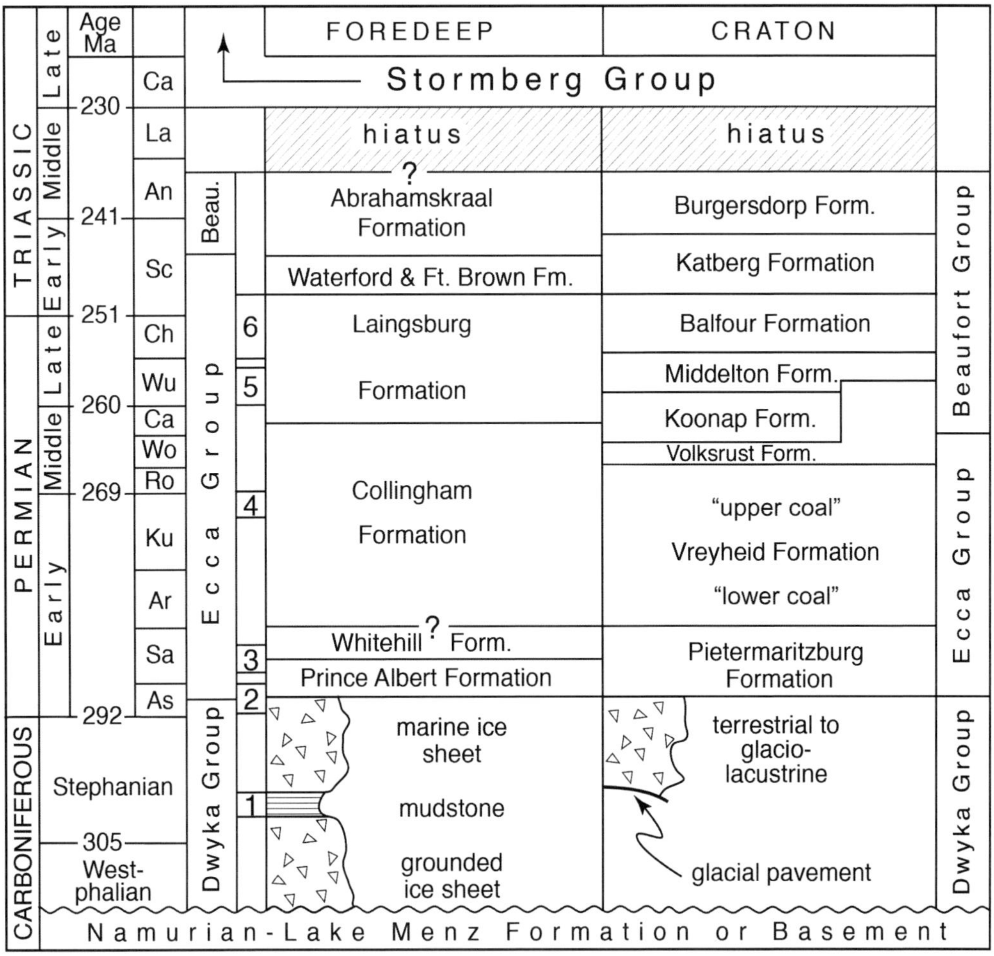

Fig. 5. Simplified stratigraphic columns of the Karoo Basin in South Africa, demonstrating the marked diachronism and facies change between the actively subsiding foredeep in the SW and the mainly terrestrial environments in the NE and north of the basin. The numbers 1–6 identify events dated by SHRIMP age determinations. (1) Tuffs in Ganigobis Shale Member, 302 ± 3.0 Ma; 299.2 ± 3.2 Ma. DS I & II of Bangert *et al.* (1999), DS 1 of Visser (1997). (2) Tuffs from base of Prince Albert Formation, 288.0 ± 3.0 Ma; 289.6 ± 3.8 Ma. (Bangert *et al.* 1999). (3) *Eurydesma* transgression, DS 2 of Visser (1997); deglaciation event (Wopfner 1999). (4) Ash beds in Collingham Formation, 270 ± 1 Ma (Catuneanu *et al.* 2005). (5) Tuff samples CVX 10 in Laingsburg Formation, 258.1 ± 1.6 Ma (Fildani *et al.* 2009). (6) Tuff samples CVX 12 in Laingsburg Formation, 252.7 ± 2 Ma (Fildani *et al.* 2009). Depositional environments of the Dwyka Group are from Visser (1997). Absolute timescale after Wardlaw (1999).

fluvio-deltaic and lacustrine deposits and thus shows a great variety of lithologies and interchanging facies (Catuneanu *et al.* 2005). An outstanding feature is the widespread occurrence of tetrapod fossils which also form the base for the subdivision of the sequence (Rubidge *et al.* 1995). The abundance of mammal-like reptiles (e.g. *Gorgonopsidae*) is indicative of a warm to hot climate which produced sufficient nutrients to sustain the reptile population. Uranium mineralizations within certain levels of these sequences are indicative of a specific groundwater chemistry which corroborates a warm to hot climate. Precipitation ranged from semi-humid at the *Eodicynodon*-zone to semi-arid in the *Dicynodon*-zone at the end of the Permian (Rubidge *et al.* 1995). The Triassic deposition is introduced by the Katberg Sandstone, an up to 1000 m thick sequence of white stacked sheet and current bedded sandstone. The Katberg Sandstone has been interpreted as a braid plain deposit of a bedload-dominated ephemeral stream system in an arid climate (see Catuneanu *et al.* 2005). The maturity of the sandstone, the sparse kaolinitic matrix and the lack of fine-grained deposits of diminishing flow regimes at the termination of flood events are however indicative of rivers with a permanent, upper flow regime. A correlation with the Kingori Sandstone in Tanzania and equivalent sandstones was suggested (see East Africa). The red and purple mudstones with mud cracks are indicative of floodplain deposits. In the succeeding Burgersdorp Formation, climate reverted to semi-arid and warm conditions as indicated by playa lakes (Rubidge *et al.* 1995). Sedimentation ceased at the end of the Anisian and, by definition, this hiatus terminated the Pangea Megasequence (Fig. 5).

Of particular significance is the occurrence of the reptile *Lystrosaurus*. It is the only group to survive the Permian–Triassic (P/T) extinction and its distribution in Antarctica, India, China, Mongolia and European Russia demonstrates available migration paths with comparable environmental conditions in the Early Triassic. A study of the changes in species composition and variation of *Lystrosaurus* across the Permian–Triassic boundary was presented by Botha & Smith (2007). A specific feeding habit, preferring *Dicroidium* as sustenance, is thought to have been responsible for their survival of the P/T crisis.

East Africa

The Permian rift basins of eastern Africa extend from Namibia to East Africa. They include the Kalahari Basin in Namibia–Botswana, the Lukasashi–Luangwa Basin in Zambia (Utting 1978), the Metangula Basin straddling Mozambique and Tanzania, the system of the Ruhuhu, Ilima, Selous and Tanga basins in Tanzania (Wopfner & Kreuser 1986; Wopfner 2002), the Duruma Basin in SE Kenya, the Morondava and Majunga basins in Madagascar (Piqué *et al.* 1999) and the Mandera–Lugh Basin in the Ogaden (Bosellini 1989). They were all parts of the complex rift system of the Malagasy Trough (Wopfner 1994), a Late Palaeozoic chasm emanating from the nascent Tethys (Fig. 2). Depositional sequences representing proglacial, periglacial and deglaciation deposits are known from all these basins. Palynological data indicate that they resulted mainly from the Stephano–Asselian glacial maximum (SAGM). The Ruhuhu Basin in southern Tanzania is described here as representative of all these rift basins.

The glacial deposits of the Ruhuhu Basin (Idusi Formation) resulted from the SAGM, except for the so-called 'long section' exposed at the Namschweia Hills and encountered in some coal exploration bores. At Namschweia, a deep U-shaped trough was gauged into the Ubendian basement. In the deepest parts of the trough, the lower 80 m of the Idusi Formation consists of thick, current bedded sandstones and conglomerates, indicative of deposits of high-volume high-energy streams (Fig. 6). The fluvial deposits are overlain by a massive diamictite, followed by periglacial deposits and a black lutite unit representing deglaciation. The succession postdating the fluvial deposits corresponds to the sequence of the SAGM elsewhere in the Ruhuhu Basin (Fig. 7; Wopfner & Diekmann 1996). The basal sandstone/conglomerate unit at the 'long section' is here interpreted as a meltwater deposit resulting from the glacier retreat at the end of the first 'Karoo glacial maximum', corresponding to DS 1 of Visser (1997).

The glacial deposits are characterized not only by specific litho-facies but also by characteristic mineralogical compositions. The matrix of the pro- and periglacial deposits consists of finely ground up quartz and feldspar and little chlorite. This 'micro-greywacke' composition is retained even in the finest lutite fraction (Diekmann & Wopfner 1996). This contrasts with the Early Sakmarian deglaciation facies of the Lilangu Member of the Idusi Formation (Fig. 7). The black organic-rich lutites consist entirely of kaolinite, except for the quartziferous silt fraction and the carbonate concretions. Even the matrix of the interbedded turbidites, deposited in the actively subsiding parts of the rift, is composed of kaolinite (Diekmann & Wopfner 1996). This deglaciation unit (which corresponds to DS 2 of Visser 1997) contains a microfloral assemblage of the *Granulatisporites confluens* Oppel-zone which is of Early Sakmarian age (Foster in Wopfner & Kreuser 1986; Foster & Waterhouse 1988; Wopfner 1999).

The succeeding coal-bearing Mchuchuma Formation (and its time-equivalents; Fig. 7) commenced with the Mpera Sandstone Member, a stacked sequence of current bedded, kaolinitic and pyritic, coarse- to medium-grained sandstones. These are typical deposits of 'blackwater' rivers, the acid waters of which contained high concentrations of organic acids. They represent the continuation of the humid phase which commenced with deglaciation. The gradients for these high-energy fluvial systems resulted from crustal rebound following the removal of the ice load from the glacial centres, modified locally by rift movements.

The overlaying coal measures of the Mchuchuma Formation were formed primarily as end-members of stacked point bar cyclothems, but near the top of the sequence back swamp environments became more prominent (Semkiwa 1992). A new high-energy fluvial event resulting in the current bedded Scarp Sandstone Member of the Mbuyura Formation introduced the next depositional cycle (Fig. 7). Kaaya (1992) demonstrated that the sandstone, which interfingers laterally with green

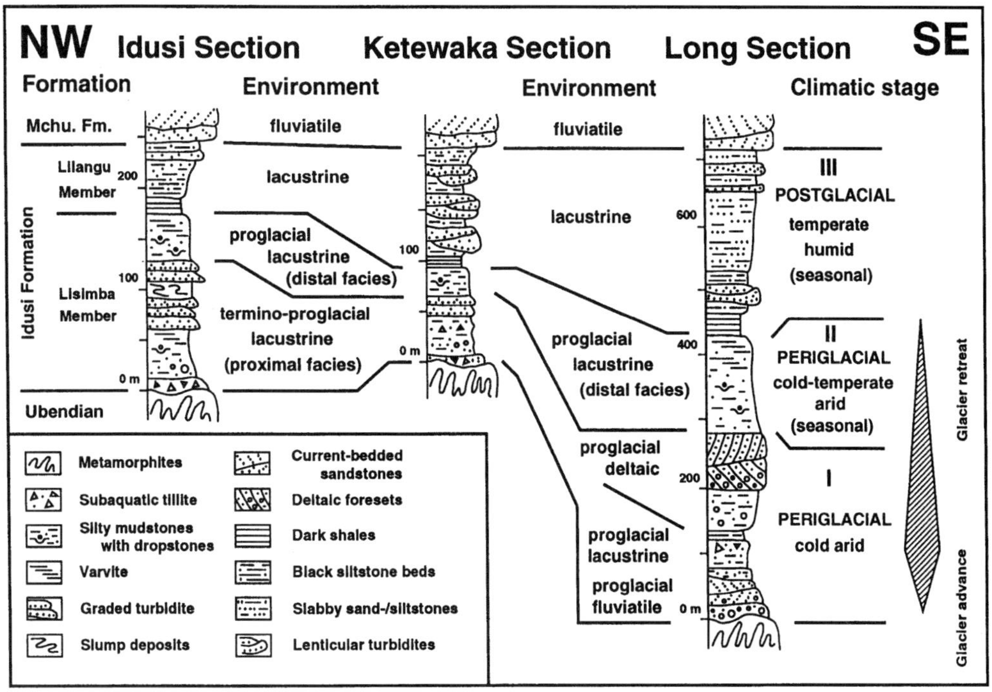

Fig. 6. Section across the Ruhuhu Basin in southern Tanzania, showing difference between the 'normal' glacigene sequence and that of the Long Section (after Wopfner & Diekmann 1996).

floodplain deposits, was deposited by braided rivers flowing off actively rising rift shoulders. An increase in precipitation was probably also involved. The following red beds of the Mbuyura Formation herald new climatic conditions. Conglomeratic units near the base may be composed of up to 60% of subangular feldspars derived from the nearby migmatites of the Palaeo-Proterozoic Ubendian basement, indicating physical disintegration of the basement rocks in an arid and hot environment prior to being transported into the basin by braided wadi and fan systems. The coal measures of the Mhukuru Formation which are dominant in the eastern rift basins (e.g. Metangulu, Mhukuru; Wopfner 2002; Catuneanu *et al.* 2005) demonstrate an increase of precipitation and a temporary lowering of temperature. In the Middle and early Late Permian hot and humid conditions created large, slightly alkaline inland lakes (Yemane & Kelts 1990; Kreuser 1994; Kaaya & Kreuser 1995; see also Utting 1978) in which the dolomitic limestones and green siltstones of the Ruhuhu Formation (and equivalents) were laid down. The end of the Permian saw a reduction of precipitation but temperatures remained high. The playa deposits of the Usili Formation with silicified wood and tetrapod remnants including *Dicynodon lacerticeps* resulted from this time (Kaaya 1992).

Deposition was interrupted around the Permian–Triassic boundary (Fig. 7). After the hiatus the stacked sequence of the Kingori Sandstone Member of the Manda Beds was laid down. Planar and asymptotic current bedding, dewatering and friction drag structures shown in Figure 8 demonstrate that the super-mature quartz sandstone originated in high-volume bedload-dominated river systems. This interpretation is supported by plots of current directions falling within narrow angled sectors (Markwort 1991). The high discharge volumes combined with low-pH waters (evidenced by homoaxial quartz overgrowth on clastic grains) are indicative of a warm, pluvial climate. The lithofacies at the top of the Kingori Sandstone consists of decimetre thick, parallel bedded sandstones with densely spaced vertical burrow tubes, demonstrating the existence of large bonds of quiet water. Comparable Early Triassic sandstone formations are present in all rift basins of southeastern Africa (Catuneanu *et al.* 2005).

Later in the Early Triassic a short evaporative event introduced the hot and semi-humid conditions which controlled the deposition of repeated fining-upwards sequences of the Olenekian–Anisian Lifua Member of the Manda Beds. Early diagenetic Laumontite concretions in some of the sandstones

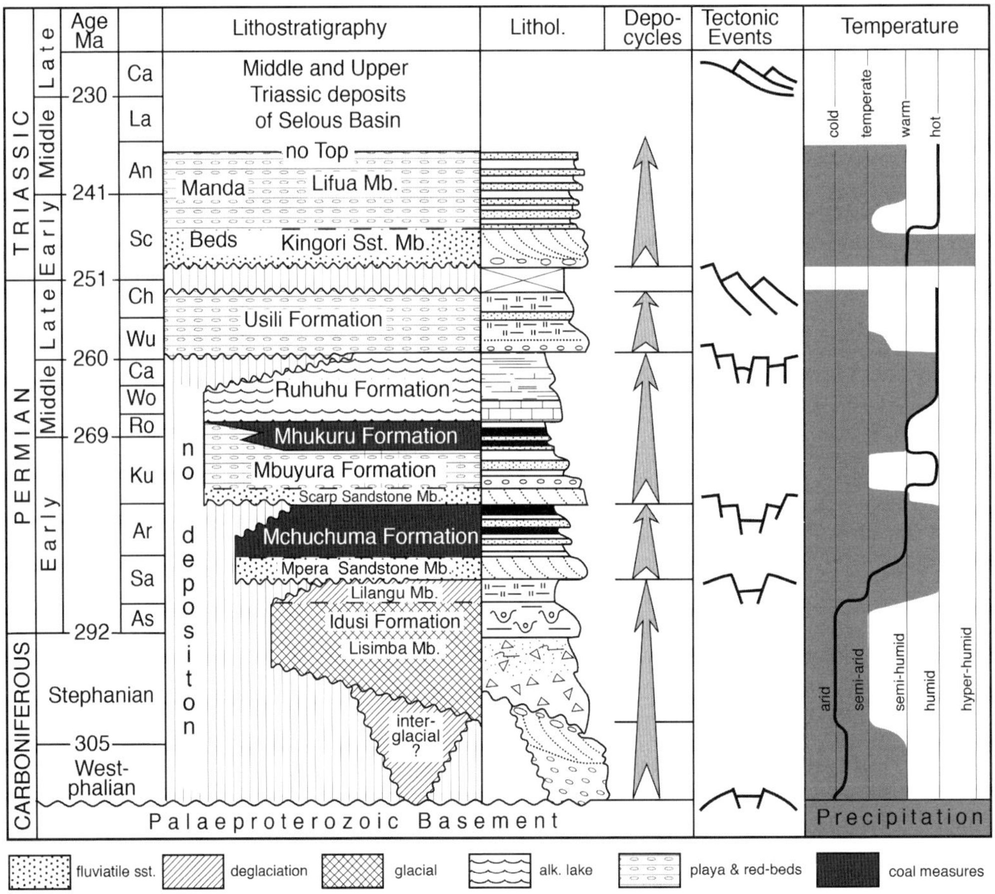

Fig. 7. Lithostratigraphy and depositional environment of the Songea Group (PMS) in southern Tanzania, modified from Wopfner & Jin (2009). The rudites and arenites at the base of the succession are here interpreted as deposits resulting from the Westphalian interglacial (DS 1 of Visser 1997). Absolute timescale after Wardlaw (1999).

(Wopfner *et al.* 1991) confirm such a climatic interpretation.

North of the Tanga–Duruma Basin which straddles the Tanzanian–Kenyan border, no exposures of Karoo deposits are known. Bosellini (1989) identified diamictites in deep boreholes in the Mandera–Lugh Basin in Somalia, and these apparently extend into the Ogaden of southern Ethiopia. The age of glacigene deposits in northern Ethiopia is uncertain (Hambrey 1981).

It should be noted that no coal deposits of any significance are known north of Tanzania.

Madagascar

Karoo deposits are exposed in the Morondava and Majunga basins along the western coast of the island. The Karoo sequences of western Madagascar mirror the sedimentary sequences of the Permian rift basins of eastern Africa (Piqué *et al.* 1999; see Fig. 2). Its glacial deposits and the succeeding deglaciation mudstones of the Early Sakmarian ‘Schistes noirs’ and the following coal beds are very similar to their counterparts on the western side of the Malagasy Trough. This also pertains to the remaining Permo-Triassic successions, except for a marine incursion in the Late Permian in the Morondava Basin and a transgression from the Tethys into the Majunga Basin in the Wuchapingian (Catuneanu *et al.* 2005).

Congo Basin

Deposits equivalent to the Karoo Supergroup of the Congo Basin are well exposed along the eastern basin margin, where they have been elevated along

Fig. 8. Planar and low-angle asymptotic current beds are typical for the extremely well-sorted and super mature Kingori Sandstone Member of the Manda Beds. Together with dewatering and friction drag structures displayed on top of the lower unit of picture, they are symptomatic of a high-volume bed-load-dominated hydraulic system.

the Cainozoic Tanganyika Rift. Boutakoff (1948) identified tillites, fluvio-glacial and post-glacial deposits within glacial valleys leading from the glaciation centre in the east down to the deeper parts of the basin in the west, where they are covered by younger rocks. Several exploration wells transected thick sequences of Karoo equivalents, their cumulative thickness exceeding 2000 m. These include the Lower Lukuga Formation (Dwyka equivalent) comprising slurried diamictites, rain-out tills, turbidites and laminites studded with dropstones (Hübner 1965; Catuneanu *et al.* 2005), the coal and sandstone sequence of the Upper Lukuga Formation (Ecca equivalent) and the red sandstones and mudstones of the Haute Luecki Formation (Kadima *et al.* 2011). In the Walikali region the glacial deposition was terminated by black shale (Boutakoff 1948), equivalent to the Lilangu Member of the Idusi Formation in Tanzania.

The Triassic comprises a comparatively thin Anisian succession of red beds, marls and bituminous mudstones followed by Late Triassic fluviatile sandstones and bituminous shale, apparently separated from the Early Triassic strata by a hiatus (Boutakoff 1948).

Gabon and Sudan

Glacial deposition extended NW towards the Gulf of Guinea, where tillites and laminites with dropstones have been described from the base of the coastal basin by Micholet *et al.* (1970). This is the northernmost occurrence of glacial deposits in western Africa. The glacigene succession is overlain by black bituminous shale with dolomite intercalations of Early Permian age, comparable to the black shale of Walikale in the Congo and which indicate the termination of glaciation. The deglaciation sequence is followed by multicoloured argillites, cherty shale, anhydrite and stromatolites, which may be attributed to a warm to hot, evaporitic climate. The Permian of Gabon represents a transition from a typical Gondwana facies to a Euramerican succession, reflected by the palynoflora. The glacial and deglaciation deposits contain a typical Gondwana microflora, whereas specific European forms such as *Lueckisporites* and *Klausipollenites* characterize the Upper Permian succession (Jardine 1974). This facies is similar to the Permian succession of the Maranhão Basin in northern Brasil (Mesner & Wooldridge 1964; Faria 1980).

The northernmost Late Palaeozoic deposits with glacial affinity are known from Abd el Malik and other localities in the central Sudan (Klitzsch & Wycisk 1987).

Arabian Peninsula

Tillites and other proglacial deposits resting on polished and striated glacial pavements are known from South Yemen (Kruck & Thiele 1983) and from the South Oman Basin (Braakman *et al.* 1982).

Oman

Lodgement tillites resting on glaciated pavement in South Oman contain Westphalian palynomorphs (Broutin & Berthelin 2007), but the main glacial

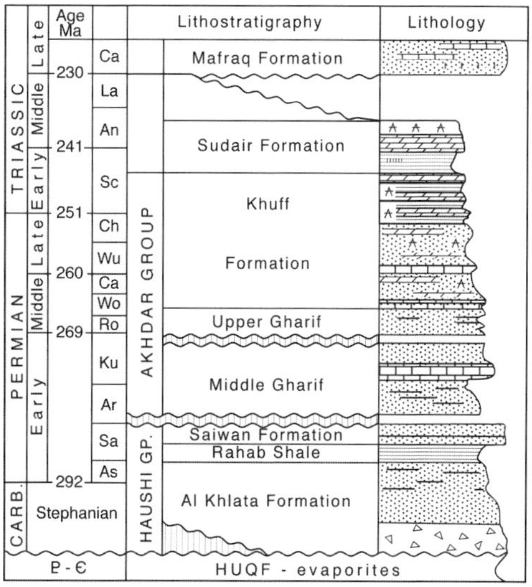

Fig. 9. Generalized stratigraphic column of Pangaea Megasequence of southern Arabian Peninsula (after Wopfner & Jin 2009).

succession of the Al Khlata Formation was deposited during the Stephanian–Asselian glaciation event (Besems & Schuurman 1987; Levell *et al.* 1988; Love 1994). The termination of the glaciation is marked by the organic-rich black lutite of the Rhahab Shale (Fig. 9). As in Tanzania, this distinguished deglaciation facies has yielded a palynomorph assemblage of the *Granulatisporites confluens* Oppel Zone of Early Sakmarian age. Wopfner (1999) and Wopfner & Jin (2009) have demonstrated that the black deglaciation unit can be correlated with time-equivalent black shale units all along the Tethyan margin of Gondwana from East Africa and Oman to SW Yunnan and Australia, where it is associated with marine fossils (Foster & Waterhouse 1988). The reliability of this correlation has been discussed by Stephenson (2008) and biological aspects of deglaciation by Stephenson *et al.* (2007). The Rahab Shale is transgressed by the marine Saiwan Formation, carrying the typical faunal assemblage of the Late Sakmarian eustatic transgressive facies (Angiolini *et al.* 1997, 2001; Archbold 2001). In Yemen the deglaciation facies is apparently represented by the Lower Permian Akbra Shale (Besems & Schuurman 1987).

There is no evidence for coal formation in Yemen or in Oman. There, the post-glacial eustatic event is succeeded by the marine Gharif Formation which in turn is overlain by the Middle–Late Permian evaporitic deposits of the Khuff Formation (Fig. 9). Deposits of the Khuff Group are known in the subsurface of the entire central and eastern Arab Peninsula, where they provide the major reservoirs for natural gas accumulations (Maurer *et al.* 2009). According to these authors a 'middle anhydrite unit' is identified, demonstrating a hot and arid interval in the late Middle Permian.

Saudi Arabia

The situation in Saudi Arabia suffers from a diverging stratigraphic terminology. Sandstones of Late Palaeozoic age resting on a lateritized post-Hercynian unconformity surface have been identified originally as Unayzah Formation by Al-Laboun (1987). The unit was regarded as part of the sander apron of Permo-Carboniferous age. Sander aprons are sand spreads peripheral to glacier fronts, deposited by meltwater streams issuing from the glaciers. Based on drilling information the formation was subsequently divided into several sub-units, restricting the name Unayzah to fluvial and aeolian sandstones and playa deposits of Roadian–Wuchiapingian age. The basal Late Carboniferous unit, comprising braided stream and frigid aeolian sandstones, was termed Haradh Formation; the equivalent to the Early Permian deglaciation unit was identified as Jawb Formation. The latter is separated from the overlaying Unayzah Formation by an (possibly Artinskian) unconformity (Senalp & Al-Duaiji 2001). More recently, Melvin *et al.* (2010) reverted to a numerical subdivision of the term Unayzah Formation, applying the name Unayzah for the whole Early–Middle Permian hydrocarbon-bearing succession. For further discussion in this review, the terminology of Senalp & Al-Duaiji (2001) is used. Whether the basal rudaceous units within the sander apron are glacial deposits is still being debated (Melvin *et al.* 2010). The fact that the base of the unconformity surface consists of a lateritic palaeosol precludes a former existence of scouring ice.

The Permian succession is concluded by the dolomite–anhydrite beds of the Khuff Formation. This shallow platform sequence is present in the subsurface of most of the Saudi Arabia Basin (Alsharan 2006). Its equivalent is exposed in NE Oman (Maurer *et al.* 2009) and it extends across the Persian Gulf into the subsurface of the coastal Fars Province of Iran. As in Arabia, a prominent anhydrite member attests to the Middle Permian evaporate event (Schlumberger 1976). Permo-Carboniferous fluvial conglomeratic sandstones at the base of this sequence are part of the sander apron, deposited by meltwater rivers. The Early Permian deposits of the south Oman Basin belong to the Gondwana phytogeographic province, whereas time-equivalent strata of north-central Saudi Arabia comprise typical Euramerican forms (Besems & Schuurman 1987; Berthelin *et al.* 2006). The boundary between the two provinces coincides roughly with the northern limit of glaciation.

Africa north of the glacial front

Accepting the northernmost exposures of glacigene sediments as control points, Wopfner (1991) and Wopfner & Casshyap (1997) identified the northern limit of glaciation as a north-convex line extending from the Gulf of Guinea to the Persian Gulf (Figs 1 & 3). The line denotes the northern limit of known glacial deposits.

Early Permian sediments laid down north of that line were influenced by their proximity to the glacier front, but they cannot be viewed as glacial deposits. Such non-glacial deposits of Early Permian age north of the glacial front are known from the Eulemmeden Basin in Nigeria and Niger, from the Sudan and from the Arabian Peninsula (see previous section). They are generally fluvial sandstones containing Late Carboniferous–Early Permian microfloras and, where known, show current structures indicative of northerly gradients. These sand-dominated sequences are interpreted as sander-type braid plain deposits in the vicinity of glaciers, developing into large, meandering fluvial systems further north.

They are akin to the large fluvial apron of Antarctica–India described by Tewari & Veevers (1993). In Niger such lower Permian sands host the Arlit and Akouta uranium deposits, indicating the presence of sufficient organic matter to facilitate accumulation and fixation of the metal. The microflora contained within these sands is characterized by an association of Euramerican taxa mixed with Gondwana forms, the composition of which stands between that of Early Permian strata of Gabon and that of SW Spain (Broutin *et al.* 1990). These authors conclude that in Afro-Arabia, floristic exchange between Gondwana and Laurasia was effective by Early Permian times. The floral differences were controlled by the climate. Further north, sedimentary deposits of the Permo-Carboniferous periods change to coaly deposits succeeded by red-bed-dominated sequences in the Permian (see following section).

Morocco

Late Carboniferous and Permian deposits generally rest unconformably on folded complexes deformed by the Variscan orogeny. Post-Variscan extensional movements commenced in the Westphalian and intensified in the Stephanian. The main unconformity is observed at the beginning of the Autunian (Hoepffner *et al.* 2005). In many instances, the receptacles for the post-orogenic deposits were extensional structures such as graben and pull-apart basins (Biron 1982; Fig. 10). The extensional regime was related to the opening of the western Tethys.

Of the diverse Stephano-Autunian and Permo-Triassic basins of Morocco, only a few typical successions can be discussed in this review. Feys & Greber (1963) presented the first comprehensive description of the Stephano-Autunian of the Souss region, south of the western High Atlas Mountains. They recognized a lower grey–black 'productive' sequence consisting of dark-coloured plant-bearing carbonaceous mudstones and siltstones with coal, generally succeeding grey fluvial sandstones. The flora is of typical Euramerican character, containing a rich assemblage of *Pecopteridae* and *Sphenophyllidae*. The 'productive' sequence rests on up to 500 m of pebble conglomerate containing some plant debris, including *Cordaites*. The 'productive' sequence and the basal conglomerate are considered to be of Middle Stephanian age. The overlying 'sterile' sequence consists of greenish-coloured silty sandstones with sparse plant fragments. It was assigned to the Upper Stephanian by Feys & Greber (1963). Red, purple and green-coloured sandstones and varicoloured mudstones follow on top of the 'sterile' sequence. Red sandstones changing laterally into black mudstones and green dolomitic sandstones conclude the succession. This is a typical Autunian lithofacies association which records increasing aridification from the bottom to the top of the section (Essamoud 1989). The age assignment is supported by an abundance of Walchia and *Estheria tenella* (Feys & Greber 1963). More recently, Broutin *et al.* (1989) reported on the discovery of a diverse macrofloral assemblage from the upper sequence of one of these basins (Ida ou Ziki), including well-preserved specimen of Ginkgophyte leafs, which firmly establishes an Early Permian age for the upper part of the succession. The Stephanian age of the lower units was confirmed by Hmich *et al.* (2006).

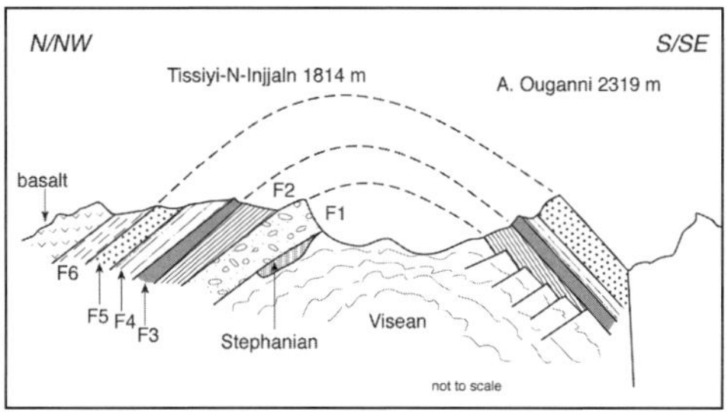

Fig 10. Evidence for rifting during early deposition of the PMS displayed in the High Atlas of Morocco (redrawn from Biron 1982).

Further exposures of Permo-Triassic red-bed sequences exist on the High Atlas south of Marakhesh (Ourika Basin) and in rift basins of Mesetian Morocco. The latter have been summarized by El Wartiti *et al.* (1990) who characterized the structural setting as follows: 'the Permian is a transitional period between compressional Hercynian phases and Mesozoic overstrechings'. This conforms to the general extensional conditions experienced along the Tethyan margin of Gondwana (Catuneanu *et al.* 2005; Wopfner & Jin 2009). These movements may have been influenced by the change of an anticlockwise rotation of Gondwana in the Carboniferous to a clockwise rotation in the Permian (Visser & Praekelt 1998) but, as the timing of the changeover is not clear (Klootwijk 2009), this model should be treated with caution.

Recent discoveries of tetrapod remnants and foot imprints as well as new floral associations in some of the Moroccan basins were presented by Hmich *et al.* (2006). These pertain among others to the Khenifra Basin, where Early Permian red–purple-coloured proximal fan deposits resting directly on Variscan basement intertongue laterally with stacked fluvial sandstones. The section is capped by thick rhyolitic ignimbrites which issued from deep-seated strike-slip fractures (El Wartiti *et al.* 1990) and accompanied the development of Early Permian post-Variscan rift basins from Morocco to Egypt (Guiraud *et al.* 2005). A striking example of

Fig. 11. Railway cut *c.* 110 km NNE of Marrakech exposes coarse clastics of proximal fan and mudflow provenance interbedded with red, silty playa deposits of the (possibly) Late Permian Mechra Ben Abbou Basin. The coarse material was dumped directly from the Devonian rim into the rapidly subsiding pull-apart basin, as evidenced by the common occurrence of fragments of fossiliferous Devonian reef limestone.

a pull-apart basin is the Mechra Ben Abbou Basin NW of Marrakech. The basin was incised into a terrain of Devonian reef limestone, eroded material of which issued directly into the rapidly subsiding trough. Proximal fan and wadi rudites, composed largely of Devonian components from the basin rim, alternate with red mudflat and playa deposits (Fig. 11). Some andesitic volcanics are interleafed near the top of the sequence. Based on lithological correlations, the deposits of the Mechra Ben Abbou Basin are placed in the Upper Permian.

A classic Permo-Triassic profile is exposed on the northern slope of the High Atlas along the road to Oukaimeden. The sequence exhibits the whole breadth of red-bed lithotypes from coarse wadi, fan and sheet flood deposits to fine-grained playa and mudflat deposits with mud cracks and other signs of aridity.

Libya and Tunisia

In the Late Carboniferous deposition in Libya was characterized by restricted marine and evaporitic conditions. Varicoloured siltstones with intercalations of gypsum and anhydrite of the Tiguentourin Formation were laid down (Tawadros 2001).

Permian sedimentary rocks are known only from the subsurface and their distribution is quite uneven. They are of terrestrial origin, comprising red- and white-coloured sandstones with interbeds of varicoloured shales with thin beds of lignite. Palynological studies show them to be of Early–early Middle Permian age (El Arnauti & Shelmani 1985). Neither Late Permian nor Early Triassic sediments appear to be present. Some Middle–Late Triassic terrestrial deposits were encountered in the south and limestones in the NE. In Tunisia, Late Permian marine sequences with biohermal complexes (Newell *et al.* 1976) change upwards to non-marine red beds, attaining a thickness of about 800 m. They witness the westernmost extension of the nascent Tethys into Pangea at that time. The Early Triassic consists mainly of sandstones and dolomites; evaporates were deposited in the Carnian.

Levante

This area contains the northernmost deposits which are reliably identified as time equivalents of the Karroo succession and are still positioned on the Afro-Arabian craton. They provide firm control points for the reconstruction of palaeoclimatic gradients.

Gulf of Suez and Sinai

Late Palaeozoic and Early Triassic deposits are known from both the Sinai Peninsula and from boreholes drilled in the course of oil exploration in the Gulf of Suez (Alsharhan & Salah 1997). According to these authors, Late Carboniferous sedimentary rocks rest with partial unconformity on fossiliferous marls and dolomites of Early Carboniferous age.

The Late Carboniferous succession commences with the Abu Durba Formation, a black shale unit with *Neospirifer* sp., *Spirifer* sp., *Productus* sp., and *Bellerophon* sp. The formation was deposited in a warm shelf sea whereas the following Rod el Hamal Formation grades from a marine sandstone, shale and limestone sequence up to fluvio-marine sandstones. The succession straddles the Carboniferous–Permian boundary and, according to Alsharhan & Salah (1997), extends to the top of the Permian. The fluvio-marine Qiseb Formation rests unconformably on the Rod el Hamal Formation and is placed in the Triassic. This contrasts with the stratigraphic position assigned to the Qiseib Formation by Hermina *et al.* (1989). Based on Triassic bivalves and gastropods found at the top of the formation and on Permian microfloral assemblages identified from its lower part, they place the Qiseib Formation in the Permo-Triassic. This view was shared by Kora (1998), who presented a correlation between the terminologies used for surface outcrops and the nomenclature applied to subsurface units. Kora (1998) recognized that NE Afro-Arabia was accessible to Tethyan waters since the Late Carboniferous, a fact frequently overlooked in Pangea reconstructions.

Negev

Middle and Upper Triasic rocks are well exposed in the breached anticline of Makhtech Ramon (Picard & Flexer 1974), but knowledge of older sedimentary successions is derived primarily from borehole data. A comprehensive description of the Permo-Triassic strata encountered in exploration wells drilled in the Negev was presented by Eshet (1990*a*). In the subsurface of the Negev, Late Palaeozoic sediments rest directly on Precambrian basement. The basal unit is the Sa'ad Formation, consisting dominantly of anhydrite cemented quartz-arenite and occasional interbeds of plant-bearing carbonaceous shale with stringers of dolomites and limestone. Earlier investigators (Glickson 1964) placed this formation in the Late Carboniferous. Eshet (1990*a*) and Eshet & Cousminer (1986) identified a palyno-assemblage which they placed within the *Potoniesporites novicus* Zone and which they considered to indicate an Autunian age. Judging from the stratigraphic column and from the fact that Eshet (1990*a*) subdivided the Permian into two stages only (i.e. Autunian and Thuringian), it may be concluded that the formation ranges from the Asselian to about the beginning of the Kungurian.

The following Arqov Formation comprises sandstones, shales and some carbonates of a nearshore marine environment. According to Eshet (1990*a*), the formation extends from the Late Autunian to the Late Thuringian, about Kungurian–Changhsingian. Sedimentation from the Permian into the Triassic was apparently continuous (Druckmann 1974), the P/T-boundary being situated within the Yamin Formation (a succession of carbonates with some interbedded sandstones). A typical feature of the P/T boundary horizon is the absence of *in situ* spores and pollen, the microflora consisting only of fungal spores (Eshet 1990*b*) (see discussion on that subject in description of Etsch Basin). The Triassic depositional sequence extends into the Carnian, comprising (in ascending order): Zafir, Ra'af, Gevanim, Saharonim and Mohilla formations. The succession up to the middle of the Ladinian (i.e. the second third of the Saharonim Formation) consists of shallow marine deposits, dominantly carbonates with some oolitic limestones and siliciclastics. The first signs of evaporitic conditions indicating the beginning of regression, such as gypsum interbeds and stromatolites, appear in the middle of the Saharonim Formation (Picard & Flexer 1974).

Continuation of this trend leads to the anhydrite, gypsum and dolomite sequence of the Carnian Mohilla Formation (Druckmann 1974; Eshet 1990*a*). The Mohilla Formation concludes the Pangea Megasequence in the Negev. It is unconformably overlain by Rhaetian–Jurassic deposits.

The climate in the Early Permian appeared to alternate between hot mildly evaporitic conditions (indicated by the anhydrite cement of the sandstones) and seasonally humid (suggested by the plant-bearing carbonaceous shale). This is supported by a pronounced alternation of characteristic palynological assemblages observed by Eshet (1990*b*). An abundance of fungal spores (or perhaps algae) in the uppermost Permian and lowermost Triassic sections may be ascribed to a similar pluvial event as that observed in Karoo successions further south. The oolithic limestones in the Early Anisian are symptomatic of a warm to hot climate. Based on statistical analyses of microfloral assemblages, Eshet (1990*a*) suggested a humid climate for the Anisian but with some 'dry niches for xerophytic elements to grow'. From the middle of the Ladinian the climate became increasingly arid, culminating in the hot conditions in which evaporates of the Mohilla Formation were laid down.

Southern Alps

The Southern Alps are separated from the main body of the Alps by the Periadriatic Line, edging the Alps from about Lake Como to northern Slowenia, a distance exceeding 600 km. Deposits corresponding to the Pangea Megasequence (PMS) are contained in three essentially rift-controlled basin complexes. From west to east they comprise, in their wider sense, the Collio, the Etsch (Adige) and the Carnia basins (Wopfner 1984; Cassinis 1986; Venturini 1990). Their basin fill represents the initiation of the Alpine depositional phase and witnesses the incursion of the Tethys from east to west. The Carnia Basin was already inundated by the sea in the Late Carboniferous, whereas transgression of the Etsch Basin did not take place until the Late Permian (Fig. 12). The Collio Basin was dominated by fluvio-lacustrine environments up to the end of the Permian (Cassinis & Ronchi 2001), except for a brief marine incursion in the Lower Permian (Sciunnach 2001).

Collio Basin

Deposition in the Collio Basin (including Orobic and Tione basins) commenced with sporadic basal conglomerates of Late Carboniferous age resting unconformably on Variscan basement (Cassinis & Ronchi 2001). They were succeeded by the Early Permian Collio Formation (Cassinis 1966*a*, *b*; Ori *et al.* 1988; Cassinis & Ronchi 2001). This unit comprises multicoloured siliciclastics of various grain size, volcanoclastics and intercalated volcanics laid down in an intracratonic rift environment under warm, seasonally sub-arid conditions. A short marine incursion in the upper Collio Formation, evidenced by benthic foraminifera (Sciunnach 2001), demonstrates the accessibility of down-faulted parts of the rift system to the Tethys in the Artinskian.

After a period of further rift movements, the Dosso dei Galli Conglomerate and the succeeding,

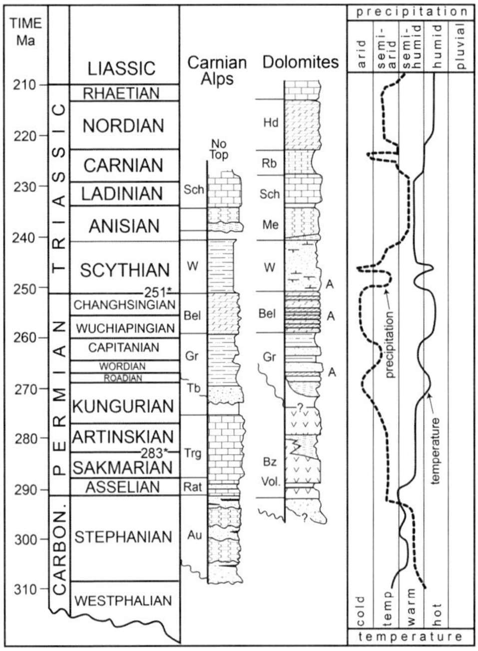

Fig. 12. Lithostratigraphic colomns of the Late Palaeozoic–Early Triassic rock sequences (PMS) of the Dolomites (Etsch Basin) and the Carnian Alps (Carnia Basin) of the Southern Alps. Generalized temperature and precipitation curves are shown in the right column. Abbreviations (in ascending order): Au: Auernig Beds; Rat: Rattendorf Beds; Bz Vol: Bozen Volcanics; Trg: Trogkofel Limestone; Tb: Tarvis Breccie; Gr: Groeden Sandstone; Bel: Bellerophon Formation; W: Werfen Formation; Me: Sarl (Mendel) Dolomite with Richthofen conglomerate at base; Sch: Schlern Dolomite; Rb: Raibl Formation.

rhyolitic Auccia Volcanics were deposited. These are followed by the red beds of the Monte Mignolo Sandstone, a thick red-bed sequence of stacked siltstone/sandstone couplets of braided to meandering stream provenance (Ori *et al.* 1988). The formation, also referred to as Verrucano Lombardo, is a facies equivalent of the Groeden Sandstone but which extends to the P/T boundary. Except for the sporadic marine ingression in the Early Permian mentioned above, the sea did not enter this rift system until the beginning of the Triassic when the partially evaporitic Servino Formation was laid down (Cassinis 1966*b*).

Etsch Basin

Within the central Etsch Basin, deposition commenced with basal conglomerates of Late Carboniferous age. These were succeeded by more than 1000 m of rhyolitic to dacitic ignimbrites, tuffs and lavas of the Bozen and Lagorai volcanic suites (Wopfner 1984) also referred to as Athesia Volcanics (Brandner & Keim 2011). Their radiometric ages range from 300 to 280 Ma (Muttoni *et al.* 2003; Visonà *et al.* 2007). Intercalations of alluvial fan deposits have yielded a macroflora comprising *cordaites*, conifers and voltzia-type branches indicating a lower Artinskian age and a warm, semi-humid climate (Fritz & Krainer 2006). Rift volcanism ceased at about the middle of the Kungurian. It was followed by a period of deep weathering which resulted in thick grus aprons containing calcrete palaeosols and an abundance of leucoxene, indicative of a hot and semi-arid climate (Wopfner & Farrokh 1986).

The overlaying Groeden Sandstone comprises purple, gypsiferous sabkha deposits, succeeded by a stacked sequence of purple floodplain argillites and grey sandstones of sheet flood or point bar origin of ephemeral streams (Masari *et al.* 1994). The ephemeral nature of these streams is demonstrated by green reduction halos extending from the sandstone base 50 cm or more into the purple argillite underneath. Syngenetic concentrations of U, Cu and Pb complete the evidence for a semi-arid warm to hot climate.

The evaporite sequence of the Bellerophon Formation heralds the encroachment of the Tethys from the east. The formation which comprises dolomites, thick gypsum beds and algal mats was laid down in a hot evaporative climate, similar to that of the present-day Persian Gulf (Bosseline & Hardie 1973). This depositional style extended to the end of the Permian period (Fig. 12).

Sedimentation was continuous across the Permian–Triassic boundary, but depositional environment experienced a dramatic change. The oolitic limestone of the Tesero Horizon at the very base of the Triassic Werfen Formation which follows on top of the dolomite/gypsum sequence of the Bellerophon Formation constitutes a fundamental change in salinity and hydrodynamic energy. This is combined with a high incidence of ‘fungal remains’ (Broglia-Loriga & Cassinis 1992) or green algae (C.B. Foster, pers. comm., 2001) and a marked negative shift of δ^{13} C and δ^{18} O at the base of the Tesero Horizon (Magaritz *et al.* 1988); this not only reflects a change in marine topography but also a substantial climatic change.

The overlying hard and resistant marls of the Mazzin Member of the Werfen Formation form a marked contrast to the underlying Bellerophon Formation (Fig. 13) and demonstrate a reduction in salinity, perhaps due to an increase in precipitation. It still has to be resolved if this was also the cause for the high incidence of fungal spores or possibly green algae about the P/T boundary.

Fig. 13. Permian–Triassic rocks exposed at the head of the famous gorge of Bletterbach in the western Dolomites near Bozen (South Tyrol). Gypsiferous red beds in left foreground form the transition from the Groeden Sandstone to well-bedded dolomites of the Bellerophon Formation at the head of the gorge. The latter is succeeded by light-coloured Tesero Oolite at the base of the Mazzin Member of the Werfen Formation. The latter is identified by its blocky outcrop appearance. The uppermost red exposures are formed by the gypsiferous Andraz Member. Upper portions of the Werfen Formation were removed by the pre-Anisian erosion event. The white peak of the Weisshorn consists of the Anisian algal reef of the Contrin Formation, earlier termed Sarl or Mendel Dolomite.

There has been considerable discussion on whether the forms identified as 'fungal spores' and 'fungal remnants' (e.g. *Reduviasporinites*, *Chordecystia*, *Tympanicysta*) are indeed of fungal origin as suggested by Eshet (1990*b*) or Visscher *et al.* (1996). This interpretation was challenged by Krassilov *et al.* (1999) who favoured an algal origin. Based on biochemical markers, Foster *et al.* (2002) and Stephenson & Foster (2002) also argued for an algal origin. Recently Visscher *et al.* (2011) observed a 'close morphological similarity of the microfossils to modern resting structures of *Rhizoctonia*, a complex of soil borne filamentous fungi'. These authors suggest that pathogens of soil-borne fungi contributed to the destruction of Late Permian forest vegetation, thus enhancing end-Permian crisis scenarios such as global anoxia or breakdown of the ionosphere. Whatever the answer is, *Reduviasporinites* is symptomatic of stressed biological systems.

The period of increased humidity was of short duration however, as demonstrated by the gypsiferous red beds of the Andraz Member of the Werfen Formation (Figs 12 & 13). The remaining succession of the Werfen Formation indicates again slight lowering of salinity and deposition in a warm shallow subtidal environment. An angular or erosional unconformity separates the succeeding Late Anisian Richthofen Conglomerate from the Werfen. The following reef formations of the Anisian Contrin Formation (Mendel; Sarl Dolomite) and the Ladinian Schlern Dolomites were built mainly by calcareous algae and reflect a warm tropical climate. The Carnian Raibl Formation was initiated by regression and by increased aridity (Fig. 12).

Carnia Basin

East of the Etsch Basin the Carnia Basin, including the Cadore Basin of Venturini (1990), extends from NE Italy into Slowenia. The eastern part of the basin was covered by a shallow sea in Stephanian times, and the marginal marine Auernig Beds with interbedded coaly beds were formed in a warm and semihumid climate. With the onset of the Permian, temperature and aridity increased as witnessed by the abundance of fusulinidae in the Trogkofel Limestone (Venturini 1990; Fig. 9). Rift movements at the beginning of the Kungurian terminated deposition and uplifted parts of the Trogkofel Limestone were eroded and transported into adjoining basins to form the rudite-fans of the Late Kungurian Tarvis Breccia. Climate was arid and hot as evidenced by marginal playas with algal nodules and small stromatolites in red beds (Kober 1984). The succeeding Middle Permian Gröden Sandstone consists largely of wadi deposits, whereas the Late Permian Bellerophon Formation is here composed of platform dolomites. The Triassic succession is comparable to the sequence in the Dolomites further west (Fig. 12).

Climate history

In the Late Palaeozoic, Africa and Arabia extended from areas proximal to the South Pole into lower latitudes so that the regions near the Tethyan margin remained free of ice cover during the Gondwana glaciation (Fig. 1). Generally the palaeo-south-pole for Gondwana in the Late Carboniferous/Early Permian (310 Ma) is placed on or near central Antarctica in the vicinity of the present South Pole

(Powell & Li 1994; Scotese & Langford 1995; Muttoni *et al.* 2003; Klootwijk 2009). Lopez-Gamundi (1997) placed the Early Permian South Pole in southeastern South Africa, which disagrees slightly with the views of the other authors. The pole position shown in Figure 1 is close to that suggested by the authors cited above.

As mentioned before, in Permo-Triassic times Afro-Arabia was only affected by intracratonic rifting and its margins did not experience any spreading (Wopfner 1994, 2002; Catuneanu *et al.* 2005). The only exception was the Middle Permian oceanization in eastern Oman (Bechennec *et al.* 1990). Thus, the regions provide excellent prerequisites for the assessment of climatic gradients during the great glaciation and for the identification of climatic zones after the Gondwana glaciers had waned. The results may be juxtaposed to the palaeoclimatic model developed for Late Palaeozoic basins of central Europe by Roscher & Schneider (2006).

The question of whether the glacial Karoo deposits resulted from a single ice sheet or from various individual glaciation centres is still under debate. Visser (1989, 1997) presented overviews of the various concepts and favoured a subpolar ice sheet emanating from Antarctic highlands. The present author envisages a large ice sheet covering Antarctica and the Samfrau Mountains and individual but coalescing centres in Africa (Wopfner & Diekmann 1996), the eastern Cimmerian terranes (Wopfner & Jin 2009) and in central Australia (Wopfner 1970, 1980).

The evidence presented in this paper indicates that the glacier tongues extended to about 40° palaeolatitude (Fig. 1). Rivers emanating from the glacial tongues deposited aprons of sand onto a foreland which extended from the glacial front to progressively warmer regions. Permo-Carboniferous fluvial sandstones in Niger as well as braid plain and frigid aeolian deposits in subsurface Saudi Arabia attest to that scenario. The situation was similar to that described for Indian basins by Tewari & Veevers (1993). In northern Africa, coal formation was still widespread at the end of the Carboniferous, but a rapid change to warm and in places more arid conditions commenced in the Permian. Similar conditions are also applicable for the northern shores of Gondwana, now represented by the Southern Alps.

A similar trend as that observed in Africa is also observed between southern Arabia (Yemen and South Oman) and the Suez-Negev region. In the latter, red sandstones and shales encountered in deep drill holes, which are time equivalent to the glacial–post-glacial succession of the Haushi Group, are indicative of a warm arid to semi-arid climate in the Early Permian (Eshet 1990*a*). Coarse, fluvial sandstones encountered in drill holes at the northern part of the Persian Gulf are also considered to be of Early Permian age (Schlumberger 1976). This suggests that a narrow belt of temperate climate existed between a warm and semi-arid belt of the sub-equatorial regions and the glaciated region.

The termination of the glaciation in Congo, Gabon, East Africa, Madagascar and southern Arabia is clearly demonstrated by the Early Sakmarian black lutites carrying a *Granulatisporites confluens* microflora as described above (Foster & Waterhouse 1988; Wopfner & Kreuser 1986; Love 1994; Wopfner 1999; Stephenson 2008) and, where present, by the succeeding transgressive marine beds resulting from the eustatic rise in sea level (Angiolini *et al.* 1997, 2001; Archbold 2001; Stephenson *et al.* 2007). The transgressive Jawb Formation (Senalp & Duaiji 2001) of subsurface Saudi Arabia was also apparently related to the deglaciation event. In South Africa, Visser (1997) placed the second deglaciation event (DS 2) in the Early Sakmarian, coinciding with the deglaciation event identified by the *Granulatisporites confluens* zone elsewhere. Samples from post-glacial mudrocks analysed for their C and S contents by Visser (1989) indicate euxinic bottom conditions and 'a higher input of organic carbon than in modern marine sedimentation'. These results are in agreement with geochemical data obtained from the black deglaciation facies in Tanzania (Lilangu Member) by Diekmann & Wopfner (1996).

The Early Sakmarian eustatic rise of sea level in the Karoo Basin is witnessed by *Eurydesma*-bearing argillaceous beds succeeding the Dwyka glacials (Dickins 1985). The evidence for a younger post-Sakmarian glacial event based on the interpretation of rudites as meltwater deposits is tenuous, considering facts known from Antarctica. In southern Argentina and on the Falkland Islands the eustatic transgression of the Early Sakmarian is documented by the Piedra Azul and the Port Sussex formations, respectively (López-Gamundi & Rossello 1998).

The black carbonaceous shales with interbedded turbidites of the Early Permian Mackellar Formation in the Transantarctic Mountains witness the deglaciation event in Antarctica (Miller & Isbell 2010). The Mackellar Formation is succeeded by the Buckley Coal Measures. Deposited at a palaeolatitude of *c.* 80°, the Mackellar Formation is a direct correlate of the Lilangu Member and its equivalents, underscoring the statement by Wopfner (1999) that deglaciation was a synchronous event in almost all parts of Gondwana. Taylor *et al.* (1992) reported trees in growth position from late Early Permian strata of Antarctica and the existence of a forest in the Early Triassic (Taylor *et al.* 2000).

The omnipresence of kaolinite and pyrite within the deglaciation facies was caused by the high influx of organic material; it is not only observed in Gondwana basins of Africa (Diekmann & Wopfner 1996) and southern Arabia (Wopfner 1999), but also in SW China (Jin 1994; Wopfner & Jin 2009) and Australia (Wopfner 1970, 1980, 1999) as well as in modern fjords of New Zealand (Bruun *et al.* 1955; Brodie 1964). The New Zealand example indicates a densely vegetated landscape in a temperate and humid climate (temperate rain forests) which shed large amounts of organic matter into stratified water bodies, where the boundary between positive and negative Eh (redox potential) is just above the water/sediment interface. In the southern half of Africa, as for most other parts of central Gondwana, the deglaciation event and the short eustatic transgression were followed by stacked fluvial sandstones, apparently related to post-glacial, crustal rebound and high precipitation. The succeeding formation of coal measures indicates further amelioration of the climate and the widening of the temperate to warm and humid climate zone. In Africa the coal formation extended to *c.* 55° Gondwana-latitude, which also approximates the limit on other parts of Gondwana (Fig. 1). Coal formation ceased in South Africa when temperature increased in the Middle Permian, ushering in the deposition of the Beaufort Group.

In eastern Africa and Madagascar coal formation ceased temporarily in the Kungurian, when thick red beds developed in a hot and arid climate. Towards the end of the Kungurian coal formation was re-established, especially in eastern basins (Semkiwa 1992), whereas red bed deposition continued further west. North of the 'coal line' (Fig. 1), climate became rapidly dryer and warmer as evidenced by the anhydrite beds in Gabon, the 'Middle Gharif Beds' of Oman and by playa and sand dune deposits of the lower Khuff Formation in central Arabia. In northern Africa, red beds dominated Permian depositional sequences and warm to hot semi-humid conditions prevailed in the South Alpine Terrane.

Rising temperatures and increased precipitation at the end of the Wordian lead to large alkaline lakes extending from South Africa to East Africa (Utting 1978; Yemane & Kelts 1990; Kreuser 1994; Kaaya & Kreuser 1995). These relatively moist conditions lasted until the end of the Capitanian, when climate became more arid again and remained so until the end of the Permian (Fig. 7).

In Arabia this trend is evident by the change from the mildly evaporitic Lower Khuff to the anhydrite formation of the Middle Khuff, which extended over most of the Arabian basin (Alsharan 2006; Maurer *et al.* 2009). The Wordian–Capitanian period of increased precipitation may be reflected by the stacked playa/flood sequence of the Groeden Sandstone of the South Alpine Terrane (see Etsch Basin).

Generally there was a distinct gradient from the warm semi-arid climate dominating the environment of the Beaufort Group in southern Africa, to the more arid and apparently also hotter conditions on the Arabian Peninsula, northern Africa and the Levante, and finally to the evaporitic conditions in the South Alpine Terrane at the end of the Permian.

The generalized climatic profile in Figure 14 shows that from the termination of the glaciation to the end of the Permian a successive widening of the warm and then hot climatic belt took place. In terms of the Köppen-Geiger climatic subdivision (Peel *et al.* 2007) the polar and tundra zone reached to about 30° south Gondwana-latitude at the beginning of the Permian. With the beginning of deglaciation it was replaced by the temperate humid zone, quickly followed by the humid and temperate to seasonally warm zone of coal swamps. A hot desert climate replaced the coal swamp climate in the Kungurian before it reverted to the seasonally moist and hot climate responsible for the mildly alkaline lakes in southern and eastern Africa and evaporitic conditions further north. After the Wuchiapingian, the savannah and hot arid zones expanded and took possession of higher latitudes. Apparently, there was no equivalent of a tropical rainforest zone.

The recently dated deepwater succession across the Permian–Triassic boundary of the Ecca Group of the South African Karoo (Fildani *et al.* 2009) would seem to require a critical review of the concept of global anoxia as reason for the extinction.

The beginning of the Triassic was characterized by a sudden increase in precipitation which fostered the development of large high-volume and generally blackwater rivers in practically all rift basins in eastern Africa (Markwort 1991). This may have been the only time when tropical rainforests existed. Whether the humid and hot conditions explain the dominance of fungal spores at the P/T boundary or if these conditions fostered the growth of algae is still open to debate, although the latter option appears more likely. Increased humidity at the beginning of the Triassic is reflected by increasing values of strontium isotope ratios (Korte *et al.* 2003), and the question has to be asked whether the wet period was a consequence of the extinction event or just a spurious effect. From the Induan to the Carnian regression climate was hot and, due to the deep embayment of the Tethys, seasonal (possibly monsoonal) conditions prevailed (Lucas 1998).

Discussion and conclusions

A shift of Afro-Arabia or of Gondwana as a whole towards the equator was assumed to explain the amelioration of the post-Sakmarian climate (e.g. Scotese & Langford 1995; Ziegler *et al.* 1997; Roscher & Schneider 2006). During the Permian

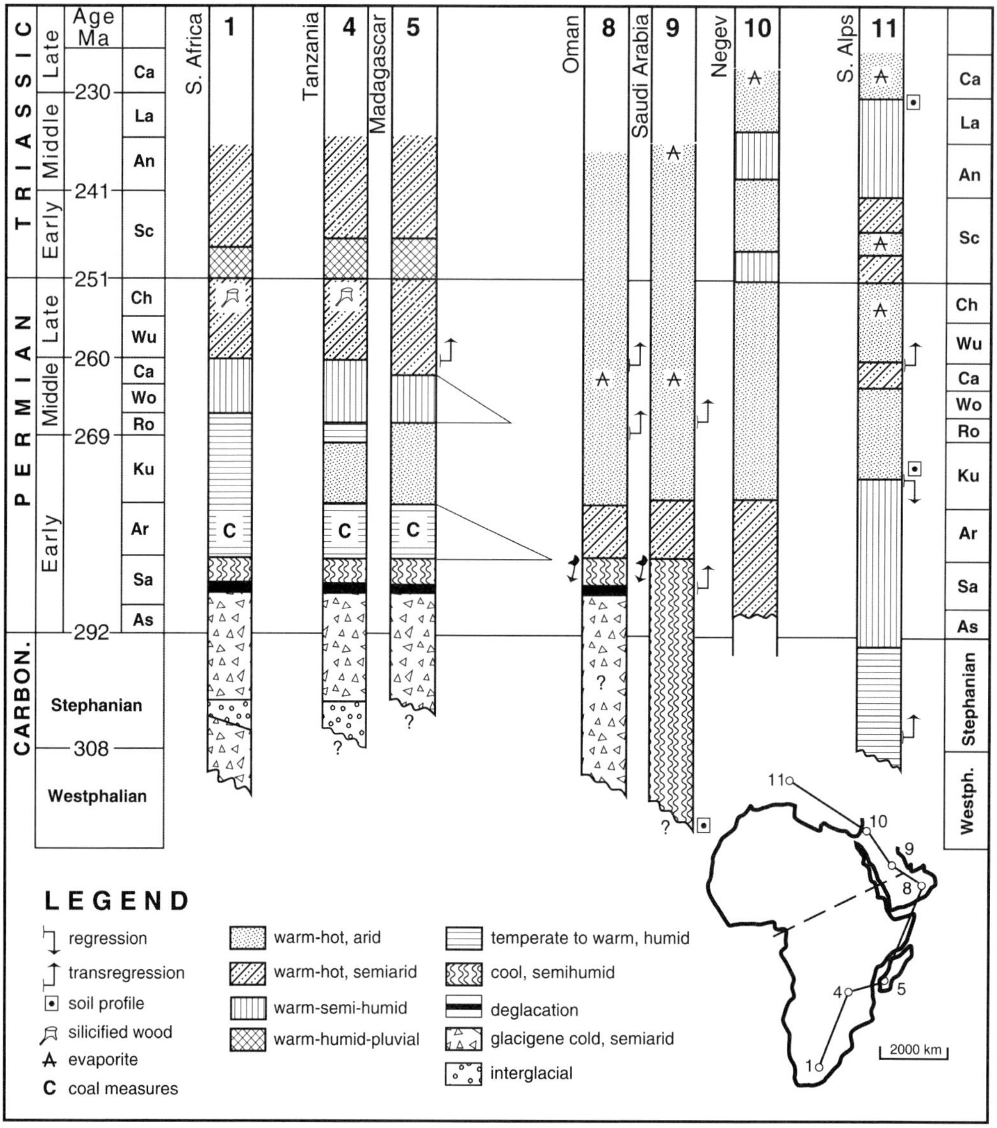

Fig. 14. Simplified climate profile from the sub-polar Karoo Basin (1) to the South Alpine Terrane (11) on the southern shores of Tethys, covering the time from Late Westphalian to the end of Anisian. Although not strictly perpendicular to palaeolatitudes, the profile demonstrates the rapid expansion of warm climate zones after deglaciation. In terms of the Köppen–Geiger climate classification (Peel *et al.* 2007) there is a dominance of warm to hot, steppe and desert climates and increased aridity towards the end of the Permian. The numerals identifying the locations are as shown in Figure 1.

however, the pole apparently moved towards Australia to reach a position in southeastern Australia (roughly in the lower reaches of the Murray River) in the Early Triassic (250 Ma) (Powell & Li 1994; Blakey 2008; see Fig. 1). Thus, the pole shift of about 20° Gondwana latitude to the 'east' effected the central-latitudinal parts Gondwana's rather than its Tethyan margin. The dominance of coal deposition in Permian sedimentary basins of Australia and the less-pronounced distribution of typical red beds support a closer position of Australia to the South Pole in Middle–Late Permian times (Fig. 1). There are other models and pole movements, but they can neither explain the synchronous retreat of the glaciers over the whole of Gondwana nor the quasi-simultaneous Gondwana-wide onset of coal formation, red-bed development or other climatic events. Reciprocate pole movements as envisaged by some authors (e.g. Melvin *et al.* 2010) are attempts to convert Gondwana to a veritable yo-yo. As Africa did not start to separate from Antarctica until the Jurassic, it is not likely that it moved on its own prior to the final break-up of Gondwana in the Jurassic. Clockwise and anticlockwise rotations of the supercontinent in relation to Euramerica (Visser & Praekelt 1998; Klootwijk 2009) influenced stress patterns but not the distribution of climate zones.

The closing and opening of ocean basins (Roscher & Schneider 2006), the formation of mountain ranges (Ziegler *et al.* 1997) or the occurrence of super-volcanic events no doubt changed marine currents, influenced local weather conditions and steered the development of specific environments, but it is difficult to accept that such events triggered simultaneous and long-lasting changes over the entire southern hemisphere.

As stressed before, the present review confirms the statement of Wopfner (1999) that the retreat of the glaciers was, within the limit of present dating capabilities, a synchronous Gondwana-wide event. Equally simultaneous was the onset of the formation of coal beds in all of Gondwana, including Antarctica and the formation of red beds (Wopfner & Casshyap 1997). Middle Permian fossil forests in Antarctica mentioned above (Taylor *et al.* 1992, 2000) and Middle Permian plant assemblages from Dronning Maud Land (McLoughlin *et al.* 2005) demonstrate that, from about the Late Artinskian onward, Antarctica was vegetated and free of glaciers as was the rest of Gondwana (Wopfner & Jin 2009), except perhaps for a few mountain glaciers in Australia.

Oxygen isotope measurements of Permian brachiopods and bivalves by Korte *et al.* (2006) support that scenario. Their statement that the 'parallel decline in $\delta^{18}O$ at low and high latitudes suggests that the warming was a global phenomenon' enforces the demand for a higher level of control than a mere shift into lower latitudes.

Timing and tendencies of the climatic changes observed during the remaining parts of the Permian and of the Early Triassic are quasi-identical in all parts of Gondwana. It shows a continuing desiccation and temperature increase right up to the end of the Permian. This was a global process (Zharkov & Chumakov 2001; Wopfner & Jin 2009) although influenced and modified by different morphologies, land configuration and ocean currents. Even onset and duration of periods of increased precipitation, recurring at intervals of *c.* 16–19 Ma in East Africa, may eventually fall into a larger pattern (compare Roscher & Schneider 2006).

To summarize, the data presented in this review result in a scenario in which changing temperatures were the driving force in controlling global climate. The climate history of the Permian was not a change from icehouse to greenhouse, but rather a change from glaciers to deserts. Or, as Melvin *et al.* (2010) put it, 'from bergs to ergs'. In other words, the Gondwana glaciers did not melt because an ocean closed, a seaway opened or a mountain rose, but simply because the global temperature increased.

This raises the question of just what governed global temperature? Variations in CO_2 concentration are not a very convincing cause. There were major periods of glaciation in the Precambrian and the Ordovician, when CO_2 concentrations were much higher than in the Permian. Furthermore, ice core data from Quaternary glaciers show that rising temperature preceded the increase of atmospheric CO_2. For instance, Caillon *et al.* (2003) analysed gas bubbles in the Vostok ice core across the Termination III-event. They found that the increase in CO_2 concentration lagged consistently behind the increasing temperature curve. Even if varying rates of CO_2 emissions from mid-oceanic ridges and volcanic eruptions due to intensified tectonic activities were responsible for variations of CO_2 concentrations in the atmosphere, hence climate change, the question of what controlled the changing intensities of tectonic activities remains. It follows that causes for global changes may have to be searched for elsewhere.

In modern climatology the cyclic recurrence of specific events such as Milankovitch cycles or other oscillations in the range of 0.01–2 Ma are commonly accepted. With that acceptance, the notion of extraterrestrial forcing of climate development is explicitly acknowledged. However, when it comes to explain cycles of larger magnitude (in the range of tens of million years or more) such as changes in sea level, biodiversity, orogenic events, Wilson cycles (Dewey & Burke 1974) or changes of depositional megasequences (Veevers 1990, 1993), our thinking is still shackled by almost pre-Galilean concepts. Golubev (2005) stated that the 'atavistic geocentric model disguised by the concept of self-development of the Earth and lack of mechanisms explaining the galactic influence' prevent us from accepting driving forces emanating from beyond our solar system. Yet, the orbit of the solar system around the core of our galaxy (black hole) constitutes the highest-order cycle of geological time. Such a cycle corresponds to a galactic year. Similarly to the varying influences to which Earth is exposed to in the course of its orbit around the Sun, the solar system is exposed to changing forces and environments in the course of its galactic orbit. Influences include, among many others, variability in cosmic ray flux (CRF), changing strength and polarity of magnetic fields and modifications of gravitational pull. For instance, Rampino & Stothers (1984) identified a 33 ± 3 Ma cycle related to the regular crossing of the sun of the galactic plane. At each crossing, the solar system encounters more interstellar dust which these authors associate with times of mass extinction. The encounter with the interstellar dust also increases the luminosity of the Sun. Fairbridge (1987) recognised the 33 Ma cycle identified by Rampino & Stothers (1984) and, like others before him (Steiner 1967; Steiner & Grillmair 1973), suggested a 'fundamental cyclicity of 200 Ma to 300 Ma' corresponding to the time the solar system takes to orbit the Milky Way. Wilson cycles (Dewey & Burke 1974) and depositional megacycles (Veevers 1990) can be placed in this category.

Medvedev & Mellot (2007) proposed that the observed 64 Ma cycle in which marine biodiversity swells and vanes is related to the 63 $\pm$ 3 Ma cycle which the solar system takes to complete a full oscillation across the galactic plane and back. During the Sun's position above the galactic plane it will be exposed to increased bombardment by cosmic rays, causing detrimental conditions for life on Earth. This is a modification of the concept proposed by Rampino & Stothers (1984). These models also indicate a correspondence between the end-Permian mass extinction and the peak of this oscillation cycle.

As far as climate change is concerned, Shaviv (2003) for instance suggested a connection between ice ages and increased cosmic ray flux (CRF). Shaviv & Veizer (2003) found that at least 66% of temperature variance in the Phanerozoic was due to changes in CRF. Their graph juxtaposing CRF, derived from Fe-meteorite records and temperature changes based on $\delta^{18}O$ ratios for the Phanerozoic, shows excellent correlation between CRF maxima and negative temperature changes.

The concept of periodic influences exerted upon the solar system during transits of spiral arms of the Milky Way galaxy (Shaviv 2003; Shaviv & Veizer 2003) was challenged by Overholt *et al.* (2009). Presenting a new model of the galaxy they suggest that, for reasons of geometry, the 140 Ma cycle of transits cannot be verified. They argue that due to the asymmetry of their new galactic picture the crossing of spiral arms would be non-periodic; they therefore reject any correlation with global cycles, especially that of 140 Ma. Even if this dogmatic view should be verified, it does not mean that the voyage of our Sun around the centre of the galaxy is a journey in a homogeneous medium.

There are many other galactic cycles which could be related to successions of specific geological or climatic events (Scherer *et al.* 2006). One such aspect pertains to magnetic polarity changes and their influence on extinction events. The time of a reversal is a time of severely reduced magnetic field strength, causing diminution or even collapse of the ionosphere. The resulting unimpeded entrance of cosmic rays will be detrimental to biodiversity and change the climate. Following this concept, diverse scenarios were evoked holding the Illawarra reversal responsible for the extinction mechanisms at the P/T boundary. The Illawarra event ended the reversed polarity of the Kiaman Superchron to a magnetic field of normal polarity in the Capitanian. However, this change was followed by a period of frequent reversals from normal to reverse and back right into the Triassic. Chang *et al.* (2006) dated a reverse to normal transition at 252.5 Ma in western Texas which they relate to the extinction event, but without discussing this event in the context of the overall situation of the magnetic field since the Illawarra reversal.

Obviously there is a large cloud of unknowns between galactic movements and Earth processes but, with rapidly improving dating methods, it should be possible to identify increasing numbers of well-constrained cycles of specific events. These could then be assigned to defined positions of our solar system on its path around the centre of the Milky Way. Just because we do not fully understand galactic mechanisms and their resulting forces should not prevent us from accepting them. There is no reason for not leaving that well-worn path of a geocentric thinking. We should accept the fact that our Earth and the solar system are parts of the galaxy.

The paper has benefited from the constructive criticism of the reviewers, C. B. Foster of Geoscience Australia, Canberra and an anonymous colleague. Their comments and contributions are gratefully acknowledged. The author records his appreciation to C. Krings of the Geology Department of the University of Cologne for the preparation of the drawings.

References

Al-Laboun, A. A. 1987. Unayzah Formation: a new Permian–Carboniferous unit in Saudi Arabia. *American Association of Petroleum Geologists Bulletin*, **71**, 29–38.

Alsharan, A. S. 2006. Sedimentological character and hydrocarbon parameters of the Middle Permian to Early Triassic Khuff Formation, United ArabEmirates. *GeoArabia*, **11**, 121–158.

Alsharhan, A. S. & Salah, M. 1997. Lithostratigraphy and hydrocarbon habitat of the Pre-Cenomanian Nubian Sandstone in the Gulf of Suez Oil Province, Egypt. *GeoArabia, Bahrain*, **2**, 385–400.

Angiolini, L., Bucher, H. *et al.* 1997. Early Permian (Sakmarian) brachiopods from south-eastern Oman. *Geobios*, **30**, 389–406.

Angiolini, L. & Balini, M. *et al.* 2001. Lower to Middle Permian sedimentation on the Arabian Platform: the succession of the Haushi-Huqf (Interior Oman). International Conference on the Geology of Oman, Sultan Qaboos University, Excursion A **02**.

ARCHBOLD, N. W. 2001. Pan-Gondwana, Early Permian (Asselian–Sakmarian–Artinskian) correlations. *In*: WEISS, R. H. (ed.) *Contributions to Geology and Palaeontology of Gondwana in Honour of Helmut Wopfner*. Geological Institute, University of Cologne, 29–39.

BANGERT, B., STOLLHOFEN, H., LORENZ, V. & ARMSTRONG, R. 1999. The geochronology and significance of ash-fall tuffs in the glaciogenic Carboniferous-Permian Dwyka Group of Namibia and South Africa. *Journal of African Earth Science*, **29**, 33–49.

BECHENNEC, F., LE METOUR, J., RABU, D., BOURDILLON-DE-GRISSAC, P., DE WEVER, P., BEURRIER, M. & VILLEY, M. 1990. The Hawasina Nappes: stratigraphy, palaeogeography and structural evolution of a fragment of the south-Tethyan passive continental margin. *In*: ROBERTSON, A. H. F., SEARLE, M. P. & RIES, A. C. (eds) *The Geology and Tectonics of the Oman Region*. Geological Society, London, Special Publication, **49**, 213–223.

BERTHELIN, M., BROUTIN, J., VASLET, D., LE NINDRE, Y. M. & HALAWANI, M. 2006. Mixed Late Permian floras and related palaeoenvironments in the Khuff Formation of central Saudi Arabia. *GeoArabia*, **11**, 65–86.

BESEMS, R. E. & SCHUURMAN, W. M. L. 1987. Palynostratigraphy of Late Palaeozoic glacial deposits of the Arabian Peninsul with special reference to Oman. *Palynology*, **11**, 37–53.

BIRON, P. E. 1982. *Le Permo-Triassic de la région de l'Ourika (Haute Atlas du Marrakech)*. Thése de Docteure de 3o cycle de gèologie applicee, Université scientifique et medical de Grenoble.

BLACKEY, R. C. 2008. Gondwana palaeogeography from assembly to breakup – A 500 m. y. odyssey. *In*: FIELDING, C. R., FRANK, T. D. & ISBELL, J. L. (eds) *Resolving the Late Palaeozoic Ice Age in Time and Space*. Geological Society of America, Boulder, Special paper, **441**, 1–28.

BOSELLINI, A. 1989. The continental margins of Somalia: their structural evolution and sequence stratigraphy. *Memorie di Science Geologiche, Università di Padova*, **41**, 373–458.

BOSELLINI, A. & HARDIE, L. 1973. Depositional theme of a marginal evaporite. *Sedimentology*, **20**, 5–27.

BOTHA, J. & SMITH, R. M. H. 2007. *Lystrosaurus* species composition across the Permo–Triassic boundary in the Karoo Basin of South Africa. *Lethaia*, **40**, 125–137, http://dx.doi.org/10.1111/j.1502-3931.2007.00011.x

BOUTAKOFF, N. 1948. Les formations glaciaires et post glaciaires fossilifères d'age permo carbonifère (Karroo inferieur) de la Région Walikale (Kivu, Congo belge). *Memoire Institute Geologie Universitè Louvain*, **9**.

BRAAKMAN, J. H., LEVELL, B. K., MARTIN, J. H., POTTER, T. L. & VAN VLIET, A. 1982. Late Palaeozoic Gondwana glaciation in Oman. *Nature*, **299**, 48–50.

BRANDNER, R. & KEIM, L. 2011. A 4-day geological field trip in the western Dolomites. *Geo Alp*, **8**, 76–118.

BRODIE, J. W. 1964. The Fiordland Shelf and Milford Sound: Studies of a Southern Fiord. *N.Z. Department of Scientific & Industrial Research Bulletin*, **257**, 15–23.

BROGLIO LORIGA, C. & CASINIS, G. 1992. The Permo-Triassic boundary in the Southern Alps (Italy) and in adjacent Periadriatic regions. *In*: SWEET, W. C., YANG, Z. Y., DICKINS, J. M. & YIN, H. F. (eds) *Permo-Triassic Events in the Eastern Tethys*. Cambridge University Press, Cambridge, 78–97.

BROUTIN, J. & BERTHELIN, M. 2007. The Permian mixed flora of the southern Tethyan margin: palaeogeographic and palaeoclimatic significance. *Abstracts, XVI International Congress on the Carboniferous and Permian*, Nanjing, China.

BROUTIN, J., FERRANDINI, J. & SABER, H. 1989. Implications stratigraphiques et paléogéographique de la découverte dune flore permienne euroaméricaine dans le Haute-Atlas occidental (Maroc). *Compte Rendu Academie Science, Paris*, **308**, Serie II, 1509–1515.

BROUTIN, J., DOUBINGER, J., EL HAMET, M. O. & LANG, J. 1990. Palynologie comparée du Permien nigérien (Afrique occidental) et Péritéthysien. Implicacions stratigraphiques et phytogéographiques. *Review of Palaeobotany and Palynology*, **66**, 243–261.

BRUUN, A. F., BRODIE, L. W. & FLEMING, C. A. 1955. Submarine geology of Milford sound. *N.Z. Journal Science & Technology, B*, **36**, 397–410.

CAILLON, N., SEVERINGHAUS, J. P., JOUZEL, J., BARNOLA, J-M., KANG, J. C. & LIPENKOV, V. Y. 2003. Timing of atmospheric CO2 and Antarctic temperature changes across Termination III. *Science*, **299**, 1728–1731.

CAIRNCROSS, B. 2001. An overview of the Permian (Karoo) coal deposits in southern Africa. *Journal of African Earth Sciences*, **33**, 529–562.

CAIRNCROSS, B. & CADLE, A. B. 1987. A genetic stratigraphy for the Permian coal-bearing Vryheid Formation in the East-Witbank Coalfield, South Africa. *South African Journal of Geology*, **90**, 219–230.

CASSINIS, G. 1966*a*. La Formazione di Collio nell'area-tipo dell' alta Val Trompia (Permiano inferiore bresciano). *Rivista Italiana di Paleontologia*, **72**, 507–588.

CASSINIS, G. 1966*b*. Rassegna delle formazioni Permiane dell' alta val Trompia (Brescia). *Atti dell' Institozo Geologico della Università die Pavia*, **17**, 50–66.

CASSINIS, G. 1986. Guide – Book, Field Conference on Permian and Permian–Triassic boundary in the South-Alpine segment of the western Tethys, 4–8. *Societá Geologica Italiana*, Brescia, 176.

CASSINIS, G. & RONCHI, A. 2001. Permian chronostratigraphy of the Southern Alps (Italy) – an update. *In*: WEISS, R. H. (ed.) *Contributions to Geology and Palaeontology of Gondwana in Honour of Helmut Wopfner*. Geological Institute, University of Cologne, Cologne, 73–88.

CATUNEANU, O., WOPFNER, H., CAIRNCROSS, B., RUBIDGE, B. S., SMITH, R. M. H. & HANCOX, P. J. 2005. The Karoo basins of south-central Africa. *In*: CATUNEANU, O., GUIRAUD, R., ERIKSSON, P., THOMAS, B., SHONE, R. & KEY, R. (eds) Phanerozoic Evolution of Africa. *Journal of African Earth Sciences*, **43**, 211–253.

CHANG, S., MUNDIL, R. & RENNE, P. R. 2006. U/Pb and Ar/Ar Dating of Latest Permian Magnetic Polarity Reversals in the Ochoan of western Texas. American Geophysical Union, Fall Meeting 2006, abstract No. V21A-0565 – Bibl. Code, 2006AGUFM.V21A 0565C.

COLLINSON, J. W. 1997. Paleoclimate of Permo-Triassic Antarctica. *In*: RICCI, C. A. (ed.) *The Antarctic Region: Geological Evolution and Processes*. Terra Antarctica Publications, Siena, 1029–1034.

CROWELL, J. C. 1978. Gondwana glaciation, cyclothemes, continental positioning and climate change. *American Journal of Sciences*, **278**, 1345–1372.

CROWELL, J. C. 1995. The ending of the Late Palaeozoic Ice Age during the Permian period. *In*: SCHOLLE, P. A., PERYT, T. M. & ULMER-SCHOLLE, D. S. (eds) *The Permian of Northern Pangea*. Springer Verlag, Berlin etc., **1**, 62–74.

CROWELL, J. C. & FRAKES, L. A. 1970. *Ancient Gondwana Glaciations*. Second Gondwana Symposium, Proceedings and Papers, 469–476, Pretoria, South Africa.

DEWEY, J. F. & BURKE, K. 1974. Hotspots and continental break up: implications for collisional orogeny. *Geology*, **2**, 57–60.

DICKINS, J. M. 1985. Late Palaeozoic glaciation. *BMR Journal of Australian Geology and Geophysics*, **9**, 163–196.

DICKINS, J. M. 1997. Some problems of the Permian (Asselian) glaciation and the subsequent climate in the Permian. *In*: MARTINI, I. P. (ed.) *Deglaciation and Global Changes: Quaternary, Permo-Carboniferous and Proterozoic*. Oxford University Press, Oxford, 243–245.

DIEKMANN, B. & WOPFNER, H. 1996. Petrographic and diagenetic signatures of climatic changes in peri- and postglacial Karoo sediments of SW Tanzania. *Palaeogeography, Palaeoclimatology, Palaeoecology*, **125**, 5–25.

DRUCKMAN, Y. 1974. The stratigraphy of the Triassic sequence in southern Israel. *Israel Geological Survey Bulletin*, **64**.

DU TOIT, A. L. 1927. *A Geological Comparison of South America with South Africa. With a Palaeontological Contribution by F.R. Cowper Reed*. Carnegie Institute Publication, Washington, **381**, 1–157.

EL ARNAUTI, A. & SHELMANI, M. 1985. Stratigraphy and structural setting. *In*: THUSU, B. & OWENS, B. (eds) *Palynostratigraphy of Northeast Libya*. British Micropalaeontological Society, London, **4**, 1–10.

EL WARTITI, M., BROUTIN, J., FREYTET, P., LARHRIB, M. & TOUTIN-MORIN, N. 1990. Continental deposits in Permian Basins of Mesetian Morocco, geodynamic history. *Journal of African Earth Sciences*, **10**, 361–368.

ESHET, Y. 1990*a*. Palynological aspects of the Permo-Triassic successions in the subsurface of Israel. *Geological Survey of Israel, Bulletin*, **81**.

ESHET, Y. 1990*b*. The palynostratigraphy of the Permo-Triassic boundary in Israel: two approaches to biostratigraphy. *Israel Journal of Earth Sciences*, **39**, 1–15.

ESHET, Y. & COUSMINER, H. L. 1986. Palynozonation and correlation of the Permo-Triassic succession in the Negev, Israel. *Micropalaeontology*, **32**, 193–214.

ESSAMOUD, R. 1989. *Serie Stéphano-Permienne de la région de Marrakech – Maroc*. Thèse de Doctorat de L'université de Bourgogne, Centre des Sciences de la Terre.

FAIRBRIDGE, R. W. 1987. Climatic variation, historical record. *In*: OLIVER, J. E. & FAIRBRIDGE, R. W. (eds) *The Encyclopedia of Climatology*. Van Nostrand Reinhold Company, New York, 305–323.

FARIA, L. E. DO C. 1980. *Estratigrafia e petrografia da Formaçâo Pedra de Fogo-Permiano da bacia do Maranhâo*. Anais do XXXI Congrsso Brasileiro de Geologia, Balneario de Camburrio, Santa Catarina, 740–753.

FEYS, R. & GREBER, C. 1963. Le Stéphanien et L'Autunien du Souss dans les Ida ou Zal (Haute Atlas occidental – Maroc). *Notes Service Géologique du Maroc*, **22**, 19–35.

FILDANI, A., WEISLOGEL, A. *ET AL*. 2009. U-Pb zircon ages from the southwestern Karoo Basin, South Africa—Implications for the Permian–Triassic boundary. *Geology*, **37**, 719–722.

FOSTER, C. B. & WATERHOUSE, J.-B. 1988. The Granulatisporites confluence Oppel-zone and eatly Permian marine faunas from the Grant Formation on the Barbwire Terrace, Canning Basin, Western Australia. *Australian Journal of Earth Sciences*, **35**, 135–157.

FOSTER, C. B., STEPHENSON, M. H., MARSHALL, C., LOGAN, G. A. & GREENWOOD, P. F. 2002. A revision of Reduviasporonites Wilson 1962: Description, illustration, comparison and biological affinities. *Palynology*, **26**, 35–58, http://dx.doi.org/10.2113/0260035

FRITZ, A. & KRAINER, K. 2006. Eine Rotliegendflora aus Sedimenten des Bozner Vulkanitkomplexes bei Sinnich (Südtirol). *Geo. Alp*, **3**, 33–45.

GENIK, G. J. 1993. Petroleum geology of Cretaceous–Tertiary rift basins in Niger, Chad and 1101 d the Central African Rpublic. *American Association of Petroleum Geologists Bulletin*, **77**, 1405–1434.

GLICKSON, M. R. 1964. Palynological investigations in Makhtesh Qatan 2 boring in the Negev, Israel. *Israel Journal of Earth Science*, **13**, 13–26.

GOLUBEV, V. M. 2005. Causes of cyclicity. *In*: MABESOONE, J. M. & NEUMANN, V. H. (eds) *Cyclic Development of Sedimentary Basins*. Development in Sedimentology, Elsevier, Amsterdam, **503**, 15–37.

GUIRAUD, R., BOSWORTH, W., THIERRY, J. & DELPLANQUE, A. 2005. Phanerozoic geological ecolution of North and Central Africa: an overview. *Journal of African Earth Sciences*, **43**, 83–143.

HAMBREY, M. J. 1981. Palaeozoic tillites in northern Ethiopia. *In*: HAMBREY, M. J. & HARLAND, W. B. (eds) *Earth's Pre-Pleistocene Glacial Record*. Cambridge University Press, Cambridge, 38–40.

HERMINA, M., KLITZSCH, E. & LIST, F. K. 1989. *Stratigraphic Lexicon and Explanatory Notes to the Geological Map of Egypt 1: 500 000*. Conoco Inc., Cairo, Egypt.

HMICH, D., SCHNEIDER, J. W., SABER, H., VOIGT, S. & EL WARTITI, M. (eds) 2006. *New Continental Carboniferous and Permian Faunas of Morocco: Implications for Biostratigraphy, Palaeobiogeography and Palaeoclimate*. Geological Society, London, Special Publication, **265**, 297–324.

HOEPFFNER, C., SOULAIMANI, A. & PIQEÉ, A. 2005. The Moroccan Hercynides. *Journal of African Earth Sciences*, **43**, 144–165.

HÜBNER, H. 1965. Permokarbonische glazigene und periglaziale Ablagerungen aus dem zentralen Teil des Kongobeckens. *Stockholm Contributions in Geology*, **13**, 41–61.

JARDINE, S. 1974. Microflores des formations du Gabon attribuées au Karroo. *Review of Palaeobotany and Palynology*, **17**, 75–112.

JIN, X. C. 1994. *Sedimentary and Palaeogeographic Significance of Permo-Carboniferous Sequences in Western Yunnan, China.* Geologisches Institut, Universität zu Köln, Sonderveröffentlichungen, **99**.

KAAYA, C. Z. 1992. *Depositional Environment of Late Permian Karoo Beds in the Ruhuhu Basin and Mikumi area of Tanzania.* Geologisches Institut, Universität zu Köln, Sonderveröffentlichungen, **83**.

KAAYA, C. Z. & KREUSER, T. 1995. Permian lakes in East Africa. *In*: GIERLOWSKI-KORDESCH, E. & KELTS, K. (eds) *Global Geological Record of Lake Basins.* Cambridge University Press, Cambridge, 83–87.

KADIMA, E., DELVAUX, D., SEBAGENZI, S., TACK, N. L. & KABEYA, S. M. 2011. Structure and geological history of the Congo Basin: an integrated interpretation of gravity, magnetic and reflection seismic data. *Basin Research*, **23**, 499–527.

KLITZSCH, E. & WYCISK, P. 1987. Geology of the sedimentary basins of northern Sudan and bordering areas. *Berliner Geowissenschaftliche Abhandlungen (A)*, **73**, 97–136.

KLOOTWIJK, C. 2009. Sedimentary basins of eastern Australia: palaeomagnetic constraints on geodynamic evolution in a global context. *Australian Journal of Earth Sciences*, **56**, 273–308.

KOBER, V. 1984. *Zur Genese der Tarviser Breccie in den Karawanken, NW-Jugoslawien.* Geologisches Institut Universität Köln, Sonderveröffentlichungen, **56**.

KORA, M. 1998. The Permo-Carboniferous outcrops of the Gulf of Suez region, Egypt: classification and correlation. *Geodiversitas*, **20**, 701–721.

KORTE, C., KOZUR, H. W., BRUCKSHEN, P. & VEIZER, J. 2003. Strontium isotope evolution of Late Permian and Triassic seawater. *Geochimica et Cosmochimica Acta*, **67**, 47–62.

KORTE, C., BRAND, U., JONES, P. J., KOZUR, H. W., MERTMANN, D. & VEIZER, J. 2006. Oxygen isotope values from Permian brachiopods and bivalves: a record of deglaciation and latitudinal sea-surface temperature gradients. *In*: ARETZ, M. & HERBIG, H.-G. (eds) *Kölner Forum Geologie-Palaeontologie.* University of Cologne, Germany, **15**, 72–73.

KRASSILOV, V. A., AFONIN, S. A. & BARINOVA, S. S. 1999. Tympanicysta and the terminal Permian events. *Permophiles*, **35**, 16–17.

KREUSER, T. 1994. Tectonic and climatic controls of lacustrine sedimenmtation in pre-rift and rift settings in the Permian–Triassic of East Africa. *Journal Palaeolimnology*, **10**, 1–18.

KRUCK, W. & THIELE, J. 1983. Late Palaeozoic glacial depositsin the Yemen. *Geologisches Jahrbuch (B)*, **46**, 3–29.

LEVELL, B. K., BRAAKMAN, J. H. & RUTTEN, K. W. 1988. Oil-bearing sediments of Gondwana glaciation in Oman. *American Association of Petroleum Geologists Bulletin*, **72**, 775–796.

LÓPEZ-GAMUNDI, O. R. 1997. Glacial-postglacial transition in the Late Paleozoic basins of southern South America. *In*: MARTINI, I. P. (ed.) *Deglaciation and Global Changes: Quaternary, Permo-Carboniferous and Proterozoic.* Oxford University Press, Oxford, 147–168.

LÓPEZ-GAMUNDI, O. R. & ROSSELLO, E. A. 1998. Basin fill evolution and paleotectonic patterns along the Samfrau geosyncline: the Sauce Grande basin-Ventana foldbelt (Argentina) and Karoo basin-Cape foldbelt South Africa) revisited. *Geologische Rundschau*, **86**, 819–834.

LOVE, C. F. 1994. The palynostratigraphy of the Haushi Group (Westphalian–Artinskian) in Oman. *In*: SIMMONS, M. D. (ed.) *Micropalaeontology and Hydrocarbon Exploration in the Middle East.* Chapman & Hall, London, 23–39.

LUCAS, S. G. 1998. The epicontinental Triassic, an overview. *Zentralblatt für Geologie und Palaeontologie, Teil 1, H.* **7–8**, 475–496, Stuttgart.

MAGARITZ, M., BÄR, R., BAUD, A. & HOLSER, W. T. 1988. The carbon-isotope shift at the Permian–Triassic boundary in the Southern Alps is gradual. *Nature*, **331**, 337–339.

MARKWORT, S. 1991. *Sedimentation and diagenese der untertriadischen Karoo Schichten des Ruhuhu Beckens, SW Tansania.* Geologisches Institut, Universität Köln, Sonderveröffentlichungen, **80**.

MASARI, F., NERI, C., PITTAU, P., FONTANA, D. & STEFANI, C. 1994. Sedimentology, palynostratigraphy and sequence stratigraphy of a continental to shallow-marine rift related succession: Upper Permian of the eastern Southern Alps (Italy). *Memorie di Scienze Geologiche*, **46**, 119–243.

MAURER, F., MARTINI, R., RETTORI, R., HILLGÄRTNER, H. & CIRILLI, S. 2009. The Geology of Khuff outcrop analogues in the Musandam Peninsula, United Arab Emirates and Oman. *GeoArabia*, **14**, 125–158.

MCLOUGHLIN, S., LARSSON, K. & LINDSTRÖM, S. 2005. Permian plant macrofossils from Fossilryggen, Vestfjella, Dronning Maud Land. *Antarctic Science*, **17**, 73–86.

MEDVEDEV, M. V. & MELOTT, A. L. 2007. Do extragalactic cosmic rays induce cycles in fossil diversity? *The Astrophysical Journal*, **664**, 879–889.

MELVIN, J., SPRAGUE, R. A. & HEINE, C. J. 2010. From bergs to ergs: the late Paleozoic Gondwana glaciation and its aftermath in Saudi Arabia. *In*: LOPEZ-GAMUNDI, O. R. & BUATOIS, L. A. (eds) *Late Palaeozoic glacial events and postglacial transgressions in Gondwana.* Geological Society of America, Boulder, Special Paper, **468**, 37–80.

MENNING, M., ALEKSEEV, A. S. ET AL. 2006. Global time scale and regional stratigraphic reference scales of Central and West Europe, East Europe, Tethys, South China, and North America as used in the Devonian–Carboniferous–Perm ian Correlation Chart 2003 (DCP 2003). *Palaeogeography, Palaeoclimatology, Palaeoecology*, **240**, 318–372.

MESNER, J. C. & WOOLDRIDGE, L. C. D. 1964. Maranhao Palaeozoic basin and Cretaceous coastal basins, north Brazil. *American Association of Petroleum Geologists Bulletin*, **48**, 1475–1512.

METCALFE, I. 1988. Origin and assembly of south-east Asian continetal terranes. *In*: AUDLEY CHARLES, M. G. & HALLAM, A. (eds) *Gondwana and Tethys.* Geological Society, London, Special Publication, **37**, 101–118.

METCALFE, I. 2002. Permian tectonic framework and palaeogeography of SE Asia. *Journal of Asian Earth Sciences*, **20**, 551–566.

MICHOLET, J., MOLINAS, E. & PENET, B. 1970. Le systeme du Karroo au Gabon. *In*: LOPEZ-GAMUNDI, O. R. & BUATOIS, L. A. (eds) *Late Palaeozoic glacial events and postglacial transgressions in Gondwana*. Proceedings and papers, Second Gondwana Symposium. CSIR, Pretoria, South Africa, 371–380.

MILLER, M. F. & ISBELL, J. L. 2010. Reconstruction of a high-latitude postglacial lake: Mackellar Formation (Permian), Transantarctic Mountains. *Geological Society of America*, Special Paper, **468**, 193–207.

MUTTONI, G., KENT, D. V., GARZANTI, E., BRACK, P., ABRAHAMSEN, N. & GAETANI, M. 2003. Early Permian Pangea 'B' to Late Permian Pangea 'A'. *Earth and Planetory Science Letters*, **215**, 379–394.

NEWELL, N. D., RIGBY, J. K., DRIGGS, A., BOYD, D. W. & STEHLI, F. G. 1976. *Permian Reef Complex, Tunesia*. Brigham Young University Geology Studies, **23**, 75–112.

ORI, G. G., DALLA, S. & CASSINIS, G. 1988. Depositional history of the Permian continental sequence in the Val Trompia – Passo Croce Domini area (Brescian Alps, Italy). *Memorie Società Geologica Italiana*, **34**, 141–154.

OVERHOLT, A. C., MELOTT, A. L. & POHL, M. 2009. Testing the link between terrestrial climate change and galactic spiral arm transit. *The Astrophysical Journal Letters*, **705**, 101–103, http://dx.doi.org/10.1088/0004-637X/705/2/L101.

PATZELT, A., LI, H. M., WANG, J. D. & APPEL, E. 1996. Palaeomagnetism of Cretaceous to Tertiary sediments from southern Tibet: evidence for the extent of the northern margin of India prior to the collision with Eurasia. *Tectonophysics*, **259**, 259–284.

PEEL, M. C., FINLAYSON, B. L. & MCMAHON, T. A. 2007. Updated world map of the Köppen-Geiger climate classification. *Hydrology and Earth System Sciences*, **11**, 1633–1644.

PICARD, L. & FLEXER, A. 1974. *Studies on the Stratigraphy of Israel: The Triassic*. The Israel Institut of Petroleum, Tel Aviv.

PIQUÉ, A., LAVILLE, E., BIGNOT, G., RABARIMANANA, M. & THOUIN, C. 1999. L'ouverture et le dévelopment du bassin de Morondava (Madagascar) car du Carbonifère supérieur au Jurassique moyen. Données stratigraphiques, sédimentaires, paléontologiques et structurales. *Journal African Earth Sciences*, **28**, 931–948.

POWELL, C. M. & LI, Z. X. 1994. Reconstruction of the Panthalassan margin of Gondwanaland. *In*: VEEVERS, J. J. & POWELL, C. McA. (eds) *Permian–Triassic Pangean Basins and Foldbelts along the Panthalassan Margin of Gondwana-land*. Geological Society of America, Boulder, Memoir **184**, 5–10.

RAMPINO, M. R. & STOTHERS, R. B. 1984. Terrestrial mass extinctions, cometary impacts and the Sun's motion perpendicular to the galactic plane. *Nature*, **308**, 709–716.

RING, U. 1995. Tectonic and lithological constraints on the evolution of the Karoo graben of northern Malawi (East Africa). *Geologische Rundschau*, **84**, 607–625.

ROSCHER, M. & SCHNEIDER, J. W. 2006. Permo-Carboniferous climate: Early Pensylvanian to Late Permian climate development of central Europe in a regional and global context. *In*: LUCAS, S. G., CASSINIS, G. & SCHNEIDER, J. W. (eds) *Non-Marine Permian Biostratigraphy and Biochronology*. Geological Society, London, Special Publications, **265**, 95–136.

RUBIDGE, B. S., JOHNSON, M. R., KITCHING, J. W., SMITH, R. M. H., KEYSER, A. W. & GROENEWALD, G. H. 1995. An introduction to the biozonation ofthe Beaufort Group. *In*: RUBIDGE, B. S. (ed.) *Biostratigraphy of the Beaufort Group (Karoo Supergroup)*. South African Committee of Stratigraphy, Biostratigraphic Series **1**.

RUST, I. C. 1975. Tectonic and sedimentary framework of Gondwana basins in southern Africa. *In*: CAMPBELL, K. S. W. (ed.) *Gondwana Geology*, Australian National University Press, Canberra, 537–564.

SCHERER, K., FICHTNER, H. ET AL. 2006. Interstellar-terrestrial relations: variable cosmic environments, the dynamic heliosphere, and their imprints on terrestrial archives and climate. *Space Science Reviews*, **127**, 327, http://dx.doi.org/10.1007/s11214-006-9126-6.

SCHLUMBERGER 1976. *Well Evaluation Conference Iran*. Schlumberger Well Logging Company, Services Techniques Schlumberger, Paris.

SCHULL, T. J. 1988. Rift basins of interior Sudan: petroleum exploration and discovery. *American Assosiation of Petroleum Geologists Bulletin*, **72**, 1128–1142.

SCIUNNACH, D. 2001. Benthic foraminifera from the upper Collio Formation (Lower Perm Ian, Lombardy Southern Alps): implications for the palaeogeography of the peri-Tethyan area. *Terra Nova*, **13**, 150–155.

SCOTESE, C. R. & LANGFORD, R. P. 1995. Pangea and the palaeogeography of the Permian. *In*: SCHOLLE, P. A., PERYT, T. M. & ULMERSCHOLLE, D. S. (eds) *The Permian of Northern Pangea, Vol 1: Palaeogeography Palaeoclimates, Stratigraphy*, Springer, Berlin, 3–19.

SEMKIWA, P. Z. 1992. *Depositional Environment and Coal Petrography of Permian Coal Deposits in Karoo Basins of SW Tanzania*. Geologisches Institut, Universität zu Köln, Sonderveröffentlichungen, **84**.

SENALP, M. & AL-DUAIJI, A. 2001. Sequence stratigraphy of the Unayzah reservoir in central Saudi Arabia. *Saudi ARAMCO Journal of Technology*, **Summer 2001**.

ŞENGÖR, A. M. C. 1998. Die Tethys: vor hundert Jahren und heute. *Mitteilungen der Österreichischen Geologischen Gesellschaft*, **89**, 5–177.

SHAVIV, N. J. 2003. The spiral structure of the Milky Way, cosmic rays, and ice age epochs on Earth. *New Astronomy*, **8**, 39–77.

SHAVIV, N. J. & VEIZER, J. 2003. Celestial driver of Phanerozoic climate? *GSA TODAY*, **July 2003**.

SHI, G. R., ARCHBOLD, N. W. & ZHAN, L. P. 1995. Distribution and characteristics of mid-Permian (Late Artinskian-Ufimian) mixed/transitional marine faunas in the Asian region and their palaeogeographical implications. *Palaeogeography, Palaeoclimatology, Palaeoecology*, **114**, 241–271.

SMITH, R. M. H., ERIKSSON, P. G. & BOTHA, W. J. 1993. A review of the stratigraphy and sedimentary environments of the Karoo-aged basins of Southern Africa. *Journal of Afrrican Earth Science*, **16**, 143–169.

STAMPFLI, G. M., MOSAR, J., DE BONO, A. & VAVASIS, I. 1998. Late Palaeozoic, early >Mesotoic plate tectonics of the Western Tethys. *Bulletin Geological Society of Greece*, **32**, 113–120.

STEINER, J. 1967. The sequence of geological events and the dynamics of the Milky Way Galaxy. The present cosmic year – a preliminary study. *Journal of the Geological Society of Australia*, **14**, 99–131.

STEINER, J. & GRILLMAIR, E. 1973. Possible galactic causes for periodic and episodic glaciarion. *Geological Society of America Bulletin*, **84**, 1003–1018.

STEPHENSON, M. H. 2008. A review of the palynostratigraphy of Gondwanan Late Carboniferous to Early Permian glacigene successions. *In*: FIELDING, C. R., FRANK, T. D. & ISBELL, J. L. (eds) *Resolving the Late Palaeozoic Ice Age in Time and Space*. The Geological Society of America, Boulder, Special Paper, **441**, 317–330.

STEPHENSON, M. H. & FOSTER, C. B. 2002. *Photomicrograph Gallery of Species of Reduviasporonites Wilson 1962: A joint initiative with Geoscience Australia*. Nottingham, UK, British Geological Survey, http://www.bgs.ac.uk/taxonomy/reduviasporonites/home.html

STEPHENSON, M. H., ANGIOLINI, L. & LENG, M. J. 2007. The Early Permian fossil record of Gondwana and its relationship to deglaciation: a review. *In*: WILLIAMS, M., HAYWOOD, A. M., GREGORY, F. J. & SCHMIDT, D. N. (eds) *Deep-Time Perspectives on Climate Change: Marrying the Signals from Computer Models and Biological Proxies*. The Micropalaeontological Society, Special Publications, The Geological Society, London, 169–189.

STRATTEN, T. 1970. *Tectonic framework of sedimentation during the Dwyka period in South Africa*. Second Gondwana Symposium, Proceedings and Papers, Pretoria, South Africa, 483–490.

TAVENER-SMITH, R. 1982. Prograding coastal facies associations in the Vryheid formation (Permian) at Effingham quarries near Durban, South Africa. *Sedimentary Geology*, **32**, 111–140.

TAWADROS, E. E. 2001. *Geology of Egypt and Libya*. Balkema Science, Rotterdam.

TAYLOR, E. L., TAYLOR, T. N. & CUNEO, N. R. 1992. The present is not the key to the past: a Permian petrified forest in Antarctica. *Science*, **257**, 1675–1677.

TAYLOR, E. L., TAYLOR, T. N. & CUNEO, N. R. 2000. Permian and Triassic high latitude palaeoclimates: evidence from fossil biotas. *In*: HUBER, B. T. & WING, S. L. (eds) *Warm Climates in Earth History*. Cambridge University Press, Cambridge, 321–350.

TEWARI, R. C. & VEEVERS, J. J. 1993. Gondwana basins of India occupy the middle of a 7500 km sector of radial valleys and lobes in central-eastern Gondwanaland. *In*: FINDLAY, R. H., UNRUG, R., BANKS, M. R. & VEEVERS, J. J. (eds) *Gondwana Eight*, Balkema, Rotterdam, 507–512.

UTTING, J. 1978. *The Karroo stratigraphy of the northern part of the Luangwa Valley*. Geological Survey of Zambia, Lusaka, Memoir 4.

VEEVERS, J. J. 1990. Tectonic-climatic supercycle in the billion-year plate-tectonic era: Permian Pangean icehouse alternates with Cretaceous dispersed continents greenhouse. *Sedimentary Geology*, **68**, 1–16.

VEEVERS, J. J. 1993. Gondwana facies of the Pangea supersequence: a review. *In*: FINDLAY, R. H., UNRUG, R., BANKS, M. R. & VEEVERS, J. J. (eds) *Gondwana Eight*, Balkema, Rotterdam, 513–520.

VEEVERS, J. J. & POWELL, C. M. (eds) 1994. *Permian–Triassic Pangean Basins and Foldbelts along the Panthalassan Margin of Gondwanaland*. Geological Society of America, Boulder, Memoir **184**.

VENTURINI, C. 1990. Geologia delle Alpi Carnichi centro orientale. *Edizioni del Museo Friulano di storia naturale, Udine, Publiccazione*, **36**.

VISONÀ, D., FIORETTI, A. M., POLI, M. E., ZANFERRARI, A. & FANNING, M. 2007. U–Pb SHRIMP zircon dating of andesite from the Dolomite area (NE Italy): geochronological evidence for the early onset of Permian Volcanism in the eastern part oft he Southern Alps. *Swiss Journal of Geoscience*, **100**, 313–324.

VISSCHER, H., BRINKHUIS, H. *ET AL*. 1996. The terminal Palaeozoic fungal event: evidence of terrestrial ecosystem destabilization and collapse. *Proceedings of the National Academy of Science*, **93**, 2155–2158.

VISSCHER, H., SEPHTON, M. A. & LOOY, C. V. 2011. Fungal virulence at the time of the end-Permian biosphere crisis? *Geology*, **39**, 883–886.

VISSER, J. N. J. 1986. Lateral lithofacies relationships in the glacigene Dwyka Formation in the western and central parts of the Karoo Basin. *Transactions of the Geological Society of South Africa*, **89**, 373–383.

VISSER, J. N. J. 1989. The Permo-Carboniferous Dwyka Formation of Southern Africa: deposition by a predominantly subpolar marine ice sheet. *Palaeogeography, Palaeoclimatology, Palaeoecology*, **70**, 377–391.

VISSER, J. N. J. 1995. Post-glacial Permian stratigraphy and geography of southern and central Africa: boundary conditions for climatic modelling. *Palaeogeography, Palaeoclimatology, Palaeoecology*, **118**, 213–243.

VISSER, J. N. J. 1997. A review of the Permo-Carboniferous glaciation in Africa. *In*: MARTINI, I. P. (ed.) *Deglaciation and Global Changes: Quaternary, Permo-Carboniferous and Proterozoic*. Oxford University Press, Oxford, 169–191.

VISSER, J. N. J. & PRAEKELT, H. E. 1998. Late Palaeozoic crustal block rotations within the Gondwana sector of Pangea. *Tectonophysics*, **287**, 201–212.

WARDLAW, B. H. 1999. Notes from the SPS Chair. *Permophiles*, **35**, 1–4.

WOPFNER, H. 1970. *Permian palaeogeography and depositional environment of the Arckaringa Basin, South Australia*. Proceedings and papers 2nd Gondwana Symposium, 273–291, Pretoria, South Africa.

WOPFNER, H. 1980. Development of Permian intracratonic basins in Australia. *In*: CREWSSWELL, M. M. & VELLA, P. (eds) *Proceedings of Fifth International Gondwana Symposium*, Wellington, New Zealand. A. A. Balkema, Rotterdam, 185–190.

WOPFNER, H. 1984. Permian deposits of the Southern Alps as product of initial alpidic taphrogenesis. *Geologische Rundschau*, **73**, 259–277, Stuttgart.

WOPFNER, H. 1991. *Extent and timing of the late Palaeozoic glaciation in Africa*. Festschrift Karl Brunnacker, Univversität zu Köln, Sonderveröffentlichungen des geologischen Instituts, **82**, 447–453.

WOPFNER, H. 1994. The Malagasy Rift, a chasm in the Tethyan margin of Gondwana. *Journal of Southeast Asian Earth Sciences*, **9**, 456–461.

WOPFNER, H. 1996. Gondwana origin of the Baoshan and Tengchong terranes of west Yunnan. *In*: HALL, R. & BLUNDELL, D. (eds) *Tectonic Evolution of Southeast Asia*. Geological Society, London, Special Publication, **106**, 539–547.

WOPFNER, H. 1999. The Early Permian deglaciation event between East Africa and northwestern Australia. *Proceedings of Tenth Gondwana Symposium, Cape Town. Journal of African Earth Sciences*, **29**, 77–90.

WOPFNER, H. 2002. Tectonic and climatic events controlling deposition in Tanzanian Karoo basins. *Journal of African Earth Sciences*, **34**, 167–177.

WOPFNER, H. & CASSHYAP, S. M. 1997. Transition from freezing to subtropical climates in the Permo-Carboniferous of Afro-Arabia and India. *In*: MARTINI, I. P. (ed.) *Deglaciation and Global Changes: Quaternary, Permo-Carboniferous and Proterozoic*. Oxford University Press, Oxford, 192–212.

WOPFNER, H. & DIEKMANN, B. 1996. The Late Palaeozoic Idusi Formation of Southwest Tanzania: a record of change from glacial to postglacial conditions. *Journal of African Earth Sciences*, **22**, 575–595.

WOPFNER, H. & FARROKH, F. 1986. Palaeosols and heavy mineral distribution in the Groeden Sandstone of the Dolomites. *Memorie Società Geologica Italiana*, **34**, 161–173.

WOPFNER, H. & JIN, X. C. 2009. Pangea Megasequences of Tethyan Gondwana-margin reflect global changes of climate and tectonism in Late Palaeozoic and Early Triassic times – a review. *Palaeoworld*, **18**, 169–192.

WOPFNER, H. & KREUSER, T. 1986. Evidence for Late Palaeozoic glaciation in southern Tanzania. *Palaeogeography, Palaeoclimatology, Palaeoecology*, **56**, 259–275.

WOPFNER, H., MARKWORT, S. & SEMKIWA, P. M. 1991. Early diagenetic laumontite in the Lower Triassic Manda Beds of the Ruhuhu Basin, southern Tanzania. *Journal of Sedimentary Petrology*, **61**, 65–72.

WORTMANN, U. G., WEISSERT, H., FUNK, H. & HAUCK, J. 2001. Alpine plate kinematics revisited: the Adria Problem. *Tectonics*, **20**, 134–147.

YEMANE, K. & KELTS, K. 1990. A short review of palaeoenvironments for Lower Beaufort (Upper Permian) Karoo sequences from southern to central Africa: a major Gondwana lacustrine episode. *Journal of African Earth Sciences*, **10**, 169–185.

ZHARKOV, M. A. & CHUMAKOV, N. M. 2001. Palaeogeography and sedimentation settings during Permian–Triassic reorganizations in biosphere. *Stratigraphy and Geological Correlation*, **9**, 340–363. (Translated from: Stratigrafya, Geologicheskaya Korrelyatsiya, 9, 29–54, Nauka/Interperiodica, Russia.)

ZIEGLER, A. M., HULVER, M. L. & ROWLEY, D. B. 1997. Permian world topography and climate. *In*: MARTINI, I. P. (ed.) *Late Glacial and Postglacial Environmental Changes–Quaternary, Carboniferous–Permian and Proterozoic*. Oxford University Press, New York.

Carboniferous and Lower Permian sedimentological cycles and biotic events of South China

XIANGDONG WANG[1]*, WENKUN QIE[1], QINGYI SHENG[1], YUPING QI[1], YUE WANG[1], ZHUOTING LIAO[1], SHUZHONG SHEN[1] & KATSUMI UENO[2]

[1]*State Key Laboratory of Palaeobiology and Stratigraphy, Nanjing Institute of Geology and Palaeontology, Chinese Academy of Sciences, Nanjing 210008, China*

[2]*Department of Earth System Science, Faculty of Science, Fukuoka University, Fukuoka 814-0180, Japan*

**Corresponding author (e-mail: xdwang@nigpas.ac.cn)*

Abstract: The sedimentary successions and four fossil groups, including rugose corals, brachiopods, fusulinaceans and conodonts, from the Carboniferous and Lower Permian of South China have been studied in order to reveal the sedimentary characteristics and evolutionary pattern of main biological groups in the East Tethyan region during the Late Palaeozoic Ice Age. The Lower Carboniferous lithology of South China is diverse, ranging from basinal and shelf carbonate rocks to coal measures and continental clastics, while the Upper Carboniferous and Lower Permian are composed mostly of shallow-marine carbonates. From uppermost Devonian to Lower Carboniferous, five major regression events are recognized at the topmost Devonian, middle and upper Tournaisian boundary, Tournasian–Viséan boundary, uppermost Viséan and the Mid-Carboniferous boundary in South China, separately. The Upper Carboniferous and Lower Permian shallow-water carbonate rocks consist of remarkable, high-frequency cyclothems. Moreover, another major sea-level fall is recognized and characterized by an extensive sedimentary hiatus at the Sakmarian–Artinskian boundary throughout South China. All of the sedimentary basins of South China were formed in extensional tectonic settings during this time; thus, multiple regressive events that occurred throughout South China should be primarily induced by glacio-eustatic drawdown. In addition, two biotic events characterized by a remarkable decline in the diversity of benthic biota and a turnover in the composition of fossil assemblages occurred, respectively, at the Mid-Carboniferous and Sakmarian–Artinskian boundaries, consistent with two major regressions, and were probably caused by the glaciations in Gondwana.

South China in terms of a palaeogeographical unit is bordered by the Tanlu Fault and the Qinlin–Dabie convergent zone in the north, the Houlongmenshan–Jinpingshan Fault in the west, and the Honghe Fault in the SW (Chen *et al.* 1994). Palaeogeographically, it formed a tropical archipelago layout during Carboniferous time (Fig. 1).

The Carboniferous in South China is typical East Tethyan type, which recorded relatively continuous sedimentary sequences and abundant marine fossils. There was no distinctive orogeny in South China during Carboniferous time; whereas the collision between West Gondwana and Laurasia in the western part of the Tethys started in the late Early Carboniferous and resulted in a major lack of marine Late Carboniferous deposition. The Carboniferous can be clearly divided into two major successions in South China. The Lower Carboniferous (the Mississippian) has a relatively limited distribution, and is composed of several major alternations between marine carbonates and nearshore siliciclastics. Each alternation reflects a major (possibly the second or third order of sequence) circulation that might be caused by global sea-level change. The Upper Carboniferous (the Pennsylvanian), which is used to include Asselian and Sakmarian deposition together as a complete sedimentary succession, has a wider distribution and is mainly composed of carbonates. The high-frequency–high-amplitude sea-level-changed cyclothems, which are expressed by the cyclic appearance of subaerial exposure surfaces occurring in the Late Carboniferous and Early Permian carbonates, can be interpreted to have been formed by alternating marine transgression and regression caused by the waxing and waning of coeval Gondwanan ice sheets (see Ueno *et al.* 2012).

The Carboniferous in South China can be roughly recognized as three depositional systems, namely: the deep basin that has a very limited distribution, and consists of cherts and thin-bedded limestones; the platform slope that comprises

From: GĄSIEWICZ, A. & SŁOWAKIEWICZ, M. (eds) 2013. *Palaeozoic Climate Cycles: Their Evolutionary and Sedimentological Impact*. Geological Society, London, Special Publications, **376**, 33–46.
First published online August 6, 2013, http://dx.doi.org/10.1144/SP376.11

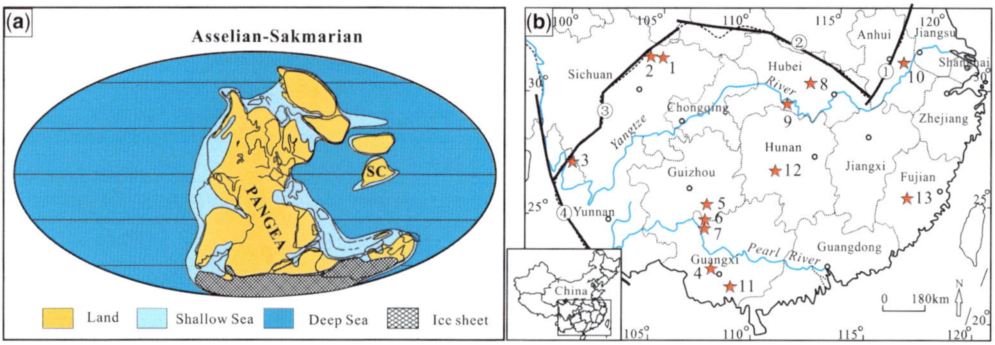

Fig. 1. (**a**) Early Permian global palaeogeographical reconstruction (modified from Metcalfe 2011): SC, South China. (**b**) Major faults bordering South China: the Tanlu Fault Zone, the Qinling-Dabie Orogen, the Houlongmenshan–Jinpingshan Fault Zone and the Honghe Fault Zone. The locations of the studied sections (also shown in Fig. 3): 1, the Majiaoba section; 2, the Shawozi section; 3, the Jinzigou section; 4, the Long'an section; 5, the Dushan section; 6, the Baping section; 7, the Luofu section; 8, the Jingshan section; 9, the Songzi section; 10, the Chaohu section; 11, the Bancheng section; 12, the Longhui section; 13, the Jiangshan section. Inset shows the location of South China on the map of China.

continuously depositional thin- to medium-bedded limestones containing abundant conodont fossils; and the platform that has a wide distribution, and contains limestones and siliciclastic lithofacies. Both the slope and the platform facies of the Carboniferous have been intensively studied. Searching for stratotype sections for the Carboniferous boundaries resulted in establishing detailed biostratigraphical sequences of conodonts in slope facies. With regard to platform facies, regional geological survey as well as synthetic stratigraphical and palaeontological studies promoted important palaeontological findings of benthic faunas, such as foraminifers, rugose corals and brachiopods, which led to the recognition of some biotic events during the Carboniferous and Early Permian (e.g. Yang *et al.* 2005; Wang *et al.* 2006).

Climatic alternation between greenhouse and icehouse conditions during the Carboniferous and Early Permian is evidenced worldwide by isotope geochemistry and cyclostratigraphy (Mii *et al.* 1999, 2001; Saltzman *et al.* 2000; Heckel *et al.* 2007; Grossman *et al.* 2008; Buggisch *et al.* 2011), which also resulted in global biotic turnovers (Jackson & Johnson 2001; Stanley & Powell 2003; Bambach *et al.* 2004; McGhee *et al.* 2012). Distinctive turnovers of biotic assemblages during the Mid-Carboniferous and post-Sakmarian have been recognized in South China (Wang *et al.* 2006). The main purpose of this paper is to present major depositional cycles of the Carboniferous and Lower Permian, to recognize biotic changeover events of various fossil groups, and to search for the possible environmental causes leading to these changes.

Distribution of the Carboniferous and Lower Permian in South China

During Carboniferous and Early Permian time, South China consisted of several subunits, including, from west to east, the Upper Yangtze Land, the Lower Yangtze platform, the Dian-Qian-Gui (Yunnan-Guizhou-Guangxi) Basin, the Jiangnan Basin and Cathaysia Land (uplift started from latest Carboniferous: Fig. 2).

South China was located within a subequatorial–equatorial position near the eastern border of the Palaeo-tethys Ocean during the Permo-Carboniferous Ice Age (Scotese 2001; Blakey 2008) (Fig. 1a). The sedimentary sequences of Carboniferous and Lower Permian sequences are quite widespread throughout South China, and crop out in structurally separated units with different settings, and which are primarily limited by some well-known fault zones (Chen *et al.* 1994) (Fig. 2). Based on the discrepancies in tectonic settings, lithofacies associations and biotic assemblages, South China could be divided into two depositional provinces:

Yangtze Province (I)

The Yangtze Province is exposed along the Yangtze River valley, and is characterized by distinctive Palaeo-Tethyan biota and diverse depositional types. The province is structurally and palaeogeographically divided into five subprovinces: Longmenshan; Ninglang-Dali; Dian-Qian-Gui; Upper Yangtze; and Lower Yangtze (Fig. 2).

Longmenshan subprovince (I_1). During the Permo-Carboniferous, a carbonate ramp occurred near

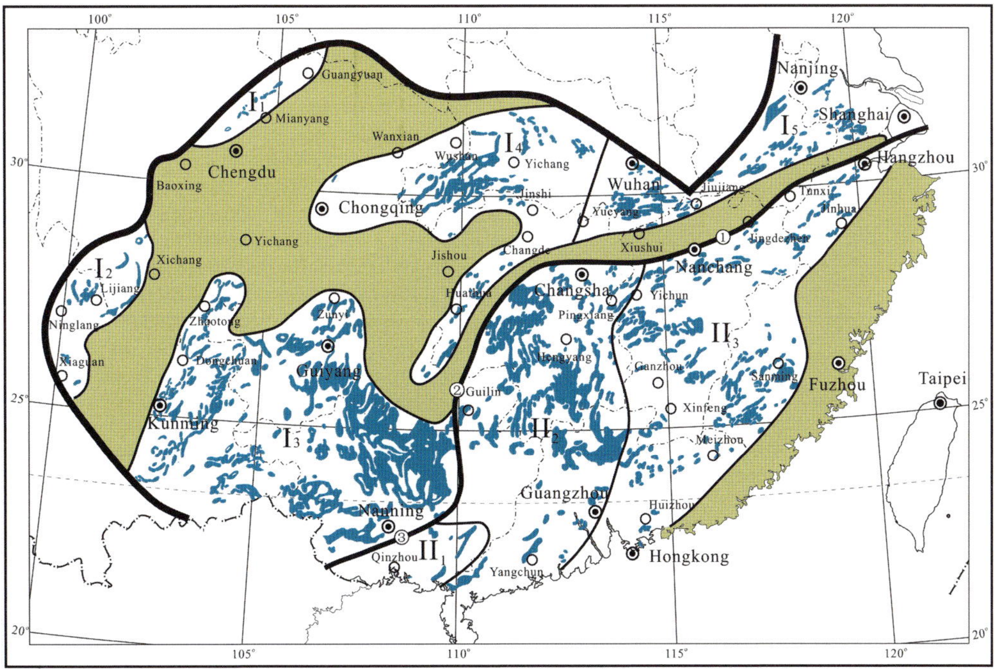

Fig. 2. Carboniferous distributions (shown in blue) and land (shown in brown) in South China. I, Yangtze province; I_1, Longmenshan subprovince; I_2, Ninglang-Dali subprovince; I_3, Dian-Qian-Gui subprovince; I_4, Upper Yangtze subprovince; I_5, Lower Yangtze subprovince; II, Southeast province; II_1, Qinzhou subprovince; II_2, Xiang-Gui subprovince; II_3, Zhe-Min-Gan-Yue subprovince. Major faults that separated the Yangtze province and the Southeast province: ① the Wuzhen-Jingdezhen-Yifeng Fault Zone; ② the Longsheng-Yongfu Fault Zone; ③ the Pingxiang-Dali Fault Zone.

the margin of the Upper Yangtze land in response to divergent tectonic background. Shallow-water carbonates dominated the sequences in this region, and were deposited in roughly NE-trending facies belts that parallel the continental margin (Fig. 2). The Lower Carboniferous comprises about 150 m of mudstones, wackestones and grainstones intercalated with small amounts of sandstone, muddy siltstones and hematite oolites. The Upper Carboniferous and the Asselian–Sakmarian of the Lower Permian consist mostly of shallow subtidal–intertidal oolitic, bioclastic, and peloidal wackestones, packstones and grainstones interbedded with dolostones, with several exposure surfaces; the contact between the Sakmarian dolostone and the Kungurian sandstones is unconformable, reflecting a regression event that led to the disappearance of the Artinskian in this region (Zhang *et al.* 2009; Liao *et al.* 2010). In ascending order, the Carboniferous and Lower Permian encompass the Majiaoba, Zongchanggou, Huanglong, Chuanshan and Liangshan formations.

Ninglang-Dali subprovince (I_2). Following the Devonian divergent plate motions, a carbonate ramp was developed on the passive continental margin in this region, lasting from Carboniferous to Early Permian time. The Lower Carboniferous comprises shallow-water bioclastic and intraclastic grainstones; the Upper Carboniferous and Lower Permian consist of bioclastic grainstones and packstones, dolomitic limestones and dolostones. In ascending order, the Carboniferous and Lower Permian sequence includes the Jinzigou, Laolongdong, Baizuo, Huashiban, Dala and Maping formations.

Dian-Qian-Gui subprovince (I_3). During Early Carboniferous time, a 'platform–basin–platform–basin' palaeogeographical framework formed along newly rifted continental margins that resulted in various depositional settings, such as attached platforms, rifting basins and isolated platforms. Tectonic controls on this palaeogeographical distribution are linked to rifting along NW- and NE-trending fault zones across this region. The Lower Carboniferous has a diverse lithofacies representing various water depths and sedimentary environments. Attached platforms are lithologically mixed, consisting of varying amounts of carbonates, siliciclastics, evaporites and coal measures, while isolated platforms are made up almost entirely of carbonates. The slope or basinal facies association includes laminated lime mudstone, thin- to medium-bedded bioclastic wackestone and siliceous mudstone. The Upper Carboniferous and Lower Permian are composed overwhelmingly of carbonates in platform settings, but have some siliceous or volcanic rocks intercalations in basinal environment. In ascending order, the sedimentary sequences of shallow-water facies consist of the Tangbagou, Xiangbai, Jiusi, Shangsi, Baizuo, Dabu, Huashiban, Dala and Maping formations, whereas that of slope or basinal facies comprise the Luzhai, Baping and Nandan formations.

Upper Yangtze subprovince (I_4). From the Late Devonian to Early Carboniferous, this subprovince was characterized by an extensive exposed landmass and a partly developed interior sag basin, in which shallow-water carbonates and clastics were deposited in a similar manner to contemporary deposits of the Lower Yangtze subprovince. The Lower Carboniferous has a very limited distribution, and is composed of shallow-marine carbonates and siliciclastics, even swamp coal-bearing strata. The lower part of the Upper Carboniferous is dominated by dolostone formed in a restricted environment; the middle part of the Upper Carboniferous, the Huanglong Formation, has a wider distribution that extensively covers this subprovince, and is underlain by Devonian or Pre-Devonian strata, reflecting a major transgression event in South China. The upper part of the Upper Carboniferous and Lower Permian consists mainly of shallow-shoal carbonates, represented by oncoid grainstones with several exposure surfaces. In ascending order, the Carboniferous and Lower Permian include the Dabu, Huanglong and Chuanshan formations.

Lower Yangtze subprovince (I_5). The Lower Yangtze region belonged to passive continental margin, and was flooded by a shallow sea that extended from eastern Hubei to Jiangsu during the Carboniferous. The Lower Carboniferous comprises mudstones, wackestones, packstones, dolostones, sandstones and siltstones intercalated with coal-bearing strata; the Upper Carboniferous and Lower Permian are mainly composed of shallow-water carbonates, such as bioclastic packstones, grainstones, wackestones and dolostones. In ascending order, the Carboniferous and Lower Permian encompass the Jinling, Gaolishan, Hezhou, Laohudong, Huanglong and Chuanshan formations.

Southeast Province (II)

This province, in the SE of South China, could be structurally divided into three subprovinces. From west to east, they are: the Qinzhou subprovince; the Xiang-Gui subprovince; and the Zhe-Min-Gan-Yue subprovince.

Qinzhou subprovince (II_1). The Qin-Fang Basin, where the most complete succession from the Silurian to Permian in China is exposed, is a remnant trough formed during the Silurian–Devonian Caledon Orogeny. It subsequently persisted to develop an oceanic sag basin following the Late Devonian–Carboniferous divergent plate separation. The Carboniferous and Lower Permian consist of deep-sea mudstones, bedded cherts and manganese claystones (Kuang *et al.* 1999). In ascending order, they include the Shijia and Bancheng formations.

Xiang-Gui subprovince (II_2). During the Carboniferous–Early Permian, a depositional basin that was bounded by NE- and NNE-trending faults persisted between the Yangtze and Cathaysia lands. The lower part of the Lower Carboniferous consists mainly of carbonates, with a few clastics intercalations, while the upper part of the Lower Carboniferous can be subdivided into three parts in terms of lithology: the lower part comprising mainly of limestone; the middle part composed mainly of coal-bearing strata; and the upper part of mainly limestone with intercalations of gypsum horizons. The Upper Carboniferous and Lower Permian consist mainly of deep-ramp, platform or shoal carbonates. In ascending order, the Carboniferous and Lower Permian comprise the Malanbian, Tian'e'ping, Shidengzi, Ceshui, Zimenqiao, Dabu, Huanglong and Chuanshan formations.

Zhe-Min-Gan-Yue subprovince (II_3). During Carboniferous and Early Permian time, the depositional basin occurred between the Jiangnan and Cathaysia lands, and covered southern Zhejiang, eastern Guangdong, and most parts of Jiangxi and Fujian. The Lower Carboniferous is characterized by clastics of great thickness, sometimes intercalated with coal-bearing strata. By contrast, the Upper Carboniferous and Lower Permian consist overwhelmingly of carbonates. In ascending order, the Carboniferous and Lower Permian include the Yejiatang (Zishan, Zhongxin), Dabu, Huanglong and Chuanshan formations.

Major sedimentary cycles during the Carboniferous and Early Permian in South China

The Permo-Carboniferous Ice Age was one of the most severe in Earth's history (Veevers & Powell 1987; Isbell *et al.* 2003, 2012; Fielding *et al.* 2008*b*; Wopfner 2012). The variations in ice-sheet volume had dramatic influences on marine temperature and oceanic circulation, and caused glacio-eustatic fluctuations of about 10–120 m (Joachimski *et al.* 2006; Rygel *et al.* 2008). As mentioned above, all of the sedimentary basins of South China were formed in extensional tectonic settings during Carboniferous and Early Permian time; thus, multiple regressive events that occurred throughout South China should be primarily induced by glacio-eustatic drawdown. It is thought to have contributed to the major lithological variations and sedimentary cycles in South China, especially in shallow-water deposition regions. The lowstand during glacial period could result in erosion of the underlying shelves, the occurrence of abundant siliciclastics and extensive reflux dolomitization (Isaacson *et al.* 2008). Subaerially exposed carbonate shelves would commonly undergo a dramatic, geomorphic modification into a karst landscape owing to weathering by dissolution.

The topmost Devonian (middle *praesulcata* zone) regression is one of the most characteristic events that could be traced worldwide, including North America, the Urals, Belgium, France and South China (Kalvoda 1989; Isaacson *et al.* 2008). The regressive series is represented by the intercalation of the Changshun shale in deep-water carbonates that was deposited in the intra-platform basin (Wang & Yin 1984; Hou *et al.* 1985; Ji *et al.* 1989), siliciclastic influx, reflux dolomites and limestone-solution breccia in shallow-water platforms throughout South China (Fig. 3). The base of the Carboniferous is characterized by microbial carbonates in the Dian-Qian-Gui (I_3) and Xiang-Gui subprovinces (II_2) (Hou *et al.* 2011*b*), and by an onlap of the Earliest Tournaisian (Tangbagouan) limestone in the Longmenshan subprovince (I_1). This sedimentary sequence indicates the beginning of transgression after the Hangenberg regression event.

A major regression associated with obvious sedimentological changes marks the boundary between the middle and upper Tournaisian sequences in South China (Wang *et al.* 1997) (Fig. 3), which could correspond to the Kinderhookian–Osagean boundary in North America, and represents a level of a conspicuous lithological as well as a parallel biotic change. The siliciclastic and coal-bearing strata mostly underlain by shallow-water limestone dominate the Longmenshan (I_1), Dian-Qian-Gui (I_3), Xiang-Gui (II_2) and Zhe-Min-Gan-Yue (II_3) subprovinces in South China. This may suggest that the increased prevalence of clastic facies possibly indicates a decline of carbonate sedimentation during a glacial or cooler climatic episode. The overlying carbonates of the Majiaoba, Tangbagou and Doulingao formations suggest a subsequent transgression.

A major regression occurred at the Tournasian–Viséan boundary in South China (Hou *et al.* 2011*a*). This regressive event resulted in the occurrences of sedimentary hiatus in the Longmenshan(I_1) and Zhe-Min-Gan-Yue (II_3) subprovinces, abundant clastic influx underlain by shallow-water limestone in the Dian-Qian-Gui (I_3), Upper Yangtze (I_4) and Lower Yangtze (I_5) subprovinces, and a series of widely distributed dolostone in Xiang-Gui subprovince (II_2), and indicates possible climatic changes during this time.

A coeval sea-level fall can be easily recognized in the uppermost Viséan deposits in the Longmenshan (I_1), Ninglang-Dali (I_2), Dian-Qian-Gui (I_3), Upper Yangtze (I_4), Lower Yangtze (I_5) as well as the Xiang-Gui (II_2) subprovinces. It is reflected by occurrences of siliciclastic and coal-bearing strata, and limestone-solution breccia and reflux dolostone. These lowstand deposits in South China coincide with the glacio-eustasy sequences in North America (Veevers & Powell 1987), glacial event (C1) in East Australia (Fielding *et al.* 2008*a*) and extensively distributed glacial deposits from Gondwana (González-Bonorino 1992; Garzanti & Sciunnach 1997), marking the onset of the Permian–Carboniferous icehouse.

A major sea-level lowstand at the Mid-Carboniferous boundary, manifested by widespread dolostone, can be traced in the Longmenshan (I_1), Dian-Qian-Gui (I_3), Upper Yangtze (I_4), and Lower Yangtze (I_5) subprovinces. This major regressive event seems to be compatible with another glacial event (C2: Fig. 3). It is recognized by the absence of upper Serpukhovian strata in many places in South China, and by several erosional surfaces in carbonate sequences in the Lower Yangtze (I_5) subprovince. Subsequent transgression occurred in the Early Bashkirian. A wide and uniform shallow-water platform developed in South China, on which tidal-flat dolostone and pure limestone were deposited.

The Upper Carboniferous and Lower Permian are composed mostly of shallow-marine carbonates, which consist of remarkable, high-frequency cyclothems that can be interpreted to have been formed from icehouse-type high-frequency–high-amplitude sea-level change caused by the cyclic waxing and waning of Gondwanan ice sheets (see Ueno *et al.* 2012). The cyclothems in South China are expressed by cyclic appearances of subaerial exposure surfaces, which were well studied in southern Guizhou (Ueno *et al.* 2012) in the Dian-Qian-Gui subprovince (I_3) and can also be observed in many areas in South China. These cyclothems can be compared with the well-studied Pennsylvanian–Early Permian Midcontinent cyclothems in North America. It has also been reported that the Maping Formation (spanning from the

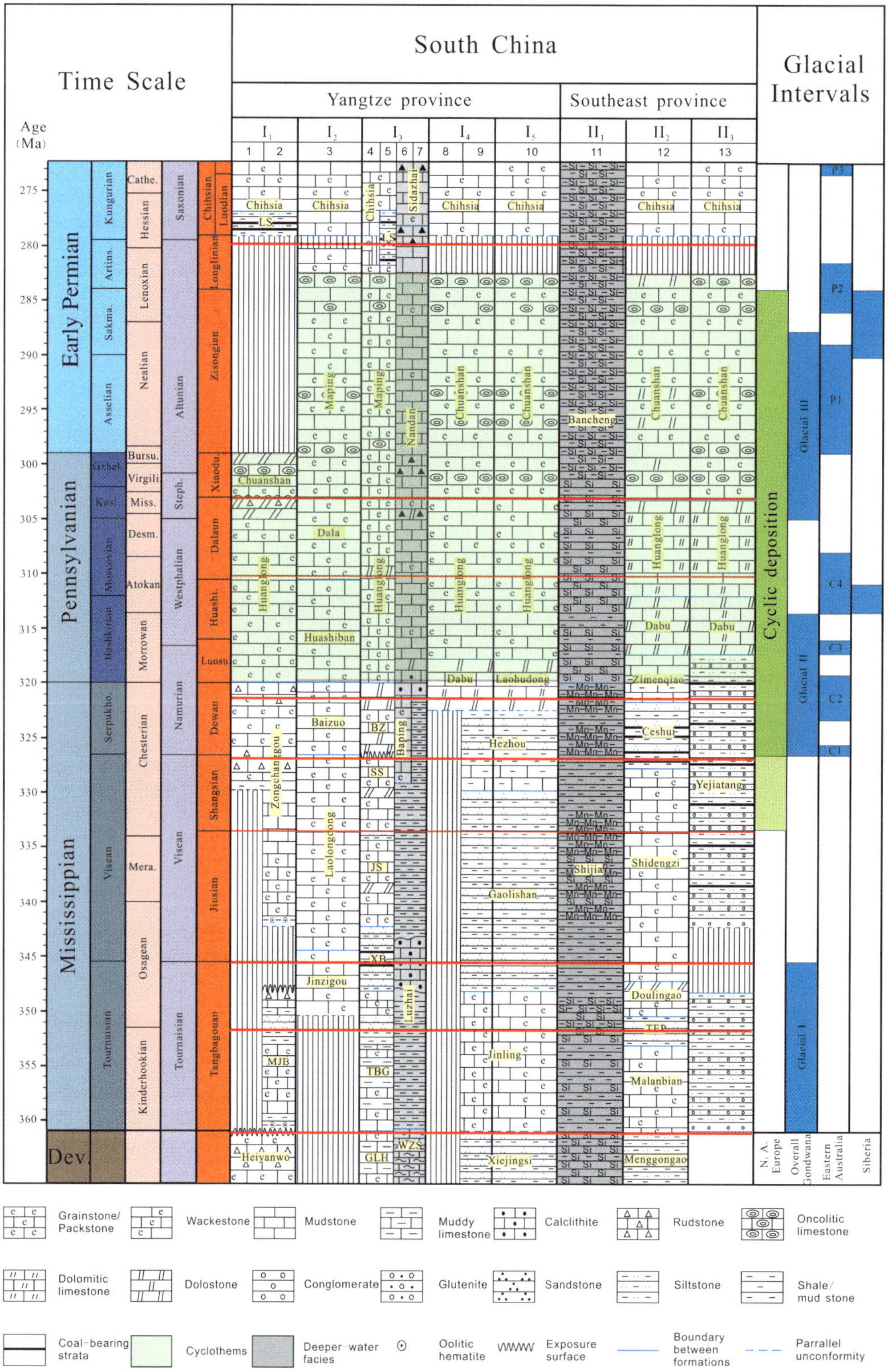
Time Scale
South China
Yangtze province
Southeast province
Glacial Intervals
Age (Ma)
Early Permian
Pennsylvanian
Mississippian
Dev.
Cyclic deposition
Glacial I
Glacial II
Glacial III
N. A. Europe
Overall Gondwana
Eastern Australia
Siberia
Grainstone/ Packstone
Wackestone
Mudstone
Muddy limestone
Calclithite
Rudstone
Oncolitic limestone
Dolomitic limestone
Dolostone
Conglomerate
Glutenite
Sandstone
Siltstone
Shale/ mud stone
Coal-bearing strata
Cyclothems
Deeper water facies
Oolitic hematite
Exposure surface
Boundary between formations
Parrallel unconformity

middle Kasimovian to the Artinskian) at Dushan, southern Guizhou includes more than 30 parasequences, which are composed of, in ascending order, bioclastic wackestone, solitary coral-bearing grainstone, fusulinid-bearing grainstone and algal boundstone, reflecting the evolution from the subtidal to peritidal zone (Li *et al.* 1997).

Another major sea-level fall occurs at the top of the Sakmarian, which produced a series of coal-bearing strata (Liangshan Formation) in the Longmenshan (I_1) and Dian-Qian-Gui(I_3) subprovinces, and an extensive sedimentary hiatus throughout South China (Fig. 3). Subsequent transgression occurred in the earliest Kungurian, when a wide, uniform shallow-water platform or ramp developed in South China. The platform was covered by a unique carbonate succession, the Chihsia Formation, rich in organic matter and chert nodules (Yan 2004).

Biotic events of the Carboniferous and Early Permian in South China

Rugose corals

The rugose corals in South China show several declines and changeovers in the diversity during the Carboniferous and Early Permian (Wang *et al.* 2006) (Fig. 4). There are two changeovers of coral assemblages developed through the Carboniferous and Permian, one at the Sepukhovian–Bashkirian boundary and another at the end of the Sarkmarian. These evolutionary changes are closely related to environmental changes.

The changeover event in rugose coral assemblages at the Sepukhovian–Bashkirian boundary. The Tournaisian and Viséan constitute an interval of radiation of rugose corals, marked by a rapid increase in the number of taxa. The rate of origination was much higher than the rate of extinction. During Serpukhovian time, however, this relationship changed, so that the rate of extinction became higher than the rate of introduction, which was expressed by the disappearance of many large, dissepimented solitary corals with axial structures such as *Kueichouphyllum*, *Heterocaninia*, *Axophyllum*, *Aspidophyllum*, *Kusbassophyllum*, *Lophophyllum*, *Spirophyllum*, *Qinghaiphyllum* and *Yuanophyllum* as well as many compound corals such as *Kwangsiphyllum*, *Orionastraea*, *Stelechophyllum* and *Stylastraea*, all of which are common genera in the Viséan coral faunas in South China (Wang *et al.* 2006).

In contrast to the high rate of extinction in the Serpukhovian, the Bashkirian was again characterized by a low rate of extinction and a high rate of origination, with increased appearance of new taxa belonging to the families Geyerophyllidae, Kepingophyllidae, Waagenophyllidae and Petalaxidae. Compound corals replaced the large dissepimented solitary corals in the Mississippian to become the dominant types in the Pennsylvanian. This compositional change in the rugose coral assemblages at the Serpukhovian–Bashkirian boundary was named as the Mid-Carboniferous rugose coral evolutionary event (Wang *et al.* 2006). This biotic change is mainly related to the change in geological settings from the siliciclastic source to the clear carbonate source in South China. Abundant compound rugose corals, especially massive colonies which prefer to live in relatively clear and well-circulated environments, replaced the large solitary dissepimented corals and were dominant from the Bashkirian to the end of the Sakmarian (Wang *et al.* 2006).

The changeover event in rugose corals at the end of the Sakmarian. The Asselian–Sakmarian interval reveals the highest diversity of rugose corals during the Carboniferous and Permian in South China (see fig. 2 of Wang *et al.* 2006). Comparatively, the endemic Tethyan family, the Kepingophyllidae, and other families such as the Petalaxidae, Waagenophyllidae, Cyathopsidae and Bothrophyllidae also show higher diversification. However, a distinctive changeover occurred in rugose coral assemblages between the Sakmarian and the Artinskian. The changeover includes a continuous decrease in the diversity starting from the Artinskian to the end of the Permian, a consistently higher rate of extinction than that of the origination, and a dramatic disappearance of typical Pennsylvanian and Early Permian families such as Bothrophyllidae, Lithostrotionidae, Cyathopsidae and Petalaxidae. In contrast to the transitional faunas of the Asselian–Sakmarian interval, the Artinskian was dominated by two uniquely typical Permian rugose families: the Kepingophyllidae and the Waagenophyllidae. This change in rugose coral faunas is also related to a major environmental change caused by the

Fig. 3. Carboniferous–Lower Permian sedimentary sequences of South China, and sea-level curves and their correlation with cyclic deposition in Laurentia and glaciation in Gondwana. I, Yangtze province; I_1, Longmenshan subprovince; I_2, Ninglang-Dali subprovince; I_3, Dian-Qian-Gui subprovince; I_4, Upper Yangtze subprovince; I_5, Lower Yangtze subprovince; II, Southeast province: II_1, Qinzhou subprovince; II_2, Xiang-Gui subprovince; II_3, Zhe-Min-Gan-Yue subprovince. Temporal distribution of cyclic deposition and ice age are from Epshteyn (1981*a*, *b*), Veevers & Powell (1987), Chumakov (1994), Smith & Read (2000), Isbell *et al.* (2003) and Fielding *et al.* (2008*a*). Timescale after Menning *et al.* (2006).

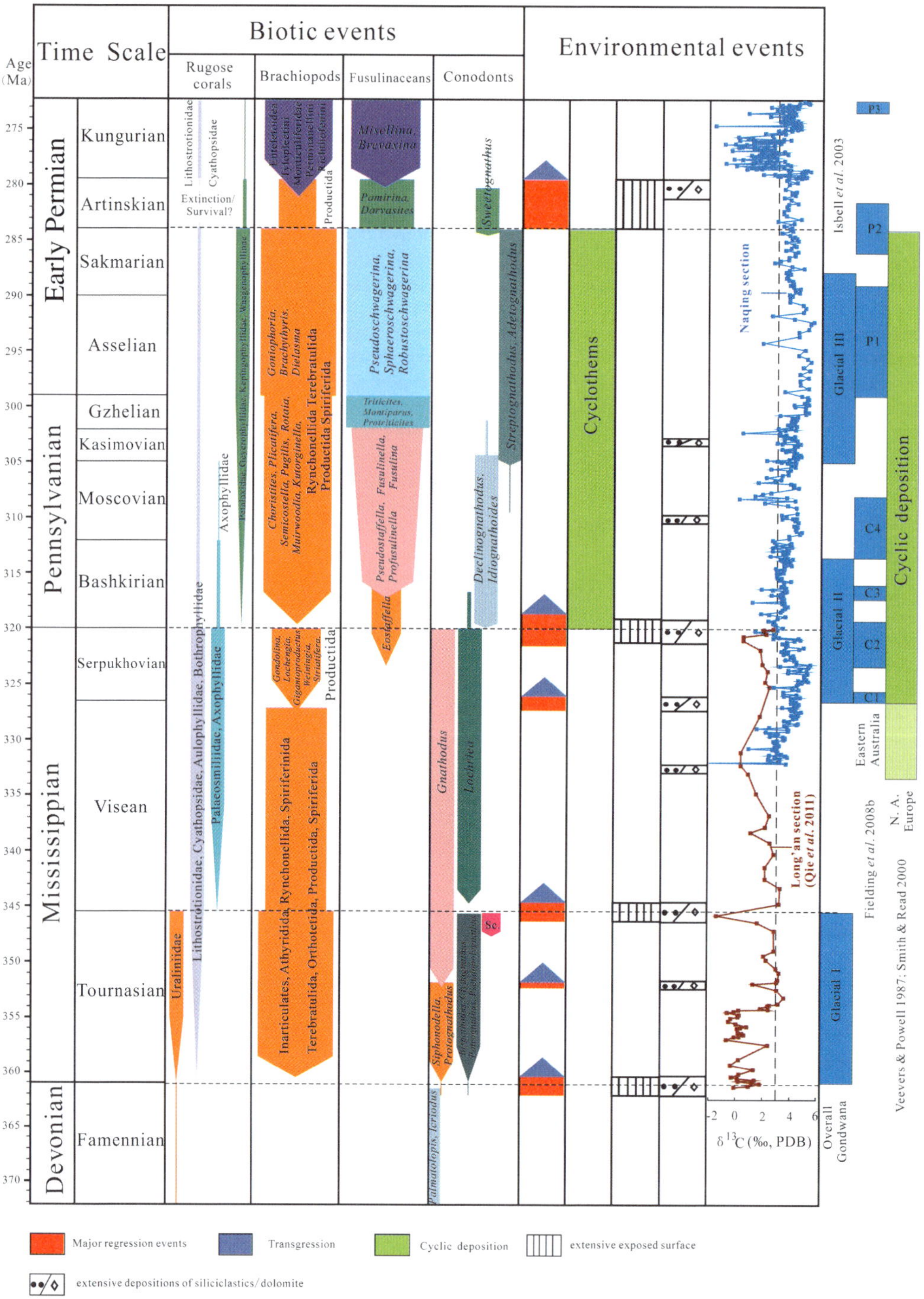

Fig. 4. Correlation of evolutionary patterns of rugose corals, brachiopods, fusulinaceans and conodonts as well as eustatic oscillations during the Carboniferous and Early Permian with composite $\delta^{13}C$ value curves from the Naqing section, Guizhou and the Long'an section, Guangxi (Qie *et al.* 2011). Temporal distribution of cyclic deposition and ice age are from Veevers & Powell (1987), Smith & Read (2000), Isbell *et al.* (2003) and Fielding *et al.* (2008*a*). Timescale after Menning *et al.* (2006).

worldwide regression event at the end of the Sakmarian (Leven 1994). This regression is marked in South China by an apparent disconformity above the Sakmarian Chuanshan (or Maping) Formation composed of massive limestone.

Brachiopods

The evolutionary event in brachiopod faunas at the mid-Carboniferous boundary. The palaeobiogeographical provinces of brachiopods are clearly exhibited in the Early Carboniferous, especially in the Serpukhovian fossil record. In South China, *Gigantoproductus*-type brachiopod, which was rarely recorded from the Early Viséan, was highly diversified in the Serpukhovian in terms of the large shell size and complex ornamentation. Highly diversified *Gigantoproductus*, such as *Gigantoproductus edelburgensis* and *G. weiningensis*, and other distinctive taxa such as *Striatifera striata*, *S. angusta*, *Gondolina* spp., *Kansuellakansuensis*, *Lochengia* spp. and *Weiningia* spp., etc., dominated the Serpukhovian strata in South China. The brachiopod assemblage could be well recognized as an indicator of the Serpukhovian. In contrast to the brachiopod assemblage of South China, those found in the coeval Boreal Realm (e.g. Junggar-Hinggan, Mongol, Siberia and North America) and Gondwana Realm (e.g. Himalaya–Western Yunnan and Australia) obviously lacked such taxa as *Gigantoproductus*, *Striatifera* and *Lochengia*. Instead, these areas were dominated by the genera of families Syringothyridae and Davisiellidae (Zhang & Jin 1976; Yang 1990; Liao 1995).

The changeover in the composition of the brachiopod faunas between the Serpukhovian and Bashkirian was reflected by dramatic variations in shell size, abundance of fossils and shell ornamentation. In contrast to large-sized brachiopods in the Serpukhovian, the Bashkirian brachiopods are mainly small to moderate sized with respect to the shells of *Choristites* (the family Spiriferidae), and to other families belonging to Productida, Rhynchonellida and Terebratulida. The most common species in the Bashkirian in South China are *Choristites mansuyi*, *C. mosquensis*, *Plicatifera chaoi*, *Semicostella panxianensis*, *Pugilis pugiliformis*, *Muirwoodia yohi*, *Rotaia subtrigona*, *Kutorginella* sp., *Goniophoria carinata*, *Brachythyris rectongula* and *Dielasma giganteum*. These Bashkirian taxa mostly persist throughout the Late Carboniferous with a few new originated taxa in the other stages (Fig. 4).

The evolutionary event in brachiopod faunas at the Sakmarian–Artinskian boundary. Brachiopods in South China as well as in other region of the world experienced a short-term flourishing period in the Asselian and Sakmarian, when distinctive brachiopod faunas were recorded from Guizhou and Guangxi (Dian-Qian-Gui subprovince). However, a drop in the diversity of brachiopods dramatically occurred at the end of the Sakmarian. The *Dictyoclostus uralicus–Neochonetes puanensis* assemblage that was present at the top of the Sakmarian in South China is composed of such common taxa as *Nantanella mapingensis*, *Alexenia* sp., *Meekella* sp. and *Pugnax* sp., most of which are genera extending from the previous intervals and showing relatively low diversity and abundance (Wu *et al.* 1974; Liao 1979; Liao 1995).

From the Artinskian, the palaeobiogeographical provinces of brachiopods gradually became distinct again. As a typical Tethyan Realm, South China possesses brachiopod faunas of several distinctive features: (1) the superfamily Enteletoidea, which had continued a low diversity since the Devonian, began their wide diversification in the Artinskian, and continued to flourish in the Middle and Late Permian; (2) the family Monticuliferidae and Tyloplectini-type taxa of the order Productida started to become the most common genera of the brachiopod faunas at this time; and (3) some new taxa belonging to Richthofenini and Perminanellini, and many earlier elements of the order Lytoniidina, started to appear or even flourished in the Artinskian (Fig. 4). Representative species of the Artinskian brachiopods include *Orthotichia chekiangensis*, *O. magnifica*, *Lyraplecta richthofeni*, *Chaoina reticulate*, *Monticulifera sinensis*, *Tyloplecta* sp. and *Pygmochonetes* sp.

Fusulinaceans

The framework of the fusuline taxonomy and classification in China was established with the pioneering work of Deprat (1912), Lee (1924, 1931) and others (e.g. Chen 1934) nearly 100 years ago. The fusuline succession of the Carboniferous and Early Permian was elaborated through the regional studies in South China (Zhang 1987; Kong *et al.* 1996). The need for refinement of the fusuline succession is realized, however, along with the growth of knowledge on the evolution of lineages.

The evolutionary pattern of the fusuline diversity for the Early Permian was analysed based on the fusuline database from South China (Yang *et al.* 2005). It reveals an increase in species diversity from the Gzhelian to the Early Sakmarian. This increase was followed by a significant loss of 40% of species diversity from the Late Sakmarian to the Early Kungurian, largely because of the sea-level changes during this period (Fig. 4). The diversity for the Late Carboniferous, however, remains unexplored. It is generally accepted that the first fusuline genus *Eostaffella* appeared in the Hezhou

Formation of the upper Lower Carboniferous (Chang 1962). Fusulines started to flourish from the late Bashkirian, and then experienced a slow and steady increase in its diversity in the Late Carboniferous (Sheng *et al.* 1988). However, the latest analysis of the fusuline database from the Arctic and Eurasia indicated a sharp drop in the diversity within Kasimovian time, followed by a rapid rebound in Gzhelian time. The diversity peak was reached at the Pennsylvanian–Permian boundary, and dropped again in the Asselian and Sakmarian (Groves & Wang 2009).

Conodonts

Several conodont biotic turnovers that were expressed as strongly reduced taxonomic diversities or extinctions coincide with the major Late Palaeozoic eustatic events during several critical intervals, such as the Devonian–Carboniferous boundary, the Mid-Carboniferous boundary and the Sakmarian–Artinskian boundary intervals.

The Devonian–Carboniferous boundary beds are well developed and exposed in Guangxi Province and Guizhou Province, South China. The Nanbiancun section in Guilin of Guangxi and the Daposhang section in Muhua of Guizhou are two of the most representative sections of the Devonian–Carboniferous boundary stratotype (Wang & Yin 1988; Ji *et al.* 1989). In both of the sections, the conodont fauna from the Lower and Middle *Siphonodella praesulcata* zones occurs below the Hangenberg Shale and was dominated by *Palmatolepis* elements. However, above the Hangenberg Shale, the conodont fauna from the Upper *Siphonodella praesulcata* Zone and the *Siphonodella sulcata* Zone was characterized by species of *Siphonodella*, *Polygnathus* and *Protognathodus*. This is a typical conodont distribution pattern from the Devonian–Carboniferous boundary beds in South China. This pattern of the same interval is recognized elsewhere in the world. *Palmatolepis* biofacies generally occurred in the deep-water slope facies or in the outer shelf where a good circulation existed. *Siphonodella–Polygnathus* biofacies represents a relatively shallow outer-shelf environment (Wang &Yin 1988).

The Naqing (Nashui) section in Luodian of Guizhou is the best-studied section for Carboniferous and Lower Permian conodonts (Wang & Higgins 1989; Wang 1996; Wang & Qi 2003; Qi 2008; Chen 2011). The abundant and high-diversity conodont faunas were obtained also from the Mid-Carboniferous boundary and the Sakmarian–Artinskian boundary intervals. The major difference in conodont assemblage across the Mid-Carboniferous boundary is the changeover of dominant genera. Below the boundary, *Gnathodus* and *Lochriea* are the dominant genera, which are the key elements for defining the upper Lower Carboniferous zones. By contrast, *Declinognathodus* and *Idiognathoides* became the key taxa to define conodont zonations of the lower Upper Carboniferous. This change can be observed in other areas over the world (Nemirovskaya & Nigmadganov 1994).

A dramatic change in conodont assemblage occurs at the Sakmarian–Artinskian boundary in the Naqing section and is characterized by: (1) the extinction of *Streptognathodus* and *Adetognathus* at the end of the Late Sakmarian; and (2) the fact that *Sweetognathus* appeared in the Late Sakmarian and became a dominant taxon in the Early Artinskian (Fig. 4).

Discussion

Possible climatic impact on the biotic events in South China

South China is well known for its complete marine Permo-Carboniferous sequences, which have been well studied for biostratigraphy, chronostratigraphy and sedimentary geology, and, thus, provide invaluable material for the investigation of the co-evolution of organisms and environment during the Palaeozoic Ice Age. As mentioned earlier, there are two major biotic events that developed through the Carboniferous–Early Permian. The first one occurred in latest Mississippian and coincided with a lowstand in South China. Analyses of oxygen isotopes of well-preserved brachiopod shells from North America at low latitudes show convincing evidence for positive shifts in $\delta^{18}O$ values at the mid-Carboniferous boundary (Grossman *et al.* 2008). The timing of this biotic event agrees much more closely with the beginning of low-latitude cooling and the expansion of glaciation just prior to the mid-Carboniferous boundary. Moreover, these new Bashkirian genera of rugose coral and brachiopod dominated the benthic fauna in South China during the ice age. Thus, it is not surprising that global cooling has contributed to the decline in the diversity and abundance of benthic biota, and to a changeover of fossil assemblages in South China. Furthermore, the changes of geological settings from the Early Carboniferous siliciclastic source to the Late Carboniferous clear carbonates source may also have played an important role in controlling the evolutionary pattern of the rugose corals.

The second major biotic event occurred near the Sakmarian–Artinskian boundary and coincided with another major sea-level fall in South China. The major regression resulted in the development of a series of coal-bearing strata and an extensive sedimentary hiatus throughout South China. The $\delta^{18}O$ values of Sakmarian–Artinskian brachiopod shells from the Urals show a decreasing trend (Korte *et al.* 2005), which is expected with an Artinskian waning of the glaciers (e.g. Isbell *et al.* 2003). However, the data from North America and the Urals by Grossman *et al.* (2008) show an increase. In South China, no comparable data have been released until now. Thus, the thermal change at low-latitude zones during this time interval is questionable. Although most reviews indicate that the Permo-Carboniferous ice age collapsed during the late Early Permian, the eastern Australian records show that glacial conditions persisted for a further 20–30 Ma and reached an acme in the late Early Permian (Fielding *et al.* 2008*a*). The Artinskian lowstand in South China, consistent with the glaciation in eastern Australia (P2: Fielding *et al.* 2008*a*), leading to the reduction of habitable area on the continental shelves for shallow-marine species, caused the changeover of the main biological groups in this area. The second biotic event may be related to cooling and the major fall in sea level, which were induced by the eastern Australian glaciation. Clearly, these important results require further confirmation.

Carbon isotopic records in South China and their palaeoclimatic implications

The major transitions of marine ecosystems represented by multiple sedimentary cycles and biotic events occurred during the Carboniferous and Early Permian, indicating the influences of the Late Palaeozoic Ice Age on the Earth system. It has been widely observed that continental glaciation and deglaciation often coincide with positive and negative $\delta^{13}C$ shifts in marine carbonates, respectively.

Lower Carboniferous carbonates from the Long'an section in the Dian-Qian-Gui subprovince have $\delta^{13}C$ values of between of −1.42 and 3.58‰ (V-PDB (Vienna-Pee Dee Belemnite)). The $\delta^{13}C$ values have a range from −0.36 to 1.41‰ in the lower Tournaisian and have an abrupt increase to a maximum of 3.56‰ in the middle Tournaisian. Relatively homogenous $\delta^{13}C$ values around 3.2‰ are observed in the upper Tournaisian and lower Viséan at the Long'an section, which was interrupted by a major negative shift to a minimum of −1.42‰ near the Tournaisian–Viséan boundary. Following peak values in the earliest Viséan, $\delta^{13}C$ values began to gradually decline in the early Viséan from 3.32 to 0.37‰. Another major positive shift in $\delta^{13}C$ values with amplitude of 2.1‰ is observed from the late Viséan to the early Serpukhovian. This shift has also been recorded in slope carbonates at the Naqing section in the Dian-Qian-Gui subprovince (Fig. 4). The variability in peak values of $\delta^{13}C$ in the middle Serpukhovian between the Long'an and Naqing sections may be due to the discrepancies in the palaeogeographical settings. The major positive excursions of $\delta^{13}C$ values in Lower Carboniferous carbonates in South China are consistent with the fall in global sea level, indicating that glaciations may have occurred during these time intervals (Qie *et al.* 2011).

Most $\delta^{13}C$ values for slope carbonates from Bashkirian to Artinskian at the Naqing section are greater than 3‰ (Fig. 4) (see also Buggisch *et al.* 2011). These persistently high isotopic values in the Pennsylvanian and the shift towards more positive $\delta^{13}C$ values in the latest Pennsylvanian and earliest Permian are consistent with results from isotopic studies of the brachiopod shells in North America (Grossman *et al.* 2008). The highest $\delta^{13}C$ values around the Gzhelian–Asselian boundary in the Naqing section coincide with the inferred Permo-Carboniferous maximum glaciations (Buggisch *et al.* 2011). The causes for this interesting relationship are still unknown. Nonetheless, carbon isotope positive shifts are generally recognized to be related to abundant organic burial or enhanced environment productivity. Since no abundant organic burial has been reported during this time interval, we assume that the high $\delta^{13}C$ values in the Pennsylvanian and Lower Permian are related to higher primary productivity, induced by plentiful nutrient element input and this, in turn, by enhanced continental weathering and current turnovers during the ice age.

Conclusions

The Carboniferous and Early Permian of South China could be structurally divided into two depositional provinces, namely the Yangtze province and the Southeast province. The Yangtze province, along the Yangtze River valley, consists of five subprovinces, including the Longmenshan, Ninglang-Dali, Dian-Qian-Gui, Upper Yangtze and Lower Yangtze subprovinces. The Southeast province, in the SE of South China, could be further divided into three subprovinces, from west to east, the Qinzhou, Xiang-Gui and Zhe-Min-Gan-Yue subprovinces. The Lower Carboniferous lithology of South China is diverse, ranging from basinal and shelf carbonates to coal measures and continental clastics, while the Upper Carboniferous and Lower Permian are composed mostly of shallow-marine carbonates.

From latest Devonian to Early Carboniferous, five major regression events are recognized separately at the topmost Devonian, middle and upper

Tournasian boundary, Tournaisian–Viséan boundary, uppermost Viséan, and the Mid-Carboniferous boundary in South China, which are marked by the occurrences of sedimentary hiatus, abundant clastic influx or widely distributed dolostone. Subsequent transgressions usually resulted in the formation of lithologically mixed or pure carbonate platform on the former exposed shelves. The Upper Carboniferous and Lower Permian consist of remarkable high-frequency cyclothems from shallow-water carbonate platforms in South China, which are expressed by cyclic appearances of subaerial exposure surfaces and can be observed in many areas in South China. Another major sea-level fall is recognized at the top of the Sakmarian, which resulted in an extensive sedimentary hiatus throughout South China. Subsequent transgression occurred in the earliest Kungurian and formed a uniform shallow-water carbonate platform or ramp in South China.

From the Carboniferous to Early Permian, two major biotic events could be identified, one at the Mid-Carboniferous boundary and the other at the Sakmarian–Artinskian boundary, respectively. A decrease in the diversity of benthic biota occurred in the Serpukhovian, followed by a changeover in the composition of fossil assemblages at the Serpukhovian–Bashkirian boundary. Compound corals replaced the large dissepimented solitary corals in the Early Carboniferous to become the dominant types in the Late Carboniferous–Early Permian, accompanied by the appearances of new taxa of the families Geyerophyllidae, Kepingophyllidae, Waagenophyllidae and Petalaxidae. The Bashkirian brachiopods show dramatic changes in shell size, shell ornamentation and abundance. The *Declinognathodus* and *Idiognathoides*, instead of *Gnathodus* and *Lochriea*, become the dominant conodont genera above the Mid-Carboniferous boundary. Another major change in the faunal assemblages, at the Sakmarian–Artinskian boundary, resulted in the absence from the Artinskian of representatives of typical Pennsylvanian and Earliest Permian families of rugose corals and fusulinids. Families uniquely typical of the Kungurian and mid-Permian, characterized by rugose coral Waagenophyllidae and Kepingophyllidae as well as by fusulinid Verbeekinidae, dominated post-Sakmarian strata. These two biotic events are associated with the major regression events at the Mid-Carboniferous and Sakmarian–Artinskian boundaries, and were probably caused by episodes of glaciations in Gondwana.

This study was financially supported by the National Natural Science Foundation of China, the Chinese Academy of Sciences, and the State Key Laboratory of Palaeobiology and Stratigraphy, Nanjing Institute of Geology and Palaeontology. The authors would like to thank the anonymous reviewers, who helped to improve this manuscript.

References

Bambach, R. K., Knoll, A. H. & Wang, S. C. 2004. Origination, extinction, and mass depletions of marine diversity. *Paleobiology*, **30**, 522–542.

Blakey, R. C. 2008. Gondwana paleogeography from assembly to breakup-A 500 m.y. odyssey. *In*: Fielding, C. R., Frank, T. D. & Isbell, J. L. (eds) *Resolving the Late Paleozoic Ice Age in Time and Space*. Geological Society of America, Special Papers, **441**, 1–28.

Buggisch, W., Wang, X. D., Alekseev, A. S. & Joachimski, M. M. 2011. Carboniferous–Permian carbon isotope stratigraphy of successions from China (Yangtze platfom), Usa(Kansas) and Russia (Moscow Basin and Urals). *Palaeogeography, Palaeoclimatology, Palaeoecology*, **301**, 18–38.

Chang, L. H. 1962. Fusulinids from the Hochow Limestone, Hohsien, Anhui. *Acta Palaeontologica Sinica*, **10**, 433–443.

Chen, H. M., Wu, X. H. et al. 1994. *Carboniferous Lithofacies, Paleogeography and Mineralization in South China*. Geological Publishing House, Beijing (in Chinese).

Chen, J. 2011. *Early Permian (Cisuralian) conodont biostratigraphy in southern Guizhou and Global correlation*. PhD thesis, Graduate University of Chinese Academy of Sciences.

Chen, S. 1934. Fusulinidae of South China, Part 1. *Palaeontologica Sinica*, **4**, 1–185.

Chumakov, N. M. 1994. Evidence of Late Permian glaciations in the Kolyma River Basin: a repercussion of the Gondwana glaciations in northeast Asia? *Stratigraphy and Geological Correlation*, **2**, 426–444.

Deprat, J. 1912. *Étude des Fusulinidés de Chine et d'Indochine et classification de calcaires à fusulines. Mémoires du Service géologique de l'Indochine*, **1**, 1–76.

Epshteyn, O. G. 1981*a*. Late Permian ice-marine deposits of northeastern U.S.S.R. *In*: Hambrey, M. J. & Harland, W. B. (eds) *Earth's Pre-Pleistocene Glacial Record*. Cambridge University Press, London, 268–269.

Epshteyn, O. G. 1981*b*. Middle Carboniferous ice-marine deposits of northeastern U.S.S.R. *In*: Hambrey, M. J. & Harland, W. B. (eds) *Earth's Pre-Pleistocene Glacial Record*. Cambridge University Press, London, 270–273.

Fielding, C. R., Frank, T. D., Birgenheier, L. P., Rygel, M. C., Jones, A. T. & Roberts, J. 2008*a*. Stratigraphic imprint of the Late Palaeozoic Ice Age in eastern Australia: a record of alternating glacial and nonglacial climate regime. *Journal of the Geological Society*, **165**, 129–140.

Fielding, C. R., Frank, T. D. & Isbell, J. L. 2008*b*. The late Paleozoic ice age – a review of current understanding and synthesis of global climate patterns. *In*: Fielding, C. R., Frank, T. D. & Isbell, J. L. (eds) *Resolving the Late Paleozoic Ice Age in Time and Space*. Geological Society of America, Special Papers, **441**, 343–354.

Garzanti, E. & Sciunnach, D. 1997. Early Carboniferous onset of Gondwanian glaciation and Neo-tethyan rifting in South Tibet. *Earth and Planetary Science Letters*, **148**, 359–365.

González-Bonorino, G. 1992. Carboniferous glaciation in Gondwana. Evidence for grounded marine ice and continental glaciation in southwestern Argentina. *Palaeogeography, Palaeoclimatology, Palaeoecology*, **91**, 363–375.

Grossman, E., Yancey, T., Jones, T., Bruckschen, P., Chuvashov, B., Mazzullo, S. & Mii, H. 2008. Glaciation, aridification, and carbon sequestration in the Permo-Carboniferous: the isotopic record from low latitudes. *Palaeogeography, Palaeoclimatology, Palaeoecology*, **268**, 222–233.

Groves, J. R. & Wang, Y. 2009. Foraminiferal diversification during the late Paleozoic ice age. *Paleobiology*, **35**, 367–392.

Heckel, P. H., Alekseev, A. S. *et al.* 2007. Cyclothem ['digital'] correlation and biostratigraphy across global Moscovian-Kasimovian-Gzhelian Stage boundary interval (Middle–Upper Pennsylvanian Series) in North America and Eurasia. *Geology*, **35**, 607–610.

Hou, H. F., Ji, Q. *et al.* 1985. *Muhua Section of Devonian–Carboniferous Boundary Beds*. Geological Publishing House, Beijing (in Chinese with English summary).

Hou, H. F., Wu, X. H. & Yin, B. A. 2011*a*. Correlation of the Tournaisian-Visean boundary beds. *Acta Geologica Sinica*, **85**, 354–365.

Hou, H. F., Zhou, H. L. & Liu, J. B. 2011*b*. Microbial sediments occurring after the end-Devonian extinction event on the Hunan platform. *Acta Geologica Sinica*, **85**, 145–156.

Isaacson, P., Díaz-Martínez, E., Grader, G., Kalvoda, J., Babek, O. & Devuyst, F. 2008. Late Devonian-earliest Mississippian glaciation in Gondwanaland and its biogeographic consequences. *Palaeogeography, Palaeoclimatology, Palaeoecology*, **268**, 126–142.

Isbell, J. L., Miller, M. F., Wolfe, K. L. & Lenaker, P. A. 2003. Timing of late Paleozoic glaciation in Gondwana: was glaciation responsible for the development of northern hemisphere cyclothems? *In*: Chan, M. A. & Archer, A. W. (eds) *Extreme Depositional Environments: Mega End Members in Geologic Time*. Geological Society of America, Special Papers, **370**, 5–24.

Isbell, J. L., Henry, L. C., Reid, C. M. & Fraiser, M. L. 2012. Sedimentology and palaeoecology of lonestone-bearing mixed clastic rocks and cold-water carbonates of the Lower Permian Basal Beds at Fossil Cliffs, Maria Island, Tasmania (Australia): insight into the initial decline of the late Palaeozoic ice age. *In*: Gąsiewicz, A. & Słowakiewicz, M. (eds) *Palaeozoic Climate Cycles: Their Evolutionary and Sedimentological Impact*. Geological Society, London, Special Publications, **376**. First published online November 26, 2012, http://dx.doi.org/10.1144/SP376.2

Jackson, J. B. C. & Johnson, K. G. 2001. Measuring past biodiversity. *Science*, **293**, 2401–2404.

Ji, Q., Wei, J. Y. *et al.* 1989. *The Dapoushang Section*. Science Press, Beijing.

Joachimski, M. M., Von Bitter, P. H. & Buggisch, W. 2006. Constraints on Pennsylvanian glacioeustatic sea-level changes using oxygen isotopes of conodont apatite. *Geology*, **34**, 277–280.

Kalvoda, J. 1989. Tournaisian events in Moravia (Czechoslovakia) and their significance. *Courier Forschungsinstitut Senckenberg*, **117**, 353–358.

Kong, X. Z., Xu, H. L. *et al.* 1996. *Late Palaeozoic Coal-bearing Strata and Biota in Shanxi, China*. Shanxi Science and Technology Press, Taiyuan.

Korte, C., Jasper, T., Kozur, H. & Veizer, J. 2005. $\delta^{18}O$ and $\delta^{13}C$ of Permian brachiopods: a record of seawater evolution and continental glaciation. *Palaeogeography, Palaeoclimatology, Palaeoecology*, **224**, 333–351.

Kuang, G. D., Li, J. X., Zhong, K., Su, Y. B. & Tao, Y. B. 1999. *Carboniferous of Guangxi, China*. University of Geosciences Press, Wuhan (in Chinese).

Lee, J. S. 1924. Grabauina, A Transitional Form between Fusulinella and Fusulina. *Bulletin of the Geological Society of China*, **3**, 51–54.

Lee, J. S. 1931. Distribution of dominant types of the fusulinid foraminifera in the Chinese seas. *Bulletin of the Geological Society of China*, **10**, 273–290.

Leven, E. J. A. 1994. The mid-Early Permian regression and transgression of the Tethys. *In*: Embry, A. F., Beauchamp, B. & Glass, D. J. (eds) *Pangea: Global Environments and Resources*. Canadian Society of Petroleum Geologists, Memoirs, **17**, 233– 233.

Li, R. F., Liu, B. P. & Zhao, C. L. 1997. Cycle-sequences, carbon isotope features and glacio-eustasy of the *Triticites* Zone in Southern Guizhou. *Acta Geologica Sinica*, **71**, 217–226.

Liao, Z. T. 1979. Uppermost Carboniferous brachiopods from western Guizhou. *Acta Palaeontologica Sinica*, **18**, 527–546.

Liao, Z. T. 1995. Faunal provinces of Carboniferous brachiopods in China and their variations across the Carboniferous boundaries. *In*: *Palaeontologia Cathayana, Volume 6*. Beijing Sciences Press, Beijing, 365–374.

Liao, Z. T., Wang, X. D. *et al.* 2010. The Carboniferous strata in Mt. Longmen. *Journal of Stratigraphy*, **34**, 349–362 (in Chinese with English abstract).

McGhee, G. R., Jr., Sheehan, P. M., Bottjer, D. J. & Droser, M. L. 2012. Ecological ranking of Phanerozoic biodiversity crises: the Serpukhovian (early Carboniferous) crisis had a greater ecological impact than the end-Ordovician. *Geology*, **40**, 147–150.

Menning, M., Alekseev, A. S. *et al.* 2006. Global time scale and regional stratigraphic reference scales of central and west Europe, east Europe, Tethys, south China, and North America as used in the Devonian-Carboniferous-Permian Correlation Chart 2003 (DCP 2003). *Palaeogeography, Palaeoclimatology, Palaeoecology*, **240**, 318–372.

Metcalfe, I. 2011. Tectonic framework and Phanerozoic evolution of Sundaland. *Gondwana Research*, **19**, 3–21.

Mii, H. S., Grossman, E. & Yancey, T. 1999. Carboniferous isotope stratigraphies of North America; implications for Carboniferous paleoceanography and Mississippian glaciation. *Bulletin of the Geological Society of America*, **111**, 960–973.

MII, H. S., GROSSMAN, E., YANCEY, T., CHUVASHOV, B. & EGOROV, A. 2001. Isotopic records of brachiopod shells from the Russian Platform – evidence for the onset of mid-Carboniferous glaciations. *Chemical Geology*, **175**, 133–147.

NEMIROVSKAYA, T. & NIGMADGANOV, I. 1994. The Mid-Carboniferous conodont event. *Courier Forschungsinstitut Senckenberg*, **168**, 319–333.

QI, Y. P. 2008. *Conodont biostratigraphy of the Naqing section: candidate GSSP for both the base of the Serpukhovian and Moscovian Stages, Luosu, Luodian, Guizhou, South China*. PhD thesis, Graduate University of Chinese Academy of Sciences.

QIE, W. K., ZHANG, X. H., DU, Y. S. & ZHANG, Y. 2011. Lower Carboniferous carbon isotope stratigraphy in South China: implications for the Late Paleozoic glaciation. *Science China Earth Sciences*, **54**, 84–92.

RYGEL, M. C., FIELDING, C. R., FRANK, T. D. & BIRGENHEIER, L. P. 2008. The magnitude of late Paleozoic glacioeustatic fluctuations: a synthesis. *Journal of Sedimentary Research*, **78**, 500–511.

SALTZMAN, M., GONZÁLEZ, L. & LOHMANN, K. 2000. Earliest Carboniferous cooling step triggered by the Antler orogeny? *Geology*, **28**, 347–350.

SCOTESE, C. R. 2001. *Atlas of Earth History*. Paleomap Project, Department of Geology, University of Texas at Arlington.

SHENG, J. Z., ZHANG, L. X. & WANG, J. H. 1988. *Fusulinids*. Science Press, Beijing.

SMITH, L. B. & READ, J. F. 2000. Rapid onset of late Paleozoic glaciation on Gondwana: evidence from Upper Mississippian strata of the Midcontinent, United States. *Geology*, **28**, 279–282.

STANLEY, S. M. & POWELL, M. G. 2003. Depressed rates of origination and extinction during the late Paleozoic ice age: a new state for the global marine ecosystem. *Geology*, **31**, 877–880.

UENO, K., HAYAKAWA, N., NAKAZAWA, T., WANG, T. & WANG, X. 2012. Pennsylvanian–Early Permian cyclothemic succession on the Yangtze Carbonate Platform, South China. *In*: GĄSIEWICZ, A. & SŁOWAKIEWICZ, M. (eds) *Palaeozoic Climate Cycles: Their Evolutionary and Sedimentological Impact*. Geological Society, London, Special Publications, **376**. First published online November 29, 2012, http://dx.doi.org/10.1144/SP376.5

VEEVERS, J. & POWELL, C. M. 1987. Late Paleozoic glacial episodes in Gondwanaland reflected in transgressive-regressive depositional sequences in Euramerica. *Bulletin of the Geological Society of America*, **98**, 475–487.

WANG, C. Y. & YIN, B. A. 1984. Conodont zonations of early Lower Carboniferous and Devonian-Carboniferous boundary in pelagic facies, South China. *Acta Palaeontologica Sinica*, **23**, 224–238 (in Chinese and English).

WANG, C. Y. & YIN, B. A. 1988. Conodonts. *In*: YU, C.-M. (ed.) *Devonian–Carboniferous Boundary in Nanbiancun, Guilin, China – Aspects and Records*. Science Press, Beijing, China, 105–148.

WANG, X. D., WANG, X. J., ZHANG, F. & ZHANG, H. 2006. Diversity patterns of Carboniferous and Permian rugose corals in South China. *Geological Journal*, **41**, 329–343.

WANG, X. L., LI, S. L. & WANG, Y. 1997. Upper Devonian and Lower Carboniferous Sequence Stratigraphy of South China. *Acta Geoscientia Sinica*, **18**, 99–105.

WANG, Z. H. 1996. Middle-Carboniferous boundary and the conodonts across this boundary in South Guizhou and North Guangxi. *Acta Micropalaeotologica Sinica*, **13**, 261–276 (in Chinese).

WANG, Z. H. & HIGGINS, A. C. 1989. Conodont zonation of the Namurian-Lower Permian strata in south Guizhou, China. *Palaeontologia Cathayana*, **4**, 261–325.

WANG, Z. H. & QI, Y. P. 2003. Upper Carboniferous (Pennsylvanian) conodonts from South Guizhou of China. *Rivista Italiana di Paleontologia e Stratigrafia*, **109**, 379–397.

WOPFNER, H. 2012. Late Palaeozoic–Early Triassic deposition and climates between Samfrau and Tethys: a review. *In*: GĄSIEWICZ, A. & SŁOWAKIEWICZ, M. (eds) *Palaeozoic Climate Cycles: Their Evolutionary and Sedimentological Impact*. Geological Society, London, Special Publications, **376**. First published online November 26, 2012, http://dx.doi.org/10.1144/SP376.4

WU, W. S., ZHANG, L. X. & JIN, Y. G. 1974. The Carboniferous system of western Guizhou. *Memoir of Nanjing Institute of Geology and Paleontology, Academia Sinica*, **6**, 72–98.

YAN, J. X. 2004. Origin of Permian Chihsian Carbonates from South China and its geological implications. *Acta Sedimentologica Sinica*, **22**, 579–587.

YANG, S. P. 1990. On the biogeographical provinces of Early Carboniferous brachiopoda in China and its adjacent regions. *In*: WANG, H.-Z., YANG, S.-P. ET AL. (eds) *Tectonopalaeogeography and Palaeobiogeography of China and Adjacent Regions*. China University of Geosciences Press, Wuhan, 317–335.

YANG, X. N., RUN, J., MING, L. & SHI, Z. G. J. 2005. Early Permian bioevent in the fusulinacean fauna of South China. *Lethaia*, **38**, 1–16.

ZHANG, L. X. 1987. *Carboniferous Stratigraphy in China*. Science Press, Beijing.

ZHANG, S. X. & JIN, Y. G. 1976. Brachiopods from the Upper Paleozoic Erathem of the Mount Qomolongma region. *In*: *A Report of a Scientific Expedition to the Mount Qomolangma Region (1966–1968), Palaeontology fasc. II*. Tibetan scientific expedition group of Chinese Academy of Sciences, Science Press, Beijing, 159–271.

ZHANG, Y., QIE, W. K., LI, Y. L. & ZHANG, X. H. 2009. Carboniferous oolitic hematite in Longmenshan area of Sichuan Province and its paleoenvironmental significance. *Acta Petrologica et Mineralogica*, **28**, 51–57 (in Chinese with English abstract).

Variability of Pennsylvanian–Permian Carbonate Associations and implications for NW Pangea Palaeogeography, east-central British Columbia, Canada

K. D. ZUBIN-STATHOPOULOS[1], B. BEAUCHAMP[1], V. I. DAVYDOV[2] & C. M. HENDERSON[1]*

[1]*Department of Geoscience, University of Calgary, 2500 University Dr. NW, Calgary, Alberta T2N 1N4, Canada*

[2]*Department of Geosciences, Boise State University, 1910 University Drive, Boise, ID, USA 83725*

**Corresponding author (e-mail: charles.henderson@ucalgary.ca)*

Abstract: Pennsylvanian–Early Permian carbonate sedimentation in east-central British Columbia records a complex history of changing environments influenced by evolving palaeogeography and climate. Newly recognized tectonically controlled features affected the distribution and variability of carbonate associations, providing new interpretations for this portion of the NW coast of Pangea. Both a heterozoan (cool-water) and photozoan (warm-water) association were identified on either side of a palaeogeographical high. Cool-water carbonates were located outboard or to the west of this high, an area influenced by upwelling waters. Inboard of this high, a warm, protected sea developed at about 20°N palaeolatitude during the Asselian and Sakmarian. This configuration and palaeolatitude is similar to that of Baja California, Mexico and the Sea of Cortéz, providing a good modern analogue for these deposits where warm-water carbonates grow at latitudes otherwise dominated by cool-water deposits. The warm sea provided a place for a photozoan association to develop during the Early Permian when the low-latitude NW coast of Pangea was dominated by cool-water carbonates.

Pennsylvanian–Permian strata in east-central British Columbia (BC), western Canada, consist of carbonate rocks with a small siliciclastic component and are predominantly skeletal wackestone and packstone (Bamber & Macqueen 1979). Localized occurrences of grainstone and boundstone that record warm-water carbonate deposition also occur in the eastern and southern portion of the area. This occurrence is unusual because it is present in an area that is otherwise dominated by cool-water limestone, dolostone and phosphatic siltstone. This aspect of western Pangean sedimentation has not been addressed in previous studies (McGugan & Rapson-McGugan 1976; Bamber & Macqueen 1979). This paper demonstrates that this anomalous occurrence of these warm-water carbonates is controlled by the emergence of a Late Pennsylvanian topographic high separating a protected, warm inland sea to the east from a significantly cooler open ocean to the west.

Pennsylvanian and Lower Permian carbonate reefs and mounds typical of tropical to subtropical settings have been well documented in the western United States and the Canadian Arctic (Davies *et al.* 1989; Morin *et al.* 1994; Beauchamp & Desrochers 1997; Wahlman 2002). The Pennsylvanian–Permian basins of the western US were located near the palaeo-equator where warm shallow water prevailed (Blakey 2008). At higher latitudes, tropical to subtropical seas also developed such as in the Sverdrup Basin of the Canadian Arctic, an area that was bathed by warm waters originating from the Tethyan Ocean prior to the closure of the Uralian seaway (Reid *et al.* 2007). The reef-building organism *Palaeoaplysina*, as well as colonial rugose corals and calcareous green algae, have been documented in east-central BC (Bamber & Macqueen 1979). These fossils form a photozoan biotic association, which is typical of shallow warm-water conditions (James 1997).

This paper documents the facies variability of Pennsylvanian–Lower Permian carbonates in east-central BC, focusing on differences in biotic associations and other sedimentologic attributes. Such differences are often attributed to climate change over time (e.g. Beauchamp 1994). However,

From: Gąsiewicz, A. & Słowakiewicz, M. (eds) 2013. *Palaeozoic Climate Cycles: Their Evolutionary and Sedimentological Impact*. Geological Society, London, Special Publications, **376**, 47–71.
First published online November 26, 2012, http://dx.doi.org/10.1144/SP376.1

this study shows that distinctive warm- and cool-water shallow-water shelf deposits accumulated at the same time while remaining unaffected by the major climatic shift that occurred across the Asselian–Sakmarian boundary associated with the significant melting of Gondwana glaciers. We present an alternative interpretation, whereby the significant difference in carbonate associations is explained by the existence of a protected sea.

Geological setting

Study area and methods

Pennsylvanian–Permian strata from the westernmost portion of the Western Canada Sedimentary Basin (WCSB) crop out in a NW–SE-trending belt in eastern BC and western Alberta. This study focuses on outcrops in map sheets 93I, P, O and 94B where the succession is relatively well exposed in a series of Laramide (Late Cretaceous–Eocene) thrust sheets of the Rocky Mountains (Fig. 1). The eight measured sections include Peck Creek, Mountain Creek, Watson Peak, Mount Palsson, Mount Crum, Fellers Creek, Mount Cornock and Ganoid Ridge. In addition to new outcrop data collected in 2009 and 2010, our study incorporates published field descriptions of Bamber & Macqueen (1979) and McGugan & Rapson-McGugan (1976). Mountain Creek and Fellers Creek, which are the most complete sections we measured, are described in greater detail. Exploration wells are also used for correlation within the eastern Peace River Embayment where the biostratigraphy and sedimentology is better understood.

This paper relies on biostratigraphic data and age interpretations outlined in Zubin-Stathopoulos (2011), in which 116 conodont samples were processed from 8 measured sections. Facies analysis was conducted using 203 thin sections, cut slabs, outcrop photographs and field notes. Standard procedures for petrographic analysis were used for identifying and imaging carbonate constituents. The carbonate classification scheme of Dunham (1962) is used as well as modifiers for carbonate associations including the terms ‘photozoan’ and ‘heterozoan’ to qualify the environmental controls (temperature, nutrients, etc.) of carbonate constituents (James 1997). In addition, terminology specific to Late Palaeozoic biota (bryonoderm, bryonoderm-extended) were used (Beauchamp 1994).

Stratigraphic setting

The Pennsylvanian–Permian succession of western Alberta and eastern BC is relatively thin (up to 300 m in Telford Basin), quite variable in composition and recorded a complex stratigraphic-sedimentological history at a time of ongoing tectonic activity in the WCSB. In east-central BC, Pennsylvanian–Permian rocks are dominated by shallow-water carbonate and chert with varying siliciclastic proportions that generally increase upwards (Bamber & Macqueen 1979). This succession comprises eight relatively thin unconformity-bounded low-order sequences that can be correlated from the Rocky Mountains in the west (east-central BC) to the subsurface areas (east-central BC and west-central Alberta) in the east (Fig. 2) (McGugan & Rapson-McGugan 1976; Bamber & Macqueen 1979; Zubin-Stathopoulos 2011). In the study area, these sequences are represented by the Kindle, Belcourt, Fantasque and Mowitch formations (Bamber & Macqueen 1979). Pennsylvanian strata in the area are equivalent to the Ksituan Formation (Henderson *et al.* 1994; Richards *et al.* 1994, fig. 14.31e) from the Peace River Embayment. The Middle Pennsylvanian–Lower Permian sequences in east-central BC are equivalent to part of the Belloy Formation in the subsurface to the east (Naqvi 1972; Dunn 2003) while the Middle Permian units are equivalent to the upper Middle and Upper Belloy Formation (Fig. 2). While they differ lithologically, these units are also equivalent in age to formations in SE BC and SW Alberta (McGugan & Rapson 1962, 1963; Rapson-McGugan 1970; MacRae & McGugan 1977).

The studied succession is part of three low-order sequences of Moscovian-, Kasimovian–Gzhelian- and Asselian–Sakmarian-aged strata, respectively (Figs 2 & 3), all contained within the Belcourt Formation. The sequence boundaries are sharp and erosive unconformable surfaces associated with intraformational conglomerates of probable nearshore origin (Fig. 4). The Belcourt Formation is a unit of fossiliferous carbonate that recorded moderately deep-water to shallow-shelf or ramp cyclic sedimentation (Bamber & Macqueen 1979). In the study area, the formation varies in thickness (Fig. 3) ranging from zero at Mt Cornock up to 127 m at Mountain Creek (Fig. 3). Southern and eastern outcrops display typical Belcourt facies, that is, grainstone (ooid and skeletal), boundstone and lesser amounts of skeletal wackestone and packstone, a suite of facies that is best preserved at Fellers Creek (Figs 4 & 5). The western section displays a different composition, which consists dominantly of lime-mudstone, skeletal wackestone and packstone and minor amounts of skeletal grainstone. This succession is best exemplified at Mountain Creek (Figs 6 & 7). We are of the opinion that a new formation could be erected to reflect this basic and mappable lithological difference within the Belcourt Formation. For the purposes of this paper, however,

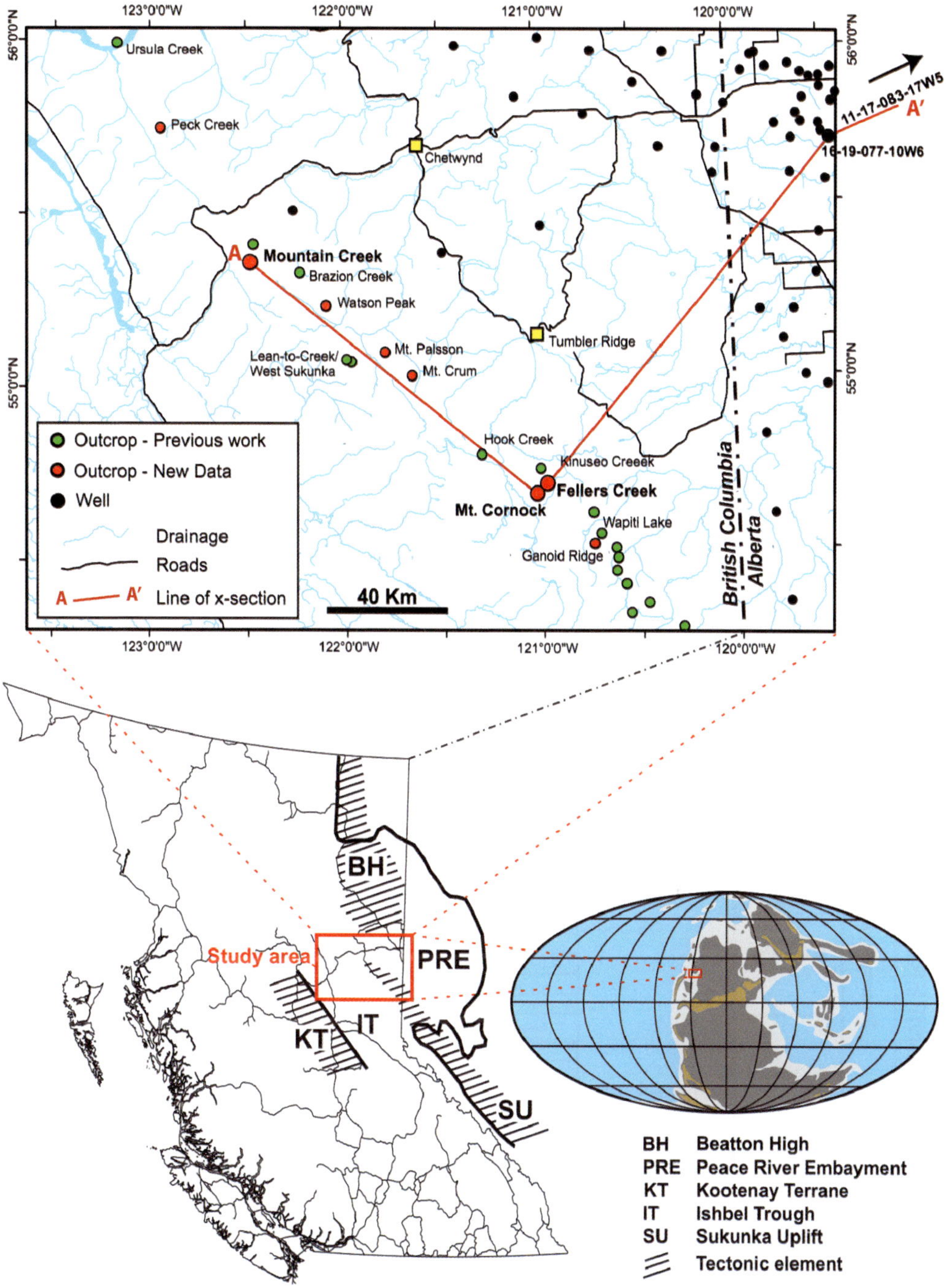

Fig. 1. Study area, east-central British Columbia. The line of cross-sections for Figures 3 and 8 are also indicated. Studied locations include Peck Creek (55.74678°N; 122.95042°W), Mountain Creek (55.35834°N; 122.48333°W), Watson Peak (55.23278°N; 122.08528°W), Mt Palsson (55.09476°N; 121.78314°W), Mt Crum (55.02655°N; 121.64192°W), Fellers Creek (54.70720°N; 120.96396°W), Mt Cornock (54.67972°N; 121.02023°W), Ganoid Range (54.52884°N; 120.73638°W), well 16-19-077-10W6 (55.69219°N; 119.52921°W) and well 11-17-083-17W5 (56.19737°N; 116.65344°W). Pangea modified from www.scotese.com; study area shown by red box. Map of British Columbia includes study area and tectonic elements discussed in text. Modified from Zubin-Stathopoulos *et al.* 2011.

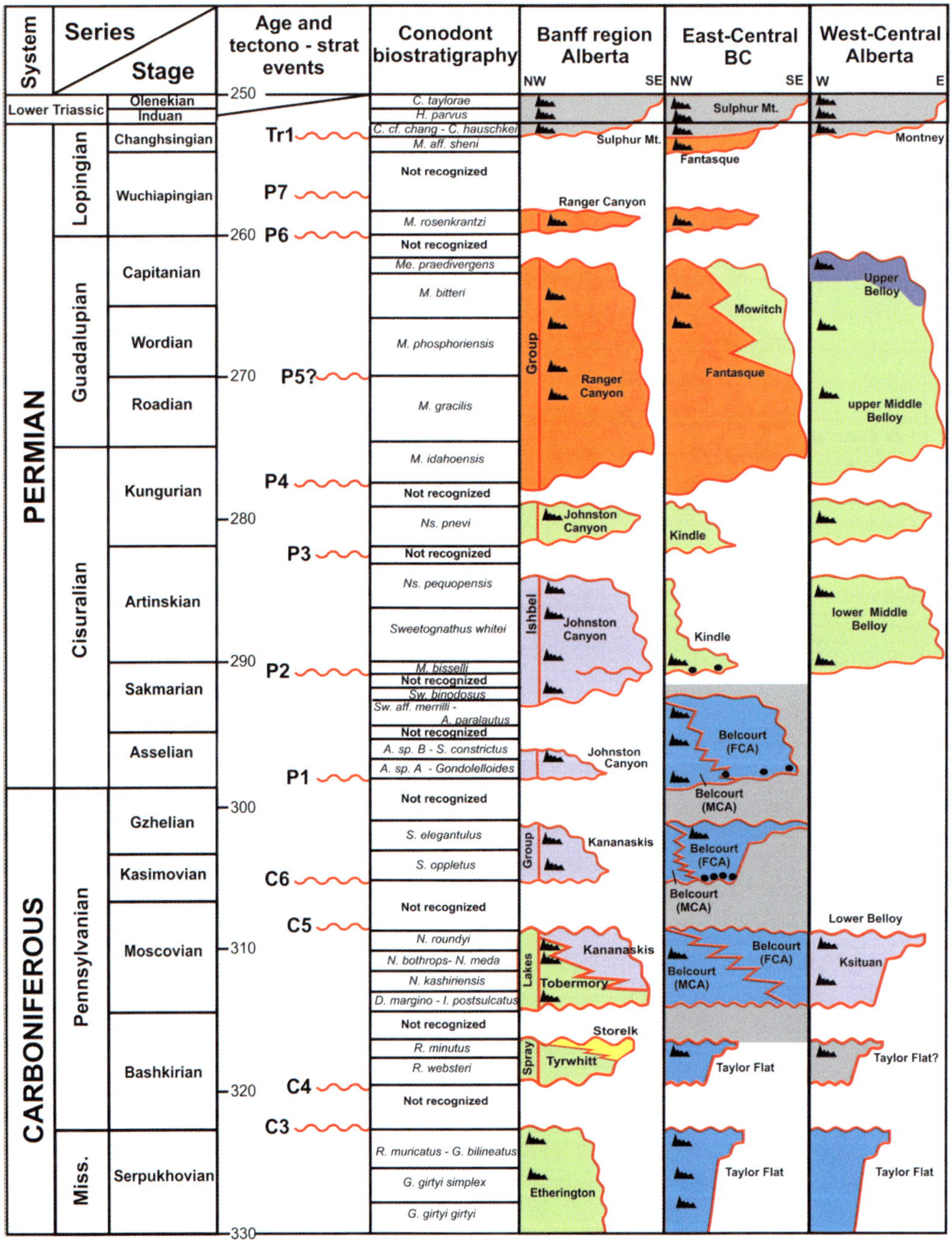

Fig. 2. Stratigraphy and tectonostratigraphic sequences of east-central British Columbia, west-central Alberta and the Banff region of the southern Alberta. The interval of focus in this study is highlighted in grey. Colours represent primary lithology. Blue: limestone; purple: dolostone; orange: chert; yellow: quartz arenite; green: bioturbated/bioclastic sandstone; grey: silty shale. C = Carboniferous, P = Permian. Tectonostratigraphic sequences modified from Snyder *et al.* 2002 and Trexler *et al.* 2004. Stratigraphy modified from Zubin-Stathopoulos *et al.* 2011.

we will refer to the Fellers Creek Assemblage (FCA; eastern and southern area) and the Mountain Creek Assemblage (MCA; western area) facies of the Belcourt Formation (Fig. 2).

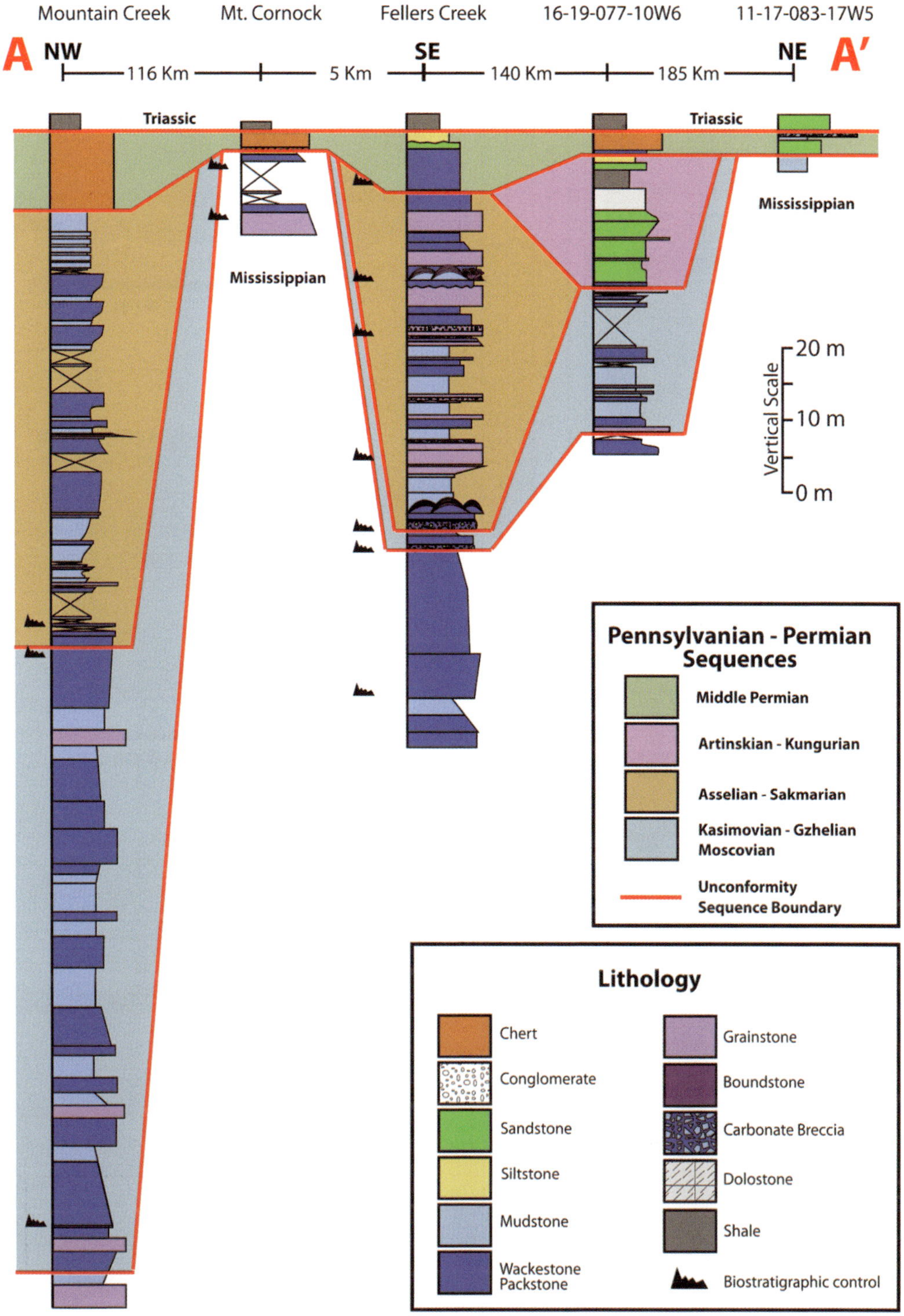

Fig. 3. Cross section A–A′ as indicated on Figures 1 and 8 that highlights how the Tipinahokan Peninsula (centred on Mt Cornock) affects stratigraphy in the area. Correlations are based on ages obtained from conodonts, foraminifers and corals.

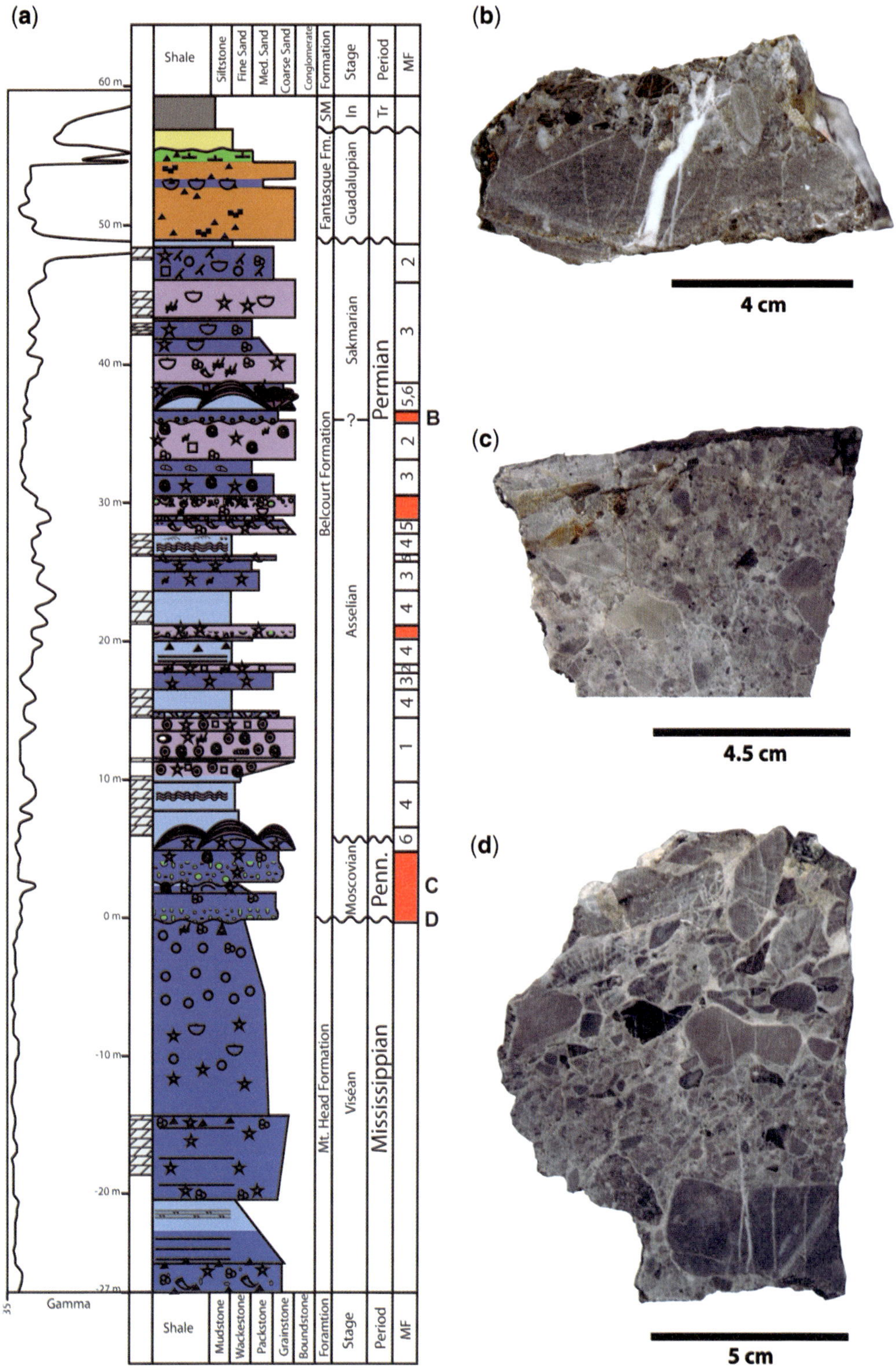
(a)
Shale
Siltstone
Fine Sand
Med. Sand
Coarse Sand
Conglomerate
Formation
Stage
Period
MF
60 m
50 m
40 m
30 m
20 m
10 m
0 m
-10 m
-20 m
-27 m
SM
Fantasque Fm.
Belcourt Formation
Mt. Head Formation
In
Guadalupian
Sakmarian
Asselian
Moscovian
Viséan
Tr
Permian
Penn.
Mississippian
-?-
2
3
5,6
2
3
5
4
3
4
4
3
2
4
1
4
6
B
C
D
35
Gamma
Shale
Mudstone
Wackestone
Packstone
Grainstone
Boundstone
Formation
Stage
Period
MF
(b)
4 cm
(c)
4.5 cm
(d)
5 cm

Palaeogeographical setting and tectonic elements

The deposits described in this study are located in the western part of the Peace River Embayment in east-central BC (Fig. 1). The Peace River Embayment is a fault-bounded indentation of the eastern margin of the Prophet (Carboniferous) and Ishbel (Permian) troughs that represents the down-warped and down-faulted western margin of the North American plate (Richards 1989; Richards *et al.* 1994). The western boundary of the Prophet and Ishbel troughs was a rim elevated by the latest Devonian Antler Orogeny and exposed today in the pericratonic Kootenay Terrane (KT of Fig. 1). The troughs and embayment were areas of carbonate and clastic deposition during the Pennsylvanian–Permian interval (Henderson *et al.* 1994). The Peace River Embayment is a complex tectonostratigraphic element at the convergence of multiple tectonic interactions and was the locus of both differential subsidence and uplift that occurred at varying rates and times in different areas. The location of the Peace River Embayment was in part determined by tectonically controlled palaeogeographical elements such as the Beatton High and Sukunka Uplift (BH and SU in Fig. 1). The Sukunka Uplift (Richards 1989) first formed in the middle Tournaisian and was episodically positive for the remaining Palaeozoic. Snyder *et al.* (2002) and Trexler *et al.* (2004) documented a series of tectonic events and corresponding unconformities that developed during the Carboniferous and Permian in Nevada; these have now been correlated into east-central BC (Fig. 2). It is now apparent that the Peace River Embayment and adjacent Ishbel Trough to the west (Richards *et al.* 1994) are divided into discrete, temporally restricted sub-basins (Henderson *et al.* 2002, 2012).

The Sukunka Uplift was transgressed during the Moscovian and was the site of open marine carbonate sedimentation (Fig. 8a). The western part of the Sukunka Uplift was elevated during the Late Pennsylvanian C6 and earliest Permian P1 events (Fig. 2), resulting in a sub-aerially exposed western high and protected interior sea to the east (Figs 8b, c). The name Kisosowin Sea (Kisosowin means 'warm' in Native American Cree language) is here ascribed to the interior sea. The Kisosowin Sea was protected by a Late Pennsylvanian–Early Permian topographic high, referred to here as the Tipinahokan Peninsula (Tipinahokan means 'shelter from the cold' in Native American Cree language), that acted as a barrier sheltering the area of warm-water sedimentation to the east from an area cooled by upwelling of cool oceanic waters to the west. Several outcrops within the study area have no Pennsylvanian–Early Permian deposits. In contrast, the Fantasque Formation (Middle Permian) is present nearly everywhere, although it is missing at Watson Peak. The Tipinahokan Peninsula, the NW–SE axis of which is intersected at Watson Peak and Mt Cornock (Figs 1 & 3), separated two distinct depositional areas to the west and east. While the Belcourt Formation is absent on the crest of the peninsula, it thickens markedly to the east and west (Fig. 3). The peninsula also constitutes the physical boundary between the area dominated by the MCA facies of the Belcourt Formation to the west and the FCA facies to the east (Fig. 3). Evidence of recurrent tectonic activity along the structure is shown by several horizons with intraformational conglomerates, some of which contain clasts derived from the immediately underlying succession (Fig. 4).

Palaeolatitudinal setting

Palaeogeographical reconstructions of Pangea place the study area in eastern British Columbia between 15 and 20°N during the Moscovian–Kasimovian and 20–25°N during the Asselian–Kungurian (Golonka & Ford 2000; Vai 2003; Blakey 2008). Contemporaneous deposits in the southwestern United States (Texas to Utah) are interpreted to be equatorial, ranging from 0 to 10°N and having migrated 10° northwards during the Kasimovian–Kungurian interval (Tabor *et al.* 2008). The Sverdrup Basin of the Canadian Arctic is interpreted as being located at *c.* 25–30°N in the latest Pennsylvanian (Gzhelian) to Early Permian (Asselian–Sakmarian) based on extensive warm-water photozoan carbonates, and to have migrated to *c.* 40°N by the Middle Permian as suggested by dominance of cool- to cold-water heterozoan carbonates (Beauchamp 1994; Bensing *et al.* 2008). This significant oceanic cooling has been associated with the closure of Uralian seaway during the Artinskian that prevented warm Tethyan-derived waters from reaching NW Pangea (Reid *et al.* 2007). East-central British Columbia may have been at slightly lower latitude than suggested by some global reconstructions, possibly 15–20°N during the Early Permian,

Fig. 4. (**a**) Fellers Creek litholog indicating age based on conodont biostratigraphy, formations, conglomerates (red areas) and microfacies (MF) occurrence (see Table 1). Key to symbols and lithologies is shown in Figure 6. (**b**) Conglomerate within the Sakmarian sequence containing reworked Pennsylvanian conodonts; 36.45 m. (**c**) Second Belcourt conglomerate; 2.85 m. (**d**) Basal Belcourt conglomerate; 0 m. Gamma values were generated by hand-held scintillometer and are useful for correlation with wells in the area.

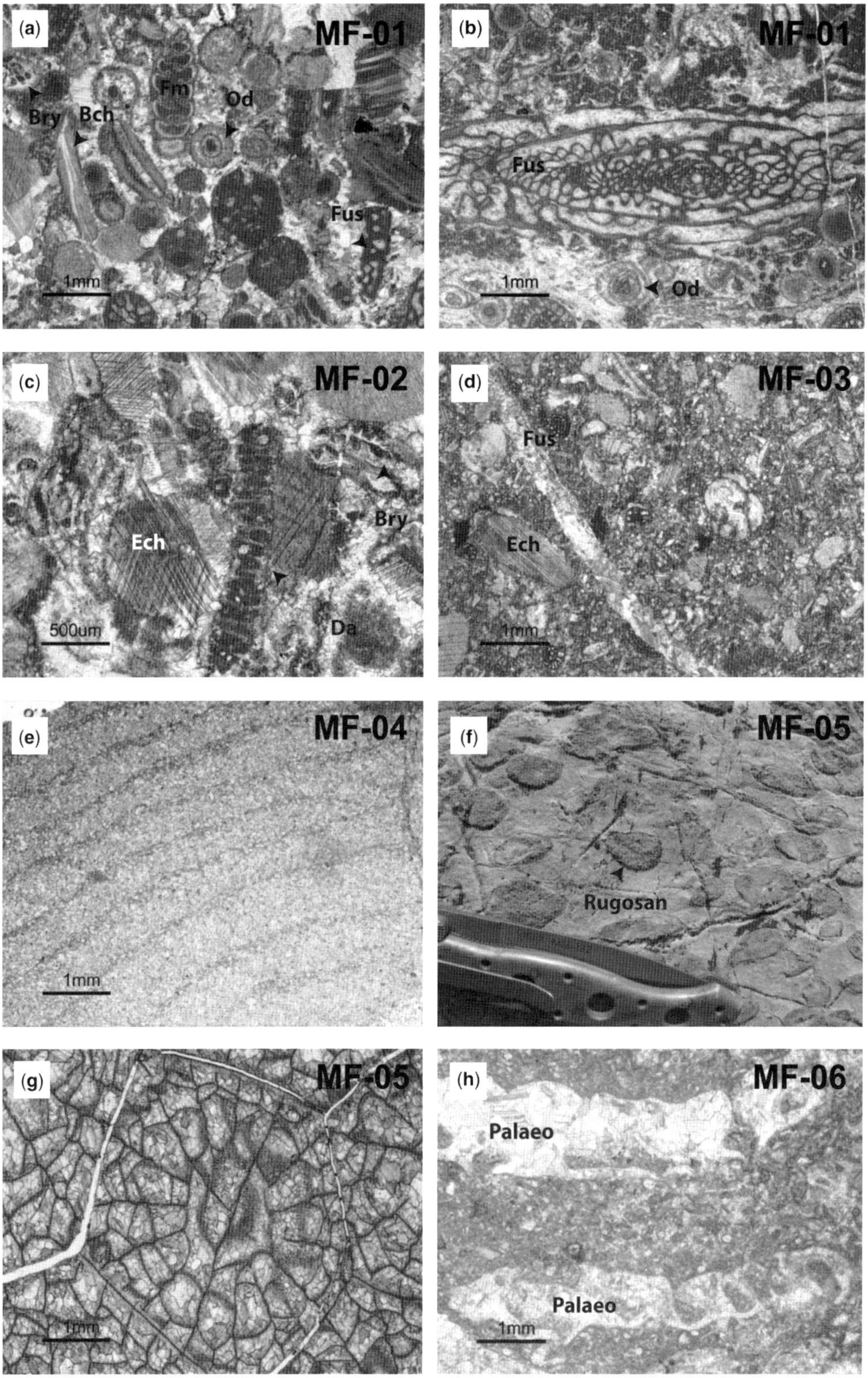
(a)
MF-01
Bry
Bch
Fm
Od
Fus
1mm
(b)
MF-01
Fus
1mm
Od
(c)
MF-02
Ech
Bry
Da
500um
(d)
MF-03
Fus
Ech
1mm
(e)
MF-04
1mm
(f)
MF-05
Rugosan
(g)
MF-05
1mm
(h)
MF-06
Palaeo
Palaeo
1mm

which would coincide well within the range of Coriolis-driven Ekman transport and upwelling along the western margin of Pangea.

Shelf cyclicity and palaeoclimatic setting

The Pennsylvanian–Early Permian interval was characterized by relatively high sea level with cyclic influence from glacial eustasy (Golonka & Ford 2000) at a time of widespread glaciation in Gondwana (Wanless & Shepard 1936). Cyclothems are well known and described from the western United States where the climate was wet-equatorial (Wanless & Shepard 1936; Heckel 1986). These cyclothems classically consist of deep marine shale, followed by regressive marine limestone and capped by shallow marine or terrestrial (coal) deposits (Heckel 2008). The transition from humid to more arid cyclothems in the Permian Basin region of southwestern US is illustrated by the disappearance of coal during the latest Pennsylvanian, red beds in the lowest Permian and evaporate capped sequences in Late Artinskian and Kungurian. Arid cyclothems are also shown to be present in higher-latitude deposits of western and northern Canada (Morin *et al.* 1994; Heckel 2002; Moore 2002; Ford *et al.* 2009). Some of these cyclothems commonly contain evaporites as their capping unit (e.g. sabkha-type dolostone and sulphate evaporites). Aeolian-sourced silt is pervasive throughout these deposits (Heckel 2002).

Arid conditions persisted throughout the Late Pennsylvanian in the western US, western Canada and parts of Russia (Francis 1994) and continued during the Permian, resulting in widespread evaporitic and desert environments. These conditions also led to abundant aeolian silt deposition within Pennsylvanian–Middle Permian marine carbonates (Francis 1994; Soreghan *et al.* 2008). The end of major Gondwana glaciation roughly coincides with the Early Sakmarian, above which high-amplitude–high-frequency sequences or cycles are not as well developed as in older Late Pennsylvanian–Early Permian sediments (Beauchamp & Henderson 1994). Sea level was at a near minimum towards the end of the Kungurian (Golonka & Ford 2000; Soreghan *et al.* 2008). During the Middle Permian, arid conditions coupled with cool-water deposition prevailed all along the northwestern margin of Pangea at a time of global warming (Beauchamp & Baud 2002; Clapham 2010).

As observed around the world, the Pennsylvanian–Lower Permian succession of east-central British Columbia also displays a series of high-order cycles as recorded by fluctuations in carbonate facies, representing environments ranging from outer shelf (or ramp) to shoreline (see descriptions below). This is shown by fluctuations from lime mudstone to packstone to grainstone in the Asselian succession at Fellers Creek (Fig. 4), and from lime mudstone to wackstone and packstone in the Moscovian–Asselian succession at Mountain Creek (Fig. 6). Cycles average 5–10 m in thickness, which is similar to contemporaneous cyclothems in the mid-continent (Heckel 2002) and in the Arctic (Morin *et al.* 1994). However, the number of observed cycles varies greatly from section to section and is considerably smaller than the number of cycles observed elsewhere. This reflects the incomplete nature and highly variable preservation of the Pennsylvanian–Permian succession in east-central BC, which attests for erosion and/or non-deposition at times of active differential tectonic uplift and subsidence. The Sakmarian succession at Fellers Creek displays only 2–3 shelf cycles, which may also reflect an incomplete stratigraphic record, at a time of widespread global transgression contemporaneous with the melting of a large proportion of Gondwana glaciers.

Facies descriptions, interpretations and depositional models

The Moscovian–Sakmarian succession of east-central British Columbia comprises 11 carbonate microfacies (MF), the content and interpretation of which is summarized in Table 1 and illustrated in Figures 5 and 7. The interpreted depositional environments range from relatively deep (below storm wave base) low-energy outer shelf or ramp (MF-08), to storm-influenced middle ramp (MF-03, MF-07), to high-energy shallow inner ramp/shoreface (above fair weather wave base) (MF-01 to MF-03; MF-05, MF-09, MF-11). Most facies represent open marine sedimentation, except for MF-05 (protected inner ramp) and MF-04 (potentially inter- to supra-tidal back-ramp deposition).

Both photozoan and heterozoan biotic associations were observed. Warm-water tropical conditions are indicated by dasycladacean algae, colonial rugose corals, *Palaeoaplysina* and ooids.

Fig. 5. Belcourt Formation microfacies (FCA facies) photomicrographs taken in plain polarized light. All measurements are from the base of the Belcourt Formation (basal conglomerate) at the Fellers Creek section. (**a**) MF-01, 12.35 m; (**b**) MF-01, 11.25 m; (**c**) MF-02, 18.15 m; (**d**) MF-03, 40.95 m; (**e**) MF-04, 27.75 m; (**f**) outcrop photograph, knife is 10 cm long, MF-05, 26.25 m; (**g**) MF-05, 39.1 m; (**h**) MF-06, 5.9 m. Ech: echinoderm; Bch: brachiopod; Bry: bryozoan; Fus: Fusulinacean; Da: Dasycladacean algae; Od: Ooid; Palaeo: *Palaeoaplysina*.

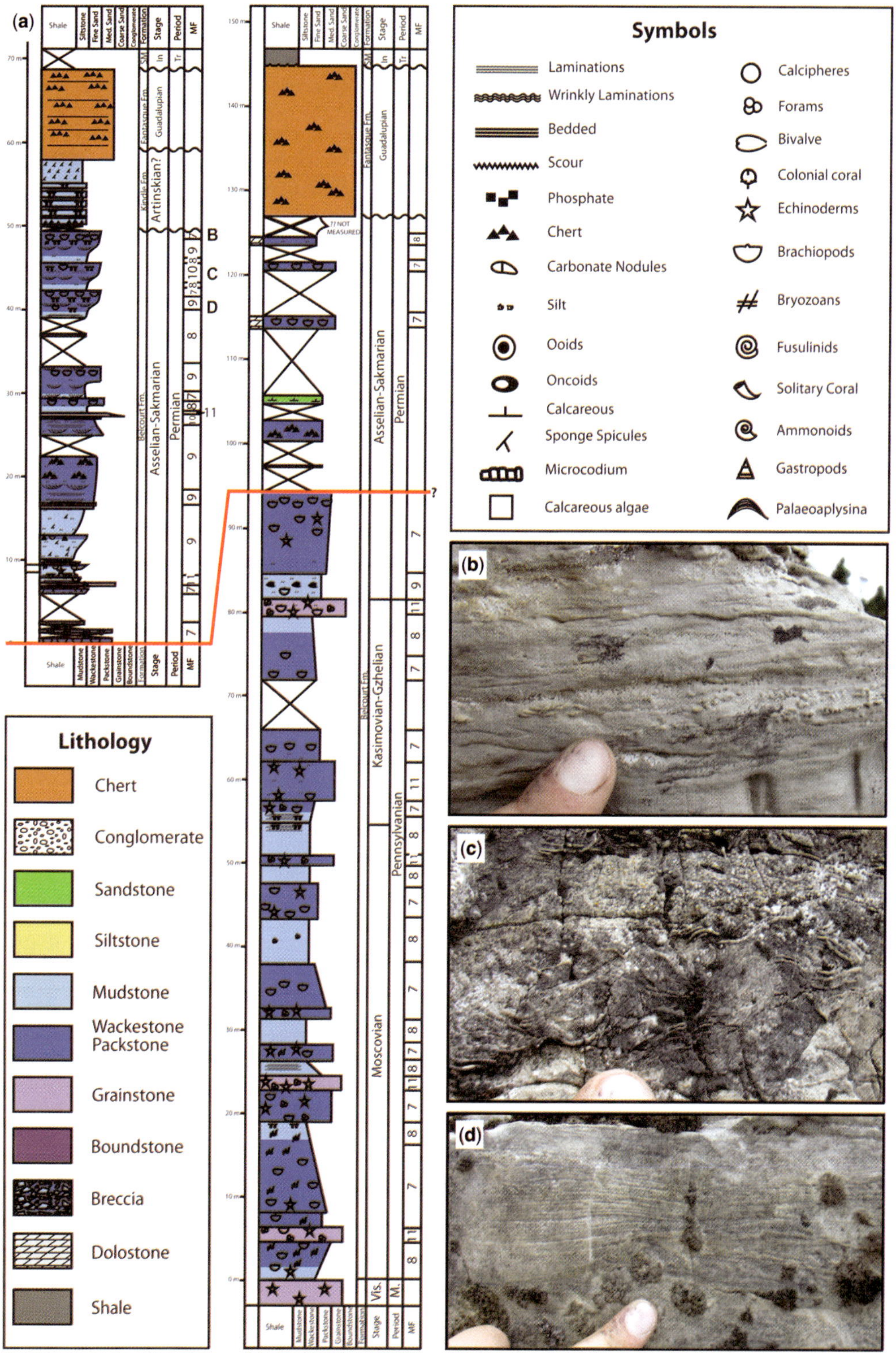
(a)
(b)
(c)
(d)
Symbols
Laminations
Wrinkly Laminations
Bedded
Scour
Phosphate
Chert
Carbonate Nodules
Silt
Ooids
Oncoids
Calcareous
Sponge Spicules
Microcodium
Calcareous algae
Calcipheres
Forams
Bivalve
Colonial coral
Echinoderms
Brachiopods
Bryozoans
Fusulinids
Solitary Coral
Ammonoids
Gastropods
Palaeoaplysina
Lithology
Chert
Conglomerate
Sandstone
Siltstone
Mudstone
Wackestone Packstone
Grainstone
Boundstone
Breccia
Dolostone
Shale
Shale
Mudstone
Wackestone
Packstone
Grainstone
Boundstone
Formation
Stage
Period
MF
Siltstone
Fine Sand
Med Sand
Coarse Sand
Conglomerate
Fantasque Fm
Guadalupian
Kindle Fm
Artinskian?
Belcourt Fm
Asselian-Sakmarian
Permian
Kasimovian-Gzhelian
Moscovian
Pennsylvanian
Vis.
M.
B
C
D
11
?? NOT MEASURED

Palaeoaplysina is an organism with unknown biological affinity that may belong to the class hydrozoa (Davies & Nassichuk 1973), although it has also been interpreted to be closely related to various calcareous algae (Watkins & Wilson 1989) including phylloid algae (Kiessling *et al.* 1999; Wahlman & Konovalova 2002). It is usually found in shallow-water high-productivity environments where photosynthesizing organisms are common (Davies & Nassichuk 1973). It is considered to be part of the photozoan association and formed bioherms in a moderate- to low-energy environment on the inner to outer ramp.

Heterozoan biota are far less diversified and dominated by sponge spicules, bryozoans, echinoderms and brachiopods, a typical Late Palaeozoic cool-water assemblage also known as Bryonoderm (Beauchamp 1994). Cool-water conditions reflect deeper depositional settings (MF-07 and MF-08) or shallow-water deposition in an area bathed by cool to cold waters (MF-09 to MF-11). In the latter case, it is not the biota that indicates shallow water deposition, but different lines of evidence such as the dominance of grainstone or presence of cross-bedding or ripples. A distinctive aspect of the studied succession is the dominance of photozoan carbonates in the eastern and southern sections, as seen at Fellers Creek (Fig. 4; FCA facies). In contrast, heterozoan carbonates dominate the western sections as exemplified at Mountain Creek (Fig. 6; MCA facies).

Various combinations of the 11 facies occur in the study area and can be found at various stratigraphic levels of the Belcourt Formation. The recurrence of facies associations attest to shifts in relative sea level in response to ongoing high-frequency glacio-eustatic fluctuations. Facies variations suggest bathymetric shifts of the order 30–50 m on average for each cycle as environments shifted from offshore, distal outer-shelf sedimentation below storm wave base to high-energy nearshore, shoreline and even supra-tidal sedimentation and erosion.

It is impossible to correlate individual cycles from section to section, but we can analyse the spectrum of microfacies within the three low-order sequences in the area: the Moscovian, Kasimovian–Gzhelian and the Asselian–Sakmarian sequences. Each of these sequences, which represent the grouping of an undetermined number of high-order cycles, has its own set of depositional characteristics as described in the following sections (Fig. 8).

Moscovian

The Moscovian portion of the Belcourt Formation consists of bioturbated silty mudstone (MF-07), bryozoan-brachiopod wackestone-packstone (MF-08) and fine-grained packstone-grainstone (MF-11) (Fig. 4). These three facies alternate in a cyclic fashion, shallowing upwards from MF-08 to MF-07 and capped by MF-11. The capping facies become progressively muddier upwards, and the mudstone portion of the cycle becomes thicker indicating overall deepening upwards succession for these cycles. The Moscovian portion of the Fellers Creek section consists of conglomerate-containing chert and carbonate clasts (Fig. 4). The bryozoans and brachiopods require normal marine salinity and circulation in order to develop, indicating that these sediments represent deposition in an open marine environment (Fig. 8a). The Moscovian found elsewhere in the Peace River Embayment is mostly assigned to the Ksituan Formation, which is predominantly composed of finely crystalline dolostone and is interpreted as shallow tidal flat deposits in sabkhas and lagoons (Dunn 2003; Wamsteeker 2007).

A Moscovian age is confirmed for this sequence by the presence of the conodont taxa *Adetognathus lautus*, *Diplognathodus edentulus*, *Neognathodus bothrops* and *Idiognathodus expansus* (Zubin-Stathopoulos 2011). This sequence starts at the basal conglomerate of the Belcourt Formation at Fellers Creek and Mountain Creek section (Figs 4 & 6) and displays extreme thickness variations ranging from zero at some outcrops (Mount Palsson, Mount Cornock, etc.) to 164 m in the subsurface in the Peace River Basin. The Moscovian portion of the Mountain Creek section coarsens upwards and contains six higher-frequency shallowing-upwards cycles (Fig. 4).

Moscovian-aged conglomerates containing Mississippian clasts at Fellers Creek were eroded from part of the Sukunka Uplift that was subsequently transgressed. The development of the Tipinahokan Peninsula resulted in the erosion of this sequence, but the preservation of Moscovian-aged conodonts (*Neognathodus bothrops*) in carbonate clasts found within a lag conglomerate indicates that deposition of Moscovian sediments was originally more widespread than seen at many outcrops (Zubin-Stathopoulos 2011).

The Moscovian sequence is characterized by overall open marine conditions (Fig. 8a) with no indication of a restricted or protected marine

Fig. 6. (**a**) Mountain Creek litholog indicating age based on conodont biostratigraphy, formations and microfacies occurrence (see Table 1). The upper portion was re-measured at a slightly different location and logged as a separate section; the equivalent level is indicated by red line. Field occurrences of (**b**) MF-10 (49.5 m), (**c**) MF-07 (43.7 m) and (**d**) MF-09 (41.0 m) are shown.

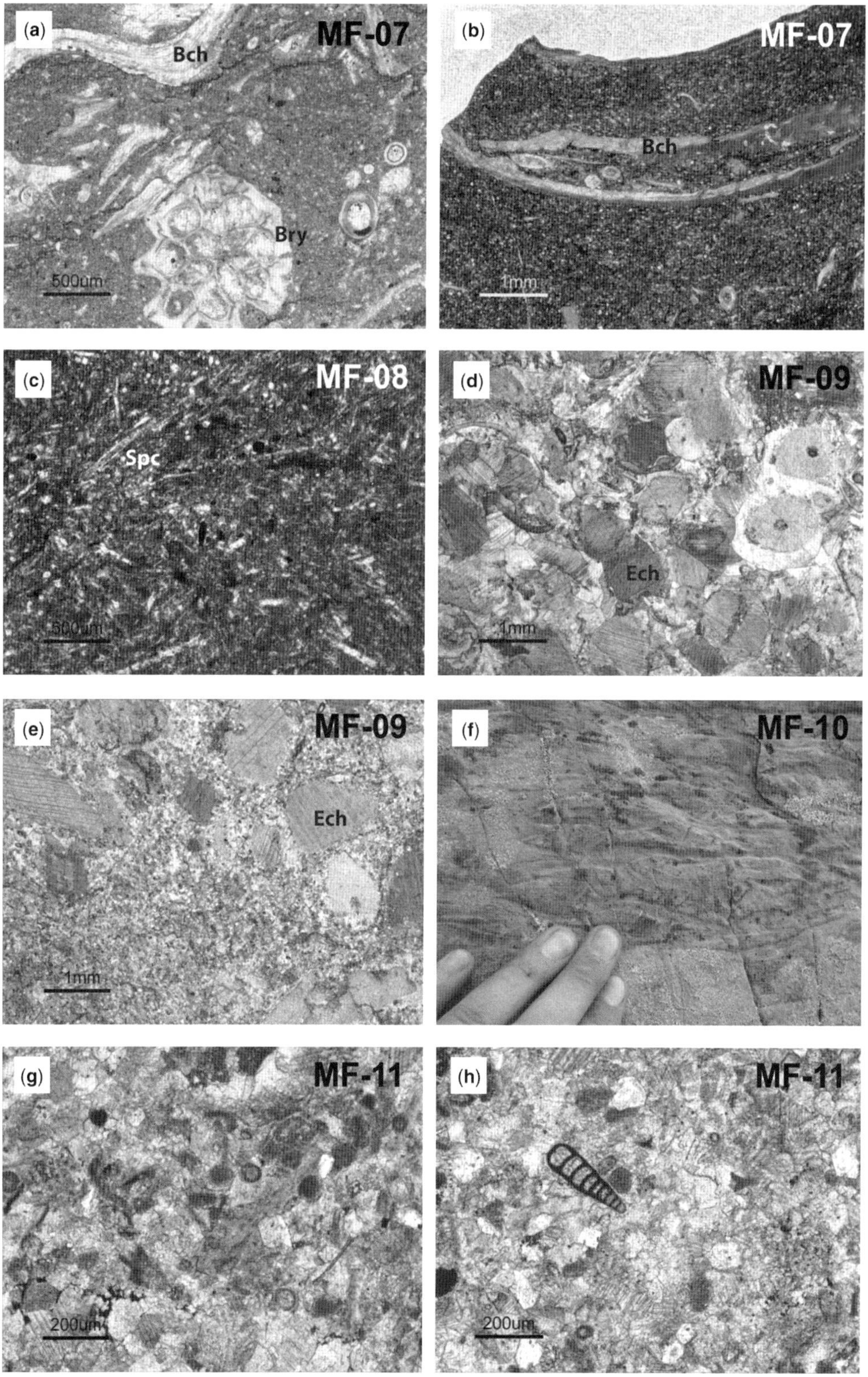
(a)
MF-07
Bch
Bry
500um
(b)
MF-07
Bch
1mm
(c)
MF-08
Spc
500um
(d)
MF-09
Ech
1mm
(e)
MF-09
Ech
1mm
(f)
MF-10
(g)
MF-11
200um
(h)
MF-11
200um

environment, except in inner ramp environments suggested by facies of the Ksituan Formation. The alternation of dominantly lime mudstone beds with periodic wackestone and packstone beds containing chaotically organized brachiopod and bryozoan fragments indicates an overall deep low-energy environment below storm wave base with shallowing-upwards cycles that end in storm-influenced beds at the tops. Mountain Creek section is located in the hanging wall of the westernmost thrust sheet with Upper Palaeozoic strata, indicating that these are the most distal facies in the study area; the carbonate association indicates deposition on a relatively deep to shallow cool-water carbonate ramp (Fig. 8a).

Kasimovian–Gzhelian

Rocks representing these two Late Pennsylvanian stages are not prominent in the study area, but are present at Mountain Creek and West Sukunka and locally in the subsurface of east-central BC (Dunn 2003). The Kasimovian–Gzhelian portion of the Belcourt Formation consists of bioturbated silty mudstone (MF-04) and bryozoan-brachiopod wackestone-packstone (MF-07). The carbonate association indicates cool, moderately deep water on the outer ramp. The location within the westernmost thrust sheet with Upper Palaeozoic strata implies that these are the most distal sediments preserved. The correlation of these stages is based on the occurrence of *Adetognathus lautus* and Idiognathodontidae *New Genus A sp.* (Kasimovian) (Zubin-Stathopoulos 2011). This succession is up to 70 m in the outcrop belt, although it is usually not present. Our limited dataset for this sequence prevents us from suggesting a sequence-specific interpretation. The range of depositional environments was likely similar to that of the Moscovian (Fig. 8a).

Asselian–Sakmarian

The Asselian–Sakmarian succession is bounded by prominent unconformities and is therefore believed to constitute a single low-order sequence. However, an additional erosion surface associated with conglomerate and potentially representing an unconformity occurs at Fellers Creek (Fig. 4); it records the boundary between two higher-order sequences, at about the Asselian–Sakmarian boundary. Because of this, we here describe the Asselian part of this sequence first and then the Sakmarian part below.

The Asselian portion of the Belcourt Formation consists of ooid-foraminifer grainstone (MF-01), *Palaeoaplysina* packstone/boundstone (MF-06), algal-bioclastic grainstone (MF-02), rugose coral wackestone-packstone-boundstone (MF-05), microbial lime- to dolomudstone (MF-04) and bryozoan-echinoderm-fusulinacean packstone-grainstone (MF-03). This Photozoan Association indicates shallow and warm-water conditions east of the Tipinahokan Peninsula. The Asselian portion of the Mountain Creek section, which was deposited west of the peninsula (Fig. 8), contains a cooler-water Heterozoan Association, consisting largely of MF-07 (bryozoan-brachiopod wackestone-packstone) with some alternation with MF-08 (silty bioturbated lime mudstone). Some levels contain abundant brachiopod fragments in a lime-mud matrix and may represent brachiopod banks.

Conodont taxa in this sequence include *Adetognathus n.sp. B*, *Streptognathodus verus* and *Streptognathodus fusus* (Zubin-Stathopoulos 2011) and indicate an Asselian age; this sequence ranges from 0 to 20 m in the outcrop belt. It is not recognized in the eastern Peace River Embayment, possibly due to low global sea level resulting in non-deposition or limited preservation (Golonka & Ford 2000). Distinct shallowing-upwards cycles are present at both the Fellers Creek (Fig. 4) and Mountain Creek sections (Fig. 6). Deposits at the Fellers Creek section represent a photozoan carbonate ramp that fostered the growth of temperature-dependent organisms such as *Palaeoaplysina* and fusulinaceans as well as abiotic constituents such as ooids (Fig. 8b). Deposits at the Mountain Creek section represent a heterozoan carbonate ramp that contained only heterozoan elements including brachiopods and bryozoans (Fig. 8b).

The Sakmarian facies (Fig. 8c) found at Fellers Creek include *Palaeoaplysina* boundstone (*sensu* original Dunham 1962 carbonate classification) (MF-06), colonial rugose coral wackestone-packstone-boundstone (MF-05) algal-bioclastic grainstone (MF-02) and bryozoan-echinoderm-fusulinacean packstone-grainstone (MF-03). These facies are part of the photozoan carbonate association (James 1997). Facies found at Mountain Creek include bryozoan-brachiopod wackestone-packstone (MF-07), cross-bedded silty packstone-grainstone (MF-10) and hummocky cross-stratified silty packstone-grainstone (MF-09).

Fig. 7. Belcourt Formation microfacies (MCA facies) photomicrographs taken in plain polarized light. All measurements are from the base of the Mountain Creek section. (**a**) MF-07, 9 m; (**b**) MF-07, 104.5 m; (**c**) MF-08, abundant sponge spicules at 107 m; (**d**) MF-09, 140 m; (**e**) MF-09, 140 m, from the same thin section indicating possible storm event; (**f**) outcrop photograph, finger tips for scale, MF-10; (**g**) MF-11, 80.7 m; (**h**) MF-11, 6.35 m. Ech: echinoderm; Bch: brachiopod; Bry: bryozoans; Spc: sponge spicules.

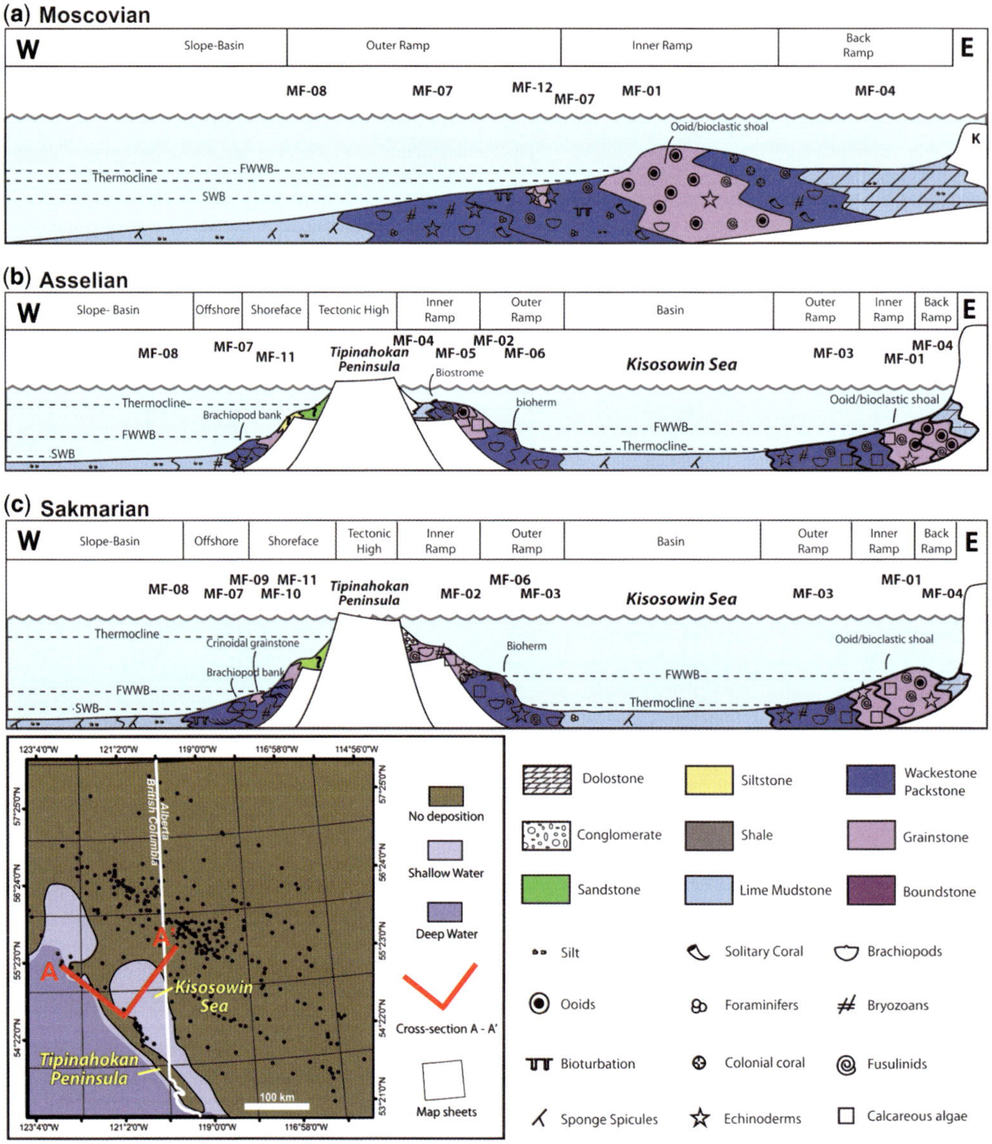

Fig. 8. Deposition model of time slices roughly based on cross-section of Figure 1 and as shown in inset map. Inset map configuration is based on the contouring function in ArcGIS and the predicted thickness distribution. FWWB: fair-weather wave base; SWB: storm wave base. Facies locations are indicated by facies number. (**a**) Moscovian profile. The occurrence and distribution of shallow-water deposits on the right side of the diagram are based on data from Wamsteeker (2007). K: extensive supratidal to shallow subtidal dolostone succession (Ksituan Formation) occurs east of the back-ramp setting. (**b**) Asselian profile showing the Tipinahokan Peninsula and Kisosowin Sea. (**c**) Sakmarian profile. Known facies are shaded in solid colours (see legend).

These belong to a heterozoan carbonate association of the bryonoderm variety (Beauchamp & Desrochers 1997).

Biostratigraphically and palaeogeographically significant fossils in this sequence include the conodont *Sweetognathus binodosus* (Zubin-Stathopoulos 2011) and the coral *Protowentzelella kunthi* (E.W. Bamber, pers. comm. 2010). Two closely spaced samples with abundant fusulinaceans at 37.5 and 39.5 m were recovered in the Fellers Creek section. Three species are identified in both samples (Fig. 10): *Schubertella* ex. gr. *kingi* Dunbar & Skinner,

Pseudofusulina attenuata Skinner & Wilde and *Ps. acuta* Skinner & Wilde. The first species is an opportunistic schubertellid that is widely distributed globally and occurs in latest Gzhelian through entire Lower Permian (Davydov 2011). The other two species were originally described from the McCloud Limestone in Shasta Lake area (Skinner & Wilde 1965). *Pseudofusulina attenuata* has also been found in Nevada in a stratigraphically very narrow horizon (Stevens *et al.* 1979; Davydov *et al.* 1997). In Nevada the horizon with *Pseudofusulina attenuata* yields the conodonts *Mesogondolella* aff. *striata* Chernykh near the bottom and *Sweetognathus* aff. *merrilli* Kozur near the top (Wang 1993; V. Chernykh, pers. comm. 2008) suggesting Late Asselian–Early Sakmarian age for this unit (Chernykh 2005). Only a 15-m-thick interval occurs at Fellers Creek (Fig. 4), which is correlated to other Sakmarian occurrences at Kinuseo Creek and Meosin Mountain. Sakmarian-aged rocks are also found at Mountain Creek. This part of the sequence developed at a time of global sea level rise and active tectonism.

The Tipinahokan Peninsula continued to separate photozoan carbonates to the east from heterozoan carbonates to the west throughout the Asselian and Sakmarian (Fig. 8b, c). Sediments on the flanks of this high were deposited on a carbonate ramp with bioherms. Sediment more distal to the flanks were deposited on a ramp that more closely resembles a siliciclastic ramp, where carbonate-producing organisms did not build mounds or wave-resistant structures (Figs 8b, c).

Discussion: Significance of distribution of carbonate associations

Western Pangean climate and oceanic currents

The occurrence of warm-water photozoan associations in east-central BC could be attributed to a climatic warming event. It has however been suggested that a southwards cool boundary current existed along the entire west coast of Pangea, creating increasing cool-water conditions starting during the Early Permian with the most pronounced effects in the Middle Permian (Beauchamp & Baud 2002; Clapham 2010). Northern and northwestern Pangea was cooling and decoupled from a broader global warming trend (Clapham 2010). The Middle Permian basin of west Texas was experiencing warmer water temperatures, while cool-water deposits prevailed north in the Phosphoria Sea (Clapham 2010). The Guadalupian of east-central BC also records a similar climatic story to that of the Sverdrup Basin and the basins of the western US. Cool-water deposits are recorded along the entire western coast of Pangea, dominantly consisting of spiculite and chert which are indicative of this cooling episode (Beauchamp & Baud 2002; Clapham 2010).

Palaeoaplysina-phylloid algae carbonate mounds that are typical of photozoan associations are well documented in the western US within tectonically controlled sub-basins (Wahlman 2002). There is an abundance of cool-water deposits in British Columbia, including the spiculitic and phosphatic siltstone of the Johnston Canyon Formation (Fig. 2) in southeastern BC and southwestern Alberta located within the southern portion of the Ishbel Trough (MacRae & McGugan 1977). This indicates that this cool boundary current also had a control on the fauna of the Early Permian in east-central BC. In addition, the palaeolatitude indicates that at least seasonal upwelling influenced these deposits, creating an environment conducive only to a heterozoan carbonate production. Despite the existence of a cool boundary current that became progressively more pronounced through the Permian, patch reefs and mounds typical of the photozoan or warm-water carbonate associations were able to develop in the protected Kisosowin Sea of northwestern Pangea.

Warm to cool carbonate deposition and palaeogeographical implications

Warm-water photozoan carbonate associations are composed of tropical non-skeletal particles such as ooids and/or light-dependent biotic elements such as green algae, fusulinaceans and warm-water corals (James 1997). The FCA facies of the Belcourt Formation predominantly consists of skeletal packstone and grainstone containing many of these constituents. It occurs in an isolated area within the outcrop belt in the central and southern portion of the study area. The Belcourt Formation can generally be characterized as deposited in a warm shallow sea, where carbonate-producing organisms were protected from cool upwelling ocean currents that prevented photozoan carbonate development on the seaward side of the Tipinahokan Peninsula. These organisms would also have required clear, oligotrophic waters in order to develop (Halfar *et al.* 2004*a*). *Microcodium* found in shallow water deposits within the Asselian and Sakmarian sequences indicates that this high may have been host to the development of soil and vegetation (Klappa 1978; Kosir 2004). It was centred approximately at 20°N palaeolatitude.

The fusulinacean assemblage with *Pseudofusulina attenuata* and *Ps. acuta* can be attributed to the McCloud province (Ross 1995), where the fusulinaceans and coral faunas at certain horizons include significant Tethyan warm-water elements (Ross 1995; Fedorowsky *et al.* 2007). The occurrences

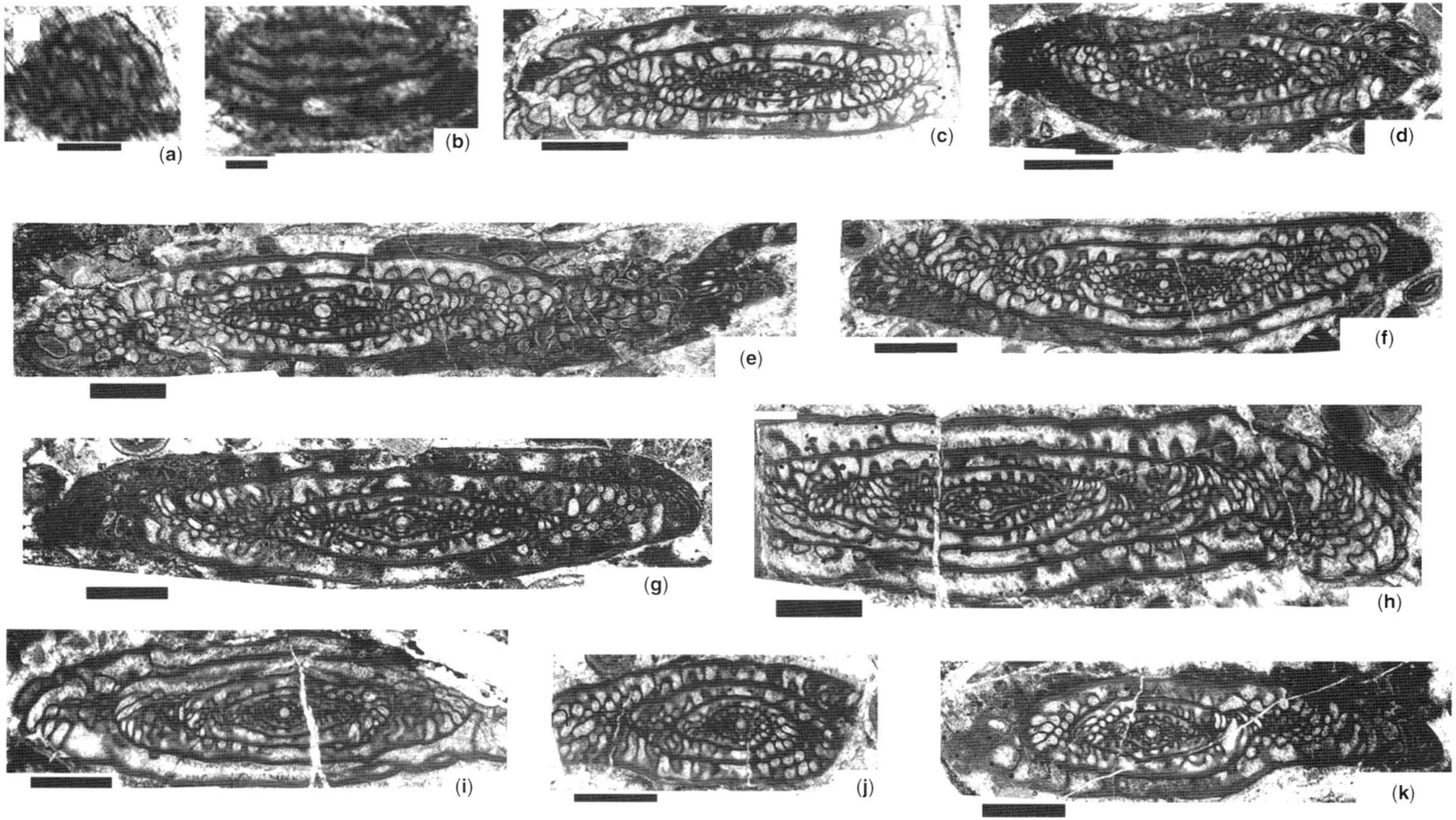

Fig. 9. Fusulinaceans from Fellers Creek section. (**a**) *Schubertella* sp., 37.5 m_6c, 0.1 mm. (**b**) *Schubertella* ex. gr. *kingi*, 39.5 m_1d, 0.1 mm. (**c**) *Pseudofusulina attenuata* Skinner and Wilde, 37.5 m_1a, 1 mm. (**d**) *Pseudofusulina attenuata* Skinner and Wilde, 37.5 m_2b, 1 mm. (**e**) *Pseudofusulina attenuata* Skinner and Wilde, 37.5 m_8a, 1 mm. (**f**) *Pseudofusulina attenuata* Skinner and Wilde, 37.5 m_1b, 1 mm. (**g**) *Pseudofusulina attenuata* Skinner and Wilde, 39.5 m_1a, 1 mm. (**h**) *Pseudofusulina attenuata* Skinner and Wilde, 37.5 m_5a, 1 mm. (**i**) *Pseudofusulina acuta* Skinner and Wilde, 37.5 m_4a, 1 mm. (**j**) *Pseudofusulina acuta* Skinner and Wilde, 37.5 m_1d, 1 mm. (**k**) *Pseudofusulina acuta* Skinner and Wilde, 37.5 m_5b, 1 mm.

Table 1. *Carbonate microfacies of Moscovian–Sakmarian Belcourt Formation, east-central British Columbia, Canada*

Name	ASSOCIATION Main biota	Carbonate grains	Petrographic attributes	Relevant field observations	Occurrence	Depositional environment
MF-01 Ooid-foraminifer grainstone (Fig. 5a, b)	PHOTOZOAN foraminifera fusulinid palaeotextularid echinoderm brachiopod	Ooids (60–90%) Bioclasts (10–40%) Broken fossils *Microcodium*	Tangential ooids Former aragonite Recrystallization		East and south outcrops (FCA) Fellers Creek Kinuseo Creek Meosin Mountain Mount Hannington Well c-52-K/93-O-8[1] Surface-subsurface NE BC[2]	INNER RAMP (SHOAL) Proximity to shoreline Warm Shallow High energy (FWB) oligotrophic Subaerial exposure
MF-02 Algal-bioclastic grainstone (Fig. 5c)	PHOTOZOAN calcareous alga dasycladacean phylloid foraminifera echinoderm brachiopod bryozoan	Broken fossils (30–70%) Bioclasts (30–70%)	Little to no mud		East and south outcrops (FCA) Fellers Creek Kinuseo Creek Meosin Mountain Mount Hanington[1]	INNER RAMP Proximity to Shoal Warm Shallow High energy (>FWB) oligotrophic
MF-03 Bryozoan-echinoderm fusulinacean packstone-grainstone (Fig. 5d)	HETEROZOAN-EXTENDED foraminifera fusulinid palaeotextularid rugose coral solitary echinoderm brachiopod bryozoan	Bioclasts (0–30%) Fossils whole and broken (0–100%)	Bryozoan branches intact <1 cm in diameter		East and south outcrops (FCA) Fellers Creek Kinuseo Creek[1,3] Meosin Mountain[1,3] Mount Hanington[1,3]	INNER TO MIDDLE RAMP Cool Shallow Moderate–high energy (<FWB) oligotrophic
MF-04 Microbial lime to dolomudstone (Fig. 5e)		Chert clasts Rare 1–2 per cm Sub-angular	Dolomitization Partial to complete Uniform Finely crystalline 5–10 μm rhombs Laminations Light and dark bands	Recessive >0.5 m units Poorly exposed Laminated	East and south outcrops (FCA) Fellers Creek Kinuseo Creek(?)	INTERTIDAL BACK-RAMP Low energy Suspension settling High-energy events Microbial stabilization[4,5] Stressed Environment Evaporative Arid climate[6,7]

(Continued)

Table 1. *Continued*

Name	ASSOCIATION Main biota	Carbonate grains	Petrographic attributes	Relevant field observations	Occurrence	Depositional environment
MF-05 Rugose coral wackestone packstone boundstone (Fig. 5f, 5g)	PHOTOZOAN rugose coral colonial[11] solitary	Whole fossils (100%)	Dark micritic matrix	30–50 cm beds Rugose corals not in life position not broken or abraded corallite diam: 1–3 cm irregular patches on top of and in sharp contact with *Palaeoaplysina* Boundstone Ceroid growth form	East and south outcrops (FCA) Fellers Creek	INNER RAMP (PROTECTED TO OPEN) Biostromes and isolated bioherms Protected areas >FWB Rugose corals Knocked over and buried quickly Photic zone limitation[8–10] Light and depth dependent because photic zone food source[7–9] Constant wave agitation Hard (lithified) substrate
MF-06 *Palaeoaplysina* packstone boundstone (Fig. 5h)	PHOTOZOAN calcareous alga *Tubiphytes* foraminifera echinoderm brachiopod bryozoan	Fossils whole and broken (100%)	Lime mudstone and wackestone matrix in between *Palaeoaplysina* plates	Massively bedded units beds: 0.2–1 m thick irregular upper and lower contacts *Palaeoaplysina* plates are 2–5 mm thick and 2–5 cm long	East and south outcrops (FCA) Fellers Creek (two levels) Kinuseo Creek[1] Western outcrops (MCA) West Sukunka[1]	OUTER RAMP Bioherms Moderate–low energy Associated with colonial rugose coral bioherms and grainstone (MF-05 and MF-03) that formed within high-energy environments
MF-07 Bryozoan-brachiopod wackestone packstone (Fig. 7a, b)	HETEROZOAN echinoderm brachiopod bryozoan foraminifera endothyrid palaeotextularid	Fossils broken (100%)	Matrix Mixed argillaceous – lime mud sometimes dolomitic	Broken fossils preserved in multiple different orientations	Western outcrops (MCA) Mountain Creek West Sukunka[1]	MIDDLE RAMP Relatively shallow Just below FWB Low energy (most of the time) Cool water

MF-08 Bioturbated silty lime mudstone (Fig. 7c)	HETEROZOAN sponge spicule brachiopod bryozoan foraminifera protonodosarid	Fossils whole and broken (100%)	Terrigenous sub-angular and coarse quartz silt up to 20% Sponge spicules commonly found in burrow fills	Variably bioturbated Feeding and dwelling traces *Chondrites* *Helminthopsis* *Palaeophycus* Planar laminae often disrupted by bioturbation	Western outcrops (MCA) Mountain Creek West Sukunka[1]	OUTER RAMP Relatively deep Below FWB and SWB Low energy Sporadic high-energy events Suspension settling Aeolian silt in arid climate[12–13] Oxic sea floor conditions Cool to cold water
MF-09 Hummocky cross-stratified silty packstone grainstone (Fig. 7d, e)	HETEROZOAN echinoderm brachiopod bryozoan ostracod	Fossils broken silt-size (100%)	Microgranular	Small-scale hummocky cross-stratified	Western outcrops (MCA) Mountain Creek West Sukunka(?)[1]	SHOREFACE TO OFFSHORE TRANSITION Relatively shallow water Below FWB, above SWB Regular storms Storm-related lag deposition Cool water Opportunistic organisms[14]
MF-10 Silty cross-bedded packstone grainstone (Fig. 7f)	HETEROZOAN brachiopod bryozoan	Fossils bioclasts silt-size (100%)	Silty (as least 20%) and argillaceous Microgranular	Ripples	Western outcrops (MCA) Mountain Creek West Sukunka(?)[1]	SHOREFACE Shallow water Above FWB Cool, eutrophic(?) water[15, 16]
MF-11 Bioclastic grainstone packstone (Fig. 7g, h)	HETEROZOAN echinoderm brachiopod (rare) bryozoan (rare) foraminifera endothyrid palaeotextularid	Fossils bioclasts silt-size (70–100%) broken (10–30%) Oeloid	Matrix is mixed argillaceous and micrite		Western outcrops (MCA) Mountain Creek (upper part) West Sukunka(?)[1]	INNER RAMP Relatively shallow water > FWB, above SWB High energy, constant agitation Periodic storms Cool water Sediment-starved environment(?)[17,18]

[1]Bamber & Macqueen 1979; [2]Wamsteeker 2007; [3]McGugan & Rapson-McGugan 1976; [4]Kepper 1966; [5]Mastandrea *et al.* 2006; [6]Frisia 1994; [7]Aretz *et al.* 2010; [8]Coates & Jackson 1987; [9]Wells 1963; [10]Mastandrea *et al.* 2006; [11]*Protowentzelella kunthi* (E.W. Bamber, pers. comm. 2010); [12]Soreghan *et al.* 2008; [13]Sun *et al.* 2002; [14]Frey 1990; [15]Boyd 2010; [16]Jones 2010; [17]James *et al.* 2009; [18]Saxena & Betzler 2003.

Note: FWB, SWB: fair-weather, storm wave base.

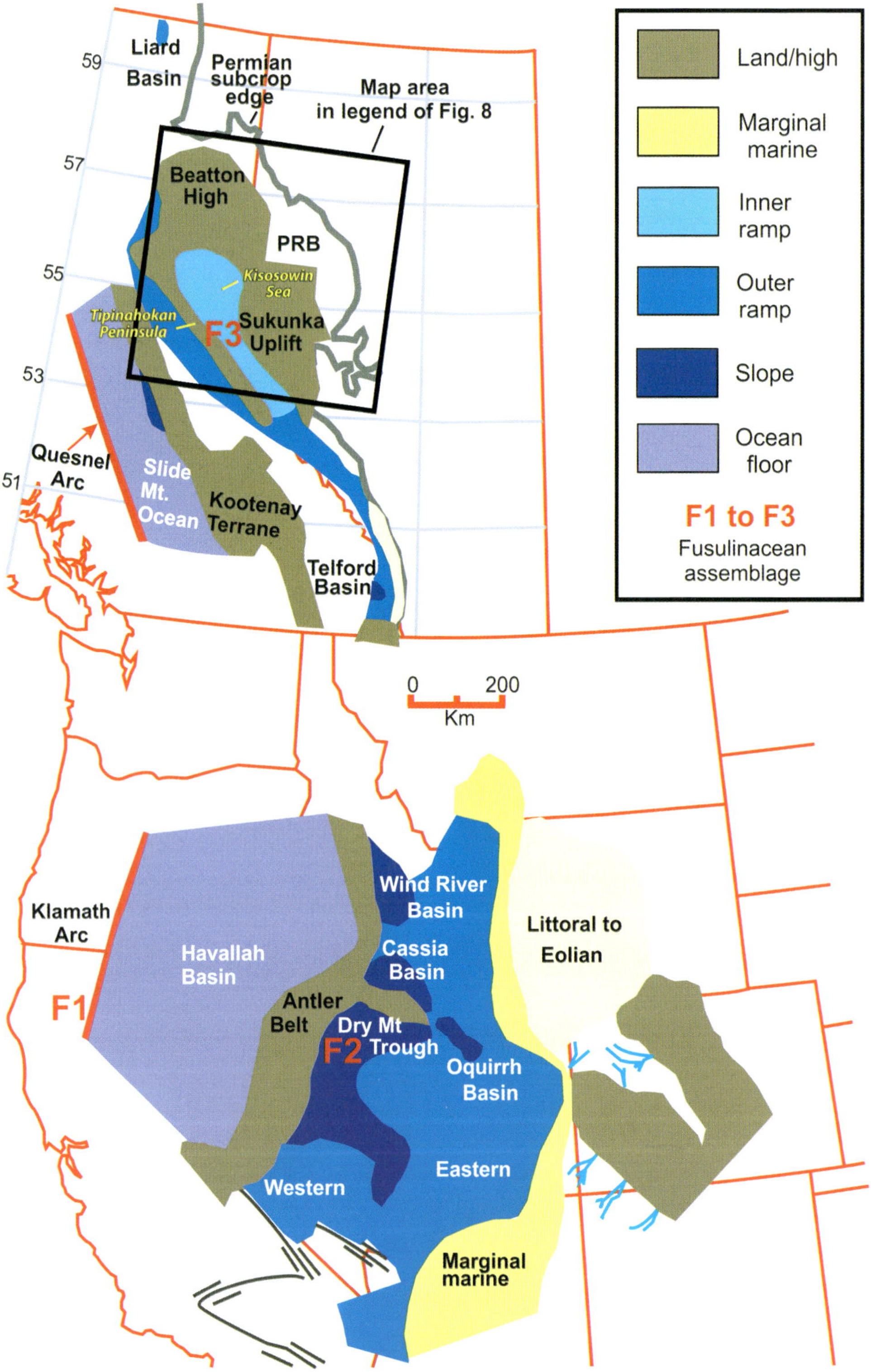
Land/high
Marginal marine
Inner ramp
Outer ramp
Slope
Ocean floor
F1 to F3
Fusulinacean assemblage
Liard Basin
Permian subcrop edge
Map area in legend of Fig. 8
59
57
55
53
51
Beatton High
PRB
Kisosowin Sea
Tipinahokan Peninsula
F3
Sukunka Uplift
Quesnel Arc
Slide Mt. Ocean
Kootenay Terrane
Telford Basin
0 200
Km
Klamath Arc
Havallah Basin
F1
Antler Belt
Wind River Basin
Cassia Basin
Dry Mt Trough
F2
Oquirrh Basin
Littoral to Eolian
Western
Eastern
Marginal marine

of this exotic assemblage on the North American (NA) craton in central Nevada and in east-central BC 1800 km to the north may suggest a warming episode along the NA margin during early Sakmarian time as well as linkage with Klamath/Quesnel arc rocks (Fig. 10) to the west. On the basis of faunal similarities, Belasky *et al.* (2002) suggested that the Quesnel and Klamath terranes must have been 2000–3000 km away from their latitudinal equivalents on the NA craton during the Early Permian. Models developed by Nelson *et al.* (2006) suggest that the Slide Mountain Ocean (and therefore also the Havallah Basin) was the locus for back-arc seafloor spreading and would have been distant from the NA craton. Henderson *et al.* (2012) highlighted the importance of timing and suggested the development of an uplifted 'peripheral bulge' during the Artinskian that closed the Kisosowin Sea and led to the development of an isolated Peace River Basin in west-central Alberta (Fig. 2). This suggests that tectonism within the Quesnel/Klamath arc terrane and Slide Mountain Ocean influenced the NA cratonic margin indicating a narrower Slide Mountain Ocean and Havallah Basin (Fig. 10), which seems to be supported by the fusulinacean assemblage (Fig. 10). It is apparent that the climatic warming suggested by the occurrence of these McCloud Tethyan warm-water elements in east-central British Columbia is insufficient by itself to account for this association, given the prevailing cool-water currents affecting the margin at these palaeolatitudes. The Kisosowin Sea clearly represents a protected embayment that was able to foster these warm-water organisms during the Early Permian (Fig. 10).

Cool-water carbonate associations or heterozoan associations are defined as a group of benthic carbonate particles produced by organisms that are non-light dependent, apart from red calcareous algae (James 1997). Common carbonate-producing organisms found within this association include brachiopods, bryozoans, mollusks, echinoderms and some smaller foraminifers. The MCA facies of the Belcourt Formation predominantly consists of wackestone and packstone with minor grainstone that are part of the Heterozoan association with no indication of photozoan elements. This assemblage occurs at outcrops in the westernmost thrust sheet in the study area, located west of outcrops that include the Tipinahokan Peninsula record.

Baja California: Modern analogue for the Tipinahokan Peninsula

Baja California is a southwards-extending peninsula on the west coast of Mexico that protects a gulf or sea (Sea of Cortéz/Gulf of California), with the opening to this embayment to the south. The peninsula is located between 22 and 32°N latitude and experiences seasonal upwelling along the Pacific coast (Walsh *et al.* 1977). Upwelling directly affects food chain dynamics, with marked changes when upwelling is slow or even at times when the current reverses and downwelling occurs along this coast (Walsh *et al.* 1977). Upwelling is at its maximum from February to June. The Sea of Cortéz is considered to be 'a mostly isolated, distinct body of water' with different biological populations to those on the Pacific coast of the Baja peninsula (Lluch-Belda *et al.* 2003). Despite this, the California Current reaches the mouth of the Sea of Cortéz, allowing some interchange between the Pacific Ocean and the opening of the gulf (Lluch-Belda *et al.* 2003). This brings not only cool water but also nutrients to the sediments at the mouth of the Sea of Cortéz.

The most northern occurrence of reef-forming hermatypic corals occur within the southern portion of the Sea of Cortéz near an area called La Paz, which is at 24°N latitude (Halfar *et al.* 2004*a*). This area is characterized as a warm-temperate carbonate realm with a mixed heterozoan–photozoan association (Halfar *et al.* 2004*b*). Mean sea-surface temperature is at 24 °C, allowing the growth of photozoan carbonates. Farther north in the Sea of Cortéz, the majority of carbonate-producing organisms consists of molluscs and rhodoliths with occasional, and often older, reworked coral indicating a heterozoan carbonate association (Halfar *et al.* 2004*b*). This occurrence of reef-forming corals is due to the protected oceanographic conditions that allow for warm water and oligotrophic to mesotrophic conditions necessary for photozoan carbonate production (Halfar *et al.* 2004*b*).

The configuration of the Sea of Cortéz with protected photozoan carbonates on the inside of the peninsula is a good analogue for the Late Palaeozoic of east-central British Columbia. It not only occurs at comparable latitude on the west coast of a continent, but the modern climate is representative of an interglacial period, probably similar to that of the

Fig. 10. Asselian–Sakmarian palaeogeography and tectonic elements for British Columbia, Alberta and western United States, modified from Henderson *et al.* (2002) and Wardlaw *et al.* (1995). This is a non-palinspastic reconstruction, so the Kisosowin Sea would be *c.* 20 km wider than shown (Richards *et al.* 1994). The width of the Slide Mountain Ocean and Havallah Basin is speculative. F1 to F3 indicates the location of fusulinacean assemblages discussed in the text. PRB: Peace River Basin (site of non-deposition during Asselian–Sakmarian, but isolated downwarped basin during Artinskian–Middle Permian).

many interglacials during the Asselian–Sakmarian. This presence of a palaeogeographical high with warm-water carbonates to the east and cool-water carbonates to the west, within latitudes that might be expected to experience at least seasonal upwelling, resembles the geographical configuration and biotic distribution of Baja California (inset Fig. 8). Photozoan carbonates within the Sea of Cortéz are characterized as warm–temperate because of the lack of green algae and extensive reef-forming carbonates (Halfar *et al.* 2004*b*). Photozoan carbonates within the Kisosowin Sea can be characterized as tropical because of the presence of calcareous green algae, coral and ooids (James *et al.* 1999). This difference in the carbonate organisms between the Sea of Cortéz and the Kisosowin Sea, despite the similarity in latitude and geography, may be due to several factors, including warmer global temperatures during the Permian and a basin configuration that would promote more oligotrophic conditions allowing photozoan carbonates to develop.

Conclusions

The emergence of the Tipinahokan Peninsula during the Late Pennsylvanian and Early Permian created a protected sea that emulated the conditions found in tropical to subtropical Pennsylvanian–Permian basins of the western US as well as the Tethyan McCloud limestone of the Klamath arc (Fig. 10). The warm-water carbonates of the Kisosowin Sea were situated in a palaeolatitude that should have experienced cool-water sedimentation from upwelling, indicating important linkages between climate, oceanic currents and tectonically controlled basins. In particular, our study demonstrates the existence of a Moscovian open ocean embayment with little restriction except in an inner ramp lagoon and sabkha environments. This was followed by the emergence of the Tipinahokan Peninsula, which began during the Late Pennsylvanian C6 event and later climaxed during the Early Permian P1 event. The Tipinahokan Peninsula was fully emergent by the Asselian–Sakmarian. This allowed a photozoan carbonate ramp to develop in the protected Kisosowin Sea to the east, adjacent to an arid and warm continental margin, comparable to the modern Sea of Cortéz and NW Mexico. A cool-water heterozoan carbonate ramp influenced by nutrient-rich upwelling waters existed to the west of the Tipinahokan Peninsula.

This study thus demonstrates the presence of a cool upwelling system along the NW margin of Pangea at a time when substantially warmer-water carbonate sedimentation occurred well over 1000 km to the north in the Sverdrup Basin (Arctic Canada) and to the south in Nevada. Finally, this study shows that the major global climatic shift across the Asselian–Sakmarian boundary, which is associated with the melting of much of the Gondwana ice sheet, did not solely affect carbonate sedimentation in our study area.

Geoscience BC and Talisman Energy Inc. provided financial support that made this research in a remote area possible. This project was also financially supported by Natural Sciences and Engineering Research Council of Canada (NSERC) Discovery grants held by C. M. Henderson and B. Beauchamp. We thank reviewers C. Reid, B. Richards and G. Wahlman for their very thoughtful comments that have greatly improved our contribution.

References

Aretz, M., Herbig, H. G., Somerville, I. D. & Cûzar, P. 2010. Rugose coral biostromes in the late Viséan (Mississippian) of NW Ireland: bioevents on an extensive carbonate platform. *Palaeogeography, Palaeoclimatology, Palaeoecology*, **292**, 488–506.

Bamber, E. W. & Macqueen, R. W. 1979. Upper Carboniferous and Permian stratigraphy of the Monkman Pass and Southern Pine Pass areas, northeastern British Columbia. *Bulletin of Geological Survey of Canada*, **301**, 27.

Beauchamp, B. 1994. Permian climatic cooling in the Canadian Arctic. *In*: Klein, G. D. (ed.) *Pangea: Paleoclimate Tectonics, and Sedimentation During Accretion, Zenith, and Breakup of a Supercontinent*, Geological Society of America, Boulder, Special Paper, **288**, 229–246.

Beauchamp, B. & Baud, A. 2002. Growth and demise of Permian biogenic chert along northwest Pangea: evidence for end-Permian collapse of thermohaline circulation. *Palaeogeography, Palaeoclimatology, Palaeoecology*, **184**, 37–63.

Beauchamp, B. & Desrochers, A. 1997. Permian warm- to very cold-water carbonates and cherts in Northwest Pangea. Special Publication, *Society for Sedimentary Geology*, **56**, 327–347.

Beauchamp, B. & Henderson, C. M. 1994. The Lower Permian Raanes, Great Bear Cape and Trappers Cove formations, Sverdrup Basin, Canadian Arctic: stratigraphy and conodont zonation. *Bulletin of Canadian Petroleum Geology*, **42**, 562–597.

Belasky, P., Stevens, C. H. & Hanger, R. A. 2002. Early Permian location of western North American terranes based on brachiopod, fusulinid and coral biogeography. *Palaeogeography, Palaeoclimatology, Palaeoecology*, **179**, 245–266.

Bensing, J. P., James, N. P. & Beauchamp, B. 2008. Carbonate deposition during a time of mid-latitude ocean cooling: early permian subtropical sedimentation in the sverdrup basin, Arctic Canada. *Journal of Sedimentary Research*, **78**, 2–15.

Blakey, R. C. 2008. Pennsylvanian–Jurassic sedimentary basins of the Colorado plateau and Southern Rocky Mountains. *In*: Andrew, D. M. (ed.) *Sedimentary Basins of the World*, Elsevier, Amsterdam, **5**, 245–296.

BOYD, R. 2010. Transgressive wave-dominated coasts. *In*: JAMES, N. P. & DALRYMPLE, R. W. (eds) *Facies Models 4*, Canadian Society of Petroleum Geologists, Calgary, 265–294.

CHERNYKH, V. V. 2005. *Zonal methods in biostratigraphy: zonal conodont scale of the Lower Permian in the Urals. (In Russian)*. Institute of Geology and Geochemistry, Uralian Branch of the Russian Academy of Sciences Ekaterinburg.

CLAPHAM, M. E. 2010. Faunal evidence for a cool boundary current and decoupled regional climate cooling in the Permian of western Laurentia. *Palaeogeography, Palaeoclimatology, Palaeoecology*, **298**, 3–4.

COATES, A. G. & JACKSON, J. B. C. 1987. Clonal growth, algal symbiosis, and reef formation by corals. *Paleobiology*, **13**, 363–378.

DAVIES, G. R. & NASSICHUK, W. W. 1973. The Hydrozoan? *Palaeoaplysina* from the Upper Paleozoic of Ellesmere Island, Arctic Canada. *Journal of Paleontology*, **47**, 251–265.

DAVIES, G. R., RICHARDS, B. C., BEAUCHAMP, B. & NASSICHUK, W. W. 1989. Carboniferous and Permian reefs in Canada and adjacent areas. *In*: GELDSEYZER, H. H. J., JAMES, N. P. & TEBBUTT, G. E. (eds) *Reefs, Canada and Adjacent areas*. Canadian Society of Petroleum Geologists, Calgary, **13**, 565–574.

DAVYDOV, V. I. 2011. Taxonomy, nomenclature and evolution of the early schubertellids (Fusulinida, Foraminifera). *Acta Palaeontologica Polonica*, **56**, 181–194.

DAVYDOV, V. I., SNYDER, W. S., SPINOSA, C., ROSS, C. A., ROSS, J. R. P. & BRENCKLE, P. L. 1997. Permian foraminiferal biostratigraphy and sequence stratigraphy of Nevada. *Special Publications – Cushman Foundation for Foraminiferal Research*, **36**, 31–34.

DUNHAM, R. J. 1962. Classification of carbonate rocks according to depositional texture. *In*: HAM, W. E. (ed.) *Classification of carbonate rocks*. American Association of Petroleum Geologists, Tulsa, Memoir, **1**, 108–121.

DUNN, L. 2003. *Sequence biostratigraphy and depositional environmental modeling of the Pennsylvanian–Permian Belloy Formation, northwest Alberta and northeast British Columbia*. PhD thesis, University of Calgary.

FEDOROWSKI, J., BAMBER, E. W. & STEVENS, C. H. 2007. *Lower Permian Colonial Rugose Corals, Western and Northwestern Pangaea; Taxonomy and Distribution*. National Research Council of Canada, Ottawa, Ont., Canada.

FORD, C. M., HENDERSON, C. M., HUBBARD, S. M., SOREGHAN, G. S., HATHAWAY, K., SOREGHAN, M. & DAVYDOV, V. I. 2009. Geologic record of arid climate cyclothems in the Upper Pennsylvanian and Lower Permian Tobermory and Kananaskis Formations of Fortress Mountain Ridge Section. *In*: *Proceedings of CSPG CSEG CWLS Convention*, Calgary, Alberta, 771–774.

FRANCIS, J. E. 1994. Paleoclimates of Pangea: Geological evidence. *In*: EMBRY, A. F., BEAUCHAMP, B. & GLASS, D. J. (eds) *Pangea: Global Environments and Resources*. Canadian Society of Petroleum Geologists, Calgary, Memoir, **17**, 265–274.

FREY, R. W. 1990. Trace fossils and hummocky cross-stratification, Upper Cretaceous of Utah. *Palaios*, **5**, 203–218.

FRISIA, S. 1994. Mechanisms of complete dolomitization in a carbonate shelf: comparison between the Norian Dolomia Principale (Italy) and the Holocene of Abu Dhabi Sabkha. *In*: PURSER, B., TUCKER, M. & ZENGER, D. (eds) *Dolomites*, Blackwell Publishing Ltd, Oxford, 55–74.

GOLONKA, J. & FORD, D. 2000. Pangean (Late Carboniferous-Middle Jurassic) paleoenvironment and lithofacies. *Palaeogeography, Palaeoclimatology, Palaeoecology*, **161**, 1–34.

HALFAR, J., GODINEZ-ORTA, L., MUTTI, M., VALDEZ-HOLGUIN, J. E. & BORGES, J. M. 2004*a*. Nutrient and temperature controls on modern carbonate production: an example from the Gulf of California, Mexico. *Geology*, **32**, 213–216.

HALFAR, J., INGLE, J. C. & GODINEZ-ORTA, L. 2004*b*. Modern non-tropical mixed carbonate-siliciclastic sediments and environments of the southwestern Gulf of California, Mexico. *Sedimentary Geology*, **165**, 93–115.

HECKEL, P. H. 1986. Sea-level curve for Pennsylvanian eustatic marine transgressive-regressive depositional cycles along Midcontinent outcrop belt, North America. *Geology*, **14**, 330–334.

HECKEL, P. H. 2002. Overview of Pennsylvanian cyclothems in Midcontinent North America and brief summary of those elsewhere in the world. *In*: HILLS, L.V., HENDERSON, C. M. & BAMBER, E. W. (eds) *Carboniferous and Permian of the World*. Canadian Society of Petroleum Geologists, Calgary, Memoir, **19**, 79–98.

HECKEL, P. H. 2008. Pennsylvanian cyclothems in Midcontinent North America as far-field effects of waxing and waning of Gondwana ice sheets. *In*: FIELDING, C.R., FRANK, T. D. & ISBELL, J. L. (eds) *Resolving the Late Paleozoic ice age in time and space*. Geological Society of America, Tulsa, Special papers, **441**, 275–289.

HENDERSON, C. M., RICHARDS, B. C. & BARCLAY, J. E. 1994. Permian strata of the Western Canada Sedimentary Basin. *In*: MOSSOP, G. D. & SHETSEN, I. (eds) *Geologic Atlas of the Western Canada Sedimentary Basin*, Canadian Society of Petroleum Geologists and Alberta Research Council, Calgary, 251–258.

HENDERSON, C. M., DUNN, L., FOSSENIER, K. & MOORE, D. 2002. Sequence biostratigraphy and paleogeography of the Pennsylvanian–Permian Belloy Formation and outcrop equivalents in Western Canada. *In*: HILLS, L.V., HENDERSON, C. M. & BAMBER, E. W. (eds) *Carboniferous and Permian of the World*. Canadian Society of Petroleum Geologists, Calgary, Memoir, **19**, 934–947.

HENDERSON, C. M., ZUBIN-STATHOPOULOS, K. D. & DEAN, G. J. 2012. *Chronostratigraphic and tectonostratigraphic summary of the Late Paleozoic and Early Triassic succession in east-central British Columbia*. Geoscience BC Summary of Activities 2011, Geoscience BC Report, **2012-1**, 115–123.

JAMES, N. P. 1997. The cool-water carbonate depositional realm. *In*: JAMES, N.P. & CLARKE, A. D. (eds) *Cool Water Carbonates*. Society for Sedimentary Geology, Tulsa, Special Publication, **56**, 1–20.

JAMES, N. P., COLLINS, L. B., BONE, Y. & HALLOCK, P. 1999. Subtropical carbonates in a temperate realm; modern sediments on the Southwest Australian shelf. *Journal of Sedimentary Research*, **69**, 1297–1321.

JAMES, N. P., FRANK, T. D. & FIELDING, C. R. 2009. Carbonate sedimentation in a Permian high-latitude, subpolar depositional realm; Queensland, Australia. *Journal of Sedimentary Research*, **79**, 125–143.

JONES, B. 2010. Warm-water neritic carbonates. *In*: JAMES, N. P. & DALRYMPLE, R. W. (eds) *Facies Models 4*, Canadian Society of Petroleum Geology, Calgary, 341–369.

KEPPER, J. C. 1966. Primary dolostone patterns in the Utah-Nevada Middle Cambrian. *Journal of Sedimentary Research*, **36**, 548–562.

KIESSLING, W., FLUEGEL, E. & GOLONKA, J. 1999. Paleoreef maps; evaluation of a comprehensive database on Phanerozoic reefs. *American Association of Petroleum Geologists Bulletin*, **83**, 1552–1587.

KLAPPA, C. F. 1978. Biolithogenesis of *Microcodium*: elucidation. *Sedimentology*, **25**, 489–522.

KOSIR, A. 2004. Microcodium revisited: root calcification products of terrestrial plants on carbonate-rich substrates. *Journal of Sedimentary Research*, **74**, 845–857.

LLUCH-BELDA, D., LLUCH-COTA, D. B. & LLUCH-COTA, S. E. 2003. Baja California's biological transition zones: refuges for the California Sardine. *Journal of Oceanography*, **59**, 503–513.

MACRAE, J. & MCGUGAN, A. 1977. Permian stratigraphy and sedimentology-southwestern Alberta and southeastern British Columbia. *Bulletin of Canadian Petroleum Geology*, **25**, 752–766.

MASTANDREA, A., PERRI, E., RUSSO, F., SPADAFORA, A. & TUCKER, M. 2006. Microbial primary dolomite from a Norian carbonate platform: northern Calabria, southern Italy. *Sedimentology*, **53**, 465–480.

MCGUGAN, A. & RAPSON, J. E. 1962. Permo-Carboniferous stratigraphy, Crowsnest area, Alberta and British Columbia. *Journal of the Alberta Society of Petroleum Geologists*, **10**, 352–368.

MCGUGAN, A. & RAPSON, J. E. 1963. Permo-Carboniferous stratigraphy between Banff and Jasper, Alberta. *Bulletin of Canadian Petroleum Geology*, **11**, 150–160.

MCGUGAN, A. & RAPSON-MCGUGAN, J. E. 1976. Permian and Carboniferous stratigraphy, Wapiti Lake area, northeastern British Columbia. *Bulletin of Canadian Petroleum Geology*, **24**, 193–210.

MOORE, D. 2002. *The Stratigraphy of the Pennsylvanian and Lower Permian Tobermory, Kananaskis and Johnston Canyon Formations of the Front Ranges of the Southern Canadian Rocky Mountains, Alberta and British Columbia*. Masters thesis, University of Calgary.

MORIN, J., DESROCHERS, A. & BEAUCHAMP, B. 1994. Facies analysis of Lower Permian platform carbonates, Sverdrup Basin, Canadian Arctic Archipelago. *Facies*, **31**, 105–130.

NAQVI, I. H. 1972. The Belloy Formation (Permian), Peace River area, northern Alberta and northeastern British Columbia. *Bulletin of Canadian Petroleum Geology*, **20**, 58–88.

NELSON, J. L., COLPRON, M., PIERCEY, S. J., DUSEL-BACON, C., MURPHY, D. C. & ROOTS, C. F. 2006. Paleozoic tectonic and metallogenic evolution of the pericratonic terranes in Yukon, northern British Columbia and eastern Alaska. *In*: COLPRON, M. & NELSON, J. (eds) *Paleozoic Evolution and Metallogeny of Pericratonic Terranes at the Ancient Pacific Margin of North America, Canadian and Alaskan Cordillera*. Geological Association of Canada, Calgary, Special Paper, **45**, 323–360.

RAPSON-MCGUGAN, J. E. 1970. The diagenesis and depositional environment of the Permian Ranger Canyon and Mowitch formations, Ishbel Group, from the southern Canadian Rocky Mountains. *Sedimentology*, **15**, 363–417.

REID, C. M., JAMES, N. P., BEAUCHAMP, B. & KYSER, T. K. 2007. Faunal turnover and changing oceanography: Late Palaeozoic warm-to-cool water carbonates, Sverdrup Basin, Canadian Arctic Archipelago. *Palaeogeography, Palaeoclimatology, Palaeoecology*, **249**, 128–159.

RICHARDS, B. C. 1989. Upper Kaskaskia Sequence; uppermost Devonian and Lower Carboniferous. *In*: RICKETTS, B. D. (ed.) *Western Canada Sedimentary Basin – A Case History*. Canadian Society of Petroleum Geologists, Calgary, Special Publication, **30**, 165–201.

RICHARDS, B. C., BARCLAY, J. E., BRYAN, D., HARTLING, A., HENDERSON, C. M., HINDS, R. C. & TROLLOPE, F. H. 1994. Carboniferous strata of the Western Canada Sedimentary Basin. *In*: MOSSOP, G. D. & SHETSEN, I. (eds) *Geologic Atlas of the Western Canada Sedimentary Basin*. Canadian Society of Petroleum Geologists and Alberta Research Council, Calgary, 221–250.

ROSS, C. A. 1995. Permian fusulinaceans. *In*: SCHOLLE, P. A., PERYT, T. M. & ULMER-SCHOLLE, D. S. (eds) *Permian of Northern Pangea. Vol. 1: Paleogeography, Paleoclimate, Stratigraphy*. Springer-Verlag, Berlin, 167–185.

SAXENA, S. & BETZLER, C. 2003. Genetic sequence stratigraphy of cool water slope carbonates (Pleistocene Eucla Shelf, southern Australia). *International Journal of Earth Sciences*, **92**, 482–493.

SKINNER, J. W. & WILDE, G. L. 1965. *Permian Biostratigraphy and Fusulinid Faunas of the Shasta Lake Area, Northern California*. University of Kansas Paleontological Contributions, Paper Article 6, 1–98.

SNYDER, W. S., TREXLER, J. H. JR, DAVYDOV, V. I., CASHMAN, P., SCHIAPPA, T. A. & SWEET, D. 2002. Upper Paleozoic Tectonostratigraphic Framework for the Western Margin of North America. *AAPG Hedberg Conference*, Vail, Colorado, 1–4.

SOREGHAN, G. S., SOREGHAN, M. J. & HAMILTON, M. A. 2008. Origin and significance of loess in late Paleozoic western Pangaea; a record of tropical cold? *Palaeogeography, Palaeoclimatology, Palaeoecology*, **268**, 234–259.

STEVENS, C. H., WAGNER, D. B. & SUMSION, S. R. 1979. Permian fusulinid biostratigraphy, central cordilleran miogeosyncline. *Journal of Paleontology*, **53**, 29–36.

SUN, D., BLOEMENDAL, J., REA, D. K., VANDENBERGHE, J., JIANG, F., AN, Z. & SU, R. 2002. Grain-size distribution function of polymodal sediments in hydraulic and aeolian environments, and numerical partitioning of the sedimentary components. *Sedimentary Geology*, **152**, 263–277.

TABOR, N. J., MONTANEZ, I. P., SCOTESE, C. R., POULSEN, C. J. & MACK, G. H. 2008. Paleosol archives of environmental and climatic history in paleotropical western Pangea during the latest Pennsylvanian through Early Permian. *In*: Fielding, C. R., Frank, T. D. & Isbell, J. L. (eds) *Resolving the Late Paleozoic ice age in time and space*. Geological Society of America, Tulsa, Special Paper, **441**, 291–303.

Trexler, J. H., Cashman, P. H., Snyder, W. S. & Davydov, V. I. 2004. Late Paleozoic tectonism in Nevada: Timing, kinematics, and tectonic significance. *Bulletin of Geological Society of America*, **116**, 525–538.

Vai, G. B. 2003. Development of the palaeogeography of Pangaea from Late Carboniferous to Early Permian. *Palaeogeography, Palaeoclimatology, Palaeoecology*, **196**, 125–155.

Wahlman, G. P. 2002. Upper Carboniferous-Lower Permian (Bashkirian-Kungurian) mounds and reefs. *In*: Kiessling, W., Flügel, E. & Golonka, J. (eds) *Phanerozoic Reef Patterns*. Society for Sedimentary Geology, Tulsa, Special Publication, **72**, 271–338.

Wahlman, G. P. & Konovalova, M. V. 2002. Upper Carboniferous-Lower Permian Kozhim carbonate bank, subpolar pre-Ural Mountains, northern Russia. *In*: Zempolich, W. G. & Cook, H. E. (eds) *Paleozoic Carbonates of the Commonwealth of Independent States (CIS): Subsurface Reservoirs and Outcrop Analogs*. Society for Sedimentary Geology, Tulsa, Special Publication, **74**, 219–241.

Walsh, J. J., Whitledge, T. E., Kelley, J. C., Huntsman, S. A. & Pillsbury, R. D. 1977. Further transition states of the baja California upwelling ecosystem. *Limnology and Oceanography*, **22**, 264–280.

Wamsteeker, M. L. 2007. *Diagenetic and geochemical characterization of the Ksituan Member of the Belloy Formation, east central British Columbia Foothills*. BSc thesis, University of Calgary.

Wang, D. 1993. *Conodont Biostratigraphy of the Carbon Ridge Formation, Secret Canyon, Fish Creek Range, Nevada*. Idaho State University, Pocatello, ID, Boise, 59.

Wanless, H. R. & Shepard, F. P. 1936. Sea level and climatic changes related to late Paleozoic cycles. *Geological Society of America Bulletin*, **47**, 1177–1206.

Wardlaw, B. R., Snyder, W. S., Spinosa, C. & Gallegos, D. M. 1995. Permian of the Western United States. *In*: Scholle, P. A., Peryt, T. M. & Ulmer-Scholle, D. S. (eds) *The Permian of Northern Pangea*, Sedimentary Basins and Economic Resources, Springer-Verlag, **2**, 23–40.

Watkins, R. & Wilson, E. C. 1989. Paleoecologic and biogeographic significance of the biostromal organism Palaeoaplysina in the Lower Permian McCloud Limestone, eastern Klamath Mountains, California. *Palaios*, **4**, 181–192.

Wells, J. W. 1963. Coral growth and geochronometry. *Nature*, **197**, 948–950.

Zubin-Stathopoulos, K. D. 2011. *Tectonic Evolution, Paleogeography and Paleoclimate of Pennsylvanian–Permian Strata in East-Central British Columbia: Implications from Conodont Biostratigraphy and Carbonate Sedimentology*. MSc thesis, University of Calgary.

Zubin-Stathopoulos, K. D., Dean, G. J., Beauchamp, B., Spratt, D. A. & Henderson, C. M. 2011. *Tectonic evolution and paleogeography of Pennsylvanian–Permian strata in east-central British Columbia (NTS 093I, O, P): implications from stratigraphy, fracture analysis and sedimentology*. Geoscience BC Summary of Activities 2010, Geoscience BC Report, **2011-1**, 209–222.

Late Permian (Zechstein) magnetostratigraphy in Western and Central Europe

MICHAEL SZURLIES

Institute of Earth and Environmental Science, University of Potsdam, Karl-Liebknecht-Straße 24–25, 14476 Potsdam-Golm, Germany (e-mail: szurlies@geo.uni-potsdam.de)

Abstract: Owing to the lack of fossils, the subdivision and correlation of the dominantly continental Permian (Rotliegend and Zechstein groups) and the overlying epicontinental Germanic Triassic (Buntsandstein, Muschelkalk and Keuper groups) from Western and Central Europe, traditionally, relies on lithological criteria. From the combination of magnetic polarity records from The Netherlands, Germany and Poland, a well-documented magnetostratigraphical composite for the Zechstein–lowermost Buntsandstein has been constructed. This interval is dominantly of normal polarity with few reversed magnetozones, allowing a reliable correlation throughout the intracontinental Central European Basin. Moreover, together with available biostratigraphical data, this integrated stratigraphical approach facilitates detailed links to the marine realm. The basal Zechstein is most probably equivalent to the uppermost lower–upper Wuchiapingian. The overlying upper Zechstein and the lowermost Buntsandstein are equivalent to the entire Changhsingian. Present bio- and magnetostratigraphical data indicate a possible duration of the Zechstein of only 2.8–3.5 Myr. A distinct climatic change accompanied by a major shift in sediment sources and a palynofloral turnover constitute important proxies for the terrestrial extinction event within the Central European Basin. A transition from a short reversed to a longer normal magnetozone within the uppermost Zechstein, which predates both the end-Permian ecological crisis and the conodont-calibrated biostratigraphical Permian–Triassic boundary (PTB), represents a prime time-marker for global correlation. Above this reverse to normal (R/N) reversal, the continental PTB is located within the lowermost Buntsandstein. The integration of magnetostratigraphy with other stratigraphical data is a prerequisite to establishing causal links between Late Permian climatic events in the sea and on land.

A precise stratigraphical framework is crucial to dating and correlating sedimentary rocks from distant localities and different depositional environments. This is of particular importance for the correlation of continental and marine sedimentary successions, mainly because a direct biostratigraphical comparison is ecologically limited by different palaeoenvironmental conditions. Furthermore, accurate time control is a prerequisite for understanding the chronology of natural processes, such as the many hotly debated hypotheses regarding mass extinction events.

The Permian–Triassic boundary (PTB) interval is characterized by the most severe mass extinction of the entire Phanerozoic, strongly affecting both marine and terrestrial biota (Erwin 1993; Retallack 1995). In Western and Central Europe, the transition from the Permian to the Triassic (i.e. the Zechstein–Buntsandstein boundary interval) is mainly developed in continental red-bed facies. Owing to the lack of fossils, the subdivision and correlation of the dominantly continental Permian (Rotliegend and Zechstein groups) and the overlying epicontinental Germanic Triassic (Buntsandstein, Muschelkalk and Keuper groups) traditionally relies on lithological criteria. In the last two decades, the combination of wireline logs (GR logs and sonic logs) with cyclo- and lithostratigraphy has provided a well-defined stratigraphical framework for the Permian and Triassic of Western and Central Europe (Geluk 2005; Doornenbal & Stevenson 2010 and references therein).

These Permian–Triassic lithostratigraphical units have also been used for chronostratigraphy; for instance, assuming the Rotliegend as Early Permian, the Zechstein as Late Permian and the overlying Buntsandstein as Early Triassic (Harland *et al.* 1990). However, because of the lack of index fossils, the correlation of the continental Permian (i.e. the Rotliegend and Zechstein groups) with the conodont-calibrated stages of the geological timescale is still under debate (Menning *et al.* 2006; Słowakiewicz *et al.* 2009; Doornenbal & Stevenson 2010). In terms of biostratigraphy, the global correlation of the Zechstein is limited to the use of palynological assemblages (e.g. Fijałkowska 1994), scarce conodonts in the lowermost Zechstein (Kozur 1994; Legler *et al.* 2005), and conchostracans in the uppermost Zechstein and overlying Buntsandstein (Kozur 1999; Bachmann & Kozur 2004; Kozur & Weems 2011). These biostratigraphical data allow, at least, an approximate

From: Gąsiewicz, A. & Słowakiewicz, M. (eds) 2013. *Palaeozoic Climate Cycles: Their Evolutionary and Sedimentological Impact.* Geological Society, London, Special Publications, **376**, 73–85.
First published online February 27, 2013, http://dx.doi.org/10.1144/SP376.7

comparison of the Zechstein with the marine early Late Permian–latest Permian (Kozur 1994, 1999; Menning *et al.* 2005, 2006).

Therefore, the key to a more detailed global correlation of the Zechstein and the overlying Buntsandstein lies with linking biostratigraphy to other less facies-sensitive methods, such as carbon isotope stratigraphy or magnetostratigraphy. The global synchronous character of magnetic polarity reversals has made magnetostratigraphy a powerful tool for precise facies-independent correlation between different depositional environments (e.g. marine to non-marine correlations).

In the last 15 years, the Zechstein and Buntsandstein have been the focus of several palaeomagnetic studies in Poland (Nawrocki 1997, 2004; Nawrocki *et al.* 2003) and in Germany (Soffel & Wippern 1998; Szurlies *et al.* 2003; Szurlies 2004, 2007). In this paper, a composite magnetostratigraphy for the Zechstein to lowermost Buntsandstein of Western and Central Europe is constructed, including recent palaeomagnetic data from the Netherlands (Szurlies *et al.* 2012). This composite provides important timelines for basin-wide correlation. Moreover, the integration of this composite with available biostratigraphical data allows a more detailed correlation with the geomagnetic polarity timescale of the Permian (Steiner 2006) and the global geological timescale (Gradstein *et al.* 2004), respectively. Such a precise temporal framework for Late Permian–Early Triassic sedimentary successions in the sea and on land is a prerequisite to establishing causal links between possible scenarios for the end-Permian ecological crisis and associated rapid and severe climate change. A precise dating of climatic events, such as the Upper Rotliegend–Zechstein wet phase (Roscher & Schneider 2006; Schneider *et al.* 2006) or the strong increase in humidity and freshwater discharge at the onset of the Buntsandstein (Bachmann & Kozur 2004; Hug & Gaupp 2006), enables detailed correlation with the climatic development in other sedimentary basins.

Geological setting

In Western and Central Europe, the mainly continental Permian and the overlying epicontinental Triassic were deposited in the intracontinental Central European Basin (CEB), which is also known as the Southern Permian Basin (Fig. 1). The CEB evolved essentially on the former Variscan foreland in latest Carboniferous–Early Permian times (Ziegler 1990; Doornenbal & Stevenson 2010). This WNW–ESE-trending depositional area extends from England to SE Poland and from offshore Norway to southern Germany. The CEB is delineated by several Variscan highs, including the London–Brabant, Rhenish and Bohemian massifs in the south, and the Mid North Sea and Ringkøbing–Fyn highs in the north (Ziegler 1990; Geluk 2005).

The southern margin of the CEB was differentiated into a complex system of NNE-trending highs and lows, such as the Ems Low, the

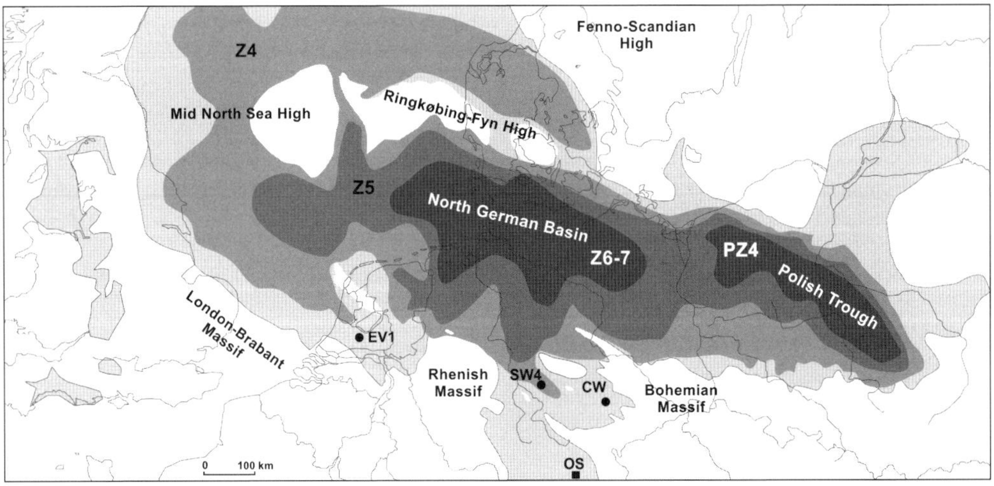

Fig. 1. Late Permian (Zechstein) palaeogeography of the Central European Basin (CEB), that is the Southern Permian Basin (after Geluk 2005). Distribution of Z4–Z7 formations. Location of wells (as circles): EV1, Everdingen-1; SW4, Schlierbachswald-4; OS, Obernsees, and outcrop (as square): CW, Caaschwitz.

Hunte Swell, the Hessian Depression and the Eichsfeld–Altmark Swell. These structures controlled the development of marginal carbonate and anhydrite platforms of the basal Zechstein (e.g. Geluk 1999). However, tectonics played only a minor role during Zechstein deposition as Late Permian evolution of the CEB continued to be dominated by thermal relaxation of the lithosphere (van Wees *et al.* 2000). During the latest Permian and, particularly, the Early Triassic (i.e. the uppermost Zechstein–Buntsandstein), tectonic pulses occasionally affected the sedimentary succession. Locally, the effects of such pulses were particularly recorded on 'palaeohighs' by unconformities, such as the minor Bröckelschiefer Unconformity within the uppermost Zechstein or the Volpriehausen Unconformity at the base of the Middle Buntsandstein (e.g. Best 1989; Röhling 1993; Geluk & Röhling 1997; Geluk 2005; Szurlies *et al.* 2012).

Rifting in the Arctic–North Atlantic region, most probably contemporaneous with a global sea-level rise, caused flooding of the CEB through an incipient Viking Graben (Ziegler 1990). This seaway already channelled previous short-lived marine incursions during the Upper Rotliegend (Legler & Schneider 2008). The deposition of the 'Kupferschiefer' (copper shale) deposits is evidence of this catastrophic and permanent flooding of the sub-sea-level intracontinental CEB (Glennie & Buller 1983; van Wees *et al.* 2000) by the Zechstein transgression. These marly Kupferschiefer black shales were deposited under basin-wide euxinic conditions (Grice *et al.* 1997; Pancost *et al.* 2002; Paul 2006) and has a relatively uniform thickness of about 0.5 m throughout the CEB. It represents an excellent time-marker for regional correlation (Geluk 2005; Doornenbal & Stevenson 2010). A Re–Os age of 257.3 ± 1.6 Ma for the Kupferschiefer in central Germany (Brauns *et al.* 2003) is consistent with the Late Permian (i.e. lower Wuchiapingian) age of scarce conodonts (*Mesogondolella britannica*, *Merrillina divergens*) from the basal Zechstein (Kozur 1994, 2007; Legler *et al.* 2005; Menning *et al.* 2006). However, it has to be stressed that this Re–Os age is not well documented and, thus, discussed controversially (see Słowakiewicz *et al.* 2009).

During the Late Palaeozoic, the CEB area of northern Pangea drifted northwards until it reached its Late Permian location in the northern hemisphere desert belt between about 10 and 30°N (Glennie 2007). Arid and semi-arid climatic conditions during deposition of the Zechstein are geologically evidenced by the development of seven stacked evaporation cycles, which are used for lithostratigraphical subdivision and basin-wide correlation (Richter-Bernburg 1955).

Zechstein and Lower Buntsandstein lithostratigraphy

Traditionally, the Zechstein and Buntsandstein of Western and Central Europe have been subdivided by lithological criteria (Richter-Bernburg 1955; Boigk 1959). The approximately 1500–2000 m-thick Zechstein follows the model of cyclic chemical precipitation in a giant saline basin (Richter-Bernburg 1955). In Germany, the Zechstein Group consists of seven such evaporation cycles (i.e. formations): the Werra (Z1), the Staßfurt (Z2), the Leine (Z3), the Aller (Z4), the Ohre (Z5), the Friesland (Z6) and the Fulda (Z7) formations (Fig. 2). The individual cycles, if ideally developed, exhibit successions of shales, carbonate rocks and evaporites (Richter-Bernburg 1955). However, the Zechstein as a whole displays an overall regressive and hypersaline trend, with the thickness of the Zechstein gradually thinning towards the margin of the CEB (Fig. 1). Therefore, only the lower three cycles have the full cyclic inventory, including marine clastics, carbonate and evaporite rocks. The upper four cycles are dominated by clay- and siltstones and, only occasionally, evaporites. Moreover, these latter cycles become more and more confined to the interior of the CEB (Best 1989). In the marginal parts of the basin, however, these upper cycles are represented by red brittle evaporitic clay- and siltstones (the so-called 'Bröckelschiefer facies') that were deposited in sabkha environments.

The base of the Zechstein is marked by the Kupferschiefer (Fig. 2), which represents one of the principal correlation units in NW European stratigraphy (cf. Geluk 2005). Two further regionally extensive correlation markers are the 'Grauer Salzton' (Grey Salt Clay) at the base of the Z3 Formation and the 'Roter Salzton' (Red Salt Clay) at the base of the Z4 Formation. These markers, which are assumed to represent nearly synchronous flooding events, strongly support a reliable basin-wide lithostratigraphical correlation (cf. Doornenbal & Stevenson 2010).

The Zechstein lithostratigraphy is well defined and can be correlated throughout the interior of the CEB using wireline logs, such as GR logs and sonic logs (Best 1989; Geluk & Röhling 1997; Käding 2000; Geluk 2005; Szurlies *et al.* 2012). These correlations reveal only a few discrepancies in the interpretation of the upper Zechstein between different European countries. First, there is a much wider definition of the Z4 Formation in Poland (PZ4a–PZ4e), where it includes correlatives of the Z5–lower Z7 formations (Fig. 2). The correlation of PZ4a–PZ4e with subdivisions in Germany is still a matter of debate (Wagner & Peryt 1997; Käding 2000). The overlying Rewal

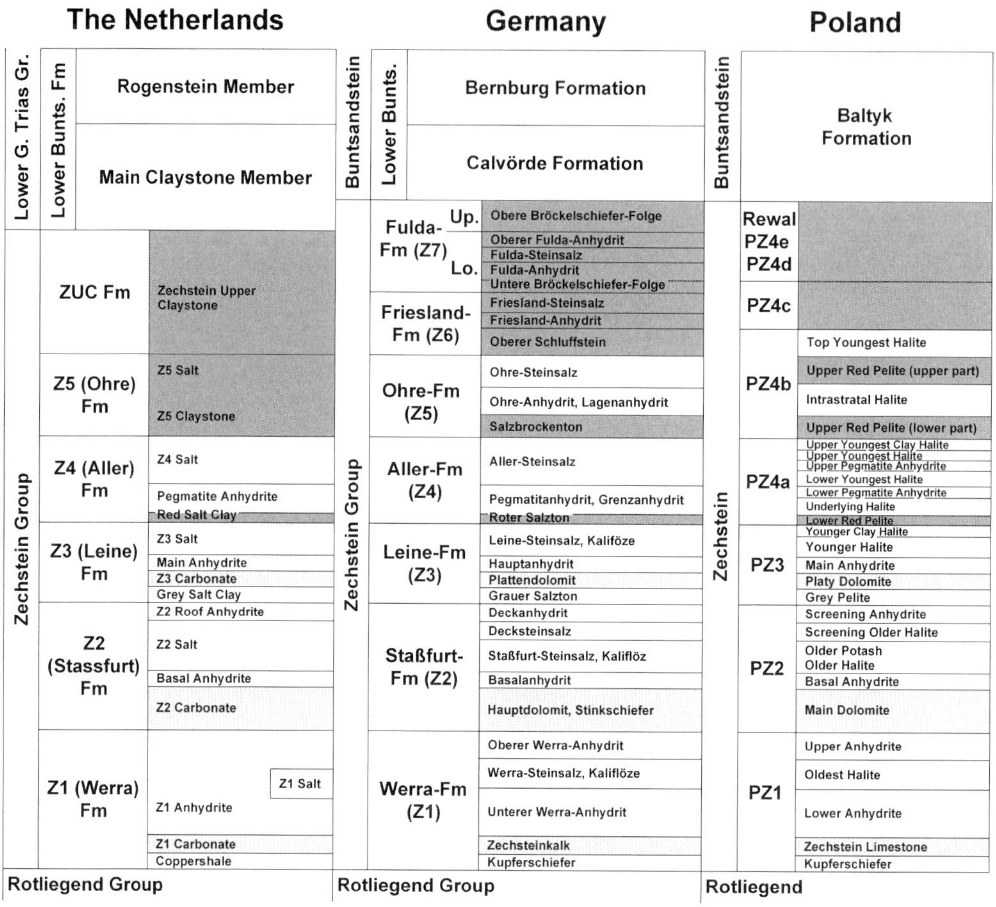

Fig. 2. Lithostratigraphical correlation chart of the Zechstein–Lower Buntsandstein in The Netherlands, Germany and Poland (after Peryt *et al.* 2010).

Formation is most probably equivalent to the Z7 Formation (Käding 2000). In The Netherlands, the Z6 and lower Z7 formations are combined into the Zechstein Upper Claystone Formation (Fig. 2). Second, the overlying correlatives of the upper Z7 are already placed in the Lower Buntsandstein Formation (i.e. the Main Claystone Member) in The Netherlands. In Germany, however, the Zechstein–Buntsandstein boundary is placed at the base of the first prominent (oolitic) sandstone of the Lower Buntsandstein directly above the Z7 Formation (Best 1989; Röhling 1993 and references therein).

The 300–400 m-thick sedimentary succession of the Lower Buntsandstein was deposited mainly in fluviolacustrine environments, with the fluvial influx originating principally from the massifs at the southern margin of the CEB (e.g. Hiete 2004; Paul *et al.* 2008). In The Netherlands and northern Germany, the Lower Buntsandstein is further subdivided into fining-upwards cycles (Paul & Klarr 1988; Röhling 1993; Geluk & Röhling 1997; Szurlies 1999; Szurlies *et al.* 2003). The most obvious cycle is the 10–20 m-thick so-called small-scale cycle, which is ascribed to changes in lake level (base level). Thus, these cycles are considered to represent nearly synchronous basin-wide events, which have recently been proven by bio- and magnetostratigraphical means (Szurlies *et al.* 2003, 2012; Bachmann & Kozur 2004). The base-level cycles are assumed to reflect variability in precipitation (wet–dry cycles) within the CEB, which is most probably related to solar-induced approximately 100 ka eccentricity cycles (Geluk & Röhling 1997; Szurlies *et al.* 2003; Bachmann & Kozur 2004; Menning *et al.* 2005; Szurlies 2007). Based on wireline logs, these cycles can be correlated with confidence throughout the interior of the

CEB (Best 1989; Röhling 1993; Geluk & Röhling 1997; Szurlies 1999; Szurlies *et al.* 2003, 2012; Geluk 2005).

Zechstein magnetostratigraphy

In the last 15 years, various studies have focused on Zechstein and Buntsandstein magnetostratigraphy (Burek 1967; Dachroth 1971, 1976; Nawrocki 1997, 2004; Soffel & Wippern 1998; Nawrocki *et al.* 2003; Szurlies *et al.* 2003, 2012; Szurlies 2004, 2007), allowing the construction of a detailed composite magnetic polarity record for the upper Zechstein (i.e. Z3–Z7 formations) (Fig. 3). The lower Zechstein (Z1 and Z2 formations), however, has only been investigated in south Germany (Soffel & Wippern 1998) and partly in Poland (PZ1n of Nawrocki 1997; Nawrocki *et al.* 2003), where it is indicated mainly by normal polarity (Fig. 4). The between-site correlation of magnetostratigraphical records from the overlying Z3–Z7 formations reveals close similarities concerning the number of magnetozones and their relative length, as well as the overall polarity pattern (Fig. 3). Moreover, the magnetostratigraphical comparison is consistent with the lithostratigraphical correlation of the Zechstein and Lower Buntsandstein (Best 1989; Geluk & Röhling 1997; Käding 2000; Szurlies *et al.* 2003, 2012; Geluk 2005).

In the Everdingen-1 core (The Netherlands) and the Caaschwitz outcrop (east Germany), the lower Z3 Formation (i.e. the lower Z3 Carbonate: Fig. 2) is characterized by normal polarity, that is the magnetozones EV1n and CG1n (Fig. 3). The uppermost Z3 Carbonate in the Everdingen-1 core belongs to the younger reversed magnetozone EV1r, which is correlated with CG1r from the Schlierbachswald-4 core in central Germany (Szurlies *et al.* 2012). Additional support for this correlation comes from the Jaworzna IG-1 core from SE Poland, where Nawrocki (1997) studied two samples from the PZ3 Carbonate (Platy Dolomite) of the PZ3 cycle (Fig. 2). The lower sample exhibits normal polarity equivalent to EV1n and CG1n, whereas the upper sample is of reversed polarity. This reversed polarity interval is most probably equivalent to EV1r, CG1r and Os1r in the Obernsees core from south Germany (Fig. 3). In the Schlierbachswald-4 core, this reversed magnetozone ranges from the upper Z3 Formation into the overlying lower Z4 Formation.

In the Schlierbachswald-4 and the Obernsees cores, the overlying Z5 and Z6 formations are

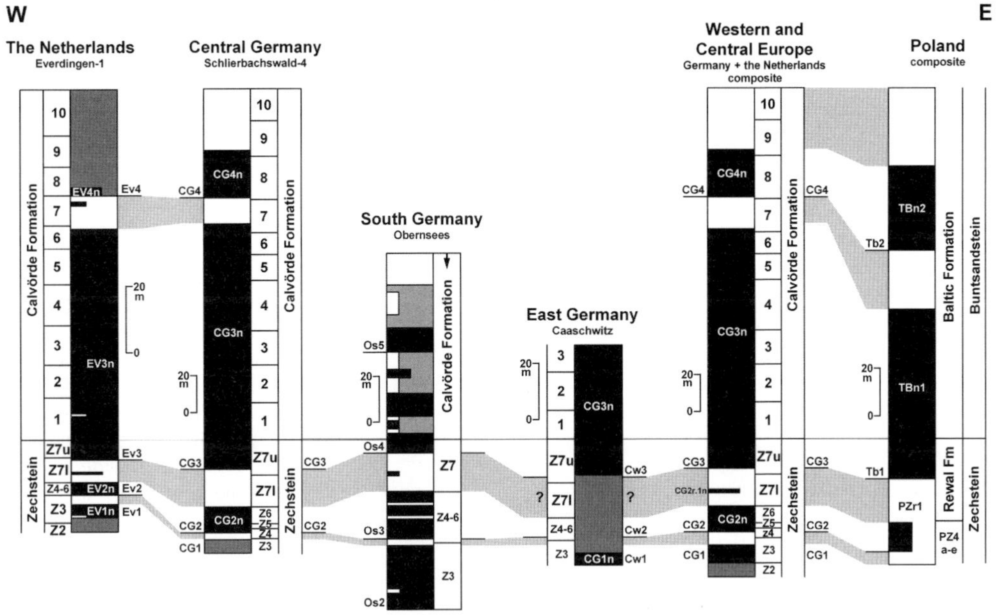

Fig. 3. Correlation of Zechstein–Buntsandstein litho- and magnetostratigraphies from The Netherlands (Everdingen-1 core; Szurlies *et al.* 2012), Central Germany (Schlierbachswald-4 core; Szurlies *et al.* 2003), south Germany (Obernsees core; Soffel & Wippern 1998), east Germany (Caaschwitz outcrop; Szurlies *et al.* 2003) and Poland (Nawrocki 1997). Composite magnetostratigraphy for Western and Central Europe (this study). Normal (reverse) polarity is displayed as black (white) bars. Please note that in the early magnetostratigraphical literature on Permian–Triassic rocks, black (white) bars have been used to indicate reverse (normal) polarity (e.g. McElhinny & Burek 1971; Menning *et al.* 1988).

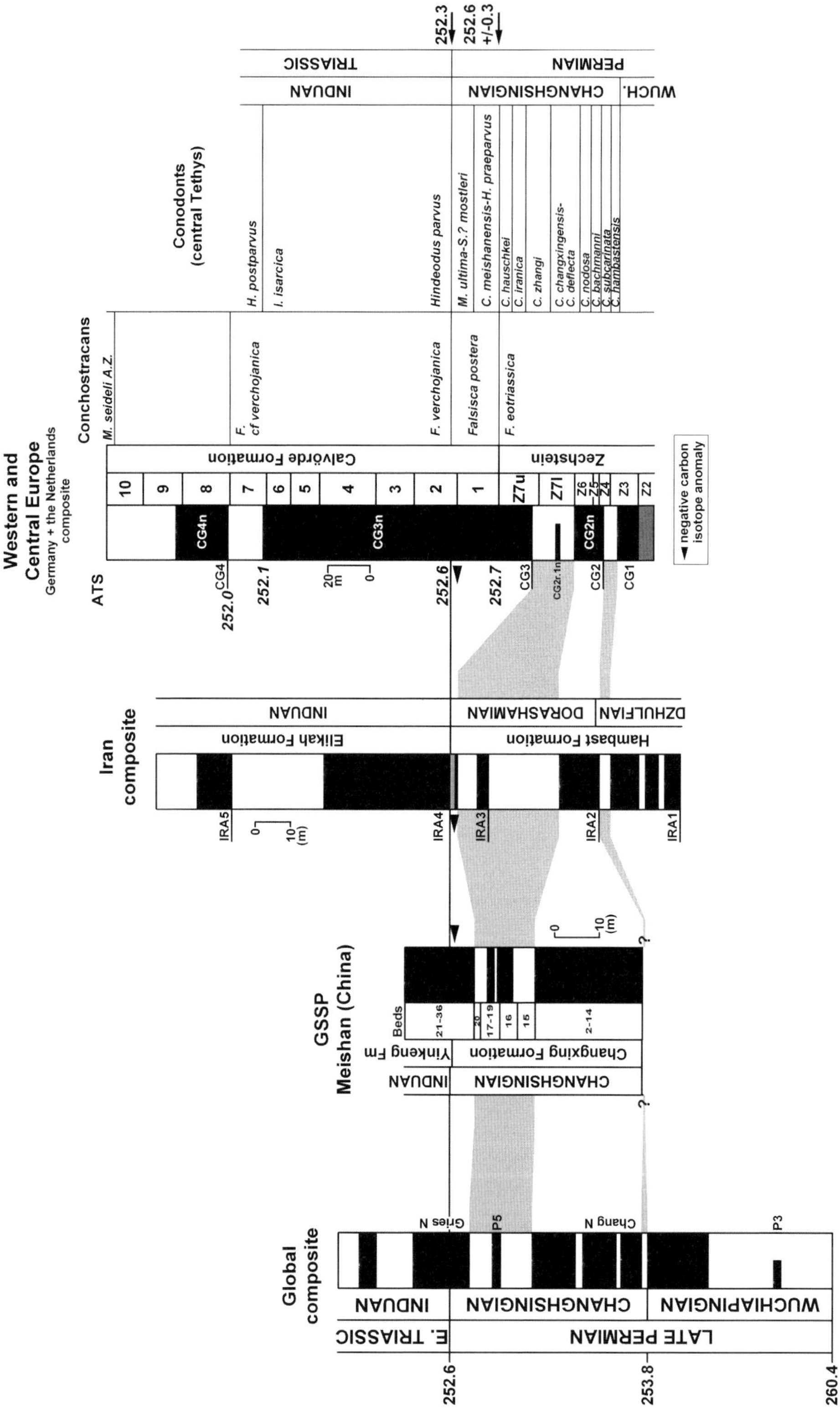
Global composite
GSSP Meishan (China)
Iran composite
Western and Central Europe
Germany + the Netherlands composite
ATS
Conchostracans
Conodonts (central Tethys)
E. TRIASSIC
LATE PERMIAN
INDUAN
CHANGHSINGIAN
WUCHIAPINGIAN
252.6
253.8
260.4
Gries N
P5
Chang N
P3
Yinkeng Fm
Changxing Formation
Beds
21-36
20
17-19
16
15
2-14
Elikah Formation
Hambast Formation
DORASHAMIAN
DZHULFIAN
IRA5
IRA4
IRA3
IRA2
IRA1
Calvörde Formation
Zechstein
CG4n
CG3n
CG2n
CG4
CG3
CG2r.1n
CG2
CG1
252.0
252.1
252.7
Z7u
Z7l
Z6
Z5
Z4
Z3
Z2
negative carbon isotope anomaly
M. seideli A.Z.
F. cf verchojanica
F. verchojanica
Falsisca postera
F. eotriassica
H. postparvus
I. isarcica
Hindeodus parvus
M. ultima-S.? mostleri
C. meishanensis-H. praeparvus
C. hauschkei
C. iranica
C. zhangi
C. changxingensis-C. deflecta
C. nodosa
C. bachmanni
C. subcarinata
C. hambastensis
TRIASSIC
PERMIAN
WUCH.
252.3
252.6 +/-0.3

characterized by normal polarity (CG2n), which is equivalent to EV2n in the Everdingen-1 core (Fig. 3). Moreover, this comparison is consistent with high-resolution cyclostratigraphical correlation (Szurlies *et al.* 2012). However, the base of magnetozones CG2n and EV2n is at different levels of the Z4 Formation. The suggested discrepancy is most probably due to the low-resolution sampling of the Z4 Formation in the Schlierbachswald-4 core (cf. Szurlies *et al.* 2012). In Poland, the PZ4a–PZ4c subcycles (Fig. 2), which correlate to the Z4–lowermost Z7 formations (Peryt *et al.* 2010), are dominantly of reversed polarity (PZr1) with one normal submagnetozone (PZr1.1n) that may be equivalent to EV2n, Os2n and CG2n, respectively (Fig. 3).

Higher in the section, magnetozone PZr1 spans the lower Rewal Formation, which is correlated with the lower Z7 Formation in Germany and The Netherlands by lithostratigraphical means (Käding 2000; Peryt *et al.* 2010). This is in full agreement with magnetostratigraphical correlation, where PZr1 is correlated with CG2r and Os2r from Germany, and EV2n from The Netherlands, respectively (Fig. 3). This interval represents the youngest reversed magnetozone of the Zechstein.

Following on from this, the correlation of the relatively long magnetozones CG3n and Os3n in Germany, EV3n in The Netherlands and Tbn1 in Poland is in full accordance with GR log- and cyclostratigraphy (Szurlies *et al.* 2012). In all of these sections, this long normal magnetic polarity interval spans the Zechstein–Buntsandstein boundary, representing a prime marker for basin-wide correlation (Szurlies *et al.* 2003, 2012). In the Schlierbachswald-4 core, CG3n ranges from the upper Z7 Formation into the lowermost cycle 7 of the Calvörde Formation (Lower Buntsandstein); whereas, in the Everdingen-1 core, it ends somewhat earlier in the uppermost cycle 6 (Fig. 3). There are two likely explanations for this discrepancy. First, the sampling resolution in the Schlierbachswald-4 core, where, for instance, the CG3n–CG3r transition is marked by a sampling gap of about 3 m (Szurlies *et al.* 2003). Second, the concept of base-level cycles postulates time intervals of non-deposition or erosion associated with the base-level fall; that is, that is the dry phase of the fluviolacustrine cycles of the Buntsandstein (Szurlies *et al.* 2012).

The uppermost part of the Everdingen-1 core is indicated by a brief interval of reversed polarity (EV3r) spanning cycle 7 and a younger normal magnetozone (EV4n) that begins in cycle 8. Both magnetozones are equivalent to polarity intervals CG3r and CG4n in Germany, and Tbr1 and Tbn2 in Poland, respectively (Fig. 3). In both the Schlierbachswald-4 and the Everdingen-1 cores, the base of EV4n and CG4n correlates with the base of cycle 8 of the Calvörde Formation.

On a regional scale, there is a good agreement of Zechstein and Buntsandstein lithostratigraphical units and magnetostratigraphy (Szurlies *et al.* 2003, 2012). In the large interior of the CEB, the good agreement of magnetozones and Buntsandstein base-level cycles strongly supports the assumption of nearly synchronous sedimentary cycles (i.e. *c.* 100 kyr eccentricity cycles) within the Lower Buntsandstein (Geluk & Röhling 1997; Szurlies *et al.* 2003, 2012; Bachmann & Kozur 2004; Menning *et al.* 2005; Szurlies 2007). Through integration with Buntsandstein cyclostratigraphy, the duration of the magnetozones can be estimated. The long normal polarity interval CG3n, for instance, that straddles the Zechstein–Buntsandstein boundary, spans seven base-level cycles (i.e. seven *c.* 100 kyr eccentricity cycles) and yields a total duration of approximately 700 kyr for CG3n (Szurlies *et al.* 2003, 2012).

Overall, the magnetostratigraphical correlation within the CEB enables the establishment of a well-defined composite magnetostratigraphy (Fig. 3), in which the long normal magnetozone CG3n provides a prime time-marker for correlation (Szurlies *et al.* 2003). Likewise, this long magnetozone is of great importance for marine to non-marine correlation (Szurlies *et al.* 2003, 2012; Szurlies 2007).

Global correlation and discussion

A sole biostratigraphical correlation of the Zechstein with the marine stages is limited owing to the lack of index fossils in the continental Permian of Central Europe. Therefore, the key to a detailed global correlation of the Zechstein lies with linking biostratigraphy to other less facies-sensitive methods, such as magnetostratigraphy. In the last two decades, many palaeomagnetic studies have focused on the Late Permian (Steiner 2006 and

Fig. 4. Correlation of magnetostratigraphies: global composite (Steiner 2006), GSSP from Meishan, China (Zhu & Liu 1999; Yin *et al.* 2001), NW and Central Iran composite (Gallet *et al.* 2000; Szurlies & Kozur 2004), and Western and Central Europe composite. Correlation of conchostracan biostratigraphy of the Zechstein–lowermost Buntsandstein with the Tethyan conodont biostratigraphy (after Kozur & Weems 2011). ATS, Buntsandstein astronomical timescale (after Szurlies 2007). For more details on carbon isotope records across the PTB see Korte & Kozur (2010 and references therein).

references therein) and the ensuing Early Triassic (Hounslow & Muttoni 2010 and references therein) from both continental and marine deposits. In Western and Central Europe, the intersection correlation of palaeomagnetic data from The Netherlands, Germany and Poland allows establishment of a composite geomagnetic polarity record for the Zechstein–lowermost Buntsandstein. It encompasses eight well-defined magnetozones; that is, magnetic polarity intervals CG1n–CG4n (Figs 3 & 4). Combined with available biostratigraphy, this composite allows a more detailed global correlation.

In terms of magnetostratigraphy, the most notable feature of the global Permian magnetic polarity record is the Kiaman Reversed Polarity Superchron, which spans the Late Carboniferous–Middle Permian. This superchron was terminated by the Illawarra Reversal (Irving & Parry 1963), which is considered a prime time-marker to anchor the continental Permian lithostratigraphy to the marine Middle Permian stages (Menning *et al.* 2005, 2006; Langereis *et al.* 2010). The Illawarra Reversal has long been regarded as earliest Capitanian in age (Menning *et al.* 2006). According to a more recent composite Permian magnetostratigraphy, it is now at the base of the Word N magnetic polarity interval within the Middle Permian (Steiner 2006). In the continental Permian of Central Europe, the Illawarra Reversal has been identified within the Upper Rotliegend (Menning *et al.* 1988), thus, clearly predating the Zechstein. Steiner (2006) suggested that the Illawarra Reversal may even be located somewhat deeper within the Upper Rotliegend. Above the Illawarra Reversal, the Middle–lower Late Permian is indicated by reversed polarity (Steiner 2006). The younger middle Late Permian–lowermost Triassic time interval (i.e. the upper Wuchiapingian–lowermost Induan) is dominantly of normal polarity (Hounslow & Muttoni 2010 and references therein) (Fig. 5).

In Western and Central Europe, the lower Zechstein in the Obernsees core is mainly of normal polarity (Soffel & Wippern 1998) (Fig. 5), which is consistent with a normal magnetozone detected within the Zechstein limestone of the PZ1 cycle of Poland (PZ1n of Nawrocki 1997). Therefore, the Z1 and Z2 formations may correlate with most of the dominantly normal upper Wuchiapingian (Fig. 5). This is consistent with the Re–Os date from the Kupferschiefer (Z1 Formation) of a mid Wuchiapingian value of 257.3 ± 1.6 Ma (Brauns *et al.* 2003). Further support for the magnetostratigraphical correlation comes from biostratigraphy. The occurrence of the conodonts *Mesogondolella britannica* and *Merrillina divergens* within the Kupferschiefer and Zechsteinkalk of the Z1 Formation (Fig. 2) (Kozur 1994; Legler *et al.* 2005)

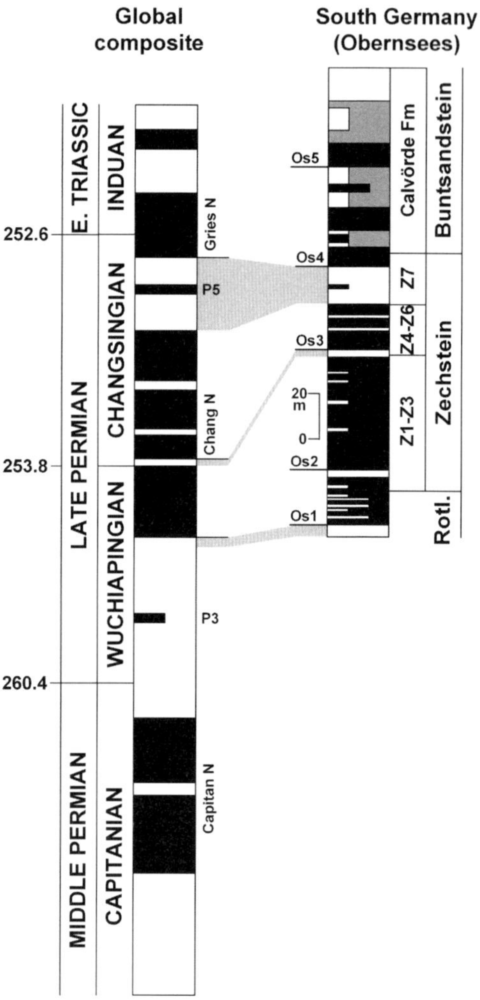

Fig. 5. Magnetostratigraphical correlation of the magnetic polarity record from the Obernsees core (Soffel & Wippern 1998) with the global composite (Steiner 2006).

allows an approximate correlation with the marine scale. According to Kozur (2007, pers. comm. 2012), *M. divergens* is known from the uppermost part of the lower Wuchiapingian (i.e. the *Clarkina leveni* Zone: Fig. 4) in Iran, indicating an early–mid Late Permian age of the basal Zechstein. This is consistent with Zechstein palynofloras belonging to the Late Permian *Lueckisporites virkkiae* Zone that spans the entire Zechstein (Fijałkowska 1994; Kürschner & Herngreen 2010 and references therein).

For the Z3–Z7 formations and the lowermost Buntsandstein, a detailed magnetostratigraphical composite for Western and Central Europe has been constructed (Fig. 3). Together with available conchostracan biostratigraphy (Kozur 1999; Bachmann & Kozur 2004; Kozur & Weems 2011), a high-resolution continental to marine correlation can be established (Fig. 4). The magnetozone CG1n from the Western and Central Europe composite is most probably equivalent to the lower part of Chang N from the global composite (Steiner 2006) and the uppermost part of IRA1n from Iran (Gallet *et al.* 2000), respectively. Thus, CG1n is the youngest normal magnetozone of the Wuchiapingian (Figs 4 & 5).

The subsequent reversed magnetozone CG1r is correlated with the first brief reversed interval within Chang N and magnetozone IRA1r, respectively (Fig. 4). It represents the first magnetozone of the Changhsingian. This is in accordance with recent correlations presented by Kozur & Weems (2011), who correlated the Changhsingian *Clarkina hambastiensis–C. hauschkei* conodont zones with the upper Z3–Z7 formations (Fig. 4). Thus, the magnetozones CG2n–CG3n (i.e. the Z4–Z7 formations and lowermost Buntsandstein) are most probably equivalent to IRA2n–IRA4n from Iran, M1n–lower M3n from the Meishan GSSP (Global Stratotype Section and Point) in China (Zhu & Liu 1999; Yin *et al.* 2001) and the middle Chang N–Gries N magnetic polarity intervals from the global composite (Steiner 2006), respectively (Fig. 4). Additional support for this correlation comes from data reported by Szurlies & Kozur (2004), who have equated CG2r with a reversed polarity interval (IRA2r) that they detected within the upper *C. changhsingensis–C. deflecta* to lower *C. zhangi* zones from the upper Changhsingian. This magnetozone represents the last reversed magnetic polarity interval of the Late Permian (Figs 4 & 5). The lower part of the following long magnetozone CG3n encompasses the Late Permian *Falsisca eotriassica* and *F. postera* zones, which are correlated with the latest Permian upper *C. zhangi* to *M. ultima–S.? mostleri* zones (Kozur & Weems 2011 and references therein). This interval corresponds to the upper Z7 Formation and the lowermost Buntsandstein (Fig. 4).

Although there is good agreement of magnetic records from marine and non-marine PTB sections, correlation details are still lively debated, mainly because of the absence of index fossils within the continental successions (Steiner 2006; Szurlies 2007; Glen *et al.* 2009; Hounslow & Muttoni 2010). However, based on a combination of magnetostratigraphy and conchostracan biostratigraphy, a detailed global correlation of the PTB interval has been achieved (Szurlies *et al.* 2003, 2012; Bachmann & Kozur 2004; Szurlies 2007; Kozur & Weems 2011).

The most distinctive feature of Late Permian palaeomagnetic data from both marine and terrestrial sections is the reversal from a short reversed to a long dominantly normal magnetic polarity interval that predates the PTB (Li & Wang 1989; Scholger *et al.* 2000; Szurlies *et al.* 2003, 2012; De Kock & Kirschvink 2004; Hounslow *et al.* 2008; Glen *et al.* 2009) (Fig. 4). In virtually all continental and marine sections, the biostratigraphical PTB falls within the lower part of the relatively long normal magnetic polarity interval that provides a further prime marker to anchor continental sections to the biostratigraphically defined marine stages (Szurlies *et al.* 2003; Metcalfe & Nicoll 2007; Glen *et al.* 2009). In Western and Central Europe, this normal polarity magnetozone (i.e. CG3n) spans the uppermost Zechstein–lowermost Buntsandstein (Szurlies *et al.* 2003, 2012). The continental PTB is located within the lower part of this long normal polarity interval (i.e. in the lowermost Buntsandstein) and is approximated by the last occurrence of the conchostracan *Falsisca postera* (Kozur 1999; Bachmann & Kozur 2004; Kozur & Weems 2011) and a negative excursion in the carbon isotope record (Korte & Kozur 2005, 2010). The magnetic polarity interval CG3n is correlated with IRA4n, M3n and Gries N, respectively (Fig. 4).

Notably, the reverse to normal (R/N) reversal predates the end-Permian ecological crisis. In the CEB, the base of the Buntsandstein is characterized by a marked palynofloral turnover, from pollen-dominated assemblages of the Zechstein, with Permian markers such as *Lueckisporites* sp. and *Vittatina*, to spore-dominated assemblages within the Buntsandstein, with typical Triassic forms, such as *Densoisporites* sp. and *Lundbladispora* sp. (Kürschner & Herngreen 2010 and references therein). Besides, the onset of the Buntsandstein is indicated by a strong increase in humidity and freshwater discharge, resulting in an increased influx of sands into the CEB (Best 1989) transported mainly by braided river systems. This distinct climatic change was accompanied by a major shift in sediment sources and transport directions (Bachmann & Kozur 2004; Hug & Gaupp 2006). Such a radical change in depositional environment towards braided river systems is also known from other possibly coeval end-Permian continental records (Ward *et al.* 2000; Michaelsen 2002), probably indicating a decrease in the coherence of riverbank sediments after a vegetation loss due to a terrestrial extinction event (Sephton *et al.* 2005). This major sediment provenance change and the palynofloral turnover at the base of the Buntsandstein constitute important proxies for the terrestrial extinction event, which falls within the lowermost part of CG3n (Szurlies *et al.* 2003). Consequently, the Zechstein–Buntsandstein boundary has already been correlated with the base of the Boundary Clay (i.e.

the main mass extinction event), at the Meishan GSSP (Bachmann & Kozur 2004).

A precise dating of climatic events, such as the Upper Rotliegend–Zechstein wet phase (Roscher & Schneider 2006; Schneider *et al.* 2006) or the strong increase in humidity and freshwater discharge at the onset of the Buntsandstein (Bachman & Kozur 2004; Hug & Gaupp 2006), enables detailed correlation with the climatic development in other sedimentary basins. Modelling of Late Permian climates has revealed a pronounced impact of monsoonal circulation patterns on Pangea (e.g. Kutzbach & Gallimore 1989; Parrish 1993; Schneider *et al.* 2006; Roscher 2009). Together with periodic glacio-eustatic sea-level change, which controlled the marine incursions from the Barents Sea (Ziegler 1990), and in combination with high evaporation rates, these so-called megamonsoons may have favoured the development of stacked evaporation cycles (cf. Stollhofen *et al.* 2008). Furthermore, a systematic change in palaeogeography during the assembly of Pangea, and a major change from Early Permian icehouse to Late Permian greenhouse conditions, may have considerably influenced sedimentation within the CEB (e.g. Roscher *et al.* 2011).

The age of the *Hindeodus parvus*-calibrated PTB is dated at around 252.3 Ma, based on a U–Pb age of 252.6 $\pm$ 0.2 Ma from an ash layer slightly below this boundary (Mundil *et al.* 2010). Assuming that the upper Z7 Formation represents one approximately 100 kyr eccentricity cycle, the long normal magnetozone CG3n is estimated to have lasted about 700 kyr (Szurlies *et al.* 2003, 2012). Above the PTB, most of CG3n is correlated with the earliest Triassic *Hindeodus parvus* and *Isarcicella isarcica* conodont zones (Fig. 4) (Kozur & Weems 2011). The base of this magnetozone predates the marine end-Permian main extinction event by about 0.1 Myr and the marine biostratigraphical PTB by about 0.2 Myr (Szurlies *et al.* 2003).

According to Gradstein *et al.* (2004) and Ogg *et al.* (2008), the base of the Changhsingian is placed at an age of 253.8 Ma, giving a duration of about 1.5 Myr for the coeval upper Z3–Z7 formations and the overlying lowermost Buntsandstein (Figs 4 & 5). Considering the Re–Os date of 257.3 $\pm$ 1.6 Ma (Brauns *et al.* 2003) for the Zechstein base, the underlying Z1–lower Z3 formations, which are correlated by bio- and magnetostratigraphical means with the upper Wuchiapingian, may have a duration of approximately 1.9 and 5.1 Myr. This obviously long duration of the lower Zechstein is consistent with the general consensus that the three predominantly marine lower Zechstein formations are of a different magnitude to the following four younger formations (cf. Peryt *et al.* 2010), during which salt deposition appears to have been driven by short-term climatic fluctuations (Wagner 1994).

However, because the underlying dominantly normal polarity Upper Rotliegend in Germany (Soffel & Wippern 1998) and Poland (Nawrocki 1997) may also be equivalent to the upper Wuchiapingian (cf. Langereis *et al.* 2010), the duration of the Zechstein is assumed to be closer to a minimum duration of about 3.4 Myr. This is close to suggestions of approximately 2.8 Myr (Menning & Szurlies in Menning *et al.* 2005). Alternatively, a duration of about 5 Myr (Käding *in* Menning *et al.* 2005) has been proposed. Recently, a duration of 8 Myr has been proposed by Słowakiewicz *et al.* (2009), who used a comparison of carbon isotope data around the base of the continental Late Permian in the USA and Central Europe. However, based on present bio- and magnetostratigraphical data, the base of the Zechstein is dated as mid Wuchiapingian. This gives a possible duration of the Zechstein of about 2.8–3.5 Myr.

Conclusions

In Western and Central Europe, regional correlation of palaeomagnetic data from The Netherlands, Germany and Poland allows the construction of a well-documented composite geomagnetic polarity record for the Zechstein–lowermost Buntsandstein (early Late Permian–earliest Triassic). This composite includes eight magnetozones; that is, the magnetic polarity intervals CG1n–CG4n. Overall, the upper Zechstein–lowermost Buntsandstein is dominantly of normal polarity.

The magnetostratigraphical composite allows a reliable basin-wide correlation within the CEB. Together with available biostratigraphical data, such as sparse conodonts in the lowermost Zechstein and conchostracans in the uppermost Zechstein–lowermost Buntsandstein, this integrated bio- and magnetostratigraphy facilitates detailed links to the marine realm, confirming a mid Late Permian–earliest Triassic age for these lithostratigraphical units. The basal Zechstein (i.e. the Z1–lower Z3 formations) is most probably equivalent to the uppermost lower–upper Wuchiapingian. The overlying upper Z3–Z7 formations and the lowermost Buntsandstein are equivalent to the entire Changhsingian. A transition from a short reversed to a long normal magnetozone within the uppermost Zechstein that predates the end-Permian mass extinction and the conodont-calibrated biostratigraphical PTB represents a prime marker for global correlation. In the CEB, the continental PTB is located within the so-called Grey Beds of the lowermost Buntsandstein, and is approximated by the last occurrence of the conchostracan *Falsisca postera* and a negative excursion in the carbon isotope record.

According to cyclostratigraphy, the long normal magnetozone CG3n is estimated to have lasted approximately 0.7 Myr. The R/N reversal at its base predates the marine end-Permian extinction event by about 0.1 Myr and the marine biostratigraphical PTB by around 0.2 Myr. Using recent ages for the PTB and the base of the Changhsingian, the upper Z3–Z7 formations to lowermost Buntsandstein lasted approximately 1.5 Myr. Present bio- and magnetostratigraphical correlation indicates a possible duration of the entire Zechstein of only 2.8–3.5 Myr. Thus, the integration of magnetostratigraphy with other stratigraphical methods, such as biostratigraphy and carbon isotope stratigraphy, is a prerequisite for establishing both a precise chronology and a detailed correlation of climatic events in the sea and on land. However, problems still remain, particularly, in properly correlating available Wuchiapingian magnetostratigraphical records from the marine realm with the continental successions, such as the Zechstein, and elsewhere in the world.

The author thanks H. W. Kozur and M. C. Geluk for valuable discussions. Helpful comments by A. Gąsiewicz, M. Słowakiewicz, J. W. Geissman and an anonymous reviewer greatly improved the manuscript.

References

Bachmann, G. H. & Kozur, H. W. 2004. The Germanic Triassic: Correlation with the international chronostratigraphic scale, numerical ages, Milankovitch cyclicity. *Hallesches Jahrbuch für Geowissenschaften*, **B26**, 17–62.

Best, G. 1989. Die Grenze Zechstein/Buntsandstein in Nordwestdeutschland nach Bohrlochmessungen. *Zeitschrift der Deutschen Geologischen Gesellschaft*, **140**, 73–85.

Boigk, H. 1959. Zur Gliederung und Fazies des Buntsandsteinsą zwischen Harz und Emsland. *Geologisches Jahrbuch*, **76**, 597–636.

Brauns, C. M., Pätzold, T. & Haack, U. 2003. A Re–Os study bearing on the age of the Kupferschiefer black shale at Sangerhausen (Germany). *In*: *Abstracts XVth International Congress on Carboniferous and Permian Stratigraphy, Utrecht, Netherlands*. TNO, Delft, 66.

Burek, P. J. 1967. Korrelation revers magnetisierter Gesteinsfolgen im Oberen Buntsandstein SW-Deutschlands. *Geologisches Jahrbuch*, **84**, 591–616.

Dachroth, W. 1971. Beispiele zur stratigraphischen Anwendung gesteinsmagnetischer Daten. *Geologische Rundschau*, **60**, 1061–1087.

Dachroth, W. 1976. Gesteinsmagnetische Marken im Perm Mitteleuropas. *Geologisches Jahrbuch*, **E10**, 3–63.

De Kock, M. O. & Kirschvink, J. L. 2004. Paleomagnetic constraints on the Permian–Triassic boundary in terrestrial strata of the Karoo Supergroup, South Africa: implications for causes of the end-Permian extinction event. *Gondwana Research*, **7**, 175–183.

Doornenbal, J. C. & Stevenson, A. G. (eds) 2010. *Petroleum Geological Atlas of the Southern Permian Basin Area*. European Association of Geoscientists & Engineers, Houten.

Erwin, D. H. 1993. *The Great Paleozoic Crisis: life and death in the Permian*. Columbia University Press, New York.

Fijałkowska, A. 1994. Palynological aspects of the Permo-Triassic succession in the Holy Cross Mountains, Poland. *Documenta Naturae*, **87**, 1–76.

Gallet, Y., Krystyn, L., Besse, J., Saidi, A. & Ricou, L. E. 2000. New constraints on the Upper Permian and Lower Triassic geomagnetic polarity timescale from the Abadeh section (central Iran). *Journal of Geophysical Research*, **105**, 2805–2815.

Geluk, M. C. 1999. Palaeogeographic and structural development of the Triassic in the Netherlands – new insights. *In*: Bachmann, G. H. & Lerche, I. (eds) *The Epicontinental Triassic, Volume 1. Zentralblatt für Geologie und Paläontologie*, **7–8**, (7–12), 727–745.

Geluk, M. C. 2005. *Stratigraphy and tectonics of Permo-Triassic basins in the Netherlands and surrounding areas*. PhD thesis, University of Utrecht.

Geluk, M. C. & Röhling, H.-G. 1997. High-resolution sequence stratigraphy of the Lower Triassic 'Buntsandstein' in the Netherlands and northwestern Germany. *Geologie en Mijnbouw*, **76**, 227–246.

Glen, J. M., Nomade, S., Lyons, J. L., Metcalfe, I., Mundil, R. & Renne, P. R. 2009. Magnetostratigraphic correlations of Permian–Triassic marine-to-terrestrial sections from China. *Journal of Asian Earth Sciences*, **36**, 521–540.

Glennie, K. W. 2007. The Permo-Carboniferous Rotliegend of NW Europe. *In*: Wong, T. E. (ed.) *Proceedings of the XVth International Congress on Carboniferous and Permian Stratigraphy*. Netherlands Academy of Arts and Sciences, Amsterdam, 10–16.

Glennie, K. W. & Buller, A. T. 1983. The Permian Weissliegend of NW Europe: the partial deformation of aeolian dune sands caused by the Zechstein transgression. *Sedimentary Geology*, **35**, 43–81.

Gradstein, F. M., Ogg, J. G. *et al.* 2004. *A Geologic Time Scale 2004*. Cambridge University Press, Cambridge.

Grice, K., Schaeffer, P., Schwark, L. & Maxwell, J. R. 1997. Changes in palaeoenvironmental conditions during deposition of the Permian Kupferschiefer (Lower Rhine Basin, N.W. Germany) from variations in isotopic compositions of biomarker components. *Organic Geochemistry*, **26**, 677–690.

Harland, W. B., Armstrong, R. L., Cox, A. V., Craig, L. E., Smith, A. G. & Smith, D. E. 1990. *A Geological Timescale 1989*. Cambridge University Press, Cambridge.

Hiete, M. 2004. *Umweltveränderungen in der Permo-Trias*. PhD thesis, University of Braunschweig.

Hounslow, M. W. & Muttoni, G. 2010. The geomagnetic polarity timescale for the Triassic: linkage to stage boundary definitions. *In*: LUCAS, S. G. (ed.) *The Triassic Timescale*. Geological Society, London, Special Publications, **334**, 61–102.

HOUNSLOW, M. W., PETERS, C., MORK, A., WEITSCHAT, W. & VIGRAN, J. O. 2008. Biomagnetostratigraphy of the Vikinghøgda Formation, Svalbard (arctic Norway) and the geomagnetic polarity timescale for the Lower Triassic. *Geological Society of America Bulletin*, **120**, 1305–1325.

HUG, N. & GAUPP, R. 2006. Palaeogeographic reconstruction in red beds by means of genetically related correlation: results from the upper part of the German Zechstein (Late Permian). *Zeitschrift der Deutschen Gesellschaft für Geowissenschaften*, **157**, 107–120.

IRVING, E. & PARRY, L. G. 1963. The magnetism of some Permian rocks from New South Wales. *Geophysical Journal of the Royal Astronomical Society*, **7**, 395–411.

KÄDING, K. C. 2000. Die Aller-, Ohre-, Friesland- und Fulda-Folge (vormals Bröckelschiefer-Folge). *Kali und Steinsalz*, **13**, 86–96.

KORTE, C. & KOZUR, H. W. 2005. Carbon isotope trends in continental lake deposits of uppermost Permian to Lower Olenekian: germanic Lower Buntsandstein (Calvörde and Bernburg Formations). *Hallesches Jahrbuch für Geowissenschaften*, **B19**, 87–94.

KORTE, C. & KOZUR, H. W. 2010. Carbon-isotope stratigraphy across the Permian–Triassic boundary: a review. *Journal of Asian Earth Sciences*, **39**, 215–235.

KOZUR, H. 1994. The correlation of the Zechstein with the marine standard. *Jahrbuch der Geologischen Bundesanstalt*, **137**, 85–103.

KOZUR, H. W. 1999. The correlation of the Germanic Buntsandstein and Muschelkalk with the Tethyan scale. *In*: BACHMANN, G. H. & LERCHE, I. (eds) *The Epicontinental Triassic, Volume 1. Zentralblatt für Geologie und Paläontologie*, **7–8** (7–12), 701–725.

KOZUR, H. W. 2007. Permophiles perspective: 30 years and 50 issues: the long journey from the globally most unsatisfactory stratigraphic subdivision of all Phanerozoic systems to one of the best elaborated international scale. *Permophiles*, **50**, 4–6.

KOZUR, H. W. & WEEMS, R. E. 2011. Detailed correlation and age of continental late Changhsingian and earliest Triassic beds: implications for the role of the Siberian Trap in the Permian–Triassic biotic crisis. *Palaeogeography, Palaeoclimatology, Palaeoecology*, **308**, 22–40.

KÜRSCHNER, W. M. & HERNGREEN, G. F. W. 2010. Triassic palynology of central and northwestern Europe: a review of palynofloral diversity patterns and biostratigraphic subdivisions. *In*: LUCAS, S. G. (ed.) *The Triassic Timescale*. Geological Society, London, Special Publications, **334**, 263–283.

KUTZBACH, J. E. & GALLIMORE, R. G. 1989. Pangaean climates: megamonsoons of the megacontinent. *Journal of Geophysical Research*, **94**, 3341–3357.

LANGEREIS, C. G., KRIJGSMAN, W., MUTTONI, G. & MENNING, M. 2010. Magnetostratigraphy – concepts, definitions, and applications. *Newsletters on Stratigraphy*, **43**, 207–233.

LEGLER, L. & SCHNEIDER, J. W. 2008. Marine ingressions into the Middle/Late Permian saline lake of the Southern Permian Basin (Rotliegend, Northern Germany) possibly linked to sea-level highstands in the Arctic rift system. *Palaeogeography, Palaeoclimatology, Palaeoecology*, **267**, 102–114.

LEGLER, B., GEBHARDT, U. & SCHNEIDER, J. W. 2005. Late Permian nonmarine–marine transitional profiles in the central Southern Permian Basin, northern Germany. *International Journal of Earth Sciences*, **94**, 851–862.

LI, H. & WANG, Z. 1989. Magnetostratigraphy of Permo-Triassic Boundary Section of Meishan of Changxing, Zhejiang. *Science in China, B*, **32**, 1401–1408.

MCELHINNY, M. W. & BUREK, P. J. 1971. Mesozoic palaeomagnetic stratigraphy. *Nature*, **232**, 98–102.

MENNING, M., KATZUNG, G. & LÜTZNER, H. 1988. Magnetostratigraphic investigations in the Rotliegendes (300–252 Ma) of Central Europe. *Zeitschrift für Geologische Wissenschaften*, **16**, 1045–1063.

MENNING, M., GAST, R., HAGDORN, H., KÄDING, K.-C., SZURLIES, M. & NITSCH, E. 2005. Die Zeitskala für die höhere Dyas und die Germanische Trias der Stratigraphischen Tabelle von Deutschland 2002. *Newsletters on Stratigraphy*, **41**, 173–210.

MENNING, M., ALEKSEEV, A. S. ET AL. 2006. Global time scale and regional stratigraphic reference scales of central and west Europe, east Europe, Tethys, south China, and North America as used in the Devonian-Carboniferous-Permian Correlation Chart 2003 (DCP 2003). *Palaeogeography, Palaeoclimatology, Palaeoecology*, **240**, 318–372.

METCALFE, I. & NICOLL, R. S. 2007. Conodont biostratigraphic control on transitional marine to non-marine Permian-Triassic boundary sequences in Yunnan-Guizhou, China. *Palaeogeography, Palaeoclimatology, Palaeoecology*, **252**, 56–65.

MICHAELSEN, P. 2002. Mass extinction of peat-forming plants and the effect on fluvial styles across the Permian-Triassic boundary, northern Bowen Basin, Australia. *Palaeogeography, Palaeoclimatology, Palaeoecology*, **179**, 173–188.

MUNDIL, R., PÁLFY, J., RENNE, P. R. & BRACK, P. 2010. The Triassic timescale: new constraints and a review of geochronological data. *In*: LUCAS, S. G. (ed.) *The Triassic Timescale*. Geological Society, London, Special Publications, **334**, 41–60.

NAWROCKI, J. 1997. Permian to Early Triassic magnetostratigraphy from the Central European Basin in Poland: Implications on regional and worldwide correlations. *Earth and Planetary Science Letters*, **152**, 37–58.

NAWROCKI, J. 2004. The Permian-Triassic boundary in the Central European Basin: magnetostratigraphic constraints. *Terra Nova*, **16**, 139–145.

NAWROCKI, J., KULETA, M. & ZBROJA, S. 2003. Buntsandstein magnetostratigraphy from the northern part of the Holy Cross Mountains. *Geological Quarterly*, **47**, 253–260.

OGG, J. G., OGG, G. & GRADSTEIN, F. M. 2008. Permian period. *In*: OGG, J. G., OGG, G. & GRADSTEIN, F. M. (eds) *The Concise Geologic Time Scale*. Cambridge University Press, Cambridge.

PANCOST, R. D., CRAWFORD, N. & MAXWELL, J. R. 2002. Molecular evidence for basin-scale photic zone euxinia in the Permian Zechstein Sea. *Chemical Geology*, **188**, 217–227.

PARRISH, J. T. 1993. Climate of the supercontinent Pangea. *Journal of Geology*, **101**, 215–233.

PAUL, J. 2006. Der Kupferschiefer: Lithologie, Stratigraphie, Fazies and Metallogenese eines Schwarzschiefers. *Zeitschrift der Deutschen Gesellschaft für Geowissenschaften*, **157**, 57–76.

PAUL, J. & KLARR, K. 1988. Feinstratigraphie und Fazies des Unteren und Mittleren Buntsandsteins in der Bohrung Remlingen 5. Gesellschaft für Strahlenforschung, Bericht, 8/87.

PAUL, J., WEMMER, K. & AHRENDT, H. 2008. Provenance of siliciclastic sediments (Permian to Jurassic) in the Central European Basin. *Zeitschrift der Deutschen Gesellschaft für Geowissenschaften*, **159**, 641–650.

PERYT, T. M., GELUK, M. C., MATHIESEN, A., PAUL, J. & SMITH, K. 2010. Zechstein. *In*: DOORNENBAL, J. C. & STEVENSON, A. G. (eds) *Petroleum Geological Atlas of the Southern Permian Basin Area*. European Association of Geoscientists & Engineers, Houten, 123–147.

RETALLACK, G. J. 1995. Permian–Triassic life crisis on land. *Science*, **267**, 77–80.

RICHTER-BERNBURG, G. 1955. Stratigraphische Gliederung des deutschen Zechsteins. *Zeitschrift der Deutschen Geologischen Gesellschaft*, **105**, 843–854.

RÖHLING, H.-G. 1993. Der Untere Buntsandstein in Nordwest- und Nordostdeutschland – Ein Beitrag zur Vereinheitlichung der stratigraphischen Nomenklatur. *Geologisches Jahrbuch, A*, **142**, 149–183.

ROSCHER, M. 2009. *Environmental reconstruction of the Late Palaeozoic – numeric modelling and geological evidences*. PhD thesis, University of Freiberg.

ROSCHER, M. & SCHNEIDER, J. W. 2006. Permo-Carboniferous climate: early Pennsylvanian to Late Permian climate development of central Europe in a regional and global context. *In*: LUCAS, S. G., CASSINIS, G. & SCHNEIDER, J. W. (eds) *Non-Marine Permian Biostratigraphy and Biochronology*. Geological Society, London, Special Publications, **265**, 95–136.

ROSCHER, M., STORDAL, F. & SVENSEN, H. 2011. The effect of global warming and global cooling on the distribution of the latest Permian climate zones. *Palaeogeography, Palaeoclimatology, Palaeoecology*, **309**, 186–200.

SCHNEIDER, J. W., KÖRNER, F., ROSCHER, M. & KRONER, U. 2006. Permian climate development in the northern peri-Tethys area – The Lodève basin, French Massif Central, compared in a European and global context. *Palaeogeography, Palaeoclimatology, Palaeoecology*, **240**, 161–183.

SCHOLGER, R., MAURITSCH, H. J. & BRANDNER, R. 2000. Permian–Triassic boundary magnetostratigraphy from the Southern Alps (Italy). *Earth and Planetary Science Letters*, **176**, 495–508.

SEPHTON, M. A., LOOY, C. V., BRINKHUIS, H., WIGNALL, P. B., DE LEEUW, J. W. & VISSCHER, H. 2005. Catastrophic soil erosion during the end-Permian biotic crisis. *Geology*, **33**, 941–944.

SŁOWAKIEWICZ, M., KIERSNOWSKI, H. & WAGNER, R. 2009. Correlation of the Middle and Upper Permian marine and terrestrial sedimentary sequences in Polish, German, and USA Western Interior Basins with reference to global time markers. *Palaeoworld*, **18**, 193–211.

SOFFEL, H. C. & WIPPERN, J. 1998. Magnetostratigraphy of Upper Permian and Lower Triassic Rocks from the Drill Site Obernsees near Bayreuth, Germany. *Geologica Bavarica*, **103**, 275–294.

STEINER, M. 2006. The magnetic polarity time scale across the Permian–Triassic boundary. *In*: LUCAS, S. G., CASSINIS, G. & SCHNEIDER, J. W. (eds) *Non-marine Permian Biostratigraphy and Biochronology*. Geological Society, London, Special Publications, **265**, 15–38.

STOLLHOFEN, H., BACHMANN, G. H. ET AL. 2008. Upper Rotliegend to Early Cretaceous basin development. *In*: LITTKE, R., BAYER, U., GAJEWSKI, D. & NELSKAMP, S. (eds) *Dynamics of complex intracontinental Basins: the Central European Basin System*. Springer, Berlin, 181–210.

SZURLIES, M. 1999. Zyklenstratigraphie und Gamma-Ray-Log-Korrelation im Unteren Buntsandstein (Untere Trias) des nördlichen Harzvorlandes. *Hallesches Jahrbuch für Geowissenschaften*, **B21**, 35–54.

SZURLIES, M. 2004. Magnetostratigraphy: the key to a global correlation of the classic Germanic Trias – case study Volpriehausen Formation (Middle Buntsandstein), Central Germany. *Earth and Planetary Science Letters*, **227**, 395–410.

SZURLIES, M. 2007. Latest Permian to Middle Triassic cyclo-magnetostratigraphy from the Central European Basin, Germany: implications for the geomagnetic polarity timescale. *Earth and Planetary Science Letters*, **261**, 602–619.

SZURLIES, M. & KOZUR, H. W. 2004. Preliminary palaeomagnetic results from the Permian–Triassic boundary interval, Central and NW Iran. *Albertiana*, **31**, 41–46.

SZURLIES, M., BACHMANN, G. H., MENNING, M., NOWACZYK, N. R. & KÄDING, K.-C. 2003. Magnetostratigraphy and high-resolution lithostratigraphy of the Permian-Triassic boundary interval in Central Germany. *Earth and Planetary Science Letters*, **212**, 263–278.

SZURLIES, M., GELUK, M. C., KRIJGSMAN, W. & KÜRSCHNER, W. M. 2012. The continental Permian-Triassic boundary in the Netherlands: Implications for the geomagnetic polarity time scale. *Earth and Planetary Science Letters*, **317–318**, 165–176.

VAN WEES, J. D., STEPHENSON, R. A. ET AL. 2000. On the origin of the Southern Permian Basin, Central Europe. *Marine and Petroleum Geology*, **17**, 43–59.

WAGNER, R. 1994. Stratigraphy and development of the Zechstein Basin in Poland. *Prace Państwowego Instytutu Geologicznego*, **146**, 1–71.

WAGNER, R. & PERYT, T. M. 1997. Possibility of sequence stratigraphic subdivision of the Zechstein in the Polish Basin. *Geological Quarterly*, **41**, 457–474.

WARD, P. D., MONTGOMERY, D. R. & SMITH, R. 2000. Altered River Morphology in South Africa Related to the Permian-Triassic Extinction. *Science*, **289**, 1740–1743.

YIN, H., ZHANG, K., TONG, J.-N., YANG, Z. & WU, S. 2001. The Global Stratotype Section and Point (GSSP) of the Permian-Triassic Boundary. *Episodes*, **24**, 102–114.

ZIEGLER, P. A. 1990. *Geological Atlas of Western and Central Europe*. 2nd edn. Geological Society, London.

ZHU, Y.-M. & LIU, Y.-Y. 1999. Magnetostratigraphy of the Permo-Triassic boundary section at Meishan, Changxing, Zhejiang Province. *In*: YIN, H. & TONG, J. (eds) *Proceedings of the International Conference on Pangea and the Paleozoic–Mesozoic Transition*. China University of Geosciences Press, Wuhan, 79–84.

Geochemical record of Late Devonian to Early Carboniferous events, Palaeozoic of Graz, Eastern Alps, Austria

ANA-VOICA BOJAR[1,2]*, FRANZ NEUBAUER[1] & CHRISTIAN KOEBERL[3,4]

[1]*Department of Geography and Geology, Paris-Lodron University, Hellbrunnerstrasse 32, A-5020, Salzburg, Austria*

[2]*Department of Mineralogy, Universalmuseum Joanneum, Weinzöttlstrasse 16, A-8045, Graz, Austria*

[3]*Center for Earth Sciences, University of Vienna, Althanstrasse 14, 1090 Vienna, Austria*

[4]*Natural History Museum, Burgring 7, 1010 Vienna, Austria*

**Corresponding author (e-mail: ana-voica.bojar@sbg.ac.at)*

Abstract: The late Famennian mass extinction has been associated with a worldwide anoxic event evidenced at many sites by the accumulation of organic-rich shales starting prior to the Devonian–Carboniferous (D/C) boundary. Environmental reconstructions are hampered by the fact that only a few D/C boundary sections exist in carbonate facies. One of these is the Palaeozoic Steinberg section, Eastern Alps, Austria. Based on detailed geochemical investigations and correlation with previously investigated sections, we propose a succession of events from the late Famennian to the Tournaisian. Both the low total rare earth element (REE) contents and the normalized REE distribution patterns indicate low terrigenous input from the *praesulcata* Zone to the *typicus* Zone. The $\delta^{13}C$, $\delta^{18}O$, $\delta^{34}S$ and total organic content (TOC) show a shift towards higher values, which correlates well with the late Famennian isotopic trends that are related to a period of active ocean circulation, cooling and increased productivity. Two isotopic shifts are present in the section: the older isotopic shift of the $\delta^{13}C$ values is situated at the level of the Lower to Middle *praesulcata* Zone, while the younger one is found at the level of the Upper *praesulcata* Zone. In the *sulcata* Zone, a sharp decrease in $\delta^{18}O$ values and a shift in the Ce anomaly follow. Concomitantly, a drop in the $\delta^{13}C$ and $\delta^{34}S$ values, highly variable TOC values, and the main and secondary element concentrations are interpreted to be related to a drop in primary productivity, with the recovery continuing in the *crenulata* Zone. In this context, the variations towards lower CaO, Sr, Rb and Ba values probably correlate with biomass variations and a transition from cooler towards warmer conditions.

The mass extinction that occurred prior to the Devonian/Carboniferous (D/C) boundary was the sixth largest extinction of the Phanerozoic (Sepkoski 1996). It mainly affected pelagic and hempelagic organisms (e.g. conodonts, ammonoids, echinoderms, trilobites, ancient agnathan and placodermian fishes), continental plant communities and, to a lesser extent, shallow-marine ostracods, foraminifera and corals (Paproth & Streel 1984; Flajs *et al.* 1988; Becker & Paproth 1993; Simakov 1993; Becker 1994; Sepkoski 1996; Streel *et al.* 2000). A short time after the extinction, there was a phase of radiation and biotic innovations (Caplan & Bustin 1999; Streel *et al.* 2000). This phase was characterized by the diversification of certain groups of organisms, such as tetrapods (Walliser 1996). Subsequent palaeontological advances have shown that changes in different groups of marine organisms and terrestrial plants are not synchronous (Streel *et al.* 2000).

The late Famennian is characterized by the occurrence of a grey shale level, and sometimes black shales, known as 'Hangenberg Shale' after a locality in the Rheinisches Schiefergebirge, Germany (Schmidt 1924). The Hangenberg Shale shows variable lithology, with the black shale not always occurring at the base (Bless *et al.* 1988). Worldwide, the D/C sections display similar litho- and biostratigraphy, consisting of a lower cephalopod limestone, shale and an upper cephalopod limestone (Walliser 1984), although there is little agreement on the estimation of the absolute age of the D/C boundary (Gradstein *et al.* 2004 and references therein). The shales appear even when the Famennian strata occur in carbonate ramp settings or are situated at lower latitudes (Walliser 1984; Caplan *et al.* 1996). The position of the D/C global boundary stratotype has been defined 'at the first appearance of the conodont *Siphonodella sulcata* within the evolutionary lineage from

From: GĄSIEWICZ, A. & SŁOWAKIEWICZ, M. (eds) 2013. *Palaeozoic Climate Cycles: Their Evolutionary and Sedimentological Impact*. Geological Society, London, Special Publications, **376**, 87–108.
First published online September 19, 2013, http://dx.doi.org/10.1144/SP376.15

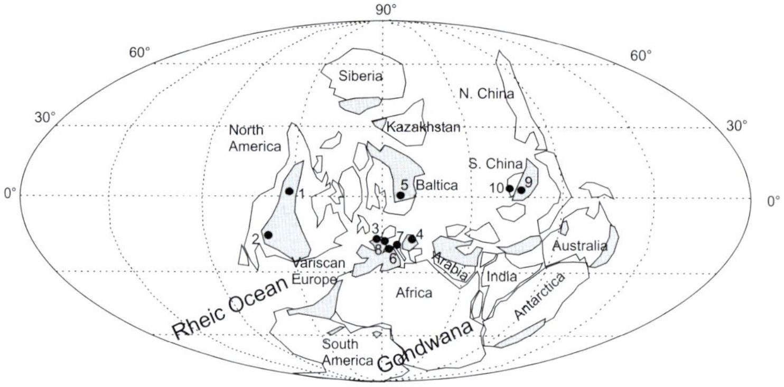

Fig. 1. Palaeogeographic reconstruction of continents during the Famennian (after Scotese & McKerrow 1990; Scotese 2002). Grey areas represent the distribution of Famennian–Tournaisian shales (after Caplan & Bustin 1999). Numbers refer to the sections described in Figure 2a.

Siphonodella praesulcata to *Siphonodella sulcata*' as found in La Serre, Montagne Noir, France (Paproth & Streel 1984; Paproth *et al.* 1991; Feist *et al.* 2000). This section was accepted as the stratotype for the D/C boundary (Paproth *et al.* 1991). Two other sections from Hasselbachtal, Germany (Becker & Paproth 1993), and Nanbiancun, China (Wang 1993), were chosen as auxiliary

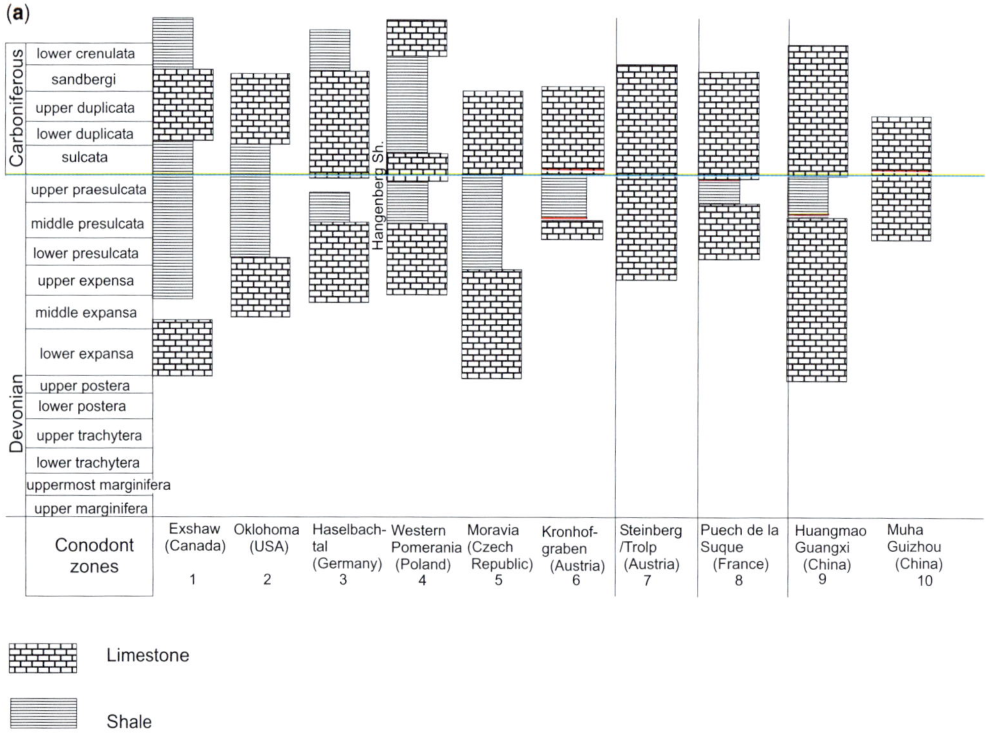

Fig. 2. (**a**) Global lithological sections at the D/C boundary. 1, Richards & Higgins (1988); 2, Orth *et al.* (1988); 3, Ziegler & Sandberg (1984), Becker & Paproth (1993); 4, Matyja & Stempien-Salek (1993); 5, Kalvoda (1986); 6, Schönlaub *et al.* (1992); 7, Nössing (1975), Ebner (1980) and this study; 8, Wang *et al.* (1993), Girard (1994); 9, Bai & Ning (1988); 10, Hou *et al.* (1984), Xu *et al.* (1989). Red lines indicate the position of the Ir anomalies.

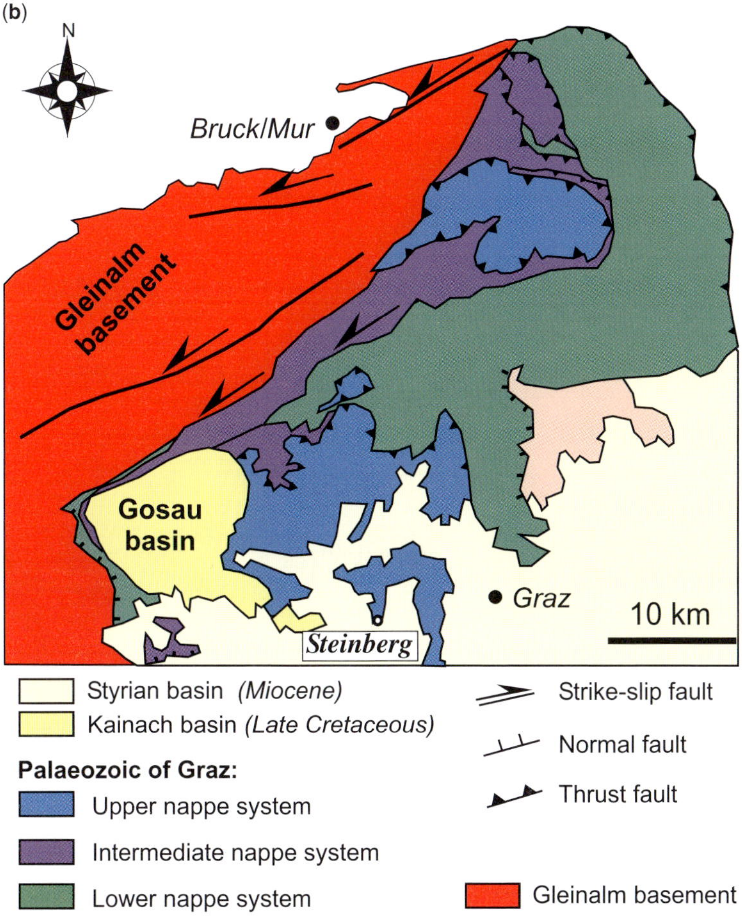

Fig. 2. (**b**) Simplified geological map of the Palaeozoic of Graz, showing the position of outcrop at the Steinberg/Trolp quarry (Figs 1 & 2a).

global boundary stratotype sections (Fig. 1). The timing of the extinction might be different among fossil groups, with a final episode occurring very close to the D/C boundary within the *Siphonodella praesulcata* to *Siphonodella sulcata* transition. In most sections, the D/C boundary is situated in the upper part, or above the black shales (Fig. 2a). Detailed stratigraphic work has shown that the mass extinction coincided with the base of the black shales (Walliser 1984), although there is no global synchronicity for its deposition as shown in Figure 2a (Streel *et al.* 2000). For the northern border of the Rhenish Massif, the base of the black shales was considered diachronic, the deposition starting from the Middle to the Upper *costatus* Zone (Bless *et al.* 1988). Further investigations in the same region concluded that the black shales were deposited in the Middle *praesulcata* Zone (Bless *et al.* 1993; Kaiser *et al.* 2008). By reinterpreting several sections from the Carnic Alps, Palaeozoic of Graz, Montagne Noir and Pyrenees, Kaiser *et al.* (2006, 2008) restricted the deposition of the black shales associated with mass extinction to the middle part of the Middle *praesulcata* Zone. In the compilation of Corradini *et al.* (2011), the Hangenberg event is also placed in the Middle *praesulcata* Zone. In this case bed 69 (La Serre section, Montagne Noir), which is situated at the base of the Middle *praesulcata* Zone (Feist *et al.* 2000) and previously considered a possible equivalent of the Hangenberg (Black) Shale from the Northern Rhenisch Rheinisches Schiefergebirge, is older. Kaiser & Corradini (2011) and Corradini *et al.* (2011) tested a new definition of the D/C boundary, using *Protognathodus* as a tool instead of *Siphonodella*, but Corradini *et al.* (2011) concluded that the use of *Protognathodus* is problematic. Kaiser (2009) did not include discussion

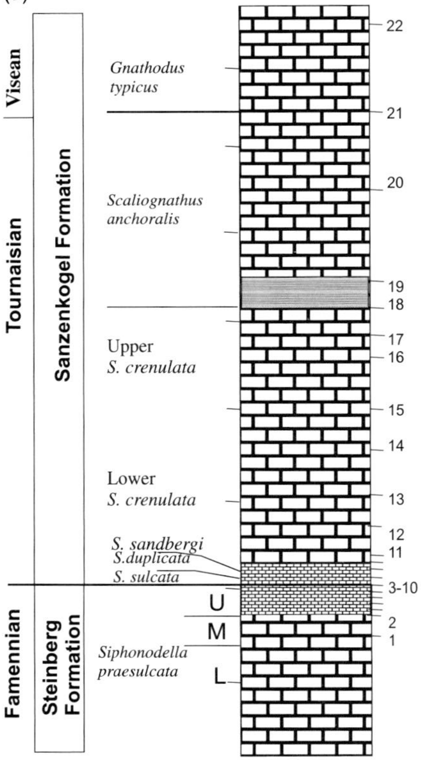

	A							B		C			D			E		F			G	
	1	2	3	4	5	6	7	8	9	10	11	12	13	14	15	16	17	18	19	20	21	22
Bispathodus costatus	x	x																				
B. ultimus	x	x																				
Palmatolepis gracilis gracilis	x																					
P. gracilis gonioclymeniae	x	x																				
P.gracilis sigmoidalis	x	x																				
Pseudop.marburgensis trigonicus	x	x																				
Siphonodella praesulcata	x				x	x																
Spathognadus supremus		x																				
Polygnathus communis communis			x		x	x						x	x	x	x	x	x	x	x			
P. communis ?			x	x		x	x															
Protognathodus collinsoni			x		x	x						x										
P. kockeli			x	x	x	x			x		x	x			x							
P. meischneri			x	x	x	x						x			x							
Polygnathus purus subplanus				x	x	x	x	x	x	x	x	x										
P. purus purus					x	x				x			x	x	x	x	x					
Pseudopolygnathus multistriatus						x																
P. dentilineatus							x			x	x	x										
P. sp. ex. gr. primus - dentilineatus							x		x								x					
P.sp.								x		x	x			x								
Siphonodella sulcata								x	x													
Polygnathus bischoffi									x										x			
Pseudopolyg. triangulus triangulus									x				x	x			x	x				
Siphonodella cooperi MT2									x		x				x		x	x				
S. duplicata MT1									x													
S.obsoleta									x						x							
Bispathodus aculeatus aculeatus										x												
Siphonodella duplicata sensu HAAS										x												
S. cf. S. Isosticha										x					x		x	x				
S.sp										x	x		x						x			
Polygnathus biconstrictus													x	x		x						
Pseudopolygnathus fusiformis													x			x						
P. marginatus															x							
P. triangulus ssp.															x	x						
Dollymae sp.																x						
Gnathodus c. antetexanus																x						
G. de licatus																x		x		x		
Siphonodella lobata																x	x					
Polygnathus cf. flabellus																	x					
Siphonodella crenulata MT1																	x	x				
Polygnathus communis carinus																		x				
Pseudopolygnathus triang. pinnatus																		x	x	x		
Siphonodella isosticha																		x	x			
S. quadruplicata																		x				
Gnathodus antetexanus																		x		x		
Siphonodella crenulata MT2																			x			
Gnathodus cuneiformis																			x	x	x	x
G. isomescens																			x			
G. semiglaber																			x	x	x	
Hindeodella segaformis																			x			
Scaliognathus anchoralis																			x	x		
Spathognathodus stabilis																			x			
Gnathodus texanus																				x	x	x
G. typicus																						x

Fig. 3. (**a**) Distribution of conodont fauna of the Steinberg quarry (after Ebner 1980). Conodont zones after Ziegler & Sandberg (1984, 1990) and Sandberg *et al.* (1978): A, *Siphonodella praesulcata*, layer no. 7 contains no siphonodellides; B, *Siphonodella sulcata–S. duplicata*; C, *Siphonodella sandbergi*; D, Lower *Siphonodella crenulata*; E, Upper *Siphonodella crenulata*; F, *Scaliognathus anchoralis*; G, *Gnathodus typicus*. U, M, L: Upper, Middle, Lower.

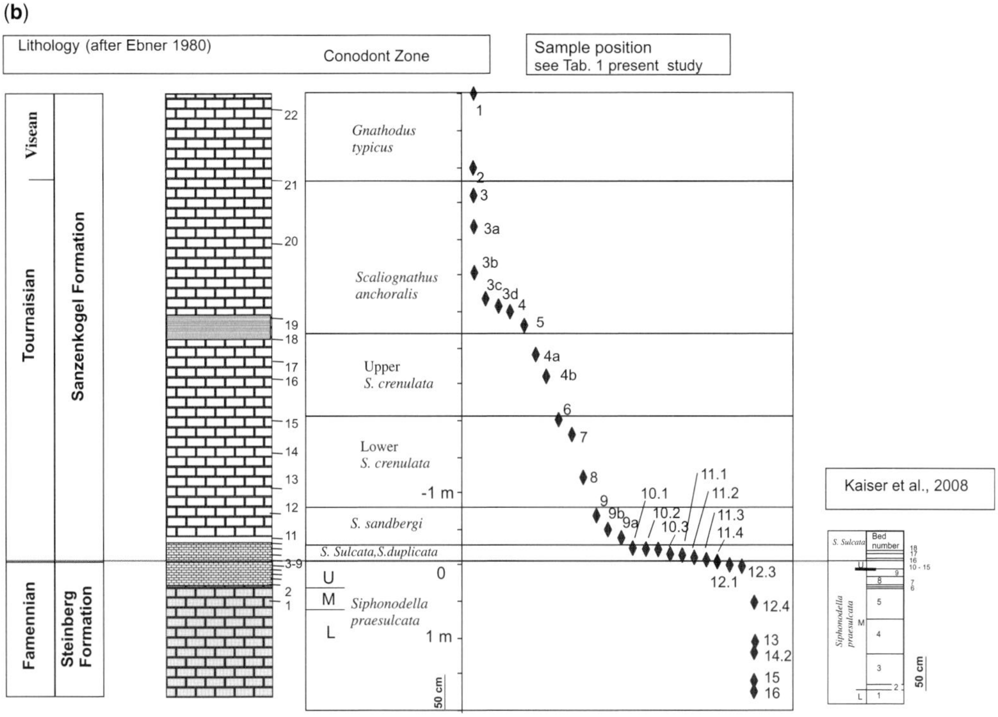

Fig. 3. (**b**) Position of samples presented in Tables 1–3. Nössing (1975) and Ebner (1980) found *S. sulcata* in the thin layered carbonates. Correlation of the Steinberg section with Kaiser *et al.* (2008). The sections are in all publications rotated with 180°, in order to display the Devonain in the lower part and the Carboniferous in the upper part.

about the great number of macrofauna and microfauna that are at La Serre being of correlative value (Paproth *et al.* 1991). The transitional morphotypes between *S. praesulcata* and *S. sulcata*, as well as the importance of other fossil groups, are discussed by Feist *et al.* (2000). In the light of the results above, we kept the previously established zonation of conodont zones after Ziegler & Sandberg (1984, 1990) and Sandberg *et al.* (1978), with the D/C boundary as defined by Paproth *et al.* (1991).

Geochemical data for this boundary are not conclusive. In this paper, we investigate the distribution of C, O and S stable isotopes, total organic carbon (TOC), and main and trace elements including rare earth element (REE) patterns in a very well-preserved, continuous D/C succession developed entirely in carbonate facies at Steinberg. This section is situated within the so-called Palaeozoic of Graz, Styria, Austria (Fig. 2b). We then correlate our data with previous conodont stratigraphy of this outcrop and with other sections and propose a succession of events during the *praesulcata* to *sulcata* Zones, raising a more general question concerning the turn-off mechanisms of anoxic events.

Geological and geochemical investigations of the late Famennian and D/C boundary

Geological evidence has shown that, during the late Famennian, generally regressive, anoxic conditions dominated, followed by an extensive transgression during the Tournaisian (Walliser 1984; Sandberg *et al.* 1988; Webster & Groessens 1990; Bless *et al.* 1993; Streel *et al.* 2000; Streel & Marshall 2006). In most of the D/C sections (Figs 1 & 2a), the carbonates are sharply to gradually overlain by organic-rich shales that indicate the transition to an anoxic environment. The reasons for the deposition of organic-rich shales, the cooling and the cause of the global biological turnover are still under debate. Walliser (1984) linked the extinction of the clymenides and goniatites to the elimination of an important part of the food chain during the ascent of the anoxic zone. The development of the black shales on top of a carbonate ramp has been interpreted to indicate surface-water eutrophication related to a period of active oceanic

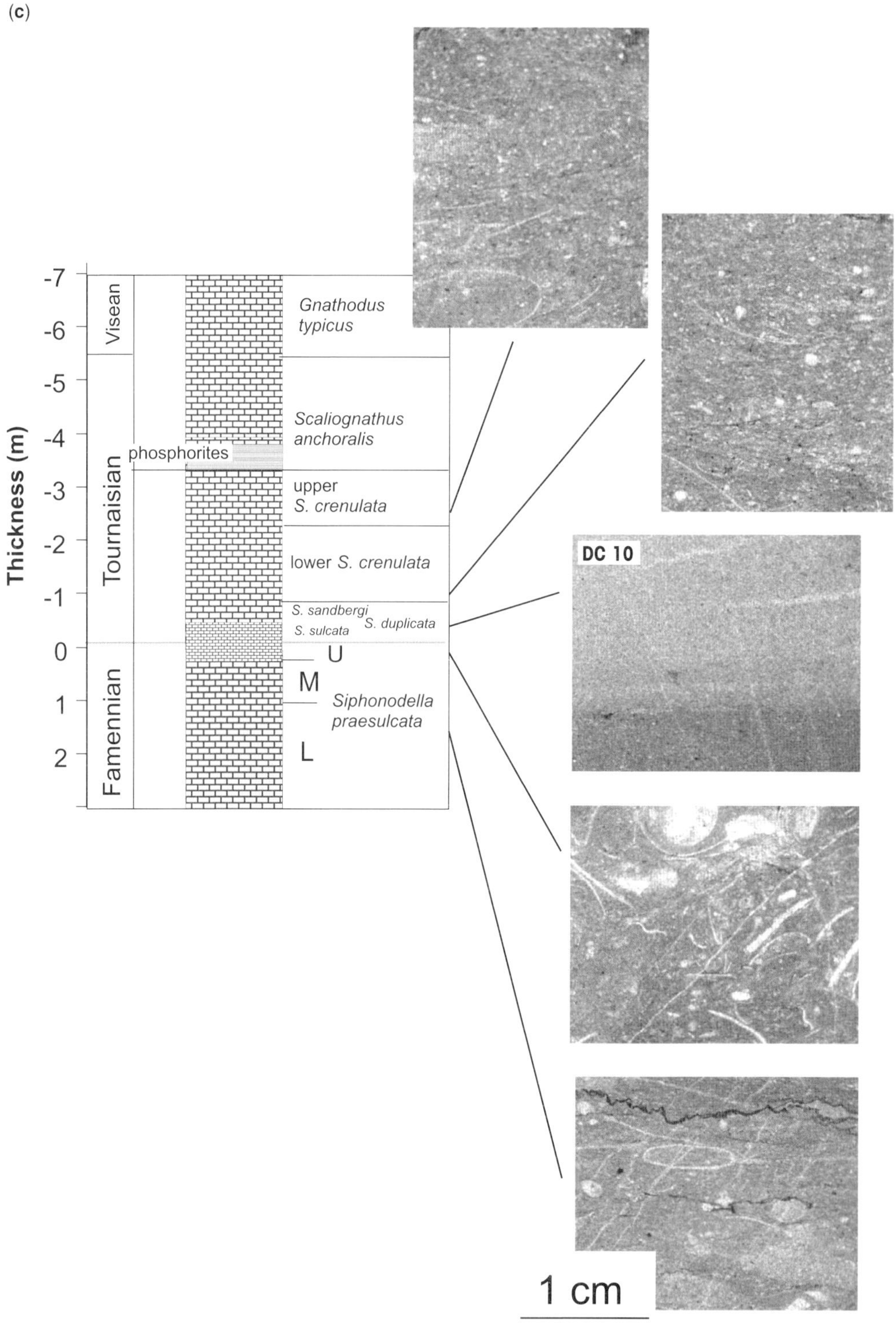

Fig. 3. (**c**) Conodont biozones and schematical lithological section of the Steinberg section. To the right, thin sections with microfacies are shown.

(d)

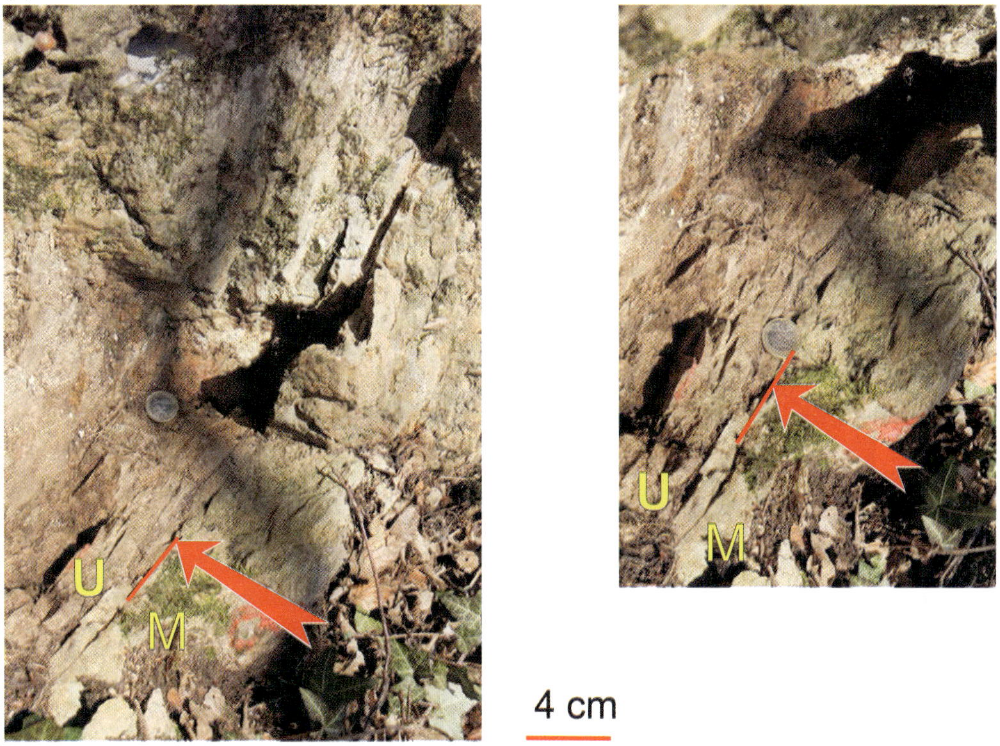

Fig. 3. (**d**) Field photograph of the thin layered carbonates (note the position of the boundary between the Middle and Upper *presulcata* Zones: there is no black shale intercalation within these intervals). The photographs are rotated through 180° in order to be easily correlated with the section presented.

circulation, upwelling, delivery of nutrients and increased productivity (Caplan *et al.* 1996; Caplan & Bustin 1999; Streel *et al.* 2000; Caplan & Bustin 2001). Isaacson *et al.* (2008) suggested that the black shales were deposited during low-stand sea-level and increased marine nutrient supply related to transport from the continent as a result of land plant diversification. Peterhänsel & Pratt (2001) linked the destruction of the tropical carbonate platform of Fammenian age to orogeny and increasing nutrient supply due to colonization of land surfaces by deep-rooting gymnosperms. Just below the D/C boundary, a glacial episode is known (Streel *et al.* 2000; Isaacson *et al.* 2008).

On a global scale, weak Ir abundance anomalies have been found at the bottom and/or at the top of the shales (Bai & Ning 1988; Wang *et al.* 1993). In the sections investigated to date, the anomalies occur near the limestone–shale transition (Fig. 2a). Such lithological transitions, which also represent redox boundaries, could have favoured syn- to post-depositional Ir concentration (Caplan & Bustin 1999) (Fig. 2a). The lithological transition from carbonates to shales, as well as the fact that a complete lithological record across the boundary has only been preserved in a few sites, has made the interpretation of the geochemical data difficult. Only a few occurrences of the D/C transitions are known with continuous sedimentation in carbonate facies. Such sections are found for the D/C boundary stratotype at La Serre in the southeastern Montagne Noire, France (Paproth *et al.* 1991), at Grüne Schneid in the Carnic Alps (Schönlaub *et al.* 1992) or at the Steinberg section (Fig. 2a, b) in the Eastern Alps (Nössing 1975; Ebner 1980).

Several geochemical studies of the late Famennian and D/C boundary have provided information concerning the environmental changes. For the Famennian, positive shifts of $\delta^{13}C$ and $\delta^{34}S$ values, in addition to TOC, have been documented (Popp *et al.* 1986; Veizer *et al.* 1986, 1999; Kürschner *et al.* 1992; Brand & Legrand-Blain 1993; Holser *et al.* 1996; Brand *et al.* 2004; Kaiser 2005; Buggisch & Joachimski 2006; Kaiser *et al.* 2006, 2008; Cramer *et al.* 2008). During the last two

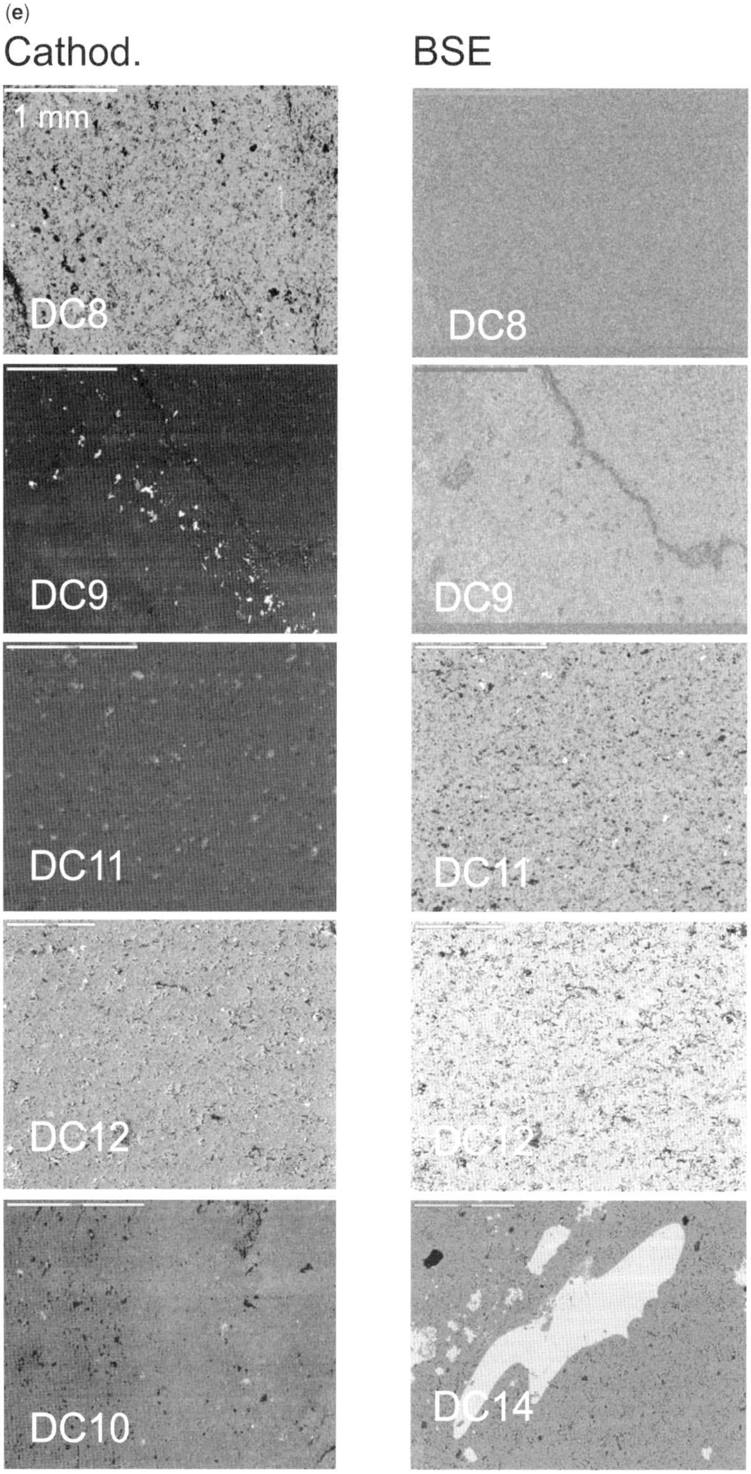

Fig. 3. (**e**) Cathodoluminescence (left) and back-scattered electron (BSE) image (right) for several samples across the boundary.

Table 1. *Stable isotope data, S content and TOC from the Steinberg section*

Sample	Position (m)	$\delta^{13}C$ (PDB) (‰)	$\delta^{18}O$ (PDB) (‰)	TOC (%)	S (ppm)	$\delta^{34}S$ CDT (‰)	Position of the D/C boundary
dc-1	−6.5	2.2	−2.7	0.3	524	14.4	
dc-2	−5.5	2.3	−3.2	0.1	653		
dc-3k	−5.1	2.6	−3.3	0.5	312	13.1	
dc-3a	−4.7	2.5	−3.3				
dc-3b	−4	1.1	−3.9				
dc-3c	−3.7	2.4	−2.8				
dc-3d	−3.55	2.2	−2.6				
dc-5	−3.5	Phosphorite					
dc-4	−3.3	1.7	−2.8	0.2	322		
dc-4a	−2.9	1.7	−3.4				
dc-4b	−2.6	1.2	−4.1				
dc-6	−2	2.3	−2.7	0.2	833		
dc-7	−1.8	1.7	−3.1	0.1	308		
dc-7a	−1.2	−0.2	−3.9				
dc-8	−0.7	−0.2	−4.2	0.1	299		
dc-9	−0.5	−0.2	−4.0	0.3	285	13.5	
dc-9b	−0.37	−2.1	−5.4				
dc-9a	−0.25	−0.5	−4.5	3.1	303	11.2	
dc-10.1	−0.23	−0.1	−5.4	0	505	16	
dc-10.2	−0.21	−0.9	−4.8	1.4	312		
dc-10.3	−0.15	−0.2	−4.7	0.3	686		
dc-11.1	−0.13	0.6	−3.9	0.2	426	13.8	
dc-11.2	−0.11	0.5	−4.4	0.1	470		
dc-11.3	−0.09	−0.8	−4.4	0.3	475	15	Tournaisian
dc-11.4	−0.05	2.3	−3.9	0.1	843	20.3	*Famennian*
dc-12.1	−0.03	1.4	−3.5	1.7	496		
dc-12.3	0	1.6	−3.8	0.1	360		
dc-12.4	0.5	1.5	−4.1	0.4	336		
dc-13	1.1	1.5	−3.2	0.8	252	16.7	
dc-14	1.2	1.7	−3.4	0.8	298		
dc-15	1.6	2.2	−3.1	0.5	503		
dc-16	1.75	2.3	−3.3	1.5	355	13.4	
dc-16a	2.75	1.9	−3.5	1.5	200	13	
dc-16b	3.1	1.4	−3.8	1.5	180	12.8	

decades, the temporal positions of the black shales and $\delta^{13}C$ shifts, as well as the presence and significance of hiatuses, have been continuously reassessed (Bless *et al.* 1988, 1993; Kaiser 2005; Kaiser *et al.* 2006, 2008; Cramer *et al.* 2008). For the Devonian of central and southern Europe, Buggisch & Joachimski (2006) suggested a positive shift for the $\delta^{13}C$ values for the *praesulcata* Zone. A similar shift was found by Cramer *et al.* (2008) for the Luisiana Limestone in the Upper *praesulcata* Zone. Using the C and O isotopic composition of unaltered brachiopod shells from the stratotype section at La Serre, Montagne Noire, Brand *et al.* (2004) observed for the Middle to Upper *praesulcata* Zone a shift toward higher $\delta^{18}O$ and $\delta^{13}C$ values of brachiopods shells related to an event clearly younger than bed 69 from La Serre. Kaiser *et al.* (2006) found a similar pattern for the C isotopic composition of the organic material and, to a lesser extent, for the micrites, in three sections from the Carnic Alps and Rhenish Massif.

Several studies also suggest the presence of an isotopic shift, clearly younger than the anoxic event covering the *expansa–praesulcata* Zones. For example, Buggisch & Joachimski (2006) and Kaiser *et al.* (2006, 2008) detected a shift toward lower $\delta^{13}C$ values of organic material/carbonate at the D/C boundary, although this trend was not discussed further. Brand *et al.* (2004) observed a decrease in the $\delta^{18}O$ values of brachiopod shells at the D/C boundary, although the scarcity of the material makes the data resolution low. The data discussed above demonstrate that there are several shifts of the $\delta^{13}C$ values within the *praesulcata* Zone, and ongoing discussions regarding their

Table 2. *Major and minor elements from the Steinberg section*

Sample	SiO_2	Al_2O_3	Fe_2O_3	MnO	MgO	CaO	Na_2O	K_2O	TiO_2	P_2O_5
	(%)	(%)	(%)	(%)	(%)	(%)	(%)	(%)	(%)	(%)
dc-1	3.19	1.21	0.59	0.067	0.64	52.44	0.18	0.31	0.060	0.09
dc-3	2.30	0.74	0.50	0.027	0.46	53.92	0.18	0.17	0.051	0.07
dc-9	4.14	1.65	0.78	0.043	0.35	51.47	0.21	0.58	0.094	0.11
dc-10.0	4.66	1.12	5.20	0.246	0.29	48.38	0.23	0.44	0.060	0.08
dc-10.1.2.3	3.47	0.80	0.50	0.046	0.38	53.02	0.21	0.33	0.037	0.05
dc-11.1	3.79	1.31	0.59	0.049	0.34	52.67	0.22	0.45	0.082	0.07
dc-11.3	4.26	1.03	0.68	0.040	0.37	52.36	0.24	0.31	0.072	0.07
dc-12.1.3.4	2.41	0.55	0.36	0.038	0.45	53.57	0.29	0.14	0.041	0.04
dc-13	2.60	0.89	0.46	0.049	0.48	53.57	0.23	0.28	0.053	0.04
dc-16	3.84	1.50	0.79	0.026	0.48	52.16	0.23	0.43	0.118	0.39

Sample	Ba (ppm)	Sr (ppm)	Y (ppm)	Sc (ppm)	Zr (ppm)	V (ppm)	Ni (ppm)	Cu (ppm)	Rb (ppm)	Nb (ppm)	Sb (ppm)	Cs (ppm)	Ir (ppb)
dc-1	33	183	17	3	21	8	74	78	12	2.0	0.2	1.1	
dc-3	26	129	11	3	26	7	68	72	9	1.4	0.4	0.5	
dc-3a	22			2.45					9.1		2.5	0.7	0.14
dc-7	8			0.97					3.0		1.44	0.39	0.14
dc-8	35			3.35					15.6		1.25	0.88	0.14
dc-9	52	95	22	5	26	12	72	69	18	2.0	0.5	0.7	0.14
dc-9a	31			3.59					16.2		1.65	0.81	
dc-10.0	135	79	19	4	28	28	96	176	11	1.5	2.3	1.1	
dc-10.1.2.3	27	133	11	3.58	16	6	67	68	15.1	0.5	1.7	0.8	
dc-10.1	32			2.11					5.0		3.12	0.54	0.14
dc-11.1	21	99	20	3.04	36	14	73	86	14.7	1.7	5.7	0.8	0.06
dc-11.3	30	110	13	5	38	9	66	73	11	1.2	0.8	0.7	
dc-12.1.3.4	19	151	11	2.59	17	−5	71	75	12.1	0.3	6.56	0.9	
dc-13	19	195	14	2.88	18	15	70	76	10.8	0.6	0.7	0.4	0.13
dc-14	26			6.91					16.5		1.84	1.85	
dc-15	49			3.53					20.1		0.72	0.75	0.08
dc-16	38	127	18	4	36	17	85	88	16	2.6	0.4	0.9	

	Cr (ppm)	Co (ppm)	Zn (ppm)	Ga (ppm)	As (ppm)	Y (ppm)	Zr (ppm)	Hf (ppm)	Ta (ppm)	W (ppm)	Tl (ppm)	Pb (ppm)	Th (ppm)	U (ppm)
dc-1				2		17.6	21	0.5	0.18	0.6	0.13	0	2.33	0.19
dc-3				1		10.5	17	0.3	0.16	1.1	0.14	0	1.30	0.18
dc-3a	4.5	4.0	21.0		14.3			0.4	0.19				1.65	0.31
dc-7	1.7	1.0	13.0		5.4			0.2	0.04				0.77	0.2
dc-8	4.1	1.2	19.0		5.3			0.7	0.23				2.19	0.4
dc-9	7.1	1.2	14.0	3	5.6	22.7	19	0.5	0.18	0.7	0.14	0	2.35	0.44
dc-9a	7.1	1.33	15		5.11			0.57	0.13				2.24	0.47
dc-10.0				2	14	18.6	18	0.4	0.12	1.4	1.10	11	1.72	0.98
dc-10.1.2.3	6.5	1.3	13.0	1	5.3	10.4	12	0.6	0.15	0.7	0.08	0	2.27	0.40
dc-10.1	6.0	2.8	23.0		7.6			0.4	0.09				1.31	0.72
dc-11.1	5.7	1.3	15.0	2	4.2	18.5	24	0.5	0.12	0.9	0.20	6	1.85	0.38
dc-11.3				2		12.3	22	0.4	0.12	0.6	0.08	0	1.38	0.22
dc-12.1.3.4	7.4	2.3	18.0	1	5.6	10.4	12	0.4	0.13	0.0	0.08	0	1.46	0.55
dc-13	4.8	0.7	10.0	1	1.6	12.4	15	0.3	0.09	0.9	0.07	0	1.14	0.32
dc-14	11.6	1.92	49		8.31			0.6	0.17				2.08	0.75
dc-15	7.6	1.4	17.0		1.9			0.6	0.21				2.01	4.16
dc-16				2		18.5	35	0.7	0.22	0.8	0.29	0	1.79	0.47

stratigraphic position and temporal extent have been published in recent years. There is a general consensus that a significant drop in atmospheric CO_2 and cool conditions correlate with the deposition of organic-rich shales and widespread anoxia, although there is no agreement on the causes that drew the atmospheric CO_2 down or the temporal extent of the event.

Table 3. *REE composition from the Steinberg section and the NASC values (in ppm)*

	La	Ce	Pr	Nd	Sm	Eu	Gd	Tb	Dy	Ho	Er	Tm	Yb	Lu
NASC	32.0	73.0	7.90	33.0	5.70	1.24	5.20	0.85	5.80	1.04	3.40	0.50	3.10	0.48
dc-1	11.11	15.0	2.34	10.0	2.0	0.48	2.26	0.36	2.09	0.44	1.25	0.1	1.13	0.17
dc-3	8.36	12.3	1.82	7.6	1.45	0.33	1.49	0.24	1.35	0.28	0.77	0.05	0.67	0.09
dc-3A	12.7	15.9		13.1	2.83	0.59	2.7	0.42					1.45	0.2
dc-7	4.84	6.55		5.5	0.91	0.21	0.9	0.13					0.47	0.07
dc-8	14.9	15.1		13.3	3.02	0.51	2.9	0.47					1.47	0.2
dc-9	19.7	21.6	4.24	16.7	3.58	0.69	3.4	0.51	3.08	0.61	1.62	0.11	1.56	0.22
dc-9a	17.1	21.2		16.8	3.11	0.64	3.1	0.44					1.51	0.21
dc-10.0	9.99	17.18	2.2	9.8	2.04	0.5	2.20	0.4	2.13	0.45	1.22	0.08	1.07	0.16
dc-10.1.2.3	16.9	21		15.7	3.25	0.64	2.9	0.45					1.59	0.21
dc-10.1	9.35	14.1		7.9	1.68	0.38	1.8	0.24					0.98	0.13
dc-11.1	15.5	13.8	2.93	12.8	2.45	0.51	2.5	0.36	2.19	0.49	1.34	0.1	1.29	0.19
dc-11.3	11.32	15.74	2.32	10.0	1.74	0.41	1.7	0.27	1.56	0.32	0.92	0.07	0.84	0.12
dc-12.1.3.4	11.2	11.5	2.58	11.3	2.06	0.38	2.1	0.31	1.38	0.27	0.77	0.06	1.11	0.16
dc-13	10.5	13.9	2.51	10.2	2.04	0.48	2.0	0.27	1.88	0.36	0.98	0.07	1.05	0.15
dc-14	24.5	20.5		21.6	4.79	1.12	5.5	0.82					3.1	0.41
dc-15	15.1	20.0		14.6	2.95	0.59	2.9	0.41					1.33	0.17
dc-16	13.9	22.13	3.08	13.23	2.63	0.63	2.67	0.42	2.29	0.47	1.32	0.09	1.13	0.16

Geological setting of the studied section and microfacies distribution

The studied outcrop is situated in the so-called Palaeozoic of Graz (at the abandoned quarry Trolp, 47°04′29″N, 15°19′15″E), which is part of the Upper Austroalpine Nappe System. The Palaeozoic of Graz is a nappe pile comprising weakly metamorphosed sediments and metavolcanic rocks of Silurian to Carboniferous age (Fig. 2b). The Palaeozoic rocks are covered by Upper Cretaceous deposits, known as the Kainach Gosau. The nappe pile consists of lower, intermediate and upper nappe systems. The upper nappe system contains metavolcanics and clastics of Silurian to Early Devonian age, as well as platform and pelagic

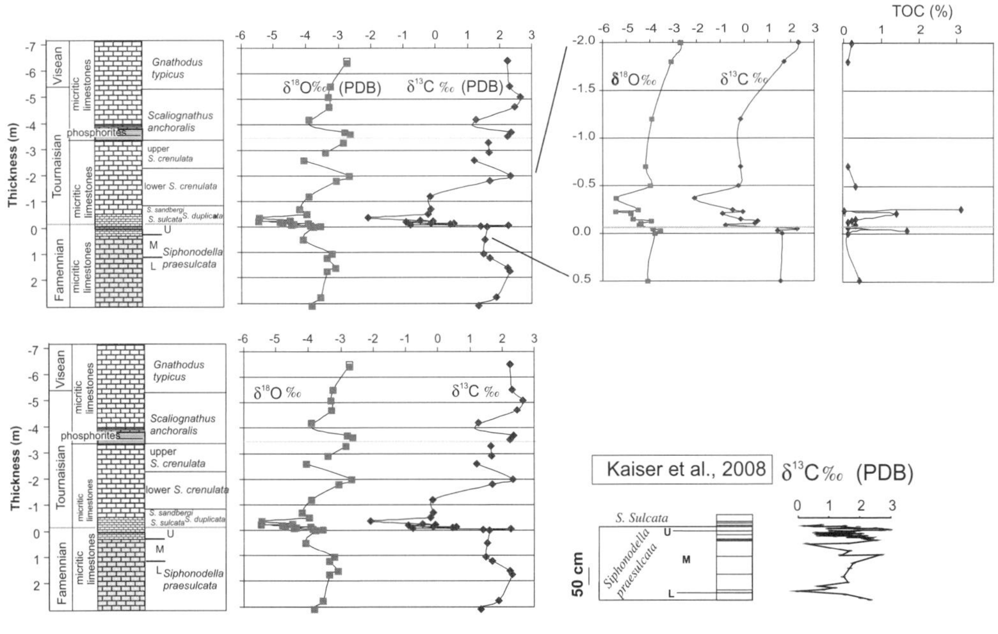

Fig. 4. Variations in O and C isotopic composition for the Steinberg section. Correlation with Kaiser *et al.* (2008).

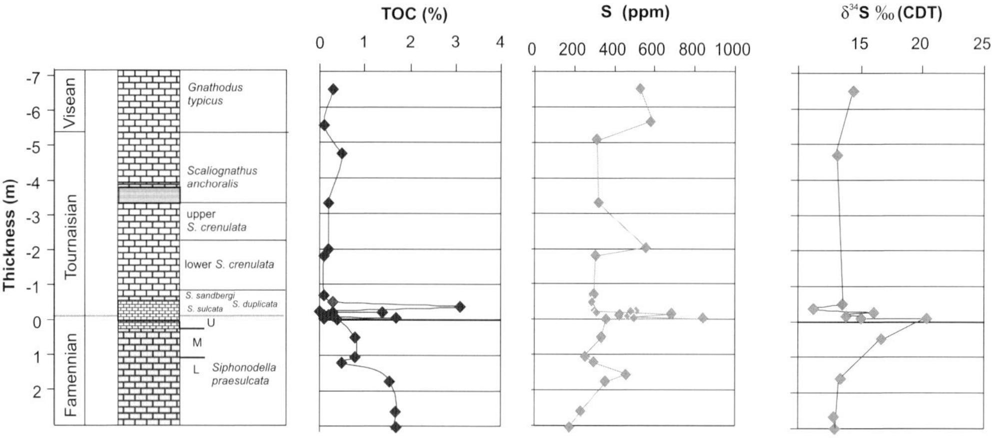

Fig. 5. Variation in TOC, S content and $\delta^{34}S$ of bulk S.

carbonates up to the Westphalian (Fritz *et al.* 1993; Kreutzer *et al.* 2000). Ebner *et al.* (2008) proposed that the Devonian evolution of the Palaeozoic of Graz was that of a rifted continental margin. The Upper Devonian Steinberg Formation, characterized by the presence of conodont- and cephalopod-bearing pelagic limestones, is followed by Lower Carboniferous pelagic deposits, known as the Sanzenkogel Formation. An overview concerning the occurrences and the facies distribution of Devonian and Carboniferous strata in the Eastern Alps is given by Ebner *et al.* (2008), Belka & Narkiewicz (2008) and McCann *et al.* (2008).

The Steinberg section presented in this paper provides a well-documented, lithologically homogenous, D/C boundary succession within pelagic limestones, possibly deposited at a water depth of *c.* 300 m (Nössing 1974; Ebner 1980). The Upper Devonian rocks cropping out below the boundary are characterized by biomicrites of the Steinberg Formation, which lie beneath the Sanzenkogel Formation of Carboniferous age. The tectonically inverted Steinberg section is the type locality for the Sanzenkogel Formation, which consists of thick-bedded limestones with a wavy, respectively flasser bedding. At Steinberg, the boundary section consists of 90–95% pure carbonates, with little organic carbon (Fig. 3). The lithology is characterized by thick-bedded micrites and biomicrites with biogenic components (molluscs, brachiopods, trilobites, ostracods and cephalopods; Fig. 3c). At the boundary, the carbonate beds are much thinner and, after a layer rich in biogenic components, layers completely devoid of such components follow (Fig. 3c). The thickness of the thin bedded carbonates is *c.* 45 cm. Nössing (1975), Ebner (1980) and our own field observations found no black shale level at the bottom or within the thin bedded micrites (Fig. 3d). In the Tournaisian strata, a 20-cm-thick chert–phosphorite intercalation can be found.

Detailed palaeontological work was carried out by Flügel & Ziegler (1957), Nössing (1974, 1975), Surenian (1977), Buchroitner *et al.* (1979), Ebner (1980) and Kaiser (2005). The conodont content helps to distinguish a complete zonation from the Lower *praesulcata* Zone to the *typicus* Zone (Fig. 3a, b). At La Serre the thickness of the *praesulcata* Zone is *c.* 160 cm, and that of the *sulcata* Zone is 20 cm. At Steinberg the thickness of *praesulcata* Zone is *c.* 100 cm, and that of the *sulcata* Zone is 10 cm. Although the sedimentation rate is lower at Steinberg than at La Serre, the present investigated section has a considerable thickness, which is well preserved and not affected by tectonics.

At Steinberg, considering (i) the last occurrence of *Palmatolepsis gracilis gonioclymeniae* and the first occurrence of *Protognathodus kockeli* (Nössing 1975; Ebner 1980) and (ii) the thickness of the Upper *costatus* Zone (10 cm, Nössing 1975), we can approximate the thickness of the Middle *praesulcata* Zone as *c.* 10–15 cm (Fig. 3a, b). The correlations of the new and old conodont zonations are both done after Ziegler & Sandberg (1984). According to Ebner (1980), between the last occurrence of the *Siphonodella praesulcata* and the first occurrence of the *Siphonodella sulcata* there are *c.* 5 cm without siphonodellides. Nössing (1974, 1975) also found *Siphonodella sulcata* in the thin bedded limestones. Figure 3b shows correlation of the layers and conodont zone from Ebner (1980) with the samples analysed in this study (Table 1).

There are several differences between the previous palaeontological works (Nössing 1975; Ebner 1980; Kaiser 2005; Kaiser *et al.* 2008). Kaiser (2005) mapped only 15 cm of thin bedded

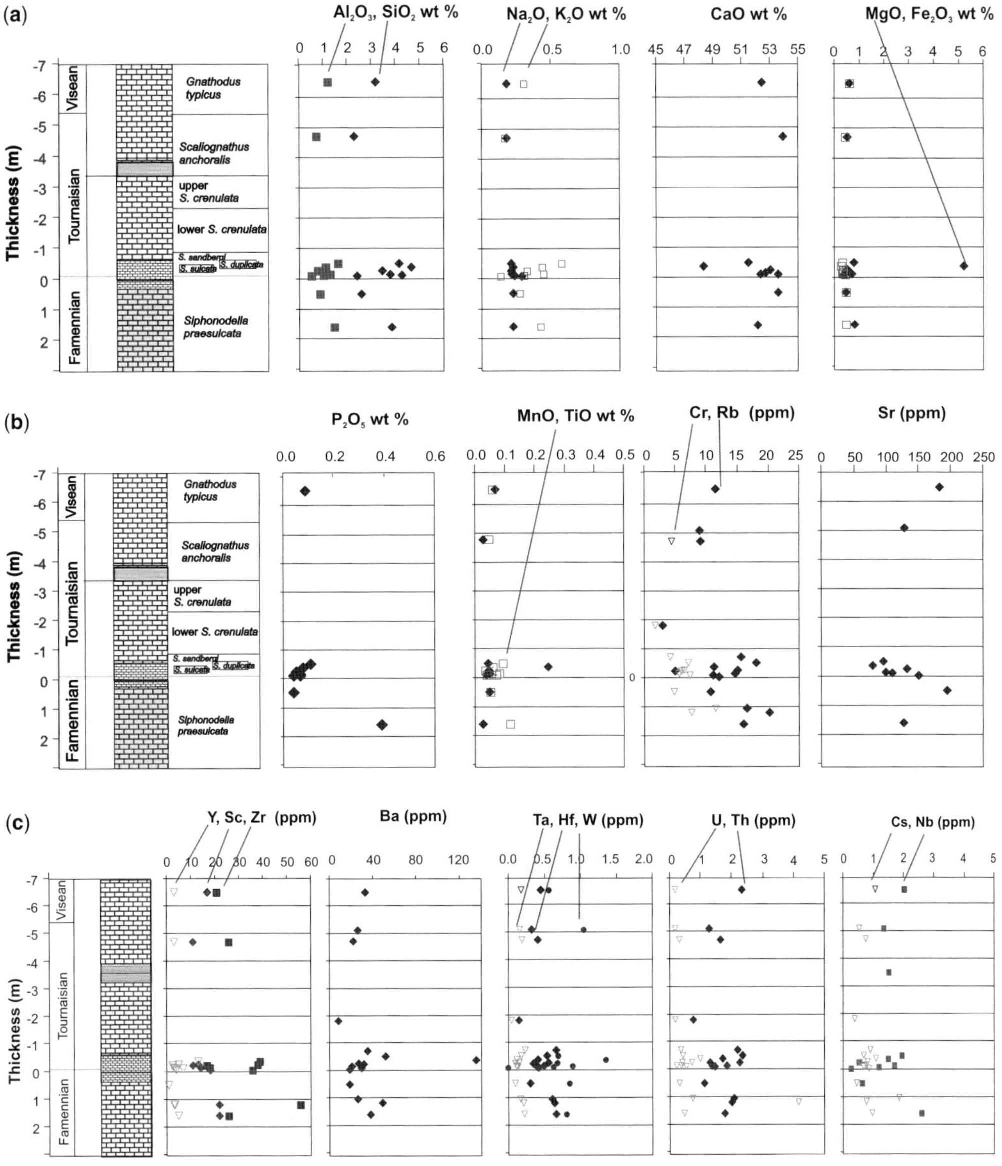

Fig. 6. (**a**) Distribution of the major elements. (**b**, **c**) Distribution of the major and lithophile elements.

limestones (below the boundary) and the D/C is not defined by the first appearance of *Siphonodella sulcata*, although the boundary has a similar position to that defined by Nössing (1975), Ebner (1980) and this study. Kaiser (2005) describes a '1 cm argillaceous layer' situated *c.* 15 cm below the D/C boundary. Although there are no further geochemical data about this layer, in a later work it is interpreted as representing a 'Hangenberg Black Shale equivalent' at the bottom of the Upper *praesulcata* Zone and inserted into the lithological sequence (Kaiser *et al.* 2008). Also according to Kaiser (2005), the Middle *praesulcata* Zone has a thickness of *c.* 180 cm, and *Palmatolepsis gracilis gonioclymeniae* was not found.

Methods and material studied

Selected limestone samples were split in two for microscope and geochemical investigations. Figure 3b and Table 1 present the names and

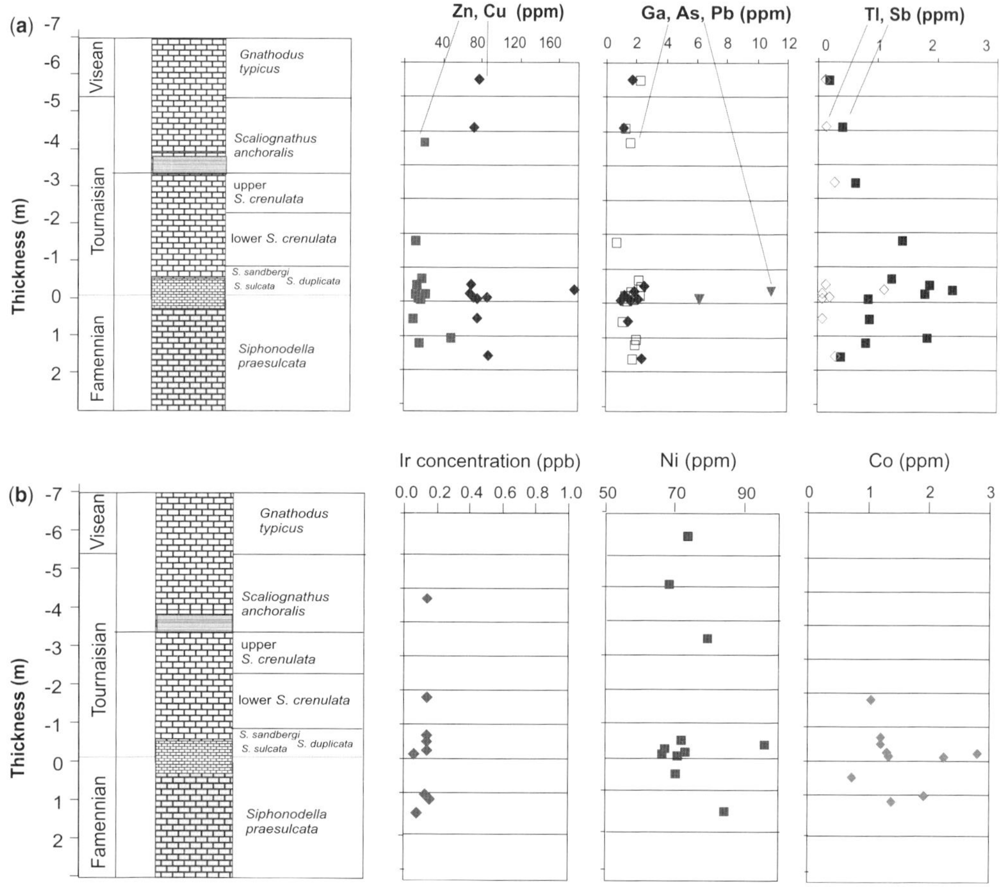

Fig. 7. (**a**) Distribution of the chalcophile elements (S distribution is shown in Fig. 5). (**b**) Distribution of siderophile elements.

positions of the samples. Table 1 contains the stable isotope data, S content and TOC. Tables 2 and 3 provide the major and minor elements (Table 2) and REE distributions (Table 3). All samples were taken from the micrites, avoiding veins, in order to minimize secondary, diagenetic effects, and were broken and ground in an agate mill. Carbonate samples were then analysed in duplicate at the Department of Geology and Palaeontology, Graz, Austria, for (i) total S and TOC using a Leco apparatus and (ii) C and O isotopic compositions using a Delta Plus mass spectrometer and Kiel II preparation line. The standard deviations are 0.06‰ for $\delta^{13}C$ and 0.1‰ for $\delta^{18}O$. Both C and O isotopic compositions are given in Pee Dee Belemnite (PDB). The rare earth and other trace element concentrations were measured at the University of Vienna and by Actlabs, Canada. The analytical procedure for Vienna is described in Koeberl (1993). The S isotopic compositions (given in Canyon Diablo Troilite, CDT) of the bulk rock were measured by Actlabs, Canada. Cathodoluminescence investigations were performed with a JEOL JSM-6310 scanning electron microscope. The operating conditions were 15 kV with a beam current of 15 nA on Al.

Results

The isotope composition of carbonate whole rock is generally very well preserved, because the rate of solid-state diffusion of C and O in carbonates at low temperatures is negligible (O'Neil 1977). The system may be altered if an input of organic or volcanic-derived CO_2 or a fluid with a different isotopic composition than the marine one occurred during early diagenesis (Allen & Matthews 1982; Lohmann 1988). Because the section studied generally contains a low amount of organic material, the presence of organic-derived CO_2 can be omitted. Marine non-luminescent micrites were largely and successfully used in order to reconstruct environmental conditions. For example, there is a wide range of literature about the preservation of $\delta^{13}C$ values in micrites, mudstones, shales, calcareous shaly siltstones, dolomitic siltstone, shaly mud-wackestone and mudstone-wackestone of Precambrian to Palaeozoic age (Wang *et al.* 1996; Kump *et al.* 1999; Joachimski *et al.* 2001; Kimura & Watanabe 2001; Brenchley *et al.* 2003; Cramer *et al.* 2008; Kaiser *et al.* 2008).

At Steinberg, further investigations of the thin sections across the boundary, as back-scattered electron (BSE) and cathodoluminescence images show no evidence for meteoric fluid input (Fig. 3e). This information allows us to assume that the primary marine isotopic signal of micrites is largely preserved and may be further used in palaeo-oceanographic considerations.

O and C isotopic composition, TOC

Across the studied section, the $\delta^{18}O$ values display several fluctuations (Fig. 4). In the Famennian, the values increase (samples dc-16b to dc-13) from the Lower to Middle *praesulcata* Zones. At the end of the Famennian, at the top of the *praesulcata* Zone, a sharp decrease to $-5.4‰$ (dc-11.3 to dc-8) is followed by a shift to heavier isotopic compositions within the Lower *crenulata* Zone, *c.* 2 m above the boundary. In the Middle *praesulcata* Zone, the $\delta^{13}C$ values show two shifts to heavier isotopic compositions. The older shift up to 2.3‰ is between the Lower and Middle *praesulcata* Zones (dc-16a to dc-13), and the second in the Upper *praesulcata* Zone (dc-11.4). At the top of the *praesulcata* Zone, the $\delta^{13}C$ values also show an abrupt decrease of *c.* 3‰ and a recovery upwards (Fig. 4, Table 1, samples dc-11.3 to dc-7). Within the *praesulcata* Zone, the TOC content is up to 2% and generally higher than in the Tournaisian (Figs 4 & 5). At the top of the *praesulcata* Zone, the TOC decreases to close to 0% and shows large variations with positive peaks within the *sulcata* and *duplicata* Zones.

S isotopic composition and S content

Within the Famennian, the S content varies between *c.* 300 ppm and 500 ppm, and the $\delta^{34}S$ composition increases from 13‰ to 18‰ (CDT) (Fig. 5). Near the top of the *praesulcata* Zone, the S content and isotopic composition increase up to 843 ppm and 20‰ (CDT), respectively. Both the S content and the isotopic composition show a drop of values starting with the *sulcata* Zone (dc-11.3 to dc-9, Table 1). Above the *sandbergi* Zone, in the Tournaisian and Visean, the S content and S isotopic composition show values similar to those of the Famennian.

Major and minor elements

Variations in the major element contents are given in Figure 6a and Table 2. The Na_2O content shows constant values across the boundary, SiO_2 increases from 2.4 wt% to 5 wt%, Al_2O_3 increases from 1 wt% to 3 wt%, and K_2O increases to 0.7 wt%. Fe_2O_3 shows one higher value at 5.5 wt%, and MnO shows one higher value at 0.25 wt%. The CaO concentration decreases within the *sulcata* and *duplicata* Zones by *c.* 4 wt% from 53 wt% to 48.4 wt%, and reaches higher values during the Tournaisian. P_2O_5 shows concentrations around 0.1 wt%, with one higher value of *c.* 0.4% during the Famennian, and values of up to 26.5% within the phosphorite layer.

Lithophile elements, such as Ta, Hf, U, Th, Cs, W, Ti and Nb, show generally low concentrations of up to a few ppm and no clear trend across the section. The contents of Rb and Cr show a similar variation pattern, with higher values for the Famennian, decreasing values at the top of the *praesulcata* Zone and lower values during the Tournaisian (Fig. 6b, c). In addition, at the top of the *praesulcata* Zone, Ba has higher values, and Sr and Zr show a variation within the *sulcata* and *duplicata* Zones.

Several chalcophile elements, such as S, Cu, Pb, Sb and Tl, show enrichments near the top of *praesulcata* Zone (Fig. 7a); others, like Zn and Ga, show relatively constant values across the section. All the measured chalcophile elements have similar values for the Famennian and Tournaisian.

The Ir concentrations are very low and show no variation across the section. Both Ni and Co display a rather scattered pattern (Fig. 7b).

REE distribution

Beside the lanthanide, REEs also include Y and Sc. Both Y and Sc behave similarly to the lanthanides in low-temperature fluids (McLennan 1989). The concentration of REEs has been normalized to North American shale composition (NASC) (Haskin *et al.* 1968; Gromet *et al.* 1984) and is plotted in Figure 8a. The data (in ppm) are given in Table 3. For the measured samples, the normalized values are between 0.1 and 1 (i.e. below NASC). For the samples where all the REEs were measured, the sum of the REEs was calculated (ΣREE), including Y and Sc, and then plotted across the section. The sum REE content decreases toward the upper part of the Famennian from a value of *c.* 80 ppm to values of *c.* 50 ppm within the *sulcata* Zone.

The Ce anomaly was calculated based on Wilde *et al.* (1996) (Fig. 8b). The Ce anomaly values

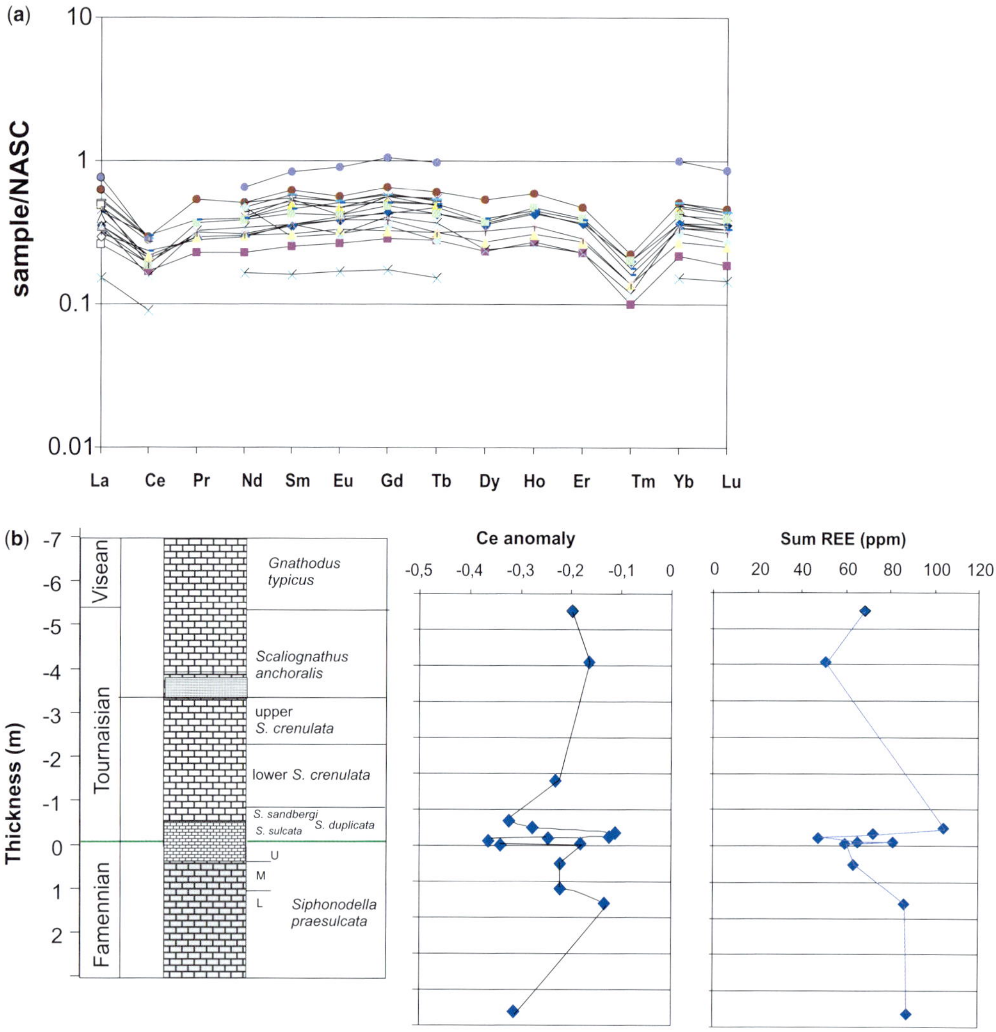

Fig. 8. (**a**) Distribution of the REE normalized to NASC (North Atlantic Shale Composition, Haskin *et al.* 1968; Gromet *et al.* 1984). (**b**) Plot of sum REE for Al_2O_3, K_2O and Zr.

calculated with this method show a negative peak within the Famennian and variations across the boundary. Within the Tournaisian, the Ce anomaly shows less variation.

Discussion

At Steinberg, in the Lower to Middle and Upper *praesulcata* Zones, positive shifts of $\delta^{18}O$ and $\delta^{13}C$, TOC and S content have been documented (sample dc-16b to dc-11.4). The $\delta^{34}S$ values show increasing values, with maximum values in the Upper *praesulcata* Zone (dc-16b to dc-11.4, Table 1). The $\delta^{13}C$ shifts can be correlated with similar trends found in the Upper *praesulcata* Zone in the 'La Serre' section, Montaigne Noir (Brand *et al.* 2004), Carnic Alps (Kaiser *et al.* 2006; Kaiser *et al.* 2008), central and southern Europe (Buggisch & Joachimski 2006) and in the North American Mid-continent region in the Louisiana Limestone (US) (Cramer *et al.* 2008). The shifts toward higher $\delta^{13}C$ values have a similar position in this study and in Kaiser *et al.* (2008) (Fig. 4), the differences being related to the extent of the conodont zones. Based on the conodont distribution of Nössing (1975) and Ebner (1980), the $\delta^{13}C$ values increase from the Lower to Middle *praesulcata* Zones. Because the conodont zones in this section are slightly condensed, the shift of the $\delta^{13}C$ values

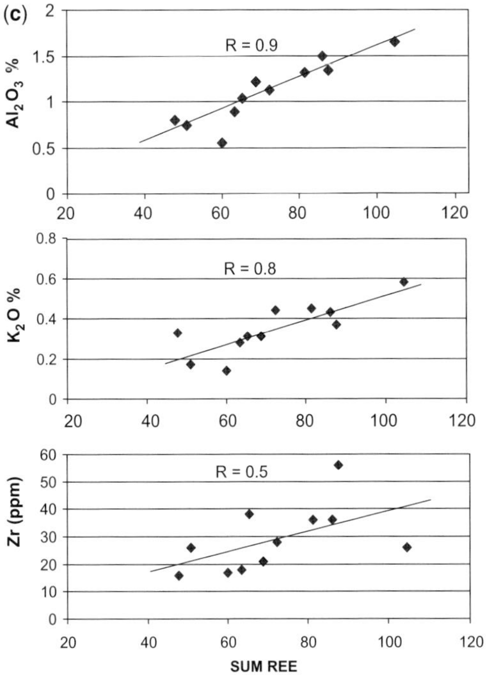

Fig. 8. (**c**) Plot of Ce anomaly and sum REE across the section.

situated in the Upper *sulcata* Zone, is less pronounced than in the Louisiana Limestone, for example, where the Upper *praesulcata* Zone has a thickness of 2.8–20 m (Cramer *et al.* 2008). For Steinberg, the higher TOC, $\delta^{13}C$ and $\delta^{34}S$ values in the Lower to Middle and Upper *praesulcata* Zones suggest higher productivity/bacterial reduction rate (Harrison & Thode 1958; Chambers & Thrudinger 1979; Kump *et al.* 1999) during the late Famennian.

Following these shifts, sharp drops in $\delta^{13}C$ and $\delta^{34}S$ values were found at Steinberg in the *sulcata* Zone. There are similarities between the CaO, $\delta^{13}C$ and TOC variation patterns across the boundary described at Steinberg and those found at the K/T boundary stratotype from El Kef (Keller & Lindinger 1989; Keller *et al.* 1995; Remane *et al.* 1999). These similarities are reflected at Steinberg in (i) a decrease in the CaO of *c.* 3%; (ii) a drop of $\delta^{13}C$ values accompanied by TOC values varying from 0 to 3%; (iii) the minimum $\delta^{13}C$ value occurring not at the boundary, but later on; and (iv) the lack of biogenic remains just above the boundary. The variations towards lower CaO, Sr, Rb and Ba values probably correlate with biomass variations (Figs 4 & 5). In addition, the CaO minimum is associated with a maximum in the abundances of Mn, Fe, Cu, Pb, Sb and Tl, but not with a maximum in S, indicating that these elements are not bound on sulphides.

We suggest that, for the investigated profile, the shift toward lower $\delta^{13}C$ values associated with a drop in TOC and subsequent variable TOC values are related to a drop in the organic carbon flux and reduction of the primary productivity of the shallow ocean as described by Hsü & McKenzie (1985), Magaritz (1989), Kump (1991) and Holser (1997). This is also consistent with the sharp decrease in $\delta^{34}S$ values found across the boundary at Steinberg, which is characteristic of other extinction boundaries and interpreted as a drop in ocean primary productivity (Strauss 1997).

In a preliminary study, Bojar & Neubauer (2003) reported a shift toward lower $\delta^{18}O$ values from the Steinberg section. Brand *et al.* (2004) observed a drop of the $\delta^{18}O$ values of the brachiopod shells in the Upper *sulcata* Zone, although the drop was not as sharp as the one we found at Steinberg. This is most probably due to the low resolution related to the occurrence of suitable brachiopod shells. Kaiser (2005) and Kaiser *et al.* (2006, 2008) did not report the $\delta^{18}O$ composition of carbonates for the sections deposited in carbonate facies at the Steinberg or Grüne Schneid, Carnic Alps.

For the Steinberg section, this shift may represent a rapid episode related to the increase in temperature and related decrease of the $\delta^{18}O$ value of seawater, or an influx of freshwater, which coincides with the transition from the *praesulcata* to *sulcata* Zones. Because no large fluctuations were found in the concentrations of Al_2O_3, SiO_2 or CaO at the studied site, there is no evidence for a significant detritic input at the boundary. Therefore, for Steinberg, we suggest that abrupt warming at the end the latest Famennian glaciation was the mechanism responsible for the negative shift of the $\delta^{18}O$ values. The shift of the $\delta^{18}O$ values is *c.* 1.5‰ (from the maximum value in sample dc-11.4 to the minimum value in dc-10.1) and similar, for example, to the variations found during Quaternary glacial–interglacial fluctuations (Labeyrie 2000) or at the end of the Palaeocene deep-sea warming (Kennett & Stott 1991). The transgression and warming starting in the *sulcata* Zone are also documented by Webster & Groessens (1990), Holser *et al.* (1996), Streel *et al.* (2000) and Brand *et al.* (2004).

Because the REEs are not easily fractionated during sedimentation, the REE sedimentary patterns are related to the average provenance composition. In sedimentary rocks, the REEs reside predominantly in clay minerals, quartz being a mineral with very low REE content (Cullers *et al.* 1979). For the studied section, the sum REE is between

50 and 100 ppm (Fig. 8b). This represents 25–50% of NASC, which has a sum REE content (including Sc and Y) of 215 ppm. The normalized REE distribution with values below 1 is typical for limestones with low detritic input (Thomson *et al.* 1984). Plotting the (ΣREE) + Sc + Y (ppm) against Al_2O_3 and K_2O and Zr, the data show a good correlation (Fig. 8c). The minimum values of Al_2O_3, SiO_2, K_2O and Zr found at the boundary correspond also to the minimum in (ΣREE) + Sc + Y, indicating a low terrigenous input (Claeys *et al.* 1996; Pattan *et al.* 2005).

Cerium exhibits multiple valences (Ce^{3+}, Ce^{4+}). Under reducing conditions, Ce undergoes reduction from the insoluble Ce^{4+} to the more soluble Ce^{3+}. Therefore, its concentration in the ocean water increases and, concomitantly, it decreases in the rocks (Wang *et al.* 1986; de Baar *et al.* 1988; Wilde *et al.* 1996). All the calculated Ce anomalies are negative, indicating widespread occurrence of anoxic water at that time. The Ce anomaly points to anoxic conditions at the top of the *praesulcata* Zone, followed by a period characterized by a lower anomaly and more oxidizing conditions. A rapid shift from anoxic to more oxic conditions around the boundary was also detected by Streel *et al.* (2000). Sulphur and Cr, which are redox-sensitive elements, show different enrichment levels, with higher concentrations during the Famennian, a drop at the boundary, and lower values during the Tournaisian. Elements such as MnO, Fe_2O_3, Ni, Cu, Sb, Tl, Pb, Cr and Ba show elevated concentrations that are associated with higher values for the Ce anomaly. The association with higher Ce anomaly indicates that shelf water turned to more oxygenated water conditions for a short time interval. We propose that, within the geological context described for Steinberg, with pelagic micrites/biomicrites, the sudden variation of the Ce anomaly towards lower values may be related to warming and transgression (Wilde *et al.* 1996), which is in agreement with the general pattern proposed for the D/C boundary (Bless *et al.* 1993; Brand *et al.* 2004). This was interrupted by an event with more oxic conditions marked by increasing values toward zero for the Ce anomaly. We can speculate that this may be related to an episode of vertical advection of deeper, more oxygenated, metal-rich waters during the general transgression episode. In the present-day ocean, it has been documented that below a zone with minimum oxygen, the waters have higher concentration of O and metals; upwelling of these waters changes the local chemistry (Wilde *et al.* 1990). As for the D/C boundary, because there are no data from comparative sections, the regional extent of this mechanism at that time cannot yet be evaluated. Further investigation is necessary.

Conclusions

(1) At Steinberg, the position of the sharp drop of the $\delta^{34}S$, $\delta^{13}C$, $\delta^{18}O$ and TOC values correlates very well with the boundary position defined as the transition from *praesulcata* to *sulcata* Zones.

(2) No large fluctuations were found in the concentrations of Al_2O_3, SiO_2 or CaO at the studied site, so there is no evidence for a significant detritic input at the boundary. The generally low sum REE and the normalized REE abundances also indicate low terrigenous input. The slight variation in the sum REE at the boundary and the very good correlation between Al_2O_3, K_2O, Zr and sum REE are evidence that terrigenous influx decreased towards the boundary for the investigated profile.

(3) Globally, the late Famennian is characterized by increased productivity and deposition of organic-rich shales, cooling and glaciation. At Steinberg, this is evidenced by increases in the $\delta^{18}O$, $\delta^{13}C$ and $\delta^{34}S$ and generally high TOC values. Two shifts of the $\delta^{13}C$ values are present: the older one situated at the level of the Lower to Middle *praesulcata* Zone, and the younger present in the Upper *praesulcata* Zone.

(4) At Steinberg, after the last occurrence of *Siphonodella praesulcata*, a sharp drop in the $\delta^{18}O$, $\delta^{13}C$ and $\delta^{34}S$ values and high variable TOC values were found. Taking into account the geochemical similarities with other extinction boundaries, we propose that this event is related to a warming episode and a drop in the relatively high productivity at the end of the anoxic event. In this context, the minimum Ce anomaly at the boundary would correspond to a transgression younger than the anoxic event, which had already started in the Late Devonian.

This paper benefited from constructive discussions with F. Ebner (Leoben University). We acknowledge interesting comments and detailed remarks from P. Wilde, two anonymous reviewers, as well as from the editors of the book. This work was partially supported by Salzburg University, Geology Division, and FWF Research Project P16258-N06.

References

Allen, J. R. & Matthews, R. K. 1982. Isotope signatures associated with early meteoric diagenesis. *Sedimentology*, **29**, 797–817.

Bai, S. & Ning, Z. 1988. Faunal change and events across the Devonian–Carboniferous boundary of Huangmao

Section, Guanxi, South China. *In*: McMillan, N. J., Embrey, A. F. & Glass, D. J. (eds) *Devonian of the World. Proceedings of the Second International Symposium on the Devonian System, CSPG Memoir* **14/ III**, 147–157.

Becker, R. T. 1994. New faunal records and holostratigraphic correlation of the Hasselbachtal D/C-boundary auxiliary stratotype (Germany). *Annales de la Societe Geologique Belgique*, **115**, 19–45.

Becker, R. T. & Paproth, E. 1993. Auxiliar stratotype section for the Global Stratotype Section and Point (GSSP) for the Devonian–Carboniferous boundary. Hasselbachtal. *Annales de la Societe Geologique Belgique*, **115**, 703–706.

Belka, Z. & Narkiewicz, M. 2008. Devonian. *In*: McCann, T. (ed.) *The Geology of Central Europe. Vol. 1: Precambrian and Palaeozoic*. Geological Society, London, 383–410.

Bless, M. J. M., Simakov, K. V. & Streel, M. 1988. Advantages and disadvantages of a conodont-based or event-stratigraphic Devonian–Carboniferous boundary. *Courier Forschungsinstitut Senckenberg*, **100**, 3–14.

Bless, M. J. M., Becker, R. T., Higgs, K., Paproth, E. & Streel, M. 1993. Eustatic cycles around the Devonian–Carboniferous boundary and the sedimentary and fossil record in Sauerland (Federal Republic of Germany). *Annales de la Societe geologique de Belgique*, **115**, 689–702.

Bojar, A.-V. & Neubauer, F. 2003. Geochemical anomaly across the Devonian Carboniferous extinction boundary, eastern Alps, Austria. *Geophysical Research Abstracts*, **5**, 14913.

Brand, U. & Legrand-Blain, M. 1993. Global perspective of Famennian–Tournaisian oceanography; geochemical analysis of brachiopods. *Annales de la Societe Geologique Belgique*, **115**, 491–496.

Brand, U., Legrand-Blain, M. & Streel, M. 2004. Biochemostratigraphy of the Devonian–Carboniferous boundary global stratotype section and point, Griotte Formation, La Serre, Montagne Noire, France. *Palaeogeography, Paleoclimatology, Palaeoecology*, **205**, 337–357.

Brenchley, P. J., Carden, G. A. *et al.* 2003. High resolution stable isotope stratigraphy of Upper Ordovician sequences: constraints on the timing of bioevents and environmental changes associated with mass extinction and glaciation. *Geological Society of America Bulletin*, **115**, 89–104.

Buchroitner, M. F., Ebner, F. & Surenian, R. 1979. Die Entwicklung der Steinbergkalke (Oberdevon, Grazer Paläozoikum) an ihrer Typuslokalität. *Mitteilungen des Naturwissenschaftlichen Vereins für Steiermark*, **109**, 71–84.

Buggisch, W. & Joachimski, M. M. 2006. Carbon isosotpe stratigraphy of the Devonian of central and Southern Europe. *Palaeogeography, Palaeoclimatology, Palaeoecology*, **240**, 68–88.

Caplan, M. L. & Bustin, R. M. 1999. Devonian–Carboniferous Hangenberg mass extinction event, widespread organic-rich mudrock and anoxia: causes and consequences. *Palaeogeography, Palaeoclimatology, Palaeoecology*, **148**, 187–207.

Caplan, M. L. & Bustin, R. M. 2001. Palaeoenvironmental and palaeoceanographic control on black, laminated mudrock deposits: example from Devonian–Carboniferous strata, Alberta, Canada. *Sedimentary Geology*, **154**, 45–72.

Caplan, M. L., Bustin, R. M. & Grimm, K. A. 1996. Demise of a Devonian–Carboniferous carbonate ramp by eutrophication. *Geology*, **24**, 715–718.

Chambers, L. A. & Trudinger, P. A. 1979. Microbiological fractionation of stable sulfur isotopes: a review and critique. *Geomicrobiology Journal*, **1**, 249–293.

Claeys, P., Kyte, F. T., Herbosch, A. & Casier, J.-G. 1996. *Geochemistry of the Frasnian–Famennian Boundary in Belgium: Mass Extinction, Anoxic Oceans and Microtektite Layer, But Not Too Much Iridium?* Geological Society America, Boulder, Special Publications, **307**, 491–504.

Corradini, C., Kaiser, S. I., Perri, M. C. & Spalletta, C. 2011. Protognathodus (Conodonta) and its potential as a tool for defining the Devonian/Carboniferous boundary. *Rivista Italiana di Paleontologia e Stratigrafia*, **117**, 15–28.

Cramer, B. D., Salzman, M. R., Day, J. E. & Witzke, B. 2008. Record of the Late Devonian Hangenberg global positive carbon-isotope excursion in an epeiric sea setting: carbonate production, organic-carbon burial and paleoceanography during the Late Famennian. *In*: Pratt, B. R. & Holmden, C. (eds) *Dynamics of Epeiric Seas*. Geological Association of Canada, St. Johns, Special Papers, **48**, 103–118.

Cullers, R. L., Chaudhuri, S., Kilbane, N. & Koch, R. 1979. Rare earths in size fractions and sedimentary rocks of Pennsylvanian–Permian age from the mid-continent of the USA. *Geochimica et Cosmochimica Acta*, **43**, 1285–1302.

de Baar, H. J. W., German, C. R., Elderfield, H. & van Gaans, P. 1988. Rare earth element distribution in anoxic waters of the Carioca trench. *Geochimica et Cosmochimica Acta*, **52**, 1203–1219.

Ebner, F. 1980. Conodont localities in the surroundings of Graz/Styria. *Abhandlungen der Geologischen Bundesanstalt Wien*, **35**, 101–127.

Ebner, F., Vozarova, A. *et al.* 2008. Devonian–Carboniferous pre-flysch and flysch environments in the Circum Pannonian Region. *Geologica Carpathica*, **59**, 159–195.

Feist, R., Flajs, F. & Girard, C. 2000. The stratotype section of the Devonian–Carboniferous Boundary. *Courier Forschungsinstitut Senckenberg*, **225**, 77–82.

Flajs, G., Feist, R. & Ziegler, W. (eds) 1988. Devonian Carboniferous boundary: results of recent studies. *Courier des Forschungs-Instituts Senckenberg*, **100**, 245.

Flügel, H. W. & Ziegler, W. 1957. Die Gliederung des Oberdevons und Unterkarbons am Steinberg westlich von Graz mit Conodonten. *Mitteilungen des Naturwissenschaftlichen Vereins für Steiermark*, **87**, 25–60.

Fritz, H., Ebner, F. & Neubauer, F. 1993. The Graz thrust-complex (Paleozoic of Graz). *In*: Neubauer, F. (ed.) *ALCAPA-Field Guide, The Eastern Central Alps of Austria*, Karl-Franzens University, Graz, 83–92.

Girard, C. 1994. Conodont biofacies and event stratigraphy across the D/C boundary in the stratotype area

(Montagne Noire, France). *Courier Forschungsinstitut Senckenberg*, **168**, 299–309.

GRADSTEIN, F., OGG, J. & SMITH, A. 2004. *A Geologic Time Scale*. Cambridge University Press, Cambridge.

GROMET, L. P., DYMEK, R. F., HASKIN, L. A. & KOROTEV, R. L. 1984. The 'North American shale composite': its compilation, major and trace element characteristics. *Geochimica et Cosmochimica Acta*, **48**, 2469–2482.

HARRISON, A. G. & THODE, H. G. 1958. The kinetic isotope effect in the chemical reduction of sulphate. *Transactions of the Faraday Society*, **53**, 1648–1651.

HASKIN, L. A., HASKIN, M. A., FREY, F. A. & WILDMAN, T. R. 1968. Relative and absolute terrestrial abundance of the rare earths. *In*: AHRENS, L. H. (ed.) *Origin and Distribution of the Elements*. Pergamon, New York, 889–912.

HOLSER, W. T. 1997. Geochemical events documented in inorganic carbon isotopes. *Palaeogeography, Palaeoclimatology, Palaeoecology*, **32**, 173–182.

HOLSER, W. T., MAGARITZ, M. & RIPPERDAN, R. L. 1996. Global isotopic events. *In*: WALLISER, O. H. (eds) *Global Events and Event Stratigraphy in the Phanerozoic*. Springer Verlag, Berlin Heidelberg, 63–88.

HOU, H.-F., JI, Q., XIONG, J.-F. & WU, X.-H. 1984. A possible stratotype of Devonian Carboniferous boundary in Guizhou Province, South China. *Courier Forschungsinstitut Senckenberg*, **67**, 193–167.

HSÜ, K. J. & MCKENZIE, J. A. 1985. A 'Strangelove' ocean in the earliest Tertiary. *In*: SUNDQUIST, E. T. & BROECKER, W. S. (eds) *The Carbon Cycle and Atmospheric CO_2: Natural Variations Archean to Present*. American Geophysical Union, Washington, DC, 487–492.

ISAACSON, P. E., DIAZ-MARTINEZ, E., GRADER, G. W., KALVODA, J., BABEK, O. & DEVUYST, F. X. 2008. Late Devonian–earliest Mississippian glaciation in Gondwanaland and its biogeographic consequences. *Palaeogeography, Palaeoclimatology, Palaeoecology*, **268**, 126–142.

JOACHIMSKI, M. M., OSTERTAG-HENNING, C. ET AL. 2001. Water column anoxia, enhanced productivity and concomitant changes in $\delta^{13}C$ and $\delta^{34}S$ across the Frasnian–Famennian boundary (Kowala–Holly Cross Mountains/Poland). *Chemical Geology*, **175**, 109–131.

KALVODA, J. 1986. Upper Frasnian–Lower Tournaisian events and evolution of calcareous foraminifera, close link to climatic changes. *In*: WALLISER, O. H. (ed.) *Global Bio-Events: A Critical Approach. Lecture Notes in Earth Sciences*, **8**, Springer-Verlag, Berlin Heidelberg, 225–236.

KAISER, S. I. 2005. *Mass Extinction, Climatic and Oceanographic Changes at the Devonian–Carboniferous Boundary*. Faculty of Natural Sciences, Ruhr University of Bochum.

KAISER, S. I. 2009. The Devonian/Carboniferous stratotype section La Serre (Montagne Noir) revisited. *Newsletter on Stratigraphy*, **43**, 195–205.

KAISER, S. I. & CORRADINI, C. 2011. The early siphonodellids (Conodonta, Late Devonian–Early Carboniferous): overview and taxonomic state. *Neues Jahrbuch für Geologie und Paläontologie Abhandlungen*, **261**, 19–35.

KAISER, S. I., STEUBER, T., BECKER, R. T. & JOACHIMSKI, M. M. 2006. Geochemical evidence for major environmental changes at the Devonian–Carboniferous boundary in the Carnic Alps. *Palaeogeography, Palaeoclimatology Palaeoecology*, **240**, 146–160.

KAISER, S. I., STEUBER, T. & BECKER, R. T. 2008. Environmental change during the Late Famennian and Early Tournaisian (Late Devonian–Early Carboniferous): implications from stable isotopes and conodont biofacies in southern Europe. *Geological Journal*, **43**, 241–260.

KELLER, G. & LINDINGER, M. 1989. Stable isotope, TOC and $CaCO_3$ record across the Cretaceous/Tertiary boundary at El Kef, Tunisia. *Palaeogeography, Palaeoclimatology, Palaeoecology*, **119**, 221–254.

KELLER, G., LI, L. & MACLEOD, N. 1995. The Cretaceous/Tertiary boundary stratotype section at ElKef, Tunisia: how catastrophic was the mass extinction? *Palaeogeography, Palaeoclimatology, Palaeoecology*, **119**, 221–254.

KENNETT, J. P. & STOTT, L. D. 1991. Abrupt deep-sea warming palaeoceanographic changes and benthic extinctions at the end of the Paleocene. *Nature*, **353**, 225–229.

KIMURA, H. & WATANABE, Y. 2001. Ocean anoxia at the Precambrian–Cambrian boundary. *Geology*, **29**, 995–998.

KOEBERL, C. 1993. Instrumental neutron activation analysis of geochemical and cosmochemical samples: a fast and reliable method for small sample analysis. *Journal of Radioanalytical and Nuclear Chemistry*, **168**, 47–60.

KREUTZER, L. H., SCHÖNLAUB, H. P. & HUBMANN, B. 2000. The Devonian of Austria. *Courier Forschungsinstitut Senkenberg*, **225**, 173–183.

KUMP, L. R. 1991. Interpreting carbon-isotope excursions: Strangelove oceans. *Geology*, **19**, 229–302.

KUMP, L. R., ATHUR, M. A., PATZKOWSKY, M. E., GIBBS, M. T., PINKUS, D. S. & SHEEHAN, P. M. 1999. A weathering hypothesis for glaciation at high atmospheric pCO_2 during the Late Ordovician. *Palaeogeography, Palaeoclimatology, Palaeoecology*, **152**, 173–187.

KÜRSCHNER, W., BECKER, R. T., BUHL, R. T. & VEIZER, J. 1992. Strontium isotopes in conodonts: Devonian–Carboniferous transition, the Northern Rhenish Slate Mountains, Germany. *Annales de la Societé Géologique de Belgique*, **115**, 595–621.

LABEYRIE, L. 2000. Glacial climate instability. *Science*, **290**, 1905–1907.

LOHMANN, K. C. 1988. Geochemical patterns of meteoric diagenetic systems and their application to studies to palaeokarst. *In*: JAMES, N. P. & CHOQUETTE, P. W. (eds) *Palaeokarst*. Springer-Verlag, Berlin, 55–80.

MAGARITZ, M. 1989. $\delta^{13}C$ minima follow extinction events: a clue to faunal radiation. *Geology*, **17**, 337–340.

MATYJA, H. & STEMPIEN-SALEK, M. 1993. Devonian/Carboniferous boundary and the associated phenomena in western Pomerania (NW Poland). *Annales de la Societé Géologique de Belgique*, **116**, 249–263.

MCCANN, T., SKOMPSKI, S. ET AL. 2008. Carboniferous. *In*: MCCANN, T. (ed.) *The Geology of Central Europe. Vol. 1: Precambrian and Palaeozoic*. Geological Society, London, 411–529.

McLennan, S. M. 1989. Rare earth elements in sedimentary rocks: influence of provenance and sedimentary processes. *In*: Lipin, B. R. & McKay, G. A. (eds) *Geochemistry and Mineralogy of Rare Earth Elements. Reviews in Mineralogy*, **21**, The Mineralogical Society of America and the Geochemical Society, Washington, 169–200.

Nössing, L. 1974. *Beitrag zur Kenntnis des Oberdevons und Unterkarbons des Grazer Paläozoikums*. Inaugural Dissertation zur Erlangung des Doktoratgrades der philosophischen, Fakultät an der Karl-Franzens, Universität in Graz.

Nössing, L. 1975. Die Sanzenkogel-Schichten (Unterkarbon), eine biostratigraphische Einheit des Grazer Paläozoikums. *Mitteilungen des Naturwissenschaftlichen Vereins für Steiermark*, **105**, 79–92.

O'Neil, J. 1977. Stable isotopes in mineralogy. *Physics and Chemistry of Minerals*, **2**, 105–123.

Orth, C. J., Quintana, L. R., Gilmore, J. S., Barrick, J. E., Haywa, J. N. & Spesshardt, S. A. 1988. Pt-group metal anomalies in the Lower Mississippian of southern Oklahoma. *Geology*, **16**, 627–630.

Paproth, E. & Streel, M. (eds) 1984. The Devonian–Carboniferous boundary. *Courier Forschungsinstitut Senckenberg*, **67**, 260.

Paproth, E., Feist, R. & Flajs, G. 1991. Decision on the Devonian–Carboniferous boundary stratotype. *Episodes*, **14**, 331–336.

Pattan, J. N., Pearcen, N. J. G. & Mislankar, R. G. 2005. Constraints in using cerium-anomaly of bulk sediments as an indicator of paleo bottom water redox environment: a case study from the Central Indian Ocean Basin. *Chemical Geology*, **221**, 260–278.

Peterhänsel, A. & Pratt, B. R. 2001. Nutrient-triggered bioerosion on a giant carbonate platform masking the postextinction Famennian benthic community. *Geology*, **29**, 1079–1082.

Popp, B. N., Anderson, T. F. & Sandberg, P. A. 1986. Brachiopods as indicators of original isotopic compositions in some Paleozoic limestones. *Geological Society of America Bulletin*, **97**, 1262–1269.

Remane, J., Keller, G., Hardenbol, J. & Ben Haj Ali, M. 1999. Report on the international workshop on Cretaceous–Paleogene transition. *Episodes*, **22**, 47–48.

Richards, B. C. & Higgins, A. C. 1988. Devonian–Carboniferous boundary beds of the Palliser and Exshaw formations at Jura Creek, Rocky Mountains, southwestern Alberta. *In*: McMillan, N. J., Embry, A. F. & Glass, D. J. (eds) *Devonian of the World. Proceedings of the Second International Symposium on the Devonian System, CSPG Memoir*, **14(II)**, 399–412.

Sandberg, C. A., Ziegler, W., Leuteritz, K. & Brill, S. M. 1978. Phylogeny, speciation, and zonation of Siphonodella (Conodonta, Upper Devonian and Lower Carboniferous). *Newsletter on Stratigraphy*, **7**, 102–120.

Sandberg, C. A., Ziegler, W., Dreesen, R. & Butler, N. 1988. Late Frasnian mass extinction: conodont event stratigaphy, global changes and possible causes. *Courier Forschungsinstitut Senckenberg*, **102**, 263–307.

Schmidt, H. 1924. Zwei Cephalopodenfaunen an der Devon-Karbon-Grenze im Sauerland. *Jahrbuch der Preussischen Geologischen Landes-Anstalt*, **44**, 98–171.

Schönlaub, H. P.,, Attrepp, M. *et al.* 1992. The Devonian/Carboniferous boundary in the Carnic Alps (Austria) – a multidisciplinary approach. *Jahrbuch der Geologischen Bundesanstalt Wien*, **135**, 57–98.

Scotese, C. R. 2002. *Paleomap Project*, http://www.scotese.com.

Scotese, C. R. & McKerrow, W. S. 1990. Revised world maps and introductions. *In*: Kerrow, W. S. & Scotese, C. R. (eds) *Paleozoic Paleogeography and Biogeography*. Geological Society, London, Memoirs, **12**, 1–21.

Sepkoski, J. J., Jr. 1996. Patterns of Phanerozoic extinctions: a perspective from global data bases. *In*: Walliser, O. H. (ed.) *Global Events and Event Stratigraphy in the Phanerozoic*. Springer Verlag, Berlin Heidelberg, 35–51.

Simakov, K. V. 1993. The dynamics and biochronological structure of the Hangenbergian bioevent. *Palaeogeography, Palaeoclimatology, Palaeoecology*, **104**, 127–137.

Strauss, H. 1997. The isotopic composition of sedimentary sulfur through time. *Palaeogeography, Palaeoclimatology, Palaeoecology*, **132**, 97–118.

Streel, M. & Marshall, J. E. A. 2006. Devonian–Carboniferous boundary global correlations and their paleogeographic implications for the Assembly of Pangaea. *In*: Wong, Th. E. (ed.) *Proceedings of the XVth International Congress on Carboniferous and Permian Stratigraphy*. Utrecht, The Netherlands, 10–16 August 2003. Royal Netherlands Academy of Arts and Sciences, Amsterdam, 481–496.

Streel, M., Caputo, M. V., Loboziak, S. & Melo, J. H. G. 2000. Late Frasian–Famennian climates based on palynomorph analysis and the question of the Late Devonian glaciations. *Earth-Science Reviews*, **52**, 121–173.

Surenian, R. 1977. *Biostratigraphie und Sedimentologie der Steinbergkalke (Oberdevon) des Steinberges westlich Graz*. PhD thesis, Faculty of Geology and Palaeontology, University of Graz.

Thompson, J., Carpenter, M. S. N., Colley, S., Wilson, T. R. S., Elderfield, H. & Kennedy, H. 1984. Metal accumulation rates in northwest Atlantic pelagic sediments. *Geochimica et Cosmochimica Acta*, **48**, 1935–1948.

Veizer, J., Fritz, P. & Jones, B. 1986. Geochemistry of the brachiopods: oxygen and carbon isotopic records of Paleozoic oceans. *Geochimica et Cosmochimica Acta*, **50**, 1679–1696.

Veizer, J., Ala, D. *et al.* 1999. $^{87}Sr/^{86}Sr$, $\delta^{13}C$ and $\delta^{18}O$ evolution of Phanerozoic seawater. *Chemical Geology*, **161**, 16159–16188.

Walliser, O. H. 1984. Pleading for a natural D/C boundary. *Courier Forschungsinstitut Senckenberg*, **67**, 241–246.

Walliser, O. H. 1996. Global events in the Devonian and Carboniferous. *In*: Walliser, O. H. (ed.) *Global Events and Event Stratigraphy in the Phanerozoic*. Springer Verlag, Berlin, 225–250.

WANG, C.-Y. 1993. Auxiliar stratotype sections for the global stratotype section and point (GSSP) for the Devonian–Carboniferous boundary: Nanbiancun. *Annales Societe Geologique Belgique*, **115**, 707–708.

WANG, Y. L., LIU, Y.-G. & SCHMITT, R. A. 1986. Rare earth element geochemistry of South Atlantic deep sea sediments: Ce anomaly change at ~54 My. *Geochimica et Cosmochimica Acta*, **50**, 1337–1355.

WANG, K., ATTREP, M., JR. & ORTH, C. J. 1993. Global iridium anomaly, mass extinction, and redox change at the Devonian–Carboniferous boundary. *Geology*, **23**, 625–628.

WANG, K., GELDSETZER, H. H. J., GOODFELLOW, W. D. & KROUSE, H. R. 1996. Carbon and sulphur isotope anomalies across the Frasnian–Famennian extinction boundary, Alberta, Canada. *Geology*, **24**, 187–191.

WEBSTER, G. D. & GROESSENS, E. 1990. Conodont subdivision of the Lower Carboniferous. *Courier Forschungsinstitut Senckenberg*, **130**, 31–40.

WILDE, P., QUINBY-HUNT, M. S. & BERRY, W. B. N. 1990. Vertical advection from oxic or anoxic water from the main pycnocline as a cause of rapid extinction or rapid radiation. *In*: KAUFMANN, E. G. & WALLISER, O. H. (eds) *Extinction Events in Earth History*. Springer Verlag, Berlin, 85–98.

WILDE, P., QUINBY-HUNT, M. S. & ERDTMANN, B. -D. 1996. The whole-rock cerium anomaly: a potential indicator of eustatic sea-level changes in shales of the anoxic facies. *Sedimentary Geology*, **101**, 43–53.

XU, D. Y., YAN, Z., ZHANG, Q. W., SHEN, Z. D., SUN, Y. Y. & YE, L. F. 1989. Significance of an anomaly near the Devonian/Carboniferous boundary at the Muha section, South China. *Nature*, **321**, 854–855.

ZIEGLER, W. & SANDBERG, C. A. 1984. Important candidate sections for stratotype of conodont based Devonian–Carboniferous boundary. *Courier Forschungsinstitut Senckenberg*, **67**, 231–239.

ZIEGLER, W. & SANDBERG, C. A. 1990. The Late Devonian Standard Conodont Zonation. *Courier Forschungsinstitut Senckenberg*, **121**, 1–115.

Palaeoecology and sedimentology of Carboniferous glacial and post-glacial successions in the Paganzo and Río Blanco basins of northwestern Argentina

A. A. DINEEN*, M. L. FRAISER & J. L. ISBELL

Department of Geosciences, University of Wisconsin – Milwaukee, Milwaukee, WI 53201, USA

**Corresponding author (e-mail: aadineen@uwm.edu)*

Abstract: Recent studies show that Late Palaeozoic Ice Age (LPIA) climate change broadly affected marine invertebrate faunas: glaciations decreased origination and extinction and long-term gradual global warming during the final deglaciation altered palaeocommunity composition. Carboniferous stratigraphy and palaeoecology in the Paganzo and Río Blanco basins of southwestern Gondwana (present-day Argentina) were studied in ordered to further determine how palaeocommunities were influenced by glacial and post-glacial processes during the LPIA. In the Paganzo Basin, the Guandacol Formation consisted of an ice-proximal to very ice-distal glaciomarine succession, while the Tupe Formation represented continental fluvial lowstand sedimentation interrupted by periodic marine incursions (estuary setting). In the Río Blanco Basin, the Río del Peñon Formation represents a shallowing-upwards coastal marine facies under the influence of wave and storm activity. Diversity and abundance data from northwestern Argentina reflects that ice-proximal environments proved physiologically stressful to organisms, limiting colonization to only opportunistic fauna. Following glaciation, the fauna of northwestern Argentina diversified and became increasingly ecologically complex during two marine transgressions in the Río Blanco Basin. It is interpreted that palaeocommunity establishment, composition and diversification in western Argentina during the LPIA was mostly dependent upon localized environmental conditions and palaeogeographic location.

The Late Palaeozoic Ice Age (LPIA), which spanned 345–260 Ma, has long been known as an important climatic event in Earth's history. The end of the LPIA is the only example in Earth's history where a vegetated and biologically complex Earth transitioned from a bipolar icehouse to a greenhouse state (Berner 1990; Frakes *et al.* 1992; Gastaldo *et al.* 1996; Isbell *et al.* 2003*a*; Montañez *et al.* 2007; Fielding *et al.* 2008*a*, *b*). Earth's climate system is currently undergoing a similar change, with glaciers and ice sheets retreating, global temperatures increasing and sea levels rising (e.g. IPCC 2007). The LPIA's transition had major impacts on Earth's physical and biological systems, and is therefore an important analogue for addressing questions regarding the nature of future global warming and its impacts upon life on Earth (Stanley & Powell 2003; Buatois *et al.* 2006; Limarino *et al.* 2006). For these reasons the LPIA is of great interest, yet many questions remain about environmental and biotic changes that occurred during this interval of time.

Most recent work on the LPIA concerns determining the extent of the glaciation (Veevers & Powell 1987; Isbell *et al.* 2003*a*, *b*, 2008*b*; Fielding *et al.* 2008*b*). It had been previously proposed that one massive long-lasting (60–80 million years) ice sheet covered Gondwana during the LPIA (Frakes 1979). Recent data do not support this idea however, but rather support the concept that glaciation occurred as multiple, short, geographically restricted glacial events that were separated by non-glacial periods of comparable duration (e.g. López-Gamundí 1997; Visser 1997; Isbell *et al.* 2003*a*, 2008*a*, *b*, 2012; Fielding *et al.* 2008*a*, *b*). These new perspectives raise many questions regarding the character of the ice age, including the magnitude of glacioeustatic fluctuations and the impact of repeated climate transitions on marine and terrestrial systems (cf. Heckel 1994, 2008; Davies 2008; Rygel *et al.* 2008; Horton & Poulson 2009).

The stratigraphic record of Gondwana in Argentina contains deposits that can be classified as 'near-field' (e.g. high-latitude such as South America, southern Africa, Antarctica and Australia during the LPIA) as opposed to 'far-field' (e.g. low-latitude such as North America, Europe and parts of Asia) due to their palaeolatitudinal location and palaeoproximity to glaciers. Studies in near-field regions such as Argentina are advantageous due to their high-latitude proximity to glaciers because they provide detailed and sensitive records of glacial–interglacial climate changes. Argentina is an ideal

From: Gąsiewicz, A. & Słowakiewicz, M. (eds) 2013. *Palaeozoic Climate Cycles: Their Evolutionary and Sedimentological Impact*. Geological Society, London, Special Publications, **376**, 109–140.
First published online November 26, 2012, http://dx.doi.org/10.1144/SP376.3

location for studying the LPIA, as it contains one of the most complete stratigraphic records of this important interval. LPIA strata are well-exposed in the Paganzo (Cuesta de Huaco locality) and Río Blanco (Río del Peñon locality) basins (Fig. 1).

Recent studies show that climate change may cause extinctions through indirect palaeoecological effects and that long-term gradual global warming can have important consequences on palaeocommunities (M. G. Powell 2005; Clapham & James 2008). However, the effects of the glacial/non-glacial fluctuations during the LPIA on the near-field marine biota have received very little attention. Much of the previous work is primarily from large global datasets (e.g. Kelley *et al.* 1990; Stanley & Powell 2003; M. G. Powell 2005) or based on taxonomic diversity (e.g. González 1997*a*), including single taxa (e.g. Pérez-Huerta & Sheldon 2006; Groves & Yue 2009; Heim 2009). As a result, what are particularly lacking are quantitative analyses of near-field marine faunas rooted in data collected in the field. This is especially important because palaeoecology and taphonomy of Carboniferous high-latitude faunas, such as those in western Argentina, are qualitatively and quantitatively different to their low-latitude counterparts (Hanger & Gabadirwe 1999; Shi & Waterhouse 2010; Waterhouse & Shi 2010); comparing patterns and processes of tropical faunas with high-latitude faunas during this period of time therefore cannot be justified. Additionally, recent data have shown that local environmental and physical factors play a significant role in diversification (i.e. Novack-Gottshall & Miller 2003), indicating that regional studies may be key to understanding complex palaeocommunity responses to climate change.

Several far-field studies have reported that ecological persistence, instead of change, is the normal biotic response to high-amplitude and high-frequency glacioeustasy during the LPIA, with broadly adapted taxa (eurytopes) increasing in importance in regional areas (Bonelli & Patzkowsky 2008, 2011; Heim 2009). Far-field marine invertebrates at this time have also been shown to have sluggish evolutionary turnover punctuated with low rates of origination and extinction (Stanley & Powell 2003; M. G. Powell 2005). As these rates did not recover until after the LPIA, it has been proposed that circumstances during the LPIA suppressed them. Several hypotheses have emerged to explain these patterns, including the hypothesis that biota with more limited environmental tolerances were unable to adapt to changing conditions following large climate fluctuations (Stanley & Powell 2003; Clapham & James 2008). Thus, it seems that short-term glacial and non-glacial fluctuations during this time in Gondwana basins had the potential to lead to ecosystem disruption.

This study uses quantitative analysis of palaeoecology to establish how regional marine palaeocommunities in western Gondwana were influenced by the waxing and waning of glaciation during the LPIA. To assess patterns of biotic change, we examined post-glacial stratigraphy and fossil assemblages in the Paganzo (Huaco locality) and Río Blanco (Río del Peñon locality) basins of western Argentina (Fig. 1). The combined sedimentologic, palaeoecological and palaeoenvironmental emphases of this project will help in understanding short-term and broad-scale biotic change in response to climate/environmental changes concomitant with varying glaciation history.

Geological setting

Statigraphy/Palaeogeography

During the Late Palaeozoic, the southwestern margin of Gondwana (present-day Argentina) underwent multiple sea level and tectonic changes, evident by the stratigraphy and sedimentology of the Mississippian–Pennsylvanian basins in which several transgressions and tectonic events are recorded. The eastern region of western Argentina (Paganzo Basin) is dominated by continental sedimentation, while the western region (Río Blanco Basin) is mostly marine sedimentation and volcanic facies. An orogenic belt, the Protoprecordillera, was an upland area that divided the two basins (Limarino *et al.* 2006). The Protoprecordillera formed as an accretionary prism during Devonian subduction of Chilenia beneath the Cuyania Terrane in western South America, which subsequently evolved into a doubly convergent foreland fold and thrust belt during collision and suture of the two terranes.

Glaciation has also been identified as a factor in the complexity of the sea level and tectonic changes. Waxing and waning of glaciers during the LPIA was driven by a number of dynamic mechanisms including Gondwana's drift over the South Pole, changing pCO_2 values in the atmosphere, Gondwana ocean and atmospheric circulation patterns and tectonism (Caputo & Crowell 1985; Powell & Veevers 1987; Eyles 1993; Montañez *et al.* 2007). Palaeomagnetic

Fig. 1. Location of study areas in western Argentina. Study Area (1) Río del Peñon locality (S 28°36.373′, W 068°43.013′), Río del Peñon Formation, Río Blanco Basin (La Rioja Province) and (2) Huaco locality, Guandacol (S 30°08.275′, W 068°32.609′) and Tupe (S 30°10.009′, W 068°34.523′) formations, Paganzo Basin (San Juan Province). Modified from Limarino *et al.* 2006.

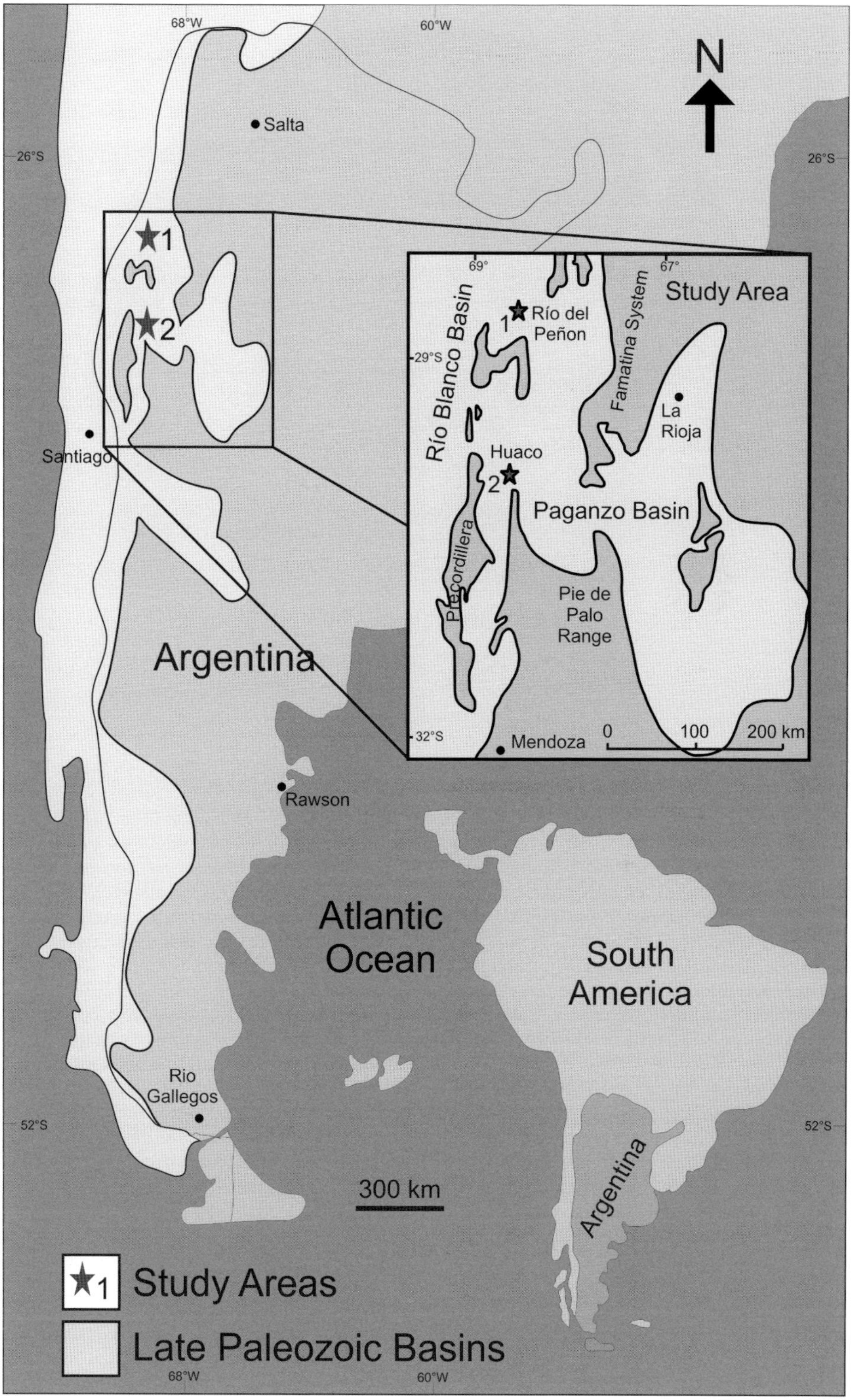
68°W
60°W
N
Salta
26°S
26°S
1
2
Santiago
69°
67°
Study Area
Río Blanco Basin
Famatina System
Río del Peñon
29°S
La Rioja
Huaco
Paganzo Basin
Precordillera
Pie de Palo Range
32°S
Mendoza
0 100 200 km
Argentina
Rawson
Atlantic Ocean
South America
Rio Gallegos
52°S
52°S
Argentina
300 km
Study Areas
Late Paleozoic Basins
68°W
60°W

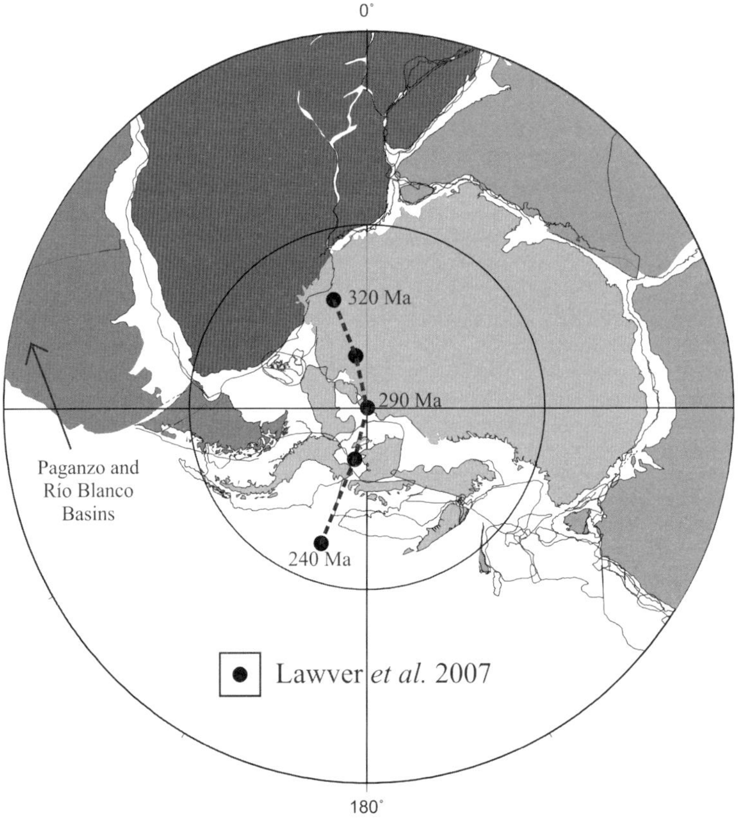

Fig. 2. Plate reconstruction with apparent polar wander path (modified from PLATES program at University of Texas, Institute for Geophysics; Lawver *et al.* 2007).

data suggest that during the Pennsylvanian (318–300 Ma), the pole was located in Antarctica (Fig. 2; Scotese & Barrett 1990; Lawver *et al.* 2007; Torsvik *et al.* 2008; Domeier *et al.* 2011).

During the Latest Devonian–Early Carboniferous, subsidence associated with the development of the Protoprecordillera mountain range (i.e. Chanic orogeny) produced the Proto-Paganzo foreland basin (Ramos *et al.* 1986). However, post-collisional extension during the Río Blanco tectonic phase (Late Carboniferous) caused the collapse of the Protoprecordillera and expansion of the basin into a much larger back-arc basin, the Paganzo Basin, during the Early–Middle Pennsylvanian (Limarino *et al.* 2006; Net & Limarino 2006; Isbell *et al.* 2012). This basin was located along the southwestern margin of Gondwana between 55 and 60°S during the Early–Middle Pennsylvanian.

The western margin of the Paganzo Basin contains the Paganzo Group, which overlies the tectonically formed Río Blanco unconformity (Fig. 3). The lower part of the Paganzo Group is divided into the Guandacol (Serpukhovian–Bashkirian) and Tupe (Late Bashkirian–Moscovian) formations (Limarino *et al.* 2002, 2006; Césari *et al.* 2011). Fine-grained marine deposits in these units correspond to three extensive post-glacial transgressions that flooded the region (Marenssi *et al.* 2005; Limarino & Spalletti 2006; Tedesco *et al.* 2010). The exact dates of these transgressions have been contested, but recent radiometric dating places the first within the Bashkirian and the second and third in the Moscovian (referred to as the Moscovian Transgression and Moscovian–Late Pennsylvanian Transgression in this study; Gulbranson *et al.* 2010). These represent marine incursions from the Panthalassan margin that breached the collapsing Protoprecordillera and flooded into the Paganzo Basin, creating a marine embayment for colonization of benthic invertebrates (López-Gamundí *et al.* 1994; López-Gamundí 1997).

The Río Blanco Basin, located to the NW of the present Precordillera, formed during the same orogenic-/basin-forming events that produced the Paganzo Basin during the Late Devonian and Mississippian. It was also located in Gondwana between

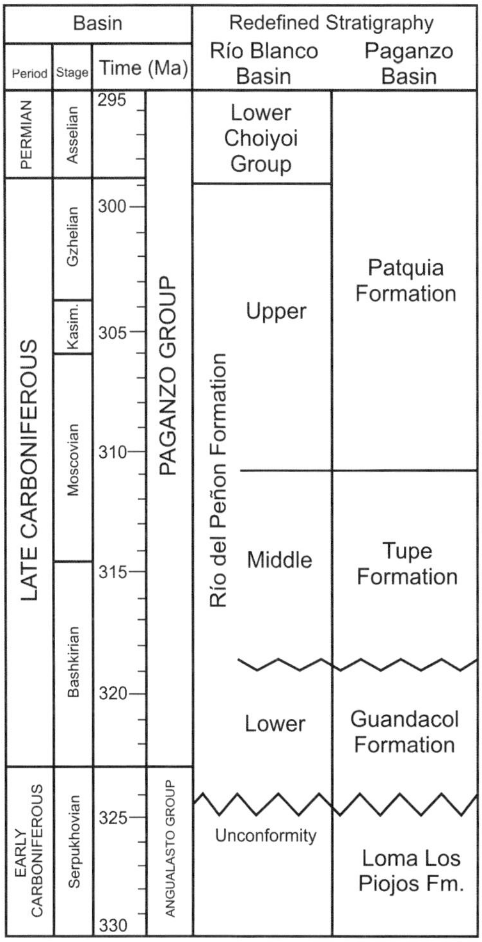

Fig. 3. Stratigraphy of the Río Blanco and Paganzo basins and their correlation. Modified from Gulbranson *et al.* 2010.

55 and 60°S during the Early–Middle Pennsylvanian and contains post-orogenic sediment related to the end of the Protoprecordillera Orogeny (Limarino *et al.* 2006). However, this back-arc basin was connected directly to the Panthalassan Ocean due to its location on the western side of the Protoprecordillera, and it is dominated by open-marine sedimentation in the Carboniferous followed by volcanic activity in the Permian. Diamictites found in both basins resulted from alpine glaciers that drained radially away from the Protoprecordillera, and advanced into the marine basin as a grounded tidewater ice front (López-Gamundí *et al.* 1994; Henry *et al.* 2008, 2010). A regional unconformity (Río Blanco tectonic phase) forms the base of a large palaeovalley in the Río Blanco Basin. The basal units of the Río del Peñon (Pennsylvanian) Formation form the initial fill of this palaeovalley, and consist of a glaciogenic succession deposited in a proximal glaciomarine setting; these strata correlate to strata in the Guandacol Formation (Paganzo Basin; Marenssi *et al.* 2005; Gulbranson *et al.* 2008, 2010; Césari *et al.* 2011). The middle and upper parts of the Río del Peñon Formation are separated from the lower glacial facies by an unconformity, which correlates to the unconformity between the Guandacol and Tupe formations in the Paganzo Basin (Fig. 3; Limarino *et al.* 2006; Henry *et al.* 2008).

Biostratigraphy

The Early Pennsylvanian of western Argentina is marked by the *Levipustula* faunal assemblage, a biozone associated with glacial and early post-glacial successions (González 1999; López-Gamundí & Martinez 2000; Azcuy *et al.* 2007; Cisterna & Sterren 2010). This fauna consists of a poor diversity assemblage of brachiopods, bivalves, bryozoans and gastropods that had a wide geographic distribution within marine Gondwana (Cisterna & Sterren 2008). However, the *Levipustula* faunal assemblage has not been identified within the Guandacol Formation, despite the strata being considered correlatable with other Early Pennsylvanian successions (e.g. Hoyada Verde Formation, Calingasta-Uspallata Basin) (Taboada 2010). Thus, local environmental conditions may be important in explaining the absence of the *Levipustula* faunal assemblage in the Guandacol Formation.

During the Late Pennsylvanian, the *Levipustula* faunal assemblage of western Argentina was succeeded by the *Tivertonia-Streptorhynchus* Biozone; the stratotype section is the Tupe Formation (Paganzo Basin) at La Herradura Creek, about 20 km to the NW of Huaco (Sabattini *et al.* 1990). The *Tivertonia-Streptohynchus* faunal assemblage is generally related to more temperate conditions in shallower waters (i.e. interglacial) (Archangelsky *et al.* 1987). Like the *Levipustula* fauna, this Late Pennsylvanian assemblage contains some cool-water genera; however, it is generally considered to be a warm or temperate water assemblage (Taboada 2010). Several studies have identified members of the *Tivertonia-Streptohynchus* faunal assemblage in the marine transgressions of the Tupe Formation, including brachiopods (*Septosyringothyris* sp., *Tivertonia* sp., *Kochiproductus* sp., *Trigonotreta* sp., *Streptorhynchus* sp.), bivalves (*Phestia* sp.), gastropods (*Peruvispira* sp.) and crinoids (Leanza 1945; Sabattini *et al.* 1990; Archbold & Simanauskas 2001; Cisterna *et al.* 2002*a*, *b*, 2006; Sterren 2004).

The middle member of the Río del Peñon Formation (Río Blanco Basin) contains several excellent fossiliferous horizons, and is considered a para-stratotype for the *Tivertonia-Streptorhynchus* Biozone (Late Pennsylvanian–Early Permian). Numerous studies have identified invertebrates in the Río del Peñon Formation including bivalves, brachiopods, gastropods, crinoids and ostracods (Lech & Aceñolaza 1987; González 1997*b*; Cisterna & Sabattini 1998; Cisterna & Simanauskas 2000; Sterren 2000). The *Tivertonia-Streptorhynchus* faunal assemblage lived on silty/sandy sea floors, probably close to shore. Bivalves such as *Schizodus* sp., *Edmondia* sp., *Modiolus* sp. and *Wilkingia* sp. appear to be the most abundant organisms, comprising roughly 60% of the total number of individuals (González 1992; Cisterna & Sterren 2007). Brachiopods, however, are the most prominent elements in the Río del Peñon Formation, particularly *Tivertonia jachalensis* (Amos), *Costatumulus* sp., *Septosyringothyris jaguelensis* (Lech) and *Trigonotreta riojanensis* (Lech and Aceñolaza). Although prominent, the fauna of *Tivertonia-Streptorhynchus* is poorly diversified and has unclear geographic affinities. These characteristics are thought to have been climatically controlled (González 1993), but these faunas require further study in order to determine the extent to which environmental conditions affected their ecological affinities.

Methods

Stratigraphic sections were measured in the field; lithology, contacts, bedding, sedimentary structures, deformation structures and sequence stratigraphy including significant bounding surfaces were identified, analysed and described. These data were used to create stratigraphic columns of the post-glacial transgressions in the Guandacol, Tupe, and Río del Peñon formations. These columns were then used to: (1) compare environments of deposition for the various transgressions within the two basins; (2) identify the changes that the various successions experienced in the transgressions; and (3) determine the variety of physical factors that influenced the palaeocommunities.

In the field, benthic marine invertebrates within shell beds were identified from whole and fragmented fossils (where identifiable) and their relative abundance and taphonomy were determined. Raw counts of bi-valved individuals were divided by two in order to achieve a more accurate number of individuals (two valves per individual). An estimation of the number of crinoid individuals was achieved by counting the number of individual ossicles, as ossicle abundance can be considered a rough measure of relative abundance (Meyer *et al.* 2002); a small crinoid is made up of 1500 ossicles (Schubert *et al.* 1992). Bryozoan abundance was measured using the Summed Combination Method (Hageman *et al.* 1997). With this method, each bryozoan's growth form was assigned an index value based on its abundance (e.g. 0 = absent; 1 = rare; 2 = common; and 3 = abundant), and the combined value of these indices represents the relative abundance for specific growth forms.

Statistical analyses were conducted to measure changes in generic diversity and abundance through the stratigraphic successions, including relative abundance, alpha diversity (S) (also sometimes called species richness), the Simpson Index of Diversity (D), the Shannon Index (H′), evenness (e), dominance (d), mean rank order and breadth of distribution. The Simpson Index of Diversity (D) takes into account the total number of species present and the abundance of each (Simpson 1949). The Shannon Index (H′) is a measure of the difficulty in predicting the species of the next individual collected. The lowest possible value is $H' = 0$, while the maximum value can be calculated as $H'_{max} = \ln S$ (Shannon 1948). Meanwhile evenness (e) is a measure of how similar the abundances of groups of species are, while dominance (d) is the conceptual inverse of this value (Hammer & Harper 2006). Both of these values can be used to determine if species and genera are evenly distributed within the shell bed in terms of abundance.

Mean rank order and breadth of distribution were used in order to determine the most ecologically abundant group in the shell bed. To determine rank order for a sample, the most abundant group receives a rank order of 1, the second most abundant group receives a 2 and so on. To determine the mean rank order, the sum of the rank orders is divided by the number of samples in which the group is present. Breadth of distribution measures the percentage of samples in which the organism was present. Additionally, trophic analysis was employed using information from the Paleobiology Database (www.paleodb.org). Analysing the trophic structure of palaeocommunities provides additional ecological information about the relative abundance of each guild as well as indicating environmental conditions.

Furthermore, similarity indices and multivariate analysis (detrended correspondence analysis or DCA and non-metric multidimensional scaling or NMS) were calculated for the Río del Peñon Formation data using the PAST software package (Hammer *et al.* 2001). DCA is an ordination method that attempts to place similar samples in similar positions on a plot via reciprocal averaging, while correcting for the arching (i.e. horseshoe effect) that typically results in correspondence analysis (Hill & Gauch 1980; Holland 2005). Despite some criticism (e.g. Wartenberg *et al.* 1987; Gotelli & Ellison 2004), multiple studies have demonstrated the ability of DCA to recover ecological gradients while overcoming problems associated with correspondence analysis (Shi 1993; Clapham & James 2008). Meanwhile, non-metric multidimensional scaling (based on Euclidian and Bray-Curtis similarity) is a technique that attempts to place the samples in a two- or three-dimensional coordinate system such that the ranked differences are preserved, and is based on a matrix of similarity coefficients for each pair of samples. The results from these techniques were then compared to ecological and sedimentological data in order to quantitatively determine the palaeoecological and palaeoenvironmental nature of community change in post-glacial transgressions during the LPIA in this area.

Paganzo Basin results

Measured sections in the Paganzo Basin (Huaco locality, San Juan Province, Argentina) include the Guandacol (Serpukhovian–Bashkirian) and Tupe (Late Bashkirian–Moscovian) formations (Fig. 1). In this paper, the measured section in the Tupe Formation is split into two transgressive units referred to here as the Moscovian and Moscovian–Late Pennsylvanian Transgressions.

Guandacol Formation

Sedimentological description. The measured Guandacol Formation section (total of 37.5 m; Huaco locality) occurs at the base of the Upper Palaeozoic succession, which is stratigraphically overlain by the Tupe Formation. Guandacol strata consist of massive and slumped diamictite and laminated siltstone/sandstone at the base of the measured section (Fig. 4). Overlying strata consist of 5.5 m of interbedded siltstone and sandstone that contains: deformational structures, including small-scale folds and slump features; lonestones which penetrate stratification (Fig. 5a, b); and rare, poorly preserved plant fossils. This unit is overlain by a 10.5-m- thick interbedded concretion-bearing (1–3 cm in diameter) interval of laminated siltstones and thin inversely graded sandstones. Lonestones are absent from this unit. These units are followed by a 14-m-thick succession of interstratified black carbonaceous mudrock and multiple (1–6 cm) graded sandstone beds. The mudrock contains abundant marcasite 'dollars' and the sandstone thickens and becomes more abundant upwards (Fig. 5c). Individual sandstone beds display sharp lower and sharp to gradational upper contacts and are laminated with small-scale cross-laminations occurring at the top of some beds. The carbonaceous mudrock abruptly becomes massive siltstone in the upper 7.5 m of the unit. This unit is overlain by a fine- to medium-grained massive sandstone body containing abundant dewatering pipes and dish structures. The base of this sandstone rests on a surface loaded into the underlying siltstone/sandstone unit (Fig. 5d). The massive sandstone represents the top of the measured section.

Sedimentological interpretation. This section of the Guandacol Formation is interpreted to represent an ice-proximal to very ice-distal glaciomarine succession that represents retreat of a temperate tidewater glacier up and out of a palaeofjord (Fig. 4). The coarse-grained massive diamictite at the base of the section most likely represents ice-proximal deposition and resedimentation on sediment along a morainal bank (cf. Cai *et al.* 1997). Such diamictite units typically represent a combination of depositional processes associated with settling from suspension from meltwater plumes, rainout of ice-rafted material including dropstones from icebergs and sea ice, and resedimentation due to mass movement and sediment gravity flows (Fig. 6). The deformation at the top of the diamictite may have formed due to: post-depositional mass movement on the morainal bank due to glacial shove during advance of the ice front; collapse of the bank as the glacier pulled away from the bank as it retreated up the fjord; or resedimentation of water-saturated sediments via slides, slumps or sediment gravity flows (e.g. Powell 1981; Powell & Domack 1995; Cai *et al.* 1997).

Lonestones contained within siltstones and shales that disrupt stratification were likely deposited as dropstones associated with iceberg and/or sea-ice-rafted debris (Fig. 5a, b). As the glacier retreated, deposition of dropstone-rich silts and muds occurred. These sediments represent deposition within the iceberg zone, indicating derivation of the sediments from a nearby glacier. Icebergs originating from the glacial termini can provide a significant volume of glaciomarine sediment from clay to boulder size (Powell & Domack 1995). The study area became more and more sediment starved as the glacier continued to retreat; finer-grained particles are therefore interpreted to have settled from suspension, possibly from a meltwater plume (Fig. 6). As the ice margin retreated, icebergs continued to deposit dropstones into the fine silts of the basin. The laminations contained in the siltstones suggest deposition from suspension due to possibly fluctuating marine or seasonal energy regimes (e.g. tides or seasonal changes in discharge; Cowan & Powell 1990; Cowan *et al.* 1997). However, they most likely represent meltwater plume deposition from the ice margin; as freshwater entered the basin, it rose to the surface as buoyant laden sediment (cf. Boulton 1990; R. D. Powell 2005).

Approximately 5 m from the base of the section, dropstones no longer occur within the siltstones; this may indicate that icebergs became grounded on a

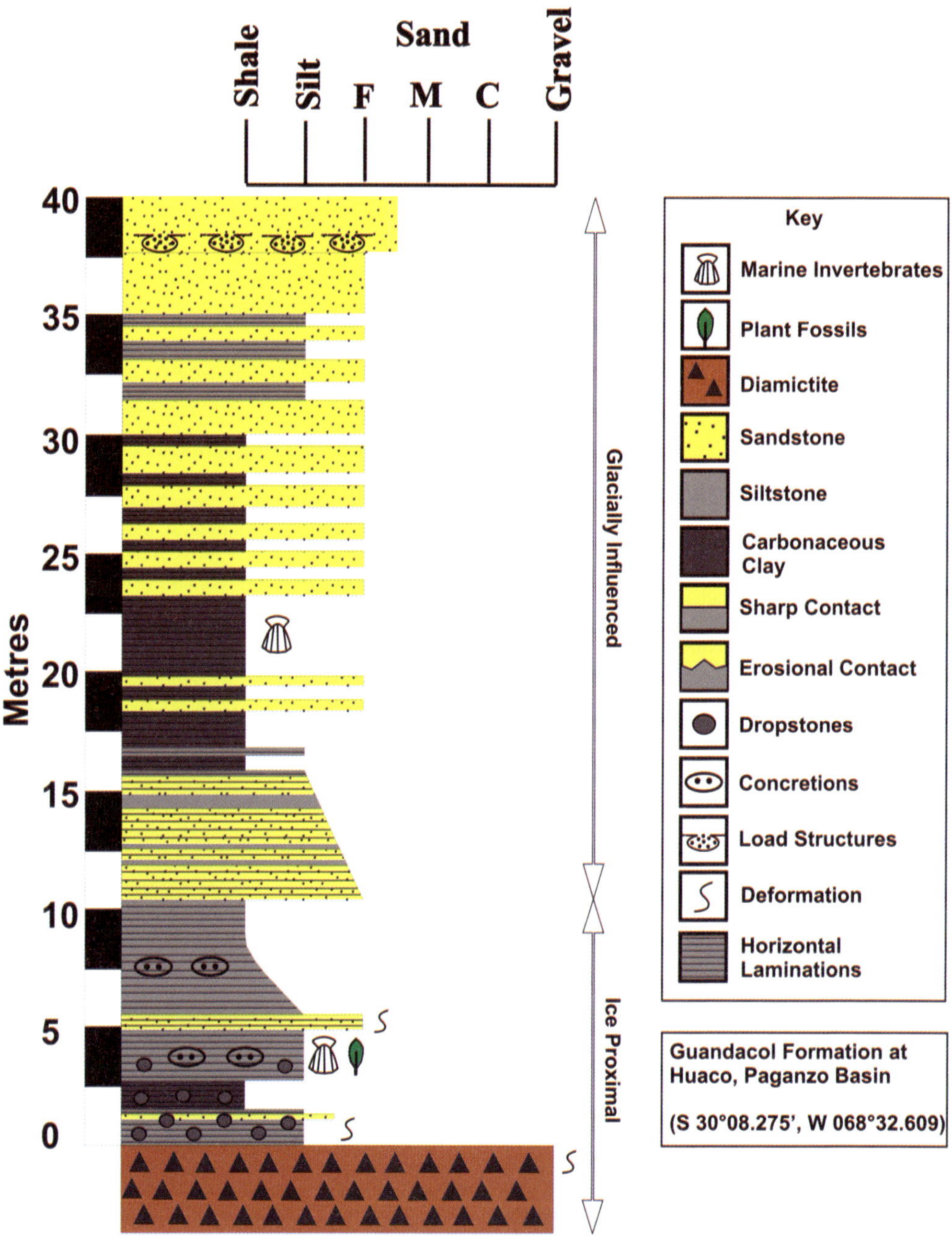

Fig. 4. Sedimentary facies and glacial sequence stratigraphy of the Guandacol Formation (0–37.5 m) from Cuesta de Huaco, Paganzo Basin (San Juan Province, Argentina).

shallow morainal bank closer to the ice front or that ice had retreated out of the sea and onto land. Concretions were precipitated post-compaction of the unit as evidenced by the lack of elongation of the concretions within these units. The laminated grading-upwards, medium- to fine-grained sandstone unit that overlies the siltstone is interpreted to be the result of underflow or turbidity currents. Hyperpycnal turbidity flows tend to include an inversely graded unit capped by a normally graded

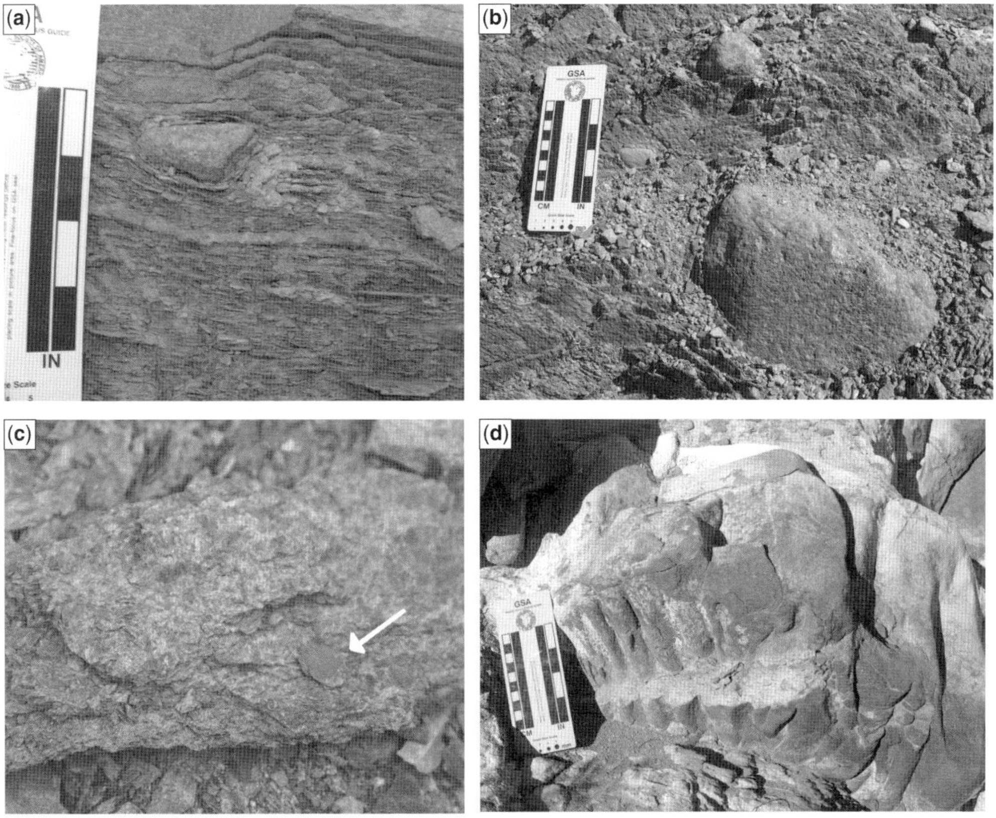

Fig. 5. (**a**) Laminated siltstone with lonestone (3 cm in diameter), which penetrates/pierces stratification (Guandacol Formation – 1 m above the base of the section). Siltstone also contains thin fine-grained sandstone layers (0.5–4 cm). (**b**) Thinly laminated black shale contained scattered lonestones (Guandacol Formation – 2 m above the base of the section). (**c**) Thinly laminated (less than 0.1 cm) black calcareous shale. Arrow points to marcasite 'dollar' (1.5 cm), abundant within the unit, 22 m above the base of the section in the Guandacol Formation. (**d**) Load structures at the base of fine- to medium-grained massive sandstone unit located 37.5 m above the base of the section in the Guandacol Formation.

unit (e.g. Mulder & Alexander 2001). However, the lower inversely graded unit may be missing due to erosion or bypass, which was probably the case with the deposit identified.

As the glacier retreated further, a silled basin formed between two morainal banks distal to the ice margin is proposed. This would help to explain the decreasing abundance of course-clastics in the area, evidence for limited circulation and isolation from glacial activity (Fig. 6; lack of diamictites and dropstones). This is evident in the deposition of carbonaceous black shales, which are punctuated by the occasional thin massive sandstone layer. These marine shales are representative of suspension settling and were most likely deposited over long periods of time. Black shales have previously been interpreted as representing anoxic bottom-water conditions as well as post-glacial facies, usually representing deposition in enclosed, strongly stratified basins that occur during transgressive episodes (Zhangerl & Richardson 1963; Arthur & Sageman 1994; Wignall 1994). The thin sandstones in the shales are interpreted to be the result of either hyperconcentrated density flows or distal turbidity currents (e.g. Mulder & Alexander 2001; Gani 2004). These types of deposits are commonly represented in the rock record by massive beds with slight grading near the top of the units, like those seen in this section of the Guandacol Formation.

These dense flows may have originated from several possible sources: (1) tributaries on the side of the fjord; (2) resedimentation of unstable sediment located on a morainal bank; or (3) the glacial margin (cf. Boulton 1990). This illustrates the possibility that fluvial systems (i.e. streams or rivers) along the side of fjords could be a source of

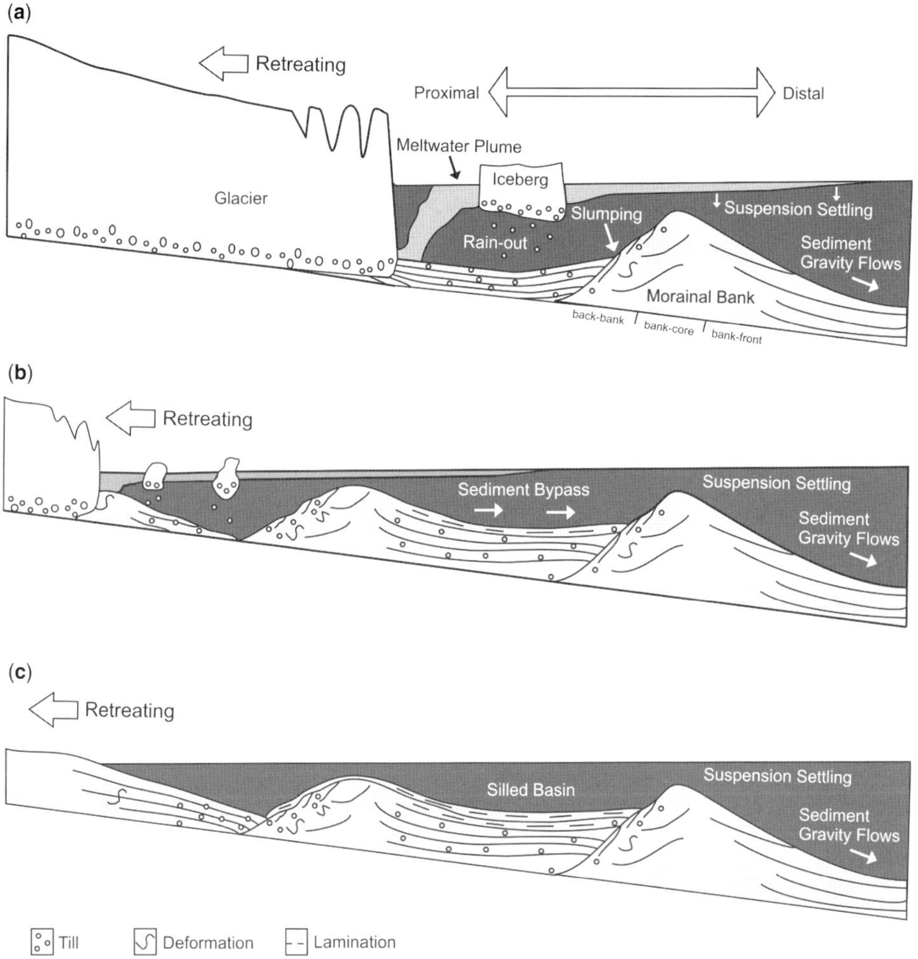

Fig. 6. Hypothetical glaciomarine schematic for the Guandacol Formation, Paganzo Basin. Models (**a**)–(**c**) represent the retreat sequence in time-steps. Modified from R. D. Powell 2005.

sediment input within glacial systems (Bennet & Glasser 1996). As these flows brought in the occasional thin layer of sand, they also likely brought in oxygen and hence created a low-oxygen (dysoxic) environment interspersed with anoxic conditions in this depositional setting. The mineral marcasite, which was found within the black shales, has recently been shown to indicate intermittent bottom-water oxygenation, hence suggesting that the black shales with marcasite were not deposited under conditions of long-term bottom-water anoxia (Schieber 2008). As sediment flows and fine-particle sedimentation slowly filled up the isolated silled basin, a shallowing-upwards succession resulted. At this time the sedimentation of silts and muds would have been exceedingly slow; it has been estimated that some deeper basins rarely exceed a net fine-grained sedimentation rate of 10 cm/1000 a (Boulton 1990). Hyperconcentrated flows become thicker, coarser and more frequent up-section, possibly representing the beginning of a distal prograding delta.

Overall, this section of the Guandacol Formation displays a large-scale coarsening-upwards succession that suggests that the ice front margin was retreating up the fjord and moving away from the study area. This succession is therefore interpreted to represent ice-proximal to ice-distal glaciomarine retreat. A glaciomarine interpretation is supported by facies and structures such as massive diamictite, deformational folding and slumping, density flows and dropstones. The ice-proximal portion includes facies deposited from resedimented portions of the morainal bank (i.e. massive diamictite), meltwater plumes (i.e. laminated sandstone and siltstone), rainout debris (i.e. dropstones penetrating stratification) and hyperpycnal turbidity flows (i.e. inversely graded sandstone). The ice-distal section comprised suspension settling (i.e. black carbonaceous shale) and multiple hyperconcentrated flow (i.e. massive sandstone bodies) deposits.

Palaeoecological description. Two species of inarticulate brachiopods were identified within the black carbonaceous shale unit, *Oehlertella* sp. and *Orbiculoidea* sp., which were very abundant and small in size (3–4 mm). Additionally, rare fragments of unidentifiable inarticulate brachiopods and gastropods (2–3 mm) were found lower in the section (Fig. 4).

Palaeoecological interpretation. As discussed in the previous section, the Guandacol Formation is interpreted to represent glaciomarine retreat facies. In the lower Guandacol Formation, a depauperate assemblage consisting of rare unidentifiable inarticulate brachiopod and gastropod fossils (2–4 mm) were found contained within a dropstone-shale unit. It is probable that the depauperate assemblage was subject to high rates of sedimentation from the glacial margin as well as brackish water conditions, both of which resulted in a highly stressful and inhospitable environment for local stable community growth. The extremely impoverished assemblage may represent opportunistic individuals trying to establish themselves in recently created marine floor space; as glaciers backed out of the fjord, space would have became available for colonization. These conditions would have been nearly inhospitable to colonizing fauna, as most brachiopods and gastropods need a firm substrate in order to settle on and feed from (Brenchley & Harper 1998).

As the glacier continued to retreat another morainal bank is interpreted to have been deposited close to the glacier margin, creating a sediment-starved silled basin. This resulted in the deposition of carbonaceous finely laminated black shale with interspersed massive siltstone layers higher up-section in the Guandacol Formation. It was in the black shale units that abundant small inarticulate brachiopods (3–4 mm) were found and identified as *Oehlertella* sp. and *Orbiculoidea* sp. This identification and occurrence supports previous work conducted in the Guandacol Formation (Martinez 1993; Pazos 2002*a*, *b*).

The low diversity of the faunal assemblage found in the upper Guandacol Formation is interpreted to be a result of a low-oxygen stressed ice-distal environment. The presence of eurytropic inarticulate brachiopods such as *Orbiculoidea* sp. has been previously interpreted to represent dysoxic conditions (Kammer *et al.* 1986). A low-oxygen environment also explains the extremely small size of the specimens (3–4 mm), as dysaerobic fauna growth tends to be stunted. In addition to low-oxygen conditions, persistent instability of bottom sediments has also been shown to reduce faunal diversity (Probert 1981). *Oehlertella* sp. and *Orbiculoidea* sp. are described as epifaunal suspension feeders, and Kammer *et al.* (1986) suggested that these types of brachiopods were tolerant to low oxygen, low salinity and muddy substrates. The broad, flat form and small body size of the inarticulates would have allowed for survival on the soft substrate surface (i.e. opportunists), where few other species would have been able to live. However, it is likely that the Guandacol local palaeocommunity existed well below normal wave base, as evident by the absence of wave and current structures at this point in the succession.

Overall, the fauna of the Guandacol Formation is interpreted to represent an opportunistic assemblage, consisting of a low-diversity brackish and dysoxic water biota. The ice-proximal nature of the fjord environment was highly stressful for fauna with high rates of sedimentation, freshwater input, unstable substrates and low-oxygen conditions. These circumstances were the result of a complex glaciomarine environment controlled by physical factors, in which only fauna able to exploit available space and resources were able to survive.

Tupe Formation (Moscovian Transgression)

Sedimentological and palaeoecological description. The Moscovian Transgression (MT; a total of 6.6 m) occurs near the base of the Tupe Formation and is in sharp contact with an underlying medium-grained sandstone (Fluvial 1) and coal succession. Within the measured section, the basal sandstone contains cross-bedding and downstream accretionary macroforms, while the coal is thin and lenticular (Fig. 7). A sharp contact separates the basal sandstone from deposits of the MT. The MT consists of 6.6 m of interbedded siltstone and fine-grained sandstone. In this unit, the sandstone contains ripple cross-laminations, silt-draped cross-laminations, climbing ripple laminations, horizontal laminations and micro-hummocky cross-stratification (Fig. 8a, b). No invertebrate fossils were found within deposits of the MT; however, rare plant impressions occur. At the top of measured section, thick concretion-bearing medium-grained cross-bedded sandstone (Fluvial 2) truncates deposits of the MT (Fig. 8c).

Sedimentological interpretation. The MT (6.6 m) of the Tupe Formation is interpreted to represent an estuarine environment with influences from both tide and wave activity (Fig. 7). The medium-grained cross-bedded sandstone at the base of the section (here labelled Fluvial 1 deposits) has been

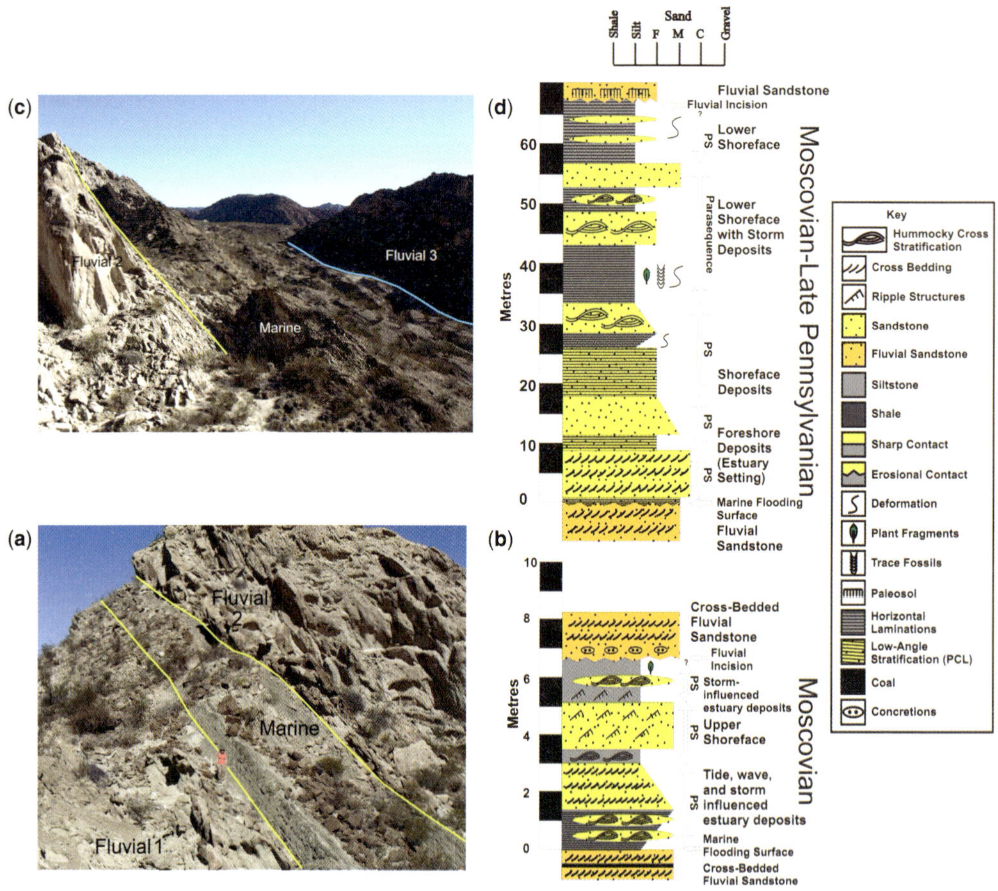

Fig. 7. Huaco locality, Paganzo Basin, San Juan Province, Argentina. (**a**) Moscovian marine transgression (6.6 m total) of the Tupe Formation, which occurs between units of fluvial sedimentation. The marine facies rests conformably on Fluvial 1; however, they are incised into by Fluvial 2. (**b**) Sedimentary facies and sequence stratigraphy of the Moscovian Transgression in the Tupe Formation (0–6.6 m). Fluvial sandstone below section not to scale. (**c**) The Moscovian–Late Pennsylvanian Transgression (labelled Marine) in the Tupe Formation. The yellow line represents the marine flooding surface, which rests on Fluvial 2 deposits. The blue line represents the Fluvial 3 incision surface, as well as the end of the marine transgression. (**d**) Sedimentary facies and sequence stratigraphy of the Moscovian–Late Pennsylvanian Transgression in the Tupe Formation (0–67.2 m).

previously interpreted to represent an anastomosed network of sandy fluvial channels (Tedesco *et al.* 2010). Downstream accretion and associated cross-bedding within the unit indicates the occurrence of prograding barforms in a fluvial environment, while the lenticular coal deposit suggests wet and temperate conditions on a floodplain. Such conditions likely occurred as the initial deposits within either an incised valley or as part of a prograding fluvial succession during an apparent sea level lowstand (lowstand systems tract).

At the base of the measured unit (0 m), laminated siltstone with sandstone bodies containing silt-draped cross-laminations, climbing ripples and micro-hummocky cross-stratification suggests deposition under a marine influence (i.e. reworking by wave, tide and storm activity). This siltstone unit is therefore interpreted to represent initial marine flooding and the beginning of a transgressive system tract. The surface (maximum regressive) separating the Fluvial 1 deposits from the overlying marine deposits is conformable, indicating lack of significant marine erosion. The change from lowstand to transgressive facies can further be recognized by an abrupt change from coarse-grained sand to fine-grained sediment with abundant tidal structures.

After initial flooding, several sedimentary structures are representative of continued deposition above storm wave base in a tidal estuary. The silt drapes within the cross-laminated sandstone bodies

Fig. 8. (**a**) Fine- to medium-grained sandstone body with trough cross-laminations and low-angle climbing ripples within siltstone unit. Cross-laminations and ripples within sandstone bodies are draped in silt, 1.3 m, Moscovian Transgression, Tupe Formation. (**b**) Fine- to medium-grained sandstone with climbing ripples below and down-climbing ripples above the low-angle truncation surface near the middle of the photograph, 5 m, Moscovian Transgression, Tupe Formation. Hammer is 22 cm. (**c**) Top of Moscovian marine transgression. Black arrow points to fluvial incision surface (from Fluvial 2), below which is rippled siltstone with micro-hummocky sandstone bodies, 6.6 m, Tupe Formation. Hammer is 22 cm long. (**d**) Locomotion and feeding traces (*Gyrochorte* isp., *Treptichnus* isp.(?) and *Diplichnites* isp.(?)) within horizontally laminated siltstone unit, 40 m, Moscovian–Late Pennsylvanian Transgression, Tupe Formation. (**e1**) Original photo. (**e2**) Vertically climbing ripples in siltstone facies, 26 m, Moscovian–Late Pennsylvanian Transgression, Tupe Formation.

contained within the siltstone unit suggest deposition from fluctuating energy conditions (drop in energy results in settling from suspension), which could signal deposition due to tidal activity or quiet water intervals between storm events. Additionally, climbing ripples within the succession indicate rapid deposition. As sediment-laden water exits a distributary mouth, the expanding unconfined outflow slows allowing ripples to migrate up the backs of slower-moving ripples. Very-fine-grained rippled sandstone that contains fine- to medium-grained micro-hummocky cross-stratified sandstone bodies also suggests deposition in slightly deeper water, possibly by storm activity.

Plant fragments found within the rippled fine-grained sandstone at the top of the unit are interpreted to have been deposited from vegetation derived further inland up the estuary and delivered to the site of deposition via fluvial-deltaic and/or storm activity. This sandstone is unconformably overlain by thick medium-grained cross-bedded sandstones. This surface, which separates hummocky sandstone below from thick cross-bedded sandstone above, is interpreted to represent fluvial incision into the underlying marine sediments (i.e. very-fine-grained sandstone) and the formation of an incised valley. The overlying fluvial sediments, here termed Fluvial 2 deposits, were previously interpreted to have formed on a wide braided alluvial plain dominated by transversal bars (Tedesco *et al.* 2010).

It is interpreted that this section of the Tupe Formation at the Huaco locality represents marine deposition within an unconformity-bounded packet of sediment, a depositional sequence. The marine deposits which represent the Moscovian Transgression in this area, and occurred between fluvial lowstand units that are filling incised valleys (Fluvial 1 and 2), possibly represent deposits of a transgressive or highstand system track (Fig. 6). A marine interpretation for these deposits at this location is supported by silt-draped cross-laminations, climbing ripples and swaley and hummocky cross-laminations, all of which suggest tidal and storm activity in a shallow marine setting. As a result, the Moscovian Transgression is interpreted as representing an estuarine environment which contains storm-, tide- and wave-reworked facies. This transgressive/highstand facies therefore represents a marine flooding of palaeovalleys in the Paganzo Basin from the west, and interrupts coastal fluvial sedimentation taking place at that time in the Huaco area (Tedesco *et al.* 2010). However, it is clear that the Huaco area was a shallow marine environment located near the head of the estuary. In such a position, sea level did not progress much farther eastwards into the basin than this point during this transgressive event.

Palaeoecological interpretation. While some plant fossil fragments were found near the top of the section, no marine invertebrate fossils or traces were discovered within this sequence at the Cuesta de Huaco site. This supports recent research from the Tupe Formation in the area (Tedesco *et al.* 2010) although other locations, such as the La Herradura Creek locality (*c.* 20 km to the NW of Huaco), have yielded the *Tivertonia-Streptorhynchus* faunal assemblage (Leanza 1945; Archbold & Simanauskas 2001; Cisterna *et al.* 2002*a*, *b*). The *Tivertonia-Streptorhynchus* fauna at La Herradura Creek includes brachiopods, gastropods and crinoids and, while the strata at this locality have not been extensively examined in terms of environmental interpretations, it is generally interpreted to have been part of a deltaic system (Ottone & Azcuy 1986). It is hypothesized here that La Herradura Creek, located 20 km to the NW, would likely have been a slightly deeper, more open marine environment than the Cuesta de Huaco locality that was proximal to estuary and shoreline environments. Environments at La Herradura Creek would have been characterized by more stable environmental conditions for organisms, and therefore may explain the occurrence of the *Tivertonia-Streptorhynchus* faunal assemblage there (compared to the lack of fossils that are found at the estuarine Cuesta de Huaco locality).

Estuarine (peritidal) environments are often subject to extreme changes in physical conditions such as temperature, salinity, sedimentation and hydrodynamic energy (currents). Subtidal environments in fjords, such as those interpreted for the Paganzo Basin at Huaco, may be under even more environmental stresses; high-latitude fjords can also have glacial meltwater and sea ice (Weslawski *et al.* 1995; Buatois *et al.* 2006). There is also the possibility of occasional periods of anoxia, which has previously been documented in the Paganzo Basin (Pazos 2002*a*). The estuary interpretation for the Cuesta de Huaco locality implies that it was proximal to freshwater input from fluvial systems, and would therefore mix with the marine saltwater of the fjord resulting in brackish conditions. Additionally, the estuary would have been subject to hydrodynamic energy such as turbidity and hyperpycnal flows, and presented a highly stressful environmental limitation to most organisms in the Paganzo Basin (MacEachern *et al.* 1999, 2005; Maughan 2001; Chou *et al.* 2004; Carmona *et al.* 2009; Lokier *et al.* 2009).

Overall, fluctuating environmental parameters that are common in estuaries, such as salinity, hydrodynamic energy and sedimentation, are interpreted to have inhibited faunal colonization at the Huaco locality during the Moscovian Transgression. Because most of the fauna could not overcome the physiological stresses represented in the Tupe Formation at Cuesta de Huaco, it is hypothesized that instead most invertebrates found in the Paganzo Basin developed communities in deeper, more offshore environments or near the mouth of the estuary/fjord (such as La Herradura Creek). A large volume of sediment is stored in estuaries, so much so that offshore sedimentation becomes markedly reduced and sediment starved during a relative rise in sea level (Brenchley & Harper 1998). Deeper, offshore areas (west of the Cuesta de Huaco locality, such as La Herradura Creek), may therefore have provided more ideal conditions (i.e. less sedimentation) for a suspension-feeding fauna at this time.

It is also a possibility that some organisms were eventually able to colonize the area but, due to fluvial incision into the Moscovian Transgressive facies, these may have been eroded away. Nevertheless, it is possible that during the first stages of sea level rise in the Cuesta de Huaco locality organisms were not able to exploit the addition of sea floor at this time. Physiologically stressful conditions are therefore interpreted to have suppressed faunal migration and colonization in the Tupe Formation during this marine incursion.

Tupe Formation (Moscovian–Late Pennsylvanian Transgression)

Sedimentological and palaeoecological description. The Moscovian–Late Pennsylvanian Transgression section (MLPT; a total of 67.2 m) measured at Huaco rests on top of the medium-grained cross-bedded sandstone (Fluvial 2) that is truncating the MT (Fig. 7). The measured section consists of siltstones and fine- to medium-grained sandstones containing varying amounts of hummocky cross-bedding, ripple cross-laminations, primary-current lineations (PCL), wave-ripple cross-laminations, climbing ripples, horizontal laminations and massive bedding (Fig. 8e).This unit is then capped by a red clay-rich crust with angular blocky structure and fine-grained red sandstone (Fluvial 3) that rests on an erosive surface cutting into the underlying siltstone. Similar to the measured section of the underlying strata, no fossils were found within the section; however, several trace fossils (*Gyrochorte isp.*, *Treptichnus isp.* and *Diplichnites isp.*) were identified, as well as rare plant fragments (Fig. 8d).

Sedimentological interpretation. The MLPT section of the Tupe Formation (67.2 m) is interpreted to represent a tidal or estuary-influenced marine environment that transitions into a more extensively flooded estuary (Fig. 7). The transgressive unit begins on top of the sandstone truncating the underlying MT (above 6.6 m), suggesting that it is an incised valley. This lowstand facies (Fluvial 2) was previously interpreted to have formed on a wide braided, alluvial plain (Tedesco *et al.* 2010) and represent initial filling of the incised palaeovalley during an interval characterized by the creation of low fluvial accommodation. Fluvial 2 deposits are unconformably overlain by several fine- to medium-grained sandstone units with low-angle laminations, cross-stratification and primary current lineations (PCL). This suggests deposition in a marine foreshore environment as base level began to rise more rapidly and flood the estuary, causing sedimentary environments to 'backstep' up the estuary. These beds are the result of the landward limit of tide and wave penetration, and are possibly due to high-velocity swash flow along the strand line in an estuary-influenced environment. The erosion surface at the base of the section is therefore interpreted to represent a marine flooding surface (transgressive ravinement surface) which scoured into the Fluvial 2 deposits. These foreshore deposits begin the next transgressive systems tract, and indicate the initial landwards shift of the shoreline during marine flooding.

The deposition of various massive, swaley and horizontally laminated fine- to medium-grained sandstones are interpreted to represent additional marine flooding, in addition to deposition in a deeper estuary setting (similar to a shoreface setting). Swaley cross-stratification suggests deposition by storms in water depths below or near fair-weather wave base, while massive bedding is possibly the result of hyperconcentrated flows initiated off of delta tops/fronts or off the walls of the still existent fjord (Mulder & Alexander 2001). Horizontally laminated sandstone deposition suggests continued deposition from the river mouth. As sediment exits the river mouth, it is most likely travelling in both traction and suspension. As freshwater enters the basin, it floats on the surface as a buoyant plume on top of the more dense seawater below. This will result in fine-grained sediment becoming entrained within this plume and depositing due to settling from suspension; material will also travel along the basin bottom in traction for short distance until the velocities decrease and buoyancy becomes the dominant force (Mulder & Alexander 2001; Strachan 2008). To deposit sand from suspension, these deposits would have had to have been located within a few hundred metres of the river mouth and close to the freshwater plumes detachment point. This is further supported by the deposition of steep-climbing ripples (including vertically climbing ripples), as well as down-climbing ripples in the succession (Fig. 8e). Steeply climbing ripples are the result of the continuous addition of sand to the bed via suspension and traction where current velocities are rapidly dropping. This occurs at the point where the flow goes from confined to unconfined and/or near the detachment point where the plume begins to float (Lucchi 1995). It is therefore likely that deposition took place at the head of the estuary, where a rapid influx of sediment occurred.

An interval of horizontally laminated siltstones containing plant fragments and trace fossils possibly indicates that the Huaco locality began to evolve into a deeper estuary environment (i.e. more extensively flooded). It is likely that, as the shoreline moved landwards, water depth at Huaco became deeper resulting in shoreface to offshore sedimentation. Plant fragments most likely originated from the backshore/foreshore environment, or were introduced into the estuary by rivers. Deposit feeding traces and burrows found throughout the siltstone are interpreted to be from the few organisms that were able to colonize the area, despite stressful physiological conditions (i.e. brackish water, limited water circulation and high sedimentation rates).

Near the top of the unit, deposition of the occasional fine- to medium-grained cross-laminated and hummocky cross-bedded sandstone possibly represents deposition in the lower shoreface. Sedimentary structures (i.e. wave ripple cross-laminations and hummocky cross-bedding) at this time are interpreted to be a result of periodic wave and storm activity, so it is likely that deposition was taking place near fair-weather wave base. These units are truncated by an erosion surface, which is overlain by a red clay-rich crust with angular blocky structure interpreted to be a palaeosol. This interpretation supports previous research (Gulbranson *et al.* 2010; Tedesco *et al.* 2010) which proposes that this surface represents a regional sub-aerial unconformity, suggesting extended exposure following a rapid drop in base level. Furthermore, the fine-grained red sandstone that overlies the palaeosol suggests fluvial deposition (Fluvial 3) due to the abrupt lithological change between units (i.e. siltstone to sandstone), as well as the unconformity that separates them. The fine-grained red sandstone is interpreted to represent initial fluvial palaeovalley fill after extended sub-aerial exposure and incised valley formation at the Huaco locality.

Overall, a marine interpretation for this section (MLPT) of the Tupe Formation at the Huaco locality is supported by the presence of several structures including hummocky and swaley cross-bedding, wave ripple cross-laminations and steeply-climbing ripples, in addition to hyperconcentrated density deposits and marine trace fossils. This marine incursion represents the second marine flooding of Huaco above the Guandacol Formation from the Panthalassan Ocean (located towards the present west), which breached the Protoprecordillera. Initial palaeovalley fill of the MLPT is thought to represent a tidal or estuary-influenced marine environment. However, as base level rose, the Huaco locality then developed into a deeper estuary environment, as indicated by the deposition of lower shoreface facies. Upper highstand facies were then removed due to sub-aerial exposure and subsequent fluvial incision, resulting in the formation of an incised valley and the deposition of Fluvial 3 deposits.

Palaeoecological interpretation. Similar to the Moscovian Transgression, no skeletonized invertebrate fossils were found within the marine facies of these transgressive deposits at the Huaco locality; however, some trace fossils and plant fragments were discovered. As previously mentioned, invertebrate fossils have been found in this stratigraphic interval by other authors (i.e. *Tivertonia-Streptorhynchus* faunal assemblage; Leanza 1945; Sterren 2004) at different localities (e.g. La Herradura Creek). Ottone & Azcuy (1986) interpreted the La Herradura Creek locality to have been a deltaic environment, although the stratigraphy has not been critically examined. A hypothesis therefore needs to be established in order to explain the lack of fossil invertebrates at this locality.

The MLPT in the Tupe Formation at this locality is interpreted to represent a tidal estuary environment. Organism distribution in estuaries is typically controlled by salinity gradients and sediment deposition rates, resulting in sporadic colonization of the substrate (Buatois *et al.* 2005). Bann *et al.* (2004) found that, in contrast to open marine settings, estuaries tend to support an impoverished ichnogenera with predominantly simple forms. Trace fossils identified within the middle of the marine succession (40 m) are interpreted to be mostly *Gyrochorte isp.*, a locomotion trace and some feeding and locomotion trails (*Treptichnus isp.* and *Diplichnites isp.*; Fig. 8d). These traces are typically found in brackish water estuaries (Buatois & Mangano 2011), and therefore support the environmental interpretation. The organisms making these traces were likely opportunistic, a common feature of brackish water and estuary biota, and thus able to take advantage of periodic colonization opportunities (e.g. between storm events). Additionally, Mangano & Buatois (2004) noted that in some restricted tide-influenced environments, such as incised valleys and embayments, marine body fossils may be notably absent such as the case here. It is therefore likely that stressful environmental parameters common in estuaries, such as high sedimentation, low salinity and high hydrodynamic energy, suppressed colonization of marine invertebrates. As previously discussed, deeper, more offshore environments closer to the mouth of the estuary/fjord (such as the La Herradurra Creek locality) may have presented less stressful conditions to colonizing invertebrates, and were therefore more ideal for palaeocommunity development.

Río Blanco Basin results

Measured sections in the Río Blanco Basin (Río del Peñon locality, La Rioja Province, Argentina) include two transgressive units in the middle Río del Peñon Formation (Moscovian; Fig. 1). These units can be correlated to the Moscovian and Moscovian–Late Pennsylvanian Transgressive units identified in the Tupe Formation at the Huaco locality (Paganzo Basin) (Limarino *et al.* 2002, 2006; Gulbranson *et al.* 2010; Fig. 3).

Río del Peñon Formation (Moscovian Transgression)

Sedimentological description. The Moscovian Transgression section (39.5 m) in the Río del Peñon Formation of the Río Blanco Basin begins as well-sorted very-coarse-grained cross-bedded sandstone containing quartz granules (Fig. 9). The majority of the measured section contains several coarsening-upwards successions consisting of alternating horizontally laminated siltstone and fine- to medium-grained sandstone. Structures identified included ripples, sigmoid-shaped cross-bedding, hummocky cross-bedding, load structures, trough cross-bedding and rare ball and pillow structures (Fig. 10a–c). Within the siltstone, units are several fine-grained sandstone layers containing invertebrate fossil material. The very top of the measured sections contains a stromatolite layer (5–8 cm thick) as well as a layer of shell debris (9 cm thick). The top unit is truncated by an erosional surface, which is then overlain by thick red-coloured, medium- to coarse-grained, cross-bedded sandstone.

Sedimentological interpretation. The Moscovian Transgression section of the middle Río del Peñon Formation is interpreted to represent wave- and storm-influenced coastal marine deposition (Fig. 9). The measured section (a total of 39.5 m) lies conformably on a unit of coarse-grained downstream-accreting trough cross-bedded sandstone, interpreted to represent coastal lowstand fluvial deposition (i.e. series of stacked bars) possibly within an incised valley. The overlying unit, fine- to medium-grained sandstone with wave ripple cross-laminations that grades into horizontally laminated siltstone, indicates marine influence (i.e. wave-reworked shoreface). This interpretation is further supported by an abrupt lithological change from coarse-grained sand to fine-grained sediment, suggesting reworking during transgression. It is therefore interpreted that the coarse-grained cross-bedded unit is representative of a fluvial incised valley deposit, conformably overlain by the start of transgressive systems tract facies. The flooding surface between the units indicates very little marine erosion. Initial flooding was followed by a rapid deepening of base level, resulting in fining of sediment up-section (i.e. rippled sandstone to horizontally laminated siltstone) and a transition into an offshore setting.

The deposition of various rippled and hummocky sandstone bodies contained within siltstone facies indicates a lower shoreface environment, with the hummocky sandstone beds representing storm deposits and the siltstone facies representing quiet periods between storm events. Ball and pillow structures, such as those found in the sandstone deposits, are load structures that usually sink into muddy water-saturated substrates (e.g. Reineck & Singh 1980; Lucchi 1995). These structures suggest rapid sedimentation of both the underlying and overlying unit, and are interpreted to be a result of storm deposition. A medium-grained sandstone body with cross-bedding and sigmoidal-shaped geometries suggests progradation of a subtidal bar form in a shallow shoreface environment.

This shallow-water environment is then overlain by alternating horizontally laminated siltstone and massive and rippled sandstone bodies, indicating that the environment is transitioning between upper and lower shoreface and deposition within storm wave base. The sandstone bodies coarsen upwards several times throughout this section, representing shallowing-upwards parasequences. Deposition of siltstone is interrupted by periodic storm events, including sandy storm beds with shell hash near the top of the section. Curiously, a discontinuous lithified stromatolite layer (9 cm thick) was identified within the siltstone succession, and is interpreted to have been washed-in via a storm event from a supratidal environment. The overlying unit, medium- to coarse-grained cross-bedded sandstone, is interpreted to be a fluvial deposit. This is apparent by the abrupt lithological change, as well as the erosion surface between the units. The erosional surface between the coarse sandstone and siltstone is indicative of fluvial incision, suggesting that some highstand (i.e. progradational) facies from this transgressive marine section have been eroded out due to a drop in base level, incision and formation of an incised valley.

It is therefore interpreted that this section of the Río del Peñon Formation represents marine deposition related to the Moscovian Transgression. The studied locality along the Río del Peñon is located on the eastern margin of the Río Blanco Basin (NW of the Precordillera). Additionally, the studied section of the middle Río del Peñon Formation is thought to be equivalent to the Tupe Formation of the Paganzo Basin located to the east. Recent radiometric U–Pb dating in the middle Río del Peñon places deposition of these facies in the Late Bashkirian–Early Moscovian (around 310 Ma), similar to the age of the Tupe Formation deposits (Gulbranson *et al.* 2010). It is therefore assumed that the transgressive facies identified in the Tupe Formation are the same transgressive facies that occur in the middle Río del Peñon Formation. However, it is important to

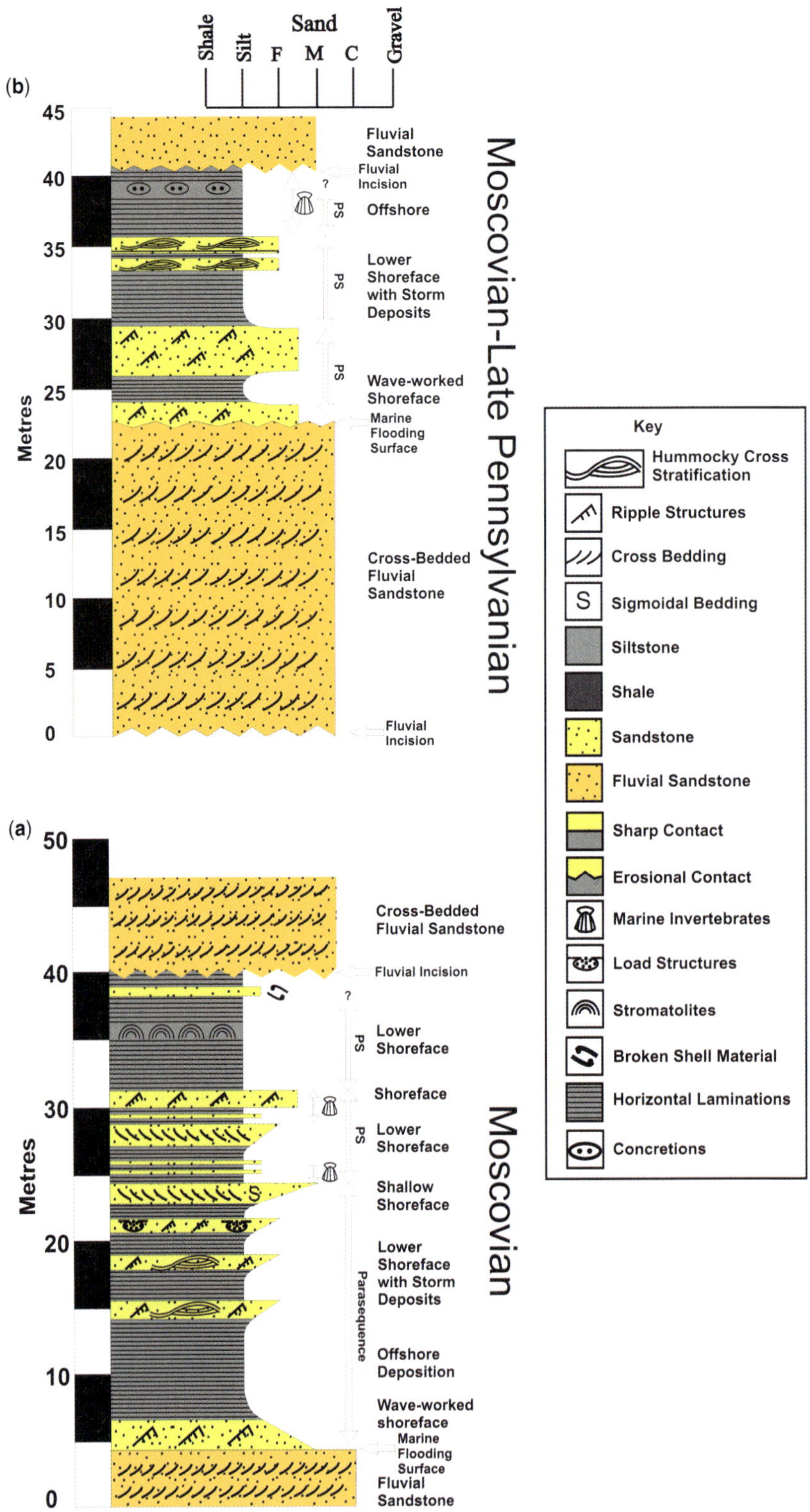
Shale
Silt
Sand
F M C
Gravel
(b)
Metres
45 40 35 30 25 20 15 10 5 0
Fluvial Sandstone
Fluvial Incision
Offshore
Lower Shoreface with Storm Deposits
Wave-worked Shoreface
Marine Flooding Surface
Cross-Bedded Fluvial Sandstone
Fluvial Incision
PS
Moscovian-Late Pennsylvanian
(a)
Metres
50 40 30 20 10 0
Cross-Bedded Fluvial Sandstone
Fluvial Incision
Lower Shoreface
Shoreface
Lower Shoreface
Shallow Shoreface
Lower Shoreface with Storm Deposits
Offshore Deposition
Wave-worked shoreface
Marine Flooding Surface
Fluvial Sandstone
Parasequence
Moscovian
Key
Hummocky Cross Stratification
Ripple Structures
Cross Bedding
Sigmoidal Bedding
Siltstone
Shale
Sandstone
Fluvial Sandstone
Sharp Contact
Erosional Contact
Marine Invertebrates
Load Structures
Stromatolites
Broken Shell Material
Horizontal Laminations
Concretions

Fig. 10. (**a**) Wave ripple structures in very-fine- to fine-grained sandstone, 15 m, Moscovian Transgression, Río del Peñon Formation. (**b**) Ball and pillow load structures within sandstone body, 21 m, Moscovian Transgression, Río del Peñon Formation. (**c**) Medium sand unit containing sigmoidal-shaped foresets, 23.5 m, Moscovian Transgression, Río del Peñon Formation. (**d**) Large (34 cm) ironstone concretion within horizontally laminated siltstone unit. Invertebrate fossils are visible on the outside of concretion, 38 m, Moscovian–Late Pennsylvanian Transgression, Río del Peñon Formation.

note that the transgressions were probably recorded first in the marine Río Blanco Basin, reaching the Paganzo Basin later as base-level rose and flooded the basin from the west.

Overall, the Moscovian Transgression in the Río del Peñon Formation is interpreted to have several coarsening-upwards parasequences representing shallowing-upwards coastal marine facies (i.e. shallow to offshore environment deposition) under the influence of wave and storm activity. These were deposited as a Transgressive Systems Tract (TST) as the rate of base-level rise increased. While the Río Blanco Basin is traditionally considered an open marine basin (Limarino *et al.* 2002, 2006), the occurrence of the measured section between fluvial deposits indicates that the Río del Peñon locality was situated closer to the palaeocoastline of the Río Blanco Basin than previously thought. This therefore suggests that this palaeovalley at the Río del Peñon locality is resting near the seaward limit of a much larger palaeovalley system (i.e. palaeovalley with several hundred metres of relief, and kilometres to possibly tens of kilometres width).

Palaeoecological description. A total of nine shell beds within the Moscovian Transgression were identified and analysed; a grouping of six are termed here #1–6 with #1 being lowest stratigraphically at *c.* 25 m from the bottom of the section, and another grouping of three termed here #7–9 with #9

Fig. 9. Río del Peñon, Río Blanco Basin (La Rioja Province, Argentina) (**a**) Sedimentary facies and sequence stratigraphy of the Moscovian Transgression in the Río del Peñon Formation (0–39.5 m). (**b**) Sedimentary facies and sequence stratigraphy of the Moscovian–Late Pennsylvanian Transgression in the Río del Peñon Formation (0–40.7 m).

Table 1. *Dominance and evenness of benthic marine invertebrates in the Rio del Peñon Formation. Shell beds #1-9 are within the MT while beds #10-12 are within the MLPT.*

	Shell bed	Alpha diversity	Simpson index of diversity (D)	Shannon index (H′)	Evenness (e)	Dominance (d)
MLPT	12	7	0.80	1.66	0.85	0.15
	11	7	0.82	1.69	0.87	0.13
	10	14	0.85	2.12	0.80	0.20
	9	12	0.58	1.46	0.59	0.41
	8	16	0.82	2.09	0.76	0.25
	7	10	0.68	1.53	0.66	0.34
MT	6	5	0.73	1.34	0.84	0.17
	5	4	0.77	1.29	0.93	0.07
	4	5	0.56	0.98	0.61	0.39
	3	3	0.52	0.77	0.70	0.30
	2	3	0.53	0.69	0.63	0.37
	1	2	0.47	0.65	0.94	0.06

being highest stratigraphically at *c.* 30 m from the bottom of the section (Fig. 9). The two groupings were separated by *c.* 3 m of sandstone and siltstone absent of fossils. We identified 420 individual invertebrate fossils belonging to 5 classes and 28 species within the 9 shell beds in the measured section. Classes identified within the Moscovian Transgression at the Río del Peñon locality include rhynchonelliform brachiopods, bivalves, gastropods, crinoids and ostracods. Most of the species identified belong to the *Tivertonia-Streptorhynchus* Biozone which, as previously discussed, is commonly identified in the Río del Peñon Formation.

Within shell beds #1–6, alpha diversity varied from 2 to 5 species (3 classes), whereas alpha diversity varied from 10 to 16 species (5 classes) for shell beds #7–9 (Table 1). The Simpson Index of Diversity (D) indicates that shell beds #1–6 showed a gradual increase in diversity throughout, while shell beds #7–9 showed some variation (Table 1). Shannon Index (H′) values show that species became more evenly distributed through time (Table 1). This is supported by evenness (e) and dominance (d) statistics, which also indicate high values (Table 1). Additionally, when all shell beds are taken into consideration (#1–9), bivalves and gastropods have the highest mean rank order with 1.8 and 1.9, respectively (Table 2). Breadth of distribution statistics indicate that only bivalves occur in 100% of all shell beds.

The overall assemblage consisted of a majority of infaunal (55% of total) suspension feeders (64% of total) in the first shell bed grouping (#1–6), while epifaunal (83% of total) algae grazing (55% of total) was the primary mode of life in the second shell bed grouping (#7–9). This change in mode of life corresponds to the relative abundance change from bivalves to gastropods between the two groupings. However, epifaunal organisms (65% of total) and suspension feeders (54%) were the most diverse and abundant of the entire succession (shell beds #1–9). A total of three gastropod species were identified, which made up the total abundance of algae grazers (46% of total); nearly all other organisms (i.e. 18 species of bivalves, brachiopods, crinoids and ostracods) were suspension feeders (54%).

Palaeoecological interpretation. In contrast to the scarcity of fossils found within the Paganzo Basin transgressive successions, the Moscovian Transgression in the Río del Peñon Formation has yielded a diverse and abundant faunal assemblage. The *Tivertonia-Streptorhynchus* Biozone is considered to be a 'post-glacial' faunal assemblage, and is commonly found in silty/sandy seafloor environments (González 1997*b*). Stratigraphic study of the Moscovian Transgression has shown that deposition took place in a shallow to offshore marine environment without the influence of ice. These identifying

Table 2. *Measures of numerical dominance of benthic marine invertebrates in the Rio del Peñon Formation*

	Mean rank order		Breadth of distribution	
	MT	MLPT	MT (%)	MLPT (%)
Brachiopods	2.88	1.00	88.9	100.0
Bivalves	1.78	2.67	100.0	100.0
Gastropods	1.88	4.00	88.9	66.7
Crinoids	16.00	3.33	22.2	100.0
Ostracods	34.00		11.1	

biozone factors, as well as individual species identification, can therefore be used to support the previous interpretation that this faunal assemblage is indeed the *Tivertonia-Streptorhynchus* faunal assemblage.

Throughout the succession, alpha diversity among benthic species increased slightly (Table 1). This diversity increase is however potentially biased because of issues from sampling in thin shell accumulations including possible taphonomic bias, time averaging and patchy individual distribution (Kidwell & Holland 1991; Bennington 2003), although Kidwell (2001) proposed that death assemblages provide reliable means to acquire palaeobiologic data. It is also possible that this pattern may have been controlled by either changing lithology or environment, but it is important to point out that all shell beds were interpreted to have been deposited within fine- to medium-grained sands in a lower shoreface environment. Nonetheless, this data is suggestive of a statistically significant increase in faunal diversity (T-test; $p = 0.0076$) in total number of species within the first interval of relative sea level rise in the Río Blanco Basin.

This is supported by the Simpson Index of Diversity (D) values, which also indicated increasing diversity throughout the succession (Table 1). It is important to note that diversity generally seems higher in the stratigraphically higher beds (i.e. #5–8) than in the lower beds (i.e. #1–4). This excludes shell bed #9, in which D appears to drop drastically from the preceding shell bed. Shell bed #9 is the last shell bed in the succession, and only a shell hash (9 cm thick) containing fragmented invertebrate fossils was deposited at the very top of the transgression. Shell bed #9 may therefore reflect changing environmental conditions within the systems track or possibly a taphonomic bias. Evenness and dominance statistics, including the Shannon Index (H′), indicate that species became more evenly distributed throughout the succession (Table 1). This is important to note because it shows that no one particular species is distributed unevenly among the community, or that one genus or species was dominant. The high evenness values for the shell beds signify that the Río del Peñon fauna was generally even in its abundance and distribution, although bivalves were the most common element (Table 2). Palaeoecological analysis indicates that conditions at this time (i.e. shoreface environment; see 'Sedimentological interpretation' above) were ideal for epifaunal suspension feeding and algae grazing, which were the dominant modes of life.

Overall, throughout the Moscovian Transgression it is apparent that diversity increased at higher and lower taxonomic levels and that taxa distribution is fairly even. As previously mentioned, the lower Río del Peñon Formation has been interpreted to have been deposited in proximal glaciomarine conditions (Marenssi *et al.* 2005; Gulbranson *et al.* 2008). The middle and upper Río del Peñon facies comprised open-marine environments, following the deglaciation in the area. It is interpreted here that the invertebrate fossils found at the Río del Peñon locality are representative of biota that gained diversity and became increasingly ecologically complex during the Moscovian Transgression, several million years after glaciation ended in this area.

Río del Peñon Formation (MLPT)

Sedimentological description. The MLPT section (40.7 m) in the Río del Peñon Formation of the Río Blanco Basin begins with the same medium- to coarse-grained sandstone overlaying the fluvial incision surface at the top of the first measured section (i.e. MT; above 39.5 m; Fig. 9). As previously discussed, this fluvial lowstand facies was interpreted to represent the initial fill after a drop in base level and the formation of an incised valley. Beginning on the fluvial incision surface, the fluvial facies consists of cross-bedded medium- to coarse-grained sandstone (23.3 m thick) and contains large feldspar and quartz grains, as well as the occasional calcite vein. This unit is scoured into by a series of alternating horizontally laminated siltstone and rippled and hummocky fine- to medium-grained sandstone. The top of the measured section consists of brown and black horizontally laminated siltstone, which contains abundant invertebrate fossils throughout as well as large ironstone concretions (5–34 cm; Fig. 10d). Invertebrate fossils are apparent on the outside and inside of the concretions. The siltstone unit is unconformably overlain by medium-grained sandstone at the top of the section.

Sedimentological interpretation. The MLPT section of the middle Río del Peñon Formation is interpreted to represent offshore to upper shoreface settings (Fig. 9). The measured section (a total of 40.7 m) begins on top of the previous fluvial incision surface identified in the Moscovian Transgression (above 39.5 m). The interpretation of this fluvial lowstand unit is supported by the abrupt lithological change between facies (i.e. siltstone to coarse sandstone) as well as the erosional incision surface separating them. As base level fell and a fluvial system eroded into the transgressive deposits, an incised valley formed. The initial base-level rise at this location resulted in fluvial lowstand deposition (i.e. 23.3 m of cross-bedded medium- to coarse-grained sandstone).

This fluvial lowstand unit is however erosionally overlain by ripple cross-laminated fine- to medium-grained sandstone. This erosion surface (i.e. transgressive ravinement surface) is interpreted to represent marine flooding and the beginning of another transgressive systems tract. Transgressive scouring is common due to tidal and wave erosion during the initial landwards shift of the shoreline (Catuneanu 2006). The fining-upwards of this sandstone and siltstone unit represents the early flooding of the fluvial environment and subsequent rise in base level, transitioning the environment in this locality into a shoreface setting. This environmental interpretation is supported by the presence of ripple cross-laminations, which suggests wave activity. The ripple cross-laminated fine- to medium-grained sandstone at the base of the sequence is gradationally overlain by a series of horizontally laminated siltstone with alternating fine- to medium-grained cross-laminated sandstone, suggesting a lower shore face environment with periodic storm activity. Lower shoreface environments are often characterized by alternating layers of waved-worked sandstone (storm deposit) and horizontally laminated siltstone (quiet period of suspension settling), similar to this part of the section. This is the result of deposition below normal wave base, but within storm wave base (Coe *et al.* 2003).

As base level continues to rise, additional siltstone is deposited (i.e. the start of another parasequence) during periods of quiet. At *c.* 34 m, this siltstone unit is periodically interrupted by multiple hummocky cross-stratified fine-grained sandstones. These sandstones indicate continued deposition in a lower shoreface environment, as well as formation due to high-energy storm waves. This unit is overlain by thick siltstone facies, indicating a period of minimal storm activity and suggests deposition in an offshore environment (i.e. below storm wave base). Large (5–34 cm) ironstone concretions found within this unit were formed pre-compaction due to the presence of invertebrates on the inside and outsides of the nodules. Ironstone (i.e. iron oxide) concretions can form in a variety of host-rock conditions, but require the mixing of Fe^{2+}- and O_2-rich waters in order to precipitate (Aubrey *et al.* 2007; Chan *et al.* 2007). This therefore indicates that ironstone concretions found within the siltstone unit are a result of an iron- and oxygen-rich marine environment.

At the top of the measured section, horizontally laminated siltstone is overlain by an erosion surface on which rests medium-grained sandstone. Due to the lithological change (i.e. siltstone to medium-grained sandstone) between these units, as well as the erosion surface, it is interpreted that the medium-grained sandstone was deposited by fluvial incision during a base-level fall followed by deposition of a lowstand systems tract during the onset of the next relative rise in sea level. Additionally, the lack of coarsening-upwards deposits at the top of the transgressive systems tract indicates that late highstand (i.e. progradational) marine facies have been eliminated by this incision.

Overall, the measured section of the middle Río del Peñon Formation is interpreted to represent a coastal marine environment, transitioning from a shoreface to offshore setting (i.e. fining-upwards sequence). In addition to the presence of ripples, which indicate deposition within normal wave base, hummocky cross-stratified sandstones deposited by storms were also identified, supporting a marine interpretation. Marine deposits (i.e. transgressive systems tract) occurring between fluvial lowstand units suggests flooding of palaeovalleys by marine incursion; it is therefore interpreted that this transgressive sequence is related to the MLPT. As previously stated, it is assumed that the transgressive facies identified in the Tupe Formation are the same transgressive facies that occur in the Río del Peñon Formation. Therefore, the MLPT represents the next flooding (following the Moscovian Transgression) of the Río Blanco Basin and is the same base-level rise that occurs in the Tupe Formation of the Paganzo Basin.

Palaeoecological description. The Río del Peñon Moscovian–Late Pennsylvanian Transgressive facies yielded a total of three shell beds termed #10–12 (#10 being lowest stratigraphically and #12 highest; Fig. 9). These beds were found within horizontally laminated siltstone at *c.* 37 m from the bottom of the succession. We identified 84 individual invertebrate fossils belonging to 4 classes and 19 species. Comparable to the Moscovian Transgressive facies, these organisms belong to the Tivertonia-Streptorhynchus faunal assemblage and include rhynchonelliform brachiopods, bivalves, gastropods and crinoids.

Analysis reveals that within the three shell beds, alpha diversity varied from 7 to 14 species (4 classes) with the most species (14) belonging to shell bed #10 (Table 1). The Simpson Index of Diversity (D) for the shell beds shows the same overall trend as alpha diversity (Table 1). The Shannon Index (H') and evenness (e) statistics indicate high values; however, they were higher in shell bed #10 than in shell beds #11 and #12 (Table 1). Additionally, brachiopods and bivalves had the highest mean rank orders with 1.0 and 2.7, respectively (Table 2). Breadth of distribution statistics indicates that brachiopods, bivalves and crinoids appear in 100% of shell beds (Table 2). Brachiopods were also the most diverse group identified in the second transgression with a total of nine species, although bivalves closely follow with seven species. Additionally, epifaunal (90% of total) suspension feeding (84% of total) was the dominant mode of life.

Palaeoecological interpretation. Similar to the Moscovian Transgressive facies, a diverse and abundant *Tivertonia-Streptorhynchus* faunal assemblage was identified in the MLPT of the Río del Peñon Formation and includes rhynchonelliform brachiopods, bivalves, gastropods and crinoids. While both transgressions contained similar faunal assemblages (i.e. *Tivertonia-Streptorhynchus*), they varied in abundance and species/class diversity. However, identifying characteristics such as deposition in an ice-free shoreface environment and species identification confirm that this is indeed a *Tivertonia-Streptorhynchus* faunal assemblage.

Within the succession, it appears that alpha diversity declines up-section but is in general fairly high (Table 1). Similarly, the Simpson Index of Diversity (D) indicates that diversity was moderately high with values from 0.80 to 0.85 (Table 1). The slight diversity decrease throughout the succession may be due to changing marine conditions, although there is potential taphonomic bias within the individual shell beds (Kidwell & Holland 1991; Bennington 2003). Regardless of this decrease, the overall high values indicate a highly diverse palaeocommunity in the MLPT. In addition, evenness statistics including Shannon Index (H′) indicate that the community was evenly distributed among groups, with no particular group dominating (Table 1). When looking at ecological dominants however, brachiopods were the most common element with a mean rank order of 1.0 (Table 2). Epifaunal suspension feeding was the dominant mode of life due to the abundance of brachiopods within this succession. The dominance of brachiopods also indicates that this environment (i.e. offshore; see 'Sedimentological interpretation' above) was likely low in turbidity (i.e. below storm wave base) and therefore ideal for epifaunal suspension feeders.

Overall, throughout the MLPT species and class diversity is generally high. As the lower Río del Peñon has been interpreted to have been deposited in proximal glaciomarine conditions (Gulbranson *et al.* 2008), it is possible to interpret that palaeocommunities diversified following these glacial conditions. In addition, it appears that diversity decreased throughout the transgression and may be related to changing environmental conditions within the succession. Regardless, faunal diversity was overall very high, and indicates that benthic communities during the MLPT had a high number of species in the Río Blanco Basin. An increase in evenness also indicates that communities were becoming increasingly complex in terms of ecology, exploiting additional niches within the environment.

Discussion

Successions measured within the Paganzo and Río Blanco basins were deposited within the same time interval (Late Serpukhovian–Moscovian) and contain equivalent transgressive deposits, although likely represent different systems tracts. Each provides different environmental perspectives of the LPIA: the Paganzo Basin contains primarily continental/estuarine-influenced deposits, while the Río Blanco Basin is dominated by marine sedimentation. Additionally, the Río del Peñon locality (Río Blanco Basin) appears to be near the seaward limit of a very large, deep palaeovalley system (i.e. several hundred metres of relief; a likely palaeofjord) containing several smaller incised palaeovalleys (Fig. 11). In northwestern Argentina, the fauna appears to have been subject to localized stresses

Fig. 11. Río del Peñon palaeovalley system (dashed line indicates incision surface of the large palaeovalley), which contains numerous depositional sequences and associated small-scale incised palaeovalleys, Río del Peñon locality, Río Blanco Basin.

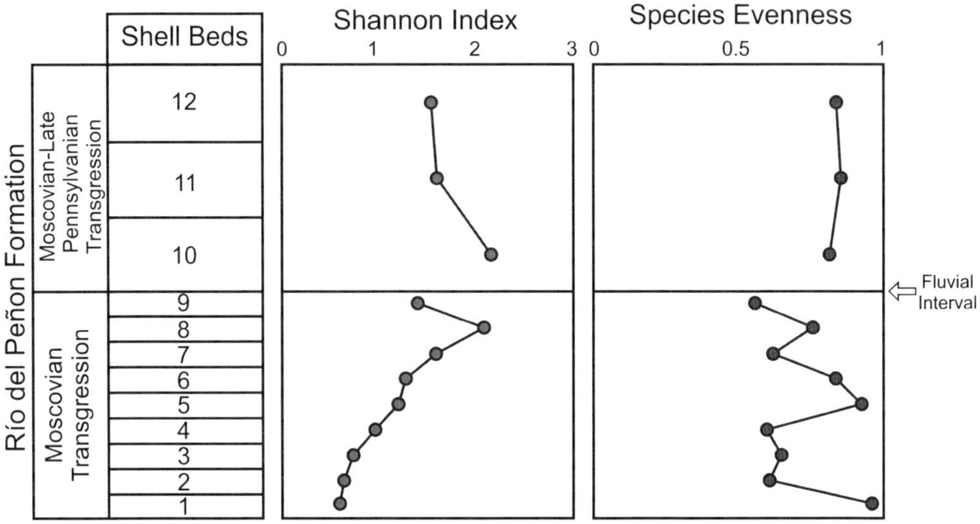

Fig. 12. Shannon Index (H′) and evenness (e) for the Moscovian and Moscovian–Late Pennsylvanian Transgressions of the middle Río del Peñon Formation, Río Blanco Basin. Stratigraphy not to scale. H′ values range from 0 to a maximum of ln (1/number of groups of organisms), with higher values indicating that individuals are evenly distributed among groups. Evenness (e) values range from 0 to 1, with higher values indicating similar individual proportion distribution in the groups. Using radiometric dating, Gulbranson *et al.* (2010) speculated that the middle Río del Peñon section was deposited 318–310 Ma; it can therefore be assumed that both transgressions occurred sometime within a minimal 8 Ma span, but it is difficult to determine exact durations of these marine incursions due to fluvial removal.

associated with either initial local alpine glaciation or later marine incursions into continental basins/narrow estuaries. Changes in relative sea level can affect the biota at different ecological levels by opening or closing seaways, changing estuary water depths and limiting the area of the habitable sea floor.

The Paganzo Basin represents a unique succession due to its proximal fjord environment and repeated marine transgressions interrupting fluvial sedimentation. A low-diversity faunal assemblage was identified within the ice-distal section of the Guandacol Formation at the Cuesta de Huaco locality, and is interpreted to be an opportunistic stressed assemblage. This depauperate palaeocommunity appeared to be able to rapidly colonize, quickly achieve high abundances and survive deglaciation stresses (i.e. low salinities, mobile substrates, dysoxic water) which led to successful short-term occupancy of this area of the Paganzo Basin. As the basin underwent continued deglaciation, the opportunistic fauna appears to disappear due to delta progradation (i.e. base-level fall) and successive loss of sea floor at the Cuesta de Huaco locality.

No marine invertebrates were discovered within the Moscovian and Moscovian–Late Pennsylvanian Transgressions (Tupe Formation) of the Paganzo Basin at the Cuesta de Huaco locality. However, a highly diverse *Tivertonia-Streptorhynchus* faunal assemblage was found in time-equivalent deposits within the Río del Peñon Formation in the Río Blanco Basin. It is speculated that this is due to the location of the basin and its connectedness to the continental margin and proximity to the Panthalassan Ocean, resulting in easier migration for biota already established elsewhere in the Río Blanco Basin. As the Paganzo Basin was mostly restricted by the Protoprecordillera during the Pennsylvanian, the fauna established in the Río Blanco basin would have had colonized adjacent marine localities first, such as Río del Peñon. Additionally, Paganzo basin localities with stronger connections to the Panthalassan margin (i.e. near the mouth of estuaries/fjords), such as La Herradura Creek, were discovered to contain an abundant *Tivertonia-Streptorhynchus* faunal assemblage (Leanza 1945; Archbold & Simanauskas 2001; Cisterna *et al.* 2002*a*, *b*). This suggests that environmental conditions were more

Fig. 13. Multivariate analysis for the Río del Peñon Formation, Río Blanco Basin. (**a**) Detrended correspondence analysis (DCA) plot showing the relationship of all 12 shell beds. (**b**) Non-metric multidimensional scaling (NMS) plot using Bray-Curtis index. (**c**) NMS plot using Euclidian index.

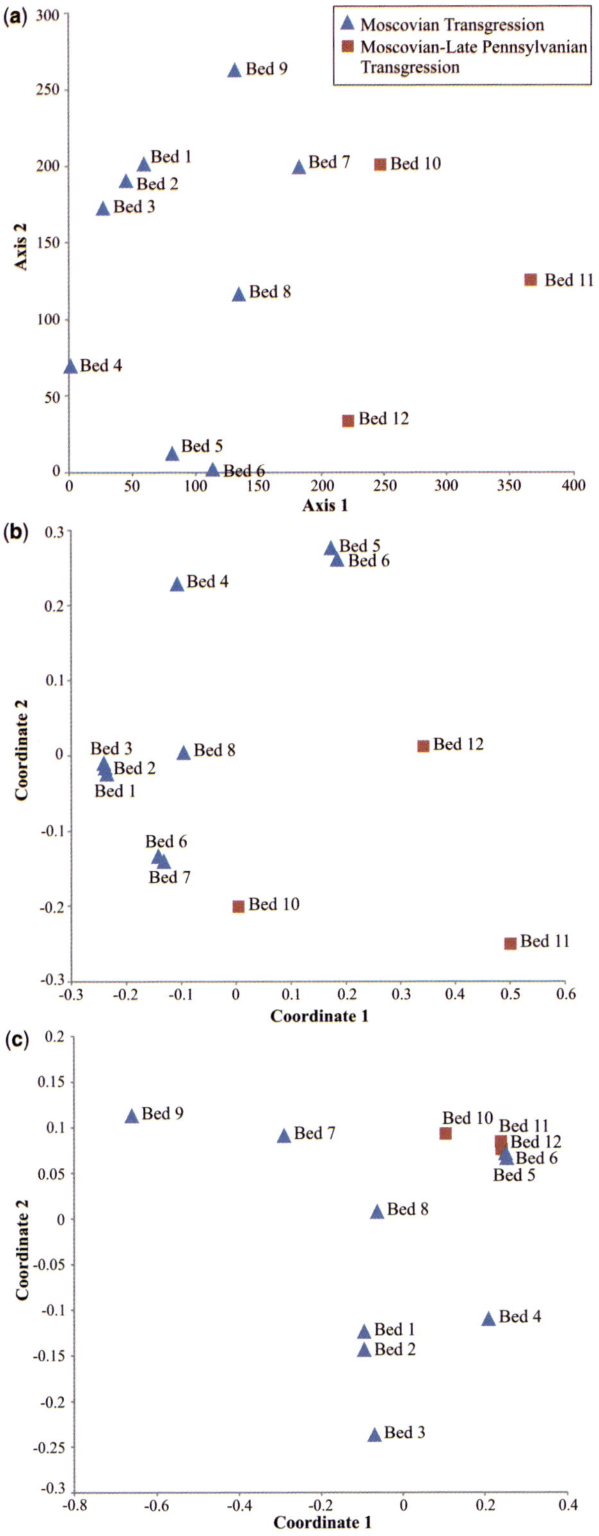
(a)
Moscovian Transgression
Moscovian-Late Pennsylvanian Transgression
Axis 1
Axis 2
Bed 1
Bed 2
Bed 3
Bed 4
Bed 5
Bed 6
Bed 7
Bed 8
Bed 9
Bed 10
Bed 11
Bed 12
(b)
Coordinate 1
Coordinate 2
(c)
Coordinate 1
Coordinate 2

stable near the mouth of estuaries and settings with close links to the Panthalassan margin localities (i.e. Río del Peñon and La Herradura Creek), and presented ideal locations for palaeocommunity growth.

In addition, proximal areas would have flooded earlier during base-level rise when compared to the distal Cuesta de Huaco locality, which has been interpreted to be the landward limit of the transgressions. This suggests that transgression would not have reached the Cuesta de Huaco locality until late during the marine incursion. The lack of an invertebrate fossil record in the Tupe Formation is therefore interpreted to be a result of physiologically stressful local estuary conditions (i.e. high sedimentation rate and freshwater input) at the Cuesta de Huaco locality, which would have been located inland near the head of estuaries. This may also be related to position within the depositional sequence, with the Huaco locality likely representing a part of the highstand systems track (lack of fossil invertebrates) and localities such as La Herradura Creek representing a part of the transgressive systems track (abundant fossil invertebrates).

Despite containing time-equivalent transgressive deposits, local environmental conditions and palaeogeography are interpreted to have resulted in the vastly different palaeontologic records of the Tupe (Paganzo Basin) and Río del Peñon (Río Blanco Basin) formations. Additionally, it appears that palaeocommunities of the Río del Peñon locality in the Río Blanco Basin diversify throughout the marine incursions. While both transgressions of the Río del Peñon Formation contained the same faunal assemblage (i.e. *Tivertonia-Streptorhynchus*), they varied in abundance, composition and species/class diversity. Species diversity is shown to have increased throughout both of the post-glacial marine transgressions, with diversity reaching its highest in the MLPT (Fig. 12; Table 1). Palaeocommunities also began to show increased evenness, indicating greater utilization of resources. This has been interpreted to signify that local palaeocommunities in northwestern Argentina were diversifying and becoming increasing ecologically complex throughout the post-glacial transgressions. This indicates that factors from the preceding glaciation, evident in the lower unit of the Río del Peñon Formation, may have suppressed local palaeocommunity diversification. These factors, such as brackish waters, high sedimentation and mobile substrates, presented a high-stress environment that proved to hinder most glaciomarine benthic invertebrates (e.g. Guandacol Formation at the Huaco locality). This suggests that the removal of these factors may have resulted in the palaeocommunity diversification that is evident in the Río del Peñon Formation. However, the proximal glaciomarine unit located in the lower Río del Peñon Formation at the Río del Peñon locality will need to be examined in order to further test this hypothesis.

Additionally, results from multivariate analysis (DCA and Bray-Curtis and Euclidean NMS) of the Río del Peñon Formation shell beds show that there is generally a separation between the transgression assemblages (Fig. 13); shell beds from the Moscovian Transgression are distinct from those of the MLPT. This separation is not as distinct when Euclidian distance is used in the NMS analysis, but still no overlap occurs. This separation is curious because while both transgressions contained species belonging to the *Tivertonia-Streptorhynchus* assemblage, it appears that a compositional and/or dominance difference existed. Thus, these results further provide evidence that ecological complexity (i.e. diversity, evenness, composition) was indeed changing throughout these two transgressions, resulting in different palaeocommunities. While it is possible that this may be a result of the difference in sample size between transgressions, with a sample size of 420 individuals for the Moscovian and 84 individuals for the MLPT, rarefaction analysis shows no significant difference ($p = 0.221$) between the number of expected species per sample size between Río del Peñon transgressions, indicating that both transgressions were sufficiently sampled.

Results from DCA show that clear partitioning between transgression assemblages is especially prominent along axis 1, which generally corresponds to the shell beds' age within the succession (with Moscovian Transgression beds at low values and MLPT beds at high values; Fig. 13a). It is therefore likely that axis 1 is representative of time as well as differences in the transgression depositional environment; this is apparent as lower shoreface and offshore transition (fine sandstone) shell beds plot at low values, while offshore (siltstone) shell beds plot at higher values. This pattern can also be seen on both NMS plots, although it is less distinct on the Euclidean plot (Fig. 13b, c). The DCA plot also demonstrates that separation along axis 2 appears to correspond to variation between shell beds within transgressions (Fig. 13a). While the reason for this is not as apparent as that for axis 1, one hypothesis is that axis 2 corresponds to the shell beds' location within smaller parasequences in each transgression; shell beds that formed during initial deposition of the parasequence have high values and shell beds that were deposited later in the parasequence have lower values. Additionally, there appears to be a distinction between beds that were gastropod-dominated (higher axis 2 values) and brachiopod/bivalve-dominated (lower axis 2 values). Together, this suggests that community composition and dominance within each transgression was strongly controlled by physical factors and marine conditions.

The variation along axis 2 may also be a factor of sampling as well as mechanic taphonomic bias within the shell beds. Chave (1964) concluded that the primary factor in controlling durability of carbonate skeletons was the micro-architecture of the shell (i.e. gastropods with dense, fine-grained skeletons were the most durable and bivalves, being more porous, were intermediate). Class taphonomy can also be dependent on other factors, such as shell hydrodynamics and thickness (Cummins *et al.* 1986; Martin 1999; Kidwell 2001). The changes in composition within the transgressions can therefore be possibly attributed to taphonomic bias. Additional sampling is needed in order to determine if this is indeed the most likely explanation. Despite this, the high degree of variability within and between each of the post-glacial transgression assemblages implies that this time period was indeed multifaceted. This suggests that local marine conditions (i.e. water depth, wave activity and sedimentation related to transgression) played an important role in community composition and dominance in post-glacial conditions.

Conclusions

Near-field glacial and post-glacial marine faunas of the Paganzo and Río Blanco basins in northwestern Argentina were identified and compared in order to gain an understanding of the effects of deglaciation on palaeocommunity establishment and diversification. Palaeontologic data from this study shows that there were distinct differences in palaeocommunity composition and diversity depending on glacial and post-glacial environments, marine setting and palaeogeographic constraints. Diversity and abundance data from northwestern Argentina reflects that ice-proximal environments proved physiologically stressful to organisms; however, opportunistic assemblages successfully exploited ice-distal settings (Paganzo Basin). Following glaciation, the fauna of northwestern Argentina diversified and became increasingly ecologically complex during two periods of marine transgression (Río del Peñon Formation). It is interpreted that palaeocommunity establishment and diversification in western Argentina during the LPIA was mostly dependent upon localized environmental conditions including ice proximity, glaciation style, marine conditions, water depth, palaeogeographic constraints and position within the systems tract.

These conclusions contribute to the palaeoecological record of western Gondwana, and provide an additional high-latitude perspective on the effects of deglaciation on near-field palaeocommunity establishment and diversification. Further research is needed in order to determine the magnitude to which glaciation affected local palaeocommunity diversification, and if current glacial marine communities will be affected in a similar manner as Earth's present climate transitions from an icehouse to greenhouse state.

The authors would like to thank A. Pagani and A. Taboada for their valuable and insightful reviews on this manuscript. Additional thanks go to O. Limarino, R. Cúneo, E. Gulbranson, I. Montañez, S. Dornbos and P. Sheehan for their clarification of LPIA ideas, stratigraphy and taxonomy. This research was supported by grants from NSF (IRES-0825617), the UW-Milwaukee Geosciences Department, the UW-Milwaukee Center for Latin American and Caribbean Studies (CLACS), the Palaeontological Society (PS) and the Wisconsin Geological Society (WGS).

References

Archangelsky, A., Azcuy, C. L., González, C. R., Sabattini, N. & Aceñolaza, G. 1987. Paleontologia, Biostratigrafia y Paleoecologia de las Cuencas Paganzo, Calingasta-Uspallata y Río Blanco. *In*: Archangelsky, A. (ed.) *El Sistema Carbonífero en la República Argentina*, Academia Nacional de Ciencias, Córdoba, 133–151.

Archbold, N. W. & Simanauskas, T. 2001. New strophalosiidae (Brachiopoda) from the early permian of Argentina. *Proceedings of the Royal Society of Victoria*, **113**, 217–227.

Arthur, M. A. & Sageman, B. B. 1994. Marine black shales: depositional mechanisms and environments of ancient deposits. *Annual Review of Earth Planetary Sciences*, **22**, 499–551.

Aubrey, A. D., Parker, E., Chalmers, J. H., Lal, D. & Bada, J. L. 2007. Ironstone concretions – analogs to martian hematite spherules. *In*: *Lunar and Planetary Science Conference*, **XXXVIII**, Proceedings of the 38th Lunar and Planetary Science Conference, League City, 2053.

Azcuy, C. L., Beri, A. et al. 2007. Bioestratigrafía del Paleozoico Superior de América del Sur: Primera Etapa de Trabajo Hacia una Nueva Propuesta Cronoestratigráfica. *Asociación Geológica Argentina Serie D: Publicación especial*, **11**, 9–64.

Bann, K. L., Fielding, C. R., MacEachern, J. A. & Tye, S. C. 2004. Differentiation of estuarine and offshore marine deposits using integrated ichnology and sedimentology: permian Pebbley Beach Formation, Sydney Basin, Australia. *In*: McIlroy, D. (ed.) *The Application of Ichnology to Palaeoenvironmental and Stratigraphic Analysis*. Geological Society, London, Special Publications, **228**, 179–211.

Bennett, M. R. & Glasser, N. F. 1996. *Glacial Geology: Ice Sheets and Landforms*. John Wiley & Sons, New York.

Bennington, J. B. 2003. Transcending patchiness in the comparative analysis of paleocommunities: a test case from the upper cretaceous in New Jersey. *PALAIOS*, **18**, 22–33.

Berner, R. A. 1990. Atmospheric carbon dioxide levels over Phanerozoic time. *Science*, **249**, 1382–1386.

Bonelli, J. R. & Patzkowsky, M. E. 2008. How are global patterns of faunal turnover expressed at regional scales? Evidence from the Upper Mississippian (Chesterian Series), Illinois basin, USA. *PALAIOS*, **23**, 760–772.

Bonelli, J. R. & Patzkowsky, M. E. 2011. Taxonomic and ecologic persistence across the onset of the late Paleozoic Ice age: evidence from the Upper Mississippian (Chesterian Series), Illinois basin, United States. *PALAIOS*, **26**, 5–17.

Boulton, G. S. 1990. Sedimentary and sea level changes during glacial cycles and their control on glacimarine facies architecture. *In*: Dowdeswell, J. A. & Scourse, J. D. (eds) *Glacimarine Environments: Processes and Sediments*. Geological Society, London, Special Publications, **53**, 15–52.

Brenchley, P. J. & Harper, D. A. 1998. *Paleoecology: Ecosystems, Environments and Evolution*. Chapman & Hall, London.

Buatois, L. A. & Mangano, M. G. 2011. *Ichnology: Organism-Substrate Interactions in Space and Time*. Cambridge University Press, Cambridge.

Buatois, L. A., Gingras, M. K., MachEachern, J. A., Mangano, M. G., Zonnefeld, J.-P., Pemberton, S. G., Netto, R. G. & Martin, A. 2005. Colonization of Brackish-Water Systems through time: evidence from the Trace-Fossil Record. *PALAIOS*, **20**, 321–347.

Buatois, L. A., Netto, R. G., Mangano, M. G. & Balistieri, P. R. 2006. Extreme freshwater release during the late Paleozoic Gondwana deglaciation and its impact on coastal ecosystems. *Geology*, **34**, 1021–1024.

Cai, J., Powell, R. D., Cowan, E. A. & Carlson, P. R. 1997. Lithofacies and seismic-reflection interpretation of temperate glacimarine sedimentation in Tarr Inlet, Glacier Bay, Alaska. *Marine Geology*, **143**, 5–37.

Caputo, M. V. & Crowell, J. C. 1985. Migration of glacial centers across Gondwana during Paleozoic Era. *Geological Society of America Bulletin*, **96**, 1020–1036.

Carmona, N. B., Buatois, L. A., Ponce, J. J. & Mangano, M. G. 2009. Ichnology and sedimentology of a tide-influenced delta, lower Miocene Chenque Formation, Patagonia, Argentina; trace-fossil distribution and response to environmental stresses. *Palaeogeography, Palaeoclimatology, Palaeoecology*, **273**, 75–86.

Catuneanu, O. 2006. *Principles of Sequence Stratigraphy*. Elsevier, Oxford.

Césari, S. N., Limarino, C. O. & Gulbranson, E. L. 2011. An Upper Paleozoic bio-chronostratigraphic scheme for the western margin of Gondwana. *Earth-Science Reviews*, **106**, 149–161.

Chan, M. A., Ormo, J., Park, A. J., Stich, M., Souza-Egipsy, V. & Komatsu, G. 2007. Models of iron oxide concretion formation: field, numerical, and laboratory comparisons. *Geofluids*, **7**, 356–368.

Chave, K. E. 1964. Skeletal durability and preservation. *In*: Imbrie, J. & Newell, N. D. (eds) *Approaches to Paleoecology*. Wiley, New York, 377–387.

Chou, L. M., Yu, J. Y. & Loh, T. L. 2004. Impacts of sedimentation on soft-bottom benthic communities in the southern islands of Singapore. *Hydrobiologia*, **515**, 91–106.

Cisterna, G. A. & Sabattini, N. 1998. Algunos Gastropoda de la Formacion Río del Peñon (Carbonifero Superior – Permico Inferior), Provincia de la Rioja, Argentina. *Revista Española de Paleontología*, **53**, 212–218.

Cisterna, G. A. & Simanauskas, T. 2000. Brachiopods from the Río del Peñon Formation, Río Blanco Basin, Upper Paleozoic of Argentina. *Revista Española de Paleontología*, **15**, 129–151.

Cisterna, G. A. & Sterren, A. F. 2007. Early Permian Marine Fauna from the Quebrada Larga Formation, San Juan Province, Argentine Precordillera: biostratigraphical implications. *Acta Geologica Lilloana*, **20**, 113–118.

Cisterna, G. A. & Sterren, A. F. 2008. Late Carboniferous Levipustula Fauna in the Leoncito Formation, San Juan Province, Argentine Precordillera: biostratigraphical and palaeoclimatological implications. *Proceedings of the Royal Society of Victoria*, **120**, 137–147.

Cisterna, G. A. & Sterren, A. F. 2010. 'Levipustula Fauna' in central-western Argentina and its relationships with the Carboniferous glacial event in the southwestern Gondwanan margin. *In*: López-Gamundí, O. R. & Buatois, L. A. (eds) *Late Paleozoic Glacial Events and Postglacial Transgressions in Gondwana*. Geological Society of America, Boulder, Special Paper, **468**, 133–147.

Cisterna, G. A., Simanauskas, T. & Archbold, N. W. 2002*a*. Permian brachiopods from the Tupe Formation, San Juan Province, Precordillera, Argentina. *Alcheringa*, **26**, 177–200.

Cisterna, G. A., Archbold, N. W. & Simanauskas, T. 2002*b*. The Permian brachiopod genus Trigonotreta Koenig and its occurrence in Argentina. *Ameghiniana*, **39**, 213–220.

Cisterna, G. A., Sterren, A. F. & Archbold, N. W. 2006. A review of the Tivertonia jachalensis-Streptorhynchus inaequiornatus Biozone in La Delfina Creek, San Juan Province, Argentina. *Ameghiniana*, **43**, 487–491.

Clapham, M. E. & James, N. P. 2008. Paleoecology of Early-Middle Permian Marine Communities in eastern Australia: response to global climate change in the aftermath of the Late Paleozoic Ice Age. *Palaios*, **23**, 738–750.

Coe, A. L., Bosence, D. W. J., Church, K. D., Flint, S. S., Howell, J. A. & Wilson, R. C. L. 2003. *The Sedimentary Record of Sea-Level Change*. Cambridge University Press, Cambridge.

Cowan, E. A. & Powell, R. D. 1990. Suspended sediment transport and deposition of cyclically interlaminated sediment in a temperate glacial fjord, Alaska, U.S.A. *In*: Dowdeswell, J. A. & Scourse, J. D. (eds) *Glacimarine Environments: Processes and Sediments*. Geological Society, London, Special Publication, **53**, 75–89.

Cowan, E. A., Cai, J., Powell, R. D., Clark, J. D. & Pitcher, J. N. 1997. Temperate glacimarine varves: an example from Disenchantment Bay, Southern Alaska. *Journal of Sedimentary Research*, **67**, 536–549.

Cummins, H., Powell, E. N., Newton, H. J., Stanton, R. J. & Staff, G. 1986. Assessing transportation by the covariance of species with comments on contagious and random distributions. *Lethaia*, **19**, 1–22.

DAVIES, S. J. 2008. The record of Carboniferous sea-level change in low-latitude sedimentary successions from Britain and Ireland during the onset of the late Paleozoic ice age. *In*: FIELDING, C. R., FRANK, T. D. & ISBELL, J. L. (eds) *Resolving the Late Paleozoic Ice Age in Time and Space*. Geological Society of America, Boulder, Special Paper, **441**, 187–204.

DOMEIER, M., VAN DER VOO, R., TOHVER, E., TOMEZZOLI, R. N., VIZAN, H., TORSVIK, T. H. & KIRSHNER, J. 2011. New Late Permian paleomagnetic data from Argentina: refinement of the apparent polar wander path of Gondwana. *Geochemistry, Geophysics, Geosystems*, **12**, 1–21.

EYLES, N. 1993. Earth's glacial record and its tectonic setting. *Earth-Science Reviews*, **35**, 1–248.

FIELDING, C. R., FRANK, T. D. & ISBELL, J. L. 2008*a*. The late Paleozoic ice age – A review of current understanding and synthesis of global climate patterns. *In*: FIELDING, C. R., FRANK, T. D. & ISBELL, J. L. (eds) *Resolving the Late Paleozoic Ice Age in Time and Space*. Geological Society of America, Boulder, Special Paper, **441**, 343–354.

FIELDING, C. R., FRANK, T. D., BIRGENHEIER, L. P., RYGEL, M. C., JONES, A. T. & ROBERTS, J. 2008*b*. Stratigraphic imprint of the late Paleozoic ice age in eastern Australia: a record of alternating glacial and nonglacial climate regime. *Journal of the Geological Society: London*, **165**, 129–140.

FRAKES, L. A. 1979. *Climates throughout Geologic Time*. Elsevier Scientific Publishing Company, New York, 129–139.

FRAKES, L. A., FRANCIS, J. E. & SYKTUS, J. I. 1992. *Climate Modes of the Phanerozoic: The History of Earth's Climate over the Past 600 Million Years*. Cambridge University Press, Cambridge.

GANI, M. R. 2004. From turbid to lucid: a straightforward approach to sediment gravity flows and their deposits. *The Sedimentary Record*, **2**, 4–8.

GASTALDO, R. A., DIMICHELE, W. A. & PFEFFERKORN, H. W. 1996. Out of the icehouse into the greenhouse: a Late Paleozoic analog for modern global vegetational change. *GSA Today*, **6**, 1–7.

GONZÁLEZ, C. R. 1992. La Fauna de Cerro Veladero, Carbónico de la Cuenca del Río Blanco, La Rioja. *Acta Geologica Lilloana*, **17**, 101–113.

GONZÁLEZ, C. R. 1993. Late Paleozoic faunal succession in Argentina. *Comptes Rendus XII*, **1**, 537–550.

GONZÁLEZ, C. R. 1997*a*. Upper Paleozoic glaciation and Carboniferous and Permian faunal changes in Argentina. *In*: DICKINS, J. M., ZUNYI, Y., HONGFU, Y., LUCAS, S. G. & ACHARYYA, S. K. (eds) *Late Paleozoic and Early Mesozoic Circum-Pacific Events and Their Global Correlation*. World and Regional Geology 10, Cambridge University Press, Cambridge, 235–241.

GONZÁLEZ, C. R. 1997*b*. Late Carboniferous Bivalvia from western Argentina. *Geologica et Palaeontologica*, **31**, 193–214.

GONZÁLEZ, C. R. 1999. Permo-Carboniferous faunas and provincialism in the South American Gondwana region. *Carboniferous Newsletter*, **17**, 12–13.

GOTELLI, N. J. & ELLISON, A. M. 2004. *A Primer of Ecological Statistics*. Sinauer Associates, Inc., Sunderland, MA.

GROVES, J. R. & YUE, W. 2009. Foraminiferal diversification during the late Paleozoic ice age. *Paleobiology*, **35**, 367–392.

GULBRANSON, E. L., LIMARINO, C. O., MARENSSI, S., MONTAÑEZ, I., TABOR, N. J., DAVYDOV, V. & COLOMBI, C. 2008. Glacial deposits in the Río del Peñon Formation (Late Carboniferous), Río Blanco Basin, Northwestern Argentina. *Latin American Journal of Sedimentology and Basin Analysis*, **15**, 37–42.

GULBRANSON, E. L., MONTAÑEZ, I., SCHMITZ, M. D., LIMARINO, C. O., ISBELL, J. L., MARENSSI, S. & CROWLEY, J. L. 2010. High-Precision U-Pb calibration of Carboniferous glaciation and climate history, Paganzo Group, NW Argentina. *GSA Bulletin*, **122**, 1480–1498.

HAGEMAN, S. J., BONE, Y., MCGOWRAN, B. & JAMES, N. P. 1997. Bryozoan Growth-forms as Paleoenvironmental Indicators: evaluation of methodology. *PALAIOS*, **12**, 405–419.

HAMMER, O. & HARPER, D. A. 2006. *Paleontological Data Analysis*. Blackwell Publishing, Oxford.

HAMMER, O., HARPER, D. A. & RYAN, P. D. 2001. PAST: Paleontological statistics software package for education and data analysis. *Palaeontologia Electronica*, **4**, 1–9.

HANGER, R. A. & GABADIRWE, M. 1999. Comparative Taphonomy and Paleoecology of a Glaciomarine Fauna, Carboniferous (Westphalian – Namurian) La Capilla Fm., Argentina. *Virginia Journal of Science*, **50**, 259–268.

HECKEL, P. H. 1994. Evaluation of evidence for glacio-eustatic control over marine Pennsylvanian cyclothems in North America and consideration of possible tectonic effects. *In*: DENNISON, M. & ETTENSOHN, F. R. (eds) *Tectonic and Eustatic Controls on Sedimentary Cycles*. SEPM (Society of Sedimentary Geology), Tulsa, Concepts in Sedimentology and Paleontology, **4**, 65–87.

HECKEL, P. H. 2008. Pennsylvanian cyclothems in Midcontinent North America as far-field effects of waxing and waning of Gondwana ice sheets. *In*: FIELDING, C. R., FRANK, T. D. & ISBELL, J. L. (eds) *Resolving the Late Paleozoic Ice Age in Time and Space*. Geological Society of America, Boulder, Special Paper, **441**, 275–289.

HEIM, N. A. 2009. Stability of regional brachiopod diversity structure across the Mississippian/Pennsylvanian boundary. *Paleobiology*, **35**, 393–412.

HENRY, L. C., ISBELL, J. L. & LIMARINO, C. O. 2008. Carboniferous glaciogenic deposits of the proto-Precordillera of west-central Argentina. *In*: FIELDING, C. R., FRANK, T. D. & ISBELL, J. L. (eds) *Resolving the Late Paleozoic Ice Age in Time and Space*. Geological Society of America, Boulder, Special Paper, **441**, 131–142.

HENRY, L. C., ISBELL, J. L., LIMARINO, C. O., MCHENRY, L. J. & FRAISER, M. L. 2010. Mid-Carboniferous deglaciation of the Protoprecordillera, Argentina recorded in the Aqua de Jaguel palaeovalley. *Palaeogeography, Palaeoclimatology, Palaeoecology*, **298**, 112–129.

HILL, M. O. & GAUCH, H. G. JR. 1980. Detrended correspondance analysis: an improved ordination technique. *Vegetatio*, **42**, 47–58.

Holland, S. M. 2005. The Signatures of Patches and Gradients in Ecological Ordination. *PALAIOS*, **20**, 573–580.

Horton, D. E. & Poulsen, C. J. 2009. Paradox of late Paleozoic glacioeustacy. *Geology*, **37**, 715–718.

IPCC. 2007. *Climate Change 2007-Mitigation of Climate Change, Contribution of Working Group III to the Fourth Assessment Report of the Intergovernmental Panel on Climate Change*, Cambridge University Press, Cambridge.

Isbell, J. L., Lenaker, P. A., Askin, R. A., Miller, M. F. & Babcock, L. E. 2003*a*. Reevaluation of the timing and extent of late Paleozoic glaciation in Gondwana: role of the Transantarctic Mountains. *Geology*, **31**, 977–980.

Isbell, J. L., Miller, M. F., Wolfe, K. L. & Lenaker, P. A. 2003*b*. Timing of late Paleozoic glaciation in Gondwana: was glaciation responsible for the development of northern hemisphere cyclothems. *In*: Chan, M. A. & Archer, A. W. (eds) *Extreme Depositional Environments: Mega end Members in Geologic Time*. Geological Society of America, Boulder, Colorado, Special Paper, **370**, 5–24.

Isbell, J. L., Fraiser, M. L. & Henry, L. C. 2008*a*. Examining the complexity of environmental change during the Late Paleozoic and early Mesozoic. *PALAIOS*, **23**, 267–269.

Isbell, J. L., Cole, D. I. & Catuneanu, O. 2008*b*. Carboniferous-Permian glaciation in the main Karoo Basin, South Africa: stratigraphy, depositional controls, and glacial dynamics. *In*: Frank, T. D. & Isbell, J. L. (eds) *Resolving the Late Paleozoic Ice Age in Time and Space*. Geological Society of America, Boulder, Colorado, Special Paper, **441**, 71–82.

Isbell, J. L., Henry, L. C. *et al.* 2012. Glacial paradoxes during the late Paleozoic ice age: evaluating the equilibrium line altitude as a control on glaciation. *Gondwana Research*, **22**, 1–19.

Kammer, T. W., Brett, C. E., Boardman, D. R. & Mapes, R. H. 1986. Ecologic Stability of the dysaerobic biofacies during the Late Paleozoic. *Lethaia*, **19**, 109–121.

Kelley, P. H., Raymond, A. & Lutken, C. B . 1990. Carboniferous brachiopod migration and latitudinal diversity: a new palaeoclimatic method. *In*: McKerrow, W. S. & Scotese, C. R. (eds) *Palaeozoic Palaeogeography and Biogeography*. Geological Society, London, Memoirs, **12**, 325–332.

Kidwell, S. M. 2001. Preservation of species abundance in marine death assemblages. *Science*, **294**, 1091–1094.

Kidwell, S. M. & Holland, S. M. 1991. Field description of coarse bioclastic fabrics. *PALAIOS*, **6**, 426–434.

Lawver, L. A., Dalziel, I. W. D., Norton, I. O. & Gahagan, L. M. 2007. *The Plates 2007 Atlas of Plate Reconstructions (750 Ma to Present Day), Plates Progress Report No. 305-0307*, Technical Report No. 195, 160, University of Texas.

Leanza, A. F. 1945. Braquiópodos Carboniferos de la Quebrada de la Herradura al NE de Jachal, San Juan. *Museo de la Plata, Sección Palaeontologia, Notas Paleontologicas*, **86**, 277–314.

Lech, R. R. & Aceñolaza, G. 1987. Braquiópodos en el Penoniano (Carbonífero superior) de la provincia de La Rioja. *IV Congreso Latinoamericano de Paleontología*, **1**, 255–256.

Limarino, C. O. & Spalletti, L. A. 2006. Paleogeography of the Upper Paleozoic basins of southern South America: an overview. *Journal of South American Earth Sciences*, **22**, 134–155.

Limarino, C. O., Cesari, S. N., Net, L. I., Marenssi, S. A., Gutierrez, R. P. & Tripaldi, A. 2002. The Upper Carboniferous Postglacial transgression in the Paganzo and Río Blanco basins (northwestern Argentina): facies and stratigraphic significance. *Journal of South American Earth Sciences*, **15**, 445–460.

Limarino, C., Tripaldi, A., Marenssi, S. & Fauque, L. 2006. Tectonic, sea-level, and climatic controls on Late Paleozoic sedimentation in the western basins of Argentina. *Journal of South American Earth Sciences*, **22**, 205–226.

Lokier, S. W., Wilson, M. E. & Burton, L. M. 2009. Marine biota response to clastic sediment influx: a quantitative approach. *Palaeogeography, Palaeoclimatology, Palaeoecology*, **281**, 25–42.

López-Gamundí, O. R. 1997. Glacial-postglacial transition in the late Paleozoic basins of southern South America. *In*: Martini, I. P. (ed.) *Late Glacial and Postglacial Environmental Changes: Quaternary, Carboniferous-Permian, and Proterozoic*. Oxford University Press, New York, Chapter 9, 147–168.

López-Gamundí, O. & Martinez, M. 2000. Evidence of glacial abrasion in the Calingasta-Uspallata and western Paganzo basins, mid-Carboniferous of western Argentina. *Palaeogeography, Palaeoclimatology, Palaeoecology*, **159**, 145–165.

López-Gamundí, O. R., Espejo, I. S., Conaghan, P. J. & Powell, C. M. 1994. Southern South America. *In*: Veevers, J. J. & Powell, C. M. (eds) *Permian-Triassic Pangean Basins and Foldbelts Along the Panthalassan Margin of Gondwanaland*. Geological Society of America, Boulder, Memoir, **184**, 281–329.

Lucchi, F. R. 1995. *Sedimentographica: Photographic Atlas of Sedimentary Structures*. Columbia University Press, New York.

MacEachern, J. A., Stelck, C. R. & Pemberton, S. G. 1999. Marine and marginal marine mudstone deposition: Paleoenvironmental interpretations based on the integration of ichnology, Palynology, and Foraminiferal Paleoecology. *In*: Bergman, K. M. & Snedden, J. W. (eds) *Isolated Shallow Marine Sand Bodies. Sequence Stratigraphy Analysis and Sedimentologic Interpretation*. SEPM, Tulsa, Special Publication, **64**, 205–225.

MacEachern, J. A., Bann, K. L., Bhattacharya, J. P. & Howell, C. D. 2005. Ichnology of deltas: organism responses to the dynamic interplay of rivers, waves, storms, and tides. *In*: Giosan, L. & Bhattacharya, J. (eds) *River Deltas – Concepts, Models, and Examples*. SEPM, Tulsa, Special Publication, **83**, 49–83.

Mangano, M. G. & Buatois, L. A. 2004. Ichnology of Carboniferous tide-influenced environments and tidal flat variability in the North American Midcontinent. *In*: McIlroy, D. (ed.) *The Application of Ichnology to Palaeoenvironmental and Stratigraphic Analysis*. Geological Society, London, Special Publications, **228**, 157–178.

Marenssi, S. A., Tripaldi, A., Limarino, C. O. & Caselli, A. T. 2005. Facies and architecture of a carboniferous grounding-line system from the guandacol formation, Paganzo Basin, Northwestern Argentina. *Gondwana Research*, **8**, 187–202.

Martin, R. E. 1999. *Taphonomy: A Process Approach*. Cambridge University Press, Cambridge.

Martínez, M. 1993. Hallazgo de Fauna Marina en La Formación Guandacol (Carbonífero) en La Localidad de Agua Hedionda, San Juan, Precordillera Nororiental, Argentina. *Comptes Rendus XII*, **2**, 291–296.

Maughan, B. C. 2001. The effects of sedimentation and light on recruitment and development of a temperate, subtidal, epifaunal community. *Journal of Experimental Marine Biology and Ecology*, **256**, 59–71.

Meyer, D. L., Miller, A. I., Holland, S. M. & Dattilo, B. E. 2002. Crinoid distribution and feeding morphology through depositional sequence: kope and fairview formations, upper Ordovician, Cincinnati arch region. *Journal of Paleontology*, **76**, 725–732.

Montañez, I. P., Tabor, N. J. *et al.* 2007. CO_2-Forced climate and vegetation instability during late paleozoic deglaciation. *Science*, **315**, 87–91.

Mulder, T. & Alexander, J. 2001. The physical character of subaqueous sedimentary density flows and their deposits. *Sedimentology*, **48**, 269–299.

Net, L. I. & Limarino, C. O. 2006. Applying sandstone petrofacies to unravel the Upper Carboniferous evolution of the Paganzo Basin, northwest Argentina. *Journal of South American Earth Sciences*, **22**, 239–254.

Novack-Gottshall, P. M. & Miller, A. I. 2003. Comparative geographic and environmental diversity dynamics of gastropods and bivalves during the Ordovician Radiation. *Paleobiology*, **29**, 576–604.

Ottone, E. G. & Azcuy, C. L. 1986. El Perfil de la Quebrada La Delfina, Provincia de San Juan. *Revista de la Asociación Geológica Argentina*, **41**, 124–136.

Pazos, P. J. 2002*a*. The Late Carboniferous Glacial to postglacial transition: facies and sequence stratigraphy, Western Paganzo Basin, Argentina. *Gondwana Research*, **5**, 467–487.

Pazos, P. J. 2002*b*. Palaeoenvironmental framework of the Glacial-Postglacial transition (Late Paleozoic) in the Paganzo-Calingasta Basin (Southern South America) and the Great Karoo-Kalahari Basin (Southern Africa): ichnological implications. *Gondwana Research*, **5**, 619–640.

Pérez-Huerta, A. & Sheldon, N. D. 2006. Pennsylvanian sea level cycles, nutrient availability and brachiopod paleoecology. *Palaeogeography, Palaeoclimatology, Palaeoecology*, **230**, 264–27.

Powell, C. M. & Veevers, J. J. 1987. Namurian uplift in Australia and South America tiggered the main Gondwana glaciation. *Nature*, **326**, 177–79.

Powell, M. G. 2005. Climate basis for sluggish macroevolution during Late Paleozoic ice age. *Geology*, **33**, 381–384.

Powell, R. D. 1981. A model for sedimentation by tidewater glaciers. *Annals of Glaciology*, **2**, 129–134.

Powell, R. D. 2005. Subaquatic landsystems: fjords. *In*: Evans, D. J. A. (ed.) *Glacial Landsystems*. Oxford University Press, New York, 313–347.

Powell, R. & Domack, E. 1995. Chapter 14: modern glaciomarine environments. *In*: Menzies, J. (ed.) *Glacial Environments, Modern Glacial Environments: Processes, Dynamics, and Sediments*, **1**, 445–486.

Probert, P. K. 1981. Changes in benthic community of china clay deposits in Mevagissey Bay following a reduction in discharges. *Journal of Marine Biological Assessment UK*, **61**, 789–804.

Ramos, V. A., Jordan, T. E., Allmendinger, R. W., Mpodozis, M. C., Kay, S. M., Cortes, J. M. & Palma, M. A. 1986. Paleozoic terranes of the central Argentine-Chilean Andes. *Tectonics*, **5**, 855–880.

Reineck, H. E. & Singh, I. B. 1980. *Depositional Sedimentary Environments*. Springer-Verlag, Berlin.

Rygel, M. C., Fielding, C. R., Frank, T. D. & Birgenheier, L. P. 2008. The magnitude of late Paleozoic glacioeustatic fluctuations: a synthesis. *Journal of Research*, **78**, 500–511.

Sabattini, N., Ottone, E. G. & Azcuy, C. L. 1990. La Zona de Lissochonetes jachalensis – Streptorhynchus inaequiornatus (Carbonifero Tardio) en La Localidad de La Delfina, Provincia de San Juan. *Ameghiniana*, **27**, 75–81.

Schieber, J. 2008. Marcasite in black shales – A mineral based indicator of ‘burndown’ in ancient sapropels? *Geochimica et Cosmochimica Acta*, **72**, A830.

Schubert, J. K., Bottjer, D. J. & Simms, M. J. 1992. Paleobiology of the oldest known articulate crinoid. *Lethaia*, **25**, 97–110.

Scotese, C. R. & Barrett, S. F. 1990. Gondwana’s movement over the South Pole during the Paleozoic: evidence from lithological indicators of climate. *In*: McKerrow, W. S. & Scotese, C. R. (eds) *Palaeozoic Palaeogeography and Biogeography*. Geological Society, London, Memoirs, **12**, 75–85.

Shannon, C. E. 1948. A mathematical theory of communication. *Bell System Technical Journal*, **27**, 466–467.

Shi, G. R. 1993. Multivariate data analysis in palaeoecology and palaeobiology – A review. *Palaeogeography, Palaeoclimatology, Palaeoecology*, **105**, 199–234.

Shi, G. R. & Waterhouse, J. B. 2010. Late Palaeozoic global changes affecting high-latitude environments and biotas: an introduction. *Palaeogeography, Palaeoclimatology, Palaeoecology*, **298**, 1–16.

Simpson, E. H. 1949. Measurement of diversity. *Nature*, **189**, 688.

Stanley, S. M. & Powell, M. G. 2003. Depressed rates of origination and extinction during the late Paleozoic ice age: a new state for the global marine ecosystem. *Geology*, **31**, 877–880.

Sterren, A. F. 2000. Moluscos bivalvos en la Formacion Río del Peñon (Carbonifero Tardio-Permico Temprano), provincia de La Rioja, Argentina. *Ameghiniana*, **37**, 421–438.

Sterren, A. F. 2004. Permian bivalves of the Tupe Formation, Herradura Gorge, San Juan Province, Argentina. *Ameghiniana*, **41**, 57–74.

Strachan, L. J. 2008. Flow transformations in slumps: a case study from the Waitemata Basin, New Zealand. *Sedimentology*, **55**, 1311–1332.

Taboada, A. C. 2010. Mississippian-Early Permian brachiopods from western Argentina: tools for middle to high-latitude correlation, paleobiogeographic and paleoclimatic reconstruction. *Palaeogeography, Palaeoclimatology, Palaeoecology*, **298**, 152–173.

Tedesco, A., Ciccoli, P., Suriano, J. & Limarino, C. O. 2010. Changes in the architecture of fluvial deposits in the Paganzo Basin (Upper Paleozoic of San Juan Province): an example of sea level and climatic controls on the development of coastal fluvial environments. *Geologica Acta*, **8**, 463–482.

Torsvik, T. H., Muller, R. D., Van der Voo, R., Steinberger, B. & Gaina, C. 2008. Global plate motion frames: toward a unified model. *Review of Geophysics*, **46**, 1–44.

Veevers, J. J. & Powell, C. M. 1987. Late Paleozoic glacial episodes in Gondwanaland reflected in transgressive-regressive depositional sequences in Euramerica. *Geological Society of America Bulletin*, **98**, 475–487.

Visser, J. N. J. 1997. Deglaciation sequences in the Permo-Carboniferous Karoo and Kalahari basins of Southern Africa: a tool in the analysis of cyclic glaciomarine basin fills. *Sedimentology*, **44**, 507–521.

Wartenberg, D., Ferson, S. & Rohlf, F. J. 1987. Putting things in order: a critique of detrended correspondance analysis. *The American Naturalist*, **129**, 434–448.

Waterhouse, J. B. & Shi, G. R. 2010. Evolution in a cold climate. *Palaeogeography, Palaeoclimatology, Palaeoecology*, **298**, 17–30.

Weslawski, J. M., Koszteyn, J., Zajackowski, M., Wiktor, J. & Kwasniewski, S . 1995. Fresh water in Svalbard fjord ecosystems. *In*: Skjoldal, H. R., Hopkins, C., Erikstad, K. E. & Leinaas, H. P. (eds) *Ecology of Fjords and Coastal Waters*. Elsevier Science B.V., Amsterdam, 229–241.

Wignall, P. B. 1994. *Black Shales*. Oxford University Press, New York.

Zhangerl, R. & Richardson, E. S. 1963. *The Paleoecological History of Two Pennsylvanian Black Shales*. Chicago Natural History Museum, Chicago.

Carboniferous coarsening-upward and non-gradational cyclothems in the Lublin Basin (SE Poland): palaeoclimatic implications

MARIA I. WAKSMUNDZKA

Polish Geological Institute-National Research Institute, 4 Rakowiecka St., 00-975 Warsaw, Poland (e-mail: maria.waksmundzka@pgi.gov.pl)

Abstract: The paper presents the characteristics of profiles consisting of coarsening-upward and non-gradational cyclothems that correspond to parasequences in the Visean, Serpukhovian and Bashkirian sections of the Lublin Basin in SE Poland. The accompanying single fining-upward cyclothems are also described. Six types of cyclothems are distinguished. The most complete types of coarsening-upward cyclothems Ic, Id and incomplete IIc and IId have been identified. The least complete are non-gradational cyclothems IIIc and IIId. The cyclothems developed within the oscillating environments of shallow carbonate and clay shelf, and in shallow-water deltas. The lowstand systems tract (LST) deposits are composed of fining-upward cyclothems, rarely of coarsening-upward IId and non-gradational IIId. Within the transgressive systems tract (TST), type IIc is dominant, and the quantity of Ic and non-gradational cyclothem IIIc is moderate. The maximum flooding surface (MFS) usually occurs within coarsening-upward cyclothems IIc. The highstand systems tract (HST) consists of similar types of cyclothems as the TST, but their quantity is less. Relative sea-level fluctuations during the Carboniferous deposition in the Lublin Basin are associated with sea-level changes generated by glacial episodes in Gondwana. The relative sea-level fluctations also affected both erosional processes related to the successive relative sea-level lowstands and the formation of stratigraphic gaps. The influence of glacio-eustatic fluctuations was modified by local factors such as the supply of sedimentary material, subsidence rate and local tectonics.

Coarsening-upward cyclothems are known from both recent deltaic environments (Scruton 1960; Coleman & Wright 1975; Tye & Coleman 1989*a*, *b*; Riegel 1991) and also ancient Carboniferous sections of the UK (Elliott 1975, 1976*a*, *b*), Ireland (Pulham 1989; Hampson *et al.* 1997) and Germany (Hampson *et al.* 1999).

The Lublin Basin (LB) belongs to the Northwest European Carboniferous Basin (NWECB), which is connected with the Variscan Orogeny. The evolution of the NWECB with plate-tectonic, subsidence and basin-fill mechanisms, as well as a discussion of stratigraphy and hydrocarbon systems, have been provided by Kombrink *et al.* (2010). The LB is located in the eastern part of the NWECB and is composed of Carboniferous rocks that occur in the form of a wide zone extending from the SE to the NW (Fig. 1). This zone is nearly parallel to the western margin of the East-European Craton. The development of the LB was probably controlled by strike–slip movements along the Teisseyre–Tornquist Zone under the pull-apart regime (Żelichowski 1987; Narkiewicz *et al.* 1998). The LB is filled with Carboniferous sediments that are not strongly tectonically deformed, and, because of their fully cored well sections, are very valuable for cyclicity investigations.

First studies on the Carboniferous in the LB area were devoted to the issues of sedimentology and stratigraphy, usually on a broad, regional scale, and the problem of cyclicity was presented very roughly. First data on the cyclicity of the Carboniferous section from south LB (SE Poland) were provided by Korejwo (1958). She presented a detailed lithological description of the Visean and Lower Namurian section, distinguishing 30 cyclothems. These were categorized according to thickness into large, medium, thin and very thin cycles. The classification does not include, however, the trend of grain-size gradation. As is apparent from her description, the cyclothems were coarsening-upward cyclothems, starting with limestones or claystones with marine, brackish or freshwater fauna, and ending with coals. In the UK, such cyclicity is recognized as Yoredale cyclothems (e.g. Elliott 1975; Leeder & Strudwick 1987; Tucker 1991). Moreover, Korejwo (1958) interprets the depositional environment as deltaic and marine. Later, Żelichowski (1964), Dembowski (1970) and Porzycki (1979, 1980) interpreted the Carboniferous of the LB cyclicity as typical for paralic basins. Skompski (1995) found a similarity between the coarsening-upward cyclothems occurring in the Carboniferous of the LB and the Yoredale cyclothems described from the Carboniferous of the UK, while Waksmundzka (1998) distinguished three main types of cyclothems: I, II and III. Further studies enabled refinement of this division

From: Gąsiewicz, A. & Słowakiewicz, M. (eds) 2013. *Palaeozoic Climate Cycles: Their Evolutionary and Sedimentological Impact*. Geological Society, London, Special Publications, **376**, 141–175.
First published online September 19, 2013, http://dx.doi.org/10.1144/SP376.18

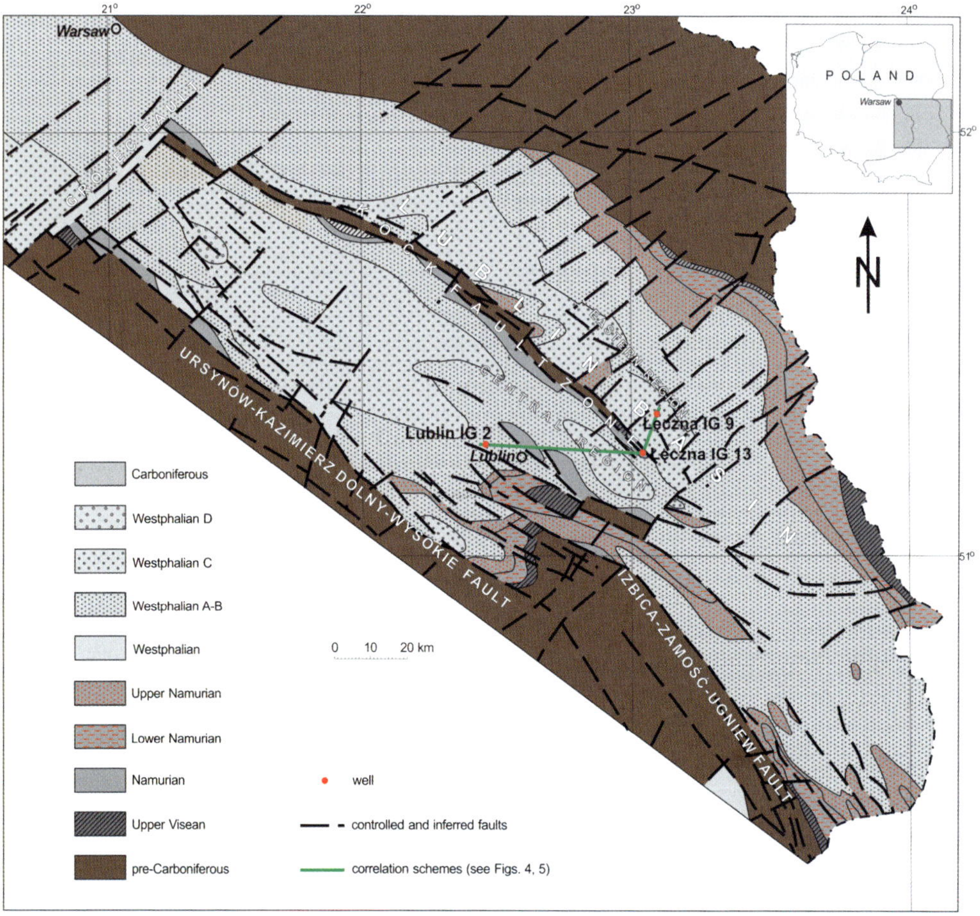

Fig. 1. Location map: extract of the structural–geological map of the Lublin Basin without strata younger than Carboniferous (modified after Żelichowski & Porzycki 1983).

and the distinction of four types of coarsening-upward cyclothems (Ic, Id, IIc, IId) and two types of non-gradational cyclothems (IIIc, IIId) (Waksmundzka 2005, 2010).

As indicated above, previous studies on the Carboniferous in the LB were devoted to the issues of sedimentology and stratigraphy, usually on a broad, regional scale, and the problem of cyclicity was not presented in a comprehensive way. The purpose of this paper is to fill this gap by presenting detailed lithofacies characterization of different types of coarsening-upward and non-gradational cyclothems in terms of sequence stratigraphy and chronostratigraphy of the Carboniferous deposits in the LB. The aim of this study is also to reconstruct the environments and subenvironments, as well as their features, based on analysis of cyclicity in relation to relative sea-level fluctuations. The results may also be helpful in sedimentological and sequence stratigraphic studies from other sedimentary basins that contain cyclothems. In addition, the analysis of cyclicity is important because it is related mainly to claystones and siltstones that occur above the potential sandstone reservoirs that formed as thick sealing horizons in incised valleys (Waksmundzka 2008*b*). Some of these horizons could be prospective for shale gas exploration.

Geological setting

The Carboniferous of the LB includes the whole area of sub-Mesozoic Carboniferous outcrops in SE Poland (Fig. 1), and its SE continuation is the the Lviv-Volyn Coal Basin in Ukraine. Although the term Carboniferous Lublin Basin suggests an approximate compliance of its boundaries with the area of deposition, in fact, its extent was wider towards the north and NE than the previously interpreted boundaries (Waksmundzka 2010). The southwestern boundary and northwestern end marked by the Grójec Fault (Żelichowski 1972), along which the basin borders the Warsaw Block, are of tectonic origin. It suggests that the Carboniferous deposition could also occur farther to the south and SW of the Ursynów–Kazimierz Dolny–Wysokie fault zone, where the basin adjoins the Radom–Kraśnik Elevation (Żelichowski 1969).

The LB is filled with Carboniferous deposits that are separated by a stratigraphic gap from the underlying Devonian, Lower Palaeozoic, Ediacaran and older crystalline basement rocks (Cebulak 1988). The Carboniferous sedimentation was preceded by erosion, which lasted in this area during the Tournaisian and Early Visean (Żelichowski 1972). The strongest denudation processes affected the sedimentary cover of the Precambrian platform, where the erosion removed up to 3000 m of older Palaeozoic deposits. The Carboniferous is unconformably overlain by Permian–Mesozoic and Cenozoic deposits. The base of this complex descends from 200 m in the east to *c.*1000 m b.s.l. in the SW.

In the LB, Żelichowski (1969) identified two major structural units: the eastern unit, and the central unit (the Lublin Trough), whose northwestern part is called the Mazowsze–Lublin Trough (Żelichowski 1972; Żelichowski & Kozłowski 1983). The development of subsidence in the basin was presumably controlled by strike–slip movements along the Teisseyre–Tornquist Zone in the pull-apart regime (Żelichowski 1987; Narkiewicz *et al.* 1998).

Carboniferous deposits are developed as claystones, siltstones, sandstones, conglomerates, limestones, marls, *Stigmaria* soils, carbonaceous claystones and coals. The lowest part of the section occasionally contains volcanites and bauxites. The thickness of the Carboniferous succession clearly increases towards the SW. The smallest thickness (*c.* 100 m) is found in the northeastern part of the basin, where an elevated part of the East European Platform occurs in the basement. Greater thicknesses, from a few hundred metres to *c.* 2000 m, are found in the remaining area of the basin, within the so-called Mazowsze–Lublin Trough.

The oldest lithostratigraphic schemes of the Carboniferous in the LB, introduced by Cebulak & Porzycki (1966), Żelichowski (1969) and Porzycki (1979), now only have historical significance (Fig. 2). Musiał & Tabor (1979, 1988) and Porzycki & Zdanowski (1995) determined the age of the Upper Carboniferous deposits as Upper Visean?–Westphalian D, based on biostratigraphic evidence. Recent results of sequence stratigraphic studies point to the earlier commencement of Carboniferous deposition, during the Middle Visean. The subdivision of the Carboniferous, based on the sequence stratigraphy approach and linked to the global scheme (Fig. 3), has been applied in this paper (Waksmundzka 2008*b*, 2010). It consists of 22 depositional sequences defined after Mitchum (1977) as generally conformable successions of genetically related strata, bounded on the base and top by unconformities and their correlative conformities. All the recognized sequences are of type I *sensu* Vail & Todd (1981), because they are bounded by type I unconformities. This type of unconformity forms due to strong fluvial erosion and subaerial exposure of the shelf during the relative sea-level lowstand. Within the sequences, the maximum regression surfaces (initial transgression) have been identified. They occur at the base of the first transgressive deposits, that is, coals found at the top of alluvial or deltaic deposits, overlying the alluvial series. The maximum flooding surface (MFS) separates the transgressive and highstand parts of sequences and the majority of the MFS has been identified on the basis of prominent peaks on gamma-ray logs. They are associated with the relatively deepest lithofacies enriched with marine organic matter. Three types of depositional systems tracts have been identified: the lowstand systems tract (LST), the transgressive systems tract (TST) and the highstand systems tract (HST). The depositional systems tracts show a cyclic structure. They include a number of coarsening-upward, non-gradational and fining-upward cyclothems corresponding to parasequences (Waksmundzka 2010, 2012).

The analysis of sequences clearly reveals a tripartition of the Carboniferous section. Its lower part comprises sequences 1–10 (Visean–Lower Bashkirian), and TST and HST deposits predominate in terms of thickness. They commonly contain thick limestone beds and shallow-shelf claystones. They developed in the foreland of retrograding delta lobes on a shallow shelf. Sedimentation of even deeper-marine shallow-shelf clay facies and the appearance of anoxic conditions during the maximum transgression are usually associated with subsequent deepening of the LB. The LST deposits of incised valleys in sequences 1–10 are of small thicknesses or missing. This part of the section was formed in a generally marine and distal delta sedimentation regime. The central part of the section, which includes sequences 11–16 (Lower and Middle Bashkirian), is generally characterized by a considerable thickness of deposits filling the incised valleys and by smaller thicknesses of the TST and HST complexes. Thin limestone beds and shallow-shelf claystones are occasionally

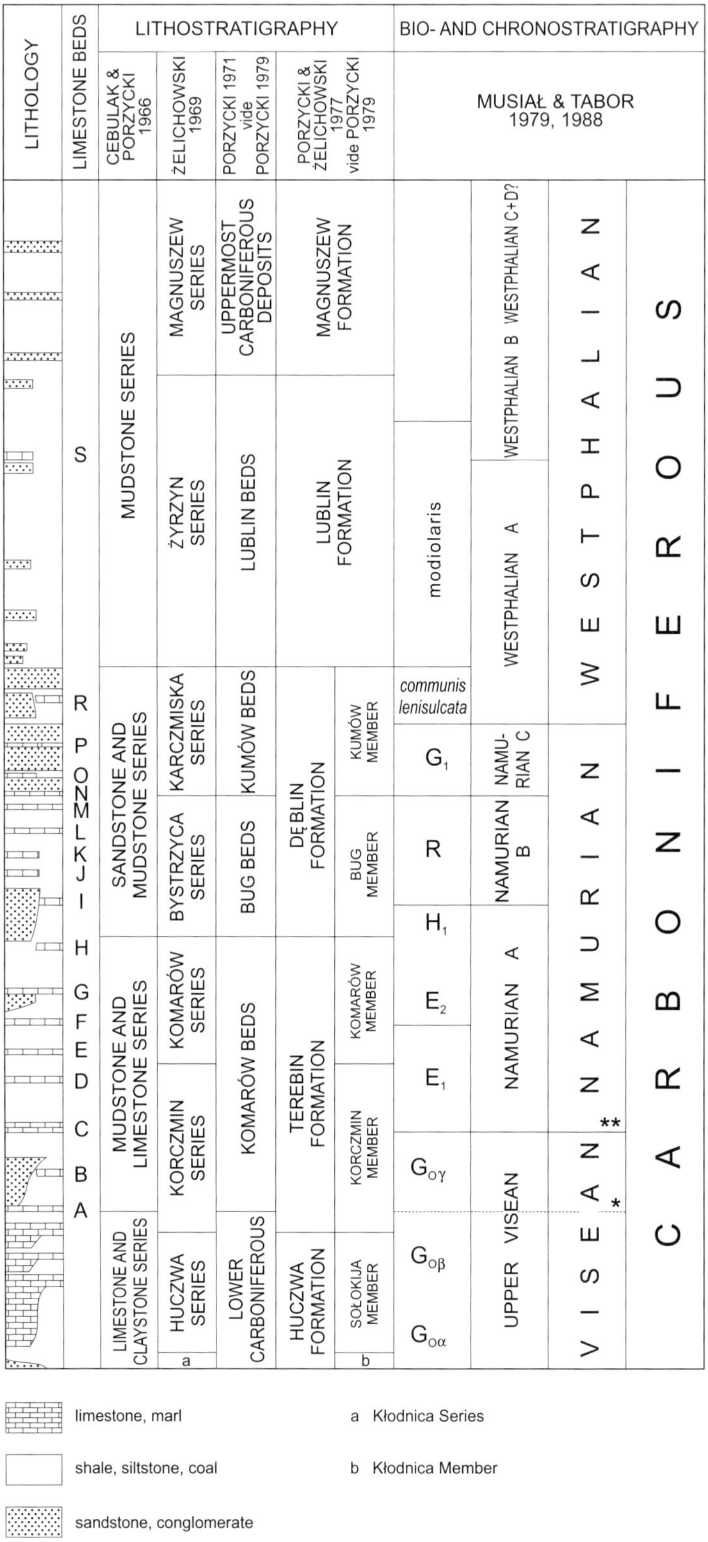
LITHOLOGY
LIMESTONE BEDS
LITHOSTRATIGRAPHY
BIO- AND CHRONOSTRATIGRAPHY
CEBULAK & PORZYCKI 1966
ŻELICHOWSKI 1969
PORZYCKI 1971 vide PORZYCKI 1979
PORZYCKI & ŻELICHOWSKI 1977 vide PORZYCKI 1979
MUSIAŁ & TABOR 1979, 1988
S
R
P
O
N
M
L
K
J
I
H
G
F
E
D
C
B
A
MUDSTONE SERIES
SANDSTONE AND MUDSTONE SERIES
MUDSTONE AND LIMESTONE SERIES
LIMESTONE AND CLAYSTONE SERIES
MAGNUSZEW SERIES
ŻYRZYN SERIES
KARCZMISKA SERIES
BYSTRZYCA SERIES
KOMARÓW SERIES
KORCZMIN SERIES
HUCZWA SERIES
a
UPPERMOST CARBONIFEROUS DEPOSITS
LUBLIN BEDS
KUMÓW BEDS
BUG BEDS
KOMARÓW BEDS
LOWER CARBONIFEROUS
MAGNUSZEW FORMATION
LUBLIN FORMATION
DĘBLIN FORMATION
TEREBIN FORMATION
HUCZWA FORMATION
KUMÓW MEMBER
BUG MEMBER
KOMARÓW MEMBER
KORCZMIN MEMBER
SOŁOKIJA MEMBER
b
modiolaris
communis lenisulcata
G1
R
H1
E2
E1
Goγ
Goβ
Goα
WESTPHALIAN C+D?
WESTPHALIAN B
WESTPHALIAN A
NAMURIAN C
NAMURIAN B
NAMURIAN A
UPPER VISEAN
WESTPHALIAN
NAMURIAN
**
VISEAN
*
CARBONIFEROUS
limestone, marl
shale, siltstone, coal
sandstone, conglomerate
a Kłodnica Series
b Kłodnica Member

found within them. The central part of the section was formed generally in a deltaic and fluvial sedimentation regime. The uppermost part of the section (sequences 17–22 of Middle Bashkirian–Lower Moscovian) is marked by terrestrial deposits, such as alluvial, lacustrine, bog and deltaic deposits.

The boundary between the paralic and terrestrial part of the Carboniferous section runs at the top of the uppermost deltaic part within the HST of sequence 16. This sequence contains the *Dunbarella papyracea* biozone, which is the last interbed of deposits with marine fauna in the Carboniferous section of the Lublin region. The boundary between the Westphalian A and B runs within this horizon (Musiał & Tabor 1988; Porzycki 1988).

The most complete Carboniferous section occurs in the central and northwestern part of the basin, whereas towards the east, NE and SW it is increasingly reduced at the top due to epigenetic erosion.

Materials and methods

Sedimentological and sequence stratigraphic studies were performed on three well cores located in the central and eastern parts of the basin (Fig. 1). They mainly represent the paralic portion of the profile and the lowest part of terrestial part with occurrence of lake delta deposits. Owing to the large thickness of the sections, it was possible to analyse a great number of cyclothems and make a statistical approach to their frequency in different stratigraphic units. Lithofacies analysis was carried out on the sections that contain coarsening-upward cyclothems and non-gradational cyclothems and single fining-upward cyclothems occurring between them (Waksmundzka 1998, 2005, 2010). Sedimentological studies were complemented by the analysis of well logging curves (Waksmundzka 1998, 2005, 2008*c*, 2010, 2012). Their age and spatial relationships in relation to the intervals studied are presented in Figures 4 and 5.

The analysed sections correspond to the Visean, Serpukhovian and Bashkirian. In the West European stratigraphic scheme, this is the ?Middle and Upper Visean, Namurian A, B and C, and Westphalian A and the lower part of Westphalian B (Musiał & Tabor 1979, 1988; Skompski 1996, 1998; Waksmundzka 2007*a*, *b*, 2008*a*, *b*, 2010). In the sequence stratigraphic scheme distinguished by Waksmundzka (2008*b*, 2010), they represent sequences 2 to 17 (Fig. 3). This part of the profile is composed of coarsening-upward cyclothems. Almost all the analysed sections correspond to the paralic part of the Carboniferous, which top runs within sequence 16. In the Łęczna IG 9 and IG 13 wells, the entire Carboniferous section has been drilled through. The Lublin IG 2 well drilled into Serpukhovian deposits only, not reaching the base of the Carboniferous (Figs 4 & 5). About 430–540 m of drill cores have been described in individual well sections. A total length of *c.* 1490 m of drill cores has been analysed. The thickness of the investigated Carboniferous deposits in the Łęczna IG 9 and Łęczna IG 13 wells is *c.* 650–800 m; in the Lublin IG 2, it is *c.* 760 m, but the section is incomplete.

Facies interpretation was based on an analysis of individual lithofacies and their succession. A total of 208 cyclothems have been analysed, of which 136 are coarsening-upward and 50 are non-gradational.

The boundaries of the coarsening-upward and non-gradational cyclothems were placed at the top of phytogenic deposits, that is, coals or carbonaceous claystones. In the absence of such deposits, they were placed at the top of *Stigmaria* soils (cf. Duff & Walton 1962; Read & Dean 1976). The top surface defined in this way is coincident with the base of the overlying cyclothem. In the case of absence of soil lithofacies, the boundaries of coarsening-upward cyclothems were placed at the base of the finest-grained lithofacies and at the top of the coarsest-grained lithofacies.

The lithofacies are described using a lithofacies coding system introduced by Miall (1977, 1978) and Rust (1978), including modifications by Zieliński (1992, 1995) (Table 1). Additional symbols (i.e. n, L, M and R) were used after Gradziński *et al.* (1995) When the lithofacies are of the same symbol, for example, Fm, its several types are distinguished as Fm1, Fm2 and Fm3, based on the colour, faunal and floral content, and the presence of siderite spherulites. In these cases, the symbol Fm is considered as a lithofacies group. By adding a digital index to the basic lithofacies symbols, distinction and characterization of different types of claystones and siltstones are possible.

The cyclothems are divided into the main types marked as I, II, III. Each type is divided into subtypes from a to d. Letter indices a and b are assigned to fining-upward cyclothems (Waksmundzka 2012), and c and d to coarsening-upward and non-gradational cyclothems (Waksmundzka 2010).

Fig. 2. Litho-, bio- and chronostratigraphic divisions of the Carboniferous succession in the Lublin Basin (after Musiał & Tabor 1979, 1988; Porzycki 1979; Porzycki & Zdanowski 1995); after Waksmundzka 2010. Top of Visean *after Musiał & Tabor (1979, 1988); **after Skompski (1996).

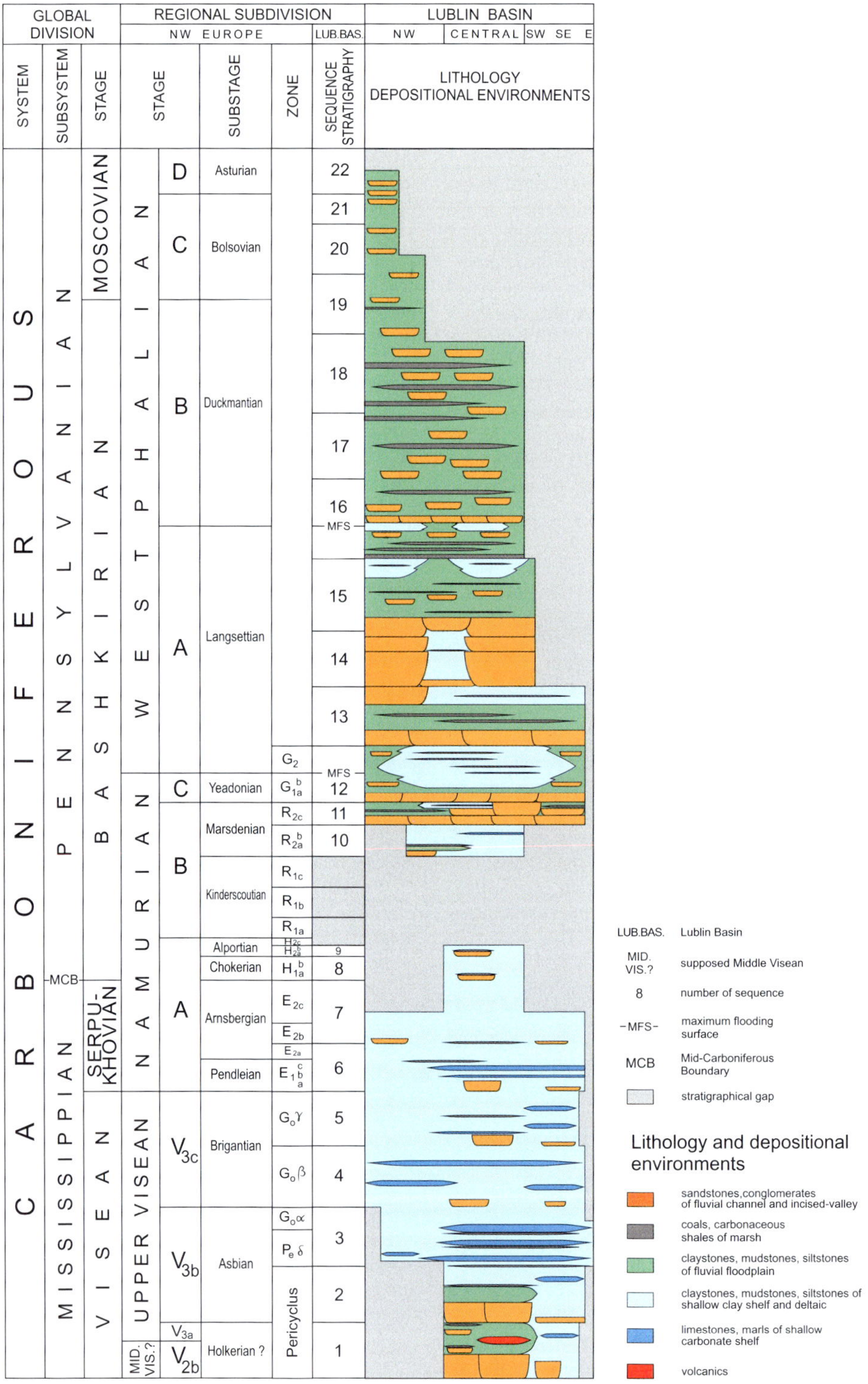

Fig. 3. Sequence stratigraphy of the Carboniferous deposits in the Lublin Basin and correlation with chronostratigraphy.

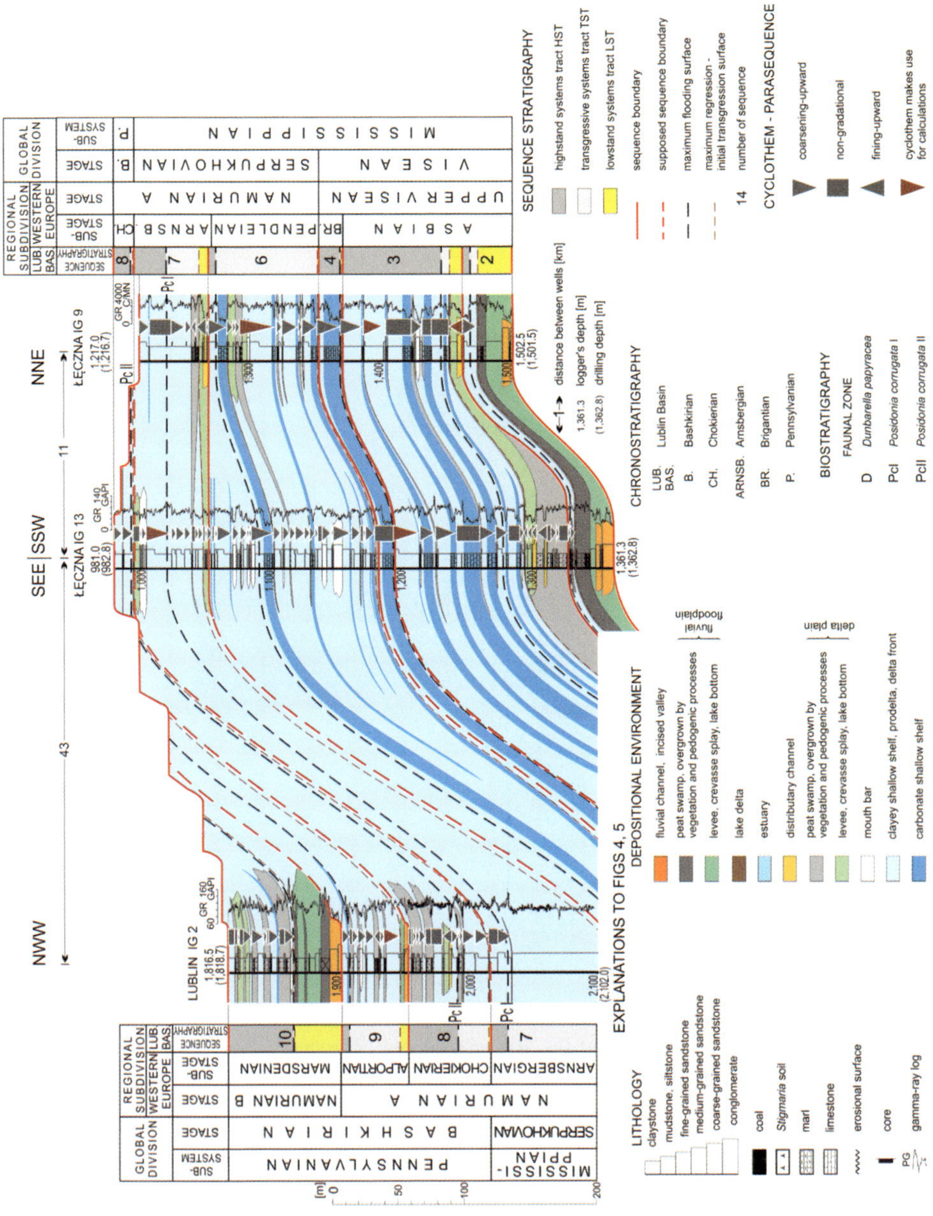

Fig. 4. Lithofacies, sequence stratigraphy and chronostratigraphy of the Visean, Serpukhovian and lower part of Bashkirian succession in investigated wells from central and east parts of the Lublin Basin (modified after Waksmundzka 2010).

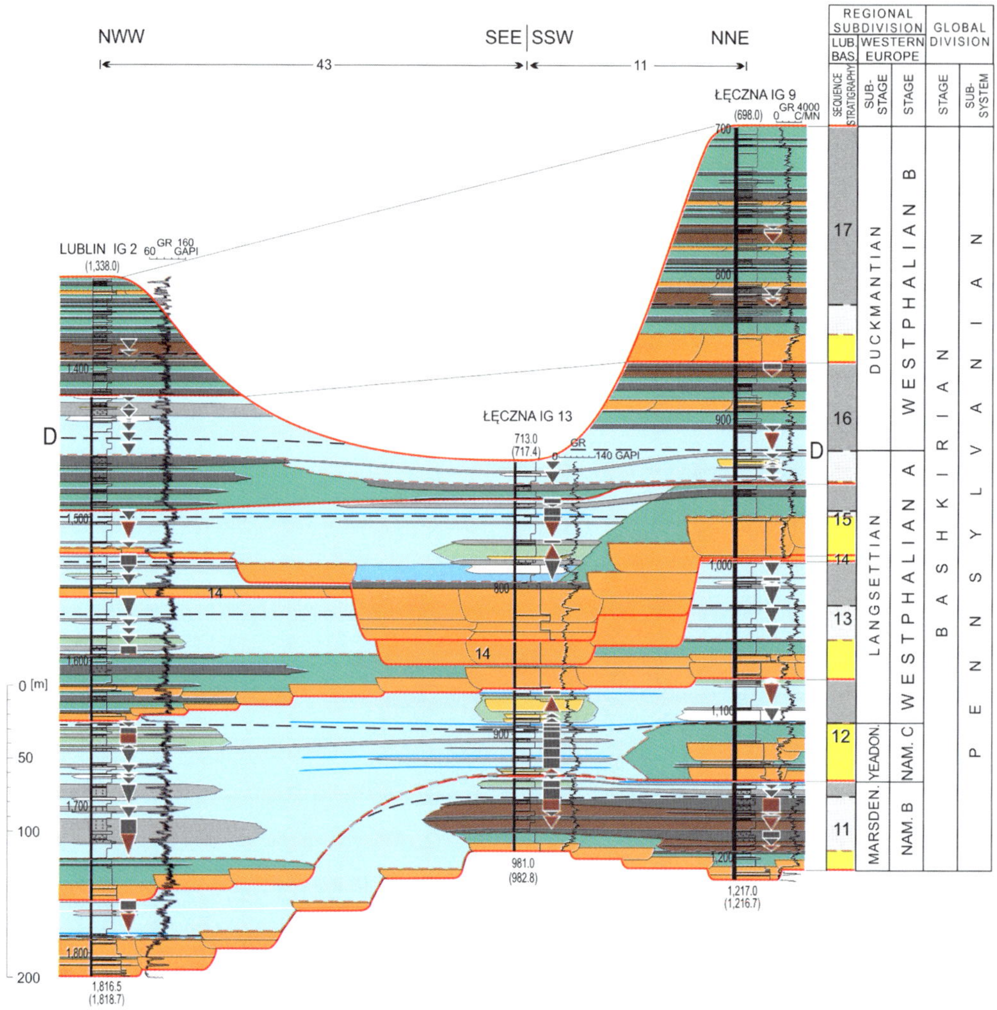

Fig. 5. Lithofacies, sequence stratigraphy and chronostratigraphy of the middle and upper parts of the Bashkirian succession in investigated wells from central and east parts of the Lublin Basin (explanations on Fig. 4) (modified after Waksmundzka 2010).

The coarsening-upward cyclothems are represented by four main types: Ic, Id, IIc and IId. It has been assumed that the completeness of the claystone-clastic part of the cyclothem, in which there are continuous transitions upward from the finest-grained lithofacies to lithofacies of successively coarser grains, is the basis for classification. In addition to the grain-size coarsening feature, the succession and types of lithofacies as well as the thickness of individual cyclothems were also analysed. In distinguishing the subtypes, the presence or absence of sediments genetically unrelated with the coarsening-upward trend, that is carbonates and phytogenic deposits such as coals, carbonaceous claystones and *Stigmaria* soils, were considered. The main diagnostic feature for the two main types of IIIc and IIId was the lack of change in grain diameter. Specific lithofacies succession and thickness were also taken into consideration (Waksmundzka 2010). Subtypes within this group are characterized by the presence or absence of the carbonate part and clay- or silt-size grain fraction.

In order to reconstruct the original thickness relations the lithofacies thickness was corrected by compaction coefficient ratio. It was calculated according to the method proposed by Baldwin & Butler (1985), by determining the so-called

Table 1. *Lithofacies code symbols (after Miall 1977, 1978; Rust 1978; Gradziński* et al. *1995; Zieliński 1992, 1995, completed).*

Code	Lithology
Textural feature	
G	Conglomerate
GS	Sandy conglomerate
S	Sandstone
SG	Gravelly sandstone
SF	Silty sandstone
FS	Sandy siltstone
F	claystone/mudstone
R	*Stigmaria* claystone/mudstone (*Stigmaria* soil)
C	Coal, carbonaceous claystone
Structural feature	
m	Massive structure
h	Horizontal lamination (stratification)
f	Flaser lamination
w	Wavy lamination
n	Lenticular lamination
r	Ripple cross-lamination
l	Low-angle cross-stratification
p	Planar cross-stratification (tabular and wedge-shaped)
t	Trough cross-stratification
x	Large-scale cross-stratification
d	Disturbed (bioturbations, unstable density stratification, by roots – *Stigmaria*)
e	Erosional scours – poorly sorted massive sandstone/conglomerate with clasts laying above erosional surface; thickness to 0.3 m
s	Scour-fill:poorly sorted massive sandstone/conglomerate with clasts laying above erosional surface; thickness over 0.3 m

thickness reduction ratio (Table 2). Based on decompacted thickness of the cyclothems, minimum palaeodepths (without estimate of basin subsidence) of the marine basin and lakes were estimated using this ratio.

The individual sections were correlated using isochronous horizons such as limestone beds C and F (Skompski 1996), *Posidonia corrugata* I, *Posidonia corrugata* II and *Dunbarella papyracea* (Musial & Tabor 1988). The basis for a sequence stratigraphic model of the Carboniferous in the LB used herein is lithofacies–stratigraphic sections presented by Waksmundzka (2010).

Characteristics of lithofacies and cyclothems

Lithofacies analysis of drill cores from the LB has allowed identification of 28 lithofacies (Table 3) based on the thickness, colour, texture, presence of organic debris and fauna, origin and interpretation of sedimentary subenvironment and environment. The most typical lithofacies are illustrated in Figures 6 and 7. The lithofacies form associations that usually correspond to coarsening-upward, non-gradational cyclothems, and also rare fining-upward cyclothems as described below.

Main types of cyclothems

Type Ic

The most complete type of coarsening-upward cyclothem, Ic, consists of parts represented by the following lithofacies, from base to top (Fig. 8, Table 4):

- part 0: limestones L and marls M;
- part 1: claystones Fm1, Fh1 and mudstones Fm3, Fh2;
- part 2: mudstones Fh2 and siltstones Fn, FSw, FSd;
- part 3: fine-grained sandstones Sh1, Sf, Sd;
- part 4: coals C, *Stigmaria* soils R and rare claystones Fm2 and mudstones Fm4, Fm5.

The feature of grain-size coarsening is observed within parts 1, 2 and 3, and is manifested as a transition from claystone into mudstone, and then into silty and sandy. Within type Ic, the most common successions from base to top occur in the order $0 \rightarrow 1 \rightarrow 2 \rightarrow 3 \rightarrow 4$ and $1 \rightarrow 2 \rightarrow 3 \rightarrow 4$ while $0 \rightarrow 1 \rightarrow 2 \rightarrow 3$ and $1 \rightarrow 2 \rightarrow 3$, are rarer. These subtypes are lacking part 4.

The current thickness of individual coarsening-upward cyclothems of type Ic is *c.* 4–24 m (average 7.4 m) (Fig. 8). In terms of thickness, the dominant is part 1, which consists of claystones Fm1, Fh1 and mudstones Fh2. The thickness correction of the parts of Ic by the decompaction factor

Table 2. *Thickness reduction ratio*

Lithology	Thickness reduction ratio	Average ratio
Conglomerate/sandstone/limestone	1.2–1.4	1.3
Claystone/mudstone/siltstone	4.3–5	4.7

Table 3. *Characteristics of lithofacies and interpretation their depositional environments*

Lithofacies	Structure	Colour	Flora	Fauna	Range of thickness in meters	Process	Shallow shelf		Subaquatic part of inner-shelf delta				Delta plain							
													Lake delta							
							Shallow clay shelf	Shallow carbonate shelf	Prodelta	Delta front	Mouth bar	Lake bottom	outer	distal	proximal	Peat swamp	Overgrown lake	Overgrown plain	Crevasse channel	Distributary channel
C Coal/carbonaceous claystone	Massive	Black	Very common unlabelled	Lack	0.05–1.0	Deposition of plant remains Coalification										■				
R *Stigmaria* claystone/mudstone	Nodular	Dark grey grey light grey	Stigmaria appendixes plant chaff	Lack	0.1–4.0	Lack of flow Deposition of clayey or muddy suspension Deposition of plant remains Pedogenic											■	■		
Gm Massive conglomerate	Massive	Light grey	Lack	Lack	0.1–0.2	Deposition in upper plain bed (upper flow regime)														■
Ge Erosional scours conglomerate	Massive	Light grey	Large plant fragments	Lack	0.1	Development of erosional surfaces with cobbles Deposition in upper plain bed (upper flow regime)														■
Se Erosional scours sandstone	Massive	Light grey	Large plant fragments	Lack	0.1															■
Gs Scour-fill conglomerate	Massive	Light grey	Large plant fragments	Lack	0.5	Erosion of large scours and their filling in upper plain bed (upper flow regime)														■
Ss Scour-fill sandstone	Massive	Light grey	Large plant fragments	Lack	0.5															■
Sm Massive sandstone	Massive	Light grey	Plant chaff	Lack	0.5–2.0	Deposition in upper plain bed (upper flow regime)														■
Sx Large-scale cross-stratified sandstone	Cross-stratification	Light grey	Plant chaff	Lack	0.3–1.5	Rhythmic bed load transport in transverse bars or megaripples (lower or upper part of lower flow regime)														■
Sl Low-angle cross-stratified sandstone	Low-angle cross-stratification	Light grey	Plant chaff/ lack	Lack	0.6	Rhythmic bed load transport in washed-out megaripples and transverse bars (transition from lower to upper flow regime)														■
St Trough cross-stratified sandstone	Trough cross-stratification	Light grey	Plant chaff/ lack	Lack	0.5	Rhythmic bed load transport in megaripples (upper part of lower flow regime)														■
Sr Ripple cross-stratified sandstone	Ripple cross-lamination	Light grey	Plant chaff	Lack	0.2–2.7	Rhythmic bed load transport in ripples (lower part of lower flow regime)													■	■
Sf Flaser laminated sandstone	Flaser lamination	Light grey	Plant chaff	Lack	0.2–1.9	Rhythmic bed load transport in ripples (lower part of lower flow regime) and lack of flow – deposition from muddy and silty suspension					■				■					

Sh	Sh2 Horizontal stratified sandstone	Horizontal stratification	Light grey	Plant chaff	Lack	0.2–0.8	Deposition in upper plain bed (upper flow regime)														■
	Sh1 Horizontal laminated sandstone	Horizontal lamination	Light grey	Plant chaff	Lack	0.2–1.5	Deposition in lower plain bed from sandy suspension					■									
Sd Disturbed sandstone		Disturbed	Light grey	Plant chaff	Lack	0.2–1.1	Roots – *Stigmaria* disturbances, Unstable density stratification, Bioturbations					■									
FSd Disturbed sandy siltstone		Disturbed	Dark grey	Plant chaff	Lack	0.2–3.5	Unstable density stratification, Bioturbations Roots – *Stigmaria* disturbances, Shear fractures				■				■						
FSw Wavy laminated sandy siltstone		Wavy lamination	Dark grey grey	Plant chaff	Lack	0.2–1.4	Rhythmic bed load transport in current and wave ripples and lack of flow – deposition from muddy and silty suspension				■				■						
Fn Lenticular laminated siltstone		Lenticular lamination	Dark grey	Plant chaff	Lack	0.2–1.6					■				■						
Fh Horizontal laminated claystone and mudstone	Fh2 Horizontal laminated mudstone	Horizontal lamination	Dark grey	Plant chaff	Lack	0.2–5	Lack of flow – deposition from clayey and muddy suspension			■	■		■	■							
	Fh1 Horizontal laminated claystone		Black dark grey	Lack	Marine	2.0–4		■		■											
Fm massive claystone and mudstone	Fm5 grey massive mudstone	Massive	Grey light grey	Lack/plant chaff	Lack	0.5–3.5							■	■							
	Fm4 Black massive mudstone		Black dark grey	Very common plant chaff, large plant fragments	Lack								■	■							
	Fm3 Dark grey massive mudstone		Dark grey black	Plant chaff	Lack	0.2–0.6				■											
						1.4–5		■		■											
	Fm2 Grey massive claystone		Light grey grey grey-red	Lack/plant chaff	Lack	0.1–5.9							■	■							
			dark grey	plant chaff/lack	Lack	0.2–2.3							■	■							
	Fm1 Black massive claystone		Black dark grey	Lack	Marine / lack	0.2–17.8				■											
						1–3		■													
M Marl		Massive	Black dark grey	Lack	Marine	0.1–2.8	Lack of flow – deposition from clayey suspension Deposition of limy marine fauna remains		■												
L Limestone		Massive nodular	Dark grey grey, beige	Lack	Marine	0.2–12	Deposition of limy marine fauna remains		■												

Fig. 6. Examples of lithofacies. Units on scale bar, 1 cm. (**a**) Lithofacies L: grey massive limestone; siderite nodule in the lower part is visible Łęczna IG 13 well; depth 1.072.7–1.072.8 m. (**b**) Lithofacies Fm1: black massive claystone with detritus of fauna; Łęczna IG 13 well; depth 1.380.75 m. (**c**) Lithofacies Fm1: black massive claystone with detritus of fauna and siderite nodules; Łęczna IG 13 well; depth 1.380.70–1.380.75 m. (**d**) Lithofacies Fm3: dark grey massive mudstone; Lublin IG 2 well; depth 2101.3 m. (**e**) Lithofacies Fm4: black massive mudstone with very common plant fragments; Łęczna IG 13 well; depth 772.0 m. (**f**) Lithofacies Fm5: grey massive mudstone with plant chaff; Łęczna IG 13 well; depth 1.541.7 m. (**g**) Lithofacies R: grey muddy *Stigmaria* soil; slickenside related compaction is observable; Lublin IG 2 well; depth 1.823.7 m. (**h**) Lithofacies C: black massive carbonaceous claystone; Łęczna IG 9 well; depth 755.6 m.

indicates that originally there was also a considerable proportion of lithofacies of part 2.

Type IIc

Type IIc is represented by incomplete coarsening-upward cyclothems that, compared with type Ic, are lacking in part 3 or rarely part 1 (Fig. 8). The lithofacies composition of the remaining parts is identical to that found in type Ic. Type IIc consists of $0 \rightarrow 1 \rightarrow 2 \rightarrow 4$, $1 \rightarrow 2$ and $2 \rightarrow 3 \rightarrow 4$ subtypes. The present-day thicknesses of cyclothems IIc is *c.* 1–17 m (average 7 m). As in type Ic, the maximum thicknesses are found in parts 1. The

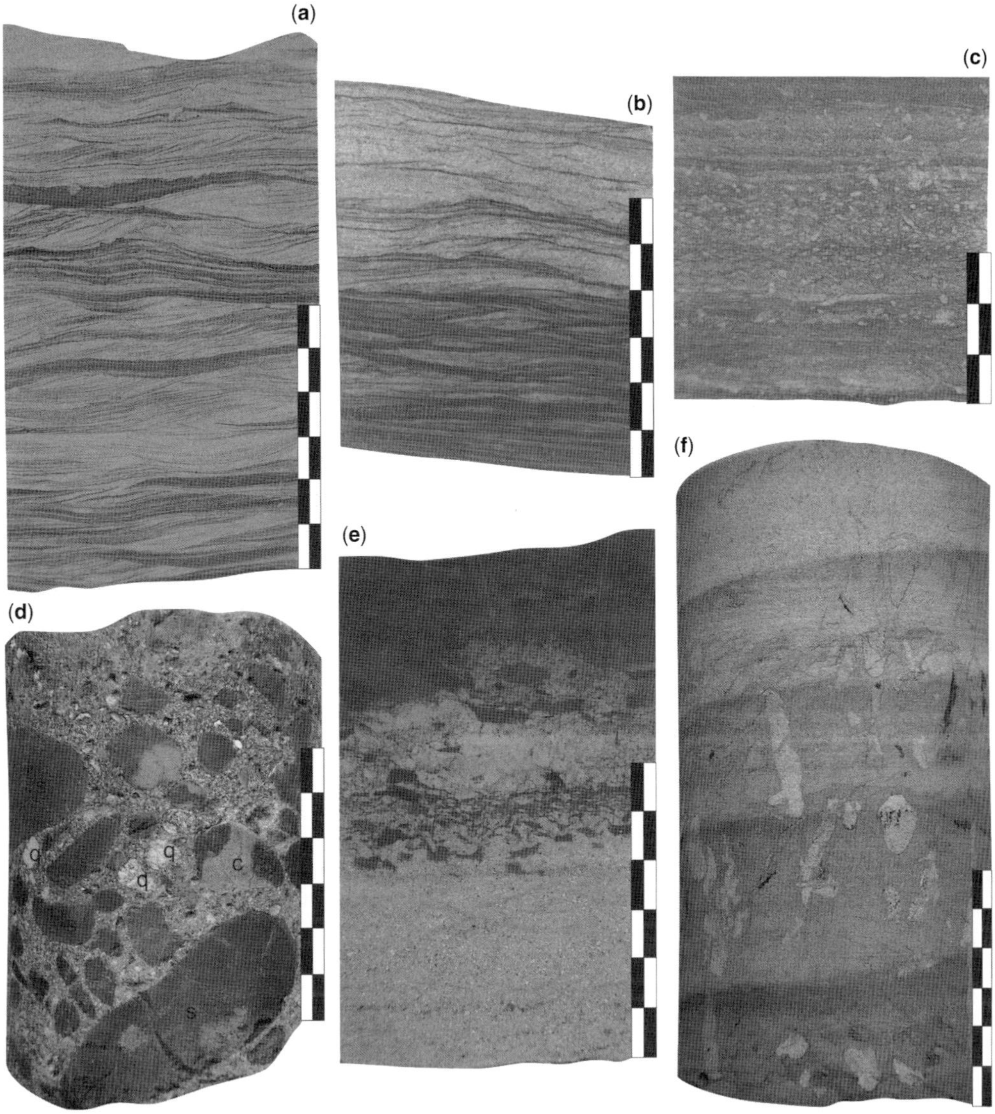

Fig. 7. Examples of lithofacies. Units on scale bar, 1 cm. (**a**) Lithofacies FSw: grey wavy laminated sandy siltstone; Łęczna IG 9 well; depth 1.084.0–1.084.1 m. (**b**) Lithofacies Fn: dark grey lenticular laminated siltstone higher up grading into light grey fine-grained flaser laminated sandstones Sf; Łęczna IG 13 well; depth 888.80–888.88 m. (**c**) Lithofacies FSd: dark grey disturbed sandy siltstones by bioturbations; Łęczna IG 9 well; depth 1.086.70–1.086.76 m. (**d**) Lithofacies Gs: grey-red massive conglomerate with siderite (s), clayey (c) and quartz (q) clasts; Łęczna IG 9 well; depth 1.266.50–1.266.59 m. (**e**) Lithofacies Sm: light grey massive medium-grained sandstone higher up into dark grey disturbed sandy siltstones FSd with unstable density stratification; Łęczna IG 9 well; depth 1.461.4–1.461.5 m. (**f**) Lithofacies FSd: dark grey disturbed sandy siltstones higher up into light grey fine-grained disturbed sandstones Sd by plant roots; quasi-horizontally rills after drill bit are observable; Lublin IG 2 well; depth 1.680.4–1.680.7 m.

thicknesses of parts 2, as well as of part 0, are also considerable. However, the decompacted thicknesses show the dominance of lithofacies of part 1, a considerable thickness proportion of part 2 and the smallest proportion of parts 0 and 4, as in type Ic.

Type IIIc

Type IIIc includes the following subtypes: $0 \rightarrow 1 \rightarrow 4$ and $1 \rightarrow 4$ (Fig. 9). The lithofacies composition of the parts of the cyclothem is the

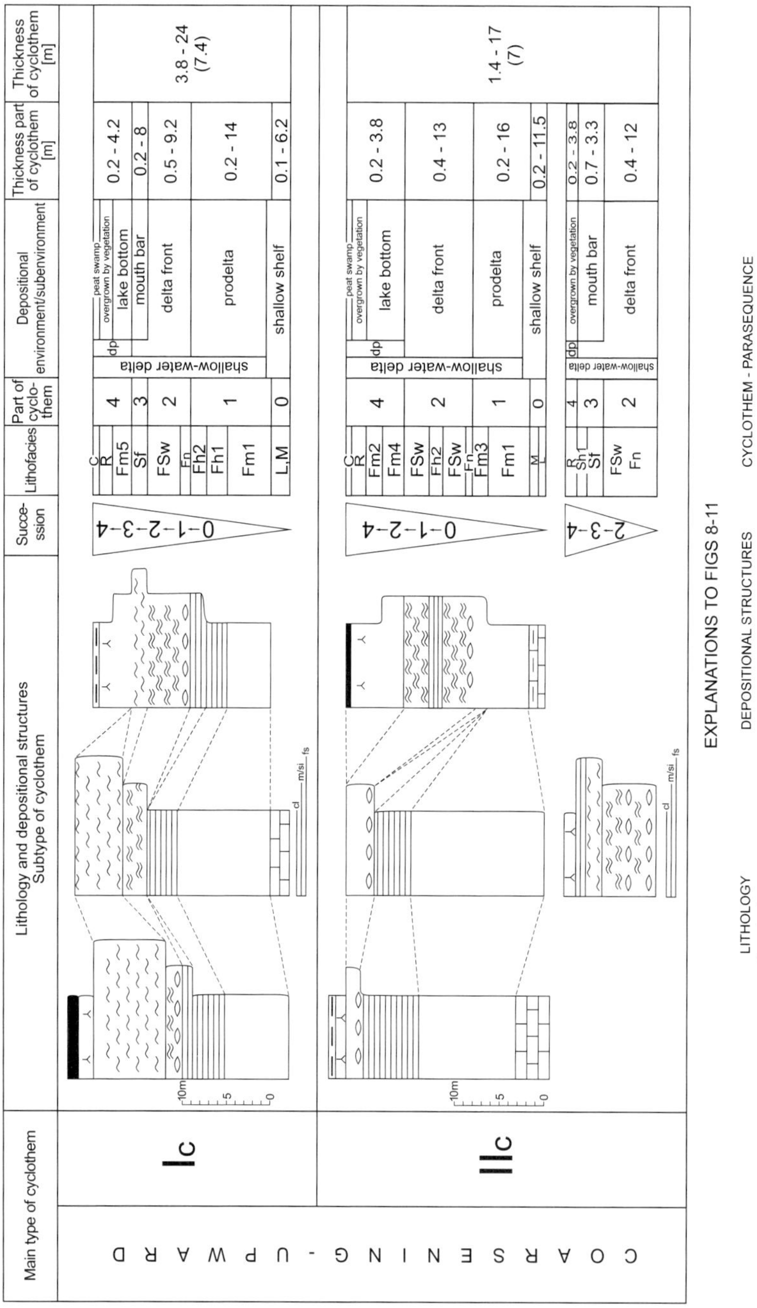

Fig. 8. Schematic example of coarsening-upward cyclothems types Ic and IIc with their depositional environments: dp, delta plain. Values of thickness reduction ratio used for calculations: claystone/mudstone/siltstone, 4.3–5; sandstone/limestone, 1.2–1.4.

Table 4. *Characteristics of coarsening-upward cyclothem type Ic and interpretation of its depositional environments and subenvironments*

Depositional environments and subenvironments		Part of cyclothem	Lithofacies	Lithology
Delta plain	Peat swamp	4	C	Coal, carbonaceous claystone
	Overgrown lake		R	*Stigmaria* claystone/mudstone (*Stigmaria* soil)
	Lake		Fm2, Fm4, Fm5, Fh2	Claystone/mudstone
	Crevasse channel		Sr	Fine-grained sandstone
	Distributary channel		Sr, Sx, St, Sl, Sh2, Sm, Se, Ss, Ge, Gs, Gm	Fine-, medium-, coarse-grained sandstone conglomerate
Subaquatic part of delta	Mouth bar	3	Sh1, Sf, Sd	Fine-grained sandstone
	Delta front	2	Fn, FSw, FSd, Fh2	Mudstone, siltstone, sandy siltstone
	Prodelta	1	Fh2 Fm3	Mudstone
			Fh1 Fm1	Claystone
Shallow shelf		0	L, M, Fm1	Limestone, marl claystone

same as in types Ic and IIc, indicating their genetic affinity. Cyclothem IIIc is a type of coarsening-upward cyclothem lacking grain-size coarsening, and hereafter is therefore separated into a single group termed 'non-gradational'. Part 1 can contain claystones or mudstones.

The current thicknesses of cyclothems IIIc are *c.* 1–20 m (average 6.7 m). The thickest part within them is part 0, although part 1 also shows considerable thickness. Correction for compaction reveals different relations within cyclothems IIIc. The dominant lithofacies were those of part 1. However, in addition to part 0, coals and *Stigmaria* soils also comprise a considerable proportion of part 4.

A comparison of the original thicknesses of Ic, IIc and IIIc cyclothems shows that the thickest parts 0 occurred in type IIIc, and the thinnest ones in type Ic. Parts 1 and 2 in type IIc were the thickest, while parts 3 and 4 had a maximum thickness in type Ic.

Type Id

Differences in the lithofacies composition and smaller thicknesses of cyclothems Id, IId and IIId were the basis for the distinction of three additional main types of cyclothems (Fig. 10). The most complete type of coarsening-upward cyclothem Id (Table 5) consists of parts represented by the following lithofacies, from base to top:

- part 1: claystones Fm2 and mudstones Fm4, Fm5;
- part 2: mudstones Fh2 and siltstones Fn, FSw;

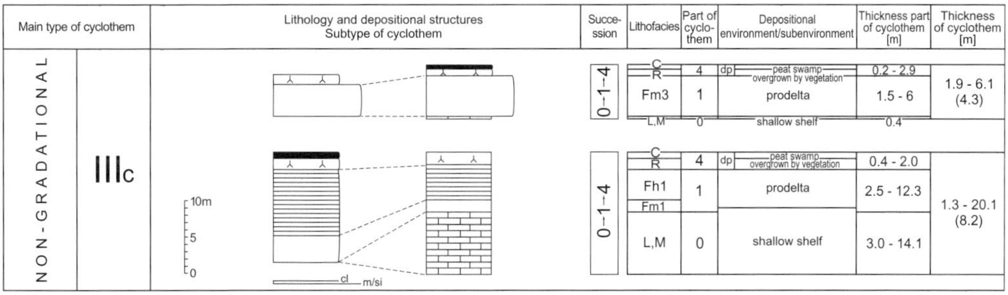

Fig. 9. Schematic example of non-gradational cyclothems types IIIc with their depositional environments: dp, delta plain. Values of thickness reduction ratio used for calculations: claystone/mudstone/siltstone, 4.3–5; limestone, 1.2–1.4.

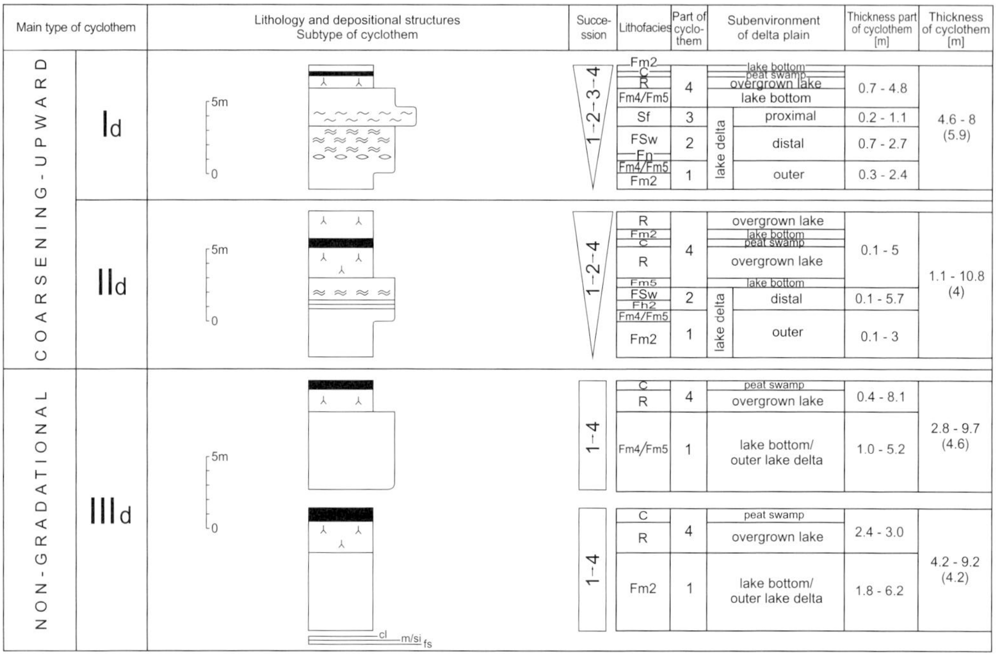

Fig. 10. Schematic example of coarsening-upward types Id, IId and non-gradational type IIId cyclothems with their depositional subenvironments (modified after Waksmundzka 2010). Values of average thickness reduction ratio used for calculations: claystone/mudstone/siltstone, 4.3–5; sandstone, 1.2–1.4.

- part 3: fine-grained sandstones Sf;
- part 4: claystones Fm2 and mudstones Fm4, Fm5; C coals and *Stigmaria* soils R.

Cyclothems Id are characterised by grain-size coarsening-upwards within parts 1, 2 and 3, and the presence of succession 1 → 2 → 3 → 4. The present-day thicknesses of the cyclothems are approximately 5–8 m (average 5.9 m). Among them, part 4 shows the greatest present-day and decompacted thicknesses. Lithofacies of parts 2 and 1 also have considerable thickness proportions.

Type IId

Cyclothem IId is an incomplete variety of cyclothem Id. The coarsening is observed in parts 1 and 2 as a transition of the claystone into the mudstone and then into silty ones (Fig. 10). Compared with cyclothem Id, cyclothem IId is lacking

Table 5. *Characteristics of coarsening-upward cyclothem type Id and interpretation of its depositional subenvironments*

Depositional subenvironments of delta plain		Part of cyclothem	Lithofacies	Lithology
Peat swamp		4	C	Coal, carbonaceous claystone
Overgrown lake or distal plain		4	R	*Stigmaria* claystone/ mudstone (*Stigmaria* soil)
Lake bottom		4	Fm2, Fm4, Fm5	Claystone/mudstone
LAKE DELTA	Proximal	3	Sf	Fine-grained sandstone
LAKE DELTA	Distal	2	Fh2, Fn, FSw	Mudstone, siltstone, sandy siltstone
LAKE DELTA	Outer/lake bottom	1	Fm2, Fm4, Fm5	Claystone/mudstone

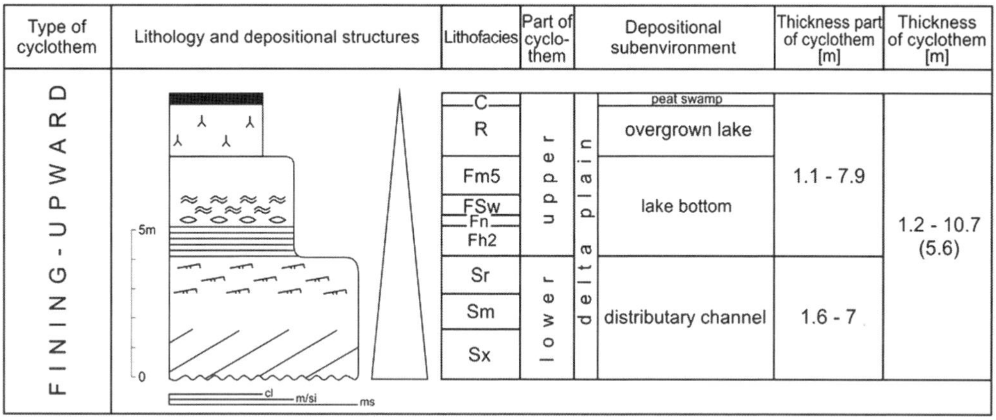

Fig. 11. Schematic example of fining-upward cyclothems with their depositional subenvironments. Values of thickness reduction ratio used for calculations: claystone/mudstone/siltstone, 4.3–5; sandstone, 1.2–1.4.

part 3 – sandstones Sf. Thus, the typical succession is 1 → 2 → 4. Recent thicknesses of the cyclothems range from 1 to 11 m (average 4 m). The greatest thicknesses, both present-day and decompacted, are observed for lithofacies of parts 2 and 4.

Type IIId

Type IIId cyclothems are non-gradational cyclothems lacking grain-size and coarsening-upward (Fig. 10). There are two subtypes of the same succession 1 → 4, which differ in lithofacies types composing part 1. In the first subtype, it is represented by mudstones Fm4 or Fm5, and in the other type by claystones Fm2. The present-day thicknesses of cyclothems IIId are *c.* 3–10 m (average 4.4 m). In the first subtype, part 4 is dominant in terms of thickness, while in the other subtype, part 1 is dominant.

Types IIc and IIIc are the most frequent. A comparison of the original thicknesses of corresponding parts within cyclothems of types Id, IId and IIId shows that the thickest parts 1 occurred in claystone subtype IIIc. The maximum thickness of part 2 is observed in type IId, and that of part 4 in mudstone subtype IIId.

Fining-upward cyclothems

The sections also contain numerous fining-upward cyclothems, mainly in thick beds of fluvial deposits (Waksmundzka 1998, 2010, 2012). However, single fining-upward cyclothems are occasionally observed within the sections composed of coarsening-upward and non-gradational cyclothems. These fining-upward cyclothems are genetically related to the above-described types of cyclothems.

The fining-upward cyclothem consists of lower and upper parts (Fig. 11). The lower parts are composed of fine-, medium- or coarse-grained sandstones represented here by lithofacies Sr, St, Sx, Sl, Sh2, Sm, Se and Ss. Locally, conglomerate lithofacies Gs, Ge or Gm are found at the base of the lower part. The upper parts are analogous to parts 4 of cyclothems Id, IId, and IIId, which are composed of siltstone, mudstone, claystone and phytogenic (coals and *Stigmaria* soils) lithofacies. Sandstones of the lower part pass upwards into siltstones FSw, Fn and FSd, and then into mudstones Fh2, Fm4 and Fm5 or into claystones Fm2. The cyclothem is topped by *Stigmaria* soils R, coals and carbonaceous claystones C. This lithofacies succession in the clastic-claystone part of the cyclothem is characterized by a fining-upward trend.

The current thickness of the fining-upward cyclothems is *c.* 1–11 m (average 5.6 m); the lower part is *c.* 1.6–7 m thick and the upper *c.* 1–8 m thick. After decompaction, lithofacies of the upper part account for 64–85% of the original thickness.

Identification of cyclothems on well logs

Sedimentological investigations of the cyclothems, made on continuous drill core intervals, were complemented by well logs analysis (gamma ray, neutron-gamma and caliper logs). The highest values on the curves correspond to claystones; mudstones and siltstones show lower values. Relatively low values are typical of sandstones, whereas very low values correspond to limestones and coals.

Boundaries of coarsening-upward cyclothems Ic (0 → 1 → 2 → 3 → 4 and 1 → 2 → 3 → 4) are readily identifiable on well logs (Figs 12 & 13).

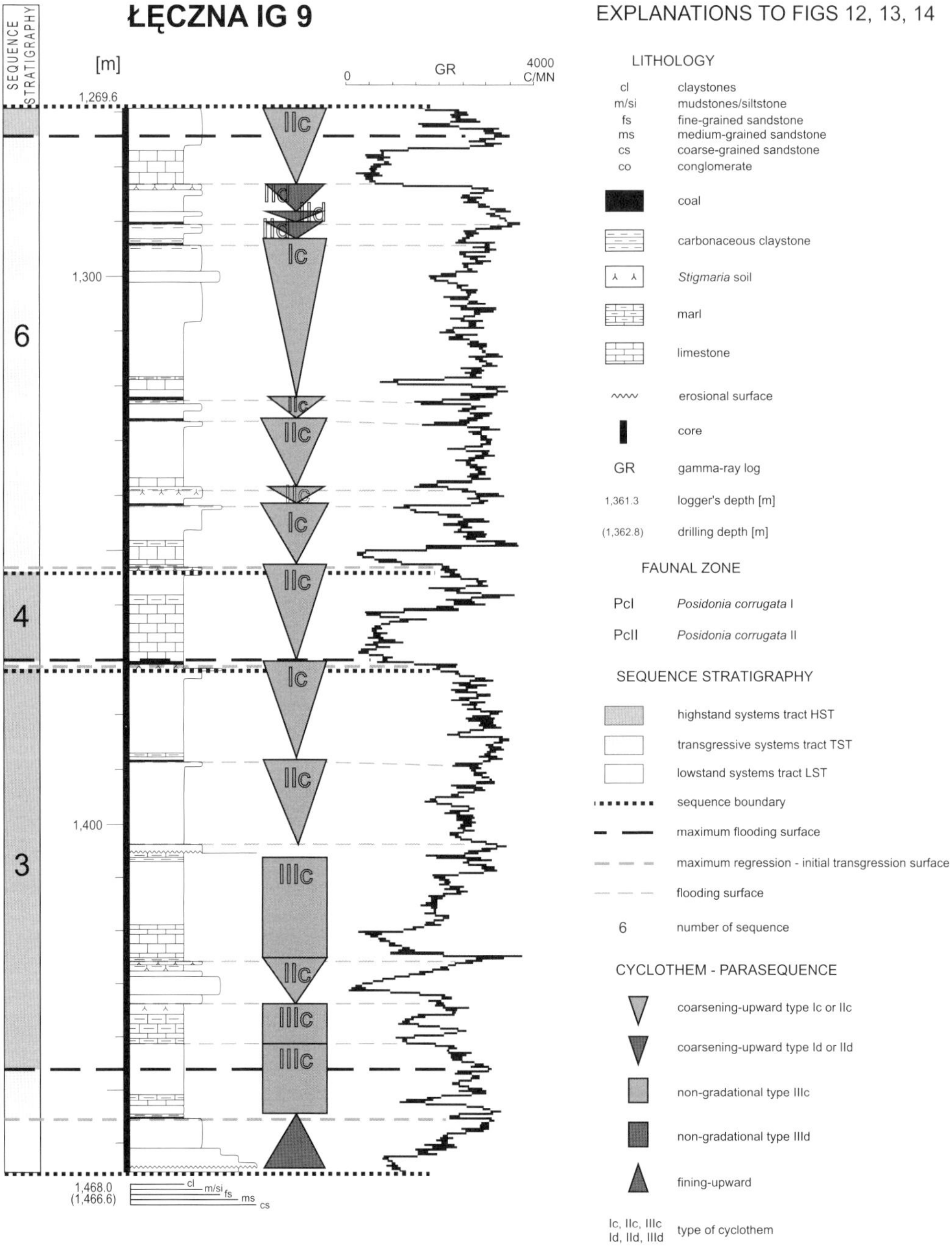

Fig. 12. Schematic example of coarsening-upward, non-gradational and fining-upward cyclothems that form the LST, TST and HST of sequences 3, 4 and 6 from the Łęczna IG 9 well section.

The type that contains part 0 begins with a block-shaped log of very low values corresponding to the limestones. The transition to claystones and mudstones of part 1 is sharp and associated with a rapid increase in radiation value. In mudstones and siltstones of part 2 and in sandstones of part 3, the values gradually decrease, giving the curve a funnel shape. In the successive parts of 1, 2 and 3,

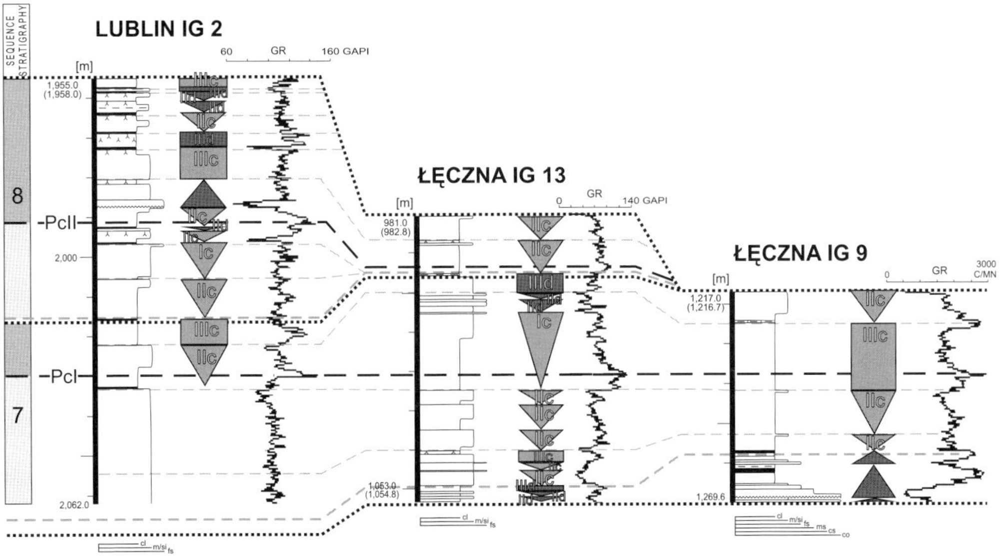

Fig. 13. Schematic example of coarsening-upward, non-gradational and fining-upward cyclothems that form the LST, TST and HST of sequences 7 and 8 from the Łęczna IG 9, Łęczna IG 13 and Lublin IG 2 well sections.

there is a simultaneous increase in grain diameter and a decrease in the content of clay minerals, which are recorded on the curve as continuously lower radiation values. Two types of curve shape may indicate the occurrence of part 4 in the cyclothems. The coal layer causes extremely low radiation values. It may overlap with the effect of low values corresponding to sandstones of part 3. The coal is not then evident. Claystones or mudstones show a block-shaped range of medium values for the curves. The beginning of cyclothems Ic, in which limestones of part 0 are absent, is marked by a sharp excursion towards high values. In the upper parts, the curve shape gradually decreases towards lower values. The cyclothems of this type are identified by the lack of high, block-shaped minima corresponding to limestones.

Within part 1 (cyclothems Ic), composed of claystones and mudstones, there is a gradual increase in the gamma values (up to the maximum) above a sharp excursion on the curve, which corresponds to the base of the cyclothem. From this point, the curve gradually decreases towards lower values, and its further course is typical for the transition from claystones to sandstones. This initial increase, up to abnormally high radiation values, is probably associated with an increase in the content of clay minerals and biogenic matter within part 1. Peaks of this type are usually associated with maximum flooding surfaces (MFSs), as described below.

Coarsening-upward cyclothems of type IIc are less clearly identified on gamma-ray curves (Figs 12 & 13). It is connected predominantly with monotonous lithology claystones, mudstones and siltstones of parts 1 and 2. The curve shape for the cyclothems showing the succession of $0 \rightarrow 1 \rightarrow 2 \rightarrow 4$ and $1 \rightarrow 2$ is similar to that of the corresponding parts observed in cyclothems Ic. There are also differences in the image of part 1. The curve may oscillate in the vertical section around high values and occasionally slightly towards higher values. Some cyclothems IIc, which are characterized by a small thickness of only a few metres, do not show a funnel shape on gamma-ray logs, and their distinction is possible only on the basis of lithofacies analysis of drill cores.

Non-gradational cyclothems IIIc (Figs 12–14) showing the succession $0 \rightarrow 1 \rightarrow 4$ start with a block shape of very low values due to the presence of limestones of part 0. Above, there may be a sharp transition to high values that oscillate within the interval corresponding to part 1. Extremely low values may occur at the top due to the presence of part 4, containing coals. In the cyclothems missing limestones, no lower effect of low values is observed on the curves. In contrast, when part 4 contains only carbonaceous claystones or *Stigmaria* soils, then the upper minimum may be absent.

Coarsening-upward cyclothems Id only occurred within sequence 17 (Fig. 5). They are not easily identifiable on gamma ray logs because of their small thicknesses and thin sandstone layers of part 1. However, in the interval consisting of segments 1, 2 and 3, there is a transition from higher to lower gamma values, with no clear funnel shape. At the top of these cyclothems, *Stigmaria* soils of

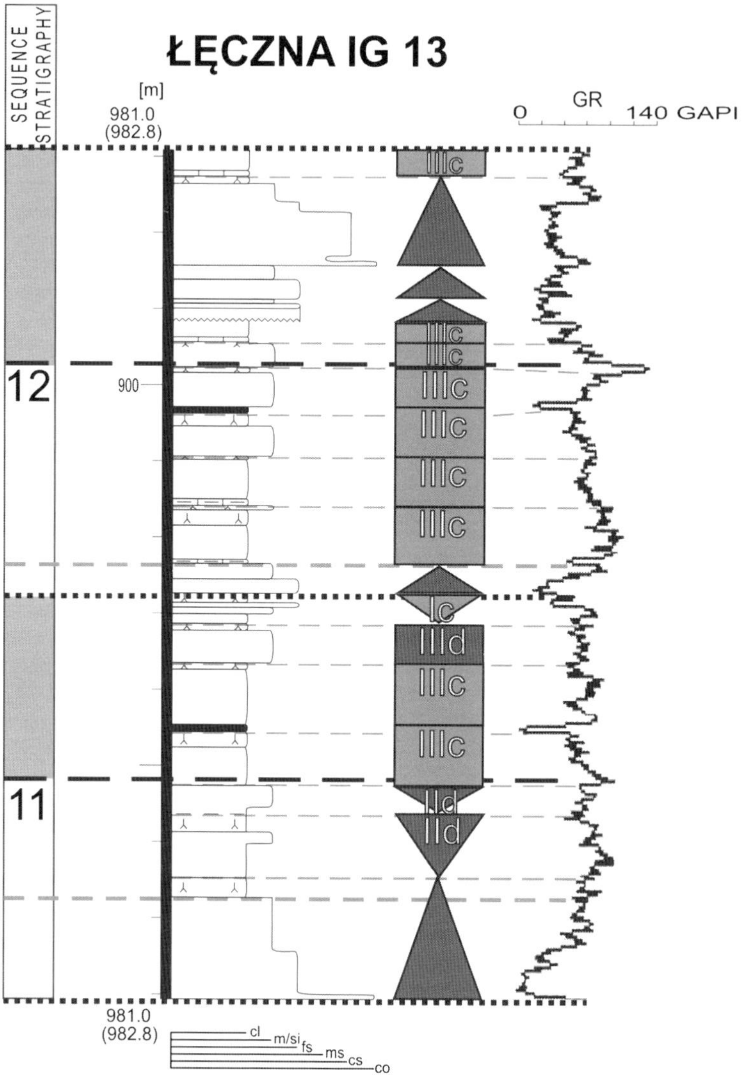

Fig. 14. Schematic example of coarsening-upward, non-gradational and fining-upward cyclothems that form the LST, TST and HST of sequences 11 and 12 from the Łęczna IG 13 well section.

part 4 are usually found, which show much higher gamma values than sandstone.

Confident identification of the top and bottom of coarsening-upward cyclothems IId and non-gradational cyclothems IIId is not possible based on well logs only (Figs 12 & 13) because of monotonous lithology and the small thicknesses of the cyclothems. Additionally, the gamma values on well logs are significantly affected by the high content of carbonized organic matter, which shifts the radioactivity of clay minerals towards the minima, preventing the interpretation of grain-size coarsening.

Interpretation of the boundaries of fining-upward cyclothems can be based on well logs (Figs 12–14). The lower parts composed of sandstone show clear block-shaped effects of low values on the curves. Above, there is a sharp transition towards higher gamma values corresponding to siltstones, mudstones and claystones of the upper parts. These values fluctuate slightly and pass at the top into anomalously low values corresponding to coal layers. In some cases, the transition from sandstones to fine-grained clay sediments is gradual and visible as increasing values on the curve, giving a bell-shaped effect.

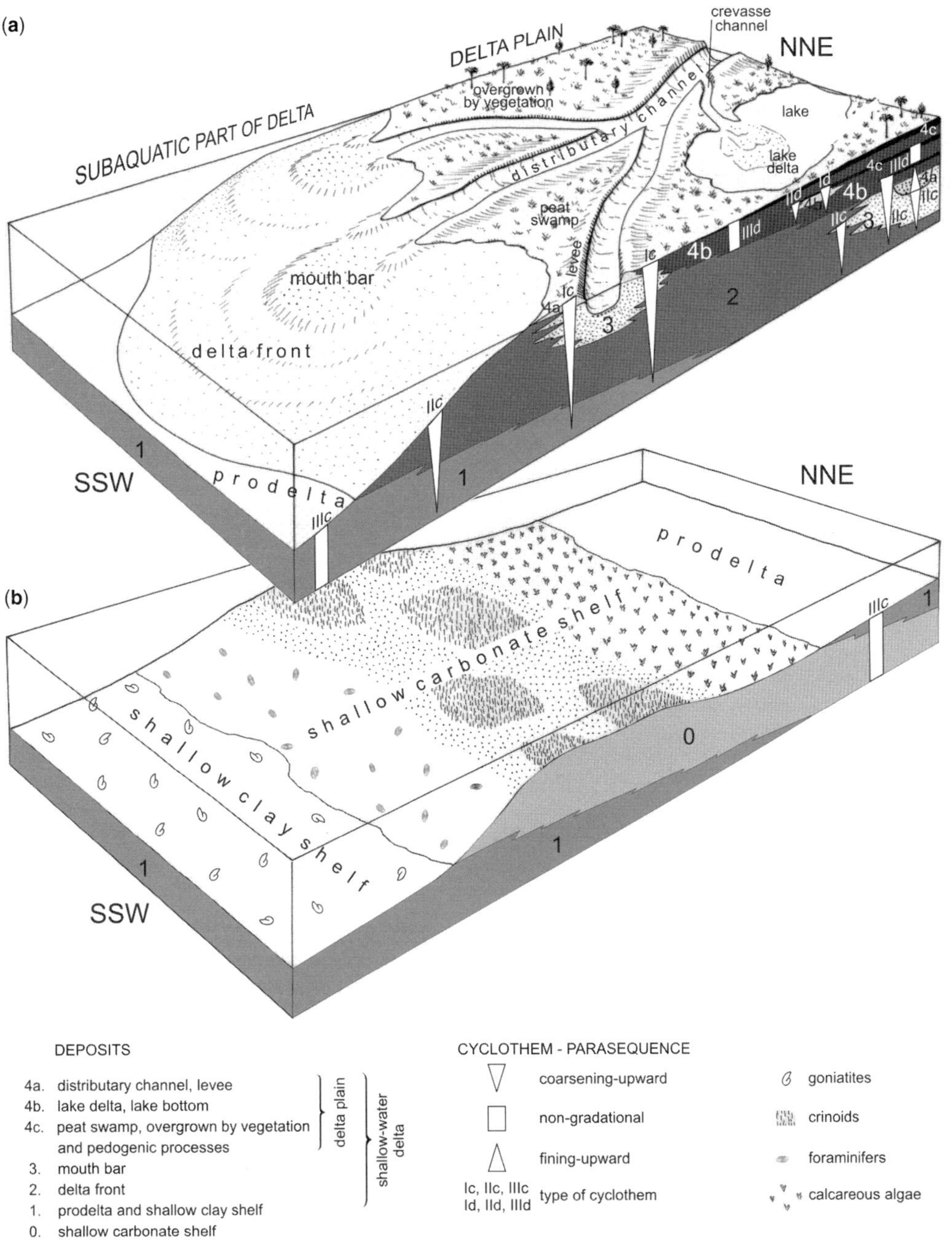

Fig. 15. Reconstruction of depositional environments and subenvironments formed during progradation phase of delta lobes: (**a**) shallow-water delta; (**b**) transition from prodelta to shallow carbonate and clay shelf.

Depositional environments and characteristics of delta systems

The presence of coarsening-upward cyclothems was the basic diagnostic feature, which pointed to their genetic relationship with a deltaic environment with subenvironments (Figs 8–10; Tables 3–5). Detailed analysis also allowed the identification of the shallow carbonate and clay shelf environments represented by lithofacies found within the cyclothems.

To categorize the delta type, within which the above-described Carboniferous deposits of the LB accumulated, the classification of Postma (1990, 1995) was used. The lithologies and facies, as well as the characteristics of the coarsening-upward cyclothems of the Carboniferous of the LB, indicate that they may have been formed within a fluvial-dominated shallow-marine delta with a delta plain of type D (deltaic system model no. 8). Its sedimentary environment is reconstructed in Figures 15a, b and 16. Deltas of this type prograded into a basin of relatively small depth ranging from *c.* 50 to 100 m, with a slightly sloping bottom at the mouths of channels and with a low-energy environment. The lithology of the deposits was highly affected by fluvial processes. A shallow depth of the marine basin was already suggested for limestones that formed further towards the open shelf relative to the deltaic deposits. Wave activity probably had little effect on the delta morphology, despite the wavy lamination observed in delta front deposits. Similarly, tides were also of low significance, as no features of tidal sedimentation have been found in the sediments. The low energy of the basin was probably related to the existence of a zone of carbonate sedimentation, further towards the open sea in the forefield of prodelta. In this zone, there were vast shallows that played the role of an energy barrier that reduced wave energy and protected the depositional areas extending closer to the shore (Skompski 1988).

Using the classification scheme of deltas developed by Porębski & Steel (2003), taking into account their position on the shelf depending on the relative sea-level position, the Carboniferous deltas of the LB are categorized into the group of inner-shelf deltas.

A recent analogue to the shallow-water deltas of the lower part of the Carboniferous section in the LB is the Mahakam Delta in Indonesia, which coexists with biogenic carbonate sedimentation (Wilson & Lokier 2002). However, for the upper part of the section, the Mississippi Delta is a good recent example (Coleman 1988). Among ancient analogues, there is a delta described by Moussavi-Harami & Brenner (1984) from the Upper Pennsylvanian of the Stanton Formation of North America. The classic examples are sedimentation models of Carboniferous deltas in the UK (Elliott 1975, 1976*a*, *b*) and Spain (Bowman 1985).

The occurrence of coarsening-upward cyclothems Id and IId and non-gradational cyclothems IIId within the Carboniferous lake-delta deposits of the LB results from the autocyclic variability of delta plain subenvironments. They migrate during progradation of lake-delta lobes, and overlap each other both vertically and laterally. The quite considerable thicknesses of the cyclothems were probably due to allogenic factors such as fluctuations in relative sea-level (Figs 16 & 17). The decompaction procedure of the present-day thicknesses of the Id, IId, IIId and fining-upward cyclothems that formed on the delta plain allowed the amount of accommodation space of this subenvironment to be estimated. This was *c.* 28–66 m during the formation of individual cyclothems. Similar environments described from the Mississippi Delta plain (Tye & Kösters 1986; Tye & Coleman 1989*a*, *b*) can be considered as a contemporary model for the Carboniferous lake deltas of the Lublin region. The above-cited sedimentation models of Carboniferous deltas from Spain (Bowman 1985) can be used as ancient examples.

Progradation of delta lobes, during which the deposits of individual subenvironments overlap each other and successively coarser fractions forming the coarsening-upward cyclothem of type Ic are deposited on prodelta sediments, is an autocyclic phenomenon. Delta lobes may develop in three phases. The first phase of progradation moves into the abandonment phase, which moves into the so-called post-abandonment phase (Elliott 1975, 1976*b*). However, complete cyclothems Ic account for only 12–22% of the studied sections. This indicates a relatively small proportion of autocyclic processes in the formation of the Carboniferous deposits of the LB. Thus, the origin of the dominant coarsening-upward cyclothems IIc and non-gradational cyclothems IIIc can be attributed to allocyclic factors. The factors influencing autocyclicity are commonly subsidence and/or eustatic sea-level changes (Turner & Whateley 1983; Cotter 1983). In examining the Carboniferous sections, a repeated alternation of successions deposited in shallow-shelf and deltaic environments and fluvial intervals was observed. A reason for such a succession is the fluctuations of relative sea-level that occurred in the Visean, Serpukhovian and Lower Bashkirian (Waksmundzka 2010). They resulted in the formation of depositional sequences including the LSTs, TSTs and HSTs. The sequences are bounded by unconformity surfaces and contain intrasequence unconformities, that is, the maximum regression–initial transgression and MFSs.

Relative sea-level fluctuations during the Carboniferous deposition in the LB can be linked to sea-level changes associated with the rhythm of successive glacial episodes (Fig. 17). They took place in the Southern Hemisphere on Gondwana Land in the latest Visean and, with shorter or longer breaks, lasted at least until the Early Permian (Veevers & Powell 1987). These fluctuations influenced the development of Carboniferous deposits in Western Europe (Ramsbottom 1977, 1979),

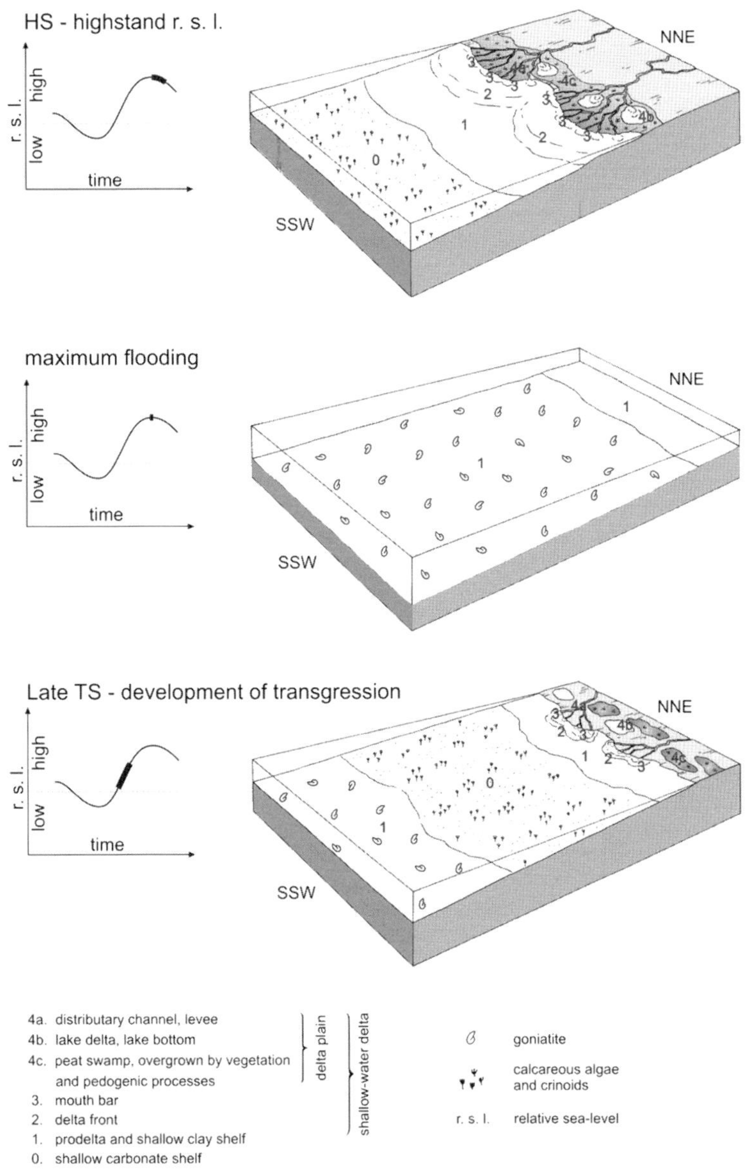

Fig. 16. Reconstruction of depositional environments and subenvironments of shallow-water delta, shallow carbonate and clay shelf that were formed during transgression, maximum flooding and highstand of relative sea-level (modified after Waksmundzka 2010).

Eastern Europe (Russian Platform; Alekseev *et al.* 1996, 2004) and North America (Saunders *et al.* 1979; Maynard & Leeder 1992; Smith & Read, 2001). The recent studies of Waters & Condon (2012) clearly show that four main glaciations took place from the Serpukhovian to the beginning of the Moscovian. Thus, glacio-eustatic fluctuations constrained by climatic changes may have been an essential factor affecting sedimentary processes in the Carboniferous of the LB (Skompski 2003). These factors significantly affected the formation of depositional sequences, parasequences – coarsening-upward and non-gradational cyclothems – and the development of lithofacies. They also greatly influenced the erosional processes associated with the successive relative sea-level lowstands and the formation of stratigraphic gaps (Saunders & Ramsbottom 1986; Waters & Condon

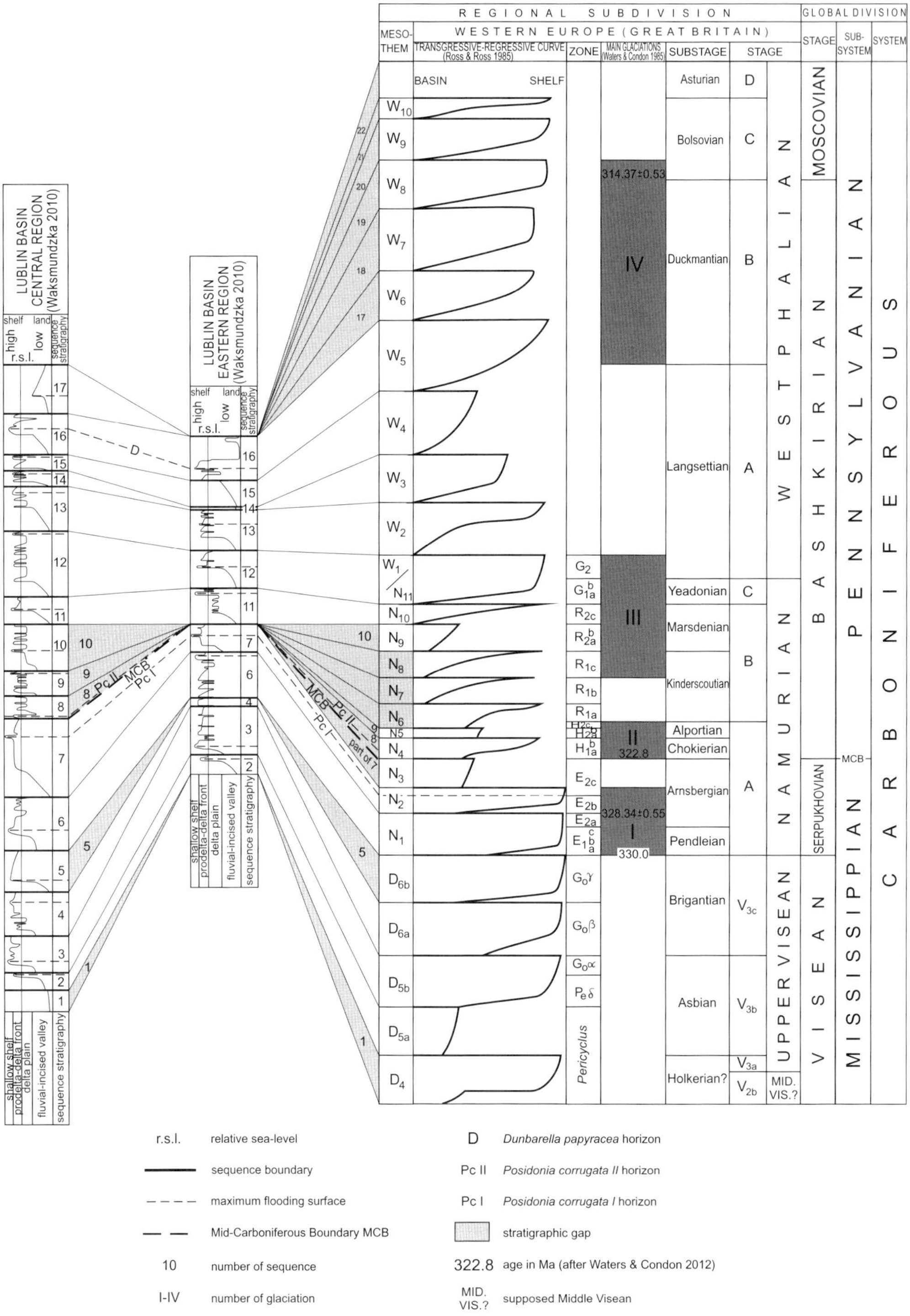

Fig. 17. Correlation of sequence stratigraphy and curves of relative sea-level oscillations of the Carboniferous in the central and eastern regions of the Lublin Basin with the transgressive–regressive curve for the Carboniferous of Western Europe (Ross & Ross 1985) and four main glaciations in Gondwana dated by Waters & Condon (2012).

2012). However, it seems that the effect of glacio-eustatic fluctuations was modified by local factors such as the supply of clastic material, subsidence rate and local tectonics. These factors significantly modified the Carboniferous sedimentary environments in the LB. The overall effect of their mutual relations with respect to the relative sea-level position is characterized later in this paper and illustrated in Figures 4 and 5.

Position of the cyclothems in the sequence stratigraphic scheme

Parasequences

The analysis of lithofacies and sedimentary cyclicity was the basis for developing a sequence stratigraphic model for the Carboniferous of the LB (Waksmundzka 2008*b*, 2010). Most of the identified types of cyclothems are parasequences that occur in parasequence sets, usually TSTs and HSTs (Figs 4, 5 & 12–14). The main issue arising from the definition of a parasequence, to be taken into account when determining its boundaries, is to locate the flooding surface. Within the Carboniferous the flooding surfaces are considered to coincide with a base of coals. This surface is related to the beginning of the formation of peat swamps, the development of which was affected by the rise of groundwater table, which was dependent, in turn, on the relative sea-level rise. The boundaries of cyclothems, composed of coal or carbonaceous claystone, coincide with the top of these lithofacies. These boundaries of cyclothems are shifted relative to the parasequence boundaries from the base to the top of coal or carbonaceous claystone. The second type of relation of boundaries is connected with the cyclothems in which there are no coals or carbonaceous claystones. In this case there is exact coincidence of the cyclothem and parasequence boundaries. In the investigated section, both types of boundary relation are observed.

Coal layers (and locally carbonaceous claystones) play a very important role in the sequence stratigraphy of the Carboniferous, because their basal surfaces correspond to stratigraphic surfaces of various types. The type of surface depends on the position of coal in the lithofacies succession of the section. Flooding surfaces representing parasequence boundaries occur at the base of most coals. The first flooding surface, overlying fluvial or delta plain lithofacies, is the initial transgression surface. The base of the coal bed can also be a sequence boundary when it is underlain and overlain by marine or deltaic lithofacies, and it is the only one to represent the terrestrial environment. In this case, it is always a correlative conformity of a typical sequence boundary running at the base of river channels or incised valleys.

Depositional sequence boundaries

Depositional sequences of type I *sensu* Vail & Todd (1981), bounded by unconformity surfaces, have been distinguished in the sequence stratigraphic scheme of the Carboniferous in the LB (Waksmundzka 2010). Most have been identified at the bases of river channels, and their belts or incised valleys developed as a result of strong fluvial erosion associated with the relative sea-level lowstand. Analysis of these deposits is beyond the scope of this paper.

In some parts of the sections, the sequence boundaries were placed at the bases of fining-upward cyclothems deposited in the delta-plain distributary channels and in the interdistributary bay (lithofacies R, C) due to the absence of deposits of river channels and incised valleys. The lower unconformities of sequences 3 and 7 in the Łęczna IG 9 well and of sequence 12 in the Łęczna IG 13 well were determined in this way. Several other sequence boundaries – that is, 4 and 6 (Łęczna IG 9 and IG 13) and 8 (Lublin IG 2) – were placed at the bases of coals, carbonaceous claystones or *Stigmaria* soils occurring in coarsening-upward cyclothems Ic and IIc, and in non-gradational cyclothem IIIc (Figs 12–14). These deposits were formed during the relative sea-level lowstand in peat swamps and vast delta-plain areas overgrown with vegetation.

Lowstand systems tracts

Above the lower boundaries of the sequences, there are relative sea-level lowstand deposits, bounded from the top by the maximum regression–initial transgression surfaces. They contain sandstone lithofacies Sx, St, Sm, Ss, SL and Sr and conglomerate lithofacies Gm, Gs and Ge, as well as fine-grained, clay and phytogenic lithofacies FSw, FSd, Fn, Fm4, Fm5, Fh2, R and C. They usually occur within fining-upward cyclothems, rarely within coarsening-upward cyclothems IId and non-gradational cyclothems IIId. These cyclothems were deposited in various delta-plain subenvironments (i.e. distributary channels, vegetation-overgrown areas) and in lakes and swamps.

Maximum regression–initial transgression surfaces

Initial transgression surfaces occur mostly at the bases of coals within coarsening-upward cyclothems that formed on the delta plain and are included in the LST (Figs 12–14). The formation of peat swamps with the deposition of organic material was associated both with a relative sea-level rise accompanied by a rise in the groundwater table level and with the flooding of the delta plain. In cases where coals are absent, the initial transgression surface coincides with the base of transgressive deposits overlying terrestrial sediments of the LST. These deposits are most commonly overlain by clay lithofacies Fm1 and Fh1 and carbonate lithofacies L and M, forming the lowest parts (0 and 1) of coarsening-upward cyclothems Ic and IIc and non-gradational cyclothems IIIc. In these types of cyclothem, corresponding to parasequences, the initial transgression surface exactly coincides with the base. The first transgressive deposits may have been deposited in prodelta, shallow clay shelf or carbonate shelf environments during a relative sea-level rise.

Transgressive systems tracts

The initial transgression surface is overlain by TST deposits bounded at the top by MFSs. They usually consist of several cyclothems – parasequences. Their lithofacies development and thicknesses vary depending on the stratigraphic position in the Carboniferous section. Clearly visible are the differences resulting from the location of the surveyed sections in different parts of the LB. The regional aspects of the Carboniferous lithologies are discussed hereafter.

The TST of sequences 6 and 12, characterized by considerable thicknesses, consists of several cyclothems. In other sequences (e.g. 2 and 10), these deposits are reduced to the lower part of cyclothem Ic, represented by lithofacies Fm1. A completely unique TST lithology is found in sequence 4, where it is represented by coals and carbonaceous claystones of lithofacies C. These deposits directly overlie the basal unconformity of the sequence, which is associated with a lack of deposition during the lowstand.

The cyclothems that constitute the TST are predominant in the studied Carboniferous sections, because they account for 39–49% of the total number of cyclothems. Among them, the most common are distal coarsening-upward cyclothems IIc. The greatest proportion of original thickness is due to claystone, mudstone and siltstone lithofacies of parts 1 and 2, deposited on a shallow clay shelf and in the distal parts of shallow-water delta lobes. Cyclothems IIIc, which are the most distal variety, account for a moderate proportion. The predominant lithofacies are claystones of part 1 and carbonates of part 0. Carbonate beds are found most often in the lower part of the Carboniferous section. Proximal coarsening-upward cyclothems Ic, in which in addition to lithofacies of parts 1 and 2 there is also mouth bar sandstone of part 3, also account for a moderate proportion. Their presence indicates that – during a relative sea-level rise – the area of their occurrence was influenced by the supply of sand sediments and may have been located closer to the source area. The TST of sequence 11 contains quite numerous coarsening-upward cyclothems IId (Fig. 14), the formation of which was associated with the development of lakes and sedimentation within small lake deltas. The development of lakes and their filling with sediments occurred at least in three stages, as indicated by a set of parasequences – coarsening-upward cyclothems IId.

The formation of TST deposits was affected mainly by aggradation of sediments on a shallow clay and carbonate shelf, and in areas of prodelta and delta plain. In some parts of the basin, the sediments prograded over the slope and mouth bar of a shallow-water delta at that time. However, the end result of these depositional processes was retrogradation of the shoreline, which was shifted landwards as the relative sea-level rose.

Maximum flooding surfaces

These intrasequence unconformities separate TST and HST deposits. Their formation is associated with the maximum sea depth and a slowdown in sedimentation. Frequently, they run within lithofacies Fm1 or Fh1 of part 1 deposited on a shallow clay shelf. These surfaces were identified based on specific lithologies of the lithofacies, indicating the presence of a great amount of dispersed organic matter, goniatite fauna, as well as the corresponding peaks on gamma-ray curves. In other cases, the surfaces run within the relatively deepest prodelta lithofacies (part 1), with relatively high radioactivity, or at the bases of carbonate beds (part 0). Lithofacies that formed during the maximum depth of the marine basin, and corresponding surfaces, most frequently occur within coarsening-upward cyclothems IIc, and quite often in cyclothems Ic and non-gradational cyclothems IIIc (Figs 12–14). They account for 8–13% of the total number of cyclothems in the analysed Carboniferous sections.

Highstand systems tracts

HSTs are bounded by MFSs and basal unconformities of the overlying sequences. Erosion, which resulted in the formation of these unconformities, contributed to a considerable reduction in thickness of the HST deposits (e.g. sequence 12) or to their complete removal (e.g. sequence 13). The cyclothems that form the HST account for 9–38%

of the total number of cyclothems in the analysed Carboniferous sections. Their lower number in present-day sections compared to the number of TST-containing cyclothems is probably due to erosional reduction of the HST deposits. By taking into account the degree of this reduction, it seems that the original number of cyclothems may have been comparable to the number of TST-containing cyclothems or greater. Thus, the characteristics of these deposits practically refer to the preserved, lower parts of the HST.

The HST deposits usually consist of several cyclothems – parasequences; rarely is their number more than 10. Coarsening-upward cyclothems IIc are quantitatively dominant. They are accompanied by cyclothems Ic and non-gradational cyclothems IIIc. Coarsening-upward cyclothems IId and non-gradational cyclothems IIId are observed in the HST sections of the Lublin IG 2 well. In this section, the number of cyclothems Ic is much smaller than in the TST. The lithologies of the lithofacies in the cyclothems are generally similar to those in the TST deposits. Thus, their formation was mainly influenced by aggradation of sediments in the areas of shallow clay and carbonate shelf, prodelta and delta plain. Progradational processes within the delta front and mouth bar of the shallow-water delta also took place, but on a smaller scale. The process of filling the accomodation space presumably occurred on a greater scale during the late HST, the deposits of which were erosionally reduced or entirely removed.

Identification of depositional systems tracts on the basis of cyclicity

Based on a statistical analysis, the following correlation between the types of cyclothem and the phase of depositional sequences within which they occur (Fig. 18a, b) has been proposed. The least diverse set of cyclothems is found in the LST deposits. They can be composed exclusively of fining-upward cyclothems of fluvial (Lublin IG 2 well) or deltaic (Łęczna IG 9 well) origin. In addition to fining-upward cyclothems, they can contain coarsening-upward cyclothems IId and non-gradational IIId. A characteristic feature of the LST deposits is the lack of coarsening-upward cyclothems Ic and IIc, and non-gradational cyclothems IIIc.

The TST and HST deposits are represented by all types of cyclothems, excluding type Id, which is found only in the HST of sequence 17. Among the TST deposits, coarsening-upward cyclothems IIc are dominant (16–23%). A considerable proportion of coarsening-upward cyclothems Ic (up to 15%) and non-gradational IIIc (12%) can also be observed. The percentage of cyclothems IId and IIId is much lower (3–6%).

Coarsening-upward cyclothems IIc dominate within the HST deposits; however, compared with the TST deposits, their proportion is smaller (13–16%). Coarsening-upward cyclothems Ic (3–4%) and non-gradational IIIc (5–6%) occur in smaller proportions. There is also a greater proportion of coarsening-upward cyclothems IId and non-gradational IIId in the HST deposits. Despite the differences between the studied sections in the frequency of occurrence of the individual types of cyclothems, the above-characterized relationships are also manifested in the calculated average percentages (Fig. 18b).

Evolution of depositional environments and subenvironments in relation to relative sea-level fluctuations: palaeoclimatic implications

Visean, Serpukhovian

Deposits of sequences 2–7, corresponding to the Visean and Serpukhovian, are characterized based on data from the Łęczna IG 13 and Łęczna IG 9 wells (Fig. 4). In this region, deposits of the oldest sequence 1 are missing because the Carboniferous deposition started later than in the centre of the basin. In this area, there is a stratigraphic gap associated with the lack of sequence 5 including the upper Brigantian. Its presence is probably related to a period of erosion and non-deposition that occurred during the relative sea-level lowstand of this sequence (Waksmundzka 2010). This succesion was not reached by Lublin IG 2 well.

The thicknesses of the Visean and Serpukhovian deposits are *c.* 150–180 m and 140–190 m, respectively. They include cyclothems – parasequences – which are more numerous in the vicinity of Łęczna IG 13 well. This suggests that the accommodation space of the LB generally increased towards the SW. By thickness, the dominant deposits are those of the TST and HST. The LST deposits are usually very thin or missing. Quantitatively, coarsening-upward cyclothems IIc are dominant. Non-gradational cyclothems IIIc and coarsening-upward cyclothems Ic are also quite common (Fig. 19b). Shallow shelf limestones, marls and claystone, prodelta claystones and mudstones and delta front mudstones and siltstones show the greatest thickness within the cyclothems. Carbonates are repeatedly interbedded with delta-lobe and shallow clay shelf deposits, which is particularly visible in the HST of sequence 3 and in the TST of sequence 6 (Fig. 12). This indicates that the development of the transgression until its maximum, as well as the filling of

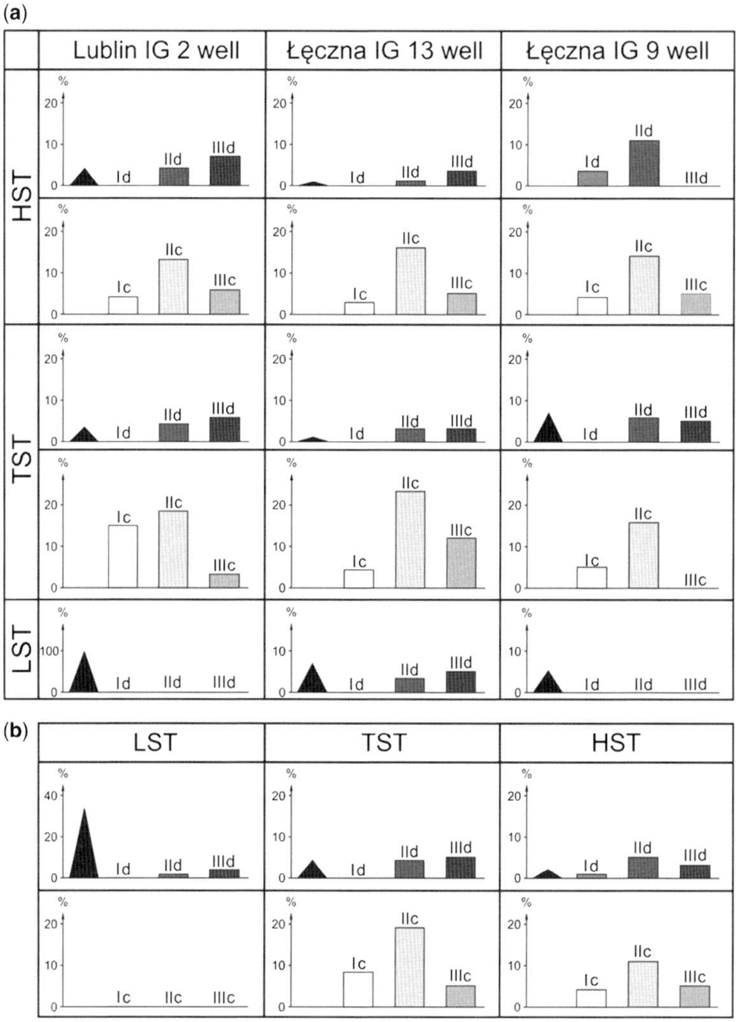

Fig. 18. (**a**) Percentage account of different types of cyclothems that form sequences 11–17 (middle and upper part of Bashkirian). (**b**) Percentage account of different types of cyclothems that form sequences 2–10 (Visean, Serpukhovian and lower part of Bashkirian).

the basin with sediments during the highstand, were associated with numerous sea-level fluctuations.

Similar Carboniferous sequences in terms of lithofacies and origin are known from different regions, including Kansas (Maynard & Leeder 1992; Archer *et al.* 1994) and Illinois (Smith & Read 1999, 2001) in the USA, as well as from the UK (Tucker 2003).

As regards thickness, the greatest proportion of shallow shelf and prodelta lithofacies suggests a generally marine and distal-delta sedimentary regime, and mainly aggradation of sediments.

Progradation of delta-lobe deposits during both the transgression and relative sea-level highstand contributed to the filling of the accommodation space to a lesser extent. It can be supposed that the present-day smaller number of cyclothems Ic than theoretically expected within the HST is associated with the preservation of their lower parts or their complete removal. Erosion may also have reduced the frequency of cyclothems IId and IIId formed in delta-plain lakes.

An estimate of the depth of the marine basin, based on the original thicknesses of coarsening-upward cyclothems Ic, indicates that it generally became greater during the consecutive transgressions. It seems that the lowest depth was during the transgression of sequence 2. It then reached

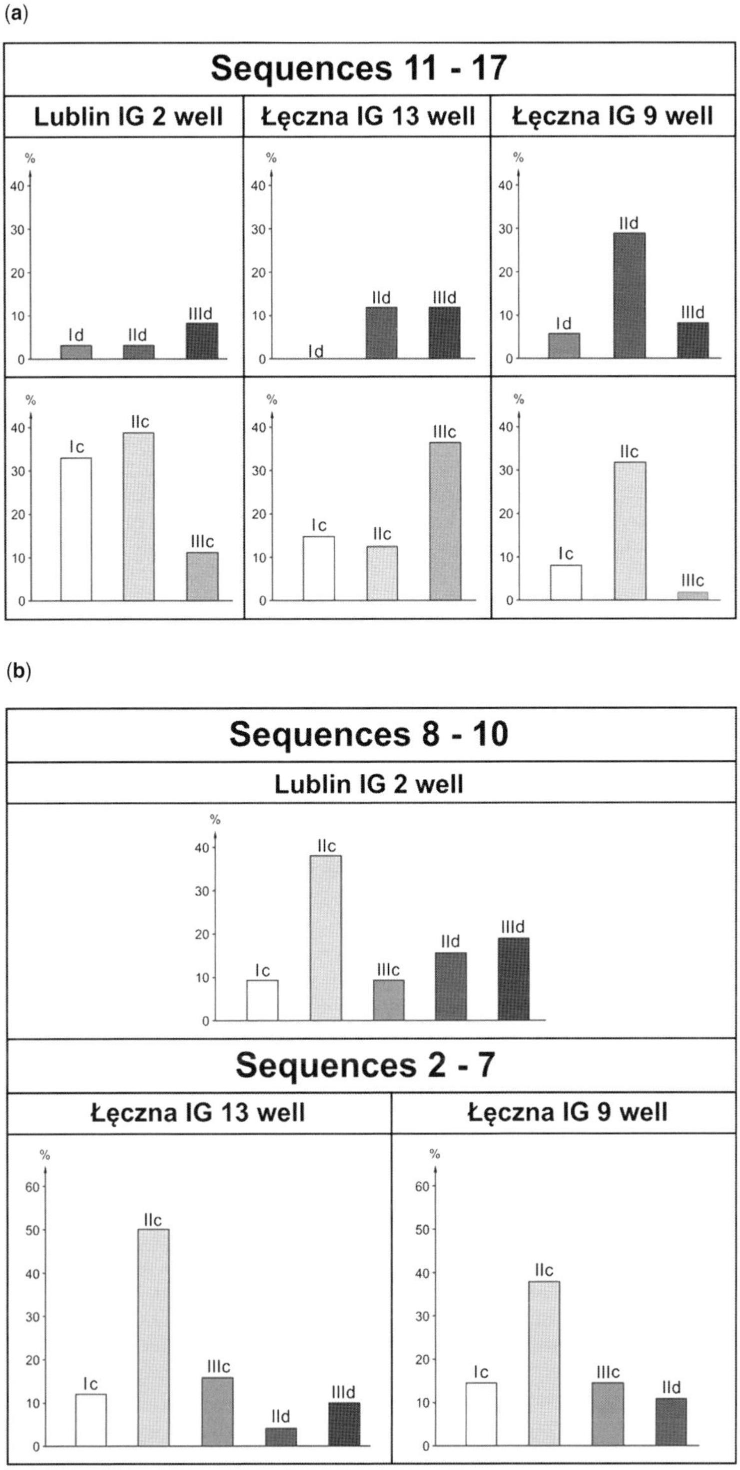

Fig. 19. (**a**) Percentages of different types of cyclothems that form the LST, TST and HST. (**b**) Average percentages of different types of cyclothems that form the LST, TST and HST.

c. 80–85 m during the deposition of sequence 3, and *c.* 90 m during deposition of sequence 6. The greatest depth of the marine basin of *c.* 100 m is estimated for sequence 7 in the Łęczna IG 13 section, which was accompanied by the maximum accommodation space.

The fluctuations occurring during the formation of sequences 6 and 7 were probably related to the first main glaciation in Gondwana, dated by Waters & Condon (2012) at the early Pendleian to late Arnsbergian. However, it did not cause any significant drop in the relative sea-level in the LB. This indicates that local factors predominantly influenced the deposition at that time. The greatest depth of the marine basin and the creation of large accommodation space are interpreted for sequences 6 and 7. These estimates confirm the earlier suggestions that the transgression of sequence 7 had the greatest extent compared to the extents of transgressions from both the lower and upper sequences in the Carboniferous of the LB (Waksmundzka 2010). It seems that the main factor responsible for the development of these phenomena was the maximum subsidence of the sedimentary basin.

Bashkirian

Deposits of sequences 8–10, corresponding to the lower part of the Bashkirian, have been found only in the Lublin IG 2 well located in the central part of the LB (Fig. 4). Their cumulative thickness is *c.* 200 m. In the other wells, this section was entirely (or almost entirely) eroded during the successive lowstands of sequences 9–11. The result of these processes was the formation of a stratigraphic gap, whose the longest range recorded among others in the eastern region, covers the interval from the late Arnsbergian to the early Marsdenian. A smaller gap (comprising latest Alportian–Kinderscoutian) is observed in the central part of the basin (Waksmundzka 2010). It seems that the formation of this gap was associated with the second and third main glaciations in Gondwana dated by Waters & Condon (2012). Because there was no relative-sea level rise during the interglacial interval between these glaciations, it can be assumed that it was due to deceleration of the subsidence rate in the LB. However, the slowdown was not uniform during the second glaciation (Chokierian–Alportian), because a stronger subsidence in the central region affected the behaviour of the deposits of sequences 8–10. Regarding thickness, the TST and HST deposits are dominant. The LST deposits are thin in sequences 8–9 and thick in sequence 10.

The section consists of coarsening-upward cyclothems – parasequences – among which type IIc is quantitatively dominant. In contrast to the Visean and Serpukhovian deposits, coarsening-upward cyclothems IId and non-gradational cyclothems IIId account for a significant proportion here (Fig. 19b). Coarsening-upward cyclothems Ic and non-gradational cyclothems IIIc occur in smaller proportions. Coarsening-upward cyclothems Ic and IIc, and non-gradational cyclothems IIIc differ substantially from those of sequences 2–7, because they are commonly lacking in part 0 (i.e. carbonates). Thus, the formation of the lower part of Lower Bashkirian sediments was dominated by aggradation of the deposits of cyclothems IIc in the environments of shallow clay shelf, prodelta and shallow-water delta front. The progradational accumulation of deposits was not significant, but it took place in the proximal parts of delta lobes in the slope and mouth bar areas. The considerable proportion of coarsening-upward cyclothems IId and non-gradational cyclothems IIId, as well as their significant thickness, indicates fairly long periods of deposition in an environment of rapidly aggrading delta plain, on which lakes and peat swamps developed.

Deposits of the lower part of the Bashkirian were deposited mainly in the deltaic sedimentation regime. The original thickness of cyclothems Ic shows that the maximum depth of the marine basin was *c.* 50 m and was less than that existing at the time of deposition of the Visean and Serpukhovian deposits. It was also associated with a smaller accommodation space of the basin. The deposition of the lower part of the Bashkirian deposits was affected by glacioeustatic sea-level fluctuations associated with the second and the beginning of the third main glaciations in Gondwana, dated by Waters & Condon (2012).

Deposits of sequences 11–17, corresponding to the middle and upper parts of the Bashkirian, are based on research performed in all the studied wells (Fig. 5). Their thickness is *c.* 280–520 m. This part of the Carboniferous section is substantially different from the older deposits of sequences 2–10 because it is characterized by a significant thickness of LST deposits. These deposits accumulated in a fluvial environment and usually filled incised valleys. They do not fall within the scope of this paper (Waksmundzka 2010, 2012). The greatest thickness proportion of these deposits is found in the Łęczna IG 9 well located farthest to the east; the relatively smallest proportion is observed in the Lublin IG well 2 located farther to the SW. The thickness proportion of TST and HST deposits is greatest in this section, too. In the other sections, their proportion is smaller. Sequences, similar in terms of lithofacies structure and thickness relations between individual depositional systems tracts, have been described from the Upper Carboniferous of the Ruhr area of Germany (Hampson *et al.* 1999; Süss *et al.* 2001),

United Kingdom (Martinsen *et al.* 1995; Waters *et al.* 2008), Ireland (Hampson *et al.* 1997), Wales (George 2001), North Sea (O'Mara *et al.* 1999), Russia (Briand *et al.* 1998), Ukraine (Izart *et al.* 1996, 2002, 2003*a*, *b*) and from the Pennsylvanian of the USA (Aitken & Flint 1994).

The sections contain coarsening-upward and non-gradational cyclothems corresponding to parasequences. The relatively greatest quantitative proportion of coarsening-upward cyclothems Ic in the whole Carboniferous section is found in the Lublin IG 2 well (Fig. 19a). This proportion decreases towards the east. Thus, the central region was under the greatest influence of progradational accumulation of shallow-water delta lobes. Cyclothems Ic, which formed during the transgression, show a very small thickness of mouth bar sandstones, while those more common, deposited during the highstand, are characterized by a greater thickness. The depths of the marine basin, estimated for some sequences from the thickness of coarsening-upward cyclothems Ic, ranged from 50 to 70 m.

Quantitatively, coarsening-upward cyclothems IIc are dominant, except in the Łęczna IG 13 well, where mostly non-gradational cyclothems IIIc occur. These cyclothems contain thin parts 0 – carbonates – deposited in a shallow carbonate shelf. In the eastern region, aggradational accumulation of prodelta and shallow clay shelf sediments was predominant. Progradation within the delta front and mouth bars was a minor process. This area was situated in a more distal zone of deltaic deposition under the influence of carbonate sedimentation. Aggradation of delta plain deposits, in lake deltas, peat swamps and areas overgrown by vegetation, took place in this region mainly during the transgression of sequence 12. Fairly numerous coarsening-upward cyclothems IId and non-gradational cyclothems IIId were deposited in these environments. Estimates of the original thickness of cyclothems IId indicate that the maximum accommodation space of the lakes may have been *c.* 66 m.

The middle Bashkirian deposits (sequences 11–16) accumulated in a deltaic sedimentary regime with a high proportion of the prograding submarine part of the delta during the highstand, and with a long aggradation of sediments of lake deltas, lakes and delta-plain peat swamps.

The last interlayer of marine deposits, which includes the *Dunbarella papyracea* Zone, lies within sequence 16. The sea depth was up to 70 m and was probably greater than in the early Bashkirian, but shallower than in the Visean and Serpukhovian. The deposition of the middle part of the Bashkirian deposits was glacioeustatically affected by sea-level fluctuations associated with the third main glaciation in Gondwana dated by Waters & Condon (2012) at the late Kinderscoutian–early Langsettian, and with the next interglacial interval. At that time, a lower subsidence rate affected deposition in the region of the Łęczna IG 9 and Łęczna IG 13 wells, where the systems of large incised valleys were formed in the Langsettian during the lowstand of sequences 13, 14 and 15. Local synsedimentary block tectonics associated with the Kock Fault Zone (Narkiewicz *et al.* 2007) may have also affected their development in the region of the Łęczna IG 13 well.

The uppermost part of the studied Carboniferous section, corresponding to the upper part of the Bashkirian (sequence 17), is represented by fluvial deposits within which intervals composed of coarsening-upward cyclothems Id and IId, deposited in lake deltas, are found. The decline of marine and deltaic sedimentation recorded in the Carboniferous section above sequence 16 may have also been caused by glacioeustatic factors related to the fourth main glaciation in Gondwana, which took place during the early Duckmantian–Bolsovian (Waters & Condon 2012). Despite this, the accommodation space of the basin developed due to local subsidence and was filled with fluvial sediments until the early Moscovian (Waksmundzka 2008*b*).

Conclusions

Depositional sequences consisting of coarsening-upward and non-gradational cyclothems in the Visean, Serpukhovian and Bashkirian sections of the LB have been characterized. The accompanying single fining-upward cyclothems are also described. The most complete coarsening-upward cyclothems Ic and Id contain the following parts, from base to top: 0, 1, 2, 3 and 4. Incomplete cyclothems IIc and IId contain parts 0, 1, 2 and 4, while the least complete non-gradational cyclothems IIIc and IIId consist of parts 0, 1 and 4. These types do not show the feature of grain-size coarsening.

The coarsening-upward and non-gradational cyclothems correspond to parasequences. The boundaries of the cyclothems that contain no coals and carbonaceous claystones coincide exactly with the boundaries of parasequences. Other boundaries of the cyclothems are shifted relative to the parasequence boundaries from the bases to the tops of coals and carbonaceous claystones.

The LST deposits may be composed of fining-upward cyclothems, rarely of coarsening-upward cyclothems IId and non-gradational cyclothems IIId. Among the TST deposits, coarsening-upward cyclothems IIc are dominant. The relative proportion of cyclothems Ic and non-gradational cyclothems IIIc is moderate. The percentage of cyclothems IId and IIId is much lower compared to the TST sections. Coarsening-upward cyclothems IIc dominate within the HST deposits; however, compared with the TST, their proportion is smaller. Cyclothems Ic and non-gradational cyclothems IIIc occur in smaller proportions and there is also a greater proportion of coarsening-upward cyclothems IId and non-gradational cyclothems IIId in the HST deposits. Complete cyclothems Ic account for only 12–22% of the Carboniferous section. This suggests a relatively small proportion of autocyclic processes in the formation of deposits of the LB. The origin of the dominant coarsening-upward cyclothems IIc and non-gradational cyclothems IIIc can be attributed to allocyclic, mainly glacio-eustatic factors.

During the Visean and Serpukhovian, a generally marine and distal-delta regime dominated. It is represented by shallow shelf and prodelta environments, in which aggradation of sediments mainly took place. High subsidence rates in the sedimentary basin contributed to the generation of the maximum accommodation space. The Lower Bashkirian deposits accumulated mainly in a deltaic sedimentary regime, at lower accommodation rates. The same environment predominated in the middle Bashkirian. However, it was characterized by a high proportion of the prograding submarine part of the delta during the HS, and by long aggradation of sediments of lake deltas, lakes and delta-plain swamps. In the upper Bashkirian, the deltaic and marine regime came to an end and terrestrial environments (i.e. fluvial and lacustrine) developed.

Relative sea-level fluctuations during the Carboniferous deposition in the LB can be associated with sea-level changes generated by glacial episodes in Gondwana connected with climatic changes. The lithology and facies development of the sections, as well as the features of the cyclothems – parasequences – indicate that the deposition was affected by the four main glaciations in Gondwana. Sea-level fluctuations also substantially affected both erosional processes related to the successive relative sea-level lowstands and the formation of stratigraphic gaps. The influence of glacio-eustatic fluctuations was modified by local factors, such as subsidence rate and local tectonics.

My gratitude is expressed to S. Skompski for supervising my PhD research. I am also deeply indebted to M. Słowakiewicz for encouraging me to write this paper. Thanks are expressed to K. Leszczyński for translating the manuscript, to the reviewers and A. Gąsiewicz for scientific and linguistic advice in making the text readable. B. Waksmundzki kindly helped in drawing the block diagrams and E. Tarka helped with the first computer version of the lithological sections of wells. The basis of this paper is the well sections and their interpretation of lithofacies and stratigraphy from my PhD thesis (Waksmundzka 2005). The presented results have been supported by two grants from the Polish Committee for Scientific Research (6 P201 026 07 and PBZ 02-03), a project supported by the National Fund for Environmental Protection and Water Management (NFOŚiGW), and a project supported from the statutory revenues of the Polish Geological Institute.

References

AITKEN, J. F. & FLINT, S. S. 1994. High-frequency sequences and the nature of incised-valley fills in fluvial systems of the Breathitt Group (Pennsylvanian), Appalachian Foreland Basin, Eastern Kentucky. *In*: DALRYMPLE, R. W., BOYD, R. & ZAITLIN, B. A. (eds) *Incised-Valley Systems*. SEPM Special Publications, Tulsa, **51**, 353–368.

ALEKSEEV, A. S., KONONOVA, L. I. & NIKISHIN, A. M. 1996. The Devonian and Carboniferous of the Moscow Syneclise (Russian Platform): stratigraphy and sea-level changes. *Tectonophysics*, **268**, 149–168.

ALEKSEEV, A. S., GOREVA, N. V., ISAKOVA, T. N. & MAKHLINA, M. KH. 2004. Biostratigraphy of the Carboniferous in the Moscow Synecline, Russia. *In*: WORK, D. M. (ed.) *Newsletter on Carboniferous Stratigraphy*. IUGS Subcommission on Carboniferous Stratigraphy, Augusta, **22**, 28–35.

ARCHER, A. W., LANIER, W. P. & FELDMAN, H. R. 1994. Stratigraphy and depositional history within incised-paleovalley fills and related facies, Douglas Group (Missourian/Virgilian; Upper Carboniferous) of Kansas, USA. *In*: DALRYMPLE, R. W., BOYD, R. & ZAITLIN, B. A. (eds) *Incised-Valley Systems*. SEPM Special Publications, Tulsa, **51**, 175–189.

BALDWIN, B. & BUTLER, C. O. 1985. Compaction Curves. *American Association of Petroleum Geologists Bulletin*, **69**, 622–626.

BOWMAN, M. B. J. 1985. The sedimentology and palaeogeographic setting of late Namurian–Westphalian a basin-fill successions in the San Emiliano and Carmenes areas of NW Leon, Cantabrian Mts, NW Spain. *Anais Da Faculdade De Ciencias*, **64**, 117–171.

BRIAND, C., IZART, A. *ET AL*. 1998. Stratigraphy and sequence stratigraphy of the Moscovian, Kasimovian and Gzhelian in the Moscow Basin. *Bulletin de la Société Géologique de France*, **169**, 35–52.

CEBULAK, S 1988. Geological outline of sub-Carboniferous basement. *In*: DEMBOWSKI, J. & PORZYCKI, J. (eds) *Carboniferous of the Lublin Coal Basin*. Prace Instytutu Geologicznego, Warsaw, **122**, 31–34, 228. [In Polish with English summary]

CEBULAK, S. & PORZYCKI, J. 1966. Lithological-petrographical characteristics of the deposits of the Lublin Carboniferous. *In*: RÜHLE, W. (ed.) *Carboniferous Sediments in the Lublin Basin*. Prace Instytutu Geologicznego, Warsaw, **44**, 21–53. [In Polish with English summary]

COLEMAN, J. M. 1988. Dynamic changes and processes in the Mississippi River delta. *Geological Society of America Bulletin*, **100**, 999–1015.

COLEMAN, J. M. & WRIGHT, L. D. 1975. Modern river deltas: variability of processes and sand bodies. *In:*

BROUSSARD, M. L. (ed.) *Deltas, models for exploration*. Houston Geological Society, 99–150.

COTTER, E. 1983. Shelf, paralic, and fluvial environments and eustatic sea-level fluctuations in the origin of the Tuscarora Formation (Lower Silurian) of Central Pennsylvania. *Journal of Sedimentary Petrology*, **53**, 25–49.

DEMBOWSKI, Z. 1970. Sedymentacja cykliczna w utworach westfalu w Lubelskim Zagłębiu Węglowym. *Kwartalnik Geologiczny*, **14**, 904–906. [In Polish with English summary].

DUFF, P. MCL. D. & WALTON, E. K. 1962. Statistical basis for cyclothems: a quantitative study of the sedimentary succession in the East Pennine Coalfield. *Sedimentology*, **1**, 235–255.

ELLIOTT, T. 1975. The sedimentary history of a delta lobe from a Yoredale (Carboniferous) cyclothem. *Proceedings of the Yorkshire Geological Society*, **40**, 505–536.

ELLIOTT, T. 1976*a*. Sedimentary sequences from the Upper Limestone Group of Northumberland. *Scottish Journal of Geology*, **12**, 115–124.

ELLIOTT, T. 1976*b*. Upper Carboniferous sedimentary cycles produced by river-dominated, elongate deltas. *Journal of the Geological Society, London*, **132**, 199–208.

GEORGE, G. T. 2001. Late Yeadonian (Upper Sandstone Group) incised valley supply and depositional systems in the South Wales peripheral foreland basin: implications for the evolution of the Culm Basin and for the Silesian hydrocarbon plays of onshore and offshore UK. *Marine and Petroleum Geology*, **18**, 671–705.

GRADZIŃSKI, R., DOKTOR, M. & SŁOMKA, T. 1995. Depositional environments of the coal-bearing Cracow Sandstone Series (upper Westphalian), Upper Silesia, Poland. *Studia Geologica Polonica*, **108**, 149–170.

HAMPSON, G. J., ELLIOTT, T. & DAVIES, S. J. 1997. The application of sequence stratigraphy to Upper Carboniferous fluvio-deltaic strata of the onshore UK and Ireland: implications for the Southern North Sea. *Journal of the Geological Society, London*, **154**, 719–733.

HAMPSON, G. J., STOLLHOFEN, H. & FLINT, S. 1999. A sequence stratigraphic model for the Lower Coal Measures (Upper Carboniferous) of the Ruhr district, north-west Germany. *Sedimentology*, **46**, 1199–1231.

IZART, A., BRIAND, C., VASLET, D., VACHARD, D., COQUEL, R. & MASLO, A. 1996. Stratigraphy and sequence stratigraphy of the Moscovian in the Donets Basin. *Tectonophysics*, **268**, 189–209.

IZART, A., VACHARD, D. & VASLET, D. 2002. Sedimentology of the Upper Carboniferous and Lower Permian in Dnieper and Donets basins. *In*: HILLS, L. V., HENDERSON, M. & BAMBER, E. W. (eds) *Carboniferous and Permian of the World, XIV International Congress on the Carboniferous and Permian*. Canadian Society of Petroleum Geologists, Alberta, Memoirs, **19**, 120–143.

IZART, A., LE NINDRE, Y., STEPHENSON, R., VASLET, D. & STOVBA, S. 2003*a*. Quantification of the control of sequences by tectonics and eustacy in the Donets Basin and on the Russian Platform during Carboniferous and Permian. *Bulletin de la Société Géologique de France*, **174**, 93–100.

IZART, A., STEPHENSON, R. ET AL. 2003*b*. Sequence stratigraphy and correlation of the Late Carboniferous and Permian in CIS, Europe, Tethyan area, North Africa, China, Gondwanaland and USA. *Palaeogeography, Palaeoclimatology, Palaeoecology*, **196**, 59–84.

KOMBRINK, H., BESLY, B. M. ET AL. 2010. Carboniferous. *In*: DOORNENBAL, J. C. & STEVENSON, A. G. (eds) *Petroleum Geological Atlas of the Southern Permian Basin Area*. EAGE Publications, Houten, 81–99.

KOREJWO, K. 1958. The Carboniferous at Strzyżów on the Bug river – Eastern Poland. *Biuletyn Instytutu Geologicznego*, **136**, 7–128. [In Polish with English abstract]

LEEDER, M. R. & STRUDWICK, A. E. 1987. Delta-marine interactions: a discussion of sedimentary models for Yoredale-type cyclicity in the Dinantian of Northern England. *In*: MILLER, J., ADAMS, A. E. & WRIGHT, V. P. (eds) *European Dinantian Environments*. 115–130.

MARTINSEN, O. J., COLLINSON, J. D. & HOLDSWORTH, B. K. 1995. Millstone Grit cyclicity revisited, II: sequence stratigraphy and sedimentary responses to changes of relative sea-level. *In*: PLINT, A. G. (ed.) *Sedimentary Facies Analysis a Tribute to the Research and Teaching of Harald G. Reading*. IAS, Oxford, Special Publications, **22**, 305–327.

MAYNARD, J. R. & LEEDER, M. R. 1992. On the periodicity and magnitude of Late Carboniferous glacio-eustatic sea-level changes. *Journal of the Geological Society, London*, **149**, 303–311.

MIALL, A. D. 1977. A review of the braided-river depositional environment. *Earth-Science Reviews*, **13**, 1–62.

MIALL, A. D. 1978. Lithofacies types and vertical profile models in braided river deposits: a summary. *In*: MIALL, A. D. (ed.) *Fluvial Sedimentology*. Canadian Society of Petroleum Geologists, Alberta, Memoirs, **5**, 597–604.

MITCHUM, R. M. JR. 1977. Seismic stratigraphy and global changes of sea level, Part 1: glossary of terms used in seismic stratigraphy. *In*: PAYTON, C. E. (ed.) *Seismic Stratigraphy – Applications to Hydrocarbon Exploration*. American Association of Petroleum Geologists, *Memoirs*, Tulsa, **26**, 205–212.

MOUSSAVI-HARAMI, R. & BRENNER, R. L. 1984. Deltaic sedimentation on a carbonate shelf: Stanton Formation (Upper Pennsylvanian), Southeastern Kansas. *American Association of Petroleum Geologists Bulletin*, **68**, 150–163.

MUSIAŁ, Ł. & TABOR, M. 1979. Stratygrafia karbonu Lubelskiego Zagłębia Węglowego na podstawie makrofauny. *In*: MIGIER, T. (ed.) *Stratygrafia Węglonośnej Formacji Karbońskiej w Polsce, II Sympozjum Sosnowiec*, 4–5 maja 1977, Wydawnictwa Geologiczne, Warsaw, 35–43. [In Polish]

MUSIAŁ, Ł. & TABOR, M. 1988. Macrofaunal stratigraphy of Carboniferous. *In*: DEMBOWSKI, Z. & PORZYCKI, J. (eds) *Carboniferous of the Lublin Coal Basin*. Prace Instytutu Geologicznego, Warsaw, **122**, 88–122, 232–233. [In Polish with English summary]

NARKIEWICZ, M., MIŁACZEWSKI, L., KRZYWIEC, P. & SZEWCZYK, J. 1998. Outline of the Devonian depositional architecture in the Radom-Lublin area. *In*: NARKIEWICZ, M. (ed.) *Sedimentary Basin Analysis of the Polish Lowlands*. Prace Państwowego Instytutu Geologicznego, Warsaw, **165**, 57–72. [In Polish with English abstract]

NARKIEWICZ, M., JAROSIŃSKI, M., KRZYWIEC, P. & WAKSMUNDZKA, M. I. 2007. Regional controls on the Lublin Basin development and inversion in the Devonian and Carboniferous. *Biuletyn Państwowego Instytutu Geologicznego*, **422**, 19–34. [In Polish with English abstract]

O'MARA, P. T., MERRYWEATHER, M., STOCKWELL, M. & BOWLER, M. M. 1999. The Trent Gas Field: correlation and reservoir quality within a complex Carboniferous stratigraphy. *In*: FLEET, A. J. & BOLDY, S. A. R. (eds) *Petroleum Geology of Northwest Europe: Proceedings of the 5th Conference*. Geological Society, London, 809–821.

PORĘBSKI, S. J. & STEEL, R. J. 2003. Shelf-margin deltas: their stratigraphic significance and relation to deepwater sands. *Earth-Science Reviews*, **62**, 283–326.

PORZYCKI, J. 1979. Litostratygrafia osadów karbonu Lubelskiego Zagłębia Węglowego. *In*: MIGIER, T. (ed.) *Stratygrafia Węglonośnej Formacji Karbońskiej w Polsce, II Sympozjum Sosnowiec, 4–5 maja 1977*, Wydawnictwa Geologiczne, Warsaw, 19–27. [In Polish]

PORZYCKI, J. 1980. *Utwory węglonośne wizenu i namuru Lubelszczyzny*. PhD thesis, Polish Geological Institut Upper Silesian Branch, Sosnowiec. [In Polish]

PORZYCKI, J. 1988. History of geological survey and discovery of the Lublin Coal Basin. Lithologic and sedimentologic characteristics of Carboniferous deposits. *In*: DEMBOWSKI, Z. & PORZYCKI, J. (eds) *Carboniferous of the Lublin Coal Basin*. Prace Instytutu Geologicznego, Warsaw, **122**, 9–18; 40–76; 226–227; 229–231. [In Polish with English summary]

PORZYCKI, J. & ZDANOWSKI, A. 1995. Southeastern Poland (Lublin Carboniferous Basin). *In*: ZDANOWSKI, A. & ŻAKOWA, H. (eds) *The Carboniferous System in Poland. Karbon w Polsce*. Prace Państwowego Instytutu Geologicznego, Warsaw, **168**, 102–109.

POSTMA, G. 1990. An analysis of the variation in delta architecture. *Terra Nova*, **2**, 124–130.

POSTMA, G. 1995. Causes of architectural variation in deltas. *Geology of Deltas*, 3–16.

PULHAM, A. J. 1989. Controls on internal structure and architecture of sandstone bodies within Upper Carboniferous fluvial-dominated deltas, County Clare, western Ireland. *In*: WHATELEY, M. K. G. & PICKERING, K. T. (eds) *Deltas: Sites and Traps for Fossil Fuels*. Geological Society, London, Special Publication, **41**, 179–203.

RAMSBOTTOM, W. H. C. 1977. Major cycles of transgression and regression (mesothems) in the Namurian. *Proceedings of the Yorkshire Geological Society*, **41**, 261–291.

RAMSBOTTOM, W. H. C. 1979. Rates of transgression and regression in the Carboniferous of NW Europe. *Journal of the Geological Society, London*, **136**, 147–154.

READ, W. A. & DEAN, J. M. 1976. Cycles and subsidence: their relationship in different sedimentary and tectonic environments in the Scottish Carboniferous. *Sedimentology*, **23**, 107–120.

RIEGEL, W. 1991. Coal cyclothems and some models for their origin. *In*: EINSELE, G., RICKEN, W. & SEILACHER, A. (eds) *Cycles and Events in Stratigraphy*, 733–750.

ROSS, C. A. & ROSS, J. R. P. 1985. Late Paleozoic depositional sequences are synchronous and worldwide. *Geology*, **13**, 194–197.

RUST, B. R. 1978. A classification of alluvial channel systems. *In*: MIALL, A. D. (ed.) *Fluvial Sedimentology*. Canadian Society of Petroleum Geologists, Calgary, *Memoirs*, **5**, 187–198.

SAUNDERS, W. B. & RAMSBOTTOM, W. H. C. 1986. The mid-Carboniferous eustatic event. *Geology*, **14**, 208–212.

SAUNDERS, W. B., RAMSBOTTOM, W. H. C. & MANGER, W. L. 1979. Mesothemic cyclicity in the mid-Carboniferous of the Ozark shelf region? *Geology*, **7**, 293–296.

SCRUTON, P. C. 1960. Delta building and the deltaic sequence. *In*: SHEPARD, F. P., PHLEGER, F. B. & ANGEL, T. H. (eds) *Recent Sediments, Northwest Gulf of Mexico*. American Association of Petroleum Geologists Bulletin, 82–102.

SKOMPSKI, S. 1988. Limestone microfacies and facies position of Upper Visean sediments in north-eastern part of the Lublin Coal Basin. *Przegląd Geologiczny*, **1**, 25–30. [In Polish with English abstract]

SKOMPSKI, S. 1995. Succession of limestone microfacies as a key to the origin of the Yoredale-type cyclicity (Viséan/Namurian, Lublin Basin, Poland). *XIII International Congres Carboniferous–Permian*, Kraków, Abstract, 133.

SKOMPSKI, S. 1996. Stratigraphic position and facies significance of the limestone bands in the subsurface Carboniferous succession of the Lublin Upland. *Acta Geologica Polonica*, **46**, 171–268.

SKOMPSKI, S. 1998. Regional and global chronostratigraphic correlation levels in the late Viséan to Westphalian succession of the Lublin Basin (SE Poland). *Geological Quarterly*, **42**, 121–130.

SKOMPSKI, S. 2003. Onset of the late Paleozoic Gondwana glaciation its sedimentary record in the Polish Carboniferous succession. *Przegląd Geologiczny*, **51**, 658–662.

SMITH, L. B. JR. & READ, J. F. 1999. Application of high-resolution sequence stratigraphy to tidally influenced Upper Mississippian carbonates, Illinois Basin. *In*: HARRIS, P. M., SALLER, A. H. & SIMO, J. A. (eds) *Advances in Carbonate Sequence Stratigraphy: Application to Reservoirs, Outcrops and Models*. SEPM Special Publications, Tulsa, **63**, 107–126.

SMITH, L. B. JR. & READ, J. F. 2001. Discrimination of local and global effects on Upper Mississippian stratigraphy, Illinois Basin, USA. *Journal of Sedimentary Research*, **71**, 985–1002.

SÜSS, M. P., SCHÄFER, A. & DROZDZEWSKI, G. 2001. A sequence stratigraphic model for the Lower Coal Measure (Upper Carboniferous) of the Ruhr district, North West Germany. *Sedimentology*, **48**, 1171–1179.

TUCKER, M. E. (ed.) 1991. *Sedimentary Petrology an Introduction to the Origin of Sedimentary Rocks*, 2nd edition, Oxford, 72–76.

TUCKER, M. E. 2003. Mixed clastic-carbonate and sequences: Quaternary of Egypt and Carboniferous of England. *Geologia Croatica*, **56**, 19–37.

TURNER, B. R. & WHATELEY, M. K. G. 1983. Structural and sedimentological controls of coal deposition in the Nongoma graben, northern Zululand, South Africa. *In*: COLLINSON, J. D. & LEWIN, J. (eds)

Modern and Ancient Fluvial Systems. International Association of Sedimentologists, Oxford, Special Publications, **6**, 457–471.

Tye, R. S. & Coleman, J. M. 1989*a*. Depositional processes and stratigraphy of fluvially dominated lacustrine deltas: Mississippi Delta plain. *Journal of Sedimentary Petrology*, **59**, 973–996.

Tye, R. S. & Coleman, J. M. 1989*b*. Evolution of Atchafalaya lacustrine deltas, south-central Louisiana. *Sedimentary Geology*, **65**, 95–112.

Tye, R. S. & Kösters, E. C. 1986. Styles of interdistributary basin sedimentation: Mississippi Delta plain, Louisiana. *Transactions of the Gulf Coast Association of Geological Societes*, **36**, 575–588.

Vail, P. R. & Todd, R. G. 1981. Northern North Sea Jurassic unconformities, chronostratigraphy and sea-level changes from seismic stratigraphy. *In*: Illing, L. V. & Hobson, G. D. (eds) *Petroleum Geology of the Continental Shelf of North West Europe*, Heyden, London, 216–235.

Veevers, J. J. & Powell, C. McA. 1987. Late Paleozoic glacial episodes in Gondwanaland reflected in transgressive–regressive depositional sequences in Euramerica. *Geological Society of America Bulletin*, **98**, 475–487.

Waksmundzka, M. I. 1998. Depositional architecture of the Carboniferous Lublin Basin. *In*: Narkiewicz, M. (ed.) *Sedimentary Basin Analysis of the Polish Lowlands*. Prace Państwowego Instytutu Geologicznego, Polish Geological Institute, Warsaw, **165**, 89–100. [In Polish with English abstract]

Waksmundzka, M. I. 2005. *Ewolucja facjalna i analiza sekwencji w paralicznych utworach karbonu z północno-zachodniej i centralnej Lubelszczyzny*. PhD thesis, Polish Geological Institut, Warsaw. [In Polish]

Waksmundzka, M. I. 2007*a*. Karbon. Wyniki badań litologicznych, sedymentologicznych i stratygraficznych. *In*: Waksmundzka, M. I. (ed.) *Profile Głębokich Otworów Wiertniczych Państwowego Instytutu Geologicznego Lublin IG 1*, Polish Geological Institute, Warsaw, **119**, 114–119.

Waksmundzka, M. I. 2007*b*. Karbon. Litologia, stratygrafia i sedymentologia. *In*: Pacześna, J. (ed.) *Profile Głębokich Otworów Wiertniczych Państwowego Instytutu Geologicznego Busówno IG 1*, Polish Geological Institute, Warsaw, **118**, 124–130. [In Polish with English summary]

Waksmundzka, M. I. 2008*a*. Karbon. Litologia, sedymentologia i stratygrafia. *In*: Pacześna, J. (ed.) *Profile Głębokich Otworów Wiertniczych Państwowego Instytutu Geologicznego Łopiennik IG 1*, Polish Geological Institute, Warsaw, **123**, 161–166. [In Polish with English summary]

Waksmundzka, M. I. 2008*b*. Correlation and origin of the Carboniferous sandstones in the light of sequence stratigraphy and their hydrocarbon potential in the NW and Central parts of the Lublin Basin. *Biuletyn Państwowego Instytutu Geologicznego*, **429**, 215–224. [In Polish with English abstract]

Waksmundzka, M. I. 2008*c*. Analiza zespołów litofacji i cykliczności w utworach rzek roztokowych, meandrujących i systemów anastomozujących, na przykładzie profilu paralicznego karbonu basenu lubelskiego (południowo-wschodnia Polska). *In*: Wojewoda, J. (ed.) *Baseny śródgórskie. Kontekst regionalny środowisk i procesów sedymentacji Kudowa Zdrój*, 15–21 September 2008 (materiały konferencyjne). WIND, Wrocław, 26–27. [In Polish]

Waksmundzka, M. I. 2010. Sequence stratigraphy of Carboniferous paralic deposits in the Lublin Basin (SE Poland). *Acta Geologica Polonica*, **60**, 557–597.

Waksmundzka, M. I. 2012. Braided-river and hyperconcentrated-flow deposits from the Carboniferous of the Lublin Basin (SE Poland) – a sedimentological study of core data. *Geologos*, **18**, 135–161.

Waters, C. N. & Condon, D. J. 2012. Nature and timing of Late Mississippian to Mid-Pennsylvanian glacioeustatic sea-level changes of the Pennine Basin, UK. *Journal of the Geological Society, London*, **169**, 37–51.

Waters, C. N., Chisholm, J. I., Benfield, A. C. & O'Beirne, A. M. 2008. Regional evolution of fluviodeltaic cyclic succession in the Marsdenian (Late Namurian Stage, Pennsylvanian) of the Central Pennine Basin, UK. *Proceedings of the Yorkshire Geological Society*, **57**, 1–28.

Wilson, M. E. J. & Lokier, S. W. 2002. Siliciclastic and volcaniclastic influences on equatorial carbonates: insights from the Neogene of Indonesia. *Sedimentology*, **49**, 583–601.

Wright, L. D. 1977. Sediment transport and deposition at river mouths: a synthesis. *Geological Society of America Bulletin*, **88**, 857–868.

Zieliński, T. 1992. Marginal moraines of NE Poland – sediments and depositional conditions. *Prace Naukowe Uniwersytetu Śląskiego*, 7–95.

Zieliński, T. 1995. Kod litofacjalny i litogenetyczny – konstrukcja i zastosowanie. *In*: Mycielska-Dowgiałło, E. & Rutkowski, J. (eds) *Badania osadów czwartorzędowych, Wybrane metody i interpretacja wyników*, Warsaw University, Warsaw, 220–235. [In Polish]

Żelichowski, A. M. 1964. Lithological and sedimentary problems of the Lower Carboniferous in Poland. *Geological Quarterly*, **8**, 524–537. [In Polish with English abstract]

Żelichowski, A. M. 1969. Karbon. *In*: Depowski, S. (ed.) *Ropo- i gazonośność obszaru lubelskiego na tle budowy geologicznej – część I: Budowa geologiczna obszaru lubelskiego*. Prace Geostrukturalne Instytutu Geologicznego, Warsaw, 70–85.

Żelichowski, A. M. 1972. Evolution of the geological structure of the area between the Góry Świętokrzyskie and the River Bug. Tectonic research in Poland. *Biuletyn Instytutu Geologicznego*, **263**, 7–97.

Żelichowski, A. M. 1987. Development of the Carboniferous of the SW margin of the East-European Platform in Poland. *Przegląd Geologiczny*, **35**, 230–237.

Żelichowski, A. M. & Kozłowski, S. 1983. *Atlas of geological structure and mineral deposits in the Lublin region*. Instytut Geologiczny Warszawa, Wydawnictwa Geologiczne, Warsaw. [In Polish with English summary]

Żelichowski, A. M. & Porzycki, J. 1983. Structural–geological map without strata younger than Carboniferous. *In*: Żelichowski, A. M. & Kozłowski, S. (eds) *Atlas of Geological Structure and Mineral Deposits in the Lublin Region*. Geological Institut, Warsaw. [In Polish with English summary]

Continental Upper Carboniferous red beds in the Variscan intermontane Saale Basin, central Germany: orbital forcing detected by wavelet analysis

UTE GEBHARDT[1]* & MICHAEL HIETE[2]

[1]*Staatliches Museum für Naturkunde Karlsruhe, Erbprinzenstraße 13, 76133 Karlsruhe, Germany*

[2]*Center for Environmental Systems Research (CESR), University of Kassel, 34109 Kassel, Germany*

**Corresponding author (e-mail: utegebhardt@gmx.de)*

Abstract: Stratigraphical correlation within continental red beds is hampered by uniformity of sediments and the lack of fossils, such that classical lithostratigraphical and biostratigraphical methods often fail. For a drilled section of Upper Carboniferous non-marine sediments of the Variscan intermontane Saale Basin, almost 800 m in thickness, an alternative approach is proposed. Wavelet-based time-series analysis is used to identify the internal organization of the cyclicity, and to distinguish cycles of different magnitude and origin as being autocyclically, tectonically or climatically controlled. Based on this distinction, basin-wide correlations of fluvial red beds are possible using a combination of high-resolution stratigraphy, biostratigraphy and classical lithostratigraphy. The analyses of well Querfurt 1/64 suggest the presence of short and long eccentricity cycles, and a duration of 5–7 Ma if the grey facies at the base of the section is to be correlated with the Grillenberg Subformation sediments. This subformation is of Stephanian A or Barruelian age, respectively, such that the well Querfurt 1/64 exposes a nearly complete section of the Mansfeld Subgroup (complete Stephanian stage).

Permo-Carboniferous non-marine sediments in Germany are the subject of intensive research owing to their great economic significance (e.g. sandstones as potential reservoir rocks for hydrocarbons, their use as underground reservoirs or geothermics, and the Upper Rotliegend clastics as the immediate footwall of the Kupferschiefer, Lower Rotliegend volcanics and Carboniferous coal seams). At present, detailed stratigraphical placement and correlation of these kinds of clastic continental red beds is associated with a number of complications, and the Saale Basin is no exception.

During the last 40 years, sedimentological research, interpretation of non-marine sections and, especially, facies analyses have improved drastically. Methods employed have included facies analysis (e.g. models of facies architecture of river systems: Miall 1985, 1996, 2006), interpretation of non-marine carbonates (Gebhardt 1988*a*, *b*; Gebhardt *et al.* 2000), sequence stratigraphy and palaeopedology (e.g. Retallack 1990, 2001; Duchaufour 1998), and series analysis of sedimentary cycles (e.g. Olsen 1990, 1994; Schwarzacher 1993; Herbert 1994; Yang & Kouwe 1995; Hinnov 2000; Hiete 2004; Hinnov & Ogg 2007; Franco *et al.* 2011).

Investigations of the Saale Basin started in the nineteenth century (e.g. von Veltheim 1824–1840; Laspeyres 1875), and the oldest geological map of this area dates back to 1815. Despite this long history of investigation, the stratigraphical problems in this area are still not satisfactorily resolved.

Since its drilling in 1964, drill hole Querfurt 1/64 has been, until now, one of the deepest within the Saale Basin. The drilling encountered a Carboniferous section with a thickness of about 800 m without reaching the basement rocks (Hagendorf & Schwahn 1969). More than two thirds of the profile are available as a solid core (Fig. 4). Using new, detailed sedimentological documentation of the core material as a strong foundation for further studies, different methods of stratigraphical analysis can now be attempted in the interpretation of the Permo-Carboniferous sediments within the Saale Basin.

Wavelet time-series methods display the internal organization of the cyclicity of profiles, and allow cycles of different magnitude and origin to be categorized as being autocyclically, tectonically or climatically controlled. Based on this distinction, basin-wide correlations of fluvial red beds are

From: GĄSIEWICZ, A. & SŁOWAKIEWICZ, M. (eds) 2013. *Palaeozoic Climate Cycles: Their Evolutionary and Sedimentological Impact*. Geological Society, London, Special Publications, **376**, 177–199.
First published online August 6, 2013, http://dx.doi.org/10.1144/SP376.12

possible using a combination of high-resolution stratigraphy, biostratigraphy and classical lithostratigraphy. In this investigation, we identified for the first time that the genetic nature of cycles in the fluvio-lacustrine Carboniferous of the Saale Basin are climatically driven and used this to solve long-standing stratigraphical problems. As cyclicity is a strong attribute of non-marine sections, it is the aim of this paper to show that wavelet time-series methods could provide a tool to help in the stratigraphical interpretation of Permo-Carboniferous continental sections.

Geological background

The Saale Basin is a SW–NE elongated continental sedimentation area that was created during Late Carboniferous time as a subcollisional structure of the Variscan Orogeny. The basin is 150 km long and 90 km wide, extending from Thuringia to the south of Brandenburg (Fig. 1a). Outcrops at the surface are found surrounding the eastern parts of the Harz Mountains, at the Kyffhäuser and in the valley of the river Saale north of Halle. Although more than 1000 wells exist in this area, being drilled for several reasons (mapping, groundwater detection, industrial purposes), most of them explored only a few metres of the Carboniferous section. Only a few drillings penetrated the Carboniferous section completely and reached the crystalline bedrock. Furthermore, these few long cores mainly date from the nineteenth century, being poorly recorded. Therefore, the database for establishing a stratigraphical concept is not really very good. Nevertheless, more than 200 years of investigation have resulted in solid knowledge about the geology of this area.

Palaeogeography and climate

The Saale Basin was palaeogeographically located in the central parts of the Variscides (Fig. 1b), having moved to the north during the Stephanian due to the general drifting of the continents. During the Westphalian, the basin was located at the equator, moving later to 10° northern latitude (Bless *et al.* 1984; Scotese 2000). The resulting climate changes led to an overall reddening of the sediments during the Stephanian, following the widely distributed coal-bearing greyish sediments of the Westphalian. At the same time, the river character changed from overall permanent and meandering rivers during the Carboniferous to mostly periodic braided river systems during Rotliegend times. These processes were superimposed by glaciations mainly occurring in the Southern Hemisphere; for example, one reaching from the Kazimovian to the Asselian (glacial time III after Lopez-Gamundi 1997; Isbell *et al.* 2003; Fielding *et al.* 2008). This glacial time can be separated into glacial episode IIIa during the early Stephanian, an interglacial episode IIIa during the late Stephanian, a glacial episode IIIb at the beginning of the Asselian and so forth (Visser 1997; Roscher & Schneider 2006). These kinds of glaciations cause strong eustatic sea-level fluctuations, on the one hand, while, on the other, affecting the position of the climate belts (e.g. Glennie *et al.*

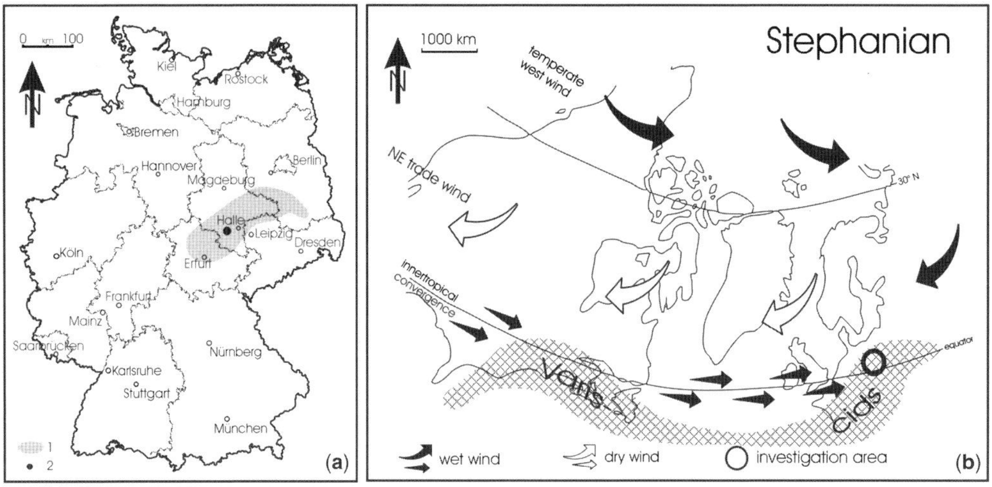

Fig. 1. (**a**) Location map: 1, distribution of Upper Carboniferous deposits of the Saale Basin, Central Germany; 2, location of the Querfurt 1/64 well. (**b**) Palaeogeographical situation of the Northern Hemisphere during the Stephanian.

1994; Horton *et al.* 2012). Within tropical latitudes, they manifest themselves by repeated changes between semi-arid and humid climate (e.g. Falcon-Lang 2003*a*, *b*, 2004; Roscher & Schneider 2006; Poulsen *et al.* 2007; DiMichele *et al.* 2009, 2010; Horton *et al.* 2012).

During the Stephanian, the Saale Basin was supposed to have been located within the tropical rain belt, which is characterized by equatorial west winds of the intertropical convergence zone (Fig. 1b). However, the Saale Basin is an elongate intramontane basin aligned perpendicular to prominent wind directions. Within such a basin, an arid–semi-arid climate is apparent even at latitudes at which a typically humid climate can be expected, as long as morphology is affecting the development of the local climate (Hay 1996). With seasonal change of the intertropical convergence zone (Kutzbach & Gallimore 1989), the Saale Basin was seasonally influenced by the dry NE trade winds, which were moving parallel to the lengthwise extension of the basin, leading to an increase of the evaporation rate within the basin (Fig. 1b). Therefore, during the Stephanian stage, the Saale Basin was located in a region that was characterized by a wet–dry (monsoon) climate with probably relatively high precipitation, but limited to individual seasons (Gebhardt *et al.* 2000).

A general increase in water flow is indicated in the sedimentation of the Mansfeld Subgroup by the transition from debris flow and braided channel to meandering channel facies, the change from calcisols and vertisols to hydromorphic soils, and the distinct increase in lacustrine–palustrine sedimentation at the centre of the basin. This increase in water flow could locally reflect the reduction in morphology within the vicinity of the basin and its influence on the inner basin climate. In a global context, overall aridization of world climate during the Stephanian following the damper Westphalian could be responsible, followed again by a more humid phase at the very end of the Carboniferous.

Lithostratigraphy and facies analyses

The general geological background and the stratigraphical placement of the Permo-Carboniferous deposits have already been clarified by Beyschlag & v.Fritsch (1899) at the end of the nineteenth century, summarizing the level of knowledge at that time in their publication.

Until the 1970s, research was focused on petrographical analyses of sediment (e.g. Nöldeke 1957; Kunert 1959; von Hoyningen-Huene 1960; Kunert 1970; Kampe & Kunert 1964; Kampe 1966; Hänel 1969; Hagendorf & Schwahn 1969; Falk *et al.* 1979, 1980). Sediment-petrographical characteristics, such as composition and maturity level of the sediments, can provide important information about source areas and palaeogeographical development, providing the foundation for lithostratigraphical classifications. The significance of the information is, however, limited if material within an intramontane basin has been reworked internally.

Sedimentation within the Saale Basin commenced with more-or-less greyish-purple and completely carbonate-free, coarse sediments (alluvial fan sediments) of the Gorenzen Formation (Fig. 4), which was emplaced discordantly on crystalline bedrock. General thicknesses range between 60 and 100 m. The material originates from the immediate substrate, followed by fluvial conglomerates with maximum pebble sizes of 15 cm in diameter, as well as pebble-rich sandstones derived from the Upper Harz Ridge. Fine clastic, grey and partly coal-bearing sediments of the Grillenberg Subformation were deposited in the depocentres of the basin (e.g. Beyschlag & von Fritsch 1899; Gothan & Schriel 1927; Hänel 1969; Schneider *et al.* 1995*b*, *c*, Schneider *et al.* 2005).

After continuous expansion of the Saale Basin to the NE, SW and SE, about 600 m of red beds of the Rothenburg Formation were deposited with an erosional unconformity at the base (Fig. 4). At the margin of the basin, deposits of fluvially dominated alluvial fans are characterized by 10–20 m-thick fining-upwards cycles. Coarse clastic parts of these cycles (Fig. 2a) include sediments of low-sinuosity bed-load channels, partly with the formation of braid plains. Grain sizes range from fine to coarse gravel, with a matrix of medium- to coarse-grained sand. Noticeable is the well rounding of quartzite pebbles up to 10 cm in diameter. The main source area is still the NW-located Upper Harz Ridge with the Acker–Bruchberg Quartzite. A rapid transition into facies of alluvial plains with meandering and braided rivers occurs towards the top of the section, as well as towards the centre of the basin. Sedimentation is mainly sandy–silty. Several metres of cross-bedded, medium- to coarse-grained sandstones deposited by meandering and braided rivers (types G and J after Miall 2006) with bed-load- to mixed-load-channels alternate with poorly bedded, fine sandy siltstone as part of overbank or floodplain deposits. Thin, mostly ripple-bedded or massive fine- to medium-grained sandstones with an erosional base have been deposited by crevasse splays or crevasse channels. The fine clastics are often penetrated by roots (Fig. 2e) and increasingly comprise calcisols in different evolutionary stages at the basin margins (Fig. 2c). Different carbonate-free palaeosols developed at the centre of the basin. Within the uppermost several metres of the Rothenburg Formation, thin, grey lacustrine carbonate

(a)
(b)
(c)
(d)
(e)
(f)

beds occur repeatedly (Fig. 2d), being precipitated by charophytes and microbes within ephemeral floodplain lakes (Gebhardt 1988*b*; Gebhardt *et al.* 2000).

With a maximum thickness of 500 m, red beds of the Siebigerode Formation, in combination with the basin-central, grey Wettin Subformation, are overlying the Rothenburg Formation (Fig. 4). Sedimentation starts with the deposition of a thick sandstone unit (Fig. 2b), representing deposits of a deep perennial braided river (type M after Miall 2006), followed by sedimentation in anastomosed rivers (type J after Miall 2006) with low-sinuosity bed-load and mixed-load channels in fining-upwards cycles. Grain sizes are, overall, smaller. Conglomerates and coarse-grained sandstones occur locally in small channel structures. Thick quartzite conglomerates, as have been observed in the Rothenburg Formation, are now missing. The large amount of kaolinized feldspars within the sandy fraction is noticeable. Source areas for these sandstones are mainly located to the south and SW (Meister 1969). It is possible that the coarse clastics at Kyffhäuser represent the marginal facies of the Siebigerode Formation, but this relationship has not been explained completely. Carbonate contents within the Siebigerode Formation are, overall, smaller. Calcisols/calcretes are generally rare and texturally immature. Lacustrine limestones are concentrated within the more than 300 m-thick Wettin Subformation; there occurring in combination with lacustrine fine clastics, hydromorphic palaeosols and coal seams (Schneider *et al.* 1984).

Well Querfurt 1/64. Well Querfurt 1/64 is located at the centre of the Saale Basin (Fig. 1a), and is therefore characterized by deposits of perennial river systems with sandy freight and comparably low transport energy. Transport energy and, connected with that, the size of river systems were decreasing upwards as expected (Fig. 4). The river system at the base of the Siebigerode Formation is different, however, indicating higher transport energies, probably tectonically or climatically driven. The second, thick fining-upwards cycle commences with the deposition of this sandstone bed (Fig. 4). Intercalated fine clastic beds are mainly overbank sediments. Crevasse splays indicate multiple flooding events, which sometimes led to the development of stable crevasse channels. Oxbow lake deposits increase towards the top, related to the development of back swamps. Lacustrine units, of extended perennial lakes, are rare and limited to the Wettin Subformation.

The lowest 24 m of the well consists of grey fine clastics (Fig. 4). They probably represent the top of another fining-upwards cycle, which was not penetrated completely, as the drilling did not reach the crystalline basement. Above these grey sediments, parts of the section have not been cored, but cuttings of grey–black fine clastics may be assigned to several grey horizons within the overall red beds. The top of the profile is characterized by grey, coal-bearing deposits of the Wettin Subformation, more than 300 m in thickness.

During the Permo-Carboniferous, the location of drill-hole Querfurt 1/64 was a flat plain in the basin centre with extensive flooding, dissected by slow flowing rivers. The intense root penetration of the fine clastics and the formation of vertisols indicate a permanently high ground-water level, leading to the development of back-swamps towards the top. Poor correlation of the coal seams confirms the interpretation of the environment being a fluvial plain. Back-swamp facies were not formed extensively, they are limited to the formation of gallery forests near active channels with a related higher ground-water level.

Hagendorf & Schwahn (1969), who first worked on the cores from well Querfurt 1/64, placed the whole Carboniferous section into the sandstone facies of the 'Mansfeld beds', the so-called 'Siebigeröder Sandstone'. Petrographical investigations of the sandstone beds by Hagendorf & Schwahn (1969) produced similar compositions for the total extent of the profile. According to their investigations, the 'Siebigeröder Sandstone' from the Querfurt borehole is an overall very-well-sorted quartzitic sandstone with feldspar concentrations of more than 10%. Feldspars are slightly sericitized, but not kaolinized. The 300 m-thick 'Wettin beds' at the top of the section, as well as the grey fine clastics at the base of the profile, are considered to represent the basin-central lacustrine facies of the 'Siebigeröder Sandstone'. Differentiation is performed lithostratigraphically, indicated by the lack of thick, coarse-grained sandstone beds (Hagendorf & Schwahn 1969).

Gebhardt (1988*b*) correlated the grey lacustrine facies at the base of the section with the maximum

Fig. 2. Lithology of the Mansfeld Subgroup: (**a**) conglomeratic depositions of braided rivers of the Rothenburg Formation, scale 1 m; (**b**) sandy deposits of the Siebigerode Formation, stacked channels of perennial braided rivers, scale 1 m; (**c**) thick calcisols of the Rothenburg Formation, scale 1 m; (**d**) thin lacustrine limestones as deposits of ephemeral ponds, Rothenburg Formation, scale in cm; (**e**) root traces (arrow) in floodplain deposits of the Siebigerode Formation, scale in cm; and (**f**) Scoyenia-burrows (arrow) as occurring in floodplain deposits of the whole Mansfeld Subgroup, scale in cm.

carbonate content (lacustrine and pedogenic) at the top of the 'Lower Mansfeld Beds' (outcrops are located in Rothenburg and Vatterode) as being a basin-wide traceable minimum of sedimentation rate. Overlying are the 'Upper Mansfeld Beds' with thick sandstone beds and low carbonate content (Kunert 1970). Schneider *et al.* (1995*b*, *c*) utilized this concept, placing the grey Querfurt Subformation at the top of the Rothenburg Formation ('Lower Mansfeld Beds') as a fine clastic, central basin sedimentary unit. Likewise, the Wettin Subformation has been placed at the top of the Siebigerode Formation ('Upper Mansfeld Beds') as a fine clastic, central basin formation.

Biostratigraphy

Except for the grey deposits of the Grillenberg and Wettin subformations, fossils are very rare throughout the Carboniferous section in the Saale Basin, as are biostratigraphical data. Therefore, attempts at biostratigraphical research have been mostly limited to frequent plant remains within grey deposits. Their significance strongly depends on facies, which does not always provide enough accuracy (Gothan & Schriel 1927; Kampe & Remy 1962; Döring & Kampe 1973; Kampe & Döring 1993). New approaches using biostratigraphical methods for the red beds were implemented in the 1980s (blattids: Schneider 1982; arthropod trackways: Walter 1983; conchostracans: Martens 1983*a*, *b*; xenacanth teeth: Schneider 1985, 1989; tetrapods: Werneburg 1989; Schneider & Werneburg 2006; Werneburg & Schneider 2006).

Deposits of the Grillenberg Subformation in the eastern foreland of the Harz Mountains can be assigned to the early Stephanian owing to the presence of palaeoflora containing *Sphenophyllum oblongifolium* (Schneider *et al.* 2005). A former separation of the Westphalian D substage by Gothan & Schriel (1927), based on the occurrence of *Imparipteris ovata* Hoffm., cannot be maintained because, according to Wagner (1966), this species is not an index fossil for Westphalian D (Schwab & Kampe 1989).

Beyschlag & von Fritsch (1899) placed the grey deposits at the base of the Carboniferous profile within the deep Schladebach 1888 borehole into the 'Grillenberg Beds'. Kampe & Döring (1993) used palynological methods to put these deposits into the upper Stephanian A–Stephanian B (spore zone ST Western Europe). Palynological floras verified by Kampe & Döring (1993) date back to Unit 104 after Beyschlag & von Fritsch (1899) from the top of the grey facies section. Furthermore, species mentioned in their descriptions allow for a slightly older age. In summary, it is possible to assign the grey deposits at the base of the Carboniferous section of the Saale Basin at the Harz border, as well as in the Schladebach borehole, to an early Stephanian age, probably the Stephanian A.

Apart from the aforementioned palaeofloras, only one fossilized tree trunk was found at Werderbruch in the Saale valley within the Rothenburg and Siebigerode formations below the Wettin Subformation (von Veltheim 1840; Laspeyres 1875). Plant fossils of no biostratigraphical use were found in lacustrine limestones (Fig. 2d) (Gebhardt 1988*b*; Gebhardt *et al.* 2000) and at Gerillgrund, a small tributary Saale valley opposite the village of Friedeburg. There, Beyschlag discovered a macroflora that enabled the sorting of the 'Mansfeld Beds' into the Stephanian (Beyschlag & v.Fritsch 1899), which was confirmed by Kampe & Remy (1962). The location is placed within the Siebigerode Formation, based on lithostratigraphical considerations. However, it is not possible to use this flora to separate the Grillenberg Subformation in the footwall from the Wettin Subformation above.

The Wettin Subformation ('Wettin Beds' in old references) has already been placed into the uppermost Stephanian by von Fritsch (1888). After multiple reclassifications during the twentieth century (e.g. Lehmann 1929; Friese 1955; Primke 1956; Kampe 1966), the uppermost Stephanian is now confirmed beyond doubt. It is supported by palaeozoological and palaeobotanical results, such as the *Syscioblatta dohrni–Sysciophlebia euglyptica* Zone (Schneider 1982; Schneider & Werneburg 1993; Schneider *et al.* 1995*a*), *Bohemiacanthus*–Ug Zone (Schneider 1988, 1996; Schneider & Zajic 1994; Schneider *et al.* 2000) and the *Branchiosaurus saalensis–Apateon intermedius* Zone (Werneburg 1996). The macroflora is typical for vegetation of the late Stephanian (Remy & Remy 1958*a*, *b*; Remy & Kampe 1961; Kampe & Remy 1962), the sporomorphs are part of the *Potonieisporites novicus–bhardwajii–Cheiledonites major* (NBM) Zone of the Stephanian C (Döring & Kampe 1973).

Well Querfurt 1/64. Fossil remains occur in several horizons in the Querfurt 1/64 borehole (Hagendorf & Schwahn 1969; Oemisch 1984). Apart from a few exceptions, fossils are concentrated in the grey beds at the base of the profile and in the coal-bearing facies of the Wettin Subformation.

The macroflora from the well Querfurt was last examined by R. Rößler (pers. comm. 2010). According to his investigations, several excellent Stephanian index forms were observed; for example, *Callipteridium pteridium* (Fig. 3b), *Sphenophyllum oblongifolium*, *Sphenophyllum angustifolium* and *Sphenophyllum verticillatum*. In addition,

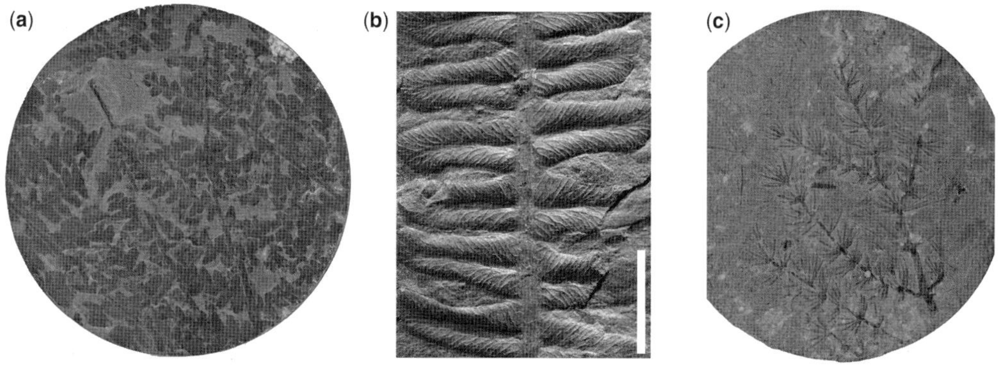

Fig. 3. Plant remains of the Querfurt 1/64 well (core diameter 9 cm): (**a**) *Dicksonites pluckenetii*; (**b**) *Callipteridium pteridium*, scale 5 mm; and (**c**) *Asterophyllites equisetiformis*.

Dicksonites pluckenetii (Fig. 3a), *Asterotheca sternbergii*, *Annularia spinulosa*, *Scolecopteris pseudobucklandii*, *Pseudomariopteris busquetii*, *Asterophyllites equisetiformis* (Fig. 3c) and *Nemejcopteris feminaeformis*, as well as *Pecopteris truncata* and further diverse species, were observed. Remains of cordaites are not useful stratigraphically; they belong to the coal-forming facies of the forest swamp (Barthel 2001). This facies existed from the beginning of the Stephanian up until nearly the end of Rotliegend times, accommodating conservative floral elements. These are swamp forest associations comprising cordaites, psaroniaceae, calamites with fern vines and callistophytaceae, as well as sphenophyllaceae with undergrowth, merging into associations consisting exclusively of cordaites. Plant remains confirm a similar floral and facies character throughout the whole length of the drilled core. Owing to the character of the macroflora, the complete section can be placed into the Stephanian B–early Stephanian C, and cannot be divided further (R. Rößler pers. comm. 2010). However, within the lowermost part of the profile, index forms are missing. Therefore nothing contradicts the placement of the grey facies at the base of the borehole into the early Stephanian.

Other fossils in the Querfurt 1/64 drill core include ostracods, *Pringlia*, *Spirorbis*, bivalves and palaeoniscid fish. In addition, hybodont sharks (confirmed by egg capsules and teeth) and insects were observed. Biostratigraphically relevant groups of insects, conchostraca and xenacanthid sharks have not been found. A general placement into both the early and late Stephanian is possible (Hagendorf & Schwahn 1969; Oemisch 1984). According to Eager (1984, 2005), the bivalve fauna of the grey deposits at the base of the Querfurt drill core (*Anthraconia prolifera* and *A. puella*) is significantly different from the fauna found in the Wettin Subformation, suggesting a placement into the Stephanian A.

Conclusions from classical lithostratigraphy and biostratigraphy

Despite a very long history of investigation and more than 1000 drillings having been completed since the nineteenth century (Ehling & Breitkreuz 2004), it is not possible to solve stratigraphic questions in the Carboniferous of the Saale Basin satisfactorily with classical lithostratigraphical and biostratigraphical methods, such as, for instance, the correct stratigraphical placement of the grey fine clastics at the final depth of the Querfurt 1/64 core. Fossils are very rare throughout the Carboniferous section in the Saale Basin. The significance of lithostratigraphy based on sediment-petrographical information is limited if material within an intramontane basin has been reworked internally. The correct placement of the basal fine clastics of the Querfurt borehole at the base or in the middle of the Stephanian, however, has severe consequences for basin-wide correlations as well as for the modelling of thickness and subsidence history of the basin. In addition, mere lithostratigraphical classifications do not allow a correlation with other areas of sedimentation.

However, analysing the internal organization of the cyclicity of profiles to categorize cycles of different magnitude and origin as being autocyclically, tectonically or climatically controlled could provide another stratigraphical tool for non-marine Late Palaeozoic sections, as has already been shown, for instance, for the Rotliegend of The Netherlands (e.g. Yang & Kouwe 1995; Nio *et al.* 2005, 2006; Minervini *et al.* 2011) or northern Germany (e.g. Gast 1995; Gast & Gebhardt 1995).

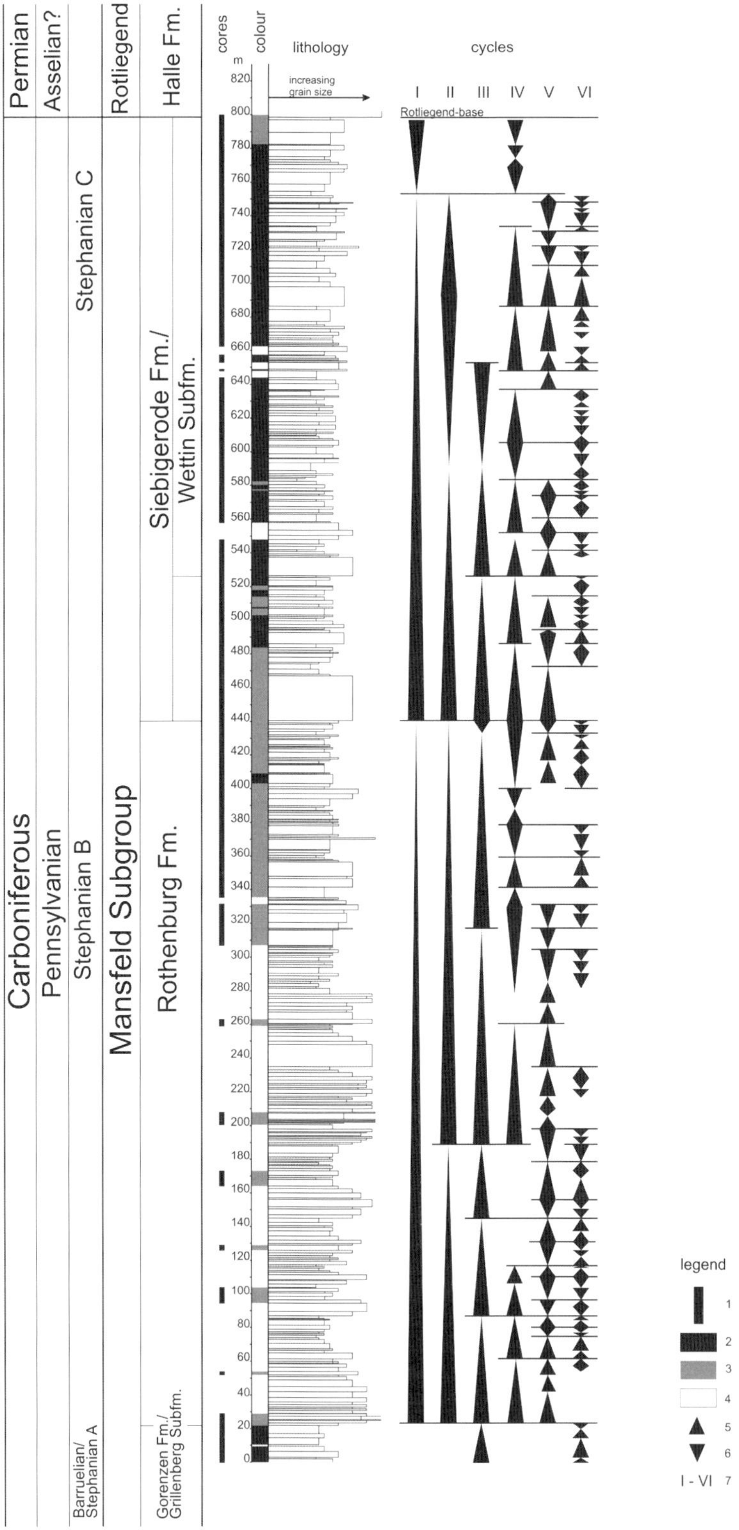
Permian
Asselian?
Rotliegend
Halle Fm.
Carboniferous
Pennsylvanian
Stephanian C
Stephanian B
Barruelian/
Stephanian A
Mansfeld Subgroup
Siebigerode Fm./
Wettin Subfm.
Rothenburg Fm.
Gorenzen Fm./
Grillenberg Subfm.
cores
colour
m
lithology
increasing
grain size
cycles
I II III IV V VI
Rotliegend-base
legend
I - VI

Analysis of cyclicity

The new detailed, lithological logging of the Querfurt core at depths of between 2998.9 m (final depth, profile metre 0) and 2200.55 m (base of Rotliegend section, profile metre 798.35) is the base for the following analyses of cyclicity within the section. Gaps in the core were completed by lithological interpretation of gamma-ray logs. Altogether, 1890 lithological units have been separated. Figure 4 shows the Carboniferous profile of the Querfurt core, which is used for visual identification of cycles of different thicknesses.

The core did not penetrate the Upper Carboniferous section. The grey fine clastics at the final depth are not basal deposits of the Upper Carboniferous section within the Saale Basin. It is more likely that they represent the fine-clastic top of a fining-upwards cycle of unknown order.

At profile metre 24, a major fining-upwards cycle starts, comprising the rest of Carboniferous section (Mansfeld Subgroup) up to the base of Rotliegend deposits (not shown in Fig. 4). In addition, cycles of at least six different categories can be separated. Columns I and II of Figure 4 show that the profile can be divided into two or three cycles above the grey-coloured section at the base. These fining-upwards cycles are definitely not autocyclically driven, but represent tectonically or climatically caused events during the development of the basin because they are traceable throughout the whole Saale Basin. Both fining-upwards cycles in column I are about 400 m thick. They each begin with massive conglomerates and sandstone, respectively, with regard to the central position in the basin. The cycles end with fine clastics, characterized by extensive root penetration in the lower cycle, and lacustrine and palustrine units in the upper cycle – both pointing to a distinct decrease in sedimentation rate. Grey deposits occur periodically in the lower cycle too, but they are partly intercalated with coarser red beds. Their distribution is probably controlled by other processes that are also triggering smaller cycles. The three fining-upwards cycles in column II tend to become thicker towards the top 165, 250 and 315 m, respectively.

Typically, cycles of all magnitudes are developed as fining-upwards cycles. Coarsening-upwards cycles are rare and generally quite thin, with thicknesses of 5–20 m. Further small cycles (down to <1 m) can be distinguished in sections if profile graphs are changed to different scales.

To what extent these cycles of varying thicknesses are tectonically, climatically or autocyclically driven needs to be determined by further investigation. Correlations between all outcrops within the Saale Basin show that, despite earlier assumptions, at least part of the larger cycles (thicknesses between 50 and 120 m) can be traced throughout the whole basin. The smaller cycles, with thicknesses of between 15 and 20 m (e.g. those exposed at the Saale slope at Rothenburg), could be autocyclically driven fluvial cycles, which cannot be correlated over long distances. Time-series analyses and specific sequence stratigraphical investigations could provide some information.

Time-series analysis

The visual identification of fining- and coarsening-upwards cycles representing different cycle hierarchies in the lithology profile is complemented by numerical time-series analysis. The aim of this analysis is to provide a second, ideally less subjective, approach that allows not only the identification of the cycles and their hierarchy, but also their frequencies and wavelengths. As astronomical cycles of Milankovitch origin, if present, may be distorted through the complex interaction of the climate system and geological processes to insolation forcing (cf. Hiete 2004 and literature cited therein), the use of appropriate time-series analysis methods is necessary. Short-term Fourier, Gabor or wavelet transform are methods of choice, as time–frequency representations of a signal are obtained (cf. Antoine 1999). Expressed in simplified terms, these methods analyse subsequentially pieces of the time series for inherent frequencies. Short-term Fourier and Gabor transform apply a moving window of fixed size to cut-out parts of the time series for analysis for inherent frequencies using sines and cosines. The choice of the window size is, therefore, a compromise between the ability to capture changes in the high frequencies over time, for which a narrow window is favourable, and the ability to capture low frequencies, for which a broad window is preferable. In contrast to infinite trigonometric functions, in the wavelet transform a small wave (therefore the name wavelet) function with compact spatial support (limited extension in time) is moved along the time series (e.g. Daubechies 1992; Farge 1992; Kumar & Foufoula-Georgiou 1997; Antoine 1999). The width and, thus, frequency of the analysing wavelet function

Fig. 4. Stratigraphy and litho-log of the Carboniferous section of the Querfurt 1/64 well, with visually identifiable cycles. Legend: 1, cores; rock colour: 2, grey; 3, red; 4, no information; cycles: 5, fining-upwards; 6, coarsening-upwards; 7, visually identifiable different hierarchy levels of cycles.

is changed optimally according to the frequencies analysed for; that is, starting with a narrow wavelet for high frequencies, the wavelet is increasingly dilated when analysing for lower frequencies (Daubechies 1992; Farge 1992) avoiding the window width problem. Therefore, the wavelet transform is regarded as better suited for geological time series that are often non-stationary in nature and containing heavily distorted periodic or quasi-periodic cycles than, for example, the Gabor transform or multi-taper method (cf. Bolton *et al.* 1995; Guyodo *et al.* 2000; Pancheva & Mukhtarov 2000). Consequently, numerous applications of the wavelet transform can be found in geological literature (e.g. Yiou *et al.* 1995; Appenzeller *et al.* 1998; Prokoph & Agterberg 1999, 2000; Guyodo *et al.* 2000; Prokoph & Thurow 2000; Rampino *et al.* 2000; Hiete 2004; Prokoph & El Bilali 2008; Saracco *et al.* 2009).

Generation of data series

The central problem when analysing geological successions is that there is – with a few exceptions in which annually laminated sediments occur (e.g. varved lake sediments) – no timescale available. Instead, the sediment depth is used. It should be taken into account that the depth scale is not proportional to time as *inter alia*: (i) time gaps due to intermittent sedimentation or an unconformity may be present; (ii) the sedimentation rate and thus the depth to time ratio may have changed (e.g. when the lowering of the erosional base level and the sedimentation were not in step); and (iii) the depth scale during the time of sedimentation is diagenetically altered if sediments are differentially compacted. For the Querfurt core, soil formation or losses of cored sections are examples of (i). Point (ii) is present in the form of probably rapidly deposited coarse conglomerates at the base of some fining-upwards cycles. Points (i) and (ii) lead to distortions at certain depth intervals that can only be accounted for by careful interpretation of the results of the wavelet transform. Diagenetic alteration of the depth scale is ubiquitous in the core and affects different lithologies to various extents, leading to subtle distortions. Therefore, the core is decompacted (backstripped) prior to wavelet analysis. For decompaction of shale, the Baldwin–Butler power-law equation was used, and for all clastics with grain sizes larger than fine sandstone the Sclater–Christie exponential curve was employed (cf. Roll 1974; Sclater & Christie 1980; Baldwin & Butler 1985; Springer 1993; Hölzel *et al.* 2008). Although Bahr *et al.* (2001) criticized the use of a power-law instead of an exponential curve, they stated that the Baldwin–Butler power-law equation nevertheless achieves a good fit. For grain sizes between shale and fine sandstone, weighted means between the two curves were used. Note that for mixed clastics, the equations might not be applicable as fine grains fill the spaces between the larger grains (cf. Bahr *et al.* 2001). The calculated decompaction factors increased from 4.83 at 2201 m to 5.07 at 2993 m, and from 1.43 to 1.53 for shale and sandstone, respectively. For coal, a constant decompaction factor of 10 was assumed.

Wavelet transform

The wavelet transform of time series $x(t)$ is defined as the integral transform:

$$W(a, b) = \int_{-\infty}^{\infty} x(t)\overline{\psi}_{a,b}(t)\,\mathrm{d}t \quad a > 0 \tag{1}$$

where $\overline{\psi}_{a,b}(t)$ is the complex conjugate of the daughter wavelet:

$$\psi_{a,b}(t) = \frac{1}{\sqrt{a}}\psi\left(\frac{t-b}{a}\right) \tag{2}$$

which is deduced from the mother wavelet, $\psi(t)$, by dilation with parameter a and translation along the time axis by parameter b (Kumar & Foufoula-Georgiou 1997). The scalogram as the representation of the time series with respect to dilation and scale parameter (analogous to the time–frequency representation offered by the spectrogram in the windowed Fourier transform) is defined by:

$$P_{a,b} = |W_{a,b}|^2. \tag{3}$$

As in Fourier transform, an inverse wavelet transform exists:

$$x(t) = \frac{1}{C_\psi}\int_{-\infty}^{\infty}\int_{0}^{\infty} W(a, b)\psi_{a,b}(t)\frac{\mathrm{d}a\,\mathrm{d}b}{a^2} \tag{4}$$

and allows reconstruction of the time series, where C_ψ is a constant wavelet-specific factor (Grossmann & Morlet 1984; Daubechies 1992; Kumar & Foufoula-Georgiou 1997). It is important to note that *local* reconstruction, locally both in time and frequency range, is possible that allows filtering (Farge 1992). For the reconstruction, another wavelet and even the delta function $\delta(t)$ can be used. Two main wavelet transforms can be distinguished: continuous wavelet transform (CWT) and discrete wavelet transform (DWT). CWT results in redundant representations that are more robust to noise and easier to interpret than DWT (cf. Kumar & Foufoula-Georgiou 1997). Accordingly, while DWT is mainly applied for, for example, compression purposes, CWT is preferred in exploratory time-series analysis. It needs to be emphasized also that CWT needs to be discretized to analyse sampled time series, such that the term discrete wavelet transform is sometimes used in the sense of discretized CWT.

Although several mother wavelets have been developed (cf. Farge 1992; Torrence & Compo 1998; Ahuja *et al.* 2005 for guidance on the choice of a mother wavelet), in exploratory time-series analysis mostly the Morlet wavelet, a plane wave modulated by a Gaussian, is preferred for several reasons (Farge 1992; Ochadlick & Kritikos 1993; Torrence & Compo 1998). First, the Morlet wavelet is non-orthogonal, resulting in redundant yet easier interpretable results and complex values, allowing the separate extraction of amplitude and phase. Then, it is quite compact (and, via the form parameter k_ψ (see below), even adjustable) allowing a good compromise between frequency and time resolution, as well as smoothly varying in shape, reflecting well the cycles looked for in the time series. Finally, the Fourier wavelength of the mother wavelet is approximately 1 for $k_\psi = 6$. A higher value for k_ψ, which increases the number of oscillations under the Gaussian and so leads to a better signal to noise ratio at the expense of time resolution, is used, for example, by Prokoph & Barthelmes (1996), Prokoph & Agterberg (1999, 2000), Prokoph & Thurow (2000), Prokoph *et al.* (2000) and Rampino *et al.* (2000). Following Torrence & Compo (1998), the Morlet wavelet can be defined (without the correction term needed to ensure admissibility if $k_\psi = 6$: cf. Farge 1992; Percival & Walden 2000) as:

$$\psi_{\mathrm{Morl}}(t) = \pi^{-0.25} \exp^{ik_\psi t} \exp^{-|t|^2/2} \tag{5}$$

A challenge in time-series analysis is to discriminate cycles against white or red (autocorrelated) noise. Hiete (2004) showed that Milankovitch signals within insolation time series distorted in several different ways are, in general, still well identifiable using wavelet analysis but that the scalogram easily becomes unclear. Therefore, the scalogram is complemented by additional information (Fig. 5), mainly following Torrence & Compo (1998). A cone of influence (COI) shows those areas of the scalogram in which the finiteness of the time series results in errors. Areas significant against an AR(1)-process (red noise) are marked in the scalogram. Ridges determined as local maxima in areas of the scalogram that are significant against red noise can be highlighted. Note that more sophisticated methods for ridge extraction exist (e.g. Carmona *et al.* 1999; Özkurt & Savaci 2005; Lin 2006). Moreover, reconstruction of a time series out of the freely selectable areas of the scalogram allows analysis of how an area of high power identified in the scalogram is related to the original time series in terms of time extension, frequency and amplitude. In practice, this check proved valuable as it avoids misinterpretation of the scalogram. In addition, a global wavelet spectrum, which is equivalent to the power spectrum, gives a fast overview of the frequencies inherent in the entire profile. Again, significance against red noise can be presented. All analyses were performed using MATLAB® and modified program code after Torrence & Compo (1998).

The approach is shown for an artificial time series in Figure 5. The scalogram in Figure 5b clearly represents the increasing wavelength of the backward chirp for $0 \leq t \leq 500$, followed by the decreasing wavelength of the second chirp and the onset of a sine of lower but constant wavelength starting at $t = 500$. Interpretation is facilitated by contours marking the significance levels for red noise and the highlighted ridges. It also becomes evident that the global wavelet power spectrum in Figure 5c is unable to resolve this non-stationary character but nevertheless provides a first overview of the main wavelengths present in the time series. Figure 5d allows a visual comparison between the time series reconstructed from the (arbitrarily) selected area of interest in the scalogram and the original time series. The reconstruction is useful both for filtering and for guidance in the interpretation of the scalogram. In the example, the selected area represents the main cycles of larger wavelength in this part of the time series.

Results of the wavelet time-series analysis

The wavelet analysis of the entire decompacted profile reveals that cycles significant at the 95% level against red noise are only present at a few positions in the scalogram (Fig. 6b). The same is true for the global wavelet power spectrum (Fig. 6c), which nevertheless shows peaks at wavelengths of around 54, 108, 188 m, and is less pronounced at 328 and 497 m. The white ridges highlighting local maxima within the scalogram in areas significant at the 50% level against red noise allow the identification of wavelengths in certain depths; for example, cycles of 20–30 m thickness in the depth range 1200–1600 m and 40–60 m in the range 200–1100 m. Reconstruction of the time series from the area marked in Figure 6b gives the impression that the time series below 1000 m is well represented, with a cycle thickness of 25–75 m, whereas above 1000 m a large part of the energy can be attributed to a thickness of around 20–40 m. This change is also reflected in a change to finer grain sizes. Possibly, a similar shift

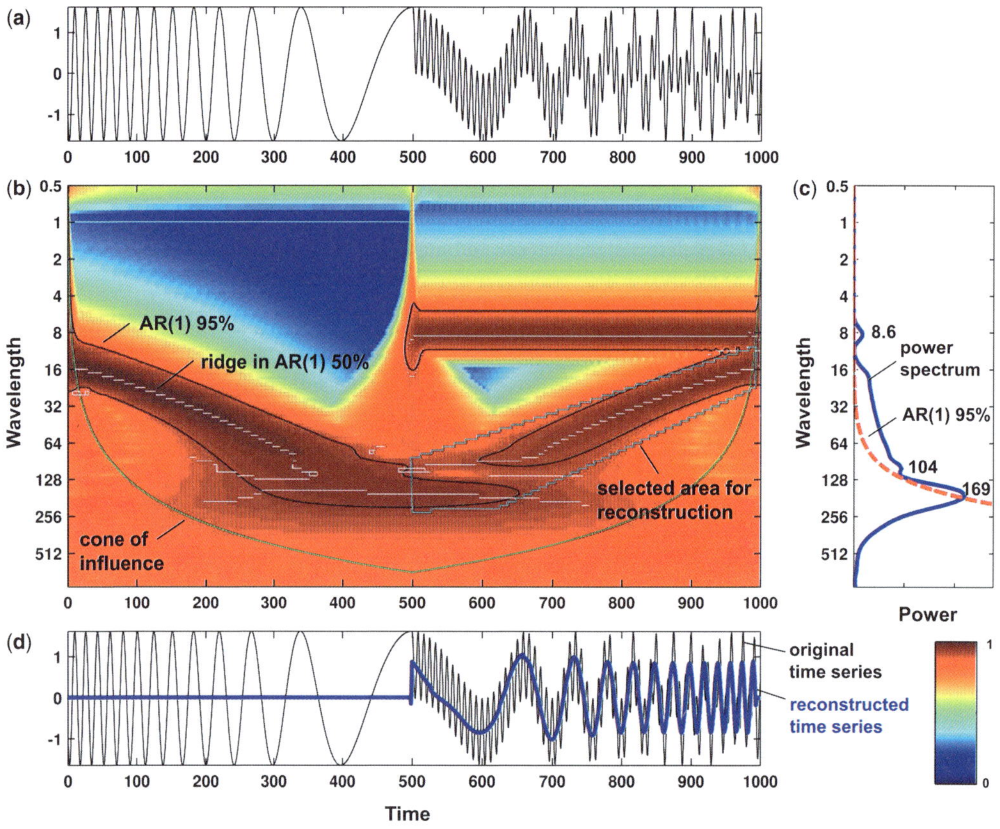

Fig. 5. (**a**) Presentation of a time series made up of a backward chirp for $0 \leq t \leq 500$, and of a chirp and a sine of wavelength 8.6 for $500 \leq t \leq 1000$. (**b**) Scalogram with log2(power) of continuous wavelet transform (Morlet-6 wavelet). The cone of influence (green line) indicates how far the edge effects extend into the wavelet power spectrum. Black framed areas mark the 95% confidence level for an AR(1) process, and the white ridges show local maxima in areas above 50% confidence level for an AR(1) process. The cyan-coloured frame marks the area selected for reconstruction of the time series. (**c**) Global wavelet power spectrum. The red dotted line marks the 95% confidence level for an AR(1) process. (**d**) Original time series (black line) and reconstructed time series (blue line) in the selected area of the scalogram.

becomes weakly apparent at greater cycle thickness, which changes from about 200 to 100 m near 1100 m. The length ratios of the cycles below and above this depth thus would be 25 to 100 m and 50 to 200 m; that is, 1:4. A comparison with the visually identified cycles of different hierarchy (Fig. 4 and presented in a decompacted manner in Fig. 6e) shows good agreement in some sections for certain hierarchy levels (e.g. for depth range 200–400 m and hierarchy level V), but the hierarchy as such is not clearly identifiable. Overall, although cycles of different thicknesses can be identified and attributed to certain bands that possibly change throughout the profile, interpretation is hampered by a strong distortion of the profile.

Analysing the finer-grained part of the profile from 1050 m upwards (Fig. 7) shows a clearer picture. A cycle thickness of around 15–30 m is present, and over long sections is also significant above the 50% and even the 95% level (Fig. 7b). Correspondingly, there is a significant peak above the 95% level in the global wavelet power spectrum at 25.6 m (Fig. 7c). Another highly significant peak is present at 109.6 m, which can also be easily traced in the scalogram in the 90–130 m band. However, below about 1250 m, the 25 m cycle splits up into 17 and 45 m cycles, and the 100 m cycle into 70 and 125 m cycles. Above 1600 m, a shift to thinner cycle thickness becomes weakly apparent, from around 100 to 60 m and from 25 to 15 m. Both the splitting-up and the shift are also visible in the reconstructed time series, where the reconstruction in the 18–34 m band (Fig. 7d) and the 80–120 m band (Fig. 7e)

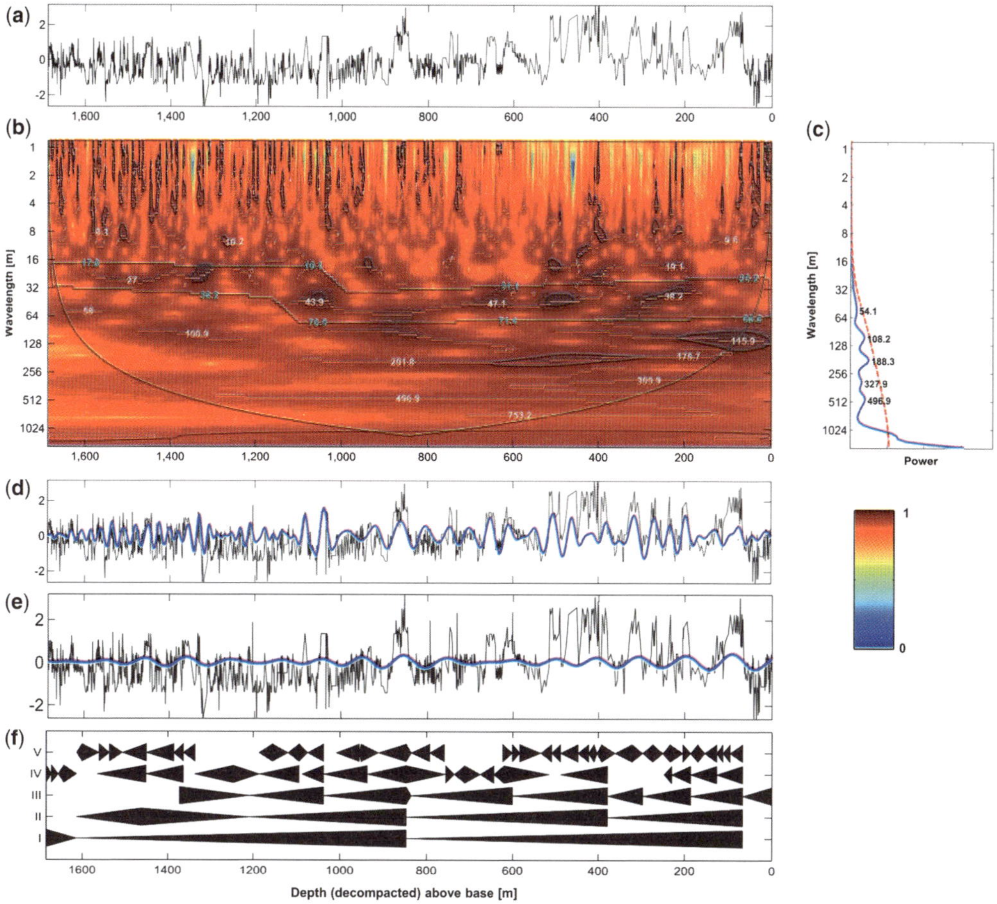

Fig. 6. (**a**) Presentation of decompacted normalized lithology profile of the Querfurt 1/64 drill core (all values above base at 2998.9 m set as zero). (**b**) Scalogram with log2(power) of lithology profile using the Morlet-6 wavelet. White numbers mark the wavelengths of ridges at these positions and the cyan-coloured numbers are the wavelengths at the borders of the area used for reconstruction of the time series. (**c**) Global wavelet power spectrum with the wavelengths of local power maxima marked. (**d**) Original (black line) and reconstructed (blue line) time series in the area marked in the scalogram. (**e**) Same as (d) but reconstructed within the wavelength band of 90–110 m. (**f**) Hierarchy levels I–V of cycles as identified from the lithology profile, decompacted for the purpose of comparison.

contains less energy in the time series visible at lower amplitudes. However, at a depth of 1200–1600 m, in particular, the time series reconstructed in the 18–34 m band covers the main cycles present in the original time series. Overall, cycles in bands of around 15–30 and 60–120 m in thickness could be identified in 1200–1600 m. Again, the cycle lengths have a ratio of 1:4.

Interpretation

Several cycle types of varying thicknesses can be observed in geological profiles, typically with a variety of reasons for cycle development (cf. House 1995). According to cycle thickness, controlling mechanisms are generally climate and tectonics; autocyclic processes are important for fluvial profiles. For geological issues, climatic and tectonic cycles with periods of a few thousand to several million years are interesting. Tectonic and climatic causes can lead to similar results within the geological record, complicating interpretations. It is possible, however, that only climatically driven cycles produce constant periods over long periods of time.

Milankovitch cycles in fluvial profiles

Marine profiles with constant sedimentation rates are well suited for the record of climatically

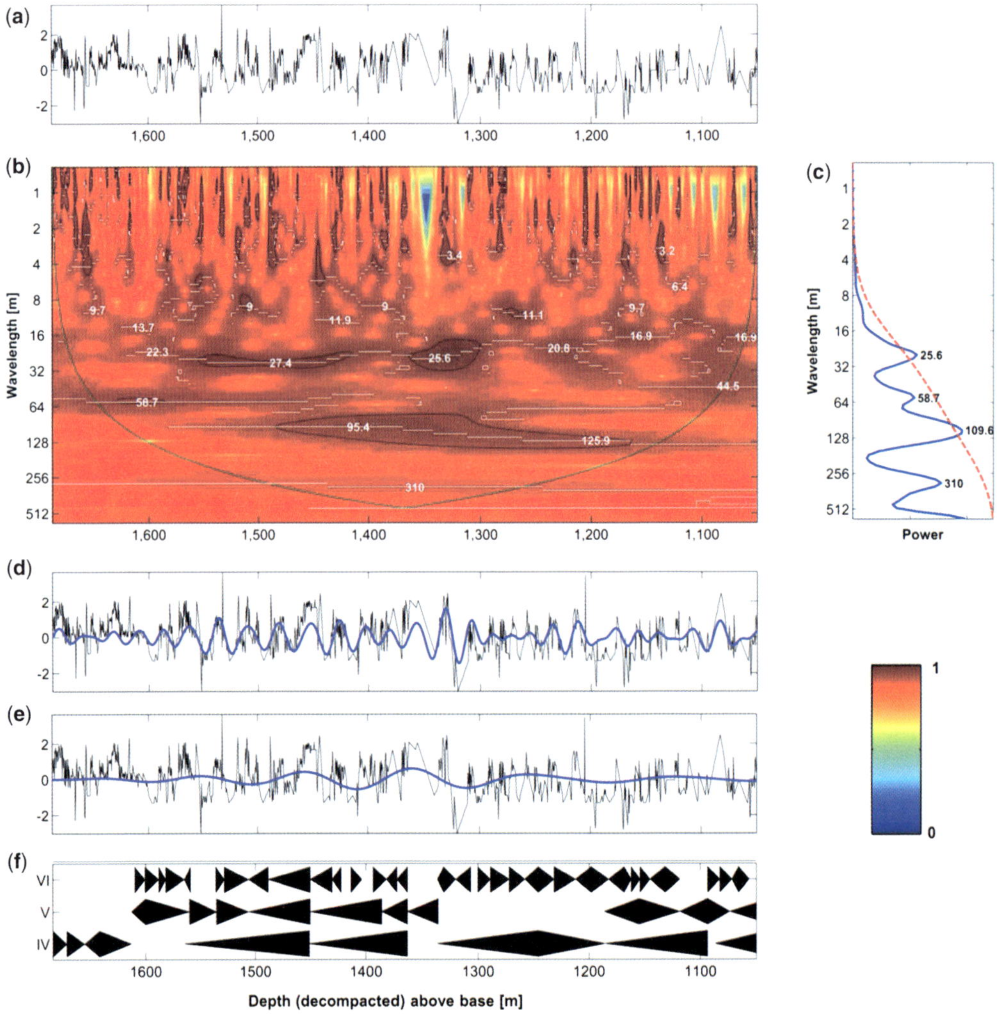

Fig. 7. (**a**) Cut-out of the decompacted normalized lithology profile of the Querfurt 1/64 core from 1050–1688 m (all values above base at 2998.9 m set as zero). (**b**) Scalogram with log2(power) of lithology profile using the Morlet-6 wavelet. White numbers mark the wavelengths of ridges at these positions. (**c**) Global wavelet power spectrum with the wavelengths of local power maxima marked. (**d**) Original (black line) and reconstructed (blue line) time series in the wavelength interval 18–34 m. (**e**) Same as (d) but with the reconstruction interval 80–120 m. (**f**) Hierarchy levels IV–VI of cycles as identified from the lithology profile, decompacted for the purpose of comparison.

driven cycles; for example, Middle Cambrian–Upper Permian sections in North America (Elrick & Hinnov 1996, 2007), the mid-Carboniferous of northern England (Tucker *et al.* 2009), the Upper Permian Zechstein of England (Mawson & Tucker 2009), the Triassic limestone succession of Latemar in the Southern Alps (Goldhammer *et al.* 1987; Schwarzacher 1993; Egenhoff *et al.* 1999; Preto *et al.* 2004; Meyers 2008) or Quaternary deposits (e.g. Broecker & van Donk 1970; Imbrie *et al.* 1984, 1989; Hilgen 1994; ten Kate *et al.* 1994; Roth & Reijmer 2005). Within extensive coastal plains, cyclic sea-level fluctuations are reflected as defined sequences of shallow-marine to continental sediments, termed cyclothems, which are well researched for the late Palaeozoic (e.g. Süss 1996; Elliot 1968; Ferm 1975; Heckel 1986, 2008; Ross & Ross 1988; Izart *et al.* 1996; Hampson *et al.* 1999; Cecil *et al.* 2003; Legler *et al.* 2005, 2011; Legler & Schneider 2008; Belt *et al.* 2011; Eros *et al.* 2012). The connection of such sea-level fluctuations with orbital driven climate cycles has been focus of research for a long time and is now generally widely accepted (e.g. Cecil 1990; Crowley & Baum 1991, 1992; Cecil *et al.* 2003; Matchen & Kammer 2006). In continental environments, lacustrine profiles are especially well suited for research, such as, for example, the Holocene lakes Tanganyika (Burnett *et al.* 2011) and Baldwin (Kirby *et al.* 2006), the Neogene Pannonian Basin (Juhász *et al.* 2007), the Eocene Green River Formation (e.g. Bradley 1929; Schwarzacher 1993; Matthews & Perlmutter 1994; Meyers 2008), the Triassic Newark Supergroup (e.g. Smoot & Olsen 1994; Olsen & Kent 1996) and the Czech Carboniferous (Lojka *et al.* 2009).

The interpretation of fluvial profiles is more complicated because of their vicinity to tectonically active environments (Etheridge *et al.* 1998; Vandenberghe 2003). The causes for cyclicity within fluvial profiles were attributed to tectonic events (e.g. Steel & Aasheim 1978; Olsen 1987; Newell *et al.* 1999; Catuneanu & Elango 2001; Juhasz *et al.* 2007; Kallmeyer *et al.* 2010) or climate changes (e.g. Clemmensen *et al.* 1989; Olsen 1990, 1994; Kelly 1992; Blum & Törnqvist 2000; Gibbard & Lewin 2002; Gao *et al.* 2007; Amorosi *et al.* 2008; Sangen *et al.* 2011). In addition, the sedimentological record is distorted by flooding events and related erosion, as well as by autocyclic displacement of active channels.

It is, however, still possible to record climatically driven cycles within fluvial profiles. Olsen (1990, 1994) reconstructed discharge rate fluctuations during the Upper Devonian (Famennian) by investigating channel geometry and the thickness of individual channel fillings of a meandering river system in East Greenland. Using three-dimensional outcrops, he could exclude the influence of changing positions within a single meander. Therefore, the observed cycles reflect a change in the discharge rate, which is climatically controlled through precipitation rates. In geological times of slight vegetation cover (Late Palaeozoic–Early Mesozoic periods), discharge rate fluctuations should be more strongly reflected within fluvial sedimentation records than nowadays. The geological framework of Olsen's investigations is identical to the conditions within the Late Carboniferous Saale Basin. It is an intramontane basin with fluvial longitudinal transport in sandy, meandering rivers, and a lateral sediment transport by alluvial fans and braided river systems. East Greenland was located at about 10° northern latitude during the Famennian, experiencing monsoon climates, whereas the global climate was controlled by glaciations in the Southern Hemisphere (Olsen 1990). Fluvial cycles display an average thickness of 101 m, with a range between 79 and 135 m; thinner cycles were also detected. The proportion of cycle lengths 5:1 could indicate a correlation with the 100 ka cycle (short eccentricity cycle) and the 21 ka cycle (combined precession cycles). Both cycles are ascribed to glaciations during the Famennian (Olsen 1990; Kelly 1992).

It has been investigated on Recent–Pleistocene rivers all over the world, that rivers respond to climate changes with changes in the internal architecture of their deposits (e.g. Blum & Törnqvist 2000; Vandenberghe 2003). Rossingol-Strick (1985) concluded that, in recent areas with monsoon climate, sedimentary cycles in fluvial profiles (e.g. the Nile) are influenced by the 21, 100 and 400 ka periods (precession and eccentricity cycles). Similar observations were made by Goodbred (2003) investigating the River Ganges. Rivers outside monsoon climates respond to climatic changes during the Pleistocene as well. So, for instance, the overall cyclic facies architecture of the Quaternary deposits of the Po river in northern Italy is controlled by the 100 ka period and correlates with sea-level fluctuations, as well as with the changing composition of the pollen flora with regard to thermophile forms (Amorosi *et al.* 2008).

Timeframe

Menning *et al.* (2005) stated a duration of 4–6 Ma for the duration of the Stephanian (Stephanian A or Barruelian, respectively, to Stephanian C) or, alternatively, 7 Ma including the Cantabrian (with a duration of *c.* 2.5 Ma). Menning *et al.* (2006) listed the duration of the Stephanian as 3–5 Ma, including the Cantabrian. Izart *et al.* (2003) listed three third-order cycles (1–2 Ma) for the Stephanian in the Saale Basin, based on research by Menning *et al.* (1997), but they did not include information on thinner cycles. In Western Europe, 400 ka cycles were interpreted as an expression of eccentricity, and 1–3 Ma cycles as a result of tectonic activity of the Variscian Mountains (Izart *et al.* 2003). Fourth-order cycles of 100 ka–1 Ma are present, particularly in East European profiles. Izart *et al.* (2003) listed seven–eight cycles for the Kasimovian (each 375–428 ka) and 16 cycles for the Gzhelian (each 375 ka) within the Donets Basin. Eros *et al.* (2012) estimated about 8 Ma for the Stephanian. Several climatically driven cycles related to ablation of the South Pole glaciation can be separated during the Carboniferous in the Southern Hemisphere (e.g. Lopez-Gamundi 1997; Visser 1997; Roscher & Schneider 2006; Schneider *et al.* 2006; Fielding *et al.* 2008). These cycles can be found again in the profiles of tropical northern continents by comparing sea-level fluctuations and global climate changes (e.g. Falcon-Lang 2003*a*, *b*, 2004; Roscher & Schneider 2006;

Poulsen *et al.* 2007; DiMichele *et al.* 2009, 2010; Eros *et al.* 2012; Horton *et al.* 2012). Climate simulations of the late Palaeozoic show linkages between ice shield coverage in the Southern Hemisphere and annual precipitation, as well as vegetation cover in low latitudes, indicating an increase in precipitation in tropical areas when the ice cover is waxing (Poulsen *et al.* 2007; Peyser & Poulsen 2008; Horton *et al.* 2012), which manifests as periodically humid periods, represented by widespread grey beds in European and North American continental sections (Roscher & Schneider 2006). Two of these humid periods can be found during the Barruelian stage, approximately Stephanian A (humid phase B), and at the threshold between uppermost Stephanian and Asselian (humid phase C). The interval between these two humid periods is 5–7 Ma (Scheffler *et al.* 2003; Roscher & Schneider 2006).

Grey facies can be found at the base of all sections within the Saale Basin, if these are deep enough. These deposits are typically classified as part of the Grillenberg Subformation. In addition, grey beds occur in several profiles as local formations at various locations, and again widely distributed at the top of the profile containing the Wettin Subformation. Provided that the grey fine clastics at the base of the Querfurt core can be correlated with the Grillenberg Subformation and be placed into the early Stephanian (i.e. Barruelian), and that the Wettin Subformation can be placed into the Stephanian C, this section records nearly the complete Mansfeld Subgroup. The major cycle of the Mansfeld Subgroup correlates then with the section of humid periods B and C after Roscher & Schneider (2006), being controlled by global climate, and continuing for between 5 and 7 Ma.

Time-series analyses showed 16 cycles with lengths of about 100 m (Fig. 6). Implying a total duration of 5–7 Ma for the complete section, each cycle is 323–453 ka long (or 329–461 ka) using the statistical average of 108 (Fig. 6c) or 110 m, respectively (Fig. 7c). Taking the number of calculated cycles into account, the duration of each cycle is 312–437 ka (16 cycles, compare Fig. 6e). The cycles are therefore part of the Milankovitch Band, probably representing changes in eccentricity of the Earth's orbit that last about 400 ka.

Distinct cycles at 26 and 110 m (Fig. 7c) show a proportion of about 1:4, becoming evident during the separate analysis of the fine clastic section above profile metre 1050 (decompacted). This result supports the interpretation that these cycles are part of the short and long eccentricity, with the 100 ka cycle being well developed, whereas the 400 ka cycle is not as distinct but can be traced though the complete profile (Figs 6e & 7d).

Conclusion

Some of the measured cycles in the Carboniferous section at the Saale Basin are most probably climatically driven. These cycles are part of the Milankovitch Band, with lengths of about 400 and 100 ka, provided that the drill hole Querfurt 1/64 shows a nearly complete record of the Mansfeld Subgroup and of the Stephanian stage, respectively (overall duration of 5–7 Ma), and that the grey beds at the base of the profile are placed into the Grillenberg Subformation.

Besides the widespread coal-bearing grey facies of the Wettin Subformation and the deposits of the Grillenberg Subformation, grey beds occur periodically within the Carboniferous section of the Saale Basin. They are typically small-scale local occurrences. Therefore, they are not suitable for basin-wide lithostratigraphical correlation.

The definition of the Querfurt Subformation by Schneider *et al.* (1995*b*, *c*) depends mainly on the correlation of the grey facies at the base of the Querfurt 1/64 borehole with grey beds of Horizon IIIb in the Schladebach core (Gebhardt 1988*b*; Beyschlag & von Fritsch 1899). Accepting the results presented here of a combination of classical lithostratigraphy and wavelet-based cyclostratigraphy, this correlation can no longer be maintained. Therefore, it is necessary to reconsider the existence and definition of a Querfurt Subformation. The deposits in question belong to the Grillenberg Subformation at the base of the Stephanian. In consequence, the maximum thickness of the Carboniferous within the Saale Basin is about 1000 m.

Provided that other methods, such as biostratigraphy and lithostratigraphy, are not able to solve stratigraphical problems, then time-series analysis can be a useful tool to use in stratigraphical interpretation, even for non-marine fossil-free red beds. Such numerical methods display the internal organization of the cyclicity of profiles, and thus provide guidance in determining whether cycles of different hierarchy and origin are autocyclically, tectonically or climatically controlled. Based on this distinction, basin-wide correlations of fluvial red beds are possible, using a combination of high-resolution stratigraphy, biostratigraphy and classical lithostratigraphy.

We would like to thank B.-C. Ehling (Halle), J. Schneider (Freiberg), E. Harms (Karlsruhe) and A. Kampe (Berlin) for their critical revisions and valuable advice. We thank E. Hanenkamp (Karlsruhe) for considerable help with the development of the English manuscript, and R. Rößler (Chemnitz) for reinvestigation of the macroflora and for providing photographs. We also are very grateful to A. Gąsiewicz and M. Słowakiewicz as well as two anonymous reviewers for their valuable contributions.

References

AHUJA, N., LERTRATTANAPANICH, S. & BOSE, N. K. 2005. Properties determining choice of mother wavelet. *IEE Proceedings: Vision, Image and Signal Processing*, **152**, 659–664.

AMOROSI, A., PAVESI, M., LUCCHI, M. R., SARTI, G. & PICCIN, A. 2008. Climatic signature of cyclic fluvial architecture from the Quaternary of the central Po Plain, Italy. *Sedimentary Geology*, **209**, 58–68.

ANTOINE, J.-P. 1999. Wavelet analysis: a new tool in physics. *In*: VAN DEN BERG, J. C. (ed.) *Wavelets in Physics*. Cambridge University Press, Cambridge, 9–22.

APPENZELLER, C., STOCKER, T. F. & ANKLIN, M. 1998. North Atlantic Oscillation dynamics recorded in Greenland ice cores. *Science*, **282**, 446–448.

BAHR, D. B., HUTTON, E. W. H., SYVITSKI, J. P. M. & PRATSON, L. F. 2001. Exponential approximations to compacted sediment porosity profiles. *Computers and Geosciences*, **27**, 691–700.

BALDWIN, B. & BUTLER, C. O. 1985. Compaction curves. *American Association of Petroleum Geologsts Bulletin*, **69**, 622–626.

BARTHEL, M. 2001. Pflanzengruppen und Vegetationseinheiten der Manebach-Formation. *Beiträge zur Geologie Thüringen N.F.*, **8**, 93–123.

BELT, E. S., HECKEL, P. H., LENTZ, L. J., BRAGONIER, W. A. & LYONS, T. W. 2011. Record of glacial–eustatic sea-level fluctuations in complex middle to late Pennsylvanian facies in the Northern Appalachian Basin and relation to similar events in the Midcontinent basin. *Sedimentary Geology*, **238**, 79–100.

BEYSCHLAG, F. & VON FRITSCH, K. 1899. Das jüngere Steinkohlengebirge und das Rotliegende in der Provinz Sachsen und den angrenzenden Gebieten. *Abhandlungen der Preußischen Geologischen Landesanstalt, N.F.*, **10**, 263S.

BLESS, M. J. M., BOUCKAERT, J. & PAPROTH, E. 1984. Migration of facies belts as a response to continental drift during the Late Devonian and Carboniferous. *Bulletin Societé belge Géologie*, **93**, 189–195.

BLUM, M. D. & TÖRNQVIST, T. E. 2000. Fluvial responses to climate and sea-level change: a review and look forward. *Sedimentology*, **47**, 2–48.

BOLTON, E. W., MAASCH, K. A. & LILLY, J. M. 1995. A wavelet analysis of Plio-Pleistocene climate indicators, a new view of periodicity evolution. *Geophysical Research Letters*, **22**, 2753–2756.

BRADLEY, W. H. 1929. *The Varves and Climate of the Green River Epoch*. United States Geological Survey, Professional Papers, **158-E**, 87–110.

BROECKER, W. S. & VAN DONK, J. 1970. Insolation changes, ice volumes, and the $\delta^{18}O$ record in deep-sea cores. *Reviews of Geophysical and Space Physics*, **8**, 169–198.

BURNETT, A. P., SOREGHAN, M. J., SCHOLZ, C. A. & BROWN, E. T. 2011. Tropical East African climate change and its relation to global climate: a record from Lake Tanganyika, Tropical East Africa, over the past 90+ kyr. *Palaeogeography, Palaeoclimatology, Palaeoecology*, **303**, 155–167.

CARMONA, R. A., HWANG, W. L. & TORRÉSANI, B. 1999. Multiridge detection and time–frequency reconstruction. *IEEE Transactions on Signal Processing*, **47**, 480–492.

CATUNEANU, O. & ELANGO, H. N. 2001. Tectonic control on fluvial styles: the Balfour Formation of the Karoo Basin, South Africa. *Sedimentary Geology*, **140**, 291–313.

CECIL, C. B. 1990. Paleoclimate controls on stratigraphic repetition of chemical and clastic rocks. *Geology*, **18**, 533–536.

CECIL, C. B., DULONG, F. T., WEST, R. R., STAMM, R., WARDLAW, B. & EDGAR, N. T. 2003. Climate controls on the stratigraphy of a Middle Pennsylvanian cyclothem in North America. *In*: CECIL, C. B. & EDGAR, N. T. (eds) *Climate Controls on Stratigraphy*. SEPM, Special Publications, **77**, 151–182.

CLEMMENSEN, L. B., OLSEN, H. & BLAKEY, R. C. 1989. Erg-margin deposits in the Lower Jurassic Moenave Formation and Wingate Sandstone, southern Utah. *Bulletin of the Geological Society of America*, **101**, 759–773.

CROWLEY, T. J. & BAUM, S. K. 1991. Estimating Carboniferous sea-level fluctuations from Gondwana ice extent. *Geology*, **19**, 507–510.

CROWLEY, T. J. & BAUM, S. K. 1992. Modelling late Paleozoic glaciation. *Geology*, **20**, 507–510.

DAUBECHIES, I. 1992. *Ten Lectures on Wavelets. (CBMS-NSF Regional Conference Series in Applied Mathematics), Volume 62*. Society for Industrial and Applied Mathematics, Philadelphia, PA.

DIMICHELE, W. A., MONTAÑEZ, I. P., POULSEN, C. J. & TABOR, N. J. 2009. Cimate and vegetational regime shiftin the late Paleozoic ice age earth. *Geobiology*, **7**, 200–226.

DIMICHELE, W. A., CECIL, C. B., MONTAÑEZ, I. P. & FALCON-LANG, H. 2010. Cyclic changes in Pennsylvanian paleoclimate and effects on floristic dynamics in tropical Pangaea. *International Journal of Coal Geology*, **83**, 329–344.

DÖRING, H. & KAMPE, A. 1973. Zur mikro- und makrofloristischen Alterseinstufung der Wettiner Schichten höchstes Stefan in der nördlichen Saalesenke. *Zeitschrift für geologische Wissenschaften*, **1**, 619–649.

DUCHAUFOUR, P. 1998. *Handbook of Pedology – Soils, Vegetation, Environment*. A.A. Balkema, Rotterdam.

EAGAR, R. M. C. 1984. Late Carboniferous–early Permian non-marine bivalve faunas of northern Europe and eastern North America. *In*: *Compte Rendu Neuvième Congrès International de Stratigraphie et de Géologie de Carbonifère, Washington and Champaign-Urbana, 1979. Compte-rendu Biostratigraphy, Volume 2*. Southern Illinois University Press, Carbondale, IL, 559–575.

EAGAR, R. M. C. 2005. Non-marine and limnic bivalves. *Courier Forschungsinstitut Senckenberg*, **254**, 55–86.

EGENHOFF, S. O., PETERHÄNSEL, A., BECHSTÄDT, T., ZÜHLKE, R. & GRÖTSCH, J. 1999. Facies architecture of an isolated carbonate platform. Tracing the cycles of the Latemàr Middle Triassic, northern Italy. *Sedimentology*, **46**, 893–912.

EHLING, B.-C. & BREITKREUZ, C. 2004. Spätpaläozoische Sedimentation und Vulkanismus im östlichen Saale-Becken. *Exkursionsführer zur GeoLeipzig2004. Schriftenreihe der Deutschen Geologischen Gesellschaft*, **35**, 133–149.

ELLIOT, R. E. 1968. Facies, sedimentation successions and cyclothems in productive coal measures in the East Midlands, Great Britain. *Mercian Geologist*, **2**, 331–372.

ELRICK, M. & HINNOV, L. A. 1996. Millennial-scale climate origins for stratification in Cambrian and Devonian deepwater rhythmites, western USA. *Palaeogeography, Palaeoclimatology, Palaeoecology*, **123**, 353–372.

ELRICK, M. & HINNOV, L. A. 2007. Millennial-scale paleoclimate cycles in widespread Paleozoic deeper water rhythmites of North America. *Palaeogeography, Palaeoclimatology, Palaeoecology*, **243**, 348–372.

EROS, J. M., MONTAÑEZ, I. P., OSLEGER, D. A., DAVYDOV, V. I., NEMYROVSKA, T. I., POLETAEV, V. I. & ZHYKALYAK, M. V. 2012. Sequence Stratigraphy and onlap history of the Donets Basin, Ukraine: insight into Carboniferous icehouse dynamics. *Palaeogeography, Palaeoclimatology, Palaeoecology*, **271**, 1–25.

ETHERIDGE, F. G., WOOD, L. J. & SCHUMM, S. A. 1998. Cyclic variables controlling fluvial sequence development: problems and perspectives. *In*: SHANLEY, K. W. & MCCABE, P. J. (eds) *Relative Role of Eustacy, Climate, and Tectonism in Continental Rocks*. SEPM, Special Publications, **59**, 17–29.

FALCON-LANG, H. J. 2003*a*. Late Carboniferous Tropical Dryland Vegetation in an Alluvial-plain Setting, Joggins, Nova Scotia, Canada. *Palaios*, **18**, 197–211.

FALCON-LANG, H. J. 2003*b*. Response of Late Carboniferous tropical vegetation to transgressive-regressive rhythms at Joggins, Nova Scotia. *Journal of the Geological Society*, **160**, 643–648.

FALCON-LANG, H. J. 2004. Pennsylvanian tropical rain forests responded to glacial-interglacial rhythms. *Geology*, **32**, 689–692.

FALK, F., ELLENBERG, J., GRUMBT, E. & LÜTZNER, H. 1979. Zur Sedimentation des Rotliegenden im Nordteil der Saale-Senke – Hallesche bis Hornburger Schichten. *Hallesches Jahrbuch für Geowissenschaften*, **4**, 3–22.

FALK, F., ELLENBERG, J., GRUMBT, E. & LÜTZNER, H. 1980. Zur Sedimentation des Rotliegenden im Nordteil der Saale-Senke–Brachwitzer und Eislebener Schichten. *Hallesches Jahrbuch für Geowissenschaften*, **5**, 77–88.

FARGE, M. 1992. Wavelet transforms and their applications to turbulence. *Annual Review Fluid Mechanics*, **24**, 395–457.

FERM, J. C. 1975. Pennsylvanian cyclothems of the Appalachian Basin, a retrospective view. *In*: MCKEE, E. D., CROSBY, E. J. ET AL. (eds) *Paleotectonic Investigations of the Pennsylvanian System in the United States, Part II: Interpretive Summary and Special Features of the Pennsylvanian System*. United States Geological Survey, Professional Papers, **853-2**, 57–64.

FIELDING, C. R., FRANK, T. D. & ISBELL, J. L. (eds). 2008. *Resolving the Late Paleozoic Ice Age in Time and Space*. Geological Society of America, Special Papers, **441**.

FRANCO, D. R., HINNOV, L. A. & ERNESTO, M. 2011. Spectral analysis and modelling of microcyclostratigraphy in late Paleozoic glaciogene rhythmites, Paraná Basin, Brazil. *Geochemistry, Geophysics, Geosystems*, **12**, Q09003, http://dx.doi.org/10.1029/2011GC003602

FRIESE, H. 1955. Zur Feingliederung der Oberen und obersten Teiles der Unteren Wettiner Schichten im Steinkohlenrevier Plötz bei Halle Saale durch Leitschichten. *Zeitschrift für angewandte Geologie*, **1**, 55–61.

GAO, C., BOREHAM, S., PREECE, R. C., GIBBARD, P. L. & BRIANT, R. M. 2007. Fluvial response to rapid climate change during the Devensian (Weichselian) Lateglacial in the River Great Ouse, southern England, UK. *Sedimentary Geology*, **202**, 193–210.

GAST, R. 1995. Sequenzstratigraphie. *In*: PLEIN, E. (ed.) *Stratigraphie von Deutschland I: Norddeutsches Rotliegendbecken; Rotliegend – Monographie Teil I*. Courier Forschungs Institut Senckenberg, **183**, 47–54.

GAST, R. & GEBHARDT, U. 1995. 4.4 Elbe-Subgruppe. *In*: PLEIN, E. (ed.) *Stratigraphie von Deutschland I: Norddeutsches Rotliegendbecken; Rotliegend – Monographie Teil I*. Courier Forschungs Institut Senckenberg, **183**, 121–145.

GEBHARDT, U. 1988*a*. Mikrofaziesanalyse und stratigraphisch-regionalgeologische Interpretation terrestrischer Karbonate der varistischen Molasse Mitteleuropas, Permokarbon. *Freiberger Forschungshefte*, **C427**, 30–59.

GEBHARDT, U. 1988*b*. Mikrofazies und Paläontologie biogener Karbonate der Unteren Mansfelder Schichten, Oberkarbon, Stefan. *Hallesches Jahrbuch für Geowissenschaften*, **13**, 5–21.

GEBHARDT, U., MERKEL, T. & SZABADOS, A. 2000. Karbonatsedimentation in siliziklastischen fluviatilen Abfolgen. *Freiberger Forschungshefte*, **C490**, 133–168.

GIBBARD, P. L. & LEWIN, J. 2002. Climate and related controls on interglacial fluvial sedimentation in lowland Britain. *Sedimentary Geology*, **151**, 187–210.

GLENNIE, K. W., PUGH, J. M. & GOODALL, T. M. 1994. Late Quaternary Arabian desert models of Permian Rotliegend reservoirs. *Shell Exploration Bulletin*, **274**, 1–19.

GOLDHAMMER, R. K., DUNN, P. A. & HARDIE, L. A. 1987. High frequency glacioeustatic sealevel oscillations with Milankovich characteristics recorded in middle Triassic platform carbonates in northern Italy. *American Journal Sciences*, **287**, 853–892.

GOODBRED, S. L. JR. 2003. Response of the Ganges dispersal system to climate change: a source-to-sink view since the last interstade. *Sedimentary Geology*, **162**, 83–104.

GOTHAN, W. & SCHRIEL, W. 1927. Die Grillenberger Schichten des Unterharzes, ein Äquivalent der oberen Saarbrücker Flammkohlenschichten. *Jahrbuch der Preußischen Geologischen Landesanstalt*, **48**, 366–375.

GROSSMANN, A. & MORLET, J. 1984. Decomposition of hardy functions into square integrable wavelets of constant shape. *SIAM Journal on Mathematical Analysis*, **15**, 723–736.

GUYODO, Y., GAILLOT, P. & CHANNELL, J. E. T. 2000. Wavelet analysis of relative geomagnetic paleointensity at ODP Site 983. *Earth Planetary Science Letters*, **184**, 109–123.

HAGENDORF, U. & SCHWAHN, H.-J. 1969. Sedimentpetrographische und paläontologische Untersuchungen des

Permosiles im Untergrund der Querfurter Mulde. *Hallesches Jahrbuch für Mitteldeutsche Erdgeschichte*, **9**, 40–74.

Hampson, G., Stollhofen, H. & Flint, S. 1999. A sequence stratigraphic model for the Lower Coal Measures (Upper Carboniferous) of the Ruhr district, North-west Germany. *Sedimentology*, **46**, 1199–1231.

Hänel, M. 1969. Fazies, Sedimentation und Paläogeographie der Grillenberger und Mansfelder Schichten am SE-Rand des Harzes. *Hallesches Jahrbuch für Mitteldeutsche Erdgeschichte*, **9**, 93–109.

Hay, W. W. 1996. Tectonics and climate. *Geologische Rundschau*, **85**, 409–437.

Heckel, P. H. 1986. Sea-level curve for Pennsylvanian eustatic marine transgressive-regressive depositional cycles along midcontinent outcrop belt, North America. *Geology*, **14**, 330–334.

Heckel, P. H. 2008. Pennsylvanian cyclothems in Midcontinent North America as far-field effects of waxing and waning of Gondwana ice sheets. *In*: Fielding, C. R., Frank, T. D. & Isbell, J.L. (eds) *Resolving the Late Paleozoic Ice Age in Time and Space*. Geological Society of America, Special Papers, **441**, 275–289.

Herbert, T. D. 1994. Reading orbital signals distorted by sedimentation. models and examples. *In*: deBoer, P. L. & Smith, G. D. (eds) *Orbital Forcing and Cyclic Sequences*. International Association of Sedimentologists, Special Publications, **19**, 483–507.

Hiete, M. 2004. *Umweltveränderungen in der Permo-Trias, Geochemische Charakterisierung, Zeitreihenanalyse, Modellierung*. Dissertation, Universität Braunschweig.

Hilgen, F. J. 1994. An astronomically calibrated polarity time scale for the Pliocene–Pleistocene. a brief review. *In*: deBoer, P. L. & Smith, G. D. (eds) *Orbital Forcing and Cyclic Sequences*. International Association of Sedimentologists, Special Publications, **19**, 109–116.

Hinnov, L. A. 2000. New perspectives on orbitally forced stratigraphy. *Annual Review of Earth and Planetary Science*, **28**, 419–475.

Hinnov, L. A. & Ogg, J. G. 2007. Cyclostratigraphy and the astronomical time scale. *Stratigraphy*, **4**, 239–251.

Hölzel, M., Faber, R. & Wagreich, M. 2008. DeCompactionTool. Software for subsidence analysis including statistical error quantification. *Computers & Geosciences*, **34**, 1454–1460.

Horton, D. E., Poulsen, C. J., Montañez, I. P. & DiMichele, W. A. 2012. Eccentricity-paced late Paleozoic climate change. *Palaeogeography, Palaeoclimatology, Palaeoecology*, **331–332**, 150–161, http://dx.doi.org/10.1016/j.palaeo.2012.03.014

House, M. R. 1995. Orbital forcing timescales. *In*: House, M. R. & Gale, A. S. (eds) *Orbital Forcing Timescales and Cyclostratigraphy*. Geological Society, London, Special Publications, **85**, 1–18.

Imbrie, J., Hays, J. D. *et al.* 1984. The orbital theory of Pleistocene climate: support from a revised chronology of the marine $\delta^{18}O$ record. *In*: Berger, A., Imbrie, J., Hays, J., Kukla, G. & Saltzman, B. (eds) *Milankovich and Climate*. NATO, ASI Series C, **126**. Kluwer, Dordrecht, 269–305.

Imbrie, J., McIntyre, A. & Mix, A. 1989. Oceanic response to orbital forcing in the late Quaternary. observational and experimental strategies. *In*: Berger, A. L., Schneider, S. & Duplessy, J. Cl. (eds) *Climate and Geosciences: A Challenge for Science and Society in the 21st Century*. NATO, ASI Series C, **285**. Kluwer, Dordrecht, 121–164.

Isbell, J. L., Miller, M. F., Wolfe, K. L. & Lenaker, P. A. 2003. Timing of Late Paleozoic glaciation in Gondwana: was glaciation responsible for the development of northern hemisphere cyclotherms? *In*: Chan, M. A. & Archer, A. W. (eds) *Extreme Depositional Environments: Mega End Members in Geologic Time*. Geological Society of America, Special Paper, **370**, 5–24.

Izart, A., Briant, C., Vaslet, D., Vachard, D., Coquel, R. & Maslo, A. 1996. Stratigraphy and sequence Stratigraphy of the Moskovian in the Donets basin. *Tectonophysics*, **268**, 189–209.

Izart, A., Stephenson, R. *et al.* 2003. Sequence Stratigraphy and correlation of late Carboniferous and Permian in the CIS, Europe, Tethyan area, North Africa, Arabia, China, Gondwanaland and the USA. *Palaeogeography, Palaeoclimatology, Palaeoecology*, **196**, 59–84.

Juhász, Gy., Pogácsás, Gy., Magyar, I. & Vakarcs, G. 2007. Tectonic v. climatic control on the evolution of fluvio-deltaic systems in a lake basin, Eastern Pannonian Basin. *Sedimentary Geology*, **202**, 72–95.

Kallmeyer, E., Breitkreuz, C., Kiersnowski, H. & Geißler, M. 2010. Issues associated with the distinction between climatic and tectonic controls on Permian alluvial fan deposits from the Kotzen and Barnim Basins (North German Basin). *Sedimentary Geology*, **223**, 15–34.

Kampe, A. 1966. *Bericht über die Stratigraphie und Fazies des Permokarbons im nordöstlichen Saaletrog*. Dissertation, Humboldt-Universität Berlin.

Kampe, A. & Döring, H. 1993. Zum mikrofloristischen Alter der tiefsten Molassesedimente in der Altbohrung Schladebach 1880 südwestlich von Leipzig. *Zeitschrift für geologische Wissenschaften*, **21**, 397–402.

Kampe, A. & Kunert, R. 1964. Die Definition der Halleschen Schichten. *Zeitschrift für angewandte Geologie*, **10**, 93–94.

Kampe, A. & Remy, W. 1962. Ausbildung und Abgrenzung des Stefanien in den Halleschen Mulde. *Monatsberichte der Deutschen Akademie der Wissenschaften*, **4**, 54–68.

Kelly, S. B. 1992. Milankovich cyclicity recorded from Devonian non-marine sediments. *Terra Nova*, **4**, 578–584.

Kirby, M. E., Lund, S. P. & Bird, B. W. 2006. Mid-Wisconsin sediment record from Baldwin Lake reveals hemispheric climate dynamics (Southern CA, USA). *Palaeogeography, Palaeoclimatology, Palaeoecology*, **241**, 267–283.

Kumar, P. & Foufoula-Georgiou, E. 1997. Wavelet analysis for geophysical applications. *Reviews of Geophysics*, **35**, 385–412.

Kunert, R. 1959. Die Zusammensetzung der permokarbonen Sandsteine im östlichen Harzvorland. *Beiträge zur Geologie*, **1**, 35–46.

KUNERT, R. 1970. Lithofazielle und tektonische Untersuchungen im Permokarbon des östlichen Harzvorlandes. *Jahrbuch für Geologie*, **3**, 57–120.

KUTZBACH, J. E. & GALLIMORE, R. G. 1989. Pangean climates. Megamonsoons of the Megacontinent. *Journal of Geophysical Research*, **94**, 3341–3357.

LASPEYRES, H. 1875. Geognostische Darstellung des Steinkohlengebirges und Rotliegenden in der Gegend nördlich von Halle an der Saale. *Abhandlungen zur geologischen Spezialkarte von Preussen und den Thüringer Staaten I*, **1**, 261–603.

LEGLER, B. & SCHNEIDER, J. W. 2008. Marine ingressions into the Middle/Late Permian saline lake of the Southern Permian Basin (Rotliegend, Northern Germany) possibly linked to sea-level highstands in the Arctic rift system. *Palaeogeography, Palaeoclimatology, Palaeoecology*, **267**, 102–114.

LEGLER, B., GEBHARDT, U. & SCHNEIDER, J. W. 2005. Late Permian non-marine–marine transitional profiles in the central Southern Permian Basin. *International Journal of Earth Sciences (Geologische Rundschau)*, **94**, 851–862.

LEGLER, B., SCHNEIDER, J. W., GEBHARDT, U., MERTEN, D. & GAUPP, R. 2011. Lake deposits of moderate salinity as sensitive indicators of lake level fluctuations: example from the Upper Rotliegend saline lake (Middle–Late Permian, Northeast Germany). *Sedimentary Geology*, **234**, 56–69.

LEHMANN, R. 1929. Das rotliegende Alter der Steinkohlen von Wettin und Löbejün. *Jahrbuch des Halleschen Verbandes zur Erforschung mitteldeutscher Bodenschätze und ihrer Verwendung, N.F.*, **8**, 172–186.

LIN, J. 2006. Ridges reconstruction based on inverse wavelet transform. *Journal of Sound and Vibration*, **294**, 916–926.

LOJKA, R., DRÁBKOVÁ, J., ZAJÍC, J., SÝKOROVÁ, I., FRANKŮ, J., BLÁHOVÁ, A. & GRYGAR, T. 2009. Climate variability in the Staphanian B based on environmental trecord of the Mšec Lake deposits (Kladno–Rakovník Basin, Czech Republik). *Palaeogeography, Palaeoclimatology, Palaeoecology*, **280**, 78–93.

LOPEZ-GAMUNDI, O. R. 1997. Glacial–postglacial transition in the late Paleozoic basins of southern south America. *In*: MARTINI, I. P. (ed.) *Late Glacial and Postglacial Environmental Changes, Quaternary, Carboniferous–Permian, and Proterozoic*. Oxford University Press, New York, 147–168.

MARTENS, T. 1983*a*. Zur Taxonomie und Biostratigraphie der Conchostraca Phyllopoda, Crustacea des Jungpaläozoikums der DDR, Teil I. *Freiberger Forschungshefte*, **C382**, 7–105.

MARTENS, T. 1983*b*. Zur Taxonomie und Biostratigraphie der Conchostraca Phyllopoda, Crustacea des Jungpaläozoikums der DDR, Teil II. *Freiberger Forschungshefte*, **C384**, 24–48.

MATCHEN, D. L. & KAMMER, T. W. 2006. Incised valley fill interpretation for Mississippian Black Hand Sandstone, Appalachian Basin, USA: implications for glacial eustacy at Kinderhookian–Osagean (Tn2–Tn3) boundary. *Sedimentary Geology*, **191**, 89–113.

MATTHEWS, M. D. & PERLMUTTER, M. A. 1994. Global cyclostratigraphy. An application to the Eocene Green River basin. *In*: DEBOER, P. L. & SMITH, G. D. (eds) *Orbital Forcing and Cyclic Sequences*. International Association of Sedimentologists, Special Publications, **19**, 459–481.

MAWSON, M. & TUCKER, M. 2009. High-frequency cyclicity (Milankovitch and millennial-scale in slope apron carbonates: Zechstein (Upper Permian), North-east England. *Sedimentology*, **56**, 1905–1936.

MEISTER, J. 1969. Sedimentpetrographische und lithologische Untersuchungen im Siles des Kyffhäusers. *Hallesches Jahrbuch für mitteldeutsche Erdgeschichte*, **9**, 75–92.

MENNING, M., WEYER, D., DROZDZEWSKI, G. & VAN AMEROM, H. W. J. 1997. Carboniferous time scales 1997 time scale A min. ages and time scale B max. ages. Use of geological time indicators. *Carboniferous Newsletter*, **15**, 26–28.

MENNING, M., WEYER, D., WENDT, I. & DROZDZEWSKI, G. 2005. 3.3 Eine numerische Zeitskala für das Pennsylvanium in Mitteleuropa. *Courier Forschungsinstitut Senckenberg*, **254**, 181–198.

MENNING, M., ALEKSEEV, A. S. *ET AL.* 2006. Global time scale and regional stratigraphic reference scales of Central and West Europe, East Europe, Tethys, South China, and North America as used in the Devonian–Carboniferous–Permian Correlation Chart 2003 DCP 2003. *Palaeogeography, Palaeoclimatology, Palaeoecology*, **240**, 318–372.

MEYERS, S. R. 2008. Resolving Milankovich controversies. The Triassic Latemar Limestone and the Eocene Green River Formation. *Geology*, **36**, 319–322.

MIALL, A. D. 1985. Architectural-element analysis. A new method of facies analysis applied to fluvial deposits. *Earth Science Review*, **22**, 261–308.

MIALL, A. D. 1996. *The Geology of Fluvial Deposits – Sedimentary Facies, Basin Analysis, and Petroleum Geology*. Springer, Berlin.

MIALL, A. D. 2006. *The Geology of Fluvial Deposits – Sedimentary Facies, Basin Analysis, and Petroleum Geology*, 4th edn. Springer, Berlin.

MINERVINI, M., ROSSI, M. & MELLERE, D. 2011. Cyclicity and facies relationships at the interaction between aeolian, fluvial, and playa depositional environments in the Upper Rotliegend: regional correlation across UK (Sole Pit Basin), The Netherlands, and Germany. *In*: GRÖTSCH, J. & GAUPP, R. (eds) *The Permian Rotliegend of The Netherlands*. SEPM, Special Publications, **98**, 119–146.

NEWELL, A. J., TVERDOKHLEBOV, V. P. & BENTON, M. J. 1999. Interplay of tectonics and climate on a transverse fluvial system, Upper Permian, Southern Uralian Foreland Basin, Russia. *Sedimentary Geology*, **127**, 11–29.

NIO, S. D., BROUWER, J., SMITH, D., DE JONG, M. G. G. & BÖHM, A. R. 2005. Spectral trend attribute analysis: application in the stratigraphic analysis of wireline logs. *First Break*, **23**, 71–75.

NIO, S. D., BÖHM, A. R., BROUWER, J., DE JONG, M. G. G. & SMITH, D. 2006. Climate stratigraphy. *In*: *Principles and Applications in Subsurface Correlations*. European Association of Geoscientists and Engineers, Short Course Series no. 1.

NÖLDECKE, W. 1957. *Schotteranalytische Untersuchungen an den Konglomerathorizonten der Unteren Mansfelder Schichten am Ostharzrand Mansfelder Mulde*. Diplomarbeit, Humboldt-Universität Berlin.

Ochadlick, A. R. & Kritikos, H. N. 1993. Variations in the period of the sunspot cycle. *Geophysical Research Letters*, **20**, 1471–1474.

Oemisch, I. 1984. *Bio- und Lithostratigraphische Bearbeitung des Permokarbon–Profils der Basisbohrung Querfurt 1*. Forschungsbericht Bergakademie Freiberg.

Özkurt, N. & Savaci, F. A. 2005. Determination of wavelet ridges of nonstationary signals by singular value decomposition. *IEEE Transactions on Circuits and Systems II: Express Briefs*, **52**, 480–485.

Olsen, H. 1987. Coarsening upward sequences – the products of laterally established river subsystems. Middle Devonian. Hornelen Basin. Norway. *Bulletin of the Geological Society of Denmark*, **36**, 203–219.

Olsen, H. 1990. Astronomical forcing of meandering river behavior. Milankovich cycles in the Devonian of East Greenland. *Palaeogeography, Paleoclimatology, Palaeoecology*, **79**, 99–115.

Olsen, H. 1994. Orbital forcing on continental depositional systems – lacustrine and fluvial cyclicity in the Devonian of East Greenland. *In*: deBoer, P. L. & Smith, G. D. (eds) *Orbital Forcing and Cyclic Sequences*. International Association of Sedimentologists, Special Publications, **19**, 429–438.

Olsen, P. E. & Kent, D. V. 1996. Milankovich climate forcing in the tropics of Pangaea during the Late Triassic. *Palaeogeography, Palaeoclimatology, Palaeoecology*, **122**, 1–26.

Pancheva, D. & Mukhtarov, P. 2000. Wavelet analysis on transient behaviour of tidal amplitude fluctuations observed by meteor radar in the lower thermosphere above Bulgaria. *Annali Geophysica*, **18**, 316–331.

Percival, D. B. & Walden, A. T. 2000. *Wavelet Methods for Time Series Analysis*. Cambridge Series in Statistical and Probabilistic Mathematics. Cambridge University Press, Cambridge.

Peyser, C. E. & Poulsen, C. J. 2008. Controls on Permo-Carboniferous precipitation over tropical Pangaea: a GCM sensitivity study. *Palaeogeography, Palaeoclimatology, Palaeoecology*, **268**, 181–192.

Poulsen, C. J., Pollard, D., Montañez, I. P. & Rowley, D. 2007. Late Paleozoic tropical climate response to Gondwana deglaciation. *Geology*, **35**, 771–774.

Preto, N., Hinnov, L. A., De Zanche, V., Mietto, P. & Hardie, L. A. 2004. The Milankovich interpretation of the Latemar platform cycles Dolomites, Italy. Implications for geochronology, biostratigraphy, and Middle Triassic carbonate accumulation. *In*: D'Argenio, B. (ed.) *Cyclostratigraphy. Approaches and Case Histories*. SEPM, Special Publications, **81**, 167–182.

Primke, G. 1956. *Stratigraphische und mikromagnetische Untersuchungen des Permokarbons der Halle-Hettstedter Gebirgsbrücke nördlich von Wettin*. Diplomarbeit, Martin-Luther-Universität Halle-Wittenberg.

Prokoph, A. & Agterberg, F. P. 1999. Detection of sedimentary cyclicity and stratigraphic completeness by wavelet analysis. An application to late Albian cyclostratigraphy of the western Canada sedimentary basin. *Journal of Sedimentary Research*, **69**, 862–875.

Prokoph, A. & Agterberg, F. P. 2000. Wavelet analysis of well-logging data from oil source rock, Egret Member, offshore eastern Canada. *American Association of Petroleum Geologists Bulletin*, **84**, 1617–1632.

Prokoph, A. & Barthelmes, F. 1996. Detection of nonstationarities in geological time series. Wavelet transform of chaotic and cyclic sequences. *Computers & Geosciences*, **22**, 1097–1108.

Prokoph, A. & El Bilali, H. 2008. Cross-wavelet analysis. a tool for detection of relationships between Paleoclimate Proxy records. *Mathematical Geosciences*, **40**, 575–586.

Prokoph, A. & Thurow, J. 2000. Diachronous pattern of Milankovitch cyclicity in late Albian pelagic marlstones of the North German basin. *Sedimentary Geology*, **134**, 287–303.

Prokoph, A., Fowler, A. D. & Patterson, R. T. 2000. Evidence for periodicity and nonlinearity in a high-resolution fossil record of long-term evolution. *Geology*, **28**, 867–870.

Rampino, M. R., Prokoph, A. & Adler, A. 2000. Tempo of the end-Permian event. High-resolution cyclostratigraphy at the Permian-Triassic boundary. *Geology*, **28**, 643–646.

Remy, W. & Kampe, A. 1961. Ausbildung und Abgrenzung des Autunien in der Halleschen Mulde. *Monatsberichte der Deutschen Akademie der Wissenschaften*, **3**, 394–408.

Remy, W. & Remy, R. 1958*a*. Über die Möglichkeit der Flözparallelisierung im Plötz–Löbejüner Raum. *Zeitschrift für angewandte Geologie*, **4**, 311–312.

Remy, W. & Remy, R. 1958*b*. Die Verteilung der Pflanzenfossilien auf die Flöze des Plötz–Löbejüner Stefan. *Zeitschrift für angewandte Geologie*, **4**, 522–524.

Retallack, G. J. 1990. *Soils of the Past*. Unwin-Hyman, London.

Retallack, G. J. 2001. *Soils of the past – An Introduction to Paleopedology*, 2nd edn. Blackwell Science, Oxford.

Roll, A. 1974. Langfristige Reduktion der Mächtigkeit von Sedimentgesteinen und ihre Auswirkung – eine Übersicht. *Geologisches Jahrbuch*, **A14**, 3–77.

Roscher, M. & Schneider, J. W. 2006. Early Pennsylvanian to Late Permian climatic development of central Europe in a regional and global context. *In*: Lucas, S. G., Schneider, J. W. & Cassinis, G. (eds) *Non-Marine Permian Chronology and Correlation*. Geological Society, London, Special Publications, **265**, 95–136.

Ross, C. A. & Ross, J. R. P. 1988. Late Paleozoic Transgressive-Regressive Deposition. *In*: Wilgus, C. K., Hastings, B. S., Posamentier, H., Van Wagoner, J., Ross, C. A. & Kendall, C. G. St C. (eds) *Sea-Level Changes – An Integrated Approach*. SEPM, Special Publications, **42**, 227–247.

Rossignol-Strick, M. 1985. Mediterranean Quaternary sapropels, an immediate response of the African monsoon to variation of insolation. *Palaeogeography, Palaeoclimatology, Palaeoecology*, **49**, 237–263.

Roth, S. & Reijmer, J. J. G. 2005. Holocene millennial to centennial carbonate cyclicity recorded in slope sediments of the great Bahama Bank and its climatic implications. *Sedimentology*, **52**, 161–181.

SANGEN, M., NEUMANN, K. & EISENBERG, J. 2011. Climate-induced fluvial dynamics in tropical Africa around the last glacial maximum? *Quaternary Research*, **76**, 417–429.

SARACCO, G., THOUVENY, N., BOURLÈS, D. L. & CARCAILLET, J. T. 2009. Extraction of non-continuous orbital frequencies from noisy insolation data and from palaeoproxy records of geomagnetic intensity using the phase of continuous wavelet transforms. *Geophysical Journal International*, **176**, 767–781.

SCHEFFLER, K., HOERNES, S. & SCHWARK, L. 2003. Global changes during Carboniferous-Permian glaciation of Gondwana. Linking polar and equatorial climate evolution by geochemical proxies. *Geology*, **31**, 605–608.

SCHNEIDER, J. 1982. Entwurf einer biostratigraphischen Zonengliederung mittels der Spiloblattinidae (Blattoidea, Insecta) für das kontinentale euramerische Permokarbon. *Freiberger Forschungshefte*, **C375**, 27–47.

SCHNEIDER, J. 1985. Elasmobranchier–Zahntypen (Pisces, Chondrichthytes) und ihre stratigraphische Verbreitung im Karbon und Perm der Saale–Senke (DDR). *Freiberger Forschungshefte*, **C400**, 90–100.

SCHNEIDER, J. 1988. Grundlagen der Morphogenie, Taxonomie und Biostratigraphie isolierter Xenacanthodier-Zähne (Elasmobranchii). *Freiberger Forschungshefte*, **C419**, 71–80.

SCHNEIDER, J. 1989. Basic problems of biogeography and biostratigraphy of the Upper Carboniferous and Rotliegendes. *Acta Musei Reginaehradecensis, S. A. scientiae naturales*, **22**, 31–44.

SCHNEIDER, J. 1996. Xenacanth teeth – a key for taxonomy and biostratigraphy. *Modern Geology*, **20**, 321–340.

SCHNEIDER, J. & WERNEBURG, R. 1993. Neue Spiloblattinidae (Insecta, Blattodea) aus dem Oberkarbon und Unterperm von Mitteleuropa sowie die Biostratigraphie des Rotliegend. *Veröffentlichungen Naturhistorisches Museum Schleusingen*, **7/8**, 31–52.

SCHNEIDER, J. W. & WERNEBURG, R. 2006. Insect biostratigraphy of the Euramerican continental Late Pennsylvanian and Early Permian. *In*: LUCAS, S. G., CASSINIS, G. & SCHNEIDER, J. W. (eds) *Non-Marine Permian Biostratigraphy and Biochronology*. Geological Society, London, Special Publications, **265**, 325–336.

SCHNEIDER, J. & ZAJIC, J. 1994. Xenacanthiden (Pisces, Chondrichthyes) des mitteleuropäischen Oberkarbon und Perm – Revision der Orginale von Goldfuss 1847, Beyrich 1848, Kner 1867 und Fritzsch 1879–1890. *Freiberger Forschungshefte*, **C452**, 101–151.

SCHNEIDER, J., SIEGESMUND, S. & GEBHARDT, U. 1984. Paläontologie und Genese limnischer Schill- und Algenkarbonate in der Randfazies der kohleführenden Wettiner Schichten Oberkarbon, Stefan C des NE-Saaletroges. *Hallesches Jahrbuch für Geowissenschaften*, **9**, 35–51.

SCHNEIDER, J., RÖßLER, R. & GAITZSCH, B. 1995*a*. Time lines of the Late Variscan volcanism – a holostratigraphic synthesis. *Zentralblatt für Geologie und Paläontologie, Teil I*, **1994**, 477–490.

SCHNEIDER, J., RÖßLER, R. & GAITZSCH, B. 1995*b*. Proposal for a combined reference section of the Central European Continental Carboniferous and Permian for correlations with marine standard sections. *Permophiles*, **26**, 26–31.

SCHNEIDER, J. W., RÖßLER, R. & GAITZSCH, B. 1995*c*. *Stratigraphy and Facies of the Middle European Continental Carboniferous and Permian: Guide to Excursion A5, August 21–August 27, 1995*. Państwowy Instytut Geologiczny, Warszawa.

SCHNEIDER, J. W., HAMPE, O. & SOLER–GIJÓN, R. 2000. The Late Carboniferous and Permian. Aquatic vertebrate zonation in southern Spain and German basins. *In*: *Palaeozoic Vertebrate Biochronology and Global Marine/Non-Marine Correlation: Final Report of IGCP 328 (1991–1996). Courier Forschungsinstitut Senckenberg*, **223**, 543–561.

SCHNEIDER, J. W., RÖßLER, R., GAITZSCH, B. G., GEBHARDT, U. & KAMPE, A. 2005. Saale-Senke. *Courier Forschungsinstitut Senckenberg*, **254**, 419–440.

SCHNEIDER, J. W., KÖRNER, F., ROSCHER, M. & KRONER, U. 2006. Permian climate development in the northern peri-Tethys area – The Lodève basin, French Massif Central, compared in a European and global context. *Palaeogeography, Palaeoclimatology, Palaeoecology*, **240**, 161–183.

SCHWAB, M. & KAMPE, A. 1989. Die Bedeutung der Tiefbohrung Schladebach 1880–1886 für die geologische Erkundung des Permosiles im nördlichen Saaletrog. *Zeitschrift für angewandte Geologie*, **35**, 350–357.

SCHWARZACHER, W. 1993. *Cyclostratigraphy and the Milankovich Theory*. Developments in Sedimentology, **52**.

SCLATER, J. G. & CHRISTIE, P. A. F. 1980. Continental stretching. An explanation of the post-mid-Cretaceous subsidence of the central North Sea basin. *Journal of Geophysical Research*, **85**, 3711–3739.

SCOTESE, C. R. 2000. *Paleomap Project*. http://www.scotese.com.

SMOOT, J. P. & OLSEN, P. E. 1994. Climatic cycles as sedimentary controls of rift basin lacustrine deposits in the early Mesozoic Newark basin based on continuous core. *In*: LOMANDO, T. & HARRIS, M. (eds) *Lacustrine Reservoirs and Depositional Systems*. SEPM, Core Workshop Notes, **19**, 201–237.

SPRINGER, J. 1993. Decompaction and backstripping with regard to erosion, salt movement, and interlayered bedding. *Computers & Geosciences*, **19**, 1115–1125.

STEEL, R. J. & AASHEIM, S. 1978. Alluvial sand deposition in a rapidly subsiding basin Devonian, Norway. *In*: MIALL, A. D. (ed.) *Fluvial Sedimentology*. Canadian Society of Petroleum Geology, Memoirs, **5**, 385–413.

SÜSS, P. 1996. *Sedimentologie und Tektonik des Ruhrbeckens. Sequenzstratigraphische Interpretation und Modellierung eines Vorlandsbeckens der Varisciden*. Bonner geowissenschaftliche Schriften, **20**.

TEN KATE, W. G. H. Z., SPRENGER, A., STEENS, T. N. F. & BEETS, C. J. 1994. Late Quaternary monsoonal variations in the western Arabian Sea based on cross-spectral analyses of geochemical and micropalaeontological data ODP Leg 117, core 728A. *In*: DEBOER, P. L. & SMITH, G. D. (eds) *Orbital Forcing and Cyclic Sequences*. International

Association of Sedimentologists, Special Publications, **19**, 109–116.

TUCKER, M. E., GALLAGHER, J. & LENG, M. 2009. Are beds in shelf carbonates millennial-scale cycles? An example from the mid-Carboniferous of NE England. *Sedimentary Geology*, **214**, 19–34.

TORRENCE, C. & COMPO, G. P. 1998. A practical guide to wavelet analysis. *Bulletin of the American Meteorological Society*, **79**, 61–78.

VANDENBERGHE, J. 2003. Climate forcing of fluvial system development: an evolution of ideas. *Quaternary Science Reviews*, **22**, 2053–2060.

VISSER, J. N. J. 1997. A review of the Permo-Carboniferous glaciation in Africa. *In*: MARTINI, I. P. (ed.) *Late Glacial and Postglacial Environmental Changes, Quaternary, Carboniferous–Permian, and Proterozoic*. Oxford University Press, New York, 169–191.

VON FRITSCH, K. 1888. Das Saalthal zwischen Wettin und Könnern. *Zeitschrift für Naturwissenschaften*, **61**, 114–142.

VON HOYNINGEN-HUENE, E. 1960. *Das Permokarbon im östlichen Harzvorland*. Freiberger Forschungshefte, **C93**. Akademie, Berlin.

VON VELTHEIM, W. 1824–40. Geognostische Betrachtung der alten Sandsteinformation am Harz und in den nördlich und östlich davon gelegenen Landstreifen. *Jahrbuch Hallescher Verband*, **18**, 15–292.

WAGNER, R. H. 1966. El significado de la flora en la Estratigrafia el Carbonifero superior. *Boletin de la Real Sociedad Espanola de Historia Natural. Seccion Geologic*, **64**, 203–208.

WALTER, H. 1983. Zur Taxonomie, Ökologie und Biostratigraphie der Ichnia limnisch–terrestrischer Arthropoden des mitteleuropäischen Jungpaläozoikums. *Freiberger Forschungshefte*, **C382**, 146–193.

WERNEBURG, R. 1989. Labyrinthodonte Amphibia aus dem Oberkarbon und Unterperm Mitteleuropas – Systematik, Phylogenie und Biostratigraphie. *Freiberger Forschungshefte*, **C436**, 7–57.

WERNEBURG, R. 1996. Temnospondyle Amphibien aus dem Karbon Mitteldeutschlands. *Veröffentlichungen Naturhistorisches Museum Schleusingen*, **11**, 23–64.

WERNEBURG, R. & SCHNEIDER, J. W. 2006. Amphibian biostratigraphy of the European Permo-Carboniferous. *In*: LUCAS, S. G., CASSINIS, G. & SCHNEIDER, J. W. (eds) *Non-Marine Permian Biostratigraphy and Biochronology*. Geological Society, London, Special Publications, **265**, 201–215.

YANG, C.-S. & KOUWE, W. F. P. 1995. Wireline log analysis as a tool for dating and correlating barren strata. An example from the Upper Rotliegend of The Netherlands. *In*: DUNAY, R. E. & HAILWOOD, E. A. (eds) *Non-Biostratigraphical Methods of Dating and Correlation*. Geological Society, London, Special Publications, **89**, 237–259.

YIOU, P., JOUZEL, J., JOHNSEN, S. & RÖGNVALDSSON, I. E. 1995. Rapid oscillations in Vostok and GRIP ice cores. *Geophysical Research Letters*, **22**, 2179–2182.

Conservatism of Late Pennsylvanian vegetational patterns during short-term cyclic and long-term directional environmental change, western equatorial Pangea

NEIL J. TABOR[1]*, CHARLES M. ROMANCHOCK[2], CYNTHIA V. LOOY[3], CAROL L. HOTTON[4,5], WILLIAM A. DIMICHELE[4] & DAN S. CHANEY[4]

[1]*Huffington Department of Earth Sciences, PO Box 750395, Dallas, TX 75275-0395, USA*

[2]*Department of Earth and Environmental Science, University of Pennsylvania, 240 South 33rd Street, Philadelphia, PA 19104, USA*

[3]*Department of Integrative Biology and Museum of Paleontology, University of California, Berkeley, CA 94720, USA*

[4]*Department of Paleobiology, NMNH Smithsonian Institution, Washington, DC 20560, USA*

[5]*National Center for Biotechnology Information, National Library of Medicine, National Institutes of Health, Bethesda, MD 20892, USA*

**Corresponding author (e-mail: ntabor@smu.edu)*

Abstract: Patterns of plant distribution by palaeoenvironment were examined across the Pennsylvanian–Permian transition in North–Central Texas. Stratigraphically recurrent packages of distinct lithofacies, representing different habitats, contain qualitatively and quantitatively different macrofloras and microfloras. The species pools demonstrate niche conservatism, remaining closely tied to specific habitats, during both short-term cyclic environmental change and a long-term trend of increasing aridity. The deposits examined principally comprise the terrestrial Markley and its approximate marine equivalent, the Harpersville Formation and parts of lower Archer City Formation. Fossiliferous deposits are lens-like, likely representing fill sequences of channels formed during abandonment phases. Palaeosols, represented by blocky mudstones, comprise a large fraction of the deposits. They suggest progressive climate change from minimally seasonal humid to seasonal subhumid to seasonal dry subhumid. Five lithofacies yielded plants: kaolinite-dominated siltstone, organic shale, mudstone beds within organic shale, coarsening upward mudstone–sandstone interbeds and channel sandstone. Both macro- and microflora were examined. Lithofacies proved compositionally distinct, with different patterns of dominance diversity. Organic shales (swamp deposits), mudstone partings (swamp drainages) and coarsening upward mudstone–sandstone interbeds (floodplains) typically contain Pennsylvanian wetland vegetation. Kaolinite-dominated siltstones and (to the extent known) sandstones contain taxa indicative of seasonally dry substrates. Some kaolinite-dominated siltstones and organic shales/coals yielded palynomorphs. Microfloras are more diverse, with greater wetland–dryland overlap than macrofloras. It appears that these two floras were coexistent at times on the regional landscape.

The fossil record offers a unique opportunity to track species pools through time and to examine their responses to long-term, directional changes in physical conditions. The ability to estimate environmental variables, such as temperature, rainfall amount (and its seasonal distribution) and atmospheric CO_2, is improving continuously. Links between proxy measures of such physical factors and quantitative data on changes in ecosystem composition and dynamics, formerly the purview of Pleistocene palaeoecology, are now available, even in the deep past, for both terrestrial (e.g. Wing *et al.* 2005; Falcon-Lang *et al.* 2006; Montañez *et al.* 2007; Secord *et al.* 2008; Currano *et al.* 2008; Tabor *et al.* 2011) and marine (Palaeozoic examples: Olszewski & Erwin 2009; Ivany *et al.* 2009; Fall & Olszewski 2010; Bonelli & Patzkowsky 2011; Holland & Zaffos 2011) systems. Plant fossils themselves can be especially useful as palaeoclimate proxies, even in the distant world of the Palaeozoic (e.g. Falcon-Lang *et al.* 2009), because of the strong sensitivities of plants to habitat and regional climate. As sessile organisms, plants are directly and inextricably linked to climate at the level of individual, population, local community and biome.

From: GĄSIEWICZ, A. & SŁOWAKIEWICZ, M. (eds) 2013. *Palaeozoic Climate Cycles: Their Evolutionary and Sedimentological Impact*. Geological Society, London, Special Publications, **376**, 201–234.
First published online September 19, 2013, http://dx.doi.org/10.1144/SP376.14

Quantitative floristic studies of vegetation from the Late Palaeozoic are numerous (e.g. Scott 1978; Pfefferkorn & Thomson 1982; Bartram 1987; Eble *et al.* 1994, 2006; Pryor & Gastaldo 2000; DiMichele *et al.* 2007; Opluštil *et al.* 2007; Šimůnek & Martínek 2009; Libertín *et al.* 2009; Bashforth *et al.* 2010), although few have had a framework with which to link plants and their associated environments to changing global climate (e.g. Phillips & Peppers 1984; Falcon-Lang 2004). It is a particular challenge to find a long, relatively uninterrupted stratigraphic section that contains both a common species pool or pools and a clear signal of an overall, directional and interpretable environmental change. The latest Pennsylvanian–earliest Permian Markley Formation of north–central Texas provides such an opportunity.

The Markley Fm was positioned in the western Pangean tropics during the Pennsylvanian–Permian transition, a time of global climate change that accompanied the onset of Permian glaciation (Fielding *et al.* 2008). Within the Markley Fm, its partial equivalent – the Harpersville Fm – and into the base of its successor – the Archer City Fm – there are distinct packets of lithofacies that repeat in the stratigraphic section. Each lithofacies contains a distinctive fossil-plant assemblage, characterized and described by Romanchock *et al.* (in press), which can be followed through the Markley Fm. The goal of this paper is to go beyond such a qualitative flora-to-facies correlation and to examine quantitatively the floristic similarity within and among these facies in space and through time in order to investigate, and in this instance document, the coherence of species assemblages in the face of short-term and long-term environmental changes, and to evaluate spatio-temporal scales on which niche versus neutral biological community assembly patterns may operate (e.g. Alonso *et al.* 2006; Weiher *et al.* 2011). In addition, we wish to examine the conformance of the Markley Fm data with interpretive models that call for climate change as the controlling variable on repeating patterns of lithofacies and plant distribution through time, in the short and long term (allocyclicity), v. models that call for changes in sedimentary environments as the driver of temporally recurrent floristic change under a relatively constant background climate (autocyclicity).

Geological setting

Geographic location

Collections from the Markley Fm, on which this study is based, were made across north–central Texas from 20 localities (Fig. 1). Eighteen of these localities are spread in a NE–SW transect across four counties: Montague, Jack, Young and Archer. Single collections come from the Harpersville Fm in Stephens County and the Archer City Fm in Clay County. Localities are lettered consecutively from A to T in Figure 1.

During the latest Pennsylvanian, the study area was located in the western Pangean tropics between 0° and 5°N (Scotese 1999).

Stratigraphic context of study

Stratigraphic correlation with the marine-equivalent Harpersville Fm indicates that the Markley Fm is composed of uppermost Pennsylvanian and lowermost Permian strata (Hentz 1988). Because of the paucity of fossiliferous marine beds in the Markley Fm, direct biostratigraphic correlation to a global standard is poor. However, Wardlaw (2005) identified the Pennsylvanian–Permian conodont boundary below the Stockwether Limestone of the Pueblo Fm, which has been lithostratigraphically correlated very close to locality Q (Fig. 1, Table 1), an exposure within the upper Markley Fm on US Highway 281 to the north of Jacksboro, Texas, USA.

The Markley Fm was deposited in a fluvial-dominated coastal plain setting on the Eastern Shelf of the Midland Basin. Source areas for Markley Fm sediments include the Ouachita fold belt to the east and the Wichita–Amarillo and Arbuckle highlands to the north. Fluvial systems emptied into open marine depositional systems to the west and south, including the proximal Midland and distal Delaware basins (Hentz 1988). In the study area, Markley Fm strata have a regional NE–SW strike and dip *c.* 0.5°NW (Hentz 1988). The fluvio-deltaic strata of the Markley Fm transition towards the SW to the more marine strata of the Harpersville Fm (Fig. 1). Immediately above the Markley Fm lie the fluvial-dominated strata of the Archer City Fm (Fig. 1).

Collections of plant fossils were made from a range of stratigraphic positions throughout the entire Markley Fm, from portions of the laterally equivalent, marine-dominated Harpersville Fm, and from the lower Archer City Fm that overlies the Markley Fm (Fig. 1, Table 1).

Sedimentology

The Markley Fm is poorly exposed in the majority of its outcrop belt. Based on the characteristics of modern soils developed upon Markley Fm strata, it is likely that the formation is predominantly composed of fine claystones and mudstones (Hentz 1988; Tabor & Montanez 2004).

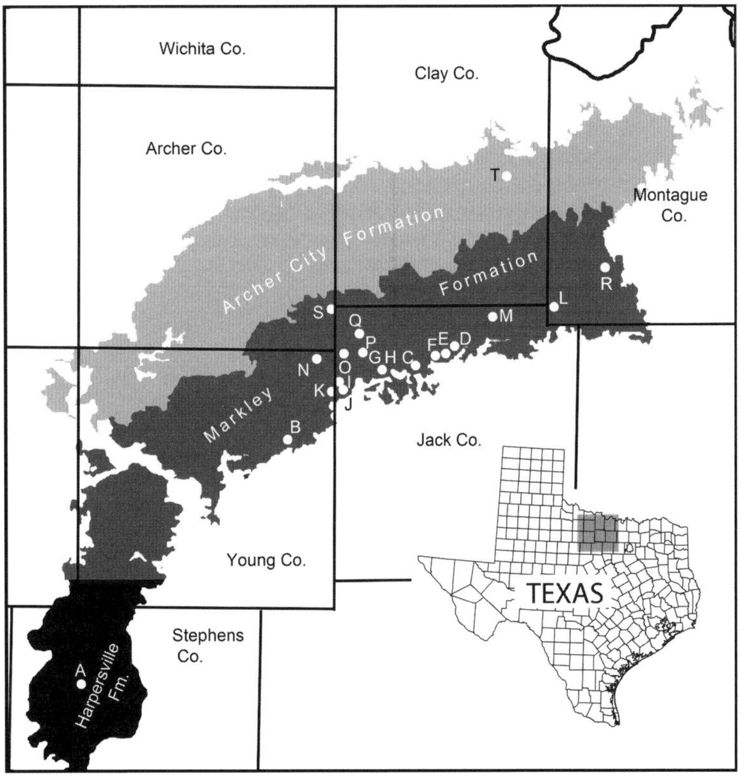

Fig. 1. Locality map of Markley Fm collecting sites, showing general collecting area and outcrop extent of Markley, Archer City and Harpersville Fm. Collecting sites are labelled by letter (A–T). The USNM or USGS locality numbers for individual collections from each site are listed after the informally given name of the general collecting site. **T** – Kola Switch USNM 40069, 40070, 40071; USGS 9998; **S** – Malone Ranch USNM 42135; **R** – Turnbow USNM 39995, 39996, 39997, 39998; **Q** – 281 Road Cut USNM 40078, 40079, 40080, 40596, 40633; **P** – Voyles USNM 39989, 39994; **O** – Maxey USNM 40602, 40603; **N** – North of Markley USNM 40006, 40064; USGS 10164; **M** – Walker USNM 40007, 40086, 40087, 40088, 40595, 43529; **L** – Scott USNM 40641, 40687, 40688; **K** – Lycopod B USNM (East) 43546, 40081, 40682; (West) 40085; **J** – Gillespie Ranch (two localities) USNM (Ironstones) 40694; (not ironstones) 40607, 40608, 40609; (North) 43534; **I** – Gas Well USNM 40610, 40611; **H** – Squaw Mountain Coal Mine USNM 40632; **G** – Squaw Mountain Road Cut USNM 39990, 40002; **F** – Bloodworth USNM (Ironstones) 40003; (not ironstones) 40004, 40005, 40692, 40693, 43547; USGS 10166; **E** – Williamson Drive USNM 40012, 40013; **D** – Cooper USNM 39991, 39992, 39999, 40000, 40001, 40068, 43528, 43530, 43531, 43532, 43533; USGS 10162; **C** – Brannon Mine USNM 40084; **B** – Loving USNM (East) 40082; (West) 40083; **A** – Coal Mountain USNM 40612, 40613.

The best outcrop exposures are usually located on the flanks of erosional cuestas that are capped by fluvial sandstones of the Markley Fm (Hentz 1988). Outcrops are typically limited in lateral extent, ranging from a few tens of metres to *c.* 200 m (Figs 2 & 3). The majority of Markley Fm exposures consist of mudstones and sandstones, which were affected by pedogenesis in Permo-Carboniferous floodplain environments (Tabor & Montanez 2004) and typified by the Markley Fm stratotype section (Hentz 1988). Fossiliferous outcrops, however, differ from the stratotype section in that such fossiliferous strata appear to have been deposited in shallow abandoned channels, or as part of backfill sequences resulting from rising groundwater table or regional base-level, and were not subject to extensive pedogenic modification. The pattern of rock units in these channel-fill sequences is repeated throughout the Markley Fm and into the basal Archer City Fm, and is similar in its lithostratigraphic content to stacking patterns described in facies models for meandering streams (e.g. Nichols 2009). A schematic diagram of a typical fossiliferous sedimentary sequence and its facies associations in the Markley Fm is shown in Figure 4. Descriptions of lithologies and

Table 1. *Relative stratigraphic positions of localities sampled in this study, located relative to sandstones as mapped and numbered by Hentz (1988)*

Archer City Fm
Sandstone 8
T-Kola Switch (Clay County)
Markley Fm
Sandstone 14
S-Malone
Sandstone 12
Q-US 281 Roadcut (Jack County)
Sandstone 11
N- North of Markley (Young County)
O-Maxey Ranch (Jack County)
P-Voyles (Jack County)
Sandstone 10
B-Loving (Young County)
D-Cooper (Jack County)
E-Williamson Drive (Jack County)
F-Bloodworth (Jack County)
H-Squaw Mountain Coal Mine (Jack County)
I-Gas Well (Jack County)
J-Gillespie Ranch (Jack County)
K-Lycopod B (Young County)
M-Walker (Jack County)
Sandstone 9
L-Scott (Montague County)
Sandstone 8
R-Turnbow (Montague County)
Sandstone 6
C-Brannon Mine (Jack County)
G-Squaw Mountain Road Cut (Jack County)
Harpersville Fm
Sandstone 5
A-Coal Mountain

sedimentary structures, as well as interpretations of depositional environments of the facies, are provided in Table 2. Figure 4 is an idealized representation, of course, and between any two localities, minor differences in lithology and relative bedding thicknesses may occur or some lithologies/facies may be entirely absent. Despite these differences, the overall bedding organization at any locality follows the sequence illustrated in Figure 4.

Material

Macrofossil collections and palynological slide preparations from the Markley Fm are housed in the Paleobotanical Collections of the United States National Museum of Natural History, Smithsonian Institution. At the request of property owners, locality information is limited to the general collecting areas rather than exact sites. The majority of the material surveyed has been collected over the past 20 years, although additional collections date back as far as the early 1900s. Selections of this older material were examined to gain a better understanding of the flora at a particular site, although this material was not analysed quantitatively.

Eighty-six separate macrofossil collections were analysed from the 20 localities shown in Figure 1, containing a total of 3544 hand samples. Collections outnumber localities because multiple collections were often made at a single locality, either from a number of stratigraphic intervals at that locality or as replicate collections in one particular bed.

Pieces of unoxidized matrix were removed from hand samples of macrofossils for palynological processing and analysis. Ten samples either did not contain palynomorphs or were too sparse for detailed floristic evaluation (Table 3). Samples from three localities (localities A, B and K; Tables 1 & 3) in the Harpersville and Markley formations between Sandstone 5 and Sandstone 10 of Hentz (1988) produced palynomorphs. Locality T in the Archer City Fm was the only other site included in this study to yield palynomorphs. Additionally, two suites of samples were collected directly from Markley Fm outcrop specifically for palynological processing, from localities A and K (Fig. 1, Table 3). Lithologies throughout the Markley Fm display a high degree of oxidation, so it is not surprising that many samples were unproductive. In addition, most of the outcrops are deeply weathered, causing post-depositional oxidation. Fresher matrix, for example, from drill cores, might prove to be more productive.

Palaeobotanical methodology

Collecting

The following is a detailed account of the collecting methods used during this study, together with the challenges, recognized biases and limitations encountered during collection of Markley Fm plant fossils. Some palynological samples were prepared from matrix removed from macrofossil samples, so the collecting methods used were the same. At localities A and K, samples were collected from distinct beds exposed in freshly dug trenches in the course of measuring outcrop sections. These details are necessary for the meaningful interpretation of our data and permit critical comparison of this study with other palaeoecological floristic analyses (Bennington *et al.* 2009).

No attempt was made to study a single outcrop exhaustively, as a major collecting goal of this study was to sample as many stratigraphic positions and lithofacies, and from as large a geographical area of Markley Fm exposures as possible. Although

Fig. 2. Markley and Harpersville formation outcrops. Key: P, basal palaeosol; K, kaolinite-dominated siltstone; O, organic shale; F, coarsening upward interbedded siltstone–sandstone; S, channel sandstone. (**a**) Maxey Ranch, locality 'O'; note thickness of basal palaeosol, prominence of quartz-kaolinitic siltstone, and weathered channel sandstone boulders. (**b**) Maxey Ranch, locality 'O', detail of erosionally resistant quartz-kaolinitic siltstone and overlying organic shale. (**c**) Maxey Ranch, locality 'O'; note lateral continuity of beds. (**d**) Lycopod B, locality 'K'; note both upper and lower coal intervals. (**e**) Squaw Mountain Coal Mine, locality 'H'; note interbeds in transition between organic shale and floodplain mudstone. (**f**) Coal Mountain, locality 'A'; note size of channel sandstone-dominated ridge in background.

Fig. 3. Markley and Harpersville formation outcrops. Key: P, basal palaeosol; K, kaolinite-dominated siltstone; O, organic shale; F, coarsening upward interbedded siltstone–sandstone. (**a**) Cooper, locality 'D'; note thick palaeosol and transition between organic shale facies and coarsening upward interbedded siltstone–sandstone facies, and blocks of channel sandstone at top of exposure. (**b**) Voyles, locality 'P'; note dipping beds in coarsening upward interbedded siltstone–sandstone facies. (**c**) Loving, locality 'B'; note both upper and lower coal intervals and channel sandstone in background. (**d**) Walker, locality 'M'; note concave shape of beds in outcrop. Kaolinite-dominated

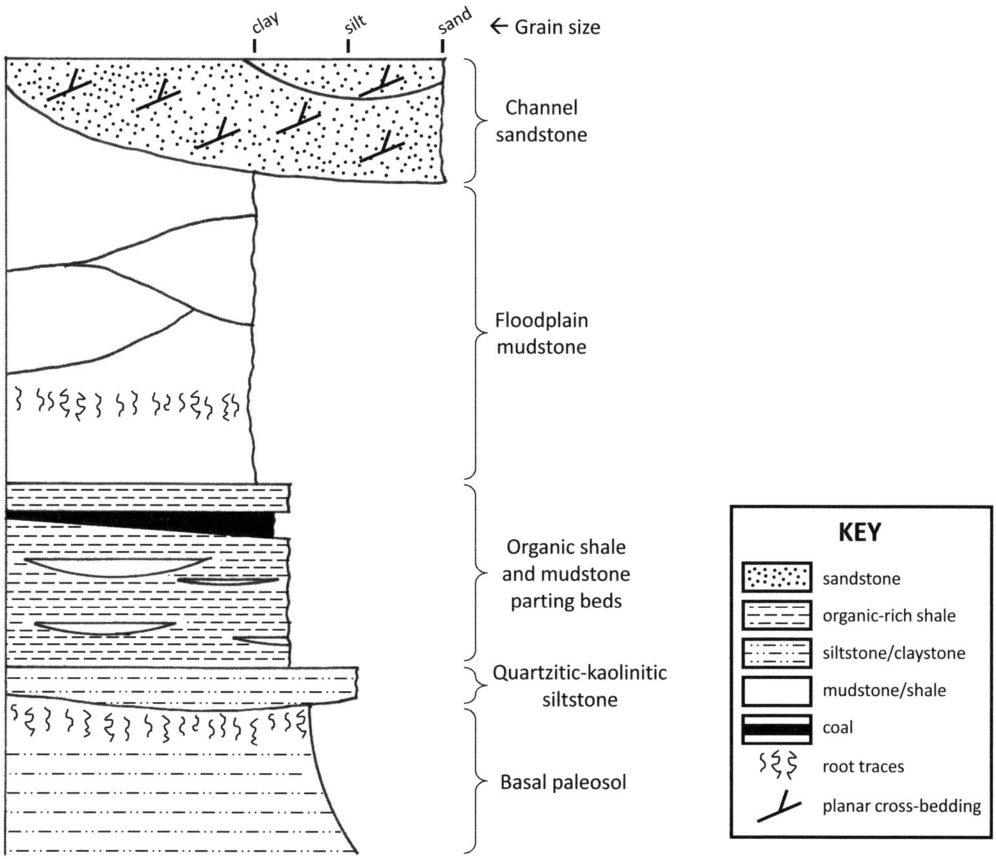

Fig. 4. Idealized vertical profile of a single Markley Fm sedimentary sequence illustrating lithology, sedimentary structures and bed geometry. Vertical scale not listed due to variability in bed thickness (see Figs 2 & 3 for comparison of bed thickness variations).

we recognize that such an exhaustive study could provide a more accurate assessment of relative abundances and diversity of taxa at that single site, practical limitations and the scope of the project prevented such an approach.

Quarry test pits were excavated into beds considered likely to contain plant fossils, followed by more extensive quarrying if warranted. Standard dimensions for a quarry were 1 m (width) × 0.5 m (depth) × 0.3 m (thickness), although these dimensions may be highly variable for a number of reasons: (i) fossiliferous deposits thinner than 0.3 m were excavated through the entire thickness of the deposit; (ii) fossiliferous deposits in some instances were highly localized and were excavated until plant fossils were exhausted; (iii) some outcrops were simply too difficult to quarry at desired dimensions and, in this case, the excavation was made as large as time and energy permitted; (iv) if several fossiliferous lithologies were distinct from each other, collections were then made on a bed-by-bed basis.

A collecting objective was to obtain a sample of plant macrofossils that accurately represented the proportional taxonomic composition of the deposit. Collections from different lithofacies were not mixed. All rocks on which plant fossils could be visually identified were examined, including those with only highly fragmentary debris. The collections housed at the NMNH usually number at least

Fig. 3. (*Continued*) siltstone bed is present directly below organic shale but not visible in photograph. (**e**) Bloodworth, locality ‘F’; note truncation of floodplain mudstone beds directly above the heads of the people in the photograph and large Fe-oxide concretions near their feet; dull orange beds at top of outcrop are Quaternary alluvium.

Table 2. *Lithological descriptions and interpretations for the six facies represented at the majority of plant bearing outcrops in the Markley, Harpersville and Archer City formations*

	Lithology	Interpretation
Channel sandstone	Lenticular bodies of tan-reddish tan, fine to very coarse sand. Can be locally conglomeratic. Cross-beds present. Individual beds often pinch and swell in thickness at outcrop scale. Overall, this unit is virtually non-fossiliferous, with only a small number of identifiable plant fossils and unidentifiable stem fragments observed at a few localities. Sharp, erosional basal contact.	Fluvial channel deposit
Floodplain mudstone	Medium grey mud/silt. Weathers to tan-rusty red. Well to fair bedded, ripple laminated, often fissile. Charcoal fragments widespread, but sparse. Plant fossils may be abundant (sometimes covering entire cleaved surfaces of a hand sample) in some intervals and absent in others, with no observed lithological differences between fossiliferous and non-fossiliferous intervals. Basal contact typically indistinct, with interbedded organic shales gradually thinning and being replaced by this unit.	Fluvial floodplain deposit
Mudstone parting beds	Medium grey, poorly bedded clay. May contain charcoal fragments. Beds are thin (<10 cm) and are deposited within organic shale beds. Plant fossils are usually present, often in abundance.	Fluvial channel deposit
Organic shale	Chocolate brown, organic-rich, fissile mud/silt. Weathers to rusty brown. Charcoal fragments (usually <2 cm in diameter and <5 mm thick) widespread, sometimes in abundance. Bedding is very thin and planar. Coal deposits locally present, ranging from 1 cm to nearly 1 m thick. Coal is sub-bituminous and impure. Thicker coal beds often contain several interbedded shaly, clastic-rich intervals. Organic beds may be rich in fossil plant foliage, but may also be dominated by stem fragments, almost to the exclusion of foliage. Basal contact gradational for *c.* 2–5 cm from underlying grey clay to more organic-rich mud.	Swamp deposit
Quartzitic-kaolinitic siltstone	Light to medium grey, quartzitic silt/fine sand clasts in a kaolinite-rich matrix. At outcrop scale, bedding is clearly present; hand samples usually show poorly developed bedding. Plant fossils locally abundant or absent. Sharp basal contact. At outcrop scale, this unit is typically lens-shaped, indicating an erosional base or deposition in a topographic depression.	Fluvial channel deposit
Basal palaeosol	Grey to reddish grey, mottled claystone, enriched in kaolinite and iron oxides. Has blocky structure. No bedding present. Colour differences are present in basal palaeosol outcrops, probably due to soil horizonation. This unit contains root traces and root casts, but no identifiable plant material was observed. This unit is identical to the 'Type B paleosols' described by Tabor & Montañez (2004).	Palaeoultisol

50 hand samples, if possible, and may reach nearly 200 hand samples. This number varies, mainly due to logistical limitations encountered in the field. If fossils were scarce, then all specimens found were collected. If fossils were abundant, then a subset of what was found was collected. A minor bias was (or could have been) introduced at this stage by favouring rare taxa. A subobjective was to capture the broadest possible range of species diversity at a site. As a consequence, the rarest of taxa at a given locality may be over-represented (but given the generally low numbers of such specimens, this 'over-representation' will have little effect on the assessment of the quantitative floristic composition).

Different methods of field preparation were used, as dictated by the nature of the rocks. Using a rock hammer, fissile shales were cleaved and non-fissile rocks were fractured. Fissile shales were sometimes forcibly fractured to expose fossils not preserved on planes of weakness in the rock (see discussion of sampling issues below).

For a given lithofacies at a given locality, excavation of one or more replicate quarries was attempted. This was done to assess the spatial and temporal variability in fossiliferous deposits in an

Table 3. *Locality numbers and codes for palynological samples processed*

Locality number	Lithofacies	DCA graph label	Locality name	Productive
USNM 42135	F		Malone Ranch	−
USNM 40069-A	K	T-13	Kola Switch 9013 Lower Bed	+
USNM 40633-A	K		281 Road Cut 9215	−
LycB90-1	K	K-1	Lycopod B East (Conifer layer)	+
LycB90-2	K/O	K-2	Lycopod B East (Cuticle layer)	+
USNM 40081	K/O	K-2.5	Lycopod B East (Bottom bed)	+
LycB90-3	O	K-3	Lycopod B East (Lower coal)	+
LycB90-7	O	K-7	Lycopod B East (Blocky mudstone)	+
LycB90-10	O	K-10	Lycopod B East (Blocky mudstone)	+
LycB90-12	O	K-12	Lycopod B East (Brown organic shale)	+
LycB90-13	O	K-13	Lycopod B East (Upper coal)	+
LycB90-14	O	K-14	Lycopod B East (Organic shale)	+
USNM 40083-A	O	B-1	Loving West (Brown organic shale)	+
USNM 40612 (1991-15-3)	O	A-3	Coal Mt. 9115A (Coal B)	+
USNM 40612 (1991-15A)	O	A-4	Coal Mt. 9115A (Organic shale)	+
USNM 39989	O		Voyles 8911	−
USNM 39994-5A	O		Voyles 9029	−
USNM 39991-A	O		Cooper 9010 (Brown organic shale)	−
USGS 5665	O		Newcastle Coal	+
USNM 40602	P		Maxey Ranch A	−
USNM 40082 (Coal)	O		Loving East Coal	−
USNM 40083 (Coal)	O		Loving West Coal	−
MTS	U		Markley type section (Grey mudstone)	−

+, productive sample; −, unproductive sample. Key to lithofacies classification: F, coarsening upward interbedded siltstone and sandstone; K, quartz–kaolinite siltstone; O, organic shale; K/O, transitional between kaolinite-dominated siltstone and organic shale; P, mudstone parting beds; U, unclassified.

outcrop. Replicate quarries may have been excavated simultaneously, or during successive field seasons. Replicate quarries were either laterally equivalent to the original quarries (usually 1–2 m from the original quarry) or vertically displaced to some degree, depending on what aspect of variability was of interest. There were no replicates in instances where the contents of a fossiliferous deposit were exhausted during initial excavation and no equivalent deposits could be located. If, after examination, the fossil contents from the replicate quarry were determined to be virtually identical in diversity and type to the original quarry, a small voucher collection was made accompanied by a field count. In some cases, no voucher collection was made but a field count accompanied by photographs served as documentation.

Macrofossil quantification method and biases

Our sampling and collecting methods were chosen for their simplicity, efficiency and applicability to the wide range of fossiliferous deposits to be analysed. Some of these deposits, either by their general lithological nature or taphonomic biases, present challenges for any attempt to study their fossil contents quantitatively. The following paragraphs list a few notes on these sampling issues with the aim of clarifying the sampling method used in this analysis.

Determination of the quantitative abundances of various macrofossil taxa in a given collection was carried out using the hand sample quadrat method of Pfefferkorn *et al.* (1975). In this method, each hand sample is considered to be a sampling quadrat; size differences among these quadrats are minor and so do not affect the overall results. Counts of taxa were made by scanning the entire surface of a hand sample for identifiable plant fossil material. Whenever two samples were found to be a part/counterpart couple, only one of the duplicate surfaces was counted. All portions of the non-mirror-image surface of the counterpart were counted as a separate quadrat. When a given taxon was identified as present on a hand sample, it was counted only once for that quadrat. The total abundance of a taxon in a given collection, therefore, is expressed first as its occurrence on a specific number of quadrats, and then normalized to a proportion or percentage of the total number of quadrats in the collection. This method produces a frequency distribution because it is theoretically possible for all taxa to be present on every quadrat. In practice, this is not the case, and in tests of the hand-sample-quadrat method against point-count methods, rank-order abundance is preserved (Lamboy & Lesnikowska 1988; DiMichele *et al.* 1991). The quadrat method tends to raise the apparent abundance of rare taxa and suppress the abundance of abundant taxa, relative to a point-count analysis (Wing & DiMichele 1995).

Fragmentary preservation, fragmentation during collection, and obscured plant characters (due to rock colour, grain size, poor fossil/matrix contrast, weathering, burial in rock matrix, etc.) rendered much of the fossil material unidentifiable. Furthermore, although some species could be identified from small fragments due to highly diagnostic morphology, others required larger surfaces for positive identification. For example, to positively identify *Cordaites* leaves (which can be easily confused with finely striated axes – stems and rachises – belonging to many groups), a sufficiently large and/or finely detailed specimen was necessary. In contrast, *Pseudomariopteris cordato-ovata* or *Nemejcopteris feminaeformis* generally could be identified from a fragment of a pinna or pinnule. In addition, some species have a natural tendency to fragment into small bits, which are inherently difficult to identify, whereas others are naturally robust and tend to be well preserved. When a positive identification to species level was not possible, identification was made to the lowest taxonomic level possible. Beyond this (usually for the most poorly preserved specimens), the fossil was ignored. In either of the latter two cases, the lack of positive identification was grounds for not including the specimen in the final floral count. It should be noted that a positive identification, either to genus or species level, was possible for the vast majority of specimens.

Because of the taphonomic factors described above, certain groups (e.g. *Cordaites*) may be unavoidably under-represented in the floral counts of some collections. However, this sampling method minimizes the amount of bias that enters into a floral count of a sufficiently large sample because some identifiable specimens are usually encountered, which provides a 'tip off' that the group is present and thus sharpens the eye of the observer. This is most likely to be the case if the group is common to abundant in the collection. If a taxon was fragmentary in its preservation, difficult to identify from fragmentary specimens and present in very low abundance, the likelihood of ultimate identification was greatly reduced. In this kind of analysis, the basic dominance patterns are likely to be robust as reconstructed. However, total species richness is likely to be underestimated due to low identifiability of the rarest elements.

The degree of ease with which fossiliferous shales could be split to plant-bearing bedding surfaces introduced bias into the analysis. Most of the rocks collected were fissile shale, which naturally cleaves along planes of weakness. These planes of weakness are most common on surfaces with large plant remains. Therefore, broad, thin plant material (e.g. various kinds of axes, large pteridosperm foliage, etc.) is likely to be the most commonly exposed during excavation or subsequent splitting of larger excavated shale blocks, if such large material is present. Conversely, small or sparse plant compressions (e.g. isolated pecopterid pinnules, sphenopsid foliage, seeds, etc.) usually will not create a significant plane of weakness and may be under-represented on the rock surfaces analysed.

Palynological preparation and biases

Samples were subjected to 10% hydrochloric and 70% hydrofluoric acid maceration, followed by heavy liquid separation in zinc bromide (Specific Gravity = 2), carried out by Global Geolab Inc (Medicine Hat, Canada). Residues were sieved through a 10 μm filter, strew-mounted in glycerine jelly on glass slides, and examined using a Nikon 80i differential interference compound microscope with transmitted light.

Fourteen samples proved productive, including the Newcastle Coal, which, due to poor preservation, was not included in this analysis. A 200-grain count was performed on each sample (Table 3). More complete analyses will be presented elsewhere.

Palynology complements data from the plant macrofossil record, but presents its own potential biases. Although more susceptible to oxidative destruction than plant macrofossils, which can yield taxonomic information even if oxidized, palynomorphs when preserved are much more abundant than plant macrofossils and capture less common or more fragile taxa not preserved in the macrofossil record. Palynomorphs, being small and readily transported by water and wind, sample a broader source area than plant macrofossils, thus capturing a larger proportion of the total regional flora, including taxa from sites undergoing active erosion that otherwise would not be preserved in the fossil record. This quality can potentially introduce noise into facies/habitat distinctions, but, in practice, in fluvial terrestrial settings, the most common palynomorphs tend to be locally derived, that is from the vegetation surrounding the site of deposition. Palynological samples can be swamped by a single or few hyperproductive plant taxa (e.g. Marattiales); conversely, taxa with low or sporadic production may be under-represented or even absent, so relative or absolute abundance data are often a poor indicator of actual abundance of parent plants on the landscape. Most palynological samples (except those processed from a single depositional event, such as a lake varve) represent some degree of time averaging, so do not offer the temporal precision of single macrofossil assemblages. Palynological samples can and often do include reworked specimens, as rivers erode their banks and release previously deposited grains back into the system for redeposition. Reworked specimens often display a higher degree of physical damage or chemical weathering, and may display a higher (or lower) degree of thermal alteration, but can be physically indistinguishable from non-reworked (or only locally reworked) grains. Finally, taxonomic resolution for palynomorphs is generally less precise compared to plant macrofossils, because many palynomorph taxa lack any known macrofossil counterpart, although almost every pollen type can be ascribed to some biologically meaningful rank, for example seed plant or fern. Fortunately, a relatively high percentage of Pennsylvanian dispersed pollen and spore morphotaxa have been linked to a parent plant. It should also be noted that palynomorph morphospecies tend to be over-split, based on minor morphological, developmental and taphonomic variation. Consequently, overall diversity based on palynomorph morphospecies probably overestimates actual species richness.

Data analysis

Based on the original quadrat counts, abundance lists of dispersed macrofossil organ types in each collection were compiled. The original count data were condensed to create, as far as possible, 'whole plants' by combining different dispersed organs or by using only those organs that are most commonly encountered and most reliable for taxonomic purposes. So, for example, where a plant typically is represented by multiple dispersed organs, such as *Sigillaria brardii*, which, in addition to the characteristic stems, is represented by distinctive leaves, reproductive organs and deeper bark adpressions, all dispersed organs were combined under that one species. Pteridosperms and ferns, on the other hand, were overwhelmingly represented by foliage, with large amounts of stem and rachial axes, and minor (almost always quantitatively insignificant) occurrences of seeds or spore-bearing organs. The unidentifiable vegetative parts often could be combined with foliage when diversity consisted of a single species, or were divided proportionately based on the proportions of identifiable foliage. Disassociated reproductive organs generally could not be combined confidently with foliage, and in most cases were kept separate and not used in the quantitative analyses of taxonomic abundance. This approach is essentially the same as that used by Gastaldo *et al.* (2004), Cleal (2007) and Cleal *et al.* (2012).

Statistical analyses in this study were based on the condensed data matrix, after normalization. Exploratory data analysis was carried out by first examining the macrofloral composition of each lithofacies, with consideration given to both diversity and quantitative abundance of different species. Composition is shown graphically with species arranged in the same sequence on each plot (Fig. 5a, c, e, g), so that changes in composition can be readily identified. The order of the taxa in these plots is based on their scores on the first axis of a Correspondence Analysis (CA), which, although not shown separately, can be seen in the species order on the *x*-axis of the plots in question. CA is a multivariate ordination method designed for categorical variables; the analysis was carried out in CANOCO for Windows 4.5 (Ter Braak & Šmilauer 2002).

In order to examine patterns in the distribution of dominants, dominance–diversity curves were prepared. These are simple bivariate plots of taxa ordered along the *x*-axis by relative abundance. Curves were created for each lithofacies type (Fig. 5b, d, f, h). 'Importance' is measured by relative percentage abundance, calculated from the original quadrat counts (number of quadrat occurrences in a sample/total number of quadrats in a sample $\times$ 100). These curves reveal with simplicity both the dominant groups in an assemblage, based on importance value, and the proportional distribution of abundance among the other taxa in a given assemblage.

Macrofossil absolute abundance data was also normalized to relative abundance data (as above). The converted data matrix was imported into R, a computer program designed for statistical computing and graphics (R Development Core Team 2010). Non-metric multidimensional scaling (NMDS) ordination was then performed on this data matrix to examine patterns of sample similarity. NMDS is a data analysis technique that uses quantitative models to describe qualitative data and display their structure as a geometrical picture (Young 1985). The technique is robust to non-normality of distribution, discontinuous data, non-linearity of relationships among variables, and questionable scales, and as such is a very robust method with which to explore data structure (McCune & Grace 2002). NMDS converts the relative abundance matrix to a dissimilarity matrix with n objects (in this case, $n = 86$ collections) on each axis. The dissimilarity between any two objects is based on their shared descriptors (in this case, species abundance). The objects are plotted as points in a multidimensional Euclidian space, the number of dimensions of which is prescribed. To

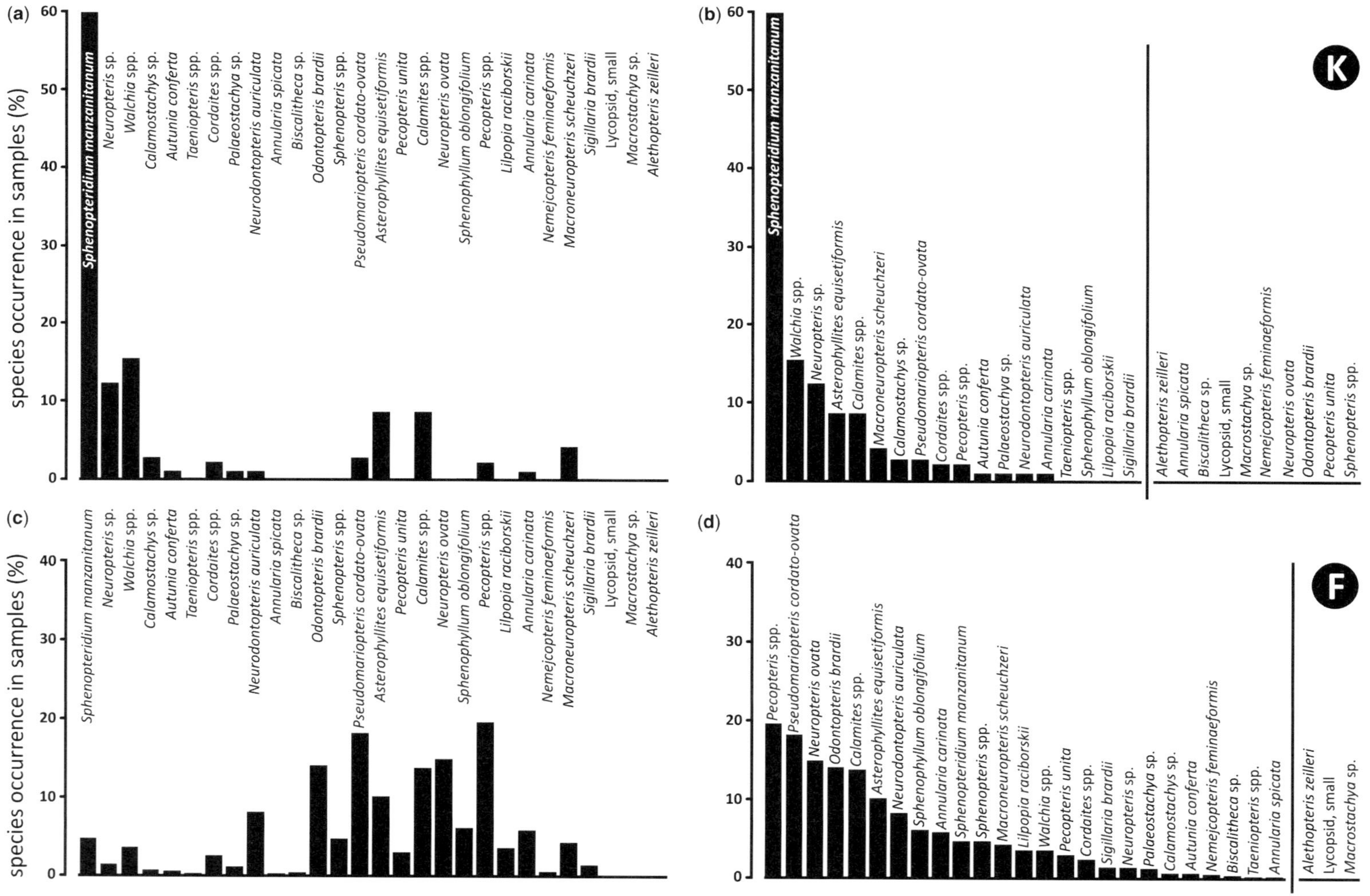

Fig. 5. Histograms illustrating the pattern of abundance and dominance in Markley Fm macrofossils by facies type. Species order on the *x*-axis is the same in (**a**), (**c**), (**e**) and (**g**). In (**b**), (**d**), (**f**) and (**h**), species are ordered by dominance along the *x*-axis. Letters to the side indicate the facies type: K, kaolinite-dominated siltstone (18 taxa, sample size 356); F, coarsening upward, interbedded siltstones and sandstones (25 taxa, sample size 1210); P, mudstone parting beds (16 taxa, sample size 696); O, organic shales (15 taxa, sample size 915).

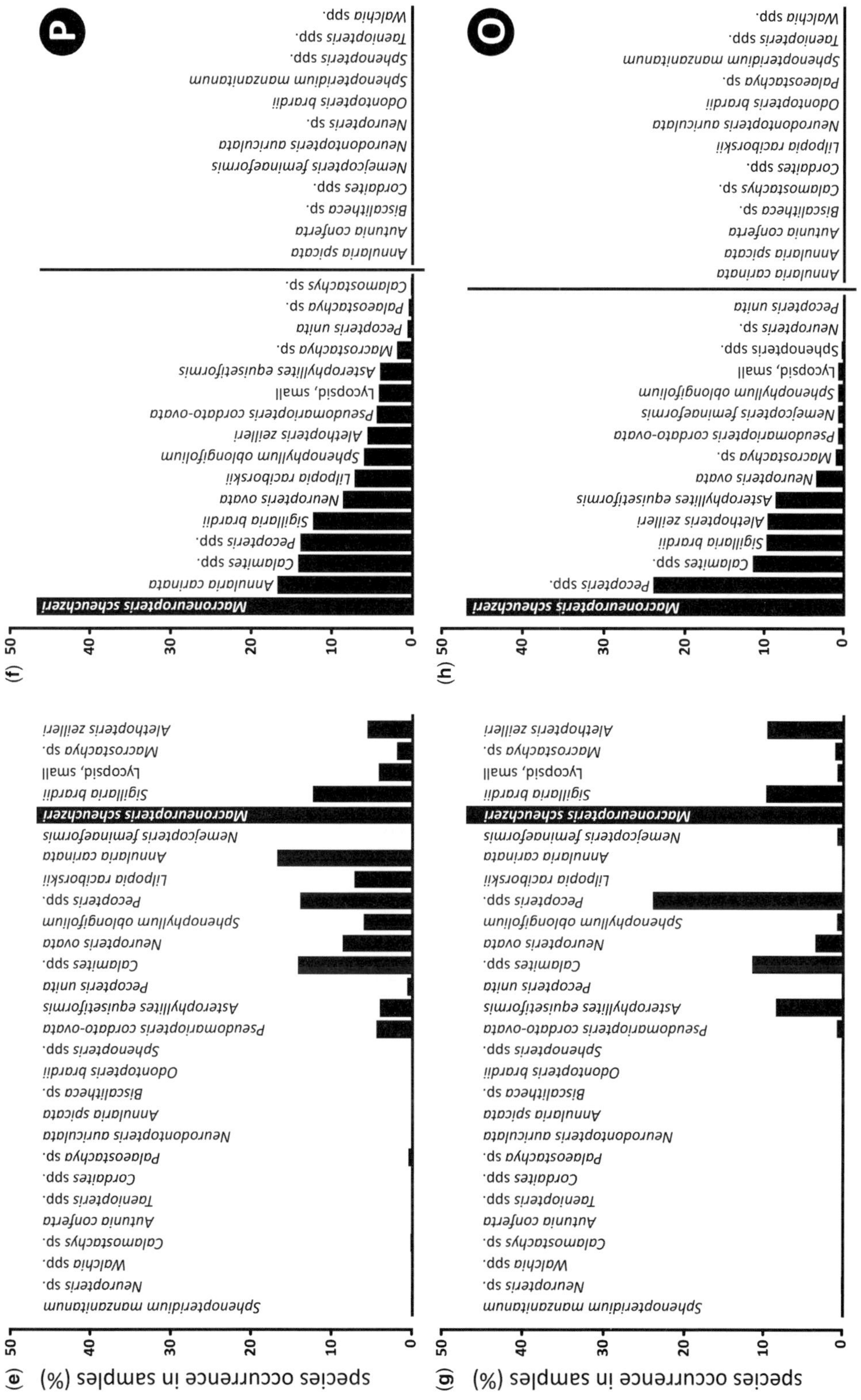

Fig. 5. (*Continued*)

simplify the display of data structure, two dimensions were used for this analysis. This was accomplished within the R program by projecting the points from a three-dimensional Euclidian space onto a two-dimensional plane. Arrangement of these points is based on their dissimilarity to each other so that proximity reflects degree of dissimilarity (i.e. highly dissimilar points will plot far from one another, and highly similar points will plot near one another). Table 4 lists the R command string used to compute the NMDS.

Confirmatory analyses were carried out on macrofossil data using the technique of analysis of similarities (ANOSIM) in R. Table 4 lists and explains the R command string used in this study to compute the ANOSIM. Further confirmatory statistical testing was not performed on palynological data because of small number of productive samples, limiting intersample comparisons. ANOSIM tests for significant dissimilarities between two or more predetermined groups of sampling units. For this analysis, fossil plant observations were divided into five groups based upon the lithology/facies in which they were found: upward coarsening mudstone–sandstone interbeds, Fe-oxide nodules (ironstones), kaolinite-dominated siltstone, organic shale and mudstone parting beds within organic shale. ANOSIM is based on a distance matrix whereby pairwise distances are first converted to rank value and then separated into within-group and between-group distances. The distance values define the level of dissimilarity between any number of objects: larger distances equate to higher dissimilarity, and *vice versa*. The dissimilarity values are displayed on a box-and-whisker diagram.

Palynological count data was the basis for a detrended correspondence analysis, carried out using PAST 2.12 software (Hammer *et al.* 2001). Correspondence analysis finds the chi-squared distances among all data points, yielding axes that account for maximum variability between points, where axis 1 accounts for the highest variance, and so on. Detrending is a form of normalization applied to data exhibiting a gradient by iteratively rescaling ordination axes to suppress 'edge' and 'arch' data artefacts generated by standard correspondence analysis (Hill & Gauch 1980; Oxanen & Minchin 1997).

Lithofacies characterization

Blocky to massive mudstone

Blocky to massive mudstones within plant-bearing successions are usually found at or near the base of outcrops and are composed of red-grey mottled claystones and siltstones (Figs 2 & 3, units labelled 'P'; Fig. 4). Thickness is often indeterminate, depending on exposure of the profile base, and the maximum exposed thickness was 6.5 m. Sedimentary bedding structure is characteristically absent. These deposits weather rapidly, producing small, low gradient slopes ($<30°$). This lithology is friable, very weakly indurated, and typically organized into angular blocky structures in the range *c.* 1–5 cm throughout the unit. The lithofacies typically has a red coloration, but may also have grey and yellow mottling that varies from rare ($<2\%$ of volume) to abundant ($>20\%$ of volume) and often includes nodular and vertically to horizontally oriented tubular Fe-oxide concretions that range from a few millimetres to several tens of centimetres across. The lithofacies can be subdivided into

Table 4. *R command strings. Commands are in standard R notation; explanations follow # symbol*

```
File Preparation
>library(vegan) # Runs vegan package
>dd=read.csv(file=file.choose(), header=T) # Choose file to analyse
>dds<- dd[,6:33] # Choose columns to analyze
>row.names(dds)<- dd[,1] # Assign cells in column "1" as headings for their
respective rows

NMDS
>w.mds<- metaMDS(dds) # Runs ordination
>w.mds # Performs NMDS using isoMDS (from MASS package)
>plot(scores(w.mds)) # Creates a plot of NMDS scores

ANOSIM
>markley.jacc=vegdist(dds, method=‘jaccard’, binary=F,diag=F,upper=F,na.rm=F)
# Calculates jaccard index
>gg<- dd$LITHOLOGY # Defines a grouping variable, in this case the grouping variable
used is lithologic type
>aa<- anosim(markley.jacc, gg) # Performs ANOSIM
>plot(aa) # Creates a box-and-whisker diagram of ANOSIM results
```

thinner layers, or horizons, based on changes in organization of the mudstone, in the amount of clay-size material, in mineral composition with depth, and in abundance of concretions. This lithofacies has been interpreted as soil profiles that formed upon the coastal plains during Late Pennsylvanian time (Tabor & Montañez 2004; pedotype B). No identifiable plant megafossils were observed in this facies (with the important exception of rooting structures). This particular sort of palaeosol morphology is restricted to the lower Markley, that is, below Sandstone 10 of Hentz (1988; Tabor & Montañez 2004).

Tabor & Montañez (2004) note an abundance of Fe-oxide nodules in the lower portions of the profiles (BCg and Cgv horizons in their fig. 4b), together with evidence for translocation of kaolinite clay and Fe- and Mn-oxide into their inferred Bt horizon. They interpreted these findings to represent a palaeoultisol that was waterlogged for prolonged periods of time near the bottom of the soil profiles (based on the Fe-oxide nodules) yet experienced relatively mild seasonal changes in soil moisture (i.e. free drainage, based on the predominance of kaolinite in the clay fraction) in the upper parts of the profiles. Taken as a whole, these observations strongly support the interpretation of these deposits as palaeosols and suggest that they likely developed on low-relief interfluves during a period of low sedimentation rates or landscape degradation. Taken a step further, it would be reasonable to assume that the climate during development of these palaeosols was predominantly humid, with periods of seasonal drying.

Whereas the blocky to massive mudstone lithofacies is typical of the base of fossiliferous stratigraphic successions and provides an appropriate representation for some non-fossiliferous successions (e.g. Markley Fm stratotype section; Hentz 1988), other palaeosol morphologies occur within different lithostratigraphic facies associations in the Markley and lower Archer City formations (Figs 2 & 4; Tabor & Montañez 2004; DiMichele *et al.* 2006). We describe several different palaeosol morphologies in the subsequent discussion and note their occurrence(s) among the lithofacies associations used in this study (Figs 2 & 4).

A frequently encountered morphology is blocky greenish-yellow mudstone up to 2 m thick with abundant yellow to reddish-brown nodules that may include a layer of brown to dusky red Fe-oxide nodules and concretions in the upper half of the profile (Tabor & Montañez 2004; Pedotype C). Below the zone of Fe-oxide cementation the mudstone is organized into wedge-shape aggregate structures and has large, low-angle, slickenplanes. Tabor & Montañez (2004) interpreted these profiles to be similar to modern inceptisols, and considered these profiles to have formed under very poorly drained conditions based on redoximorphic features such as gley coloration and Fe-oxide nodules and concretions. However, the presence of slickensides and Fe-oxide nodules does suggest that profiles saw at least a small amount of seasonal drying and aeration. This particular morphology is limited to the upward coarsening mudstone–sandstone interbed facies. Such palaeosols likely formed within floodplain deposits proximal to fluvial channels, as evidenced by their close lateral and stratigraphic proximity to channel-filling and crevasse-splay sandstones (Tabor & Montañez 2004).

Another common palaeosol that occurs in abundance through the Markley and lower Archer City formations (Pedotype E; Tabor & Montañez 2004) ranges from 0.2 to 2 m thick and is composed of variable lithologies ranging from claystone, to muddy fine-grained sandstone, to sandstone, with massive to very weakly developed blocky structure. Fine-grained profiles commonly preserve carbonized fossil root compressions, as well as abundant and prominent colour mottling. In contrast, sand-rich lithologies have no carbonized root compressions and typically preserve only faint mottling and rare Fe-oxide nodules. Tabor & Montañez (2004) considered this morphologically variable group of palaeosol profiles to be similar to modern entisols, and noted that they are abundant throughout the Permo-Pennsylvanian continental strata of the eastern shelf of the Midland Basin. This particular palaeosol morphology has not been observed in the stratigraphic position of the blocky to massive mudstone lithofacies association. It does, however, occur commonly in the organic shale and mudstone parting beds, coarsening upward mudstone–sandstone interbeds, at the tops of the channel sandstone lithofacies and at the tops of kaolinite-dominated siltstone associations.

Pedotype D of Tabor & Montañez (2004) is common in the upper Markley (above Sandstone 11 of Hentz 1988) and lower Archer City formations. Such profiles are composed primarily of red to purple mudstone up to 1.3 m thick with common to abundant, coarse and irregular grey and yellowish mottles. The mudstone is organized into wedge-shaped aggregates ranging from 5 cm to several tens of centimetres across. Fe-oxide nodules and concretions commonly occur in the upper half of the palaeosol profiles. These palaeosols preserve slickenplanes in the lower half of the profile, and often preserve V-shaped dykes that extend from the top, down to the middle, of the profile. These dykes are typically filled with material that is similar in composition to the material in the stratum that overlies the profile. Tabor & Montanez (2004) considered these profiles to be morphologically and mineralogically similar to modern vertisols. The wedge-shaped aggregate structures, slickenplanes and clastic dykes are all indications of seasonal shrink–swell processes in soil profiles. This kind of morphology in modern tropical soil systems occurs in seasonally moist climates with four to eight dry months annually (Dudal & Eswaran 1988). Furthermore, Nordt *et al.* (2006) found that modern vertisols are non-calcareous (similar to the palaeosols considered here) only when mean annual precipitation exceeds 1400 mm a^{-1}. Such high mean annual precipitation seems consistent with the abundant mottling and redoximorphic features in these profiles and suggests extended periods of soil saturation and relatively short dry intervals. These palaeovertisols are interpreted to have formed upon interfluvial surfaces of the coastal plain, and can occur either within the blocky to massive mudstone facies association (Fig. 2) or the coarsening upward mudstone–sandstone interbed facies association.

A final palaeosol morphology that we have not observed in the Markley or lower Archer City formations includes all of the morphological and mineralogical characteristics of a palaeovertisol, but also includes horizons with calcium carbonate nodules. The presence of calcite in this palaeosol type indicates lesser amounts of rainfall (Nordt *et al.* 2006) and longer and more severe intervals of soil moisture deficit than non-calcareous palaeosols, so the absence of these profiles is important from a palaeoclimate and palaeoecological perspective. Although Hentz (1988) noted calcareous palaeovertisols above Sandstone 12 of the Markley Fm, in 20 years of fieldwork we have not identified calcareous palaeovertisols below Sandstone 3 in the Archer City Fm, nor anywhere in the Markley Fm. This suggests that significant drying and greater seasonality across landscapes existed in the Eastern Midland Basin after deposition of Archer City Fm Sandstone 3, and importantly, subsequent to the interval in which the fossiliferous stratigraphic sequences discussed herein were deposited.

Kaolinite-dominated siltstone lithofacies

One or more beds of siltstone composed mainly of kaolinite and quartz, with variable amounts of other clays, mainly chlorite and illite, are present in nearly all of the fossiliferous outcrops studied, most often positioned immediately above the basal palaeosol (Figs 2 & 3, labelled 'K'; Fig. 4). Kaolinite-dominated siltstone beds also occurred at the base of scours within the stacked mudstone–sandstone lithofacies at a few sites. Such beds are laterally extensive on outcrop, variable in thickness from 10 to 80 cm, and have trough-shaped erosional basal contacts with the underlying facies. In lithological composition the kaolinite-dominated siltstone beds are very similar to the basal palaeosol facies. The most striking difference between these two facies is that the kaolinite-dominated siltstone deposits lack the pedogenic features found in the underlying palaeosols and instead show evidence of fluvial deposition, such as an erosional and scoured basal contact, poor, yet distinct sedimentary bedding, and localized deposits of identifiable plant fossils that often cross-cut bedding surfaces.

The plant fossils in this facies are xeromorphic and belong to evolutionary lineages broadly interpreted as centred in subhumid to semi-arid habitats. The only evidence for soil formation in this facies is found in some instances near the top of the unit, where bedding disruption or bioturbation by rooting rarely occurs.

Overall, the kaolinite-dominated siltstone facies appears to have been formed by rapid deposition of fine-grained, highly weathered sediments, possibly as a slurry of mud and fossil plant fragments. Because this facies is so lithologically and mineralogically similar to the underlying basal palaeosol facies, it is reasonable to speculate that its sediments were derived from local erosion of the surface layers of the underlying palaeosols from an adjacent, slightly elevated source, although this cannot be confirmed. Romanchock *et al.* (in press) compared these deposits to the so-called 'ball clays' of Eocene age from the southeastern USA (Elsik & Dilcher 1974; Potter 1976; Potter & Dilcher 1980), which are typically confined to abandoned floodplain channels, and are thought to form in hot, humid climates.

Organic shale lithofacies

Organic-rich lithofacies range from finely laminated shales with a high organic content to locally developed coal beds with high ash content (Figs 2 & 3, labelled 'O'; Fig. 4). The shale beds may contain extremely dense deposits of fossil plant material and coaly/charcoal fragments. Shales are fissile, low in density and extremely friable, yet are erosionally resistant, often forming low escarpments on outcrop. Overall, the organic shale deposits typically range between 2 and 3 m thick, with individual organic-rich beds ranging from 10 cm to nearly 2 m in thickness. At a few localities an impure coal up to 0.8 m thick was observed within the organic shale deposit. Thinner (5–20 cm), laterally discontinuous coals are present at many other localities. Fe-oxide nodules of various sizes occur within this lithofacies at a few localities, interspersed throughout the organic shale deposits in varying concentrations. These organic-rich deposits probably formed in a low-energy environment with low clastic input, such as a swamp.

Thin (<10 cm), medium-grey, clay-rich beds are locally present in the organic shales. These beds, which are referred to in this paper as mudstone parting beds, are lens-shaped on outcrop. The mudstone parting beds, based on their small size, limited extent and fine-grained composition, are interpreted as channels that periodically carried clastic sediments into the swamp environment.

Organic accumulations of this nature, especially trending to impure coals, require a humid climate and persistent standing water conditions for most of the year to retard decay and permit the accumulation and preservation of peat deposits (Cecil & Dulong 2003; DiMichele *et al.* 2006; Greb *et al.* 2006). Furthermore, a humid climate promotes the thick growth of vegetation necessary to stabilize the landscape and severely limit the erosion and transport of sediments into the fluvial system. Vegetationally controlled inhibition of sediment flux, as a secondary consequence of humid climatic conditions, helps explain the relatively low quantities of clastic sediment in the organic shale facies, permitting the concentration of organic matter. This is especially relevant considering the lack of evidence for change in landscape elevation or topographic slope, and the repetitive appearance of this facies at different stratigraphic levels, but in the same position relative to the other facies.

Coarsening upward mudstone–sandstone interbed lithofacies

The organic shale facies is directly overlain by a coarsening upward sequence of interbedded mudstone and sandstone (Figs 2 & 3, labelled 'F'; Fig. 4). The contact between these two facies may be sharp and erosional or gradational. In most of the instances we observed, this facies association can be artificially separated into three sections: a 'bottom' transitional section consisting of interbedded thin layers of organic-rich shale and mudstone, a 'middle' section consisting of mudstone and interspersed weakly developed palaeosols (types C, D and E described above), and an 'upper' section that is composed of interbedded mudstones and sandstones that collectively define an upward coarsening sequence of sedimentary strata beneath the channel sandstone lithofacies. The predominant lithology is grey-to-tan coloured, clay-and-silt rich, mudstone and fissile shale with low-to-moderate organic content, sometimes containing charcoal fragments. Beds are thin (0.5–10 cm) and discontinuous, lenticular in shape, and often ripple-laminated, giving them an undulating surface. At outcrop scale, packages of beds frequently truncate one another.

The various bed geometries and sedimentary features mentioned above suggest that the lower part of the upward coarsening mudstone–sandstone interbed facies was deposited in shallow channels within broad, low-profile channel belts.

This depositional pattern continues upward into the middle section where the organic shales disappear, the mudstones take on a more uniform lithological character, and palaeosols begin to appear within the unit. These palaeosols are angular and blocky in texture and exhibit various levels of development (Pedotypes C, D and E, discussed above). Most are thin and only slightly obscure the original sedimentary structure of the mudstones, resembling immature palaeoentisols (Pedotype E discussed above). Others are thicker and more mature, sometimes including blocky or wedge-shaped mudstone clasts and containing a horizon of densely packed Fe-oxide nodules, resembling palaeoinceptisols (Pedotype E discussed above) and palaeovertisols (Pedotype D discussed above). The upper surface of any of these immature palaeosols represents a hiatus in deposition and implies the presence of significant diastems. This middle section also contains plant fossils, which are found in small, localized lenses of sediment that are often trough-shaped and appear to have erosional basal contacts, possibly originating as scour surfaces. The enclosed plant fossils are sometimes preserved along bedding planes but often crosscut bedding planes suggesting rapid rates of sediment deposition.

Within the upper section, the floodplain mudstone facies is replaced by channel sandstone facies. This transition may happen at a single lithological contact or as a series of interbeds of mudstone and sandstone of up to 3 m in thickness that is similar in appearance to crevasse-splay deposits emplaced on floodplains by overbank flooding of meandering streams (e.g. Nichols 2009).

Based on the fine-grained nature of the sediments, frequent depositional hiatuses and overall coarsening-upward profile, deposition of the upward coarsening mudstone–sandstone interbed facies likely occurred in a low-energy, aggrading (i.e. regressive), coastal plain setting. Furthermore, the overall coarsening upward pattern in this lithofacies is consistent with a temporal transition from an initially distal floodplain site to one that was proximal to meandering river channels (e.g. Walker & James 1992).

Channel sandstone lithofacies

Sandstones are present at the uppermost portion of nearly all outcrops studied, either as intact beds or weathered blocks of talus material that are now scattered across the outcrops (Fig. 2a, labelled 'S'; Fig. 4). The sandstones cap outcrops because they are more resistant to erosion than the fine-grained mudstones. The channel sandstone lithofacies is characterized by erosional basal contacts with the coarsening upward mudstone–sandstone interbeds, which, where visible, are broadly U-shaped in cross-section. Overall unit thickness varies from a few metres to a few tens of metres, although accurate thickness measurements are complicated by modern weathering that destroyed or buried the uppermost portions of sandstones. Their internal geometries exhibit extensive pinching and swelling of individual beds, with bed thickness ranging from 0.5 m to less than 5 cm, and trough cross-bedding. Grain size ranges from fine to very coarse sand, with chert, quartz and Fe-oxide pebble conglomerates locally present at the base of trough-cross bedded units.

Plant fossils were very sparse or absent from this facies at most localities examined. In the few instances where plant fossils were present, *Cordaites* sp., unidentifiable axes and rare *Pecopteris* sp. were most common. Perhaps most interestingly within the scope of this study, the conifer *Walchia* was also observed at one locality (locality K); the only other occurrences of this taxon within Pennsylvanian strata are in the kaolinite-dominated siltstone facies (*Walchia* becomes common in younger Permian strata). Because of the scarcity of fossils, a meaningful quantitative analysis of the channel sandstone facies flora is not possible.

It is likely that these deposits fill channels that were once part of larger stream networks. Channel cutting and sand deposition were possibly contemporaneous with the development of soils on interfluve surfaces and may account for the floristic similarity between the channel sandstone and kaolinite-dominated siltstone lithofacies, if the kaolinite-dominated siltstones were derived from erosion of soil surfaces.

Temporal trends in depositional patterns

At the temporal scale of the entire Markley Fm deposition, facies recur in distinct packages, as discussed above. Yet there is a long-term trend of differential facies abundance through the Markley Fm most marked by the diminishment, and ultimately the disappearance, of organic-rich beds. In the lower Markley Fm, for example, virtually every locality from which plant fossils were collected contains deposits of organic shale, often of considerable thickness (up to 1.5 m). Progressing stratigraphically upward through the Markley Fm, these organic shale deposits decrease in number, extent and thickness until they become virtually absent from the uppermost Markley Fm and the overlying Archer City Fm. In the basal Archer City Fm, most plant-bearing shales occur in channel-fill facies on outcrops of less lithological complexity than typical of the Markley Fm. In fact, the characteristic sedimentary cycles of the Markley Fm disappear almost entirely and are replaced by a new sedimentary pattern consisting mostly of well-developed palaeosols, floodplain or channel-fill mudstones, and fluvial sandstones.

The disappearance of the organic shale–coaly facies, while at the same time the other typical Markley facies persist in sequence, strongly suggests a climatic trend toward long-term drying, or at least increased seasonality of precipitation whereby productivity and preservation of plant organic matter was increasingly limited. While one may argue that this lithological shift could also have resulted from a steepening of the topographic gradient that would have prohibited the formation of organic-rich swamps, it is inconsistent with a long period of tectonic quiescence in this region during and after deposition of the Markley Fm (Hentz 1988). Although basin subsidence did continue during this time, the sediment influx was sufficient to maintain a nearly level coastal plain with a shoreline well to the west. Further support of the inferred long-term drying trend comes from other sedimentological indicators of dry climate in formations younger than the Markley. These include widespread development of soil carbonate and gypsum nodules in the upper Bowie and Wichita groups (Tabor & Montañez 2004; Tabor *et al.* 2008), a complete lack of both organic-rich palaeosol profiles (Histosols; coal) and deeply weathered palaeoultisols indicative of humid climates, as well as marine beds of gypsum in the Wichita Group ranging from 3 to 10 m thick (Jones 1971; Jones & Hentz 1988).

Results of macrofossil palaeobotanical analysis

Dominance–diversity curves

Patterns of composition and dominance and diversity for each of the four lithofacies are illustrated in Figure 5. Each of these curves shows the full list of 28 taxa identified in the Markley Fm. However, the number of taxa identified in each facies differs considerably. The mudstone–sandstone interbed lithofacies includes 25 taxa, whereas the other facies encompass between 15 and 18. In addition, the mudstone–sandstone interbed lithofacies includes many relatively abundant taxa that are rare to absent in the other facies, such as *Pseudomariopteris cordato-ovata*, *Neuropteris ovata*, *Asterophyllites equisetiformis*, *Neurodontopteris auriculata* and *Sphenophyllum oblongifolium*.

Species dominance patterns in the coarsening upward mudstone–sandstone interbed facies are different from those of the other three facies. In the mudstones of the coarsening upward mudstone–sandstone interbed facies, no species occurs on more than 20% of the sampling quadrats, giving this facies the most even distribution of dominance among the four fossiliferous lithofacies types. Even so, eight or nine species are present at a frequency of 5% or more of the sampling quadrats in each of the four fossiliferous facies except for the organic shales, at 6 species. However, the organic shale, parting bed and kaolinite-dominated siltstone facies were more uniform from site to site than the coarsening upward mudstone–sandstone interbed facies. Furthermore, the dominance–diversity curves for each facies, which are composites of all sample sites from a given facies, are generally similar from site to site in all but the coarsening upward mudstone–sandstone interbed facies, where very high dominance by one species may occur within any local assemblage despite an overall average uniformity of species abundances among the sampling sites. These observations, together with field observations of the rocks, suggest that the coarsening upward mudstone–sandstone interbed facies was environmentally heterogeneous. Much like today, Late Pennsylvanian floodplains, which we presume were the basic depositional settings of the coarsening upward mudstone–sandstone interbed facies, were subject to high levels of disturbance and supported spatially heterogeneous, riparian plant communities.

The three remaining lithofacies are all clearly dominated by a single species. The organic shale and mudstone parting bed facies are dominated by *Macroneuropteris scheuchzeri* (present on 47% and 46% of quadrats, respectively), whereas the kaolinite-dominated siltstone facies is dominated by *Sphenopteridium manzanitanum* (present on 57% of quadrats). In each of these three facies, the second ranked species was identified on 14–23% of quadrats. Although the shapes of the dominance–diversity curves are similar for these three facies, the quartz kaolinitic siltstones have a very different taxonomic composition from the organic shale and mudstone parting bed facies. High dominance by a single species is consistent with a standing-water swamp interpretation for the physical habitat of the organic shales and mudstone parting beds. Relatively low species richness and high dominance by one or a few species are well documented for clastic and peat-swamp settings in the Pennsylvanian (e.g. Gastaldo *et al.* 2004; DiMichele *et al.* 2007; Opluštil *et al.* 2007, 2009*a*, *b*). Conversely, the limited floral diversity in the kaolinite-dominated siltstone facies may reflect the fact that this flora is transported to some extent. Low species richness may reflect a low diversity of trees on the landscape immediately surrounding the environment of deposition from which the flora is most likely drawn, a conclusion supported by modern actualistic studies of plant taphonomic patterns (e.g. Scott 1978; Scheihing & Pfefferkorn 1984; Burnham & Spicer 1986; Burnham *et al.* 1992). Significant taphonomic selection against plants farther from the edge of the deposit and against ground cover plants has been documented. In addition, low abundances and diversity of small, ground cover plants in fossiliferous assemblages is expected in most allochthonous depositional settings, even when the environment of deposition is immediately adjacent to the source vegetation (Scheihing 1980).

Non-metric multidimensional scaling (NMDS)

NMDS ordination of all 86 macrofossil collections (Fig. 6) reveals patterns consistent with expectations based on observations of species occurrence and dominance within the different lithofacies. Points in Figure 6 are labelled by their respective lithofacies. Recall that each point is a single excavation characterized by its quantitative taxonomic composition, determined to the lowest taxonomic rank possible.

The most obvious and significant pattern is the clear emergence of clusters, based on species composition, that conform to lithofacies groups. The pattern indicates that species composition has the strongest effect on the distribution of points (samples) within this ordination, and that floral composition is tied strongly to lithology/habitat of growth. Conversely, in this two-dimensional plot there is no clearly interpretable signal derived from the stratigraphic position or geographic location of a given point. In fact, there is no stratigraphically detectable pattern in the data. The ordination pattern is, then, in agreement with field and laboratory observations that the major pattern in the data is one of species by lithology, and supports niche-based models of species distribution.

The kaolinite-dominated siltstone facies sample suite does not overlap with samples of the organic shale facies (Fig. 6), consistent with the strong differences in species content of these two facies. In contrast, the suite of organic shale samples and that of the mudstone parting bed facies overlap one another nearly entirely (Fig. 6). This indicates a common species pool and is consistent with field

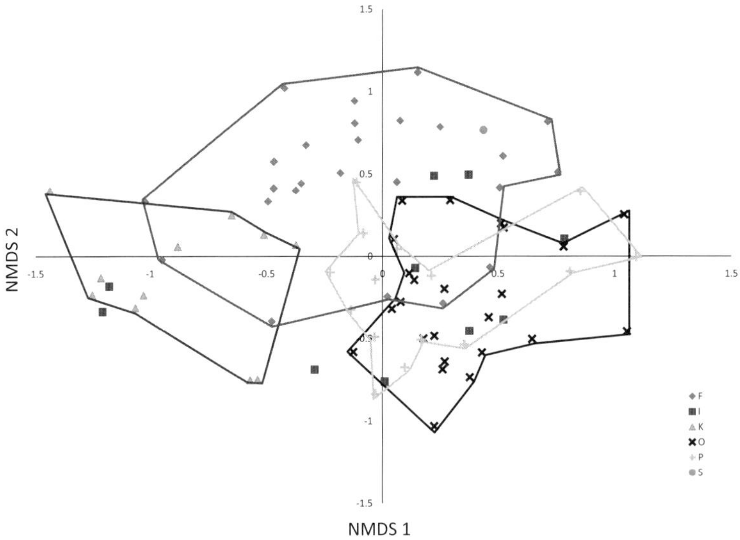

Fig. 6. Non-metric multidimensional scaling (NMDS) ordination of all 86 macrofossil samples. Points are labelled by facies type. Each facies type is outlined by a continuous line. Axis orientation is arbitrary and axis units represent NMDS scores. Key: F, coarsening upward interbedded siltstone–sandstone; I, Fe-oxide nodules (ironstones); K, kaolinite-dominated siltstone; O, organic shale; P, mudstone parting beds; S, channel sandstone.

observations of the position of the mudstone parting beds within organic shale deposits.

The area of the ordination encompassed by any given facies cluster reflects the degree of similarity among the samples of that facies. For example, the cluster of organic shale samples ($n = 24$) covers a relatively small area, consistent with *Macroneuropteris scheuchzeri* and/or *Pecopteris* abundance in all of these samples. In contrast, the coarsening upward mudstone–sandstone interbed facies samples ($n = 25$) are nearly twice as dispersed as the organic shale sample suite. Qualitative observations made during sampling, and dominance–diversity curves based on quantitative analysis, indicate that the coarsening upward mudstone–sandstone interbed facies flora is often diverse at any one site and spatiotemporally variable among sites. In contrast, organic shale samples have low richness at any given site and vary little from site to site.

A sample suite of Fe-oxide nodules (ironstones) is included in Figure 6 (red squares (grey in print version), not encompassed by a circumferential line) to show the floral variability of this collection. These samples do not form a coherent cluster but instead overlap with all other facies clouds. These nodules formed in – and were collected from – various facies, indicating that they are a taphonomic feature without vegetational or floristic significance. Locally, the ironstones often preserve plant impressions in exquisite detail and so are of great value in characterizing the flora of those deposits in which they occur.

A single sample representing the only quantifiable channel sandstone facies collection is also included in Figure 6. Whereas a single collection from this facies is perhaps of limited significance, it is interesting to note that this point is located among peripheral samples (filled orange dot (grey in print version) on the right side) of the coarsening upward mudstone–sandstone interbed facies cloud (marked by blue (grey in print version) diamonds), which are interpreted to have been deposited upon floodplain sites.

The influence exerted by five of the most abundant species on the NMDS cluster pattern was assessed by plotting the abundances of these taxa on five separate NMDS-based ordinations (Fig. 7). These subsets reveal those collecting localities (points) that contain an abundance of that particular taxon. The cutoff for abundance varied for each taxon, and was chosen to enhance the visual aspect of the pattern by maximizing the number of points while simultaneously maximizing their concentration on the NMDS plot.

These plots (Fig. 7) demonstrate the strong effect of dominant taxa on the NMDS pattern. For example, the *Macroneuropteris scheuchzeri* cluster ($n = 24$) is the main influence on the overlap between floodplain mudstone, organic shale and mudstone parting bed facies. Similarly, *Sphenopteridium manzanitanum* ($n = 11$) and *Walchia* sp.

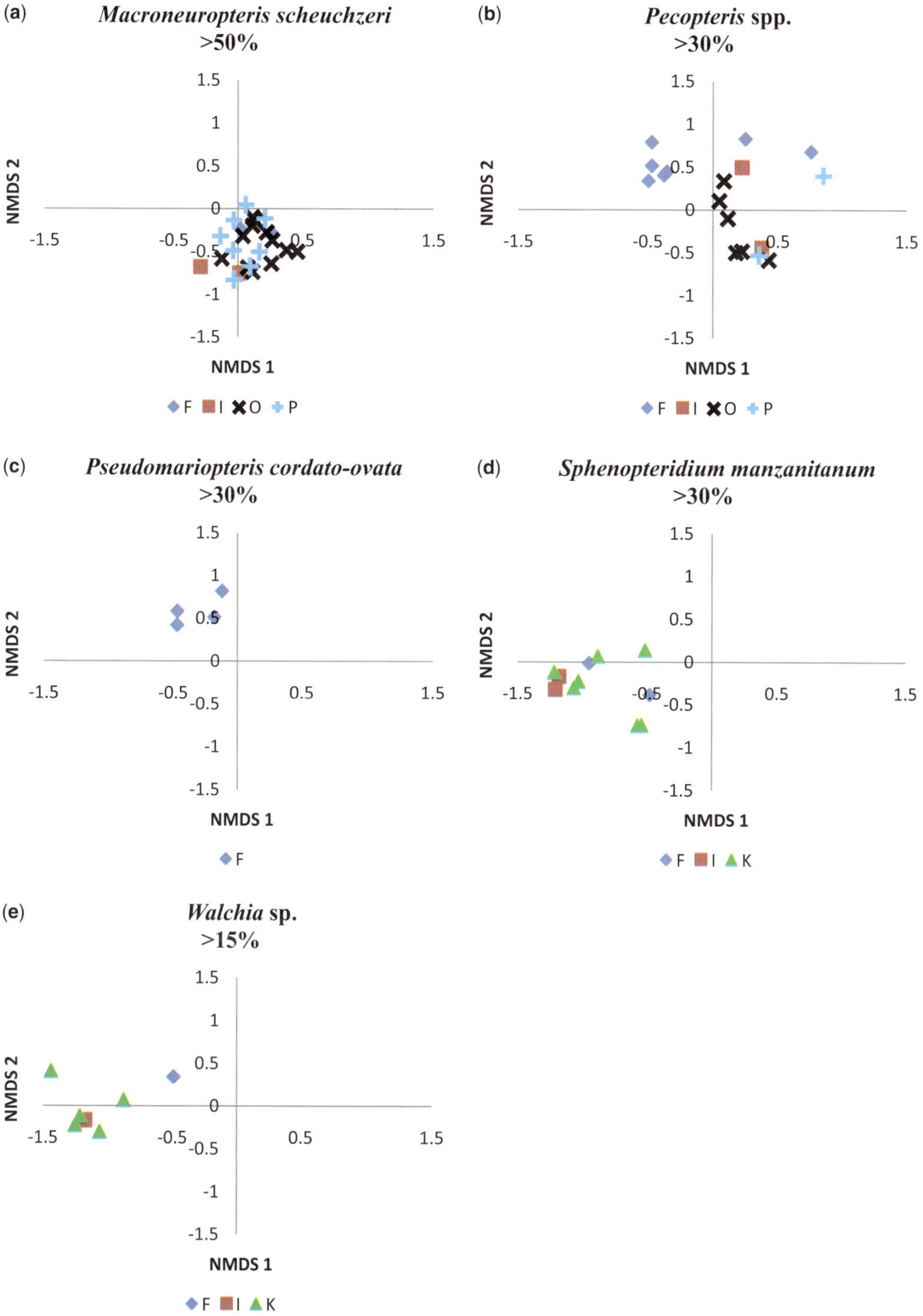

Fig. 7. NMDS-based plots for selected dominant macrofossil taxa, indicated by taxon label. Percentage values indicate relative abundance cutoff values used for that particular plot. Orientation of axes is arbitrary and axis units represent NMDS scores. Key: F, coarsening upward interbedded siltstone–sandstone; I, Fe-oxide nodules (ironstones); K, kaolinite-dominated siltstone; O, organic shale; P, mudstone parting beds.

($n = 7$) account for the distinctiveness of the kaolinite-dominated siltstones, but their rare occurrences in deposits of the coarsening upward mudstone–sandstone interbed facies are major factors in creating a zone of overlap between these two facies clusters. *Pseudomariopteris cordato-ovata* ($n = 4$), in contrast, is conspicuously abundant only in a subset of samples from the coarsening upward mudstone–sandstone interbed facies. High abundance in each of these four taxonomic groups is relatively tightly clustered in the NMDS. The opposite pattern is found in pecopterid ferns ($n = 17$), which do not form a discrete cluster and have a high abundance in a variety of facies types, occurring in many collections of the coarsening upward mudstone–sandstone interbed, organic shale and mudstone parting bed facies. These findings are consistent with a high degree of habitat specificity for many of these taxa. Species such as *M. scheuchzeri*, *S. manzanitanum* and *P. cordato-ovata* appear to have high habitat specificity towards wetland swamps, well-drained soils and wet flood plains, respectively. Pecopterid ferns, on the other hand, are common in species pools of both floodplain mudstones of the coarsening upward mudstone–sandstone interbed facies and organic shale–mudstone parting bed facies. Pecopterid ferns may be a very complex group in terms of both species diversity and ecological tolerances, but this is difficult to clarify because the taxonomy of pecopterids is refractory (e.g. Knight 1985; Pšenička *et al.* 2009), particularly where specimens are preserved mainly as small fragments, as in the Markley Fm samples (Romanchock *et al.* in press). As a group, pecopterid marattialean ferns appear to have been able to grow in a wide range of landscapes as long as they were able to locate even small areas of sufficiently wet substrates on those landscapes (DiMichele & Phillips 2002; DiMichele *et al.* 2006).

Analysis of similarities

ANOSIM (Fig. 8) compares the degree of dissimilarity within prespecified groups to the amount of variability between those groups. It calculates the statistic 'R' to determine if these dissimilarities are significantly different. In the case of the Markley Fm data, it is a test of the hypothesis that there are distinct lithofacies-based assemblages, not an exploratory analysis.

In this analysis, the lithofacies-specific floras formed the predefined groups. The most significant aspect of this test is the between-group dissimilarity. The test was carried out with and without the Ironstone samples and a similar result was obtained in each analysis; only the with-Ironstones results are reported here.

The median between-group dissimilarity is higher than the dissimilarity within any individual lithofacies group, except for ironstones (which are not facies-specific and represent floras from several different lithofacies). This indicates that the dissimilarity within any facies is significantly less than that among facies. Thus, the ANOSIM is consistent with the patterns seen in the NMDS and other qualitative observations. The kaolinite-dominated siltstone, organic shale and coarsening upward mudstone–sandstone interbed facies have relatively low within-group dissimilarities. In contrast, the coarsening upward mudstone–sandstone interbed facies has a much higher level of dissimilarity, again consistent with NMDS and visual inspection of collections, indicating high site-to-site variability.

Results of palynological analyses

Of the five lithofacies, only two – organic shale and kaolinite-dominated siltstone – were palynologically productive (Table 3). These two lithofacies contrast in palynomorph composition, largely but not completely matching the pattern observed in the plant macrofossil record.

Composition and dominance patterns

The organic shales, broadly speaking, are dominated by spores of Marattiales, *Sigillaria* and other lycopsids (e.g. *Chaloneria)*, as well as sphenophylls and calamites (Fig. 9). Microspores of *Sigillaria* appear more abundant in the organic-rich clastic units, whereas the coals appear to be dominated by spores of Marattiales and sphenophylls; however, sample size was insufficient to support this impression statistically. The palynological assemblage of the organic shale facies displays a greater degree of evenness of species in quantitative composition when compared to the dominance of *Macroneuropteris* in the macrofossil records. The spores of non-marattialian ferns display the greatest species richness, although not necessarily abundance, in this facies. Significantly, pollen of walchian conifers and seed ferns is rare but consistent in samples observed from the organic shale lithofacies (abundance, 0.5%).

Certain macrofossil taxa are poorly represented in the microfossil record and, conversely, many microfossil taxa lack obvious counterparts in the macrofossil record. The most notable incongruence is the very low abundance of *Schopfipollenites*, the pre-pollen of *Macroneuropteris* and other medullosan seed ferns (*Alethopteris*, *Neurodontopteris*, *Neuropteris*). Reasons for this are unclear, but may include size sorting in the depositional

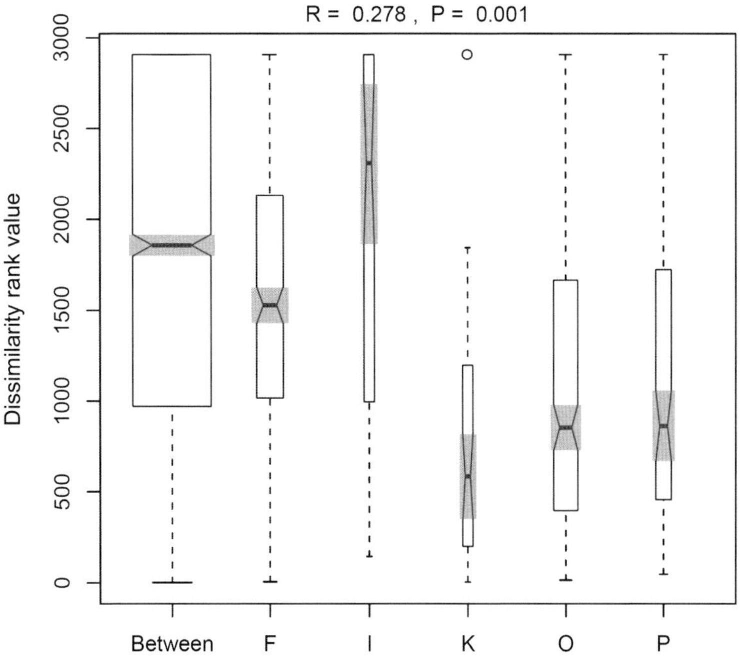

Fig. 8. Analysis of similarities (ANOSIM) plot showing dissimilarity between and within facies. Bold horizontal bar in box indicates median; bottom of box indicates 25th percentile; top of box indicates 75th percentile; whiskers extend to the most extreme data point, which is no more than the range (i.e. 1.5) times the interquartile range from the box; width of bar directly proportional to sample size; grey-shaded boxes highlight notch width; if the notches of any two plots do not overlap this is 'strong evidence' that the two medians differ (Chambers *et al.* 1983, p. 62). The ANOSIM statistic R is based on the difference of mean ranks between groups (r_B) and within groups (r_W), such that: $R = (r\_B - r\_W)/(N^*(N-1)/4)$. Key: Between, between-group dissimilarity; F, coarsening upward interbedded. siltstone–sandstone; I, Fe-oxide nodules (ironstones); K, kaolinite-dominated siltstone; O, organic shale; P, mudstone parting beds.

process, because of the large size of *Schopfipollenites* pre-pollen (up to 500 μm in length). Breakage during deposition or sample processing may be another factor, because *Schopfipollenites* sometimes occurs as fragments in strew mounts, as well as lower pollen output. In contrast to the scarcity of medullosan pre-pollen, cordaitalean pollen is abundant in the organic shale lithofacies, especially in the clastic phases, compared to the rarity of *Cordaites* leaves. This discrepancy is likely due to the difficulty of recognizing *Cordaites* leaf fragments in the macrofossil record, as noted above.

The palynological assemblage composition of the kaolinite-dominated siltstone lithofacies is also largely congruent with the macrofossil data. The most abundant pollen type in this facies is one or more species of non-taeniate bisaccate, or pseudobisaccate forms such as *Falcisporites*, *Alisporites* or *Pseudovesicaspora*. It is tempting to attribute one or more of these forms to the seed fern *Sphenopteridium manzanitum*, which is dominant in this lithofacies; however, there is no direct evidence for this. Pollen of walchian conifers (*Potonieisporites* and *Nuskoisporites*) is subdominant, along with spores of sphenophylls and marattialian ferns. Dominance–diversity curves for palynomorphs largely follow those of the macrofossils, with the exception of the virtual absence of medullosan pre-pollen (Fig. 5). Lycopsid spores and cordaitalean pollen are rare in this lithofacies, as are spores of *Sigillaria*. Marattialian fern spores display low to moderate abundance, consistent with the presence of macrofossils of this clade in many environments. A few non-marattialian ferns are sometimes abundant in this lithofacies, but their overall species richness is much lower than observed in the organic shale lithofacies.

The palynofloras from both facies are notably more species-rich than the respective macrofossil assemblages. Roughly 110 palynomorph morphospecies were recognized in the 13 productive samples, compared to 28 in the macrofossil flora. Per sample diversity is also much higher in palynological assemblages, with a mean of 30 (ranging from 19 to 42). Some of this apparent diversity may be due to taxonomic oversplitting as noted

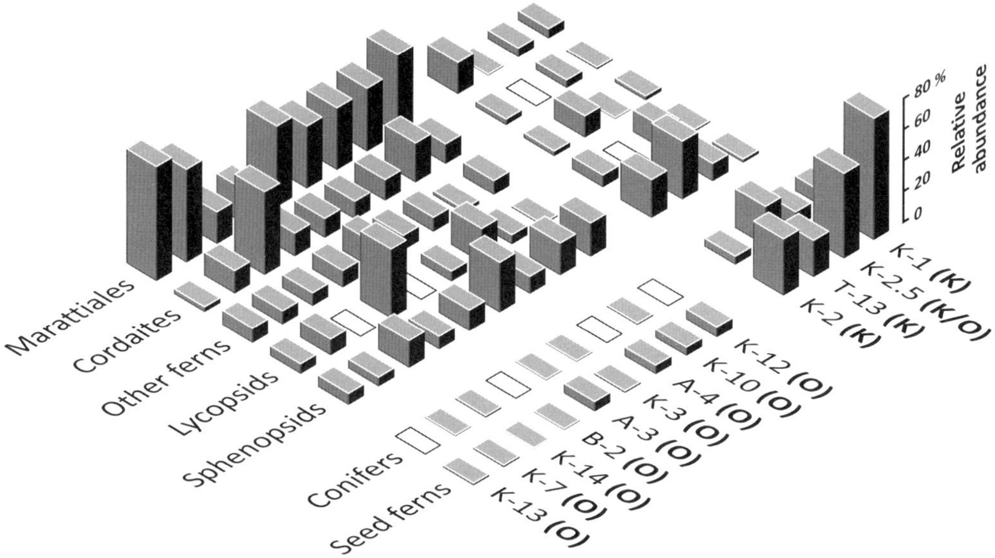

Fig. 9. Taxonomic composition and relative abundance of major taxonomic groups in palynological samples. Two lithofacies are represented: kaolinite-dominated siltstones (K) and organic shales (O), with two transitional samples (K/O). Labels of palynological samples as in Table 3.

above, but the greater part of this signal is probably real. The greatest diversity occurs within the non-marattialian ferns, with at least 20 distinct morphotypes represented, many of unknown affinity. These ferns, including members of the Tedeleaceae, Zygopteridaceae, Gleicheniaceae, among others, never comprise more than 15% of the total flora, more usually ranging from 2% to 5%, but are always present at some level. The palynological assemblages may be more diverse, in part because of time averaging, reflecting a temporal succession of floral elements represented in the macrofossil record by a single time slice, or because they are drawing from a broader source area, or probably both.

Ordination analysis

Detrended correspondence analysis of palynological abundance data (Fig. 10) reveals more overlap between the organic shale and kaolinite-dominated siltstone lithofacies than is apparent in the macrofossil samples. Axis 1 of the graph is interpreted as a wet-dry gradient. Axis 2, on the other hand, manifests primarily variation within the organic-shale facies, and appears to reflect abundance of *Sigillaria* microspores; this may indicate a subtle wet–dry gradient within that facies. Samples from the organic shale lithofacies (palynological samples B-2, A-3, A-4, K-3, K-7, K-10, K-12, K-13 and K-14) group to the left of the graph, presumably on the wet end of the spectrum, whereas the two macrofossil-bearing kaolinite-dominated siltstone beds (palynological samples K-1 and T-13) group to the right, showing a clear disjunction in palynoflora composition between these two lithofacies types. However, two samples between the kaolinite-dominated siltstone bed and the basal coal at locality K (palynological samples K-2 and K-2.5) are transitional in palynofloristic composition between the wetland and dryland floras (Fig. 10). The lithology of samples K-2 and K-2.5 becomes increasingly rich in organics and plant debris upsection. Both units contain abundant conifer and non-taeniate bisaccate pollen, together with an almost equal abundance of marattialian and non-marattialian fern, sphenopsid and lycopsid spores (Figs 9 & 10). Significantly, walchian twigs occur just below the basal coal in one of these layers. There is no evidence of a sedimentological break or hiatus between kaolinite-dominated siltstone lithofacies and the organic shale beds at locality K.

Discussion

Evidence of niche-based assembly dynamics

Perhaps the most significant specific finding of this study is that species content and relative abundance in a given facies are remarkably similar regardless of geographic location or stratigraphic position within the Markley or basal Archer City formations. These patterns of niche conservatism

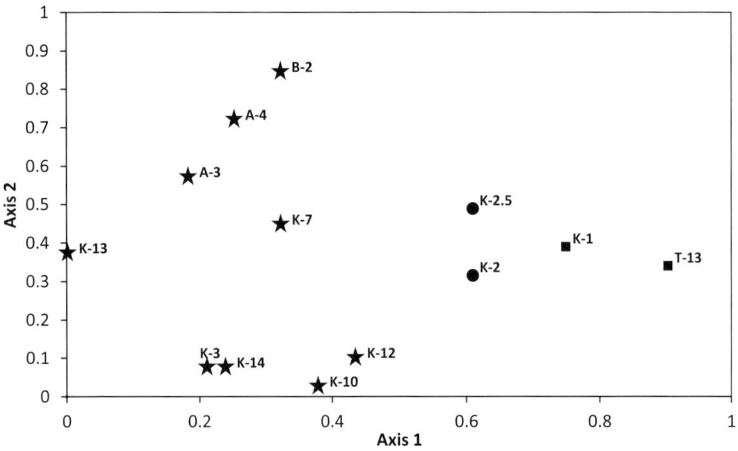

Fig. 10. Palynological site detrended correspondence analysis (DCA) based on count data of 109 taxa from 13 localities. Lithofacies are either organic shales (stars, $n = 9$), kaolinite-dominated siltstones (squares, $n = 2$) or transitional between these facies (circles, $n = 2$) (see also Fig. 4). Locality letter designations as in Table 3. Eigenvalues for Axis 1 and Axis 2 are 0.7599 and 0.3258, respectively.

are superimposed on a stratigraphic interval of uncertain duration, but likely encompassing several hundred thousand to perhaps as many as a million years. This duration is estimated based on the several distinct intervals of peat formation separated by periods of palaeosol development and marine incursion. This cyclic repetition of lithofacies packages is evidence of recurrent base level and likely climate change, in addition to the documented ecological changes. This means that a high degree of 'niche assembly' is called for in the face of changing landscape conditions within which certain very specific habitats (represented by the lithofacies types) periodically reappear.

Thus the species that comprise these different habitats do not appear to be randomly distributed but instead have strong preferences for physical conditions that reflect amount and duration of soil wetness, degree of disturbance, and possibly amount of light availability. Although a few species are present in more than one habitat, each habitat contains a unique suite of plants and dominance distribution, whereas species richness is roughly comparable. So strong are these habitat relationships that they permit statistical differentiation among the facies solely based on taxonomic composition and quantitative abundance. The strength of these conclusions about flora/facies relationships can certainly be tested in other parts of the Late Pennsylvanian (Stephanian) tropics, given the widespread distribution of these basic species pools (e.g. Wagner 1983; Doubinger *et al.* 1995; Castro 2005; Wagner & Álvarez-Vázquez 2010; Bashforth *et al.* 2010).

Vegetational patterns

Plant assemblages of the Markley Fm can be grouped into two basic vegetational types: those of periodically flooded and generally wet substrates and those of rarely or never flooded, periodically dry substrates. At the wet end of this environmental spectrum are the organic shale beds, interpreted to have formed in swamps, and containing a typical Late Pennsylvanian wetland flora (Crookall 1959, 1966; Darrah 1969; Wagner & Álvarez-Vázquez 2010; Wagner & Castro 2010; Bashforth *et al.* 2010), most commonly dominated by *Sigillaria brardii*, *Macroneuropteris scheuchzeri* and marattialean ferns. Floodplain assemblages, part of the coarsening upward mudstone–sandstone interbed facies, also fall broadly into this category, and are dominated by typically Pennsylvanian wetland plants. At the drier end of the spectrum are floras from the kaolinite-dominated siltstone facies, dominated by *Sphenopteridium manzanitanum* and *Walchia* spp., with various types of *Pecopteris* as accessory taxa. The kaolinite-dominated siltstone beds are interpreted to be composed of clay-rich sediment formed under a hot, humid climate, and later eroded from palaeosol surfaces and transported into the deposystem, perhaps in the early phases of landscape degradation and filling of abandoned channels. The flora is the key to the environmental interpretation: *Sphenopteridium* is found most commonly in association with other indicators of seasonally dry climates, both in the Mississippian and in the later Pennsylvanian and earlier Permian (Mamay 1992). Similarly, conifers are typically found in strata that include evidence for seasonality of moisture availability at the time of deposition, probably in more dry to wet subhumid climates than in semi-arid conditions (e.g. Rothwell *et al.* 1997; Feldman *et al.* 2005; Looy 2007; DiMichele *et al.* 2008; Hernandez-Castillo *et al.* 2009).

A more complex picture of the vegetated landscape appears with the addition of less abundant accessory taxa. Usually common to rare within any particular collection, these taxa may occur in abundance locally and may be the dominant species within a narrow stratigraphic interval or at a single collecting site. Some of these accessory taxa are widespread and appear in a broad range of lithologies (e.g. Equisetalean sphenopsids, *Cordaites* and some types of *Pecopteris*) whereas others are lithologically specific (e.g. *Pseudomariopteris cordato-ovata* and *Odontopteris brardii* in floodplain mudstones of the coarsening upward mudstone–sandstone interbed facies or *Alethopteris zeilleri* in organic shales). Among many possibilities, one of the most likely explanations for this pattern is habitat heterogeneity combined with niche specificity, some taxa being tolerant of a greater range of physical conditions than others. Dispersal limitation could also play a major role within a pool of species with otherwise very similar physical tolerances, particularly within the context of a heavily vegetated wetland landscape in which incumbency and standing population size would strongly affect the likelihood of colonizing available sites (e.g. Belyea & Lancaster 1999). Such a process may account for taxa that exhibit a very restricted distribution in the Markley outcrop belt, geographically and stratigraphically, such as (?)*Lepidodendron*, *Neurodontopteris*, *Alethopteris* and *Odontopteris*.

Space v. *time and lithofacies/habitat relationships: differentiating allocyclicity from autocyclicity*

The clear differentiation of vegetation by lithofacies, and the ecological implications thereof pose a question that arises in any temporal study of Pennsylvanian-age vegetation, in which the ability to assess the spatial scope of a landscape is limited (e.g. compare Dimitrova *et al.* 2011 with Dolby *et al.* 2011, examining and interpreting differently the same dataset, and published back to back in the same journal), namely, 'Did these different vegetation types coexist in relatively close proximity side by side on a landscape, shifting location through time under a fluctuating, but generally consistent climate ('autocyclic coexistence model')?' Or did they succeed one another in time, expanding and contracting within the region as physical, mainly climatic, conditions changed significantly over the course of time ('allocyclic succession model')? Or some combination of these? Clearly, this is a question of spatial scale because the different vegetation types did coexist on Earth, and in the case of the Pennsylvanian equatorial coal belts, perhaps even in the same basic latitudinal zones. We use the terms autocyclic and allocyclic from Beerbower (1964) as modifiers for reasons that will become clear below.

Autocyclic coexistence models

In the strictest construction of the autocyclic coexistence model, widespread regional climate is assumed to be relatively constant, lithological changes reflecting only shifting loci of sedimentary dynamics. During most of the Pennsylvanian Period, in those areas where coal is abundant and well developed, this calls for wet conditions with high rainfall for most of the year, exceeding evapotranspiration (e.g. Dimitrova *et al.* 2010). On such a landscape, vegetational zones are thought to be controlled by topography and consequent differences in soil drainage (for another example, see Cazzulo-Klepzig *et al.* 2007). Under this model, changes in the distribution of vegetation through time reflect changes in loci of sedimentation and erosion intrinsic to the local setting and region. In terms of a regional sedimentary system applicable to the Pennsylvanian tropics, the deposystem might look much like the deltaic settings proposed for Pennsylvanian coal-bearing sequences (e.g. Ferm & Cavaroc 1968; Esterle & Ferm 1986; Ferm & Weisenfluh 1989), in which the driver of peat accumulation is envisioned to have been the simple cutoff of wetland areas from clastic input, thus preventing dilution of accumulating organic matter, with no consideration given to climate. In this model, significant siliciclastic transport and deposition continues within the delta elsewhere, contemporaneously with peat accumulation in such areas as abandoned, sinking delta lobes. In the Markley Fm, this model would need to be reformulated such that the prevailing climate was seasonally dry, given the record of floras (Romanchock *et al.* in press) and palaeosols (Tabor & Montañez 2004), but still with high rainfall for much of the year. As discussed above, the climate also seems to have been becoming more seasonally dry through time, culminating in the restriction of the wetland flora to limited areas in the Archer City Fm, deposited during the Early Permian.

A tectonic variant of the autocyclic coexistence model has recently been proposed to explain major changes in Pennsylvanian-age equatorial vegetation. In this model, mountain building events affect such things as drainage patterns, elevation and other factors that control plant biogeographic distribution (see Cleal *et al.* 2011 for a summary). In the Pennsylvanian, for example, the growth of the European Variscan mountains has been suggested to be the cause of widespread and rapid changes in vegetation during the Middle–Late Pennsylvanian transition, in contrast to widespread climatic changes. This model strongly downplays, though does not rule out, climatic effects on vegetational changes in time and space. This is not the place to critique in detail this particular application of such a model. Its most significant weaknesses are the great disparity in the rates at which vegetation is known to change its spatial distribution v. the rates of tectonically induced changes in elevation and drainage patterns, and the somewhat geographically local, and temporally irregular effects a tectonic model can offer when examining sequences in which there are repeated patterns of facies and vegetation through time over vast areas, such as equatorial Euramerica. Tectonic effects on the landscape of any region are important long-term modifiers of many physical variables, including climate, and are perhaps best considered as an explanatory variable for biological changes that take place on such long-term timescales.

Another version of the coexistence model is found in those sequence stratigraphic interpretations of Pennsylvanian peat-forming systems in which climate is not integral to the model (e.g. Flint *et al.* 1995; Heckel 1995, 2008; Bisnett & Heckel 1996; Bohacs & Suter 1997). Such models generally refer explicitly to eustatically driven sea-level fluctuations as the proximate driver of sedimentary sequences on the shelf-edge and cratonic surface and, as such, may be referred to as 'allocyclic coexistence' models. Peat formation is envisioned to have been driven by landscape paludification caused by rising sea-level, and is part of a system of coexistent, terrestrial and marine, sedimentary belts, pushed across the craton during sea-level rise and fall, thus making each environment time transgressive. A 'default' climate is generally not specified, but must be somewhere in the range of wet subhumid to humid, in the terminology of Cecil & Dulong (2003), meaning rainfall exceeding evapotranspiration for 8–10 months of the year if there is to be peat (coal) accumulation as part of the system. We can find no examples where this particular aclimatic sequence stratigraphic model has been used as a framework for detailed interpretation of terrestrial palaeoecological patterns on Pennsylvanian landscapes, although such interpretations are implicit in the model and have been alluded to broadly, particularly as they apply to peat formation, palaeosols and non-marine limestones (e.g. Jerrett *et al.* 2011; Belt *et al.* 2011). We do not consider this model further.

Allocyclic succession model

The allocyclic succession model requires the imposition of climatic changes on the local and regional sedimentary systems and biota, thus strongly affecting both (e.g. Cecil 1990; Gastaldo *et al.* 1993; Falcon-Lang 2004; DiMichele *et al.* 2008, 2010; Falcon-Lang & DiMichele 2010). When placed in a sequence stratigraphic framework (e.g. Ziegler *et al.* 1987; Gastaldo *et al.* 1993; Cecil *et al.* 2003; Falcon-Lang 2004; Peyser & Poulsen 2008; Gastaldo *et al.* 2009), this model calls for strong linkages between eustacy, climate and sediment transport dynamics in order to account for lithofacies patterns. Such models suppose a significant element of climate control on the formation of peat (deposited under humid to perhumid climates), clastics (transport being highest under seasonal climate regimes) and carbonates (marine and non-marine, forming mainly under the more strongly seasonal to arid climates). Because plants are so sensitive to climate (see discussion and citations in DiMichele *et al.* 2008, 2009), these climatic changes are expected to have had spatially extensive effects on vegetation (Falcon-Lang & DiMichele 2010; Falcon-Lang *et al.* 2009; Dolby *et al.* 2011), effectively centring different species pools, or biomes, in different parts of the palaeogeographic landscape at any one time, with minimal zones of overlap. Thus, lithofacies–biofacies patterns, in the context of the Pennsylvanian ice house world, would be expected to be a reflection of changing climate and depositional environment, driven by glacial–interglacial fluctuations. Different vegetation types through time are interpreted to reflect climatic shifts to which vegetation responded, similar to that seen during the Pleistocene and Holocene in response to glacial–interglacial climatic cycles (e.g. Falcon-Lang 2004).

Autocyclicity v. *allocyclicity in the Markley Fm*

Markley Fm vegetational and sedimentary analyses provide clear evidence of differences in physical conditions on the landscape. At question, however, is whether these differences reflect widespread, cyclical patterns of fluctuation in climate, particularly precipitation, or whether the Markley was deposited under a fairly stable climatic background within which autocyclicity dominated. The physical data, that is, the facies sequences and their spatial expression, to the degree they are known, do not lead to a clear resolution of this problem. Outcrop exposures are limited in areal extent and are widely spaced geographically. At a local, spatial scale, at all sampled sites through time and space, the environmental sequence is the same; habitats shift from palaeosols (evidencing well-drained conditions with seasonally high rainfall) and associated kaolinite-rich fluvial deposits dominated by a relatively homogeneous association of seed ferns and walchian conifers, to monotonous medullosan pteridosperm and sigillarian lycopsid-rich swamp settings, to floodplains with a diverse, spatially heterogeneous wetland flora.

Macrofossil analyses of the kaolinite-dominated siltstone and organic shale lithofacies indicate little overlap of the two species pools: (i) the 'dry' flora, dominated by the xeromorphic pteridosperm *Sphenopteridium manzanitanum* and conifers and (ii) the 'wet' flora, dominated by a diversity of pteridosperms, lycopsids and ferns. However, floodplain lithofacies include rare occurrences of dryland elements, within an overall wetland floristic framework. The compositional differences indicate, at a minimum, strong taphonomic segregration of the floras. The processes by which the organic remains are shed, transported and buried, and the depositional conditions under which these processes operate, are at least segregated spatially or differ in terms of the major operative sedimentary processes.

The palynological data also support the existence of two distinct floras, one a typical Late Pennsylvanian wetland flora dominated by lycopsids, sphenopsids, tree and herbaceous ferns and cordaitaleans, and the other a dryland flora dominated by seed ferns, walchian conifers and more tolerant pteridophyte taxa. However, palynology introduces nuances that are not apparent in the macrofossil record, and suggests that the allocyclic succession and autocyclic coexistence models are not necessarily mutually exclusive.

That all of the palynologically productive samples from the Markley occur within a restricted stratigraphic interval characterized by regionally developed coals suggests that these beds were deposited during an overall more humid phase. Absence of pollen (due presumably to syndepositional oxidative destruction) from stratigraphically higher intervals within the Markley is consistent with a pattern of increasing aridity in younger strata; this agrees with the disappearance of Histosols and Ultisols above Sandstone 10 (Tabor & Montañez 2004). Nevertheless, palynological data indicate a spatial overlap between dryland and wetland floras, for example, the presence of rare *Sigillaria* microspores in kaolinite-dominated siltstones, and the presence of rare pollen grains of walchian conifers and seed ferns in organic shales. The transitional palynofloras preserved at locality K contain a mixture of wetland and dryland elements, in addition to *Sigillaria* macrofossils in the same beds from which the palynologically mixed floras are derived, suggesting that they may represent an ecotonal boundary. This boundary between wetland and dryland was probably quite narrow spatially, evidenced by walchian conifer twigs occurring just below a transitional kaolinite-dominated siltstone bed with *Sigillaria* in its upper portions and several centimetres below an organic shale also containing *Sigillaria*. This pattern suggests that, at least during all or perhaps the earlier parts of this wetter phase on this particular landscape, the latest Pennsylvanian wetland and dryland floras existed in relatively close proximity.

Most or all of the organic shale and coal deposits examined in this study of the Markley and lower Archer City formations are underlain by kaolinite-dominated siltstones in abandoned channel fills, suggesting the existence of a genetic relationship between the two types of lithofacies. We propose that kaolin-rich sediments were deposited during the abandonment phases of shallow channels active during the late stages of soil formation, lining channel bottoms. These clays were stripped from surrounding interfluves, sometimes carrying plant debris derived from the well-drained interfluve floras, dominated by walchian conifers and seed ferns. There, the clays formed impediments to drainage (aquicludes), creating poorly drained depressions and the formation of local wetlands that were dominated by typical Pennsylvanian wetland species. A modern analogue might be that observed today in Borneo (Gastaldo 2010). Clay-rich deposits there form under a hot, humid climate through intensive chemical leaching, suggesting that the emplaced flora within the kaolinite-dominated siltstone lithofacies may similarly have existed under hot, humid conditions. If well-drained interfluves were dominated by 'dryland' floras, particularly during more intensely seasonal periods of time, wetland elements would not be expected in the kaolinite-dominated siltstone lithofacies; thus, the apparent separation of these two types of floras could be due largely to taphonomic biases operative during a transitional period from a more seasonally dry to a more seasonally wet portion of any given glacial–interglacial climatic and sedimentary cycle. This model proposes that at times, all observed floristic elements within the Markley coexisted temporally, and were spatially rather than temporally separated.

Integration of allocyclic and autocyclic succession models still allows for high-frequency climate fluctuations between humid and more seasonal precipitation regimes as the underlying driver of the habitat changes that resulted in vegetation shifts. These shifts would most likely have been driven by glacial–interglacial cycles and the attendant changes in tropical climate and sea-level accompanying them, which occurred during the Late Pennsylvanian (Olszewski & Patzkowsky 2003).

At a coarser level of resolution, entire habitats linked to humid climates disappear and are replaced through time by those of drier and more seasonal climates. Moving upward into younger portions of the stratigraphic section, this long-term drying trend becomes more pronounced, with the development of calcrete-bearing palaeosols and evaporite deposits in the Archer City Fm (Jones 1971; Tabor & Montañez 2004).

Still, a question remains: 'What was responsible for these changes in climate at any and all scales?' To answer this, the patterns of floral and sedimentary dynamics described by this study must be viewed in the context of Late Palaeozoic climate in the western Pangean tropics. It is now well established that this region experienced a long-term drying trend accompanied by increased seasonality in precipitation beginning in latest Pennsylvanian time (Cecil 1990; Tabor & Montañez 2004; Tabor & Poulsen 2008 and references therein). This climate shift occurred at different temporal scales and was influenced by multiple drivers operating at each of these scales. For example, tectonic drivers active during the formation of Pangea (e.g. orographic uplift and erosion, change in continental position through time) likely operated at scales of 10^6–10^7 years (Tabor & Poulsen 2008), whereas glacial–interglacial cycles operated at scales of *c.* 10^5 year (Heckel 1986; Davydov *et al.* 2010; Eros *et al.* 2011). Although tectonic drivers cannot account for the repetitive, high-frequency sedimentary facies and climate shifts present in the Markley Fm, ice-volume fluctuations and coincident changes in atmospheric CO_2 are viable primary drivers (Poulsen *et al.* 2007; Peyser & Poulsen 2008). Evidence for massive expansion of multiple glacial-spreading centres during the latest Pennsylvanian has been reported from various Gondwanan basins (Isbell *et al.* 2003; Fielding *et al.* 2008). Depending on the number of spreading centres and amount of ice ablation during interglacial periods, estimates of attendant eustatic changes range from 50 m to 115 m (Isbell *et al.* 2003). Eustatic changes of this magnitude would have been sufficient to account for the observed orderly repetition of facies in the Markley Fm, given the lowland coastal plain on which deposition took place. Furthermore, conceptual (Cecil *et al.* 2003) and numerical (Peyser & Poulsen, 2008) climate models suggest that ice-volume fluctuations would have shifted the position of the Intertropical Convergence Zone (ITCZ; via the conceptual model) or altered Hadley Cell strength and, consequently, amounts of precipitation near the equator (via the numerical model). Although the mechanism is different in each model, the end result is that mean annual precipitation increases and seasonality decreases at low latitudes during glacial periods and, in contrast, mean annual precipitation decreases during interglacial periods and the seasonality of its distribution increases. When this high-frequency cycle is superimposed on the tectonically driven long-term drying trend, the average trend within the high-frequency cycles is for the wetter intervals to become less wet and the drier intervals to become more dry over the longer spans of time. This overall trend matches the precipitation signal inferred from outcrop- and regional-scale floral, pedogenic and sedimentological patterns in the Markley Fm.

Suppose, in contrast to the allocyclic succession interpretation, that the Markley Fm lithofacies–biofacies patterns reflect strictly small-scale spatial variation in the coastal plain landscape, in accordance with an autocyclic coexistence interpretation. In this case, the background climate would most likely have fluctuated no further than between humid and wet–subhumid, at least during the early and middle intervals of Markley Fm deposition, such that peat accumulation could continue on the landscape at all times, anywhere clastic input to persistently wet areas was diminished. These swamp deposits, and the buried remains of the constituent floras, are among the most likely to be preserved on a landscape with slow, tectonic creation of accommodation space, because they are low-lying, sediment-accumulating locations. The vast majority of the lowlands, in contrast, would have consisted of moderate to well drained surfaces supporting xeromorphic vegetation. This vegetation would have grown on the soils that became the palaeosol profiles that are the predominant lithology of the Markley Fm, and, as such, even though relatively poorly represented in the Markley fossil record, would have been the persistently dominant vegetation type on the coastal plain at all times. Preservation of the dryland flora occurs mainly in reworked kaolinite-dominated siltstones that seem to have been deposited in shallow channels during the final phases of abandonment. The sedimentary matrix in which the fossils are buried would have been stripped from the soil surface, over an extended time, by repeated events of erosional runoff, accounting for the variable and generally fragmentary preservation of the plant fossils. Floodplain interbedded siltstones and sandstones likely represent the encroachment phases of channels, although we cannot differentiate a fluvial system dominated by meandering processes from one in which aggradation and avulsion are the predominant processes. Such landscapes would have had relatively high soil moisture, been subjected to high levels of disturbance, and would have had complex drainage networks, consistent with the scour and fill nature of these deposits on outcrop, the weakly developed palaeosols, and the patchiness and variability of the vegetation.

Under the autocyclic coexistence scenario, the longer-term pattern of diminishment, and ultimately disappearance, of the organic-rich facies and the shifts in palaeosol characteristics from ultisols to vertisols can be accounted for by the same processes as described above for the allocyclic succession model.

Whereas it may seem somewhat unsatisfying to leave the matter of the environmental framework unresolved, we are, as a group of authors, unresolved on which of these scenarios is most likely. The increasing refinement of sequence stratigraphic models for Pennsylvanian tropical, coal-bearing sequences points ever more strongly to the need for a significant climatic element in these models. Sea-level alone is an insufficient explanation to account for the diversity of lithologies and their repeated, stereotypical successions with a sequence of Pennsylvanian cyclothems. Furthermore, an aclimatic sequence model cannot account for the widely varied climatic signals represented by, say, coal beds and vertic, sometimes calcic palaeosols, which often lie in immediate successional juxtaposition.

Palaeobotanical patterns create still further dilemmas. Repeated temporal fluctuations occur within basins, thus in the same biogeographical space, between typical wetland assemblages and so-called 'extrabasinal' floras (if they are extrabasinal, what are they doing in the basins?). Both vegetation types occur in autochthonous to parautochthonous preservation, which indicates growth in areas close to the sites of deposition. Yet there is only minimal compositional overlap between these different floras, suggesting that they represent different species pools, likely reflecting different climatic backgrounds. Extension of such findings, made elsewhere, to the Markley Fm is stymied by its limited marine influence, indicating a relatively inland position on the coastal plain, its position in the western parts of the Pangean equatorial zone, placing it under a relatively drier climate than the eastern North American coal basins, and the limitations of surface exposure. Thus, at present, we believe that while many data favour an allocyclic successional interpretation, the autocyclic, coexistence interpretation cannot be ruled out, and it is most likely that the two play important complementary roles, although perhaps in different spatial and temporal frames of reference.

We thank the numerous property owners on whose land these exposures are found. Without their permission and material assistance this study would not have been possible. We are indebted to R. W. Hook and K. Craddock for their advice and participation in various aspects with this project, to the late S. Mamay and N. Hotton III for their mentorship and collaboration, and to J. Nelson for helpful discussion. The fieldwork on which this project is based was carried out over a 20-year period with various grants from the Smithsonian Institution, including the Scholarly Studies Program, Endowed Funds, and NMNH Small Grants to W.D. and to the late N. Hotton III. The research of C.L.H. was supported in part by the Intramural Research Program of the National Institutes of Health, National Library of Medicine. The paper was improved by the thoughtful comments of two anonymous reviewers, whom we thank.

References

Alonso, D., Etienne, R. S. & McKane, A. J. 2006. The merits of neutral theory. *Trends in Ecology and Evolution*, **21**, 451–457.

Bartram, K. M. 1987. Lycopod succession in coals: an example from the Low Barnsley Seam (Westphalian B), Yorkshire, England. *In*: Scott, A. C. (ed.) *Coal and Coal-bearing Strata: Recent Advances*. Geological Society, London, Special Publications, **32**, 187–199.

Bashforth, A. R., Falcon-Lang, H. J. & Gibling, M. R. 2010. Vegetation heterogeneity on a Late Pennsylvanian braided-river plain draining the Variscan Mountains, La Magdalena Coalfield, northwestern Spain. *Palaeogeography, Palaeoclimatology, Palaeoecology*, **292**, 367–390.

Beerbower, J. R. 1964. Cyclothems and cyclic deposition mechanisms in alluvial plain sedimentation. *Kansas Geological Survey Bulletin*, **169**, 31–42.

Belt, E. S., Heckel, P. H., Lentz, L. L., Bragonier, W. A. & Lyons, T. W. 2011. Record of glacial–eustatic sea-level fluctuations in complex middle to late Pennsylvanian facies in the Northern Appalachian Basin and relation to similar events in the Midcontinent basin. *Sedimentary Geology*, **238**, 79–100.

Belyea, L. R. & Lancaster, J. 1999. Assembly rules within a contingent ecology. *Oikos*, **86**, 402–416.

Bennington, J. B., DiMichele, W. A. *et al.* 2009. Critical issues of scale in paleoecology. *Palaios*, **24**, 1–4.

Bisnett, A. J. & Heckel, P. H. 1996. Sequence stratigraphy helps to distinguish offshore from nearshore black shales in the Midcontinent Pennsylvanian succession. *In*: Witzke, B. J., Ludvigson, G. A. & Day, J. (eds) *Paleozoic Sequence Stratigraphy: Views from the North American Craton*. Geological Society of America, Special Papers, **306**, 341–350.

Bohacs, K. & Suter, J. 1997. Sequence stratigraphic distribution of coaly rocks: fundamental controls and paralic examples. *American Association of Petroleum Geologists Bulletin*, **81**, 1612–1639.

Bonelli, J. R., Jr. & Patzkowsky, M. E. 2011. Taxonomic and ecologic persistence across the onset of the Late Paleozoic Ice Age: evidence from the Upper Mississippian (Chesterian Series), Illinois Basin, United States. *Palaios*, **26**, 5–17.

Burnham, R. J. & Spicer, R. A. 1986. Forest litter preserved by volcanic activity at El Chichon, Mexico; a potentially accurate record of the pre-eruption vegetation. *Palaios*, **1**, 158–161.

Burnham, R. J., Wing, S. L. & Parker, G. G. 1992. The reflection of deciduous forest communities in leaf litter: implications for autochthonous litter assemblages from the fossil record. *Paleobiology*, **18**, 30–49.

Castro Martínez, M. P. 2005. *La flora estefaniense B de La Magdalena (Léon, España), un referente europeo*. Instituto Geológico y Minero de España, Madrid, Vols 1 & 2.

Cazzulo-Klepzig, M., Guerra-Sommer, M. & Menegar, R. 2007. Peat-forming environment of Permian coal seams from the Faxinal coalfield (Paraná Basin) in southern Brazil, based on palynology and palaeobotany. *Revista brasileira de Paleontologia*, **10**, 117–127.

Cecil, C. B. 1990. Paleoclimate controls on stratigraphic repetition of chemical and siliciclastic rocks. *Geology*, **18**, 533–536.

Cecil, C. B. & Dulong, F. T. 2003. Precipitation models for sediment supply in warm climates. *In*: Cecil, C. B. & Edgar, N. T. (eds) *Climate Controls on Stratigraphy*. SEPM Special Publications, Tulsa, **77**, 21–27.

Cecil, C. B., Dulong, F. T., West, R. R., Stamm, R., Wardlaw, B. & Edgar, N. T. 2003. Climate controls on the stratigraphy of a middle Pennsylvanian cyclothem in North America. *In*: Cecil, C. B. & Edgar, N. T. (eds) *Climate Controls on Stratigraphy*. SEPM Special Publications, Tulsa, **77**, 151–180.

Chambers, J. M., Cleveland, W. S., Kleiner, B. & Tukey, P. A. 1983. *Graphical Methods for Data Analysis*. Wadsworth, Boston.

Cleal, C. J. 2007. The Westphalian–Stephanian macrofloral record from the South Wales Coalfield. *Geological Magazine*, **144**, 465–486.

Cleal, C. J., Opluštil, S., Thomas, B. A. & Tenchov, Y. 2011. Pennsylvanian vegetation and climate in tropical Variscan Euramerica. *Episodes*, **34**, 3–12.

Cleal, C. J., Uhl, D., Cascales-Miñana, B., Thomas, B. A., Bashforth, A. R., King, S. C. & Zodrow, E. L. 2012. Plant biodiversity changes in Carboniferous tropical wetlands. *Earth-Science Reviews*, **114**, 124–155.

Crookall, R. 1959. Fossil plants of the Carboniferous rocks of Great Britain, 2nd section, Neuropteridae. *Memoirs of the Geological Survey of Great Britain, London, Palaeontology*, **4**, 85–216.

Crookall, R. 1966. Fossil plants of the Carboniferous rocks of Great Britain, 2nd section. *Memoirs of the Geological Survey of Great Britain, Palaeontology*, **4**, 355–571.

Currano, E. D., Wilf, P., Wing, S. L., Labandeira, C. C., Lovelock, E. C. & Royer, D. L. 2008. Sharply increased insect herbivory during the Paleocene-Eocene thermal maximum. *Proceedings of the National Academy of Sciences of the United States of America*, **105**, 1960–1964.

Darrah, W. C. 1969. *A critical review of the Upper Pennsylvanian floras of eastern United States with notes on the Mazon Creek flora of Illinois*. Privately printed, Gettysburg, Pennsylvania.

Davydov, V. I., Crowley, J. L., Schmitz, M. D. & Poletaev, V. I. 2010. High-precision U–Pb zircon age calibration of the global Carboniferous time scale and Milankovitch band cyclicity in the Dontz Basin, eastern Ukraine. *Geochemistry, Geophysics, Geosystems*, **11**, http://dx.doi.org/10.1029/2009GC002736

DiMichele, W. A. & Phillips, T. L. 2002. The ecology of Paleozoic ferns. *Review of Palaeobotany and Palynology*, **119**, 143–159.

DiMichele, W. A., Phillips, T. L. & McBrinn, G. E. 1991. Quantitative analysis and paleoecology of the Secor coal and roof-shale floras (Middle Pennsylvanian, Oklahoma). *Palaios*, **6**, 390–409.

DiMichele, W. A., Tabor, N. J., Chaney, D. S. & Nelson, W. J. 2006. From wetlands to wetspots: environmental tracking and the fate of Carboniferous elements in Early Permian tropical floras. *In*: Greb, S. F. & DiMichele, W. A. (eds) *Wetlands Through Time*. Geological Society of America Special Papers, **399**, 223–248.

DiMichele, W. A., Falcon-Lang, H. J., Nelson, W. J., Elrick, S. D. & Ames, P. R. 2007. Ecological gradients within a Pennsylvanian forest mire. *Geology*, **35**, 415–418.

DiMichele, W. A., Kerp, H., Tabor, N. J. & Looy, C. V. 2008. The so-called 'Paleophytic–Mesophytic' transition in equatorial Pangea: multiple biomes and vegetational tracking of climate through geological time. *Palaeogeography, Palaeoclimatology, Palaeoecology*, **268**, 152–163.

DiMichele, W. A., Montañez, I. P., Poulsen, C. J. & Tabor, N. J. 2009. Climate and vegetational regime shifts in the late Paleozoic ice age earth. *Geobiology*, **7**, 220–226.

DiMichele, W. A., Cecil, C. B., Montañez, I. P. & Falcon-Lang, H. J. 2010. Cyclic changes in Pennsylvanian paleoclimate and effects on floristic dynamics in tropical Pangaea. *International Journal of Coal Geology*, **83**, 329–344.

Dimitrova, T. K., Zodrow, E. L., Cleal, C. J. & Thomas, B. A. 2010. Palynological evidence for Pennsylvanian (Late Carboniferous) vegetation change in the Sydney Coalfield, eastern Canada. *Geological Journal*, **45**, 388–396.

Dimitrova, T. K., Cleal, C. J. & Thomas, B. A. 2011. Palynological evidence for Pennsylvanian extra-basinal vegetation in Atlantic Canada. *Journal of the Geological Society, London*, **168**, 559–569.

Dolby, G., Falcon-Lang, H. J. & Gibling, M. R. 2011. A conifer-dominated palynological assemblage from Pennsylvanian (late Moscovian) alluvial drylands in Atlantic Canada: implications for the vegetation of tropical lowlands during glacial phases. *Journal of the Geological Society, London*, **168**, 571–584.

Doubinger, J., Vetter, P., Langiaux, J., Galtier, J. & Broutin, J. 1995. The flore fossile du basin houiller de Saint-Étienne. *Mémoires du Muséum National D'Histoire Naturelle*, **164**, 1–355.

Dudal, R. & Eswaran, H. 1988. *Distribution, Properties and Classification of Vertisols*. Publication Soil Management Support Services, US Department of Agriculture, Natural Resources Conservation Service, Washington DC, 1–22.

Eble, C. F., Gastaldo, R. A., Demko, T. M. & Liu, Y. 1994. Coal compositional changes along a mire interior to mire margin transect in the Mary Lee coal bed, Warrior Basin, Alabama, USA. *International Journal of Coal Geology*, **26**, 43–62.

Eble, C. F., Grady, W. C. & Pierce, B. S. 2006. Compositional characteristics and inferred origln of three Late

Pennsylvanian coal beds from the northern Appalachian Basin. *In*: Greb, S. F. & DiMichele, W. A. (eds) *Wetlands Through Time*. Geological Society of America Special Papers, **399**, 191–222.

Elsik, W. C. & Dilcher, D. L. 1974. Palynology and age of clays exposed in Lawrence Clay Pit, Henry County, Tennessee. *Palaeontographica*, **146B**, 65–87.

Eros, J. M., Montañez, I. P., Osleger, D., Davydov, V., Nemyrovska, T., Poletaev, V. & Zhykalyak, M. 2011. Upper Carboniferous sequence stratigraphy and relative sea level history, Donets Basin, Ukraine. *Palaeogeography, Palaeoclimatology, Palaeoecology*, **313–314**, 1–25.

Esterle, J. S. & Ferm, J. C. 1986. Relationship between petrographic and chemical properties and coal seam geometry, Hance seam, Breathitt Fm, southeastern Kentucky. *International Journal of Coal Geology*, **6**, 199–214.

Falcon-Lang, H. J. 2004. Pennsylvanian tropical rain forests responded to glacial–interglacial rhythms. *Geology*, **32**, 689–692.

Falcon-Lang, H. J. & DiMichele, W. A. 2010. What happened to the coal forests during Pennsylvanian glacial phases? *Palaios*, **25**, 611–617.

Falcon-Lang, H. J., Benton, M. J., Braddy, S. J. & Davies, S. J. 2006. The Pennsylvanian tropical biome reconstructed from the Joggins Fm of Nova Scotia, Canada. *Journal of the Geological Society, London*, **163**, 561–576.

Falcon-Lang, H. J., Nelson, W. J., Elrick, S. D., Looy, C. V., Ames, P. R. & DiMichele, W. A. 2009. Incised channel-fills containing conifers imply that seasonally-dry vegetation dominated Pennsylvanian tropical lowlands. *Geology*, **37**, 923–926.

Fall, L. M. & Olszewski, T. D. 2010. Environmental disruptions influence taxonomic composition of brachiopod paleocommunities in the Middle Permian Bell Canyon Fm (Delaware Basin, West Texas). *Palaios*, **25**, 247–259.

Feldman, H. R., Franseen, E. K., Joeckel, R. M. & Heckel, P. H. 2005. Impact of longer-term modest climate shifts on architecture of high-frequency sequences (cyclothems), Pennsylvanian of midcontinent U.S.A. *Journal of Sedimentary Research*, **75**, 350–368.

Ferm, J. C. & Cavaroc, V. V., Jr. 1968. A nonmarine sedimentary model for the Allegheny rocks of West Virginia. *In*: Klein, G. D. (ed.) *Late Paleozoic and Mesozoic Continental Sedimentation, Northeastern North America*. Geological Society of America Special Papers, **106**, 1–19.

Ferm, J. C. & Weisenfluh, G. A. 1989. Evolution of some depositional models in late Carboniferous rocks of the Appalachian coal fields. *International Journal of Coal Geology*, **12**, 259–292.

Fielding, C. R., Frank, T. D. & Isbell, J. L. 2008. The late Paleozoic ice age – a review of current understanding and synthesis of global climate patterns. *In*: Fielding, C. R., Frank, T. D. & Isbell, J. L. (eds) *Resolving the Late Paleozoic Ice Age in Time and Space*. Geological Society of America Special Papers, **441**, 343–354.

Flint, S., Aitken, J. & Hampson, G. 1995. Application of sequence stratigraphy to coal-bearing coastal plain successions: implications for the UK Coal Measures. *In*: Whateley, M. K. G. & Spears, D. A. (eds) *European Coal Geology*. Geological Society, London, Special Publications, **82**, 1–16.

Gastaldo, R. A. 2010. Peat or no peat: why do the Rajang and Mahakam deltas differ? *International Journal of Coal Geology*, **83**, 162–172.

Gastaldo, R. A., Demko, T. M. & Liu, Y. 1993. Application of sequence and genetic stratigraphic concepts to Carboniferous coal-bearing strata: an example from the Black Warrior basin, USA. *Geologische Rundschau*, **82**, 212–226.

Gastaldo, R. A., Stevanovic-Walls, I., Ware, W. N. & Greb, S. F. 2004. Community heterogeneity of Early Pennsylvanian peat mires. *Geology*, **32**, 693–696.

Gastaldo, R. A., Purkyňová, E. & Šimůnek, Z. 2009. Megafloral perturbation across the Enna Marine Zone in the Upper Silesian Basin attests to Late Mississippian (Serpukhovian) deglaciation and climate change. *Palaios*, **24**, 351–366.

Greb, S. F., DiMichele, W. A. & Gastaldo, R. A. 2006. Evolution and importance of wetlands in Earth history. *In*: Greb, S. F. & DiMichele, W. A. (eds) *Wetlands Through Time*. Geological Society of America Special Papers, **399**, 1–40.

Hammer, Ø., Harper, D. A. T. & Ryan, P. D. 2001. PAST: paleontological statistics software package for education and data analysis. *Palaeontologia Electronica*, **4**, 1–9.

Heckel, P. H. 1986. Sea-level curve for Pennsylvanian eustatic marine transgressive-regressive depositional cycles along mid-continent outcrop belt, North America. *Geology*, **14**, 330–334.

Heckel, P. H. 1995. Glacial-eustatic base-level – climatic model for late Middle to Late Pennsylvanian coal-bed formation in the Appalachian Basin. *Journal of Sedimentary Research*, **B65**, 348–356.

Heckel, P. H. 2008. Pennsylvanian cyclothems in Midcontinent North America as far-field effects of waxing and waning of Gondwana ice sheets. *In*: Fielding, C. R., Frank, T. D. & Isbell, J. L. (eds) *Resolving the Late Paleozoic Ice Age in Time and Space*. Geological Society of America Special Papers, **441**, 275–289.

Hentz, T. F. 1988. *Lithostratigraphy and paleoenvironments of Upper Paleozoic Continental Red Beds, North-Central Texas–Bowie (new) and Wichita (revised) Groups*. Bureau of Economic Geology Report of Investigations, Austin, TX, Report 170.

Hernandez-Castillo, G. R., Stockey, R. A., Rothwell, G. W. & Mapes, G. 2009. Reconstructing *Emporia lockardii* (Voltziales: Emporiaceae) and initial thoughts on Paleozoic conifer ecology. *International Journal of Plant Sciences*, **170**, 1056–1074.

Hill, M. O. & Gauch, H. G. Jr. 1980. Detrended correspondence analysis: an improved ordination technique. *Vegetatio*, **42**, 47–58.

Holland, S. M. & Zaffos, A. 2011. Niche conservatism along an onshore–offshore gradient. *Paleobiology*, **37**, 270–286.

Isbell, J. L., Miller, M. F., Wolfe, K. L. & Lenaker, P. A. 2003. Timing of late Paleozoic glaciation in Gondwana: was glaciation responsible for the development of northern hemisphere cyclothems?. *In*: Chan,

M. A. & Archer, A. W. (eds) *Extreme Depositional Environments: Mega End Members in Geologic Time*. Geological Society of America Special Papers, **370**, 5–24.

Ivany, L. C., Brett, C. E., Wall, H. L. B., Wall, P. D. & Handley, J. C. 2009. Relative taxonomic and ecologic stability in Devonian marine faunas of New York State: a test of coordinated stasis. *Paleobiology*, **35**, 499–524.

Jerrett, R. M., Hodgson, D. M., Flint, S. S. & Davies, R. C. 2011. Control of relative sea level and climate on coal character in the Westphalian C (Atokan) Four Corners Fm, Central Appalachian Basin, U.S.A. *Journal of Sedimentary Research*, **81**, 420–445.

Jones, J. O. 1971. *The Blaine Fm of North Texas*. Unpublished PhD dissertation, University of Iowa, Iowa City.

Jones, J. O. & Hentz, T. R. 1988. Permian strata of North-Central Texas. *In*: *Geological Society of America Centennial Field Guide – South-Central Section*. Geological Society of America, Boulder, Colorado, 309–316.

Knight, J. 1985. The stratigraphy of the Stephanian rocks of the Sabero coalfield, León (NW Spain) and an investigation of the fossil flora. *Palaeontographica*, **197B**, 1–80.

Lamboy, W. & Lesnikowska, A. 1988. Some statistical methods useful in the analysis of plant paleoecological data. *Palaios*, **3**, 86–94.

Libertín, M., Dašková, J., Opluštil, S., Bek, J. & Edress, N. 2009. A palaeoecological model for a vegetated early Westphalian intramontane valley (Intra-Sudetic Basin, Czech Republic). *Review of Palaeobotany and Palynology*, **155**, 175–203.

Looy, C. V. 2007. Extending the range of derived late Paleozoic conifers: *Lebowskia* gen. nov. (Majonicaceae). *International Journal of Plant Sciences*, **168**, 957–972.

Mamay, S. H. 1992. Sphenopteridium and Telangiopsis in a Diplopteridium-like association from the Virgilian (Upper Pennsylvanian) of New Mexico. *American Journal of Botany*, **79**, 1092–1101.

McCune, B. & Grace, J. B. 2002. *Analysis of Ecological Communities*. MJM Software Design, Gleneden Beach, OR.

Montañez, I. P., Tabor, N. J. et al. 2007. CO_2-forced climate and vegetation instability during late Paleozoic deglaciation. *Science*, **315**, 87–91.

Nichols, G. 2009. *Sedimentology and Stratigraphy*, 2nd ed. Wiley-Blackwell, London.

Nordt, L., Orosz, M., Driese, S. & Tubbs, J. 2006. Vertisol carbonate properties in relation to mean annual precipitation: Implications for paleoprecipitation estimates. *The Journal of Geology*, **114**, 501–510.

Olszewski, T. D. & Erwin, D. H. 2009. Change and stability in the Permian brachiopod communities of western Texas. *Palaios*, **24**, 27–40.

Olszewski, T. D. & Patzkowsky, M. E. 2003. From cyclothems to sequences: the record of eustasy and climate on an icehouse epeiric platform (Pennsylvanian–Permian, North American Midcontinent). *Journal of Sedimentary Research*, **73**, 15–30.

Opluštil, S., Pšenička, J., Libertín, M. & Šimůnek, Z. 2007. Vegetation patterns of Westphalian and lower Stephanian mire assemblages preserved in tuff beds of the continental basins of Czech Republic. *Review of Palaeobotany and Palynology*, **143**, 107–154.

Opluštil, S., Pšenička, J., Libertín, M., Bashforth, A. R., Šimůnek, Z. & Daškov, J. 2009*a*. A Middle Pennsylvanian (Bolsovian) peat-forming forest preserved *in situ* in volcanic ash of the Whetstone Horizon in the Radnice Basin, Czech Republic. *Review of Palaeobotany and Palynology*, **155**, 234–274.

Opluštil, S., Pšenička, J., Libertín, M., Bek, J., Daškov, J. & Šimůnek, Z. 2009*b*. Composition and structure of an *in situ* Middle Pennsylvanian peat-forming plant assemblage buried in volcanic ash, Radnice Basin (Czech Republic). *Palaios*, **24**, 726–746.

Oxanen, J. & Minchin, P. R. 1997. Instability of ordination results under changes in input data order: explanations and remedies. *Journal of Vegetation Science*, **8**, 447–454.

Peyser, C. E. & Poulsen, C. J. 2008. Controls on Permo-Carboniferous precipitation over tropical Pangaea: a GCM sensitivity study. *In*: Soreghan, G. S. & Montañez, I. P. (eds) *Special Issue on the Late Paleozoic Earth System. Palaeogeography, Palaeoclimatology, Palaeoecology*, **268**, 181–192.

Pfefferkorn, H. W. & Thomson, M. 1982. Changes in dominance patterns in Upper Carboniferous plant-fossil assemblages. *Geology*, **10**, 641–644.

Pfefferkorn, H. W., Mustafa, H. & Hass, H. 1975. Quantitative Charakterisierung ober–karboner Abdruckfloren (quantitative characterization of compression–impression floras of Upper Carboniferous age). *Neues Jahrbuch Geologie Palaeontologie, Abhandlungen*, **150**, 253–269.

Phillips, T. L. & Peppers, R. A. 1984. Changing patterns of Pennsylvanian coal-swamp vegetation and implications of climatic control on coal occurrence. *International Journal of Coal Geology*, **3**, 205–255.

Potter, F. W. Jr. 1976. Investigations of angiosperms from the Eocene of southeastern North America: pollen assemblages from Miller Pit, Henry County, Tennessee. *Palaeontographica*, **157B**, 44–96.

Potter, F. W. Jr. & Dilcher, D. L. 1980. Biostratigraphic analysis of Eocene clay deposits in Henry County, Tennessee. *In*: Dilcher, D. L. & Taylor, T. N. (eds) *Biostratigraphy of Fossil Plants*. Dowden, Hutchinson, and Ross, Stroudsburg, PA, 211–225.

Poulsen, C. J., Pollard, D., Montañez, I. P. & Rowley, D. 2007. Late Paleozoic tropical climate response to Gondwanan deglaciation. *Geology*, **35**, 771–774.

Pryor, J. S. & Gastaldo, R. A. 2000. Paleoecological analysis of two Early Pennsylvanian mineral-substrate wetlands. *Palaios*, **15**, 3–13.

Pšenička, J., Bek, J., Cleal, C. J., Wittry, J. & Zodrow, E. L. 2009. Description of synangia and spores of the holotype of the Carboniferous fern *Lobatopteris miltoni*, with taxonomic comments. *Review of Palaeobotany and Palynology*, **155**, 133–144.

R DEVELOPMENT CORE TEAM 2010. *R: A Language and Environment for Statistical Computing*. R Foundation for Statistical Computing, Vienna, Austria, http://www.R-project.org

Romanchock, C. M., DiMichele, W. A., Chaney, D. & Hook, R. W. in press. A latest Pennsylvanian fossil flora from the Markley Fm, north-central Texas. *Review of Paleobotany and Palynology*.

ROTHWELL, G. W., MAPES, G. & MAPES, R. H. 1997. Late Paleozoic conifers of North America: structure, diversity and occurrences. *Review of Palaeobotany and Palynology*, **95**, 95–113.

SCHEIHING, M. H. 1980. Reduction of wind velocity by the forest canopy and the rarity of non-arborescent plants in the Upper Carboniferous fossil record. *Argumenta Palaeobotanica*, **6**, 133–138.

SCHEIHING, M. H. & PFEFFERKORN, H. W. 1984. The taphonomy of land plants in the Orinoco Delta: a model for the incorporation of plant parts in clastic sediments of Late Carboniferous age of Euramerica. *Review of Palaeobotany and Palynology*, **41**, 205–240.

SCOTESE, C. R. 1999. *PALEOMAP Animations 'Paleogeography'*. PALEOMAP Project, Department of Geology, University of Texas, Arlington, TX.

SCOTT, A. C. 1978. Sedimentological and ecological control of Westphalian B plant assemblages from West Yorkshire. *Proceedings of the Yorkshire Geological Society*, **41**, 461–508.

SECORD, R., WING, S. L. & CHEW, A. 2008. Stable isotopes in early Eocene mammals as indicators of forest canopy structure and resource partitioning. *Paleobiology*, **34**, 282–300.

ŠIMŮNEK, Z. & MARTÍNEK, K. 2009. A study of Late Carboniferous and Early Permian plant assemblages from the Boskovice Basin, Czech Republic. *Review of Palaeobotany and Palynology*, **155**, 275–307.

TABOR, N. J. & MONTAÑEZ, I. P. 2004. Morphology and distribution of fossil soils in the Permo-Pennsylvanian Wichita and Bowie Groups, north-central Texas, USA: implications for western equatorial Pangean palaeoclimate during icehouse–greenhouse transition. *Sedimentology*, **51**, 851–884.

TABOR, N. J. & POULSEN, C. J. 2008. Palaeoclimate across the Late Pennsylvanian–Early Permian tropical palaeolatitudes: a review of climate indicators, their distribution, and relation to palaeophysiographic climate factors. *In*: SOREGHAN, G. S. & MONTAÑEZ, I. P. (eds) *Special Issue on the late Paleozoic Earth System. Palaeogeography, Palaeoclimatology, Palaeoecology*, **268**, 293–310.

TABOR, N. J., MONTAÑEZ, I. P., SCOTESE, C. R., POULSEN, C. J. & MACK, G. H. 2008. Paleosol archives of environmental and climatic history in paleotropical western Pangea during the latest Pennsylvanian through Early Permian. *In*: FIELDING, C. R., FRANK, T. D. & ISBELL, J. L. (eds) *Resolving the Late Paleozoic Ice Age in Time and Space*. Geological Society of America Special Papers, **441**, 291–303.

TABOR, N. J., SMITH, R. M. H., STEYER, J. S., SIDOR, C. A. & POULSEN, C. J. 2011. The Permian Moradi Fm of northern Niger: paleosol morphology, petrography and mineralogy. *Palaeogeography, Palaeoclimatology, Palaeoecology*, **299**, 200–213.

TER BRAAK, C. J. F. & ŠMILAUER, P. 2002. *CANOCO Reference manual and CanoDraw for Windows User's Guide: Software for Canonical Community Ordination (version 4.5)*. Microcomputer Power, Ithaca.

WAGNER, R. H. 1983. Upper Stephanian stratigraphy and palaeontology of the Puertollano Basin, Ciudad Real, Spain. *Anais da Faculdade de Ciências do Universidade Porto*, **64**, 171–231.

WAGNER, R. H. & ÁLVAREZ-VÁZQUEZ, C. 2010. The Carboniferous floras of the Iberian Peninsula: a synthesis with geological connotations. *Review of Palaeobotany and Palynology*, **162**, 239–324.

WAGNER, R. H. & CASTRO, M. P. 2010. Compositional changes in a mid-Stephanian (Kasimovian) flora in relation to alluvial plain deposits derived from westward-receding mountains and bordered by the paleotethys: La Magdalena Coalfield, northwestern Spain. *Palaios*, **26**, 33–54.

WALKER, R. G. & JAMES, N. P. 1992. *Facies Models: Response to Sea Level Changes*. Geological Association of Canada, St. John's.

WARDLAW, B. R. 2005. Age assignment of the Pennsylvanian-Early Permian succession of North Central Texas. *Permophiles*, **46**, 21–22.

WEIHER, E., FREUND, D., BUNTON, T., STEFANSKI, A., LEE, T. & BENTIVENGA, S. 2011. Advances, challenges and a developing synthesis of ecology community assembly theory. *Philosophical Transactions of the Royal Society of London B*, **366**, 2403–2413.

WING, S. L. & DIMICHELE, W. A. 1995. Conflict between local and global changes in plant diversity through geological time. *Palaios*, **10**, 551–564.

WING, S. L., HARRINGTON, G. J., SMITH, F. A., BLOCH, J. I., BOYER, D. M. & FREEMAN, K. H. 2005. Transient floral change and rapid global warming at the Paleocene–Eocene boundary. *Science*, **310**, 993–996.

YOUNG, F. W. 1985. Multidimensional Scaling. *In*: KOTZ, S. & JOHNSON, N. L. (eds) *Encyclopedia of Statistical Sciences*. Wiley, New York, **5**, 649–659.

ZIEGLER, A. M., RAYMOND, A., GIERLOWSKI, T. C., HORRELL, M. A., ROWLEY, D. B. & LOTTES, A. L. 1987. Coal, climate and terrestrial productivity: the present and Early Cretaceous compared. *In*: SCOTT, A. C. (ed.) *Coal and Coal-Bearing Strata: Recent Advances*. Geological Society, London, Special Publications, **32**, 25–49.

Pennsylvanian–Early Permian cyclothemic succession on the Yangtze Carbonate Platform, South China

KATSUMI UENO[1]*, NAOKI HAYAKAWA[1], TSUTOMU NAKAZAWA[2], YUE WANG[3] & XIANGDONG WANG[3]

[1]*Department of Earth System Science, Fukuoka University, Fukuoka 814-0180, Japan*

[2]*Geological Survey of Japan, National Institute of Advanced Industrial Science and Technology (AIST), Tsukuba 305-8567, Japan*

[3]*State Key Laboratory of Palaeobiology and Stratigraphy, Nanjing Institute of Geology and Palaeontology, Chinese Academy of Sciences, Nanjing 210008, P. R. China*

**Corresponding author (e-mail: katsumi@fukuoka-u.ac.jp)*

Abstract: Pennsylvanian (Late Carboniferous)–Early Permian cyclothems are documented for the first time from the shallow-marine carbonate succession on the Yangtze Carbonate Platform, South China. The Zongdi section in southern Guizhou Province comprises bioclast-rich grey limestones with dark-coloured intervals yielding macro- and microfabrics related to emergent events. We recognized 26 depositional sequences (consisting essentially of basal peritidal sediments, middle shallow subtidal sediments and upper peritidal sediments affected by meteoric diagenesis and subaerial exposure) separated by 25 lowstand events. These are interpreted as having been formed by alternating marine transgression and regression resulting from variations in the Gondwanan ice sheets. They are compared with the Midcontinent (Kansas-type) cyclothems in North America. However, the Zongdi cyclothems are relatively thin and consist entirely of very shallow-marine sediments formed above wave base, whereas the Midcontinent cyclothems reflect deposition far below wave base. The Zongdi depositional environment was an upper shelf area, on a platform within the epeiric sea on the South China Craton, where sedimentation occurred in a relatively small (shallow) accommodation space during the highest stands of sea level. They were formed by major (large-scale) transgressions that submerged high-shelf areas, and most of them correspond to the major cyclothems of the North American Midcontinent Basin. The fusuline biostratigraphical framework reveals an uneven stratigraphical pattern of the Zongdi cyclothems, reflecting subtle fluctuations of tectonic subsidence on the epeiric Yangtze Carbonate Platform. Less erosion and karstification on the tops of cyclothems in the Zongdi section, together with the presence of partly calcareous palaeosols, indicates that a dry climate prevailed on the South China Block during the Pennsylvanian–Early Permian when the Zongdi cyclothems were formed.

During the Pennsylvanian (Late Carboniferous) and Cisuralian (Early Permian), a large mass of continental ice sheets covered wide parts of Gondwanaland at higher latitudes of the then southern hemisphere (e.g. Veevers & Powell 1987; Blakey 2008). The waxing and waning of the Gondwanan glaciation led to a remarkable icehouse-type of high-frequency/high-amplitude glacio-eustatic sea-level change during the late Palaeozoic (e.g. Ross & Ross 1987, 1988; Haq & Schutter 2008; Heckel 2008; Fielding *et al.* 2008*b*), which can essentially be compared to that observed throughout the Quaternary. The late Palaeozoic icehouse sea-level changes resulted in the formation of typical cyclic sedimentary successions of alternating marine and continental sedimentation, generally referred to as cyclothems (a term coined by Wanless & Weller 1932), in major Pennsylvanian–Cisuralian sedimentary basins of the world (Heckel 2002) such as the North American Midcontinent Basin (Heckel 1986, 1995, 1999, 2008), Moscow Syneclise of Russia (Alekseev *et al.* 1996; Kabanov 2003; Kabanov *et al.* 2006; Kabanov & Baranova 2007), Donets Basin of Ukraine (Izart *et al.* 1996, 2002; Eros *et al.* 2012) and British coal measure basins (Calver 1968; Ramsbottom 1977, 1979). Recent progress on sequence-stratigraphical analyses of cyclothems as well as conodont and fusuline biostratigraphy in Pennsylvanian successions makes it possible to examine not only the intra-continental but also inter-continental correlation of each depositional cycle based on frequency and magnitude of cyclothems and high-resolution biostratigraphical framework. This idea was recently synthesized as a cyclothem-based 'digital' correlation, which resulted in a better, more reliable, correlation of

From: GĄSIEWICZ, A. & SŁOWAKIEWICZ, M. (eds) 2013. *Palaeozoic Climate Cycles: Their Evolutionary and Sedimentological Impact*. Geological Society, London, Special Publications, **376**, 235–267.
First published online November 29, 2012, http://dx.doi.org/10.1144/SP376.5

Pennsylvanian successions among the North American Midcontinent Basin, Moscow Basin and Donets Basin (Heckel *et al.* 2007).

In the South China (Yangtze) Block, a large carbonate platform had developed during the Late Palaeozoic–Early Mesozoic between the Upper Yangtze Land in the NW and the Cathaysia Land in the east. This carbonate platform is referred to as the Yangtze Carbonate Platform (or often simply the Yangtze Platform) and was associated with some intra-continental deeper basins collectively called the Dian-Qian-Gui Basin in the southwestern part of the platform during the Carboniferous–Permian (Feng *et al.* 1994, 1998; Wang & Jin 2000) (Fig. 1). Shallow-marine sediments that must have been strongly controlled by coeval icehouse-type eustatic sea-level change were dominant on that platform. Although intensive regional biostratigraphical research has been performed on fusulines and conodonts in this area over the past decades (e.g. Sheng 1963; Xiao *et al.* 1986; Wang & Higgins 1989; Lin *et al.* 1990; Wang 1993; Zhang *et al.* 2010), a sequence-stratigraphical analysis of these strata has not been carried out until now, except for some studies dealing only with very broad outlines of sea-level change of the relevant ages in this area (Chen *et al.* 1998). To say that cyclothems have not been reported from this region in the readily available literature (Heckel 2002) is not an overstatement.

We recently investigated shallow-marine Pennsylvanian–Early Permian strata in southern Guizhou Province of South China and found for the first time on the Yangtze Carbonate Platform a carbonate succession in which clear sedimentary cycles can be recognized, based mainly on cyclically intervening subaerial exposure surfaces (disconformities). They can be compared to cyclothems described in other coeval major sedimentary basins of the world. In this paper, we first describe the subaerial exposure surfaces and their microfabrics, which are the clues to defining sequence-bounding unconformity events, and then reconstruct depositional sequences and sedimentary cycles in the studied Zongdi section. Based on comparison with cyclothems in the North American Midcontinent Basin, we propose a model for the depositional processes that controlled these sedimentary cycles. Moreover, we discuss three other implications in detecting epeiric tectonic subsidence from the Pennsylvanian Zongdi succession, explaining differences in sedimentary cycles between the Pennsylvanian and Early Permian and establishing the palaeoclimate of the South China Block at that time.

Geological setting and the Zongdi section

The study section is located at Zongdi, *c.* 30 km SE of Ziyun in southern Guizhou Province, South China. In this area, mildly faulted and folded Devonian–Triassic carbonates are widely distributed on the Yangtze Carbonate Platform (Bureau of Geology

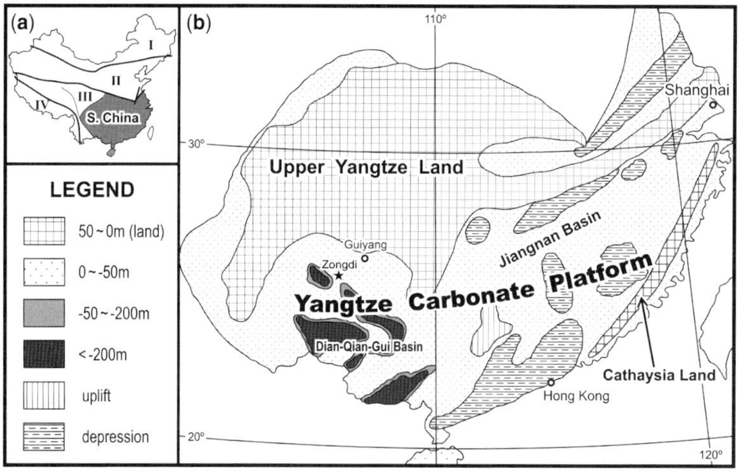

Fig. 1. Palaeogeography of South China during Pennsylvanian (Late Carboniferous)–Cisuralian (Early Permian) (modified from Buggisch *et al.* 2011). (**a**) Four stratigraphical provinces of China (after Wang & Jin 2003) and location of South China. I: Junggar–Hinggan Province; II: North China–Tarim Province; III: South China–Qiangtang Province; and IV: Himalayan Province. (**b**) Palaeogeographical reconstruction map of South China showing distribution of water depth during highstand across Yangtze Carbonate Platform (modified from Wang & Jin 2000). Zongdi section (star mark) studied in this article is located in western part of Yangtze Carbonate Platform.

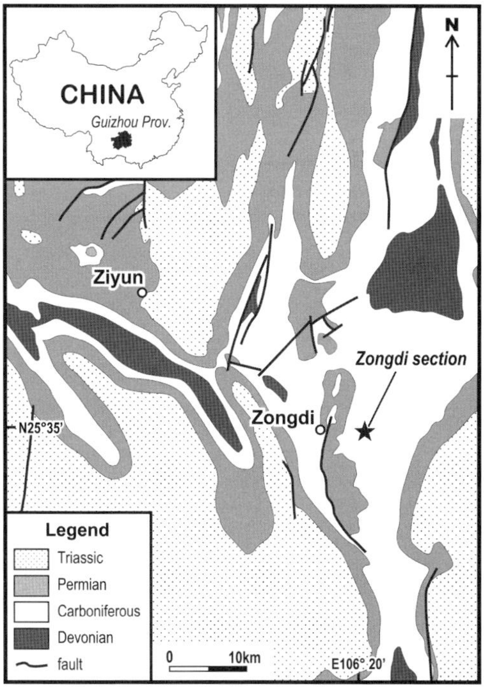

Fig. 2. Geological map of Ziyun area, showing location of Zongdi section (star) in southern Guizhou Province, South China. Geological map after Bureau of Geology and Mineral Resources of Guizhou Province (1987).

and Mineral Resources of Guizhou Province 1987; Fig. 2). Among these strata, Pennsylvanian–Cisuralian (Weiningian–Chuanshanian in regional chronostratigraphical subdivisions of China; Wang & Jin 2003) shallow-marine platform sediments are dominant in the Ziyun–Zongdi area, which abruptly change facies southwards into platform slope to deeper basinal sediments in an intra-platform basin (the Luodian Basin of Feng *et al.* 1998) of the greater Dian-Qian-Gui Basin (Fig. 1).

The Zongdi section was measured along a county road connecting Ziyun to Luodian, with coordinates N25°34′43″ and E106°21′06″ at the beginning of the study section. It provides almost continuous roadside cuts and is about 360 m thick in the measured part, ranging from the Bashkirian (Early Pennsylvanian) up to the Asselian (earliest Permian) (Fig. 3) although the strata continue further in both directions. The section is composed essentially of bioclast-dominated grey shallow-marine carbonates with frequent development of dolomitic limestone and dolostone in the lower *c.* 100 m, while the overlying part consists entirely of pure limestone. The strata generally strike north–south to NE–SW, and dip relatively gently (20–30°) to the W to NW. In this study, we focused on the upper *c.* 150-m-thick succession, which corresponds to 210–360 m interval of the measured section (Figs 3 & 4). This part as a whole exhibits very shallow-marine sedimentation above wave base with regularly intervening levels of disconformity with related macroscopic sedimentary fabrics readily recognized in the field (Fig. 3). Figure 5 shows the geological log of the study section.

Subaerial exposure and its microfabrics

One of the highlights on the study of the Zongdi section is to show cyclically appearing subaerial exposure surfaces that provide the basis for reconstructing high-frequency sedimentary cycles that are recognized for the first time in the Pennsylvanian–Early Permian succession on the South China Craton (Fig. 5). We first describe in detail their sedimentary features. These exposure surfaces can be defined by a number of macroscopic ‘palaeokarstic’ features in the field such as palaeosols, karst breccias and organically coloured (‘polluted’) dark-grey–black limestone horizons, as well as microscopic fabrics related to emergent events including rhizoliths, *Microcodium*, alveolar texture, pendant (gravitational) cement, grain leaching and others. They are conspicuous as intermittently intervening thin intervals within essentially homogeneous grey shallow-marine bioclastic limestones (Fig. 5). The sedimentological features of subaerial exposure surfaces provide good bases for recognizing major lowstand horizons (disconformities/sequence boundaries) and thus establishing depositional sequences. Together with the fusuline biostratigraphical framework, these features help to reconstruct high-frequency sedimentary cycles during the Pennsylvanian–Early Permian and interpret their origin in the Zongdi section.

As a result, we recognized a total of 35 subaerial exposure surfaces in the section, among which we defined 25 major lowstand horizons as sequence boundaries (SB1–SB25) separating 26 depositional sequences (DS1–DS26; Fig. 5). Below, we describe the diagnoses of macroscopic and microscopic fabrics related to subaerial exposure, which are the primary tools to discriminate the sequence boundaries of depositional sequences at Zongdi.

Subaerial exposure surfaces: recognition in the field

In the Zongdi section, a subaerial exposure surface can often be readily recognized on outcrop as a horizon showing a sharp colour change from dark-grey to black limestone below into grey limestone above (Fig. 6c–f; see also column showing colour

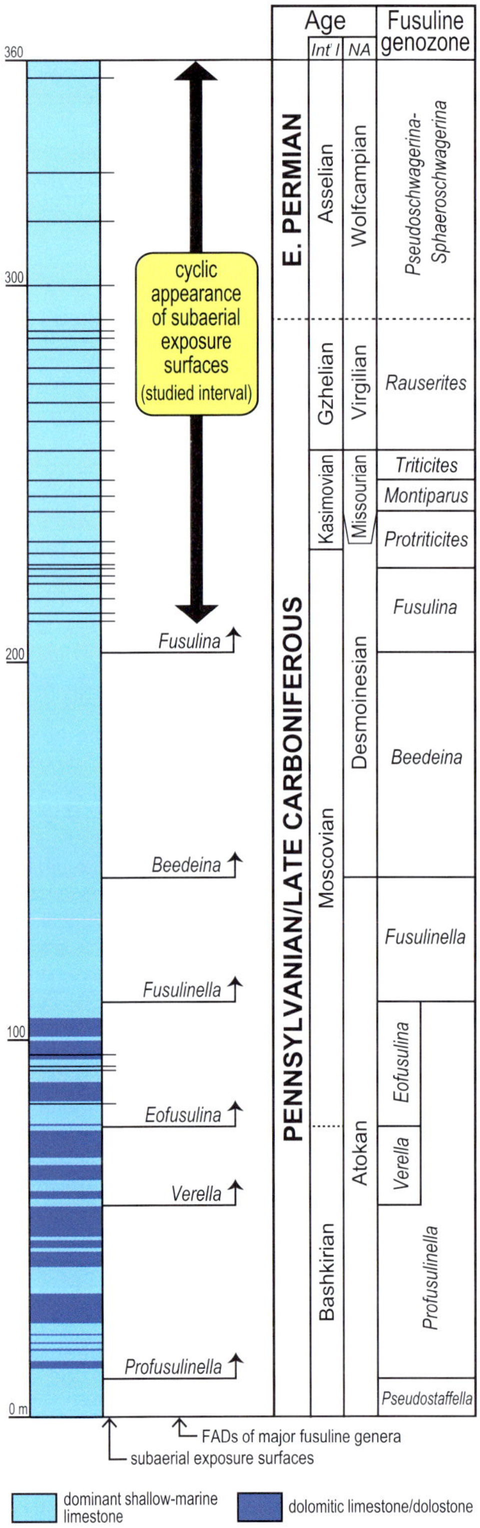

Fig. 3. Stratigraphical overview of Zongdi section. *Int'l*: International stages, *NA*: North American regional stages.

fabric on right side of lithic column in Fig. 5). The origin of the dark colour in these levels is still equivocal but strong possibilities include pigmentation by organic matter during emergence and/or the development of terrestrial cement that contains dark pedogenic materials (Fig. 7e). The former is likely similar to the formation process of black pebbles (Strasser 1984), and the latter would have resulted in giving the overall limestone a dark colour tone. Recently, a similar kind of dark-pigmented limestone just below unconformities (probably disconformities) was also documented in Middle Permian successions of Panthalassan seamount carbonates in SW Japan (Nakazawa & Ueno 2004; Hayakawa *et al.* 2008). Whichever process contributed to their formation, the dark-coloured limestones in the Zongdi section are very likely related to emergent events.

In some of these levels, other macroscopic fabrics related to subaerial exposure such as palaeosols (Fig. 6a–d, f) and karst breccia (Fig. 6f) are also associated with the dark-coloured intervals (Fig. 5). Palaeosols in the Zongdi section are yellowish–greyish brown argillaceous mudstones. They are semi-consolidated but subject to fragmentation by weathering, and somewhat calcareous in some parts. Karst breccias consist of pebble-sized limestone clasts derived from underlying beds and their distribution is irregular and patchy. Moreover, plant root casts (rhizoliths) (Figs 6c & 7b) are also observable in these dark-coloured limestone levels in the field. They are a good marker of emergent events because rhizoliths are related to terrestrial vegetation. In addition to these macroscopic examples, microscopic fabrics related to subaerial exposure are also recognized in the dark-coloured limestone intervals. Apart from coloured cements and rhizoliths mentioned above, they also include alveolar texture, *Microcodium*, pendant (gravitational) cement, crystal silts and leaching of grains by meteoric (freshwater) diagenesis.

In some works dealing with shallow-marine carbonates that underwent emergence, rhizoliths, alveolar texture and *Microcodium* are often utilized as diagnostic features in a palaeocaliche setting (e.g. Esteban & Klappa 1983; Flügel 2004). In the case of the Zongdi section, however, they are not found within actual palaeosol profiles, but are directly developed within host marine limestone substrates with marine bioclasts (mostly represented by oncoidal rudstone, fine-grained bioclastic grainstone and lime-mudstone) deposited in a peritidal environment just before subaerial exposure. They are

Fig. 4. Overview of studied 210–360 m interval of Zongdi section. (**a**) Panorama of section, viewed from SW. Strata gently dip to NW (to left in this photograph). (**b**) Abandoned small quarry, exposing *c.* 295–320 m stratigraphical interval. (**c**) Uppermost part of section, viewed from NW.

therefore better expressed as being developed by penetrating, infilling and/or replacing the host limestone in various ways. These important microfabrics observed in the Zongdi section are briefly described in the following subsection.

Rhizoliths

In the Zongdi section, rhizoliths (Figs 7a–d & 8b) can be observed in most of the dark-coloured limestone intervals just below subaerial exposure surfaces (Fig. 5). Particularly in lime-mudstones that have a characteristically monotonous occurrence on rock surface (Fig. 6c), rhizoliths are easily recognized in the field as darker, small, circular or irregular spots that are filled with sparry calcite (Fig. 7b). In thin sections, rhizoliths are observed as simple tubular cavities, often branching and/or anastomosing and containing alveolar texture (Fig. 7d). They sometimes coexist with *Microcodium* (Figs 7a & 8b). In some cases, they are represented by a tubular structure coated with micritic envelopes (Figs 7a & 8b).

Rhizoliths are root-related organosedimentary structures produced by accumulation and/or cementation within and around, or replacement of, higher plant roots by mineral matter (Klappa 1980; Flügel 2004). As pointed out by Esteban & Klappa (1983), rhizoliths are a reliable indicator of subaerial exposure since evidence of marine plant roots is rarely (if ever) preserved.

Alveolar texture

This microfabric (Figs 7a, d, e & 8a, c, d) is also common in most of the dark-coloured limestone intervals just below subaerial exposure surfaces in the Zongdi section, and is generally associated with rhizoliths (Fig. 5). It is a structure consisting of thin micritic septations that variously divide pores, forming an irregular web-like network in thin section.

The term alveolar texture was originally introduced by Esteban (1974) for describing cylindrical to irregular pores, which may be calcite-filled, separated by a network of anastomosing micrite walls. This microstructure is interpreted as resulting from rootlets penetrating sediments causing micritization of adjacent channel walls, and commonly occurs adjacent to rhizoliths (Bain & Foos 1993). Esteban & Klappa (1983) considered that alveolar texture is the product of coalesced millimetre-sized

Fig. 6. Outcrop photographs of Zongdi section, showing typical subaerial exposure surfaces (disconformities/sequence boundaries). (**a**) Thin palaeosol layer (yellow-brown portion) denotes SB10 (239.7 m level). Below boundary is lime-mudstone of probable mud-flat origin (a). Just above boundary, limestone consists of fine-grained bioclastic grainstone (b), overlain by well-sorted densely-packed bioclastic (crinoidal) grainstone with low-angle cross-stratification (c) (see also Fig. 13a). (**b**) SB14 at 263.7 m level with intervening thin palaeosol layer (P). (**c**) Thin palaeosol layer (P) at SB11 (243.8 m level). Just below palaeosol, 10-cm-thick grey bed (b) is composed of lime-mudstone of probable mud-flat origin and yields abundant rhizoliths (see also Fig. 7b). Further down, bed (a) is characteristically dark-coloured (organically 'polluted') while bed just above palaeosol at base of next depositional cycle (DS12), consisting of fine-grained bioclastic grainstone, is pale grey. (**d**) Uneven (patchy) palaeosol (P) at SB 17 (277.9 m level) and dark-coloured limestone (a) just below it. Dark-grey limestone is strongly affected by meteoric (freshwater) diagenesis with various kinds of microfabrics related to subaerial exposure (such as *Microcodium* and

rhizoliths. Our observations also support the strong genetic relation to rhizoliths.

Microcodium

In the Zongdi section, *Microcodium* (Figs 7a, 8b & 9a–c) is found not only in the dark-coloured limestone intervals just below subaerial exposure surfaces but also in the middle part of some depositional sequences (Fig. 5). Although it is essentially a microscopic structure, *Microcodium* is sometimes recognizable in the field as stain-like, irregular-shaped, slightly darker portions on rock surfaces. In thin section, it exhibits a characteristic tannish colour and is observed as a series of superposed small discs that have small central tubes and consist of a radial bundle of saccate chamberlets.

Microcodium was previously variously explained as being algal, inorganic, bacterial or actinomycete in origin, but was later accepted as being formed by the calcification of mycorrhizae, which are symbiotic associations between soil fungi and cortical cells of higher plant roots (Klappa 1978). Bain & Foos (1993) noted that this interpretation on the origin of *Microcodium* is supported by its mutual association with alveolar textures and rhizoliths in a Pleistocene example from the Bahamas. Subsequent to this long-standing view, Kabanov *et al.* (2008) recently proposed an alternative model for the formation of this microfabric. They concluded that, because non-recrystallized *Microcodium* grains preserve occasional hyphae-like morphology and hypha-like canals that imply its actinobacterial origin, the formation of *Microcodium* is likely due to a mycelial saprotrophic organism responsible for substrate corrosion and associated bacteria capable of consuming acidic metabolites and $CaCO_3$ re-precipitation into the *Microcodium* structures. Whatever its origin, *Microcodium* is a good indicator of subaerial exposure and meteoric diagenesis.

In the Zongdi section, *Microcodium* is also generally associated with other root-related microfabrics (rhizoliths and alveolar texture) (Figs 7a & 8b), providing support to the hypotheses mentioned above. As noted already, however, it is not always seen in strata just below subaerial exposure. In the example illustrated on Fig. 9c from sample H27 (245.7 m level in DS12), *Microcodium* is developed in strata about 2 m below the next higher recognized subaerial exposure (SE15). In contrast, rhizoliths are almost always found in the dark-coloured intervals just below subaerial exposure surfaces. These lines of evidence imply that *Microcodium* (calcification of mycorrhizae/corrosion of substrates by a mycelial organism and $CaCO_3$ re-precipitation by some bacteria) and rhizoliths (simple plant roots) can be formed at slightly different levels below the ground surface.

Pendant (gravitational) cement

This type of vadose cement (Figs 7e, f & 9d, e) hangs from the underside of grains, reflecting the direction of gravity at the time of cement formation. In the Zongdi section, pendant cement is common in the dark-coloured intervals just below subaerial exposure surfaces, but is also found in some cases at much lower levels below the upper-bounding subaerial exposure surface within a single depositional sequence (e.g. 336–337 m interval in DS25, which is *c.* 19 m below upper-bounding SB25; Fig. 5). The formation of pendant cement was primarily more dependent on the position of the water table (groundwater level) at that time and the preservation or development of interstitial pores; it is not necessarily associated with other root-related microfabrics, which are generally restricted to the dark-coloured levels just below subaerial exposure surfaces. Nevertheless, pendant cement is also common in the dark-coloured limestones. In thin section, it can be seen as semicircular- to crescent-shaped radial cement at the bottoms of grains and often shows a characteristic brown colour by contamination from pedogenic materials.

Pendant cement is sometimes associated with limpid isopachous circumgranular cement around grains of shallow-marine origin (Fig. 7f), which suggests meteoric phreatic cementation (Flügel 2004). In this example, it implies a complex diagenetic history of the limestone: shallow-marine sedimentation of bioclasts and oncoids was followed by emergence, during which the outer parts of those grains were slightly dissolved thus expanding interstitial spaces. They were then coated by isopachous cement in a meteoric phreatic environment, followed by the formation of pendant cement in a vadose environment.

Coloured cement

Cement of a translucent brown colour (Fig. 7e) is often observed in bioclastic limestone within the

Fig. 6. (*Continued*) rhizoliths). (**e**) Sharp colour change from dark-toned limestone (a) upwards to pale-grey normal-coloured limestone (b) at SB22 (300.0 m level). (**f**) Locally developed karst breccia (K) and palaeosol (P) at SB13 (256.4 m level). Note patchy distribution of karst breccias and palaeosol, and dark-coloured limestone below (lower right).

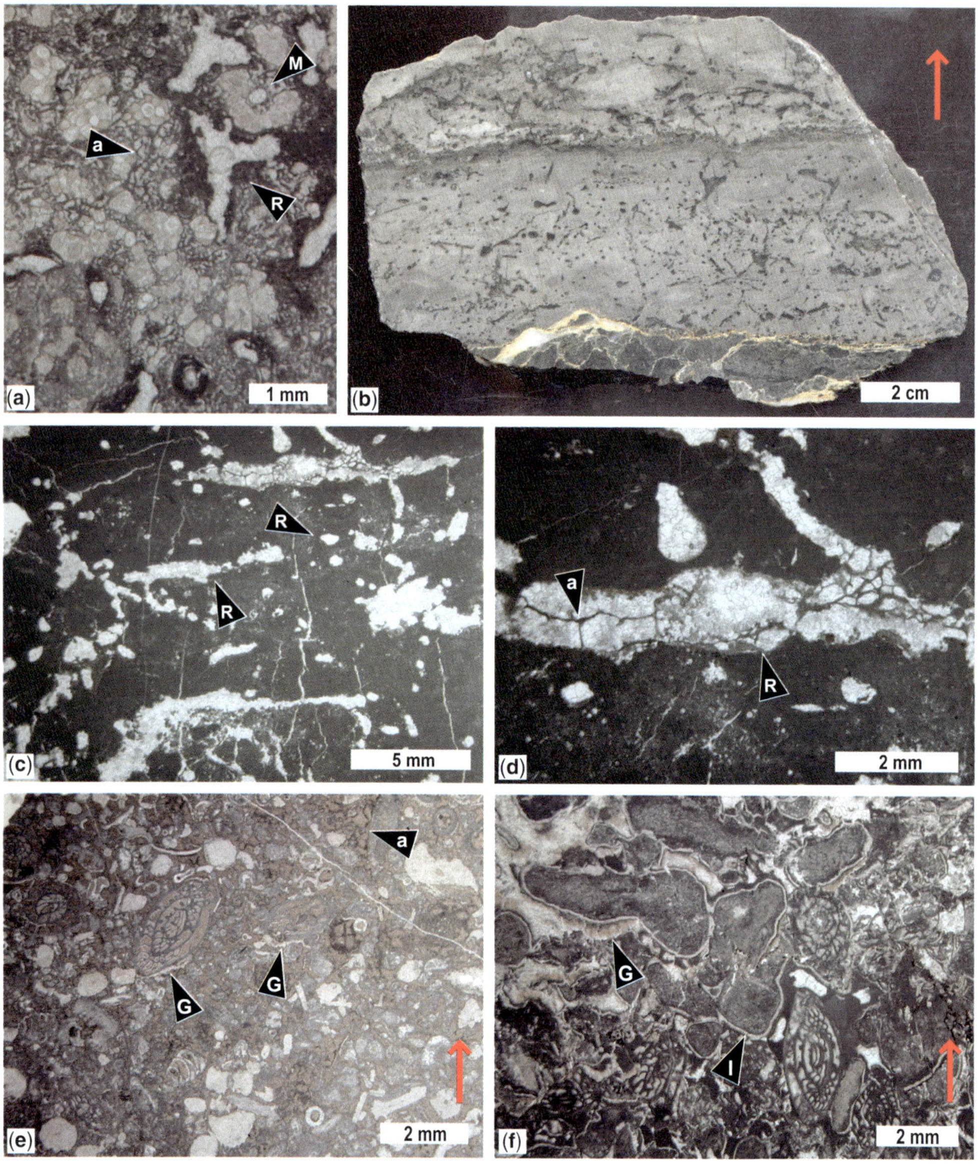

Fig. 7. Thin-section photomicrographs (**a, c–f**) and polished slab photograph (**b**) showing microfabrics related to subaerial exposure. Red arrow indicates upward direction of strata. (**a**) *Microcodium* (M), rhizolith (R) and alveolar texture (a) in sample H69 (274.6 m level in DS16). In this sample, original microfacies is hard to recognize because these microfabrics have entirely replaced the original sediments. (**b**) Polished slab of sample H20 (243.7 m level in DS11; see also Fig. 6c), with abundant rhizoliths (dark-coloured, circular to elongate spots) in weakly laminated lime-mudstone of probable mud-flat origin. (**c**) Thin-section photomicrograph of slab sample illustrated in (**b**), showing microscopic features of rhizoliths (R). (**d**) Enlarged part of (**c**). Alveolar texture (a) is shown as membrane-like partitions in rhizolith (R). (**e**) Bioclastic grainstone with brown-coloured calcite cement from sample H75 (277.9 m in DS17). This colour is due probably to pedogenic pigmentation. In places, alveolar texture (a) and pendant (gravitational) cement (G) are observed in interstitial spaces, suggesting that this limestone had emerged in a vadose environment probably just after deposition of bioclasts in a shallow-marine environment. Coloured cement was then precipitated in a meteoric phreatic zone. (**f**) Bioclastic-oncoidal grainstone from sample F128 (287.8 m level in DS20). Grains are surrounded by thin isopachous cement (I) formed in a meteoric phreatic environment. Pendant (gravitational) cement (G) with a pedogenic brown-colour then precipitated on underside of some grains in vadose zone.

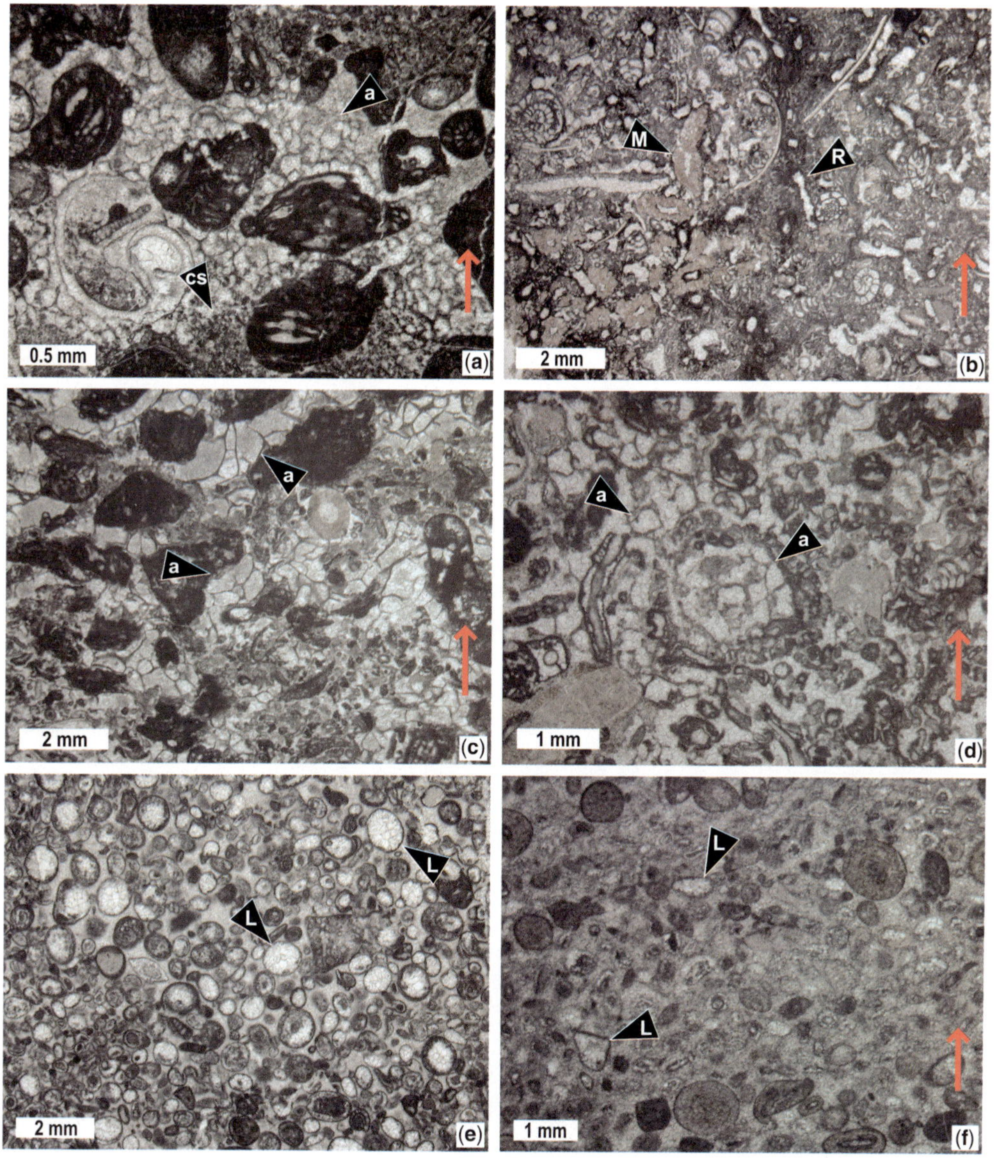

Fig. 8. Thin-section photomicrographs showing microfabrics related to subaerial exposure. Red arrow indicates upward direction of strata. (**a**) Poorly-preserved bioclastic grainstone from sample M14 (228.6 m level in DS8). Alveolar texture (a) is well developed in interstitial spaces of bioclasts, whose outer parts are more or less affected by dissolution. In the lower part, space is partly filled by crystal silts (CS). (**b**) *Microcodium* (M) and rhizoliths (R) developed in bioclastic grainstone from sample H19 (243.5 m level in DS 11). (**c**) Well-developed alveolar texture (a) within interstitial spaces of small-oncoidal rudstone from sample M24 (222.5 m level in DS5). (**d**) Bioclastic grainstone with well-developed alveolar texture (a) from sample H134 (354.7 m level in DS25). (**e**) Oolitic-bioclastic grainstone from sample F103 (253.0 m level in DS13) which has undergone leaching (particularly within ooids), producing many grain pores that were later infilled by coarse sparry calcite (L). (**f**) Ooid-bearing bioclastic grainstone from sample F90 (235.2 m level in DS10). Some grains of probable bioclasts (L) were affected by leaching followed by spar filling. This sample has also undergone slight recrystallization, making its microfacies vague.

dark-coloured intervals just below major subaerial exposure surfaces. It seemingly consists of fine-grained sparry calcite, although the calcite crystals are often somewhat obscure. In places within

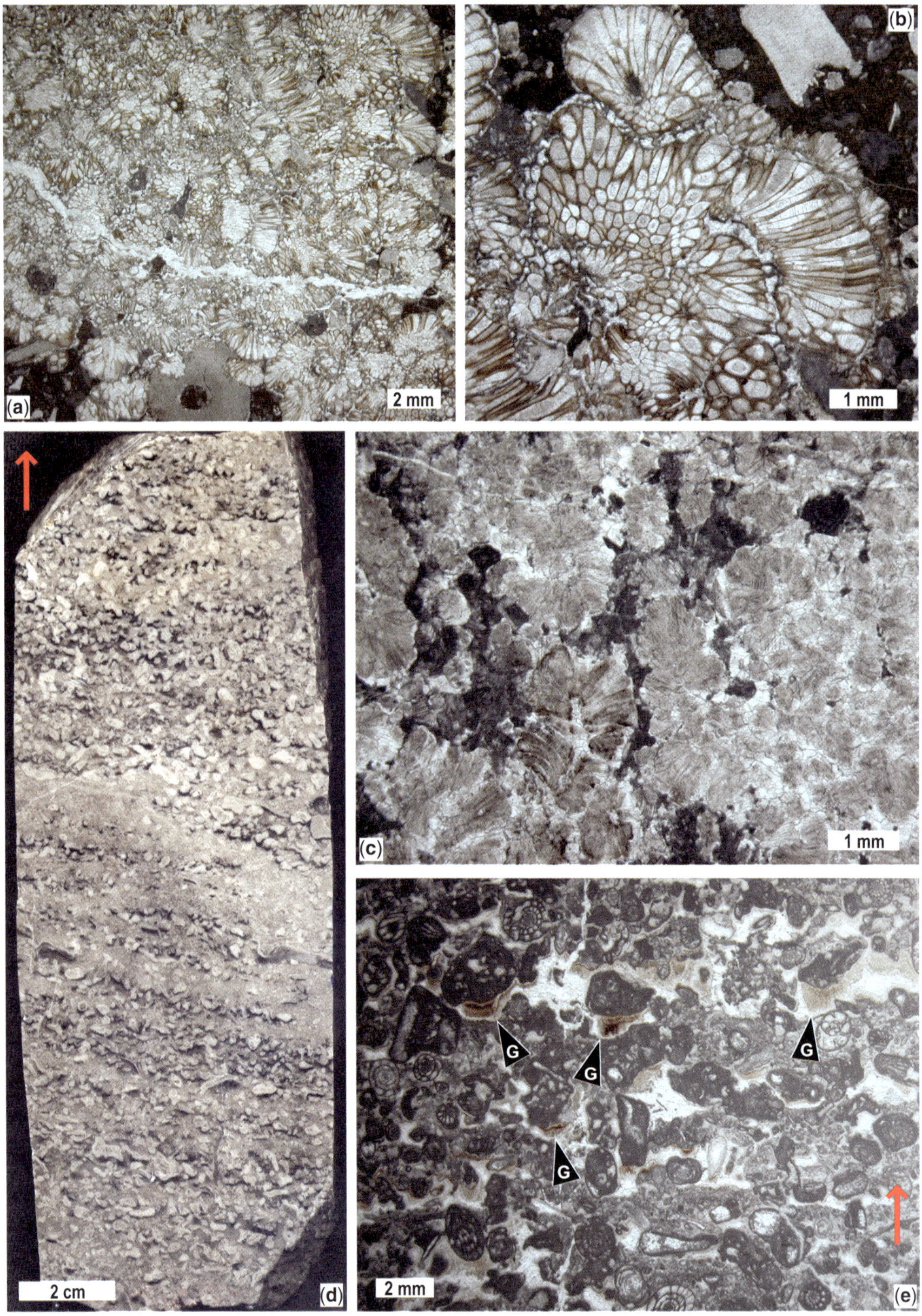

Fig. 9. Thin-section photomicrographs (**a–c, e**) and polished slab photograph (**d**) showing microfabrics related to subaerial exposure. Red arrow indicates upward direction of strata. (**a**) Well-preserved *Microcodium* in sample F120 (277.8 m level in DS17), partly replacing crinoid ossicle in lower centre. (**b**) Enlargement of *Microcodium* in sample F120 (277.8 m level in DS17), showing its detailed internal structure. (**c**) *Microcodium* in sample H27 (245.7 m level in

interstitial spaces where this type of cement is developed, remnants of alveolar texture and pendant cements are sometimes observed (Fig. 7e).

The colour of the cement is probably due to pedogenic pigmentation as it is similar to what can be seen in some pendant cement (Figs 7f & 9e), although the former is generally darker in colour. The coloured cementation probably occurred in a meteoric phreatic environment with abundant soil input, and it appeared following a meteoric vadose condition in which pendant cement and alveolar texture were formed.

Leaching

Leaching (Fig. 8e, f) is often observed in oolitic grainstone and fine-grained bioclastic grainstone in the middle of depositional sequences. Leaching (or corrosion) of grains is also sometimes seen in fine-grained bioclastic grainstone in the dark-coloured intervals just below subaerial exposure surfaces. It is indicative of dissolution in a meteoric diagenetic environment (Scholle & Ulmer-Scholle 2003), and is therefore a good indicator of subaerial exposure.

Ooids affected by leaching are filled by blocky calcite cement, with only the outermost few cortices remaining in many cases (Fig. 8e). This may suggest that they were originally superficial ooids with larger nuclei, but of unstable mineralogy different to that of the coatings. In fine-grained bioclastic grainstone, fragments of calcareous algae with probable original aragonitic mineralogy are often leached and replaced by coarse calcite cement. In that example, other bioclasts and cement are also often microcrystallized which makes the outlines of grains vague, giving them a poorly-preserved appearance on limestones under the microscope.

Crystal silts

Crystal silts (Fig. 8a) are observed as relatively well-sorted silt-sized micritic grains associated with other fabrics described above such as pendant cement, alveolar texture and leached grains. They are often found as fine-grained deposits partly infilling voids in coarse-grained grainstone in the dark-coloured limestone intervals. Crystal silts are interpreted as indicative of subaerial exposure and the vadose diagenetic environment (Dunham 1969), although its marine formation is also suggested in certain cases (Scholle & Ulmer-Scholle 2003).

Zongdi depositional sequences

Recognition and overview

Using both macro- and microscopic features/fabrics as basic criteria, primarily the palaeosols and dark-coloured intervals that typically contain rhizoliths and pendant cement (all of which are easy to recognize in the field), we identified a succession of subaerial exposure surfaces in the 150-m-thick studied section (Fig. 5). All of the recognized 35 subaerial exposure surfaces (SE1–SE35) do not necessarily define sequence boundaries however, which represent major lowstands of sea level. There are two variations in which two or more subaerial exposure surfaces are recognized within a very short stratigraphical interval (typically less than 1 m), of which only one is strong enough to be considered a sequence boundary.

Figure 10 illustrates two examples of two successive subaerial exposure surfaces within a short stratigraphical interval. One in each case is interpreted as representing only a minor fluctuation (or stillstand) of sea level. In the case illustrated in Fig. 10a, SE12 is immediately succeeded by SE13, which is accompanied with palaeosol, with only a 70 cm interval between them. Moreover, rhizoliths and alveolar textures are well developed in samples collected between SE12 and SE13, but these microfabrics are not found below SE12. Although both subaerial exposure surfaces are clearly characterized by dark-coloured limestones, only SE13 exhibits more conspicuous development of macro- and microscopic features related to subaerial exposure just below it. In addition, sediments between these two exposure surfaces are represented only by lime-mudstone and fine-grained bioclastic grainstone with a few oncoids. As discussed in the next section, they suggest very shallow-marine peritidal deposition, generally constituting the uppermost part of a depositional sequence in the Zongdi section. All these lines of evidence indicate that SE12 can be interpreted as representing only a small-scale fluctuation (or stillstand) of sea level that resulted in a minor subaerial exposure surface during late regression just before the major lowstand of sea level represented by SE13. We therefore interpret that only SE13 can be defined as a sequence-bounding disconformity (SB10).

Those shown in Fig. 10b express the other case formed by the combination of a major sea-level drop

Fig. 9. (*Continued*) DS12), here developed within densely packed crinoidal grainstone/packstone in middle part of depositional sequence. (**d**) Polished slab of sample F88 (228.5 m level in DS8), showing aggregate grain-bioclastic grainstone with conspicuous pendant (gravitational) cement (dark-coloured portions aligned roughly parallel to lamination). (**e**) Thin-section photomicrograph of part of sample illustrated in (**d**). Pendant cement (G) with pedogenic brown-colour tinge is developed only on underside of grains, suggesting direction of gravity at time of cement formation in a vadose environment (i.e. opposite to upwards direction of strata; red arrow).

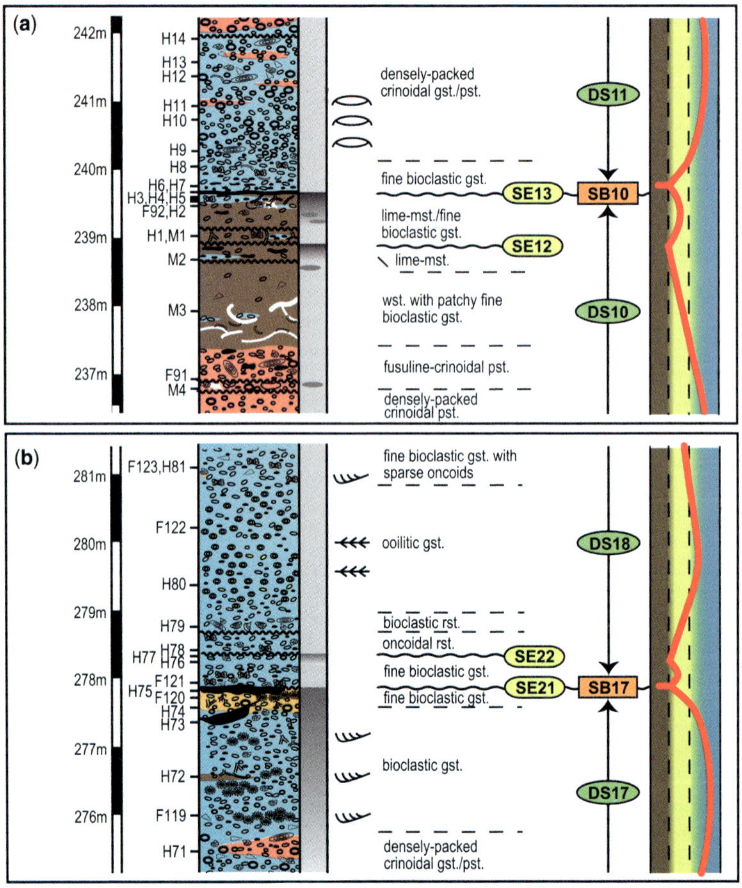

Fig. 10. Examples of two closely adjacent successive subaerial exposures in Zongdi section, presumably illustrating minor fluctuations (or stillstands) of sea level. See Figure 5 for legend. (**a**) Small-scale fluctuation of sea level in latest stage of regression (exemplified by SE12) just below sequence boundary SB10 (represented by SE13). Minor emergence represented by SE12 occurred just before major sea-level lowstand (SE13). (**b**) Stillstand of sea level in earliest stage of transgression (exemplified by SE22) just above sequence boundary SB17 (represented by SE21). This minor emergence (SE22) probably reflects minor fluctuation of sea level during general time of rising sea-level.

followed by a minor fluctuation of sea level during the early stage of transgression. In this example, SE21 has stronger signatures of macro- and microscopic fabrics related to subaerial exposure such as palaeosol, *Microcodium*, rhizoliths and alveolar textures, with thicker and more conspicuous development of the dark-coloured interval in the field up to 1.9 m below it. In contrast, overlying SE22 (separated from SE21 by 50 cm of section) developed only a thin dark-coloured horizon below it, with no evidence for long-term exposure and palaeosol formation. The limestone between SE21 and SE22 is fine-grained bioclastic grainstone, which generally constitutes the lower part of a depositional sequence interpreted as shallow wave-affected peritidal facies in the Zongdi section. SE21 can therefore be interpreted as representing a level of major sea-level lowstand, namely sequence boundary (SB17), whereas the next higher SE22 represents only a minor subaerial exposure surface formed during a small-scale drop or stillstand of sea level during transgression.

Internal architecture of depositional sequences

Based on major subaerial exposure surfaces that are interpreted as sequence-bounding disconformities, we recognized a total of 26 depositional sequences (DS1–DS26) in the Zongdi section (Fig. 5). As illustrated in Figure 11, each depositional sequence can be generally subdivided into three parts: basal, middle (core) and upper units.

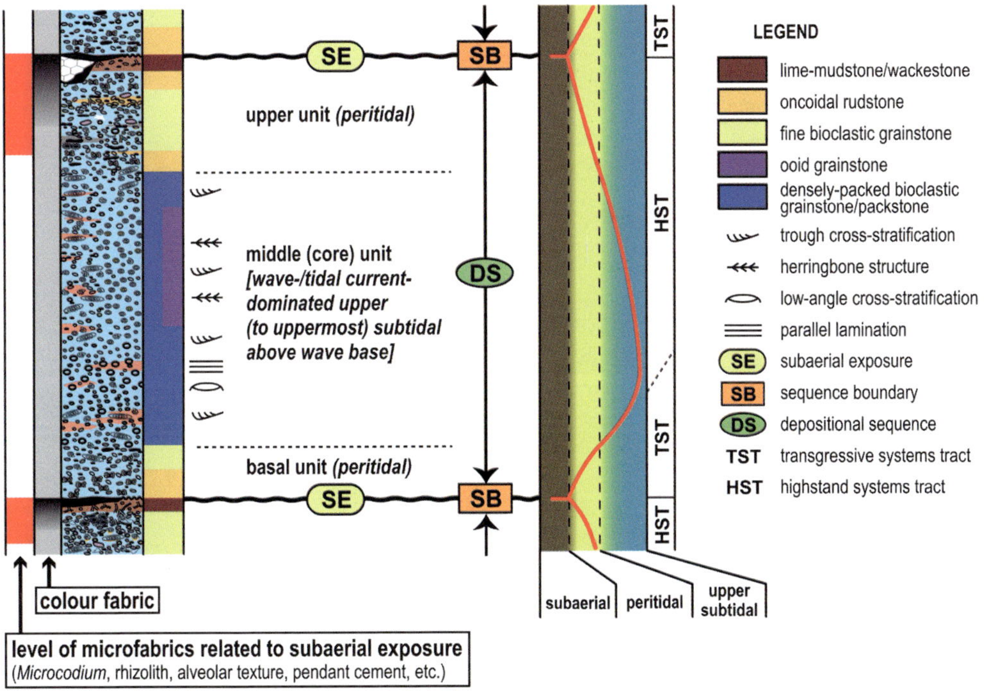

Fig. 11. Geological log and internal architecture of schematic single depositional sequence at Zongdi with interpretations of sea-level history and systems tracts. Each depositional sequence can be divided into a succession of basal (peritidal), middle (upper subtidal) and upper (peritidal) units. For legend for geological log, see Figure 5.

The basal unit consists generally of fine-grained bioclastic grainstone and oncoidal rudstone (Fig. 12g, h). This part is usually thin and even obscure or absent in some depositional sequences (e.g. DS23). Sometimes, small-scale trough cross-stratification is observed (Fig. 13d). The middle (core) unit is well developed and thickest in most depositional sequences. Conglomeratic deposits consisting of coarse-grained biomorphs (gastropods, corals, fusulines) and/or oncoids sometimes mark the base of the middle unit (Fig. 13c). The middle unit is composed essentially of densely packed, coarse-grained bioclastic (generally fusuline-crinoidal) grainstone (Fig. 12e) and well-sorted ooid grainstone (Fig. 12f). In this unit, low-angle cross-stratification, trough cross-stratification and herringbone structure are generally observed (Fig. 13a, b, e–g). Strata with low-angle cross-stratification are dominated by crinoidal grainstone. The upper unit is also well recognized in most depositional sequences and consists of fine-grained grainstone (partly packstone) with some oncoids, oncoidal rudstone and bioclast-poor lime-mudstone/wackestone (Fig. 12a–d). The lime-mudstone exhibits characteristic monotonous appearance and is often developed at the top of sequences (Fig. 6c). Moreover, various kinds of sedimentary fabrics related to subaerial exposure (Figs 6–9) are essentially restricted to this unit although, as noted already, pendant cement is sometimes observed within the basal and middle units and *Microcodium* and grain leaching are occasionally found in the middle unit.

Based on the characters of sedimentary facies and structures, the fine-grained limestones in the basal and upper units of a depositional sequence can be interpreted as a range of peritidal environments. In particular, the lime-mudstone often observed in the uppermost part of sequences can be interpreted as mud-flat deposits formed just before emergence above sea level. In contrast, the middle unit is considered to represent wave- and/or tidal-current-dominated upper to uppermost subtidal sediments. The large biomorph-rich conglomeratic beds sometimes observed at the base of the middle unit (Fig. 13c) are interpreted as transgressive lag deposits. Moreover, the low-angle cross-stratification in this unit (Fig. 13a) has a similar bedform as hummocky or swaley cross-stratification, except it is often developed in coarser-grained sediments. We interpret the densely packed crinoidal grainstone with low-angle cross-stratification seen in some depositional sequences as exhibiting the deepest-water facies in a single sequence; it probably represents deposition slightly below fair-weather

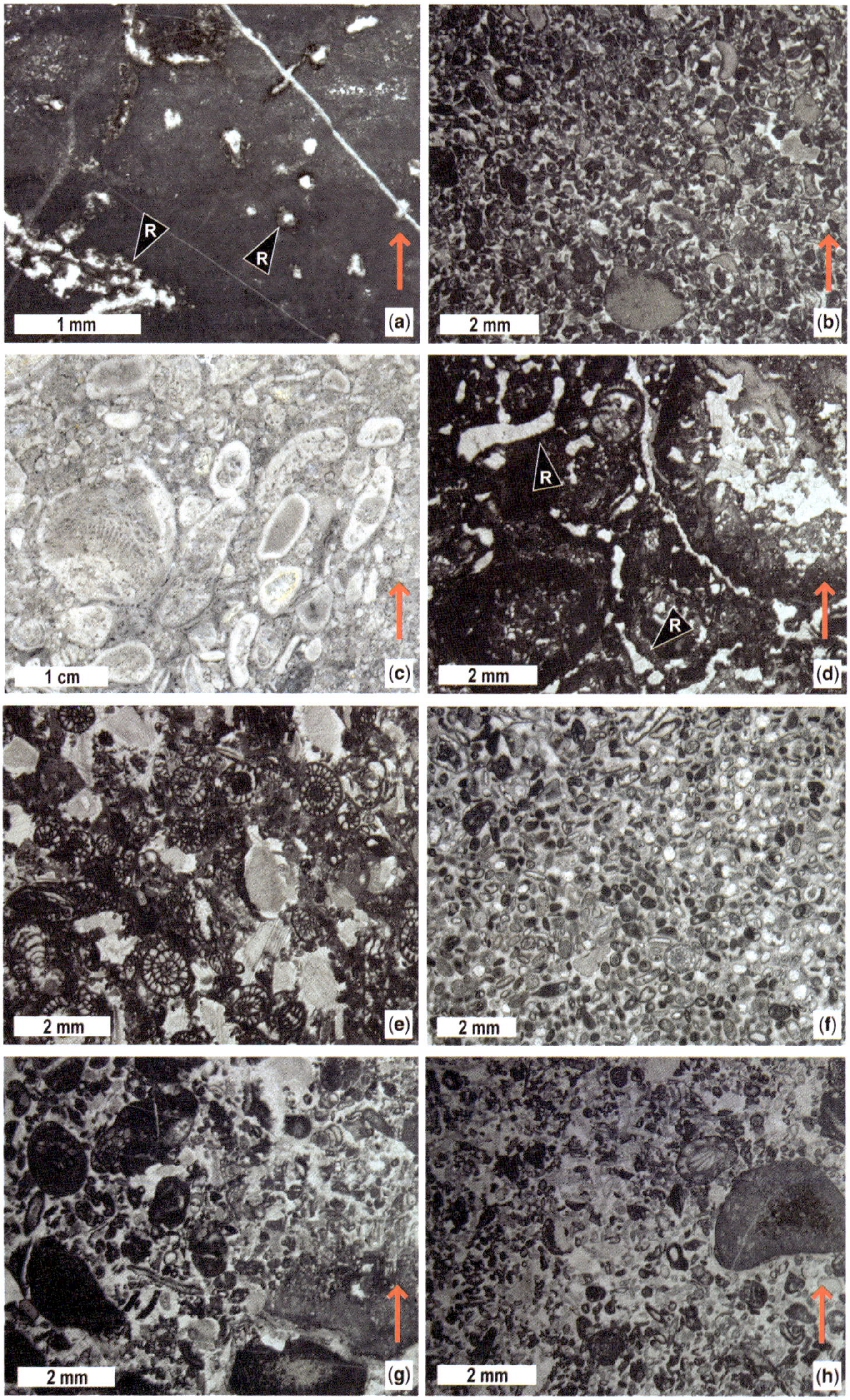
R
R
1 mm
(a)
2 mm
(b)
1 cm
(c)
R
R
2 mm
(d)
2 mm
(e)
2 mm
(f)
2 mm
(g)
2 mm
(h)

Fig. 13. Outcrop photographs showing several sedimentary structures observed in basal and middle units of Zongdi depositional sequences. (**a**) Low-angle cross stratification in densely packed crinoidal grainstone in middle unit of DS11 (241 m level). (**b**) Small-scale herringbone structure and trough cross-stratification in oolitic grainstone in middle unit of DS10 (234 m level). (**c**) Lag deposits consisting of gastropods, corals, fusulines and lithoclasts at base of middle unit of DS18 (278.9 m level). (**d**) Small-scale trough cross-stratification in bioclastic grainstone in basal unit of DS13 (248 m level). (**e**) Large-scale trough cross-stratification in coarse bioclastic grainstone in middle unit of DS24 (328 m level). (**f**) Herringbone structure in well-sorted oolitic grainstone in middle unit of DS18 (280 m level). (**g**) Herringbone structure and trough cross-stratification in oolitic grainstone in middle unit of DS13 (253 m level).

Fig. 12. Thin-section photomicrographs (**a, b, d–h**) and polished slab photograph (**c**) showing typical microfacies in Zongdi depositional sequences. Red arrow indicates upward direction of strata. (**a–d**), (**e, f**) and (**g, h**) represent typical microfacies of upper, middle and basal units of sequences, respectively. (**a**) Lime-mudstone of probably mud-flat origin with rhizoliths (R) from F95 (243.8 m level in DS11). (**b**) Fine bioclastic grainstone from M18 (226.1 m level in DS7). (**c, d**) Oncoidal rudstone with well-developed rhizoliths (R) from M10 (231.7 m level in DS9). (**e**) Densely packed crinoid-fusuline grainstone/packstone from H95 (290.0 m level in DS21). Note crinoid ossicles with syntaxial cement overgrowths. (**f**) Ooid grainstone from F122 (280.2 m level in DS18). (**g**) Oncoidal rudstone from H64 (268.8 m level in DS16). (**h**) Fine bioclastic grainstone with sparse oncoids from M39 (210.6 m in DS2).

wave base, but above storm wave base. One depositional sequence therefore corresponds well to a single transgressive-regressive cycle (Figs 5 & 11), except for the existence of small-scale fluctuations of sea level below and above some sequence boundaries (Fig. 10). Furthermore, we interpret that the lower and the lower portion of the middle units below the crinoidal grainstone with low-angle cross-stratification represent the transgressive systems tract, and the overlying remaining middle and upper units correspond to the highstand systems tract (Fig. 11). In a case that lacks this deepest-water crinoidal facies, densely packed bioclastic grainstone/packstone with trough cross-stratification would correspond to the highest stand of sea level in a single sequence, representing deposition above fair-weather wave base. It is important to note that the Zongdi depositional sequences consist entirely of very shallow-marine limestone suggesting upper subtidal to peritidal environments (Figs 5 & 11–13). In each depositional sequence, even the sediments representing the deepest facies in the middle unit are considered to be formed above general (storm plus fair-weather) wave base.

Figure 14 summarizes the general facies-stacking pattern in all depositional sequences of the Zongdi section. The threefold unit subdivisions of facies described above are well recognized in

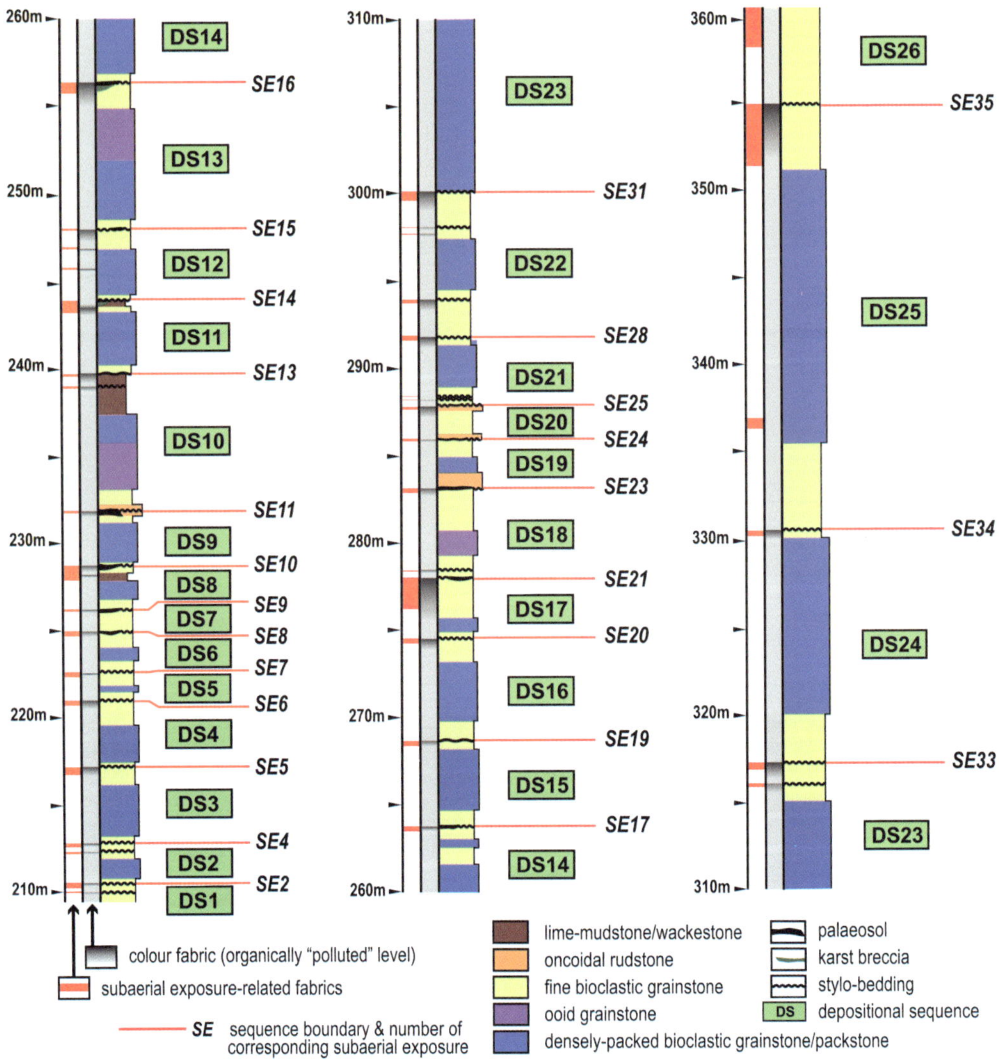

Fig. 14. Summary of sedimentary facies within all Zongdi depositional sequences.

most of the sequences. Although most depositional sequences are only 3–8 m thick, they become thicker up-section to as much as 24 m thick.

Chronostratigraphical framework by fusuline biostratigraphy

The examined part of the Zongdi section is rich in fusuline foraminifers throughout, as it is composed essentially of shallow open-marine limestones. In contrast, our preliminary conodont examination revealed that they are too few to utilize for biostratigraphy. Fusulines are therefore the primary tool to establish biostratigraphical and biochronological frameworks of the section. Figure 15 shows the stratigraphical distributions of the biostratigraphically important fusuline taxa from the Zongdi section, along with its chronostratigraphical framework. Figures 16 and 17 illustrate representative fusuline taxa from the section.

Among the Zongdi depositional sequences, the lower six (DS1–DS6) have a similar fusuline assemblage characterized by *Fusulinella*, large *Neostaffella*, *Beedeina* and *Fusulina*. This generic composition, as well as the fact that the first occurrence of the genus *Beedeina* (which first appearance datum is recorded in the late Kashirian of the late Early Moscovian) is at about 140 m level from the base in the Zongdi section (Fig. 3), indicates that these depositional sequences are correlative to the late Podolskian–early Myachkovian substages of the Late Moscovian Stage (e.g. Isakova 2001). The next sequences, DS7 and DS8, contain *Protriticites* sp. A, whose spirotheca in the outermost volution looks finely porous with vague diaphanotheca in its inner volutions. We consider that these characteristics of this form are primitive among the genus and thus conclude that DS7 and DS8 are probably referable to the late Myachkovian (latest Moscovian).

The overlying three depositional sequences (DS9–DS11) are characterized by more advanced *Protriticites* (*P.* sp. B) with more definite (coarser) pores in the spirotheca. *Quasifusulinoides* also occurs in this interval. Above, sequence DS12 also yields *Protriticites* sp. B but, in addition, records the first occurrences of *Montiparus* and *Triticites*. This implies the correlation of DS9–DS11 with the Krevyakinian Substage of the early Kasimovian and DS12 with the Khamovnikian Substage of the middle Kasimovian. However, we cannot completely exclude the possibility that DS11 should be assigned to the lower part of the Khamovnikian, judging from the fact that the *Montiparus* species from DS12 (coexisting with *Triticites*) is not considered primitive among the relevant genus. Putting the base of the Kasimovian within the *Protriticites* genozone in this study is based on the traditional definition of the Moscovian–Kasimovian boundary (Davydov 1997; also Villa & Task Group 2008). The next sequence, DS13, is not very diagnostic in fusuline composition as *Triticites* is its major element. Based on its stratigraphical position between *Montiparus*-bearing levels below and *Rauserites*-bearing levels above, DS13 is probably assignable to the Dorogomilovian Substage of the late Kasimovian. This sequence also yields '*Carbonoschwagerina*' sp. A which is smaller and less inflated, thus considered more primitive among the genus, compared to two other '*Carbonoschwagerina*' species discussed below.

In the base of DS14 *Rauserites* first appears, with short-ranging but characteristic occurrences of two '*Carbonoschwagerina*' species ('*C.*' sp. B and '*C.*' sp. C). Of these, the *Rauserites* found in the basal part of DS14 is more or less comparable morphologically with those reported from the lower Gzhelian in some areas of Eurasia, and includes *R.* ex gr. *rossicus* (e.g. Rozovskaya 1958; Leven & Davydov 2001). '*Carbonoschwagerina*' sp. B somewhat resembles *C. morikawai* reported from the middle Gzhelian of Japan (Ozawa *et al.* 1992). On the other hand, more inflated '*Carbonoschwagerina*' sp. C is quite similar to *C. minatoi*, which was conventionally subsumed in the Early Permian genus *Pseudoschwagerina* in earlier studies (e.g. Kanmera 1958) but is now reassigned to the late Gzhelian (Ozawa *et al.* 1992). These considerations, particularly based on '*Carbonoschwagerina*' sp. C, could potentially lead to the conclusion that DS14 is already in the later part of the Gzhelian. However, this assumption is not supported by the facts that: (1) the *Rauserites* species from DS14 is more similar to those known from the lower Gzhelian in other areas; (2) there is little evidence to support such a large hiatus (time gap) as the early–middle Gzhelian between late Kasimovian DS13 and the next DS14; and (3) there is no other fusuline species that strongly suggests a late Gzhelian age, not only from DS14 but also from the overlying several depositional sequences. It therefore seems more reasonable to refer DS14 to the basal Gzhelian. The occurrence of '*Carbonoschwagerina*', which would seemingly suggest younger ages due to its distinctive shell inflation, in the base of the Gzhelian would more likely show that they are better interpreted as *Carbonoschwagerina*- mimics. In a broader sense, this example shows that shell inflation itself is not necessarily age-dependent but could appear several times in a homeomorphic fashion and could evolve rather rapidly at different times in different places. This morphological character may therefore not be a diagnostic basis for detailed biochronology in some cases. Furthermore, the '*Carbonoschwagerina*' from the Zongdi section is probably not included in the appropriate genus.

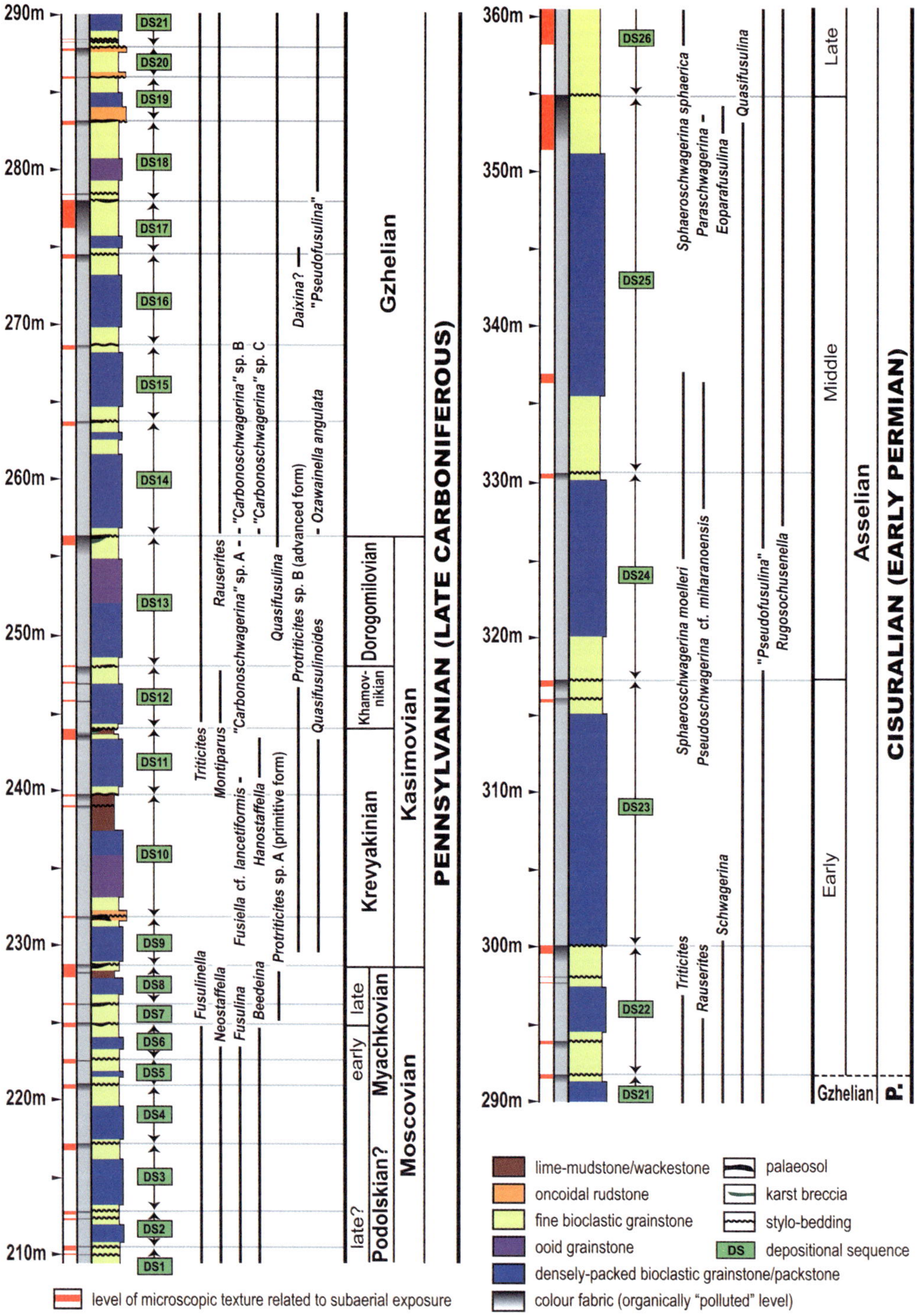

Fig. 15. Range chart showing stratigraphical distribution of major fusuline taxa and biochronology of Zongdi section.

The next several depositional sequences (DS15–DS20) are poor in strongly age-diagnostic forms, and contain only *Quasifusulina*, *Triticites*, *Rauserites*, *Daixina*? and '*Pseudofusulina*' (several forms). These genera broadly suggest a Gzhelian age for this interval. *Schwagerina* occurs in overlying sequences DS21–DS23, followed by two sequences (DS24 and DS25) yielding *Sphaeroschwagerina moelleri* and *Pseudoschwagerina* cf. *miharanoensis* which suggest a Middle Asselian age (e.g. Leven & Scherbovich 1978). Although it is not completely identical, *Schwagerina* sp. from DS21 to DS23 is somewhat reminiscent of globose *Schwagerina* such as *S. versabilis* and *S. parva*, which are known to occur generally in Gzhelian–Asselian (thus Carboniferous–Permian) boundary strata (e.g. Rauzer-Chernousova & Scherbovich 1958; Forke 2002). In the Zongdi section, this stage (and also systemic) boundary can probably be placed somewhere within this *Schwagerina*-bearing interval, and is here provisionally put at the base of DS22. DS22 and DS23 are therefore presumed to be Early Asselian, although unfortunately no diagnostic fusuline conclusively supporting this age assessment has been found from these two depositional sequences. The next two depositional sequences, DS24 and DS25, are referable to the Middle Asselian as already noted. The uppermost depositional sequence, DS26, is considered to be Late Asselian because of the occurrence of *Sphaeroschwagerina sphaerica* (e.g. Leven & Scherbovich 1978).

Summarizing the above-mentioned fusuline biostratigraphy (Fig. 15), the studied Zongdi section ranges from the Late Moscovian to the Asselian without any obvious time gap that is larger than one substage (Fig. 18) although, as we discuss in the following section, probably somewhat greater depositional gaps could be assumed in middle–late Kasimovian time judging from the number of depositional sequences formed. Moreover, the Moscovian–Kasimovian and Kasimovian–Gzhelian boundaries are very likely accurately placed at the bases of DS9 and DS14, respectively. The Carboniferous–Permian boundary in this section is not very clear at this time, however, but is provisionally placed at the DS21–DS22 boundary and thus at SE28/SB21.

Discussion

Depositional model for the Zongdi cyclothems

Judging from the facies-stacking pattern within each depositional sequence, cyclic appearances of subaerial exposure surfaces and their Pennsylvanian–Early Permian age, the depositional sequences observed in the upper part of the Zongdi section are best interpreted to have been formed by alternating marine transgression and regression, which resulted from icehouse-type high-frequency/high-amplitude sea-level change caused by the cyclic waxing and waning of Gondwanan ice sheets. We have therefore recognized for the first time in Late Palaeozoic platform carbonates on the South China Craton different cyclothems composed entirely of carbonates, which presumably correspond genetically to the well-known Midcontinent (Kansas-type) cyclothems in North America (e.g. Heckel 1986; Fig. 19) and more recently with those in the Moscow Basin of Russia (e.g. Kabanov & Baranova 2007). The sediments observed in the Zongdi section are composed entirely of peritidal and upper subtidal, very shallow-marine carbonates with wave- and/or tidal-current-generated sedimentary structures clearly formed above average wave base. The thickness of each cyclothem (depositional sequence) is relatively thin, particularly in the Pennsylvanian. This is due presumably to marine sedimentation having occurred in a relatively small (shallow) accommodation space. It further suggests that the Zongdi section was probably located during the Pennsylvanian–Early Permian in a high (upper) shelf area (platform) within an epeiric sea, where sedimentation occurred in such small accommodation space only during the highest stands of sea level in relatively large (major) sea-level cycles (Fig. 20). Moreover, this model implies that the period of time for the deposition of a single Zongdi cyclothem must be considerably shorter than that shown by an intervening unconformity (hiatus). Furthermore, the position of this part of the epeiric platform was very distant from detrital sources, unlike the high-shelf Appalachian region of North America where the cyclic marine units are also conspicuously thin but in a thicker succession, because most sediment is terrigenous and deposited in terrestrial and paralic environments and in incised palaeovalleys formed during lower stands of sea level (Belt *et al.* 2011).

Heckel (2002) subdivided Pennsylvanian cyclothems of the North American Midcontinent into major, intermediate and minor cyclothems (Fig. 19) based on the scale of transgression (which approximates the scale of sea-level rise). Among cyclothems of the three informal orders of magnitude, the major cyclothems correspond to major (large-scale) transgressions, during which even high (upper) shelf areas were submerged and marine sedimentation took place across the entire shelf. During that time, shallow-marine sedimentation would occur in a small accommodation space in high-shelf areas, which remained generally emergent during lesser sea-level rises that formed the intermediate and minor cyclothems. He recognized some 20 major cyclothems from the Middle Desmoinesian

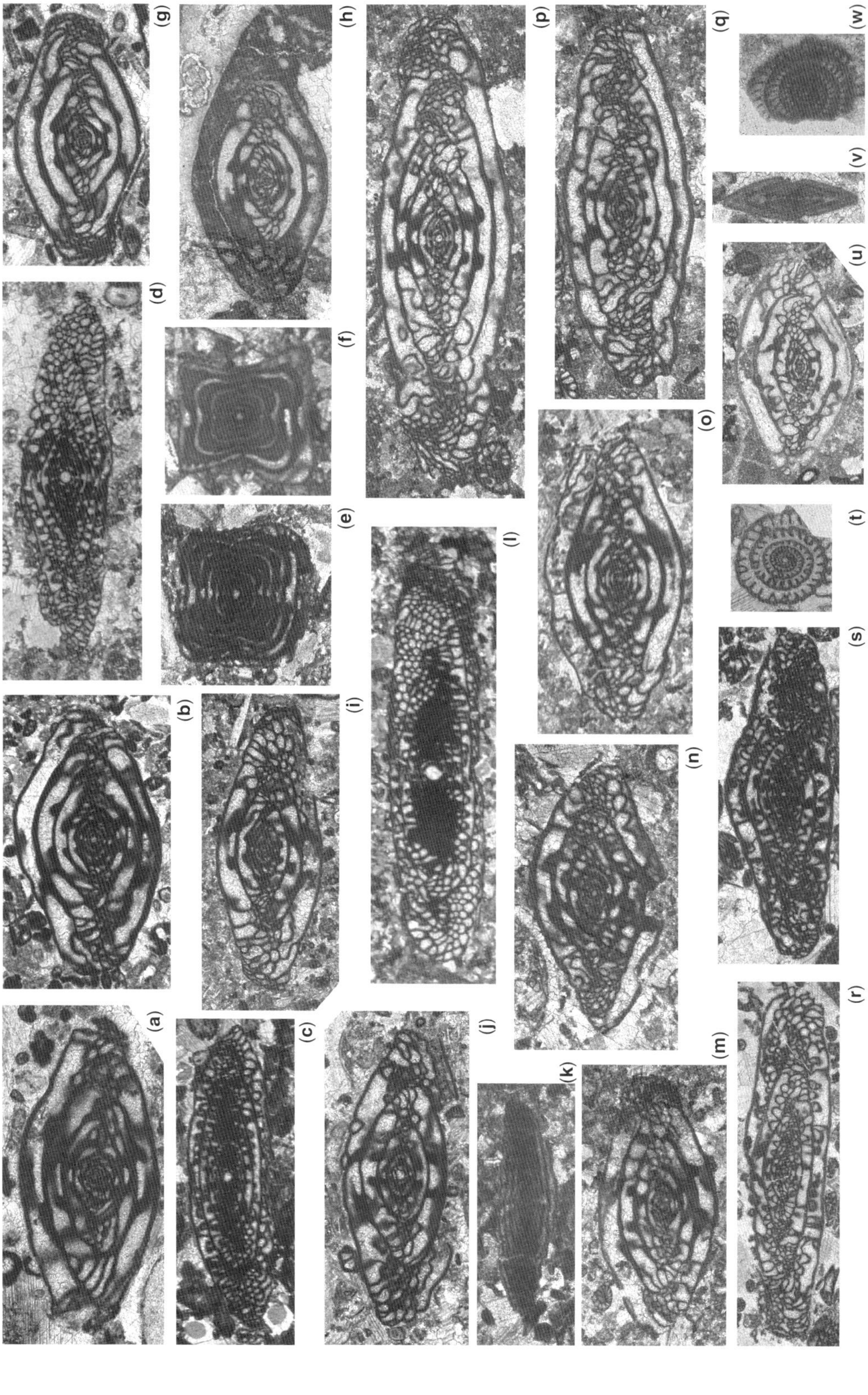
(a)
(b)
(c)
(d)
(e)
(f)
(g)
(h)
(i)
(j)
(k)
(l)
(m)
(n)
(o)
(p)
(q)
(r)
(s)
(t)
(u)
(v)
(w)

to Virgilian, which almost coincides with the number of Zongdi cyclothems of corresponding ages (Late Moscovian–Gzhelian). Because the Zongdi cyclothems are presumed to have been formed during only the highest stands of sea-level in relatively large glacio-eustatic sea-level cycles, they can be interpreted to correlate essentially with the major cyclothems in the North American Midcontinent Basin. As we examine in detail in the next subsection however, some intermediate cyclothems were probably recorded during Late Moscovian–early Kasimovian and late Gzhelian times, and some major cyclothems are presumably missing in the middle–late Kasimovian in the Zongdi succession.

Detecting possible epeiric tectonic signal from the Pennsylvanian Zongdi cyclothems

In a recent insightful study by Heckel *et al.* (2007), high-resolution inter-continental correlation of Pennsylvanian cyclothems based on their scale as well as biostratigraphical framework was attempted. This study, which they called the 'digital' correlation, demonstrated that major cyclothems in the North American Midcontinent Basin during late Desmoinesian–early Virgilian time, which approximates the global Moscovian–Kasimovian–Gzhelian stage boundary interval, are well correlated with coeval sedimentary cycles in the Moscow Basin of Russia using these criteria for identifying individual cyclothems. In the Moscow Basin, nine of the correlatable twelve cyclothems were regarded to be formed by major cycles, although the remaining three are of lesser degree (Fig. 21). Although there are some differences in cyclothem scales between these two epeiric basins, it can be suggested that the major cyclothems in the Midcontinent Basin essentially represent the global (eustatic) signal of major sea-level rises (transgressions) during the relevant time interval.

As noted in the preceding section, the number of cyclothems recognized in the Late Moscovian–Gzhelian succession at Zongdi is apparently comparable to that of the major cyclothems recognized in the Midcontinent Basin during late Desmoinesian–Virgilian time (Fig. 21). However, fusuline biostratigraphy proposed a potentially more detailed comparison of the number of cyclothems formed by larger transgressions in shorter chronological segments between Zongdi and the Midcontinent. By combining this information with the stratigraphical pattern of the Zongdi cyclothems and the postulated secular change of global sea-level, we can further estimate relative epeiric tectonic control for their formation on a high-shelf area of the South China Block.

In the Midcontinent Basin, ten major cyclothems are identified in the Desmoinesian; eight cyclothems (from the McCurtain to Lower Pawnee cyclothems) of the ten are correlated with the Late Moscovian (Heckel 2008; Fig. 21). At Zongdi, there are also eight cyclothems (DS1–DS8) recognized in the Late Moscovian succession (Figs 15 & 18). However, the time during which these eight Zongdi cyclothems were formed presumably only corresponds to a younger part of the Desmoinesian (Fig. 21) because the first *Beedeina*, which defines the base of the Desmoinesian in North America (Dunbar & Henbest 1942; Waddell 1966; Douglass 1987; Ross & Ross 1987), was found from a level further (*c.* 70 m) below the base of the studied interval (Fig. 3). These data suggest that some of the Late Moscovian cyclothems in the Zongdi section are potentially equivalent to late Desmoinesian intermediate cyclothems in the Midcontinent Basin. This further implies a large relative sea-level rise on the epeiric platform at Zongdi during the Late Moscovian, because it necessitates an increasing accommodation space for their appearance. Haq & Schutter (2008) illustrated a slight rise of long-term eustatic sea-level in the Late Moscovian (Fig. 21), but the accommodation space formed by this change of sea level would not be large enough to help the appearance of intermediate cycles in the Zongdi section. We therefore interpret that epeiric tectonic subsidence is the primary control for the formation of Late Moscovian cyclothems in this section (Fig. 21), and that DS1–DS8 are represented by a stacking of not only major cyclothems but also intervening intermediate cyclothems, formed during the time of more tectonic subsidence (Fig. 22a). It is also important to note that, further down in the section, marine sedimentation without subaerial

Fig. 16. Fusulines from Zongdi section (DS1–DS14). (**a, b**) *Fusulinella* sp. (**a**) from M37 (DS2); (**b**) from M40 (DS1), Late Moscovian, ×12.5. (**c**) *Fusulina* sp. from F82 (DS3), Late Moscovian, ×7. (**d**) *Beedeina* sp. from F86 (DS6), Late Moscovian, ×7. (**e**) *Neostaffella* sp. from F86 (DS6), Late Moscovian, ×12.5. (**f**) *Hanostaffella* sp. from F93 (DS11), early Kasimovian, ×20. (**g, h**) *Protriticites* sp. A, both from F88 (DS8), latest Moscovian, ×12.5. (**i, j**) *Protriticites* sp. B, both from M4 (DS10), early Kasimovian, ×12.5. (**k**) *Fusiella* cf. *lancetiformis* Putrja, from H10 (DS11), early Kasimovian, ×35. (**l**) *Quasifusulinoides* sp. from H16 (DS11), early Kasimovian, ×8. (**m, o**) *Montiparus* sp. (**m**) from H26 (DS12); (**n**) from H24 (DS12); (**o**) from H34 (DS12); middle Kasimovian, ×12.5. (**p, q**) *Triticites* sp. (**p**) from F101 (DS13), late Kasimovian; (**q**) from H34 (DS12), middle Kasimovian, ×12.5. (**r–t**) *Rauserites* ex gr. *rossicus* (Schellwien), all from H54 (DS14), early Gzhelian, ×8. (**u**) '*Carbonoschwagerina*' sp. A from H50 (DS13), late Kasimovian, ×8. (**v, w**) *Ozawainella angulata* (Colani), both from H54 (DS14), early Gzhelian, ×20.

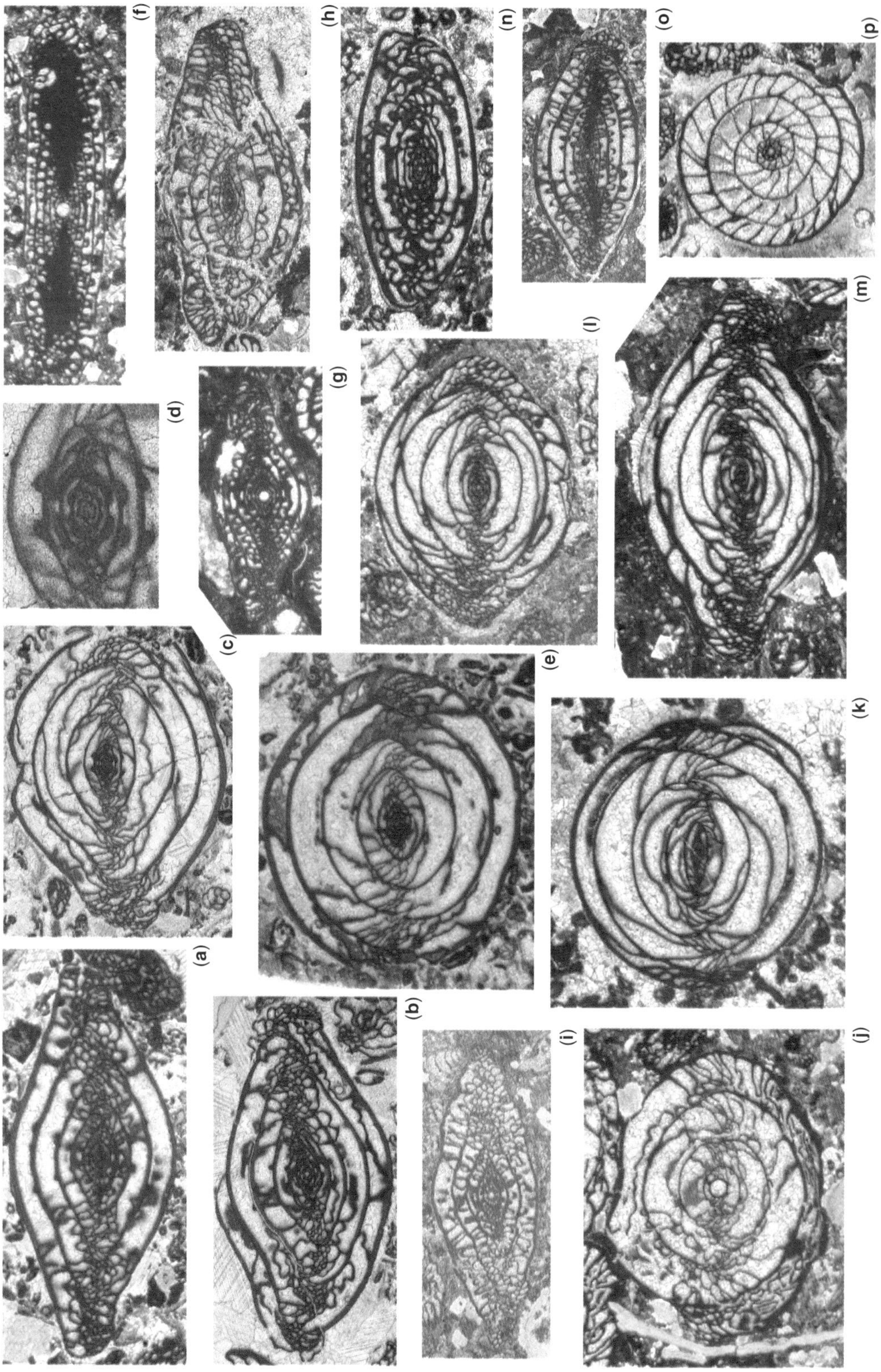
(a)
(b)
(c)
(d)
(e)
(f)
(g)
(h)
(i)
(j)
(k)
(l)
(m)
(n)
(o)
(p)

Geological age				Zongdi depositional sequences (cyclothems)
PERMIAN	Cisuralian	Sakmarian		
		Asselian	Late	DS26
			Middle	DS25 DS24
			Early	DS23 DS22
CARBONIFEROUS	Pennsylvanian	Gzhelian		DS21 DS20 DS19 DS18 DS17 DS16 DS15 DS14
		Kasimovian	Dorogomilovian	DS13
			Khamovnikian	DS12
			Krevyakinian	DS11 DS10 DS9
		Moscovian	Myachkovian	DS8 DS7 DS6 DS5 DS4 DS3 DS2
			Podolskian	DS1

Fig. 18. Age of Zongdi cyclothems based on fusuline succession.

exposure prevailed in the lower Desmoinesian part of the Zongdi succession (Fig. 3). It can also be deduced that larger continuous epeiric subsidence completely submerged the Yangtze Carbonate Platform at Zongdi during that time interval (Fig. 21).

In the Kasimovian, the Zongdi cyclothems show a different tectonic signal to the Late Moscovian. Fusuline biostratigraphy revealed that three cyclothems were formed during the early Kasimovian (Krevyakinian) and one cyclothem was formed during both the middle Kasimovian (Khamovnikian) and the late Kasimovian (Dorogomilovian) (Figs 15 & 18). In the corresponding time intervals, two (Altamont and Lost Branch), three (Hertha, Swope and Dennis) and four (Dewey, Iola, Stanton and Cass) major cyclothems are recognized in the Midcontinent Basin (Fig. 21). One of the three early Kasimovian cyclothems recognized in the Zongdi section should therefore be equivalent to one of intermediate cyclothems within this interval in the Midcontinent Basin. In the middle–late Kasimovian, five major cyclothems out of seven seen in the Midcontinent Basin are missing in the Zongdi succession. This suggests that the Late Moscovian trend of more tectonic subsidence continued to the Early Kasimovian. On the other hand, there are greater gaps among cyclothems in the younger half of the Kasimovian than in the early half at Zongdi. In other words, there was no accommodation space for five of seven major Midcontinent sea-level rises on

Fig. 17. Fusulines from Zongdi section (DS14–DS26). (**a, b**) '*Carbonoschwagerina*' sp. B, both from H54 (DS14), early Gzhelian, ×8. (**c–e**) '*Carbonoschwagerina*' sp. C. (**c, e**) from H54 (DS14), early Gzhelian, ×8, (**d**) enlarged part of (**c**), showing tightly coiled juvenile volutions with thick epithecal deposits, ×30. (**f**) *Quasifusulina* sp. from H99 (DS22), probably Early Asselian, ×8. (**g**) *Schwagerina* sp. from F135 (DS23), Early Asselian, ×8. (**h**) *Paraschwagerina* sp. from H131 (DS25), Middle Asselian, ×8. (**i**) '*Pseudofusulina*' sp. from H114 (DS24), Middle Asselian, ×8. (**j**) *Pseudoschwagerina* cf. *miharanoensis* Akagi, from H114 (DS24), Middle Asselian, ×8. (**k**) *Sphaeroschwagerina sphaerica* (Scherbovich) from H138 (DS26), Late Asselian, ×8. (**l, m, p**) *Sphaeroschwagerina moelleri* (Rauzer-Chernousova). (**l**) from H122 (DS25); (**m**) from H121 (DS25); (**p**) from H116 (DS24), Middle Asselian, ×8. (**n**) *Eoparafusulina* sp. from H130 (DS25), Middle Asselian, ×12.5. (**o**) *Rugosochusenella* sp. from H121 (DS25), Middle Asselian, ×8.

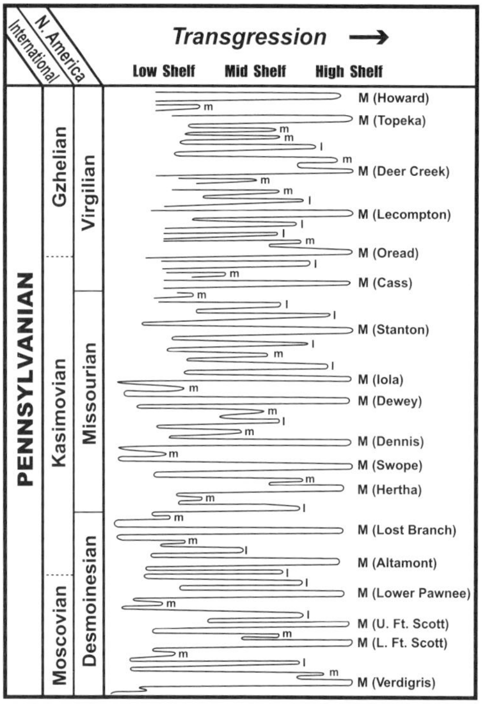

Fig. 19. Pennsylvanian glacio-eustatic sea-level curve which shows cyclothems that should be coeval on global scale, based on data from Midcontinent outcrop belt in North America (redrawn from Heckel 2008). M, I, m: major, intermediate and minor cyclothems, respectively.

this part of the Yangtze Carbonate Platform on the South China Block during the middle–late Kasimovian. As there is no large long-term sea-level change postulated during the relevant time interval (Haq & Schutter 2008), this decreasing accommodation space is primarily attributed to less tectonic subsidence of the Yangtze Carbonate Platform at Zongdi at that time (Fig. 21). In the middle–late Kasimovian, the Zongdi cyclothems are therefore interpreted to represent only of some of the major transgressions that formed the major cyclothems in the Midcontinent Basin, under less control of tectonic subsidence (Fig. 22b).

In the Gzhelian, six major cyclothems (Oread, Lecompton, Deer Creek, Topeka, Howard and Upper Hughes Creek) are recorded in the Midcontinent Basin, with a relatively long upper Virgilian interval which lacked any major cyclothems but demonstrated intermediate cyclothems (Heckel 2008; Fig. 21). The Zongdi section has eight cyclothems in the Gzhelian. This suggests that some of them are equivalent to (upper?) Virgilian intermediate cyclothems in the Midcontinent Basin. If this is the case, then we would suggest increasing epeiric subsidence on this part of the Yangtze Carbonate Platform under a small sea-level rise in long-term eus-tacy during the very latest Carboniferous (Fig. 21). Ross & Ross (1987) earlier depicted a rather sharp long-term (second-order) sea-level drop (estimated at *c.* 50 m in amplitude) at the end of the Gzhelian. This simply requires a Late Gzhelian large hiatus in the Zongdi section under the assumption of a moderate constant subsidence, which seems unlikely based on our fusuline biostratigraphical data.

The average rate of sedimentation also gives another insight into the estimation of tectonic subsidence during the Kasimovian and Gzhelian in the Zongdi section. Using 3.8 myr and 4.4 myr in Ogg *et al.* (2008) for the time interval of the Kasimovian and Gzhelian stages and thicknesses of the Zongdi cyclothems shown in Figure 5, their sedimentation rates per million years are calculated at 7.3 m for the Kasimovian and 8.0 m for the Gzhelian. This broad estimation yields a more or less constant rate of sedimentation during the Late Pennsylvanian. Taking the unwavering long-term sea-level of this time interval (Fig. 21) into account, these values further imply that there was rather steady subsidence as a long-term perspective in the Late Pennsylvanian on the Yangtze Carbonate Platform at Zongdi. Thus, an early Kasimovian trend of more subsidence which lasted from the Late Moscovian (Fig. 21), would be compensated by that of less subsidence during the middle–late Kasimovian. This resulted in the Kasimovian average sedimentation rate of 7.3 m/myr, essentially comparable to that of the Gzhelian.

Heckel (1994) has already demonstrated that the aforementioned analytical concept and method are useful for detecting regional tectonic signals in the Middle–Upper Pennsylvanian (Desmoinesian–Virgilian) of the Appalachian Basin, which was located in a high-shelf area of Midcontinent Sea on the North American Craton. He recognized an interesting pattern of major cyclothem appearance in the Appalachian Basin succession, in that the relative extents of marine incursions in the basin delineated nearly three complete long-term cycles of presence and absence of bundles of widespread marine units in that region. In that study, the Appalachian marine units appear to be bundled in the middle–late Desmoinesian (approximately equal to the latest Moscovian), the mid-Missourian (approximately equal to the middle–late Kasimovian) and the early Virgilian (corresponding to the early Gzhelian). Heckel (1994) interpreted that, during these three time periods, the Appalachian Basin was low enough to receive widespread glacio-eustatic marine inundations due to larger relative subsidence. In contrast, predominantly terrestrial intervals with only local minor marine units, commonly represented by a blackish-water or shoreline facies and more

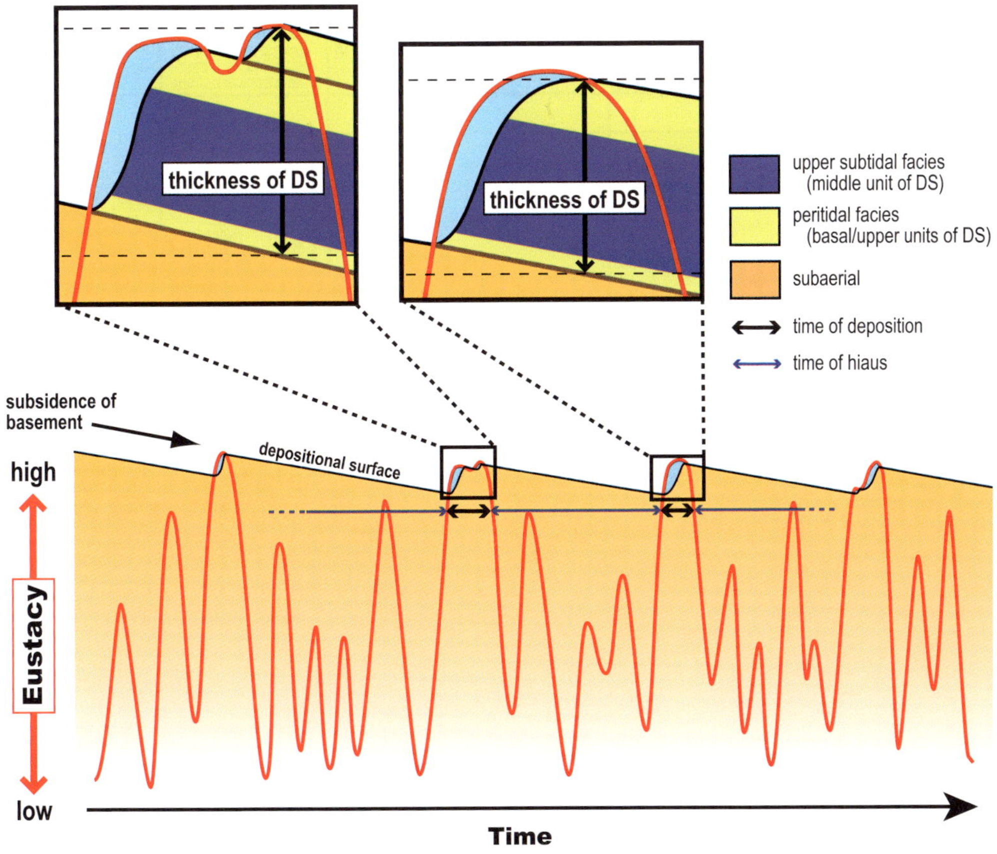

Fig. 20. Model illustrating depositional processes of Zongdi cyclothems within framework of high-frequency/high-amplitude (icehouse-type) transgressive-regressive sea-level cycles during late Palaeozoic glacial period, assuming constant rate of subsidence. Note that their deposition occurred within a relatively small (shallow) accommodation space only during highest stands of sea level in larger sea-level cycles (corresponding to those that produced major cyclothems in North American Midcontinent Basin).

complex palaeosol development in a succession, are observed between these marine-bundled intervals, suggesting smaller amount of accommodation space being formed due to less relative subsidence control. He estimated that these cycles of alternating marine-bundled and terrestrial-dominant intervals have a period of about 3 million years, and attributed their origin to epeiric tectonism. Our data for the Zongdi section indicate a differential subsidence signal from the Midcontinent (and Moscow Basin) region, and also from the Appalachian region (Fig. 21). The differences among the appearances of major cyclothems on high shelves between the Zongdi section and the Appalachians Basin suggest that their appearances are more likely related to regional differences in the rates of subsidence between Zongdi and the Appalachians, rather than from differences in levels of highstand among the individual major cyclothems. This is because they are expected to be nearly the same in different high-shelf regions if subsidence rates were constant elsewhere. Epeiric tectonism has seldom been considered and discussed in previous studies as a controlling factor of Palaeozoic sedimentation on the South China Block. We have detected for the first time a subtle tectonic control for the deposition of Pennsylvanian shallow-marine sediments on the Yangtze Carbonate Platform of South China Craton.

Cause of the formation of earliest Permian thicker sequences

The Asselian (earliest Permian) cyclothems in the Zongdi section clearly exhibit different features from the Pennsylvanian cyclothems in that they have larger thicknesses, although their internal facies composition is similar to that of the Pennsylvanian

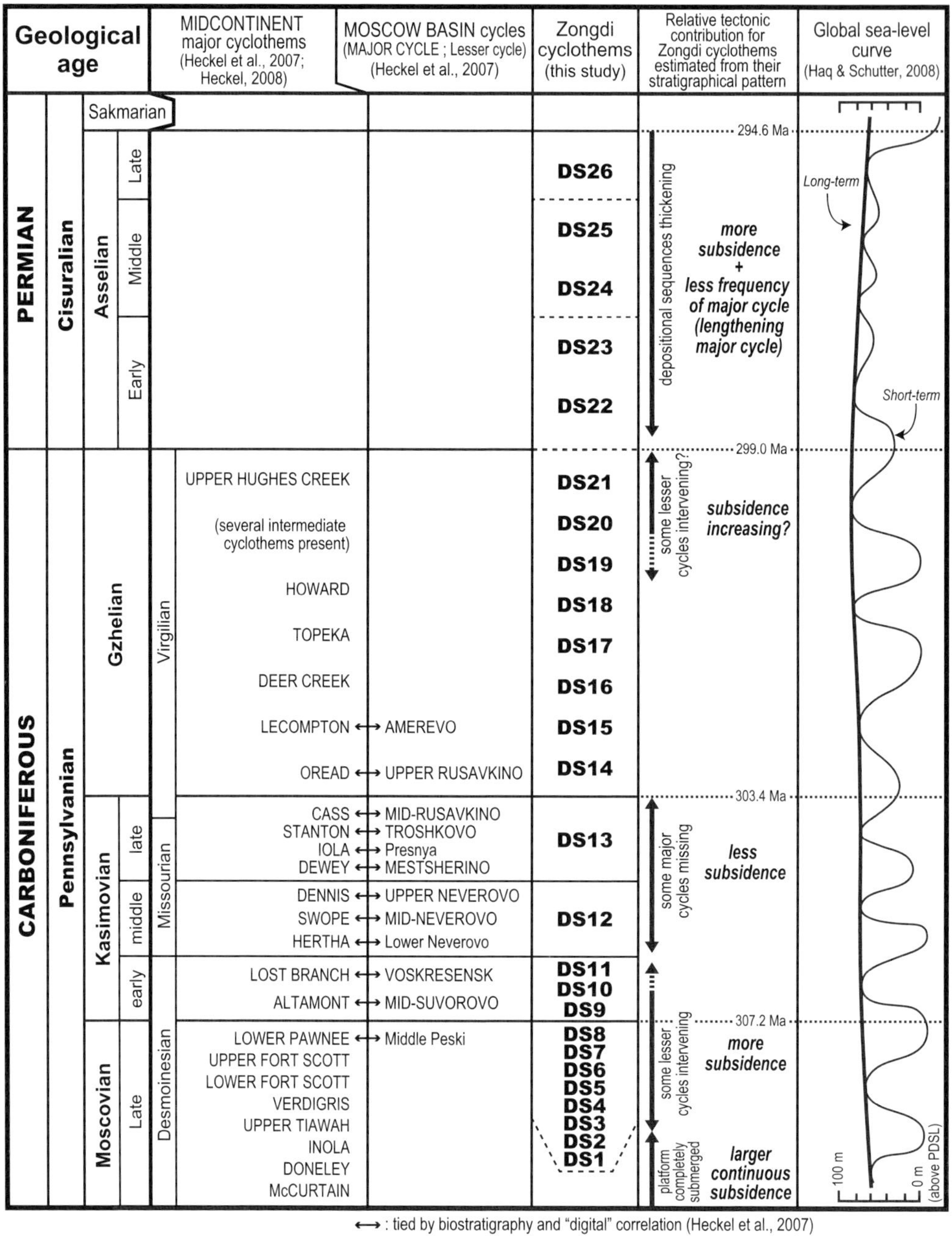

Fig. 21. Estimation of relative tectonic control for formation of Zongdi cyclothems based on stratigraphical appearances of Zongdi cyclothems and coeval cyclothems in Midcontinent Basin of North America and Moscow Basin of Russia (Heckel *et al.* 2007), and change of long-term sea-level (Haq & Schutter 2008). Numerical ages for stage boundaries are from Ogg *et al.* (2008). PDSL: present-day sea-level.

cyclothems (Figs 5 & 14). Their thickness varies from 8.5 m (DS22) to 24.5 m (DS25) (Fig. 5), averaging 15.9 m. They are 2–6 times thicker than the Pennsylvanian cyclothems, which have an average thickness of *c.* 4.1 m. This observation requires an increasing rate of formation of accommodation

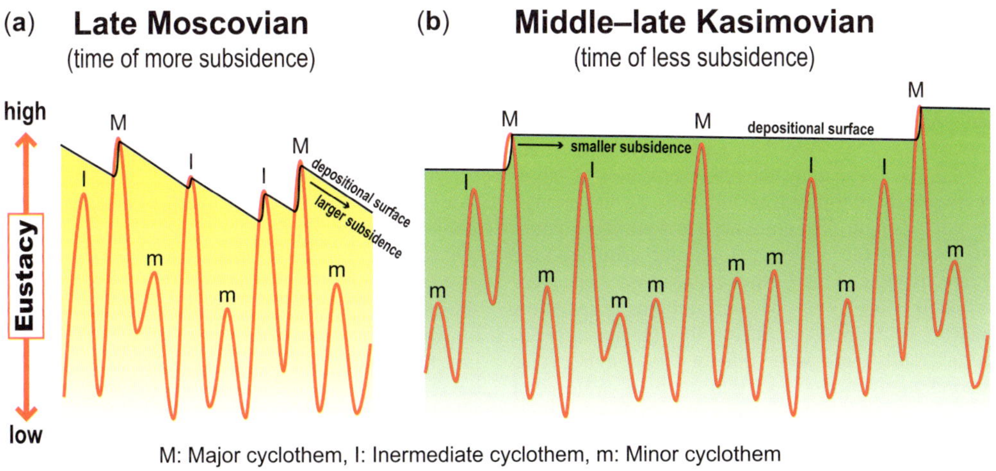

Fig. 22. Schematic diagrams accounting for differential stratigraphical patterns of Zongdi cyclothems during two Pennsylvanian time intervals that have differing rates of subsidence. (**a**) Late Moscovian cyclothems formed during a period of more subsidence, with a higher rate of subsidence than is shown in Figure 20. In this case, succession is represented by a stacking of not only major cyclothems but also some intervening intermediate cyclothems. (**b**) Middle–late Kasimovian cyclothems formed during a period of less subsidence, with a lower rate of subsidence than is shown in Figure 20. During that time interval succession consists of major cyclothems; even some major cycles that are seen in other epeiric basins (e.g. North American Midcontinent) are potentially missing.

space per unit time; in other words it concludes the trend of larger relative sea-level rise in the earliest Permian in the Zongdi succession. In such conditions, at least all sedimentary cycles that are comparable in the magnitude of sea-level rise to the major cyclothems in Pennsylvanian time must be recorded in the high-shelf area of the Yangtze Carbonate Platform, such as Zongdi; there are even possibilities of some cycles of intermediate scale being recorded in it.

On this premise, we estimate the cyclicity of these Asselian thicker cyclothems. Adopting 4.4 myr in Ogg *et al.* (2008) for the time interval of the Asselian Stage, five Asselian Zongdi cyclothems (but possibly varying four–six, taking uncertainty of the positions of the Carboniferous–Permian and Asselian–Sakmarian boundaries in the Zongdi section into account) are considered to have an average cycle period of *c.* 880 kyr (varying over the range of 730 kyr–1.1 myr). In Middle–Late Pennsylvanian successions, 400- to 450-kyr units that correspond to the Milankovitch Period (longest eccentricity period) and are probably related to the formation of major cyclothems are known to prevail (Heckel 2008; Peterson 2011). Even considering some existing uncertainties that the Asselian Zongdi cyclothems could include those made by intermediate-scale transgressions, as in the case of the Late Moscovian, and our lack of success in delineating the base and top of the Asselian in the Zongdi section, it is likely probable that the appearing time-interval of major cycles in the Asselian became about twice that of the Pennsylvanian. Because the thicknesses of cyclothems have increased (by a factor of 2–6) more than the rise in cyclothem interval (by a factor of 2) in the Asselian compared to those of the Pennsylvanian, and because the long-term sea-level is estimated to be slightly lowered (Haq & Schutter 2008; Fig. 21), the increasing thickness of the Asselian cyclothem is best interpreted to be a result of increasing tectonic subsidence, coupled with the lengthening appearance intervals of the Asselian major cyclothems (Fig. 21).

The concrete rate of subsidence can be estimated using the average rate of sedimentation for the Asselian. Employing the same method presented in the preceding subsection, the rate of sedimentation per million years is calculated at 15.7 m for the Asselian. As only a small sea-level fall is postulated in its long-term fluctuation during this time interval (Fig. 21), this value is broadly regarded to be equivalent to the average rate of subsidence at that time. The Asselian value of 15.7 m/myr is about twice as large as both that of the Kasimovian (7.3 m/myr) and Gzhelian (8.0 m/myr), implying double the rate of subsidence in the Asselian compared to the Late Pennsylvanian.

The doubling of the interval of appearance of the major cyclothems, thus the formation of fewer numbers of major cycles in a definite time interval, under subsidence-increasing conditions during the

Asselian in the Zongdi section requires explanation. The fewer larger transgressions that extended onto the high shelf at Zongdi during the Asselian are presumably related to the fact that the earliest Permian (Asselian and early Sakmarian) was a more intensely glacial time than earlier (Fielding *et al.* 2008*a*, *b*). There were far fewer episodes of large-scale glacial melting (transgression) strong enough to reach Zongdi than compared to a similar amount of time in the generally warmer Middle–Late Pennsylvanian. This further means that there were probably more times of lesser amounts of ice-sheet melting on Gondwana during this glacial interval. As a result, there were only a few major cycles among many minor and intermediate cycles that never extended onto the Yangtze Carbonate Platform to Zongdi in the Asselian.

By combining all these lines of evidence, and comparing with a hypothetical average cyclothem during the Pennsylvanian, we propose a model explaining the formation processes of the Early Permian thicker Zongdi cyclothems (Fig. 23). Under 400–450 kyr average cyclicity of major transgressions during the Pennsylvanian, relatively thin major cyclothems with moderate time intervals of their appearance are expected to stack over on a high-shelf carbonate platform area that has a moderate constant rate of subsidence comparable to those of the North American Midcontinent and the Moscow Basin of Russia (Fig. 23a). The stratigraphical pattern of cyclothems formed in such conditions would resemble that of the (early) Gzhelian in the Zongdi section. During the Early Permian on the other hand, more intense glaciation decreased chances of large-scale melting of Gondwanan ice sheets; this resulted in lengthening of the interval of two successive major cyclothems, while forming many more minor cycles of transgression that never extended high enough onto the shelf to reach Zongdi. As a result, a bigger number of lesser transgressions that could form intermediate and minor cyclothems occurred between two successive major cyclothems at lengthening appearance intervals. This, coupled with probably larger subsidence effects, would contribute to produce a larger accommodation space for the next major sea-level highstand, making an Asselian thicker cyclothem in the Zongdi section (Fig. 23b).

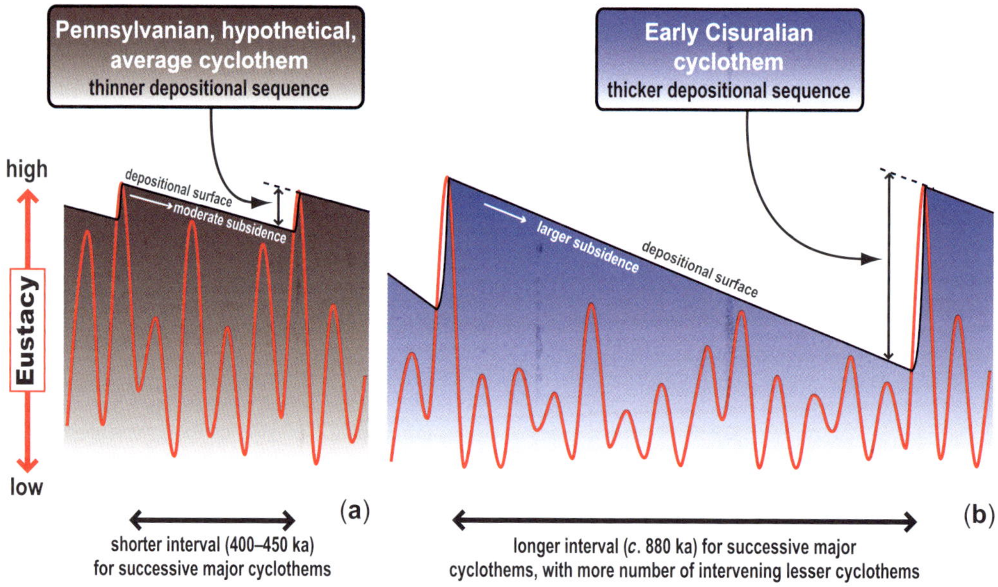

Fig. 23. Hypothesis for explaining formation processes of Early Permian (Asselian) thicker cyclothems in Zongdi section. (**a**) Formation processes of Pennsylvanian, hypothetical, average cyclothems in high-shelf area, under controls of 400–450 kyr average cyclicity of major cycles and moderate (average) rate of subsidence comparable to that of North American Midcontinent (Fig. 20). A similar frequency of 400–450 kyr average cyclicity of relatively thin major cyclothems is expected under these conditions. In Zongdi section, Gzhelian cyclothems are somewhat similar to those formed in this model. (**b**) During the Early Permian, more intense glaciation meant that there was a smaller probability of a large amount of melting of ice sheets on Gondwana, which lengthened the interval of two successive major cyclothems. As a result, a larger number of intervening lesser (intermediate and minor) cyclothems were formed between them. Together with probable, larger epeiric subsidence, this would have produced a larger accommodation space for the next major sea-level highstand, which resultantly made a thicker cyclothem at Zongdi.

Implication to the palaeoclimate of the South China Block

Another interesting implication of the Zongdi cyclothems involves their application to interpreting the climatic condition of the South China Craton during their formation. Increasing information on global palaeogeography during the Late Palaeozoic has generally located the South China Block in a palaeo-equatorial area (e.g. Golonka 2000; Blakey 2007, 2008). Moreover, major lithofacies distribution and/or environmental subdivision on the block have been documented in some works (Feng *et al.* 1991, 1994, 1998; Golonka & Ford 2000; Wang & Jin 2000; Cui 2004). Little is known, however, regarding its palaeoclimatic conditions during the time of the Gondwanan glaciation, although Chen *et al.* (2001) interpreted a relatively dry climate for the South China Craton at that time based on the formation/deposition of dolomite and oolitic limestone in the Jiangnan Basin as well as that of gypsum in places (such as eastern Sichuan Province). In contrast, Rowley *et al.* (1985) earlier suggested a wet climate for South China during Westphalian C–D time of the Pennsylvanian, which is slightly earlier than the time of the Zongdi cyclothems, based on palaeogeograpical and phytogeographical information.

In the first part of this section, we suggested that the time gap between two successive depositional sequences, represented by a sequence boundary (disconformity), would be longer than the entire depositional period of a single sequence (Fig. 20). In particular, we estimated that the middle–late Kasimovian and the Asselian had larger hiatuses between two successive cyclothems (Figs 22 & 23). These arguments provide a significant basis for the consideration of climatic conditions on the Yangtze Carbonate Platform during the Pennsylvanian–Early Permian. Despite these estimates of relatively longer emergence in the Zongdi cyclothems, the amount of erosion during each subaerial exposure is estimated to be relatively small; peritidal sediments formed just before emergent events are well preserved immediately below sequence boundaries in most depositional sequences (Fig. 14), even including mud-flat sediments corresponding to the latest stage of regression in some cases. This evidence strongly suggests that erosion/dissolution was not very effective on exposed surfaces of the Yangtze Carbonate Platform during each emergent event, and further implies that its climatic condition was probably not humid enough to promote the formation of deeply karstified ground surfaces when the platform was broadly emergent.

This conclusion is also supported by the nature of the palaeosols. In a broad perspective, it is known that soil chemistry is primarily controlled by climate. Soils in arid or semi-arid condition are dominated by calcium carbonate whereas those formed in humid climate are poor in that component, as it is easily dissolved and transported away (Retallack 2001). The former type of soils corresponds to what was called pedocal by Marbut (1935). In the Zongdi section, many of the sequence boundaries during the Pennsylvanian have thin palaeosol profiles that are more or less consolidated with partly calcareous materials. Meanwhile, a tropical climate is inevitably anticipated for the South China Block of this age based on its reconstructed palaeo-equatorial position. Moreover, pendant cement is often recognized in the Zongdi succession, suggesting at least brief times of rainfall and percolation through the vadose zone. However, neither the well-preserved peritidal sediments just before major emergent events nor the palaeosol character indicate a very humid (wet) climate with a high rate of rainfall. More likely, a less humid (rather, somewhat dry) climate would have prevailed on the South China Block during the Pennsylvanian–Early Permian when the Zongdi cyclothems were formed, particularly during those more glacial times when the Yangtze Carbonate Platform was largely emergent. This argument substantiates the dry climate proposed by Chen *et al.* (2001) based on more general features of the shallow-marine sediments deposited on the South China Block during this time. These dry conditions strongly support the location of the South China Block in the dry subtropical to tropical zone on either side of the palaeo-equator, rather than in the equatorial zone where daily rainfall prevails during much of the year.

Conclusions

Cyclothems are one of the most significant stratigraphical features in Pennsylvanian–Early Permian successions of shallow-marine (to paralic) origin. Cyclothemic successions have been reported from many of the major late Palaeozoic sedimentary basins of the world. On the South China Craton, the large Yangtze Carbonate Platform developed during the Late Palaeozoic–Early Mesozoic. Despite increasing stratigraphical and biostratigraphical information on Pennsylvanian–Early Permian strata in this area, however, cyclothemic successions have not been reported there until now.

In this paper, we have documented for the first time icehouse-type high-frequency sedimentary cycles (cyclothems) in the Pennsylvanian–Early Permian shallow-marine carbonate succession of the Yangtze Carbonate Platform at the Zongdi section in southern Guizhou Province, South China. The sedimentary cycles are primarily established by defining sequence-bounding major lowstand horizons (disconformities), which can be recognized by dark-coloured limestones and associated macro- and microfabrics related to subaerial exposure such as palaeosols, karst breccias, rhizoliths, *Microcodium*, alveolar texture and pendant cement. A total of 26 depositional sequences (DS1–DS26) were identified in the *c.* 150-m-thick Zongdi section. They range from the Late Moscovian (late Middle Pennsylvanian) to the Asselian (early Early Permian) based on fusuline biostratigraphy. Moreover, it suggests that the Moscovian–Kasimovian and Kasimovian–Gzhelian boundaries are very likely accurately placed at the base of DS9 and DS14 and the Gzhelian–Asselian (thus Carboniferous–Permian) boundary is provisionally placed at the DS21–DS22 boundary.

Each depositional sequence has a similar internal facies-stacking pattern that is subdivided into three units, consisting of peritidal sediments in the basal unit, wave- or current-dominated upper (to uppermost) subtidal sediments in the middle unit, and again peritidal sediments in the upper unit. Judging from the facies-stacking pattern within each depositional sequence, the cyclic appearances of subaerial exposure surfaces and their Pennsylvanian–Early Permian age, the Zongdi depositional sequences are best interpreted to have been formed by alternating marine transgression and regression, which resulted from icehouse-type high-frequency/high-amplitude sea-level change caused by the cyclic waxing and waning of Gondwanan ice sheets. They can be compared genetically to the well-studied Pennsylvanian–Early Permian Midcontinent cyclothems in North America and to those in the Moscow Basin of Russia.

Cyclothems in the Zongdi section consist entirely of very shallow-marine carbonates and each of them is relatively thin, lacking the deeper-facies sediments further below wave base that would be comparable to the offshore core shale of the Midcontinent (Kansas-type) cyclothems. These differences between the Zongdi and Midcontinent cyclothems can be explained by interpreting the former as having formed in a high (upper) shelf area within a thinly submerged epeiric sea on the South China Craton. In this setting, sedimentation occurred in a relatively small (shallow) accommodation space that appeared only during the highest stands of sea level in relatively larger sea-level cycles, which presumably formed the major cyclothems in the North American Midcontinent Basin. This interpretation further suggests that in Zongdi, a period of deposition for a single cyclothem was much shorter than the time represented by the hiatus between two successive cyclothems, and the position of Zongdi on the Yangtze Carbonate Platform was very distant from the detrital source.

Based on fusuline biostratigraphical framework in combination with the comparison of stratigraphical appearance of cyclothems in the North American Midcontinent Basin, we clarified an uneven stratigraphical pattern of the Zongdi cyclothems. Furthermore, we detected subtle epeiric tectonism on the Yangtze Carbonate Platform. During the Late Moscovian–early Kasimovian and possibly the late Gzhelian, major cyclothems were likely associated with some intermediate cyclothems in the Zongdi section. This suggests that epeiric tectonic subsidence was effective in making more accommodation space to form some intermediate cyclothems in these time intervals on the Yangtze Carbonate Platform at Zongdi. In the middle–late Kasimovian, on the other hand, there was a smaller number of major cyclothems formed in the Zongdi section compared to the Midcontinent Basin. This is due probably to less tectonic subsidence on the Yangtze Carbonate Platform at Zongdi. In the Asselian, the Zongdi cyclothems clearly exhibit different features from the Pennsylvanian cylothems in that they have larger thicknesses with about twice the average interval of appearance of successive major cyclothems, although their internal facies composition is similar to that of the Pennsylvanian cyclothems. The Asselian thicker cyclothems were probably produced under a larger rate of subsidence with decreased frequency of appearance of major cyclothems. This is presumably related to the development of more intensely glacial, colder climates in the earliest Permian, which resulted in the formation of far fewer episodes of large-scale glacial melting (transgression) that developed major cyclothems. Moreover, it is important to note that the tops of cyclothems were little affected by erosion and dissolution, as the highest sediments that formed just before major emergent events are relatively well-preserved in Zongdi. This evidence, coupled with the slightly calcareous nature in part of their palaeosols, implies that a less humid climate prevailed on the South China Block during the Pennsylvanian–Early Permian when the Zongdi cyclothems were formed. This helps to locate the South China Block in either of the two dry climatic zones on either side of the palaeo-equator, rather than on the palaeo-equator where humid conditions would have prevailed.

We greatly appreciate the thorough review of our manuscript and many valuable comments and suggestions by P.H. Heckel (University of Iowa, USA). Our manuscript also benefited from comments by O. Kossovaya (VSEGEI, Russia). Fieldwork in southern Guizhou Province of China and subsequent laboratory work by the first author were financially supported by Grant-in-Aids for Scientific Research (C) from the Japan Society for the Promotion of Science (Nos. 16540429 and 21540486). Two co-authors (YW and XW) acknowledge the National Natural Science Foundation of China for financial support of their fieldworks.

References

ALEKSEEV, A. S., KONONOVA, L. I. & NIKISHIN, A. M. 1996. The Devonian and Carboniferous of the Moscow Syneclise (Russian Platform): stratigraphy and sea-level changes. *Tectonophysics*, **268**, 149–168.

BAIN, R. J. & FOOS, A. M. 1993. Carbonate microfabrics related to subaerial exposure and paleosol formation. *In*: REZAK, R. & LAVOIE, D. L. (eds) *Carbonate Microfabrics*. Springer-Verlag, New York, 19–27.

BELT, E. S., HECKEL, P. H., LENTZ, L. J., BRAGONIER, W. A. & LYONS, T. W. 2011. Record of glacial-eustatic sea-level fluctuations in complex middle to late Pennsylvanian facies in the Northern Appalachian basin and relation to similar events in the Midcontinent basin. *Sedimentary Geology*, **238**, 79–100.

BLAKEY, R. C. 2007. Carboniferous–Permian paleogeography of the assembly of Pangaea. *In*: WONG, TH. E. (ed.) *Proceedings of the XVth International Congress on Carboniferous and Permian Stratigraphy*. Royal Netherlands Academy of Arts and Sciences, Amsterdam, 443–456.

BLAKEY, R. C. 2008. Gondwana paleogeography from assembly to breakup – A 500 m.y. odyssey. *In*: FIELDING, C. R., FRANK, T. D. & ISBELL, J. L. (eds) *Resolving the Late Paleozoic Ice Age in Time and Space*. The Geological Society of America, Boulder, Special Paper, **441**, 1–28.

BUGGISCH, W., WANG, X. D., ALEKSEEV, A. S. & JOACHIMSKI, M. M. 2011. Carboniferous–Permian carbon isotope stratigraphy of successions from China (Yangtze platform), USA (Kansas) and Russia (Moscow Basin and Urals). *Palaeogeography, Palaeoclimatology, Palaeoecology*, **301**, 18–38.

BUREAU OF GEOLOGY AND MINERAL RESOURCES OF GUIZHOU PROVINCE. 1987. *Regional Geology of Guizhou Province (Geological Memoirs, Series 1, Number 7)*. Geological Publishing House, Beijing (in Chinese with English summary).

CALVER, M. A. 1968. Distribution of Westphalian marine faunas in northern England and adjoining areas. *Proceedings of the Yorkshire Geological Society*, **37**, 1–72.

CHEN, X., YUAN, Y. P. & BOUCOT, A. J. (eds) 2001. *Evolution of the Paleozoic Climate of China*. Science Press, Beijing (in Chinese).

CHEN, Z. Q., JIN, Y. G. & SHI, G. R. 1998. Permian transgression-regression sequences and sea-level changes of South China. *In*: SHI, G. R., ARCHBOLD, N. W. & GROVER, M. (eds) *The Permian System: Stratigraphy, Palaeogeography & Resources. Proceedings of the Royal Society of Victoria*, **110**, 345–367.

CUI, K. X. 2004. *Atlas of the Paleogeography and Its Evolution of Southwest China*. Seismological Press, Beijing (in Chinese).

DAVYDOV, V. I. 1997. Middle/Upper Carboniferous boundary: problems of definition and correlation. *In*: PODEMSKI, M., DYBOVA-JACHOWICZ, S., JAWOROWSKI, K., JURECZKA, J. & WAGNER, R. (eds) *Proceedings of the XIII International Congress on the Carboniferous and Permian, Part 1*. Prace Państwowego Instytutu Geologicznego, **157**, 113–122.

DOUGLASS, R. C. 1987. Fusulinid biostratigraphy and correlations between the Appalachian and Eastern Interior basins. *U. S. Geological Survey, Professional Paper*, **1451**, 1–95.

DUNBAR, C. O. & HENBEST, L. G. 1942. Pennsylvanian Fusulinidae of Illinois. *Illinois State Geological Survey, Bulletin*, **67**, 1–218.

DUNHAM, R. J. 1969. Early vadose silt in Townsend mound (reef), New Mexico. *In*: FRIEDMAN, G. M. (ed.) *Depositional Environments in Carbonate Rocks*. Society of Economic Paleontologists and Mineralogists, Tulsa, Special Publication, **14**, 139–182.

EROS, J. M., MONTAÑEZ, I. P., OSLEGER, D. A., DAVYDOV, V. I., NEMYROVSKA, T. I., POLETAEV, V. I. & ZHYKALYAK, M. V. 2012. Sequence stratigraphy and onlap history of the Donets Basin, Ukraine: insight into Carboniferous icehouse dynamics. *Palaeogeography, Palaeoclimatology, Palaeoecology*, **313–314**, 1–25.

ESTEBAN, M. 1974. Caliche textures and Microcodium. *Bollettino della Societa Geologica Italiana*, **92**(Supplement), 105–125.

ESTEBAN, M. & KLAPPA, C. F. 1983. Subaerial exposure environment. *In*: SCHOLLE, P. A., BEBOUT, D. G. & MOORE, C. H. (eds) *Carbonate Depositional Environments*. American Association of Petroleum Geologists, Tulsa, Memoir, **33**, 1–95.

FENG, Z. Z., HE, Y. B. ET AL. 1991. *Lithofacies Paleogeography of Permian of Middle and Lower Yangtze Region*. Geological Publishing House, Beijing (in Chinese with English summary).

FENG, Z. Z., JIN, Z. K. ET AL. 1994. *Lithofacies Paleogeography of Permian of Yunnan–Guizhou–Guangxi Region*. Geological Publishing House, Beijing (in Chinese with English abstract).

FENG, Z. Z., YANG, Y. Q. ET AL. 1998. *Lithofacies Paleogeography of Carboniferous in South China*. Geological Publishing House, Beijing (in Chinese with English summary).

FIELDING, C. R., FRANK, T. D., BIRGENHEIER, L. P., RYGEL, M. C., JONES, A. T. & ROBERTS, J. 2008*a*. Stratigraphic record and facies associations of the late Paleozoic ice age in eastern Australia (New South Wales and Queensland). *In*: FIELDING, C. R., FRANK, T. D. & ISBELL, J. L. (eds) *Resolving the Late Paleozoic Ice Age in Time and Space*. Geological Society of America, Boulder, Special Paper, **441**, 41–57.

FIELDING, C. R., FRANK, T. D. & ISBELL, J. L. 2008*b*. The late Paleozoic ice age – A review of current understanding and synthesis of global climate patterns. *In*: FIELDING, C. R., FRANK, T. D. & ISBELL, J. L. (eds) *Resolving the Late Paleozoic Ice Age in Time and Space*. Geological Society of America, Boulder, Special Paper, **441**, 343–354.

FLÜGEL, E. 2004. *Microfacies of Carbonate Rocks: Analysis, Interpretation and Application*. Springer, Berlin.

FORKE, H. C. 2002. Biostratigraphic subdivision and correlation of uppermost Carboniferous/Lower Permian sediments in the Southern Alps: Fusulinoidean and conodont faunas from the Carnic Alps (Austria/Italy), Karavanke Mountains (Slovenia), and Southern Urals (Russia). *Facies*, **47**, 201–276.

GOLONKA, J. 2000. *Cambrian–Neogene Plate Tectonic Maps*. Wydawnictwa Uniwersytetu Jagiellonskiego, Kraków.

GOLONKA, J. & FORD, D. 2000. Pangean (Late Carboniferous–Middle Jurassic) paleoenvironment and lithofacies. *Palaeogeography, Palaeoclimatology, Palaeoecology*, **161**, 1–34.

HAQ, B. U. & SCHUTTER, S. R. 2008. A chronology of Paleozoic sea-level changes. *Science*, **322**, 64–68.

HAYAKAWA, N., NAKAZAWA, T. & UENO, K. 2008. A Middle Permian subaerial exposure surface and the overlying tidal-flat deposits in seamount-capping shallow-marine carbonates of the Omi Limestone, Niigata Prefecture, Central Japan. *Fukuoka University Science Reports*, **38**, 59–73 (in Japanese with English abstract).

HECKEL, P. H. 1986. Sea-level curve for Pennsylvanian eustatic marine transgressive-regressive depositional cycles along midcontinent outcrop belt, North America. *Geology*, **14**, 330–334.

HECKEL, P. H. 1994. Evaluation of evidence for glacio-eustatic control over marine Pennsylvanian cyclothems in North America and consideration of possible tectonic effects. *In*: DENNISON, J. M. & ETTENSOHN, F. R. (eds) *Tectonic and Eustatic Controls on Sedimentary Cycles*. SEPM (Society for Sedimentary Geology), Tulsa, Concepts in Sedimentology and Paleontology, **4**, 65–87.

HECKEL, P. H. 1995. Glacial-eustatic sea-level – Climatic model for late Middle to Late Pennsylvanian coal-bed formation in the Appalachian Basin. *Journal of Sedimentary Research*, **B65**, 348–356.

HECKEL, P. H. (ed.). 1999. Middle and Upper Pennsylvanian (Upper Carboniferous) cyclothem succession in Midcontinent Basin, U.S.A. *Kansas Geological Survey Open-file Report*, **99–27**, 1–236.

HECKEL, P. H. 2002. Overview of Pennsylvanian cyclothems in mid-continent North America and brief summary of those elsewhere in the world. *In*: HILLS, L. V., HENDERSON, C. M. & BAMBER, E. W. (eds) *Carboniferous and Permian of the World*. Canadian Society of Petroleum Geologists, Calgary, Memoir, **19**, 79–98.

HECKEL, P. H. 2008. Pennsylvanian cyclothems in Midcontinent North America as far-field effects of waxing and waning of Gondwana ice sheets. *In*: FIELDING, C. R., FRANK, T. D. & ISBELL, J. L. (eds) *Resolving the Late Paleozoic Ice Age in Time and Space*. Geological Society of America, Boulder, Special Paper, **441**, 275–289.

HECKEL, P. H., ALEKSEEV, A. S. *ET AL*. 2007. Cyclothem ['digital'] correlation and biostratigraphy across global Moscovian–Kasimovian–Gzhelian Stage boundary interval (Middle–Upper Pennsylvanian Series) in North America and Eurasia. *Geology*, **35**, 607–610.

ISAKOVA, T. N. 2001. Fuzulinidy. *In*: MAKHLINA, M. Kh., ALEKSEEV, A. S. *ET AL*. (eds) *Sredniy Karbon Moskovskoy Sineklizy (Yuzhnaya Chast') Tom 2, Paleontologicheskaya Kharakteristika*. Nauchnyy Mir, Moskva, 10–32 (in Russian).

IZART, A., BRIAND, C., VASLET, D., VACHARD, D., COQUEL, R. & MASLO, A. 1996. Stratigraphy and sequence stratigraphy of the Moscovian in the Donets basin. *Tectonophysics* **268**, 189–209.

IZART, A., VACHARD, D., VASLET, D. & MASLO, A. 2002. Sedimentology of the Upper Carboniferous and Lower Permian in the Dniepr and Donets basins. *In*: HILLS, L. V., HENDERSON, C. M. & BAMBER, E. W. (eds) *Carboniferous and Permian of the World*. Canadian Society of Petroleum Geologists, Calgary, Memoir, **19**, 120–143.

KABANOV, P. 2003. The Upper Moscovian and basal Kasimovian (Pennsylvanian) of Central European Russia: facies, subaerial exposures and depositional model. *Facies*, **49**, 243–270.

KABANOV, P. & BARANOVA, D. 2007. Cyclothems and stratigraphy of the Upper Moscovian–basal Kasimovian (Pennsylvanian) succession of central and northern European Russia. *In*: WONG, TH. E. (ed.) *Proceedings of the XVth International Congress on Carboniferous and Permian Stratigraphy*. Royal Netherlands Academy of Arts and Sciences, Amsterdam, 147–160.

KABANOV, P. B., ALEKSEEV, A. S., BARANOVA, D. V., GORJUNOVA, R. V., LAZAREV, S. S. & MALKOV, V. G. 2006. Biotic changes in a eustatic cyclothem: Domodedovo Formation (Moscovian, Carboniferous) of Peski Quarries, Moscow Region. *Paleontological Journal*, **40**, 351–368.

KABANOV, P., ANADÓN, P. & KRUMBEIN, W. E. 2008. *Microcodium*: an extensive review and a proposed nonrhizogenic biologically induced origin for its formation. *Sedimentary Geology*, **205**, 79–99.

KANMERA, K. 1958. Fusulinids from the Yayamadake Limestone of the Hikawa Valley, Kumamoto Prefecture, Kyushu, Japan. Part 3, Fusulinids of the Lower Permian. *Memoirs of the Faculty of Science, Kyushu University, Series D*, **6**, 153–215.

KLAPPA, C. F. 1978. Biolithogenesis of Microcodium; elucidation. *Sedimentology*, **25**, 489–522.

KLAPPA, C. F. 1980. Brecciation textures and tepee structures in Quaternary calcrete (caliche) profiles from eastern Spain; The plant factor in their formation. *Geological Journal*, **15**, 81–89.

LEVEN, E. JA. & DAVYDOV, V. I. 2001. Stratigraphy and fusulinids of the Kasimovian and upper Gzhelian (Upper Carboniferous) in the southwestern Darvaz (Pamir). *Rivista Italiana di Paleontologia e Stratigrafia*, **107**, 3–46.

LEVEN, E. JA. & SCHERBOVICH, S. F. 1978. *Fuzulinidy i Stratigrafiya Assel'skogo Yarusa Darvaza*. Akademiya Nauk SSSR, Moskovskoe Obschestvo Ispytateley Prirody, Izdatel'stvo Nauka, Moskva (in Russian).

LIN, J. X., LI, J. X. & SUN, Q. Y. 1990. *Late Paleozoic Foraminifers of South China*. Science Press, Beijing (in Chinese).

MARBUT, C. F. 1935. *Atlas of American Agriculture. Part III, Soils of the United States*. U.S. Government Printing Office, Washington.

NAKAZAWA, T. & UENO, K. 2004. Sequence boundary and related sedimentary and diagenetic facies formed on Middle Permian mid-oceanic carbonate platform: core observation of Akiyoshi Limestone, Southwest Japan. *Facies*, **50**, 301–311.

OGG, J. G., OGG, G. & GRADSTEIN, F. 2008. *The Concise Geologic Time Scale*. Cambridge University Press, Cambridge.

OZAWA, T., WATANABE, K. & KOBAYASHI, F. 1992. Morphologic evolution in some schwagerinid and schubertellid lineages and the definition of the Carboniferous–Permian Boundary. *In*: TAKAYANAGI, Y. & SAITO, T. (eds) *Studies in Benthic Foraminifera (Proceedings of the Fourth International Symposium on Benthic*

Foraminifera, Sendai, 1990). Tokai University Press, Tokyo, 389–401.

PETERSON, J. A. 2011. Better mathematical constraints on ages of Carboniferous stage boundaries using radiometric tuff dates and cyclostratigraphy. *Geochemistry, Geophysics, Geosystems*, **12**, 7, http://dx.doi.org/10.1029/2010GC003467

RAMSBOTTOM, W. H. C. 1977. Major cycles of transgression and regression (mesothems) in the Namurian. *Proceedings of the Yorkshire Geological Society*, **41**, 261–291.

RAMSBOTTOM, W. H. C. 1979. Rates of transgression and regression in the Carboniferous of NW Europe. *Journal of the Geological Society*, **136**, 147–153.

RAUZER-CHERNOUSOVA, D. M. & SCHERBOVICH, S. F. 1958. O shvagerinovom gorizonte tsentral'noy chasti Russkoy Platformy. *Akademiya Nauk SSSR, Trudy Geologicheskogo Instituta*, **13**, 3–56 (in Russian).

RETALLACK, G. 2001. *Soils in the Past: An Introduction to Paleopedology*. 2nd edn. Blackwell Science, Oxford.

ROSS, C. A. & ROSS, J. R. P. 1987. Late Paleozoic sea levels and depositional sequences. *In*: ROSS, C. A. & HAMAN, D. (eds) *Timing and Depositional History of Eustatic Sequences: Constraints on Seismic Stratigraphy*. Cushman Foundation for Foraminiferal Research, Washington DC, Special Publication, **24**, 137–149.

ROSS, C. A. & ROSS, J. R. P. 1988. Late Paleozoic transgressive–regressive deposition. *In*: WILGUS, C. K., HASTINGS, B. S., KENDALL, C. G. ST. C., POSAMENTIER, H. W., ROSS, C. A. & VAN WAGONER, J. C. (eds) *Sea-Level Changes: An Integrated Approach*. Society of Economic Paleontologists and Mineralogists, Tulsa, Special Publication, **42**, 227–247.

ROWLEY, D. B., RAYMOND, A., PARRISH, J. T., LOTTES, A. L., SCOTESE, C. R. & ZIEGLER, A. M. 1985. Carboniferous paleogeographic, phytogeographic, and paleoclimatic reconstructions. *International Journal of Coal Geology*, **5**, 7–42.

ROZOVSKAYA, S. E. 1958. Fuzulinidy i biostratigraficheskoe raschlenenie verkhnekamennougol'nykh otlozheniy Samarskoy Luki. *Akademiya Nauk SSSR, Trudy Geologicheskogo Instituta*, **13**, 57–120 (in Russian).

SCHOLLE, P. A. & ULMER-SCHOLLE, D. S. 2003. *A color guide to the petrography of carbonate rocks: grains, textures, porosity, diagenesis*. American Association of Petroleum Geologists, Tulsa, Memoir, **77**, 1–474.

SHENG, J. C. 1963. Permian fusulinids of Kwangsi, Kueichow and Szechuan. *Palaeontologia Sinica, New Series B*, **10**, 1–247.

STRASSER, A. 1984. Black pebbles occurrence and genesis in Holocene carbonate sediments (Florida Keys, Bahamas, and Tunisia). *Journal of Sedimentary Petrology*, **54**, 1097–1109.

VEEVERS, J. J. & POWELL, C. MCA. 1987. Late Paleozoic glacial episodes in Gondwanaland reflected in transgressive-regressive depositional sequences in Euroamerica. *Geological Society of America Bulletin*, **98**, 475–487.

VILLA, E. & TASK GROUP. 2008. Progress report of the task group to establish the Moscovian–Kasimovian and Kasimovian–Gzhelian boundaries. *Newsletter on Carboniferous Stratigraphy*, **26**, 12–13.

WADDEL, D. E. 1966. Ardmore Basin fusulinids. *Oklahoma Geological Survey, Bulletin*, **113**, 1–128.

WANG, C. Y. (ed.) 1993. *Conodonts of Lower Yangtze Valley – An Indexes to Biostratigraphy and Organic Metamorphic Maturity*. Science Press, Beijing (in Chinese with English abstract).

WANG, Z. H. & HIGGINS, A. C. 1989. Conodont zonation of the Namurian–Lower Permian strata in South Guizhou, China. *Palaeontologia Cathayana*, **4**, 261–325.

WANG, Y. & JIN, Y. G. 2000. Permian palaeogeographic evolution of the Jiangnan Basin, South China. *Palaeogeography, Palaeoclimatology, Palaeoecology*, **160**, 35–44.

WANG, X. D. & JIN, Y. G. 2003. Carboniferous biostratigraphy of China. *In*: ZHANG, W. T., CHEN, P. J. & PALMER, A. R. (eds) *Biostratigraphy of China*. Science Press, Beijing, 281–329.

WANLESS, H. R. & WELLER, J. M. 1932. Correlation and extent of Pennsylvanian cyclothems. *Bulletin of the Geological Society of America*, **43**, 1003–1016.

XIAO, W. M., WANG, H. D., ZHANG, L. X. & DONG, W. L. 1986. *Early Permian Stratigraphy and Faunas in Southern Guizhou*. The People's Publishing House of Guizhou, Guiyang (in Chinese with English summary).

ZHANG, L. X., ZHOU, J. P. & SHENG, J. Z. 2010. Upper Carboniferous and Lower Permian fusulinids from Western Guizhou. *Palaeontologia Sinica, New Series B*, **34**, 1–296 (in Chinese with English summary).

Upper Carboniferous–Lower Permian *Palaeoaplysina* build-ups on Svalbard: the influence of climate, salinity and sea-level

NILS-MARTIN HANKEN[1] & JESPER KRESTEN NIELSEN[2]*

[1]*Department of Geology, University of Tromsø, Dramsveien 201, NO-9037 Tromsø, Norway*

[2]*North Energy ASA, New Area Exploration, PO Box 1243, NO-9504 Alta, Norway*

**Corresponding author (e-mail: jesperknielsen@gmail.com)*

Abstract: Upper Carboniferous–Lower Permian *Palaeoaplysina* build-ups on Svalbard developed on a partly restricted marine warm-water carbonate platform at low palaeo-latitudes. The *Palaeoaplysina* build-ups mostly have a flat-topped tabular geometry, representing the mature stage of build-up growth. With decreasing accommodation space, continued vertical growth was hampered and lateral growth became dominant. The tabular growth form of the build-ups is probably a result of lateral amalgamation of several minor mounds. In central Spitsbergen many of the tabular build-ups form parts of continuous ridges or ring-shaped to polygonal build-up complexes encircling a lagoon. Lateral interfingering between bedded fusulinid dolostones along the build-up margins and barren dolostones further out in the lagoon may indicate increasing salinity from the lagoon margin to the central part. These lateral facies changes are often associated with evaporites (gypsum) and low biodiversity, especially abundant euryhaline fusulinids or even barren beds. On the sea side of the *Palaeoaplysina* build-ups, the frequent renewal of normal saline seawater allowed the highest biodiversity (e.g. stenohaline crinoids). The uppermost part of some build-ups contain *Microcodium*, indicating intermittent subaerial exposure and vegetation cover during lowstand, probably caused by eustatic sea-level changes and/or local tectonics. The biota of the build-ups and their lateral sediments is typical for a shallow marine chloroforam association, indicating a tropical climate during deposition.

Palaeoaplysina is an important constituent of Upper Carboniferous and Lower Permian build-ups found along the northern margin of the Late Palaeozoic supercontinent of Pangea (Scotese *et al.* 1979; Davies *et al.* 1989; Stemmerik 2000), forming a belt that can be traced from the western part of the Ural Mountains, through Timan-Pechora, across Spitsbergen and the western Barents Sea, to Ellesmere Island and the Yukon, and then southwards to British Columbia, Idaho and California (Fig. 1) (Watkins & Wilson 1989; Fossum *et al.* 2001; Henriksen *et al.* 2011). However, little is known about the palaeoenvironmental causes of the development of these build-ups, which are unique in Palaeozoic history, in terms of their composition and palaeoecology and the depositional geometry they are forming on a larger scale. Moreover, investigations based on exploration wells and seismic data (e.g. Bruce & Toomey 1993; Cecchi 1993; Bugge *et al.* 1995) give somewhat limited information about composition, geological setting and facies relationships.

Palaeoaplysina build-ups were confined to between 10°N and 40°N of the Upper Palaeozoic equator in a warm and arid climate belt; however, they were also under the influence of sea-level fluctuations during the Carboniferous–Permian glaciations of Gondwana (Stemmerik 2008) and local tectonic events (Anisimow *et al.* 1998; Davydov *et al.* 2012). To unravel the influence of the palaeo-climate, among other factors, on the depositional evolution of *Palaeoaplysina* build-ups, we have investigated the sedimentology and the floral and faunal composition of the well-exposed *Palaeoaplysina* build-ups and the associated sediments on central Spitsbergen. The lateral extent of the *Palaeoaplysina* build-up complexes on Spitsbergen is also compared with seismic data from the Norwegian Barents Sea.

Geological setting

On Spitsbergen, Carboniferous and Permian strata can be traced as a narrow belt from Sørkapp in the south along the west coast to Brøggerhalvøya (Fig. 2). However, the largest and most continuous exposures occur from Brøggerhalvøya to Sabine Land, and there are more scattered outcrops between Sabine Land and northwards to Lomfjordhalvøya and the southern part of Nordaustlandet. The Upper Palaeozoic strata are also well exposed on the island of Bjørnøya in the southernmost part of the Svalbard archipelago (Steel & Worsley

From: Gąsiewicz, A. & Słowakiewicz, M. (eds) 2013. *Palaeozoic Climate Cycles: Their Evolutionary and Sedimentological Impact*. Geological Society, London, Special Publications, **376**, 269–305.
First published online September 19, 2013, http://dx.doi.org/10.1144/SP376.17

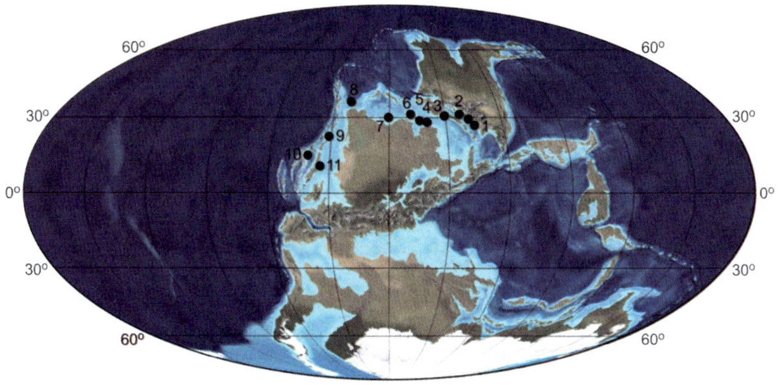

Fig. 1. Known occurrences of Upper Carboniferous and Lower Permian *Palaeoaplysina* build-ups. 1–2, western Urals (Tchuvaschov 1983); 3, Timan (Riabinin 1955); 4, Loppa High and Stappen High (Lønøy 1988); 5, Bjørnøya (Lønøy 1988); 6, Spitsbergen (Skaug *et al.* 1982; Kano 1992; Bergh *et al.* 2000); 7, Ellesmere Island and Axel Heiberg Island (Davies & Nassichuk 1973, 1986); 8, northern Yukon (Davies 1971); 9, northeastern British Columbia (Bamber & Macqueen 1979); 10, California (Watkins & Wilson 1989); 11, central Idaho (Breuninger 1976; Skipp *et al.* 1979). Approximate locations of *Palaeoaplysina* on the palaeogeographic map after Ron Blakey (http://cpgeosystems.com/index.html), showing landscape in green to pale brownish colour, shallow sea in light blue and deep ocean in dark blue.

1984). Cutbill & Challinor (1965) erected a lithostratigraphic framework for the Upper Carboniferous and Permian strata of Spitsbergen that, with slight modifications by Pickard *et al.* (1996), is still valid today (Fig. 3) (Dallmann *et al.* 1999).

The distribution of the Upper Carboniferous and Lower Permian deposits in central Spitsbergen was largely fault-controlled. The most important structural units in Spitsbergen are the Nordfjorden Block and the Billefjorden Trough (the downfaulted part of the more extensive Ny Friesland Block), which are separated by the NNW–SSE-trending Billefjorden Fault Zone. The main fault block movement occurred in the Bashkirian and Moscovian, resulting in half grabens with a syntectonic sedimentary record developed along the Billefjorden Fault Zone and other NNW–SSE lineaments (Gjelberg & Steel 1981; Steel & Worsley 1984; Mann & Townsend 1989; Braathen *et al.* 2011). The Billefjorden Fault Zone seems to have been active at least up to Asselian times, because Harland *et al.* (1974) showed an increase in thickness from 30 m to 300 m east of the fault zone in the upper Minkinfjellet and Cadellfjellet members. This view has been supported by Kleinspehn *et al.* (1984) based on a Late Carboniferous–Early Permian fan-delta complex in southwestern Spitsbergen, which was explained to reflect sporadic fault movements along the extensional margin of the Inner Hornsund Trough.

During the Carboniferous and Early Permian, both Svalbard and the Barents Sea were periodically submerged or subaerially exposed due to a combination of tectonic movements and eustasy (Stemmerik & Worsley 2005). The depositional successions were influenced by high-frequency, high-amplitude glacioeustatic sea-level fluctuations in excess of 50 m, which led to the formation of subaerially exposure-capped shelf carbonates and, most likely, growth control on *Palaeoaplysina* build-ups (Stemmerik 2008) of the Kapp Duner, Ørn and Wordiekammen formations, Gipsdalen Group (as defined in Larssen *et al.* 2002). This view has also been supported by Anisimow *et al.* (1998) and Davydov *et al.* (2012), who have shown the presence of several hiatuses in the Tyrrellfjellet Member, which appear to have been controlled by local tectonic events superimposed on eustatic sea-level changes.

Local horst structures were gradually transgressed in the Gzhelian with the deposition of widespread shallow marine carbonates in the Tyrrellfjellet Member (Skaug *et al.* 1982; Steel & Worsley 1984; Samuelsberg & Pickard 1999; Hüneke *et al.* 2001; Blomeier *et al.* 2009). With waning tectonic movements in the Late Carboniferous, most of Svalbard developed into a carbonate platform. Regions with carbonate build-ups are well known from the platform, where they occur at three different stratigraphic levels within the Minkinfjellet and Wordiekammen formations (Fig. 3). The oldest build-ups are dominated by bryozoans and are so far only known from two localities within the Minkinfjellet Formation in Bünsow Land (Samuelsberg & Pickard 1999; Nakrem 2002). The well-known phylloid algae bioherms in the Cadellfjellet Member have a much larger regional extent (Dallmann *et al.* 1999; Samuelsberg & Pickard 1999; Hüneke *et al.* 2001; Blomeier *et al.* 2009). The youngest build-ups

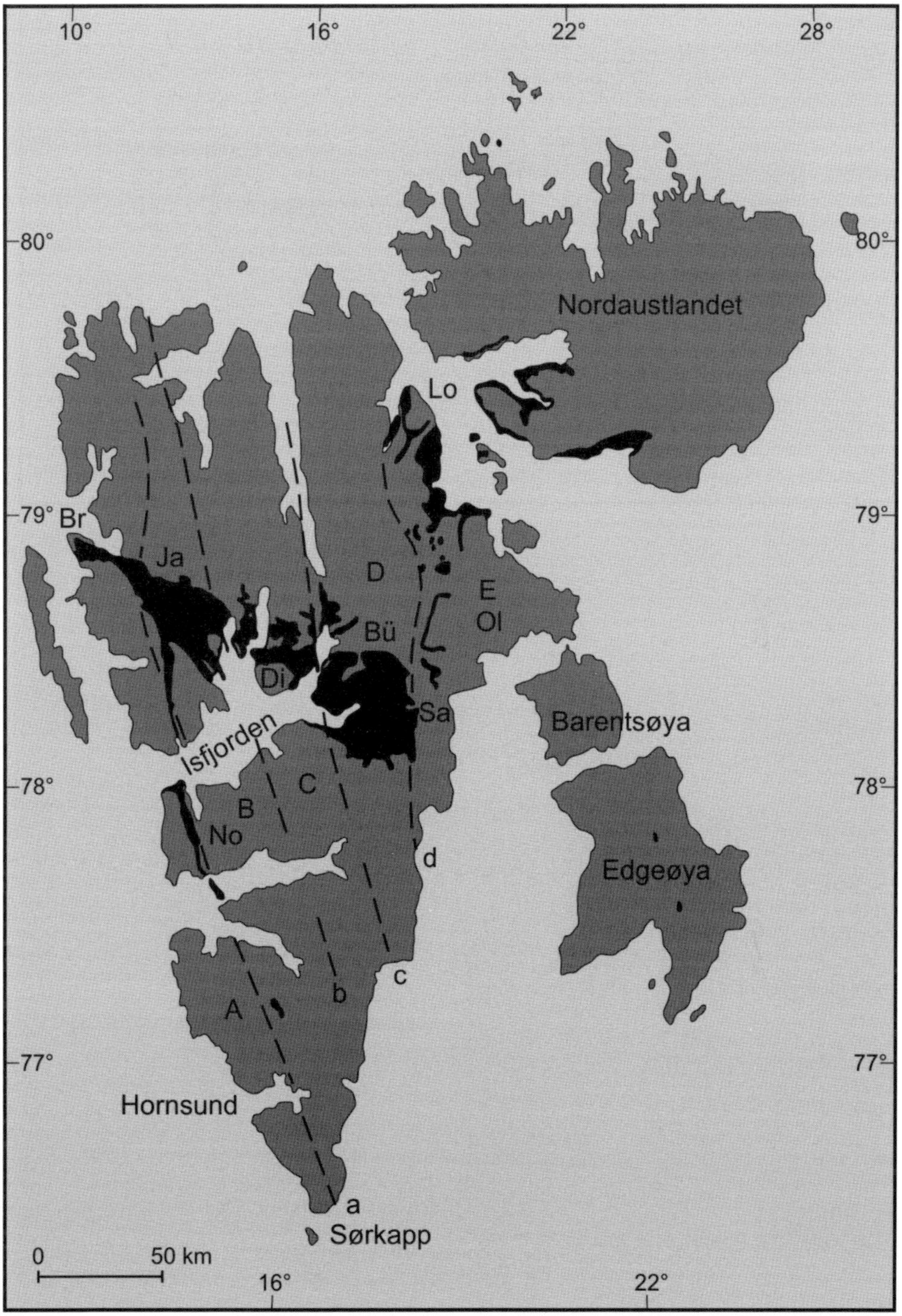

Fig. 2. Map of Svalbard, with Carboniferous–Permian deposits in black. Structural elements: A, Sørkapp–Hornsund High; B, Central Trough (St. Jonsfjorden Trough); C, Nordfjorden Block; D, Billefjorden Trough; E, Ny Friesland Block. a–d, fault lineaments; d, Billefjorden Fault Zone. *Abbreviations*: Br, Brøggerhalvøya; Bü, Bünsow Land; Di, Dickson Land; Ja, James 1 Land; Lo, Lomfjordhalvøya; No, Nordenskiöld Land; Ol, Olav V Land; Sa, Sabine Land. Modified from Pickard *et al.* (1996).

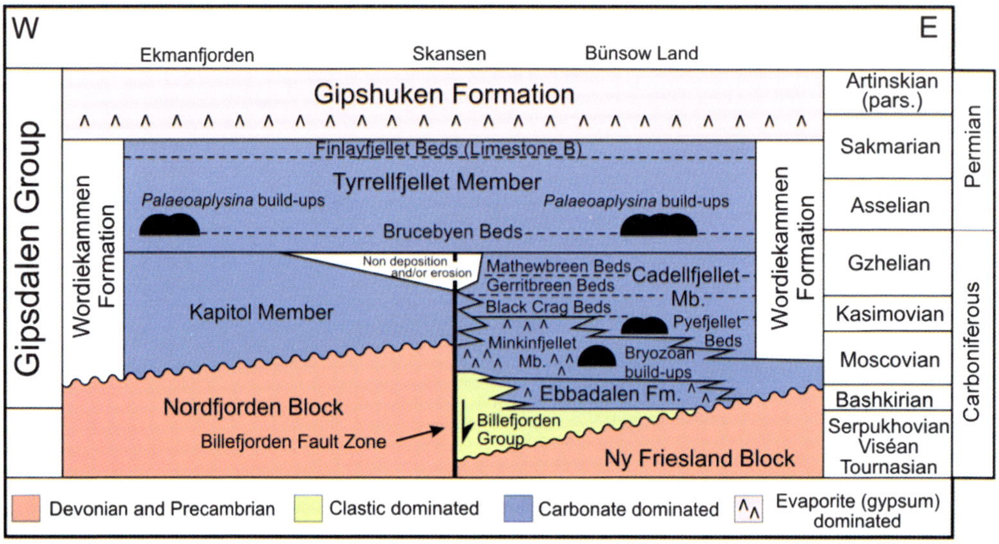

Fig. 3. Diagram illustrating the Upper Carboniferous–Lower Permian stratigraphy of central Spitsbergen. The stratigraphic levels and relative geographic positions (projected onto an east–west section) of the build-ups are indicated. Modified from Pickard *et al.* (1996).

are dominated by *Palaeoaplysina* build-ups, and these occur in some areas in the lower part of the Tyrrellfjellet Member (Skaug *et al.* 1982; Johansen 1994; Dallmann *et al.* 1999; Samuelsberg & Pickard 1999; Hüneke *et al.* 2001; Blomeier *et al.* 2009).

The lower part of the Tyrrellfjellet Member consists of organic-rich fusulinid packstone and wackestone deposits. This distinctive unit, the Brucebyen Beds, was originally defined by Cutbill & Challinor (1965). The total organic content (TOC) is fairly variable, but Bjorøy *et al.* (1982) recorded a maximum TOC of 25% at Kapitol. The organic material consists of kerogen with a composition between Type II and Type III, indicating that a significant part of the material has been derived from terrestrial vegetation. The Brucebyen Beds have a relatively extensive lateral extension although their thickness is variable, but a maximum of *c.* 11 m has been recorded at Tyrrellfjellet (Dallmann *et al.* 1999). They interfinger with the *Palaeoaplysina* build-ups, but are absent or very thin in areas with thick build-ups.

The top of the Tyrrellfjellet Member is characterized by a 35- to 40-m-thick cliff-forming limestone unit in Bünsow Land, which Gee *et al.* (1952) informally named 'Limestone B'.

The Spitsbergen and Barents Sea build-ups occur within a series of cyclic sedimentation. Each cycle of sediment deposition is separated by an exposure surface, and the primary control in cyclic development is considered to have had changes in sea-level driven by glacio-eustasy due to the Late Carboniferous and Early Permian glaciation in southern Gondwanaland (cf. Ross & Ross 1985; Veevers & Powell 1987; Ishell *et al.* 2003; Chen *et al.* 2013). A relative rise in sea-level resulted in the flooding of a widespread, relatively shallow-water carbonate platform centred on the Nordfjorden and Ny Friesland structural blocks of central Spitsbergen. Sediment accretion led to an ever-decreasing water depth and, when coupled with a subsequent fall in sea-level, resulted in at least partial exposure of the platform.

Age of the *Palaeoaplysina* build-ups

All the *Palaeoaplysina* build-ups have been found within the upper part of the Cadellfjellet Member and the lower–middle part of the overlying Tyrrellfjellet Member, in the Wordiekammen Formation. The biostratigraphic correlation and dating have been based largely on fusulinids. Cutbill & Challinor (1965) showed that the Nordenskiöldbreen Formation (equivalent to the Minkinfjellet and Wordiekammen formations, according to Pickard *et al.* (1996)) was of Moscovian to Sakmarian age, therefore agreeing with the fusulinid studies of Ross (1965), which showed that the Cyathophyllum limestone (Cadellfjellet and Tyrrellfjellet members) was no younger than the Sakmarian. Ross (1965) also assumed that the Tyrrellfjellet Member was Asselian to Sakmarian in age, which is in accordance with the biostratigraphic correlation by Keilen (1991) and Buggisch *et al.* (2001).

The Svalbard fusulinid fauna forms part of a relatively uniform fauna that occurs along the northern part of Pangaea from Arctic Canada in the west to the Timan–Pechora Basin in the east, and has therefore been extensively used for biostratigraphic dating and correlation (e.g. Nakrem *et al.* 1992; Pickard *et al.* 1996; Davydov 1997; Davydov *et al.* 2001). Nilsson (1988, 1993) and Nilsson & Davydov (1992) recognized a series of breaks in the sedimentary succession, but the precise duration was not established. However, Anisimow *et al.* (1998) and Davydov *et al.* (2012) have shown how graphic correlation (Mann & Lane 1995) can be used for correlating unconformities and estimating the duration of hiatuses in the Carboniferous–Permian sequence at Svalbard. In this way, they were able to show the presence of several hiatuses in the lower part of the Tyrrellfjellet Member.

Rasmussen & Hanken (1991, 1992) used conodonts to correlate the Tyrrellfjellet Member in Bünsow Land with equivalent successions in Arctic Canada, China, the USA and Russia. They showed that the Tyrrellfjellet Member spans the interval from a level close to the Gzhelian–Asselian boundary to within the Artinskian. The ages are based on relatively frequent occurrences of *Streptognathodus elongates* and *Cavusgnathus latus* in the lower part of the member and the reported appearance of *Sweetognathus inornatus* and *Sweetognathus whitei* in the uppermost part of the member, Limestone B (Nakrem 1991*a*). In the zonation of Arctic Canada, the Tyrrellfjellet Member would equate with the *C. latus* zone (P1) to the Artinskian *N. biselli–S. whitei* (P7) zones (Beauchamp *et al.* 1989).

Further biostratigraphic work on the Late Palaeozoic succession on Spitsbergen has also been performed by Nilsson (1988, 1993), Nakrem *et al.* (1992) and Sommerville (1997).

Controversy persists regarding the position of the Carboniferous–Permian boundary in the Wordiekammen Formation. Unfortunately, the boundary has often been based on different criteria. For instance, Forbes *et al.* (1958) placed it within the Brucebyen Beds at the contact between an Upper Carboniferous *Triticites* zone and a Lower Permian *Pseudoschwagerina* zone, whereas Winsnes (1966) considered that it occurred at the base of the Brucebyen Beds, at the first appearance of the fusulinid *Schwagerina* sp. Cutbill & Challinor (1965), however, placed the boundary at a disconformity represented by a thin conglomerate located *c.* 10 m beneath the Brucebyen Beds, and therefore considered it to be both a biostratigraphic and a lithostratigraphic boundary. Hanken *et al.* (1993) found a matrix-supported polymictic conglomerate (consisting of detrital chert and megaquartz) beneath the Brucebyen Beds on Tyrrellfjellet, but it was not identified in the Pyefjellet sections further to the east. This indicates that the conglomerate only has a local distribution and cannot be used as a general marker horizon on Spitsbergen.

Because of the variable development of the Brucebyen Beds and the presence of similar lithological units in the upper part of the Cadellfjellet Member, Samuelsberg & Pickard (1999) proposed that the base of the Tyrrellfjellet Member should be placed at a lower level than the Brucebyen Beds. According to them, the boundary should coincide with the base of a series of limestones containing an open-marine fauna characterized by colonial rugose corals, brachiopods, fenestrate bryozoans, crinoid debris and, locally, *Palaeoaplysina* build-ups. These limestones mark a distinct transgressive episode across central Spitsbergen, although the contact with the underlying rocks of the Cadellfjellet and Kapitol members is variable. However, within the Billefjorden Trough in Bünsow Land these open-marine limestones directly overlie dolostones and thin evaporites of the Matthewbreen Beds.

Terminology

The definition and usage of the term reef, and related terms such as bioherm and mound, have long been a matter of debate (e.g. James 1980; James & Bourque 1992; Bosence & Bridges 1995). One of the main problems is determining whether the structures observed actually possessed depositional topography on the sea floor. The *Palaeoaplysina* accumulations in central Spitsbergen possessed depositional topography and these structures are herein termed build-ups following the definition of Wilson (1975, p. 20): ‘A body of locally formed (laterally restricted) carbonate sediment which possesses topographic relief’. This generic term is very useful because it carries no inference about internal composition and encompasses all carbonate bodies that most workers would describe as reefs, bioherms, mounds or banks.

In small outcrops where no topography can be demonstrated, the sediments containing *Palaeoaplysina* are termed horizons. The term bioherm is used to describe any horizon composed of *Palaeoaplysina* and possessing a fine-grained carbonate matrix, while the term bank (Bank A, B, C etc.) is used for discrete bodies of sediments that contain *Palaeoaplysina* and form mappable lithosomes (Fig. 4). In some instances, it is possible to discern several banks within a single build-up (cf. Pickard 1992).

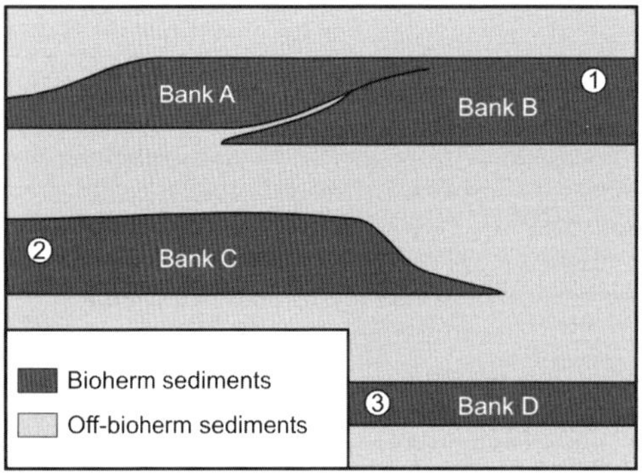

Fig. 4. Diagrammatic sketch illustrating the usage of terminology in this paper. The sketch shows three bioherms (1–3). The first two have depositional relief and are termed build-ups, but the third cannot be demonstrated to have had primary depositional relief and is called a *Palaeoaplysina* horizon. Build-up 1 is composed of two distinct, mappable units or banks (A and B).

Introduction to *Palaeoaplysina* build-ups

Despite numerous earlier studies of the Permian rocks of Spitsbergen, *Palaeoaplysina* build-ups were first recognized in the early 1980s (Lønøy 1981; Skaug 1982; Skaug *et al.* 1982; Sundsbø 1982; Dons 1983). Since then, they have been treated by Samuelsberg & Pickard (1999) and Hüneke *et al.* (2001). *Palaeoaplysina* build-ups normally occur in the lower part of the Tyrrellfjellet Member and may intercalate with the Brucebyen Beds. Most of the build-ups are located in Bünsow Land, Dickson Land and James 1 Land. More recent investigations NE of these areas (Olav V Land and southwestern Nordaustlandet) have not revealed the presence of build-ups (Fig. 5). However, on the western margin of Spitsbergen, build-ups have been described from Haroldfjellet in Nordenskiöld Land (Hjelle *et al.* 1986) and on Brøggerhalvøya (Bergh *et al.* 2000). These build-ups had previously been overlooked. Consequently, *Palaeoaplysina* build-ups may have a wider distribution along the western margin of Spitsbergen than has hitherto been assumed.

Sundsbø (1982) showed that the *Palaeoaplysina* build-ups in the Billefjord area are more or less dome-shaped, wedging out laterally after some tens of metres. They are typically 2–8 m thick and are thought to be equidimensional as they appear to wedge out equally quickly in all directions. In a few localities, the build-ups are stacked but show a lateral shift in their maximum thickness.

In a regional overview of the *Palaeoaplysina* build-ups in central Spitsbergen, Skaug *et al.* (1982) divided them into three different types:

(1) small, isolated dome-shaped structures (up to 6 m in height);
(2) tabular units with flat bases and tops – they may be several tens of metres long and up to 15 m thick;
(3) offlap sequences in which one bioherm drapes laterally over an adjacent one – individual build-ups may be *c.* 10 m thick.

However, Hanken *et al.* (1993) showed that the last two types are synonymous; the offlapping relationship simply represents the progradational margin of a laterally accreting, tabular-shaped build-up.

Palaeoaplysina build-ups have also been recorded from the Kapp Dunér Member (Asselian) on the island of Bjørnøya (Lønøy 1988; Stemmerik & Larsen 1993; Stemmerik *et al.* 1994; Worsley *et al.* 2001). On both Spitsbergen and Bjørnøya, *Palaeoaplysina* build-ups occur as isolated domes or tabular (disc-shaped?) bodies of considerable lateral extent and are commonly stacked; individual build-ups sit directly on top of one another or are separated by thin sequences of inter-build-up sediments. Individual build-ups have a relatively modest thickness (*c.* 1–20 m, but most are less than 10 m), although they may have a great lateral extent and some have been traced over an area of *c.* 10 km^2. Although some build-ups can be traced for several kilometres, others wedge out after a few tens of metres. Consequently, the number of build-ups in any given section of the Tyrrellfjellet Member varies from one locality to another; in some sections build-ups are absent, whereas nearby localities may have several horizons with build-ups. Figure 5 gives an impression of this variable occurrence of build-ups in the Tyrrellfjellet Member and shows their regional distribution in Spitsbergen. Lateral mapping of the build-ups indicates that there are well-defined, continuous areas containing build-ups, whereas intervening areas contain none.

Well data from the subsurface of the Norwegian Barents Sea have documented *Palaeoaplysina* build-ups on the Loppa High (Lønøy 1988; Stemmerik *et al.* 1999) and Finnmark Platform (Bugge *et al.* 1995; Ehrenberg *et al.* 1998). Three-dimensional seismic investigations in the Norwegian Barents Sea have shown that similar structures form well-defined, continuous but somewhat sinuous ridges and semi-enclosed polygonal basins/lagoons (Elvebakk *et al.* 2002; Rafaelsen *et al.* 2003; Samuelsberg *et al.* 2003; Colpaert *et al.* 2007, 2010; Hunt *et al.* 2010; see Fig. 6). Owing to a combination of overburden and lack of adequate exposure, it is not possible to show the exact configuration of the ridges and basins in Svalbard, but the somewhat limited data strongly indicate that the Spitsbergen build-ups also form the same type

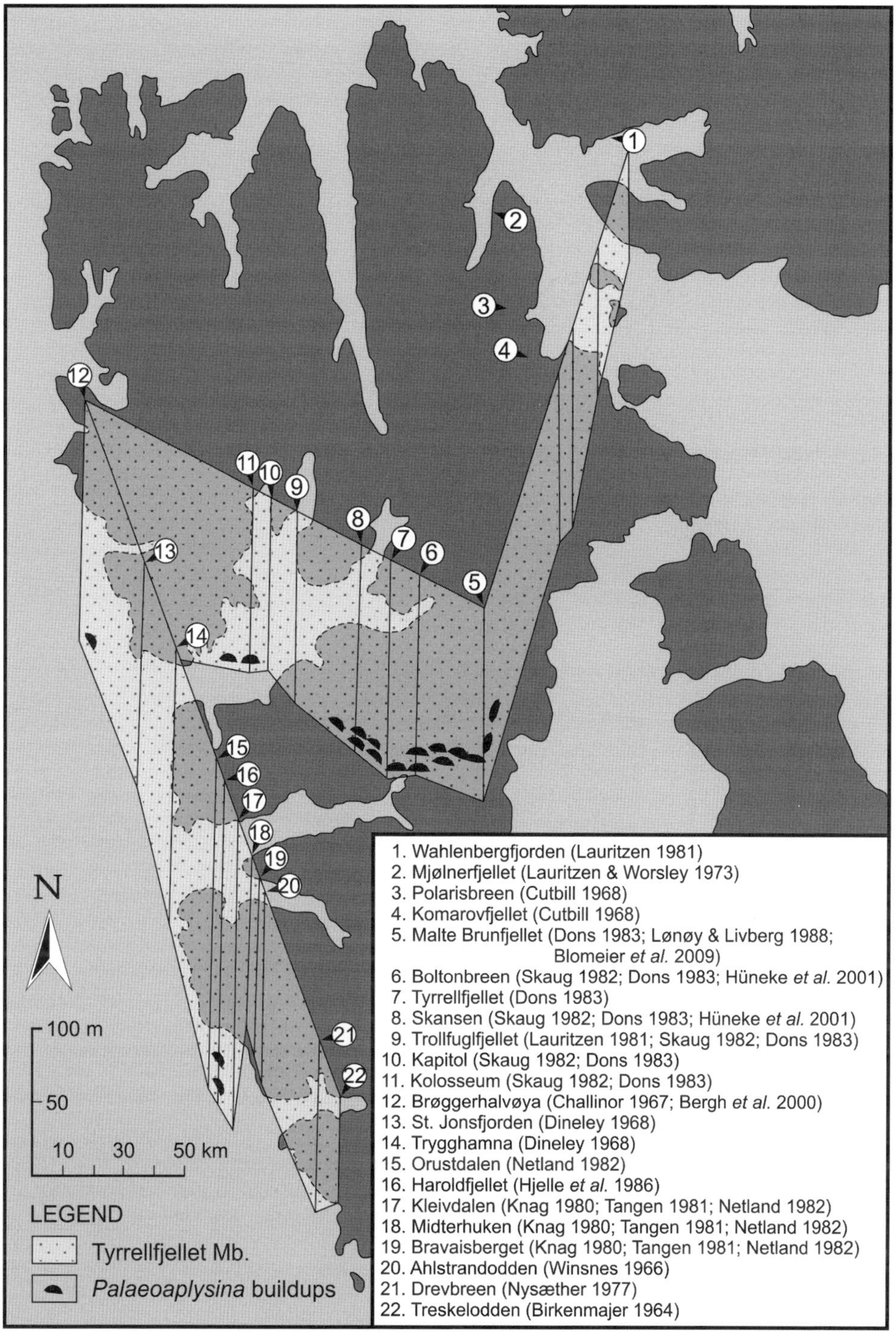

Fig. 5. Fence diagram showing the regional variation in the thickness of the Tyrrellfjellet Member on Spitsbergen. *Palaeoaplysina* build-ups are confined to three areas (Brøggerhalvøya, north of Isfjord between Bünsow Land and James I Land, and the western part of Nordenskiöld Land). 1–22, localities under study. Citations given in parentheses refer to descriptions of the localities. Modified from Dons (1983) and Hanken *et al.* (1993).

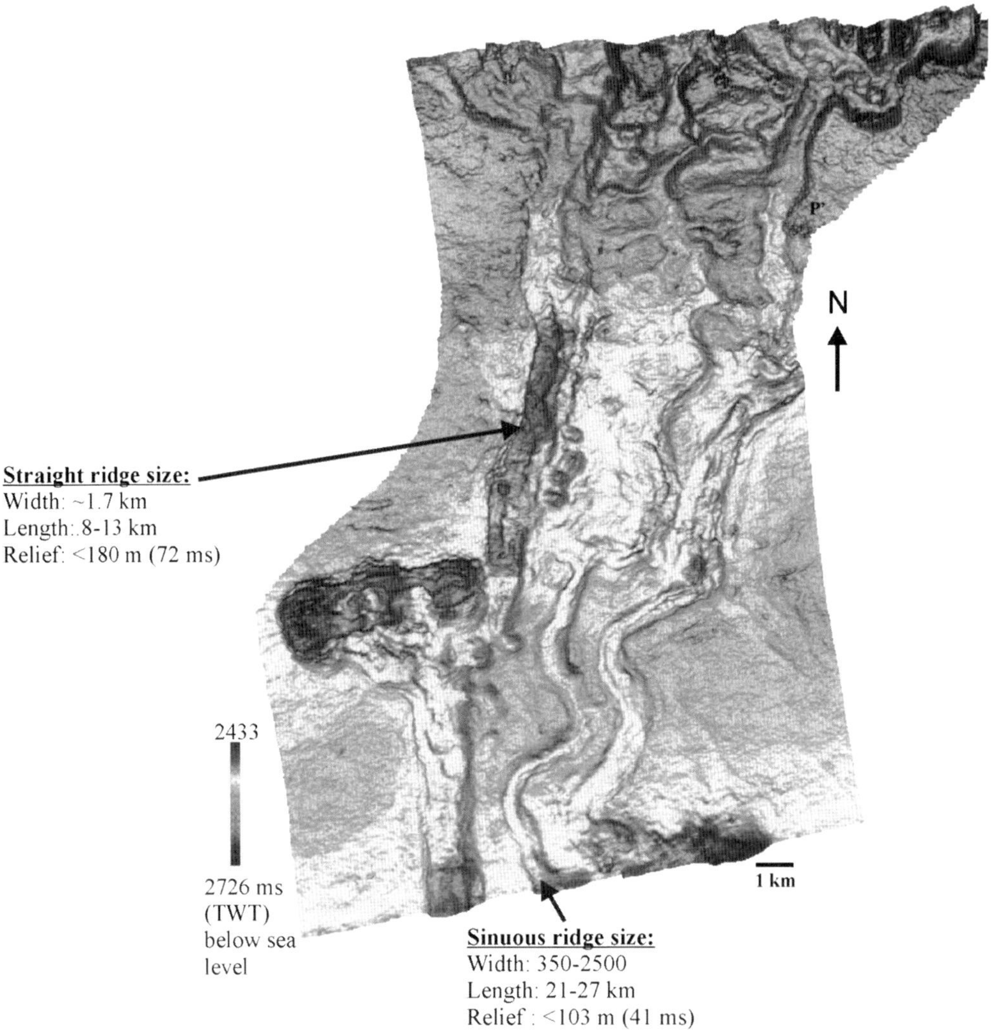

Fig. 6. Illuminated time–structure map of the Base Tatar horizon, displaying sinuous and straight ridges of carbonate build-ups. The ridges to the far left and right are located on the edge of a topographical high. These continuous ridges are tentatively interpreted to be fault-controlled and/or controlled by topography. The vertical exaggeration on the figure is ×8. From Rafaelsen *et al.* (2003) and Rafaelsen *et al.* (2008).

of structures as seen in the Barents Sea (Fig. 7). Accordingly, the Spitsbergen outcrops offer a unique opportunity to study this type of facies relationship in great detail.

Flora and fauna in the Tyrrellfjellet Member

The *Palaeoaplysina* build-ups on Spitsbergen contain fairly sparse fauna and flora compared with Recent reef structures. Except for the large plate-shaped *Palaeoaplysina*, the carbonate build-ups were constructed by relatively small organisms such as calcareous algae, *Tubiphytes* and bryozoans, with contributions of variable significance from calcareous sponges, brachiopods, encrusting foraminifera and tabulate and rugose corals. These low-growing, partly enigmatic bioherm builders became extinct later in Earth history. This community therefore differs greatly from the reef communities developed in Mesozoic and Cenozoic times, and analogues with recent reef facies must therefore not be drawn uncritically.

The *Palaeoaplysina* build-ups contain a muddy carbonate matrix with a clotted fabric between the

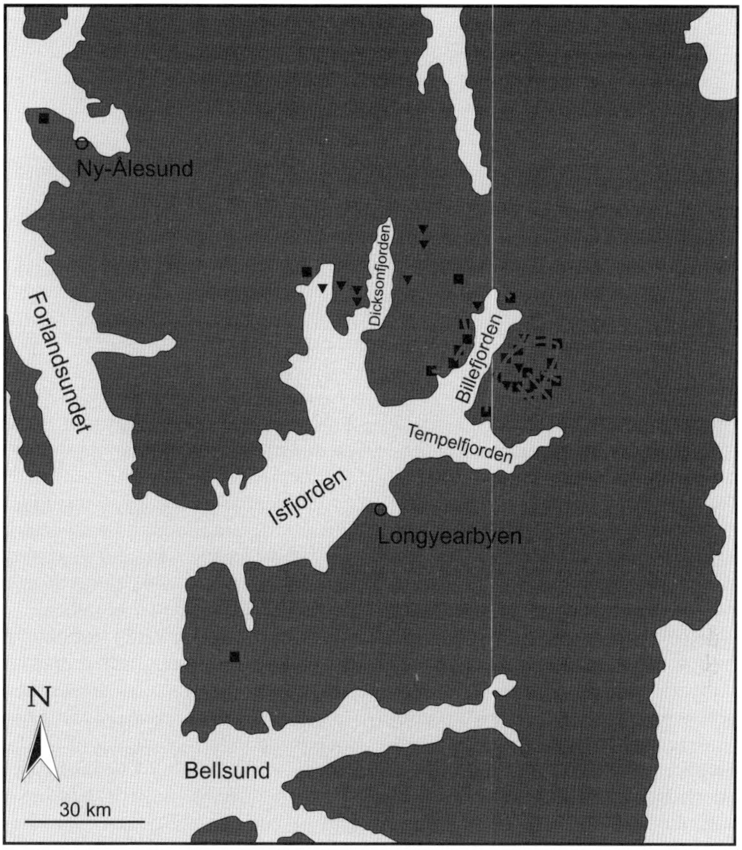

Fig. 7. Geographical distribution of *Palaeoaplysina* build-ups in central Spitsbergen. Sections through the Tyrrellfjellet Member that contain build-ups are marked with a square, and sections lacking build-ups are marked with a triangle. A red line (black in hardcopy) has been drawn connecting *Palaeoaplysina* build-ups lying close to one another. The *Palaeoaplysina* build-ups in Bünsow Land seem to form an oval structure surrounding an inner area with no build-ups. Further to the north and west, the build-ups seem to form elongate, ridge-like structures, but this may be an oversimplification because of limited outcrop data. Modified after Dons (1983).

skeletal fossils. Both in the build-ups themselves and in nearby lateral deposits it is common to find separate *Palaeoaplysina* fragments and angular to subrounded intraclasts derived from the *Palaeoaplysina* build-ups. These intraclasts are characterized by consisting of *Palaeoaplysina* plates together with micritic matrix between the plates. This observation indicates that the micritic matrix represents early marine cement.

Analogous micritic matrix is well known from Phanerozoic build-ups (e.g. Wahlman 2002), and there is growing evidence to suggest that most of the micrite in muddy carbonate build-ups is autochthonous and microbial in origin (Pratt 1982; Wahlman 1985, 1988, 2002; Monty *et al.* 1995; Blendinger *et al.* 1997). Weidlich (2002) proposed that microbial carbonate production was crucial for the stabilization of Phanerozoic reefs and mounds, and Grotzinger & Knoll (1995) claimed that many Permian reefs would resemble bioclastic sediments without the presence of this early marine cement.

Calcareous algae

The Tyrrellfjellet Member contains a relatively rich algal flora, but the algae are generally poorly preserved, rendering identification difficult. Thin-section studies have revealed that calcareous algae constitute a significant portion of the skeletal material in some facies, implying that they were important producers of carbonate particles, at least locally. Calcareous algae are important in the reconstruction of ancient carbonate environments because they provide an important complement to other benthonic organisms in delimiting marine biofacies. Calcareous algae are widespread and

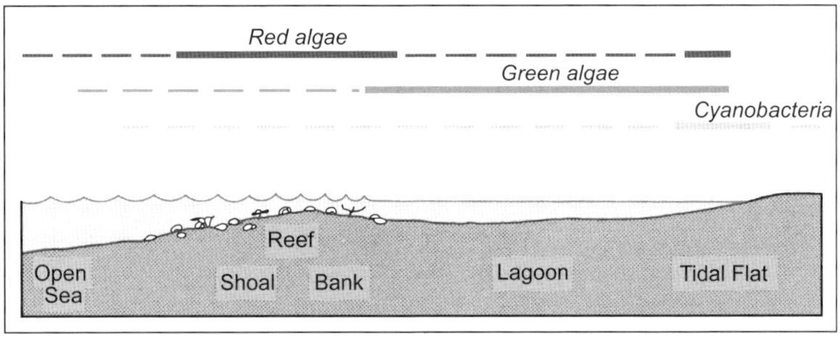

Fig. 8. Generalized distribution of Recent skeletal and non-skeletal algae across an idealized profile of a calcareous shelf in a low-latitude region. Modified from Wray (1977).

abundant in shallow shelf environments, and many species are restricted to narrow ecological niches. Figure 8 shows a generalized distribution of Recent algae on a low-latitude shelf.

Johansen (1994) described several algae belonging to cyanophyta (blue-green algae), rhodophyta (red algae) and chlorophyta (green algae) from the Tyrrellfjellet Member. There is only sporadic evidence of Girvanella from wackestone facies. Evidence of red algae (Squamariacea and Corallinacea) suggests a very shallow marine (subtidal) environment, where some of their growth forms suggest a fairly strong agitation in build-up environments. These are generally regarded as less tolerant of salinity variations than green algae and cyanobacteria. Furthermore, the coralline alga *Archaeolithophyllum* is known from wackestone–packstone facies, but is fairly rare.

In the Tyrrellfjellet Member, *Eugonophyllum* is found as a leaf-shaped phylloid alga in the wackestone–packstone and build-up facies (Fig. 9). Toomey & Babcock (1983) pointed out that the leaf-like thalli baffled debris contributed both to stabilizing sediments and increasing sedimentation rates.

Tubiphytes is well known as one of the most significant encrusters in the Late Carboniferous–Lower Permian bioherm communities (Mamet *et al.* 1987; Mamet & Stemmerik 2000; Wahlman 2002). This possibly red alga is also abundantly encountered in the *Palaeoaplysina* build-ups on Spitsbergen (Fig. 10).

Problematical red algae such as *Aoujgalia* and *Ungdarella* are found in facies lateral to the build-ups, while *Stacheoides* and *Fourstonella* are found both in build-ups and lateral facies in the lower part of the Tyrrellfjellet Member.

Green algae are fairly common in the Tyrrellfjellet Member. Single-celled prasinophytous algae are found sporadically preserved in magnetite (Fig. 11) (Hanken *et al.* 1993). *Epimastopora* and *Epimastoporella*, belonging to the family Dasycladacea, are sparsely distributed in wackestones, while *Dvinella* has only been observed in the lower part of the Tyrrellfjellet Member. Calcispheres are widely distributed throughout the Tyrrellfjellet Member and are most commonly observed in wackestone and packstone deposits and *Palaeoaplysina* build-ups. Analogous to most Recent green algae, these may have lived in shallow (depths of 5–10 m) marine tropical waters of normal salinity. However, some dasycladaceans tolerate hypersaline or brackish water (Wray 1977).

Palaeoaplysina was initially compared to the Recent sponge *Aplysina* (Krotov 1888) but was allocated to the class Hydrozoa (see Davies 1971; Davies & Nassichuk 1973). The similarity of the canal system in *Palaeoaplysina* (Fig. 12) is

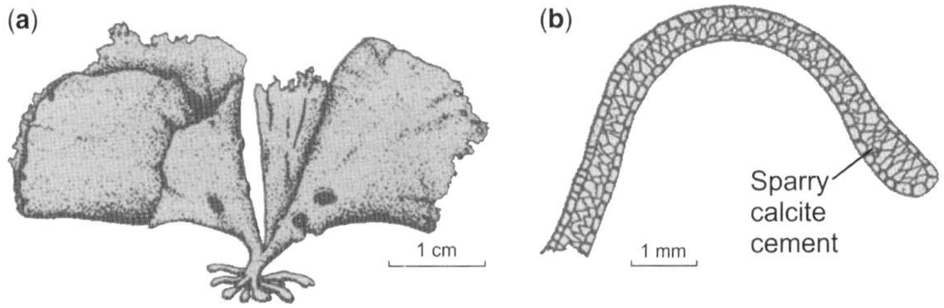

Fig. 9. (**a**) Suggested growth form of the phylloid alga *Eugonophyllum* sp. Modified from Wray (1977). (**b**) Example of the preservation of a phylloid alga. The original microstructure is normally poorly preserved.

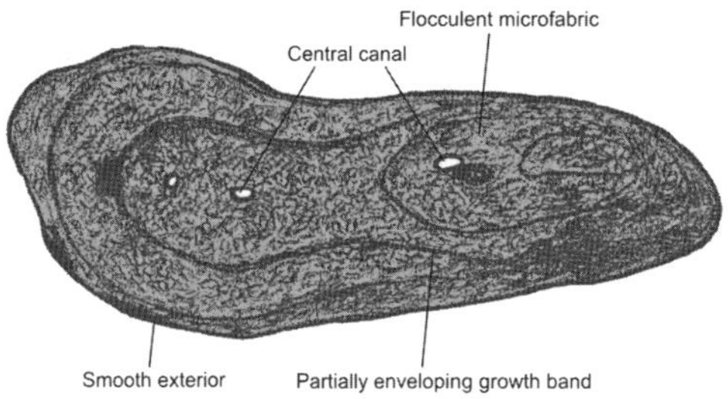

Fig. 10. Characteristic morphological features of *Tubiphytes* exemplified by the Upper Permian *Tubiphytes obscurus*. Modified from Riding & Guo (1992).

particularly striking when compared with the plate-shaped segments in the green algae *Halimeda*. This led Watkins & Wilson (1989) to consider the *Palaeoaplysina* to be enormous carbonate-secreting green algae. It is therefore tempting to compare the Permian *Palaeoaplysina* bioherms with the Holocene *Halimeda* bioherms described from the Java Sea, the Great Barrier Reef and the Nicaraguan Rise (Roberts *et al.* 1987; Hine *et al.* 1988; Marshall & Davies 1988).

Palaeoaplysina from Spitsbergen and their associated fauna and flora

As *Palaeoaplysina* build-ups are generally exposed only in vertical sections, the plate-shaped *Palaeoaplysina* is generally seen in cross-section. Consequently, difficulties arise when trying to obtain information about its three-dimensional shape. Measurements of maximum plate length are generally unreliable because they tend to be made on fortuitous sections and (from experience) on fragmented plates. As a consequence, plate lengths must be considered minimum values. Nonetheless, Skaug *et al.* (1982) reported plate lengths of at least 40 cm for *Palaeoaplysina* from central Spitsbergen. Sundsbø (1982) gave a length of 20 cm for plates from the Billefjord district, whereas Kano (1992) found an average plate length of 130 cm at Skansen. It is not known whether these differences in size represent differing degrees of fragmentation, actual size differences or different species.

The density *of Palaeoaplysina* can vary enormously within a single build-up. A cross-section through one build-up may be characterized by alternations of densely crowded *Palaeoaplysina* horizons and horizons containing more scattered *Palaeoaplysina* plates (Fig. 13a). Skaug *et al.* (1982) stated that the build-ups have a matrix-to-binder ratio ranging from 1:1 to 2:1. In this calculation, both *Palaeoaplysina* and the encrusting fauna (bryozoans and tubular foraminifera) were counted as binders. Observations by Hanken *et al.* (1993) from Pyefjellet and Tyrrellfjellet showed that the relationship between matrix and *Palaeoaplysina* varies a great deal both laterally and vertically within a single build-up. In the parts of a build-up richest in *Palaeoaplysina*, the

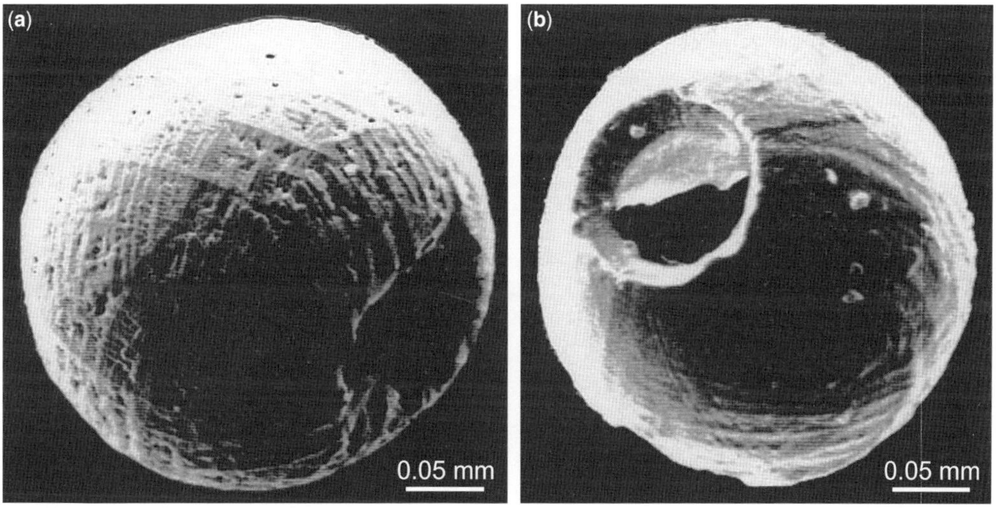

Fig. 11. Prasinophytous algae preserved in magnetite. (**a**) and (**b**) show two different specimens with a circular opening. The opening shows that the algae had 'hatched', indicating that the fossils represent empty algal skeletons. Tyrrellfjellet, level 19.6 m.

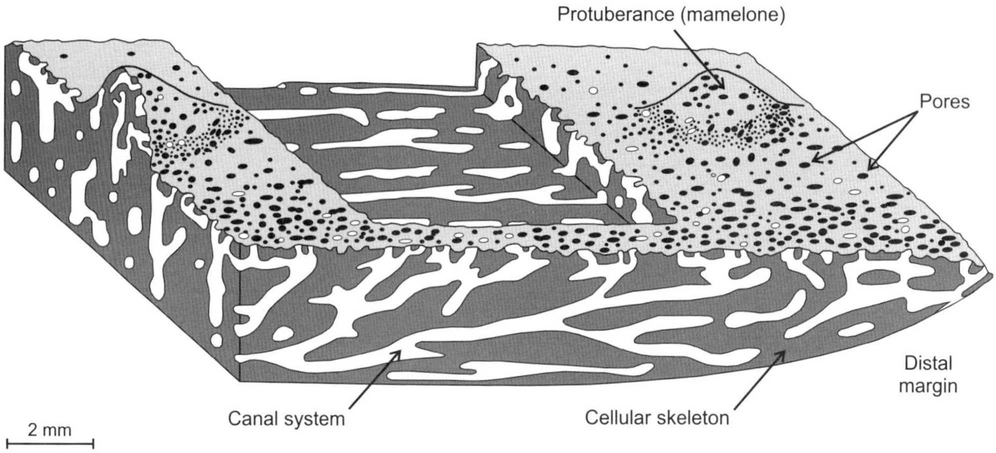

Fig. 12. Schematic reconstruction of a *Palaeoaplysina* plate illustrating the distribution of canals within the organism. Modified after Beauchamp *et al.* (1989).

matrix-to-*Palaeoaplysina* relationship may vary from 1:10 to 1:1, whereas it may be as high as 20:1 in mud-dominated sequences.

Immediately adjacent plates have the same growth form, as growth irregularities in underlying plates are commonly repeated in overlying ones. Topographical irregularities in the substrate are thus reflected in the growth form. This is clearly seen where *Palaeoaplysina* has overgrown a hard substrate, such as reworked gypsum nodules (Fig. 13b). These observations suggest that either (i) *Palaeoaplysina* has grown as an encruster on the sea floor or (ii) it was flexible and moulded itself to the underlying structure.

Foraminifera

The foraminifera in the Cadellfjellet and Tyrrellfjellet members are dominated by fusulinids, which have been described by Forbes *et al.* (1958), Forbes (1960), Cutbill & Challinor (1965), Ross (1965), Sosipatrova (1967), Igo & Okimura (1992), Nilsson (1988, 1993), Nilsson & Davydov (1992, 1997) and Anisimow *et al.* (1998).

The lower part of the Tyrrellfjellet Member containing *Palaeoaplysina* build-ups interfingers with the fusulinid wackestone and packstone of the Brucebyen Beds. The change between facies is abrupt and the marginal parts of the build-ups may

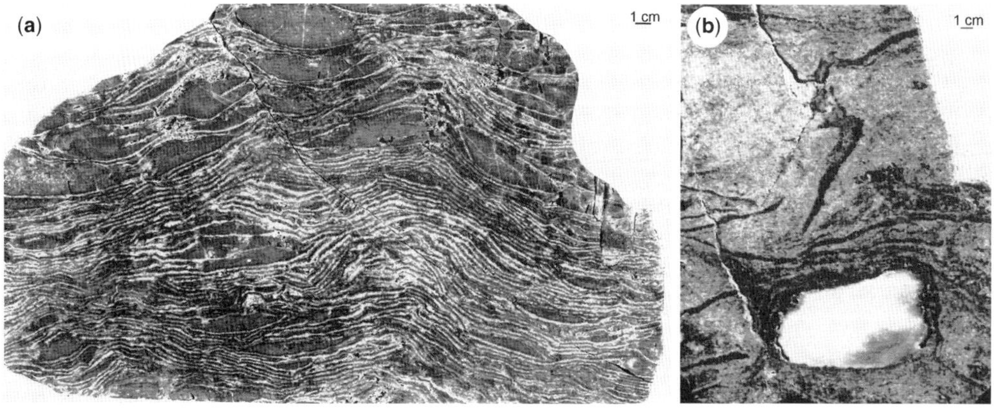

Fig. 13. (**a**) Polished slab showing stacked *Palaeoaplysina* plates from a horizon of *Palaeoaplysina* boundstone. Note that the plates sit directly on one another and an overlying one often possesses a similar growth form to the one beneath. Tyrrellfjellet Member, Pyefjellet. (**b**) Polished slab showing *Palaeoaplysina* plates encrusting a reworked clast of gypsum at the base of a build-up. Tyrrellfjellet Member, Pyefjellet. From Hanken *et al.* (1993).

contain numerous fusulinids in the matrix between *Palaeoaplysina* plates. Consequently, the modest fusulinid content of the build-ups contrasts sharply with the unusually rich foraminifera fauna in the Brucebyen Beds.

The ecology of fusulinids is not very well known, but Ross (1982*a*) pointed out that a few species can have mass occurrences in back-reef lagoons. To explain the mass occurrences in the sediments intercalated with the *Palaeoaplysina* build-ups, Ross (1982*b*) drew an analogy with the Holocene foraminifera *Calcarina*. *Calcarina* is often extremely abundant in depressions between reef structures and may constitute more than 90% of the sand fraction. These foraminifera are able to live there and withstand being washed out, because they are attached to the substrate by pseudopodia that may extend 15–20 cm down into the sediment. In this way, a cohesive mat of living foraminifera is formed. Individual foraminifera in the mat are constantly moving, thereby preventing any other organisms from being able to grow on the mobile substrate. For this reason, these areas are characterized by an extremely low faunal diversity but an unusually high population density. A similar mode of life for the Permian fusulinids may account for their equally high abundance between the *Palaeoaplysina* build-ups.

The inter-build-up sediments often contain gypsum nodules or gypsum crystals, pointing to hypersalinity. These parts of the succession contain a rather homogeneous fauna dominated by abundant fusulinids, indicating that they were adapted to this harsh environment. It is well known that the solubility of oxygen in brine decreases progressively as salinity increases, and this affects the survival of many organisms (e.g. Copeland 1967; Hite 1970; Kinsman *et al.* 1973). Because of the decreased diversity in the biota, the adapted species have little or no competition from other species for nutrients, space or any other requirements for existence, and will flourish. The unusually rich fusulinid fauna in otherwise nearly barren sequences may be an example of this adaptation.

The diversity of fusulinid faunas has been correlated to seawater temperature, where the higher diversity of shallow-water assemblages is related to optimal conditions between 20 °C and 30 °C, while the lower-diversity fauna is linked to cooler or warmer seawater, with water temperature never below *c.* 15 °C (Davydov & Arefifard 2007). Silicified fusulinids from the Tyrrellfjellet Formation show the presence of a characteristic honeycomb-like surface (Hanken *et al.* 1993), and Ross (1982*a*) pointed out that several large, Recent foraminifera possess an analogous surface pattern resulting from a symbiotic relationship with photosynthesizing zooxanthellae. The algae are essential to the metabolism of the foraminifera, supplying it with nutrients and oxygen. A similar symbiotic relationship with algae could explain the characteristic surface pattern of fusulinids (e.g. Ross 1972, 1974, 1979; Yancey & Kennedy 1984; Lee & Hallock 1987; Beavington-Penney & Racey 2004; Vachard *et al.* 2004; Leppig *et al.* 2005). This extra contribution of oxygen from the symbiotic algae may have been very important for the fusulinids as an adaptation to oxygen-poor saline water.

Limpet-like tetrataxids (Fig. 14) are fairly common in the *Palaeoaplysina* build-ups (Hanken *et al.* 1993). Cossey & Mundy (1990) indicated that their growth form was an adaptation for grazing on benthic algae and a variety of marine vertebrates in search of food and shelter. Our material is based on silicified specimens from build-up samples dissolved in diluted HCl, so we have no information

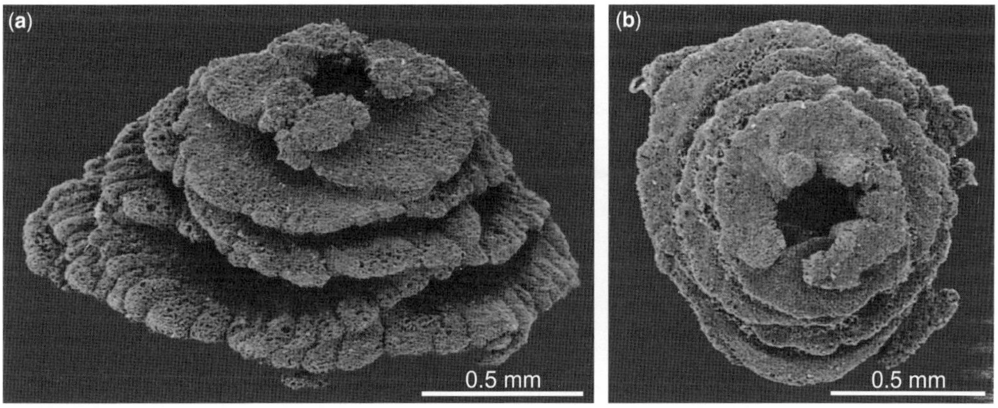

Fig. 14. Silicified tetrataxid foraminifera. (**a**) Side view showing the conical form. (**b**) View from above. Tyrrellfjellet, level 19.6 m.

about the association of *Tetrataxis* and the biogenic substrate.

Corals

Ezaki & Kawamura (1992) published a preliminary species list of the coral fauna in the Carboniferous–Permian deposits of central Spitsbergen. The Wordiekammen Formation has a relatively meagre coral fauna, and there are strikingly few colonial rugose corals. Most of the corals occur in bioclastic limestones, and they found no corals in the *Palaeoaplysina* build-ups. This is very surprising as Sommerville (1997) reported the presence of scattered, large, fasciculate coral colonies in the lower part of the larger build-ups, while small, solitary rugose corals occur in the basal parts. The coral fauna constitutes an extremely small proportion (less than 1%) of the build-up biomass, and this may explain why Ezaki & Kawamura (1992) found no corals in the build-ups.

The richest coral fauna in the Tyrrellfjellet Member has been found in the inter-build-up facies above the Brucebyen Beds, where certain horizons may be extremely rich in solitary rugose corals. These often show evidence of mechanical abrasion and have commonly suffered mechanical compaction during burial and are severely crushed. Early Permian tabulate corals from the Hornsund area, situated in the southern part of the basin, have been described by Nowiński (1982, 1991).

Gastropods

The Permian gastropod fauna on Spitsbergen is poorly known. Dons (1983) showed that gastropods are common at the base of the Kapitol Member, where the transgressive conglomerate overlying the Nordfjord Block may be entirely dominated by gastropod fragments. Shelly gravels rich in gastropod fragments also occur in the lower part of the Tyrrellfjellet Member, especially in the Brucebyen Beds. Large gastropods are common at the top of some *Palaeoaplysina* build-ups.

Brachiopods

Skaug (1982) showed the presence of two characteristic brachiopod associations in the Tyrrellfjellet Member of the Nordenskiöldbreen Formation at Boltonbreen, named, respectively, the *Septacamera*

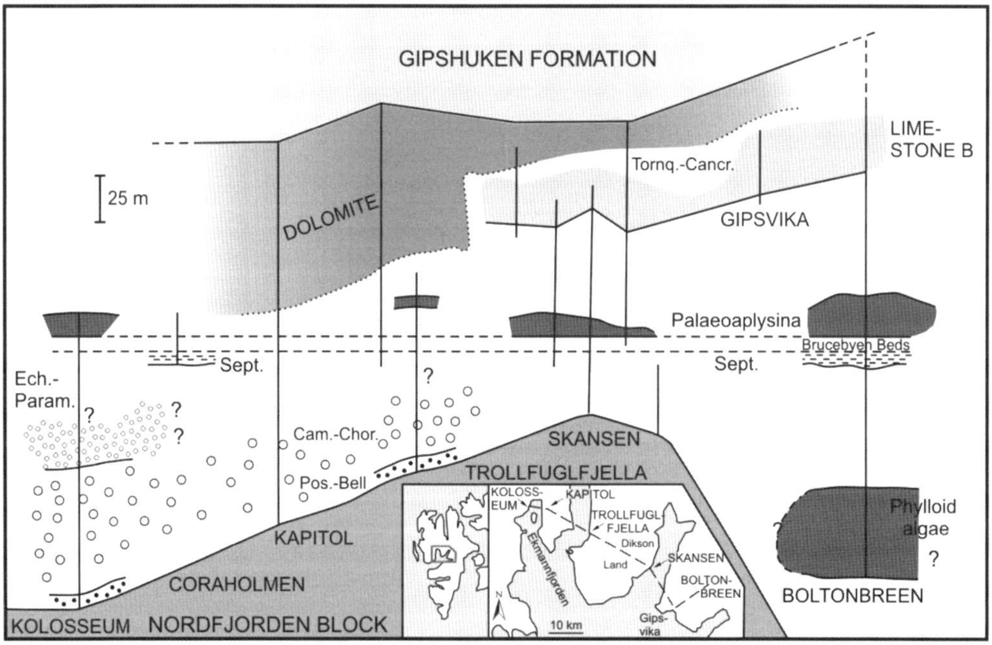

Fig. 15. Stratigraphic distribution of brachiopod associations within the Nordenskiöldbreen Formation in central Spitsbergen. Tornq.–Cancr., *Tornquistia–Cancrinella* association; Sept., *Septacamera* association; Ech.–Param., *Echinoconchus–Paramarginifera* association; Cam.–Chor., *Camerisma–Choristites* association; Pos.–Bell., *Posidoniella–Bellerophontocea* association. Modified after Skaug (1982).

association, found below the Brucebyen Beds, and the *Tornquistia–Cancrinella* association, found in Limestone B (Fig. 15). Most of the brachiopods described by Skaug (1982) are disarticulated, and the ratio between the pedicle valve and the brachial valve often deviates from one. Likewise, the degree of shell fragmentation and abrasion varies across a broad range, indicating that a significant proportion of the skeletal debris was mechanically destroyed before final deposition. The post-mortem transport may have resulted in mixing of assemblages, but Skaug (1982) considered that, although this was likely, it did not overshadow the dominant taxa in any given association.

A few brachiopods occur in life position, the most common being *Cancrinella*, which is frequently found in the wackestones from the upper part of the Tyrrellfjellet Member. The productids in the same part of the sequence have their pedicle valve facing downwards and long spines are still attached and unbroken, suggesting that part of the brachiopod fauna is autochthonous (or at least parautochthonous).

According to Skaug (1982), the build-up/inter-build-up facies contains an extremely meagre brachiopod fauna. However, an investigation of the silicified fauna of this part of the sequence showed that the brachiopod fauna in the Tyrrellfjellet Member is significantly richer than previously thought. Hanken *et al.* (1993) showed that small brachiopod species (less than 5 mm in length) are relatively common in both the Brucebyen Beds and the *Palaeoaplysin*a build-ups. The silicification is relatively coarse and patchy, so finer details have often been destroyed, and only fragments of the skeletal material were recovered. For this reason, no systematic study was undertaken, but Matveev (1996, pers. comm.) suggested that the silicified brachiopod fauna was stunted.

Bryozoans

Nakrem (1991*b*, 1994*a*, *b*, 2008) showed that the lower part of the Tyrrellfjellet Member is characterized by a meagre bryozoan fauna with delicate vinculariform zoarians, fenestrates and pinnate–fenestrate bryozoans. Vinculariform zoarians and fenestrate bryozoans have been recognized in the Brucebyen Beds, but only fenestrate bryozoans (both delicate and robust growth forms) have been found in the *Palaeoaplysina* build-ups. The low diversity of the bryozoan fauna in the Brucebyen Beds indicates, according to Skaug *et al.* (1982), that this part of the succession was deposited in a quiet environment under low oxic conditions not favouring bryozoan growth. The bryozoan content in the build-ups is very meagre, and this may explain why Sommerville (1997) found no bryozoans in the build-ups. However, there is a marked increase in bryozoan diversity in the upper part of the Tyrrellfjellet Member. In this part of the sequence, the bryozoan fauna is dominated by robust and delicate vinculariform and fenestrate bryozoans, which, according to Nakrem (1994*a*), indicates stronger currents during deposition of the upper part of the sequence.

Echinoderms

Skaug (1982) found that within the Nordenskiöldbreen Formation echinoderm fragments occurred most frequently in the Kapitol Member. Here, crinoid fragments are the dominant bioclasts in most facies. In both the Cadellfjellet and Kapitol Members, echinoderms mostly occur as fragmented *Arhaeocidaris* and echinoid spines (Dons 1983).

Echinoderms are not common in the *Palaeoaplysina* build-ups, but silicified, disarticulated crinoid ossicle debris may constitute a significant proportion of the biomass in laterally equivalent wackestone–packstone facies.

Trace fossils

With the exception of Limestone B, bioturbation is not commonly encountered in the Tyrrellfjellet Member. Only a few specimens of *Zoophycos* and *Teichichnus* have been observed in wackestone–packstone facies, while *Planolites* has a somewhat broader distribution with regard to facies.

No endolithic borings or substrate excavating grazing traces have been observed in the build-ups or the enclosing lateral sediments. However, this might be due to the fact that the surface of the *Palaeoaplysina* plates might be too badly exposed or preserved. The lack of boring organisms, and probably also substrate-excavating grazers, indicates some sort of stress in the sedimentary environment because both bioeroders and grazers are well known from Palaeozoic reefs growing in water with normal salinity (cf. Vogel 1993; Bromley 2004).

James (1997) showed that carbonate-associated biota can be broadly divided by water temperature into warm-water photozoan carbonates and cool- to cold-water heterozoan carbonates. In modern environments, warm-water carbonates are those accumulating in water warmer than *c.* 20 °C. The biota of the *Palaeoaplysina* build-ups and their lateral sediments are characterized by calcareous algae and foraminifera-dominated assemblages that are typical for the photozoan carbonates, indicating a tropical climate. Reid *et al.* (2007) have given a biogeographic model showing how warm water from the west might have supported a photozoan fauna at Svalbard.

Field observations of Palaeoaplysina build-ups

Hanken *et al.* (1993) described well-exposed *Palaeoaplysina* build-ups in the Tyrrellfjellet Member from the southern face of Pyefjellet, Gipsdalen (Figs 16–18). Four levels with *Palaeoaplysina* build-ups are present in this sequence. The bulk of these build-ups occur in the lower part of the Tyrrellfjellet Member, either in the upper part of the Brucebyen Beds or just above them. However, one horizon was found much higher in the sequence, some 50 m above the main build-up horizon. Another section was measured on the northern side of Tyrrellfjellet (Figs 19 & 20). There are only a few small *Palaeoaplysina* build-ups in the lowermost part of the Tyrrellfjellet Formation in this area.

The thickest *Palaeoaplysina* build-up forms a pronounced cliff, *c.* 10 m high, which, despite sporadic exposure to the west, can be traced laterally for nearly 2 km along the southern face of Pyefjellet. This build-up is also exposed on the southern side of Boltonbreen and, although its southern and eastern margins are unknown, it covers an area of a least 2 km^2. Exposure is less frequent beneath the thick *Palaeoaplysina* build-up, but it was sufficient to ascertain that at least four *Palaeoaplysina* build-ups are present. Of these, the upper three (Banks C, D & E in Fig. 21) show primary depositional relief and may therefore be termed build-ups *sensu stricto*. The lateral extent of these build-ups is uncertain; some appear to pinch out laterally so that the exact number of *Palaeoaplysina* build-ups beneath the main build-up remains variable (Fig. 21). The lowermost *Palaeoaplysina* level occurs within the Brucebyen Beds and is overlain by dolostones containing *Microcodium*. The only other occurrence of *Microcodium* noted in the Pyefjellet section occurs in the topmost build-up of the main sequence (Bank A and possibly the very top of Bank B).

The *Palaeoaplysina* build-ups on the southern side of Pyefjellet have tabular geometries and vary from *c.* 1 m to 10 m in thickness. Build-up margins are typically wedge-shaped; the build-ups thin rapidly over a distance of *c.* 10–15 m before passing transitionally into level-bedded, inter-build-up sediments. Crude bedding can occasionally be discerned near the build-up margin and the beds typically thicken into the build-up before amalgamating into massive build-up sediments. Build-up margins are exposed at several locations on Pyefjellet. The relief measured between the margins and contemporaneous inter-build-up sediments varies from *c.* 1 m to 8 m (locality PH and PG, see Fig. 21). Geopetal structures, although rare, are occasionally preserved near the margins of the build-ups and indicate that depositional slopes of up to *c.* 20° were attained. These features suggest that, in plan view, build-up margins were irregular and comprised a series of 'spurs' or 'promontories'

Fig. 16. (**a**, **b**) Photographs showing the Pyefjellet locality on the north side of Boltonbreen. Lines 1–4 show the locations of the profiles, which comprise the composite log through the upper part of the Cadellfjellet Member and the Tyrrellfjellet Member (see Fig. 18). Line 5 shows the position of the lateral profile shown in Figure 21.

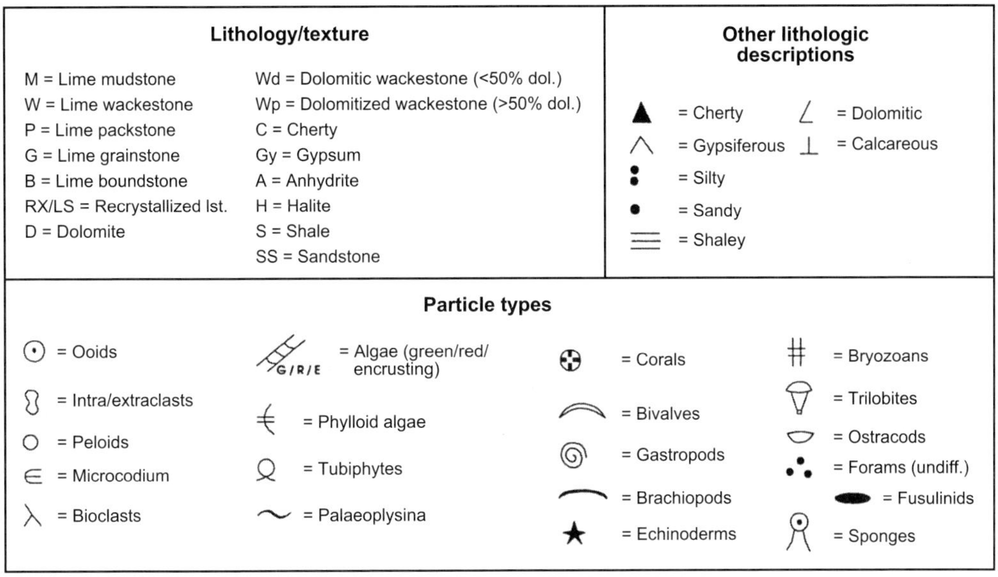

Fig. 17. Legend for the sedimentary logs.

separated by 're-entrants' or 'grooves'. This is best shown by the thickest build-up in the Pyefjellet sequence (Bank B/A), the northern margin of which is exposed in a series of gullies that cut northwards into the mountainside (Figs 22 & 23).

Although the overall geometry of the build-ups is tabular, the upper surface of some build-ups is not always planar but consists of a series of 'hummocks' or 'domes' separated by depressions in which inter-build-up sediments accumulated. This is best observed by the upper Bank D (Fig. 24). Taking the base of the succeeding build-up (Bank C) as a datum, the thickness of inter-build-up sediments between Bank D and Bank C varies from 0.2 m to 2.0 m, and the top of Bank D is seen to consist of a series of domes or hummocks spaced between 25 m and 50 m apart. The maximum height of an individual dome approaches 1.8 m. However, as build-up sediments on the flanks of the hummocks can be traced into contemporaneous inter-build-up sediments in the intervening depressions, depositional relief would have been less than 1.8 m.

The tabular geometry shown by the build-ups on Pyefjellet suggests that build-up facies rapidly established themselves over a wide area. However, lateral progradation of the larger build-ups did occur and may be observed in the thick build-up exposed on Pyefjellet (Bank A–B). At the base of Bank B, successive build-up margins are seen to overstep each other and adjacent inter-build-up sediments as the build-up margin prograded towards the north and west (Fig. 25). The build-up margins are preserved by a thin drape of inter-build-up sediments deposited during a temporary standstill in the lateral accretion of the build-up. A similar situation occurs at the top of the main build-up, where renewed progradation of the build-up led to the development of Bank A, which offlaps and partly infills the topography developed on the margin of Bank B.

Build-up facies

Build-up sediments are readily distinguished as grey or brown, massive to crudely bedded wackestone and boundstone. Packstone fabrics are also locally developed. Partial to complete dolomitization is common, but in the central Spitsbergen build-ups it is generally confined to the margins and top of the structures. The faunal content of the build-ups is dominated by *Palaeoaplysina* plates and fusulinid foraminifera together with small, indeterminate, tubular microfossils, possibly encrusting foraminifera and the problematicum *Tubiphytes*. However, faunal diversity is generally low and the distribution of biota listed earlier is patchy and, in the case of some organisms, is related to certain areas within the build-up or to specific stratigraphical levels.

Four facies are recognized within build-ups, based on varying proportions of the major faunal components and the preservation of *Palaeoaplysina* plates:

(1) *Palaeoaplysina* wackestone;
(2) *Palaeoaplysina* boundstone;

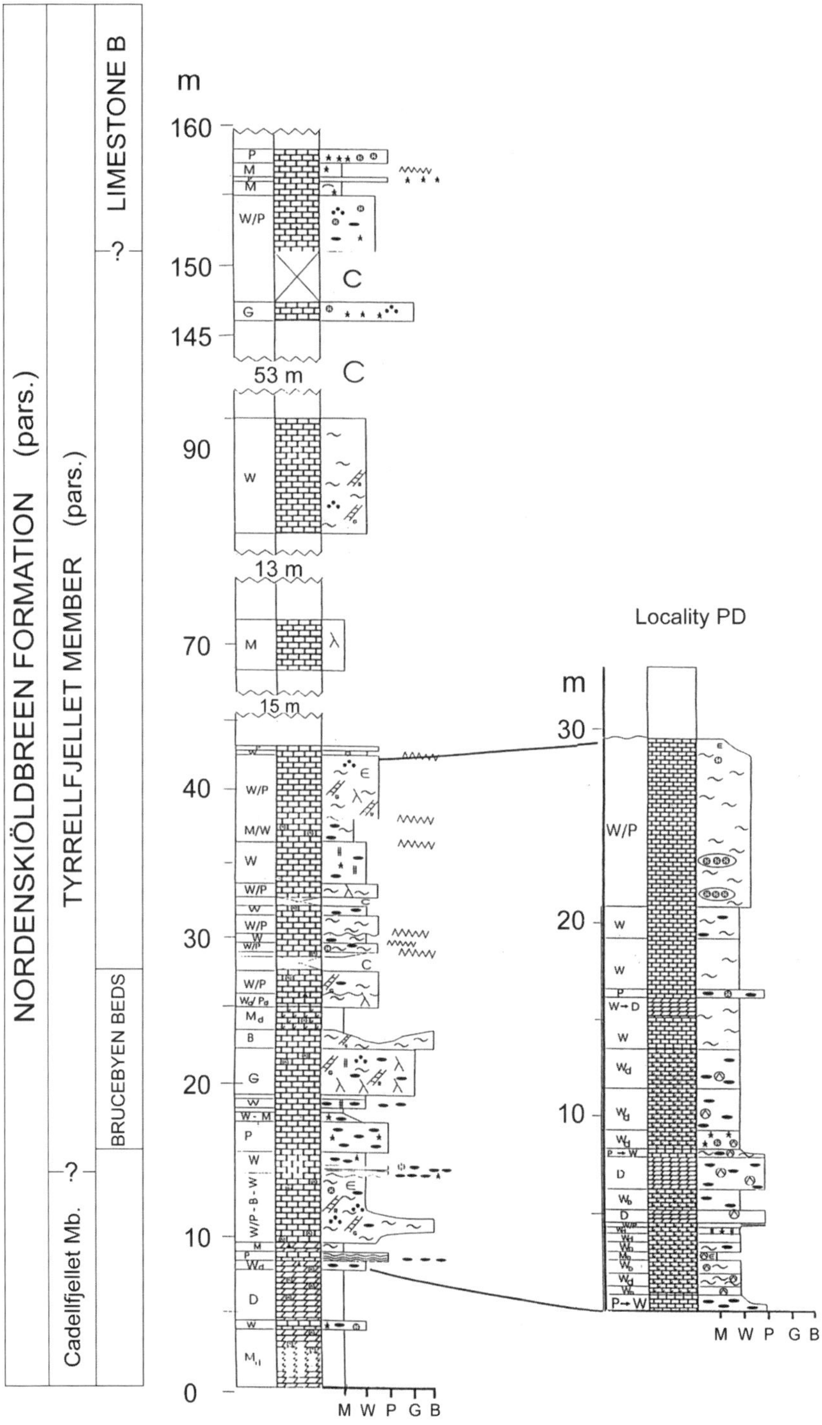

Fig. 18. Sedimentary log from the Pyefjellet locality. Level 0–35 m is represented in line 1 on Figure 16, while line 2 represents level 35–70 m on the log. Level 70–85 m is represented in line 3. The uppermost part of the log is represented in line 4 on Figure 16. The location of locality PD is shown in Figure 21. For the legend, see Figure 17. From Hanken *et al.* (1993) and Johansen (1994).

Fig. 19. (**a**) The northern slope of Tyrrellfjellet seen from Mathiesondalen. (**b**) Close-up showing the location of the two profiles investigated. 'Limestone B' constitutes the cliff-forming limestone unit at the upper end of profile 2.

(3) *Palaeoaplysina*–fusulinid wackestone and packstone;
(4) Crinoid–*Palaeoaplysina* wackestone.

Palaeoaplysina wackestone is the predominant lithology in the build-ups and frequently alternates with *Palaeoaplysina* boundstone horizons. Each facies is described below.

Palaeoaplysina wackestone

Palaeoaplysina wackestone is volumetrically the most important type in the build-ups. It is characterized by scattered *Palaeoaplysina* plates separated by a micritic matrix (Fig. 26). Given the irregular plate-like morphology of *Palaeoaplysina*, many of the plates may be in contact, giving a

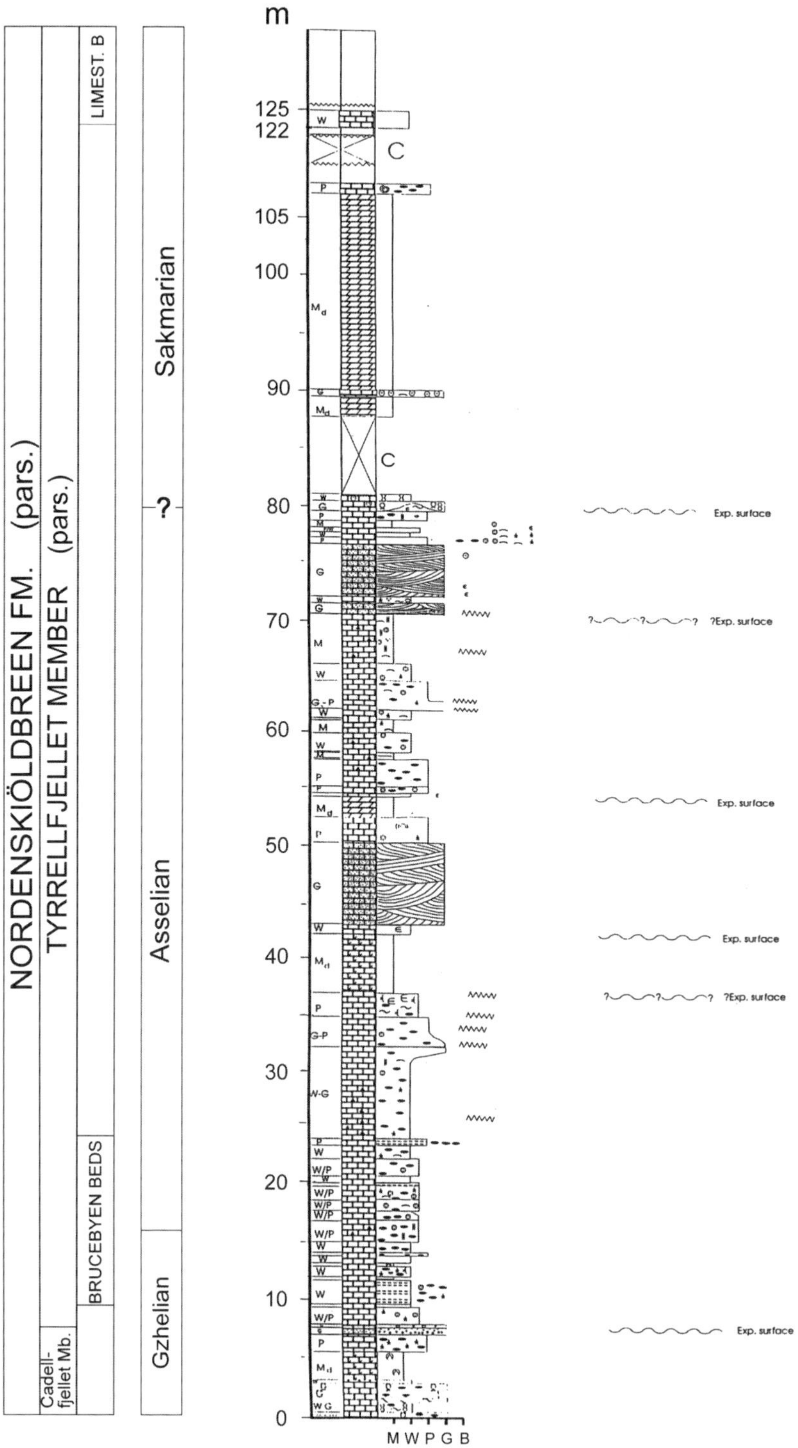

Fig. 20. Sedimentary log from the Tyrrellfjellet locality. Line 1 in Figure 19 shows where the lowermost 55 m were logged. Line 2 shows where the uppermost part was logged. For the legend, see Figure 17. Pars., an incomplete section of a member or formation; Limest. B, Limestone B; C, covered section. From Johansen (1994).

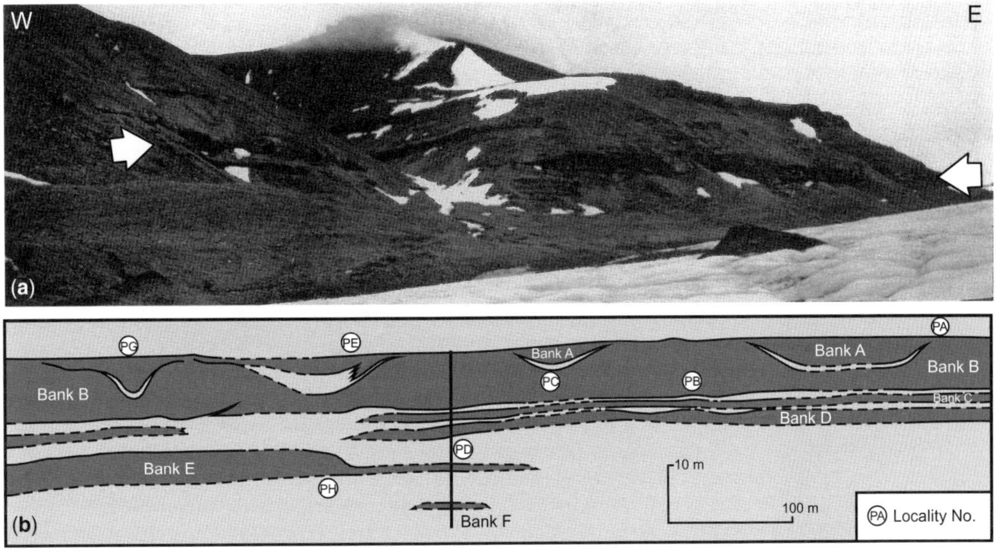

Fig. 21. (**a**) Photograph showing the Pyefjellet locality on the north side of Boltonbreen. The two arrows show the area from which the lateral profile was taken (line 5 in Fig. 16 shows the location). (**b**) Drawing of the lateral profile showing the distribution of *Palaeoaplysina* build-ups. Note the vertical exaggeration. Modified from Hanken *et al.* (1993).

grain-supported fabric. The *Palaeoaplysina* plates are often curved lying at high angles to the upper surface of the build-up. Given the seemingly random orientation of many of the *Palaeoaplysina* plates, there is a total absence of shelter cavities, even beneath curved plates. Other faunal elements present in this facies include fusulinids, colonial and solitary rugose corals, brachiopods (particularly Rhynchonellids), large gastropods and crinoidal debris. Large, fasciculate, colonial rugose corals are locally important and occasionally form small biostromes at the base of the build-ups (Skaug *et al.* 1982). Horizons containing large gastropods (up to 6 cm in length) have been observed near the top of some build-ups.

Interpretation

The fine-grained nature of the matrix does not imply that this facies developed in a low-energy depositional environment; rather, it is a result of early micritic cementation. The seemingly chaotic orientation of many *Palaeoaplysina* plates probably indicates material eroded from other parts of the build-up during storms. When a storm is declining, rapid deposition from the highly sediment-laden water can trap large fragments in chaotic orientations (cf. Jeffery & Aigner 1982). It is unlikely that the oblique orientation was caused by bioturbation (cf. Bromley 1996) because no trace fossils have been observed in the build-ups.

Palaeoaplysina boundstone

The sediment consists almost entirely of stacked, elongate *Palaeoaplysina* plates. These are commonly large, reaching thicknesses of 0.5 cm, with individual plates attaining lengths of more than 1 m (Skaug *et al.* 1982; Sundsbø 1982; Kano 1992). However, the majority of plates observed in the Tyrrellfjellet sequences rarely exceeded 30 cm in length. Thin layers and/or lenses of carbonate mud may occur between the plates but are a volumetrically minor component of this facies. Stacked plates commonly have similar growth forms, such that irregularities in underlying plates are commonly repeated in overlying ones (Fig. 13a). Topographical irregularities in the substrate are also reflected in the growth form. This is clearly seen where *Palaeoaplysina* has overgrown a hard substrate, such as reworked gypsum clasts (Fig. 13b).

Palaeoaplysina boundstone occurs as a distinct horizon (5–12 cm in thickness), which can be traced laterally for tens of metres (*c.* 20 m for Bank D at locality PB, Fig. 21) before gradually pinching out, normally close to the margin of a build-up.

Interpretation

The abundance of *Palaeoaplysina* plates indicates dense colonization of the sea floor and implies either (i) low sedimentation rates or (ii) rapid

Fig. 22. (**a**) Contact between Bank A and Bank B on the western side of the scree-filled gully at locality PA (see Fig. 21 for location). The blocky weathering inter-build-up dolostones overlying Bank B (large arrow) pass laterally into massive build-up sediments. Note how the beds in the lower part of Bank A thicken towards the ENE and infill the topography on the margin of Bank B. These beds are overlain by a pronounced disconformity (small arrows). (**b**) Photograph illustrating the nature of the contact between Bank A and Bank B at locality PE (see Fig. 21 for location). Thinly bedded, fusulinid dolostones form the recess between the two banks and wedge out towards the SSE. Crude beds within Bank A thicken towards the NNW as they infill the topography developed on the margin of Bank B and also appear to be overlain by a disconformity (small arrows, see also (a)). From Hanken *et al.* (1993).

growth rates, which allowed *Palaeoaplysina* mats to develop across part (or the whole?) of an individual build-up. If *Palaeoaplysina* was encrusting the sediment surface, the growth forms in the *Palaeoaplysina* boundstones indicate that the build-ups possessed significant micro-topography (up to 15 cm) and would have been capable of withstanding increased levels of turbulence.

Palaeoaplysina–fusulinid wackestone and packstone

Comprised predominantly of *Palaeoaplysina*–fusulinid wackestone and packstone, this facies is present at the margins and tops of build-ups, where it forms part of a transitional sequence between the *Palaeoaplysina* wackestone and boundstone of the build-up interior and level-bedded, off-build-up sediments (Fig. 27). *Palaeoaplysina* plates are commonly in contact with each other and appear to form an open framework 'filled' with a fusulinid wackestone matrix. Locally, the fusulinids are preserved in small packstone lenses (up to 5 cm thick and 20 cm long), which are commonly overlain by a single *Palaeoaplysina* plate. Many of the *Palaeoaplysina* plates are broken and they may be chaotically oriented.

Interpretation

Increased numbers of broken plates and the presence of packstone textures suggest higher levels

Fig. 23. Photograph illustrating locality PG (see Fig. 21 for location). Banks A, B and E are labelled. Note how Bank A infills the topography developed on Bank B and how the two banks appear to merge towards the west and east near the top of the build-up. The contact between these banks is marked by a thin sequence of dolomitized mudstones which wedge out to the east and west (arrowed). From Hanken *et al.* (1993).

of water turbulence and reworking than the *Palaeoaplysina* wackestone. Fusulinids are common, both in this facies and in immediately adjacent or overlying inter-build-up sediments. Discrete packstone lenses of fusulinids (sand ripples?) are again indicative of wave or current reworking of sediment.

Crinoid-Palaeoaplysina wackestone

Build-up facies rich in crinoids are relatively rare, being confined to the uppermost part of the main build-up sequence at Pyefjellet (Bank A/B, see Fig. 21). The crinoids are generally preserved as single ossicles (up to 1 cm in diameter), but small stem fragments are locally present. Fusulinids are also a common constituent of this facies. *Palaeoaplysina* plates are small and may appear fragmented, suggesting some reworking has occurred.

This facies has been identified at several localities on Pyefjellet (locs. PA, PC and PE, Fig. 21) and in fallen blocks believed to have originated from the same stratigraphical level (top of the vertical cliffs at loc. PD). Crinoidal-rich build-up facies occur in Bank A, which is itself a lateral development of Bank B. Near the margin of Bank A and at the top of Bank B it intercalates with well-sorted crinoidal-fusulinid grainstone. Fallen blocks (loc. PD) from the top of Bank B show crinoidal-fusulinid grainstone directly overlying an erosion surface cut into the crinoidal build-up facies (Fig. 28).

Interpretation

The abundance of fragmented stenohaline organisms, such as crinoids, is indicative of a more open marine setting. There is no sign of any crinoid holdfasts in the build-ups, which indicates that the crinoids were living laterally to the build-ups and the crinoid ossicles found together with *Palaeoaplysina* plates represent a mixing of parautochthonous and/or allochthonous material.

Inter-build-up sediments

Inter-build-up sediments include those which are known to have been, or from their stratigraphical position are likely to have been, contemporaneous with the *Palaeoaplysina* build-ups. They are typically fine-grained dolostones that possess a sparse and restricted fauna dominated by fusulinids. However, towards the top of the main Pyefjellet build-up sequence, inter-build-up sediments are more variable, and several lithologies are recognized. As the *Palaeoaplysina* horizons on Tyrrellfjellet are small and no depositional margins have been observed, the nature of the contemporaneous off-build-up sediments is not known. However, fusulinid dolostone is present above the lowermost *Palaeoaplysina* horizon and fusulinid wackestone and dolostone also bound the middle horizon.

Fine-grained dolostone

Fine-grained, bedded dolostone is volumetrically the most important inter-build-up sediment in the Pyefjellet sequence. The bedded dolostone is subdivided here based on the presence or absence of fauna, notably fusulinids. At outcrop, it is typically weathered to an orange-brown colour and may possess small laths or nodules of gypsum (Fig. 29).

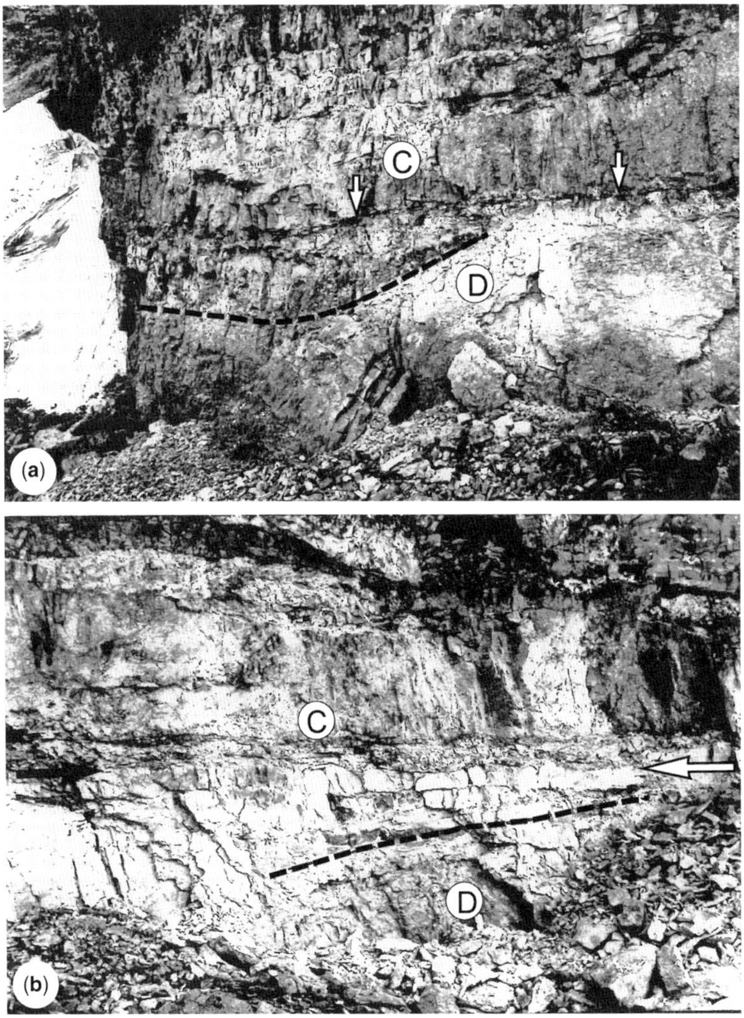

Fig. 24. (**a**, **b**) Photographs showing the irregular topography developed on the top of Bank D at locality PB (see Fig. 21 for location). The contact between Bank D and overlying dolostones is dashed. The base of Bank C is arrowed. From Hanken *et al.* (1993).

Fusulinid dolostone

Fusulinids are the main biota in these sediments. Small, often fragmented *Palaeoaplysina* plates are also present, together with rare solitary corals and brachiopods (generally Rhynchonellida). In many instances, the fusulinids have been dissolved and are preserved as mouldic porosity, ribbing of the void surface being the only means of identifying the former presence of a fusulinid. This facies commonly possesses a laminated fabric formed by the alternation of very thin (<1 mm), undulating, dark brown or black (organic-rich?) laminae with thicker (1–2 mm) laminae of orange-brown weathering dolostone. The undulose, dark brown laminae are generally impersistent and can only be traced for several tens of centimetres at the most.

Thin horizons (generally <10 cm) of broken *Palaeoaplysina* plates are occasionally preserved in this facies and appear to represent debris reworked from the adjacent build-ups. Bedded fusulinid dolostone passes transitionally into *Palaeoaplysina*–fusulinid wackestone at the margins of the build-ups (Fig. 27). The fusulinid dolostone has also been observed to pass, with increasing distance from the build-ups, into faunally barren dolostones possessing numerous gypsum laths and nodules.

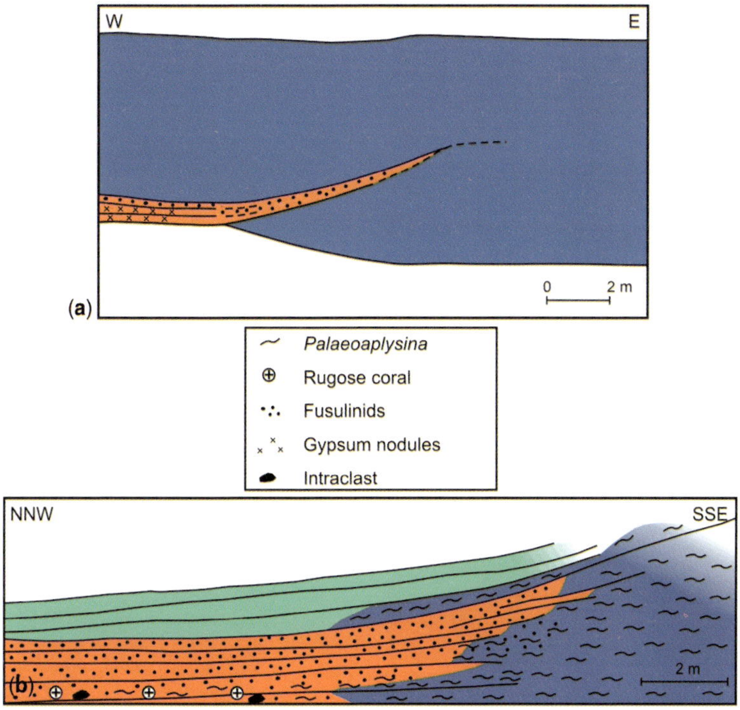

Fig. 25. (**a**) Field sketch of the base of Bank B just east of locality PG (see Fig. 21 for location) illustrating a 'palaeo-build-up margin'. Massive build-up sediments thin rapidly to the west and pass into level-bedded fusulinid-rich dolostones with numerous gypsum nodules. The dolostone sediments are themselves overlain by build-up facies. The margin of the build-up appears to have prograded laterally across the dolostones following a hiatus in build-up growth. (**b**) Field sketch of the margin of Bank B as exposed at locality P1, which is situated just west of the lateral profile shown in Figure 21. The lower sketch is a close-up of the build-up margin showing the transition between *Palaeoaplysina*-rich build-up facies and level-bedded, fusulinid-rich inter-build-up sediments. The build-up margin is dominated by fusulinid-rich *Palaeoaplysina* sediments. *Palaeoaplysina* plates in the margin beds commonly lie parallel to the depositional slope. Modified from Hanken *et al.* (1993).

Fig. 26. Polished slab of *Palaeoaplysina* wackestone with several *Palaeoaplysina* plates (arrowed) that appear to fold back on themselves.

Fossil barren dolostone

This facies resembles the fusulinid dolostone in all respects except that there is no recognizable fossil debris. It is commonly laminated, with numerous dark brown, undulose laminae. Small laths of gypsum are common, and gypsum nodules are locally preserved. The nodules are irregular and may be aligned parallel to the bedding.

Interpretation of fine-grained dolostone

The bedded dolostone represents dolomitized mudstone and wackestone, implying a low-energy depositional setting. As lateral equivalents of build-up facies, they must have been deposited in slightly deeper water than the build-ups. The low diversity or absence of fauna suggests a restricted environment, while the abundance of gypsum indicates that hypersaline conditions pertained for

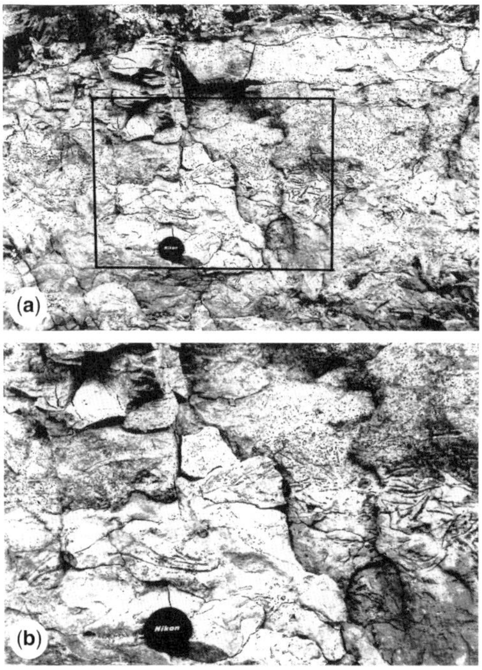

Fig. 27. (**a**) Transitional contact at the top of Bank D, Pyefjellet section. *Palaeoaplysina* wackestone at the base of the photograph passes up into fusulinid-rich build-up sediments and then into fusulinid dolostones at the top of the picture. (**b**) Enlargement of the centre of A showing the fusulinid-rich build-up facies. From Hanken *et al.* (1993).

Fig. 28. Polished slab of a crinoidal–fusulinid grainstone containing large intraclasts of crinoid-rich build-up facies and reworked *Palaeoaplysina* plates. Note (i) the *Palaeoaplysina* plate folding back on itself (large arrow) and (ii) *Microcodium* replacing *Palaeoaplysina* plates (small arrows). Crinoidal–fusulinid grainstones overlie an erosive contact with the crinoid-rich build-up facies. Sample collected from the top of a large block at locality PD (see Fig. 21 for location) that has probably fallen from the top of Bank D. Scale in centimetres. From Hanken *et al.* (1993).

part of the time. There is no sign of halite in the Tyrrellfjellet Member, and this indicates that the brine never became concentrated enough for precipitation of the more soluble salts, which would have required a concentration to at least one-tenth of the original seawater.

The ring-shaped *Palaeoaplysina* complex found in Bünsow Land (Fig. 7) may have formed a barrier preventing the unrestricted access of seawater into the basin (lagoon), thus allowing evaporation to concentrate seawater into a brine and to reduce the outflow of the evaporation-concentrated brines (Fig. 30). Any change in the rate of influx, for instance through changes in the size or shape of inlet(s), tectonic movements, glacio-eustatic sea-level changes (Stemmerik *et al.* 1999; Stemmerik 2008) or precipitation, may have altered the fine balance between the rate of inflow, reflux and evaporation. If, for example, the ring-shaped bank complex was subaerially exposed during a regression, the lagoon might be disconnected and only be fed by seawater derived from seepage reflux through the barrier. As the build-ups had suffered early cementation and the intervening sediments were mostly fairly fine-grained, the rates of inflow must have been very much smaller than those for surface influx (cf. Maiklem 1971). The permeability in the barrier has probably been so low that it has been insufficient to offset all the water lost by evaporation. During this evaporative drawdown in the lagoon, the salinity increased to a value high enough for gypsum precipitation, but the brine never became sufficiently concentrated for halite precipitation.

It is significant that nearly all gypsum observed in the Tyrrellfjellet Member consists of nodules enclosed within the carbonate matrix, whereas horizons of dispersed gypsum crystals are uncommon and no continuous layers of gypsum have been

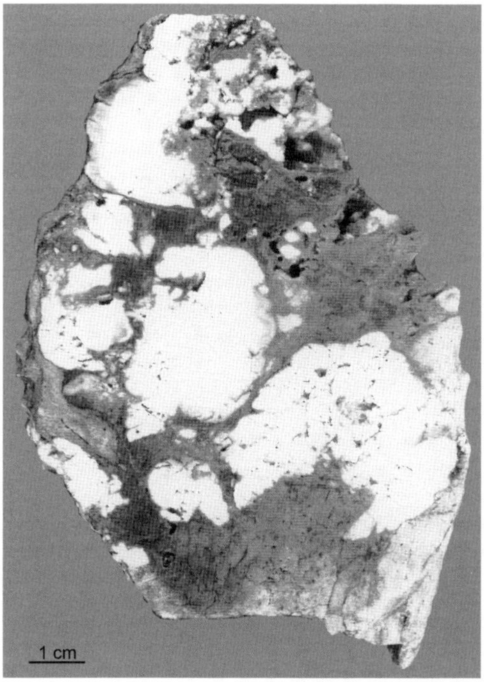

Fig. 29. Polished slab showing gypsum nodules in fine-grained dolostones. Tyrrellfjellet Member, Pyefjellet section.

observed. These observations indicate that the gypsum was precipitated in a sabkha facies and not from a standing body of water (cf. Kendall & Warren 1988).

The gypsum precipitation in the lagoon sediments implies at least partial subaerial exposure of the lagoons. This can only be achieved by a significant reduction in inflow followed by evaporative drawdown. This process is most probably connected with the subaerial exposure of the surrounding build-up complex, thus forming a barrier to the surrounding sea. This implies that the gypsiferous parts of the lagoon sediments may be correlated with hiati in the surrounding build-up complex (Fig. 31). The modest deposition of gypsum in the lagoon sediments indicates that gypsum precipitation was episodic and the lagoon water was mostly only saline enough for carbonate precipitation.

Crinoidal–fusulinid grainstone

Crinoidal–fusulinid grainstone has only been observed near the top of the main build-up sequence at the Pyefjellet locality. There, this facies makes up moderately to thickly bedded inter-build-up sediments that can be traced laterally into the crinoid-rich build-up facies of Bank A. The grainstone is moderately to well-sorted and contains fragments of *Palaeoaplysina* plates and intraclasts of *Palaeoaplysina* wackestone (Fig. 28). This facies is believed to overlie the top of Bank B at locality PD (Fig. 21), where it has been observed in fallen blocks.

Interpretation

The moderately to well-sorted nature of this facies and the highly fragmented state of large bioclasts such as *Palaeoaplysina* suggest deposition in a relatively high-energy environment that was probably above the normal wave base.

Ecology of the lagoon

Extensive fieldwork on Spitsbergen has shown that lagoon sediments surrounded by the *Palaeoaplysina* build-up complex often contain a low-diversity fauna dominated by fusulinids. Study of Recent hypersaline lagoons has established that such environments are highly productive but poor in species (e.g. Friedman & Krumbein 1985). It is well known that the diversity of species endemic to a particular ecological niche diminishes as the harshness of the environment increases. This is well known from Recent hypersaline locales where Zenkevitch (1963), for instance, showed that diversity decreases with increasing salinity. Concurrent with the impoverishment in the number of species, such as very abundant fusulinids, adapted to hypersaline environments, the number of individuals is enriched in the Tyrrellfjellet Member. Despite few taxa, the organic production of the lagoonal facies may therefore have reached a maximum of biomass at about two or three times the concentration of seawater (cf. Raymont 1963). The bottom sediments of Recent hypersaline lagoons typically have a high organic content, as is also the case for lagoonal facies in the Tyrrellfjellet Member.

To explain some of our observations from the lagoonal facies, we will stress the following factors:

(1) An increase in salinity involves an increase in alkalinity. The highest pH recorded in Recent hypersaline lagoons is *c.* 9.0 and occurs when the salinity is about twice that of normal seawater (Copeland 1967).
(2) The concentration of CO_2 in the brine is nearly zero at a pH of *c.* 9.0 (Moberg *et al.* 1934). Hence, algae that depend upon carbon dioxide for photosynthesis will perish.
(3) The solubility of oxygen in the brine decreases with increasing salinity and temperature

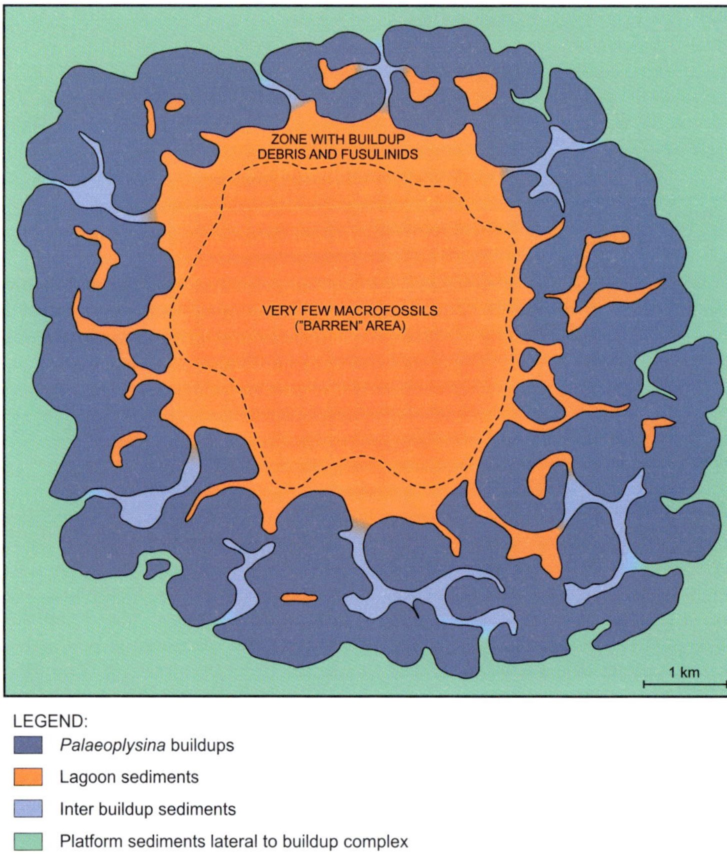

Fig. 30. Reconstruction of a ring-shaped to polygonal *Palaeoaplysina* build-up complex encircling a lagoon. A well-developed ring complex has only small, winding channels leading into the lagoon between the broad build-ups. This hampers the influx of water into the lagoon, and in a dry climate evaporation produces hypersaline conditions in the lagoon. Because few plants and animals are adapted to hypersaline environments, there are significantly fewer species in the lagoon compared with the much richer biota outside the barrier. Because salinity is highest in the middle of the lagoon and decreases somewhat towards the inlets, there is a richer biota near the margins than in the middle of the lagoon.

(Copeland 1967; Hite 1970; Kinsman *et al.* 1973), and this affects the survival of some organisms.

The lagoonal facies of the Tyrrellfjellet Member often contains a flourishing fusulinid fauna with very few other fossil taxa present. We interpret this low-diversity assemblage as a response to the harsh environment in hypersaline lagoons where only symbiotic fusulinids have been adapted.

Facies distribution and facies associations

The *Palaeoaplysina* build-ups in the area investigated mostly have the shape of tabular banks. However, the great lateral extent of these build-ups strongly indicates that there must have been several growth centres that amalgamated; the large build-ups originated as many small mounded structures resembling patch reefs (Fig. 32). By a combination of vertical and lateral growth, the incipient patch reefs coalesced to form wide, tabular build-ups. The flat, even bases of the build-ups indicate that lateral growth was high compared with the inter-reef sedimentation. The presence of stenohaline organisms such as corals in the lower part of the build-ups indicates that this part of the build-ups grew in more or less normally saline water.

In a few cases, we have observed build-ups with a hummocky surface. These 'abnormal' occurrences can be regarded as build-ups that never attained full equilibrium with their sedimentary environment and were buried during an incipient stage of their development.

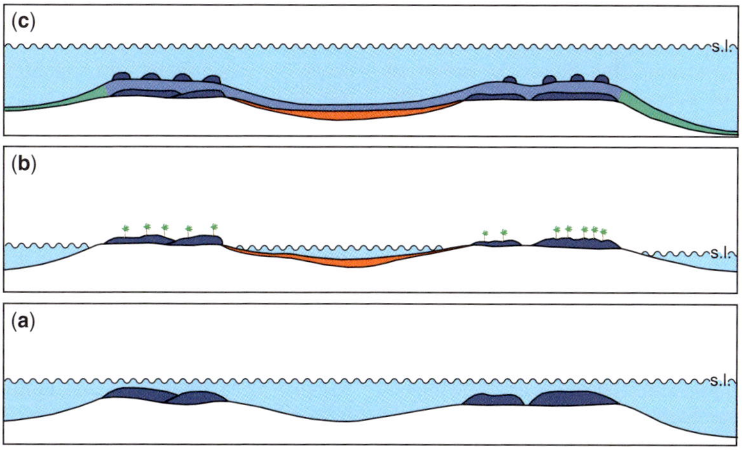

Fig. 31. Schematic cross-sections through a polygonal lagoon with surrounding *Palaeoaplysina* build-ups. (**a**) *Palaeoaplysina* build-ups (dark blue) are growing on former highs, but are submerged due to little restriction in the flow of water across the barrier. (**b**) Regression and subaerial exposure of the *Palaeoaplysina* complex. The lagoon water is cut off from the seawater and gypsum is precipitated in the hypersaline pool. (**c**) During transgression, both the partly weathered *Palaeoaplysina* complex and the lagoon sediments are flooded and buried by marine sediments characterized by a more normal marine fauna and flora. New *Palaeoaplysina* build-ups start to grow as small disconnected structures on the palaeo-high. See Figure 30 for legend.

The flat-topped tabular build-ups can be viewed as the mature stage of the build-up growth. With decreasing accommodation space, continued vertical growth was hampered and lateral growth became dominant. This stage in the build-up development is equal to the 'keep-up stage' as defined by Kendall & Schlager (1981). *Palaeoaplysina* plates lying parallel to the depositional slope at the build-up margins indicate that part of the lateral accretion was due to *in situ* growth as well as mechanical transportation of build-up material that originated on the top of the build-up.

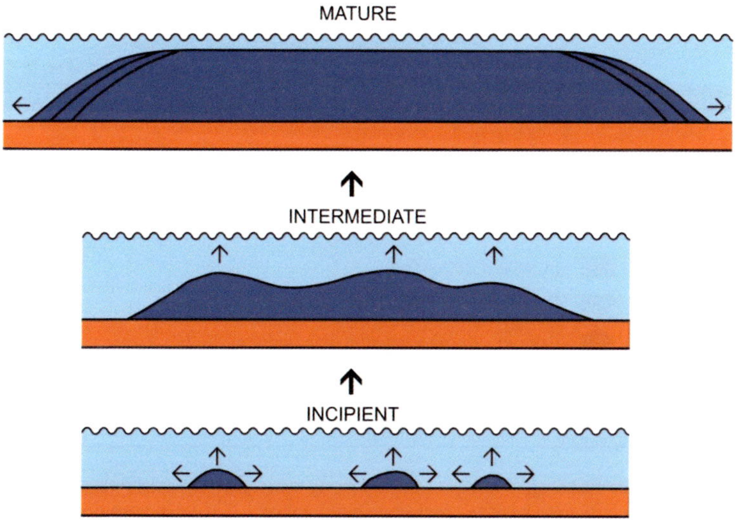

Fig. 32. Diagram showing how a tabular *Palaeoaplysina* build-up evolved from an incipient patch reef stage with several isolated mounds through an intermediate stage where the mounds have grown both vertically and laterally, giving a large build-up characterized with an undulating topography. In the mature stage, with decreasing accommodation space, continued vertical growth was hampered and lateral growth became dominant. Modified from Beauchamp (1993).

The build-ups are associated with distinct facies belts. The development of the different facies varies from locality to locality, but Figure 33 shows an interpretation of the various facies associations. The main volume of the build-ups is dominated by *Palaeoaplysina* wackestone and boundstone. The uppermost part of the build-ups is characterized by a thin veneer of *Palaeoaplysina*–fusulinid dolostone on the lagoon side and crinoid–*Palaeoaplysina* wackestone on the sea side. This lateral facies shift is probably due to differences in salinity between the isolated lagoon and the open sea. The stenohaline crinoids could only live on the sea side, but the euryhaline fusulinids were well adapted to the hypersaline lagoon water. The uppermost part of the build-ups may contain *Microcodium*, indicating subaerial exposure. The lateral interfingering between the bedded fusulinid dolostone along the build-up margins and barren dolostone further out in the lagoon may indicate an increasing salinity from the lagoon margin to the central part.

Subaerial exposures

Freytet & Plaziat (1982) showed that the presence of *Microcodium* in marine sediments is evidence of subaerial exposure because it is a pedological feature that forms in a vadose environment, that is, in well-drained, aerobic parts that are subaerially exposed. *Microcodium* has often been used as evidence for precipitation of calcium carbonate in root cells of terrestrial plants inhabiting carbonate-rich substrates (see discussion by Košir 2004). However, Kabanov *et al.* (2008) suggested a non-rhizogenic biologically induced origin where saprotrophs decomposed dead terrestrial and connate organic matter in carbonate deposits. Disregarding the reaction mechanism, there is a general agreement that the presence of *Microcodium* is due to an organo–mineral interaction in the subaerial vadose environment.

At Svalbard, *Microcodium* occurs in the upper parts of some *Palaeoaplysina* build-ups, where it is found in the centre of individual *Palaeoaplysina* plates and in the micritic matrix (Fig. 34). In the upper part of the Tyrrellfjellet Member, it has been observed in grainstone where it occurs in units up to 1.5 m thick. In the first case, *Microcodium* is autochthonous, whereas it has been transported and redeposited as single grains in the second. However, in both cases the presence of *Microcodium* is taken as evidence of subaerial exposure of the build-ups themselves or erosion and redeposition in marine sediments during transgression of formerly subaerially exposed parts of the sequence.

The restriction of *Microcodium* to the upper part of *Palaeoaplysina* build-ups also indicates that

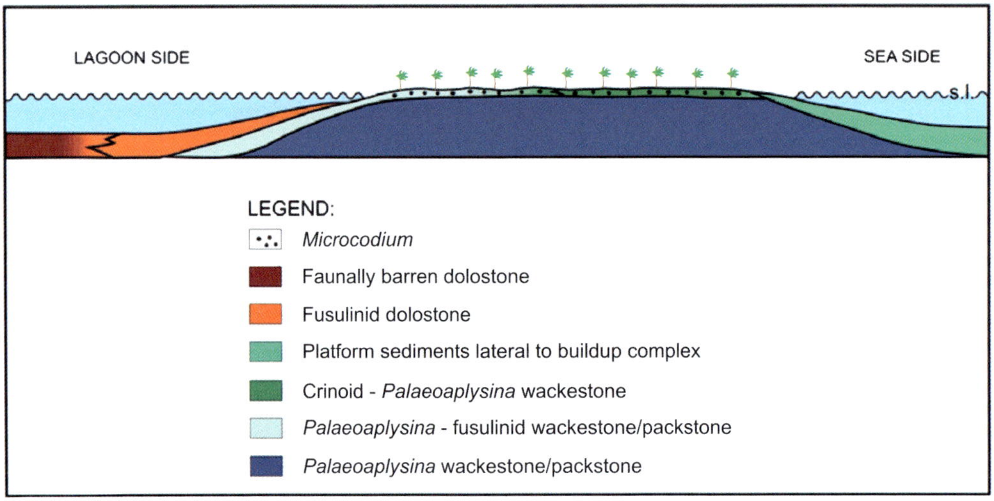

Fig. 33. Diagram showing the facies relationships around a partly subaerially exposed *Palaeoaplysina* build-up. The topmost part of the *Palaeoaplysina* wackestone–packstone build-up has a fairly thin transition zone consisting of *Palaeoaplysina*–fusulinid wackestone and packstone on the lagoon side and, in some places, crinoid–*Palaeoaplysina* wackestone towards the sea side where the seawater has near-normal salinity. The lagoon water is hypersaline, but, because of some inflow of seawater over or through the barrier, the water along the periphery is less saline than in the middle of the lagoon. Because fusulinids flourish in the least saline parts of the lagoon water, the lagoon sediments consist of fusulinid dolostone along the margin and faunally barren dolostone in the central part. The subaerially exposed part of the build-up is karstified and contains *Microcodium*, indicating vegetation on the island.

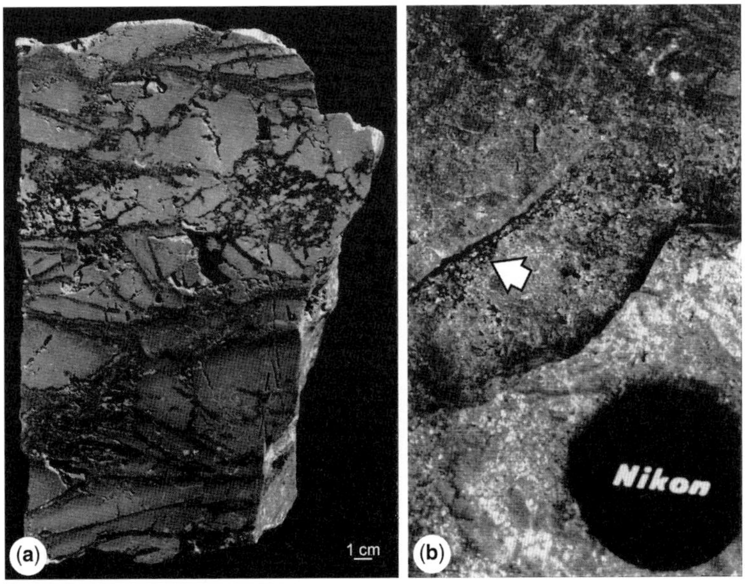

Fig. 34. (**a**) *Microcodium* (small black grains) developed in the micritic matrix of the uppermost *Palaeoaplysina* build-up in the Tyrrellfjellet section (see Fig. 20). (**b**) *Microcodium* completely replacing a *Palaeoaplysina* plate (arrowed). Note the 'beard' of *Microcodium* extending down into the sediment from the replaced plate. Top of Bank B at Locality PD (see Fig. 21 for location). From Hanken *et al.* (1993).

during lowstands these were islands surrounded by a shallow sea. Although plant fossils are very rare in this part of the sequence, the islands must have been covered by vegetation because both the presence of *Microcodium* in the top of the build-ups and kerogen with a composition of both Type II and Type III in the lateral facies indicate a terrestrial contribution.

The subaerial exposure of limestone deposits results in karstification with the development of both positive and negative karst forms (i.e. areas with positive and negative relief, respectively). However, subaerially exposed *Palaeoaplysina* build-ups with *in situ Microcodium* only have a flat or slightly undulating surface. The lack of positive karst forms might be explained as selective erosion of topographical highs due to rapid erosion in the tidal zone by a combination of bioerosion and wave-induced erosion (cf. Longman & Brownlee 1980). In the case of relatively slow transgression, shallow-water erosion may thus result in the levelling of positive karst forms so that the original karst surface will appear as a relatively smooth erosion surface where only negative karst forms have been preserved. However, negative karst forms such as solution-enlarged fissures, cracks and caves have not been observed. This indicates that the karstification has been relatively brief, the karstification processes having ceased in an early phase of the modification of the subaerially exposed sequence. This hypothesis is also indirectly supported by studies of fusulinids by Nilsson (1993), which indicate that no fusulinid zones are missing in the Tyrrellfjellet Member. The individual fusulinid zones are believed to have a resolution of *c.* 1.5 million years (Nilsson 1993), which indicates that the individual subaerial exposure events cannot represent hiati of any significant duration.

Summary

During the Late Carboniferous and Early Permian, the northern margin of Pangaea was situated at relatively low palaeo-latitudes where a warm dry climate with strong evaporation prevailed. It appears that *Palaeoaplysina* accumulations were confined to between 10°N and 40°N of the Upper Palaeozoic equator. The biota were characterized by a photozoan assemblage, which indicates deposition in a tropical climate.

The continental glaciation over the Gondwana south pole during the Late Carboniferous–Early Permian had discrete glacial intervals separated by warmer periods, which led to glacioeustatic worldwide sea-level variations. During lowstands, parts of the *Palaeoaplysina* build-ups could be subaerially exposed, karstified and covered with vegetation.

The *Palaeoaplysina* build-ups started as separate mounds. In an early stage, vertical growth dominated, but with decreasing accommodation space, continued vertical growth was hampered and lateral growth became dominant. The flat-topped build-ups could in several cases be situated close together (only separated by canals), forming elongated ridges or big ring-shaped to polygonal structures surrounding a lagoon. During lowstands, the lagoon could be more or less cut off from normal seawater outside the build-ups. In dry climates, evaporation caused hypersalinity. Only symbiotic fusulinids were well adapted to this harsh environment.

We wish to thank I. Nilsson, J. A. Rasmussen, T. Samuelsberg, I. Sommerville and N. Pickard for constructive discussions in the field. Amoco, British Gas, Elf NORGE A/S, Norwegian Petroleum Directorate, Norsk Hydro ASA, Norske Shell, Saga Petroleum and Statoil funded the fieldwork. Most of the drawings were created by T. Midtun and J. P. Holm. Reviewers G. Elvebakk and T. A. Svånå and editors A. Gąsiewicz and M. Słowakiewicz are acknowledged for valuable contributions that improved the manuscript.

References

ANISIMOW, R. M., DAVYDOV, V. I. & NILSSON, I. 1998. Graphic correlation of Upper Carboniferous–Lower Permian strata of Spitsbergen and the Loppa High, Barents Sea shelf. *Permophiles*, **32**, 32–37.

BAMBER, E. W. & MACQUEEN, R. W. 1979. Upper Carboniferous and Permian stratigraphy of the Monkman Pass and southern Pine Pass areas, northeastern Columbia. *Canadian Geological Survey Bulletin*, **301**, 27.

BEAUCHAMP, B. 1993. Carboniferous and Permian reefs of Sverdrup Basin, Canadian Arctic: an aid to Barents Sea exploration. *In*: VORREN, T. O., BERGSAGER, E., DAHL-STAMNES, Ø. A., HOLTER, E., JOHANSEN, B., LIE, E. & LUND, T. B. (eds) *Arctic Geology and Petroleum Potential*. Norwegian Petroleum Society (NPF), Elsevier, Amsterdam, Special Publications, **2**, 217–241.

BEAUCHAMP, B., DAVIES, G. R. & NASSICHUK, W. W. 1989. Upper Carboniferous to Lower Permian Palaeoaplysina-phylloid algal buildups, Canadian Arctic Archipelago. *In*: GELDSETZER, H. H. J., JAMES, N. P. & TEBBUTT, G. E. (eds) *Reefs: Canada and Adjacent Areas*. Canadian Society of Petroleum Geology, Calgary, Memoirs, **13**, 631–637.

BEAVINGTON-PENNEY, S. J. & RACEY, A. 2004. Ecology of extant nummulitids and other larger benthic foraminifera: application in palaeoenvironmental analysis. *Earth-Science Reviews*, **67**, 219–265.

BERGH, S. G., MAHER, H. D. & BRAATHEN, A. 2000. Tertiary divergent thrust directions from partitioned transpression, Brøggerhalvøya, Spitsbergen. *Norsk Geologisk Tidsskrift*, **80**, 63–82.

BIRKENMAJER, K. 1964. Devonian, Carboniferous and Permian formations of Hornsund, Vestspitsbergen (Utwory dewonu, karbonu i permu we fiordzie Hornsund na Spitsbergenie). *Studia Geol. Polon.*, **XI**, 47–123.

BJORØY, M., HELLEM, T. A., MILLS, N. & VIGRAN, J. O. 1982. *Organic geochemical analysis on Paleozoic rocks from Svalbard*. IKU-report **P-160/1/82**.

BLENDINGER, W., BOWLIN, B., ZIJP, F. R., DARKE, G. & EKROLL, M. 1997. Carbonate buildup flank deposits: an example from the Permian (Barents Sea, northern Norway) challenges classical facies models. *Sedimentary Geology*, **112**, 89–103.

BLOMEIER, D., SCHEIBNER, C. & FORKE, H. 2009. Facies arrangement and cyclostratigraphic architecture of a shallow-marine, warm-water carbonate platform: the Late Carboniferous Ny Friesland Platform in eastern Spitsbergen (Pyefjellet Beds, Wordiekammen Formation, Gipsdalen Group). *Facies*, **55**, 291–324.

BOSENCE, D. W. J. & BRIDGES, P. H. 1995. A review of the origin and evolution of carbonate mud-mounds. *In*: MONTY, C. L. V., BRIDGES, P. H. & PRATT, B. R. (eds) *Carbonate Mud-Mounds, their Origin and Evolution*. Blackwell Science for the International Association of Sedimentologists, Oxford, Special Publications, **23**, 3–10.

BRAATHEN, A., BÆLUM, K., MAHER, H. JR. & BUCKLEY, S. 2011. Growth of extensional faults and folds during deposition of an evaporite dominated half-graben basin: the Carboniferous Billefjorden Trough, Svalbard. *Norwegian Journal of Geology*, **91**, 137–160.

BREUNINGER, R. H. 1976. Palaeoaplysina (Hydrozoan?) carbonate buildups from Upper Paleozoic of Idaho. *American Association of Petroleum Geologists Bulletin*, **60**, 584–607.

BROMLEY, R. G. 1996. *Trace Fossils. Biology, Taphonomy and Applications*. Chapman & Hall, London.

BROMLEY, R. G. 2004. A stratigraphy of marine bioerosion. *In*: MCILROY, D. (ed.) *The Application of Ichnology to Palaeoenvironmental and Stratigraphic Analysis*. Geological Society, London, Special Publications, **228**, 455–481.

BRUCE, J. R. & TOOMEY, D. F. 1993. Late Paleozoic bioherm occurrences of the Finnmark Shelf, Norwegian Barents Sea: analogues and regional significance. *In*: VORREN, T. O., BERGSAGER, E., DAHL-STAMNES, Ø. A., HOLTER, E., JOHANSEN, B., LIE, E. & LUND, T. B. (eds) *Arctic Geology and Petroleum Potential*. Norwegian Petroleum Society, Elsevier, Amsterdam, Special Publications, **2**, 377–392.

BUGGE, T., MANGERUD, G., ELVEBAKK, G., MØRK, A., NILSSON, I., FANAVOLL, S. & VIGRAN, J. O. 1995. The Upper Palaeozoic succession on the Finnmark Platform, Barents Sea. *Norsk Geologisk Tidsskrift*, **75**, 3–30.

BUGGISCH, W., JOACHIMSKI, M., LÜTZNER, H., THIEDIG, F. & HÜNEKE, H. 2001. Conodont stratigraphy of the Carboniferous and Permian strata from Brøggerhalvøya and the Billefjorden Trough (Svalbard). *Geologische Jahrbuch*, **B91**, 637–689.

CECCHI, M. 1993. Carbonate sequence stratigraphy: application to the determination of play-models in the Upper Paleozoic succession of the Barents Sea, offshore northern Norway. *In*: VORREN, T. O., BERGSAGER, E., DAHL-STAMNES, Ø. A., HOLTER, E., JOHANSEN, B., LIE, E. & LUND, T. B. (eds) *Arctic Geology and Petroleum Potential*. Norwegian Petroleum

Society, Elsevier, Amsterdam, Special Publications, **2**, 419–438.

CHALLINOR, A. 1967. The structure of Brøggerhalvøya, Spitsbergen. *Geological Magazine*, **104**, 322–336.

CHEN, B., JOACHIMSKI, M. M. *ET AL.* 2013. Permian ice volume and palaeoclimate history: oxygen isotope proxies revisited. *Gondwana Research*, **24**, 77–89.

COLPAERT, A., PICKARD, N., MIENERT, J., HENRIKSEN, L. B., RAFAELSEN, B. & ANDREASSEN, K. 2007. 3D seismic analysis of an Upper Palaeozoic carbonate succession of Eastern Finnmark Platform area, Norwegian Barents Sea. *Sedimentary Geology*, **197**, 79–98.

COLPAERT, A., HUNT, D., MIQUELIS, F., BROCHERAY, S., GRAHAM-WALL, B., PATTON, G. & AVU, A. 2010. Subsurface identification of a Palaeozoic carbonate system in the Barents Sea: buildup distribution, geometry and development. *In*: WOOD, L. J., SIMO, T. T. & ROSEN, N. C. (eds) *Seismic Imaging of Depositional and Geomorphic Systems. 30th Annual Gulf Coast Section SEPM Foundation Bob F. Perkins Research Conference*, Houston, Texas, 5–8 December 2010, 11

COPELAND, B. J. 1967. Environmental characteristics of hypersaline lagoons. *Contribution to Marine Sciences*, **12**, 207–218.

COSSEY, P. J. & MUNDY, D. J. C. 1990. Tetrataxis: a loosely attached limpet-like foraminifer from the Upper Palaeozoic. *Lethaia*, **23**, 311–322.

CUTBILL, J. L. 1968. Carboniferous and Permian stratigraphy of Ny Friesland, Spitsbergen. *Norsk Polarinstitutt Årbok*, **1966**, 12–24.

CUTBILL, J. L. & CHALLINOR, A. 1965. Revision of the stratigraphical scheme for the Carboniferous and Permian rocks of Spitsbergen and Bjørnøya. *Geological Magazine*, **102**, 418–439.

DALLMANN, W. K., GJELBERG, J. G. *ET AL.* 1999. Upper Palaeozoic lithostratigraphy. *In*: DALLMANN, W. K. (ed.) *Lithostratigraphic Lexicon of Svalbard. Upper Palaeozoic to Quaternary Bedrock. Review and Recommendations for Nomenclature Use*. Committee on the Stratigraphy of Svalbard. Gjøvik Trykkeri As, Gjøvik, Norway, for Norsk Polarinstitutt, Tromsø, Norway, 25–126.

DAVIES, G. R. 1971. A Permian hydrozoan mound, Yukon Territory. *Canadian Journal of Earth Sciences*, **8**, 971–988.

DAVIES, G. R. & NASSICHUK, W. W. 1973. The hydrozoan? *Palaeoaplysina* from the upper Paleozoic of Ellesmere Island, Arctic Canada. *Journal of Paleontology*, **47**, 251–265.

DAVIES, G. R. & NASSICHUK, W. W. 1986. Ancient reefs in the high Arctic. *Geos*, **15**, 1–5.

DAVIES, G. R., RICHARDS, B. C., BEAUCHAMP, B. & NASSICHUK, W. W. 1989. Carboniferous and Permian reefs in Canada and adjacent areas. *In*: GELDSETZER, H. H. J., JAMES, N. P. & TEBBUTT, G. E. (eds) *Reefs, Canada and Adjacent Areas*. Canadian Society of Petroleum Geologists, Calgary, Memoirs, **13**, 565–574.

DAVYDOV, V. I. 1997. Fusulinid biostratigraphy of Upper Paleozoic of the Kolguev Island and Franz Josef Land Archipelago. *In*: KIRICHKOVA, A. I. & BELONIN, M. D. (eds) *Biostratigraphy of Oil-Bearing Basins. Transactions of first International Symposium*, Sankt Petersburg, 40–59.

DAVYDOV, V. I. & AREFIFARD, S. 2007. Tectonics and paleogeographic applications of Peri-Gondwanan Permian fusulinid fauna from Kalmard Region, East-Central Iran. *Permophiles Issue*, **49**, 13–24.

DAVYDOV, V. I., NILSSON, I. & STEMMERIK, L. 2001. Fusulinid zonation of the Upper Carboniferous Kap Jungersen and Foldedal Formations, southern Amdrup Land, eastern North Greenland. *Bulletin of the Geological Society of Denmark*, **48**, 31–77.

DAVYDOV, V. I., SCHIAPPA, T. A. & SNYDER, W. S. 2012. Testing the sequence stratigraphy model: response of fusulinacean fauna to sea-level fluctuations (examples from Pennsylvanian and Cisuralian of the Pricaspian–southern Urals region). *Society for Sedimentary Geology (SEPM)*, **75**, 361–378.

DINELEY, D. L. 1958. A review of the Carboniferous and Permian rocks of the west coast of Vestspitsbergen. *Norsk geologisk Tidsskrift*, **38**, 197–217.

DONS, C. E. 1983. *Fasies og paleostrømanalyse av Nordenskiöldbreen formasjonen (overkarbon-underperm), sentrale Spitsbergen.* Unpublished cand. real. thesis, University of Oslo.

EHRENBERG, S. N., NIELSEN, E. B., SVÅNÅ, T. A. & STEMMERIK, L. 1998. Depositional evolution of the Finnmark carbonate platform, Barents Sea: results from wells 7128/6-1 and 7128/4-1. *Norsk Geologisk Tidsskrift*, **78**, 185–224.

ELVEBAKK, G., HUNT, D. W. & STEMMERIK, L. 2002. From isolated buildups to buildup mosaics: 3D seismic sheds new light on Upper Carboniferous–Permian fault controlled carbonate buildups, Norwegian Barents Sea. *Sedimentary Geology*, **152**, 7–17.

EZAKI, Y. & KAWAMURA, T. 1992. Carboniferous–Permian corals from Skansen and Festningen, Central Spitsbergen: their faunal characteristics. *In*: NAKAMURA, K. (ed.) *Investigation on the Upper Carboniferous–Upper Permian Succession of West Spitsbergen 1989–1991.* Hokkaido University, Sapporo, Japan, 59–75.

FORBES, C. L. 1960. Carboniferous and Permian fusulinids from Spitsbergen. *Palaeontology*, **2**, 210–225.

FORBES, C. L., HARLAND, W. B. & HUGHES, N. F. 1958. Paleontological evidence for the age of the Carboniferous and Permian rocks of central Spitsbergen. *Geological Magazine*, **95**, 465–490.

FOSSUM, B. J., SCHMIDT, W. J., JENKINS, D. A., BOGATSKY, V. L. & RAPPOPORT, B. I. 2001. New frontiers for hydrocarbon production in the Timan–Pechora Basin, Russia. *In*: DOWNEY, M. W., THREET, J. C. & MORGANS, W. A. (eds) *Petroleum Provinces of the Twenty-First Century.* American Association of Petroleum Geologists, Tulsa, OK, Memoirs, **74**, 259–279.

FREYTET, P. & PLAZIAT, J. C. 1982. Continental carbonate sedimentation and pedogenesis – Late Cretaceous and Early Tertiary of Southern France. *Contribution to Sedimentology*, **12**, 1–213.

FRIEDMAN, G. M. & KRUMBEIN, W. E. 1985. *Hypersaline Ecosystems: The Gavish Sabkha.* Springer-Verlag, Berlin.

GEE, E. R., HARLAND, W. B. & MCWHAE, J. R. H. 1952. Geology of central Vestspitsbergen. Part 1: Review of the geology of Spitsbergen with special reference to central Vestspitsbergen. *Transactions of the Royal Society of Edinburgh*, **52**, 299–356.

GJELBERG, J. G. & STEEL, R. J. 1981. An outline of Lower–Middle Carboniferous sedimentation on Svalbard: effects on climatic, tectonic and sea level changes in rift basin sequences. *In*: KERR, J. W. & FERGUSSON, A. J. (eds) *Geology of the North Atlantic Borderlands*. Canadian Society of Petroleum Geologist, Calgary, Memoirs, **7**, 543–561.

GROTZINGER, J. P. & KNOLL, A. H. 1995. Anomalous carbonate precipitates: Is the Precambrian the key to the Permian? *Palaios*, **10**, 578–596.

HANKEN, N.-M., PICKARD, N. A. H., JOHANSEN, T. & RASMUSSEN, J. A. 1993. *Sedimentology of the Palaeoaplysina bioherms of central Spitsbergen, with notes on the palaeontology of the Tyrrellfjellet Member*. Unpublished report to Norsk Hydro AS.

HARLAND, W. B., CUTBILL, J. I. *ET AL*. 1974. The Billefjorden Fault zone, Spitsbergen; the long history of a major tectonic lineament. *Norsk Polarinstitutt Skrifter*, **161**, 72.

HENRIKSEN, H., RYSETH, A. E., LARSSEN, G. B., HEIDE, T., RØNNING, K., SOLLID, K. & STOUPAKOVA, A. V. 2011. Tectonostratigraphy of the greater Barents Sea: implications for petroleum systems. *In*: SPENCER, A. M., EMBRY, A. F., GAUTIER, D. L., STOUPAKOVA, A. V. & SØRENSEN, K. (eds) *Arctic Petroleum Geology*. Geological Society, London, Memoirs, **35**, 163–195.

HINE, A. C., HALLOCK, P., HARRIS, M. W., MULLINS, H. T.,, BELKNAP, D. F. & JAPP, W. C. 1988. Halimeda bioherms along an open seaway: Miskito Channel, Nicaraguan Rise, SW Caribbean Sea. *Coral Reefs*, **6**, 173–178.

HITE, R. J. 1970. Shelf carbonate sedimentation controlled by salinity in the Paradox Basin, southeast Utah. *In*: RAU, J. L. & DELLWIG, L. F. (eds) *Third Symposium on Salt*. Northern Ohio Geological Society, Cleveland, **1**, 48–66.

HJELLE, A., LAURITZEN, Ø., SALVIGSEN, O. & WINSNES, T. 1986. *Geological map of Svalbard 1:100 000*. Sheet B10G Van Mijenfjorden. Norsk Polarinstitutt, Temakart, no. 2.

HÜNEKE, H., JOACHIMSKI, M., BAGGISCH, W. & LÜTZNER, H. 2001. Marine carbonate facies in response to climate and nutrient level: the Upper Carboniferous and Permian of Central Spitsbergen (Svalbard). *Facies*, **45**, 93–136.

HUNT, D., COLPAERT, A. *ET AL*. 2010. Seismic geomorphology of paleokarst systems in Paleozoic carbonates, Norwegian Barents Sea. *In*: WOOD, L. J., SIMO, T. T. & ROSEN, N. C. (eds) *Seismic Imaging of Depositional and Geomorphic Systems. 30th Annual Gulf Coast Section SEPM Foundation Bob F. Perkins Research Conference*, Houston, TX, 5–8 December 2010, Bergan *et al*., inc., Rockport, Texas.

IGO, H. & OKIMURA, Y. 1992. Carboniferous–Permian foraminifers of west Spitsbergen. *In*: NAKAMURA, K. (ed.) *Investigations on the Upper Carboniferous–Upper Permian Succession of West Spitsbergen 1989–1991*. Department of Geology and Mineralogy Faculty of Sciences, Hokkaido University, Japan, 97–118.

ISHELL, J. I., MILLER, M. F., WOLFE, K. I. & LENAKER, P. A. 2003. Timing of late Paleozoic glaciation in Gondwana: Was glaciation responsible for the development of northern hemisphere cyclothems? *In*: CHAN, M. A. & ARCHER, A. W. (eds) *Extreme depositional environments: Mega end members in geologic time* Geological Society of America, Boulder, Colorado, Special Paper, **370**, 5–24.

JAMES, N. P. 1980. Reef environment. *In*: SCHOLLE, P. A., BEBEOUT, D. G. & MOORE, C. H. (eds) *Carbonate Depositional Environments*. American Association of Petroleum Geologists, Tulsa, Memoirs, **33**, 346–444.

JAMES, N. P. 1997. The cool-water carbonate depositional realm. *In*: JAMES, N. P. & CLARKE, J. A. D. (eds) *Cool-Water Carbonates*. SEPM Special Publications, Tulsa, **56**, 1–20.

JAMES, N. P. & BOURQUE, P. A. 1992. Reefs and mounds. *In*: WALKER, R. G. & JAMES, N. P. (eds) *Facies Models: Response to Sea Level Change*. Geological Association of Canada, St. John, 323–347.

JEFFERY, D. & AIGNER, T. 1982. Storm sedimentation in the Carboniferous Limestones near Weston-Super-Mare (Dinantian, SW-England). *In*: EINSELE, G. & SEILACHER, A. (eds) *Cyclic and Event Stratification*. Springer-Verlag, Berlin, 240–247.

JOHANSEN, T. 1994. *Sedimentology and diagenesis of Tyrrellfjellet Member, Bünsow Land, Central Spitsbergen*. Unpublished cand. scient. thesis, University of Tromsø.

KABANOV, P., ANADÓN, P. & KRUMBEIN, W. E. 2008. *Microcodium*: an extensive review and a proposed non-rhizogenic biologically induced origin for its formation. *Sedimentary Geology*, **205**, 79–99.

KANO, A. 1992. Paleoecology of palaeoaplysinid bioherms of the Lower Permian in central Spitsbergen. *In*: NAKAMURA, K. (ed.) *Investigations on the Upper Carboniferous–Upper Permian succession of West Spitsbergen 1989–1991*. Department of Geology and Mineralogy Faculty of Sciences, Hokkaido University, Japan, 119–126.

KEILEN, H. B. 1991. Lower Permian sedimentary sequences in Central Spitsbergen, Svalbard. *In*: NAKAMURA, K. (ed.) *Investigations on the Upper Carboniferous–Upper Permian Succession of West Spitsbergen 1989–1991*. Department of Geology and Mineralogy Faculty of Sciences, Hokkaido University, Japan, 127–134.

KENDALL, C. G. ST. C. & SCHLAGER, W. 1981. Carbonates and relative changes in sea level. *Marine Geology*, **44**, 181–212.

KENDALL, C. G. & WARREN, J. K. 1988. Peritidal evaporites and their sedimentary assemblages. *In*: SCHREIBER, B. C. (ed.) *Evaporites and Hydrocarbons*. Columbia University Press, New York, 66–138.

KINSMAN, D. J. J., BOARDMAN, M. & BORCSIK, M. 1973. An experimental determination of the solubility of oxygen in marine brines. *In*: COOGAN, A. (ed.) *Fourth Symposium on Salt*. Northern Ohio Geological Society, Cleveland, **1**, 325–327.

KLEINSPEHN, K. I., STEEL, R. J., JOHANNESEN, E. & NETLAND, A. 1984. Conglomeratic fan-delta sequences, Late Carboniferous -Early Permian, western Spitsbergen. *In*: KOSTER, E. H. & STEEL, R. J. (eds) *Sedimentology of Gravels and Conglomerates*. Canadian Society of Petroleum Geologists, Calgary, Memoirs, **10**, 279–294.

KNAG, G. 1980. *Gipshuk- og Kapp Starostinformasjonen, mellom til øvre perm, langs vestkysten av Svalbard.* Cand. real. thesis, University of Bergen.

KOŠIR, A. 2004. Microcodium revisited: root calcification products of terrestrial plants on carbonate-rich substrates. *Journal of Sedimentary Research*, **74**, 845–857.

KROTOV, P. 1888. Geologicheskiya izsledovaniya na Zapadnom sklon Solikamskogo i Cherdynskago Urala (Geologiccal investigation on the west slope of the Solikamskogo and Cherdyn Urals). *Trudy Geologicheskago Komiteta*, **6**, 431–434. (in Russian; German summary)

LARSSEN, G. B., ELVEBAKK, G. *ET AL.* 2002. Upper Palaeozoic lithostratigraphy of the southern Norwegian Barents Sea. *NPD Bulletin*, **9**, http://www.npd.no/en/Publications/NPD-bulletins/259-Bulletin-9/. (Also published in *NGU-Bulletin* 444).

LAURITZEN, Ø. & WORSLEY, D. 1973. Observations on the Upper Paleozoic stratigraphy of the Ny Friesland area. *Norsk Polarinstitutt Årbok*, **1973**, 41–51.

LEE, J. J. & HALLOCK, P. 1987. Algal symbiosis as the driving force in the evolution of larger foraminifera. *Annals of the New York Academy of Sciences*, **503**, 330–347.

LEPPIG, U., FORKE, H. C., MONTENARI, M. & FOHRER, B. 2005. A three- and two-dimensional documentation of structural elements in schwagerinids (superfamily Fusulinoidea) exemplified by silicified material from the upper Carboniferous of the Carnic Alps (Austria/Italy): a comparison with verbeekinoideans and alveolinids. *Facies*, **51**, 541–553.

LONGMAN, M. V. & BROWNLEE, D. N. 1980. Characteristics of karst topography, Palawan, Philippines. *Zeitschrift für Geomorphologie*, **24**, 299–317.

LØNØY, A. 1981. *Fasies analyse av undre permiske karbonater i Tyrrellfjellet led av Nordenskiöldbreen formasjonen, Billefjordområdet, Spitsbergen.* Unpublished cand. real. thesis, University of Bergen.

LØNØY, A. 1988. Environmental setting and diagenesis of Lower Permian Palaeoaplysinid buildups and associated sediments from Bjørnøya: implications for the exploration of the Barents Sea. *Journal of Petroleum Geology*, **11**, 141–156.

MAIKLEM, W. R. 1971. Evaporative drawdown – a mechanism for water-level lowering and diagenesis in the Elk Point Basin. *Bulletin of Canadian Petroleum Geology*, **19**, 487–501.

MAMET, B. L. & STEMMERIK, L. 2000. Carboniferous algal microflora, Kap Jungersen and Foldedal Formations, Holm Land and Amdrup Land, eastern North Greenland. *Geology of Greenland Survey Bulletin*, **187**, 79–100.

MAMET, B. L., ROUX, A. & NASSICHUK, W. W. 1987. Algues carbonifères et permiennes de l'Arctique canadien. *Geological Survey of Canada Bulletin*, **342**, 1–143.

MANN, K. & LANE, R. (eds) 1995. *Graphic Correlation.* Society of Economic Paleontologists and Mineralogists, Tulsa, Special Publications, **53**.

MANN, A. & TOWNSEND, C. 1989. The post-Devonian tectonic evolution of southern Spitsbergen illustrated by cross-sections through Bellsund and Hornsund. *Geological Magazine*, **126**, 549–566.

MARSHALL, J. F. & DAVIES, P. J. 1988. Halimeda bioherms of the northern Great Barrier Reef. *Coral Reefs*, **6**, 139–148.

MOBERG, E. G., GREENBERG, D. M., REVELLE, R. & ALLEN, E. C. 1934. The buffer mechanism of sea water. *Scripps Institution of Oceanography Technical Series Bulletin*, **3**, 231–278.

MONTY, C. L. V., BOSENCE, D. W. J., BRIDGES, P. H. & PRATT, B. R. (eds) 1995. *Carbonate Mud-Mounds – Their Origin and Evolution.* Blackwell Science for the International Association of Sedimentologists, Oxford, Special Publications, **23**.

NAKREM, H. A. 1991*a*. Distribution of conodonts through the Permian succession of Svalbard. *Geonytt*, **1**, 38–39.

NAKREM, H. A. 1991*b*. Distribution of bryozoans in the Permian succession of Svalbard (preliminary data). *In*: BIGEY, F. P. & D'HONDT, J.-L. (eds) *Bryzoaires Actuels et fossiles: Bryozoa Living and Fossil.* Bulletin de la Société des Sciences Naturelles de l'Ouest de la France, Nantes, Mémoire HS, **1**, 291–298.

NAKREM, H. A. 1994*a*. Middle Carboniferous–Lower Permian bryozoans from Spitsbergen. *Acta Palaeontologia Polonica*, **39**, 45–116.

NAKREM, H. A. 1994*b*. Environmental distribution of bryozoans in the Permian of Spitsbergen. *In*: HAYWARD, P. J., RYLAND, J. S. & TAYLOR, P. D. (eds) *Biology and Palaeobiology of Bryozoans.* Olsen & Olsen, Fredensborg, 133–137.

NAKREM, H. A. 2002. A Moscovian (Carboniferous) bryozoan buildup from Svalbard. *In*: WYSE JACKSON, P. N., BUTTLER, C. J. & SPENCER JONES, M. E. (eds) *Bryozoan Studies 2001.* A. A. Balkema, Lisse, 239–245.

NAKREM, H. A. 2008. Fossil bryozoa from Svalbard (arctic Norway): a research history. *In*: WYSE JACKSON, P. N. & SPENCER JONES, M. (eds) *Annals of Bryozoology 2: Aspects of the History of Research on Bryozoans.* International Bryozoology Association, Dublin, 71–88.

NAKREM, H. A., NILSSON, I. & MANGERUD, G. 1992. Permian biostratigraphy of Svalbard (Arctic Norway) – a review. *International Geology Review*, **34**, 933–959.

NETLAND, A. 1982. *Fasies analyse av Drevbreen beds og Nordenskiöldbreen Formasjonen, øvre karbon til undre perm, Bellsund området, Svalbard.* Cand. real. thesis, University of Bergen.

NILSSON, I. 1993. *Upper Palaeozoic fusulinid stratigraphy of the Barents shelf and surrounding areas.* Unpublished PhD thesis, University of Tromsø.

NILSSON, I. 1988. *Fusulinids on the Nordfjorden Block, Central Spitsbergen, Svalbard.* IKU Report **23.1252.06/01/88**.

NILSSON, I. & DAVYDOV, V. I. 1992. Upper Carboniferous (Kasimovian–Gzhelian) and Lower Permian (Asselian–Sakmarian) fusulinid stratigraphy in central Spitsbergen, Arctic Norway. IKU Petroleumsforskning a.s. Report **No. 23. 1356.00/07/92**.

NILSSON, I. & DAVYDOV, V. I. 1997. Fusulinid biostratigraphy in Upper Carboniferous (Gzhelian) and Lower Permian (Asselian–Sakmarian) successions of Spitsbergen, Arctic Norway. *Permophiles*, **30**, 18–24.

NOWIŃSKI, A. 1982. Some new species of tabulata from the Lower Permian of Hornsund, Spitsbergen. *In*: BIERNAT, G. & SZYMAŃSKA, W. (eds) *Palaeontological Spitsbergen Studies, Part 1*. Palaeontologia Polonica, Warszawa-Krakow, **43**, 83–96.

NOWIŃSKI, A. 1991. Carboniferous to Early Permian tabulata from Spitsbergen. *Palaeontologia Polonica*, **51**, 3–74.

NYSAETHER, E. 1977. Investigations on the Carboniferous and Permian stratigraphy of the Torell Land area, Spitsbergen. *Norsk Polarinst. Arbok*, **1976**, 22–41.

PICKARD, N. A. H. 1992. Depositional controls on Lower Carboniferous microbial buildups, eastern Midland Valley of Scotland. *Sedimentology*, **39**, 1081–1100.

PICKARD, N. A. H., EILERTSEN, F. *ET AL*. 1996. Stratigraphic framework of Upper Carboniferous (Moscovian–Kasimovian) strata in Bünsow Land, central Spitsbergen: palaeogeographic implications. *Norsk Geologisk Tidsskrift*, **76**, 169–185.

PRATT, B. R. 1982. Stromatolitic framework of carbonate mudmounds. *Journal of Sedimentary Petrology*, **52**, 1203–1227.

RAFAELSEN, B., ANDREASSEN, K., SAMUELSBERG, T. J., HOGSTAD, K. & RANDEN, T. 2003. Upper Paleozoic buildups in the Norwegian Barents Sea – new insights from 3-D seismic and automated facies mapping. *65th EAGE Conference & Exhibition*, EAGE Business Office, Houten, Stavanger, Norway, 2–5 June 2003. Extended abstract.

RAFAELSEN, B., ELVEBAKK, G., ANDREASSEN, K., STEMMERIK, L., COLPAERT, A. & SAMUELSBERG, T. J. 2008. From detached to attached carbonate buildup complexes – 3D seismic data from the upper Palaeozoic, Finnmark Platform, southwestern Barents Sea. *Sedimentary Geology*, **206**, 17–32.

RASMUSSEN, J. A. & HANKEN, N.-M. 1991. *Conodont stratigraphy and thermal maturation of the Upper Carboniferous and Lower Permian Nordenskiöldbreen Formation at Bünsow Land, Central Spitsbergen*. Report to Norsk Hydro A/S.

RASMUSSEN, J. A. & HANKEN, N.-M. 1992. *Conodont biostratigraphy and thermal maturation of the Permo-Carboniferous Nordenskiöldbreen Formation at Bünsow Land, Central Spitsbergen*. Report to Norsk Hydro A/S.

RAYMONT, J. E. G. 1963. *Plankton and Productivity in the Oceans*. Macmillian, New York.

REID, C. M., JAMES, N. P., BEAUCHAMP, B. & KYSER, T. K. 2007. Faunal turnover and changing oceanography: Late Palaeozoic warm-to-cool water carbonates, Sverdrup Basin, Canadian Arctic Archipelago. *Palaeogeography, Palaeoclimatology, Palaeoecology*, **249**, 128–159.

RIABININ, V. N. 1955. *O Kamennougolnykhi Permskikh Paleoaplyixinakh Uralai Timana*. Vsesoiuznyi Neftianoy Nauchno-Issledovatelskii Geologorazvedochnyi Institut, Trudy, **90**, 331–337.

RIDING, R. & GUO, L. 1992. Affinity of *Tubiphytes*. *Palaeontology*, **35**, 37–49.

ROBERTS, H. H., PHIPPS, C. V. & EFFENDI, L. 1987. Halimeda bioherms of the eastern Java Sea, Indonesia. *Geology*, **15**, 371–374.

ROSS, C. A. 1965. Fusulinids from the Cyathophyllum Limestone, Central Westspitsbergen. *Contribution from the Cushman Foundation for Foraminiferal Research*, **15**, 74–86.

ROSS, C. A. 1972. Paleobiological analysis of fusulinacean (Foraminiferida) shell morphology. *Journal of Paleontology*, **46**, 719–728.

ROSS, C. A. 1974. Evolutionary and ecological significance of large, calcareous Foraminiferida (Protozoa), Great Barrier Reef. *In*: CAMERON, A. M. & CAMBELL, B. M. (eds) *Proceedings of the Second International Coral Reef Symposium*, Vol. 1. The Great Barrier Reef Committee, Brisbane, Australia, 327–334.

ROSS, C. A. 1979. Ecology of large, shallow-water, tropical foraminifera. *In*: LIPPS, J. H., BERGER, W. H., BUZAS, M. A., DOUGLAS, R. G. & ROSS, C. A. (eds) *Foraminiferal Ecology and Paleoecology*. SEPM, Tulsa, Short Course, **6**, 43–61.

ROSS, C. A. 1982*a*. Paleozoic foraminifera – fusulinids. *In*: BROADHEAD, T. W. (ed.) *Foraminifera – Notes for a Short Course* (organized by Buzas, M. A. & Sen Gupta, B. K), *Studies in Geology*. University of Tennessee. Department of Geological Sciences, Tennessee, **6**, 163–176.

ROSS, C. A. 1982*b*. Paleobiology of fusulinaceans. *Proceedings of the North American Paleontological Convention*, **2**, 442–445.

ROSS, C. A. & ROSS, J. R. P. 1985. Late Palaeozoic depositional sequences are synchronous and worldwide. *Geology*, **13**, 27–30.

SAMUELSBERG, T. J. & PICKARD, N. A. H. 1999. Upper Carboniferous to Lower Permian transgressive–regressive sequences of central Spitsbergen, Arctic Norway. *Geological Journal*, **34**, 393–411.

SAMUELSBERG, T. J., ELVEBAKK, G. & STEMMERIK, L. 2003. Late Palaeozoic evolution of the Finnmark Platform, southern Barents Sea. *Norwegian Journal of Geology*, **83**, 351–362.

SCOTESE, C. R., BAMBACH, R. K., BARTON, C., VANDERVOO, R. & ZIEGLER, A. M. 1979. Palaeozoic base maps. *Journal of Geology*, **87**, 217–277.

SKAUG, M. 1982. *Bentiske fossile assosiasjoner og faciesvariasjoner i Nordenskiöldbreen formasjonen (overkarbon–underperm)*, sentrale Spitsbergen. Unpublished cand. real thesis, University of Oslo.

SKAUG, M., DONS, C. E., LAURITZEN, Ø. & WORSLEY, D. 1982. Lower Permian Palaeoaplysinid bioherms and associated sediments from central Spitsbergen. *Polar Research*, **2**, 57–75.

SKIPP, B., HOGGAN, R. D., SCHLEICHER, D. L. & DOUGLASS, R. C. 1979. Upper Paleozoic carbonate bank in east-central Idaho–Snaky Canyon, Bluebird Mountain, and Arco Hills Formations, and their paleotectonic significance. *US Geological Survey Bulletin*, **1486**, 1–78.

SOMMERVILLE, I. D. 1997. Biostratigraphy and biofacies of Upper Carboniferous–Lower Permian rugose coral assemblages from the Isfjorden area, central Spitsbergen. *In*: PEREJON, A. & COMAS, R. M. J. (eds) *Proceedings of the VII International Symposium on Fossil Cnidaria and Porifera. Boletin de la Real Sociedad Espanola de Historia Natural, Seccion Geologica*, **92**, 1–4, 365–380.

SOSIPATROVA, G. P. 1967. Upper Palaeozoic foraminifera of Spitsbergen. *In*: SOKOLOV, V. N. (ed.) *Stratigraphy of Spitsbergen*. Institut Geologii Arktiki, Leningrad (In Russian, English translation available from the

Lending Division of the British Library Board, Boston Spa, 1977: 125–163).

STEEL, R. J. & WORSLEY, D. 1984. Svalbard's post-Caledonian strata – an atlas of sedimentational patterns and paleogeographic evolution. *In*: SPENCER, A. M. *ET AL.* (eds) *Petroleum Geology of the Northern European Margin*. Graham and Trotman, London, for the Norwegian Petroleum Society, 109–135.

STEMMERIK, L. 2000. Late Palaeozoic evolution of the North Atlantic margin of Pangea. *Palaeogeography, Palaeoclimatology, Palaeoecology*, **161**, 95–126.

STEMMERIK, L. 2008. Influence of late Paleozoic Gondwana glaciations on the depositional evolution of the northern Pangean shelf, North Greenland, Svalbard, and the Barents Sea. *In*: FIELDING, C. R., FRANK, T. D. & ISBELL, J. L. (eds) *Resolving the Late Paleozoic ice age in Time and Space*. Geological Society of America, Boulder, CO, Special Papers, **441**, 205–217.

STEMMERIK, L. & LARSSEN, G. B. 1993. Diagenesis and porosity of Lower Permian *Palaeoaplysina* buildups, Bjørnøya, Barents Sea: an example of diagenetic response to high frequency sea level fluctuations in an arid climate. *In*: HORBURY, A. D. & ROBINSON, A. G. (eds) *Diagenesis and Basin Development*. American Association of Petroleum Geologists Studies in Geology, Tulsa, **36**, 199–211.

STEMMERIK, L. & WORSLEY, D. 2005. 30 years on – Arctic Upper Palaeozoic stratigraphy, depositional evolution and hydrocarbon prospectivity. *Norwegian Journal of Geology*, **50**, 151–168.

STEMMERIK, L., LARSON, P. A., LARSSEN, G. B., MØRK, A. & SIMONSEN, B. T. 1994. Depositional evolution of Lower Permian *Palaeoaplysina* buildups, Kapp Duner Formation, Bjørnøya, Arctic Norway. *Sedimentary Geology*, **92**, 161–174.

STEMMERIK, L., ELVEBAKK, G. & WORSLEY, D. 1999. Upper Palaeozoic carbonate reservoirs on the Norwegian Arctic Shelf: delineation of reservoir models with application to the Loppa High. *Petroleum Geoscience*, **5**, 173–187.

SUNDSBØ, G. 1982. *Facies analysis of Late Carboniferous and Early Permian carbonates in the Billefjorden area, Spitsbergen*. Unpublished cand. real. thesis, University of Bergen.

TANGEN, O. 1981. *A sedimentological and environmental interpretation of the upper part of the Nordenskiöldbreen Formation and the Gipshuken Formation on western Spitsbergen*. Cand. real. thesis, University of Bergen.

TCHUVASCHOV, B. I. 1983. Permian reefs of the Urals. *Facies*, **8**, 191–212.

TOOMEY, D. F. & BABCOCK, D. F. 1983. *Precambrian and Paleozoic Algal Carbonates, West Texas – Southern New Mexico*. Colorado School of Mines, Mines Press, Golden, Colorado, **11**, 343.

VACHARD, D., MUNNECKE, A. & SERVAIS, T. 2004. New SEM observations of keriothecal walls: implications for the evolution of Fusulinida. *Journal of Foraminiferal Research*, **34**, 232–242.

VEEVERS, J. J. & POWELL, C. MCA. 1987. Late Palaeozoic glacial episodes in Gondwanaland reflected in transgressive–regressive depositional sequences in Euramerica. *Geological Society of America Bulletin*, **98**, 475–487.

VOGEL, K. 1993. Bioeroders in fossil reefs. *Facies*, **28**, 109–114.

WAHLMAN, G. P. 1985. Lower Permian (Wolfcampian) *Archaeolithoporella-Tubiphytes*-sponge boundstone from the subsurface of West Texas. *In*: TOOMEY, D. F. & NITECKI, M. H. (eds) *Paleoalgology, Contemporary Research and Applications*. Springer-Verlag, Berlin, 208–215.

WAHLMAN, G. P. 1988. Subsurface Wolfcampian (Lower Permian) shelf margin reefs in the Permian basin of west Texas and southeastern New Mexico. *In*: MORGAN, W. P. & BABCOCK, J. A. (eds) *The Permian of the Midcontinent*. SEPM, Midcontinent Section, Special Publications, Tulsa, **1**, 177–204.

WAHLMAN, G. P. 2002. Upper Carboniferous–Lower Permian (Bashkirian–Kungurian) mounds and reefs. *In*: KIESSLING, W., FLÜGEL, E. & GOLONKA, J. (eds) *Phanerozoic Reef Patterns*. SEPM Special Publications, Tulsa, **72**, 271–338.

WATKINS, R. & WILSON, E. C. 1989. Paleoecologic and biogeographic significance of the biostromal organism Palaeoaplysina in the Lower Permian McCloud Limestone, eastern Klamath Mountains, California. *Palaios*, **4**, 181–192.

WEIDLICH, O. 2002. Middle and Late Permian Reefs – distributional patterns and reservoir potential. *In*: KIESSLING, W., FLÜGEL, E. & GOLONKA, J. (eds) *Phanerozoic Reef Patterns*. SEPM Special Publications, Tulsa, **72**, 339–390.

WILSON, J. L. 1975. *Carbonate Facies in Geologic History*. Springer Verlag, New York.

WINSNES, T. S. 1966. Observations on the Carboniferous and Permian rocks of Vestspitsbergen. *Norsk Polarinstitutt Årbok*, **1964**, 7–29.

WORSLEY, D., AGDESTEIN, T. *ET AL.* 2001. The geological evolution of Bjørnøya, Arctic Norway: implications for the Barents Shelf. *Norwegian Journal of Geology*, **81**, 195–234.

WRAY, J. L. 1977. *Calcareous Algae*. Elsevier Scientific, Amsterdam, Oxford, New York.

YANCEY, T. E. & KENNEDY, N. L. 1984. Relative depth position of fusulinids and ammonoides in Late Pennsylvanian regressive sequences in north Texas. *American Association of Petroleum Geologists Bulletin*, **68**, 541.

ZENKEVITCH, L. 1963. *Biology of the Seas of U.S.S.R.* Interscience, New York.

Sedimentology and palaeoecology of lonestone-bearing mixed clastic rocks and cold-water carbonates of the Lower Permian Basal Beds at Fossil Cliffs, Maria Island, Tasmania (Australia): Insight into the initial decline of the late Palaeozoic ice age

J. L. ISBELL[1]*, L. C. HENRY[1], C. M. REID[2] & M. L. FRAISER[1]

[1]*Department of Geosciences, University of Wisconsin – Milwaukee, PO Box 413, Milwaukee, WI 53201, USA*

[2]*Department of Geological Sciences, University of Canterbury, Private Bag 4800, Christchurch 8140, New Zealand*

**Corresponding author (e-mail: jisbell@uwm.edu)*

Abstract: The middle Sakmarian Basal Beds on Maria Island were deposited during the initial decline of the Late Palaeozoic Ice Age following Late Pennsylvanian–Early Sakmarian maximum glaciation. At that time, Tasmania was located within the South Polar Circle between an apparent ice-free pole (Antarctica) and the mid- to high-latitude Sydney-Bowen-Gunnedah glacigenic basins in eastern Australia. The dropstone-bearing Basal Beds consist of: interstratified siltstone, conglomerate and cold-water limestones of the Lower Erratic Zone; siltstone and *Eurydesma*-rich cold-water carbonates of the Darlington Limestone; and siltstone and conglomerate of the Upper Erratic Zone. Interstratification of the coarse-clastic strata, siltstones and limestones within these units were previously attributed to glacial/non-glacial cycles. However, the interfingering of beds within each of these units and the occurrence of large, fossil bryozoans crossing and abutting lithological boundaries indicate that cyclicity was of shorter duration than that of Milankovitch-driven cycles. Within these intertidal and subtidal deposits, the occurrence of rounded dropstones derived from local basement rocks exposed along a rocky coastline and an absence of glacial indicators other than dropstones, along with other evidence, suggest that ice rafting was by sea ice rather than by icebergs. Study results confirm the spatial restriction of Middle Sakmarian to earliest Wuchiapingian glaciation.

The Late Palaeozoic Ice Age (LPIA) was the most significant icehouse of the Phanerozoic, as this interval marks the only complete transition from an icehouse to a greenhouse/hothouse state for a biologically complex Earth (Gastaldo *et al.* 1996). Because of similarities between the LPIA and the Cenozoic Ice Age (i.e. long-duration glaciation interspersed with non-glacial epochs, low atmospheric pCO_2, complex fauna and flora and possible bipolar glaciation), the transition out of the LPIA and into the Late Permian–Mesozoic greenhouse/hothouse serves as an important deep-time proxy for understanding future climate change due to global warming (Frakes *et al.* 1992; Gastaldo *et al.* 1996; Montañez & Soreghan 2006).

The LPIA was previously assumed to be a single long duration event with ice covering much of Gondwana from the Mississippian until the Late Permian (Fig. 1; Veevers & Powell 1987; Frakes & Francis 1988; Frakes *et al.* 1992; Crowell 1999). However, recent work suggests that the LPIA consisted of a series of shorter glacial epochs that shifted diachronously across Gondwana with each epoch supporting multiple ice sheets, ice caps and/or alpine glaciers (Fig. 1; Crowell 1995; López-Gamundí 1997; Visser 1997; Isbell *et al.* 2003, 2008*a*; Fielding *et al.* 2008*a*, *b*; Gulbranson *et al.* 2010). Glacial acme occurred during the latest Pennsylvanian–Early Permian with the growth of numerous ice sheets scattered across Gondwana. Ensuing deglaciation occurred during the Asselian–Early Sakmarian in eastern and southern South America, Africa, Antarctica, India and the Arabian Peninsula (Veevers & Tewari 1995; López-Gamundí 1997; Wopfner & Casshyap 1997; Isbell *et al.* 2003, 2008*a*, *b*; Fielding *et al.* 2008*c*; Holz *et al.* 2008; Martin *et al.* 2008; Stollhofen *et al.* 2008). However, glaciation continued in parts of eastern Australia (Queensland, New South Wales and possibly Western Australia) until the latest Capitanian–earliest Wuchiapingian (the P2–P4 events of Fielding *et al.* 2008*a*, *b*; Mory *et al.* 2008). During that interval, the Sydney, Bowen and Gunnedah basins of eastern Australia were located between *c.* 52 and

From: Gąsiewicz, A. & Słowakiewicz, M. (eds) 2013. *Palaeozoic Climate Cycles: Their Evolutionary and Sedimentological Impact*. Geological Society, London, Special Publications, **376**, 307–341.
First published online November 26, 2012, http://dx.doi.org/10.1144/SP376.2

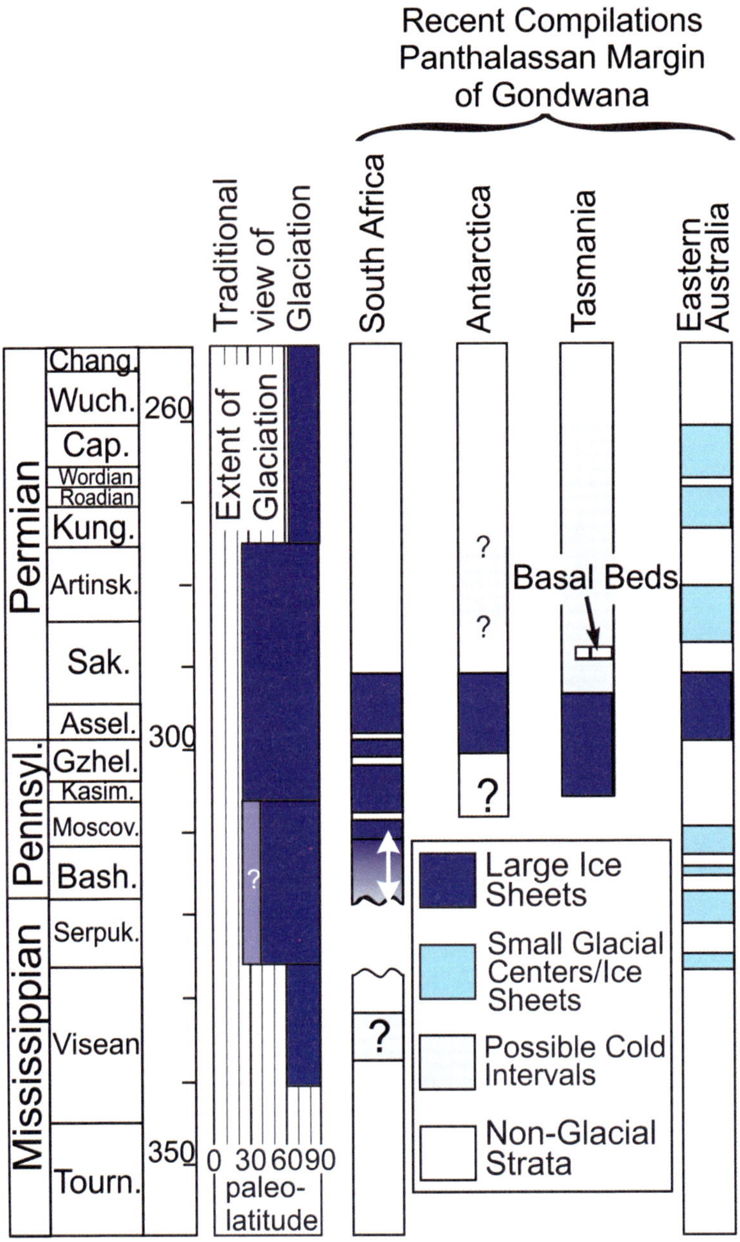

Fig. 1. Traditional and recent syntheses on the timing, duration, extent and style of glaciation during the Late Palaeozoic Ice Age (LPIA) using the timescale of Gradstein *et al.* (2004). Previous syntheses of LPIA are from Frakes & Francis (1988). Recent compilations for glaciation in South Africa are from Visser (1997), Isbell *et al.* (2008*b*) and Stollhofen *et al.* (2008); Antarctica from Askin (1998) and Isbell *et al.* (2008*a*, 2012); Tasmania from Fielding *et al.* (2010) and Henry *et al.* (2012); and Eastern Australia from Fielding *et al.* (2008*a*, *b*).

70°S latitude (Domeier *et al.* 2011; Isbell *et al.* 2012). In contrast, glacial deposits are unknown from Antarctica during that time, even though Antarctica resided over the South Pole and composed most of the landmass within the South Polar Circle (Dalziel 1997; Lawver *et al.* 2008; Domeier *et al.* 2011; Isbell *et al.* 2012). Following the Early Sakmarian deglaciation, Antarctica was characterized by post-glacial mudrock and coal-bearing fluvial-deltaic deposition. Despite the importance of understanding the causes of icehouse to greenhouse/hothouse transitions, the occurrence of mid-latitude glaciers and an ice-free pole during the decline of the LPIA (Middle Sakmarian–Early Wuchiapingian) has only recently been identified (Jones *et al.* 2006; Henry *et al.* 2012; Isbell *et al.* 2012). As of yet, this paradox has received little attention.

During the Pennsylvanian and Permian, Tasmania was situated within the South Polar Circle and was located between Antarctica to the south and eastern Australia to the north (Fig. 2; Domeier *et al.* 2011). Strata of the Parmeener Supergroup, a succession of marine to terrestrial strata, display lithological and stratigraphical elements that are transitional between strata found in Antarctica and eastern Australia (Figs 1–3; Fielding *et al.* 2010; Henry *et al.* 2012; Isbell *et al.* 2012). In Tasmania, Pennsylvanian–Lower Permian (Asselian) glacigenic diamictites at the top of the Wynyard, Truro and Stockers formations are roughly equivalent to glacigenic rocks deposited across Gondwana during peak glaciation, which in eastern Australia is referred to as the P1 glacial event (Fig. 1; Fielding *et al.* 2008*a*, *b*, 2010; Henry *et al.* 2012). Like glacigenic strata in the Transantarctic Mountains (Antarctica), the Wynyard and equivalent units contain the last direct evidence for Permian ice-contact deposits in Tasmania (Figs 1 & 3; e.g. Hand 1993; Isbell *et al.* 2008*a*; Fielding *et al.* 2010; Isbell 2010; Henry *et al.* 2012). However, unlike Antarctica, lonestones deposited as ice-rafted dropstones continued sporadically in Tasmania until the end of the Capitanian (Banks & Clarke 1987; Rogala *et al.* 2007; Fielding *et al.* 2010). In Antarctica, dropstones continue upwards only several metres above the early Sakmarian glacial/post-glacial contact (Isbell *et al.* 2008). Dropstone-bearing strata in Tasmania are roughly correlative with the P2–P4 glacial events in eastern Australia (Fielding *et al.* 2010). Perhaps the best example of ice-rafted debris in the Permian

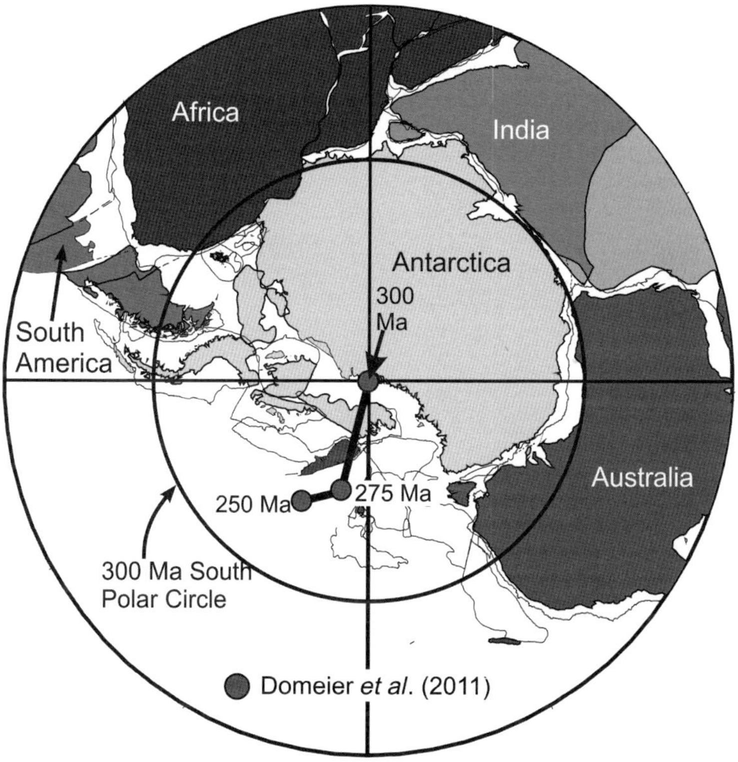

Fig. 2. Polar view of Gondwana at 300 Ma and the apparent polar wander path of Domeier *et al.* (2011), which is used in this paper. The diagram is modified from Isbell *et al.* (2012). The reconstruction of Gondwana is from Dalziel (1997) and Lawver *et al.* (2008). To construct the curve for Domeier *et al.* (2011), their poles were attached to South America then rotated into the Gondwana configuration presented by Lawver *et al.* (2008) with the view centred over the 300 Ma pole of Domeier *et al.* (2011). The reconstruction and rotations are courtesy of the Institute of Geophysics' PLATES Project at the University of Texas at Austin.

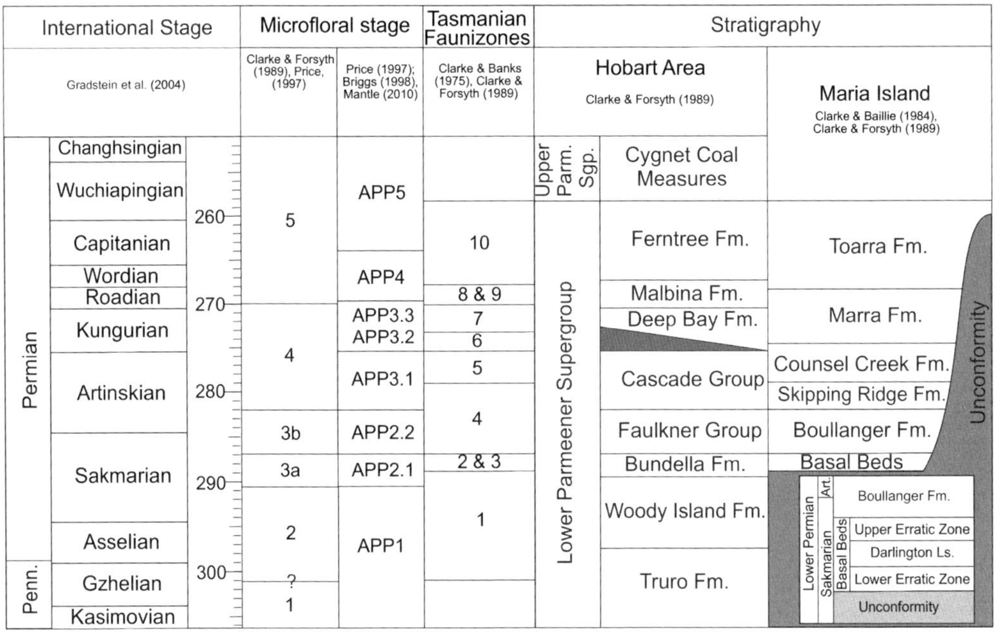

Fig. 3. Stratigraphic column for the Lower Parmeener Supergroup on Maria Island and in the adjacent Hobart area of Tasmania. Correlation between the two areas is based on Clarke & Forsyth (1989). The geological timescale is that of Gradstein *et al.* (2004). Correlations with the geological timescale are based on Truswell (1978), Clarke & Forsyth (1989), Price (1997), Briggs (1998) and Mantle *et al.* (2010).

of Tasmania occurs in the Basal Beds on Maria Island (Fig. 4), which were deposited several million years after termination of the Wynyard Formation and coeval deposition in Tasmania (Fig. 3; cf. Clarke & Forsyth 1989; Veevers *et al.* 1994; Fielding *et al.* 2010; Henry *et al.* 2012).

The Lower Permian Basal Beds exposed in sea cliffs and intertidal platforms along the east coast of Maria Island (Fig. 4) consist of siltstones, sandstones, conglomerates and cold-water limestones (Fig. 1). The origin of these beds is unresolved, in part, because the abundant lonestones within these units have led to conflicting depositional and environmental interpretations with regard to depositional processes (iceberg v. sea-ice rafting), temporal and spatial proximity to glaciation and climatic setting (glacial v. non-glacial v. interglacial; cf. Crowell & Frakes 1971; Brill 1982; Rao & Green 1982; Rao 1983; Clarke & Baillie 1984; Banks & Clarke 1987; Rogala *et al.* 2007; Fielding *et al.* 2010; Clarke & Forsyth 1989). Considering the plethora of interpretations of the origin of the Basal Beds and their significance in interpreting climatic conditions during the LPIA, detailed sedimentology of these units is warranted. One area where the study of glaciations is hindered is in the non-specific use of terminology. This practice creates much confusion in the literature. This paper uses the term glacial and glacial conditions to refer to an environment or setting that was directly influenced by glaciers and glacial processes; areas or time intervals that were not directly related to glaciers but experienced freezing conditions are referred to as cold/cold-climate areas or intervals. Polar is used to refer to the area that resides within the polar circle. This paper addresses the role that glaciation played in the deposition of the Basal Beds on Maria Island in an attempt to better understand the nature and geographical extent of Gondwana glaciation following the Early Permian LPIA glacial maximum. Such studies are important as they better define and constrain the timing, extent and causes of individual glacial events during the LPIA which, in turn, help to clarify environmental changes and their causes during Earth's transition from icehouse to greenhouse/hothouse conditions.

Geological setting

Deposition of strata in the lower portion of the Lower Parmeener Supergroup occurred within the high-relief intracratonic Tasmanian Basin. Relief of as much as 1000 m occurred along the basin margins (Fig. 5; Clarke & Forsyth 1989; Veevers *et al.* 1994; Fielding *et al.* 2010; Clarke & Forsyth

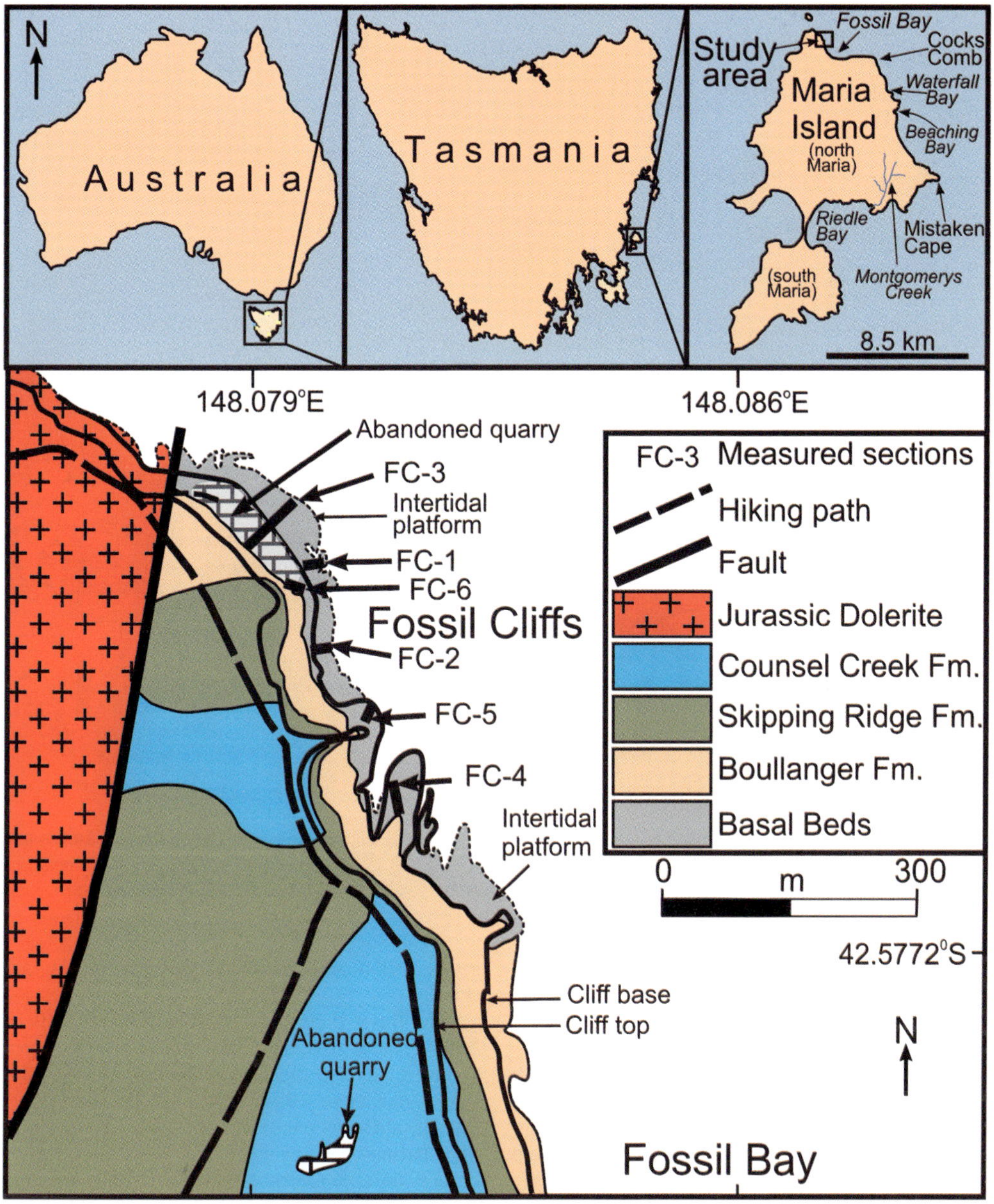

Fig. 4. Geological map of the Fossil Cliffs on Maria Island (modified from Clarke & Baillie 1984 and Reid 2004). Map shows the locations of sedimentological logs displayed in Figure 6.

1989). Pennsylvanian and Permian strata progressively onlap basement rocks, finally filling relief on the sub-Parmeener unconformity during the Kungurian–Wordian. Strata of the Wynyard, Stockers and Truro formations occupy the deeper portions of the Tasmanian Basin with deposition having occurred in glacial, periglacial and glacimarine settings (Fig. 5; Banks 1981; Powell 1990; Hand 1993; Henry *et al.* 2012). These glacigenic strata contain Pennsylvanian (Kasimovian–Gzhelian; Fig. 3) and Lower Permian (Asselian–Sakmarian) fossil plants (*Rhacopteris ovata*, *Botrychiopsis plantiana and Aphlebia sp.*) and microflora (Stage 1 and 2 of Truswell 1978; APP1 of Price 1997; cf. Clarke & Banks 1975; Veevers *et al.* 1994; Briggs 1998; Mantle *et al.* 2010). An abrupt transition from

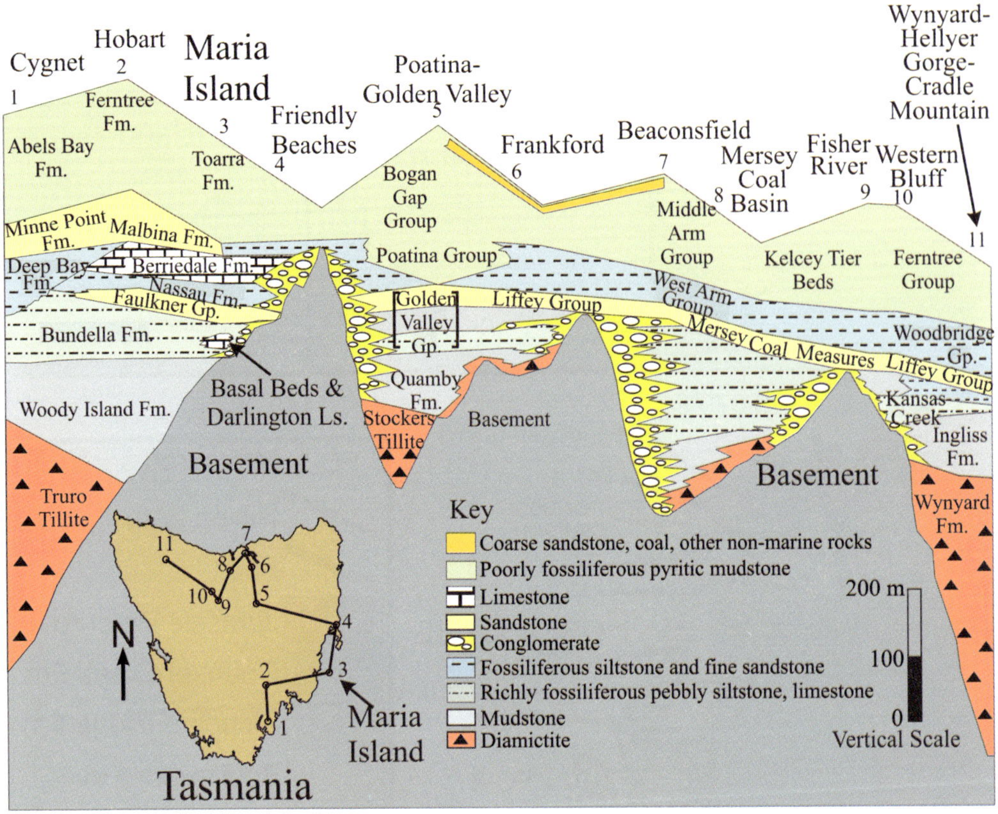

Fig. 5. Cross-section showing Parmeener Supergroup strata in Tasmania (modified from Clarke & Forsyth 1989 and Henry *et al.* 2012).

diamictite to overlying glendonite-bearing pyritic and carbonaceous mudrocks and siltstones of the Quamby Mudstone, the Woody Island Siltstone and the Inglis Formation marks the change from glacial to post-glacial deposition throughout much of the basin (Banks 1962; Domack *et al.* 1993). Ice-contact deposits and glacially eroded surfaces are absent above the Wynyard/Inglis formational (and correlative units) contact (Banks & Clarke 1987; Clarke & Forsythe 1989; Fielding *et al.* 2010; Henry *et al.* 2012). Although present, lonestones are rare in these post-Wynyard units. The occurrence of palynomorphs in the APP1.2 and APP2.1 zone suggests an Early–Middle Sakmarian age for the Woody Island Formation and equivalent units (Fig. 3; cf. Truswell 1978; Clarke & Forsyth 1989; Domack *et al.* 1993; Price 1997; Briggs 1998; Mantle *et al.* 2010). These strata are overlain by fossiliferous lonestone-bearing siltstones of the Bundella Formation (Clarke & Forsyth 1989), which occur in the APP2.1 microfloral stage (Clarke & Forsythe 1989; Price 1997; Briggs 1998; Mantle *et al.* 2010) and are Middle Sakmarian in age (Fig. 3).

The Basal Beds exposed at the Fossil Cliffs on Maria Island, which are the subject of this paper, are correlative to the Bundella Formation (Fig. 3; Banks & Clarke 1987; Clarke & Forsyth 1989). The Basal Beds consist of the Lower Erratic Zone, the Darlington Limestone and the Upper Erratic Zone (Clarke & Baillie 1984). These units accumulated next to a palaeogeographic high, as they rest directly on and onlap Devonian and older basement rocks (cf. Clarke & Baillie 1984; Rogala *et al.* 2007). The beds contain abundant lonestones, fossil-bearing siltstones and conglomerates (Lower and Upper Erratic Zones) and fossiliferous cold-water limestones (Lower Erratic Zone and Darlington Limestone). Lonestones and gravels in the Basal Beds have been interpreted as dropstones resulting from ice rafting, but the precise mechanism of rafting (iceberg or sea ice) has not been determined (Crowell & Frakes 1971; Brill 1982; Rao & Green 1982; Rao 1983; Clarke & Baillie 1984; Rogala *et al.* 2007; Fielding *et al.* 2010). In a paper on the Wasp Head Formation from the Early Sakmarian of the Sydney Basin, Eyles *et al.* (1997) suggested that lonestone-bearing beds in New South Wales and on Maria Island were analogous to Pleistocene Heinrich events and that the lonestones resulted from accelerated iceberg calving during the collapse of a marine-based ice sheet in East Antarctica. However, there is no evidence for ice sheets in Antarctica at that time (Isbell *et al.* 2008*a*, 2012), and Rygel *et al.* (2008) provided evidence that Heinrich events were not the cause for lonestone deposition in the Wasp Head Formation. The gravel-bearing Basal Beds have also been interpreted as the result of individual glacial cycles (Rao & Green 1982; Fielding *et al.* 2010). Alternatively, potential rafting of clasts by river or sea ice has also been suggested (Brill 1982; Banks & Clarke 1987; Rogala *et al.* 2007; Fielding *et al.* 2010). The Basal Beds have been correlated with late P1 glaciation, the ensuing non-glacial interval and the early stages of P2 glaciation in eastern Australia (Fielding *et al.* 2010). However, more recent analysis of available biostratigraphic data, which assign the beds to the upper portion of the APP2.1 Microfloral Zone (Mantle *et al.* 2010), suggest that deposition occurred during the non-glacial interval between the P1 and P2 glacial events in eastern Australia (Fielding *et al.* 2008*a*, *b*) and the post-glacial interval within neighbouring Antarctica (Collinson *et al.* 1994; Isbell *et al.* 1997, 2003, 2008*a*).

Within the Basal Beds at the Fossil Cliffs (Fig. 3), the bioclastic Darlington Limestone consists primarily of argillaceous rudstone and grainstone. Interstratification of siltstone and argillaceous carbonate occurs at the base of the unit with thicker and non-argillaceous carbonate occurring upwards (Brill 1982; Clarke & Baillie 1984). These strata are reported to grade shorewards into conglomeratic facies and offshore into calcareous mudrock (Rao & Green 1982; Clarke & Baillie 1984). Previous workers noted the occurrence of lonestones and an absence of iceberg keel marks within these strata (Brill 1982; Rao & Green 1982; Clarke & Baillie 1984; Rogala *et al.* 2007). Filter-feeding invertebrate fossils (large bivalves, brachiopods and bryozoans) dominate the low-diversity fauna. The *Eurydesma*-bearing beds in the Darlington Limestone have been interpreted as intertidal barrier bars or shell banks that formed adjacent to an exposed headland or rocky coastline (cf. Brill 1982; Clarke & Baillie 1984; Rogala *et al.* 2007). Fielding *et al.* (2010) suggested that the change in facies from interbedded siltstone, *Eurydesma* rudstone and conglomerate in the Lower Erratic Zone to the argillaceous limestone and siltstone at the base of the Darlington Limestone to purer carbonates at the top of the Darlington Limestone may have resulted from deepening and increased isolation from clastic sediment sources through time. They also reported a decrease in lonestone abundance upwards within the limestone and interpreted this change to represent decreasing ice rafting and decreasing glacial influence with time. Such a scenario is similar to the James *et al.* (2009) model for the stacking patterns of younger Permian carbonates in the Bowen Basin of eastern Australia. There, the limestones were deposited during post-glacial rises in sea level.

On Maria Island, carbonaceous and pyritic siltstone of the Boullanger Formation rest conformably on the Basal Beds (Fig. 3). Along the east coast of Maria Island, both the Basal Beds and the Boullanger Formation lap onto the underlying unconformity (Clarke & Baillie 1984). Dispersed gravel and detrital plant material are common in these units, while marine macro- and trace fossils are rare. Fielding *et al.* (2010) interpret these strata to have been broadly deposited in a deltaic setting with gravel introduced as iceberg-, sea-ice- or anchor-ice-rafted debris. Elsewhere in Tasmania, strata correlative to the Boullanger Formation are assigned to the non-marine Faulkner and Liffey groups or the marginal marine and marine Hickman Formation (Reid & Burrett 2004).

Upwards in the Maria Island succession, a relatively conformable succession of Permian strata overlies the Boullanger Formation and laps southwards onto basement rocks (Clarke & Baillie 1984). This succession consists of: (1) fossiliferous, thin- to medium-bedded, calcareous siltstone and argillaceous limestone of the Skipping Ridge Formation; (2) medium- to thick-bedded bioclastic (crinoidal grainstone and bryozoan floatstone) limestone of the Counsel Creek Formation; (3) fossiliferous sandstone and subordinate siltstone of the Marra Formation; and (4) siltstone and minor conglomerate of the Toarra Formation (Fig. 3). Randomly scattered granule- to cobble-sized clasts composed primarily of quartzite occur in these formations (Clarke & Baillie 1984).

Sedimentological analysis of the Sakmarian Basal Beds

Strata along the Fossil Cliffs (Figs 4 & 6) were examined and logged using a Jacob Staff and Abney Level. Samples from key beds were collected, and the succession was extensively photographed. The Basal Beds along the east coast of Maria Island (north Maria) south of the Fossil Cliffs are exposed in sea cliffs that are accessible only by boat. South of Riedle Bay (south Maria), cliff exposures are slightly more accessible and, with some difficulty, can be accessed by land. These cliff exposures

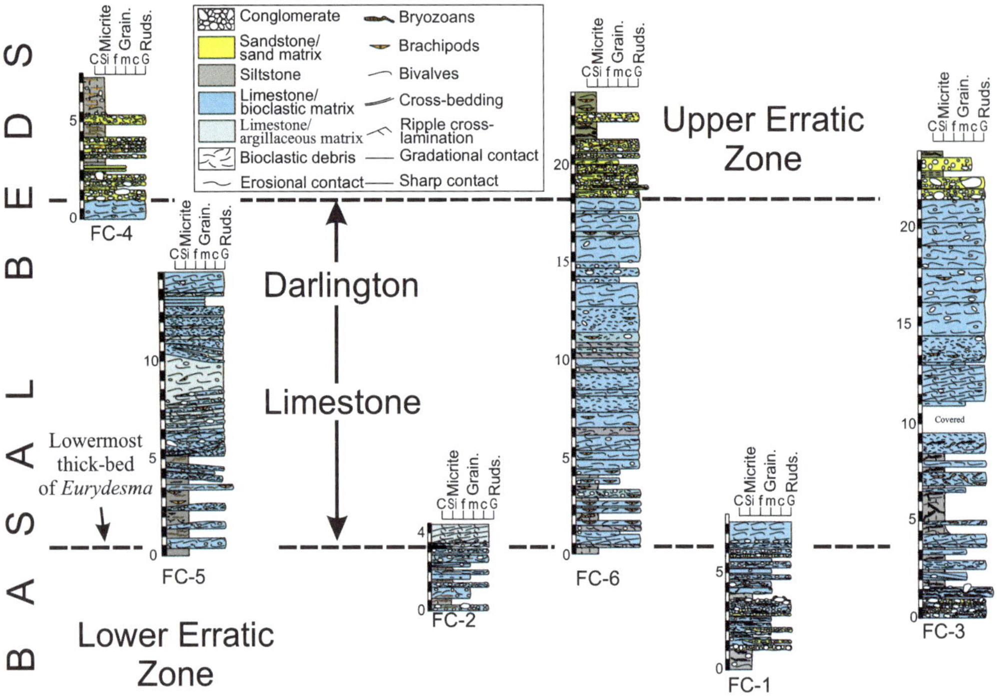

Fig. 6. Sedimentological logs of strata in the Basal Beds exposed at Fossil Cliffs on Maria Island. Section locations are shown in Figure 4.

were examined and photographed in June 2011. The bulk of the sedimentology is based on accessible strata at the Fossil Cliffs. However, the unconformity beneath the Basal Beds is below sea level in this area. The description of the unconformity and the beds immediately above this surface are therefore from exposures further south. To establish a better understanding of environmental change through time, the unconformity, Lower Erratic Zone, Darlington Limestone and Upper Erratic Zone will be described and interpreted separately.

Pre-Parmeener unconformity

Description. On Maria Island, Upper Palaeozoic strata rest on an unconformity cut into the Silurian(?)–Devonian metasedimentary rocks of the Mathinna Beds and Devonian granite (Everard 1983; Clarke & Baillie 1984; Figs 7 & 8). The Mathinna Beds are folded and consist of quartzites and pelites. These rocks are intruded by porphyritic fine- to coarse-grained granite, which produced low-temperature contact metamorphism in the Mathinna Beds (Everard 1983). At several places on south Maria Island, tor-like structures of granite (Clarke & Baillie 1984) and deeply weathered granite with a gruss-like appearance occur beneath the unconformity. On Maria Island, the pre-Permian unconformity displays a relief of greater than 260 m over a distance of 7 km. The unconformity is well exposed in sea cliffs along the east coast of Maria Island, and is draped and onlapped by discontinuous massive to bedded, clast-supported conglomerates that are a few metres to a few tens of metres thick and laterally continuous for tens to hundreds of metres (Figs 7 & 8; Banks 1962; Clarke & Baillie 1984). Conglomerates up to 100 m thick also fill depressions on the unconformity (Clarke & Baillie 1984). In places, conglomerate beds up to several metres thick extend as sheets and wedge-shaped bodies tens to hundreds of metres away from basement highs. In these wedge-shaped bodies, metre-scale boulders occur along the wedge's distal margin where the units thin abruptly and wedge-out (Fig. 8c). Smaller boulders occur behind these large boulders. Imbricated clasts occur in the conglomerate, and the clasts are predominantly subrounded to rounded quartzite with subordinate clasts of subrounded to rounded granite (Clarke & Baillie 1984). Clasts have the same composition as adjacent and underlying basement rocks. Striated pavements or striated clasts have not been observed or reported from either the unconformity or from the conglomerates (Clarke & Baillie 1984), and diamictites were not observed within the Permian succession on Maria Island. Pebble- to boulder-sized lonestones, some of which pierce bedding, are abundant within the overlying Basal Beds (Figs 7 & 8). Some boulder-scale lonestones occur just a few tens of centimetres above the wedge-shaped beds. The lonestones display the same maximum clast size, particle roundness and composition as clasts in the underlying conglomerates. However, the lonestones decrease in abundance away from the dipping unconformity in both vertical and lateral directions. Discrete, laterally extensive, lonestone-rich horizons, which onlap the underlying conglomerate-lined unconformity, also occur. To the SSE of the Fossil Cliffs, the unconformity rises and the unconformity and/or the conglomerate are progressively onlapped by undifferentiated Permian strata equivalent to the Basal Beds (including the Darlington Limestone) and the overlying Boullanger, Skipping Ridge and Counsel Creek formations (Banks 1962; Clarke & Baillie 1984). Along Montgomerys Creek, the Marra and most of the Toarra formations also lap onto an exposed granite ridge (Clarke & Baillie 1984).

Interpretation. Clarke & Baillie (1984) interpreted the unconformity on Maria Island to be the product of weathering and mass wasting with subsequent reworking along a high-relief high-energy coastline. This interpretation is supported by the occurrence of weathered granite and tor-like structures beneath the unconformity, the occurrence of rounded, imbricated and clast-supported conglomerates draping the unconformity and the onlapping of marine strata onto the high-relief unconformity. The absence of glacially modified clasts, striated basement surfaces and diamictites suggest that the coastline and overlying conglomerate were not produced or modified by glacial activity, and the composition of the clasts indicates that the conglomerates were derived directly from the underlying basement rocks. Onlap of Lower–Upper Permian marine and deltaic strata onto the unconformity indicate that basement rocks along the shoreline were exposed and progressively submerged during the Early–Late Permian.

While most of the conglomerates suggest reworking along a high-energy coastline, some of the conglomerate bodies with metre-scale clasts near their margins are similar to clast distributions in modern boulder barricades deposited due to sea-ice rafting in shallow-water settings (Rosen 1979, 1980, 2005; McCann *et al.* 1981). During spring melt, gravel (including boulders) can be incorporated into sea ice by basal freezing, lifting of the clasts from the sea floor by tidal action and upward ascent of particles through the ice due to surface melting. Break-out of shore ice then allows rafting of clasts offshore, where they are deposited when the ice floes encounter pack ice, which tends to persist longer into the melt season and occurs in localities where there is an abrupt change in slope from shallow to deeper water. Boulder-sized clasts are preferentially deposited along this slope forming clast-supported boulder barricades (Rosen 1979). Clast-strewn shallow-water platforms occur shorewards of the barricades. Lonestones in the Basal Beds deforming bedding were deposited from floating ice as dropstones. The occurrence of large lonestone boulders in the strata immediately above the basal conglomerates suggests that ice rafting may have played a part in the formation of some of the basal conglomerates; clast roundness however suggests that the gravels record a history of high-energy abrasion as part of their origin. Lonestone composition and the onlapping of clast-rich lonestone horizons onto the basal conglomerate indicate local derivation of the gravels directly from clasts in the conglomerate. The local derivation of clasts from a rocky shoreline and an absence of glacial features (other than dropstones) within the deposit are more easily explained by sea-ice rafting than by rafting of clasts by far-travelled icebergs.

Lower Erratic Zone (LEZ)

Description. The Lower Erratic Zone at the Fossil Cliffs is characterized by interbedded calcareous, fossiliferous siltstones; thin limestone beds (<1 m thick) of grainstone and rudstone; sandstones; and conglomerates (Figs 6–11). Facies in the Lower Erratic Zone are shown in Table 1. The various facies in the Lower Erratic Zone are interbedded and alternate within a vertical succession. In places, the facies wedge-out and interfinger with each other (Fig. 9).

Facies LEZ-1 at the Fossil Cliffs consist of fossiliferous siltstone, 0.01–3 m in thickness. The siltstone is faintly laminated, with the laminations defined by scattered bioclastic debris composed primarily of fenestrate bryozoans (Figs 9 & 10). Fossils consist predominantly of bryozoans (fenestellids and foliose and branching trepostomes) and rare bivalves (*Deltopecten* and *Eurydesma*; cf. Clarke & Baillie 1984; Reid 2010; Reid & James 2010). Some *Stenopora* with thin erect-rigid foliose fronds are intact and extend as much as 20 cm upwards through the siltstone. In places, these *in situ Stenopora* colonies extend from the siltstone into overlying limestone and conglomerate, and from the limestone or conglomerate into overlying siltstone (Figs 10i & 11). Scattered siliciclastic granules and pebbles occur throughout the siltstone; however, rare granite clasts up to 2.6 m in diameter also occur. SSE of Fossil Cove, interbedded lonestone-bearing siltstones and limestones are horizonally bedded (decimetres to metres thick) and comprise equal

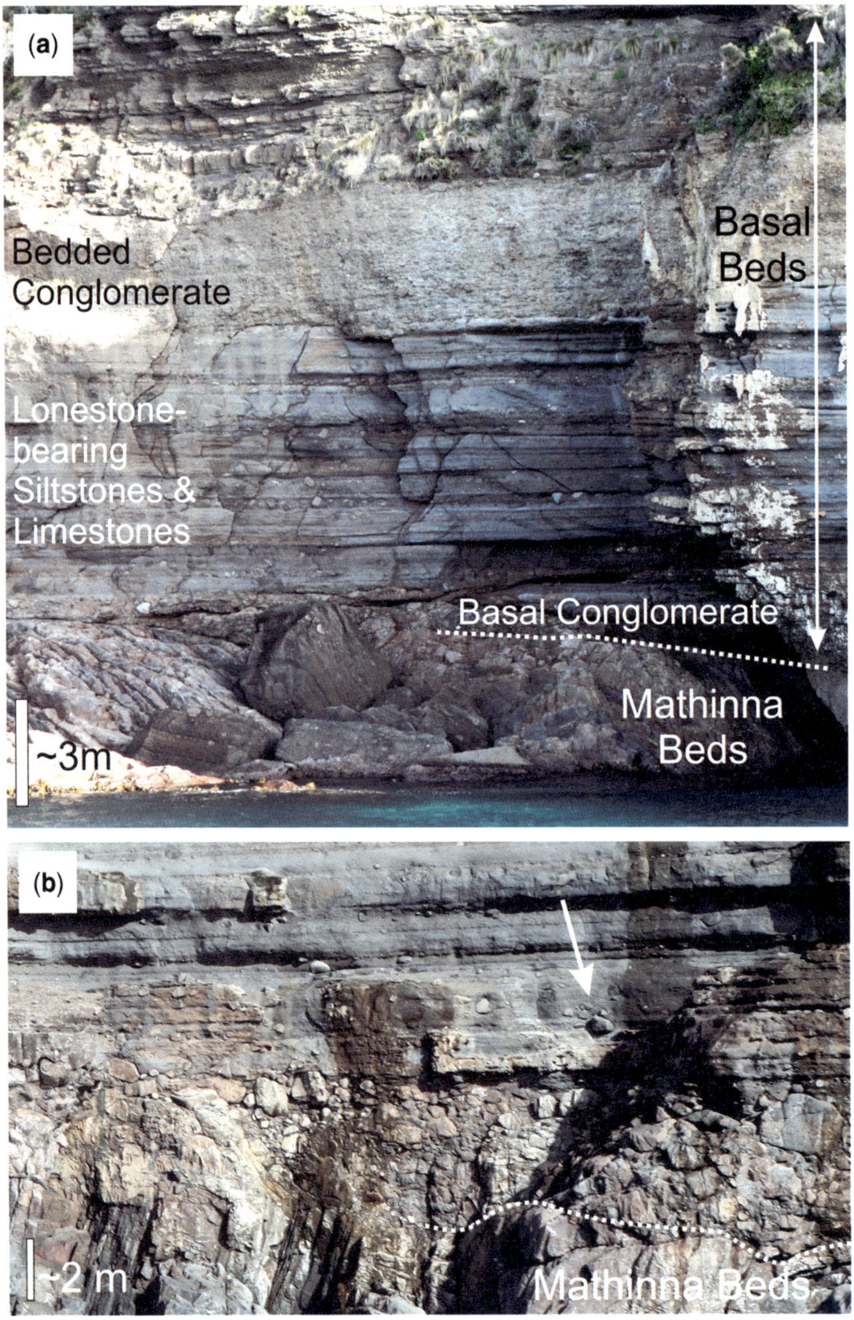

Fig. 7. Photographs showing the unconformable relationship between metamorphic rocks of the Mathinna Beds and Basal Bed conglomerates that line the pre-Parmeener unconformity along the eastern side of north Maria Island. The dashed white lines mark the position of the unconformity. (**a**) Folded quartzites and pelites in the Mathinna Beds overlain by Basal Bed conglomerates, lonestone-bearing siltstones and limestones and bedded conglomerate. The cliff face is located *c.* 0.45 km SSE of Cocks Comb (Fig. 4). (**b**) Steeply dipping rocks in the Mathinna Beds overlain by a quartzite boulder conglomerate in the Basal Beds and by lonestone-bearing siltstones and limestones in the Basal Beds. The exposure is located *c.* 8 km SSE of Cocks Comb. Note the large quartzite lonestones (arrow) in the overlying fine-grained unit.

Fig. 8. Photographs showing the unconformable relationship between metamorphic rocks of the Mathinna Beds, Devonian Granite and Basal Bed conglomerates that line the pre-Parmeener unconformity along the eastern side of Maria Island. (**a**) Granite conglomerate lining the pre-Parmeener unconformity (dashed white line) cut on Devonian granite SSE of Riedle Bay. (**b**) Titled quartzites and pelites in the Mathinna Beds unconformably (solid white line) overlain by a wedge-shaped basal conglomerate body, lonestone-bearing siltstones and bedded conglomerate of the Basal Beds. Cliff face located *c.* 0.45 km SSE of Cocks Comb. (**c**) Quartzite boulder conglomerate at the base of the Basal Beds lining the pre-Parmeener unconformity located *c.* 8 km SSE of Cocks Comb. Lonestone-rich horizons overlie and lap onto the conglomerate. Large boulders occur along the margin of the body with finer boulders occurring to the left behind the larger clasts.

percentages of the non-conglomeratic portion of the Basal Beds (Figs 7 & 8).

Facies LEZ-2 consist of limestone units up to 75 cm thick which occur as sheets, lenses and as channel-form bodies of grainstone, rudstone and bryozoan bafflestone (Fig. 9). These bodies interfinger with (or are interstratified with), and are surrounded by, siltstone units (0.01–1 m thick). Thin (centimetre to decimetre scale) limestone sheets occur low in the LEZ succession at Fossil Cliffs while thicker sheets, lenses and channels (decimetre scale) occur upwards near the top of the LEZ (Fig. 9). The limestone sheets have sharp, erosional and/or gradational bases and sharp to slightly gradational upper contacts (Figs 8 & 9a). Grainstones and rudstones may form separate sheets or may grade laterally within an individual sheet from grainstone into rudstone; alternatively, they may grade laterally into bryozoan bafflestone over a few tens of metres. Low-angle centimetre-scale grainstone foreset beds downlap onto the base of some sheets (Fig. 9), while some forest beds have thin discontinuous siltstone drapes that interfinger with underlying siltstones (Fig. 9a). Channel-like structures are up to 0.5 m thick and up to a few tens of metres wide (Fig. 9a, b). Channel fills consist primarily of grainstone. However, rudstone forms the final fill at the centre and tops of channels in some places. Rudstone also occurs as levee-like sheets that overstep the channel margins. These sheets and channels contain broken and abraded bivalve shells *(Deltopecten* and *Eurydesma*) and bryozoan debris, *in situ* encrusting and foliose bryozoans (*Stenopora*) and articulated semi-infaunal and epifaunal bivalves in life position (*Eurydesma* and *Megadesmus*). Bafflestones consist dominantly of (*in situ*) colonies of foliose *Stenopora* with sand-sized and larger shell debris and scattered pebbles occurring between the ‘fronds’. Large foliose *Stenopora* bryozoans are the dominant fossils in the bafflestones, but branching trepostomes and broken fenestellids are also common. Individual foliose bryozoan colonies are as large as 1.5 m in

Table 1. *Lower Erratic Zone facies associations*

Facies	Lithology and bed thickness	Sedimentary structures	Biota	Interpretation
LEZ-1 Fossiliferous siltstone	Fossiliferous siltstone with quartz and granite clasts. Beds 0.1 m to several metres thick. Siltstones near the top of the unit are interbedded/interfinger with limestone, conglomerate and pebbly sandstone	Laminated siltstone. Contains scattered bioclastic debris and rounded to well-rounded quartz and granite lonestones up to 2.6 m in diameter. Clasts are not striated or faceted.	Trepostome and stenoporid bryozoans with some Stenopora in growth position. Rare bivalves (*Eurydesma*, and *Deltopecten*).	Settling from suspension with lonestones (dropstones) rafted by sea ice into the subtidal zone.
LEZ-2 Carbonate grainstone and rudstone	Beds are up to 75 cm thick and are interbedded with siltstone, conglomerate and pebbly sandstone. Some beds grade into calcirudite and bryozoan bafflestones, and some grade into conglomerate/sandstone. Units contain quartzite and granite clasts.	Units occur as sheets and channels. Some beds contain foreset beds with siltstone drapes. Units contain scattered rounded to well-rounded clasts up to 70 cm in diameter with some clasts crushing bryozoans. Clasts are abundant on bedding plane surfaces. Clasts are not faceted or striated.	Stenoporid bryozoans, bivalves (eurydesmids, some deltopectens) with some bivalves in growth position.	Deposition as debris fans and channels draining bioclastic debris from shallow water into deeper subtidal areas. Clasts (dropstones) were rafted by sea ice.
LEZ-3 Conglomerate and pebbly sandstone	Fine- to coarse-grained pebbly sandstone to boulder conglomerate with quartzite and granite clasts. Unit is interbedded with siltstone and may grade laterally into pebbly sandstone and/or grainstone/rudstone. Conglomerate unit(s) are centimetres to a few metres thick.	Structureless. Beds near the base of the unit may display normal or reverse grading. Clasts are rounded to well-rounded and are granule to cobble in size with the largest observed clasts being 1.1 m in diameter. Clasts occur as both single clasts or in clast clusters. Gravels are not faceted or striated.	Bryozoans (*Stenopora* and *Fenestellids*), bivalves (*Deltopecten* and *Eurydesma*) and rare brachiopods occur as bioclastic debris. *Stenopora* occurs encrusting clasts and encrusting the top of individual conglomerate layers. Rare bivalves occur in growth position.	Gravel-sized particles rafted by sea ice into intertidal to shallow subtidal zone. Possible reworking by storm events.

diameter. Some of these bryozoans extend upwards into the bases of overlying siltstone and limestone units (Figs 10i & 11). The limestones contain scattered quartzite and granite pebbles, cobbles and boulders (Figs 9b & 10f, g). These clasts are typically more abundant on bedding plane surfaces and often project from the top of the beds. Rare boulders up to 70 cm in diameter also occur. In several cases, large clasts pierce bedding and are seen to have crushed or crumpled underlying bryozoans (Fig. 10f). In turn, many clasts are encrusted and overgrown by bryozoans. SSE of Fossil Cove, limestones occur as laterally continuous sheets that can be traced hundreds of metres across cliff exposures (Figs 7 & 8). In this region, lenses and channels are rare.

Facies LEZ-3 comprises conglomerate, which occurs as discontinuous–continuous centimetre- to decimetre-scale beds along the Fossil Cliffs or within a laterally continuous metre-thick interval that caps a resistant intertidal platform located a few metres below the contact with the overlying Darlington Limestone (Fig. 9b). Conglomeratic horizons also occur locally on the tops of the limestone beds as described in Facies LEZ-2. The thin discontinuous units of pebble conglomerate and/or conglomeratic calcarenite near the base of the exposure display either normal or inverse grading with the long axes of most clasts oriented parallel to subparallel to bedding. In places, pebbles and cobbles stick out of the top of these units (Fig. 10e, h) with some clasts having A-axis orientations at high angles to bedding.

The metre-thick conglomerate consists of multiple sheets (centimetre- to decimetre-thick beds) of pebble to boulder conglomerate, fine- to coarse-grained sandstone and granule- to boulder-bearing grainstone to rudstone (Fig. 10a–d). These lithologies grade into one another within individual sheets (Fig. 10a). Clasts are composed predominantly of quartzite (84%) with lesser amounts of granite (12%) and pelites (4%; Fig. 10). Although Everard (1983) and Clarke & Baillie (1984) reported the occurrence of rare mica schist and limestone clasts, none were observed during fieldwork for this paper. Clasts are supported by both siliciclastic and bioclastic sand/grainstone to gravel-size/rudstone grains. Silt and clay particles are absent. The clast abundances vary widely across bedding plane exposures (Fig. 10a, b, e, h). Large clasts occur in clusters with some clasts penetrating bedding. No striated or faceted clasts were observed. The average clast is granule to cobble in size with the largest observed clasts being 1.1 m in diameter. The clasts are angular to well-rounded with the average clast falling within the rounded to subrounded range (0.64); the average large clast (>30 cm) is subrounded (0.54, Fig. 10a–c, e, h). Bioclastic debris consists of fragments of bryozoans (trepostomes and fenestellids), bivalves (*Deltopecten* and *Eurydesma*) and rare brachiopods. *In situ* bryozoans (*Stenopora*) and bivalves (*Euydesma* and *Megadesmus*) also occur (Fig. 10c–e, h). Individual conglomeratic units often have encrusting and foliose trepostome bryozoans attached to clasts and bioclastic debris on their upper surfaces (Fig. 10a–c, e, h), thus allowing separation of the conglomerate into 7–9 individual beds each ranging from 5–10 cm in thickness (Fig. 10a). In places, thin siltstone beds up to 10 cm in thickness separate individual conglomerate sheets. Numerous examples of *in situ* foliose *Stenopora* bryozoans crossing from one conglomerate unit into another or into an overlying siltstone occur. A 1–2-m-thick horizontally bedded conglomerate located just below the Darlington Limestone, which is correlative to the metre-thick conglomerate at Fossil Cliffs, is exposed along the SE side of Fossil Bay (Fig. 4). This unit can be traced SSE for 4.3 km along the sea cliffs on the east side of Maria Island to Beaching Bay (Figs 7a & 8b, c). There, the conglomerate attains a thickness of >30 m and laps on to Devonian granite and steeply dipping quartzites and pelites of the underlying Mathinna Beds.

Interpretation. A wide range of processes, including settling from suspension, ice rafting and biogenic production, occurred during deposition of strata in the Lower Erratic Zone (cf. Fielding *et al.* 2010). The occurrence of bryozoan-bearing siltstones (Facies LEZ-1) suggests deposition from suspension and biological activity in a low-energy subtidal zone (Rogala *et al.* 2007; Fielding *et al.* 2010; Reid 2010). Fossils of such filter-feeding organisms indicate that conditions within the water column were seldom turbid (Reid 2010). These conditions suggest relatively slow sedimentation rates (Fielding *et al.* 2010; Reid 2010). However, the occurrence of *in situ* foliose *Stenopora* bryozoans crossing lithological boundaries between units from siltstone into conglomerate bases (facies LEZ-1 to LEZ-3) or siltstone into limestone bases (facies LEZ-1 to LEZ-2) suggest that deposition of some lithological couplets occurred within the life of individual bryozoan colonies, which may have been up to 100 years (cf. Reid 2010). Such changes in sedimentation during bryozoan growth suggest variable and episodic sedimentation rates during deposition of the various lithological units.

Most of the limestones (Facies LEZ-2) in the Lower Erratic Zone are composed of bioclastic debris re-deposited as sheets and channel bodies. Interbedding and interfingering of these units with siltstone (facies LEZ-1) suggest erosion and transport of shallow-water debris by rip currents and storm-generated currents through channels into a subtidal setting with deposition occurring as channel

(a)
Eurydesma-rich beds
Darlington Ls.
Lower Erratic Zone
Channel
Downlapping beds
(b)
Eurydesma-rich beds
Darlington Ls
Wedge-shaped downlapping beds
Dropstone crushing bryozoans
Lower Erratic Zone
(c)
Darlington Ls.
Lower Erratic Zone
Metre-thick, bedded conglomerate
(d)
Darlington Ls.
Wedge-shaped downlapping sheets
Lower Erratic Zone

fills, levee-like deposits, sheets and debris fans (cf. Fielding *et al.* 2010). Downlapping of individual beds and low-angle cross-stratification with fine-grained sediment drapes within sheets indicate episodic progradation of the carbonate bodies. Bryozoan bafflestones make up parts of some sheets. These represent trapping of bioclastic debris by *in situ* bryozoans as the sheets prograded. Examining environmental controls on the distribution of bryozoans at the Fossil Cliffs, Reid (2010) determined that foliose colonies of bryozoans that grew up above the sea floor could tolerate higher sedimentation rates than encrusting forms or bryozoans that grew closer to the sea floor. Clasts piercing stratification and/or crushing bryozoan colonies represent dropstones transported by ice rafting. Reworking of the tops of the carbonates may also have increased the abundance of the lonestones, leaving behind a lag deposit (cf. Eyles *et al.* 1997; Rygel *et al.* 2008; Fielding *et al.* 2010). However, exposure of the clast-rich horizons on the sea floor was limited to the amount of time it took colonies of bryozoans to recover and encrust some of the boulder-sized lonestones. An upward progression from siltstone to carbonate sheets to channels suggests shallowing upwards during deposition of strata in the Lower Erratic Zone.

Conglomerates (facies LEZ-3) in the Lower Erratic Zone display a wide variety of lithologies that range from granule to boulder conglomerate, fine- to coarse-grained sandstone and granule- to boulder-bearing grainstone and rudstone; they were therefore likely deposited by a range of processes (cf. Fielding *et al.* 2010). Centimetre-thick conglomerates and conglomeratic grainstones with normal or reverse grading suggest wave or current reworking, or resedimentation by sediment gravity flows, within subtidal settings. Grainstones and rudstones containing broken and abraded shell fragments within the thicker conglomeratic sheets on the main bench may represent transport of bioclastic debris into deeper waters similar to that described above for other limestones beds. *In situ* foliose *Stenopora* bryozoans likely served to trap both clastic and carbonate debris, forming bafflestones. The occurrence of sandstone and conglomerate in close association with limestones, especially within the same sheet body, are anomalous due to an absence of current- or wave-generated structures. Some reworking of the deposits may have occurred (Fielding *et al.* 2010), and primary sedimentary structures may have been destroyed by organisms. However, neither sediment gravity flows, transport of sediment due to bottom currents or winnowing by wave action fully explains the occurrence of coarse clastic particles in this setting. Many of the particles, including sand, were likely introduced as ice-rafted debris (cf. Brill 1982; Clarke & Baillie 1984; Rogala *et al.* 2007; Fielding *et al.* 2010). Evidence for ice rafting includes the occurrence of clast penetration structures indicating deposition as dropstones and the clustering of clasts, which suggests accumulation due to ice dumping (cf. Rogala *et al.* 2007; Fielding *et al.* 2010). Sand grains could also have been dropped by floating ice (Gilbert 1990), which provides an explanation of why the sheets vary in composition laterally. The conglomerate units are composed of multiple thin beds of conglomerate separated by layers of either encrusting bryozoans or thin siltstones, which indicate that conglomerate deposition occurred as multiple episodic events. The occurrence of filter-feeding organisms (bryozoans and rare bivalves) suggests an absence of turbid waters between events (Reid 2010), and the occurrence of *in situ* foliose *Stenopora* bryozoans spanning the gap from one conglomerate unit into another suggest that both conglomerate depositional events and breaks in sedimentation between events were of geologically short duration.

The composition of the clasts in the Lower Erratic Zone indicates derivation of particles from basement rocks currently exposed on Maria Island (Clarke & Baillie 1984). Roundness values for the clasts are similar to pebbles, cobbles and boulders transported and abraded in high-energy environments, suggesting that the clasts originated along the adjacent palaeocoastline of Maria Island prior to ice rafting. Because of their local derivation from a non-glaciated coastline, the clasts are easiest to explain by sea-ice rafting. Extremely rare clasts of schist and limestone, which were not observed during this study, appear to be exotic to the area as

Fig. 9. Limestones, siltstones and conglomerates near the top of the Lower Erratic Zone and the lowermost portion of the Darlington Limestone at the Fossil Cliffs. Note the occurrence of thin *Eurydesma*-rich beds near the top of the Lower Erratic Zone. (**a**) Limestone beds displaying sheet- and channel-form geometries exposed at Section FC-1. Beds within the sheets show internal downlapping stratal geometries. The lower downlapping grainstone unit has thin siltstone drapes between the low-angle foreset beds. (**b**) Interfingering of grainstone and rudstone sheets with siltstones near the top of the Lower Erratic Zone at section FC-3. Note the downlapping stratal geometries and the large lonestone (see Fig. 10f). (**c**) Interfingering deposits (small arrows) in the Lower Erratic Zone overlain by the basal *Eurydesma*-rich bed at the base of the Darlington Limestone in section FC-3. (**d**) Downlapping and wedge-shaped grainstone, rudstone and conglomerate beds at the top of the Lower Erratic Zone in Section FC-5. The wedge-shaped beds thin and pinch out within a thick siltstone unit tens of metres to the south (left).

(a)
(b)
(c)
(d)
(e)
(f)
(g)
(h)
(i)
Colony 2
Colony 1

such lithologies are not present below the sub-Parmeener unconformity on Maria Island. These clasts may represent rafting from distant sources, possibly by rare icebergs, or during a previous interval (Asselian) when icebergs from Tasmanian glaciers may have transited the Maria Island area that were later eroded and reworked during the Middle Sakmarian. It is also possible that the limestone clast may have been the result of erosion and reworking of older limestones in the Basal Beds.

Darlington Limestone facies assemblage

Description. Along the Fossil Cliffs, the Darlington Limestone is 17–20 m thick and consists of beds of *Eurydesma* rudstone, bryozoan siltstone and spiriferid rudstone and grainstone (Figs 6, 11–13; Table 2). The base of the Darlington Limestone is defined as the lowest occurrence of thick-bedded *Eurydesma*-rich limestone (Figs 6, 9 & 13a, e), and the contact between the Lower Erratic Zone and the Darlington Limestone is an abrupt, slightly undulatory contact that can be traced over several kilometres along the east coast of Maria Island. Interbeds of rudstone, siltstone and grainstone occur at the base of the unit, while rudstone dominates the upper portion of the unit (Figs 12 & 13; Clarke & Baillie 1984; Reid 2004). A sharp contact separates the Darlington Limestone from conglomerates in the overlying Upper Erratic Zone. The Darlington Limestone is exposed in an abandoned quarry along the northern end of the Fossil Cliffs, in exposures along the southern end of the Fossil Cliffs and in vertical cliff faces along the east side of Maria Island south of Fossil Bay (Fig. 4). The Darlington Limestone can be traced at least 1 km SSE of Cocks Comb where it consists of thin interbeds of siltstone and limestone (Fig. 4). However, a similar succession occurs on the north side of Waterfall Bay, suggesting that the Darlington Limestone may be traceable at least that far along the east coast of Maria Island.

Facies DL-1 consists of interbedded rudstone, siltstone and grainstone and occurs in the lower portion of the Darlington Limestone (Figs 9 & 12b, d). The *Eurydesma*-bearing units typically consist of a 2–10-cm-thick inversely graded bed at the base, composed of small, broken shells set in a sandy grainstone matrix, overlain by *Eurydesma* rudstone with a sandy to argillaceous grainstone matrix (Fig. 11a, e). Fossils in the rudstone consist almost exclusively of disarticulated valves of large bivalves, mainly *Eurydesma*, with convex-upwards orientations; however, *Deltopecten* and *Trigonotreta* are also present. Low-angle cross-beds occur in these units, but are not abundant. The top of the *Eurydesma*-bearing beds are in sharp to gradational contact with overlying siltstone. Decimetre- to metre-thick siltstone units in the lower portion of the Darlington Limestone are laminated and contain widely scattered small clasts and abundant fenestellid bryozoans (Figs 12b, d & 13f). Ramose trepostome bryozoans are locally abundant in the siltstones while *Eurydesma* and *Trigonotreta* are rare. The siltstones contain thin grainstone lenses and sheets similar to the grainstone units in the Lower Erratic Zone. Fossils in these units included broken and abraded *Eurydesma*, spirifers and trepostome bryozoans. Traced laterally, the grainstone sheets thin and pinch out into siltstone (Fig. 12b, d). Scattered lonestones up to 10 cm in diameter occur throughout this facies, while granule- to pebble-size clasts are common in both grainstone and rudstone units (Fig. 13e, f). However, rare clasts up to 50 cm in diameter also occur. The clasts are rounded to well rounded (average clast roundness 0.76), and are dominantly composed of quartzite (97%) with lesser amounts of granite and rare pelites.

Facies DL-2 is characterized by several spiriferid-bearing rudstone and grainstone units 0.5–2 m thick contained within both the lower and upper portions of the Darlington Limestone (Figs 12a & 13c, d). The bases of these units are sharp to erosional with channelized bases in places. The tops of these units are sharp to slightly gradational with either siltstone or thick-bedded *Eurydesma*-bearing rudstone. Thicker beds contain metre-scale

Fig. 10. Photographs showing representative strata in the Lower Erratic Zone at the Fossil Cliffs. (**a**) Conglomeratic facies overlain by siltstones and wedge-shaped grainstone sheets (section FC-1). Note the complex interfingering of beds at the top of the conglomeratic unit. (**b**) Conglomeratic facies (section FC-1) composed of rounded clasts, sand and bioclastic debris. Broken foliose bryozoans (arrows) separate the deposit into multiple units. (**c**) Conglomeratic facies at Section FC-1 composed of clasts, bioclastic debris and granule-sized particles. (**d**) *In situ* bivalves (arrows) within a grainstone/rudstone unit near the top of the Lower Erratic Zone (section FC-1). This bed grades laterally into the metre-thick conglomeratic unit shown in Figure 10a–c. (**e**) Quartzite boulder at the top of the metre-thick conglomeratic facies encrusted and overgrown by foliose bryozoans (section FC-1). (**f**) Granitic lonestone interpreted as a dropstone crushing a foliose *Stenopora* colony (section FC-3). (**g**) Granitic lonestone interpreted as a dropstone penetrating both grainstone and siltstone units (section FC-1). (**h**) Quartzite boulder at the top of the metre-thick conglomeratic facies encrusted by bryozoans (arrows; section FC-1). (**i**) Interbedded grainstone and siltstone units showing two bryozoan colonies in growth positions extending upwards from grainstone units and through overlying siltstone units (section FC-1).

(a)
Top
Base
(b)
Top
Top
Base
(c)
Clast-rich layer
top
Base
Second Colony

cross-stratification. The uppermost spiriferid-bearing unit in the Darlington Limestone at the Fossil Cliffs can be traced across the length of the exposure. Abundant disarticulated and slightly abraded to broken spirifers (*Trigonotreta*) and *Eurydesma* and *Deltopecten* valves occur. Rounded to well-rounded quartzite clasts up to 20 cm in diameter are present within these units and, in places, clasts are abundant on the bedding planes at the top of the facies. Some clasts and bioclastic debris are encrusted by bryozoans.

Facies DL-3 is composed of stacked, shell-supported, thick-bedded, *Eurydesma*-bearing rudstones lacking interbedded siltstone units. This facies dominates the upper 8–12 m of the Darlington Limestone at the Fossil Cliffs (Fig. 12a). These strata are only broken by a 1–2-m-thick unit of spiriferid-bearing grainstone/rudstone (Figs 12a & 13d; Facies DL-2) that break the *Eurydesma* calcirudite into a lower unit and an upper unit of approximately equal thicknesses. In the area of the quarry, broken and abraded fenestellid, crinoid and brachiopod bioclastic debris and calcarenite fill the interstitial spaces between shells within the *Eurydesma*-bearing beds. These beds range from 75% to greater than 90% calcium carbonate (Clarke & Baillie 1984). Southwards, bioclastic debris and fine-grained siliciclastic sediment occur between the larger shells. However, relatively pure calcium carbonate dominates the upper 4 m of the Darlington Limestone throughout the length of the Fossil Cliffs. *Eurydesma*, *Trigonotreta* and *Deltopecten* are abundant within these *Eurydesma*-bearing rudstones. Disarticulated valves dominate at the base of the two units while articulated valves, possibly in growth position, occur upwards. Within each unit, individual beds are sharp based and display undulatory subhorizontal bedding; however, both high- and low-angle cross-stratification are also present. Some beds interfinger with, or grade into, grainstone beds. Channel-like features on the southern end of the Fossil Cliffs, 1–2 m deep and up to 10 m wide, are filled with bioclastic debris and granule- to pebble-sized clasts are common (Fig. 12d). Siliciclastic clasts up to 50 cm in diameter occur throughout the two units but are common only on bedding planes on the southern end of the exposure where granules, pebbles and cobbles are common. Although present, clasts appear to have a lower abundance in the clean carbonate in the upper 4 m of the Darlington Limestone.

Interpretation. The Darlington Limestone at the Fossil Cliffs was previously interpreted as intertidal barrier bars or shell banks that formed adjacent to an exposed headland or high-energy coastline as suggested by the *in situ* and transported nature of the fossils and the palaeoecology of the bivalve *Eurydesma* (Runnegar 1979; Brill 1982; Clarke & Baillie 1984; Banks & Clarke 1987; Rogala *et al.* 2007; Reid 2010). Findings here support this interpretation. The contact between the Lower Erratic Zone and the Darlington Limestone is demarcated by the base of the first thick bed of *Eurydesma*-bearing rudstone (Facies DL-1). This contact is somewhat arbitrary as limestone beds, including thin beds of *Eurydesma*-bearing rudstone, first occur near the top of the Lower Erratic Zone and increase in abundance and thickness into the overlying Darlington Limestone. The interstratification of grainstones, rudstones and siltstones in both units suggests that the top of the Lower Erratic Zone may grade upwards into the base of the Darlington Limestone, thus suggesting a relationship between the depositional environments of these two units in space and time.

Facies DL-1 is interpreted to have accumulated in a low-energy subtidal zone. Siltstones in the lower portion of the Darlington Limestone are laminated and contain abundant fenestrate bryozoans. An absence of evidence for reworking suggests that siltstone units formed from settling from suspension and the palaeoecology of modern bryozoans suggests that these units accumulated slowly in lower-energy waters along the margins of the shell bank (Reid 2010). Bioclastic shell debris deposited as grainstone and rudstone sheets, debris fans and as graded beds at the base of the lowest thick-bedded *Eurydesma*-bearing calcirudite were likely derived from the shell bank(s) and transported into deeper water where they were deposited along bank margins. Interbedding and interfingering of grainstones and siltstone beds suggest progradation of the shell bank.

Fig. 11. Photographs showing foliose bryozoans in growth positions crossing lithologic contacts in the Lower Erratic Zone. (**a**) Bryozoans in growth position (arrows) near section FC-3. The *in situ* bryozoans cross from the top of a thick carbonate unit through siltstone, a decimetre-thick limestone, and into the base of a gravel-rich unit (scale is 0.25 m long). (**b**) Bryozoans in growth position extending from the top of a bryozoan bafflestone/grainstone upwards through a siltstone, a grainstone/bryozoan bafflestone and into the base of a clast-rich rudstone (scale is 0.5 m long; near section FC-1). (**c**) Multiple *in situ* bryozoans extending upwards through multiple units. The large bryozoan (white arrows) extends from a grainstone/rudstone unit upwards through multiple grainstone beds, a siltstone unit and terminates at a gravel-rich rudstone bed (scale is 0.5 m long). A second colony with a similar vertical range appears to the right of the main bryozoan colony (near section FC-3).

Fig. 12. Photographs showing features of the Darlington Limestone at the Fossil Cliffs. (**a**) Cliff exposure of the Darlington Limestone (DL) between section FC-1 and FC-5 showing the three facies identified in the limestone. The Lower Erratic Zone (LEZ) is exposed in the intertidal platform and the Upper Erratic Zone (UEZ) and the Boullanger Formation (BF) are exposed above the limestone cliff (person for scale; arrow). (**b**) Interbedded siltstone, grainstone and rudstone in the lower and middle portions of the Darlington Limestone at section FC-5. Jacob staff for scale (arrow). (**c**) Channel-form structure filled with bioclastic calcirudite within facies DL-3 in the Darlington Limestone at section FC-5. (**d**) Interbedded siltstone, grainstone and rudstone in the lower and middle portions of the Darlington Limestone at section FC-5. Visible section is *c.* 8 m thick above the waterline.

Spiriferid-rich grainstones and rudstones of facies DL-2 that occur as single beds within Facies DL-1 (lower portion of the Darlington Limestone) and between stacked beds of Facies DL-3 higher in the Darlington Limestone suggest deposition in shallow agitated water on and adjacent to the shell bank. Evidence that suggests these beds were reworked include: (1) the occurrence of erosional channelized bases; (2) the occurrence of grainstones and/or rudstones; (3) the occurrence of disarticulated and broken shells; and (4) the occurrence of cross-bedding. However, some spiriferid-rich rudstone beds are interstratified with siltstone, which suggests that the bioclastic debris and shell hash in these beds were transported into deeper water and deposited along the flanks of the shell bank.

Thicker *Eurydesma* rudstones of Facies DL-3 at the top of the Darlington Limestone suggest deposition in intertidal to subtidal shell banks. Evidence for the reworking of grains in these units includes: (1) broken and abraded bioclastic debris; (2) grain support of disarticulated bivalve and brachiopod shells; (3) the occurrence of broken and abraded

Fig. 13. Strata of the Darlington Limestone at the Fossil Cliffs. (**a**) Argillaceous rudstone at the base of the Darlington Limestone in section FC-2 consisting of *Eurydesma*-rich bioclastic debris. Note the clast-rich grainstone at the base of the unit. Increments on the Jacob staff are 10 cm. (**b**) Rudstone bed near the middle of the Darlington Limestone at Section FC-5. Note the abundant rounded quartzite clasts on the bedding-plane exposure (mechanical pencil for scale). (**c**) Low-angle cross-stratified spiriferid-rich rudstone exposed at section FC-5 in the middle of the Darlington Limestone (Jacob staff for scale). (**d**) Cross-bedded spiriferid-bearing grainstone of facies DL-2 in the upper portion of the Darlington Limestone in section FC-6. (**e**) Clast-rich *Eurydesma* rudstone near the base of the Darlington Limestone at section FC-1. (**f**) Interbedded clast-bearing *Eurydesma* rudstone and siltstone near the middle of the Darlington Limestone between sections FC-2 and FC-5 (Jacob staff for scale).

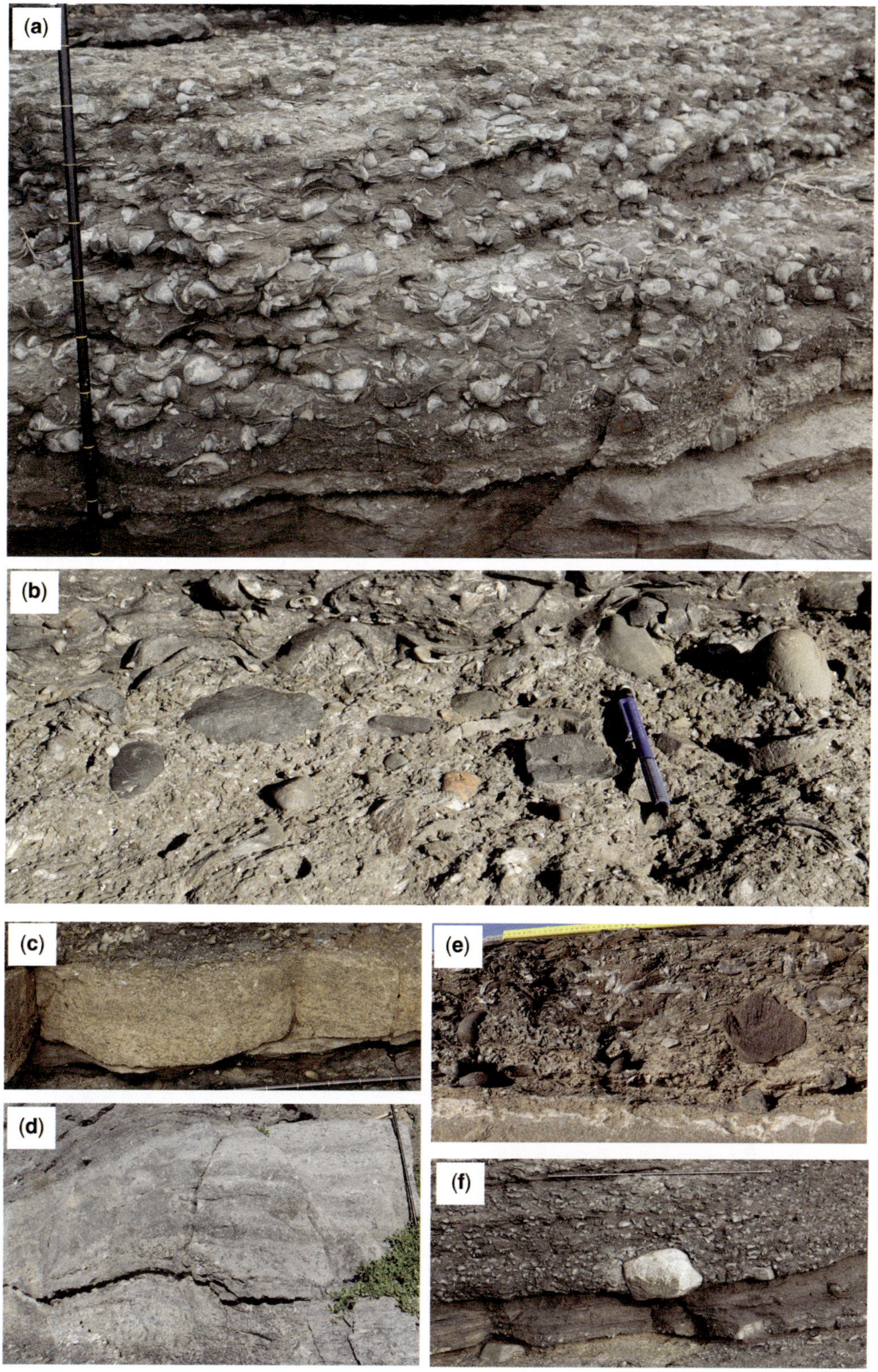
(a)
(b)
(c)
(d)
(e)
(f)

Table 2. *Darlington Limestone facies associations*

Facies	Lithology and bed thickness	Sedimentary structures	Biota	Interpretation
DL-1 Interstratified fossiliferous siltstone, grainstones and thin rudstone beds	Occurs near the base of the Darlington Limestone and consists of interstratified fossiliferous siltstone (0.2–3 m thick), grainstone (0.2–1 m thick) and rudstone (0.3–1 m thick) sheets. Traced laterally over tens to a few hundred of metres, the carbonate units thin and pinch out into thick siltstone units. Quartzite and granite lonestones are present and may be locally abundant within and at the tops of the limestone units.	Some grainstones and rudstones may contain cross-bedding. Some bryozoans occur *in situ*, while most biotic material is disarticulated, broken and abraded.	Abundant Bryozonas (fenestellids and *Stenopora*) in siltstones with scattered bivalves (*Eurydesma* and *Deltopecten*) bivalves. Calcarenites contain brachiopods (*Trigonotreta*) and bivalves (*Eurydesma* and *Deltopecten*), while rudstones contain abundant bivalves (*Eurydesma*).	Subtidal zone with transport of bioclastic debris by wave or storm activity and deposited as debris fans. Lonestones (dropstones) rafted by sea ice into the subtidal zone.
DL-2 Spirifer calcirudite	Spirifer rudstone occurs as sheets 0.5–2 m thick. The rudstone contains scattered quartzite and granite lonestones.	Unit consists of broken and abraded bioclastic debris. Some beds contain cross-bedding. Base of the unit may be channelized. Rounded to well-rounded quartzite clasts up to 20 cm in diameter occur. Clasts are abundant in and on the tops of these beds, and some clasts are encrusted by bryozoans.	Abundant spirifers (*Trigonotreta*) with scattered bivalves (*Eurydesma* and *Deltopecten*).	Shallow water adjacent to the shell bank.
DL-3 *Eurydesma* calcirudite	*Eurydesma* rudstone occurs as sheets and channels. Stacked sheets are up to 7 m thick. Channels contain coarse bioclastic debris. Unit also contains scattered quartzite and granite lonestones.	Bivalves are disarticulated with convex-side-up throughout most of the limestone except at the top where the valves occur in growth position. Rare cross-bedding is present. Channels are filled with rudstone-sized bioclastic debris. Lonestones up to 50 cm in diameter are rounded to well rounded.	*Eurydesma* dominates but other bivalves (*Deltopecten*) and spirifier brachiopods also occur. Fenestellid bryozoan, crinoid and brachiopod bioclastic debris occurs between the larger fossil grains.	Intertidal shell banks with storm and rip current channels.

shell material filling interstitial spaces between the larger shells; (4) convex side up orientation of *Eurydesma* shells suggesting transport; (5) reverse grading from calcarenite to calcirudite at the base of thick shell beds near the base of the Darlington Limestone; and (6) the scattered occurrence of cross-beds. Near the top of the Darlington Limestone, these strata form two *Eurydesma*-rich units (4–6 m thick each) separated by an erosional-based spiriferid-bearing grainstone/rudstone. The spiriferid-bearing unit suggests possible fluctuations in water depth late in the deposition of the Darlington Limestone. Nevertheless, *Eurydesma*-rich units are dominant in the upper 4–6 m of the unit, suggesting that these beds were deposited in an intertidal to subtidal setting. Reworked material in the form of disarticulated valves predominates at the base of the two units, while articulated valves, possibly in growth position, occur near the top of the units. Rudstone channels suggest reworking and transport of bioclastic debris away from the crest of the bank.

In this shallow intertidal to subtidal setting, the occurrence of lonestones within the Darlington Limestone suggests deposition as ice-rafted dropstones. These lonestones vary in abundance both vertically and laterally within the Darlington Limestone, but are more abundant along the southern end of the Fossil Cliffs than in areas to the north. Brill (1982) also noted this distribution and, using stretch line and quadrant methods, his clast counts show that lonestones occur throughout the Darlington Limestone and are highest in beds rich in bioclastic debris. Brill's (1982) counts also showed a decreased abundance at the top of the Darlington Limestone as observed here. This lonestone distribution suggests that few lonestones were deposited on the crest of the bank, but increased in abundance in reworked materials in deeper water along the margins of the bank. One interpretation is that ice would not have been able to invade shallower waters near the crest of the bars as easily as in deeper waters in off-bank areas. Clast composition and rounding suggest derivation of most of the clasts from a high-energy shoreline associated with exposures of the Mathinna Beds.

Upper Erratic Zone

Description. Strata and facies in the Upper Erratic Zone are similar to strata in the Lower Erratic Zone and consist of a 4–5-m-thick succession of thin interbedded pebble to boulder conglomerate/sandstone, massive pebble to cobble conglomerate/sandstone and fossiliferous bioturbated siltstone (Figs 6 & 14; Table 3). However, siltstones in the Upper Erratic Zone are more intensely bioturbated and fossils are less abundant than in the Lower Erratic Zone. A boulder-bearing unit resting on limestone (Darlington Limestone?) similar to that of the Upper Erratic Zone can be traced 4.5 km SSE along the east coast of Maria Island from the south side of Fossil Bay to at least Waterfall Bay (Fig. 4).

Facies UEZ-1 is characterized by units of boulder conglomerate/sandstone that rest directly on the underlying Darlington Limestone, and by overlying coarse-grained clastic units that consist of pebble to cobble conglomerates/sandstones (Fig. 14a, b, f). Boulders of granite and quartzite in the conglomeratic units are up to 2.5 m in diameter and are typically angular to subangular, while smaller clasts are rounded to subrounded (Fig. 12a–c, f). Clasts greater than 0.5 m in diameter tend to be box-shaped with *c.* 90° corners (Fig. 14c). Large clasts often occur in clusters, protrude from the tops of beds and have their long axes randomly oriented in the horizontal plane. Some smaller clasts have steeply dipping A axes or AB planes that cut underlying stratification. Strata overlying these particles onlap and drape over the clasts. Clasts are composed predominantly of quartzite. However, granite clasts are common. The basal boulder conglomerate/sandstone unit fines upwards, whereas both normal and reverse grading occurs in the overlying pebble to cobble conglomerates/sandstones. Scattered bioclastic debris occurs throughout individual units and consists of broken trepostome bryozoans, spirifers, *Deltopecten* and *Eurydesma.* Branching and foliose trepostome bryozoans also occur in growth position (Fig. 14e).

Siltstone of Facies UWZ-2, which consist of fossiliferous siltstone and sandy siltstone, are interbedded with Facies UEZ-1. These siltstones are 2–50 cm thick and increase in thickness upwards within the Upper Erratic Zone. The siltstones are bioturbated and contain *Planolites*, *Palaeophycus*, *Rhizocorallium* and *Diplocraterion* (Fig. 14d; Fielding *et al.* 2010). Rare fossils include bryozoans, spirifers and *Deltopecten.* Scattered granules to cobble-sized clasts also occur within the siltstone. The top of the Upper Erratic Zone is marked by a 3–4-m-thick fossiliferous, calcareous siltstone, known locally as the spiriferid zone (Clarke & Farmer 1976). The siltstone contains abundant spirifers (*Trigonotreta*) with occasional *Eurydesma* and *Deltopecten.* Scattered granule and pebble-sized clasts also occur. Clarke & Baillie (1984) noted a horizon of *in situ Eurydesma* near the top of this siltstone. The Basal Beds are conformably overlain by thin-bedded, carbonaceous, pyritic siltstone of the Boullanger Formation.

Interpretation. Conglomerates of Facies UEZ-1 are abundant at the base of the Upper Erratic Zone and consist of boulder conglomerates to pebbly sandstones. The occurrence of clast clusters, lonestones penetrating bedding and lonestones protruding from the tops of beds suggest that the gravels were deposited by ice rafting. However, the occurrence of bioclastic debris and both reverse and normal grading suggest, in part, that the units were reworked by wave/storm activity and/or resedimentation by sediment gravity flows in subtidal waters above storm wave base. Siltstone lamina and beds of Facies UEZ-2 contained between the gravel-rich layers suggest that deposition of the conglomerate beds was episodic. Clast compositions of quartzite and granite indicate that the gravels were derived locally from the nearby rocky coastline.

The occurrence of bioturbated siltstones containing an offshore trace fossil assemblage and a fauna consisting of trepsotome bryozoans and spiriferid brachiopods (*Trigonotreta*) suggest that deposition of Facies UEZ-2 occurred in subtidal waters below normal wave base or during long intervals between storm events (Clapham & James 2008;

Fig. 14. Photographs of the Upper Erratic Zone at or near section FC-4 at the Fossil Cliffs. (**a**) General view of section FC-4 showing large boulders resting on the contact with the underlying Darlington Limestone, interbedded conglomerates and bioturbated siltstones. (**b**) Thin bedded conglomerate with rounded to angular clasts. Jacob staff for scale. (**c**) Large 2.5-m-long granitic clasts exposed on an intertidal platform 100 m south of sections FC-4 (hammer for scale). (**d**) Bioturbated and laminated siltstone overlain by conglomerate at section FC-4 (hammer for scale). (**e**) *In situ* foliose bryozoan colony on the top of a conglomeratic layer on an intertidal platform 200 m south of sections FC-4 (ruler for scale). (**f**) Boulder within alternating conglomeratic and pebbly sandstone/siltstone layers (hammer for scale).

Table 3. *Upper Erratic Zone facies associations*

Facies	Lithology and bed thickness	Sedimentary structures	Biota	Interpretation
UEZ-1 Conglomerate and pebbly sandstone	Fine- to coarse-grained pebbly sandstone to boulder conglomerate with quartzite and granite clasts. Unit is interbedded with siltstone. Conglomerate units are centimetres to a metre thick.	Structureless. Beds display normal, reverse or no grading. Clasts are angular to well-rounded and are typically granule to cobble in size with the largest observed clasts being 2.5 m in diameter. Clasts occur as both single clasts or in clast clusters. Clasts are not faceted or striated.	Bryozoans (*Stenopora*), bivalves (*Deltopecten* and *Eurydesma*) and rare brachiopods occur as bioclastic debris.	Gravel-sized particles rafted by sea ice into subtidal/offshore zone.
UEZ-2 Fossiliferous siltstone	Fossiliferous siltstone with quartz and granite clasts. Beds centimetres to several metres thick.	Bioturbated to faintly laminated siltstone. Contains scattered bioclastic debris and rounded to well-rounded quartz and granite lonestones. Clasts are not striated or faceted.	Stenoporid bryozoans with some *Stenopora* in growth position. Rare bivalves (*Eurydesma*, and *Deltopecten*) and brachiopods also occur with some in growth position. *Planolites, Palaeophycus, Rhyzocorallium and Diplocraterion* ichnogenera occur.	Settling from suspension with lonestones (dropstones) rafted by sea ice into the subtidal/offshore zone.

Fielding *et al.* 2010; Reid 2010). Sea-ice cover may also have played a role in siltstone deposition by dampening wave energy and limiting sediment reworking. The upward increase in the abundance of siltstone and in spiriferid brachiopods suggest gradual deepening of waters in the area of the present Maria Island and the reduction in episodic gravel sedimentation through deposition of the Upper Erratic Zone.

Depositional setting of the Basal Beds

Permian strata on Maria Island were deposited across a high relief (+260 m) erosional surface cut on basement rocks consisting of metasedimentary rocks of the Mathinna Beds and Devonian porphyritic granites. No evidence of direct glacial activity (striated and grooved surfaces, faceted and striated clasts, sheared sedimentary rocks, massive or stratified diamictites) has been found on the unconformity or within any Permian strata on Maria Island (Clarke & Baillie 1984; this paper). The occurrence of tor-like structures (Clarke & Baillie 1984) and the occurrence of a granite gruss below the pre-Parmeener unconformity indicate that the basement rocks were subjected to deep weathering and mass wasting. Such features suggest that the area of present-day Maria Island was beyond the limit of glaciation (Clarke & Baillie 1984; Banks & Clarke 1987). Rounded pebbles, cobbles and boulders of quartzite, granite and other metasedimentary rocks line the unconformity and compose gravel-sized clasts in the Basal Beds. The nature of these clasts indicates their derivation from the underlying basement rocks and rounding in a high-energy nearshore environment along a rocky coastline (Clarke & Baillie 1984; Rogala *et al.* 2007). Lonestones in strata overlying the unconformity that pierce bedding, occur as clast clusters and crush bryozoans suggest deposition due to ice rafting (Thomas & Connell 1985). The progressive onlap of shallow marine strata of the Lower Parmeener Supergroup (Basal Beds through the Toarra Formation) onto the unconformity and the occurrence of imbricated clast-supported conglomerates that lines this surface suggest submergence of a rocky high-energy coastline during the Early–Late Permian (Clarke & Baillie 1984; Rogala *et al.* 2007).

In the Basal Beds, lonestones were deposited as ice-rafted dropstones which occur throughout the succession. Robust hypotheses on the origin of these dropstones must explain the following observations:

(a) an absence of glacial indicators on the pre-Parmeener unconformity and the occurrence of deeply weathered basement rocks and mass-wasting features, which suggest that the basement rocks on Maria Island were never truncated by glaciation;
(b) an absence of iceberg keel marks, which suggest a lack of ice grounding in shallow water;
(c) the occurrence of clast-supported conglomerate composed of rounded gravel lining the unconformity, which suggests derivation of the clast from the underlying basement rocks and deposition in a high-energy setting prior to ice rafting;
(d) onlap of the bedded conglomerate (lateral equivalent of the metre-thick conglomerate at Fossil Cliffs) onto and the thickening of the unit in the area where it onlaps the pre-Parmeener unconformity, which suggest derivation of the clasts from local basement rocks and from the basal conglomerate;
(e) the onlap of lonestone-bearing horizons onto the basal conglomerate, which suggests that the clasts were derived and rafted away from the conglomerate;
(f) the occurrence of quartzite, granite and pelite clasts in Basal Bed strata that have the same composition as basement rocks located along the coast of Maria Island, which indicate local derivation of the clasts;
(g) the occurrence of high roundness values and an absence of faceted and striated lonestones, which suggest that the clasts were derived from a high-energy environment rather than from a glacial source;
(h) the occurrence of clast clusters and lonestones that pierce bedding and crush bryozoans, which suggests that the clast were released or dumped from floating ice;
(i) an absence of diamictites; and
(j) deposition in subtidal to intertidal settings.

Specific features characterize iceberg-rafted clasts and deposits. These features typically include: clast composition (in part, exotic to that of the local basement geology); clast roundness (low roundness values), facets and striations that reflect glacial transport and abrasion mechanisms; lenses of diamictites and clasts clusters (iceberg dump structures); and erosional grounding stuctures (keel marks in shallow water; Thomas & Connell 1985; Gilbert 1990; Lisitzin 2002; Benn & Evans 2010). Icebergs of temperate glaciers are typically relatively small and debris rich (bergy bits, growlers and brash ice), while icebergs derived from polar glaciers are large and nearly void of debris (Powell 1987; Powell & Domack 2002; Benn & Evans 2010). If deposited by icebergs, the abundance of clasts in the Basal Beds suggests deposition occurred from temperate icebergs due to the abundance of the dropstones. Bergs from such glaciers typically remain close to the glacier front with far-travelled bergs typically travelling less than a few tens of kilometres. Therefore, most ice-rafted debris from temperate glaciers is deposited close to the glacial front (Powell 1983). If present, local glaciers would have typically entered the ocean through deep troughs or fjords and they would have been associated with meltwater plumes and the deposition of both massive and stratified diamictites, similar to deposits of the Yakataga Formation in Alaska (Eyles & Lagoe 1990; James *et al.* 2009). Deposition from far-travelled icebergs would not be associated with meltwater plumes, but temperate bergs would have been smaller and the resultant lonestone abundance within a given deposit would likely be low due to the limiting factors such as iceberg size and melt-out of debris during long transport distances. Although deposition in shallow water can occur, and it is common for small bergs (e.g. bergy bits) to collect on tidal flats in a proximal glacial position, deep-water deposition of clasts is more likely for larger far-travelled icebergs due to the draft/size of bergs required to raft clasts over long distances (Powell 1983; Powell & Domack 2002).

Sea ice is capable of abrading and striating, pushing, rolling, freezing onto, lifting and transporting particles of various sizes. Gravel-sized material can be incorporated into or on surface ice by: rock fall; freezing onto and lifting of clasts in the intertidal zone; freezing and flotation of clasts by anchor ice in rivers, lakes and the sea; freezing in of suspended material; river flow across sea or lake ice during spring melt; eolian processes; 'bulldozing' of clasts by river, lake or sea ice; as well as several other processes (Rosen 1979; Lauriol & Gray 1980; Dionne 1981, 1988; Syvitski *et al.* 1987; Kempema *et al.* 1989, 2001; Gilbert 1990; Reimnitz *et al.* 1998; Dethleff *et al.* 2000; Lisitzin 2002). Incorporation of gravel into sea ice is most effective due to downwards freezing and tidal lifting of particles in the upper intertidal zone, especially during Spring tides just prior to and during Spring break-up (Rosen 1979). Boulders up to 6.6 m in diameter have been reported to have been moved by sea ice (Lauriol & Gray 1980). However, most of those boulders were likely pushed or dragged by the ice (Drake & McCann 1982), as sea ice is known to have pushed a boulder weighing 176 tons (Dionne 1988). Although a 2-m-thick ice floe can lift and raft boulders up to 6 m in diameter (Drake & McCann 1982; Syvitski *et al.* 1987), boulders up to 2 m in diameter are commonly lifted and rafted on the top of ice floes (Rosen 1979; Fig. 15a). Anchor ice can also lift and transport boulders (Kempema *et al.* 2001), as Dayton *et al.* (1969) reported sediment masses of as much as 25 kg being lifted from the sea floor by this process. Clasts rafted by sea ice are typically locally derived (Dionne 1981). However, clasts may be transported several kilometres before being deposited in clast clusters (dump structures), boulder fields, boulder barricades or as isolated boulders and gravel (lonestones/dropstones), on tidal flats and in deeper subtidal waters (Rosen 1979; Dionne 1981; McCann *et al.* 1981; Aitkens *et al.* 1988; Fig. 15a, b). Diamictites are not typically associated with sea-ice-rafted materials.

The occurrence of mass-wasting features and an absence of glacial indicators on the pre-Parmeener unconformity imply that Maria Island was never glaciated and therefore that the clasts, which were locally derived, were not eroded or transported by glaciers. However, clast composition (primarily quartzite and granite), clast size and roundness (identical to underlying basal conglomerate), the decrease in lonestone abundance both vertically and laterally away from the dipping unconformity and the onlap of lonestone-rich horizons onto the clast-supported conglomerate-lined unconformity are more easily explained due to derivation from a nearby high-energy rocky shoreline and rafting of the gravels by sea ice rather than transport and deposition from far-travelled icebergs. Based on the characteristics of iceberg- and sea-ice-rafted deposits described above, we conclude that deposition from sea ice more accurately explains the occurrence of Basal Bed lonestones than does rafting by icebergs.

In the Lower Erratic Zone, the interbedding of laminated siltstones with conglomerates, sandstones and grainstones indicates settling from suspension, biotic production and ice rafting of gravels interspersed with protracted periods of wave winnowing and reworking during episodic high-energy events (e.g. storms or currents) in subtidal, lower shoreface or offshore settings. Such high-energy events may be related to an absence of sea ice which, when present, would have prevented reworking by waves (Syvitski *et al.* 1987). A fossil biota of *in situ* filter-feeding organisms including bryozoans, bivalves and brachiopods suggest low background sedimentation rates and that low-energy conditions occurred between these episodic high-energy events (Reid 2010).

Although clasts occur throughout the Lower Erratic Zone, their greater abundance within conglomerate, sandstone and grainstone beds than in siltstone units has led several authors to suggest that the deposition of these clast-rich beds resulted from climatic cycles (e.g. Heinrich events or glacial v. interglacial periods; Rao & Green 1982; Eyles *et al.* 1997; Fielding *et al.* 2010). However, the occurrences of *in situ* foliose *Stenopora* bryozoans crossing or abutting lithological boundaries constrains the duration between depositional cycles to the lifetime of the bryozoan colony (*c.* 100 years). Therefore, the apparent cyclicity at the Fossil Cliffs was on a much shorter timescale than that of Milankovitch cyclicity. Conglomerate, limestone

Fig. 15. Tidal flat in Ungava Bay, Canada at low tide showing debris transported by sea ice. Ungava Bay experiences a tidal range in excess of 9 m (Archer & Hubbard 2003). (**a**) Granite boulder and gravel deposited on the middle portion of the tidal flat (person for scale). (**b**) Gravel-covered flat near the base of the intertidal zone. Boulder with the side covered in barnacles is 1 m in diameter. Photos courtesy of Allen Archer.

and siltstone beds also interfinger in the Lower Erratic Zone and also occur as alternating beds within larger low-angle downlapping sediment bodies. Such relationships suggest that they could not have been the result of Milankovitch cyclicity but that the units were laterally related and part of the same progradational sediment body.

The Lower Erratic Zone at the Fossil Cliffs is dominated by siltstone at the base of the exposure. Upwards, conglomeratic units and carbonate (grainstones and rudstones) sheets and channels increase in abundance. This change suggests shallowing where storms and rip currents reworked the deposits near the top of the Lower Erratic Zone.

Rogala *et al.* (2007) interpreted the Darlington Limestone to be the result of the formation of intertidal *Eurydesma* barrier bars adjacent to a rocky coastline. Rudstone (coquina) and grainstone beds composed of *Eurydesma* and spiriferid brachiopods at the top of the Darlington Limestone were deposited as intertidal to subtidal shell banks. The occurrence of decimetre- to metre-scale cross-stratification, the occurrence of disarticulated valves and the occurrence of broken and abraded bioclastic debris suggest deposition in shallow wave-agitated waters. Beds containing *in situ* articulated shells also occur. The interpreted palaeoecology of *Eurydesma* supports carbonate deposition in a shallow-water high-energy setting (Runnegar 1979). Channel-form bodies of rudstone and grainstone occur in the middle portion of the Darlington Limestone, while grainstone and rudstone sheets and wedge-shaped bodies are interbedded and interfinger with siltstone at the base of the unit. Channel-form bodies represent debris and rip-current channels that transported bioclastic debris off of the crest of the shell bank, whereas grainstone sheets represent debris fans that accumulated along bank margins where they grade into and interfinger with siltstone deposited in deeper subtidal waters (Fielding *et al.* 2010). This upwards change in facies coupled with a vertical change in fossils from a quiet-water bryozoan-dominated fauna at the base of the Darlington Limestone to a higher-energy eurydesmid-dominated fauna at the top of the limestone suggests shoaling through time as the bank developed (Clarke & Baillie 1984; Rogala *et al.* 2007). Based on bryozoan, bivalve and brachiopod palaeoecology, Reid (2010) suggested that the change at the Fossil Cliffs from bryozoan-bearing siltstones upwards into *Euyrdesma*-rich limestones represents a low-energy offshore to high-energy onshore depositional trend.

Clast counts reported by Brill (1982) indicate that ice-rafted debris occurs throughout the Darlington Limestone. A higher percentage of gravel on bedding plane surfaces suggests that the clasts were concentrated due to wave activity and the removal of smaller bioclastic debris. Although clasts are less abundant to the north in the area of the quarry, there is no systematic reduction in the number of clasts upwards within the limestone, which suggests that ice rafting occurred throughout deposition of the limestone. Although clasts are common in the Darlington Limestone, they are less abundant than in strata of the Lower and Upper Erratic Zones. These clasts are typically rounded and derived from local bedrock, suggesting that they were not transported by icebergs but rather by sea ice. Sea ice is more efficient in transporting clasts into shallow water than icebergs due to the shallower draft of an ice flow than that of an iceberg.

The sharp contact between the Darlington Limestone and strata in the overlying Upper Erratic Zone, which includes fine-grained siliciclastic units, bryozoans and spiriferid brachiopods, suggests the occurrence of an abrupt deepening event. The upward increase in bioturbated siltstone with an offshore trace fossil assemblage suggest deepening continued to the top of the Basal Beds. The interbedding of conglomerate, conglomeratic sandstone and siltstone in the Upper Erratic Zone is similar to that of strata in the Lower Erratic Zone, and suggests deposition was due to similar processes of slow settling from suspension interspersed with ice rafting and sediment reworking. An offshore trace fossil assemblage within the siltstones provides further evidence for deposition in deeper waters than the underlying Darlington Limestone. The derivation of clasts from local sources, an absence of striated and faceted clasts and the high roundness values of the clasts suggest that gravel was rafted away from a local, unglaciated, rocky shoreline, which is more easily explained by sea-ice rafting than rafting by icebergs.

Although lonestones occur throughout the Basal Beds, they display varying abundance vertically within the succession. Three prominent horizons occur which consist of: the basal conglomerate; the 1- to >30-m-thick bedded conglomerate (LEZ-3); and the boulder-bearing unit at the base of the Upper Erratic Zone. It is tempting to relate such beds to intense cold intervals. However, cold intervals are often characterized by multiyear shorefast sea ice or sikussak. As such, development of sikussaks would correspond to a reduction in the amount of material rafted from the littoral zone. In contrast, increased sea-ice rafting should occur during slightly warmer intervals within a cold climatic setting when shorefast ice breaks up on a regular basis. Local physiographic conditions (e.g. fjords, silled basins or fringing banks) and weather patterns (e.g. absence of katabatic winds or storms) may also favour formation of sikussaks (Dowdeswell *et al.* 2000; Barry & Gan 2011). The relationship between lonestone abundance within the Basal Beds and global climate change is therefore unknown at this time.

Discussion

During the Late Pennsylvanian and Early Permian LPIA acme, extensive glacial and glacimarine deposits occurred across Gondwana (Fig. 1; López-Gamundí 1997; Visser 1997; Isbell *et al.* 2003, 2008*a*, *b*, 2012; Fielding *et al.* 2008*a*, *b*; Holz *et al.* 2008; Rocha Campos *et al.* 2008; Stollhofen *et al.* 2008). Such a scenario corresponded to major global cooling and the latitudinal lowering of the equilibrium line altitude across the globe (Isbell *et al.* 2012). The equilibrium line altitude is the altitude of the line on a glacier that separates areas of annual net ablation from annual net accumulation. The land surface must reside above the equilibrium line altitude for glaciers to form. As for the Late Cenozoic, maximum glaciation during the LPIA would have seen glaciers forming at sea level within the South Polar Circle. This can be seen by the occurrence of extensive glacimarine deposits in Antarctica, South Africa and Tasmania. Outside of the polar circle, ice centres formed on uplands above the northwards-rising local equilibrium line altitude (e.g. Windhoek and Cargonian Highlands, southern Africa). Some of these uplands rose to greater than 3000 m above sea level (Visser 1987; Stollhofen *et al.* 2008). Ice sheets centred over these uplands fed ice into the adjacent basins (Visser 1997; Holz *et al.* 2008; Rocha Campos *et al.* 2008).

Strata of the Wynyard Formation and equivalent units represent the youngest evidence of ice contact deposition in Tasmania during the LPIA (Figs 1, 3 & 5; Henry *et al.* 2012). These strata were deposited during maximum LPIA glaciation. The Wynyard deposits are correlative, in part, to the Upper Palaeozoic glacial deposits in Antarctica and South Africa (Isbell *et al.* 2008*a*, *b*, 2012; Stollhofen *et al.* 2008; Isbell 2010) and to the P1 glacial deposits in eastern Australia (Fielding *et al.* 2008*a*, *b*, 2010). However, Wynyard deposition may have ended several million years before the end of P1 glaciation and prior to the end of LPIA glaciation in Antarctica and South Africa. Ice disappeared in these regions at *c.* 291 Ma during the Sakmarian (Fielding *et al.* 2008*a*, *b*; Isbell *et al.* 2008*a*, *b*, 2012; Stollhofen *et al.* 2008).

The occurrence of an abrupt contact between glacial and post-glacial deposits across southern Gondwana in the Sakmarian records retreat of the glaciers out of the basins (e.g. Karoo Basin, Transantarctic Basin, Tasmania Basin). This retreat was likely triggered by global warming and was likely related to a major increase in atmospheric pCO_2 values (Montañez *et al.* 2007). The absence of Permian glacial deposits younger than Early Sakmarian in these areas suggests that most of southern and Polar Gondwana was never re-glaciated following the Late Pennsylvanian–Early Permian glacial episode (Isbell *et al.* 2003, 2008*a*, *b*, 2012; Catuneanu *et al.* 2005; Miller & Isbell 2010; Henry *et al.* 2012). Strata in the central Transantarctic Mountains, Antarctica, consist of post-glacial shale containing dropstones only in the basal metre of the unit, deltaic sandstones and coal- and fossil-plant-bearing fluvial deposits (Collinson *et al.* 1994; Isbell & Cúneo 1996; Isbell *et al.* 1997, 2008*a*, 2012). Permian fossil plants in Antarctica consist of compressions and impressions of the *Glossopteris-Gangamopteris* flora (Cúneo *et al.* 1993) and permineralized fossil stumps and logs with growth rings (Taylor & Taylor 1990; Gulbranson *et al.* 2012). The fossilized wood contains no evidence of frost damage during growth (Taylor *et al.* 1992). During the Early–Late Permian, the crustal blocks of Antarctica and Tasmania resided within the South Polar Circle with Antarctica comprising most of the polar landmass (Fig. 2; Domeier *et al.* 2011; Isbell *et al.* 2012). Like Antarctica, Tasmanian strata show no evidence of direct glacial deposition from the Early Sakmarian until the end of the Permian (Fig. 1; Fielding *et al.* 2010). The deposits and climate for Antarctica, and possibly Tasmania, contrast with the Bowen, Gunnedah and Sydney basins of eastern Australia located between 52 and 70°S latitude, which were episodically glaciated until the latest Capitanian/earliest Wuchiapingian (Figs 1 & 2; Fielding *et al.* 2008*a*, *b*). This spatial disparity between non-glacial polar and glacial subpolar localities raises questions as to what role global climate cooling played in initiating and terminating glacial events during the Middle–Late Permian (Jones *et al.* 2006; Isbell *et al.* 2012; Henry *et al.* 2012).

The Basal Beds are younger than the P1 glacial deposits and occur within the APP2.1 Palynomorph Zone (Fig. 3; Clarke & Forsyth 1989; Price 1997; Briggs 1998; Mantle *et al.* 2010), which correlates with the non-glacial interval between the P1 and P2 glaciations in eastern Australia (Fig. 1; Fielding *et al.* 2008*a*). Clearly, the occurrence of ice-rafted debris in Tasmania following the end of the Wynyard glaciation requires cold-climatic conditions. Defining when glaciation occurred on Earth during the LPIA is important, but equally important is defining the limits of glaciation and the distribution of other non-glacial cold-climate indicators in space and time. Currently, sea ice is a major transporting agent along the polar, subpolar and temperate coasts of northern and eastern Russia, Alaska and much of Canada, and occurs as far south as the Saint Lawrence estuary in North America. In these examples, the coastlines are separated from the nearest glaciers by hundreds to thousands of kilometres. Many of the characteristics of iceberg-rafted and -deposited debris (e.g. dropstones, clast clusters, boulder pavements, dump structures, keel marks, striated grain surfaces, large range of grain sizes including large clasts and clast abundance) are shared by sea-ice deposits in these areas (Hansom 1983; Syvitski *et al.* 1987; Gilbert 1990; Lisitzin 2002). Although the occurrence of dropstones combined with other lines of evidence is a powerful tool for identifying glacial episodes, such as those in the Bowen, Gunnedah and Sydney basins of eastern Australia, the ability of sea ice to transport clasts raises questions about the validity of using dropstones alone as a glacial indicator. Although cold-water conditions are required for the formation of sea ice or anchor ice, glacial conditions are not required to explain dropstones within the Basal Beds because they can be more easily explained using sea-ice rafting and because they are correlated to a non-glacial interval between the P1 and P2 glacial events in eastern Australia (APP2.1 Palynomorph Zone; Briggs 1998; Fielding *et al.* 2008*a*).

The magnitude of global cooling during the three additional sub-polar glacial events that occurred in eastern Australia (P2, P3 and P4; Fielding *et al.* 2008*a*) during the Permian is unknown. Although changing atmospheric pCO_2 values, the occurrence of uplands and changing atmospheric/oceanographic circulation patterns contributed to the development of these local ice centres (Fielding *et al.* 2001, 2008*a*; Royer *et al.* 2004; Jones *et al.* 2006; Royer 2006; Montañez *et al.* 2007; Henry *et al.*

2012; Isbell *et al.* 2012), global cooling was not severe enough for the re-advance of glaciers into the polar basins during the Middle–Late Permian with the exception of the southernmost portion of the Sydney Basin, which straddled the South Polar Circle.

Conclusions

The Basal Beds are an important part of the Late Palaeozoic climatic history of Tasmania and have been used to suggest glacial conditions in or adjacent to the Tasmanian sector of Gondwana during the Sakmarian. However, a detailed facies analysis of the Basal Beds has been lacking. Biostratigraphic data indicate that the strata exposed along the Fossil Cliffs on Maria Island were deposited following the latest Pennsyvlanian–Early Permian acme of Gondwanan glaciation. These strata record the actions of sea ice and wave/current activity on and along the margins of a *Eurydesma* shell bank located offshore from a rocky storm-influenced shoreline. The presence of deeply weathered basement rock (quartzite, pelite and granite) beneath the pre-Parmeener unconformity, an absence of evidence for glacial truncation of the basement rocks, the occurrence of conglomerate-containing rounded quartzite and granite gravels that line the unconformity, and the onlap of dropstone-bearing marine strata onto the conglomerate-lined unconformity with dropstones that display the same composition, size and roundness values as the conglomerate clasts all suggest that the shoreline was not glaciated at the time of deposition of the Basal Beds. Locally derived lonestones occur throughout these units and were eroded, transported and deposited from sea ice. Sea ice dampens wave activity and is capable of transporting clasts up to 6 m in diameter. Although the Basal Beds contain abundant dropstones, glacial conditions are not required to produce such successions.

This work was supported by the National Science Foundation grants OPP-0635537, OISE-0825617, OPP-0943935 and OPP-0944532 to the University of Wisconsin-Milwaukee. The work was also supported by the American Association of Petroleum Geologists Student Grants-in-Aid, the University of Wisconsin-Milwaukee and a University of Canterbury Geological Sciences Internal Research Grant. A. Archer provided photos of sea-ice-rafted debris in Ungava Bay, Canada. L. Gahagan from the Institute of Geophysics' PLATES Project at the University of Texas at Austin provided plate reconstructions and plotted the polar wander path of Domeier *et al.* (2011) on reconstructions of Gondwana shown in Figure 2 that were constructed from the Institute's geophysical database. Special thanks are extended to J. and A. Cooke, owners of the Maria Island Ferry, for making it possible to visit the eastern coast of Maria Island. Work on Maria Island was conducted through The Tasmania National Parks Service under permits issued to J. Isbell and C. Reid.

References

AITKEN, A. E., RISK, M. J. & HOWARD, J. D. 1988. Animal-sediment relationships on a subarctic intertidal flat, Pangnirtung Fiord, Baffin Island, Canada. *Journal of Sedimentary Petrology*, **58**, 969–978.

ARCHER, A. W. & HUBBARD, M. S. 2003. Highest tides in the world. *In*: CHAN, M. A. & ARCHER, A. W. (eds) *Extreme Depositional Environments: Mega end Members in Geologic Time*. Geological Society of America, Boulder, Special Paper, **370**, 151–73.

ASKIN, R. A. 1998. Floral trends in the Gondwana high latitudes: palynological evidence from the Transantarctic Mountains. *Journal of African Earth Sciences*, **27**, 12–13.

BANKS, M. R. 1962. The geology of Tasmania: Permian. *Journal of the Geological Society of Australia*, **9**, 189–215.

BANKS, M. R. 1981. Late Palaeozoic tillites of Tasmania. *In*: HAMBREY, M. J. & HARLAND, W. B. (eds) *Earth's Pre-Pleistone Glacial Record*. Cambridge University Press, Cambridge, 495–501.

BANKS, M. R. & CLARKE, M. J. 1987. Changes in the geography of the Tasmania Basin in the late Paleozoic. *In*: MCKENZIE, G. D. (ed.) *Gondwana Six: Stratigraphy, Sedimentology, and Paleontology*. American Geophysical Union, Washington, DC, 1–14.

BARRY, R. & GAN, T. Y. 2011. *The Global Cryosphere: Past, Present and Future*. Cambridge University Press, Cambridge, 472.

BENN, D. I. & EVANS, D. J. A. 2010. *Glaciers and Glaciation*. Hodder Education, Oxon.

BRIGGS, D. J. C. 1998. Permian Productidina and Strophalosiidina from the Sydney-Bowen Basin and New England Orogen; systematics and biostratigraphic significance. *Memoir of the Association of Australasian Palaeontologists*, **19**.

BRILL, K. G. 1982. Palaeoenvironment of the Darlington Limestone (early Permian), Tasmania. *Papers and Proceedings of the Royal Society of Tasmania*, **116**, 67–84.

CATUNEANU, O., WOPFNER, H., ERIKSSON, P. G., CAIRNCROSS, B., RUBIDGE, B. S., SMITH, R. M. H. & HANCOX, P. J. 2005. The Karoo basins of south-central Africa. *Journal of African Earth Sciences*, **43**, 211–253.

CLAPHAM, M. E. & JAMES, N. P. 2008. Paleoecology of Early-Middle Permian marine communities in eastern Australia; response to global climate change in the aftermath of the late Paleozoic ice age. *Palaios*, **23**, 738–750.

CLARKE, M. J. & BAILLIE, P. W. 1984. *Maria*. Tasmania Department of Mines, Geological Survey Explanatory Report & Geological Atlas 1:50 000 Series, Sheet 77, 32.

CLARKE, M. J. & BANKS, M. R. 1975. The stratigraphy of the lower (Permo-Carboniferous) parts of the Parmeener Super-Group, Tasmania. *In*: CAMPBELL, K. S. W. (ed.) *Gondwana Geology: Papers presented at the Third Gondwana Symposium*. Australian National University Press, Canberra, 453–467.

Clarke, M. J. & Farmer, N. 1976. Biostratigraphic Nomenclature for late Palaeozoic rocks in Tasmania. *Royal Society of Tasmania, Papers and Proceedings*, **110**, 91–8.

Clarke, M. J. & Forsyth, S. M. 1989. Late Carboniferous-Triassic. *In*: Burrett, C. F. & Martin, E. L. (eds) *Geology and Mineral Resources of Tasmania*. Geological Society of Australia, Sydney, Special Publication, **15**, 293–338.

Collinson, J. W., Isbell, J. L., Elliot, D. H., Miller, M. F. & Miller, J. M. G. 1994. Permian–Triassic Transantarctic basin. *In*: Veevers, J. J. & Powell, C. M. (eds) *Permian-Triassic Pangean basins and foldbelts along the Panthalassan Margin of Gondwanaland*. Geological Society of America, Boulder, Memoir, **184**, 173–222.

Crowell, J. C. 1995. The ending of the late Paleozoic ice age during the Permian period. *In*: Scholle, P. A., Peryt, T. M. & Ulmer-Scholle, D. S. (eds) *The Permian of Northern Pangea*. Springer-Verlag, Berlin, 62–74.

Crowell, J. C. 1999. *Pre-Mesozoic Ice Ages: Their Bearing on Understanding the Climate System*. Geological Society of America, Boulder, Memoir, **192**, 1–106.

Crowell, J. C. & Frakes, L. A. 1971. Late Paleozoic glaciation: Part IV, Australia. *Geological Society of America Bulletin*, **82**, 2515–2540.

Cúneo, N. R., Isbell, J. L., Taylor, T. N. & Taylor, E. L. 1993. The Glossopteris Flora in Antarctica: taphonomy and paleoecology. *Comptes Rendus XII International Congress for the Carboniferous and Permian, Buenos Aires*, **2**, 13–40.

Dalziel, I. W. D. 1997. Neoproterozoic-Paleozoic geography and tectonics: review, hypothesis, environmental speculation. *Geological Society of America Bulletin*, **109**, 16–42.

Dayton, P. K., Robilliard, G. A. & Devries, A. L. 1969. Anchor ice formation in McMurdo Sound, Antarctica, and its biological effects. *Science*, **163**, 273–274.

Dethleff, D., Rachold, V., Tintelnot, M. & Antonow, M. 2000. Sea-ice transport of riverine particles from the Laptev Sea to Fram Strait based on clay mineral studies. *International Journal of Earth Sciences*, **89**, 496–502.

Dionne, J. C. 1981. A boulder-strewn tidal flat, north shore of the Gulf of St. Lawrence, Quebec. *Geographie Physique et Quaternaire*, **35**, 261–267.

Dionne, J. C. 1988. Ploughing boulders along shorelines, with particular reference to the St. Lawrence estuary. *Geomorphology*, **1**, 297–308.

Domack, E. W., Burkley, L. A., Domack, C. R. & Banks, M. R. 1993. Facies analysis of glacial marine pebbly mudstones in the Tasmania Basin: implications for regional paleoclimates during the late Paleozoic. *In*: Findlay, R. H., Unrug, R., Banks, M. R. & Veevers, J. J. (eds) *Gondwana Eight: Assembly, Evolution and Dispersal*. A.A. Balkema, Rotterdam, 471–484.

Domeier, M., Van Der Voo, R., Tohver, E., Tomezzoli, R. N., Vizan, H. & Kirshner, J. 2011. New Late Permian paleomagnetic data from Argentina: Refinement of the apparent polar wander path of Gondwana. *Geochemistry, Geophysics, Geosystems*, **12**, 1–21.

Dowdeswell, J. A., Whittington, R. J., Jennings, A. E., Andrews, J. T., Mackensen, A. & Marienfeld, P. 2000. An origin for laminated glacimarine sediments through sea-ice build-up and suppressed iceberg rafting. *Sedimentology*, **47**, 557–76.

Drake, J. J. & Mccann, S. B. 1982. The movement of isolated boulders on tidal flats by ice floes. *Canadian Journal of Earth Sciences*, **19**, 748–754.

Everard, J. L. 1983. Petrology of contact metamorphosed Mathinna Beds, near Trigonia Corner, Maria Island. *In*: Clarke, M. J. & Baillie, (eds) Maria. Tasmania Department of Mines, Geological Survey Explanatory Report & Geological Atlas 1:50 000 Series, Sheet 77, 33–39.

Eyles, C. H. & Lagoe, M. B. 1990. Sedimentation patterns and facies geometry on a temperate glacially-influenced continental shelf: the Yakataga Formation, Middleton Island, Alaska. *In*: Dowdeswell, J. A. & Scourse, J. D. (eds) *Glacimarine Environments: Processes and Sediments*. Geological Society, London, Special Publication, **53**, 363–386.

Eyles, N., Eyles, C. H. & Gostin, V. A. 1997. Iceberg rafting and scouring in the Early Permian Shoalhaven Group of New South Wales, Australia: evidence of Heinrich-like events? *Palaeogeography, Palaeoclimatology, Palaeoecology*, **136**, 1–17.

Fielding, C. R., Sliwa, R., Holcombe, R. J. & Jones, A. T. 2001. A new Palaeogeographic synthesis for the Bowen, Gunnedah and Sydney basins of Eastern Australia. *In*: Hill, K. C. & Bernecker, T. (eds) *Eastern Australasian Basins Symposium: A Refocused Energy Perspective for the Future*. Petroleum Exploration Society of Australia, Melbourne, Special Publication, **1**, 269–278.

Fielding, C. R., Frank, T. D., Birgenheier, L. P., Rygel, M. C., Jones, A. T. & Roberts, J. 2008 *a*. Stratigraphic imprint of the late Palaeozoic ice age in eastern Australia: a record of alternating glacial and nonglacial climate regime. *Journal of the Geological Society, London*, **165**, 129–140.

Fielding, C. R., Frank, T. D., Birgenheier, L. P., Rygel, M. C., Jones, A. T. & Roberts, J. 2008*b*. Stratigraphic record and facies associations of the late Paleozoic ice age in eastern Australia (New South Wales and Queensland). *In*: Fielding, C. R., Frank, T. & Isbell, J. L. (eds) *Resolving the Late Paleozoic Ice Age in Time and Space*. Geological Society of America, Boulder, Special Paper, **441**, 41–57.

Fielding, C. R., Frank, T. D. & Isbell, J. L. 2008*c*. The late Paleozoic ice age – a review of current understanding and synthesis of global climate patterns. *In*: Fielding, C. R., Frank, T. D. & Isbell Boulder, J. L. (eds) *Resolving the Late Paleozoic Ice Age in Time and Space*. Geological Society of America, Boulder, Colorado, Special Publication, **441**, 343–354.

Fielding, C. R., Frank, T. D., Isbell, J. L., Henry, L. C. & Domack, E. W. 2010. Stratigraphic signature of the late Palaeozoic Ice Age in the Parmeener Supergroup of Tasmania, SE Australia, and inter-regional comparisons. *Palaeogeography, Palaeoclimatology, Palaeoecology*, **298**, 70–90.

Frakes, L. A. & Francis, J. E. 1988. A guide to Phanerozoic cold polar climates from high-latitude ice-rafting in the Cretaceous. *Nature*, **333**, 547–549.

FRAKES, L. A., FRANCIS, J. E. & SYKTUS, J. I. 1992. *Climate Modes of the Phanerozoic*. Cambridge University Press, Cambridge.

GASTALDO, R. A., DIMICHELE, W. A. & PFEFFERKORN, H. W. 1996. Out of the icehouse into the greenhouse; a late Paleozoic analog for modern global vegetational change. *GSA Today*, **6**, 1–7.

GILBERT, R. 1990. Rafting in glacimarine environments. *In*: DOWDESWELL, J. A. & SCOURSE, J. D. (eds) *Glacimarine Environments: Processes and Sediments*. Geological Society, London, Special Publication, **53**, 105–120.

GRADSTEIN, F. M., OGG, J. G. ET AL. 2004. *A Geologic Time Scale 2004*. Cambridge University Press, Cambridge.

GULBRANSON, E. L., MONTAÑEZ, I. P., SCHMITZ, M. D., LIMARINO, C. O., ISBELL, J. L., MARENSSI, S. A. & CROWLEY, J. L. 2010. High-precision U-Pb calibration of Carboniferous glaciation and climate history, Paganzo Group, NW Argentina. *Geological Society of America Bulletin*, **122**, 1480–1498.

GULBRANSON, E. L., ISBELL, J. L., TAYLOR, E. L., RYBERG, P. E., TAYLOR, T. N. & FLAIG, P. P. 2012. Permian polar forests: deciduousness and environmental variation. *Geobiology*, DOI: 10.1111/j.1472-4669.2012.00338.x

HAND, S. J. 1993. Palaeogeography of Tasmania's Permo-Carboniferous glacigenic sediments. *In*: FINDLAY, R. H., UNRUG, R., BANKS, M. R. & VEEVERS, J. J. (eds) *Gondwana Eight: Assembly, Evolution and Dispersal*. A.A. Balkema, Rotterdam, 459–469.

HANSOM, J. D. 1983. Ice-formed intertidal boulder pavements in the sub-Antarctic. *Journal of Sedimentary Petrology*, **53**, 135–145.

HENRY, L. C., ISBELL, J. L., FIELDING, C. R., DOMACK, E. W., FRANK, T. D. & FRAISER, M. L. 2012. Proglacial deposition and deformation in the Upper Carboniferous to Lower Permian Wynyard Formation, Tasmania: A process analysis. *Palaeogeography, Palaeoclimatology, Palaeoecology*, **315–316**, 142–157.

HOLZ, M., SOUZA, P. A. & IANNUZZI, R. 2008. Sequence stratigraphy and biostratigraphy of the Late Carboniferous to Early Permian glacial succession (Itararé subgroup) at the eastern-southeastern margin of the Paraná Basin, Brazil. *In*: FIELDING, C. R., FRANK, T. & ISBELL, J. L. (eds) *Resolving the Late Paleozoic Ice Age in Time and Space*. Geological Society of America, Boulder, Special Paper, **441**, 115–129.

ISBELL, J. L. 2010. Environmental and paleogeographic implications of glaciotectonic deformation of glaciomarine deposits within Permian strata of the Metschel Tillite, southern Victoria Land, Antarctica. *In*: LÓPEZ-GAMUNDÍ, O. R. & BUATOIS, L. A. (eds) *Late Paleozoic Glacial Events and Postglacial Transgressions in Gondwana*. Geological Society of America, Boulder, Special Paper, **468**, 81–100.

ISBELL, J. L. & CÚNEO, N. R. 1996. Depositional framework of Permian coal-bearing strata, southern Victoria Land, Antarctica. *Palaeogeography, Palaeoclimatology, Palaeoecology.*, **125**, 217–238.

ISBELL, J. L., SEEGERS, G. M. & GELHAR, G. A. 1997. Upper Paleozoic glacial and postglacial deposits, central Transantarctic Mountains, Antarctica. *In*: MARTINI, I. P. (ed.) *Late Glacial and Postglacial Environmental Changes: Quaternary, Carboniferous-Permian, and Proterozoic*. Oxford University Press, Oxford, 230–242.

ISBELL, J. L., MILLER, M. F., WOLFE, K. L. & LENAKER, P. A. 2003. Timing of late Paleozoic glaciation in Gondwana: was glaciation responsible for the development of northern hemisphere cyclothems? *In*: CHAN, M. A. & ARCHER, A. W. (eds) *Extreme Depositional Environments: Mega end Members in Geologic Time*. Geological Society of America, Boulder, Special Paper, **370**, 5–24.

ISBELL, J. L., KOCH, Z. J., SZABLEWSKI, G. M. & LENAKER, P. A. 2008*a*. Permian glacigenic deposits in the Transantarctic Mountains, Antarctica. *In*: FIELDING, C. R., FRANK, T. & ISBELL, J. L. (eds) *Resolving the Late Paleozoic Ice Age in Time and Space*. Geological Society of America, Boulder, Special Paper, **441**, 59–70.

ISBELL, J. L., COLE, D. I. & CATUNEANU, O. 2008*b*. Carboniferous-Permian glaciation in the main Karoo Basin, South Africa: stratigraphy, depositional controls, and glacial dynamics. *In*: FIELDING, C. R., FRANK, T. & ISBELL, J. L. (eds) *Resolving the Late Paleozoic Ice Age in Time and Space*. Geological Society of America, Boulder, Special Paper, **441**, 71–82.

ISBELL, J. L. & HENRY, L. C. ET AL. 2012. Glacial paradoxes during the late Paleozoic ice age: Evaluating the equilibrium line altitude as a control on glaciation. *Gondwana Research*, **22**, 1–19, http://dx.doi.org/10.1016/j.gr.2011.11.005

JAMES, N. P., FRANK, T. D. & FIELDING, C. R. 2009. Carbonate sedimentation in a Permian high-latitude, subpolar depostional realm: Queensland, Australia. *Journal of Sedimentary Research*, **79**, 125–143.

JONES, A. T., FRANK, T. D. & FIELDING, C. R. 2006. Cold climate in the eastern Australian mid to late Permian may reflect cold upwelling waters. *Palaeogeography, Palaeoclimatology, Palaeoecology*, **237**, 370–377.

KEMPEMA, E. W., REIMNITZ, E. & BARNES, P. W. 1989. Sea ice sediment entrainment and rafting in the Arctic. *Journal of Sedimentary Petrology*, **59**, 308–317.

KEMPEMA, E. W., REIMNITZ, E. & BARNES, P. W. 2001. Anchor-ice formation and ice rafting in southwestern Lake Michigan, U.S.A. *Journal of Sedimentary Research*, **71**, 346–354.

LAURIOL, B. & GRAY, J. T. 1980. Processes responsible for the concentration of boulders in the intertidal zone in Leaf Basin, Ungava. *In*: MCCANN, S. B. (ed.) *The Coastline of Canada*. Geological Survey of Canada, Ottowa, Paper 80–10, 281–292.

LAWVER, L. A., DALZIEL, I. W. D., NORTON, I. O. & GAHAGAN, L. M. 2008. *The Plates 2007 Atlas of Plate Reconstructions (750 Ma to Present Day)*. Technical Report, University of Texas, **195**.

LISITZIN, A. P. 2002. *Sea-Ice and Iceberg Sedimentation in the Ocean*. Springer-Verlag, Berlin.

LÓPEZ-GAMUNDÍ, O. R. 1997. Glacial-postglacial transition in the Late Paleozoic basins of southern South America. *In*: MARTINI, I. P. (ed.) *Late Glacial and Postglacial Environmental Changes: Quaternary, Carboniferous-Permian, and Proterozoic*. Oxford University Press, Oxford, 147–168.

MANTLE, D. J., KELMAN, A. P., NICOLL, R. S. & LAURIE, J. R. 2010. *Australian Biozonation Chart*. Geoscience Australia, Canberra, 2.

MARTIN, J. R., REDFERN, J. & AITKEN, J. F. 2008. A regional overview of the late Paleozoic glaciation in Oman. *In*: FIELDING, C. R., FRANK, T. & ISBELL, J. L. (eds) *Resolving the Late Paleozoic Ice Age in Time and Space*. Geological Society of America, Boulder, Special Paper, **441**, 175–186.

MCCANN, S. B., DALE, J. E. & HALE, P. B. 1981. Subarctic tidal flats in areas of large tidal range, southern Baffin Island, eastern Canada. *Geographie Physique et Quaternaire*, **35**, 183–204.

MILLER, M. F. & ISBELL, J. L. 2010. Reconstruction of a high-latitude, postglacial lake: Mackellar Formation (Permian), Transantarctic Mountains. *In*: LÓPEZ-GAMUNDÍ, O. R. & BUATOIS, L. A. (eds) *Late Paleozoic Glacial Events and Postglacial Transgressions in Gondwana*. Geological Society of America, Boulder, Special Paper, **468**, 193–207.

MONTAÑEZ, I. & SOREGHAN, G. S. 2006. Earth's fickle climate; lessons learned from deep-time ice ages. *Geotimes*, **51**, 24–27.

MONTAÑEZ, I., TABOR, N. J. ET AL. 2007. CO_2-forced climate and vegetation instability during late Paleozoic deglaciation. *Science*, **315**, 87–91.

MORY, A. J., MARTIN, J. R. & REDFERN, J. 2008. A review of Permian-Carboniferous glacial deposits in Western Australia. *In*: FIELDING, C. R., FRANK, T. & ISBELL, J. L. (eds) *Resolving the Late Paleozoic Ice Age in Time and Space*. Geological Society of America, Boulder, Special Paper, **441**, 29–40.

POWELL, R. D. 1987. Glacial-marine sedimentation processes and lithofacies of temperate tidewater glaciers, Glacier Bay, Alaska. *In*: MOLINA, B. F. (ed.) *Glacial-Marine Sedimentation*. Plenum Press, New York, 185–232.

POWELL, R. D. 1990. Glacimarine processes at grounding-line fans and their growth to ice contact deltas. *In*: DOWDESWELL, J. A. & SCOURSE, J. D. (eds) *Glacimarine Environments: Processes and Sediments*. Geological Society, London, Special Publication, **53**, 53–73.

POWELL, R. D. & DOMACK, E. 2002. Modern glaciomarine environments. *In*: MENZIES, J. (ed.) *Modern and Past Glacial Environments*. Butterworth-Heinemann Ltd., Oxford, 361–389.

PRICE, P. L. 1997. Permian to Jurassic palynostratigraphic nomenclature of the Bowen and Surat Basins. *Queensland Minerals and Energy Review Series*, **1997**, 137–178.

RAO, C. P. 1983. Geochemistry of Early Permian cold-water carbonates (Tasmania, Australia). *Chemical Geology*, **38**, 307–319.

RAO, C. P. & GREEN, D. C. 1982. Oxygen and carbon isotopes of Early Permian cold-water carbonates, Tasmania, Australia. *Journal of Sedimentary Petrology*, **52**, 1111–1125.

REID, C. M. 2004. *Maria Island*. Geological Society of Australia, Sydney, Field Guide A6.

REID, C. M. 2010. Environmental controls on the distribution of late Paleozoic bryozoan colony morphotypes: an example from the Permian of Tasmania, Australia. *Palaios*, **25**, 692–702.

REID, C. M. & BURRETT, C. F. 2004. The geology and hydrocarbon potential of the glaciomarine Lower Parmeener Supergroup, Tasmania Basin. *Petroleum Exploration Society of Australia Special Publication*, **2**, 265–275.

REID, C. M. & JAMES, N. P. 2010. Permian higher latitude bryozoan biogeography. *Palaeogeography, Palaeoclimatology, Palaeoecology*, **298**, 31–41.

REIMNITZ, E., MCCORMICK, M., BISCHOF, J. & DARBY, D. A. 1998. Comparing sea-ice sediment load with Beaufort sea shelf deposits: Is entrainment selective? *Journal of Sedimentary Research*, **68**, 777–787.

ROCHA CAMPOS, A. C., DOS SANTOS, P. R. & CANUTO, J. R. 2008. Late Paleozoic glacial deposits of Brasil: Paraná Basin. *In*: FIELDING, C. R., FRANK, T. & ISBELL, J. L. (eds) *Resolving the Late Paleozoic Ice Age in Time and Space*. Geological Society of America, Boulder, Special Paper, **441**, 97–114.

ROGALA, B., JAMES, N. P. & REID, C. M. 2007. Deposition of Polar carbonates during interglacial highstands on an Early Permian shelf, Tasmania. *Journal of Sedimentary Research*, **77**, 587–606.

ROSEN, P. S. 1979. Boulder barricades in central Labrador. *Journal of Sedimentary Petrology*, **49**, 1113–1124.

ROSEN, P. S. 1980. Coastal environments of the Makkovik Region, Labrador. *In*: MCCANN, S. B. (ed.) *The Coastline of Canada*. Geological Survey of Canada, Ottowa, Paper, 80–10, 267–280.

ROSEN, P. S. 2005. Boulder Barricades. *In*: SCHWARTZ NORWELL, M. L. (ed.) *Encyclopedia of Coastal Science*. Kluwer Academic Publishers, MA, 204–206.

ROYER, D. L. 2006. CO_2-forced climate thresholds during the Phanerozoic. *Geochimica et Cosmochimica*, **70**, 5665–5675.

ROYER, D. L., BERNER, R. A., MONTAÑEZ, I. P., TABOR, N. J. & BEERLING, D. J. 2004. CO_2 as a primary driver of Phanerozoic climate. *GSA Today*, **14**, 4–10.

RUNNEGAR, B. 1979. Ecology of Eurydesma and the Eurydesma fauna, Permian of eastern Australia. *Alcheringa*, **3**, 261–285.

RYGEL, M. C., FIELDING, C. R., BANN, K. L., FRANK, T. D., BIRGENHEIER, L. & TYE, S. C. 2008. The Lower Permian Wasp Head Formation, Sydney Basin: high-latitude, shallow marine sedimentation following the late Asselian to early Sakmarian glacial event in eastern Australia. *Sedimentology*, **55**, 1517–1540.

STOLLHOFEN, H., WERNER, M., STANISTREET, I. G. & ARMSTRONG, R. A. 2008. Single-zircon U-Pb dating of Carboniferous-Permian tuffs, Namibia, and the intercontinental deglaciation cycle framework. *In*: FIELDING, C. R., FRANK, T. & ISBELL, J. L. (eds) *Resolving the Late Paleozoic Ice Age in Time and Space*. Geological Society of America, Boulder, Special Paper, **441**, 83–96.

SYVITSKI, J. M., BURRELL, D. C. & SKEI, J. M. 1987. *Fjords: Processes and Products*. Springer-Verlag, New York.

TAYLOR, E. L. & TAYLOR, T. N. 1990. Structurally preserved Permian and Triassic floras from Antarctica. *In*: TAYLOR, T. N. & TAYLOR, E. L. (ed.) *Antarctic Paleobiology: Its Role in the Reconstruction of Gondwana*. Springer-Verlag, New York, 149–163.

TAYLOR, E. L., TAYLOR, T. N. & CÚNEO, N. R. 1992. The present is not the key to the past: a polar forest from the Permian of Antarctica. *Science*, **257**, 1675–1677.

THOMAS, G. S. P. & CONNELL, R. J. 1985. Iceberg drop, dump, and grounding structures from Pleistocene glacio-lacustrine sediments, Scotland. *Journal of Sedimentary Petrology*, **55**, 243–249.

TRUSWELL, E. W. 1978. Palynology of the Permo-Carboniferous in Tasmania: an interim report. *Bulletin of the Geological Survey of Tasmania*, **56**, 1–37.

VEEVERS, J. J. & POWELL, C. M. 1987. Late Paleozoic glacial episodes in Gondwanaland reflected in transgressive-regressive depositional sequences in Euramerica. *Geological Society of America Bulletin*, **98**, 475–487.

VEEVERS, J. J. & TEWARI, R. C. 1995. *Gondwana Master Basin of Peninsular India between Tethys and the Interior of the Gondwanaland Province of Pangea.* Geological Society of America, Boulder, Memoir, **187**, 1–72.

VEEVERS, J. J., CONAGHAN, P. J., POWELL, C. M., COWAN, E. J., MCDONNELL, K. L. & SHAW, S. E. 1994. Eastern Australia. *In*: VEEVERS, J. J. & POWELL, C. M. (eds) *Permian-Triassic Pangean Basins and Foldbelts along the Panthalassan Margin of Gondwanaland.* Geological Society of America, Boulder, Memoir, **184**, 11–171.

VISSER, J. N. J. 1987. The palaeogeography of part of southwestern Gondwana during the Permo-Carboniferous glaciation. *Palaeogeography, Palaeoclimatology, Palaeoecology*, **61**, 205–219.

VISSER, J. N. J. 1997. Deglaciation sequences in the Permo-Carboniferous Karoo and Kalahari basins of southern Africa: a tool in the analysis of cyclic glaciomarine basin fills. *Sedimentology*, **44**, 507–521.

WOPFNER, H. & CASSHYAP, S. M. 1997. Transition from freezing to subtropical climates in the Permo-Carboniferous of Afro-Arabia and India. *In*: MARTINI, I. P. (ed.) *Late Glacial and Postglacial Environmental Changes: Quaternary, Carboniferous-Permian, and Proterozoic*. Oxford University Press, Oxford, 192–212.

Climatic impact on the reef biota in the Cisuralian and Guadalupian (Permian), East European Platform

OLGA KOSSOVAYA[1]*, DANIEL VACHARD[2] & ALAIN IZART[3]

[1]*All Russian Geological Research Institute, Sredny pr., 74, St. Petersburg, 199106, Russia*

[2]*University of Lille 1, UMR 8217 Géosystèmes, Bâtiment SN 5, 59655 Villeneuve d'Ascq Cédex, France*

[3]*University of Lorraine, UMR 7566 Géoressources, BP239, 54506 Vandoeuvre les Nancy, France*

**Corresponding author (e-mail: Olga_Kossovaya@vsegei.ru)*

Abstract: As important components of extinct as well as modern reefs, the measurement of changes in the composition of reef biota is crucial in order to evaluate the influence of extrinsic factors on the marine communities. The dramatic abiotic changes in the eastern and northern Pangea was mirrored by a gradual transition from a photozoan to a heterozoan association and the appearance of cool- (temperate-) water carbonates both in reef and carbonate ramp environments. Analyses of large skeletal and microbiotic components as well as microfacies succession were used for detailed explanation of reef structure, especially for lesser-known heterozoan skeletal mounds. The youngest skeletal mounds are recognized in the Roadian. The stable isotope data demonstrated a negative oxygen shift between latest Sakmarian and late early Artinskian coinciding with the end of P2 glaciation. The multiplicative nature of the event included the series of successive changes of extrinsic factors such as ice melting in interglacial episodes, eustatic ocean level rise, change of oceanic circulation and decrease of water temperature. The late Artinskian and subsequent Kungurian climatic impacts in the Northern Hemisphere were irreversible for the photozoan biota and prevented its further development. Roadian (Guadalupian) bryonoderm extended skeletal mounds developed in rather warm-water environments.

The study of extinct and modern reefs allows us to understand the influence of the extrinsic factors on biota composition, type of reef-builders and the type of bioconstructions (Kiessling 2001). The Asselian–Kungurian interval is characterized by periodical waxing and waning of ice sheets in the Southern Hemisphere (Fielding *et al.* 2008). The wide distribution of Permian photozoan reefs and mounds which evolved in the late Asselian–early Sakmarian and late Sakmarian–early Artinskian in the East European Platform (EEP) is therefore an exceptional example of long-living bioconstructions which existed during the glacial age (P1–2) (Fielding *et al.* 2008).

The Lower Permian reefs of the eastern EEP have been studied systematically since the twentieth century (Rauzer-Chernousova 1950; Shamov 1957) with most significant works published in the 1970s and 1980s (Maslov 1973; Chuvashov 1983, 1995, 2011; Korolyuk 1985; Antoshkina 2003). More recent syntheses were published by Flügel (1994), Vennin (1997), Weidlich (2002), Vachard & Vennin (2007) and Krainer (2007). Some studies of limestone structure can be found in the classical Russian literature devoted to carbonate-producing organisms (facies) (Maslov 1973). Study of the reef mound of Shakhtau and the surrounding subsidence reefs of the Ishimbai region were promoted by oil exploration (Korolyuk 1985). The barrier reef complex and reef mounds of the eastern EEP margin exposed in the Uralian range in their middle and southern parts were briefly observed (Chuvashov 1983, 1995). The Asselian–Artinskian ramp-type carbonate sections of the Middle Urals were studied by Izart *et al.* (1999). The reefs of the northern part of the Urals were observed by Antoshkina (2003). Microfacies of the reef mounds of Tratau, Shakhtau and Nizhnya Irgina and their connection to the sedimentary regime were discussed by Vennin (1997, 2007*a*, *b*). The Permian reef facies studied in the EEP contain fusulinids, corals, *Tubiphytes* and *Archaeolithoporella*; some of which are probably of microbial origin. Green algae are abundant in the Lower Permian of the EEP (Kulik 1978; Korolyuk 1985; Vennin 1997) as well as in the West European Platform (Pia 1937; Flügel & Flügel-Kahler 1980; Samankassou 1998; Krainer & Vachard 2007*a*, *b*; Vachard & Vennin 2007). Despite the large number of publications, analyses were usually focused on the

From: GĄSIEWICZ, A. & SŁOWAKIEWICZ, M. (eds) 2013. *Palaeozoic Climate Cycles: Their Evolutionary and Sedimentological Impact*. Geological Society, London, Special Publications, **376**, 343–366.
First published online September 25, 2013, updated November 11, 2013, http://dx.doi.org/10.1144/SP376.19

morphology and biota composition. The spatial evolution of the Permian East European reefs was only briefly mentioned by Antoshkina (2003) and reviewed by Chuvashov (1983, 1995) and Chuvashov *et al.* (1996).

The intriguing fact that the origin and subsequent evolution of the photozoan biota and its bioconstructions occurred during the glacial age is still under discussion. The proposed ideas embraced the adaptation to high latitude (Kiessling 2001) similar to a modern distribution of high-latitude reefs in rather cold temperatures. It was also explained by a high input of nutrients in the cold currents (Clapham 2010). We maintain that the photozoan biota were adapted to warm-water conditions, contradicting the hypothesis of cold-water environments in the Northern and Eastern Pangea before the late Artinskian.

The objectives of this paper are to: (1) recognize the environmental changes that triggered the great and final replacement of the photozoan reef builders in the EEP; (2) trace the microfacies succession in the bioconstructions to understand the depositional dynamics; and (3) compile data on the extrinsic factors to show the origin of the extinction of colonial corals and other photozoan biota. Microfacies and taxonomical analyses are applied to trace a temporal facies change and reconstruct the lateral profiles of the selected mounds. Data on biotic and microfacies change in addition to stable isotope data known from the studied region and adjacent areas were applied for the environmental analyses of the crucial intervals.

Materials and methods

The typical photozoan and heterozoan carbonate associations were collected in several reefs and mounds of the Permian. Samples from the Bashkirian reefs (Tratau, Uraktau and Shakhtau), Northern Timan section, Belaya Gora (Most) section, Divia section and Roadian skeletal mound were collected in different years. The faunal sampling was performed during this work, and previous collections were also used. The microfacies were studied in thin-sections orientated across the bedding plain. The standard and large thin-sections (50×70 mm) were prepared and studied via polarizing microscope. Several thin-sections were coloured by alizarin. The large skeletal fragments of corals, algae and forams were taxonomically defined. Thin-sections from the Lower Permian deposits of Kolguev Island were also available for analyses. Some published results were analysed to obtain more complete characteristics and to trace the reef distribution in the Timan–Pechora district. To the east, the marginal position is occupied by the isolated Kozhim reef in the Polar Urals (Antoshkina 1998, 2003).

Isotope investigation

For the isotope study, 48 specimens of the colonial corals dominating the Belaya Gora (Most) were selected. In each specimen the inner part of the colony was extracted after a visual investigation. The microstructure of all specimens was studied using scanning electron microscopy (SEM) to avoid secondary alteration. Spectral analyses performed for each specimen shows the normal level of calcium. The taxonomical affiliation was carried out for all specimens and representatives of cerioid *Protowentzelella*, *Permastraea* and *Protolondaleiastraea* were selected. Isotope data were obtained by digesting the calcium carbonate samples under vacuum with 100% H_3PO_4 (ortho-phosphoric acid) for 24 h at 25 °C (McCrea 1950). After freezing it out from the separation line, released CO_2 was analysed using a Finnigan Mat delta plus gas mass-spectrometer in the Light Stable Isotopes Laboratory of the Institute of Geological Sciences and Institute of Palaeobiology, Polish Academy of Sciences, Warsaw by Dr K. Malkowski. The gas sample was introduced into the spectrometer in dual inlet mode and a mean value was calculated from eight measurements. The standard error of the spectrometer measurements was 0.02‰. The results are reported in per mil deviation from the Vienna Peedee Belemnite (VPDB) standard using δ (delta) notation. The precision (reproducibility of replicate analyses) of both carbon and oxygen isotope analysis was usually better than ± 0.1‰. Data are normalized to the VPDB scale using National Bureau of Standards NBS-19 ($\delta^{18}O = -2.20$‰ and $\delta^{13}C = 1.95$).

Biostratigraphical framework

The biostratigraphical framework for the region and its correlation with the International Chronostratigraphic Chart (Geological Time Scale 2012) and range of the sections studied are shown in Figure 1. The biostratigraphy is based on rugose corals and fusulinid assemblages and zones. The conodont data which are used for the stage boundary definition in the International Stratigraphic Chart (ISC) in many cases are inapplicable for studied shallow-water carbonate sections. The lower boundary of the studied interval is in the *Sphaeroschwagerina sphaerica* Zone, correlated with the appearance of massive colonial corals in the sections under consideration. The middle Artinskian boundary is correlated by an ostracod zone and the *Skinerella (Paraskinerella) solidissima* Zone. The late Artinskian is also defined by solitary rugose

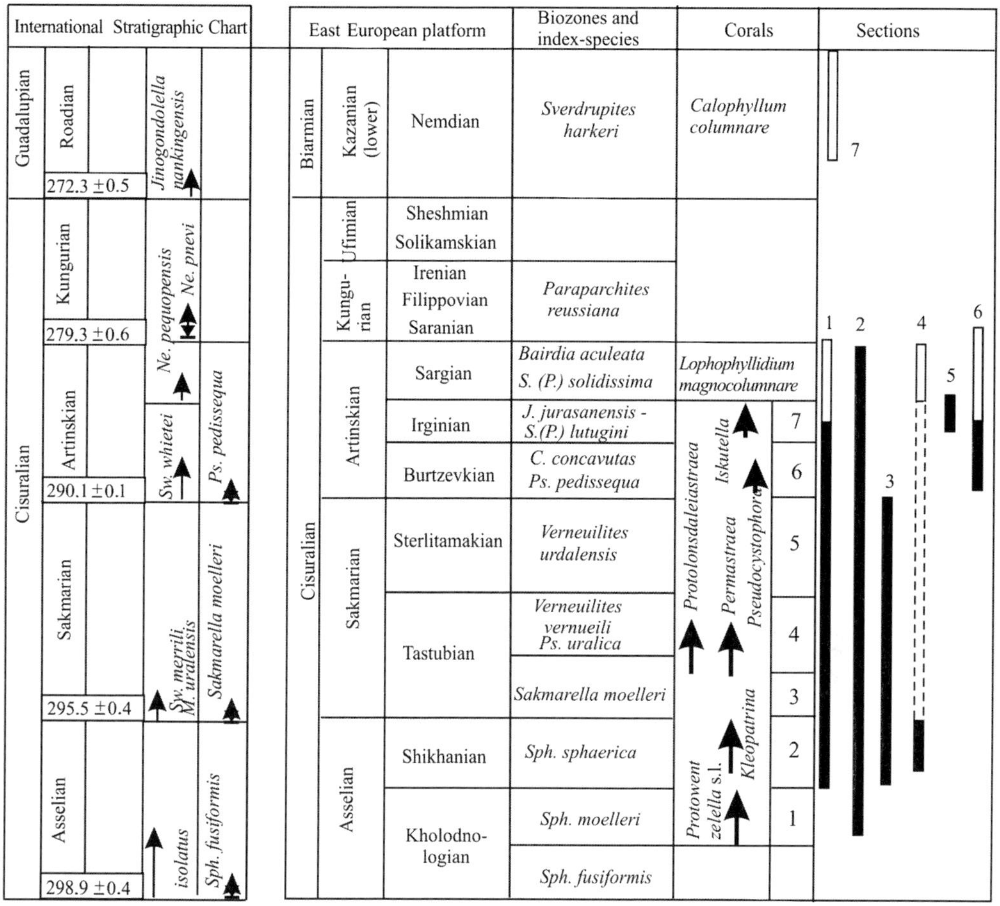

Fig. 1. Stratigraphic range of the section studied. International Stratigraphical Chart 2012 (Gradstein *et al.* 2012). Sections: 1, Belaya Gora (Most); 2, Shakhtau; 3, Uraktau; 4, Tratau; 5, Schuchya River; 6, Divia section; 7, Nemda River. Coral zonation: 1, *Protowentzelella noinskyi*; 2, *Kleopatrina pseudoelegans*; 3, *Kleopatrina magnifica*; 4, *Pseudocystophora complexa*; 5, *Permastraea pseudowisherianum*; 6, *Protolonsdaleiastraea juresanensis*; 7, *Iskutella sulensis*. Abbreviations: Sph., Sphaeroshwagerina; J., Jurasanella; C., Cancavutella; Ne., Neostreptognathodus; M., Mesogondolella; S., Skinerella; P., Paraskinerella; Ps., Pseudofusulina (Rauser-Chernousova *et al.* 1996).

corals and ammonoids (Kossovaya *et al.* 2001, 2002). In the ISC this level is defined by *Neostreptognathodus pequopensis* FAD. Only one element of *Neostreptognathodus* sp. was distinguished (Kossovaya *et al.* 2002). The Kungurian boundary is shown in the ISC by *N. pnevi* FAD (Gradstein *et al.* 2012). In the Divia reef it was defined by shallowing and disappearance of the late Artinskian brachiopods. The correlation with the Roadian is based on ammonoid data and correlation with the Canadian Arctic and the USA (Leonova & Shilovsky 2007; Shilovsky 2007).

The data on oxygen and carbon stable isotopes from biogenic carbonates of Belaya Gora (Most section) (Kossovaya *et al.* 2010) were used for environmental analyses and comparison with previously known measurement for the lower latitude and the Urals (Korte *et al.* 2008; Zeng *et al.* 2012).

Geological and palaeogeographical setting

The Uralian basin is a typical foreland basin formed in the Palaeozoic. The Uralian fold belt appeared as a result of oblique collision between the East European (Laurussia) passive margin and the active margin on the Kazakhstanian continent. Collision began south of the Urals and moved, wave-like, to the north. At the Early Permian the subduction processes were finished and the filling of the foredeep started from the east. Facies changes of the foredeep sediments are complex, but the main steps in the foredeep development are the same at all latitudes. It began with the establishment of a 'preflysch' deep-water basin on shelf sediments, west of the orogenic front. The basin was filled then with flysch and molasse sediments grading westwards into deep-water facies and, still further west, into reefs and biostromes (Puchkov 2009). The siliciclastic deposits accumulated along the eastern part of the basin during Asselian–Kungurian time and gradually changed from marine to lagoon and terrestrial.

A part of the former passive margin of the East European (Laurussian) block (forebulge) is considered as two western zones of the Uralian foreland: the West Uralian and the Central Uralian (Fig. 2). The marginal position of the Early Permian reefs along the eastern margin of the EEP determined their dependence on the destruction of healthy carbonate platform systems caused by a collision of continents and closure of the Pre-Uralian basin. The Lower Permian lateral facies profile embraces the permanent carbonate sedimentation in the platform part and predominate siliciclastic contemporaneous deposition in the eastern slope of the Uralian foredeep influenced by orogenic processes. The maximal deepening is fixed in the middle Artinskian. The western part of the platform is separated by the Baltica continental area characterized by an epeiric regime with stable shallow-marine and lagoon environments, deposition of evaporites and several gaps in the Asselian–Artinskian interval.

The Early Permian barrier reef is represented in the modern relief by a chain of isolated hills located along a meridian from south to north (Figs 3, 4a) and a series of the buried reefs marks the western boundary of the Uralian foredeep. Reefs are characterized by different thicknesses, structures and composition. They include barrier reefs (Chuvashov 2011) and isolated mounds. The barrier reef embraced the single mounds of Shakhtau and Tratau (Figs 3a, 5a, c, d, h), the reef system of Uraktau (southern Urals) (Fig. 5e, f) and a number of subsurface reefs (e.g. the Ishimbai reef described by Maslov 1973).

Isolated mounds are known along the Kos'va River in the Kholodny Log section (Izart *et al.* 1999), Kozhim River (Antoshkina 1998, 2003) and Podcherem River in the Northern Urals. The coral and *Palaeoalysina* bioherms were studied from the Belaya Gora (Most) section in the slope deposits (Izart *et al.* 1999). Isolated carbonate banks and some frame reefs and skeletal mounds also occur in the subsurface Permian deposits of the Timan–Pechora province (Menner 1989), in the Kolguev Island in the Barents Sea shelf (Preobrazhenskaya *et al.* 1993, 1995) and in Northern Timan (Goreva *et al.* 1997; Kossovaya *et al.* 2002). In boreholes in Pechora Sea they were found in the Denisov depression and in the Dolginskyi and Sorokin uplifts. Based on seismic data the zone of reefs is also traced in the Kolva mega-uplift and in the Khoreiverskaya depression. They are

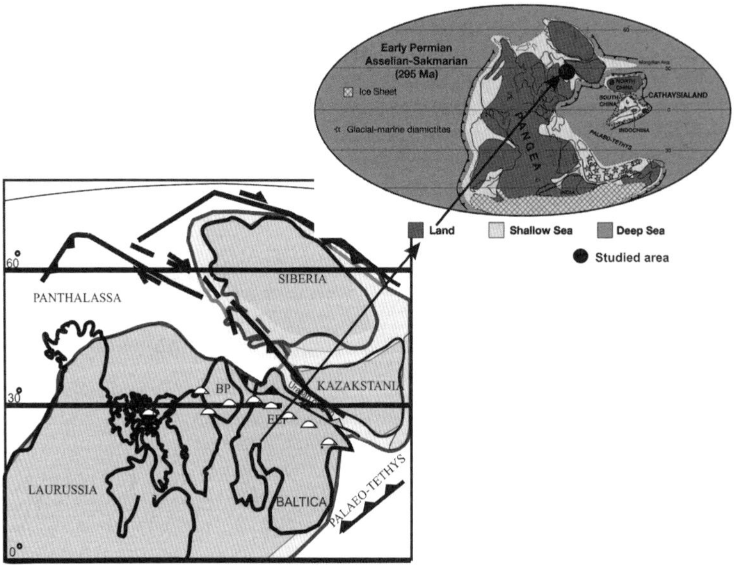

Fig. 2. Location of the area studied in schematic palaeotectonic map (based on Colpron & Nelson 2011; Metcalfe 2013). EEP, East European Platform; BP, Barents Platform.

interpreted as appearring in middle-outer shelf environments (Suvorova 2012). To the west an isolated reef is known in the eastern slope of the Northern Timan and a reef chain is shown along the Kola Peninsula.

A few bioconstructions were selected for this study (from north to south):

(a) The Schuchya frame reef, situated in the eastern slope of the Northern Timan (the western margin of the Timan–Pechora basin) at the mouth of the Schuchya River to the Sula River (Figs 4b, 5g).
(b) The Belaya Gora (Most) section, located on the left bank of the Kos'va River in the vicinity of the town of Gubakha in Permian administrative district. The frame reefs and reef system are situated in the vicinity of the town of Sterlitamak on the western slope of the Urals (Bashkortostan, Southern Urals) (Fig. 3a).
(c) The Divia skeletal mound (Figs 3b, 4d, e) is located in the vicinity of the town of Krasnoufimsk in the northern part of Bashkortostan at its boundary with the Chelyabinsk administrative region. The Roadian skeletal mound found in the Kremeshki quarry is situated in the Volga–Vyatka interfluves in the vicinity of the town of Sovetsk (Figs 3e, 4f).

General terminology of reefs and mounds

Riding (2002) proposed classification of reefs as a combination of the three main components: matrix (M), skeleton (S) and cavity/cement (C). Based on composition and structure, he recognized matrix-supported reefs (agglutinated microbial reef, cluster reef, segment reef), skeleton-supported reef (frame reef), cement-supported reef and carbonate mud-mound. Contrary to this, Antoshkina (2003) proposed a different classification where all constructions, regardless of their reef-builder type, shape and geographical position and so on, were joined in bio-constructions subdivided into organic build-ups and reefs. Each of these types was subdivided into smaller units. Organic build-ups were therefore subdivided into frame-organic build-ups (kaliptra, biogenic bank, biostrome, bioherm) and biogenic-accumulative build-ups (carbonate bank and mound; mud-mound, skeletal mound and microbial mound). Reefs are classified as either patch, isolated, fringing or barrier reefs. We have partly adopted this terminology, but emphasize that the many transitions between the various types do not allow us to strictly follow one classification (Dunham 1970). Some of the very variably built Early Permian organic build-ups of the Urals demonstrate a combination of skeletal growth and frame structures within the same construction. In addition, gradual transition between skeletal and mud-mound is widely known in late Artinskian constructions. Mud mounds were defined as either 'buildups with depositional relief composed dominantly of carbonate mud, peloidal mud, or micrite' (Bosence & Bridges 1995) or 'as reefs dominated by microcrystalline calcium carbonate' (Pratt 1995). There are no typical examples of mud-mound without participation of biotic components in our material, but some of the constructions with a dominant role of mud can be classified as mud-mounds.

To emphasize the importance of the biogenic nature of a given construction structure, we used the terms isolated frame reef (considered by Antoshkina 2003 as bioherms with maximum thickness of 25 m), barrier reef, reef system, skeletal mound, shoal, cemented mound for constructions in which cemented organisms have a prevailing role and mud-mound. According to the accepted classification, the focus is the construction role of the living organisms or the post-mortem accumulation via the destruction and cementation. This argument means it is necessary to adopt the definitions of the cement reef–reef created by cementation of essentially in-place organisms (Riding 2002, Appendix B).

Permian bioconstructions are classified as either: (1) a reef system; (2) an isolated frame reef with undeveloped/developed flanks or undifferentiated structure; or (3) a skeletal mound (probably close to the segment or skeleton segment reefs of Riding 2002) and its lateral facies, depending on the extent of accumulation of reef detritus. The change from frame-reef structures to skeletal mounds, with increasing significance of incrusting organisms, demonstrates the general evolution of bioconstruction in the Permian of the EEP.

Principal benthic carbonate associations

The principal benthic carbonate association is considered as the main key to understanding the climatic affinities of reef dwellings. To interpret the facies of the EEP, the difference between cool-water and warm-water associations must be defined. Lees & Buller (1972) defined a cool-water association, the foramol association (molluscs, foraminifera, crinoids and bryozoans) and a warm-water association, the chlorozoan association (the same bioclasts plus corals and calcareous green algae). James (1997) defined the photozoan association for shallow warm-water benthic calcareous communities and resultant sediments. The chlorozoan association is composed of green calcareous algae, invertebrates with corals and active abiotic precipitation (mud, ooids and cements). For the photozoan association, James (1997) proposed the chloralgal facies,

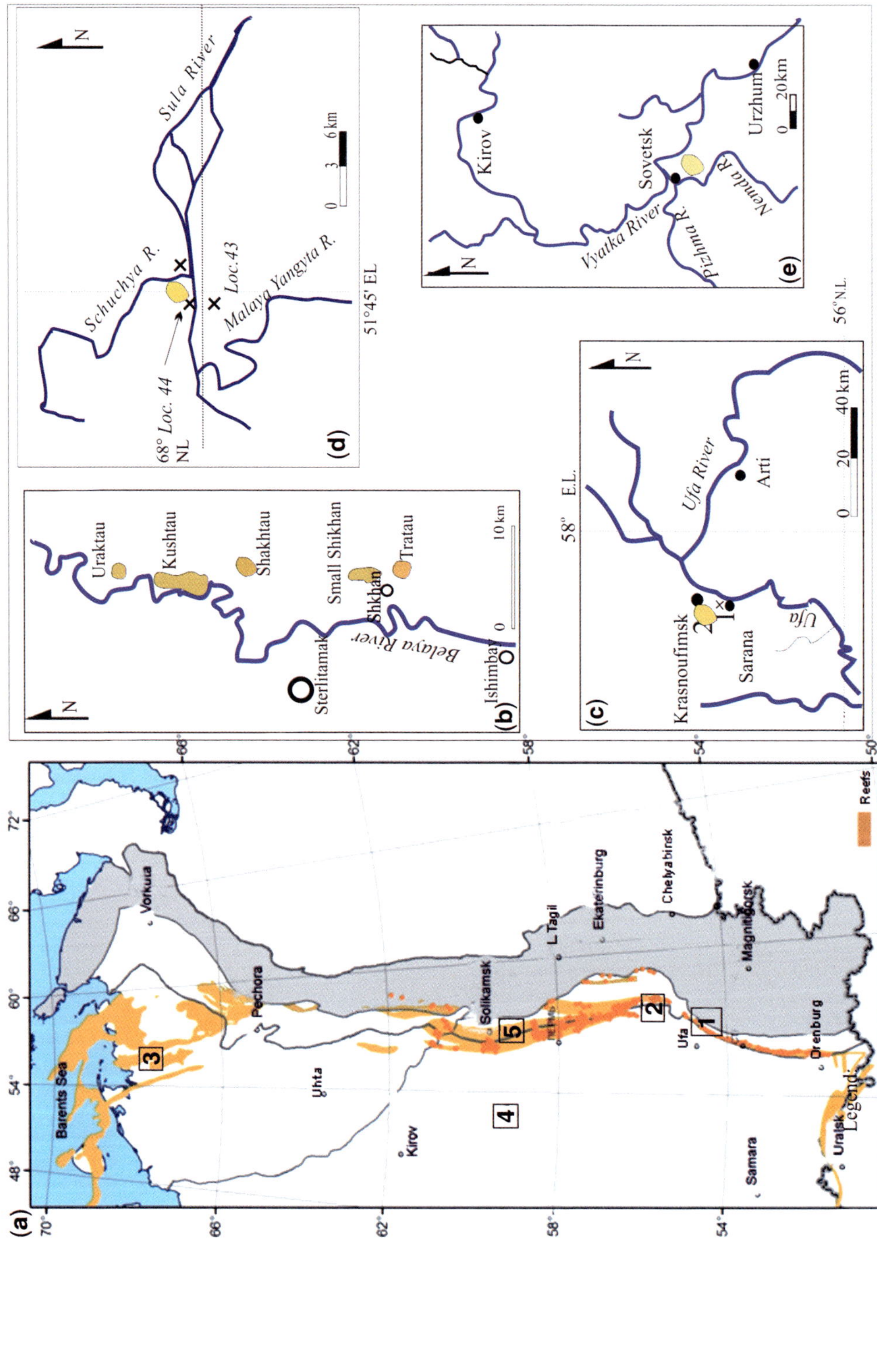
(a)
Barents Sea
Vorkuta
Pechora
Uhta
Kirov
Solikamsk
L.Tagil
Ekaterinburg
Chelyabinsk
Magnitogorsk
Ufa
Orenburg
Samara
Uralsk
Legend:
Reefs
1
2
3
4
5
(b)
Uraktau
Kushtau
Shakhtau
Small Shikhan
Shkhan
Tratau
Sterlitamak
Ishimbay
Belaya River
0
10 km
N
(c)
Krasnoufimsk
Sarana
Ufa
Ufa River
Arti
58° E.L.
0
20
40 km
N
(d)
Schuchya R.
Sula River
Malaya Yangyta R.
Loc.43
68° Loc. 44
NL
51°45' EL
0
3
6 km
N
(e)
Kirov
Sovetsk
Urzhum
Vyatka River
Pizhma R.
Nemda R.
0
20km
N
56°N.L.

oopeloid facies, fusulinid-algal facies and coral facies. Beauchamp (1994) defined a chloroforam association composed of green algae, fusulinids and small foraminifera observed from the Late Carboniferous, Asselian and Sakmarian of the Canadian Arctic and Middle Urals (Izart *et al.* 1999). For the cool-water association considered as heterozoan, James (1997) proposed the foramol facies (foraminifera + molluscs), the bryomol facies (bryozoans + molluscs) and the bryonoderm (bryozoans) facies. Beauchamp (1994) suggested bryonoderm-extended, bryonoderm and hyalosponge associations. The bryonoderm-extended association, composed of bryozoans, echinoderms, brachiopods and fusulinids, is observed in the Artinskian in the Canadian Arctic and at the top of the Sakmarian and the base of the Artinskian in the Middle Urals (Izart *et al.* 1999; Beauchamp & Baud 2002). The bryonoderm association, composed of bryozoans, brachiopods, echinoderms and solitary corals, is observed in the Artinskian–Kungurian in the Canadian Arctic and possibly the base of the Sargian in the Middle Urals (Izart *et al.* 1999). The hyalosponge association, composed of spicules of sponges, is observed in the Middle and Late Permian of the Canadian Arctic (Beauchamp *et al.* 2009). Reid *et al.* (2007) also included a heterozoan association without reference in their map of the Middle Permian in the Middle Urals. Samankassou (2002) described a cool-water carbonate in a palaeoequatorial shallow-water environment in the Permian of the Carnic Alps (Austria).

The heterozoan association is more widely distributed than cool-water only (e.g. Samankassou 2002); it is also typical for deep-water deposits and occurs in areas of upwelling. In a few cases the principal benthic carbonate associations are subdivided into the specific associations based on the predomination of the skeletal fragments.

Temperature and bathymetric control on large skeletal organisms

The composition of the benthic fauna and its mosaic distribution depend on the biotope bathymetry and temperature regime of the basin, but also play a substantial role in the formation of reef mound relief and development (Wilson 1980; Korolyuk 1985). In modern environments, massive colonial corals are mostly distributed in the photic zone of warm (tropical) zones. They sporadically occur in subtropical areas together with bryozoans, representing a mixed photozoan/heterozoan association of subtropical carbonates (Michel *et al.* 2011). Bryozoans are known from temperate waters in the modern seas off Tasmania (Rao 1993). The comparison made by James (1997) demonstrates a contrasting development of bryozoans and corals for the Cenozoic (James 1997, fig. 2). Late Artinskian bryozoans found in the western Urals are associated with a '*Cyathaxonia*' coral group characteristic of deep-water or cool-water facies (Kossovaya 2007). The co-occurrence of *Tubiphytes* with bryozoans and rather rare and small colonies of cerioid corals have been reported from the Tastubian reef crest (Korolyuk 1985), but most colonies were developed in the margin of the reef plateau where bryozoans are extremely rare. We consider that the association of colonial corals with bryozoans is not typical of the Lower Permian, and this regularity is useful for subdivision of heterozoan/photozoan associations. Although corals were not included in the list of Pre-Cenozoic characteristic bioclasts by James (1997), they were emphasized by Flügel (2004) and Kabanov (2009).

A wide distribution of brachiopods, fusulinids and colonial corals, typical of the warm-water basins referred to the photozoan biota and representing part of the thermophilic biota (Beauchamp *et al.* 2009), is limited by a pre-late Artinskian event (called the early Late Cisuralian turnover by Weidlich 2002 and Vachard & Vennin 2007). The warm-water environments include the following specific associations: (1) colonial corals and fusulinids; (2) fusulinids with *Tubiphytes*; (3) *Tubiphytes* with bryozoans; and (4) *Palaeoaplysina* in association with fragments of all mentioned fauna.

Temperate environments are characterized by bryozoans, solitary corals and (particularly for the Permian) solitary corals without dissepiments, poriferans and small foraminifers. Some faunal elements such as bryozoans, *Tubiphytes*, pelmatozoans and brachiopods are found in both warm and temperate environments. Microbial fabrics are known in a wide spectrum of environments but are rather rare in cool water. The hyalosponge association

Fig. 3. (**a**) Position of the reefs and mounds in the East European Platform: 1, Asselian–early Artinskian frame and skeletal-cemented mound in the southern Bashkortostan; 2, late Artinskian Divia skeletal mound, Krasnoufimsk district; 3, early Artinskian isolated frame reef in the Sula River valley, Northern Timan; 4, Roadian skeletal-mud mound, Nemda River, Kirov district; and 5, bioherms and tempestite in the Most section, Kos'va River. Location maps for (**b**) Shakhtau, Uraktau, Tratau, vicinity of the town of Sterlitamak; (**c**) Divia mountain – skeletal mound in the vicinity of Krasnoufimsk; (**d**) frame reef in the mouth of Schuchya River, eastern slope of the Northern Timan; (**e**) Kremeshki quarry with skeletal-mud mound in the vicinity of the town of Sovetsk.

(a)
reef-core
reef-breccia
(b)
(c)
(d)
(e)
(f)
reef-core

(Beauchamp 1994) is typical of cool-water environments in the middle and outer shelf of the carbonate platform.

Photozoan bioconstructions of the EEP: facies and biota

Middle Asselian – early Artinskian reef mounds of the EEP are mostly represented by frameworks constructed by different organisms referred to photozoan associations. The cement-producing organisms are *Tubiphytes* and some microbials. According to our interpretation, *Tubiphytes* is not just a cyanobacterium but an association between miliolid foraminifers and cyanobacteria (Vachard & Krainer 2001; Gaillot & Vachard 2007; Vachard *et al.* 2010).

Tratau isolated frame – cemented mound

The stratotype of the Shikhanian Regional Stage (Kotlyar & Stepanov 1984) in present relief is a hill with an altitude of 402 m (Figs 3a, b, 5a). It rises 270 m above the Belaya River valley and extends 1 km from north to south. It consists of massive, white and yellow limestone with abundant brachiopods, crinoids, rare ammonoids, fusulinids, the solitary coral *Amplexocarinia ruzhenzevi* Soshkina (Fig. 5b) and the cerioid rugose coral *Kleopatrina pseudoelegans* (Dobr.). A Tastubian age for the uppermost part of the section is still under discussion, but is supported by data obtained from Neptunian dykes. Clayey-carbonate deposits of Tastubian–Sterlitamakian (Sakmarian) age were reported from the Neptunian dykes (Chuvashov *et al.* 1996; Vennin 1997, 2007*a*) indicating that this reef was submerged in the Sakmarian, that is, at a time when neighbouring reefs were developing. The reason for this could be the deeper setting of the reef on the carbonate ramp which was ended by submersion during the subsequent transgression.

The lower part of the section consists of thick-bedded crinoid carbonate tempestite with ungraded brachiopod shells (Kossovaya 1987). The middle part of the reef is completed by massive boundstone (Fig. 5a) whereas the slope facies is characterized by packstone with *Tubiphytes* (Vennin 1997; Vennin *et al.* 1997), solitary corals and crinoids (Fig. 5b). The *Palaeoaplysina* and colonial rugose corals are mostly distributed on the outer reef slope. Some cyanobacterial(?) consortium such as *Archaeolithoporella* and *Tubiphytes* encrusted brachiopods and corals, and filled inner parts of microcaverns and other cavities (Fig. 5a). Radial calcite cement is also characteristic of the reef mound.

The mound is covered by red–brown marls with embryonic ammonoid shells, radiolarians and conodonts (e.g. *Neostreptognathodus pequopensis* Behnken) of Sargian age (Chuvashov *et al.* 1996), showing the abrupt submersion of the bioconstruction and change of biota.

Shakhtau frame reef

The Shakhtau reef is situated to the north of Tratau (Fig. 3a). The reef deposits form a hill which rises 210 m above the Belaya River. It differs from Tratau by the variability of the frame builders and facies, as described by Rauzer-Chernousova (1950), Shamov (1957) and Korolyuk (1985). The change of dominant facies has been traced from the middle Asselian to the Sakmarian sequentially. Maximal facies diversity is characteristic of the Tastubian (Sakmarian) and includes reef flats with maximal distribution of *Tubiphytes*, slope deposits with *Palaeoaplysina*-coral bioherms, becoming an island at the beginning of the Tastubian, and sandy shoals and brachiopod banks in the backreef. One of the most typical facies is a bryozoan framestone with bryozoan branches covered by cyanobacterial encrustations, some encrusting foraminifers and sparite (Fig. 5c). The other typical facies are *Palaeoaplysina* floatstone (Korolyuk 1985, plate VI, fig. 1) and *Tubiphytes*–bryozoan boundstone (Korolyuk 1985, plate V). Bryozoan–*Tubiphytes* boundstone similar to that of Shakhtau was reported from the Ishimbai buried reef massif (Maslov 1973, plate 2). Vennin (2007*b*, figs 32–40) demonstrated the variety of frame structures such as coral framestone or *Archaelithoporella* and *Tubiphytes* bindstone. Fasciculate and cerioid corals are rather abundant in Shakhtau (Fig. 5d). They occur as single colonies of up to 2–3 m, more often in the slope facies together with *Palaeoaplysina* (Korolyuk 1985). In the upper Asselian–Sakmarian, the assemblage also includes cerioid *Protowentzelella* sp., *Pr. noinskyi* (Porfiriev) and *Pr. simplex* Porfiriev.

Fig. 4. Frame reefs and skeletal mounds in the Central and South Urals: (**a**) General view of the Shikhan frame reef (one in the barrier reef in Bashkortostan); (**b**) isolated frame reef in the Sula River valley; (**c**) distal tempestite of the Irgina Formation and overlaying unnamed formation (Sargian) in the Most section, Central Urals; (**d**) Sargian reef deposits, Sarga Formation, Divia Mountains; (**e**) Divia Formation; and (**f**) Roadian deposits in the Kremeshki quarry (right) and the upper surface of the skeletal-mound at the top of the first floor (left).

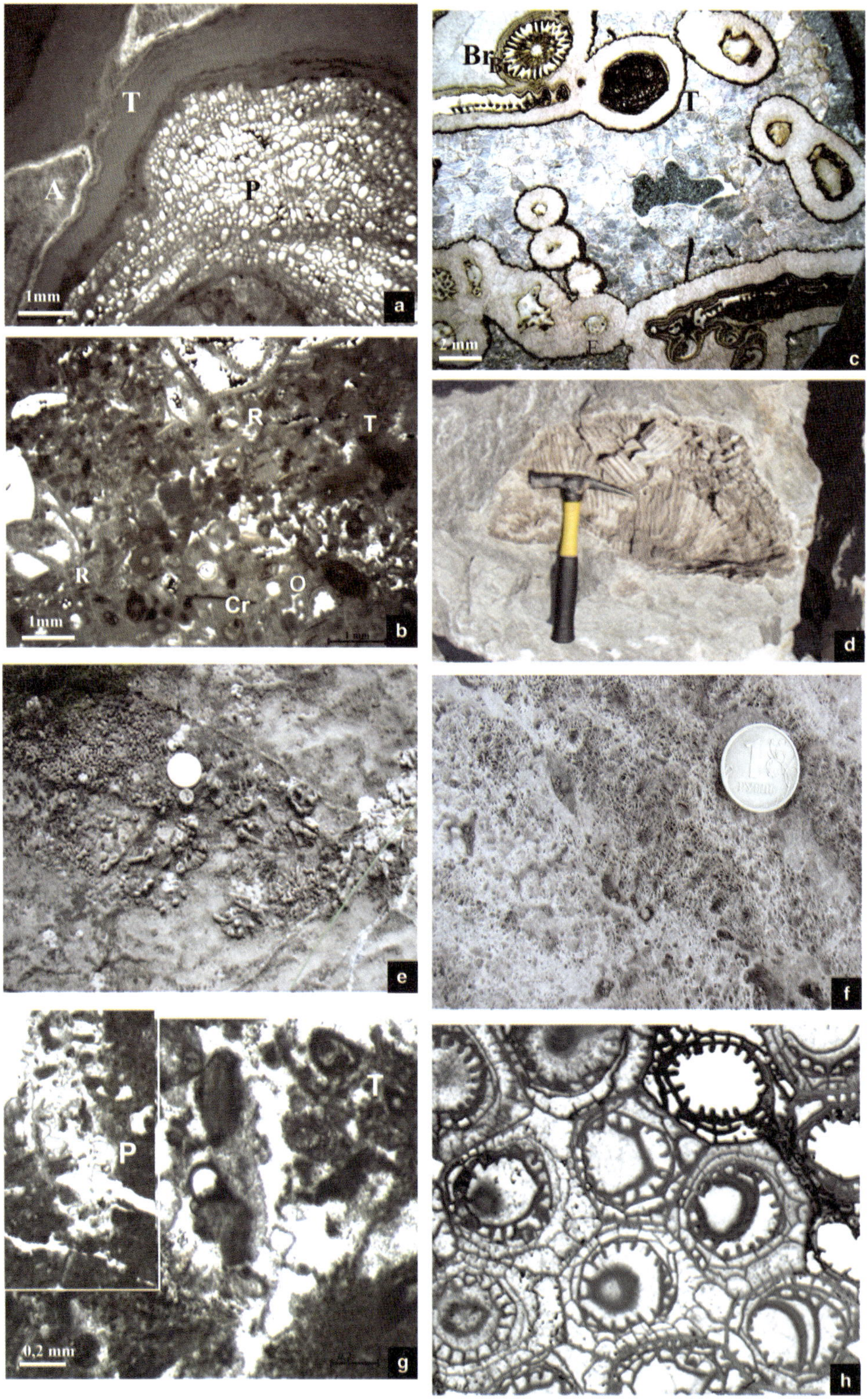
T
A
P
1mm
a
Br
T
2 mm
c
R
T
R
O
Cr
1mm
b
d
e
f
P
T
0,2 mm
g
h

Juraktau frame-reef system

The reef core is composed of massive white or yellow-white limestone including recrystallized crinoids and probably *Tubiphytes*. The transition from reef plateau to reef slope is well preserved in the upper flat part of the modern relief, where the framework consists of coral colonies overgrowing each other and preserved in their living position (Fig. 5e, f). Corals are represented by *Protowentzelella aseptatum* (Dobr.), *Pseudocystophora complexa* (Dobr.) and *Protolonsdaleiastraea juresanensis* (Dobr.). The bryozoan–foram packstone includes crinoids, brachiopod fragments, fenestellid bryozoan fragments, small forams and fragments of coral septa. The assemblage of small forams consists of *Calcitornella* sp., *Globivalvulina* ex gr. *bulloides* (Brady), *Palaeotextularia* sp., *Boultonia* ex gr. *willsi* Lee, *Nodosinelloides longa* (Lipina) and *Schubertella* sp. Fusulinids are represented by *Sakmarella* ex gr. *moelleri* of the Tastubian Regional Stage (Sakmarian). Astraeoid corals are characteristic of the upper part of this *Sakmarella moelleri* Zone. The back-reef deposits are represented by bedded yellowish limestone with numerous brachiopods.

Timan–Pechora Basin: carbonate banks, skeletal mounds and frame reefs

Build-ups considered mostly as carbonate banks are widely distributed in the Timan–Pechora basin. They are represented by *Tubiphytes* mounds, microbial or bryozoan banks in the Asselian and *Palaeoaplysina* mounds in the Sakmarian (Vazhenin & Ioffe 1986; Menner 1989; Larionova *et al.* 2002). The reefs and mounds are distributed along the depressions and long-living folds (Fig. 6). Asselian bioconstructions are known in Vuktyl and the Verkhnyaya Pechora Depression, and younger bioconstructions occur in the North Mylva area and to the west (Vazhenin & Ioffe 1986; Menner 1989). The continuation of the reef facies into the Barents Sea is demonstrated by the occurrence of reef structures in the eastern, western and northern parts of Kolguev Island. Frame builders are represented by *Tubiphytes*, *Palaeoaplysina* (less abundant) and bryozoa. The reef limestone and the reef breccia found in the core of boreholes in this island are assigned to the late Asselian–Sakmarian (Preobrazhenskaya *et al.* 1993, 1995). The age of reef deposits becomes younger to the west, where an early Artinskian isolated frame–skeletal reef has been studied on the eastern slope of Northern Timan (Schuchya River) (Figs 3c, 4b, 6).

One of the Asselian–Sakmarian mounds is situated on the eastern margin of the Timan–Pechora basin on the Kozhim River (Fig. 6, Antoshkina 1998). The Asselian rocks are lensoidal and consist of coarse-grained packstone, crinoid grainstone, fine-grained wackestone, fenestrate bryozoan–*Tubiphytes* boundstone and brachiopod-rich limestone. *Palaeoaplysina* is only present in a thin bed containing oriented fragments. Encrusting fibrous calcite cement is commonly associated with packstone containing *Palaeoaplysina* plates, (?)phylloid algae and fenestrate bryozoan fragments. Near the top of the mound there are thick beds of well-sorted skeletal packstone. The mound rocks are overlain by upper Sakmarian–Artinskian black shale, siltstone and sandstone (Antoshkina 1998).

Schuchya River frame–skeletal reef

The frame–skeletal reef is located in the estuary of the Schuchya River, a left tributary of the Sula River (Figs 3c, 4b, 6). The mound has a flattened shape with a length of *c.* 50 m and maximum thickness of 15 m. Stratigraphically, it is assigned to the upper part of the Nerma Formation of the early Artinskian (Goreva *et al.* 1997). It is overlain by mudstone and wackestone of the lower part of the Komicha Formation. The reef consists of the central reef core and surrounding brecciated flanks. The reef core is represented by *Palaeoaplysina* boundstone including small (up to 15 cm) flattened isometric colonies of *Iskutella sulensis* (Kossovaya 1997). The massive framestone is laterally changed by the breccia (thickness of the latter is *c.* 3 m). The breccia includes *Tubiphytes*, pellets, fragments of *Palaeoaplysina* and crinoids (Figs 5g, 6). The abrupt change of facies from reef deposits to clay-marl-clayey limestone suggests an unconformity within the Artinskian, corresponding to the final regressive part of the lower Artinskian cycle, seemingly absent from the section. The appearance of clayey limestones, marl and clays indicates the beginning of the new transgressive short-time cycle of the upper Artinskian.

Fig. 5. Main components of the photozoan biota: (**a**) *Tubiphytes–Palaeoaplysina* boundstone, Shikhanian, Tratau, southern Bashkortostan; (**b**) *Tubiphytes*–crinoid packstone with ostracods and rugose corals; (**c**) bryozoans–*Tubiphytes* boundstone, Shakhtau, polarized light; (**d**) colony of rugose corals in the Shikhanian, Shakhtau; (**e, f**) Sakmarian rugose corals forming bioherms in the upper reef slope facies of Uraktau frame reef; (**g**) reef-breccia in the frame reef in Sula River valley, Northern Timan; and (**h**) *Protowentzelella aseptatum* (Dobr.), magnification × 4, Juraktau. *Abbreviations*: A, Archaelithoporella; Br, bryozoans; T, *Tubiphytes*; P, *Palaeoaplysina*; Cr, crinoids; O, ostracods.

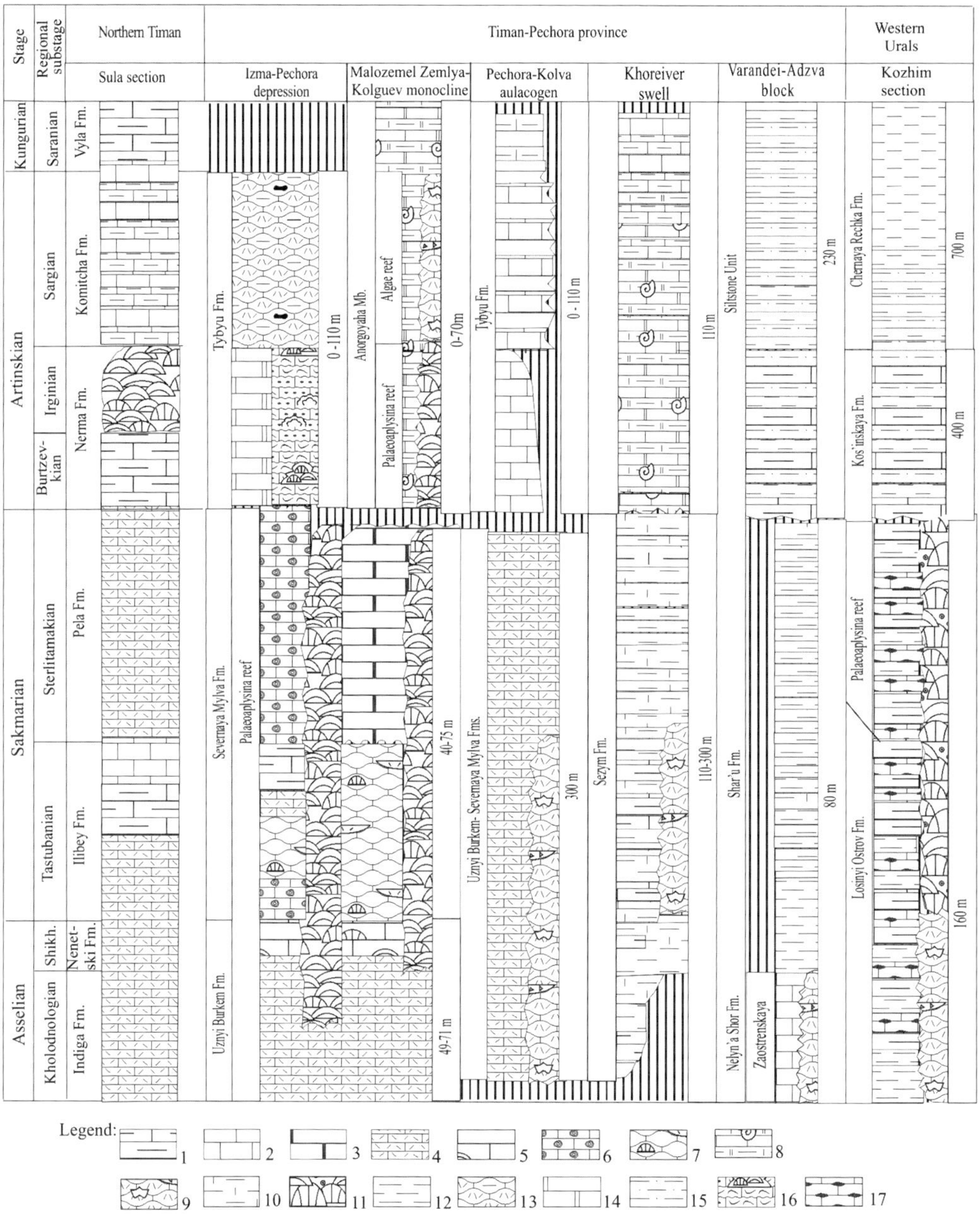

Fig. 6. Lithostratigraphic chart of the Lower Permian deposits of the Timan–Pechora basin (based on Preobrazhenskaya *et al.* 1993, 1995; Antoshkina 1998, 2003; Goreva *et al.* 1997; Larionova *et al.* 2000, 2002). Legend: 1, clayey limestone; 2, limestone; 3, thin-bedded limestone; 4, bioclastic limestone; 5, limestone with single *Palaeoaplysina*; 6, algae or *Tubiphytes* mound; 7, bedded limestone with colonial corals; 8, wacke-packstone with small foraminferans; 9, packstone with brachiopods; 10, siltstone; 11, *Palaeoaplysina* reef; 12, marl; 13, intercalation of bioclastic limestone and wackestone; 14, dolostone; 15, sandy mudstone; 16, intercalation of shelly limestone and coral-bearing limestone; 17, limestone with siliceous nodules or siliceous limestone.

Heterozoan association of the EEP skeletal mounds: detailed analysis

Late Artinskian skeletal mound and ramp facies

The extensive spreading of the late Artinskian transgression and the submersion of the shallow margins is reflected by abrupt changes of ramp and reef facies and replacement of biota.

Belaya Gora (Most) section

The replacement of biota in the slope deposits is demonstrated by the overlap of the Irginian siliceous tempestites (Figs 3, 4c) on the fusulinid-colonial coral facies of the late Sakmarian–early Artinskian (Burtzevskian) deposits of the Juresan' Formation in the Most section (Perm district, Central Urals). The proximal tempestite is characterized by a rather rare accumulation of fusulinids on the wavy bed surface and overturned small coral colonies. The deposits progressively changed to pack-wackestone with abundant solitary corals, bryozoans and brachiopods of the distal tempestite. The subsequent deepening and development of the maximal flooding surface are mirrored by the more condensed sedimentation of dark-grey or black spiculites with bryozoans, brachiopods and solitary Rugosa without dissepiments characteristic of the 'Cyathaxonia facies' (Izart *et al.* 1999; Kossovaya *et al.* 2001, 2002; Kossovaya 2007).

Divia skeletal reef mound

The replacement of photozoan by heterozoan biota in the reef facies has also been studied in the Divia Mountain (Krasnoufimsk region, Middle Urals, Figs 3d, 4d, 4e, 7; Table 1). Prior to the inception of reef-building origination the sedimentation process was influenced by an increasing transgression, probably starting in the late early Artinskian. The tempestite is subdivided into a proximal part with accumulation of fusulinid–bryozoan packstone (Irgina Formation) and a distal part with thin laminations of greenish silicified marls bearing abundant spicules (Divia Formation, early Sargian). The skeletal reef deposits were formed later in the late Sargian. From another point of view, the Sargian–Saranian age has been considered and the Divia Member was interpreted as a facies analogue of the Sargian reef deposits (Chuvashov 1995). Comprising two outcrops, the faunal list of the Irgina Formation includes *Pseudofusulina* ex gr. *pedissequa*, *Ps. uralensis*, *Skinerella (P.)* aff. *lutugini* and *Skinerella (P.)* ex gr. *solidissima* (Kossovaya *et al.* 2002). Sporadic brachiopods of the Divia Formation are represented by *Uraloproductus stuckenbergi*, *Reticulatia uralica*, *Ovatia ufensis* and *Anidantus aagardi*, typical of the Sargian (Kotlyar & Stepanov 1984). No difference was found in the brachiopod assemblages collected in the Divia and Sarga reef formations.

The bryonoderm principal association includes transitional biofacies (Fig. 7, beds 2–4) in the Irgina Formation, represented by fusulinid–bryozoan packstone with remains of brachiopod shells and small *Tubiphytes* (Fig. 7a, b), which are characteristic of the middle shelf. The higher diversity of bioclasts composed of large foraminifers, numerous bryozoans, small foraminifers and others is similar in the Raanes and Great Bear formations (Sverdrup Basin) of the Sakmarian–Artinskian. The latter has been assigned to the mixed photozoan–heterozoan association, characteristic of the transition from the subtropical to temperate zone (Bensing *et al.* 2008). The comparison between the overlying deposits of the Sverdrup and Uralian basins allows the absence of spicules from the Sverdrup basin and the great number of them in reef, outer ramp and platform facies in the EEP to be emphasized (Kossovaya *et al.* 2002).

The onset of a deepening phase led to the disappearance of fusulinids and deposition of spiculite–bryozoan packstone with small hyalosponge networks (Fig. 7c) assigned to the bryozoan–hyalosponge biofacies with predominant bryozoans. Further deepening of the basin led to deposition of spiculite deposits in the Divia Formation (Sargian substage) (Fig. 7d) of average thickness 15 m. The contemporaneous deposits of the Belaya Gora (Most) section are not as thick (*c.* 6 m) because of the incompleteness of this part of the section. Strong silicification was triggered by an input of silica from the hyalosponges. The skeletal reef mound in the upper Sargian is represented by fenestellid floatstone and abundant ramified bryozoans. Large productid brachiopods are typical. The appearance of *Microcodium* is characteristic of the uppermost part of the reef mound and is considered as bacterial activity in subaerial exposures (e.g. Blomeier *at al.* 2011) (Fig. 7j). These deposits are assigned to the Sarana Formation. One example of the late Artinskian 'Nizhne-Irginsk' bioherms (Vennin 2007*c*) shows widespread *Tubiphytes* together with fenestellid bryozoans with spicule-dominated packstone and wackestone in the inter-reef and off-reef facies.

The consequent development of the heterozoan reefs was interrupted by the post-Saranian (Kungurian) regression, when normal marine sedimentation was confined to the eastern part of the Timan–Pechora basin in the Kozhim and Pay–Khoy sections (Kotlyar *et al.* 2004).

Roadian skeletal mounds

Reefs and mounds appeared once more during the Roadian (Kazanian), probably via a transgression

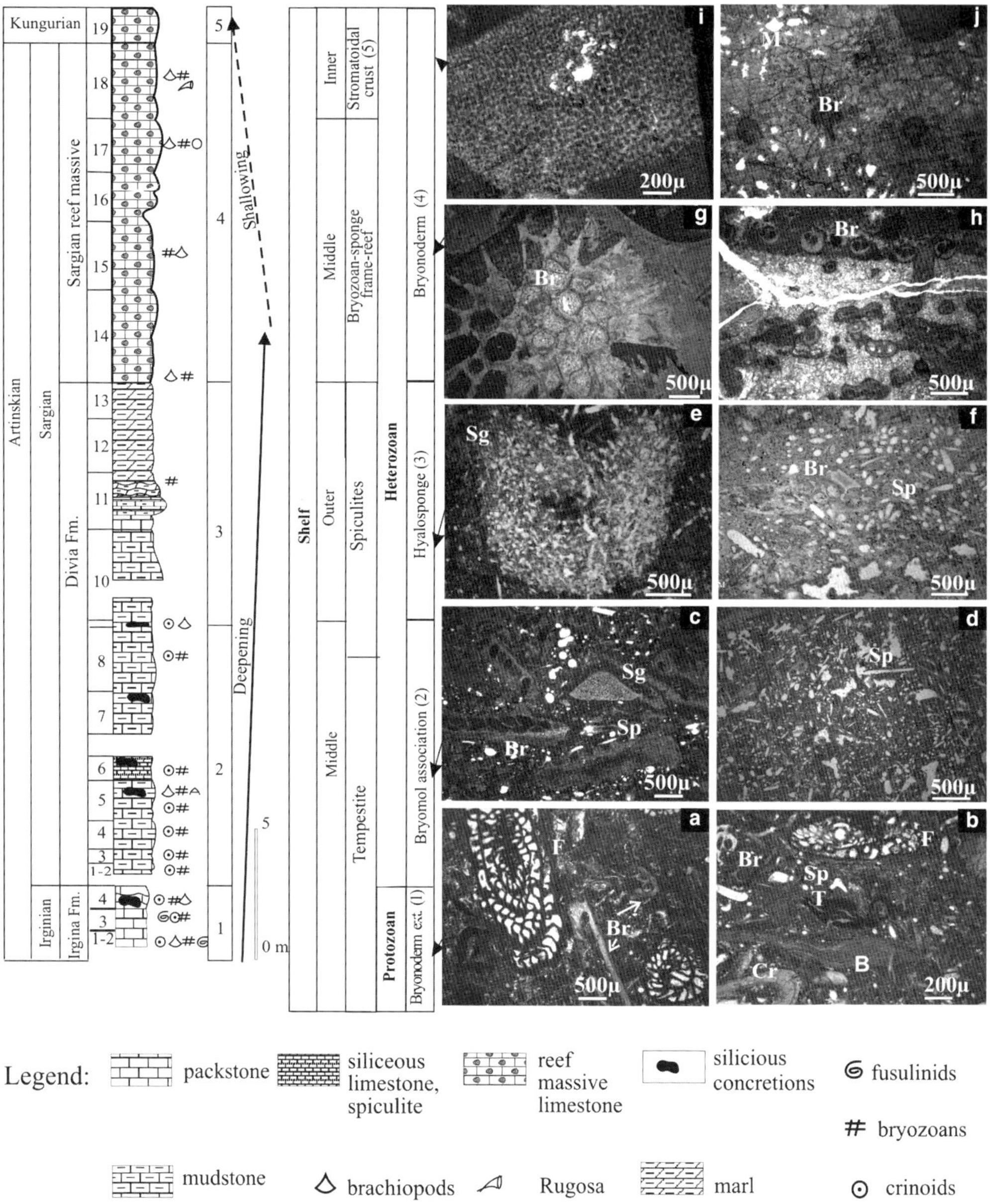

Fig. 7. Stratigraphic log and microfacies succession in the Divia Mountain. Beds 1–4: Irgina Formation; beds 1–18: Divia, Sargian reef formations; bed 19: Sarana Formation (lower part). (**a–f**) microfacies of the Irgina, Divia, Sargian reef and Sarana formations: photozoan association; Irgina Formation: bryonoderm association. (**a**, **b**) Bryozoans–fusulinids packstone with fragments of fusulinids (F), fenestellid bryozoa (Br), brachiopods (B), crinoids (C) and spicules of silicisponge (S); Heterozoan association, Divia Formation (2): Briomol biofacies (2). (**c**) Spiculate–bryozoan packstone of the lower part of formation in the Divia Mountain forming distal tempestite with graded lamination; fragment of the siliceous sponge (Sg); Hyalosponge biofacies (3). (**d**) Spiculitic packstone. (**e**) Accumulation of spicules considered as silicisponge; bryozoans are rare or absent; microfacies correspond to the deepest part of the basin and maximum of transgression. (**f**) Spicula–bryozoan packstone; bryozoans are rare or absent; bryonoderm association. (**g**) Fragment of fenestellid Bryozoa; (**h**) bryozoan packstone with zone of sparite in a birdseye vug; (**i**) secondary micritization of bioclasts; and (**j**) radial *Microcodium* around fenestellid bryozoans.

Table 1. *Facies of the Artinskian deposits of the Divia Mountains*

Formation	Type of facies	Description	Interpreted environments
Irgina (IF)	Bryonoderm-extended association. Fusulinid-bryozoan specific association (Fig. 7a, b)	Gray limestone; middle- to thick-bedded; fusulinids, pelmatozoans, bryozoans, spiriferid brachiopods, calcisponge spicules, micrite filling the space between bioclasts	Middle shelf subtropical to temperate, normal marine salinity, photic zone, middle sedimentation rate
Divia (DF)	Bryonoderm association (Fig. 7c, d)	Grey limestone with some brachiopods, small spicules, numerous bryozoans and rare hyalosponge	Outer shelf environments; cool temperature, normal salinity, below photic zone; high energy
	Hyalosponge association (Fig. 7e, f)	Dark green-grey silicified marls with numerous spicules and their association (hyalosponge)	Temperate or cool-water, normal salinity, middle sedimentation rate
Sargian reef (SF)	Bryonoderm association (Fig. 7g, h)	Light-grey–white fenestellid bryozoans floatstone with cavities filled with sparite; abundant brachiopods, lamellate and branching bryozoans	Temperate, normal salinity, middle or inner shelf
Kungurian	Stromatoidal crusts (Fig. 7i, j)	Light yellow and white boundstone with secondary micritization; radial *Microcodium* surrounding fenestellid bryozoans	Subaerial exposure

from the northern part of the Panthalassa Ocean. Marine carbonate deposits are known in the northern part of the EEP, to the south near the Onega River and in the Novgorod district and near the town of Sovetsk. The central parts of the Roadian skeletal mounds are constructed of bryozoan fragments. The section studied is situated in the SW of the Kirov district in the interfluves of the Vyatka and Nemda rivers, 10 km south of the town of Sovetsk (Fig. 3d). The stratigraphy of the section was studied recently (Zhuravlev *et al.* 2006; Kotlyar *et al.* 2007). The age of the deposits is supported by the occurrence of the ammonoids *Sverdrupites amundseni* Nassichuk, *Sv. harkeri* (Ruzhencev), *Biarmiceras esaulovae* Leonova, Kutygin & Shilovsky, *B. kremeshkense* Leonova, Kutygin & Shilovsky, *B. barskovi* Leonova, Kutygin & Shilovsky and *S. bogoslovskayae* Leonova & Shilovsky (Shilovsky 2007; Leonova & Shilovsky 2007) and therefore correlates with the deposits of the Assistance Formation of Arctic Canada. The Assistance Formation, characterized by the co-occurrence of *Sverdrupites amundseni and Sverdrupites harkeri* with the conodonts *Mesogondolella gracilis* and *M. idahoensis*, is correlated with the Roadian stratotype (Mei & Henderson 2002; Beauchamp *et al.* 2009). A few reef mounds are known in the Nemda River valley; one is situated on the right bank of the Nemda River (Solodukho 1954). The Chimbulat reef mound located nearby exhibits a dominant bryozoan component (Kotlyar *et al.* 2007).

The Kremeshki skeletal mound is an isolated composite structure located between shoal deposits (Fig. 8a). The boundstone of the central part includes a large proportion of skeletal organisms such as bryozoans and, more rarely, brachiopod fragments. The cementing organisms are encrusting foraminifera: *Palaeonubecularia* and *Tubiphytes* (Fig. 8b, f–h). Macrofauna, such as rather large crinoid fragments, solitary corals and some nautilids, occur in irregular accumulations. In some parts, marl lenses (20–50 cm thick) with abundant brachiopods were observed. The sediments in the upper part of the skeletal mound have been affected by cyanobacteria and are considered to be microbial crusts bearing *Microcodium*, small gastropods and bivalves. Coated grains are absent from this part of the section. The presence of *Microcodium* together with karst development is traditionally considered as evidence of periodical aerial exposure (e.g. Klappa 1978; Kabanov *et al.* 2008; Blomeier *et al.* 2011).

The lateral change is represented by cross-laminated oncoid grainstone. The wavy-surface-graded lamination is emphasized by accumulation of the larger oncoids in the lower part and smaller oncoids in the upper part of the bed. The differentiation was triggered by agitated wave conditions. Oncoids are of spherical shape, or more often

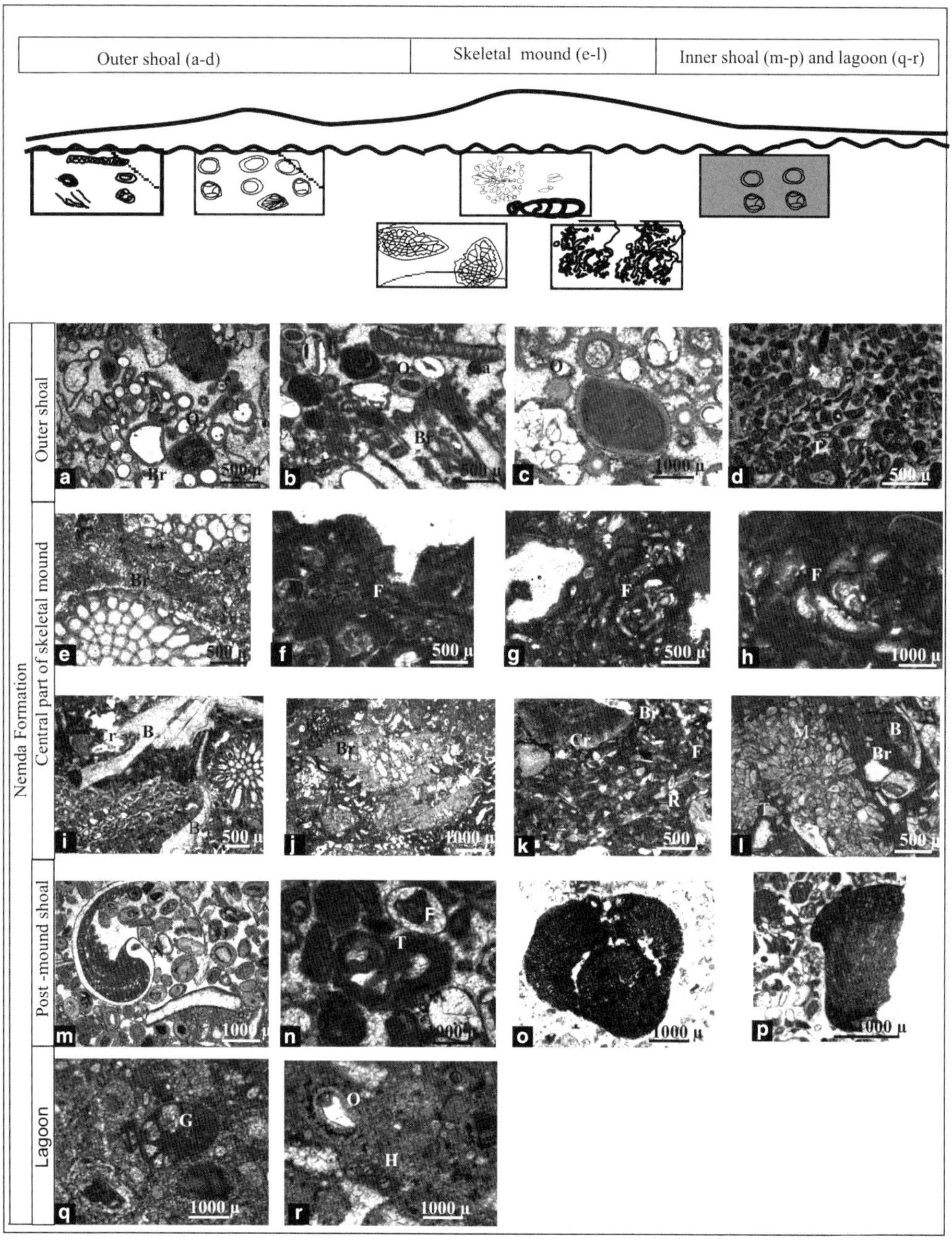

Fig. 8. Model of the facies distribution in Kremeshki skeletal-mound. (**a**–**d**) Microfacies of the outer shoal. Oncoid grainstone with bryozoans, algae, *Tubiphytes* fragments. The easily visible origination of oncoids around bioclasts-forams and bryozoans in the centre are surrounding by microbial film. (**e**–**j**) Central part of the skeletal mound consisting of bryozoan fragments incrusted by *Palaeonubecularia*. (**k**–**l**) Uppermost part of the skeletal-mound with temporal aerial exposure shown by the appearance of *Microcodium*. (**m**–**p**) Upper shoal covering mud-mound with an input of nautilids, red algae and *Tubiphytes*. (**q**, **r**) Second pulse of transgression; inner shoal and lagoon. Secondary micritization of the oncoid grainstone. B, brachiopod shell fragment; N, embryonal section of nautilidae shell; Br, bryozoans fragment; O, oncoids; M, *Microcodium*; F, forams; T, *Tubiphytes*; H, *Hemigordius*; G, *Globivalvulina*; Cr, crinoids; R, rugose fragment.

Table 2. *Facies of the Roadian deposits of the skeletal mound in the Kremeshki quarry*

Type of facies	Description	Interpreted environments
Outer shoal	Light-grey limestone with abundant brachiopods. Oncoid rudstone and grainstone. Oncoids formed around fauna fragments such as bryozoans and small foraminifers. Cyanobacterial crust is well developed. Cement: sparite (Fig. 8a–d).	Upper part of shelf with high wave energy, temperate, normal marine salinity, photic zone
Central part of skeletal mound	White limestone. bryozoan packstone and rudstone, bryozoan-foraminifera boundstone. Encrusting forams upon bryozoa (Fig. 8e–h).	Upper part of the shelf, middle wave energy, photic zone, temperate
Uppermost part of skeletal mound	White crinoid – bryozoan limestone in the upper part of the skeletal mound. Microbial facies with abundant *Microcodium* (Fig. 8i–l).	Shallow or subaerial exposure, temperate
Inner shoal and post reef environments	Light-yellow limestone. Grain- or packstone with juvenile nautilids, *Tubiphytes*, *Parachaetetes,* algae, etc. (Fig. 8m–p).	Shallow, warm-temperate, normal salinity, middle or inner shelf
Lagoon	Light yellow and grey pack-wackestone with foraminifers and secondary micritization (Fig. 8q, r).	Shallow, warm, low sedimentation rate

anisometric and tangential. The environment is interpreted as an outer shoal (Fig. 8a). The thickness of microbial crust is variable in oncoids and reaches a maximum of 200 μm in cross-section; *Tubiphytes* also occur (Fig. 8b). Weakly developed lagoon facies are characterized by secondary micritization and its presence in foraminifers and brachiopods (Fig. 8b, q, r).

The deposits overlying the reef mounds consist of coarse oncoid grainstone with pebbles and new input of typical marine fauna including fragments of juvenile nautilids, *Tubiphytes*, brachiopods and the red alga *Parachaetetes* (Fig. 8m–p). The abundant macrofauna consists of ammonoids, nautilids (Shilovsky 2007) and brachiopods. This post-reef upper shoal was formed during a new pulse of transgression, after the regression triggered the appearance of the microbial crusts on the skeletal mound surface. The replacement of facies in the reef mound and surrounding facies of the Kremeshki skeletal mound is represented in Table 2.

By the fauna components facies, this skeletal mound is assigned to the bryonoderm extended association. A few features allow the consideration that the Kremeshki skeletal mound and surrounding facies appeared in rather warm conditions.

Sedimentological and biotic events in the adjacent areas

Contemporaneous reefs are distributed along the Norwegian Sea coast (Elvebakk *et al.* 2002) and were described from Spitsbergen and Arctic Canada (Beauchamp 1989). In the late Asselian–Sakmarian they are principally constructed of photozoan organisms, whereas the heterozoan biota is prevalent in the construction of the upper Artinskian and Roadian skeletal mounds.

Sakmarian–Kungurian

The temperature gradient of the reefs and mounds located in the EEP can be compared to a gradual transition from a tropical (warm) to a subtropical and temperate (cool) environment, or temperature fall trigged by deepening of the basin. To estimate the scale of this phenomenon, the sedimentation in the transitional interval has to be considered. The temperature change in the Early Permian is described for Svalbard (Spitsbergen) by Hüneke *et al.* (2001) and Blomeier *et al.* (2011) and for the Canadian Arctic Archipelago by Morin *et al.* (1994) and Reid *et al.* (2007).

Contemporaneous deposits of the uppermost part of the Great Bear Cape Formation (GBC Formation, Sverdrup Basin) demonstrate the transition from subtropical (warm) to temperate (cool) environments (Bensing *et al.* 2008, table 1), corresponding to the *Mesogondolella bisselli – Sweetognathus whitei* Zone (late Sakmarian–early Artinskian according to the Permian Timescale, 2012). The typical *Neostreptognathodus pequopensis* is known from the GBC Formation in the Raanes Peninsula (Beauchamp & Henderson 1994), which is assigned to the Sargian. The transition was found in the Great Bear Cape Formation between levels four (GBC4) and five (GBC5) and is characterized by the change from the rugose–fusulinid-specific association to the pelmatozoan–bryozoan-specific association in shallowing-upwards facies. The cycle is ended by a *Microcodium* development in a semiarid environment (Bensing *et al.* 2008). Near the basinward extent of the GBC Formation, the sequence boundary subdivides the relatively shallow-water wackestone and packstone and much deeper-water silty and cherty spiculitic carbonates above. This upper sequence boundary is dated as late Artinskian (Beauchamp & Henderson 1994). The replacement in the Urals deposits and the change from photozoan biota to heterozoan took place near the Sargian boundary, a little later than in the Sverdrup Basin. The depositional succession seems to be different. The Uralian section represents a transition from a Sakmarian inner (or middle) shelf to a late Artinskian outer shelf, and a late Artinskian middle shelf to a Kungurian (Saranian) inner shelf.

Sponge–bryozoan build-ups are also known in the Artinskian of southwestern Ellesmere Island (Arctic Canada) in the lower part of the GBC Formation (Beauchamp 1989; Beauchamp & Henderson 1994).

Cool-water carbonates were reported from Greenland, Spitsbergen, the Finnmark Platform and Bjornoya (Stemmerik 1997) in the upper Artinskian–Kazanian interval. The Artinskian–Kazanian succession is divided into two second-order depositional sequences of Artinskian–early Kungurian and late Kungurian–Kazanian age, respectively. The Artinskian–early Kungurian second-order sequence is separated from the underlying sequence by an erosional unconformity with evidence of subaerial exposure in most areas. The hiatus separating the two sequences spans much of Early Permian time in the onshore areas of North Greenland and Svalbard, whereas it is of very short duration in more basinal settings.

On the Finnmark Platform and in the offshore areas the oldest sediments above the sequence boundary are of latest Sakmarian age. The early Artinskian platforms are composed of limestones with a bryoderm extended fauna dominated by bryozoans and crinoids, and include thick submarine cemented bryozoan–*Tubiphytes* build-ups. The basal parts of the Kim Fjelde Formation in North Greenland, the Kapp Starostin Formation on Spitsbergen and the Hambergfjellet Formation on Bjornoya are latest Artinskian in age, based on the occurrence of the conodont *Neostreptognathodus pequopensis*. The fauna, represented by the bryonoderm association (Stemmerik 2000), is known from the Sargian Substage in the Urals. The Kapp Starostin Formation was deposited in deep water. Its carbonate deposits, represented by bioclastic limestone banks with brachiopod and bryozoans, occur at the base of the Vøringen Member in the Festningen section. The other large bryozoan-dominated build-ups are known along the northern margin of the Finnmark platform on the Loppa Highland and elsewhere in the Barents Sea during the sea-level rise (Stemmerik 1997).

Roadian

The non-skeletal deposits of the Roadian found their equivalents in the deposition of ooids and oncoids and shallow submarine algae-bryozoan build-ups in East Greenland. Deposits belong to the Karstryggen and Wegener Halvø formations (Beauchamp & Baud 2002). Carbonate build-ups that formed in a more offshore setting in the Wegener Halvø Formation comprise a typical bryonoderm biota which probably indicates a deeper-water or cooler environment. In the Sverdrup Basin facies, the shallowest part of the Assistance Formation contains a bryonoderm association with locally abundant sponge spicules. Some similarity occurs for the shallow facies, but the faunal association with encrusting foraminifers and bryozoans in the Roadian deposits in the Nemda Formation is unique.

The Roadian mounds appeared during the last marine impulse fixed in the EEP after the Kungurian continental regime.

Extrinsic factors: climatic control, palaeolatitudinal position, bathymetry and circulation

The glacial period at the Asselian and early Sakmarian (P1) and the late Sakmarian–early Artinskian (P2; Isbell *et al.* 2012) coincides with the appearance and following radiation of the massive coral diversity and diversification of the photozoan bioconstructions in the Northern Hemisphere. To explain this paradox, the upwelling influence was used to interpret the distribution of so-called cool-water colonial corals in Western Laurentia; such mechanisms were also considered for the Uralian fauna (Clapham 2010). This assumption contradicts the warm-water photozoan association distributed in the late Asselian–early Artinskian reefs and mounds of the eastern margin of the EEP and epeiric sea of the central part of the EEP. The high latitudinal position of *Palaeoaplysina* and other photozoan reefs was discussed by Kiessling (2001). The lowermost position of *Palaeoaplysina* reefs found in North America was explained by the cool current along the eastern margin. The latitudinal position of bioconstructions was reported for corals (22°N), bryozoans (27°N) and microbes (17.6°N). Some space disunity between *Tubiphytes* and *Palaeoaplysina* reefs was suggested, but the South Urals frame reefs demonstrate the co-occurrence of these groups (Shakhtau) in the late Asselian–Sakmarian/early Artinskian; for the Artinskian bryozoan reefs however, the higher latitudinal position seems to be possible via the northern drift of Pangea.

The development of reefs and the replacement of reef builders were forced by several factors and processes, as proposed by combining various regional controlling factors. The distribution of coral faunas in the North American terranes and Spitsbergen display a great similarity on generic and species level to the coeval coral faunas in the Urals and Northern Timan forming the coral belt along the Pangea margin (Belasky *et al.* 2002). Northern latitudinal limitation of this coral belt and photozoan reefs was *c.* 30° or 40°N, occupying the subtropical zone of the Early Permian Uralian basin (situated within 30–60°N; Torsvik *et al.* 2012). Pangea drifted northwards at a rate of 2–3 mm a^{-1} from *c.* 20°N to 45°N during the Carboniferous and Permian (Scotese & McKerrow 1990; Golonka & Ford 2000), affecting not only the regions adjoining the Barents Sea but also the eastern shelf of Pangea settled by photozoan biota. Considering the palaeolatitudinal position of the photozoan reef belt within the subtropical zone, the appearance of massive colonial corals can be connected to the spreading of warm surface currents related to the eustatic fall of ocean level. The stabilization of warm environments over a wide area with rather high energy on the shelf was optimal for the reef growth.

Biogenic carbonate stable isotope results

The biogenic carbonate was extracted from septa of late Asselian–early Artinskian colonial corals from the Belaya Gora (Most) section. Material is limited by the last occurrence in bed 131 (Fig. 9). The fluctuation of the $\delta^{13}C$ values obtained from the biogenic carbonate in the Belaya Gora (Most) section shows an absence of the significant shifts until the middle Sakmarian (Kossovaya *et al.* 2010). The mean $\delta^{13}C$ value along the Shikhanian (late Asselian) section is in the range of 4–5‰ VPDB (Fig. 9). The reduction is demonstrated in the early Artinskian. A negative shift of $\delta^{18}O$ coincides with some transgression impulse (Izart *et al.* 1999, fig. 16) associated with the appearance of bioherms in the Belaya Gora (Most) section. The value of $\delta^{18}O$ in the Shikhanian–Tastubian fluctuated between −2‰ to −4.5‰ VPDB.

Results show a positive shift of $\delta^{18}O$ values in the Asselian-Sakmarian transition from −4.3‰ to −3.1‰ (between samples 30 and 32, Fig. 9), followed by a constant value. Increase of about 2‰ $\delta^{18}O$ at the Sakmarian boundary was reported recently for the Kondurovka sections (South Urals, slope facies) (Zeng *et al.* 2012, fig. 3). It coincides with depletion of $\delta^{13}C$ about 4‰ in the same section. Two positive duplets of $\delta^{18}O$ were found in the Sakmarian in Belaya Gora (Most): from −4.9‰ VPDB (bed 85) to −2.4‰ VPDB (bed 94 lower part) and from −3.3‰ (bed 103) to −1.7‰ VPDB (bed 105) (for bed numbers see Izart *et al.* 1999). The negative shift was found at the end of the Sterlitamakian (uppermost Sakmarian) coinciding with the gradual deepening of the basin and transition to outer ramp facies. The depletion of $\delta^{13}C$ is unremarkable (*c.* 1‰), but the $\delta^{18}O$ negative shift is *c.* 5‰ in the interval from

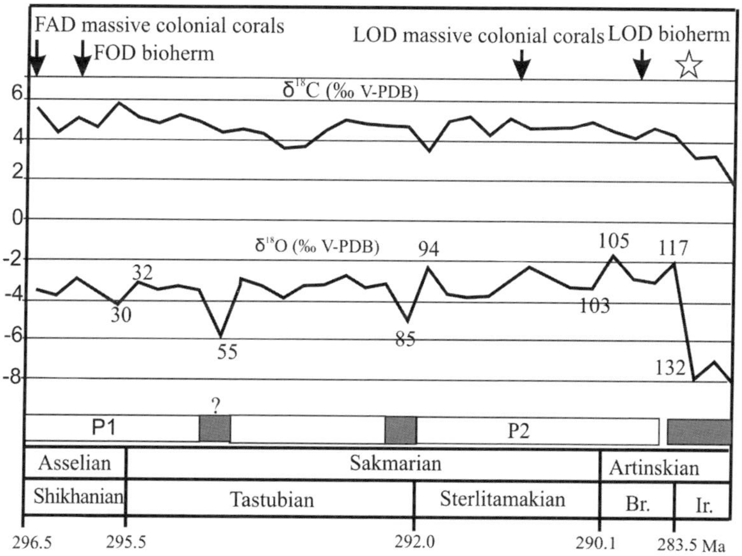

Fig. 9. Isotope data from the Belaya Gora (Most) section. The glaciation record from Fielding *et al.* (2008). Numbers on the curve are the bed numbers according Izart *et al.* (1999). Age is based on Gradstein *et al.* (2012).

bed 117 to 131. Similar tendencies (although different in absolute value) were reported from Dal'nyi Tulkas and Kondurovka sections (South Ural) (Zheng *et al.* 2012). The negative shift of δ^{18}O in the lower part of the Artinskian is comparable with data from the transition to an interglacial episode at the end of P2 glacial.

The stable isotope data for the Kremeshki section demonstrates a fluctuation from +2‰ to +4‰ VPDB of δ^{13}C for the Roadian (=lower Kazanian) part of section (Kuleshov & Sedaeva 2003). The 28–30‰ VSMOW of δ^{18}O value is permanent along the section with one deviation up to 32‰ VSMOW in the lower part of the section, which is higher than 21‰ VSMOW known from the Guadalupe Mountains (Texas) (Korte *et al.* 2005; Chen *et al.* 2013). δ^{13}C values of *c.* +2.8‰ VPDB and δ^{18}O values of −3‰ VPDB were obtained from the borehole bulk samples collected in the lower Kazanian deposits (Nurgalieva 2008). The δ^{13}C data correspond to the value obtained from the Guadalupe Mountains (Korte *et al.* 2005). The rather high values of δ^{18}O found in the carbonate samples from the Kazanian are correlated with the P3 glaciation in Eastern Australia (Fielding *et al.* 2008). It coincides with an increase in δ^{18}O shown for lower latitudes (Korte *et al.* 2005).

The hypothesis

The consequences of melting ice have to be considered as one of the reasons for eustatic rising of the ocean that could have caused the change of circulation and input of cold water to the basin, triggering the replacement of a warm surface current by a cold one. Another (and probably associated) factor is an upwelling triggered by the subsidence of the sea floor. One example of the widening of the range of heterozoan associations is their occurrence in modern cool-water carbonates located in a zone of oceanic upwelling along the African coastline in tropical latitudes (12–33°N) (Michel *et al.* 2011). Here, the main bioclasts within temperate zones are bivalves, foraminifers and echinoderms. The coastal cold current along the margin of the EEP may have been induced by an invasion of cold water from Northern Panthalassa. Upwelling leads to a decrease of general water temperature within the basin. This assumption contradicts the fact that the Coriolis force should direct the upwelling towards the eastern margin of the basin, which is characterized by turbidite or outer shelf accumulation depending on the position relative to the sea cost. An additional factor is connected to the closing of the gateway with the Tethys.

The input of cold water could be a reason for the regional cooling in the middle latitudes of Northern Panthalassa which was considered as one of the crucial factors of biota replacement (Beauchamp 1994; Bensing *et al.* 2008). The climatic change in the Early Permian was also documented in Svalbard (Hüneke *et al.* 2001; Blomeier *et al.* 2011; Hanken & Nielsen 2013), in the Canadian Arctic Archipelago (e.g. Morin *et al.* 1994; Reid *et al.* 2007) and in western Canada (Zubin-Stathopoulos *et al.* 2013). The abrupt eustatic rise of sea level was catastrophic for the shallow-water biota, but favourable for the ecological opportunists such as bryozoans playing the leading role as reef-builders in the late Artinskian. The subsequent northwards drift of Pangea seems to be one of the reasons for the facies changes that led to photozoan biota extinction.

Conclusions

(1) The evolution of photozoan reef biota in Northern and Eastern Pangea was interrupted during the middle–late Artinskian by global climatic impact trigged by indirect influence of ice melting in the Southern Hemisphere after the P2 glacial (Isbell *et al.* 2008, 2012). Modelling of the possible extrinsic factors using isotope data from the Urals has demonstrated the complex succession of abiotic changes including circulation changes and penetration of cold water from Northern Panthalassa. The invasion of cold water into the Uralian Basin led to disarray of the coastal circulation and rising of cold water via upwelling.

(2) The detailed facies succession reconstructed for the late Artinskian, Lower Permian and Roadian (Middle Permian) includes heterozoan skeletal mounds in both intervals. The essential difference was found in accumulative processes of constructions and biotic composition. The former, considered as deep-water constructions, appeared in the transgressive development of the basin and the latter developed in the short-term marine influx in the epeiric basin. The Roadian heterozoan skeletal mound contains the bryozoan-encrusting foraminifers association which was not observed in late Artinskian reef mounds. The absence of the hyalosponge association in the Roadian and development of the bryonoderm association indicates a shallower depth and warmer temperature than reported for the deeper and cooler environments of the distant northern part of the Panthalassa Ocean.

(3) The depletion of δ^{18}O during the early Artinskian was demonstrated by analyses of the biogenic carbonates from Belaya Gora (Most)

section. This coincides with the previously known trend for $\delta^{18}O$ shown for low latitudes from the Sakmarian to early Artinskian with a minimum during the middle Artinskian (Korte *et al.* 2008), and is in accordance with data from the South Urals (Zeng *et al.* 2012). The high value of $\delta^{18}O$ and value of $\delta^{13}C$ typical of marine water found for specimens from the Roadian in EEP are interpreted as a normal marine temperate-warm environment with higher evaporation formed during the last transgressive impulse in a semi-closed basin.

The participation of O. Kossovaya was supported by RFBR grants 11-05-00053a and 11-05-01162-a. We are very grateful to J. Pickett (Australia) for his comments and revision of English text. We thank A. Antoshkina (Russia) for her recommendation. We are also grateful to B. Beauchamp (Canada) for very attentive review and the valuable hard work of the editors. The authors are also very grateful to S. Schroeder (Germany) for his help with the final preparation of the manuscript.

References

ANTOSHKINA, A. I. 1998. Organic buildups and reefs on the Palaeozoic carbonate platform margin, Pechora Urals, Russia. *Sedimentary Geology*, **118**, 187–211.

ANTOSHKINA, A. I. 2003. *Reef Formation in Palaeozoic (Northern Part of the Urals and Adjacent Areas).* Uralian Branch of Russian Academy of Science, Ekaterinburg, Ekaterinburg, 1–302 (in Russian).

BEAUCHAMP, B. 1989. Lower Permian (Artinskian) sponge-bryozoan build-ups, Southern Ellesmere Island, Canadian Arctic Archipelago. *In*: GELDSETZER, H. J., JAMES, N. P. & TEBBUT, G. E. (eds) *Reefs: Canada and Adjacent Area.* Canadian Society of Petroleum Geologists, Calgary, **13**, 575–584.

BEAUCHAMP, B. 1994. Permian climatic cooling in the Canadian Arctic. *In*: KLEIN, G. D. (ed.) *Pangea: Palaeoclimate, Tectonics, and Sedimentation during Accretion, Zenith, and Breakup of a Supercontinent.* Geological Society of America, Boulder, Colorado, Special Paper, **288**, 229–245.

BEAUCHAMP, B. & BAUD, A. 2002. Growth and demise of Permian biogenic chert along northwest Pangaea: evidence for end-Permian collapse of thermohaline circulation. *Palaeogeography, Palaeoclimatology, Palaeoecology*, **184**, 37–63.

BEAUCHAMP, B. & HENDERSON, C. M. 1994. The Lower Permian Raanes. Great Bear Cape and Trappers Cove formations, Sverdrup Basin, Canadian Arctic: stratigraphy and conodont zonation. *Bulletin of Canadian Petroleum Geology*, **42**, 562–597.

BEAUCHAMP, B., HENDERSON, C. M, GRASBY, S. E., GATES, L. T., BEATTY, T. W., UTTING, J. & JAMES, N. P. 2009. Late Permian sedimentation in the Sverdrup Basin, Canadian Arctic: the Lindstrom and Black Stripe formations. *Bulletin of Canadian Petroleum Geology*, **57**, 167–191.

BELASKY, P., STEVENS, C. H. & HANGER, R. A. 2002. Early Permian location of western North American terranes based on brachiopod, fusulinid, and coral biogeography. *Palaeogeography, Palaeoclimatology, Palaeoecology*, **179**, 245–266.

BENSING, J. P., JAMES, N. P. & BEAUCHAMP, B. T. 2008. Carbonate deposition during a time of mid-latitude ocean cooling: Early Permian subtropical sedimentation in the Sverdrup basin, Arctic Canada. *Journal of Sedimentary Research*, **78**, 2–15.

BLOMEIER, D., DUSTIRA, A., FORKE, H. & SCHEIBNER, C. 2011. Environmental change in the Early Permian of NE Svalbard: from a warm-water carbonate platform (Gipshuken Formation) to a temperate, mixed siliciclastic-carbonate ramp (Kapp Starostin Formation). *Facies*, **57**, 493–523.

BOSENCE, D. W. J. & BRIDGES, P. H. 1995. A review of the origin and evolution of carbonate mud-mounds. *In*: MONTY, C. L. V., BOSENCE, D. W. J., BRIDGES, P. H. & PRATT, B. R. (eds) *Carbonate Mud Mounds, Their Origin and Evolution.* IAS Special Publications, Oxford, **23**, 3–9.

CHEN, B., JOACHIMSKI, M. M. *ET AL.* 2013. Permian ice volume and palaeoclimate history: oxygen isotope proxies revisited. *Gondwana Research*, **24**, 77–89.

CHUVASHOV, B. I. 1983. Permian reefs of the Urals. *Facies*, **8**, 191–212.

CHUVASHOV, B. I. 1995. Permian deposits of the Urals and Preduralye. *In*: SCHOLLE, P. A., PERYT, T. M. & ULMER-SCHOLL, D. S (eds) *The Permian of Northern Pangaea.* Springer-Verlag, Berlin, Heidelberg, New York, **2**, 158–183.

CHUVASHOV, B. I. 2011. Reef systems of the Uralian mobile belt and evolution of the reef building biotas. *In*: ROZHNOV, S. V. (ed.) *Reef Formations and Reefs in the Biosphere Evolution.* Palaeontological Institute of Russian Academy of Science, Moscow, 73–115 (in Russian).

CHUVASHOV, B. I., PRUST, J.-H., BUASSO, T., VENNIN, E. & CHERNYKH, V. V. 1996. To the history of origination of the Sterlitamakian shikhans (Early Permian reef massifs of the southern Cis-Urals). *Ezhegodnik-1995.* Zavaritsky Institute of the Geology and Geochemistry, Uralian Subdivision of RAS, Ekaterinburg, 25–37 (in Russian).

CLAPHAM, M. E. 2010. Faunal evidence for a cool boundary current and decoupled regional climate cooling in the Permian of western Laurentia. *Palaeogeography, Palaeoclimatology, Palaeoecology*, **298**, 348–359.

COLPRON, M. & NELSON, J. L. 2011. A Palaeozoic NW Passage and the Timanian, Caledonian and Uralian connections of some exotic terranes in the North American Cordillera. *In*: SPENCER, A. M., EMBRY, A. F., GAUTIER, D. L., STOUPAKOVA, A. V. & SØRENSEN, K. (eds) *Arctic Petroleum Geology.* Geological Society, London, Memoirs, **35**, 463–484.

DUNHAM, R. J. 1970. Stratigraphic reefs v. ecologic reefs. *AAPG Bulletin*, **54**, 1931–1932.

ELVEBAKK, G., HUNT, D. W. & STEMMERIK, L. 2002. From isolated buildups to buildup mosaics; 3D seismic sheds new light on upper Carboniferous-Permian fault controlled carbonate buildups, Norwegian Barents Sea. *Sedimentary Geology*, **152**, 7–17.

Fielding, C. R., Frank, T. D., Birgenheiler, L. P., Rygel, M. C., Jones, A. T. & Roberts, J. 2008. Stratigraphic imprint of the Late Palaeozoic Ice Age in eastern Australia: a record of alternating glacial and nonglacial climate regime. *Journal of the Geological Society, London*, **165**, 129–140.

Flügel, E. 1994. Pangaean shelf carbonates: controls and paleoclimatic significance of Permian and Triassic reefs. *In*: Klein, G. D. (ed.) *Pangea: Paleoclimate, Tectonics and Sedimentation during Accretion, Zenith and Breakup of a Supercontinent*. Geological Society of America, Boulder, Special Paper, **288**, 247–266.

Flügel, E. 2004. *Microfacies of Carbonate Rocks. Analysis, Interpretation, Application*. Springer, Berlin-Heidelberg-New York.

Flügel, E. & Flügel-Kahler, E. 1980. Algen aus den Kalken der Trogkofel-Schichten der Karnischen Alpen. *Carinthia II, Sonderheft*, **36**, 113–182.

Gaillot, J. & Vachard, D. 2007. The Khuff Formation (Middle East) and time-equivalents in Turkey and South China: biostratigraphy from Capitanian to Changhsingian times (Permian), new foraminiferal taxa, and palaeogeographical implications. *Coloquios de Paleontología*, **57**, 37–223.

Golonka, J. & Ford, D. 2000. Pangaean (Late Carboniferous–Middle Jurassic) paleoenvironment and lithofacies. *Palaeogeography, Palaeoclimatology, Palaeoecology*, **161**, 1–34.

Goreva, N. V., Kashik, D. S. *et al.* 1997. *The Key Section of the Carboniferous and Permian of the Northern Timan*. Transactions of All-Russian Research Geological Institute, Saint Petersburg. Nauka, **24** (in Russian).

Gradstein, F. M., Ogg, J. G., Schmitz, M. D. & Ogg, G. M. 2012. *The Geological Time Scale*. Elsevier, Oxford.

Hanken, N. -M. & Nielsen, J. E. 2013. Upper Carboniferous–Lower Permian Palaeoplysina buildups on Svalbard: the influence of climate, salinity and sea level. *In*: Gąsiewicz, A. & Słowakiewicz, M. (eds) *Palaeozoic Climate Cycles: Their Evolutionary and Sedimentological Impact*. Geological Society, London, Special Publications, **376**. First published online September 19, 2013, http://dx.doi.org/10.1144/SP376.17

Hüneke, H., Joachimski, M. & Buggish, W. 2001. Marine carbonate facies in response to climate and nutrient level: the Upper Carboniferous and Permian of Central Spitsbergen (Svalbard). *Facies*, **45**, 93–136.

Isbell, J., Fraiser, M. L. & Henry, L. C. 2008. Examining the complexity of environmental change during the Late Palaeozoic and Early Mesozoic. *Palaios*, **23**, 267–269.

Isbell, J. L., Henry, L. C. *et al.* 2012. Glacial paradoxes during the late Paleozoic ice age: evaluating the equilibrium line altitude as a control on glaciation. *Gondwana Research*, **22**, 1–19.

Izart, A., Kossovaya, O. L., Vachard, D. & Vaslet, D. 1999. Stratigraphy, sedimentology and sequence stratigraphy of the Early Permian along the Kosva River (Gubakha area, Central Urals, Russia). *Bulletin de la Société Géologique de France*, **170**, 799–820.

James, N. P. 1983. Reefs. *In*: Scholle, P. A., Bebout, D. G. & Moore, C. H. (eds) *Carbonate Depositional Environments*. AAPG, Tulsa, Memoir, **33**, 345–462.

James, N. P. 1997. The cool-water carbonate depositional realm. *In*: *Cool-Water Carbonate*, James, N. J. & Clarke, J (eds) SEPM, Tulsa, Special Publications, **56**, 1–20.

Kabanov, P. 2009. Benthic carbonate facies of the Phanerozoic: review and example from the Carboniferous of the Russian Platform. *Stratigraphy and Geological Correlation*, **17**, 5, 493–509.

Kabanov, P., Anadon, P. & Krumbein, W. E. 2008. *Microcodium*: an extensive review and a proposed non-rhizogenic biologically induced origin for its formation. *Sedimentary Geology*, **205**, 79–99.

Kiessling, W. 2001. Paleoclimatic significance of Phanerozoic reefs. *Geology*, **29**, 751–754.

Klappa, C. F. 1978. Biolithogenesis of *Microcodium*: elucidation. *Sedimentology*, **25**, 489–522.

Korolyuk, I. K. 1985. *Methods and Results of the Study of the Permian Reef Massif Shakhtau (Bashkirian Priural'e)*. AN SSSR, Nauka, Moscow (in Russian).

Korte, C. H., Jaspera, T. & Kozur, H. 2005. $\delta^{18}O$ and $\delta^{13}C$ of Permian brachiopods: a record of seawater evolution and continental glaciation. *Palaeogeography, Palaeoclimatology, Palaeoecology*, **224**, 333–351.

Korte, C. H., Jones, P. J., Brand, U., Mertmann, D. & Veizer, J. 2008. Oxygen isotope values from high-latitudes: clues for Permian sea-surface temperature gradient and Late Palaeozoic deglaciation. *Palaeogeography, Palaeoclimatology, Palaeoecology*, **269**, 1–16.

Kossovaya, O. L. 1987. Palaeogeographical and palaeoecological features of the Permian reef massives. *In*: *Modern and Ancient Reef Systems. Abstracts of the VI Symposium on Fossil Cnidaria and Reefs*. Vladivostok: 39–40 (in Russian).

Kossovaya, O. L. 1997. Rugosa of the type sections of the Gzhelian–Artinskian stages of Northern Timan and western slope of the Urals. *In*: *Atlas etalonnykh kompleksov paleozoiskoi bentosnoi fauny severo-vostoka Evropeiskoi Rossii*. VSEGEI, Saint-Petersburg, 34–53 (in Russian).

Kossovaya, O. L. 2007. Ecological aspects of Upper Carboniferous-Lower Permian 'Cyathaxonia Fauna' taxonomical diversity (the Urals). *In*: *Proceedings of the 9th International Symposium on Fossil Cnidaria and Porifera 2003*, Verlag der Österreichische Akademie der Wissenschaften, **17**, 383–406.

Kossovaya, O. L., Guseva, E. A., Lukin, A. E. & Zhuravlev, A. V. 2001. Middle Artinskian (Early Permian) ecological event: a case study of the Urals and Northern Timan. *Proceedings of the Estonian Academy of Science, Geology*, **50**, 95–113.

Kossovaya, O. L., Kotlyar, G. V., Shishlov, S. B. & Zhuravlev, A. V. 2002. Integrated approach to mid-Artinskian correlation. *In*: Hills, L. V., Henderson, C. M. & Bamber, E. W. (eds) *Carboniferous and Permian of the World*. Canadian Society of Petroleum Geologists, **19**, 753–775.

Kossovaya, O., Izart, A. & Malkovski, K. 2010. Permian reefs of the East European platform (evolution, crisises, biota). *In*: Rozhnov, S. V. (ed.) *Biota as Factor of Geomorphology and Geochemistry: Reef Formations and Reefs in the Evolution of Bioshere*. Palaeontological Institute of Russian Academy of Science, Moscow, 50–55 (in Russian).

KOTLYAR, G. V. & STEPANOV, D. L. (eds) 1984. *The Main Features of the Permian System.* Trudy VSEGEI, Leningrad, **286**, 1–273 (in Russian).

KOTLYAR, G. V., KOSSOVAYA, O. L., SHISHLOV, S. B. & ZHURAVLEV, A. V. 2004. Boundary between Permian series in variable sedimentary facies of north European Russia: constraints of event stratigraphy. *Stratigraphy and Geological Correlation*, **12**, 29–54.

KOTLYAR, G. V., ZHURAVLEV, A. V., KOSSOVAYA, O. L. & SHISHLOV, S. B. 2007. Sections of the Kazanian stage of the Nemda River (Volga-Vyatka region). *In*: SILANTIEV, V. V. (ed.) *Upper Palaeozoic of Russia: Stratigraphy and Palaeogeography*. Kazan State University, Kazan, 162–170 (in Russian).

KRAINER, K. 2007. *Late Paleozoic reef mounds of the Carnic Alps (Austria/Italy).* Geobios, **40**, 625–643.

KRAINER, K. & VACHARD, D. 2007*a*. Gzhelian *Anthracoporella* mounds, Carnic Alps Austria. *In*: VENNIN, E., ARETZ, M., BOULVAIN, F. & MUNNEKE, A. (eds) *Facies from Palaeozoic Reefs and Bioaccumulations.* Mémoires du Muséum National d'Histoire Naturelle, Paris, **195**, 283–286.

KRAINER, K. & VACHARD, D. 2007*b*. Sakmarian–Artinskian *Shamovella* (=*Tubiphytes*) *Archaeolithoporella* mounds, Carnic Alps Austria. *In*: VENNIN, E., ARETZ, M., BOULVAIN, F. & MUNNEKE, A. (eds) *Facies from Palaeozoic Reefs and Bioaccumulations.* Mémoires du Muséum National d'Histoire Naturelle, Paris, **195**, 291–294.

KULESHOV, V. N. & SEDAEVA, K. M. 2003. Isotopic composition ($\delta^{13}C$, $\delta^{18}O$) and depositional setting of Upper Kazanian carbonate sediments in the area between the Volga and Vjatka rivets (Russia). *In*: WONG, TH. E. (ed.) *Proceedings of the XVth International Congress on Carboniferous and Permian Stratigraphy*. Utrecht, The Netherlands, 2003. Royal Netherlands Academy of Arts and Sciences, 125–127.

KULIK, E. L. 1978. Izvestkovye zelenye (sifonovye) vodorosli asselskogo i sakmarskogo yarusov biogermnogo massiva Shakhtau (Bashkiriya). (Green calcareous algae (Siphonales) from the Asselian and Sakmarian stages of the biohermal massif of Shakhtau (Bashkiria)). *Voprosy Mikropaleontologii*, **21**, 182–215 (in Russian).

LARIONOVA, E. V., BOGATSKYI, V. I. & DOVZHIKOVA, E. G. 2000. *Timan-Pechora Sedimentation Basin. Atlas of the Geological Maps.* Timan-Pechora Scientific Centre, Uchra, Komi Republic.

LARIONOVA, E. V., BOGATSKYI, V. I. ET AL. 2002. *Timan-Pechora Sedimentation Basin. The Explanatory Record to the Atlas of the Geological Maps.* Kirov Regional Publishing House, 1–120 (in Russian).

LEES, A. & BULLER, A. T. 1972. Modern temperate-water and warm-water shelf carbonate sediments contrasted. *Marine Geology*, **13**, 67–73.

LEONOVA, T. B. & SHILOVSKY, O. P. 2007. The evolution of the Permian family Pirolegoceratidae (Goniatitida, Ammonoidea). *Paleontological Journal*, **1**, 27–36 (in Russian).

MASLOV, V. P. 1973. *Atlas of Tock-Building Organisms (Calcareous and Siliceous Organisms).* Nauka, Moscow (in Russian).

MCCREA, J. M. 1950. On the isotopic chemistry of carbonates and a paleotemperature scale. *Journal of Chemical Physics*, **18**, 849–857.

MEI, S. & HENDERSON, C. M. 2002. Conodont definition of the Kungurian (Cisuralian) and Roadian (Guadalupian) boundary. *In*: HILLS, L. V., HENDERSON, C. M. & BAMBER, E. W. (eds) *Carboniferous and Permian of the World.* Canadian Society of Petroleum Geologists, **19**, 529–551.

MENNER, V. V. 1989. *Lithological Criteria of the Oil-and-Gas Bearing Palaeozoic Deposits of the North-East of the Russian Platform.* Nauka, Moscow (in Russian).

METCALFE, I. 2013. Gondwana dispersion and Asian accretion: tectonic and palaeogeographic evolution of eastern Tethys. *Journal of Asian Earth Sciences*, **66**, 1–33.

MICHEL, J., VICENS, G. M. & WESTPHAL, H. 2011. Modern Heterozoan Carbonates from eutrophical shelf (Mauritania). *Journal of Sedimentary Research*, **81**, 641–655.

MORIN, J., DESROCHERS, A. & BEAUCHAMP, B. 1994. Facies analysis of Lower Permian platform carbonates, Sverdrup Basin, Canadian Arctic Archipelago. *Facies*, **31**, 105–130.

NURGALIEVA, N. G. 2008. Lithologicheskie aspecty issedovania struktury stratigraficheskoi zapisi permi vostoka Russkoi plity Avtoreferat dissertacii. Kazanian State University, 1–25. (in Russian).

PIA, J. V. 1937. Die wichtigsten Kalkalgen des Jungpaläozoikums und ihre geologische Bedeutung. *Compte Rendu du 2e Congrès Avancement Etudes de Stratigraphie du Carbonifère, Heerlen, 1935*, **2**, 765–856.

PRATT, B. R. 1995. The origin, biota and evolution of deep-water mudmounds. *In*: MONTY, C. L. V., BOSENCE, D. W. J., BRIDGES, P. H. & PRATT, B. R. (eds) *Carbonate Mud-Mounds. Their Origin and Evolution.* IAS Special Publications, Blackwell Science, Oxford, **23**, 49–123.

PREOBRAZHENSKAYA, E. N., USTRITSKY, B. N. & CHUVASHOV, B. I. 1993. Lower Permian reefs of the Kolguev Island (based on the deep-hole drilling). *Reports of Russian Academy of Science*, **329**, 1, 71–75 (in Russian).

PREOBRAZHENSKAYA, E. N., USTRITSKY, B. N. & BRO, E. G. 1995. Palaeozoic deposits of the Kolguev Island (Barents Sea). *Stratigraphy, Geological Correlation*, **3**, 75–85 (in Russian).

PUCHKOV, V. N. 2009. Structure of the Urals (with a special reference to the Carboniferous complexes). *Carboniferous type sections in Russia and potential global stratotypes. Southern Urals excursion. Ufa*, **2009**, 7–11.

RAO PRASADA, C. 1993. Carbonate minerals, oxygen and carbon isotopes in modern temperate bryozoa, eastern Tasmania, Australia. *Sedimentary Geology*, **88**, 123–135.

RAUSER-CHERNOUSOVA, D. M. 1950. Facies of the Upper Carboniferous and Artinskain deposits of Sterlitamak - Ishimbai Cisurals (based on fusulinids study). *Transactions of UGN AN SSSR*, **119**, 43, 1–109.

RAUSER-CHERNOUSOVA, D. M., BENSH, F. R. ET AL. 1996. *Reference Book on Systematic of Palaeozoic Foraminifera.* Nauka, Moscow (in Russian).

REID, C. M., JAMES, N. P., BEAUCHAMP, B. & KYSER, T. K. 2007. Faunal turnover and changing oceanography: Late Palaeozoic warm-to-cool water carbonates,

Sverdrup Basin, Canadian Arctic Archipelago. *Palaeogeography, Palaeoclimatology, Palaeoecology*, **249**, 128–159.

RIDING, R. 2002. Structure and composition of organic reefs and carbonate mud mounds: concepts and categories. *Earth-Science Reviews*, **58**, 163–231.

SAMANKASSOU, E. 1998. Skeletal framework mounds of dasycladacean alga *Anthracoporella*, Upper Paleozoic, Carnic Alps, Austria. *Palaios*, **13**, 297–300.

SAMANKASSOU, E. 2002. Cool-water carbonates in a palaeoequatorial shallow-water environment: the paradox of the Auernig cyclic sediments (Upper Pennsylvanian, Carnic Alps, Austria-Italy) and its implications. *Geology*, **30**, 655–658.

SCOTESE, C. R. & MCKERROW, W. S. 1990. Revised world maps and introduction *In*: MCKERROW, W. S. & SCOTESE, C. R. (eds) *Palaeozoic Palaeogeography and Biogeography*. Geological Society, London, Memoirs, **12**, 1–21.

SHAMOV, D. F. 1957. Facies of the Sakmarian–Artinskian deposits of the Ishimbai Cisurals. *Trudy of the Ufimian Oil Institute*, **11**, 3–77 (in Russian).

SHILOVSKY, O. P. 2007. *Kazanian Cephalopods of the Volga-Uralian region*. Autoreferat PhD Thesis, Kazanian State University, 1–16 (in Russian).

SOLODUKHO, M. G. 1954. Kazanian deposits of the Nemda River basin. *Memoir of Kazan University*, **114**, 127–143 (in Russian).

STEMMERIK, L. 1997. Permian (Artinskian-Kazanian) cool-water carbonates in North Greenland, Svalbard and the Western Barents Sea. Geological Survey of Denmark and Greenland. *In*: JAMES, N. P. & CLARKE, J. A. D. (eds) *Cool Water Carbonates*. SEPM, Tulsa, Special Publication, **56**, 349–364.

STEMMERIK, L. 2000. Late Palaeozoic evolution of the North Atlantic margin of Pangea. *Palaeogeography, Palaeoclimatology, Palaeoecology*, **161**, 95–126.

SUVOROVA, E. B. 2012. Lithology and sedimentary environments of the Visean-Lower Permian deposits of the Petchora off shore. *Neftegazovaya geologia. Teoria i practica*, **7**, 1–26.

TORSVIK, T. H., VAN DER VOO, R. *ET AL.* 2012. Phanerozoic polar wander, palaeogeography and dynamics. *Earth-Science Reviews*, **114**, 325–368.

VACHARD, D. & KRAINER, K. 2001. Smaller foraminifers, characteristic algae and pseudo-algae of the latest Carboniferous/Early Permian Rattendorf Group, Carnic Alps (Austria/Italy). *Rivista Italiana di Paleontologia e Stratigrafia*, **107**, 169–195.

VACHARD, D. & VENNIN, E. 2007. Permian. *In*: VENNIN, E., ARETZ, M., BOULVAIN, F. & MUNNEKE, A. (eds) *Facies from Palaeozoic Reefs and Bioaccumulations*. Mémoires du Muséum National d'Histoire Naturelle, Paris, **195**, 279–282.

VACHARD, D., PILLE, L. & GAILLOT, J. 2010. Palaeozoic Foraminifera: systematics, palaeoecology and responses to the global changes. *Revue de Micropaléontologie*, **53**, 209–254.

VAZHENIN, G. V. & IOFFE, G. A. 1986. Regularities of the localization of the organic build-ups in the Lower Permian carbonate complex of the Timan-Pechora province. *In*: BOGATSKYI, V. I (ed.) *Regularities in the Localization of Zones of the Oil-Bearing Deposits in Timan-Pechora Province*. VNIGRI, Leningrad, 80–86 (in Russian).

VENNIN, E. 1997. Architecture sédimentaire des Bioconstructions permo-carbonifères de l'Oural Méridional (Russie). *Mémoires de la Société Géologique du Nord*, Lille, **26**, 1–350.

VENNIN, E. 2007*a*. Coelobiontic communities in neptunian fissures of synsedimentary tectonic origin in Permian reef, southern Urals, Russia. *In*: *Palaeozoic Reefs and Bioaccumulations: Climatic and Evolutionary Controls*. Geological Society, London, Special Publications, **275**, 211–227.

VENNIN, E. 2007*b*. Archaeolithoporella-Shamovella (=Tubiphytes)–bryozoan reef-mound complexes, Russia. *In*: VENNIN, E., ARETZ, M., BOULVAIN, F. & MUNNEKE, A. (eds) *Facies from Palaeozoic Reefs and Bioaccumulations*. Mémoires du Muséum National d'Histoire Naturelle, Paris, **195**, 295–297.

VENNIN, E. 2007*c*. Artinskian bryozoan-phylloid algal-*Shamovella* bioherms, Russia. *In*: VENNIN, E., ARETZ, M., BOULVAIN, F. & MUNNEKE, A. (eds) *Facies from Palaeozoic Reefs and Bioaccumulations*. Mémoires du Muséum National d'Histoire Naturelle, Paris, **195**, 299–301.

VENNIN, E., VACHARD, D. & PROUST, J. N. 1997. Taphonomic and synecologic study of Tubiphytes. Examples from Tratau and Nizhni-Irginsk buildups from Lower Permian of the Urals Mountains. *Geobios*, **30**, 635–649.

WEIDLICH, O. 2002. Permian reefs re-examined: extrinsic control mechanisms of gradual and abrupt changes during 40 my of reef evolution. *Géobios*, **24**, 287–294.

WILSON, J. L. 1980. *Carbonate Facies in Geologic History*. Moscow, Nedra, 1–462.

ZENG, J., CAO, C., DAVYDOV, V. I. & SHEN, S. 2012. Carbon isotope chemostratigraphy and implications of palaeoclimatic changes during the Cisuralian (Early Permian) in the southern Urals, Russia. *Gondwana Research*, **21**, 601–610.

ZHURAVLEV, A. V., KOTLYAR, G. V. & SHISHLOV, S. B. 2006. Palaeobiogeographical and biostratigraphical analysis of the Kazanian (Middle Permian) conodonts of the east Russian platform – preliminary results. *Permophiles*, **48**, 15–20 (in Russian).

ZUBIN-STATHOPOULOS, K., BEAUCHAMP, B., DAVYDOV, V. I. & HENDERSON, C. 2013. Variability of Pennsylvanian–Permian Carbonate associations and implications for Northwest Pangea Palaeogeography, East-Central British Columbia, Canada. *In*: GĄSIEWICZ, A. & SŁOWAKIEWICZ, M. (eds) *Palaeozoic Climate Cycles: Their Evolutionary and Sedimentological Impact*. Geological Society, London, Special Publications, **376**. First published online November 26, 2012, http://dx.doi.org/10.1144/SP376.1

Carbon, oxygen and strontium isotopic composition of low-latitude Permian carbonates (Venezuelan Andes): climate proxies of tropical Pangea

JUAN CARLOS LAYA[1,2]*, MAURICE E. TUCKER[1,3], DARREN R. GRÖCKE[1] & ALBERTO PEREZ-HUERTA[4]

[1]*Department of Earth Sciences, Durham University, Durham DH1 3LE, UK*

[2]*Grupo de investigaciones de ciencias de la tierra – TERRA, Universidad de Los Andes, Mérida, Venezuela*

[3]*School of Earth Sciences, University of Bristol, Bristol BS8 1RJ, UK*

[4]*Department of Geological Sciences, University of Alabama, Tuscaloosa, AL 35487, USA*

**Corresponding author (e-mail: layajc@ula.ve)*

Abstract: The Late Palaeozoic period was an important time in Earth history, primarily as a result of the formation of Pangea and the Gondwana glaciations that had direct influences in high latitudes and indirect, but no less important, effects in low latitudes. This study documents and interprets the isotopic record for the Early–Middle Permian in northern Gondwana from detailed fieldwork in the Venezuelan Andes. Sr-isotope dating of well-preserved brachiopod shells suggests a Kungurian age from samples located within the middle part of the Palmarito Formation. $\delta^{13}C$ data show a long-term trend through the section towards more positive values, but with some very negative $\delta^{13}C$ values suggesting carbonates affected by pedogenic processes. In the middle part of the succession, major excursions in $\delta^{13}C$ and $\delta^{18}O$ show a strong connection with glacial–deglacial events during the Early–Middle Permian. Overall results presented here fill an important gap in the regional palaeogeography and, therefore, have significant implications for the palaeoclimate and palaeoceanography of this Late Palaeozoic time.

The assembly of Pangea during the Permian triggered important geological, oceanographical and climatic events, which had a direct influence on the geological record. Many isotope studies of Permian rocks from the equatorial region of Pangea have been undertaken during the last two decades, primarily focusing on understanding palaeoceanography and palaeoclimate (e.g. Tabor & Montañez 2002; Isbell *et al.* 2003; Brand 2004; Korte *et al.* 2006; Montañez *et al.* 2007; Tabor & Poulsen 2008; Fielding *et al.* 2008; Peyser & Poulsen 2008). As a result, palaeoclimate models have been developed showing reconstructions of palaeoprecipitation, palaeotemperature and the variation in position of the Intertropical Convergence Zone (ITCZ) during the Permian. However, there is still insufficient information about regions around northern Gondwana, particularly for Central and South America.

In northern South America, Upper Palaeozoic strata were deposited extensively over pericratonic areas but in many places they are rarely exposed and so are poorly documented. This is largely the result of extensive weathering and dense vegetation in the tropical Andes, as well as the remoteness of the outcrop areas. However, these strata do have significant information for the palaeogeographical and palaeoclimatic reconstructions, and for the evolution of northern Gondwana, at this critical time in Earth history. The existence of a seaway connecting the Palaeotethys and Panthalassa oceans, which lasted until the late Early Permian, has been suggested by just a few authors (Davydov & Leven 2003; Izart *et al.* 2003; Vai 2003). This deduction has been based on palaeontological evidence, including Tethyan brachiopod and fusulinid distributions. An east–west seaway across northern Gondwana accounts for the shallow-marine sedimentary succession across western Venezuela during the Permian. These earlier publications have not described the depositional setting of this Venezuelan basin or how it was connected and related to the neighbouring oceans. Figure 1, modified from the map for 280 Ma of Blakey (2007), shows the palaeogeography following the reinterpretations of Laya & Tucker (2012). Permian rocks in Venezuela have previously been described in several studies (Arnold 1966; García 1972; Odreman & Wagner 1979; Bellizia 1992; Viscarret & Laya 2007). The

From: GĄSIEWICZ, A. & SŁOWAKIEWICZ, M. (eds) 2013. *Palaeozoic Climate Cycles: Their Evolutionary and Sedimentological Impact*. Geological Society, London, Special Publications, **376**, 367–385.
First published online February 28, 2013, http://dx.doi.org/10.1144/SP376.10

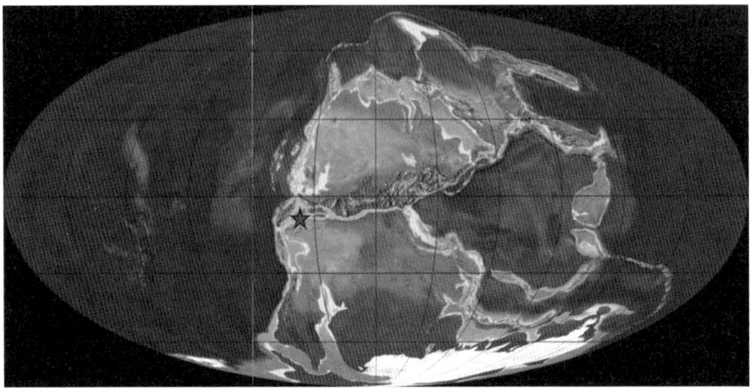

Fig. 1. Mollweide projection for the Early Permian Artinskian stage (280 Ma) modified from Blakey (2007). The red star (black in print version) represents the palaeogeographical location of the Palmarito Formation.

facies and microfacies of the Palmarito Formation, which crops out throughout the Venezuelan Andes, suggest that there was a seaway connecting central Pangea with basins to the NW, and also a connection to the SW with basins in the Bolivian–Peruvian region (Laya & Tucker 2012). This would have implications for the regional climate in terms of land–sea distribution, supercontinentality, uplift–collapse of the Central Pangean Mountains and the source of moisture.

The Palmarito sediments were deposited on a carbonate ramp that dipped basinwards towards the north facing the open ocean. The formation evolved from the underlying fluvial deposits (Sabaneta Formation) through to tidal-flat and to mid-outer ramp deposits, with all facies recording a well-developed cyclicity. The aims of this study are to present and discuss the stable isotope record in the context of the sedimentology of the Palmarito Formation located in western equatorial Pangea, and to consider these data in terms of climatic and oceanographical regimes in these low latitudes during the Permian. The Permian succession in Venezuela covers a poorly known region between Gondwana and Laurentia, before the final assembly of Pangea.

Geological background

The Palmarito Formation is part of the Palaeozoic tectonostratigraphical unit in western Venezuela that has been referred to as a supraterrane by Bellizia (1992), with sediments deposited within the Mucuchachí Basin, a name introduced by Arnold (1966). Sedimentation of Permian carbonates seems to have been a widespread phenomenon in NW Venezuela, with likely extensions and correlations through Colombia to the SW, then to Peru and Bolivia farther south, but also northwards to Guatemala, Mexico and to the southern USA (Laya & Tucker 2012).

Lithostratigraphy of the Late Palaeozoic in the Venezuelan Andes

During the 1950s–1970s, fieldwork and biostratigraphical studies connected with petroleum exploration by González De Juana (1951), Pierce *et al.* (1961), Shell de Venezuela & Creole Petroleum Corporation (1964) and Shagam (1972) resulted in different lithostratigraphical and palaeogeographical interpretations. However, these studies did not provide an accurate chronostratigraphy to establish the age of the strata to the stage level.

The Mucuchachí Formation is the oldest unit in the area that has been proposed to date from the Devonian to Lower Pennsylvanian by González De Juana (1951), Pierce *et al.* (1961), Shell de Venezuela & Creole Petroleum Corporation (1964) and Arnold (1966) on the basis of field relationships and regional lithostratigraphical correlations across the area. However, there was no strong biostratigraphical evidence to support this interpretation. Later, Odreman & Useche (1997) proposed that the age is restricted to the Pennsylvanian based on palaeontological evidence.

The Mucuchachí Formation consists of slate and fine-grained quartzite that was affected by regional metamorphism, perhaps in the late Pennsylvanian (Laya & Tucker 2012). These rocks form the basement in the region and are overlain by the sedimentary cover of the Sabaneta and Palmarito formations.

The Sabaneta Formation (Middle Pennsylvanian) is composed of conglomerates and reddish sandstones with channel structures, and these grade into thinly bedded red mudrocks with palaeosoils of floodplain origin (Arnold 1966; Viscarret 2002).

These continental deposits of the Sabaneta Formation pass up into marine sediments of the Palmarito Formation, which are attributed to the latest Carboniferous through early Middle Permian age. This lithostratigraphical unit, which is the focus of the present study, begins with a package of heterolithic calcareous sandstone, mudstone and siltstone with ostracods and plant debris of a tidal-flat environment. This grades upwards to dark-grey calcareous deeper-ramp deposits. Further up, the succession becomes more shallow marine with calcareous algal packstone and fusulinid grainstone, that represent inner-ramp deposits, and then crinoidal–bryozoan wackestone of the middle to deep ramp with local buildups. Deeper-ramp facies are interbedded with dolomites, and the formation finishes with lime mudstones and evaporites. (Laya & Tucker 2012).

The chronostratigraphy of the Palmarito Formation is based on fusulinid species that were identified by González De Juana (1951) and Sellier de Civrieux (1951), and include *Parafusulina* and *Globivalvulina*. These species were considered characteristic of the Upper Guadalupian, probably Capitanian, in west Texas. Shell de Venezuela & Creole Petroleum Corporation (1964) provided additional information about the rocks that contained the fusulinids and proposed one lithostratigraphical unit but with a split into two members that were formally described by Arnold (1966). Furthermore, some correlations were made by Hoover (1981), who described several species of brachiopod from the Portachuelo area. Several sections from here were used in this study. On the basis of these palaeontological data, Hoover (1981) proposed a relationship between Venezuela, Colombia, Ecuador, Bolivia, México, Guatemala, the USA and Argentina. However, Hoover (1981) and Rigby (1984) did not present any stratigraphical logs of the sections nor show the stratigraphical positions of the fossils described. In most cases the samples were collected from exotic blocks on a mule trail.

Sections

For this study, four sections in the Portachuelo area were sampled. These are located to the south of Merida city. Figure 2 shows the geographical location of the succession in the context of Venezuelan Andes.

- *Santa Cruz de Palmarito* is located 100 m south of a small chapel called 'Santa Cruz de Palmarito' along a mule trail; the section has been described by Arnold (1966) and Hoover (1981). The succession consists of approximately 100 m of fossiliferous massive and nodular limestones

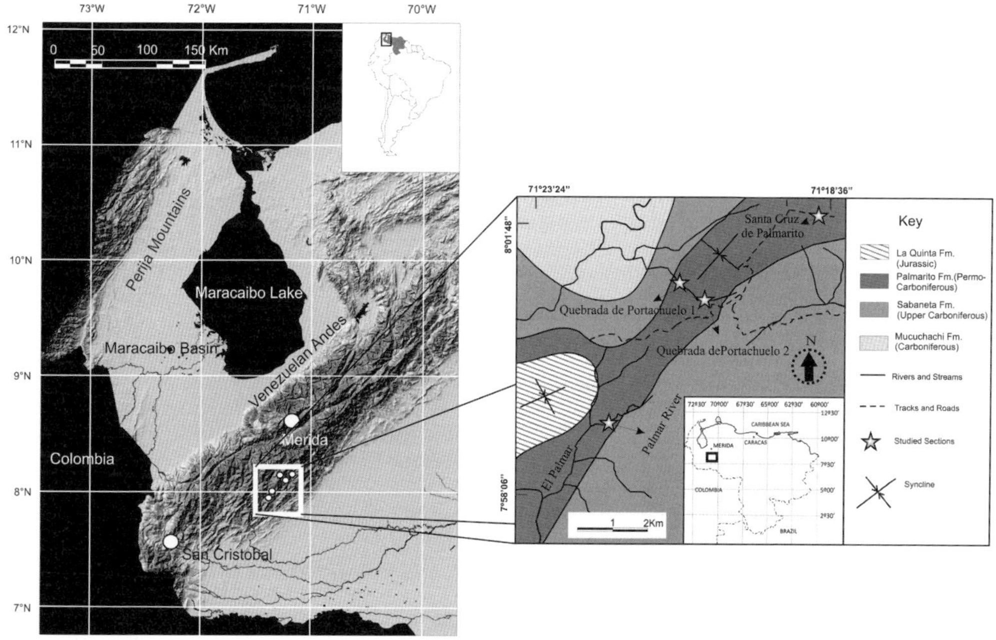

Fig. 2. Digital shaded relief image of western Venezuela showing the location of the main area of study. Modified from Garrity *et al.* (2010) and a detailed geological map of the Portachuelo area of the Venezuelan Andes.

that are assigned to the middle–upper part of the formation.

- *Quebrada de Portachuelo 2* comprises outcrops from the head of the stream towards the west, which is located on the eastern flank of a syncline. The succession shows a younging-direction that is parallel to the stream flow and opposite to that in the Quebrada de Portachuelo 1 Section. The middle and upper part of the Palmarito Formation is exposed showing massive grey-bluish, highly fossiliferous limestones and crystalline dolomite, dipping westwards.
- *Quebrada de Portachuelo 1* exposes rocks in the stream that are on the western flank of a syncline, which is the main structural feature of the area. The 175 m-thick section consists of lime mudstones and sandstones that grade into thick massive limestones, resting on reddish sandstone and mudstone of the Sabaneta Formation.
- The *Palmar River* exposes Palmarito strata consisting of a thick succession of calcareous sandstones and mudstones with ostracods that pass upwards to calcareous algal packstone, fusulinid grainstone, crinoidal–bryozoan wackestone and, finally, dolomites interbedded with boundstone containing sponges. This is the most complete and best-exposed section where it is possible to find the base and top of the Palmarito Formation.

Methods

Samples of whole rock and brachiopods were taken from the successions recorded in sections located in the Portachuelo area (Fig. 2). Carbonate powder was collected using a hand-drill on brachiopods, and pestle and mortar for matrix–whole-rock samples. Strontium, carbon and oxygen isotopes were determined on unaltered brachiopod shells of low-Mg calcite that should retain original sea-water signatures. For matrix–whole-rock material only carbon and oxygen isotopes were determined.

Sample screening

Brachiopod shells are composed of low-Mg calcite that is normally stable and resistant to diagenetic alteration. According to Brand (2004), brachiopod shell material is the most suitable for the study of ancient sea-water composition. However, a suite of tests was undertaken to discard material that had been altered; these included microtextural analyses using petrography, scanning electron microscope (SEM), cathodoluminescence (CL) and trace element analyses. The primary, outer layer of brachiopod shells is commonly altered and for this reason the secondary internal layer was sampled (Scholle & Ulmer-Scholle 2003) after examination with the SEM, as Korte *et al.* (2005) proposed, for any sign of alteration.

Trace-element content. The internal layers of 15 brachiopod shells were drilled and analysed for trace elements (Ca, Si, Na, Fe, Al, Mg, K, Ba, Sr, Mn, Zn, P, Ti and S) in a Perkin Elmer Optima 3300RL ICP OES (Inductively Coupled Plasma Optical Emission Spectrometer). In terms of this study just strontium (Sr), iron (Fe) and manganese (Mn) were used to assess any shell alteration.

Different authors have established cut-off values above which the shell material is considered to have been affected by diagenesis. Brand (2004) followed the values proposed by Denison *et al.* (1994), namely suitable material has less than 300 ppm Mn and 3000 ppm Fe. Korte *et al.* (2005) considered concentrations no greater than 250 ppm Mn and 300–3400 ppm strontium as acceptable, and Popp *et al.* (1986) supported this threshold.

Carbon and oxygen isotopes

The carbon and oxygen analyses were performed at Birmingham University using the following procedure:

> 100–140 μg of powdered carbonate were placed into 4 ml glass vials, and then sealed by a lid and pierceable septum. The vials were placed in a heated sample rack (90 °C) where the vial head space was replaced by pure helium via an automated needle system as part of an Isoprime Multiflow preparation system. Samples were then manually injected with approximately 200 μl of phosphoric acid and left to react for at least 1 h before the headspace gas was sampled by automated needle and introduced into a continuous-flow Isoprime mass-spectrometer. Duplicate samples were extracted from each vial, and a mean value obtained for both $\delta^{13}C$ and $\delta^{18}O$. Samples were calibrated using IAEA standards NBS-18 and NBS-19, and reported as ‰ on the VPDB scale. An external precision of better than 0.1‰ is typically achieved for both $\delta^{13}C$ and $\delta^{18}O$ (I. Boomer 2011 pers. comm.).

In addition, some of the samples were analysed in the Stable Isotope Biogeochemistry Laboratory in NCIET (Northern Centre for Isotopic and Elemental Tracing) at Durham University. The procedure was the same but was carried out using an Isotope Ratio Mass Spectrometer Thermo Finnigan Gasbench II connected to a Thermo Finnigan MAT 253.

Strontium isotopes

Using hand-drill and micromill techniques, powders were collected and analysed for their $^{87}Sr/^{86}Sr$ isotope ratios in the Arthur Holmes Isotope Geology Laboratory in NCIET. Samples were extracted from low-Mg calcite shells by micromill sampling using a New Wave Instruments™ Micromill™ system. This technique was developed by Charlier *et al.* (2006), and used by Font *et al.* (2008) and Deegan *et al.* (2010). The New Wave Instruments™ Micromill™ system consists of a stage, drill and binocular microscope connecting to a computer workstation. The system is driven by software that projects a microscopic image of a thick section of the sample, in this case the brachiopod shells. The drilling point location can be placed in an $X - Y$ co-ordinate system with accuracy of up to 1 μm. A hand-drill was also used directly on fragments of the brachiopod shells with less accuracy using a drill-bit up to 1 mm in diameter. Afterwards, all powders were placed in 3 ml beakers and subjected to a dissolution procedure that consisted of dissolving powders in ultra-pure distilled acid. The strontium fraction was then separated by micro-Sr column chemistry described by Charlier *et al.* (2006). Finally, samples from the micromill technique were placed in a thermal ionization mass spectrometer (TIMS: Triton), and the samples from the hand-drilling procedure were put in a plasma ionization multi-collector mass spectrometry (PIMMS: Neptune). Results were reported in ratios, with a range of standard deviation of ± 0.000012 to ± 0.000016 (2σ). The samples were analysed in a single session during which the average $^{87}Sr/^{86}Sr$ value for NBS-987 was 0.710262 ± 0.000016 (2σ) ($n = 4$) for the TIMS, which agrees extremely well with the accepted NBS-987 $^{87}Sr/^{86}Sr$ value reported by Thirlwall (1991) of 0.710248 ± 0.000023 (2σ) ($n = 427$) and the average $^{87}Sr/^{86}Sr$ value for NBS-987 was 0.710259 ± 0.000012 (2σ) ($n = 8$) for the PIMMS.

Results

Results from screening samples

Microtextural preservation. In thin section, Palmarito brachiopod samples are composed of fibrous calcite crystals with no signs of recrystallization (Fig. 3a, b). However, some of the shells do have puncti that seem to be recrystallized; in other

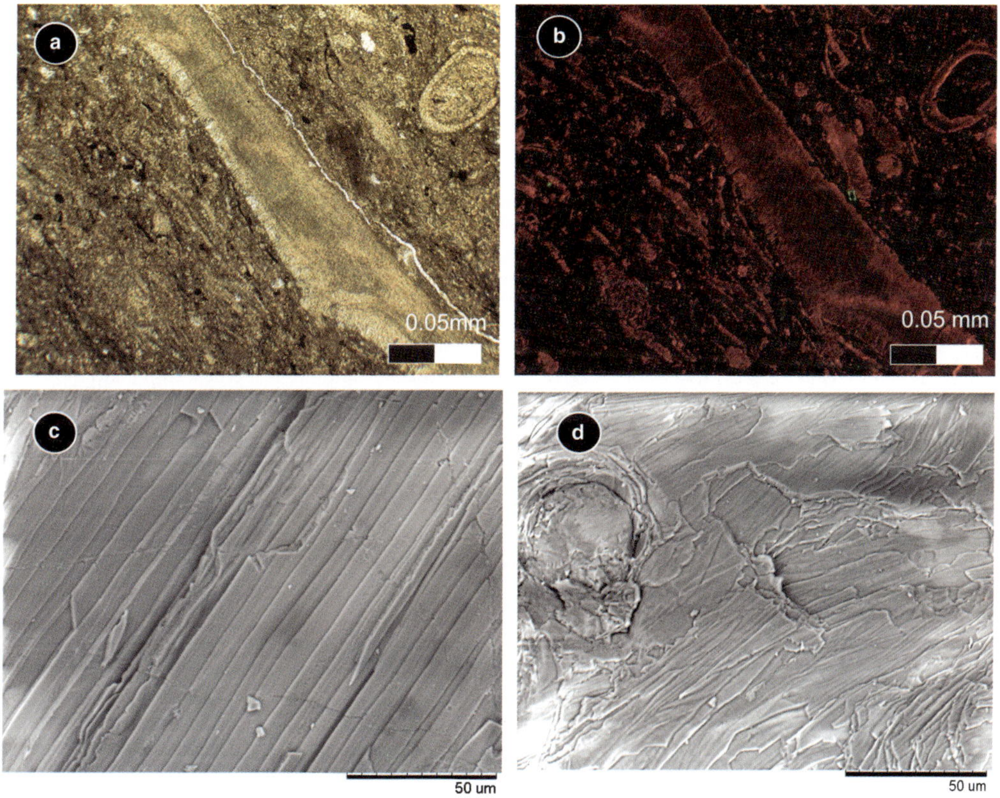

Fig. 3. (**a**) Photomicrograph (plane polarized light), (**b**) photomicrograph (cathodoluminescence, CL) and (**c**) SEM micrograph showing well-preserved brachiopod shells. (**d**) SEM micrograph showing puncti within a brachiopod shell, which were avoided for this analysis.

cases, the puncti are holes filled by diagenetic cements (Scholle & Ulmer-Scholle 2003). For this reason, these areas were avoided. Under the SEM, the fibrous structure is again well preserved with no micro- or nano-scale evidence of neomorphism or borings (Fig. 3c, d). In addition, under cathodoluminescence (CL, Fig. 3b), all samples show very dull to almost no luminescence.

Trace-element content. With the Palmarito secondary layer of the brachiopod shells, the concentration of Mn is slightly high (average 283 ppm) and Fe is also quite high (9589 ppm on average) (Table 1), but they do show very good preservation in CL, petrography and SEM (Fig. 3). For these reasons, the shells were considered suitable for Sr isotope analysis and it is believed that the values should reflect the composition of Permian sea water. Any effect of diagenesis is likely to give an older age for the samples as the general trend of diagenesis is to result in an enrichment of ^{87}Sr and the trend of the slope of the ^{87}Sr/^{86}Sr sea-water curve is decreasing through this part of the Permian.

Strontium isotopes

Ten seemingly unaltered brachiopods from the Portachuelo and Palmar sections were analysed for their strontium isotopes. The values vary between 0.707223 and 0.707435. However, sample 138-A shows a much higher value, up 0.708537 (Table 2). This could have been the result of contamination from the clay-rich matrix during the microdrilling process – clay minerals can be enriched in more radiogenic Sr, thus increasing the ^{87}Sr/^{86}Sr value – or from the effect of diagenesis, as noted above. Figure 4 shows the distribution of the analysed samples in the representative successions.

Table 1. *Trace elements of the brachiopod secondary layer*

Sample	Fe (ppm)	Sr (ppm)	Mn (ppm)
67-A	9351	903	287
N1-A	13120	978	405
N1-B	7322	1423	578
BQ-1	967	927	130
BQ-2	26154	869	323
BQ-3	12398	867	212
BQ-4	5080	874	526
BQ-5	25633	896	386
BQ-6	3361	750	116
BQ-7	9129	747	118
BQ-8	16902	532	130
BQ-9	9746	826	92
BQ-10	4050	1704	110
129	–	–	–
138-A	–	–	–
141-shell	3171	740	610
143-shell	6442	1085	488
144	–	–	–
152	–	–	–
154-shell	367	970	22
109-B	–	–	–
109-A	–	–	–

Table 2. *^{87}Sr/^{86}Sr values and ages based on Gradstein* et al. *(2004)*

Sample	^{87}Sr/^{86}Sr	Age (Ma)
67-A	0.707286	271
N1-B	0.707223	270.44
N1-A	0.707324	271.72
138-A	0.708537	–
109-B	0.707435	273.32
109-A	0.707331	271.88
152	0.707272	270.5
144	0.707319	271.6
143	0.707325	271.71
129	0.707366	272

Long-term trends in carbon isotope values

There is a long-term trend through the succession in δ^{13}C of progressively decreasingly negative and then increasingly positive values from −9‰ to +5.3‰ (Table 3). The most negative values of the carbon isotope ratios are present in the pedogenic facies within the tidal-flat facies of the lowest part of the carbonate ramp succession. The long-term positive trend is typical of the Permian time generally and it has been attributed to several factors including a long-term increase in nutrient supply to the oceans leading to a submesotrophic to mesotrophic regime during the upper part of the Permian (Martin 1996). Consequent elevated rates of organic productivity and increased burial would have led to an ocean carbon reservoir that is more enriched in ^{13}C. Cárdenas & Harries (2010) suggested that this phenomenon triggered evolutionary opportunities due to the improvement of the food-web complexity created by an increased abundance of primary producers.

The lower part of the Palmarito succession is attributed to the Artinskian stage by Laya & Tucker (2012) and this is where the δ^{13}C values of the palaeosoil facies are very negative as a result of light soil CO_2 from the decomposition of organic matter being incorporated into the pedogenic carbonate. Light δ^{13}C signatures are characteristic of early diagenesis in a subaerial–pedogenic environment. The succeeding strata are more open-marine deposits and here the $\delta^{13}C_{carb}$ rises to positive values

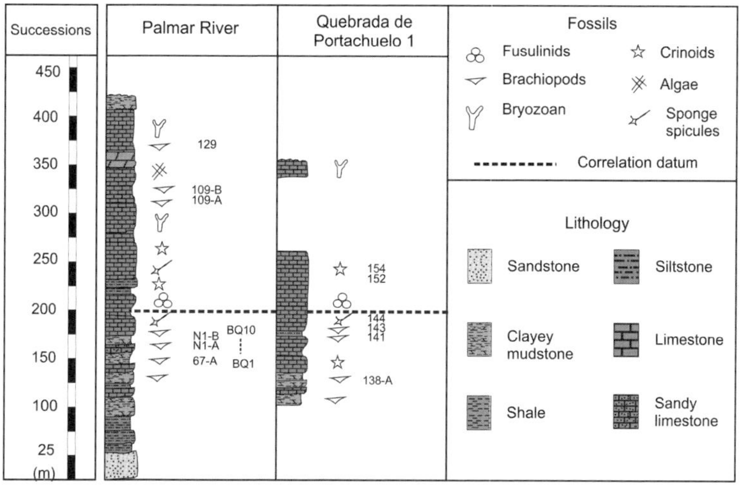

Fig. 4. Location of samples for strontium isotope analysis, and for trace elements to assess the degree of alteration and any effects of diagenesis.

reaching +2‰; this is a typical normal-marine signal (Tucker & Wright 1990). The increasingly positive trend continues through the Kungurian and then there is a zone where the data show much scatter, with values varying from −1‰ to +5‰ (Fig. 5). During the Roadian–Capitanian, the general pattern is one of less scatter and an increase in values to between +3‰ and +5‰.

In terms of comparison of this isotopic record with strata deposited in similar environments in other basins, the intervals with tidal-flat facies reflect local early diagenesis/pedogenesis and so need to be separated off from the higher, more normal-marine facies that may reflect a more global oceanic signature.

Marine carbon isotope trends

The four sections of the Palmarito Formation analysed show the same broad overall positive trend. However, each succession presents its own particular pattern of which the most common feature is a strong scatter of data-points from the cyclic deeper subtidal strata. In contrast, the data from the Portachuelo 2 section shows a smooth, increasingly positive pattern from +1.6‰ to +5.6‰.

A total of 103 samples were collected from different facies of the Palmar River succession and of these 82 correspond to normal-marine carbonates (see Fig. 6). The vertical pattern in Figure 6 shows values up to 5‰ of scatter where the curve starts with values slightly below 0 to +1‰ and a range of variation smaller than about 1‰ during the first 50 m of the succession in the marine facies (85.5–127.8 m above the base) (mean $\delta^{13}C = +1‰$). Upwards, more scatter in the data is present, with a dispersion of 5‰. Within the interval 127–162 m above the base, it is noteworthy to highlight the negative values in the deep subtidal facies that reach −1.4‰ (mean $\delta^{13}C = +2.24‰$). Above this part, the $\delta^{13}C$ values become heavier and vary between +2.8 and +4.7‰ for the next 50 m (mean $\delta^{13}C = +3.4‰$). Values for the next interval of 35 m reveal a decrease in $\delta^{13}C$ with a scatter of 2‰ (mean $\delta^{13}C = +2.4‰$). For the next 40 m above (heights of 210–250 m), the isotopes become heavier up to the maximum values for the whole succession, increasing from +1 and reaching +5.1‰ (mean $\delta^{13}C = +4.4$). Continuing up, there is a less positive excursion with values that fall almost to 3‰. Above this, there is an upwards increase with values showing a variation of 1‰ (mean $\delta^{13}C = +3.9‰$) in the next 100 m, up to the top of the succession.

Seventeen samples were analysed in the marine carbonate facies from the Quebrada de Portachuelo 1 section (Fig. 6). The succession starts with low positive values close to 0‰ and these values increase to 1‰ in first 25 m of the section (27.8–58.8 m above the base). Subsequently, the interval of 58.8–109.5 m exhibits much scatter towards more positive values where the variation is between +0.7 and +3.1‰. Upwards, $\delta^{13}C$ becomes a little more positive, reaching above +4‰ for last part of the succession.

33 samples were taken from the Quebrada de Portachuelo 2 section (Fig. 6). In contrast to the last two sections presented in Fig. 6, the data show less scatter, rising smoothly from values higher than +2‰ with a small variation of no more than

Table 3. *Stable isotope δ13C (‰ V-PDB) and δ18O (‰ V-PDB) of four main sections*

Samples	Metres above base	$\delta^{13}C$ (‰)	$\delta^{18}O$ (‰)	Samples	Metres above base	$\delta^{13}C$ (‰)	$\delta^{18}O$ (‰)
Palmar River				72-A	160.1	2.6	−6.7
14	10.2	−9.2	−5.1	72-B	160.3	3.0	−6.8
16	14	−5.3	−10.6	72-C	160.5	2.4	−6.6
17	14.8	−8.0	−6.6	72-D	160.7	2.3	−7.0
19	17.2	−3.0	−10.8	Brach BQ-9	161.00	2.6	−7.1
20	18.1	−3.2	−10.7	Brach BQ-10	161.50	4.0	−6.1
21	19	−2.7	−10.8	72	162.2	1.1	−7.4
24	26.1	−1.0	−11.8	73	163	3.6	−6.6
25	30.8	−0.6	−10.7	78	179.5	3.1	−5.0
27	36	−1.1	−11.0	79	180.4	4.2	−5.0
30	47	0.5	−9.1	80-A	188.6	2.8	−4.6
31	48	−1.1	−9.9	80	189.1	3.3	−6.1
32	49.1	−1.9	−10.0	80-B	189.5	3.2	−7.5
33	50.5	−2.1	−11.5	80-C	190.1	3.2	−6.6
34	55	−3.3	−7.7	80-D	190.3	3.1	−6.2
35	55.5	−0.8	−11.5	80-E	190.5	3.3	−5.2
38	61.2	−1.2	−9.9	80-F	190.8	3.7	−4.5
39	63	−1.3	−10.2	81	191	3.2	−4.8
40	63.5	0.6	−8.9	82	195	2.9	−2.9
43	72	−0.3	−5.9	83	196.5	3.5	−5.4
44	74.3	−0.7	−10.3	85	209.4	4.7	−4.2
47	79.6	−2.2	−2.5	85A	210.5	2.8	−6.0
48	85.5	0.9	−9.0	86	214.5	3.6	−5.6
49-A	96.4	2.1	−7.9	87	215.8	3.6	−7.5
49	99.5	0.7	−9.4	88	217.1	2.2	−6.4
51-A	106.2	0.8	−4.2	89	221.5	1.2	−5.5
52	112	1.0	−7.5	92	233.8	3.1	−5.3
53	113.5	1.0	−7.8	94	241.2	1.2	−5.6
55	127.8	0.5	−7.0	95	245.3	2.0	−6.5
56	128.3	3.8	−4.2	96	245.8	2.8	−5.9
57	131.1	3.5	−5.2	98	256.7	4.5	−4.7
58	132.8	1.7	−7.1	101-A	265.5	4.8	−4.9
59	134.5	−0.2	−6.7	101-B	267	5.0	−5.1
60	137.7	3.8	−5.8	101-C	268	4.9	−5.3
61	140.8	2.9	−6.1	101-D	269	5.0	−5.0
62	141.5	1.9	−7.4	103-A	274.5	4.7	−5.2
Brach BQ-1	141.50	3.0	−5.7	103-B	275	5.1	−5.1
Brach BQ-2	143.20	3.7	−7.2	104-A	279	4.5	−5.2
63	143.4	1.6	−7.7	104-B	287.4	2.2	−6.0
64	145	1.9	−7.4	105	292	4.1	−4.9
Brach BQ-3	145.40	2.6	−6.9	127-shell	300	4.4	−4.4
65	146.2	3.5	−6.0	107-A	301.8	4.2	−5.0
Brach BQ-4	147.50	2.5	−6.1	107-B	305.5	3.9	−4.7
66	148	−1.4	−6.9	108-A	309.5	4.1	−4.9
67-A	150.4	2.5	−6.8	108	310.5	4.4	−4.4
67	151.5	2.1	−6.9	109-A	315.2	3.7	−4.9
Brach BQ-5	152.20	2.5	−7.0	109-B	315.8	3.6	−4.9
68	153.6	2.1	−6.4	113	340.6	3.3	−7.5
Brach N1-A	154.4	4.1	−7.3	114	343.5	3.1	−7.2
69	155	2.8	−6.3	116-A	351.5	3.9	−5.8
70	155.5	2.2	−6.9	116	352.5	4.1	−5.3
71	156.7	3.1	−6.3	117-A	358.7	4.6	−6.2
Brach BQ-6	157.50	4.2	−5.9	128	359.7	4.2	−4.3
Brach N1-B	159	4.1	−7.0	129	360.7	3.4	−5.0
Brach BQ-7	159.50	4.1	−7.2	129-A	361.7	4.1	−4.3
Brach BQ-8	160.00	4.5	−6.6	129-C	362.7	3.9	−4.6

(*Continued*)

Table 3. *Continued*

Samples	Metres above base	$\delta^{13}C$ (‰)	$\delta^{18}O$ (‰)	Samples	Metres above base	$\delta^{13}C$ (‰)	$\delta^{18}O$ (‰)
129-E	363.7	3.5	−5.3	288	108.5	2.2	−7.1
129-G	365.7	3.8	−5.6	289	118.4	2.7	−6.0
129-K	366.7	4.5	−5.2	158	140.1	4.4	−4.6
130	375.8	3.6	−6.1	159	143.4	3.6	−4.0
131	381	4.3	−6.8	162	165.1	4.7	−5.7
Quebrada de Portachuelo 1				163	168.5	4.9	−5.1
133	5	−2.3	−2.1	164	171	5.0	−6.1
134	8.4	−0.3	−8.6	165	181.2	5.3	−5.6
135	27	−3.6	−7.7	166-B	196.4	5.6	−5.6
136	27.8	0.3	−2.9	166	198.8	4.6	−5.8
138	33	0.4	−8.0	166-A	204.4	4.6	−6.4
140	58.8	1.1	−5.6	167-B	214	3.8	−5.6
141-brach	37.00	2.85	−4.50	167-A	217.8	5.0	−5.0
141	60	0.9	−7.2	167	225.5	4.9	−5.6
142	74.5	3.0	−7.5	168-A	234.8	4.7	−3.7
143	78	0.7	−6.9	168	238.7	2.9	−4.5
143-brach	78.00	2.30	−6.71	169	239.5	3.9	−4.3
144	78.5	2.9	−7.1	170	241.5	3.9	−3.7
147	85.5	1.1	−7.3	171	249.9	3.9	−4.8
148	92.5	1.2	−7.4	*Santa Cruz de Palmarito*			
149	99	1.6	−6.3	236	0.5	4.0	−5.9
150	100.5	3.1	−5.5	237	0.8	3.9	−5.0
151	109.5	0.7	−6.1	238	6.2	2.5	−6.6
153	148	2.3	−6.2	239	7	2.4	−6.2
154	167.2	2.9	−3.7	240	20.5	4.2	−5.5
154-brach	167.2	4.18	−3.98	241	26	4.7	−5.9
155	341	3.8	−5.8	243	31.2	4.2	−5.7
156	344.5	4.3	−5.2	244	33.7	4.8	−5.2
157	347.2	3.7	−5.2	245	35	4.8	−5.5
Quebrada de Portachuelo 2				246	36.9	4.2	−6.0
265	3.5	2.4	−8.0	248	40.3	5.0	−4.1
267	14.4	2.1	−7.7	249	41.5	4.9	−5.1
268	18.2	2.5	−7.5	251	43.5	4.4	−5.3
269	21.6	2.3	−8.1	252	47.2	3.1	−6.6
270	24.2	2.5	−8.7	253	49.5	4.3	−5.5
271	28.5	2.3	−7.9	255	55.5	0.0	−6.1
272	32.4	2.8	−8.6	256	57.3	2.7	−6.3
273	39.4	1.6	−7.8	257	59.3	4.4	−5.6
276	52.6	2.1	−7.0	258	69.3	3.2	−5.8
277	54.6	2.7	−7.0	259	71.2	5.0	−4.2
278	61.7	2.3	−7.4	260	71.8	5.0	−4.5
280	69.5	2.2	−7.8	261	80.5	4.5	−4.6
285	88.7	2.1	−7.0	262	85.5	4.5	−5.5
287	104.2	2.5	−6.3	263	99	1.7	−7.2

0.5‰, except for one sample at 39.4 m above the base, which exhibits a signature of +1.6‰. This trend stays constant to 108.5 m where the values increase dramatically to +5.6 ‰ in the 100 m above. Afterwards, a less positive excursion is present, with values decreasing to +3.8‰. Subsequently, another interval shows $\delta^{13}C$ values increasing to +5‰. Finally, a drop from +5 to +3‰ occurs in the last part of the section.

Twenty-six samples of marine carbonate were collected from the Santa Cruz de Palmarito section (Fig. 6). The first values recorded in that succession are close to +4‰ and in the initial 10 m above the base a less positive excursion is seen. Afterwards, a heavier isotopic composition is present and this stays relatively constant between +4 and +5‰ for the next 45 m above the base. Then, the curve shows a dramatic drop from +4 to 0‰ at 55.5 m above the

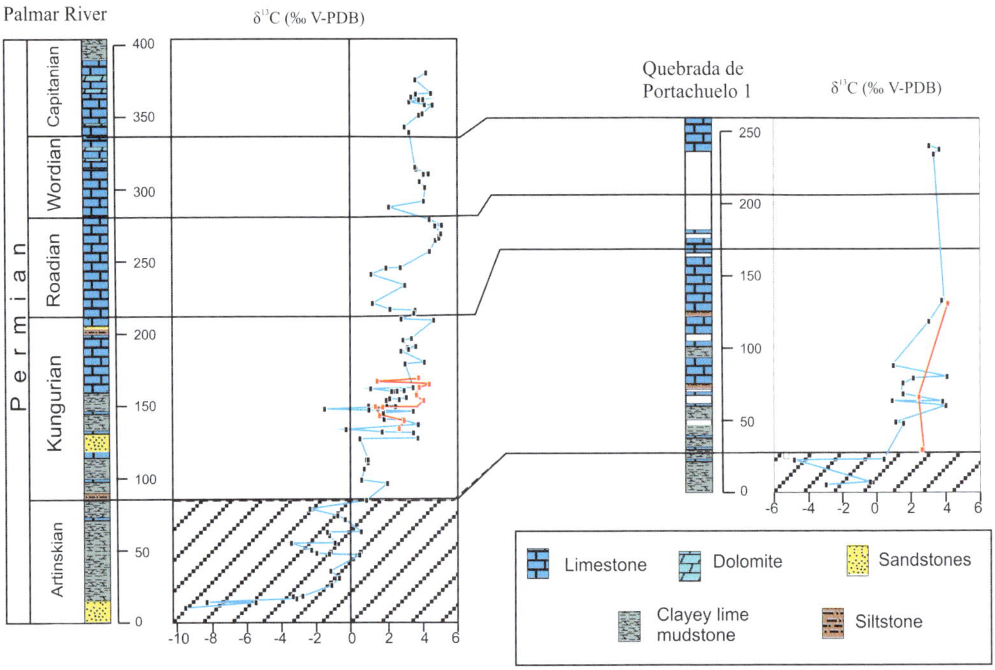

Fig. 5. Carbon isotopes of whole-rock carbonate (blue line) and well-preserved brachiopods (red line) from the Palmar River and Quebrada de Portachuelo 1 successions.

base. Subsequently, the values show a moderate scatter between +3 and +5‰ for the next 30 m.

The Naqing section in China (Buggisch *et al.* 2011), which is coeval with the Palmarito Formation, has also been analysed for its isotope stratigraphy. The trend in $\delta^{13}C$ is very similar to the Palmarito, with positive values between 0 and 5‰. Minimum values of around 1‰ occur in the Kungurian, and values increase upwards towards the Roadian and Wordian, reaching approximately 5‰ in the Capitanian (Fig. 6).

Oxygen isotopes

In general, values of $\delta^{18}O$ in the matrix–whole-rock carbonates from the Palmarito strata show a wide scatter and range from −11.8 to −2.1‰ with an overall average of −6.3‰. However, the oxygen isotopic signatures from the tidal-flat facies exhibit lighter values with an average of −8‰. In addition, these facies preserve a $\delta^{18}O$ isotopic signal with a high dispersion, which includes the maximum and minimum values of the whole succession. Following this, oxygen isotope values of the marine facies lie between −2.9 and −9.4‰ (mean $\delta^{18}O$ = −5.8‰). The range of scatter is smaller than for the tidal-flat facies. The last two groups of values exhibit a slightly lighter isotopic composition than the well-preserved low-Mg calcite brachiopod shells, which have an average of −6.24‰ (Table 3).

Fifteen low-Mg calcite brachiopod shells were collected from two sections, namely the Palmar River (n = 13), and the Portachuelo 1 sections; brachiopods are not present throughout the sections. Values vary between −7.3 and −4‰ (Table 3). These occur within the deeper subtidal facies so that their isotopic signatures may reflect the sea water of the Mucuchachí Basin.

Discussion

Age determination from strontium isotopes and chronostratigraphy

The trend of the $^{87}Sr/^{86}Sr$ ratios from Asselian to Capitanian in the mid-Permian is a decreasing pattern (see Fig. 7). This has been related to the decline of continental input as a result of the gradual aridification of Gondwana associated with the deglaciation of the southern polar region; this reduced the continental run-off from central and northern Pangea (Korte *et al.* 2006; Grossman *et al.* 2008). Sedimentological evidence of the inferred aridification trend can be seen in the Palmarito strata in the form of decreasing clastic input up

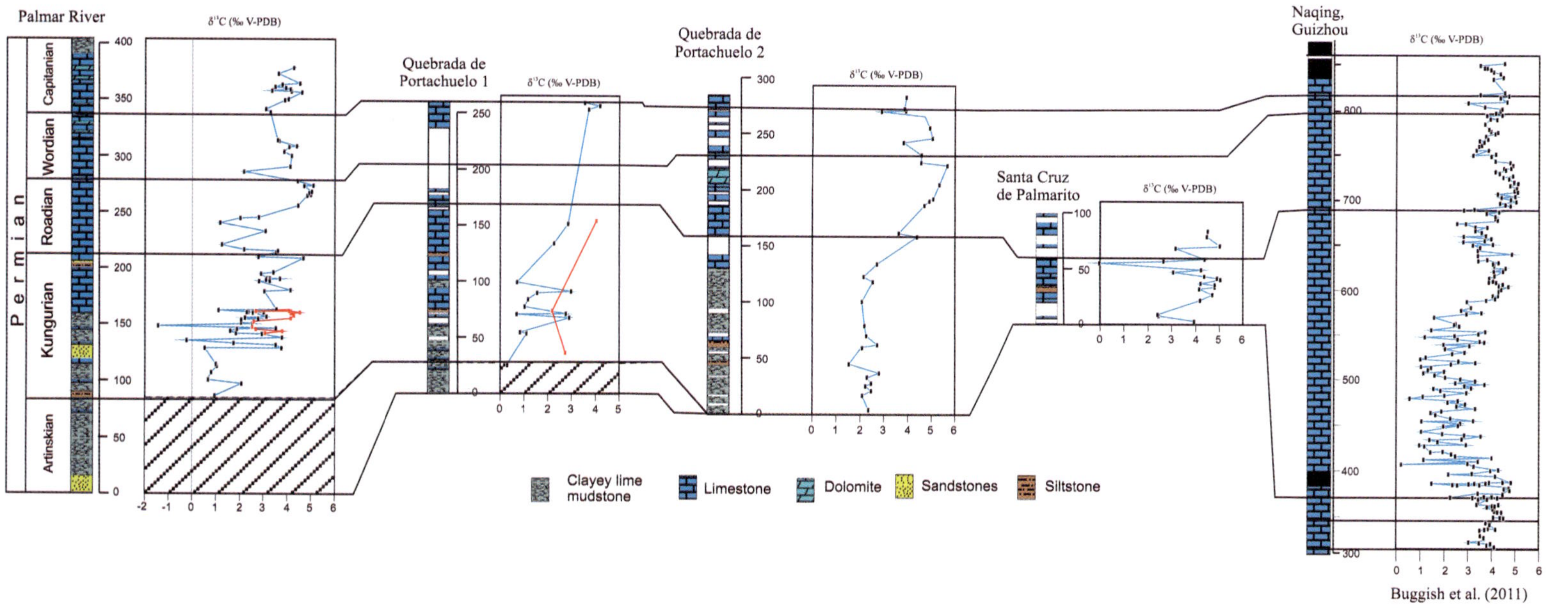

Fig. 6. Carbon isotopes of whole-rock carbonate (blue line) and well-preserved brachiopods (red line) from four successions in the Portachuelo area, Venezuelan Andes and the Naqing succession from China (Buggisch *et al.* 2011).

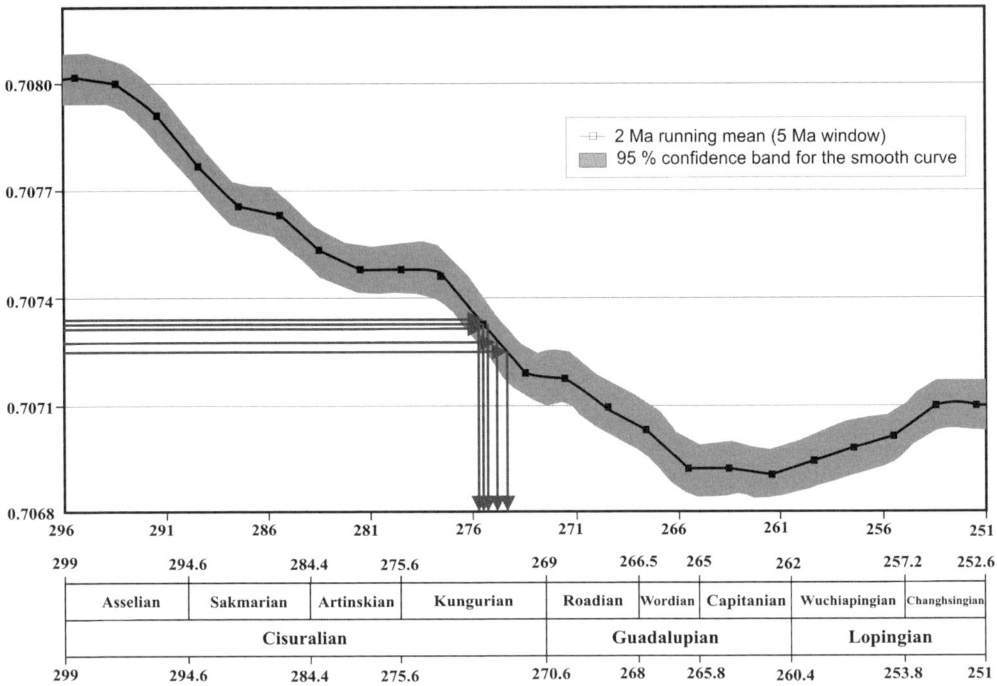

Fig. 7. $^{87}Sr/^{86}Sr$ mean (2 Ma steps, 5 Ma windows) for well-preserved Permian brachiopod shells from Korte *et al.* (2006). The orange lines and arrows (grey in print version) represent $^{87}Sr/^{86}Sr$ data from well-preserved Palmarito brachiopods.

through the succession and evaporite precipitation towards the top. Since this is a global phenomenon, it would be expected that the $^{87}Sr/^{86}Sr$ ratios from the Palmarito should show the same behaviour as that of the global trend proposed by Korte *et al.* (2006) for the Permian. Thus, although brachiopods are restricted to one part of the Palmarito Formation, their $^{87}Sr/^{86}Sr$ values should lie on the global Sr curve (Fig. 7).

The strong change in the strontium isotope curve does allow the Palmarito values to be used for absolute dating for the horizons where the brachiopods are located. $^{87}Sr/^{86}Sr$ ratios for the Permian have been refined by Korte *et al.* (2006) (see Fig. 7).

The ages obtained from the $^{87}Sr/^{86}Sr$ data (see Table 2) occur between 270.5 and 273.5 Ma, and these ages correspond to the Kungurian stage according to Gradstein *et al.* (2004). However, there is an inconsistency with the stratigraphical position of one sample, which provided an older date than expected. The sample is from a higher stratigraphical horizon and as the curve at this Permian time is showing a sharp decrease in values, the sample should have been showing a lower value. The higher $^{87}Sr/^{86}Sr$ ratio could be due to reworking of the brachiopod shell into a younger horizon or, more likely, it could be a result of diagenetic alteration, which commonly leads to a higher value because radiogenic Sr is released as a result of low-grade heating.

Brachiopods (Hoover 1981) and sponges (Rigby 1984) have been well studied for the time frame of the Palmarito Formation; however, there is poor biostratigraphical control on the horizons where these fossils are located. Thus, the precise age range of these fossils has not been easy to determine.

Localities 1–4 (described earlier) are stratigraphically located in the middle and upper part of the formation, and were assigned a Roadian–Wordian age by Hoover (1981). The $^{87}Sr/^{86}Sr$ ratio reported here does indicate a Kungurian age for the strata below the horizons studied by Hoover (1981). In addition, cephalopods were found in the Quebrada de Portachuelo section, which the same author suggested belong to the late Artinskian. However, the vertical range of the cephalopods is not known, and they could be associated with a level below that from where the samples for strontium dating were collected.

The evidence suggests that the Palmarito Formation could be restricted to the Artinskian, but possibly it could extend up to the early Capitanian. However, this time frame needs further investigation.

Carbon isotopes in Palmarito strata

The lowermost part of the formation, composed of inner-ramp deposits, was subjected to exposure and early meteoric diagenesis. Thus, the carbon isotope values show very low positive or negative values. The overlying marine facies exhibit low positive values, close to 0‰, and one possible explanation is that these could be the result of weathering and influx of lighter carbon from terrestrial organic matter into shallow waters, reducing a normal-marine $\delta^{13}C$ signature (typically 0 to +4‰: Tucker & Wright 1990). An increase in continental organic matter input into the basin, with its light carbon isotopic composition, could be expected during this late phase of deglaciation. This is supported by the interpretation of relatively humid conditions for this period proposed by Ricardi (2008). Also, it is likely that vegetation was relatively dense in terrestrial areas of Venezuela at this time.

Within the succeeding marine facies, the data are quite scattered and the more negative values are probably due to the periodic influx of terrestrial organic matter. The more positive values probably relate to the marine bioclastic sediments composed of brachiopod, crinoid and bryozoan debris, which would have a heavier isotopic composition. The strong seasonality of a humid and arid climate at this stage due to the deglaciation would be a time when the ITCZ had a strong north–south migration. However, low positive values are dominant until the middle Kungurian and this could be a reflection of the last phase of deglaciation (Figs 6 & 8). This excursion can be correlated with the isotope stratigraphy from the Naqing section in China (Buggisch *et al.* 2011), shown in Fig. 6, that shows a similar fall to less positive values. In addition, the same trend can be seen in the $\delta^{13}C$ data from brachiopods collected from the US craton, compiled by Grossman *et al.* (2008). For the late Kungurian, the curve exhibits strong dispersion but with a dominance of high positive values of around 5‰. This phenomenon is, perhaps, related to high primary productivity resulting from an increased influx of nutrients to the carbonate ramp.

Furthermore, the increased deposition of organic matter could have led to a reduction in atmospheric pCO_2 (CO_2 partial pressure) and this could have triggered a fall in temperature. As a result, ice caps may have increased in size during this time.

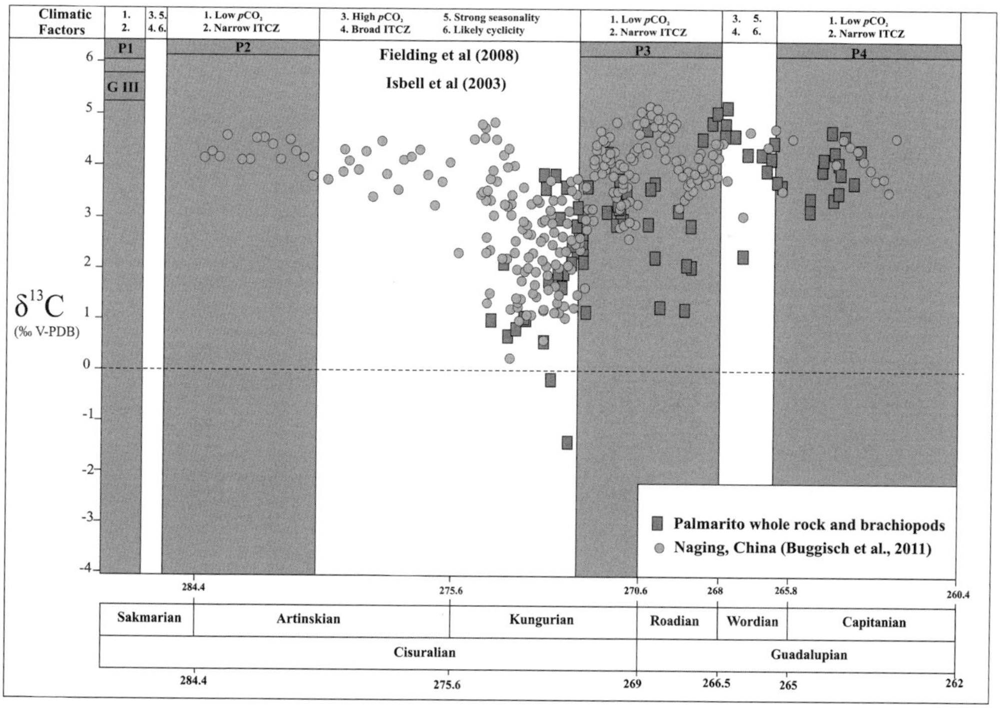

Fig. 8. Whole-rock and brachiopod carbon isotope values from Permian carbonates in the Palmar succession, Venezuelan Andes (orange squares, grey in print version) and whole-rock carbon isotopes of Permian carbonates from the Naqing succession in China (yellow circles, grey in print version). The blue areas (grey in print version) represent periods of glaciation proposed by Isbell *et al.* (2003, 2008) and Fielding *et al.* (2008).

In association with this phenomenon, Fielding *et al.* (2008) suggested a short period of glaciation during the late Kungurian and also in the Capitanian (glacial phases P3 and P4); these could correspond to the positive carbon excursions that are repeated twice in the four successions (Figs 6 & 8). Finally, the $\delta^{13}C$ trend shows a stabilization to values around +4‰ during the Capitanian stage. A similar pattern is seen in values from brachiopods reported by Grossman *et al.* (2008) from the Guadalupe Mountains and by Buggisch *et al.* (2011) from the Naqing section in China (Fig. 6). Although, in general, carbon isotope values are much more stable and less prone to alteration by diagenesis compared with oxygen isotope data, some effects of diagenesis cannot be discounted (Fig. 8).

Oxygen isotopes

Original oxygen isotope signatures in limestones are susceptible to alteration by diagenesis. For this reason, grains and cements originally (and usually still) of low-Mg calcite are preferred in isotope studies seeking the isotopic composition of sea water as they are the most stable form of the $CaCO_3$ minerals. Well-preserved low-Mg calcite brachiopods in the Palmarito Formation were analysed, as well as matrix–whole-rock samples; in general, their results exhibit close similarities. Thus, there are two possibilities: first, the values could be reflecting sea-water composition and temperature; or, secondly, if ratios have been affected by diagenesis, then it is possible that all components, fossils and matrix, could have been shifted uniformly by the alteration, normally to more negative values.

During the early–middle Kungurian, the $\delta^{18}O$ signatures show a decrease towards more negative values. This could indicate the development of a warmer period that is probably associated with the deglaciation interval between glacial stages P2 and P3 of Fielding *et al.* (2008) (Fig. 9). After this, the values become less negative during the late Kungurian–early Roadian and this trend coincides with the positive excursion of carbon isotopes as has been mentioned before; the enhanced deposition of organic matter would have resulted in a reduction of atmospheric pCO_2 and this could have triggered

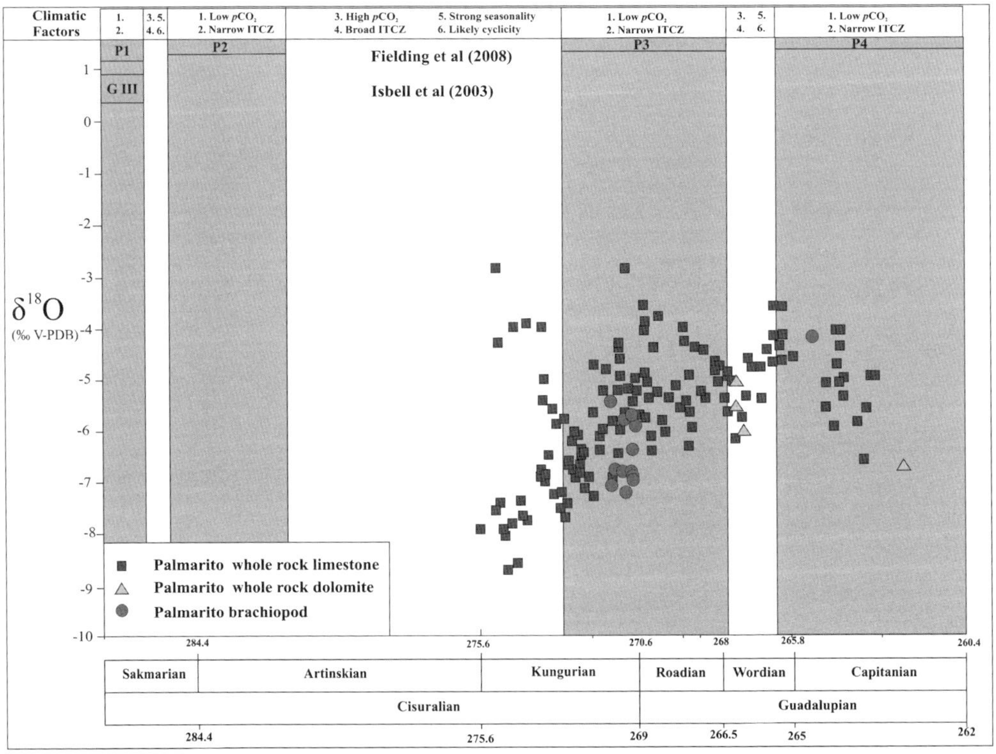

Fig. 9. Oxygen isotope data from brachiopods and whole-rock limestones and dolomites from the Palmarito Formation, Venezuelan Andes. The blue areas (grey in print version) represent periods of glaciation proposed by Isbell *et al.* (2003, 2008) and Fielding *et al.* (2008).

the temperature fall recorded by the oxygen isotope signatures. Fielding *et al.* (2008) proposed a smaller glaciation (P3) at this time. Subsequently, results from oxygen isotopes for the Roadian stage show evidence of a warming that could correlate with a deglaciation (Fig. 9). Although it is known from the Quaternary record that the temperature variation between glaciation and deglaciation in low latitudes is really quite small, Grossman *et al.* (2008) showed a quite large range of oxygen isotope signatures for low-latitude Permian samples. This could be a diagenetic effect or simply indicate quite large temperature variations during this time of unstable climate.

During the Wordian, the $\delta^{18}O$ values of the limestones become less negative in a quite smooth trend and this could be the result of a developing glacial period, correlating with stage P4 of Fielding *et al.* (2008) (Fig. 9). However, Grossman *et al.* (2008) have interpreted the heavier composition of $\delta^{18}O$ as being related to the aridification process, inferred from facies analysis, based on data from Russia and the USA. Finally, the more negative trend in the $\delta^{18}O$ in the limestones in the top part of the succession could indicate a temperature rise towards the early Capitanian (Fig. 9).

Relationship of the Palmarito strata with palaeoclimatic factors

Plate movements (tectonic drift of Tabor & Poulsen 2008) have been considered as one of the factors that could have brought about climate change in the Late Palaeozoic. Western and central Pangean plate movements were modelled using Point Tracker software developed by Scotese, and the results show that the huge Pangean land mass gradually moved northwards from 10° to 14°N (Tabor & Poulsen 2008). A consequence of this was the migration of the Pennsylvanian equatorial basins from the humid climate associated with the ITCZ (high precipitation conditions) into the adjacent northern subtropical arid zone (Witzke 1990; Gibbs *et al.* 2002; Rees *et al.* 2002; Tabor *et al.* 2008). If this is correct, then the long-term aridification of Pangea could have been caused by plate movement rather than by a change in atmospheric conditions. However, palaeomagnetic studies (May & Butler 1986; Bazard & Butler 1991; Steiner & Lucas 2000; Steiner 2003) have suggested that western Pangea remained at the same latitude until the Late Triassic. The rapid change from humid to arid conditions and the persistent humid climate in eastern tropical Pangea are arguments against this hypothesis of plate movements. Thus, plate movements can be disregarded as an important control on climate change in the low latitudes of Pangea during the Permian. Tropical humid conditions during deposition of the Palmarito have been suggested by Ricardi (2008) for the Early Permian. Venezuelan carbonates were located slightly south of the Equator within central Pangea and may well have remained in this palaeographic position for some time.

Land–sea distribution seems to have been an important factor that could have had a significant influence on the Permian climate (Tabor & Poulsen 2008). For Pennsylvanian–Permian times, an increase in land area over equatorial Pangea has been suggested by Blakey (2007) in his palaeogeographical maps due to the withdrawal of the inland epeiric sea. However, a different land–sea distribution is proposed in this study on the basis of the occurrence of Permian transgressive carbonates over central Pangea that extended until the middle Permian, and perhaps up to the Capitanian. The increase in continentality could have pushed away the source of moisture and triggered the aridification process. Nevertheless, the Palmarito strata do show evidence of humid conditions in the early Permian. In addition, the climate models of Peyser & Poulsen (2008) and Tabor & Poulsen (2008) show a decrease in precipitation, but still remaining quite high, during the early Permian, and this matches with the Palmarito evidence. Also, from the Palmarito carbon isotope curve, the humid conditions could have controlled the input of carbon into the basin; this is also the case in the north China and Turpan basins located in eastern equatorial Pangea (Rees *et al.* 2002; Ziegler *et al.* 2003; Tabor & Poulsen 2008).

The climate of tropical Pangea was characterized by monsoonal circulation (Tabor & Montañez 2002; Peyser & Poulsen 2008; Tabor & Poulsen 2008). Perlmutter & Plotnick (2003) and Tabor & Poulsen (2008) proposed that monsoon intensity over equatorial Pangea might have changed over shorter periods of time than a tectonic timescale. In addition, monsoon intensity has been more intense during glacial periods and lower during interglacial stages, as a result of the change in temperature difference between the poles and Equator (Perlmutter & Plotnick 2003).

Moist tropical air masses of the ITCZ, which came from the Panthalassa Ocean and inland seas, were affected by the presence of the Central Pangean Mountains (Parrish 1993; Peyser & Poulsen 2008; Tabor & Poulsen 2008). The Palmarito area was located south of this physiographical high where the ITCZ could migrate freely during the seasons. The migrations of the ITCZ were strong in the interglacial periods and extended in a broad band over the tropics, in contrast to times of maximum glaciation when the ITCZ would have been quite narrow and focused around the Equator, according to Tabor & Poulsen (2008). However, in a unipolar glaciation model (that is the case for the Late Palaeozoic) it would be expected that the ITCZ moved farther to the north. In any case, strong seasonality during interglacial periods, as shown by the strong scatter of $\delta^{18}O$ values in the Palmarito pedogenic facies, can be interpreted according to Tabor & Montañez (2002) as a proxy of palaeo-precipitation in equatorial Pangea. These results suggest that there could have been an alternating near and far source of rainfall for this region owing to the wide movements of the ITCZ, which could have triggered arid and sub-arid conditions alternating with a humid climate. In addition, during the deposition of the tidal-flat facies (lower part of the Palmarito Formation) evidence of exposure was found, including calcrete and thin coal horizons. These facies indicate an alternation of humid and arid climates. Furthermore, intense cyclicity in the marine facies of the Palmarito strata was mainly developed during the period when seasonality was strongly expressed as a result of a high-frequency climatic control on sedimentation on the Venezuelan carbonate ramp.

According to Peyser & Poulsen (2008), atmospheric pCO_2 concentrations are the most important climatic factor to explain long-term aridification. However, Tabor & Poulsen (2008) supported the thesis that not only atmospheric pCO_2 was responsible for climate change during the Late Palaeozoic but also latitudinal plate movements, the assembly of Pangea and orogenesis. Yet, the evidence for these processes is weak. Furthermore, atmospheric pCO_2 variations during the Late Palaeozoic have been discussed by many authors (e.g. Parrish 1993; Tabor & Montañez 2002; Saltzman 2003; Montañez *et al.* 2007; Peyser & Poulsen 2008; Tabor & Poulsen 2008); all of them suggested that CO_2 was a significant control in surface temperature, evapotranspiration and palaeo-precipitation in the short and long term. In addition, the effect on expansion and contraction of the ice caps, which controlled the migration of the ITCZ, would have caused strong seasonality and so affected sedimentation and the geochemical signals.

Although, the long-term trend towards aridity in western and central Pangea was the dominant process, short-lived wet periods were present, associated with possible intensification of monsoonal circulation during the periods of glaciation. In the Palmarito strata these were associated with the P3 and P4 glaciations proposed by Fielding *et al.* (2008) (Figs 7 & 8).

Marine circulation

Permian ocean circulation has been studied by several authors (Grossman *et al.* 2002; Winguth *et al.* 2002; Winguth & Maier-Reimer 2005). Their models suggest that there were strong global changes in atmospheric circulation, CO_2 levels and carbon sequestration at this time. However, Grossman *et al.* (2008) have emphasized the potential danger in misinterpreting regional changes as being on a global scale. As an example, they noted the misconception of sea-water ^{18}O enrichment being only due to glaciation and they suggested that excess evaporation was an equally plausible alternative explanation.

Palmarito strata were deposited in an epicontinental sea that could potentially have been restricted as a result of the partly enclosed palaeogeography; this may have given rise to distinctive geochemical signals. However, the carbon and oxygen isotope data from the Venezuelan Permian carbonates exhibit similar trends to those from other

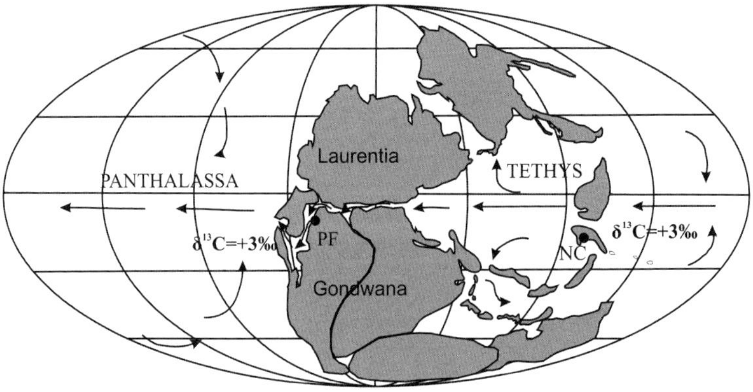

Fig. 10. Schematic palaeogeographical projection showing simplified ocean circulation and mean carbon values of the Palmarito Formation (PF) in the Venezuelan Andes and carbonates of the Naqing succession (NC) in China (Buggisch *et al.* 2011).

areas, notably the Naqing section in China (Buggisch *et al.* 2011) (Fig. 6), where deposition took place in the open Tethys Ocean. This argument supports the suggestion that the area of Palmarito deposition had free circulation with the open ocean so that the isotopic data from there reflect global isotopic trends (Fig. 10).

Summary and conclusions

The Palmarito strata in Venezuela are located in the west-central Pangea region, which is a poorly known area between Gondwana and Laurentia at a time just before the final assembly of Pangea. The data presented here thus fill an important gap in the regional palaeogeography, and also have significant implications for the palaeoclimate and palaeoceanography over equatorial Pangea at that time.

Values of $^{87}Sr/^{86}Sr$ ratios from low-Mg calcite brachiopod shells in the central part of the Palmarito Formation have provided useful dates for the strata, placing them within the Kungurian stage at 270.5–273.5 Ma. The improvement of the dating has allowed better correlations and comparison with global climate and palaeoceanographical changes at the time.

Carbon isotope data show a long-term trend towards more positive values. However, values in the lower part of the succession are more negative and this is attributed to the effects of pedogenesis and weathering during early diagenesis. Excursions of $\delta^{13}C$ higher up in the formation show a strong relationship to the inferred glaciation–deglaciation events that have been described by Isbell *et al.* (2003, 2008) and Fielding *et al.* (2008) for the early and middle Permian.

Although original oxygen isotope signatures of limestones are prone to diagenetic alteration, the similarities of the results from low-Mg calcite brachiopods (generally resistant to alteration) and whole-rock samples, and from different sections, allow the interpretation of the relative trends within the strata. The trends of oxygen values were correlated with glacial events, P3 and P4, proposed by Fielding *et al.* (2008). Changes in Permian climate related to plate movements, land–sea distributions, monsoonal circulation, the Intertropical Convergence Zone and atmospheric pCO_2 levels have been discussed with reference to the Palmarito succession in central Pangea. As a result, it has been concluded that pCO_2 was the most significant factor that controlled not only long-term climate change but also short-term variations.

Marine circulation within the Mucuchachí Basin potentially could have been restricted, as a result of which distinctive geochemical signals might have been expected. However, carbon and oxygen isotope data from the Venezuelan Permian carbonates exhibit trends similar to those from other successions, such as the Naqing section in China, which was deposited in an open-marine realm of the Panthalassa Ocean. Thus, an open-marine basin was probably established in northern Venezuela at this time.

We would like to thank, from Durham University, C. McLeod, F. Laiginhas, J. Peterkin, C. Otley, G. Nowell, C. Macpherson and G. Pearson, and also A. Finlay for his constructive comments. We are also grateful to Dr I. Boomer and Prof. A. Baker.

This research was funded by a grant from Consejo Cientifico, Humanistico y Tecnologico de la Universidad de los Andes, Venezuela (CDCHT-ULA), which provided funding with the project code I-1092-08-02-B. Also, the AAPG foundation is acknowledged for a grant from the Horst & Jessie von Bandat memorial fund.

References

Arnold, H. C. 1966. Upper Paleozoic Sabaneta-Palmarito sequence of Mérida Andes, Venezuela. *AAPG Bulletin*, **50**, 2366–2387.

Bazard, D. R. & Butler, R. F. 1991. Paleomagnetism of the Chinle and Kayenta Formations, New Mexico and Arizona. *Journal of Geophysical Research*, **96**, 9847–9871.

Bellizia, A. 1992. Terreno Mérida y Bloque Caparo. Cordillera de Los Andes de Venezuela. *In*: *Actas de las sesiones científicas: III Congreso Geológico de España, Volume 4*. Universidad de Salamanca, Facultad de Ciencias, Salamanca, 112–116.

Blakey, R. C. 2007. Carboniferous–Permian palaeogeography of the assembly of Pangea. *In*: Wong, T. E. (ed.) *XVth International Congress on Carboniferous and Permian Stratigraphy, Utrecht*, 10–16 August. Royal Netherlands Academy of Arts and Sciences, Amsterdam, 443–456.

Brand, U. 2004. Carbon, oxygen and strontium isotopes in Paleozoic carbonate components: an evaluation of original seawater-chemistry proxies. *Chemical Geology*, **204**, 23–44.

Buggisch, W., Wang, X., Alekseev, A. S. & Joachimski, M. M. 2011. Carboniferous-Permian carbon isotope stratigraphy of successions from China (Yangtze platform), USA (Kansas) and Russia (Moscow Basin and Urals). *Palaeogeography, Palaeoclimatology, Palaeoecology*, **301**, 18–38.

Cárdenas, A. L. & Harries, P. J. 2010. Effect of nutrient availability on marine origination rates throughout the Phanerozoic eon. *Nature Geoscience*, **3**, 430–434.

Charlier, B. L. A., Ginibre, C., Morgan, D., Nowell, G. M., Pearson, D. G., Davidson, J. P. & Ottley, C. J. 2006. Methods for microsampling and high-precision analysis of strontium and rubidium isotopes at single crystal scale for petrological and geochronological applications. *Chemical Geology*, **232**, 114–133.

Davydov, V. I. & Leven, E. Y. A. 2003. Correlation of Upper Carboniferous (Pennsylvanian) and

Lower Permian (Cisuralian) marine deposits of the peri-Tethys. *Palaeogeography, Palaeoclimatology, Palaeoecology*, **196**, 125–155.

Deegan, F. M., Troll, V. R., Freda, C., Misiti, V., Chadwick, J. P., Mclead, C. L. & Davidson, J. P. 2010. Magma-carbonate interaction processes and associated CO_2 release at Merapi Volcano, Indonesia: insights from experimental petrology. *Journal of Petrology*, **51**, 1027–1051.

Denison, R. E., Koepnick, R. B., Fletcher, A., Howell, M. W. & Calloway, W. S. 1994. Criteria for the retention of original seawater $^{87}Sr/^{86}Sr$ in ancient limestones. *Chemical Geology*, **112**, 132–143.

Fielding, C. R., Frank, T. D., Birgenheier, L. P., Rygel, M. C., Jones, A. T. & Roberts, J. 2008. Stratigraphic imprint of the Late Palaeozoic Ice Age in eastern Australia: a record of alternating glacial and non-glacial climate regime. *Journal of Geological Society*, **165**, 129–140.

Font, L., Davidson, J. P., Pearson, D. G., Nowell, G. M., Jerram, D. A. & Ottley, C. J. 2008. Sr and Pb isotope micro-analysis of plagioclase crystals from Skye lavas: an insight into open-system processes in a flood basalt province. *Journal of Petrology*, **49**, 1449–1471.

García, R. 1972. El Permo-Carbonífero en Venezuela. *Boletín de la Sociedad Venezolana de Geólogos*, **VII**, 203–214.

Garrity, C. P., Hackley, P. C. & Urbani, F. 2010. *Digital shaded relief map of Venezuela (v. 2.0)*. United States Geological Survey, Open-File Report, **2004-1322**. World Wide Web Address: http://pubs.usgs.gov/of/2004/1322/

Gibbs, M. T., Rees, P. M., Kutzbach, J. E., Ziegler, A. M., Behling, P. J. & Rowley, D. B. 2002. Simulations of Permian climate and comparisons with climate-sensitive sediments. *Journal of Geology*, **110**, 33–55.

González De Juana, C. 1951. Introducción al estudio de la geología en Venezuela. *Venezuela Boletín de Geología*, **1**, 195–216.

Gradstein, F., Ogg, J. & Smith, A. 2004. *A Geologic Time Scale*. Cambridge University Press, Cambridge.

Grossman, E. L., Bruckschen, P., Mii, H., Chuvashov, B. I., Yancey, T. E. & Veizer, J. 2002. Carboniferous paleoclimate and global change: isotopic evidence from the Russian Platform. *In*: Chuvashov, B. I. & Amon, E. A. (eds) *Carboniferous stratigraphy and Paleogeography in Eurasia*. Institute of Geology and Geochemistry, Russian Academy of Sciences, Ekaterinburg, 61–71. World Wide Web Address: http://geoweb1.tamu.edu/faculty/grossman/Grossman02.pdf

Grossman, E. L., Yancey, T. E., Jones, T. E., Bruckschen, P., Chuvashov, B., Mazzullo, S. J. & Mii, H. 2008. Glaciation, aridification, and carbon sequestration in the Permo-Carboniferous: The isotopic record from low latitudes. *Palaeogeography, Palaeoclimatology, Palaeoecology*, **268**, 222–233.

Hoover, P. R. 1981. Paleontology, taphonomy and paleoecology of the Palmarito Formation (Permian of Venezuela). *Bulletin of American Paleontology*, **80**, 1–128.

Isbell, J. L., Miller, M. F., Wolfe, K. L. & Lenaker, P. A. 2003. Timing of late Paleozoic glaciation in Gondwana: was glaciation responsible for the development of northern hemisphere cyclothems? *In*: Chan, M. A. & Archer, A. W. (eds) *Extreme Depositional Environments: Mega End Members in Geologic Time*. Geological Society of America, Special Papers, **370**, 5–24.

Isbell, J. L., Cole, D. I. & Catuneanu, O. 2008. Carboniferous-Permian glaciation in the main Karoo Basin, South Africa: Stratigraphy, depositional controls, and glacial dynamics. *In*: Fielding, C. R., Frank, T. D. & Isbell, J. L. (eds) *Resolving the Late Paleozoic Ice Age in Time and Space*. Geological Society of America, Special Papers, **441**, 71–82.

Izart, A., Stephenson, R., Vai, G. B. *et al.* 2003. Sequence stratigraphy and correlation of late Carboniferous and Permian in the CIS, Europe, Tethyan area, North Africa, Arabia, China, Gondwanaland and the USA. *Palaeogeography, Palaeoclimatology, Palaeoecology*, **196**, 59–84.

Korte, C., Jasper, T., Kozur, H. & Veizer, J. 2005. $\delta^{18}O$ and $\delta^{13}C$ of Permian brachiopods: a record of seawater evolution and continental glaciation. *Palaeogeography, Palaeoclimatology, Palaeoecology*, **224**, 333–351.

Korte, C., Jasper, T., Kozur, H. W. & Veizer, J. 2006. $^{87}Sr/^{86}Sr$ record of Permian seawater. *Palaeogeography, Palaeoclimatology, Palaeoecology*, **240**, 89–107.

Laya, J. C. & Tucker, M. E. 2012. Facies analysis and depositional environments of Permian carbonates of the Venezuelan Andes: Palaeogeographic implications for Northern Gondwana. *Palaeogeography, Palaeoclimatology, Palaeoecology*, **331–332**, 1–26.

May, S. R. & Butler, R. F. 1986. North American Jurassic apparent polar wander – implications for plate motions, palaeogeography and Cordilleran tectonics. *Journal of Geophysical Research*, **91**, 11519–11544.

Martin, R. E. 1996. Secular increase in nutrients levels through the Phanerozoic: Implications for productivity, biomass, and diversity of the marine biosphere. *PALAIOS*, **11**, 209–219.

Montañez, I. P., Tabor, N. J., Niemeier, D. *et al.* 2007. CO_2-forced climate and vegetation instability during Late Paleozoic deglaciation. *Science*, **315**, 87–91.

Odreman, O. & Useche, A. 1997. *Formación Palmarito. Léxico Estratigráfico de Venezuela*. Ministerio de Energía y Minas. Comisión de Estratigrafía y Terminología, Caracas.

Odreman, O. & Wagner, Y. R. 1979. Precisiones sobre algunas floras carboníferas y pérmicas se los Andes venezolanos. *Boletin de Geología Ministerio de Energía y Minas, Caracas*, **13**, 77–81.

Parrish, J. T. 1993. Climate of the supercontinent Pangaea. *Journal of Geology*, **101**, 215–233.

Peyser, C. E. & Poulsen, C. J. 2008. Controls on Permo-Carboniferous precipitation over tropical Pangaea: a GCM sensitivity study. *Palaeogeography, Palaeoclimatology, Palaeoecology*, **268**, 181–192.

Perlmutter, M. A. & Plotnick, R. E. 2003. Hemispheric asymmetry of the marine stratigraphic record: conceptual proof of a unipolar ice cap. *In*: Cecil, C. B. & Edgar, T. N. (eds) *Climate Controls on Stratigraphy*. SEPM, Special Publications, **77**, 51–66.

Pierce, G. R., Jefferson, C. C. & Smith, W. R. 1961. Fossiliferous Paleozoic localities in Mérida Andes, Venezuela. *AAPG Bulletin*, **45**, 342–375.

Popp, B. N., Anderson, T. F. & Sandberg, P. A. 1986. Brachiopods as indicators of original isotopic compositions in some Paleozoic limestones. *Geological Society of America Bulletin*, **97**, 1262–1269.

Rees, P. M., Ziegler, A. M., Gibbs, M. T., Kutzbach, J. E., Behling, P. J. & Rowley, D. B. 2002. Permian phytogeographic patterns and climate data/model comparisons. *Journal of Geology*, **110**, 1–31.

Rigby, J. K. 1984. Permian Sponges from Western Venezuela. *Journal of Paleontology*, **58**, 1436–1462.

Ricardi, T. F. 2008. Venezuelan paleoflora of the Pennsylvanian-Early Permian: Paleobiogeographical relationships to central and western equatorial Pangea. *Gondwana Research*, **14**, 297–305.

Saltzman, M. R. 2003. Late Paleozoic ice age: ocean gateway or pCO_2? *Geology*, **31**, 151–154.

Scholle, P. & Ulmer-Scholle, D. 2003. *A Color Guide to the Petrography of Carbonate Rocks, Grains, Textures, Porosity, Diagenesis*. AAPG, Memoirs, **77**, 1–460.

Sellier De Civrieux, J. 1951. Ocurrencia del género Glovobalbulina en el Permiano de Venezuela. *Boletín de Geología*, **1**, 141–146.

Shagam, R. 1972. Geología de los Andes Venezolanos. (resum.) *Boletin de Geologia Publicacion Especial*, **5**, 935–938.

Shell De Venezuela & Creole Petroleum Corporation 1964. Paleozoic rocks of Mérida Andes, Venezuela. *AAPG Bulletin*, **48**, 70–84.

Steiner, M. B. 2003. A cratonic middle Jurassic paleopole: Callovian–Oxfordian pole stillstand (J-2 cusp), rotation of the Colorado Plateau, and Jurassic North American apparent polar wander. *Tectonics*, **22**, 1020–1041.

Steiner, M. B. & Lucas, S. G. 2000. Palaeomagnetism of the Late Triassic Petrified Forest Formation, Chinle Group, western United States: further evidence of 'large' rotation of the Colorado Plateau. *Journal of Geophysical Research*, **105**, 25791–25808.

Tabor, N. J. & Montañez, I. P. 2002. Shifts in late Paleozoic atmospheric circulation over western equatorial Pangea: Insights from pedogenic mineral $\delta^{18}O$ compositions. *Geology*, **30**, 1127–1130.

Tabor, N. J. & Poulsen, C. J. 2008. Palaeoclimate across the late Pennsylvanian-Early Permian tropical palaeolatitudes: a review of climate indicators, their distribution, and relation to palaeophysiographic climate factors. *Palaeogeography, Palaeoclimatology, Palaeoecology*, **268**, 293–310.

Tabor, N. J., Montañez, I. P., Scotese, C. R., Poulsen, C. J. & Mack, G. H. 2008. Paleosol archives of environmental and climatic history in paleotropical western Pangea during the latest Pennsylvanian through Early Permian. *In*: Fielding, C. R., Frank, T. D. & Isbell, J. L. (eds) *Resolving the Late Paleozoic Ice Age in Time and Space*. Geological Society of America, Special Papers, **441**, 291–304.

Thirlwall, M. 1991. High-precision multicollector isotopic analysis of low levels of Nd as oxide. *Chemical Geology*, **94**, 13–22.

Tucker, M. E. & Wright, V. P. 1990. *Carbonate Sedimentology*. Blackwell, Oxford.

Vai, G. B. 2003. Development of the palaeogeography of Pangaea from Late Carboniferous to Early Permian. *Palaeogeography, Palaeoclimatology, Palaeoecology*, **196**, 125–155.

Viscarret, P. 2002. *Estudio integral del Paleozoico no metamorfizado (formaciones Caparo, el Horno, Sabaneta y Palmarito) en los Andes de Mérida, Venezuela*. Diploma Dissertation, Universidad de Los Andes.

Viscarret, P. & Laya, J. C. 2007. Facies sedimentarias del pérmico en el flanco surandino de Venezuela. *Ciencia e Ingeniería*, **28**, 55–60.

Winguth, A. M. E. & Maier-Reimer, E. 2005. Causes of the marine productivity and oxygen changes associated with the Permian-Triassic boundary: a re-evaluation with ocean general circulation models. *Marine Geology*, **217**, 283–304.

Winguth, A. M. E., Heinze, C., Kutzbach, J. *et al.* 2002. Simulated warm polar currents during the middle Permian. *Paleoceanography*, **17**, 1057–1075.

Witzke, B. J. 1990. Palaeoclimatic constraints for Palaeozoic palaeolatitudes of Laurentia and Euramerica. *In*: McKerrow, W. S., Scotese, C. R. (eds) *Palaeozoic Palaeogeography and Biogeography*. Geological Society, London, Memoirs, **12**, 57–74.

Ziegler, A. M., Eshel, G., Rees, P. M., Rothfus, T. A., Rowley, D. B. & Sunderlin, D. 2003. Tracing the tropics across land and sea: Permian to present. *Lethaia*, **36**, 227–254.

Carbon and oxygen isotope records of Permian brachiopods from relatively low and high palaeolatitudes: climatic seasonality and evaporation

JESPER KRESTEN NIELSEN[1,2]*, BŁAŻEJ BŁAŻEJOWSKI[3], PIOTR GIESZCZ[4] & JAN KRESTEN NIELSEN[5]

[1]*North Energy ASA, Kunnskapsparken, Markveien 38 B, NO-9504 Alta, Norway*

[2]*Department of Geology, University of Tromsø, Dramsveien 201, NO-9037 Tromsø, Norway*

[3]*Institute of Paleobiology, Polish Academy of Sciences, Twarda 51/55, PL 00-818 Warszawa, Poland*

[4]*Association of Polish Climatologists, ul. Międzynarodowa 58/36, PL03-922 Warszawa, Poland*

[5]*Statoil ASA, Field Development North, P.O. 273, NO-7501 Stjørdal, Norway*

**Corresponding author (e-mail: jesperknielsen@gmail.com)*

Abstract: Several brachiopods which belong to the same genus *Horridonia* of late Early and early Late Permian age in Spitsbergen and Poland, respectively, have been petrographically and geochemically analysed to verify seasonal variation in stable carbon and oxygen isotope values for palaeoclimatological implications. The specimens of *Horridonia timanica* from Spitsbergen show distinct cyclicity reflective of seasonal pattern, while those of *Horridonia horrida* from Poland do not. These differences are explained by the fact that the former lived at high palaeolatitudes at the northern margin of the supercontinent Pangaea where the seawater temperature differences between winter and summer seasons were greater, as expressed in the isotopic composition of the skeletal materials. In contrast, the shell growth of *Horridonia horrida* was subjected to strong evaporative influence by climatic variations in the eastern area of Pangaea.

Carbon, oxygen and strontium isotopes of brachiopods of a wide range of ages are often used to determine seawater isotopic composition chemostratigraphic correlation and characteristics of their living habitat such as seawater temperature and thereby palaeoclimate. Studies of brachiopods include: the Wegener Halvø Formation of Permian age in East Greenland (Scholle *et al.* 1991; Scholle 1995); the Carboniferous and Permian of Kansas, Texas and New Mexico (Magaritz *et al.* 1983; Given & Lohmann 1986; Grossman *et al.* 1991, 1993, 1996; Mii & Grossman 1994); Silurian brachiopods from Gotland in Sweden (Bickert *et al.* 1997); Upper Pleistocene brachiopods from New Zealand (Curry & Fallick 2002); Late Palaeozoic brachiopods from North America (Popp *et al.* 1986; Brand 1989); Early–Middle Permian bivalves and brachiopods from Australia (Korte *et al.* 2008; Ivany & Runnegar 2010; Mii *et al.* 2012) and the Kapp Starostin Formation on Spitsbergen of the archipelago Svalbard (Gruszczyński *et al.* 1992; Mii *et al.* 1997); and more geographically widespread studies (Grossman 1994; Veizer *et al.* 1997; Korte *et al.* 2005; Korte & Kozur 2010) and studies of modern brachiopods (Lowenstam 1961; Barbin & Gaspard 1995; Carpenter & Lohmann 1995; James *et al.* 1997; Auclair *et al.* 2003; Brand *et al.* 2003). The majority of brachiopod species have shells that are composed of low-magnesium calcite, which is the most stable skeletal carbonate over geological timescales; such brachiopod shells are therefore less susceptible to diagenetic alteration. Advanced petrography such as cathodoluminescence and scanning electron microscopy of shell material and its microstructure is crucial for exploring the possibility of fabric-retentive and non-luminescent materials, that is, well-preserved primary shell materials (e.g. Rush & Chafetz 1990; Barbin & Gaspard 1995; Mii *et al.* 1997). Isotopic compositions of brachiopods have often shown to be comparable to those of marine cements and thereby seawater (Given & Lohmann 1985; Scholle *et al.* 1991). A detailed sub-sampling of growth bands of brachiopods can be utilized to obtain the carbon and oxygen (and strontium) isotope values of past seawater, especially for chemostratigraphic correlations. Such isotope

From: Gąsiewicz, A. & Słowakiewicz, M. (eds) 2013. *Palaeozoic Climate Cycles: Their Evolutionary and Sedimentological Impact*. Geological Society, London, Special Publications, **376**, 387–405.
First published online November 26, 2012, http://dx.doi.org/10.1144/SP376.6

studies of brachiopods can be placed in a biogeographic and palaeoclimatic context (e.g. Shen & Shi 2000, 2004) to unravel seasonality during growth of the brachiopods (Grossman *et al.* 1996).

To investigate whether the isotopic compositions of late Early and early Late Permian brachiopods of the same genus from Spitsbergen and Poland can be used to determine seasonal variation, we have carried out a series of carbon and oxygen isotope analyses on increment level. It is shown that the analysis of preserved microstructures of the productid brachiopods of *Horridonia horrida* (Sowerby 1823) and *H. timanica* (Stuckenberg 1875) are useful in determining the palaeoclimate and, to some extent, the palaeoceanographic conditions during the Permian. The specimens from the northern margin of Pangea show distinct cyclicity (seasonal pattern), while those from Poland do not.

Geological setting

A marked change in the lithofacies and biofacies encountered for the Tempelfjorden Group (late Artinskian–Wuchiapingian, possibly Changhsingian) and the underlying Gipsdalen Group (late Serpukhovian–early Artinskian) and Bjarmeland Group (late Sakmarian–early Kungurian) on Svalbard and in the Norwegian Barents Sea (Figs 1 & 2) characterizes a major climatic change as the supercontinent Pangea drifted northwards and the oceanic current systems underwent change related to rifting (Beauchamp 1994; Stemmerik & Worsley 2005; Stemmerik 2008; Worsley 2008). Intracratonic rift systems developed to the west of Svalbard and the Boreal Realm became connected via the Greenland–Norway Sea, forming the Zechstein seaway to the central European Zechstein basins (the Northern Permian Basin and Southern Permian Basin; Ziegler 1990; Stemmerik *et al.* 1991; Ziegler *et al.* 1997; Figs 3–5). During the Late Carboniferous–Late Permian, the present-day Spitsbergen drifted approximately from 20°N to 45°N (Stemmerik 2000; Golonka 2011; Fig. 1) while the Southern Permian Basin was located *c.* 30° further to the south. Along this south–north-trending seaway, the climate changed from arid evaporative subtropical environment at low palaeolatitudes to temperate colder environment further north (Stemmerik 1995, 2000).

Along the Norwegian–Greenland seaway, the Ravnefjeld Formation (onshore East Greenland) and the offshore Mid Norway of Wuchiapingian age are regarded to be time-equivalent to the Kupferschiefer (Stemmerik 1995; Stemmerik *et al.* 2001; Bugge *et al.* 2002; Menning *et al.* 2006) and, further north, roughly equivalent to the Kapp Starostin, Røye and Ørret formations (Stemmerik & Worsley 2005; Figs 2 & 5). As for the Kupferschiefer, Ravnefjeld, Kapp Starostin and Ørret Formations, dysoxic to anoxic sulphidic bottom-water conditions existed in deeper waters of the European Zechstein basins (Pancost *et al.* 2002), the East Greenland Basin (Nielsen & Shen 2004; Nielsen *et al.* 2010) and the Barents Shelf and Spitsbergen (Nielsen *et al.* 2012).

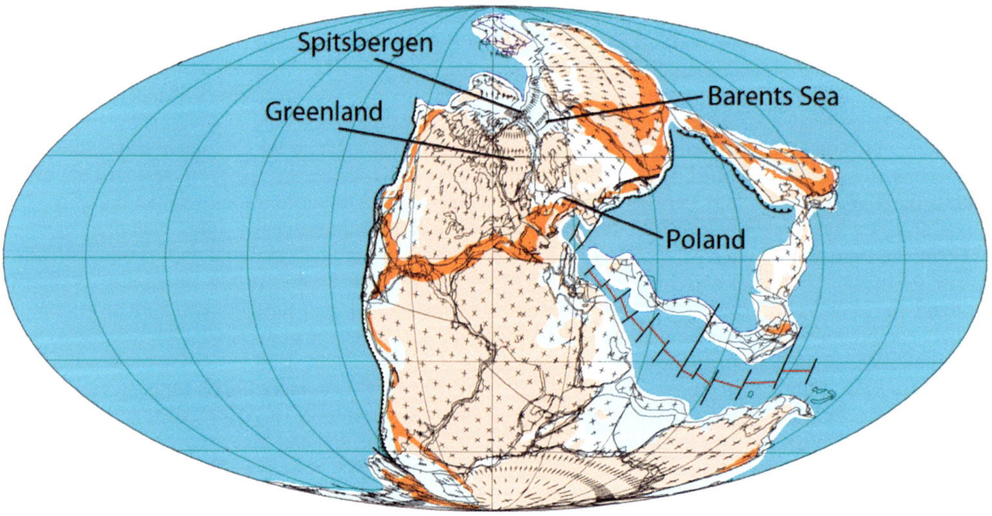

Fig. 1. Palaeogeographic map of the Guadalupian time. Modified from Golonka (2011). Blue: deep basin; pale blue: shelf area; orange: topographic highs, inactive tectonically; pale brownish: topographic medium to low, inactive tectonically.

Boreal realm (Spitsbergen)

Cutbill & Challinor (1965) introduced the term Tempelfjorden Group (equivalent to the Russian name *Starostinskaya Svita*) for a suite of spiculites (spiculitic chert), siliceous (spiculitic) shales and siltstones with intercalated minor glauconitic sandstones and silicified skeletal limestones of late Early–Late Permian age (late Artinskian–Kazanian/?Tatarian) in the onshore areas (Stemmerik 1988; Nakrem 1991, 2005; Mangerud 1994; Fig. 2). The carbonates contain a fauna often dominated by brachiopods, sponges, bryozoans and crinoids. The type area for the Tempelfjorden Group is in the innermost part of Isfjorden in central Spitsbergen (Cutbill & Challinor 1965; Dallmann *et al.* 1999). The Kapp Starostin Formation of the Tempelfjorden Group has its type section in the Festningen section (Frebold 1937; Orvin 1940; Dallmann *et al.* 1999). The lowermost part of the Kapp Starostin Formation is dominated by sandy bioclastic limestone with a rich brachiopod and bryozoan fauna in the up to 20 m thick Vøringen Member (the so-called 'Spirifer limestone'; late Artinskian–early Kungurian age), overlaid by alternating shales, siltstones and cherts and siliceous limestones of the Svenskeegga Member (Kungurian–Kazanian). The uppermost unit, the *c.* 50 m thick Hovtinden Member (Kazanian–Tatarian), is composed of shales and siltstones with few fossils (Nakamura *et al.* 1987; Małkowski 1988; Stemmerik 1988; Mangerud & Konieczny 1993; Nakrem 1995; Dallmann *et al.* 1999). The age of the Vøringen Member is based on conodonts (Szaniawski & Małkowski 1979), non-fusulinid foraminiferans in the underlying Gipsdalen Group (Sosipatrova 1967, 1969; Błażejowski 2009), palynomorphs (Mangerud & Konieczny 1993) and bryozoans (Nakrem 1995). Although the age of the upper part of the Kapp Starostin Formation is somewhat problematic, Nakrem *et al.* (1992) indicate a Kazanian age (see also Nakrem 2005). Based on the similarities of the brachiopod faunas between the upper part of the Kapp Starostin Formation and the Foldvik Creek Group in East Greenland, the

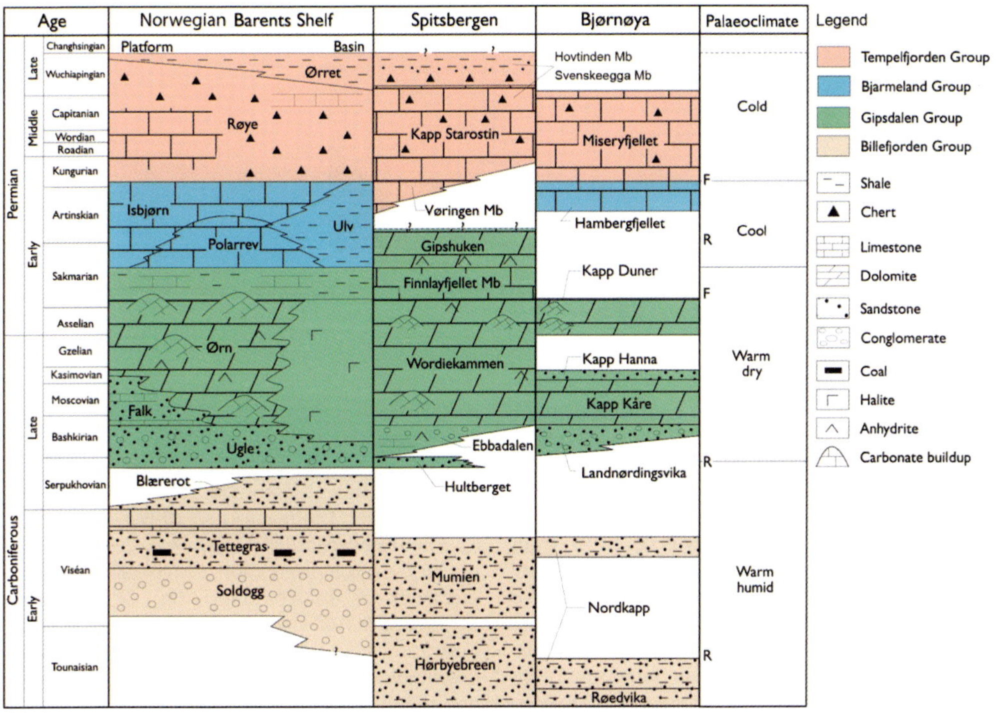

Fig. 2. Stratigraphic overview of the Norwegian Barents Sea and Svalbard. Stratigraphic scheme showing the distribution of siliciclastic deposits with coal seams (Early Carboniferous), warm-water carbonates (Late Carboniferous–Early Permian) and cool- and cold-water carbonates and cherts (late Early Permian–Late Permian) related to the northern drift of Pangea and changes of the palaeoclimate. Slightly modified from Stemmerik & Worsley (2005). The investigated *Horridonia timanica* (Stuckenberg 1875) are from the latest Early Permian Vøringen Member of the Kapp Starostin Formation.

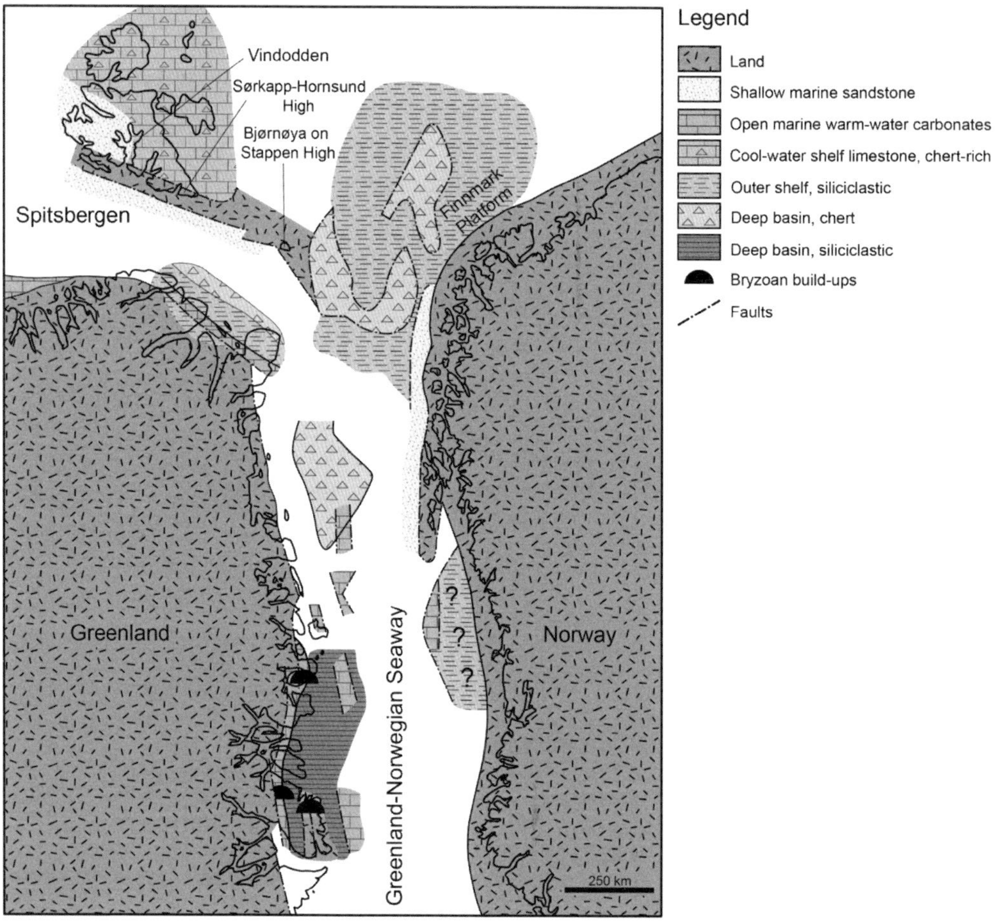

Fig. 3. Palaeogeographic map of the Greenland–Norwegian seaway, Norwegian Barents Sea and Svalbard (Kazanian time). Slightly modified from Stemmerik (2000).

upper part of the Kapp Starostin Formation may be of Kazanian–early Tatarian age (Stemmerik 1988; Fig. 2). Time-equivalent deposits are found offshore in the Norwegian Barents Sea where the spiculitic deposits of the Røye Formation and shales of the Ørret Formation are especially well described from the eastern Finnmark Platform (southeastern Norwegian Barents Sea; Henriksen *et al.* 2011). The Tempelfjorden Group usually thins over local structural highs, where the exposures on Bjørnøya on the Stappen High (western Norwegian Barents Sea), for example, show an extremely condensed development of *c.* 115 m in thickness development. Highly condensed exposures on the margins of the Sørkapp-Hornsund High of Spitsbergen are only a few metres thick and pinch out over the structure (Hellem & Worsley 1978). The thickness of the Kapp Starostin Formation reaches up to 460 m; it is 380 m in the type section at Isfjorden (the Festningen section) and becomes highly condensed and pinches out on the eastern margin of the Sørkapp-Hornsund High (Małkowski 1982, 1988; Steel & Worsley 1984; Dallmann *et al.* 1999).

The Tempelfjorden Group was deposited during an overall transgression accompanied by retrogradation of the coastline (Fig. 2). The main accumulations of spiculites of the Kapp Starostin Formation are related to transgressive periods when favourable environmental conditions for sponges prevailed over most of the shelf. Cool-water bryozoan carbonates formed along the margins and formed low-relief platforms during sea level highstand (Larssen *et al.* 2005). Above the Vøringen Member, abundant trace fossils such as *Zoophycos* and *Chondrites* indicate low-energy as well as oxic to dysoxic bottom waters well below normal

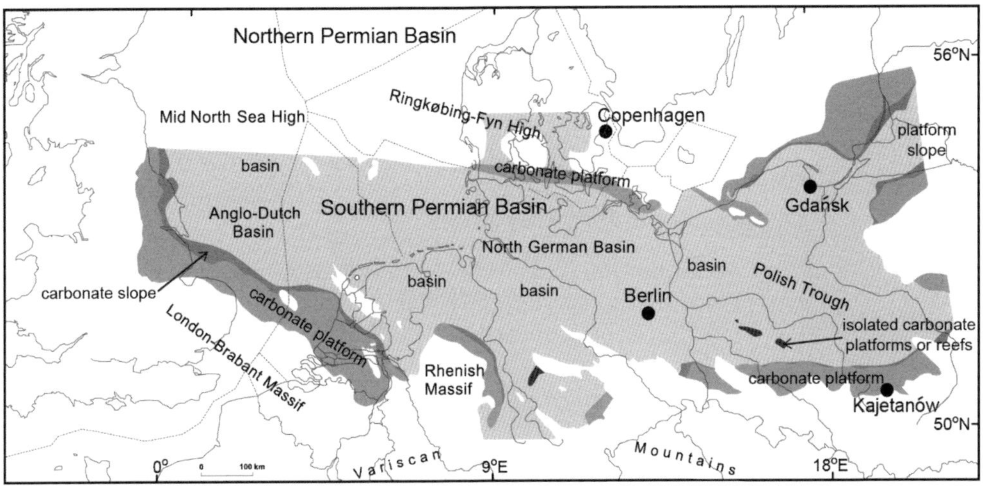

Fig. 4. Palaeogeography of the Zechstein Limestone in the Southern Permian Basin. Modified from Peryt *et al.* (2010).

wave base (Małkowski 1988; Nakrem 2005; Nielsen *et al.* 2012). Such depositional conditions were punctuated by the development of a stratified water column where anoxic and sulphidic bottom waters persisted for a shorter time, based on sedimentological, palaeoecological and geochemical evidence (Małkowski & Hoffman 1979; Małkowski 1982, 1988; Nielsen *et al.* 2012). Other factors controlling the distribution of the brachiopods may also be substrate mobility, particularly of the spiculitic deposits found above the Vøringen Member (Małkowski 1988) where brachiopods are only occasionally found in coquinas which may represent storm rework (Nakrem 2005).

The main groups of fossils in the Kapp Starostin Formation comprise brachiopods, sponges (mainly siliceous), bryozoans, foraminifers, ostracods, echinoderms (crinoids), bivalves, gastropods, corals and algae (e.g. Siedlecka 1970; Małkowski 1988; Nakrem 2005; Błażejowski 2008). The fine-grained sediments are dominated by abundant delicate bryozoans, trepostomids, rhabdomesids, cryptostomids and fenestellids, while the more robust trepostomids, *Polypora*, *Acanthocladia* and *Reteporidra*, dominate the partially silicified limestones (Nakrem 1994, 1995). Most of the fossils are typical cool-to cold-water forms while the occurrence of temperate to warm-water forms of corals, trilobites and ammonoids are rare, according to Nakrem (2005). In comparison, the underlying deposits of the Gipsdalen Group contain warm-water forms such as abundant *Palaeoaplysina*, phylloid algae and fusulinid foraminifers (Stemmerik 1997; Błażejowski *et al.* 2006; Hanken & Nielsen, this volume, in review; Fig. 2).

Brachiopods from Permian limestones, central Spitsbergen

Assemblages of brachiopods from ten outcrops of the Kapp Starostin Formation in central and southern Spitsbergen have been comprehensively described by Małkowski (1988) (Figs 2 & 6). The majority of the over 2500 specimens are well preserved, often with clearly visible morphology, growth bands, inner structure and microstructure of the shell (Figs 6–8), except for a few partially silicified (Małkowski 1988) and/or partially replaced with blocky low-Mg calcite (Figs 6 & 7) specimens. At the Vindodden section (Fig. 9), the Vøringen Member (*c.* 40 m thick) is composed of fossiliferous biocalcarenites and biomicrites interbedded with non-fossiliferous marly sediments with some gravel. The fossil assemblage is dominated by abundant productids, such as *Horridonia timanica*, and other brachiopods such as spiriferids including mostly broken bryozoans, crinoids and bivalves. The Svenskeegga Member (*c.* 180 m thick) is dominated by spiculitic deposits and some marly limestones containing bryozoans, cystoid crinoids, pectenid bivalves, solitary corals and rare gastropods, but somewhat fewer brachiopods (Małkowski 1988). The latter are mainly confined to the storm-induced coquinas and platform edge bars above normal wave base during lower sea level (Nakrem 2005). Dark shales and locally developed glauconitic porous sandstones of the Hovtinden Member (*c.* 110 m thick) contain some productids, spiriferids, bryozoans, bivalves and solitary corals. Size of the *Horridonia timanica* varies somewhat throughout the formation, often well preserved in or close

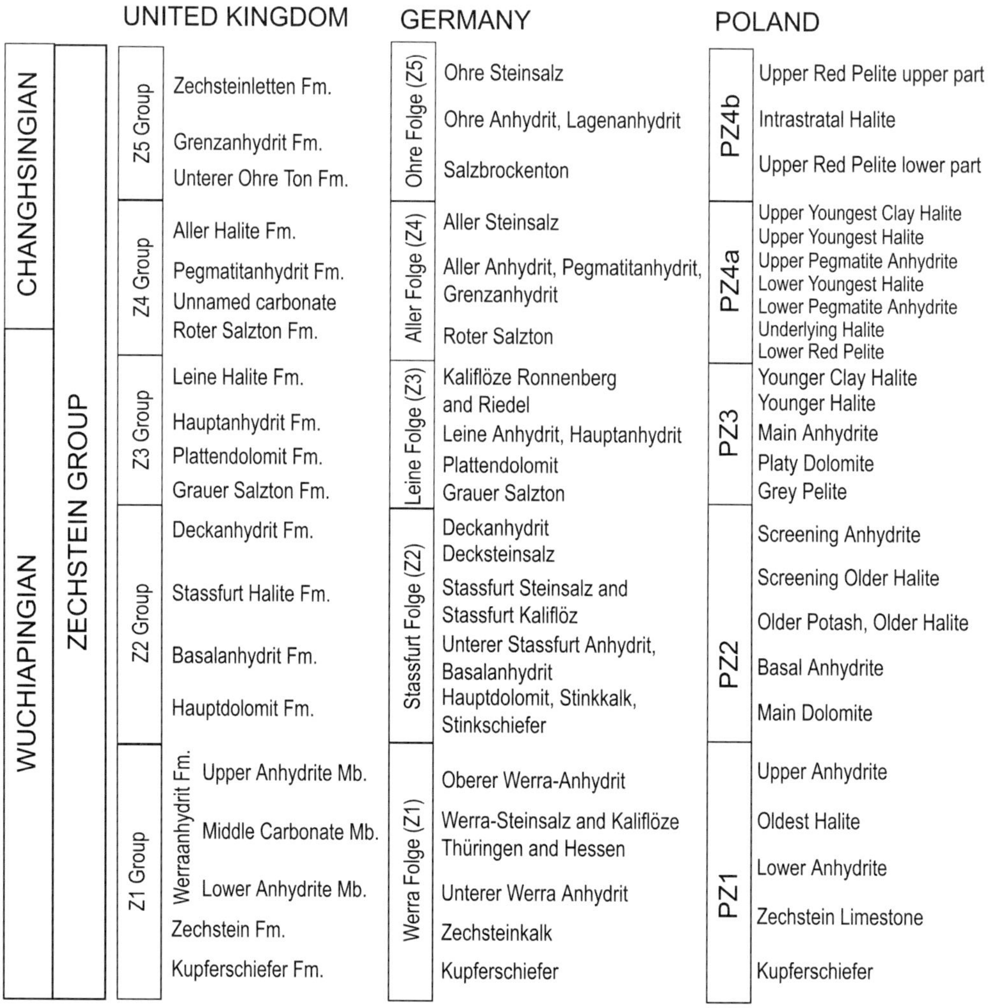

Fig. 5. Classification and correlation of the Zechstein Group in the UK, southern North Sea, North Germany and Poland according to Peryt *et al.* (2010). See also Wagner (1994), Johnson *et al.* (1994), Kiersnowski *et al.* (1995), McCann *et al.* (2008), Mawson & Tucker (2009) and Słowakiewicz *et al.* (2009). The Zechstein limestone of PZ1 (Wuchiapingian) includes the Lower Zechstein black limestones from which the investigated *Horridonia horrida* (Sowerby, 1823) comes from.

to life position. Their disappearances and reappearances are related to facies zones and most likely to fluctuations in oxygen deficiency and substrate properties (most likely substrate nature), as they have been found in northeastern Spitsbergen (Małkowski 1988). The lifestyle of *Horridonia timanica*, *Svalbardoproductus arctius* and *Anemonaria pseudohorrida* was stabilization in the sediment, commonly on coarse-grained loose substrates rich in skeletal debris, by anchoring and supporting it with their long and straight spines. The spines grow mainly on the margins of the ventral valve and on the auricle edges (Małkowski 1988). Małkowski (1988) also indicates that the *Horridonia timanica* is valuable as an indicator of the environment.

The mode of life of brachiopods such as *Waagenoconcha irginae* and *Kochiproductus porrectus* was floating at the surface of the soft substrate and such as *Linoproductus dorotheevi* that was velum-bearing forms within the sediment. *Waagenoconcha irginae* lived between the skeletal banks dominated by *Horridonia timanica* and *Svalbardoproductus arctius*. Other brachiopods achieve stability of the shell by the resistance of the ventral valve, but lack spines (Małkowski 1988).

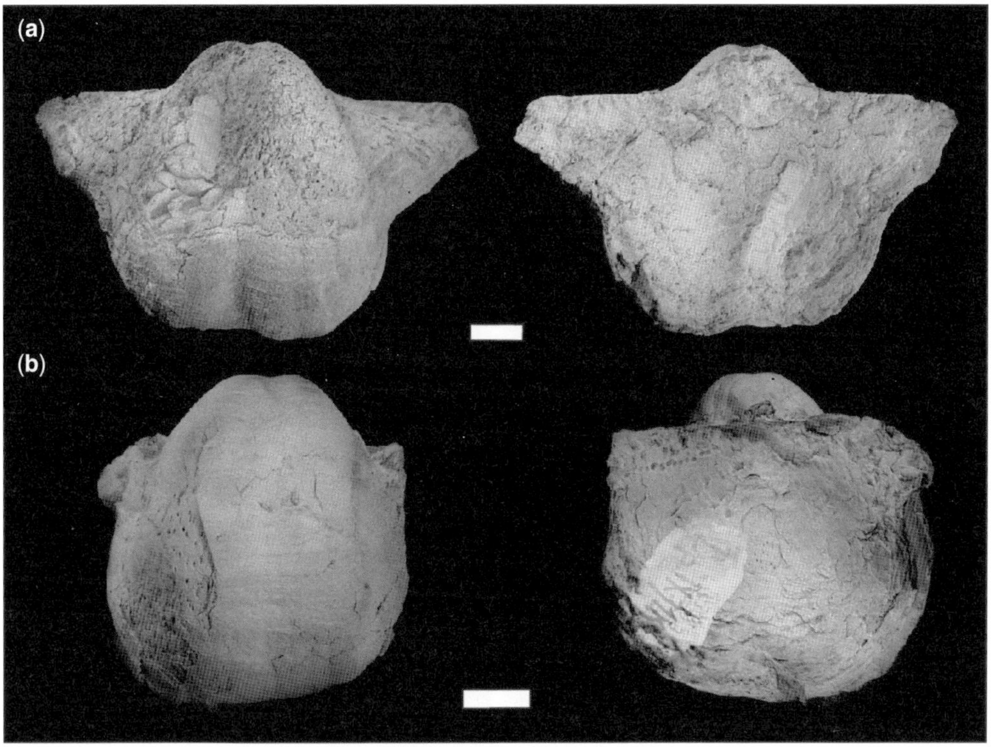

Fig. 6. Overview images of the investigated brachiopods as examples of the two species. (**a**) *Horridonia timanica* (Stuckenberg 1875), Vindodden section; viewed ventrally and dorsally (ZPAL Bp. XXX/V-414). (**b**) *Horridonia horrida* (Sowerby 1823), Kajetanów quarry; adult specimens, viewed ventrally and dorsally (ZPAL Bp. IX/1 K). The length of each scale bar is 1 cm.

Southern Permian Basin (Poland)

The Rotliegend of Germany, Netherlands and Poland begins in the latest Carboniferous Gzhelian Stage and continues into the Late Permian Wuchiapingian Stage, while the overlying Zechstein can be correlated with the Middle Wuchiapingian and most of the Changhsingian Stage (Figs 4 & 5) (Wagner 1994; Johnson *et al.* 1994; Kiersnowski *et al.* 1995; Geluk 1999; Brauns *et al.* 2003; McCann *et al.* 2008). The Zechstein–Buntsandstein boundary does not coincide with the Permian–Triassic boundary but is still latest Permian in age based on conchostrachans, palaeomagnetic and palynological data (e.g. Geluk & Röhling 1997; Kozur 1999; Szurlies *et al.* 2003; Menning *et al.* 2006). Megaspores, conchostrachans and magnetostratigraphy indicate that the Permian–Triassic boundary is located *c.* 30 m above the base of the Lower Buntsandstein (in the M2 cycle; Kozur 1999; Szurlies *et al.* 2003), that is, the Zechstein–Buntsandstein boundary is about 0.1 Ma older than the Changhsingian–Indusian (i.e. Permo-Triassic?) boundary (Menning *et al.* 2006).

During the Permian, the Norwegian–Greenland seaway was established by southwards transgression preceded by intra-Kungurian pulses of rifting which, in combination with glacio-eustatic sea level rise due to the deglaciation in Gondwana, reached the Northern Permian and Southern Permian basins (Surlyk *et al.* 1986; Ziegler 1990; Stemmerik 1995; Ziegler *et al.* 1997). These basins were topographic lows and became rapidly flooded (Glennie & Buller 1983). An arid climate and a repeated tectonic and glacio-eustatic restriction on renewal supply of marine waters from the north resulted in deposition of the cyclical Zechstein Group (Fig. 5). The Zechstein Group is very variable in thickness, partly as a result of post-depositional salt movements (halokinesis). The cycles typically comprise transgressive anoxic shales of the Kupferschiefer followed by carbonates and thick evaporites (mostly anhydrite and halite), occasionally intercalated with local clastic rocks. Evaporites such as halite

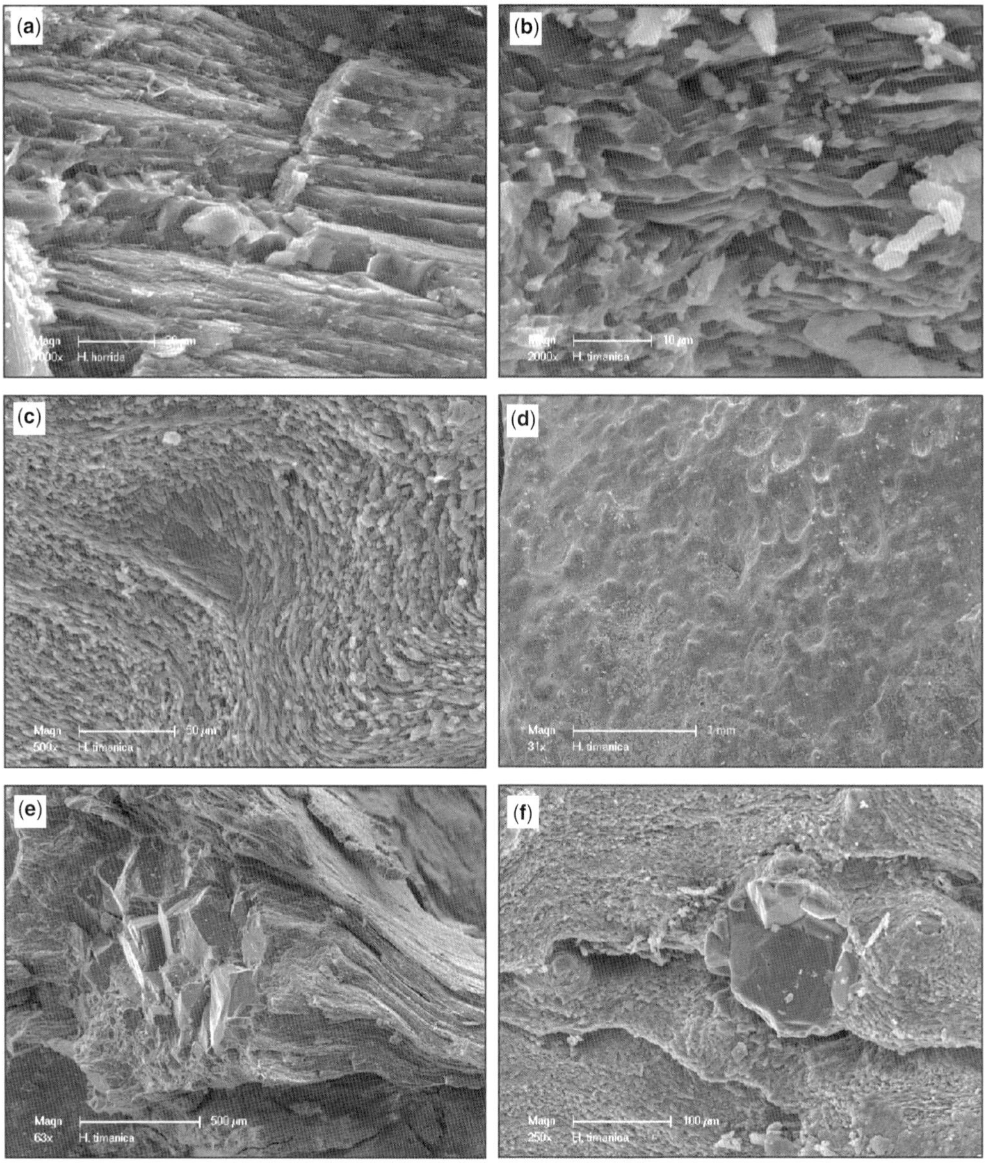

Fig. 7. Photomicrographs of the microstructures of the brachiopods. (**a**) Primary microstructure of *Horridonia horrida* (Sowerby 1823), Kajetanów quarry, Poland (ZPAL Bp. IX/1 K). (b–f) Microstructures of the *Horridonia timanica* (Stuckenberg 1875), Vindodden section, Spitsbergen (ZPAL Bp. XXX/V-414). (**b**, **c**) Well-preserved primary microstructure. (**d**) Surface structures. (**e**, **f**) Blocky calcite has replaced smaller parts of the shell.

dominate the basin-centre sequences, while limestones, dolomites and some anhydrites prevail along the basin margins (Fig. 4; Wagner 1994; Kiersnowski *et al.* 1995; Geluk 2007; McCann *et al.* 2008; Peryt *et al.* 2010). After the initial transgression, the Southern Permian Basin reached 200–300 m of water depth (Ziegler 1990). Each cycle becomes progressively more evaporitic, finally ending with potash deposits. We have investigated brachiopods from limestones of the oldest cyclothem (PZ1, Poland), which is time-equivalent to the Z1 Werra cycle in NW Germany (Fig. 5). The limestones are confined to shallow waters in basin margin areas and on intra-basinal highs, where a

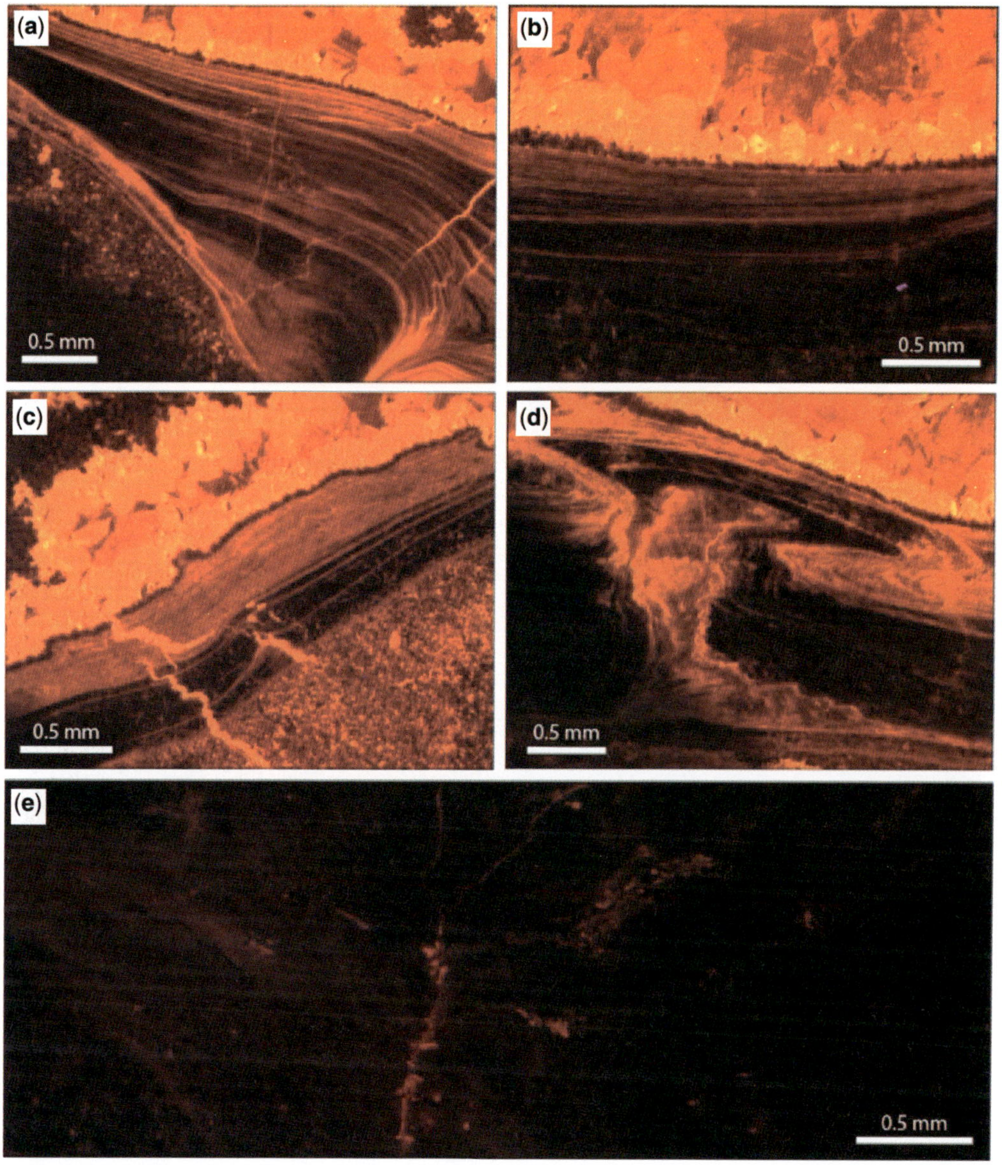

Fig. 8. Cathodoluminescence of the brachiopods. The non-luminescent to slightly luminescent (dark grey to black) is that part of the skeleton which holds its origin. (**a–d**) *Horridonia horrida* (Sowerby 1823), Kajetanów quarry, Poland. Sample ZPAL Bp. IX/65 K. The shells are slightly to non-luminescence except for a very few yellowish luminescing calcite infilled fractures. Some of the shells have worked as umbrella for early diagenetic yellow to brownish luminescing calcite cements. Examples of (a) umbo, (b) central portion of valve, (c) non-luminescing outer layer of central portion of valve and (d) hinge area. (**e**) *Horridonia timanica* (Stuckenberg 1875), Vindodden section, Spitsbergen. Sample ZPAL Bp. XXX/V-5. The shell is non-luminescing except for a slightly luminescing calcite infilled fracture. Samples for isotope analysis were taken from non-luminescing parts containing a seawater isotope signature (avoiding sampling of the luminescing diagenetic-influenced parts).

rich fauna with a close affinity to the Norwegian–Greenland seaway and the Boreal Realm lived. The deposits sequence of Z1 is considered as being formed in an arid climate within a shallow intra-continental sea (Wagner & Peryt 1997; Słowakiewicz *et al.* 2009).

Fig. 9. Field photographs of the Vøringen Member limestones, Vindodden section, central Spitsbergen. (**a**) The limestone is typically thick bedded with some vague internal stratification. As scale S.-A. Grundvåg. (**b**) A close-up of the deposits shown in (a). Most of the brachiopod shells have their convex side upwards. The deposition occurred in a high-energy environment above fair-weather wave base.

Brachiopods from the Lower Zechstein black limestones, Kajetanów, Poland

Investigated material was collected from abandoned quarry in Kajetanów village which is situated *c.* 10 km northward of Kielce (Holly Cross Mountains region, central Poland) (Figs 4 & 5). Dark and black limestones and marls bedded rhythmically dipping northwards at 12–15° are cut by minor faults and appear in the western wall of the Kajetanów quarry. The name Kajetanów Limestones was introduced by Pawłowska (1978); at present they form a succession of up to 8 m thick. The limestones are medium- and thin-bedded, and show discrete horizontal and wavy lamination. An admixture of quartz grains, mica flakes and scattered pyrite as well as small sandstone clasts accompanied by pyritized bioclasts and fragments of plants are visible in thin sections. The lower part of the profile – the Productus Limestones – consists of medium- and thick-bedded limestones intercalated by marls (*c.* 1.5 m thick), dated by the occurrence of brachiopod *Horridonia horrida* (Sowerby 1823). The upper part of the profile pass into stratified medium- and thin-bedded limestones and marls (*c.* 4 m thick, the Strophalosia Marls) with abundant macro- and microfossils that are represented by *Horridonia horrida* (Sowerby), *Strophalosia morrisiana* King, *Dielasma elongatum* Schlotheim, *Lingula credneri* Geinitz, *Bakewella ceratophaga* Schlotheim, *B. antiqua* Verneull, *Nucula beyrichi* Schlotheim, *Stenopora columnaria* Schlotheim, *Acanthocladia anceps* Schlotheim and *Agathamina pusilla* Geinitz (after Fijałkowska-Mader 2010; see also Jurkiewicz 1962; Kaźmierczak 1967).

Materials and methods

For the purposes of this study, we have chosen to investigate two species *Horridonia horrida* (Sowerby 1823) and *Horridonia timanica* (Stuckenberg 1875) based on their abundance and preservation, including the evidence of autochthonous (i.e. isotopically analysed) specimens which were collected in their life position.

The specimens of *Horridonia horrida* come from the Lower Zechstein black (bituminous) limestones from Kajetanów, Holly Cross Mountains, Poland (Kaźmierczak 1967; Figs 4, 5 & 10). The material was collected by J. Kaźmierczak during summer 1964 in the Kajetanów quarry and in several outcrops in the western part of Gałęzice-Kowala syncline in the Holy Cross Mountains, southern Poland. The collection is housed in the Institute of Palaeobiology, Polish Academy of Sciences, and has unique value since the outcrops no longer exist. The collection, which numbers a total

Fig. 10. Kajetanów no longer exists as a quarry, and it is impossible to get new samples from there. The picture of the outcrop was taken by Prof. W. Trela when it was still accessible.

of 89 specimens as well as a few separate valves, is marked with the catalogue numbers ZPAL Bp. IX/ 1–137 (Kaźmierczak 1967).

The brachiopods *Horridonia timanica* are from the Vøringen Member of the Kapp Starostin Formation (Tempelfjorden Group, late Early–Late Permian), Vindodden, central Spitsbergen (Figs 2 & 9). The material examined in this paper is from the Vindodden section, one of 10 sections of the Kapp Starostin Formation from where K. Małkowski collected his brachiopods (described in detail in 1988). Małkowski's collection is housed in the Institute of Palaeobiology of the Polish Academy of Sciences, Warsaw, labelled ZPAL Bp. 1–2500 (Małkowski 1988).

To exclude the possibility of diagenetic overprint in the investigated fossil shells (e.g. Popp *et al.* 1986; Brand *et al.* 2003), we examined the chosen samples using scanning electron microscopy (SEM) with an energy-dispersive spectrometer (EDS). Carbon-coated fragments and polished thin sections of the samples were examined in secondary electron (SE) and backscatter electron (BSE) operational modes, respectively. In that way, we could examine microstructure and composition in two and three dimensions.

Thick polished sections of *H. horrida* and *H. timanica* shell were examined using plane light and cathodoluminescence (CL) microscopy. CL analyses were carried out on a Technosyn Model

Table 1. *Carbon and oxygen isotopic composition of* Horridonia timanica *(Stuckenberg 1875) from Spitsbergen*

Horidonia timanica (Stuckenberg 1875)					
Sample ZPAL Bp. XXX/V-42			Sample ZPAL Bp. XXX/V-5		
Subsample No.	$\delta^{13}C$ (‰)	$\delta^{18}O$ (‰)	Subsample No.	$\delta^{13}C$ (‰)	$\delta^{18}O$ (‰)
1	3.40	−6.14	1	5.38	−3.68
2	3.78	−6.35	2	5.63	−3.43
3	3.71	−6.71	3	5.66	−3.50
4	3.84	−4.76	4	5.63	−3.77
5	4.20	−5.36	5	5.50	−4.28
6	4.46	−4.73	6	5.20	−4.64
7	4.05	−6.22	7	5.06	−4.63
8	3.81	−7.49	8	5.26	−4.59
9	4.08	−6.85	9	5.26	−4.91
10	4.42	−4.83	10	5.34	−4.37
11	4.81	−4.37	11	5.26	−3.98
12	3.92	−7.45	12	5.33	−4.48
13	3.79	−7.27	13	5.21	−4.29
14	4.34	−5.46	14	5.25	−4.12
15	4.33	−5.66	15	4.67	−4.29
16	4.25	−5.38	16	4.96	−3.89
17	3.88	−8.19	17	5.13	−3.95
18	4.06	−6.90	18	5.06	−4.23
19	4.46	−4.51	19	5.14	−4.32
20	4.17	−5.25	20	–	–
21	4.48	−4.65	21	4.34	−4.63
22	4.59	−5.06	22	4.91	−3.71
23	4.34	−5.73	23	4.68	−4.08
24	3.98	−6.49	24	4.91	−4.02
25	4.03	−5.80	25	5.02	−4.35
26	4.18	−4.87	26	5.07	−4.52
27	4.03	−5.50	27	5.21	−4.54
28	3.77	−6.57	28	4.79	−5.18
29	4.13	−5.67	29	4.41	−4.92
30	4.19	−5.19	30	4.51	−4.21
31	3.87	−5.17			
32	3.72	−6.23			
33	3.38	−6.78			
34	3.54	−6.69			
35	3.16	−7.29			

8200 MK II CL stage coupled to an Olympus microscope (Grossman *et al.* 1993). The digital images were taken with Nikon DS-Ri1 camera. Any diagenetically altered shell materials were eliminated from further analyses.

Prior to taking carbonate samples, the shell was cleaned to remove any surface contamination. Individual carbonate powder samples (>50 μg) were millcut under the microscope sequentially from the outer layer along the axis of maximum growth from the interior area less influenced by diagenesis. Spatial resolution was *c.* 1 mm.

For stable isotope analysis, the carbonate powder was reacted with 100% orthophosphoric acid under vacuum at 70 °C in a KIEL IV carbonate device, which was coupled to a Finnigan MAT Delta Plus isotope ratio mass spectrometer. Isotope ratios are reported in per mil (‰) in the usual delta notation relative to the VPDB (Vienna Peedee Belemnite) standard (defined via NBS 19). The spectrometer external error amounts to less than ±0.08 ‰. Experiments carried out on sample replicates showed that the average difference between replicates was less than ±0.15 ‰ for $\delta^{13}C$ and $\delta^{18}O$.

Shell morphology

Horridonia timanica (Stuckenberg 1875) has a large trapezoid shell with auricles, where the ornamentation is absent or granular (Figs 6 & 7). There is a narrow umbo and a distinct sinus. The spines are thick and straight according to Małkowski (1988). As shown in Małkowski's figure (1988, fig. 7), the phenotypic variation of *H. timanica* in the Kapp Starostin Formation may include the following variable features: shell size, shell coiling, auricle size, ornamentation density, spine distribution and the presence or absence of growth lines. The size of *H. timanica* ranges from 18.2 to 83.3 mm in width and from 14.8 to 74.0 mm in length (Małkowski 1988).

Horridonia horrida (Sowerby 1823) is generally sub-pentagonal to sometimes sub-oval in outline, with the largest convexity occurring in the middle part and large triangular ears (Fig. 6). The anterior margin is bent and forms a single fold. The ventral valve is strongly convex. The umbo is large and broad and the sulcus is broad and shallow. The dorsal valve is slightly concave and almost flat, but with a slightly outlined broad fold. The ornamentation is smooth, although concentric irregular wrinkles are clearly visible. Growth lines are often rather densely distributed and may show disturbances along the anterior margin. The ventral valves are thickest at umbo and the areas of muscle fields. On both valves, spines are distributed along the hinge margin and thick long auricular spines occur on the ears. The microstructure of hollow spines is similar to the valves. The external morphology may vary by differences in shell coiling and in width of the ears related to the length of hinge margin. The size of *H. horrida* ranges from 26 to 56 mm in width and from 16 to 47 mm in length, according to Kaźmierczak (1967).

Horridonia horrida is strongly characterized by the costae on the anterior part of the ventral valve (Dunbar 1955, 1961; Kaźmierczak 1967). *Horridonia timanica* differs from *H. horrida* in the lack of hinge and aurical spines on the ventral valve, granular ornamentation, more massive hinge and generally larger dimensions (Kaźmierczak 1967).

Horridonia horrida from the Holy Cross Mountains, Poland is identical to those described from: the Lower Zechstein (Werra cyclothem) of the Outer

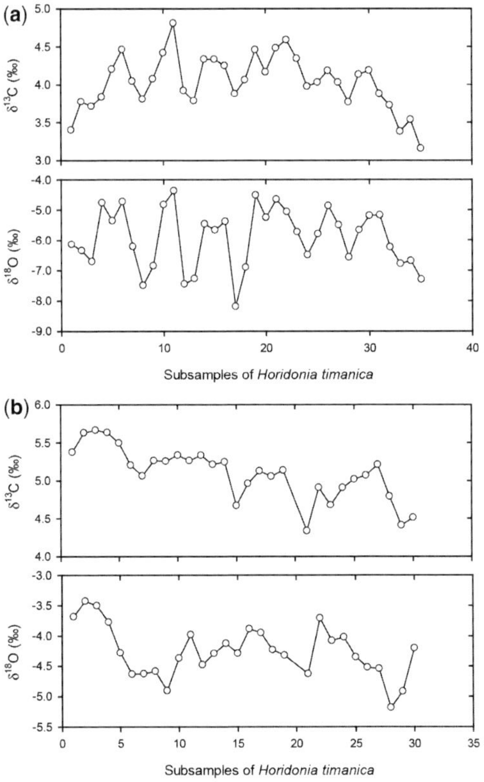

Fig. 11. Carbon and oxygen isotopic composition of *Horridonia timanica* (Stuckenberg 1875) from the Vindodden section, Spitsbergen. Sample numbers represent sampling sequence (X-axis) towards ventral margin, with *c.* 1 mm resolution. (**a**) This specimen shows a strong co-variation between the carbon and oxygen isotope curves. Sample ZPAL Bp. XXX/V-42. (**b**) A similar co-variation in isotopic composition (although somewhat less strong) is found for this specimen. Sample ZPAL Bp. XXX/V-5.

Table 2. *Carbon and oxygen isotopic composition of* Horridonia horrida *(Sowerby 1823) from Poland*

Horidonia horrida (Sowerby 1823)					
Sample ZPAL Bp. IX/1K			Sample ZPAL Bp. IX/7K		
Subsample No.	$\delta^{13}C$ (‰)	$\delta^{18}O$ (‰)	Subsample No.	$\delta^{13}C$ (‰)	$\delta^{18}O$ (‰)
1	4.41	−4.30	1	3.33	−3.48
2	4.83	−4.45	2	2.91	−3.75
3	4.84	−4.21	3	2.63	−4.33
4	4.93	−4.09	4	3.57	−4.67
5	4.94	−3.76	5	3.89	−4.54
6	4.79	−3.67	6	3.63	−4.32
7	4.47	−3.35	7	3.59	−3.28
8	4.33	−3.17	8	3.97	−2.72
9	4.29	−3.25	9	4.32	−2.48
10	4.41	−2.95	10	4.54	−2.51
11	4.44	−2.81	11	4.48	−3.00
12	4.28	−2.75	12	4.34	−3.39
13	4.40	−2.35	13	3.70	−2.22
14	3.89	−2.49	14	4.68	−2.47
15	4.59	−1.94	15	4.30	−2.65
16	3.84	−2.15	16	4.04	−1.97
17	3.98	−2.14	17	4.37	−2.41
18	4.27	−1.62			
19	4.26	−1.44			
20	4.38	−1.90			
21	4.65	−1.89			
22	4.08	−2.45			
23	2.62	−2.75			
24	3.36	−2.79			
25	3.17	−3.90			

Sudetic Basin and the Fore-Sudetic Monocline, Poland; the Zechstein of Middle England (Marl Slate, Magnesian Limestone); the Lower Zechstein of Germany (Kupferschiefer shales and Zechsteinkalk limestones); the Zechstein of Lithuania (Kaźmierczak 1967); and Late Permian of East Greenland (Dunbar 1955, 1961). *H. horrida* has however not been found in the Arctic Europe and Russia (Kaźmierczak 1967). *H. timanica* has been described from the Permian of Spitsbergen

Table 3. *Main statistical parameters and Pearson correlation coefficients of the isotopic compositions shown in Tables 1 and 2*

	Horridonia timanica		*Horridonia timanica*		*Horridonia horrida*		*Horridonia horrida*	
	Sample ZPAL Bp. XXX/V-42		Sample ZPAL Bp. XXX/V-5		Sample ZPAL Bp. IX/1K		Sample ZPAL Bp. IX/7K	
	$n = 35$		$n = 30$		$n = 25$		$n = 17$	
	$\delta^{13}C$ (‰)	$\delta^{18}O$ (‰)	$\delta^{13}C$ (‰)	$\delta^{18}O$ (‰)	$\delta^{13}C$ (‰)	$\delta^{18}O$ (‰)	$\delta^{13}C$ (‰)	$\delta^{18}O$ (‰)
Maximum	4.81	−4.37	5.66	−3.43	4.94	−1.44	4.68	−1.97
Mean	4.03	−5.93	5.10	−4.26	4.26	−2.90	3.90	−3.19
Minimum	3.16	−8.19	4.34	−5.18	2.62	−4.45	2.63	−4.67
Correlation coefficient	0.69		0.44		−0.24		0.61	

(Nakamura *et al.* 1987; Małkowski 1988), northern Russia (Kalashnikov 1993), Arctic Canada (Tschernyschew & Stepanow 1916) and Alaska (Brabb & Grant 1971). Both forms have been reported from the Pennsylvanian–Permian of Yukon Territory, Canada (Nelson 1962).

Stable isotopic composition

Carbon and oxygen isotopes have been analysed from two valves of *Horridonia timanica* (Table 1). The subsamples have $\delta^{13}C$ values ranging from 3.16 to 5.66‰. The $\delta^{18}O$ values are between −8.19 and −3.43‰. The correlation coefficients between $\delta^{13}C$ and $\delta^{18}O$ are 0.69 and 0.44 in the valves ZPAL Bp. XXX/V-42 and ZPAL Bp. XXX/V-5, respectively. The correlation is moderately positive, and the curves of $\delta^{13}C$ and $\delta^{18}O$ indicate some cyclical pattern (Fig. 11).

Two valves of *Horridonia horrida* have also been analysed isotopically (Table 2). The subsamples show $\delta^{13}C$ values between 2.62 and 4.94 ‰ (Table 3). $\delta^{18}O$ values vary between −4.67 and −1.44 ‰. Concerning the valve ZPAL Bp. IX/1 K, the correlation coefficient between $\delta^{13}C$ and $\delta^{18}O$ is −0.24 and indicates no significant relationship. In contrast, the correlation coefficient is moderately positive (0.61) for the valve ZPAL Bp. IX/7 K. There is no distinct cyclical nature in the sets of isotope values from *H. horrida* (Fig. 12).

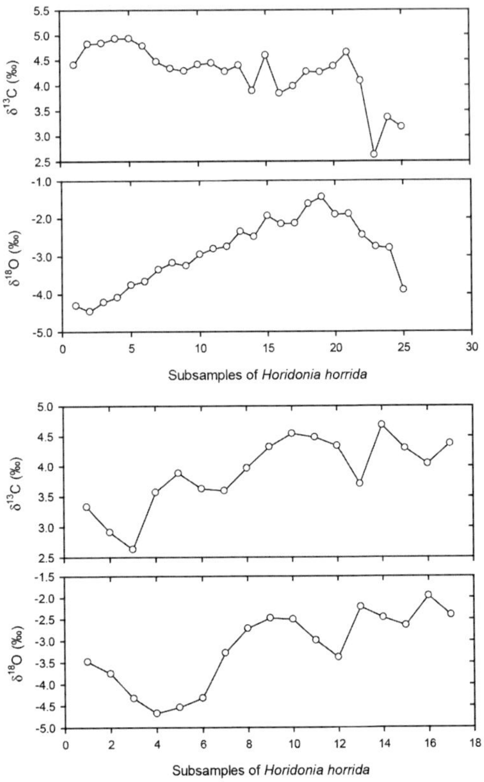

Fig. 12. Carbon and oxygen isotopic composition of *Horridonia horrida* (Sowerby 1823) from the Kajetanów quarry, Poland. Sample numbers represent sampling sequence (X-axis) towards ventral margin, with *c.* 1 mm resolution. (**a**) This specimen shows a strong independence between the carbon and oxygen isotope curves, where the oxygen isotope curve could be related to long-term cyclic evaporative variations. Sample ZPAL Bp. IX/1 K. (**b**) Overall increase in isotope values; however, not in phase. Sample ZPAL Bp. IX/7 K.

Seasonality and evaporation

The $\delta^{18}O$ values within *Horridonia timanica* show a recurring pattern that may be interpreted in terms of seawater temperature in the habitat. It assumes that the salinity was constant during the life duration of the sampled growth increments. This method of interpretation requires that the so-called vital effect upon oxygen isotopic fractionation was insignificant or even absent. Recent articulated brachiopods have nearly no vital effect and the calcite phase is in isotopic equilibrium with ambient water (Wefer & Berger 1991, fig. 30; Carpenter & Lohmann 1995; James *et al.* 1997); we assume that this was the case for the *Horridonia* specimens. The $\delta^{18}O$ cycles are therefore interpreted as reflective of seasonal changes in the isotopic composition caused by temperature variations. Previous studies have interpreted seasonal cycles in the oxygen isotopic composition of brachiopods. For instance, Mii & Grossman (1994) recognized seasonality in the brachiopod *Neospirifer dunbari* from the tropical epicontinental sea of Kansas during Late Pennsylvanian. Annual seasonal changes have also been recognized in other invertebrate groups such as in the Atlantic surf clam *Spisula solidissima* (Jones *et al.* 1983) and the geoduck *Panopea generosa* (Nielsen & Nielsen 2009), corresponding to instrumental records of sea surface temperature.

The above-mentioned $\delta^{18}O$ values from *Horridonia horrida* do not show such cyclicity; the curves are more irregular in course (Fig. 12). The $\delta^{18}O$ mean values for *H. horrida* are higher than those from *H. timanica* (Table 3), which preliminarily suggest that the former species lived in lower-temperature water in the Southern Permian Basin. Because the Southern Permian Basin was located at lower latitude than the northern entrance of the Greenland–Norway seaway, the interpretation solely in terms of temperature seems less likely. The Southern Permian Basin was an epicontinental sea characterized by a narrow entrance that restricted rejuvenation with normal marine water from the Greenland–Norway seaway (Figs 1 & 4). The salinity and therefore the oxygen isotopic composition varied with shell crystallization/precipitation, river outlets and evaporation in the region. Increased evaporation would have resulted in higher salinity as well as higher $\delta^{18}O$ values. Changes in the salinity could have blurred any seasonal pattern. The fact that the evaporative process can be reflected in the shell isotopic ratio was demonstrated by, for example, Bickert *et al.* (1997). Silurian brachiopods from Gotland (Sweden) have isotopic ratios indicating palaeoenvironmental changes between a humid and an arid climate. The arid conditions yielded higher salinity as well as $\delta^{13}C$ and $\delta^{18}O$ values (Bickert *et al.* 1997).

The oxygen isotope ratio in *Horridonia timanica* correlates moderately with $\delta^{13}C$ values, showing a partial in-phase relation (Fig. 11; Table 3). The latter was probably related to the isotopic composition of the available food such as phytoplankton that was dependent upon the water temperature. The $\delta^{13}C$ ratio in *H. horrida* was irregular during growth time (Fig. 12) and could have been distorted because of changes in the isotopic fractionation within the food items. The phytoplankton and zooplankton may have been sensitive to salinity changes so that the floral composition and its isotopic ratios altered gradually over the years. The minimum of the $\delta^{13}C$ values is higher in *H. horrida* than in *H. timanica* (Table 3). The cause of this is speculative and needs further investigation.

Conclusions

Comparative analysis of the results gives interesting clues about the differences in the record of environmental parameters at distant places and at slightly different times. We have performed precise and careful analysis, which is consistent with the large palaeolatitudinal distance between Boreal realm (Spitsbergen) and the Southern Permian Basin (Poland) during the Permian (Fig. 1). On Spitsbergen there is clearly a seasonal pattern, both on carbon and oxygen isotopes, supporting the fact that the habitat of the brachiopods was located at relatively high palaeolatitudes.

In the brachiopod shells from Poland however, we do not observe such seasonal regularities. The stable isotope pattern appears to be related to the effect of regional evaporative processes at much lower palaeolatitudes. Most likely, the somewhat distorted carbon isotope pattern was caused by available food sources and their sensitivity to salinity changes over time. Any diagenetically altered shells materials were eliminated prior to analysis; diagenetic processes therefore cannot explain the recorded isotope patterns.

The Institute of Palaeobiology, Polish Academy of Sciences (PAS), Warszawa kindly granted permission to carry out the analysis on the brachiopods collected by J. Kaźmierczak (1967) and K. Małkowski (1988). We would like to sincerely thank K. Małkowski for many helpful suggestions during the early phase of this investigation. We are grateful to W. Trela for fruitful discussions and the permission to use the picture of the Kajetanów quarry. N.-M. Hanken is thanked for fruitful discussions during fieldwork on Spitsbergen. We thank the editors, M. Sone and an anonymous reviewer for constructive comments which improved the manuscript.

References

Auclair, A.-C., Joachimski, M. M. & Lécuyer, C. 2003. Deciphering kinetic, metabolic and environmental controls on stable isotope fractionations between seawater and the shell of *Terebratalia transversa* (Brachiopoda). *Chemical Geology*, **202**, 59–78.

Barbin, V. & Gaspard, D. 1995. Cathodoluminescence of recent articulate brachiopod shells. Implications for growth stages and diagenesis evaluation. *Geobios, Memoir Special*, **18**, 39–45.

Beauchamp, B. 1994. Permian climatic cooling in the Canadian Arctic. *In*: Klein, G. D. (ed.) *Pangea: Paleoclimate, Tectonics and Sedimentation During Accretion, Zenith and Breakup of a Supercontinent.* Geological Society of America, Boulder, Special Paper, **288**, 229–246.

Bickert, T., Pätzold, J., Samtleben, C. & Munnecke, A. 1997. Paleoenvironmental changes in the Silurian, indicated by stable isotopes in brachiopod shells from Gotland (Sweden). *Geochimica et Cosmochimica Acta*, **61**, 2717–2730.

Błażejowski, B. 2008. *Otwornice późnego paleozoiku południowego Spitsbergenu.* Unpublished PhD thesis, Institute of Paleobiology, Polish Academy of Sciences, Warszawa.

Błażejowski, B. 2009. Foraminifers from Treskelodden Formation (Carboniferous–Permian) of south Spitsbergen. *Polish Polar Research*, **30**, 193–230.

Błażejowski, B., Hołda–Michalska, A. & Michalski, K. 2006. *Schellwienia arctica* (Fusulinidae), from the Carboniferous–?Permian strata of the Treskelodden Formation in south Spitsbergen. *Polish Polar Research*, **27**, 91–103.

Brabb, E. E. & Grant, R. E. 1971. Stratigraphy and paleontology of the revised type section for the Takhandit Limestone (Permian) in east-central Alaska. *United States Geological Survey Professional Paper*, **703**, 1–25.

Brand, U. 1989. Biogeochemistry of late Paleozoic North American brachiopods and secular variation of seawater composition. *Biogeochemistry*, **7**, 159–193.

Brand, U., Logan, A., Hiller, N. & Richardson, J. 2003. Geochemistry of modern brachiopods: applications and implications for oceanography and paleoceanography. *Chemical Geology*, **198**, 305–334.

Brauns, C. M., Pätzold, T. & Hacck, U. 2003. A Re–Os study bearing on the age of the Kupferschiefer black shale at Sangerhausen (Germany). *Abstracts of the*

XVth International Congress on Carboniferous and Permian Stratigraphy, Utrecht.

Bugge, T., Ringås, J. E., Leith, D. A., Mangerud, G., Weiss, H. M. & Leith, T. L. 2002. Upper Permian as a new play model on the mid-Norwegian continental shelf: investigated by shallow stratigraphic drilling. *American Association of Petroleum Geologists Bulletin*, **86**, 107–127.

Carpenter, S. J. & Lohmann, K. C. 1995. $\delta^{18}O$ and $\delta^{13}C$ values of modern brachiopod shells. *Geochimica et Cosmochimica Acta*, **59**, 3749–3764.

Curry, G. B. & Fallick, A. E. 2002. Use of stable oxygen isotope determinations from brachiopod shells in palaeoenvironmental reconstruction. *Palaeogeography, Palaeoclimatology, Palaeoecology*, **182**, 133–143.

Cutbill, J. L. & Challinor, A. 1965. Revision of the stratigraphical scheme for the Carboniferous and Permian rocks of Spitsbergen and Bjørnøya. *Geological Magazine*, **102**, 418–439.

Dallmann, W. K., Gjelberg, J. G. *et al.* 1999. Upper Palaeozoic lithostratigraphy. *In*: Dallmann, W. K. (ed.) *Lithostratigraphic Lexicon of Svalbard. Review and Recommendations for Nomenclature Use*. Upper Palaeozoic to Quaternary bedrocks. Committee on the Stratigraphy of Svalbard. Gjøvik Trykkeri AS, 25–126.

Dunbar, C. O. 1955. Permian brachiopod faunas of central East Greenland. *Meddelelser om Grønland, København*, **110**, 1–169.

Dunbar, C. O. 1961. Permian invertebrate faunas of Central East Greenland. *In*: Raasch, C. O. (ed.) *Geology of the Arctic*, University of Toronoto Press, Toronto, 224–230.

Fijałkowska-Mader, A. 2010. Stop 2. Kajetanów – The Lower Zechstein black limestones. *In*: *CIMP 2010 Field Trip Guidebook*. Institute of Geological Sciences, Polish Academy of Sciences (PAS), Warszawa, 13–17. http://www.ing.pan.pl/CIMP-2010/Graph_Attach/CIMP-2010_Guidebook.pdf

Frebold, H. 1937. Das Festungsprofil auf Spitzbergen. IV Die Brachiopoden- und Lamellibranchiatenfauna und die Stratigraphie des Oberkarbons und Unterperms. *Skrifter om Svalbard og Ishavet*, **69**.

Geluk, M. C. 1999. Late Permian (Zechstein) rifting in the Netherlands: models and implications for petroleum geology. *Petroleum Geoscience*, **5**, 189–199.

Geluk, M. C. 2007. Permian. *In*: Wong, T. E., Blatjes, D. A. J. & de Jager, J. (eds) *Geology of the Netherlands*. Royal Netherlands Academy of Arts and Sciences, Amsterdam, 63–83.

Geluk, M. C. & Röhling, H.-G. 1997. High-resolution sequence stratigraphy of the Lower Triassic 'Buntsandstein' in the Netherlands and northwestern Germany. *Geologie en Mijnbouw*, **76**, 227–246.

Given, R. K. & Lohmann, K. C. 1985. Derivation of the original isotopic composition of Permian marine cements. *Journal of Sedimentary Petrology*, **55**, 430–439.

Given, R. K. & Lohmann, K. C. 1986. Isotopic evidence for the early meteoric diagenesis of the reef facies, Permian Reef Complex of West Texas and New Mexico. *Journal of Sedimentary Petrology*, **56**, 183–193.

Glennie, K. W. & Buller, A. T. 1983. The Permian Weissliegend of N.W. Europe: the partial deformation of aeolian dune sands caused by the Zechstein transgression. *Sedimentary Geology*, **35**, 43–81.

Golonka, J. 2011. Phanerozoic palaeoenvironment and palaeolithofacies maps of the Arctic region. *In*: Spencer, A. M., Embry, A. F., Gautier, D. L., Stoupakova, A. V. & Sørensen, K. (eds) *Arctic Petroleum Geology*. Geological Society, London, Memoirs, **35**, 79–129.

Grossman, E. L. 1994. The carbon and oxygen isotope record during the evolution of Pangea: Carboniferous to Triassic. *In*: Klein, G. D. (ed.) *Pangea: Paleoclimate, Tectonics, and Sedimentation During Accretion, Zenith, and Breakup of a Supercontinent*. Geological Society of America, Boulder, Special Paper, **288**, 207–228.

Grossman, E. L., Zhang, C. & Yancey, T. E. 1991. Stable-isotope stratigraphy of brachiopods from Pennsylvanian shales in Texas. *Geological Society of America Bulletin*, **103**, 953–965.

Grossman, E. L., Mii, H.-S. & Yancey, T. E. 1993. Stable isotopes in late Pennsylvanian brachiopods from the United States: implications for Carboniferous paleoceanography. *Geological Society of America Bulletin*, **105**, 1284–1296.

Grossman, E. L., Mii, H.-S., Zhang, C. & Yancey, T. E. 1996. Chemical variation in Pennsylvanian brachiopod shells—diagenetic, taxonomic, microstructural, and seasonal effects. *Journal of Sedimentary Research*, **66**, 1011–1022.

Gruszczyński, M., Hoffman, A., Małkowski, K. & Veizer, J. 1992. Seawater strontium isotopic perturbation at the Permian-Triassic boundary, West Spitsbergen, and its implications for the interpretation of strontium isotopic data. *Geology*, **20**, 779–782.

Hanken, N.-M. & Nielsen, J. K. In review. Sedimentological investigations of Lower Permian *Palaeoaplysina* buildups on Spitsbergen. *In*: Gąsiewicz, A. & Słowakiewicz, M. (eds) *Palaeozoic Climate Cycles: Their Evolutionary and Sedimentological Impact*. Geological Society, London, Special Publications, **376**.

Hellem, T. & Worsley, D. 1978. *An Outcrop of the Kapp Starostin Formation at Austjøkeltinden, Sørkapplandet*. Norsk Polarinstitutt, Årbok 1977, 340–343.

Henriksen, E., Ryseth, A. E., Larssen, G. B., Heide, T., Rønning, K., Sollid, K. & Stoupakova, A. V. 2011. Tectonostratigraphy of the greater Barents Sea: implications for petroleum systems. *In*: Spencer, A. M., Embry, A. F., Gautier, D. L., Stoupakova, A. V. & Sørensen, K. (eds) *Arctic Petroleum Geology*. Geological Society, London, Memoirs, **35**, 163–195.

Ivany, L. C. & Runnegar, B. 2010. Early Permian seasonality from bivalve $\delta^{18}O$ and implications for the oxygen isotopic composition of seawater. *Geology*, **38**, 1027–1030.

James, N. P., Bone, Y. & Kyser, T. K. 1997. Brachiopod $\delta^{18}O$ values do reflect ambient oceanography: Lacepede Shelf, southern Australia. *Geology*, **25**, 551–554.

Johnson, H., Warrington, G. & Stoker, S. J. 1994. Permian and Triassic of the Southern North Sea. *In*: O'B Knox, R. W. & Cordey, W. G. (eds) *Lithostratigraphic Nomenclature of the North Sea*, British Geological Survey, Nottingham.

Jones, D. S., Williams, D. F. & Arthur, M. A. 1983. Growth history and ecology of the Atlantic surf clam,

Spisula solidissima (Dillwyn), as revealed by stable isotopes and annual shell increments. *Journal of Experimental Marine Biology and Ecology*, **73**, 225–242.

JURKIEWICZ, H. 1962. Fauna otwornicowa cechsztynu w rejonie Kajetonowa, Tumlina i Gałęzic. *Kwart Geologiczny*, **6**, 446–447.

KALASHNIKOV, N. V. 1993. Brakhiopody Permi Evropeiskogo Severa Rossii. (Permian brachiopods from European northern Russia). Nauka, St. Petersburg. (in Russian). http://dro.deakin.edu.au/eserv/DU:30001334/shi-latecarboniferous-2001.pdf

KAŹMIERCZAK, J. 1967. Morphology and palaeoecology of the Productid *Horridonia horrida* (Sowerby) from Zechstein of Poland. *Acta Palaeontologica Polonica*, **12**, 239–260.

KIERSNOWSKI, H., PAUL, J., PERYT, T. M. & SMITH, D. B. 1995. Facies, paleogeography, and sedimentary history of the Southern Permian Basin in Europe. *In*: SCHOLLE, P., PERYT, T. M. & ULMER-SCHOLLE, D. S. (eds) *The Permian of Northern Pangea*, Paleogeography, Paleoclimates, Stratigraphy. Springer-Verlag, New York, **1**, 119–136.

KORTE, C. & KOZUR, H. W. 2010. Carbon-isotopes stratigraphy across the Permian-Triassic boundary: a review. *Journal of Asian Earth Sciences*, **39**, 215–235.

KORTE, C., JASPER, T., KOZUR, H. W. & VEIZER, J. 2005. $\delta^{18}O$ and $\delta^{13}C$ of Permian brachiopods: a record of seawater evolution and continental glaciation. *Palaeogeography, Palaeoclimatology, Palaeoecology*, **224**, 333–351.

KORTE, C., JONES, P. J., BRAND, U., MERTMANN, D. & VEIZER, J. 2008. Oxygen isotope values from high-latitudes: clues for Permian sea-surface temperature gradients and Late Palaeozoic deglaciation. *Palaeogeography, Palaeoclimatology, Palaeoecology*, **269**, 1–16.

KOZUR, H. W. 1999. The correlation of the Germanic Buntsandstein and Muschelkalk with the Tethyan scale. *Zentralblatt für Geologie und Paläontologie Teil*, **I, 1998**, 701–725.

LARSSEN, G. B., ELVEBAKK, G. ET AL. 2005. Upper Palaeozoic lithostratigraphy of the southern part of the Norwegian Barents Sea. *Norges Geologiske Undersøkelse Bulletin*, **444**, 3–45.

LOWENSTAM, H. A. 1961. Mineralogy, O^{18}/O^{16} ratios, and strontium and magnesium contents of recent and fossil brachiopods and their bearing on the history of the oceans. *Journal of Geology*, **69**, 241–260.

MAGARITZ, M., ANDERSON, R. Y., HOLSER, W. T., SALTZMAN, E. S. & GARBER, J. 1983. Isotope shifts in the Late Permian of the Delaware Basin, Texas, precisely timed by varved sediments. *Earth and Planetary Science Letters*, **66**, 111–124.

MAŁKOWSKI, K. 1982. Development and stratigraphy of the Kapp Starostin Formation (Permian) of Spitsbergen. *Palaeont. Polonica*, **43**, 69–81.

MAŁKOWSKI, K. 1988. Paleoecology of Productacea (Brachiopoda) from the Permian Kapp Starostin Formation, Spitsbergen. *Polish Polar Research*, **9**, 3–60.

MAŁKOWSKI, K. & HOFFMAN, A. 1979. Semi-quantitative facies model for the Kapp Starostin Formation (Permian) Spitsbergen. *Acta Palaeotologica Polonica*, **24**, 217–230.

MANGERUD, G. 1994. Palynostratigraphy of the Permian and lowermost Triassic succession, Finnmark Platform, Barents Sea. *Review of Palaeobotany and Palynology*, **82**, 317–349.

MANGERUD, G. & KONIECZNY, R. M. 1993. Palynology of the Permian succession of Spitsbergen, Svalbard. *Polar Research*, **12**, 65–93.

MAWSON, M. & TUCKER, M. 2009. High-frequency cyclicity (Milankovitch and millennial-scale) in slope-apron carbonates: Zechstein (Upper Permian), North-east England. *Sedimentology*, **56**, 1905–1936.

MCCANN, T., KIERSNOWSKI, H. ET AL. 2008. Permian. *In*: MCCANN, T. (ed.) *The Geology of Central Europe, Vol. 1: Precambrian and Palaeozoic*. Geological Society, London, 531–597.

MENNING, M., ALEKSEEV, A. S. ET AL. 2006. Global time scale and regional stratigraphic reference scales of Central and West Europe, East Europe, Tethys, South China, and North America as used in the Devonian–Carboniferous–Permian Correlation Chart 2003 (DCP 2003). *Palaeogeography, Palaeoclimatology, Palaeoecology*, **240**, 318–372.

MII, H. & GROSSMAN, E. L. 1994. Late Pennsylvanian seasonality reflected in the ^{18}O and elemental composition of a brachiopod shell. *Geology*, **22**, 661–664.

MII, H.-S., GROSSMAN, E. L. & YANCEY, T. E. 1997. Stable carbon and oxygen isotope shifts in Permian seas of West Spitsbergen—global change or diagenetic artifact? *Geology*, **25**, 227–230.

MII, H.-S., SHI, G. R., CHENG, C.-J. & CHEN, Y.-Y. 2012. Permian Gondwanaland palaeoenvironment inferred from carbon and oxygen isotope records of brachiopod fossils from Sydney Basin, southeast Australia. *Chemical Geology*, **291**, 87–103.

NAKAMURA, K., KIMURA, G. & WINSNES, T. S. 1987. Brachiopod zonation and age of the Permian Kapp Starostin Formation (Central Spitsbergen). *Polar Research*, **5**, 207–219.

NAKREM, H. A. 1991. Conodonts from the Permian succession of Bjørnøya (Svalbard). *Norsk Geologisk Tidsskrift*, **71**, 235–248.

NAKREM, H. A. 1994. Environmental distribution of bryozoans in the Permian of Spitsbergen. *In*: HAYWARD, P. J., RYLAND, J. S. & TAYLOR, P. D. (eds) *Biology and Palaeobiology of Bryozoans*. Olsen & Olsen, Denmark, 133–137.

NAKREM, H. A. 1995. Bryozoans from the Lower Permian Vøringen Member (Kapp Starostin Formation), Spitsbergen (Svalbard). *Norsk Polarinstitutt Skrifter*, **196**, 92.

NAKREM, H. A. 2005. Some middle Permian bryozoans from Svalbard, Arctic Norway. *In*: MOYANO, H. I., CANCINO, G. J. M. & WYSE JACKSON, P. N. (eds) *Bryozoan Studies 2004*. Taylor & Francis Group, London, 197–205.

NAKREM, H. A., NILSSON, I. & MANGERUD, G. 1992. Permian biostratigraphy of Svalbard (Arctic Norway) – a review. *International Geological Review*, **34**, 933–959.

NELSON, S. J. 1962. Horridonid brachiopods as horizon indicators, Permo-Pennsylvanian of the Yukon Territory. *Journal of Alberta Society of Petroleum Geology*, **10**, 192–197.

Nielsen, J. K. & Nielsen, J. K. 2009. Geoducks (*Panopea abrupta*) as isotopic bioarchives of seasonality in the climate of British Columbia. *Ecological Research*, **24**, 987–995.

Nielsen, J. K. & Shen, Y. 2004. Evidence for sulfidic deep water during the Late Permian in the East Greenland Basin. *Geology*, **32**, 1037–1040.

Nielsen, J. K., Shen, Y., Piasecki, S. & Stemmerik, L. 2010. No abrupt change in redox condition caused the end-Permian marine ecosystem collapse in the East Greenland Basin. *Earth and Planetary Science Letters*, **291**, 32–38.

Nielsen, J. K., Nielsen, J. K. & Hanken, N.-M. 2012. Anoxicity controlling the formation of potential source rocks of Permian age: examples from East Greenland, Spitsbergen and the Norwegian Barents Sea. *Arctic Frontiers 2012 Conference – Energies of the High North, Part I: Arctic geology, hydrocarbon reservoirs & gas hydrates*, 25–27 January 2012, University of Tromsø, Norway.

Orvin, A. K. 1940. Outline of the geological history of Spitsbergen. *Skrifter om Svalbard og Ishavet*, **78**.

Pancost, R. D., Crawford, N. & Maxwell, J. R. 2002. Molecular evidence for basin-scale photic zone euxinia in the Permian Zechstein Sea. *Chemical Geology*, **188**, 217–227.

Pawłowska, K. 1978. Zechstein in the Holy Cross Mts. *In*: Symposium on Central European Permian. Quide of Excursions. Part 2 Zechstein of the Holy Cross Mts, Instytut Geologiczny, Warszawa, 9–19.

Peryt, T. M., Geluk, M. C., Mathiesen, A., Paul, J. & Smith, K. 2010. Zechstein. *In*: Doornenbal, J. C. & Stevenson, A. G. (eds) *Petroleum Geological Atlas of the Southern Permian Basin Area*. EAGE Publications, Houten, 123–147.

Popp, B. N., Anderson, T. F. & Sandberg, P. A. 1986. Brachiopods as indicators of original isotopic compositions in some Paleozoic limestones. *Geological Society of America Bulletin*, **97**, 1262–1269.

Rush, P. F. & Chafetz, H. S. 1990. Fabric-retentive, non-luminescent brachiopods as indicators of original delta ^{13}C and delta ^{18}O composition; a test. *Journal of Sedimentary Petrology*, **60**, 968–981.

Scholle, P. A. 1995. Carbon and sulfur isotope stratigraphy of the Permian and adjacent intervals. *In*: Scholle, P. A., Peryt, T. M. & Ulmer-Scholle, D. S. (eds) *The Permian of Northern Pangea—Volume 1: Paleogeography, Paleoclimates, Stratigraphy*. Springer, Berlin-Heidelberg, 133–149.

Scholle, P. A., Stemmerik, L. & Ulmer, D. S. 1991. Diagenetic history and hydrocarbon potential of Upper Permian carbonate buildups, Wegener Halvø area, Jameson Land Basin, East Greenland. *The American Association of Petroleum Geologists Bulletin*, **75**, 701–725.

Shen, S. & Shi, G. R. 2000. Wuchiapingian (early Lopingian, Permian) global brachiopod palaeobiogeography: a quantitative approach. *Palaeogeography, Palaeoclimatology, Palaeoecology*, **162**, 299–318.

Shen, S.-Z. & Shi, G. R. 2004. Capitanian (Late Guadalupian, Permian) global brachiopod palaeobiogeography and latitudinal diversity pattern. *Palaeogeography, Palaeoclimatology, Palaeoecology*, **208**, 235–262.

Siedlecka, A. 1970. Investigations of Permian cherts and associated rocks in southern Spitsbergen. Parts I and II. *Norsk Polarinstitutt Skrifter*, **147**, 86.

Słowakiewicz, M., Kiersnowski, H. & Wagner, R. 2009. Correlation of the Middle and Upper Permian marine and terrestrial sedimentary sequences in Polish, German, and USA Western Interior Basins with reference to global time markers. *Palaeoworld*, **18**, 193–211.

Sosipatrova, G. P. 1967. Upper Paleozoic Foraminifera of Spitsbergen. *In*: Sokolov, V. N. (ed.) *Stratigraphy of Spitsbergen*. Institut Geologii Arktiki, Leningrad, 1–238 (in Russian).

Sosipatrova, G. P. 1969. Foraminifers from the Starostinskaya Suite of Spitsbergen. *Nauchnoissledovatel's-kiy Institut Geologii Arktiki, Ministerstvo Geologii SSSR, Uchenye Zapiski, Paleontologiya i Biostratigrafiya*, **26**, 46–79 (in Russian).

Sowerby, J. 1823. *The Mineral Conchology of Great Britain: or Coloured Figures and Descriptions of those Remains of Testaceous Animals or Shells, which have been Preserved at Various Times and Depths in the Earth*. **Volume 4**. W. Arding, London.

Steel, R. J. & Worsley, D. 1984. Svalbard's post-Caledonian strata. An atlas of sedimentational patterns and palaeogeographic evolution. *In*: Spencer, A. M. et al. (eds) *Petroleum Geology of the North European Margin*. Norwegian Petroleum Society, Graham & Trotman, London, 109–135.

Stemmerik, L. 1988. Discussion. Brachiopod zonation and age of the Permian Kapp Starostin Formation (Central Spitsbergen). *Polar Research*, **6**, 179–180.

Stemmerik, L. 1995. Permian history of the Norwegian-Greenland Sea area. *In*: Scholle, P. A., Peryt, T. M. & Ulmer-Scholle, D. S. (eds) *The Permian of Northern Pangea. Volume 2: Sedimentary Basins and Economic Resources*. Springer-Verlag, Berlin, Heidelberg, 98–118.

Stemmerik, L. 1997. Permian (Artinskian–Kazanian) cool-water carbonates in North Greenland, Svalbard and the western Barents Sea. *In*: James, N. P. & Clarke, J. A. D. (eds) *Cool-Water Carbonates*. Society of Economic Paleontologists and Mineralogists, Tulsa, Special Publication, **56**, 349–364.

Stemmerik, L. 2000. Late Palaeozoic evolution of the North Atlantic margin of Pangea. *Palaeogeography, Palaeoclimatology, Palaeoecology*, **161**, 95–126.

Stemmerik, L. 2008. Influence of late Paleozoic Gondwana glaciations on the depositional evolution of the northern Pangean shelf, North Greenland, Svalbard, and the Barents Sea. *In*: Fielding, C. R., Frank, T. D. & Isbell, J. L. (eds) *Resolving the Late Paleozoic Ice Age in Time and Space*. Geological Society of America, Boulder, Special Publication, **441**, 205–217.

Stemmerik, L. & Worsley, D. 2005. 30 years on – Arctic Upper Palaeozoic stratigraphy, depositional evolution and hydrocarbon prospectivity. *Norwegian Journal of Geology*, **85**, 151–168.

Stemmerik, L., Vigran, J. O. & Piasecki, S. 1991. Dating of late Palaeozoic rifting events in the North Atlantic: new biostratigraphic data from the uppermost Devonian and Carboniferous of East Greenland. *Geology*, **19**, 218–221.

Stemmerik, L., Bendix-Almgreen, S. E. & Piasecki, S. 2001. The Permian-Triassic boundary in central East

Greenland: past and present views. *Geological Society of Denmark Bulletin*, **48**, 159–167.

Stuckenberg, A. 1875. Report on Geological Journey to Pechora region and Timan tundra. *Materials for the Geology of Russia*, **6**, 1–125 (in Russian).

Surlyk, F., Hurst, J. M., Piasecki, S., Rolle, F., Scholle, P. A., Stemmerik, L. & Thomsen, E. 1986. The Permian of the western margin of the Greenland sea – a future exploration target. *In*: Halbouty, M. T. (ed.) *Future Petroleum Provinces of the World*. American Association of Petroleum Geologists, Tulsa, Memoir, **40**, 629–659.

Szaniawski, H. & Małkowski, K. 1979. Conodonts from the Kapp Starostin Formation (Permian) of Spitsbergen. *Acta Palaeontologica Polonica*, **24**, 231–264.

Szurlies, M., Bachmann, G. H., Menning, M., Nowaczyk, N. R. & Käding, K.-C. 2003. Magnetostratigraphy and high-resolution lithostratigraphy of the Permian–Triassic boundary interval in Central Germany. *Earth and Planetary Science Letters*, **212**, 263–278.

Tschernyschew, T. & Stepanow, P. 1916. *Obercarbonfauna von Konig Oscars und Heibergs Land*. Report of the Second Norwegian Arctic Expedition in the 'Fram' 1898–1902, **34**, 1–67.

Veizer, J., Bruckschen, P. *et al.* 1997. Oxygen isotope evolution of Phanerozoic seawater. *Palaeogeography, Palaeoclimatology, Palaeoecology*, **132**, 159–172.

Wagner, R. 1994. Stratigraphy and evolution of the Zechstein Basin in the Polish Lowland. *Prace Panstwowego Instytutu Geologicznego*, **146**, 1–71 (in Polish with English summary).

Wagner, R. & Peryt, T. M. 1997. Possibility of sequence stratigraphic subdivision of the Zechstein in the Polish Basin. *Geological Quarterly*, **41**, 457–474.

Wefer, G. & Berger, W. H. 1991. Isotope paleontology: growth and composition of extant calcareous species. *Marine Geology*, **100**, 207–248.

Worsley, D. 2008. The post-Caledonian development of Svalbard and the western Barents Sea. *Polar Research*, **27**, 298–317.

Ziegler, P. A. 1990. *Geological Atlas of Western and Central Europe*. 2nd edn. Shell International Petroleum Mij. B.V.

Ziegler, A. M., Hulver, M. L. & Rowley, D. B. 1997. Permian world topography and climate. *In*: Martini, I. P. (ed.) *Late Glacial and Postglacial Environmental Changes*. Oxford University Press, New York, 111–146.

Extinction of Guadalupian rugose corals: an example of biotic response to the Kamura event (southern Primorye, Russia)

OLGA L. KOSSOVAYA* & GALINA S. KROPATCHEVA

All-Russian Research Geological Institute, Sredny Prospect 74, St Petersburg 199106, Russia

**Corresponding author (e-mail: olga_kossovaya@vsegei.ru)*

Abstract: Permian Rugosa of southern Primorye (Russian Far East) occur in a series of terranes of different tectonic origin. The taxonomical composition of Guadalupian (Wordian–Capitanian) rugose corals distributed in southern Primorye changed progressively in an ecological succession, starting with primitive persistent cosmopolitan taxa (assemblage I). This was replaced by assemblage II with a marked invasion of Peri-Gondwanian genera such as *Timorphyllum* and *Verbeekiella*, and, in the Late Capitanian, by typical Cathyasian colonial waagenophyllids (assemblage III). Subsequent communities have nearly no transitional forms. The latter assemblage was abruptly replaced by small fasciculate and solitary primitive forms (assemblage IV). Species peculiarities of assemblages of the *Parafusulina stricta* Zone and the specific lithological features of coral-bearing deposits indicate a palaeogeographical differentiation of terrane positions. The terranes are of various origins: oceanic arcs, passive margin and guyot. The extinction of massive colonial Rugosa in southern Primorye (western Palaeo-Pacific) is identified as a biotic event that is recognizable in Japan, and probably in Inner Mongolia. It also coincides with a general reduction in coral diversity at the end of the Capitanian in China. The present study of the taxonomical distribution of corals has enabled the level of massive waagenophyllid extinction to be defined precisely and to correlate it to a similar pattern in adjacent territories of the Palaeo-Pacific below the Lopingian boundary. The diversity decrease coincides with the last phase of the Kamura event, and the effects of the Emeishan volcanism is correlated with a second positive shift in $\delta^{13}C$ (South China). The assumed cooling impacted on the fauna in the western Palaeo-Pacific. The stepwise elimination of massive colonial waagenophyllids and large fusulinids documented in the southern Primorye at the end of the Capitanian is considered to be the first step in the Permian–Triassic extinction, about 8 Myr prior to the final Permian–Triassic (P–T) global event.

The extinction of Rugosa at the Permian–Triassic (P–T) boundary has been discussed widely, using examples of different faunal groups. A stepwise scenario and two-phase pattern of rugose coral extinction at the end of the Permian have been recognized (Ezaki 1994; Kossovaya 1995; Wang & Sugiyama 2000). Faunal extinction in Permian basins of the Palaeo-Pacific (including the territory of the present-day southern Primorye) was equated with the Lopingian–Wuchiapingian boundary (Kotlyar *et al.* 2004*a*). The first step in P–T extinction has been documented at the end of the Guadalupian, illustrated by the extinction of typical Tethyan assemblages inclusive of the tropical trio: the giant alatoconchid bivalves; waagenophyllid corals; and large fusulinids of the family Verbeekinidae (Isozaki & Aljinović 2009). Recently, the Kamura signal in the *Yabeina* Zone has been discovered in the European Tethys (Isozaki *et al.* 2011). The similarity amongst assemblages of different fauna distributed in the eastern Palaeo-Pacific (Inner Mongolia, Japan and southern Primorye in Russia) (Wang *et al.* 2006; Kotlyar *et al.* 2007) is based on shared generic and specific compositions, but the number of endemic species is fairly high. The coral fauna of southern Primorye is one of the richest in the eastern Palaeo-Pacific. The similar coeval fauna occurred in a variety of reef facies referred to the Iwaizaki Limestone of the Kanokura Formation in the stratotype section at Iwaizaki Cape (Kawamura & Machiyama 1995). Overlying the reef facies is deep-water shale, which proves the submersion of the reef; a similar depositional history has been recorded for the Nakhodka section in southern Primorye (Zakharov & Pavlov 1987). The Kamura event embraces the interval between the end of the *Yabeina* Zone, correlated with the Lower Capitanian, the *Lepidolina* Zone and the lower part of the barren interval coinciding with the Upper Capitanian (Ota & Isozaki 2006). The extinction of large fusulinids took place at the end of the *Lepidolina* Zone (Ota & Isozaki 2006). Geochemical data obtained from the atoll carbonates of the Kamura section (after which the event is named) show a positive $\delta^{13}C$ anomaly in the Early Capitanian interval, with an extremely high

From: GĄSIEWICZ, A. & SŁOWAKIEWICZ, M. (eds) 2013. *Palaeozoic Climate Cycles: Their Evolutionary and Sedimentological Impact*. Geological Society, London, Special Publications, **376**, 407–429.
First published online August 6, 2013, http://dx.doi.org/10.1144/SP376.13

$\delta^{13}C$ value of +7‰ PDB (Pee Dee Belemnite) in the middle part of the *Yabeina* Zone (Ota & Isozaki 2006). Following their data, most extinctions occurred at the 'high positive plateau' of $\delta^{13}C$ values ranging between +5 and +6‰ PDB. Data recorded are restricted by a $\delta^{13}C$ value of +3.8‰ PDB in the Sen'kina Shapka section (Zakharov *et al.* 2001) and by +4.1‰ PDB in the Iwaizaki Limestone (Japan). Recently, the Capitanian extinction amongst fusulinids and algae has been noted in the Maokou Formation in China (Bond *et al.* 2010*a*, *b*). By correlation, in South China the 'positive plateau' of the Kamura event coincides with a positive trend during the *Jinogondolella postserrata–Jinogondolella altudaensis* conodont zones. Later it is changed by a lighter carbon isotope signal in the Late Capitanian, at the level of the *Jinogondolella prexuanhanensis–xuanhanensis* conodont Zone coinciding with the uppermost part of the *Lepidolina–Yabeina gubieri* Zone (fig. 2 in Bond *et al.* 2010*a*). In spite of some discrepancy in geochemical data, diversity changes and extinction pattern seem to be coeval. Reliable biostratigraphical data are used as framework for ecological reconstructions.

The extinction of modern-day scleractinians through increasing CO_2 and ocean acidification is amongst widely discussed hypotheses (see e.g. Veron 2008). However, data on the Kamura cooling as a possible trigger for the elimination of massive Rugosa during the latest Guadalupian requires an analysis of other temperature impacts. The reason behind periodic extinction of Phanerozoic corals is widely debated, mostly within the scope of reef elimination, which occurred several times during the Earth's history. Tropical reefs and one of the most typical constructors of the reef systems – colonial corals – are considered, in view of their modern distribution, to be indicators of climate and, in particular, of its temperature component (Kiessling 2001). The mechanism of the extinction of Palaeozoic reef communities used to document the major role of cooling was also considered as follows: (1) non-tropical marine taxa suffered a less severe extinction than the tropical group, presumably because many of their species were able to migrate equatorwards; (2) during the protracted Permian crisis biogeographical provinces contracted towards the equator; (3) for each crisis the disproportionally large percentage of surviving taxa were cosmopolitan in distribution; (4) in the aftermath of each event carbonate production was reduced; (5) reef building was suppressed for several millions of years after each crisis event, although potential reef-builders survived; and (6) each mass extinction approximately coincides with the initiation of a glacial interval in the polar region (Stanley 1988). All of these features coincide with the Permian–Triassic event and, in part, are recognizable as the Guadalupian–Lopingian extinction (see later).

Cooper (1994) also considered cooling as the main reason behind coral extirpation at global extinction events. Twitchett (2006) emphasized the significance of change rapidity, as most organisms should be able to cope with slow change over millions of years. This assumption is important for the Guadalupian–Lopingian extinction, because the Kamura event is considered to have been a phase of rather long-lasting cooling. Contrary to that, Veron (2008) found other reasons for global events. Thus, for the Permian–Triassic, the reasons for stress amongst Rugosa and tabulate corals were wildly fluctuating temperatures in response to sharp increases in atmospheric CO_2, acid rain and lack of oxygen in shallow waters after protracted times of plenty, all of which have been held responsible. The oceans are widely believed to have turned anoxic and to have contained free hydrogen sulphide. Another study suggests that a sudden release of methane occurred at this time. Recently, a comprehensive palaeophysiological study clearly implicates CO_2, perhaps associated with the Siberian Traps, the largest volcanic outpouring of all time (Veron 2008). The richness of coral reefs in the tropical zone leads us to the logical conclusion that low temperatures were one of the factors that restricted their latitudinal distribution and, thus, might have been fatal for these animals. One example from the Phanerozoic is coral extinction connected with an abrupt drop of 5 °C in sea-water temperature at the Eocene–Oligocene boundary, connected with the onset of Antarctic glaciation. The hypothermal effect led to the extirpation of cosmopolitan taxa, with only the highly tolerant *Actinacis* dominating in low-diversity build-ups (Montaggioni & Braithwaite 2009). For modern corals, in latitude communities, mostly in the Caribbean, the Persian Gulf and in Japan, very few hermatypic species are known to tolerate temperatures below 11 °C under natural conditions. Contrariwise observations on low-temperature mortality in tropical reefs are few and far between (Veron & Minchin 1992); however, throughout the past 2 Myr of the Pleistocene, they survived at least 17 glacial–interglacial cycles of dramatic climate change and sea-level fluctuations (Pandolfi 1999). Some evidence can be found in modern experiments conducted at Scripps' Experimental Aquarium, documenting the mortal effect of abrupt, short-term cooling and adaptation of scleractinian corals to a gradual temperature decrease (Rath *et al.* 2012).

Estimating the Guadalupian–Lopingian extinction, on the basis of the material studied, we can conclude that there was a total annihilation of massive tropical Rugosa in the Palaeo-Pacific in the so-called 'northern transitional gateway' (Shi 2006).

The aim of the present paper is to compare the successive changes in corals from southern Primorye with assemblages from adjacent areas, and to consider a possible model of massive colonial coral extinction using results on biodiversity, palaeoclimatic and palaeoenvironmental changes.

Historical review of coral collections

The first data on rugose corals from southern Primorye were published by Fomichev in 1953 (Fig. 1), based on a study completed in the 1940s (Fomichev 1953). The collection included specimens obtained during geological mapping carried out by Maslennikov and others in those years (Maslennikov 1948). The occurrence was reported merely as the Doliolinovaya Formation, considered to be of early Late Permian (Partizanskaya River = Suchan and Shkotovo district). The collection was housed at the Central Research Geological Museum (CNIGR Museum), St Petersburg, number 7184, and later was in part restudied by Iljina (1977, 1984, 1987). All attempts to trace this collection have so far been unsuccessful. The whereabouts of the holotypes mentioned by Iljina (1984), chosen from Fomichev's collection, are unknown. Fomichev (1953) described 14 taxa (the ones with * being new): *Tachylasma columbinum**, *T. vanganense* *, *T.* cf. *densa* Hill, *T. voinovae**, *Sochkineophyllum lophoplylloides* (Soshkina) var. *orientalis**, *S. artiense* (Soshkina) var. *densa**, *S. intabulatum**, *Polycoelia* cf. *simplex* (Soshkina), *Polycoelia* sp. nov.*, *Verbeekia ussurica**, *Lophocarinophyllum* cf. *acanthiseptatum* Grabau, *V. ussurica** var. *elongate**, *Timorphyllum maichense** and one new genus, *Maichelasma*, for *M. magnum**. Subsequently, Iljina (1984) included a number of the species described in her revision of the Suborder Polycoelina and transferred most species of *Tachylasma* described by Fomichev to *Ufimia*, the exception being *Tachylasma columbinum* Fomichev, which was assigned to *Paracaninia*. In addition, *Calophyllum simplex* (Fomichev) was recorded from the *Monodiexodina sutchanica* fusulinid Zone, and *Calophylloides gemmatum* Iljina and *Calophyllum columnare* (Schlotheim) (Iljina 1984, p. 140) from the *Parafusulina stricta* fusulinid Zone (Fig. 1; Table 1). Now we consider the material redescribed by Iljina (1984) to be topotypical. Some topotypes for species not redescribed by Iljina (1984) were selected from an unpublished collection, 2873 PIN RAS, for the present study. Kropatcheva (1978) described some species, mostly colonial ones, from the *Metadoliolina lepida–Lepidolina kumaensis* fusulinid Zone of southern Primorye, and a single colonial form was described by Ivanovskiy (1969).

Materials

Data on the geology of southern Primorye (Fig. 1) were obtained from mapping reports (Burago *et al.* 1968; Kotlyar *et al.* 1975) and the results of our fieldwork. In addition to collections obtained during these field trips, the unpublished collection number 2873 at the Palaeontological Museum of the Russian Academy of Science, Moscow has been investigated. In addition, topotypical material from Iljina (1984) was restudied (collection number 2376, PIN RAS).

Geological setting and age

The region studied is located between the Sino-Korean craton in the south and the Sikhote–Alin folded belt in the east (Fig. 2). According to their tectonic position, the Laoelin–Grodekovo and Voznesenka terranes belong to the Khanka superterrane – one of the tectonic blocks of the Burea–Khanka–Yamusi Palaeozoic orogenic belt (Figs 2 & 3a) (Khanchuk 2006). The Permian marine and non-marine volcanic and siliciclastic deposits that overlie most of the various terranes of the Khanka superterrane are interpreted as having formed in a back-arc basin adjacent to the Permian island arc. The Sergeevka terrane belongs to the Sikhote-Alin' folded belt, which is separated by faults from the Burea–Khanka orogenic belt. They acted separately during the Late Palaeozoic and were accreted to the Siberian Platform in Mesozoic (Figs 2 & 3a).

The Laoelin-Grodekovo terrane is situated to the south of Khanka Lake. It made up a fragment of the Late Palaeozoic island arc and was accreted to the Khanka superterrane during the Late Permian. The NE part of the Laoelin–Grodekovo terrane (western part of the area studied) is characterized by siliciclastic and carbonate Permian deposits containing a large amount of volcaniclastic and volcanic formations penetrated by Upper Permian and Jurassic granitoids. The Laoelin–Grodekovo terrane in its tectonic setting is similar to that of the modern Sea of Japan (Khanchuk 2006). Carbonate deposits, mainly of Middle Permian age, are referred to the Barabashevka Formation of the Guadalupian (Wordian–Capitanian: Fig. 4).

The Voznesenka terrane is considered to be a fragment of the Early Palaeozoic passive margin and forms the SW part of the Khanka superterrane (Khanchuk 2006). The coral-bearing deposits are

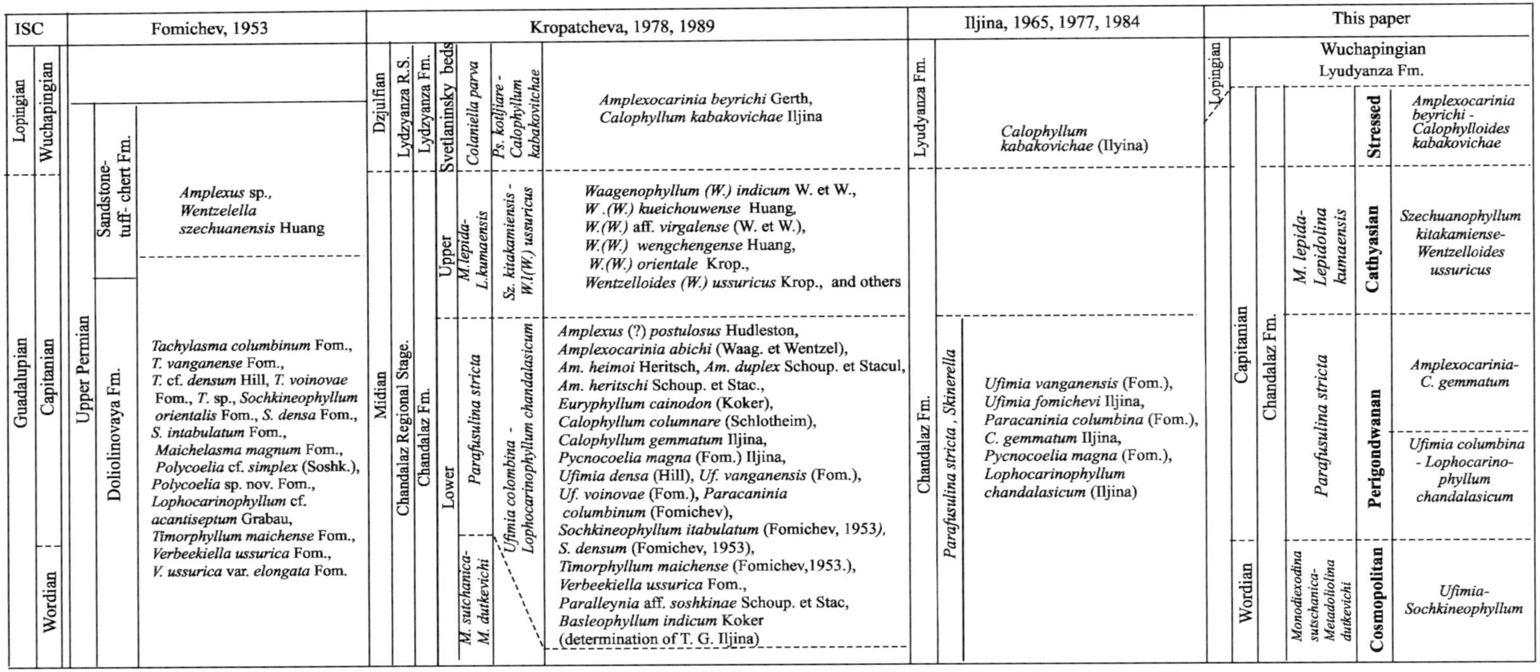

Fig. 1. The history of the coral study in southern Primorye.

Table 1. *The occurrences and location of some holotypes and topotypes*

	Holotype (coll. Fomichev)	Topotype (Iljina 1984)
Ufimia vanganensis (Fomichev 1953)	Specimen 137b, basin of the Artemovka River, Shumilov Spring, Chandalazian Horizon, *Parafusulina stricta* Zone (Iljina 1984)	Specimen 2873/240 PM RAS 3 km upstream from the mouth of the Shumilov Spring, Artemovka River, *Parafusulina stricta* Zone (p. 109, plate XV, figs 1–3, Iljina 1984)
Ufimia fomitchevi Iljina (1984)		Holotype 2873/88 PM RAS right bank of the Artemovka River, Golubinaya Mountain, *Parafusulina stricta* Zone (p. 110, plate XVI, fig. 1, Iljina 1984)
Paracaninia columbina (Fomichev 1953)	Specimen 28, the basin of the Artemovka River, Chandalazian Horizon, *Parafusulina stricta* Zone (Iljina 1984)	Topotype 2873/113, right bank of Artemovka River, Golubinaya mountain, Chandalaz Formation, *Parafusulina stricta* Zone (Iljina 1984)
Calophylloides gemmatum (Iljina 1977)		Holotype 2873/356, Porganichnyi District, Nesterovka River, Chandalaz Horizon, *Parafusulina stricta* Zone
Calophylloides kabakovitchae (Iljina 1977)		Holotype 2873/486, NE from Voltchanets Lake, the uppermost Capitanian
Calophyllum magnum (Fomichev 1953)	Specimen 137_2, basin of the Artemovka River, Shumilov Spring, Chandalaz Horizon, *Parafusulina stricta* Zone (Iljina 1984)	Topotype 2873/169, 3 km upstream from the mouth of the Shumilov Spring, Chandalaz Formation, Artemovka River, *Parafusulina stricta* Zone
Lophocarinophyllum chandalazicum (Iljina 1987)	*Lophocarinophyllum* cf. *acantiseptum* Grabau	Holotype 2873/19 left bank of the Pad'Shirokaya, Barabashevka Formation, *Parafusulina stricta* Zone (Iljina 1987)
Lophocarinophyllum inaequale (Iljina 1987)		Holotype 2873/564, right bank of the Partizanskaya River, Sen'kina Shapka section, Chandalaz Formation, *Metadoliolina lepida* Zone (Iljina 1987)

represented by the carbonate Chandalaz Formation of Wordian–Capitanian age, and by the mostly siliciclastic Lyadyanza Formation of Lopingian age.

Permian deposits of the Sergeevka terrane are assigned to the Palaeozoic–Mesozoic passive margin, which was later included in the structures of the Jurassic accretion prism. It is situated in the southern coastal part of southern Primorye. The carbonate Chandalaz Formation here makes up the entire Wordian–Capitanian (Figs 1 & 3).

The biostratigraphy, zonal subdivisions and correlation of stratigraphic units have been discussed previously (Kotlyar *et al.* 1983, 1989, 1990, 2004*b*, 2006).

Coral occurrences

The Laoelin–Grodekovo terrane

Corals of the key section Pad' Shirokaya belong to the Barabashevka Formation (Fig. 4). The early Wordian date of the underlying Vladivostok Formation is based on the occurrence of the ammonoids *Daubichites orientalis* in the lower part of that formation and *Roadoceras subroadense* in its upper part (Zakharov & Pavlov 1986). The overlying deposits are represented by the Barabashevka Formation, which consists of tuffs, sandstones, conglomerates and mudstones with benthic fauna. The basal layer of the Barabashevka Formation is composed of a conglomerate (Unit V: Fig. 4). The overlying part of the formation includes an intercalation of tuff, sandstone and thin beds of limestone (Unit VI), with rare brachiopods, foraminifers, bryozoans, crinoids and corals. Two assemblages of solitary corals were distinguished within the Barabashevka Formation at a few outcrops in the Barabashevka River Basin. The first assemblage in the Pad' Shirokaya section within the interval of *Monodiexodina sutchanica* Zone (Unit VI) is represented by *Ufimia* sp., *Ufimia densa* Hill, *Cyathocarinia multituberculata* Soshkina, *Amplexocarinia* sp. and *Sochkineophyllum lophophylloides* (Soshkina) (Fig. 4). The second assemblage is

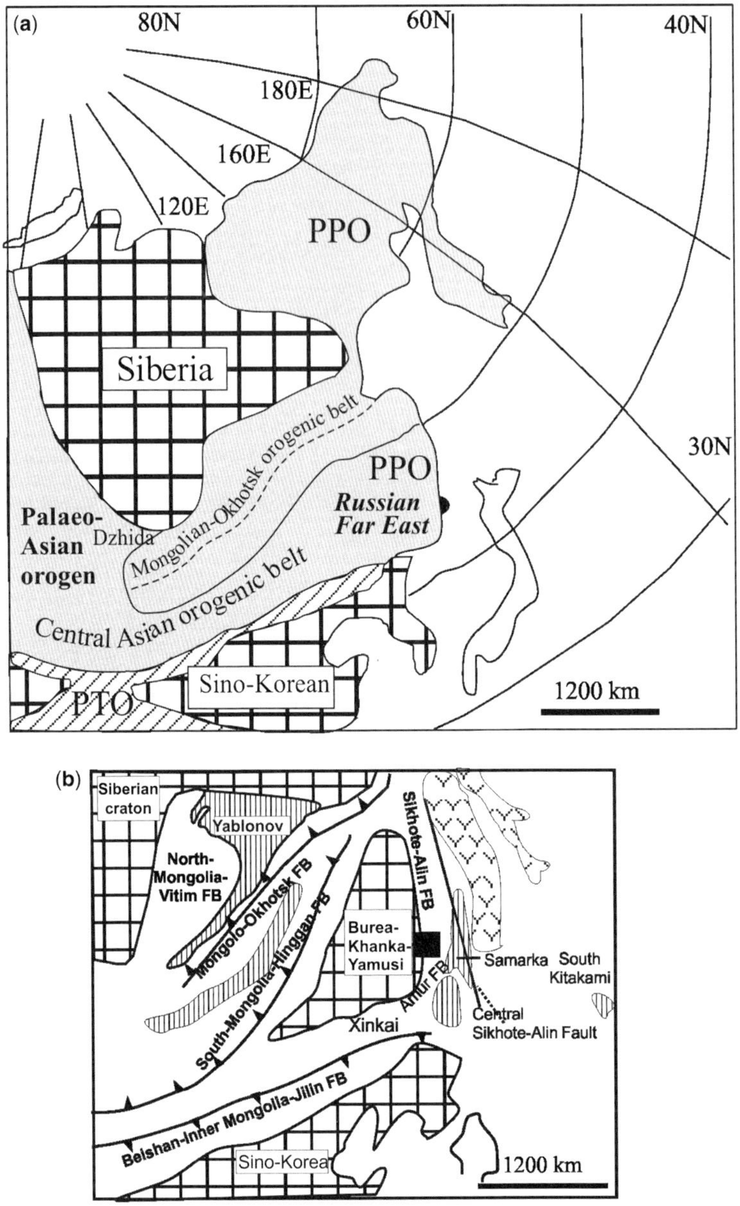

Fig. 2. (**a**) General tectonic scheme of Asia showing main orogens formed in place of former oceans: PPO, Palaeo-Pacific (modified from Safonova 2009). (**b**) Detailed tectonic chart of the NE Pacific (modified from Shi 2006). ■, the region studied.

located at 7.2 m above the lower boundary of the *Parafusulina stricta* Zone in dark-grey clayey limestone of the Barabashevka Formation (Fig. 4). It co-occurs with remains of *Skinnerella*, *Codonofusiella* and other taxa, corresponding to the *Parafusulina stricta* Zone of Midian age (Capitanian) (Burago *et al.* 1968). The other outcrop of the Barabashevka Formation occurs in the basin of the Nesterovka River (=Takheyazh River, Pogranichnyi District: Fig. 3). There the sequence consists of an intercalation of effusive rocks with lenses of siliciclastic and carbonate rocks. The coral-bearing part of the formation is also exposed at a limestone quarry nearby.

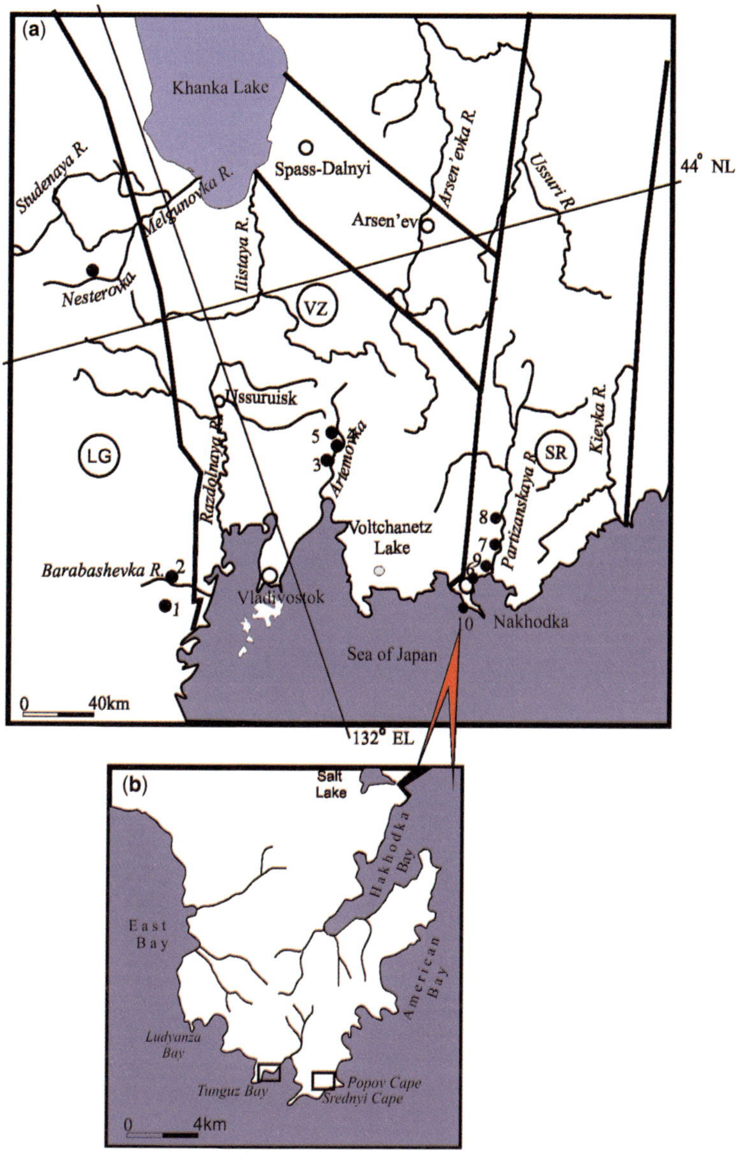

Fig. 3. (**a**) Position of terranes and coral occurrences in southern Primorye (Russian Far East). LG, Laoelin–Gradekovo terrane; 1, Pad' Shirokaya section; 2, Barabashevka (-Mongugay); VZ, Voznesensk terraine; 3, Golubinaya Mountain; 4 & 5, Vangan, Flangovaya, Artemovka River; SR, Sergeevka terraine; 6, Nakhodka; 7 & 8, Sen'kina Shapka; 9; Sestra Mountain; 10; Trudnyi Peninsula. (**b**) Location of outcrops in the Trudnyi Peninsula.

The Sergeevka terrane

Material came from outcrops situated in the Partizanskaya River Basin, Trudnyi Peninsula (Fig 3b), and in the vicinity of the town of Nakhodka. Middle–Upper Permian deposits are represented by shallow-marine limestone and siltstone up to 1500 m thick, and are characterized by abundant and diverse fusulinids, small foraminifers, bryozoans, corals and brachiopods (Taschi *et al.* 1992). Carbonate deposits in the region include bedded bioclastic limestone and some carbonate build-ups – Sestra and Brat mountains (the estuary of Partizanskaya River: Fig. 2a) of the Chandalaz Formation and siliciclastic deposits of the Lyadyanza Formation. The latter is cropped out in the Trudnyi

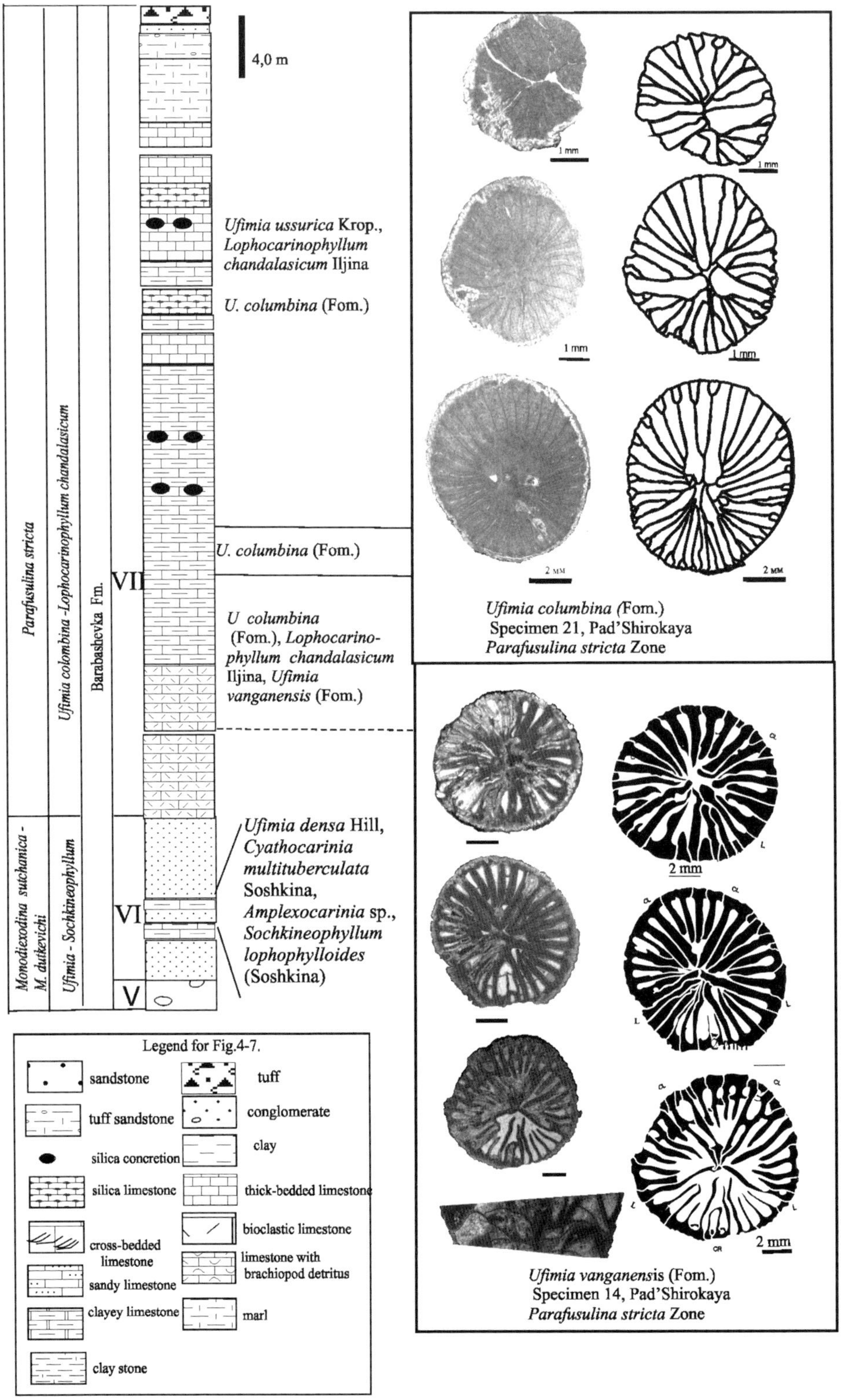
4,0 m
Parafusulina stricta
Ufimia colombina -Lophocarinophyllum chandalasicum
Barabashevka Fm.
VII
Monodiexodina sutchanica - M. dutkevichi
Ufimia - Sochkineophyllum
VI
V
Ufimia ussurica Krop., Lophocarinophyllum chandalasicum Iljina
U. columbina (Fom.)
U. columbina (Fom.)
U columbina (Fom.), Lophocarino-phyllum chandalasicum Iljina, Ufimia vanganensis (Fom.)
Ufimia densa Hill, Cyathocarinia multituberculata Soshkina, Amplexocarinia sp., Sochkineophyllum lophophylloides (Soshkina)
1 mm
1 mm
1 mm
1 mm
2 мм
2 мм
Ufimia columbina (Fom.)
Specimen 21, Pad'Shirokaya
Parafusulina stricta Zone
2 mm
2 mm
Ufimia vanganensis (Fom.)
Specimen 14, Pad'Shirokaya
Parafusulina stricta Zone
Legend for Fig.4-7.
sandstone
tuff
tuff sandstone
conglomerate
clay
silica concretion
silica limestone
thick-bedded limestone
bioclastic limestone
cross-bedded limestone
limestone with brachiopod detritus
sandy limestone
clayey limestone
marl
clay stone

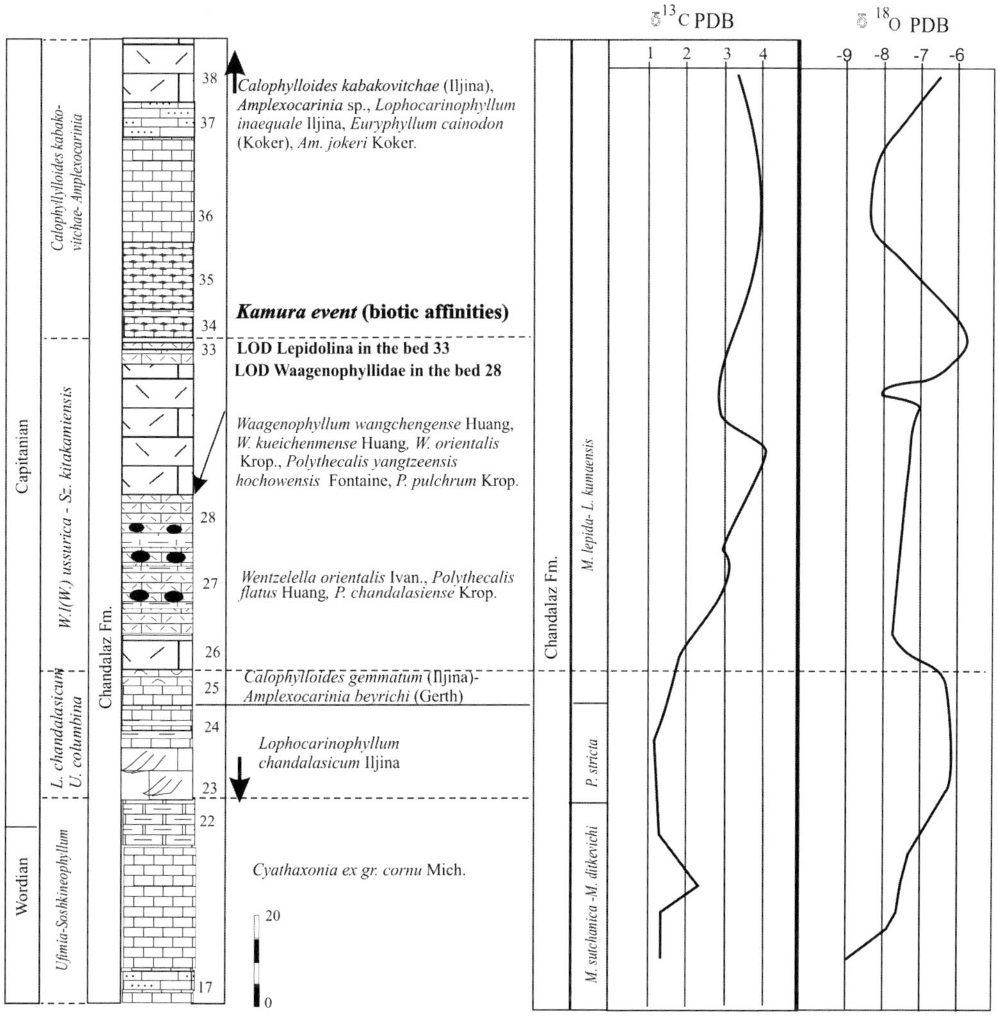

Fig. 5. Distribution of rugose corals and stable isotopes in the Sen'kina Shapka section, Sergeevka terrane (isotope data based on Zakharov *et al.* 2001).

Peninsula and Nakhodka. One of the key sections, Sen'kina Shapka (Fig. 5), is situated in the Partizanskaya River Basin where the coral unit is well correlated with coral-bearing deposits of the Sestra and Brat sections (Figs 3b, 5 & 6). The Trudnyi Peninsula composite section is represented by a series of outcrops situated between Popov Cape and Srednyi Cape (Fig. 3b). *Ufimia vanganensis* (Fomichev), *Verbeekiella*? sp. and *Timorphyllum maichense* (Fomichev), characteristic of the *Parafusulina stricta* Zone, were identified (Figs 3b & 6). At the Nakhodka quarry (Fig. 7) and the Verblud Mountain, the reef limestone is overlain by black mudstones of the Lyudyanza Formation bearing ammonoids (*Cyclolobus kiselevi*: Zakharov & Pavlov 1986) and the small foraminifers *Hemigordius* sp. and *Kamurana* sp., assigned to the Lyadyanza Formation of the Wuchiapingian

Fig. 4. Distribution of corals in the Barabashevka Formation of the Pad'Shirokaya section (Laoelin–Grodekovo terrane). The right tributary of the Barabashevka River. (The log is based on Burago *et al.* 1968 and Kotlyar *et al.* 1975). Thickness of Unit VII, out of scale.

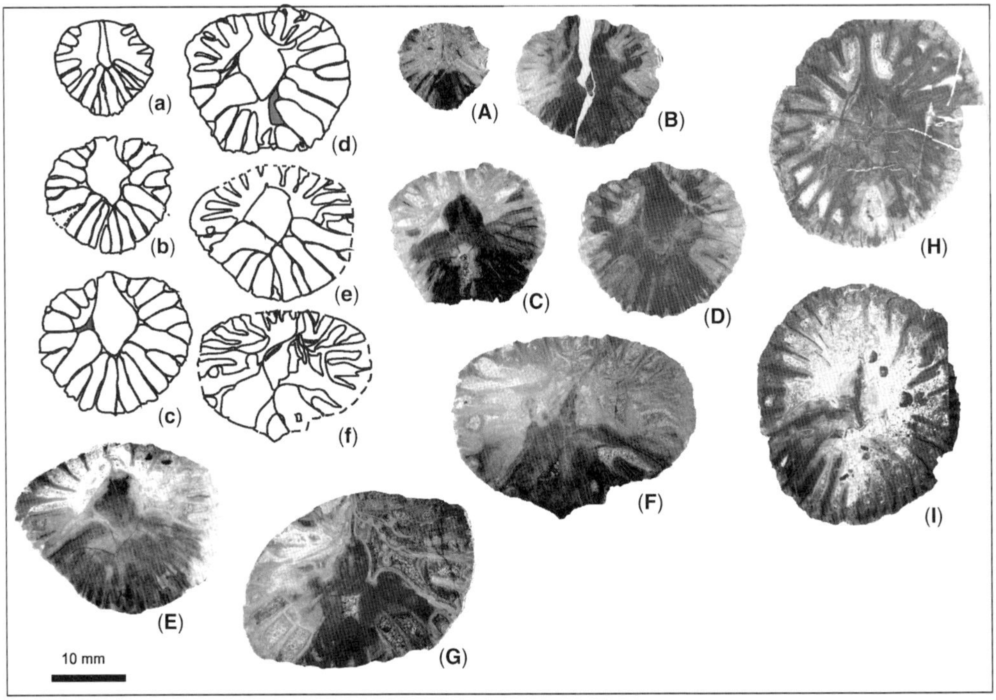

Fig. 6. Coral fauna from Trudnyi Peninsula. *Timorphyllum maichense* Fomichev. Drawings (**a**)–(**f**) The successive ontogenetic stages, specimen 967/2873, Cape Popov, Chandalaz Formation, *Parafusulina stricta* Zone. Photographs (**A**)–(**G**) The transverse sections of the same specimen. The drawings correspond to the photograph of the specimen marked by the capital letter. (**H**)–(**I**) *Timorphyllum maichense* Fomichev, specimen 962/2873, Cape Popov, *Parafusulina stricta* Zone.

(Figs 1 & 7). Fusulinids are rare in the uppermost part of the Chandalaz and Lyadyanza formations. The stratotype of the Lyadyanza Formation, situated along the shore of Lyudyanza (Progulochnaya) Bay in the Trudnyi Peninsula (Fig. 3b), is represented by black siltstone with lenses of crinoid limestone and calcareous sandstone. The underlying reef or reef-breccia is referred to the uppermost part of the Chandalaz Formation (Zakharov & Pavlov 1986). It includes fasciculate *Calophylloides* and *Amplexocarinia*, considered here to be as young as the Capitanian. The lower boundary of the Lyudyanza Formation marks the termination of carbonate deposition. The upper part of the Chandalaz Formation includes an ammonoid assemblage with *Stacheoceras orientale* and *Xenodiscus subcarbonarius* appearing sequentially (Figs 1 & 7). Earlier, the Lopingian boundary was correlated with the appearance of the latter species (Kotlyar *et al.* 2006). According to Zahkarov & Pavlov, the position of the Guadalupian–Lopingian boundary coincides with the base of the Lyudyanza Formation and is dated by the appearance of the ammonoid *Cyclolobus kiselevi*, which was found 10 m above the base of the formation (Zakharov & Pavlov 1987, p. 26).

The Voznesenka terrane (Artemovka River Basin)

Permian deposits are widely distributed in the Artemovka River Basin (outcrops in the Vangan, Golubinaya, Flangovaya and Izvestnyakovaya mountains, on the left bank of the Artemovka River, in Murav'inaya, and near the settlement of Krasnoarmeyskyi). Coral-bearing deposits characteristic of the Guadalupian are assigned to the Chandalaz Formation. The Chandalaz Formation, overlying the Vladivostok Formation in the Artemovka River Basin in the Golubinaya and Vangan mountains, is about 1000 m thick, and consists of limestones, sandy limestones and tuffaceous sandstones (Fig. 8). The lowest part of the Chandalaz Formation belongs to the *Monodiexodina sutchanica* Zone. In the Voznesenka terrane, corals of the *Monodiexodina* Zone were found in Tabayza (Murav'inaya) Bay, in the Golubinaya and

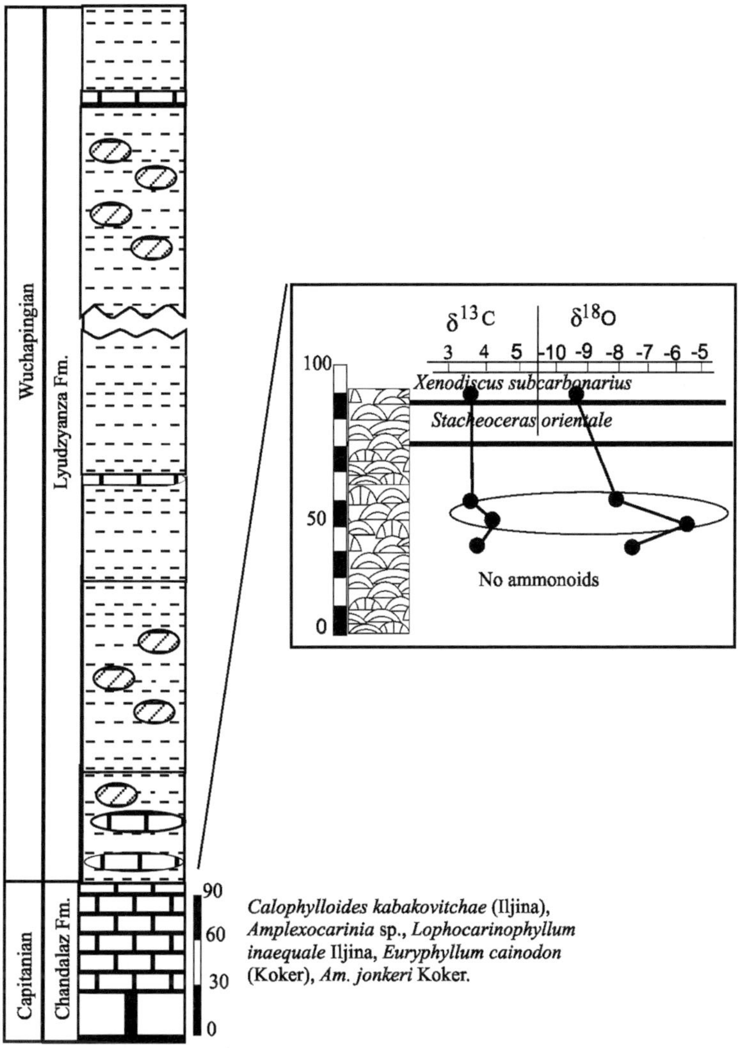

Fig. 7. Chandalaz and Lyudyanza formations, Sergeevka terrane, Nakhodka outcrop, Guadalupian–Lopingian boundary. (Isotope and ammonoid data based on Zakharov *et al.* 2001.)

Flangovaya mountains, near the dam of the Artemovka River and at Shumilov Spring. The assemblage includes *Ufimia densa* (Hill), *Amplexocarinia* sp. and *Lophophyllidium* sp. Much more abundant is the assemblage in the *Parafusulina stricta* Zone, which consists of numerous *Ufimia* such as *U. vanganensis* Fomichev, *U. densa* Hill, *U. fomichevi* Iljina, *Calophyllum magnum* (Fomichev), *Calophylloides gemmatus* (Iljina), *Timorphyllum maichense* Fomichev, *Timorphyllum* sp. and *Verbeekiella*. The coral-bearing deposits are referred to as the *Lophocarinophyllum chandalasicum–Ufimia columbina* beds. Some indeterminate massive colonial corals occur above the previous coral unit in the Vangan Mountain (collection 2873, PIN RAS).

Analyses of coral assemblages

Four coral assemblages (I–IV) are recognized from the Wordian to the Capitanian (Guadalupian) (Figs 9 & 10). They are associated with four foraminiferal zones: the *Monodiexodina sutchanica–Metadoliolina dutkevichi* Zone, the *Parafusulina stricta* Zone, the *Metadoliolina lepida–Lepidolina kumaensis* Zone and the small foram zone of *Bradyina ussurica–Dagmarita cuneata* (Kotlyar

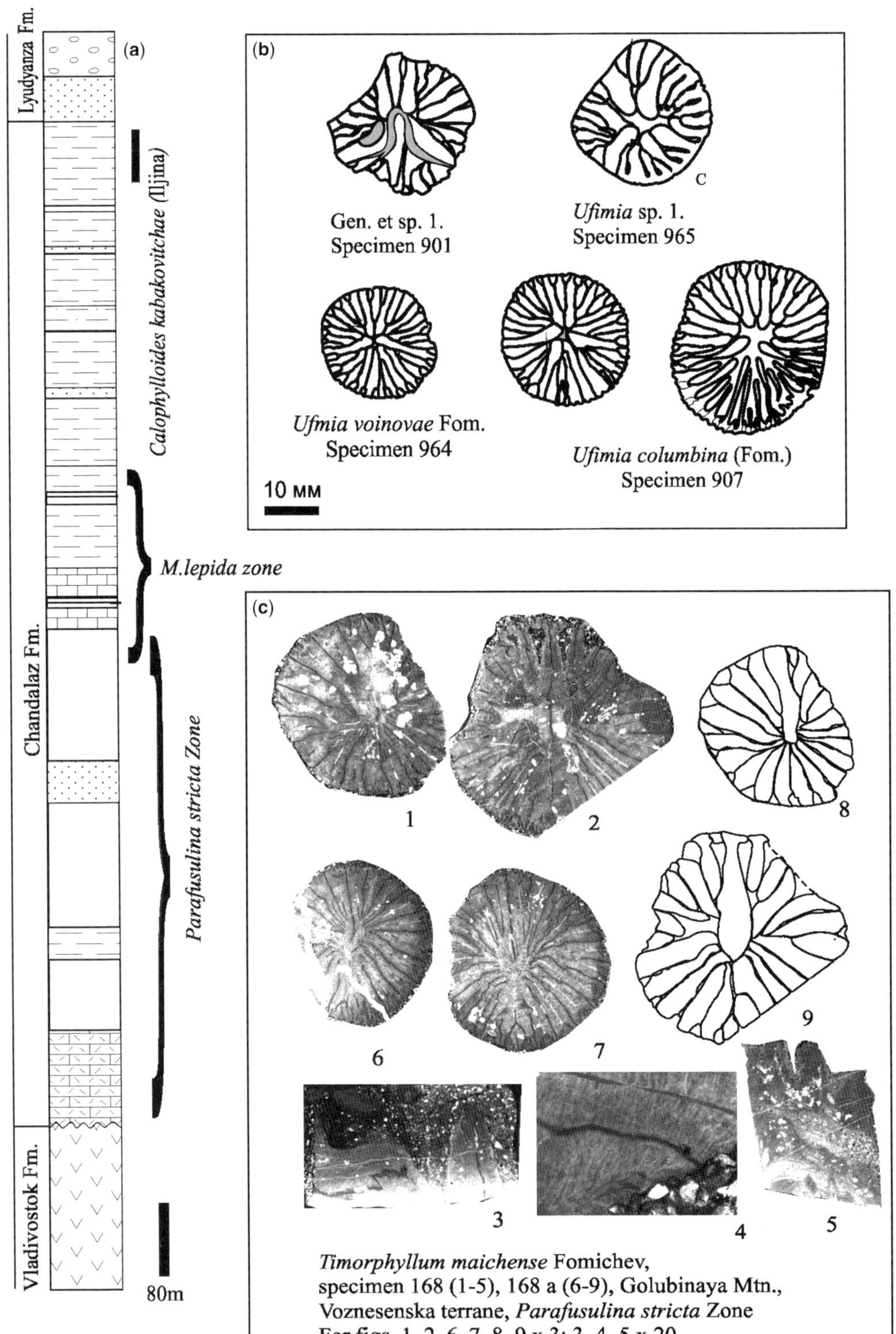

Timorphyllum maichense Fomichev,
specimen 168 (1-5), 168 a (6-9), Golubinaya Mtn.,
Voznesenska terrane, *Parafusulina stricta* Zone
For figs. 1, 2, 6, 7, 8, 9 x 3; 3, 4, 5 x 20

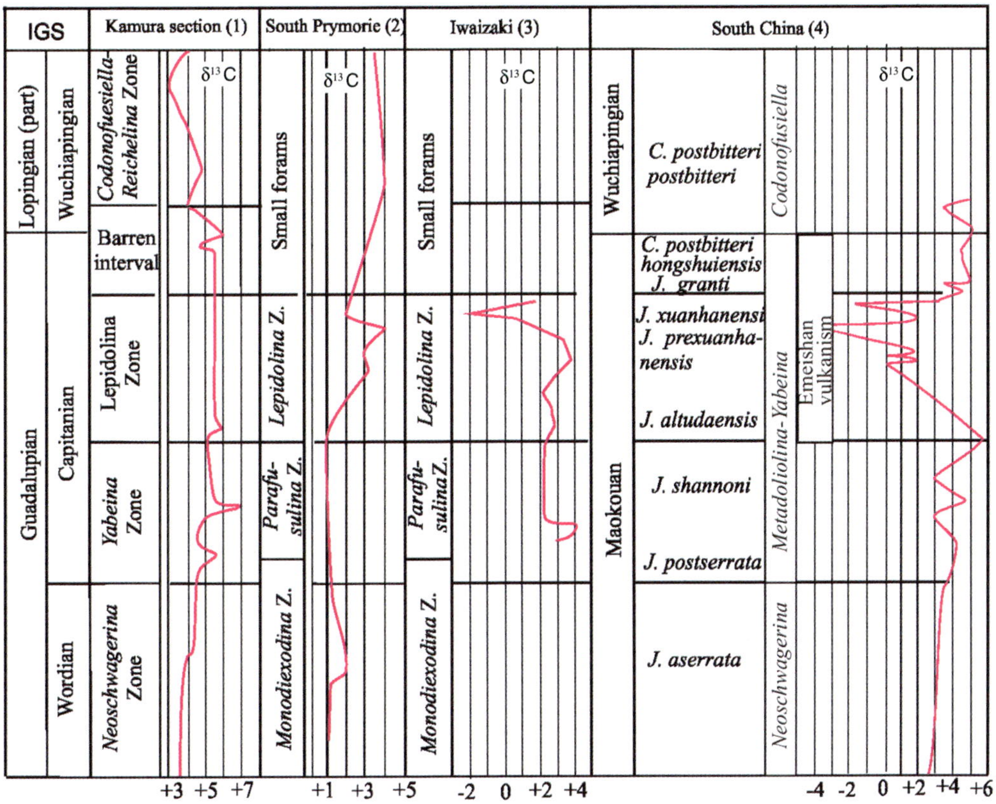

Fig. 9. Isotope data in the Wordian–Capitanian in the Palaeo-Pacific region (summarized data from Zakharov *et al.* 2001; Isozaki *et al.* 2007; Bond *et al.* 2010*a*, *b*).

et al. 2004*a*, *b*). Coral assemblages demonstrate a gradual ecological replacement from the cosmopolitan type with non-specific palaeogeographical affinities to solitary corals bearing a strong similarity to the Peri-Gondwanian type, and then to colonial fauna of the typical Cathyasian type. The last youngest assemblage of the Capitanian consists of rather primitive forms. This assemblage is considered as a stressed one, despite the fact that there is no substantial lithological change within the coral-bearing rocks (Figs 1 & 7).

Assemblage I: Ufimia *and* Sochkineophyllum

Corals occurring in the *Monodiexodina sutchanica* Zone are represented by *Calophyllum simplex* (Soshkina), *Ufimia* sp. and *Sochkineophyllum* sp. *Calophyllum simplex* was described by Fomichev as *Tachylasma simplex* Soshkina. He also included in the synonymy '*T. angustum* Rothpletz' based on Soshkina (1928). The latter had been described by Soshkina as *Tachylasma gerthi* Soshkina and is now considered a junior synonym of *C. angustum* (Rothpletz) (D. Weyer pers. comm.). Thus, *C. simplex* (Soshkina) has to be attributed to *C. angustum* (Rothpletz). The assemblage also includes *Cyathocarinia* sp. Numerous *Ufimia* and *Sochkineophyllum* are widely distributed in the Upper Artinskian and, in restricted numbers, occur in the Kungurian of the Ural Basin (Kossovaya 2007, 2009). All aforementioned genera are found in the Kungurian–Capitanian of SE Pamir (Iljina 1997). Their palaeogeographical position is considered here to be originally in the Southern Hemisphere. Dating

Fig. 8. (**a**) Vangan Mountain, Artemovka River Basin (Voznesenska terrane), *Parafusulina stricta* Zone. (**b**) Fauna from the outcrop near the dam: Gen. et sp. 1, specimen 901; *Ufimia* sp.1, specimen 965; *Ufmia voinovae* Fomichev, specimen 964; two views of *Ufimia columbina* (Fom.), specimen 907. (**c**) Corals from the Golubinaya Mountain: *Timorphyllum maichense* Fomichev, specimens 168 (1–5) and 168a (6–9). For (1, 2, 6–9) ×3; for (3)–(5) ×20.

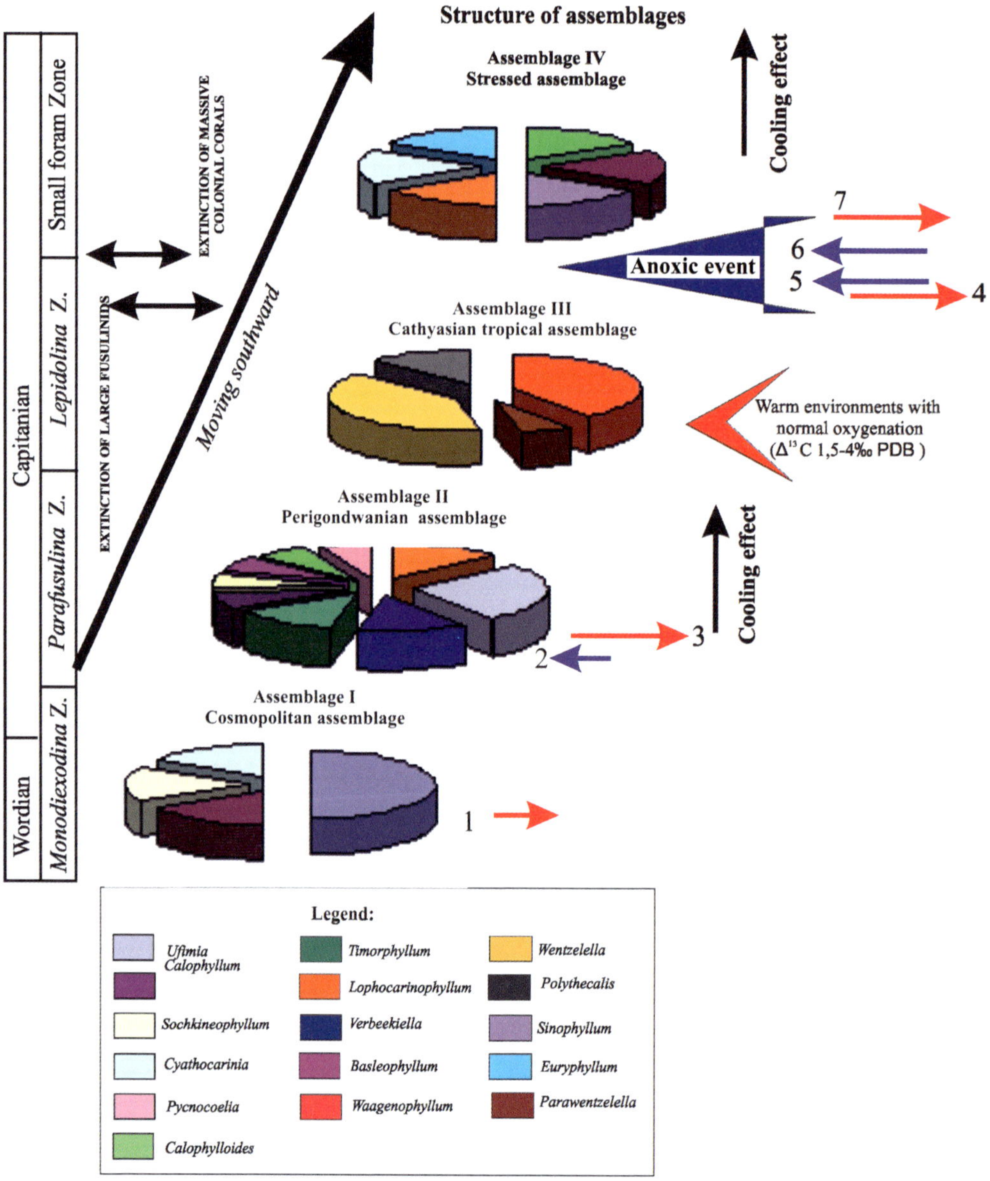

Fig. 10. Proposed model of the extinction event in the southern Primorye (western Palaeo-Pacific): 1, positive shift δ^{13}C 1‰ PDB (southern Primorye); 2, negative shift δ^{13}C 1.5‰ PDB (southern Primorye); 3, positive shift δ^{13}C 4‰ PDB (Japan); 4, positive shift δ^{13}C 3‰ PDB (southern Primorye); 5, negative shift δ^{13}C 1.5‰ PDB (southern Primorye); 6, negative shift δ^{13}C 7‰ PDB (South China); 7, positive shift δ^{13}C 6‰ PDB (South China) (based on Zakharov *et al.* 2001; Isozaki *et al.* 2007; Bond *et al.* 2010*a*, *b*).

by fusulinids of Assemblage I as Wordian–Capitanian is different in taxonomical composition to the Roadian (Kazanian) or Wordian–Capitanian of the Boreal Province (Kossovaya 2009). However, it is similar to coeval assemblages of the Hugete Formation of Inner Mongolia (Ding *et al.* 1985).

Assemblage II: Ufimia columbina–Lophocarinophyllum chandalasicum

Ufimia columbina–Lophocarinophyllum chandalasicum was demonstrated in the lower part of the Chandalazian Regional Stage. The first corals of this assemblage appear a little above the lower boundary of the *Parafusulina stricta* fusulinid Zone. The stratotype of zone Unit VII (Fig. 4) is situated in the lower part of limestones of the Barabashevka Formation on the left bank of the Pad' Shirokaya section near the settlement of Barabash. The index species, *Ufimia columbina*, was originally described by Fomichev (1953) as *Tachylasma columbinum*. In consideration of the taxonomical value of early ontogenetic features it is referred here to *Ufimia*. This species was found in many outcrops: along the right bank of the Artemovka River, Golubinaya Mountain, Suvorovka River, Devil's Spring, along the left bank of the Barabashevka River and across the mouth of Bogatyi Spring. We can also include in the synonymy of this species '*Plerophyllum clavatum* Iljina' (Ding *et al.* 1985, plate 8, fig. 3) reported from the Zhesi Honguer Formation in the Zhesi Hill section (Inner Mongolia) (Ding *et al.* 1985).

Lophocarinophyllum chandalasicum was originally described as *Lophocarinophyllum* cf. *acanthiseptatum* Grabau (Fomichev 1953), and then redescribed and renamed on the basis of the new collection (Iljina 1987). This species was found along the left bank of the Shirokaya Pad' ravine, along the right bank of the River Nesterovka, in the Golubinaya Mountain, in the upper course of the Artemovka River (Devil's Spring) (Fig. 8), near the mouth of Murav'inaya River and elsewhere. The type level of both index species is Unit VII in the lower–middle part of the Barabashevka Formation.

Assemblage II includes *Calophylloides gemmatus* (Iljina), *Pycnocoelia magna* (Fomichev) emend. Iljina (1984), *Ufimia densa* (Hill), *Uf. vanganensis* (Fomichev), *Uf. voinovae* (Fomichev) (Fig. 8), *Ufimia columbina* (Fomichev), *Sochkineophyllum intabulatum* (Fomichev 1953), *S. densum* (Fomichev 1953), *Timorphyllum maichense* (Fomichev) (Figs 5 & 8), *Verbeekiella ussurica* Fomichev, *Basleophyllum indicum* Koker and others (Figs 1, 4 & 6). In addition to *Ufimia columbina*, which enables discussion on the similarity between assemblages of the Zhesi and Barabashevka formations and the lower part of the Chandalaz Formation, *Ufimia vanganensis* (Fomichev) seems to be similar to *Tachylasma znense* Ding (Ding *et al.* 1985).

Part of the coral assemblage represented by endemic forms

Calophyllum magnum (Fomichev 1953), which occurs in the Shumilovka Spring, the right tributary of the Artemovka River, the Izvestkovaya Mountain and near the village of Pogranichnyi, was redescribed by Iljina (1984). *Ufimia voinovae* (Fomichev) and pseudofasciculate forms of *Calophylloides gemmatus* Iljina (Artemovka River) are also considered here to be endemic species. Columellate forms of Peri-Gondwanian origin found in the Chandalaz Formation are represented by *Timorphyllum* (Figs 5, 6 & 8), *Lophocarinophyllum*, *Verbeekiella*, *Lophophyllidium* (*Lophophyllidium*) and *Wannerophyllum* (Kossovaya 2009). They are also known from the Basleo Formation of Timor (Schouppé & Stacul 1955; Fedorowski 1974, 1986). Palaeogeographically, the Timor region is situated in the temperate zone of the Himalaya Province of the Southern Hemisphere (Shen & Shi 2004).The list of genera is to be amended by the addition of *Lophophyllidium*, *Ufimia*, *Timorphyllum* and *Verbeekiella*, described from Xilingolmeng, central Inner Mongolia (Wang *et al.* 2006). Two species, *Tachylasma maichense* Fomichev and *Verbeekiella ussurica* Fomichev, from this occurrence were originally described from southern Primorye. The coral fauna of the fusulinid *Parafusulina stricta* Zone of southern Primorye is generically, as well as specifically, very closely related to the fauna known from the Jisu Honguer Formation in Inner Mongolia (Ding *et al.* 1985). Three common species were mentioned from the Zhesi Formation in Wulancilao of Sonid Zuoqi, Xilin Gol Meng, Inner Mongolia, namely *Pleophullum magnum* (Fomichev), *Timorphyllum maichense* Fomichev and *V. ussurica* Fomichev (Ding 1983). Corals of the Jisu Honguer Formation document the *Plerophyllum crassoseptatum–Tachylasma zhesiense–T. variabile* Zone (Ding *et al.* 1985), which is comparable with the *Ufimia colimbina–Lophocarinophyllum chandalasicum* beds in southern Primorye, and both are characterized by solitary forms without dissepiments.

Assemblage IIa (Fig. 1), characterized by corals without dissepiments, includes a few species: *Amplexocarinia* sp., *Calophylloides gemmatum* Iljina found in the Sen'kina Shapka section in the upper part of the *Parafusulina stricta* interval.

Assemblage III: Szechuanophyllum kitakamiense–Wentzelloides (Wentzelloides) ussuricus

This assemblage, typical of the upper part of the Chandalaz Formation, was noted in strata of the *Metadoliolina lepida–Lepidolina kumaensis* fusulinid Zone. The stratotype of this zone was established in the Sen'kina Shapka section, where corals of Assemblage III occur in the interval of beds 27–28 assigned to the upper part of the Chandalaz Formation (Figs 5 & 11). The assemblage is widely distributed and was found in the Sestra and

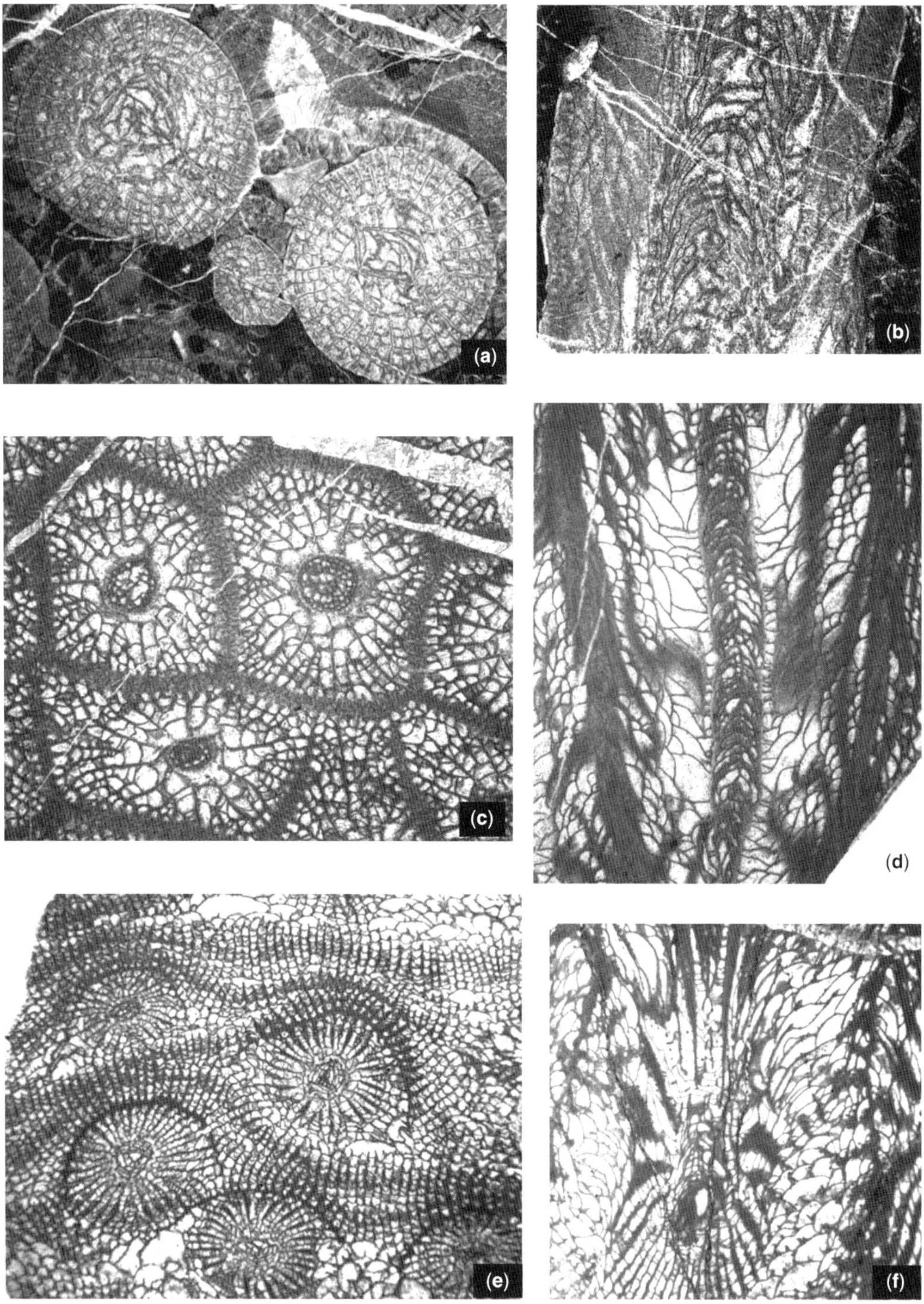

Fig. 11. Waagenophyllidae corals from the Chandalaz Formation., Sergeevka terrane. (**a**) & (**b**) *Waagenophyllum indicum* (Waagen et Wentzel); (**c**) & (**d**) *Szechuanophyllum szechuanense orientale* Ivan.; (**e**) & (**f**) *Wentzelloides* sp.1, ×10. Sestra Mountains, *Metadoliolina lepida–Lepidolina kumaensis* Zone, Capitanian.

Brat Mountains, the Yastrebovka River and others. It includes mostly colonial corals such as *Waagenophyllum (Waagenophyllum) indicum* Waagen et Wentzel, *W. (W.) kueichouwense* Huang, *W. (W.)* aff. *virgalense* (Waagen et Wentzel), *W. (W.) wenghengense* Huang, *W. (W.)* ex gr. *novozelandiae* Leed, *W. (W.) orientale* Kropatcheva, *Parawentzelella* sp., *Wentzelella (Wentzelella) maichensis* Iljina, *W. (W.)* ex. gr. *winneri* Waagen et Wentzel, *W. (W.) golubinae* Kropatcheva, *W. (W.) ussurica* Kropatcheva, *W. (W.) simplex* Yu et Lin, *Szechuanophyllum kitakamiense* Yabe et Minato, *Wentzelloides (Wentzelloides) maiyanensis* Yabe et Minato, *W. (W.) primoricum* Kropatcheva, *W. (W.) grandicolumellatus* Kropatcheva, *Polythecalis flatus* Huang and *P. yangtzeensis hochowensis* Huang. One species, *Wentzelella orientale* Ivanovskiy (Ivanovskiy 1969), was described from the Chandalaz Formation in the Partizanskaya River valley near the settlement of Ekaterinovka, Gold Hill. A similar assemblage of massive colonial corals is known from the Yihewusu Formation of the Jisu Honguer, Inner Mongolia (Ding *et al.* 1985), where two zones with waagenophyllid corals were recognized: *Carnoverbeekiella sinensis–Pseudowaagenophyllum vesiculosum–Diphycarinophyllum* and *Waagenophyllum virgalense mongoliense–W. stereoseptatum-Wentzelella*. According Ding *et al.* (1985), colonial Waagenophyllidae extend to the top of the Yihewusu Formation.

Assemblage IV: Amplexocarinia beyrichi–Calophylloides kabakovitchae

The assemblage with predominant colonial Waagenophyllidae changes into an assemblage with solitary corals, restricted to the uppermost part of the Chandalaz Formation in the Sen'kina Shapka Mountain (Sergeevka terrane): *Lophocarinophyllum inaequale* Iljina, *Amplexocarinia* aff. *heimoi* Heritsch, *Am. jorkeri* Koker, *Cyathocarinia* sp., *Sinophyllum pendulum* Grabau and *Euryphyllum cainodon* (Koker), which correspond to the uppermost Capitanian. A similar fauna was also reported from the Nakhodka and Volchanetz occurrences in the algal–bryozoan reef limestone, separated as Svetlanka beds. The taxonomical diversity here is very poor and the assemblage comprises only two species of solitary corals: *Amplexocarinia beyrichi* (Gerth) and *Calophylloides kabakovitchae* (Iljina). Iljina (1977) described the fasciculate colonies *C. kabakovitchae* with intracalical and lateral increase from the contemporaneous deposits near Volchanetz Lake (Primorye) (Fig. 3a) and near the town of Nakhodka (Figs 7 & 12). One of the typical features of the latter species is the small corallite diameter and their fasciculate growth. In addition, the list of coral occurrences from the Nakhodka sphinctozoan reef includes *Paracaninia subtilis* (Niermann) and *P. simplex* Zhao (Kotlyar *et al.* 1987).

Palaeogeographical affinities of coral associations

Considering the palaeogeographical affinities of the non-tropical assemblage of solitary forms found in the *Parafusulina stricta* Zone in both the Chandalaz and Barabash formations, similarities between faunas from Boreal and Peri-Gondwanian provinces are recognizable. The Peri-Gondwanian component of the assemblage involves species of polyprovincial genera, such as *Euryphyllum cainodon* (Koker), *Kabakovitchiella duplex* (Schouppe et Stacul), *Am. heritschi* Schouppe et Stacul, and species and genera unknown in the Boreal province, yet distributed in the Basleo Formation in Timor: *Basleophyllum indicum* (Koker), *Verbeekiella* and *Timorphyllum* (Charlton *et al.* 2002). The palaeogeographical evolution of the Permian terranes of southern Primorye and the territories of Inner Mongolia and North China all belong to the 'northern transitional belt' of the western Palaeo-Pacific (Shen & Shi 2004), assumed to be the sequential movement to the south. It resulted in the invasion of Peri-Gondwanian elements followed by typical Cathyasian forms represented by colonial warm-water tropical corals (Figs 1 & 11). The subsequent, and last, assemblage composed of small *Calophylloides* fauna is comparable by morphological similarity (connected possibly with palaeotemperature) with solitary *Calophyllum* distributed in the Middle–Upper Permian in the Boreal province of the Northern Hemisphere, such as in Greenland (Flügel 1973; Fedorowski 1982), Arctic Canada (Fedorowski & Bamber 2001, 2002), the Russian Platform (Weyer 1997), the UK (Howse 1848; King 1948, 1950), Poland (Sarnecka 2001) and Germany (Zechstein Basin) (Weyer 1979, 1984). Species of *Calophyllum* are typical of the Basleo Formation in Timor (Schouppé & Stacul 1959; Niermann 1975). The bi-polar position of *Calophyllum* provinces in the Roadian–Capitanian was considered to coincide with temperate belts formed in the Southern and Northern Hemisphere (Kossovaya 2009). It can be indirectly related to the long-lasting cooling effect found in the Early Capitanian (Isozaki *et al.* 2007). The appearance of tropical massive colonies could be a result of the following southward movement of terranes situated in the area studied. The biotic change coincides with the replacement of typical tropical reefs (the Sestra and Brat mountains) – characterized by the essential role of massive colonial Waagenophyllidae, by microbial or by sphinctozoan

(a)
(b)
(c)
(d)
(e)
(f)
(g)
(h)
(i)
(j)
(k)
(l)
(m)
T
S
R
A

reef mounds – with microfossils such as *Tubiphytes* and *Antracoporella* as binders (Fig. 12), and sphinctozoan constructing the core of the reef mound studied in the Nakhodka outcrop (Belyaeva 1987). A small *Calophylloides* fauna was documented in the Nakhodka reef in the probably microbial upper part of the mound with small sphinctozoans (Fig. 12).

Geochemical data

Our consideration is aimed at comparing the dynamics of stable isotope and coral diversity in the Capitanian in order to recognize the possible influence of the Kamura event, a cooling effect established in the Palaeo-Pacific (Isozaki 1997; Isozaki *et al.* 2007), and data of the isotope dynamics from South China (Bond *et al.* 2010*a*, *b*). Data published by Zakharov *et al.* (2001) show an increase in $\delta^{13}C$ from +1.2 to 4‰ PDB at the end of the *Metadoliolina lepida–Lepidolina kumaensis* Zone in the Sen'kina Shapka section, which coincides with the beginning of the interval of 'stressed' fauna (Figs 5 & 9). The isotope data from the stratotype section of the Iwaizaki Limestone (Japan) show a minimum value in the *Pseudofusulina paramotohashii* Zone ($\delta^{13}C$ of +2–4‰ PDB), comparable in stratigraphic position with the *Parafusulina stricta* Zone. The lower and middle part of the *Lepidolina multiseptata* Zone (with positive trend of $\delta^{13}C$ to 4.1‰ PDB) (Zakharov *et al.* 2001) approximately coincides with the maximal value in the Lepidolina Zone of the Iwaizaki Formation, $\delta^{13}C$ of about 4.0‰ PDB. Comparison of $\delta^{13}C$ isotopic data from the Iwaizaki Limestone and the Chandalaz Formation (Sen'kina Shapka section) demonstrates a similar trend to the progressive increase in *Lepidolina* Zone deposits. The isotopic ratio known from the Iwato Formation, Kamura (Isozaki *et al.* 2007), shows a higher value of $\delta^{13}C$ in the *Yabeina* Zone (coeval with the *Parafusulina* Zone: Fig. 9) followed by a high value plateau. In the Iwato Formation, the high value plateau is correlated with the *Yabeina* Zone + *Lepidolina* Zone + barren interval (Isozaki *et al.* 2007). The equivalent of the fauna of the barren interval, the stressed fauna developed at the end of the Capitanian, was reported from the Nakhodka section (Fig. 6) and some others (see earlier). The isotope composition shows about a 2‰ PDB drop and then a positive trend to 4‰ PDB up to the Guadalupian–Lopingian boundary. The maximum $\delta^{18}O$ value is documented for the *Parafusulina* Zone. The positive shift in $\delta^{18}O$ is characteristic of the onset of the 'stressed' fauna and can be interpreted as a second phase of the cooling effect. At the end of the stressed interval near the Lopingian boundary, $\delta^{18}O$ values changed progressively from −8 to −9‰ PDB (Fig. 5); this coincides with the appearance of microbial-sphinctozoan reefs. Then the following decrease in $\delta^{18}O$ values up to a very low figure of about −10‰ PDB is considered as an anoxic event. This rather approximate correlation of the geochemical (from Zakharov *et al.* 2001) and biotic data the impact of the Kamura cooling event followed by Emeishan volcanism to be linked with the extinction of the waagenophyllid corals in southern Primorye, an abrupt replacement of reef-builders and the disappearance of large fusulinids.

Extinction of Guadalupian corals: the mechanism and possible reasons

In the interval of the long-lasting Kamura event, nearly all fauna of the Capitanian had been influenced by cooling (Isozaki *et al.* 2007). However, the cooling effect was, in part, disguised by tectonic movement of the blocks towards the equator, and the migration of fauna from the Cathasyan and Peri-Gondwanian realms. The extinction of massive colonial 'tropical' corals at the end of the Capitanian and the general replacement of reef biota, together with the extirpation of other groups of typical tropical fauna, can be considered as the main biotic features of the 'small' event preceding the Permian–Triassic global extinction.

Assemblages I and II are typical of the rather deep-slope depositional environments, with some influx of siliciclastic material (Kossovaya 2009), and seem to be tolerant to some volcanic activity.

Fig. 12. Corals of the uppermost part of the Chandalaz Formation (Svetlanka beds), Voznesenka terrane, Artemovka River Basin, collection 2373, Palaeontological museum PAS. (**a**) *Kabakovitchiella duplex* (Schoup. et Stacul.), Golubinaya Mountain, Artemovka River, specimen 179 (×7); (**b**) & (**c**) *Amplexocarinia* sp., specimen 191 (b, ×8; c, ×7); (**d**) & (**e**) *Calophylloides kabakovitchae* (Iljina 1977) (×7), specimen 192, Vangan, Artemovka River; (**f**) *Amplexocarinia beyrichi* (Gerth), specimen 197 (×7), Artemovka River, Mnogoudobnoe village; (**g**) early stage of *Calophylloides kabakovitchiae* (Iljina 1977) (×10), specimen 767, Artemovka River, quarry; (**h**) *Calophylloides kabakovitchae* (Iljina 1977), specimen 797, Artemovka River, quarry (×10); (**i**) & (**j**) *Calophylloides kabakovitchiae* Iljina 1977, specimen 767, quarry; (**k**) fasciculate form of *Calophylloides kabakovitchae* (Iljina), specimen 800 (×7); Artemovka River, quarry; (**l**) & (**m**) microphotographs: packstone with *Anthracoporella* (A), *Tubiphytes* (T) and sphinctozoa (S).

Judging from their palaeogeographical position, the diverse and abundant fauna of the Laoelin–Grodekovo terrane was situated in a back-arc basin. This level coincides with the first positive excursion of ^{13}C. The absence of colonial corals from Assemblage II in remnants of carbonate platform (Sergeevka and Voznesenka terranes) and also in adjacent territories is indirect evidence of the Kamura cooling. The assemblage is predominated by Peri-Gondwanan immigrant taxa. It shows a great affinity with other simultaneous faunas of the 'northern transitional belt' (Fig. 10). The absence of colonial forms can be explained by exposure to cooling. Consequently, the appearance of colonial forms has to be connected with the progressive southward movement of all compared blocks of the 'northern transitional belt'. Abundant massive colonial corals (Assemblage III) in the middle Capitanian are also widely distributed in coeval assemblages in Maokouan in South China, situated in the equatorial zone (Ezaki 1994). According to the reconstruction of the Kamura event, this embraced the whole *Lepidolina* Zone, and the anomaly between the long duration of cooling and the appearance of Cathyasian fauna is evident. We do not have any critical biochemical evidence of isotopic disturbances at this level. Thus, by analogy with modern corals, we assume that the adaptation of colonial fauna to decreased temperatures was, in part, compensated by the southward movement of terranes. Considering the palaeogeographical position of southern Primorye along the northern limit of the subtropical zone, the change in temperature could be rather crucial. The following event, which halted the development of colonial corals, seems to be Emeishan volcanism (Bond *et al.* 2010*a*, *b*) (Fig. 10). This event is behind the simultaneous restriction of massive colonies in the equatorial zone (South China), and coincides with a negative $\delta^{13}C$ shift in southern Prymorie and a giant leap in South China (Fig. 9). The disappearance of colonial corals matches approximately the middle of the high $\delta^{13}C$ plateau in Japan (Isozaki *et al.* 2007). The Emeishan exposure triggered some cooling and the following anoxia is supported by a positive excursion in $\delta^{13}C$. The extinction of Assemblage III can presumably be explained by a cooling in low latitudes and associated eutrophication, which probably was responsible for the extinction of tropically adapted faunas in the Palaeo-Pacific at the end of the *Lepidolina* Zone (late Capitanian), not only in the eastern but also in the western Palaeo-Pacific. Assemblage IV could exist in a fairly cool-water basin, and the replacement of reef-builders by sphinctozoan and microbial structures is also evidence of environmental changes. The terminal exposure of the Guadalupian–Lopingian event at 260.4 Ma was connected with the onset of superanoxia. Thus, based on the progressive change of the Rugosa assemblages and supported by geochemical data, we propose a two-phase pattern of the Kamura event with regional differentiation of its features.

The event at the end of the Capitanian shows only a weak coincidence with typical features of major events (Stanley 1988) such as the reappearance of a restricted number of polyprovincial genera (*Amplexocarinia*, *Calophylloides*) after the extinction of widespread and abundant tropical taxa. Contrary to the proposed model (Stanley 1988), carbonate sedimentation in microbial-sphinctozoan reefal facies in southern Primorye lasted some time after the extinction of tropical massive corals and was accompanied by the spreading of the 'stressed coral community'. The carbonate sedimentation ceased later, during the Lopingian. Following anoxia at the Guadalupian–Lopingian boundary, unfavourable conditions for corals occurred, even for the subsequent development of the 'stressed' assemblage. This minimized changes of survival and acted as a barrier for migration from the south, where fasciculate and simple solitary waagenophylliids survived.

Conclusion

The coral fauna of the Capitanian in southern Primorye is represented by four distinct assemblages. The absence of transitional forms between assemblages allows them to be considered as an ecological succession triggered by the gradual movement of terranes towards the south. The different palaeogeographical gradient of terranes is demonstrated by a differentiation in coral composition. The extinction of massive colonial Waagenophylliidae in southern Primorye, assigned to western Palaeo-Pacific, is well correlated with a synchronous biotic event recognized in Japan, where waagenophylliid corals disappeared in the uppermost part of the *Lepidolina multiseptata* Zone and coeval deposits of Inner Mongolia, and is also well correlated with a general reduction in coral diversity at the end of the Capitanian in China (Wang & Sugiyama 2000). The disappearance of massive colonial corals occurs slightly prior to the extinction of large fusulinids. The rugose event coincides with replacement of coral-reef facies, including reef mounds of sphinctozoans. This mechanism was well documented in the Iwaizaki reef where the massive colonies, as reef constructors, were present in different reef facies, but then disappeared in the uppermost reef lagoon inhabited by tubiphytes, calcisponges, bryozoans, brachiopods and crinoids (Fig. 8) (Kawamura & Machiyama 1995). Thus, some ecological changes of biota in the uppermost part of the *Lepidolina* Zone are traceable.

The effect of the Kamura event is demonstrated in the western Palaeo-Pacific (southern Primorye region) by isotope trends, massive colonial coral extinction, and the following extinction of large fusulinids and tropical reef biota. It is considered as a complex event and the cooling period influenced the extinction of massive colonial corals.

The proposed pattern of the event documented in the western Palaeo-Pacific (southern Primorye region) embracing massive colonial coral extinction, the later elimination of large fusulinids and the replacement of reef biota was a result of two phases of the abiotic pressure: (a) the temperature control of the Kamura event; and (b) hypostatized influence of the Emeishan volcanism.

The authors are very grateful to Dr G. Kotlyar for advice in the geological and stratigraphical part of the work. In addition, we appreciate Dr O. Erlanger's help in arranging our studies of collections in the Palaeontological Museum of the Russian Academy of Science, and the administration of the Palaeontological Institute RAS for giving us permission to work and photograph the museum collections. We are grateful to Dr D. Weyer for carefully reading an earlier typescript and for his comments, and to Prof. I, Somerville and Dr J. Yagt for important comments and linguistic corrections. Also I am very grateful to editors. This study was supported by RFBR grants 11-05-00950-a and 11-05-00053-a.

References

BELYAEVA, G. V. 1987. Some late Permian Sphinctozoa of the South Primorye. *In*: ZAKHAROV, YU. D. & ONORPIENKO, YU. I. (eds) *Problems of the Permian and Triassic Biostratigraphy of East USSR*. Far East Scientific Centre of the Academy of Sciences of the USSR, Vladivostok, 49–53 (in Russian).

BOND, D. P. G., HILTON, J., WIGNALL, P. B., ALI, J. R., STEVENS, L. G., SUN, Y. & LAI, X. 2010*a*. The Middle Permian (Capitanian) mass extinction on land and in the oceans. *Earth-Science Reviews*, **102**, 100–116.

BOND, D. P. G., WIGNALL, P. B. *ET AL*. 2010*b*. The mid-Capitanian (Middle Permian) mass extinction and carbon isotope record of South China. *Palaeogeography, Palaeoclimatology, Palaeoecology*, **292**, 282–294.

BURAGO, V. I., KISILEVA, A. V., KOTLYAR, G. V., NIKITINA, A. P. & CHERNYSH, I. P. 1968. *Biostratigraphy of the Upper Permian of the South-Western Primorye.* Unpublished Report of Primorsk Geological Department, Vladivistok.

CHARLTON, T. R., BARBER, A. J. *ET AL*. 2002. The Permian of Timor: stratigraphy, palaeontology and palaeogeography. *Journal of Asian Earth Sciences, 20*, **7**, 19–774.

COPPER, P. 1994. Ancient reef ecosystem expansion and collapse. *Coral Reefs*, **13**, 3–11.

DING, Y. 1983. On some new species of Early Permian Tetracoralla from Xilin Gol Meng, Inner Mongolia. *Bulletin of the Tianjin Institute of Geology and Mineral Resources*. Chinese Academy of Geological Sciences, Beijing, **8**, 105–122.

DING, Y. I., XIA, G. Y., DUAN, C. H., LI, W. G., LIU, X. L. & LIANG, Z. F. 1985. Study of the early Permian Stratigraphy and fauna in Zhesi district, Nei Mongol, Zizhiqu (Inner Mongolia). *Bulletin Tianjin Institute of Geology Mineral Resources*, Chinese Academy of Geological Sciences, Beijing, **10**, 160–244.

EZAKI, E. 1994. Patterns and paleoenvironmental implication of end-Permian extinction of Rugosa in South China. *Palaeogeography, Palaeoclimatology, Palaeoecology*, **107**, 165–177.

FEDOROWSKI, J. 1974. The Upper Palaeozoic tetracoral genera *Lophophyllidium* and *Timorphyllum*. *Palaeontology*, **17**, 441–473.

FEDOROWSKI, J. 1982. Some rugose corals from the Upper Permian of East Greenland. *Rapport – Grønlands Geologiske Undersogelse*, **108**, 71–91.

FEDOROWSKI, J. 1986. Permian rugose corals from Timor (remarks on Schouppé and Stacul's collections and publications from 1955 and 1959). *Palaeontographica A*, **191**, 173–226.

FEDOROWSKI, J. & BAMBER, E. W. 2001. Guadalupian (Middle Permian) solitary rugose corals from Degerböls and Trold Fiord Formations, Ellesmere and Melville islands, Canadian Arctic Archipelago. *Acta Geologica Polonica, 51*, **1**, 31–79.

FEDOROWSKI, J. & BAMBER, E. W. 2002. Palaegeographic and biostratigraphic significance of Guadalupian (Middle Permian) solitary corals, Sverdrup Basin, Arctic Canada. *In*: HILLS, L. V., HENDERSON, C. M. & BAMBER, E. W. (eds) *Carboniferous and Permian of the World: Proceedings of the 14th International Congress of the Carboniferous and Permian.* Canadian Society of Petroleum Geologists, Memoirs, **19**, 427–436.

FLÜGEL, H. W. 1973. Rugose Korallen aus dem Perm Ost-Grönlands. *Verhandlungen der Geologischen Bundesanstalt*, **1**, 1–57.

FOMICHEV, V. D. 1953. *Permian Rugose Corals from Far East.* Transactions of the All-Union Research Geological Institute, St Petersburg, 1–55 (in Russian).

HOWSE, R. 1848. A catalogue of the fossils of the Permian system of the counties of Northumberland and Durham, drawn up at the request of the Tyneside Naturalist's Field Club. *Transactions of the Tyneside Naturalist's Field Club*, **1**, 219–265.

ILJINA, T. G. 1977. Development of septa in rugose corals of the superfamily Polycoeliaceae. *Mémoire du Bureau de recherches géologiques et minières*, **89**, 78–86.

ILJINA, T. G. 1984. The historical development of corals. Suborder Polycoeliina. *Transactions of the Palaeontological Institute of Academy of Science of USSR*, **198**, 1–184 (in Russian).

ILJINA, T. G. 1987. Morphogenesis and microstructure of carina in the corals of the genus *Lophocarinophyllum*. *Paleontological Journal*, **2**, 41–51 (in Russian).

ILJINA, T. G. 1997. Distribution, taxonomy and morphology of Permian Rugosa of Southeastern Pamir (Tadzhikistan). *Boletín de la Real Sociedad Espanolã de Historia Natural*, **91**, 127–141.

ISOZAKI, Y. 1997. Permo-Triassic boundary Superanoxia and stratified superocean: records from lost deep-sea. *Science*, **276**, 235–238.

ISOZAKI, Y. & ALJINOVIC, D. 2009. Extinction of the Permian large bivalve Alatoconchidae: End of gigantism in tropical seas at the end-Guadalupian. *Palaeogeography, Palaeoclimatology, Palaeoecology*, **284**, 11–21.

ISOZAKI, Y., KAWAHATA, H. & OTA, A. 2007. A unique carbon isotope record across the Guadalupian–Lopingian (Middle–Upper Permian) boundary in mid-oceanic paleo-atoll carbonates: The high-productivity "Kamura event" and its collapse in Panthalassa. *Global and Planetary Change*, **55**, 21–38.

ISOZAKI, Y., ALJINOVIĆ, D. & KAWAHATA, H. 2011. The Guadalupian (Permian) Kamura event in European Tethys. *Palaeogeography, Palaeoclimatology, Palaeoecology*, **308**, 12–21.

IVANOVSKIY, A. B. 1969. Wentzelella orientalis sp.n. from the Permian of the Far East. *Proceedings of the Geology and Geophysics of the Sibirian Branch of the AS USSR*, **68**, 120–126.

KAWAMURA, T. & MACHIYAMA, H. 1995. A Late Permian coral reef complex, South Kitakami Terrane, Japan. *Sedimentary Geology*, **99**, 135–150.

KHANCHUK, A. I. (ed.) 2006. *Geodynamics, Magmatism and Metallogeny of the Eastern Russia*. Far East Scientific Centre of the Academy of Sciences of the USSR, Vladivostok, **1**, 1–577 (in Russian).

KIESSLING, W. 2001. Paleoclimatic significance of Phanerozoic reefs. *Geology*, **29**, 751–754.

KING, W. 1848. *A Catalogue of the Organic Remains of the Permian Rocks of Northumberland and Durham*. Newcastle-upon-Tyne, 1–16.

KING, W. 1850. *A Monograph of the Permian Fossils of England*. The Palaeontographical Society, London, Monographs, **3**.

KOSSOVAYA, O. L. 1995. The morphogenetic differenciation of the Rugosa in the Late Palaeozoic. *Paleontological Journal*, **28**, 11–19.

KOSSOVAYA, O. L. 2007. Ecological aspect of upper Carboniferous–lower Permian 'Cyathaxonia fauna' taxonomical diversity (the Urals). *In*: HUBMANN, B. & PILLER, W. E. (eds) *Fossil Corals and Sponges. Proceedings of the 9th International Symposium on Fossil Cnidaria and Porifera, Graz 2003. Österreichische Akademie der Wissenschaften , Schriftenreihe der Erdwissenschaftlichen Kommissionen*, **19**, 383–406.

KOSSOVAYA, O. L. 2009. Artinskian–Wordian antitropical rugose coral associations: a palaeogeographical approach. *Palaeoworld*, **18**, 136–151.

KOTLYAR, G. V., BURAGO, V. I., KISILEVA, A. V. & NIKITINA, A. P. 1975. *The Stage Subdivisions of the Permian of the Central Siberia and Southern Primorye*. Unpublished Report of the All-Union Research Geological Institute, Leningrad, 1–373.

KOTLYAR, G. V., ZAKHAROV, YU. D. ET AL. 1983. Late Permian phase of the evolution of the organic world. *In*: GRAMM, M. N. & POSTOVTZEV, K. O. (eds) *Dzhulfian and Dorashamian stages in the USSR*. Nauka, Leningrad, 1–200 (in Russian).

KOTLYAR, G. V., VUKS, G. P., KROPACHEVA, G. S. & KUSHNAR', L. V. 1987. Nakhodka reef-rock massif and position of Ludianza horizon within the stage scale of Permian in the Tethys. *In*: ZAKHAROV, YU. D. & ONOPRIENKO, YU. I. (eds) *Problems of the Permian and Triassic Biostratigraphy of East USSR*. Far-Eastern Scientific Centre, USSR, Academy of Science, Vladivostok, 54–63 (in Russian).

KOTLYAR, G. V., ZAKHAROV, YU. D., KROPACHEVA, G. S., PRONINA, G. P., CHEDIA, I. O. & BURAGO, V. P. 1989. Late Permian phase of the evolution of the organic world. Midian Stage of USSR. *In*: KOTLYAR, G. V. & ZAKHAROV, YU. D. (eds) *Evolution of the latest Permian biota. Midian Stage*. Nauka, Leningrad, 1–145 (in Russian).

KOTLYAR, G. V., KROPATCHEVA, G. S., SOSNINA, M. I., PRONINA, G. P. & CHEDIA, I. O. 1990. Zonal subdivisions of the marine Upper Permian deposits of the South Primorye. *In*: ZAKHAROV, YU. D. & ONOPRIENKO, YU. I. (eds) *New data on biostratigraphy of Palaeozoic and Mesozoic in the south of Far East of Russia. Late Palaeozoic and early Mesozoic events in Circum Pacific and their global correlation*. AN USSR, Dalnevostochnoe otdelenie, Dalnevostochnyi Geologicheskyi Institut, Vladivostok, 104–116 (in Russian).

KOTLYAR, G. V., KOSSOVAYA, O. L. & ZHURAVLEV, A. V. 2004*a*. Interregional correlation of the main event levels of the Permian system. *Pacific Geology*, **23**, 25–42.

KOTLYAR, G. V., NIKITINA, A. P., ZHURAVLEV, A. V. & KOSSOVAYA, O. L. 2004*b*. Midian (Upper Wordian–Capitanian) mixed Fauna of Southeastern Asia. *Bulleten Moskovskogo obschestva ispytatelei prirody, otdel geologicheskyi*, **78, 1**, 33–48 (in Russian).

KOTLYAR, G. V., BELYANSKY, G. C., BURAGO, V. I., NIKITINA, A. P., ZAKHAROV, YU. D. & ZHURAVLEV, A. V. 2006. South Primorye, Far East Russia – a key region for global Permian correlation. *Journal of Asian Earth Sciences*, **26**, 280–293.

KOTLYAR, G. V., SHEN, S.-Z., KOSSOVAYA, O. L. & ZHURAVLEV, A. V. 2007. Middle Permian (Guadalupian) biostratigraphy in South Primorye, Russian Far East and correlation with Northern China. *Palaeoworld*, **16**, 173–189.

KROPATCHEVA, G. S. 1978. The late Permian tetracorals of the South Primorye. *In*: POPEKO, L. I. (ed.) *Upper Palaeozoic of the Northeastern Asia*. Institute of Tectonics and Geophysics, Far East Scientific Centre of the Academy of Sciences of the USSR, Vladivostok, 44–95 (in Russian).

MASLENNIKOV, D. F. 1948. New data on the stratigraphy of the Upper Palaeozoic deposits of the southern part of the Far East. *In*: *Recodes of the All-Union Scientific Research Geological Institute (Vsegei)*. Ministry of Geology of USSR, Moscow, General Series, **8**, 79–84.

MONTAGGIONI, C. J. R. & BRAITHWAITE, C. J. R. (eds) 2009. *Quaternary Coral Reef Systems: History, Development Processes and Controlling Factors. Developments in Marine Geology*, **5**, 1–532.

NIERMANN, H. T. 1975. Polycoeliidae aus dem Oberperm von Basleo auf Timor. *Münstersche Forschungen zur Geologie und Paläontologie*, **27**, 131–225.

OTA, A. & ISOZAKI, Y. 2006. Fusuline biotic turnover across the Guadalupian–Lopingian (Middle–Upper Permian) boundary in mid-oceanic carbonate buildups:

biostratigraphy of accreted limestone in Japan. *Journal of Asian Earth Sciences*, **26**, 353–368.

PANDOLFI, J. M. 1999. Response of Pleistocene coral reefs to environmental change over long temporal scales. *American Zoologist*, **39**, 113–130.

RATH, M. S., GOERICKE, R. & DEHEYN, D. D. 2012. Cold induces acute stress but heat is ultimately more deleterious for the reef-building coral *Acropora yongei*. *Scientific Reports*, **2**, 1–5.

SAFONOVA, I. YU. 2009. Intraplate magmatism and oceanic plate stratigraphy of the Paleo-Asian and Paleo-Pacific Oceans from 600 to 140 Ma. *Ore Geology Reviews*, **35**, 137–154.

SARNECKA, E. 2001. Gromada Anthozoa Ehrenberg, 1834. *In*: PAJCHLOWA, M. & WAGNER, R. (eds) *Budowa geologiczna Polski. Atlas skamieniałości przewodnich i charakterysticznych*. Młodszy Paleozoik, Państwowy Instytut Geologiczny, Warszawa, **III**, 1–224.

SCHOUPPÉ, A. & STACUL, P. 1955. Die Genera *Verbeekiella* Penecke, *Timorphyllum* Gerth, Wannerophyllum n. gen., Lophophyllidium Grabau aus dem Perm von Timor. *Palaeontographica*, **3**, 95–196.

SCHOUPPÉ, A. & STACUL, P. 1959. Säulchenlose Pterocorallia aus dem Perm von Indonesisch Timor (mit Ausnahme der Polycoeliidae). *Palaeontographica*, **4**, 197–359.

SHEN, S. Z. & SHI, G. R. 2004. Capitanian (Late Guadalupian, Permian) global brachiopod palaeobiogeography and latitudinal diversity pattern. *Palaeogeography, Palaeoclimatology, Palaeoecology*, **208**, 235–262.

SHI, G. R. 2006. The marine Permian of East and Northeast Asia: an overview of biostratigraphy, palaeobiogeography and palaeogeographical implications. *Journal of Asian Earth Sciences*, **265**, 175–206.

SOSHKINA, E. D. 1928. Nizhnepermskie (artinskie) korally zapadnogo sklona Severnogo Urala. *Byulleten' Moskovskogo Obshchestva Ispytateley Prirody*, **36**, 339–393 (in Russian).

STANLEY, S. M. 1988. Climatic cooling and mass extinction of Paleozoic Reef communities. *Palaios*, **3**, 228–232.

TASCHI, S. M., KOTLYAR, G. V., NIKITINA, A. P. & KISELEVA, A. V. 1992. The marine Upper Permian of the Trudnyi Peninsula. Senkina Shapka Mountain. *In*: *A Field Guide to the Late Palaeozoic and Early Mesozoic Circum-Pacific Bio Geological Events*. Far East Scientific Centre of the Academy of Sciences of the USSR, Vladivostok, 51–59.

TWITCHETT, R. J. 2006. The palaeoclimatology, palaeoecology and palaeoenvironmental analysis of mass extinction events. *Palaeogeography, Palaeoclimatology, Palaeoecology*, **232**, 190–213.

VERON, J. E. N. 2008. Mass extinctions and ocean acidification: biological constraints on geological dilemmas. *Coral Reefs*, **27**, 459–472.

VERON, J. E. N. & MINCHIN, P. R. 1992. Correlations between sea surface temperature, circulation patterns and the distribution of hermatypic corals of Japan. *Continental Shelf Research*, **12**, 835–857.

WANG, X. D. & SUGIYAMA, T. 2000. Diversity and extinction patterns of Permian coral faunas of China. *Lethaia*, **33**, 285–294.

WANG, X. D., SUGIYAMA, T., KIDO, E. & WANG, X. J. 2006. Permian rugose faunas of Inner Mongolia-Northern China and Japan: paleobiogeographical implications. *Journal of Asian Earth Sciences*, **26**, 369–379.

WEYER, D. 1979. Korallen-Funde im europäischen Zechstein-Meer. *Zeitschrift für Geologische Wissenschaften*, **7**, 981–1021.

WEYER, D. 1984. Korallen im Paläozoikum von Thüringen. *Hallesches Jahrbuch für Geowissenschaften*, **9**, 5–33.

WEYER, D. 1997. Korallen im Oberpermischen Kazan-Meer von Russland. *Freiberger Forschungshefte*, **466**, 85–107.

ZAKHAROV, YU. D. & PAVLOV, A. M. 1986. Permian cephalopods of the Primorye and the problem of the Permian zonal stratigraphy of Tethys. *In*: ZAKHAROV, YU. D. & ONOPRIENCO, Y. I. (eds) *Correlation of the Permian-Triassic sediments of East USSR*. Far East Scientific Centre of the Academy of Sciences of the USSR, Vladivostok, 5–32 (in Russian).

ZAKHAROV, YU. D. & PAVLOV, A. M. 1987. A standard for the Permian-Triassic in the Tethys. *In*: ZAKHAROV, Y. D. & ONOPROENKO, Y. I. (eds) *Problems of the Permian and Triassic Biostratigraphy of East USSR*. Far East Scientific Centre of the Academy of Sciences of the USSR, Vladivostok, 6–48 (in Russian).

ZAKHAROV, YU. D., BORISKINA, N. G. & POPOV, A. M. 2001. *Reconstruction of the Marine Environment of the Late Palaeozoic and Early Mezozoic by the Isotope Data*. Far East Scientific Centre of the Academy of Sciences of the USSR, Vladivostok, 1–110 (in Russian).

Late Permian aeolian sand seas from the Polish Upper Rotliegend Basin in the context of palaeoclimatic periodicity

HUBERT KIERSNOWSKI

Polish Geological Institute – National Research Institute, ul. Rakowiecka 4, 00-975 Warszawa, Poland (e-mail: Hubert.Kiersnowski@pgi.gov.pl)

Abstract: A study of a Permian aeolian depositional system in the Polish Upper Rotliegend Basin (PURB) is described and is placed in the wider context of similar studies of the German, Dutch and UK parts of the Southern Permian Basin (SPB). Aeolian complexes in the PURB consist of several stacked erg units inter-bedded with fluvial and playa deposits. These deposits are grouped into seven main units: six predominantly aeolian and one fluvial. The separate aeolian units were considered as representative of the changing pattern of palaeowinds directions. The correlations of wind mean directions assisted in the development of a new depositional model for part of the Eastern Erg aeolian complex within the PURB.

Palaeowind directions from the Polish Upper Rotliegend Basin (PURB) are compared with those recorded from various parts of the Southern Permian Basin (SPB) and discussed in the context of Rotliegend cross-basin lithostratigraphic correlations over a time span related to particular lithostratigraphic units. An established and revised model for wind directions in the entire SPB during deposition of Upper Rotliegend deposits provides new information on the SPB palaeotopography and course of climatic changes. Orographic barriers influenced local climatic conditions within the SPB area. A rain shadow desert with lower precipitation rates developed in the western and central parts of the SPB. Consequently, the western part of the SPB (Anglo-Dutch and partly North German basins) was drier in comparison to the PURB in the east. Aeolian deposits in the PURB form the thickest and most complex system in the SPB. Lower aeolian units record climatic fluctuations (arid–humid cycles in Late Permian time, *c.* Guadalupian). The upper aeolian units attributed to the whole SPB, including the Polish Rotliegend Basin, formed in more constant and dry climatic conditions. The levelling of the SPB topography through incision and erosion together with basin expansion may have contributed to the change in the wind patterns through time from prevailing monsoon circulation winds (seasonal) to prevailing NE trade winds (more constant). This change could have occurred due to SPB migration northwards over a long time interval. This assumption forces a hypothesis for a longer time equivalent for Upper Rotliegend (UR II) aeolian deposition than is assumed on lithostratigraphic scales adjustable to the geologic time scale.

Late Permian aeolian depositional systems in northern Pangea are the largest known from the entire Phanerozoic. Climatic conditions (characterized by significant climate periodicity; Yang & Baumfalk 1994) for the development of such extensive aeolian environments need to be analysed over the time span of the Late Permian. Alternating arid and humid conditions controlling the origin or disappearance of sand deserts in the Permian Basins of the United States and Europe are described by many authors (Glennie 1983*a*, *b*; Blakey *et al.* 1988; Parrish & Peterson 1988; George & Berry 1993, 1997).

After Roscher & Schneider (2006) and Schneider *et al.* (2006), Roscher *et al.* (2008) re-evaluated palaeoclimatic models for Pangea and stressed the importance of the equatorial Variscan mountain chain as a high or significant topographic barrier during Carboniferous and Permian times, influencing the variability of climatic indicators and hence wind patterns. These palaeogeographic and hyper arid conditions were crucial for initiation, development and demise of Late Permian (mostly Capitanian or even Wordian Stage) aeolian sand 'seas' of Europe.

The Permian sand seas of the PURB formed a stable, thick sedimentary cover similar to that of the west part of the SPB. Aeolian deposits from the PURB are different to those which occurred in the Southern and Northern Permian basins of Europe. In the PURB they developed during Upper Rotliegend time (Drawa and Noteć formations), whereas in the western part of the SPB they occurred only during deposition of the Leman and Slochteren formations (lithostratigraphic equivalent of the Noteć Formation in the PURB). A general and simplified top Rotliegend lithofacies pattern across the SPB (W–E axis) was desribed by Kiersnowski *et al.* (1995), McCann *et al.* (2008) and Gast *et al.* (2010).

From: GĄSIEWICZ, A. & SŁOWAKIEWICZ, M. (eds) 2013. *Palaeozoic Climate Cycles: Their Evolutionary and Sedimentological Impact*. Geological Society, London, Special Publications, **376**, 431–456.
First published online September 26, 2013, updated October 9, 2013, http://dx.doi.org/10.1144/SP376.20

Data describing aeolian sand sources, transport routes and their seasonal variability in delivery, the fabric of the stacked ergs of the sand seas, playa-lake fluctuations and fluvial events from common or separate climatic, tectonic and hydrologic factors are discussed. However, as detailed sedimentological analyses show, aeolian ergs developed under the rhythm of palaeoclimatic changes. Their aerial spread was controlled by distance from sand sources, sand supply and wind directions. Late Permian palaeowind directions and the fluvial drainage pattern within the PURB do not appear to be linked to similar data from the whole SPB. Gibbs *et al.* (2002) showed how measured average wind directions should be used for reconstructions of principal wind directions, controlled for basin-wide scales and focused on main climatic periods. Wind directions calculated from dune set azimuths are used here, adjusted for palaeoclimatic models showing wind direction patterns embracing the SPB area. The aim of this paper is to adjust such data for existing palaeoclimatic models and lithofacies and palaeogegraphic maps from western SPB.

Geological setting

Sand seas in the Southern Permian Basin (SPB) developed on a structurally complex bedrock of Caledonian and Variscan structural elements (Ziegler & Dèzes 2006; Kroner *et al.* 2008). Sutures and orogenic belts (Fig. 1) segmented the SPB area into parts with higher subsidence occupied by playas and salt lakes and uplifted basin margins occupied by aeolian and fluvial deposits.

The PURB is the eastern extension of the Dutch, British and German portions of the Southern Rotliegend Basin and thus contains very similar lithofacies patterns and depositional systems (Kiersnowski *et al.* 1995). Regional palaeoclimatic changes were therefore important across the entire continent. The SPB had an asymmetric lithofacies pattern (Verdier 1996) during Upper Rotliegend time (including roughly Capitanian and Wordian stages). The southern part was occupied by broad aeolian sand deserts, while the central and northern portions were represented by vast saline mudflats surrounding a giant salt lake (Gast 1991; Gaupp *et al.* 2000; Legler *et al.* 2010). The PURB is also asymmetric in shape with permanent aeolian sand deserts in the SW to southern area and a long-lasting playa environment to the NE, a pattern was caused by increased tectonic subsidence to the NE. The strongest subsidence was associated with the tectonic belt (Teisseyre–Tornquist Zone or TTZ) running along the East European platform and the Variscan foredeep of the Carboniferous–Permian basins (Figs 1, 2a–d). The lowest level of subsidence occurred on the southwestern side of the basin, as part of the Variscan Externides (Dadlez *et al.* 1995).

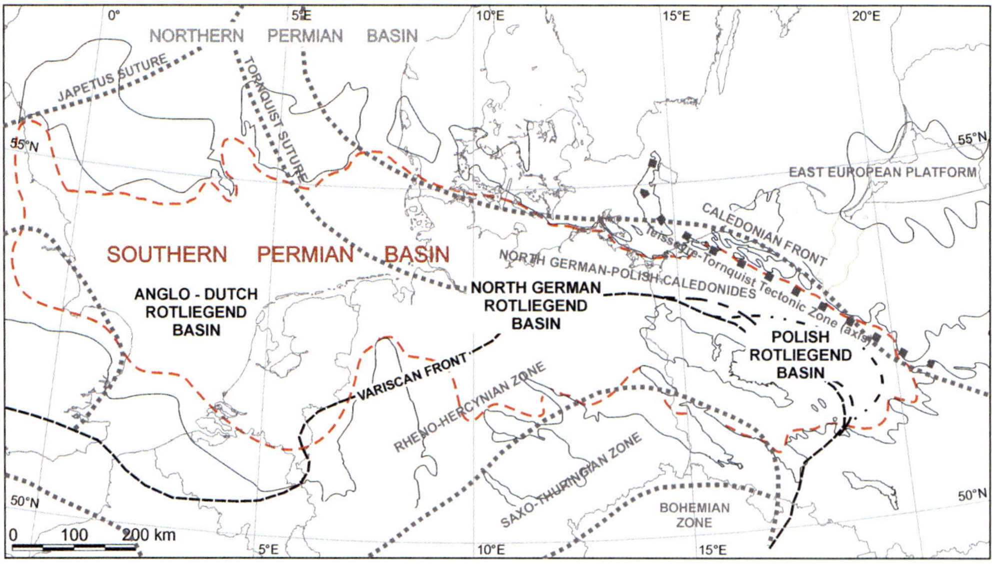

Fig. 1. Southern Permian Basin in Europe on a simplified sketch of Caledonian and Variscan structural elements (Ziegler & Dèzes 2006). Sutures and orogenic fronts are shown as broken lines.

Aeolian characteristics

Roscher & Schneider (2006) recognized two features that controlled the occurrence of aeolian deposits within the SPB: (1) increasing aridity within northern Pangea, reaching a maximum in late Wordian to Capitanian time; and (2) a directionally constant wind system. Roscher & Schneider (2006) argued that the ongoing aridization produced large areas of arid to semi-arid environments, enabling the production and movement of a large amount of sand. Permian aeolian dunes in the SPB are mostly deeply buried and data are derived from well cores. Some outcrops exist in uplifted tectonic blocks however, for example, Flechtingen block in Germany (Ellenberg *et al.* 1976) and along the western SPB margin in east England (i.e. Yellow Sands, Clemmensen 1989).

Aeolian deposits from the Lower Rotliegend (roughly Asselian and Sakmarian stages) and lower part of the Upper Rotliegend (Rotliegend I, roughly Artinskian and Kungurian stages) basins in Poland are unknown. It is only assumed that these may exist, as in the French and German basins. For example in eastern Germany, the Lower Rotliegend is recognized in the Thuringian Forest Basin (Eisenach Formation) and the Saale Basin (upper Hornburg Formation) (Roscher & Schneider 2006; Schneider *et al.* 2006). In the Polish Basin, initial aeolian sand occurrences affiliated with the Upper Rotliegend (UR II) basin are poorly recognized. The onset of the early aeolian sand deposition in the PURB is practically unknown due to the scarcity of preserved well cores. In some wells log data indicate sand units, interpreted as possible aeolian in origin and interbedded with alluvial-fluvial deposits.

Aeolian sandstones within the PURB belong to the Upper Rotliegend (Drawa and Noteć formations) (Pokorski 1989). They occur in three distinct areas: the Southern (Silesian) Erg, the Eastern Erg (Kiersnowski 1997) and the Northern Erg (Kiersnowski & Buniak 2006) (Fig. 2a–d). The Southern Erg was separated from the Eastern Erg by the Wolsztyn–Pogorzela Palaeo-High (Kiersnowski *et al.* 2010*b*). The Eastern Erg was separated from the Northern Erg by the central playa-lake deposits (Fig. 2a). Several additional less-prominent dune fields are poorly recognized (Łeba Erg, Podlasie Erg) although there may be southern smaller ergs (Fig. 2d).

The aeolian sand cover within the best-recognized Eastern Erg varies from several metres to 700 m in thickness and reaches a maximum of 1000 m in higher-subsidence areas as tectonic grabens. The complete sand cover also comprises fluvial sands and can therefore be partitioned into several units that represent stacked ergs and fluvial deposits. This complex was divided into six stacked erg units and one fluvial unit, whereby each unit reflects climatic and tectonic (subsidence) changes on a basin-wide scale during Upper Rotliegend deposition. The whole aeolian–fluvial complex from the Eastern Erg is separated by erosional surfaces, interpreted as sequence-stratigraphic boundary surfaces (Kocurek 1988; Kocurek & Havholm 1993), into seven depositional main units with four of the unit names based on colours designated for palaeowind directions (Fig. 2a–d).

Most of the analysed dune cross-bed sets are preserved only in bottom sets, although there are significant exceptions where thick dune units are preserved. Bottom dune sets can demonstrate a slightly different dune form migration azimuth than upper dune sets due to changeable wind directions and strengths. Interpreted dune forms are barchans, barchanoid ridges to transversal forms and, in some cases, more complex forms. Parts of longitudinal dunes or low-angle dome dunes are rarely recognized. Interdune deposits, in the form of both dry and damp sand-sheets, and deposits of local temporary ponds are common. In some profiles large stacked transverse dunes with uniform dips and azimuth dips are recognized, exceeding over 40 m in thickness in some cases.

The origin of large dune forms is well described (e.g. Andreotti *et al.* 2009), but preservation in the sedimentary record is unusual and may be due to special physical conditions: high tectonic subsidence rates in accordance with a constant or sudden sand supply. This would decrease the possibility of deflation and keep palaeowater table levels high enough to stabilize aeolian sand bodies, especially with capillary water effects from high evaporation rates. Especially important in the PURB with its thick preserved units is the occurrence of ergs associated with depositional lows. These lows are interpreted as localized high-subsidence areas, deeply deflated areas, large interdraa corridors (Langford *et al.* 2008) or, in some cases, incised valleys related to large wadi systems (Mijnlieff & Geluk 2011). Almost fully preserved dunes are only found at the top of the Rotliegend within the PURB and Anglo-Dutch basin (Steele 1983; Karnkowski 1986; Strömbäck & Howell 2002; Kiersnowski & Tomaszczyk 2010), where they were preserved during Zechstein sea inundation. Only the topmost dune units were eroded with reworked sand commonly deposited as mass flow deposits on dune slopes and within interdune areas. These are covered by Weissliegend shallow marine siliciclastic deposits or directly by Zechstein marine carbonates and evaporites (Glennie & Buller 1983). Such geomorphologic forms (dune fields or yardangs) are known from other aeolian systems (Fryberger 1986); however, in the case of the SPB, they are mostly found in the Polish part of the SPB.

Aeolian ergs are interbedded with the basin-centred playa-lake deposits (Fig. 2a, b). The areal extent of these playas was controlled by fluctuations of playa-lake levels and these changes probably strongly influenced erg development. However, within the Eastern Erg deposits of the PURB, there is a lack of well log data showing the relationships of erg deposits to the central playa-lake deposits, as in the case of the Anglo-Dutch Basin (Sweet 1999; Bailey & Lloyd 2001). In this case, it is unclear how central playa-lake hydrology was affected by basin subsidence and major climatic changes, and thus how playa-lake level variations changed groundwater levels. These changes (relative water table rise or fall) may have controlled the base levels within the erg body, being key for possible wind erosion or erg stabilization.

In models of erg development it is assumed that erg margin sandstones are interbedded with playa deposits (Sweet 1999). In some Rotliegend profiles from the central parts of the PURB, aeolian dunes prograded rapidly onto playa deposits during dry periods, leaving thick aeolian sequences. In some cases, aeolian dunes crossed playa mudflats leaving thin sand units or sand-drift surfaces (Clemmensen & Tirsgaard 1990).

Wind direction analyses

The measurements of sedimentary dips and the azimuths of lamina sets allow the reconstruction of dune migration direction with high accuracy. Stacked dune forms show changes of dune types and of migration directions resulting from the changes in palaeowind directions and strengths. Averaged directions in individual well cores show gradual or rapid changes in prevailing winds over time. Rapid changes in directions of prevailing winds may indicate seasonal winds as well as erosional gaps in the sedimentary record. Gradual changes of wind directions across aeolian deposits hundreds of metres thick may indicate long-lasting climatic cycles and progressive changes in topography. Aeolian dunes, interdune deposits and playa strata from the PURB sections are recognized in well cores and wireline logs. Dunes can be identified using image logs and dipmeter data, where dune cross-sets and directional dip variability (directional azimuths) are distinguishable (Aleksandrowski & Kiersnowski 1998).

All wind directions interpreted from measured dip azimuths from dune sets were fitted to palaeonorth on Pangean northern hemisphere during Late Permian time (*c.* Capitanian Stage). For the SPB, the rotation of the Permian north pole is 18–20° eastwards from present magnetic north because the Late Permian dipole field for Europe is oriented 18° clockwise from the present day (Cogne *et al.* 1990; Besse *et al.* 1996). Corrected wind directions are grouped in four sectors, each 90° quadrant with a different colour. The blue sector represents wind blowing from the NE and East, red from the SE and South, green from the SW and West and yellow from the NW and North (Figs 2a–d, 3).

Correlation of sand sea deposits

Dipmeter and core data from the Eastern Erg of the PURB show that on the scale of whole erg occurrences it is almost impossible to recognize correlatable super-bounding surfaces (SBS) based on stacked erg fabric. Such surfaces can be traced on shorter well-to-well distances. In the case of the Anglo-Dutch Rotliegend basin, such SBSs (George & Berry 1993, 1997) are related to sabkha–saline lake lithostratigraphy (Gast 1991; Legler 2006); sequence-stratigraphy models used (Kocurek & Havholm 1993); the recognition of maximum wetting surfaces (MWS) (Van Ojik *et al.* 2011); and salt lake level maxima in the North German Rotliegend Basin (Legler *et al.* 2010).

For more complexity, aeolian deposits are partly faulted and segmented due to Late Permian syndepositional tectonic movements related to older differentiated Carboniferous horsts and grabens in the eastern Variscan Externides area (Kiersnowski 1997; Kiersnowski *et al.* 2010*a*). However, regional tectonic subsidence within the Eastern Erg area allows the preservation of aeolian cover with an average thickness exceeding 600 m (in grabens even up to 1000 m). In some places, the thickness was reduced by the uplift of tectonic blocks, resulting in stratigraphic condensation and significant erosion (Fig. 3). Inversely, in more rapidly subsiding areas it was possible to preserve thicker units with individual large dune forms. Consequently, aeolian deposits have variable thicknesses across an area and cannot be easily correlated across longer distances using roughly similar and constant thickness of erg units bounded by flat-topped erosional surfaces (Loope 1985; Moscariello 2011).

Another complexity in attempting erg-wide correlations is due to strongly variable wind directions. This does not support the simplified assumption on correlation based on time-related prevailing wind directions (i.e. winter and summer dominant winds) or wind direction coherency across the erg area. In the Anglo-Dutch Basin, a variety of dune types developed during the same time period (Glennie 1983*a*, *b*; Steele 1983; George & Berry 1997). The different dune forms and related variability of dip azimuths in the PURB ergs also may be attributed to the particular time periods.

The general inventory of dune types in the Upper Rotliegend II (including roughly Capitanian and Wordian stages) ergs consists of single and complex dunes: barchan dune; barchanoid ridge or draa (with superimposed oblique crescentic and linear dunes); transverse dunes (Karpeta 1990; George & Berry 1997; Howell & Mountney 1997); star dunes (George & Berry 1993); longitudinal draa (Steele 1983); longitudinal or seif dunes (Glennie 1985); complex linear draa (Clemmensen 1989; Chrinz & Clemmensen 1993); crescent-shaped dunes and dome dunes (Karpeta 1990); and sand sheets.

However, George & Berry (1993) show dependence between wind directions and changing dune forms, showing the predicted lateral succession of aeolian bedforms originating within a variable NE trade wind regime (Glennie 1986). It is still not easy to predict what kind of dune forms can be expected in a particular erg area.

Sneh (1988) reported Rotliegend dune forms from several places in Anglo-Dutch and German basins as well as from the Isle of Arran and the English Midlands. He interprets dunes as oblique forms and notes that dunes which he describes as having a transverse pattern should be reinterpreted as oblique. This statement has far-reaching consequence for wind direction interpretation, stating that: 'in most cases the prevailing winds were northerly and not northeasterly and easterly' (Sneh 1988; Fig. 4b).

Description of aeolian and fluvial units of the PURB

Aeolian deposits rest directly on Lower Rotliegend (roughly Asselian and Sakmarian stages) volcanic or sedimentary rocks or on erosional surfaces of Carboniferous bedrock (Kiersnowski *et al.* 2010*a*). Seven sedimentary units (ERU, ERBU 1, EBYU, ERBU 2, FIU, ECU 1 and ECU 2), described from bottom to the top in the following sections, are recognized in the PURB Eastern Erg area.

Aeolian red unit (ERU). This mostly aeolian complex features palaeowinds prevailing from the SE and represents a stage of abundant and constant sand supply. It is attributed to the Lower Drawa Formation and correlated with the Parchim Formation of the North German Basin. This unit is characterized by abundant and constant sand supply from the south and SE. The sand source was in the southern basin fringe, where a broad alluvial plain was sourced by weathered Cambrian and Carboniferous sandstones. The complex is characterized by large transversal dunes migrating to the north and NE. During periods of very dry climatic conditions, aeolian dunes migrated sporadically across the early playa-lake to the northern basin slopes, where the Northern Erg in western Pomerania (NW Poland) was accumulated. This shows that the wind direction from south to north was dominant over a long period. Other wind directions are not excluded as possible input sources for sand for the Northern Erg, but there is no sufficient data to confirm this. A northern sand source is therefore less probable. A long-lasting southern sand source allowing transport of sand across the PURB and forming a thick (up to 150 m) aeolian sequence in Rønne Graben (Lindert *et al.* 1993), located north of the Northern Erg area (present southern Baltic Sea area), is postulated (Kiersnowski & Buniak 2006; Fig. 2a). This assumption is based on correlation using fabric and bounding surfaces, but not supported by dipmeter data.

Aeolian red-blue unit 1 (ERBU 1), blue-yellow unit (EBYU) and red-blue unit 2 (ERBU 2). These aeolian units are attributed to the lower and upper Drawa Formation and correlated with the Parchim and Mirow formations of the North German Basin (Parchim Graben: Gast *et al.* 1998; Schneverdingen Graben: Drong *et al.* 1982). ERBU 1, EBYU and ERBU 2 are mostly aeolian complexes with a stage of abundant but variable sand supply from bimodal directions. Palaeowinds prevailed from the SE and NE, NE and NW, and SE and NE, respectively.

Interpreted erosional gaps are located between the lower and upper Drawa Formation (Parchim and Mirow formations) and between the upper Drawa and lower Noteć formations (Parchim and Detchlingen formations) (Gebhardt *et al.* 1991; Gast 1993; Helmuth & Süssmuth 1993; Plein 1993). These deposits are not recorded in the western part of the SPB.

These three aeolian units are characterized by differentiated wind directions, forming distinctively stacked thick sand units. These units are grouped together but each unit could be treated as separately stacked ergs with characteristic bimodal wind directions. The prevailing wind directions were from the east and south and secondary winds were from north and NW. These southeasterly winds, as in the ERU, are interpreted as transporting sand completely across the Polish Rotliegend basin. Sand migration across the basin was repeated several times. The aeolian sand source was, as for the ERU, from the southern basin fringe but probably also from the east. The dune types were barchan or transverse dunes. The ERBU 2 unit deposits are limited on the top by erosional surfaces and the fluvial deposits of the fluvial intermediate unit (FIU). Between ERBU 2 and the underlying ERBU 1 and EBYU is an interpreted erosional surface.

These three aeolian units developed simultaneously with playa-lake deposits (Fig. 2b), characterized by thick red claystones which are more

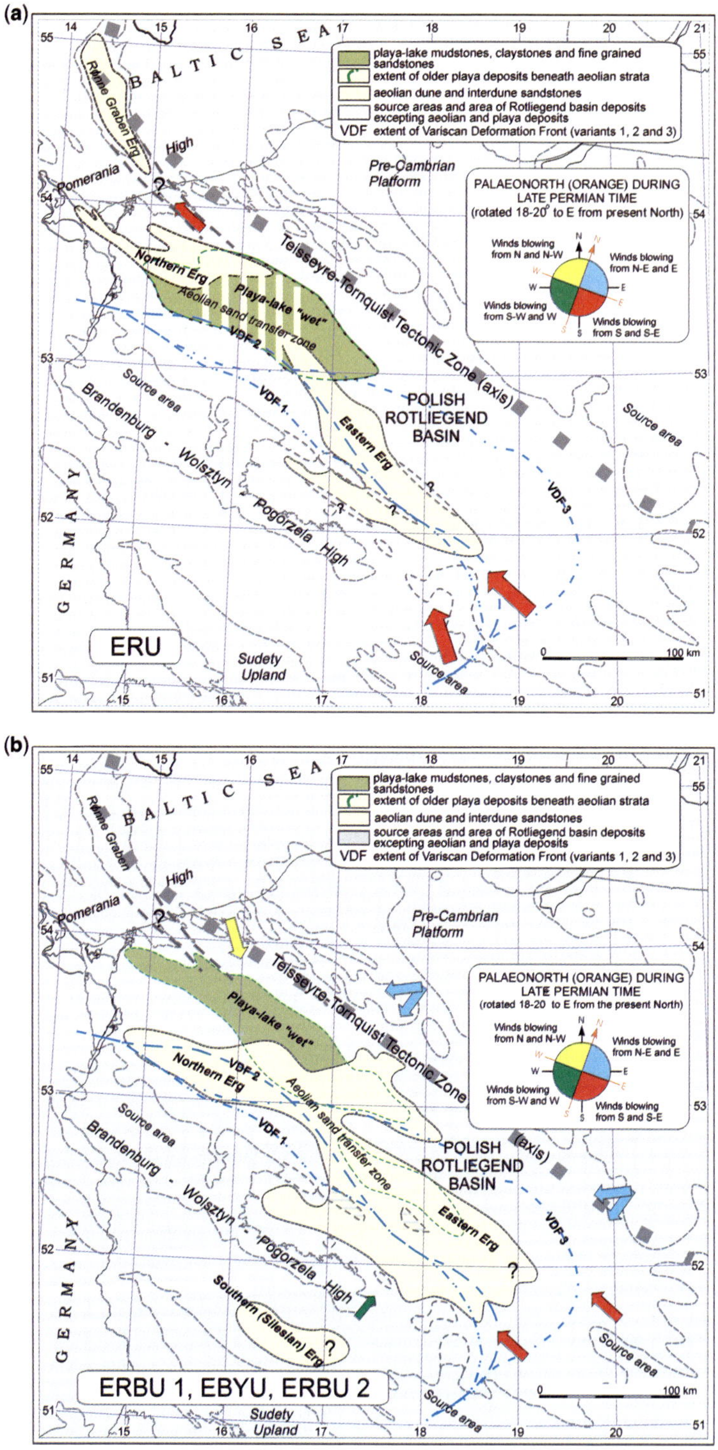

Fig. 2. Evolution and distribution of aeolian sand cover within the Polish Upper Rotliegend Basin. Palaeography: (**a**) during ERU deposition; (**b**) during ERBU 1, EBYU and ERBU 2 deposition; (**c**) during ECU 1 deposition; and (**d**) of topmost ECU 2. Sedimentary environments and dominant lithofacies are described on each figure.

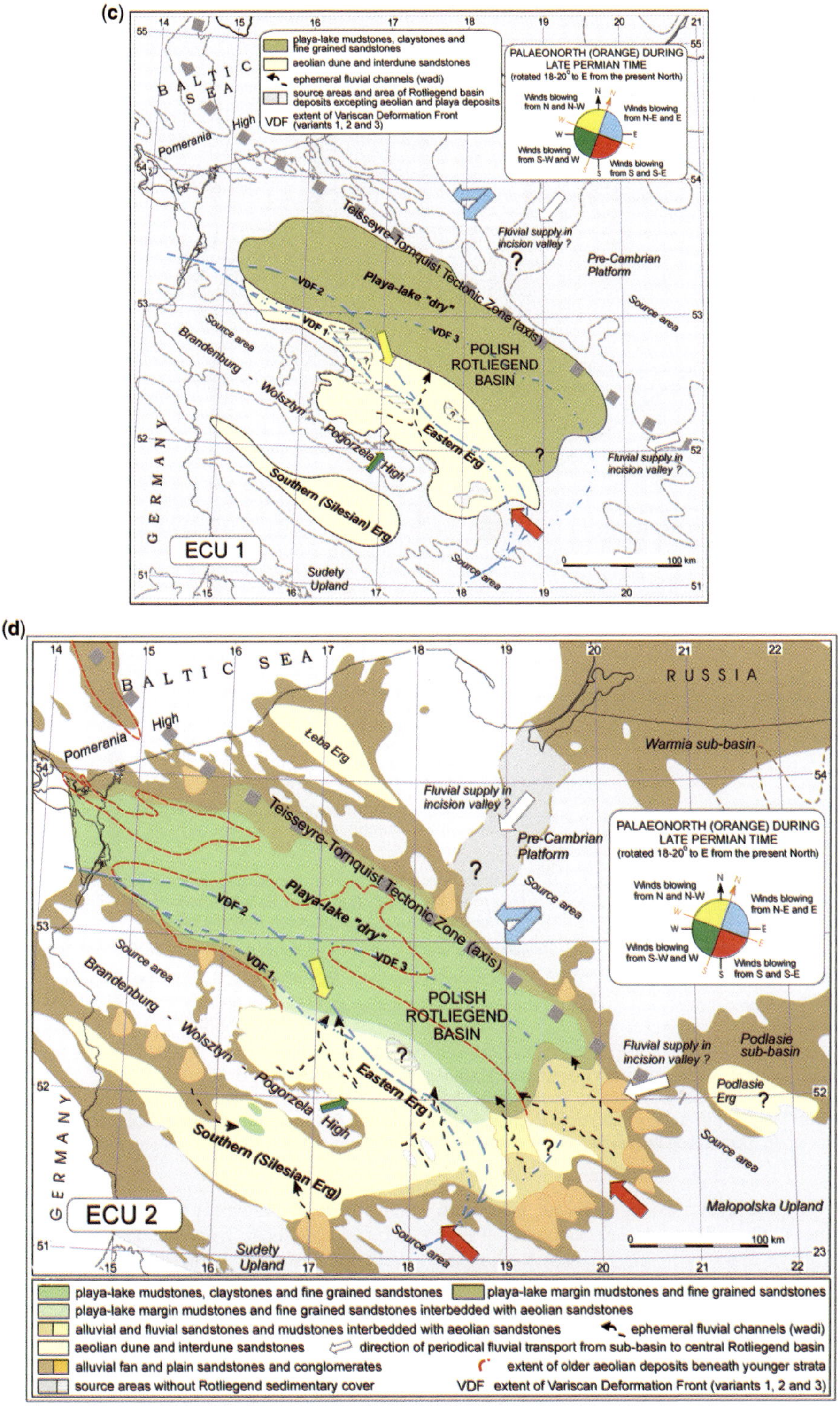

Fig. 2. *Continued.*

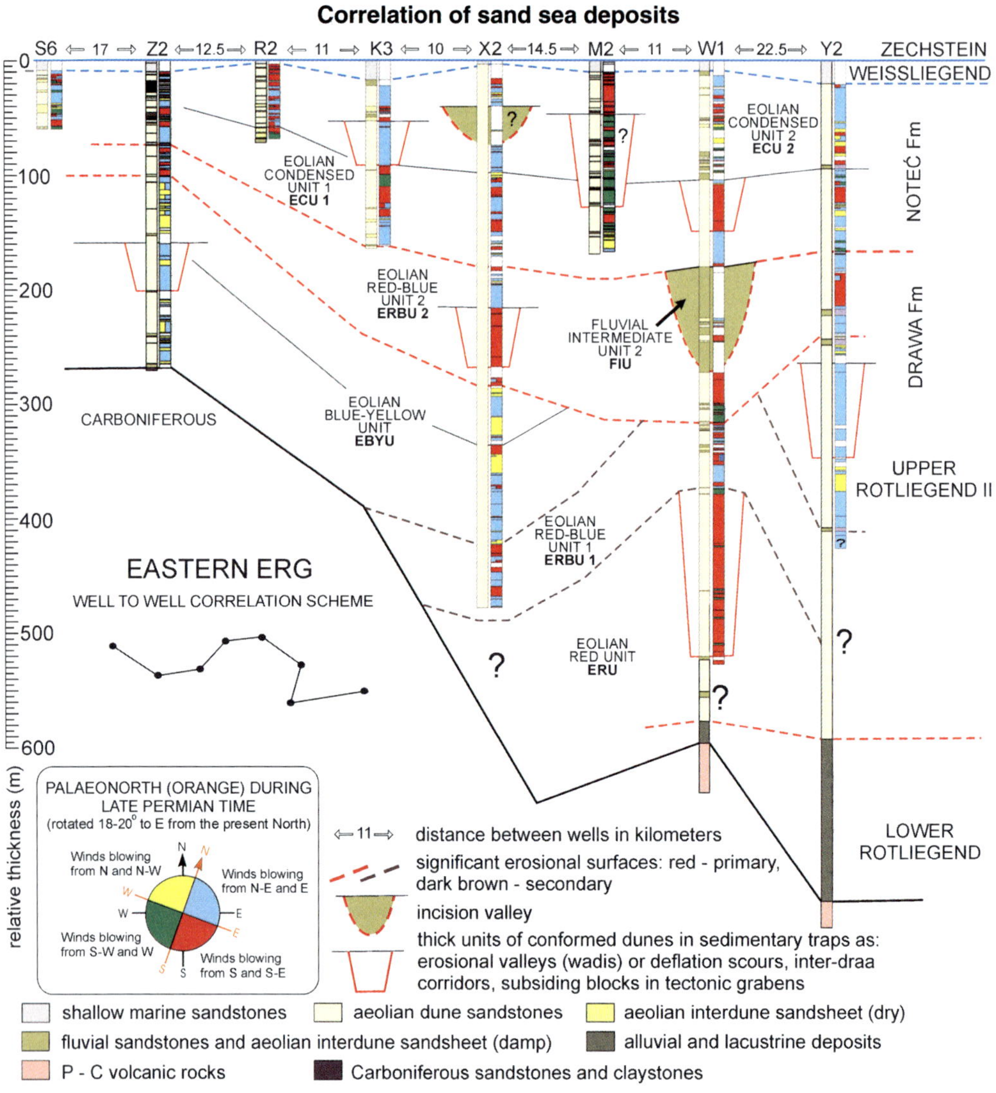

Fig. 3. Correlation diagram of wells with aeolian deposits from the Eastern Erg of the PURB. Wells are located in *c.* West–East direction. Sedimentary environment and lithology are indicated on left bar in each well profile. The right bar indicates colours attributed to four main sectors with wind directions. Wind directions are interpreted from dipmeter data (aeolian laminae dip and azimuth) from Rotliegend aeolian sands. White colour of the right column indicates a lack of aeolian strata (occurrence of fluvial? and Weissliegend marine deposits) or a lack of sufficient dipmeter data.

similar to those described in Variscan hinterland lacustrine basins (Schneider & Gebhardt 1993) than to playa deposits in the topmost Rotliegend units. The claystones were deposited under more freshwater conditions with rare occurrence of limnic fauna, abundant in Variscan lacustrine basins. For these playas, the term 'wet' lower playa is used.

The aeolian deposits from the Parchim Graben (North German Basin) originated from an abundant and constant sand supply from the south and SE. The aeolian sequences from the Parchim and Mirow wells of Germany form thick non-differentiated sand units similar to those in the PURB. It can perhaps be postulated that wind directions were similar to those in the PURB at this time. However, dipmeter data from the aeolian Schneverdingen Sandstones from Schneverdingen Graben (Drong *et al.* 1982) in western Germany (Figs 5 & 6), which are correlated to the Parchim and Mirow formations, show mostly winds from eastern directions which rotated into southeasterly winds in the upper portion of sandstones attributed to the Mirow Formation (Gast 1993; Gast & Gundlach 2006). It is also assumed that the tectonic origins of the Schneverdingen Graben was earlier, and the lower part of aeolian deposits in the SOZ well belong to the pre-Parchim Formation.

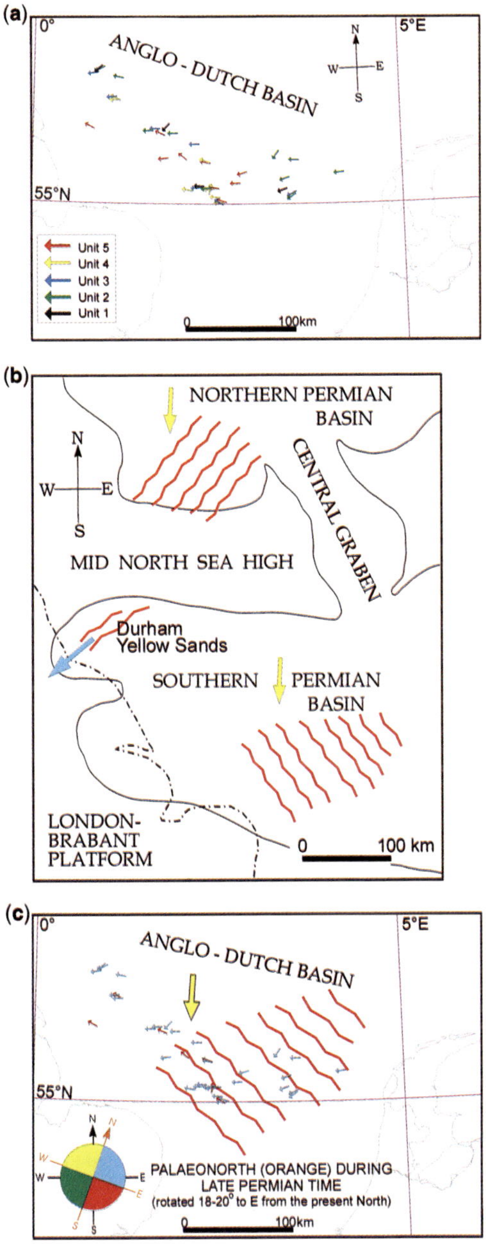

Fluvial Intermediate Unit (FIU). This fluvial and aeolian complex is attributed to the lower part of the lower Noteć Formation, being an equivalent of the lower Dethlingen Formation in the North German Basin. In addition, the interpreted incised valleys may be within the aeolian sandstones of the upper Drawa Formation.

This unit consists of fluvial deposits with interbeds of aeolian deposits, characterized by rapid changes in sedimentary environment from aeolian to fluvial. These changes were caused by both climatic and increased tectonic subsidence, lowering the basin-wide erosional base and rapid progradation of fluvial deposits. The fluvial deposits are in some place tens of metres thick. Fluvial sandstones and gravels are even deposited in the central portion of the basin. In other places, fluvial features are observed only as erosional surfaces within aeolian deposits since fluvial incision produced large valleys in the erg area. However, some intercalations of aeolian deposits are present within the FIU; this may indicate changing conditions for sediment input (whether from tectonic or climatic influences).

The set of stacked, small fluvial sandy channels is interpreted as the product of a long-lasting wadi system which intruded into aeolian habitat and crossed the Eastern Erg towards the playa-lake on the NE. The repeating fluvial courses and water streams direction (dipmeter data) indicate relatively large water inflow from the area of limited drainage system. In this case, it is instead assumed that water abundance resulted from frequently repeating rainfalls, indicating a more humid period. In case of drier periods, wadi deposits were limited to source area and water vanished into neighbouring dune fields. In case of northern source areas (Kiersnowski & Buniak 2006) with large drainage area, periodical

Fig. 4. Winds directions interpreted from aeolian sandstones from the Rotliegend Anglo-Dutch Basin (data from George & Berry 1997, adapted mostly from Glennie 1983*a* and Sneh 1988). (**a**) Arrows indicate mean wind directions. Arrow colours are attributed to five depositional units recognized from bottom to top by George & Berry (1997). Most of the wind directions are constant during the entire Upper Rotliegend deposition. (**b**) Dune patterns (oblique draas) and prevailing wind direction inferred by Sneh (1988) (redrawn from Sneh 1988, fig. 6b). Note the good correlation with Durham Yellow Sands dune pattern (Clemmensen 1989; Chrinz & Clemmensen 1993) with different assumptions of prevailing winds directions. (**c**) The same arrows with colours attributed to main sectors of wind directions during Late Permian time. These data show only winds blowing from the east and NE as primary, and winds blowing from the south and SE as secondary. Note the prevailing wind direction (yellow arrow) inferred by Sneh (1988) from the some data as used by George & Berry (1997) for the Anglo-Dutch Basin.

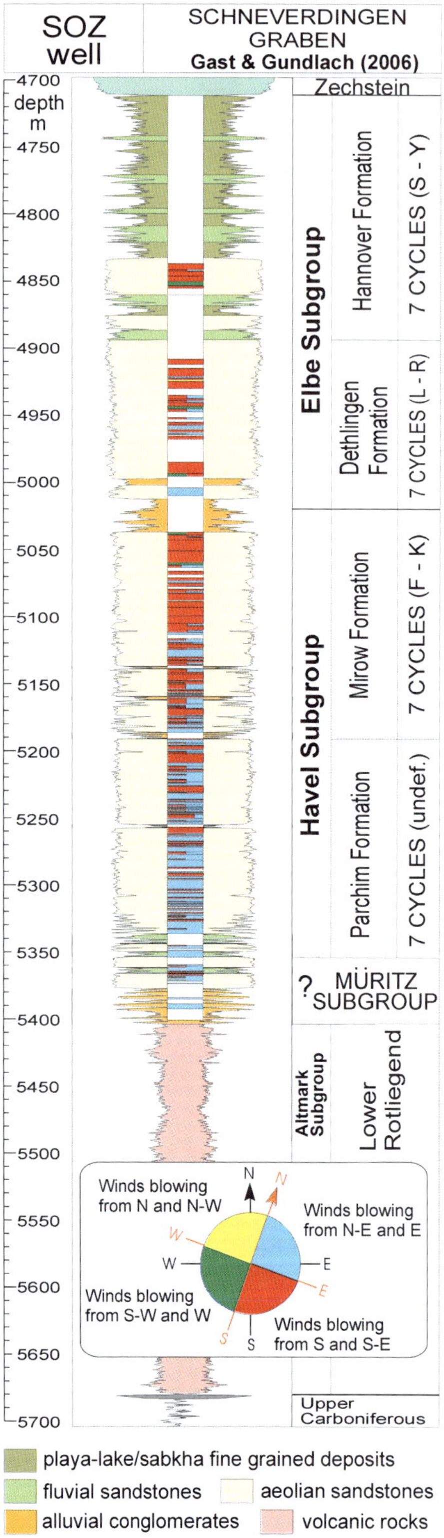

water courses (braided fluvial system) brought more coarse deposits, even gravel dominated.

It is assumed that the FIU represents a significant time period, including several deeply erosional and stratigraphic gaps. The erosion surfaces even extend into the central parts of the PURB. The top of this complex is also characterized by rapid change from fluvial to aeolian deposition. Palaeocurrent directions from fluvial channels (well W1 on Fig. 3) and the provenance of gravel clasts from the NE part of the PURB exhibit southern and northern sediment sources.

Discussion regarding entire SPB in Rotliegend time. Comparison to the Permian Anglo-Dutch Basin reveals similar features to the PURB. Using core and wire-line log correlation, Sweet (1999) identified up to 70 m deep incised valleys in the Rotliegend aeolian deposits of the UK sector of the southern North Sea Basin (NSB). Valleys are backfilled with interlayered fluvial and aeolian deposits. These valleys are present within the entire Rotliegend section in the NSB, in association with probably active tectonic faults. Similarly, Newell (2001) reported such mixed aeolian-fluvial systems and incised valleys from the Rotliegend Dawlish Sandstone Formation in the Wessex Basin, UK where 'aeolian sands were deposited by winds blowing from the SE to NW' (Laming 1966). Newell (2001) stated that 'the decimeter-scale alternation of aeolian and fluvial facies may record short term (20 ka) arid to humid climatic cycles'. However, such facies alternations may represent less-significant reasons than climatic cycles such as periodic rainfalls in arid and semi-arid climate, leaving channel fill and flood deposits.

Such alternation of aeolian-fluvial facies is recorded in some sections of the PURB, that is, the northern flanks of the Wolsztyn–Pogorzela High (Kiersnowski *et al.* 2010*b*) and southern flanks of the Pomerania High (Kiersnowski & Buniak 2006) (Fig. 2a–d). In the case of stratigraphically condensed profiles with a large number of erosional surfaces (characteristic of the topmost aeolian units from the Eastern and Southern ergs), recorded alternations of dune and dray interdune deposits interbedding with fluvial or damp interdune deposits resulted not only from climatic cycles but instead represents sedimentary units condensed

Fig. 5. Rotliegend stratigraphy from the North German Basin and gamma ray log from SOZ well from Schneverdingen Graben in Germany (Gast 1993; Gast & Gundlach 2006) with schematic interpretation of main wind directions. Bar indicates colours attributed to sectors with corroborated wind directions. White colour indicates fluvial and alluvial deposits or a lack of sufficient data.

due to slow and weak regional subsidence. In this case, such stacked units may represent significant time intervals with a number of aeolian and fluvial sedimentary events. In the case of incised valley aeolian and fluvial deposits, they represent event sedimentary units unrelated to neighbouring lateral aeolian and fluvial sequences.

Analogous Rotliegend deposits in the UK sector of the southern North Sea Basin record some fluvial-aeolian units (Sweet 1999). Similar fluvial units are also reported from well cores and wire-log sections from examined well profiles from the Eastern Erg of the PURB (Fig. 3). For instance, the reported W1 well profile is unusual because of the scale of assumed incision, ranging up to 90 m. According to Sweet (1999), incised valleys into aeolian deposits can form in response to rapid events such as changes in the water table or regional base levels coeval with a rapid fall of playa or salt lake levels driven by a combination of climatic change and syndepositional tectonics. A rapid fall of playa-lake levels can indicate a very dry period while fluvial system development can indicate a shift to a humid climate or concomitant change in drainage due to tectonic movements. Simultaneously, changes of relative playa-lake base levels could force strong erosion and rejuvenation of fluvial drainage systems.

There may be a possible link based on the origin of incised valleys between the Upper Rotliegend Warmia sub-basin (Pokorski 1989) and the central portion of the PURB (Kiersnowski 1997), as well as between the Podlasie sub-basin and southeastern part of the PURB (Fig. 2d). Palaeogeographic reconstructions indicate that both sub-basins were at higher elevations than the central playa-lake area of the PURB, so periodic fluvial transport of detrital material from their catchment areas fed into the lower playa basin. In the case of a significant fall of playa-lake level (forming a local lowermost base level), induced erosion through older Palaeozoic strata promoted the deep incision of valleys. Perhaps this event could be correlated with the origin of the incised valley (W1 well profile, Fig. 3) that shows a significant change of base level from climatic change and/or coeval with basin subsidence. A rapid change of base level would enable the spread of the fluvial system towards the basin centre.

This base level fall event is also recorded in the North German Basin by the lowering of the salt lake level (Gast 1991) along with the aridization of the area and the origin of the Wustrow Member (Hannover Formation). Exposure of a large marginal mudflat allowed the encroachment of dune systems and the accumulation of the aeolian sand cover. Simultaneously, the gradual lake fall generated coastal barriers. Most importantly, the Wustrow sandstone barriers are gas-bearing hydrocarbon reservoir rocks with potential subtle gas traps (Maynard & Gibson 2001). The lowering of water levels also promoted the progradation of fluvial-deltaic shaped systems (McKie 2011), terminated within the playa margin area in the form of pseudo-lobe stacked sand sheets or flood deposits.

Aeolian Condensed Unit 1 (ECU 1). This aeolian and fluvial complex is characterized by variable wind directions (prevailing winds from the SE and NE), forming distinctively thick sand units. These deposits are stratigraphically condensed due to low accommodation space. The dune types were barchan or transverse dunes. The sand sources for this complex are disputed. The winds coming from the south and SE were most important on the southern alluvial plains, this is, the same area as the sand source for older aeolian ergs. The winds coming from the east and NE have a sand source from the eroded and exposed Pre-Cambrian platform (see palaeogeographic map in Fig. 2d). However, the sand source for the SW margin of the PURB must be different due to the morphologic barrier of the central playa-lake. Sections through the playa-lake deposits contain no evidence of aeolian sand transport with only thin fine sandstone layers from subaqueous bedload deposition. In such circumstances, the most probable sand source was from the south and SE basin margin. Southern and easterly winds removed sand gradually and alternatively to north and west direction. This process was probably associated with erosion (cannibalization) of older aeolian beds.

Aeolian Condensed Unit 2 (ECU 2). This unit is composed of tens of metres of aeolian and fluvial beds representing a multi-stage development of ergs. The variable palaeowind directions prevailed from the NE and SE (possibly occasionally from the SW). The aeolian complex thickness depends on local tectonic subsidence and the frequency of repeated erosion and accumulation processes. These deposits are stratigraphically condensed due to low accommodation space. The dune type was barchan or an unknown type, due to the common preservation of only bottom dune sets. Aeolian deposits of this complex are limited on the top by an erosional surface and deposits of Zechstein transgression (Weissliegend). However, marine erosion is possible, especially in places with high-angle slopes. In many cases palaeotopography (Kiersnowski *et al.* 2010*b*) and large dune fields were preserved (Kiersnowski & Tomaszczyk 2010). These forms are described as barchanoid ridges, star dunes and even large yardangs. In the Southern (Silesian) Erg of the PURB in western Poland (Fig. 2d) large dune forms exist with possible

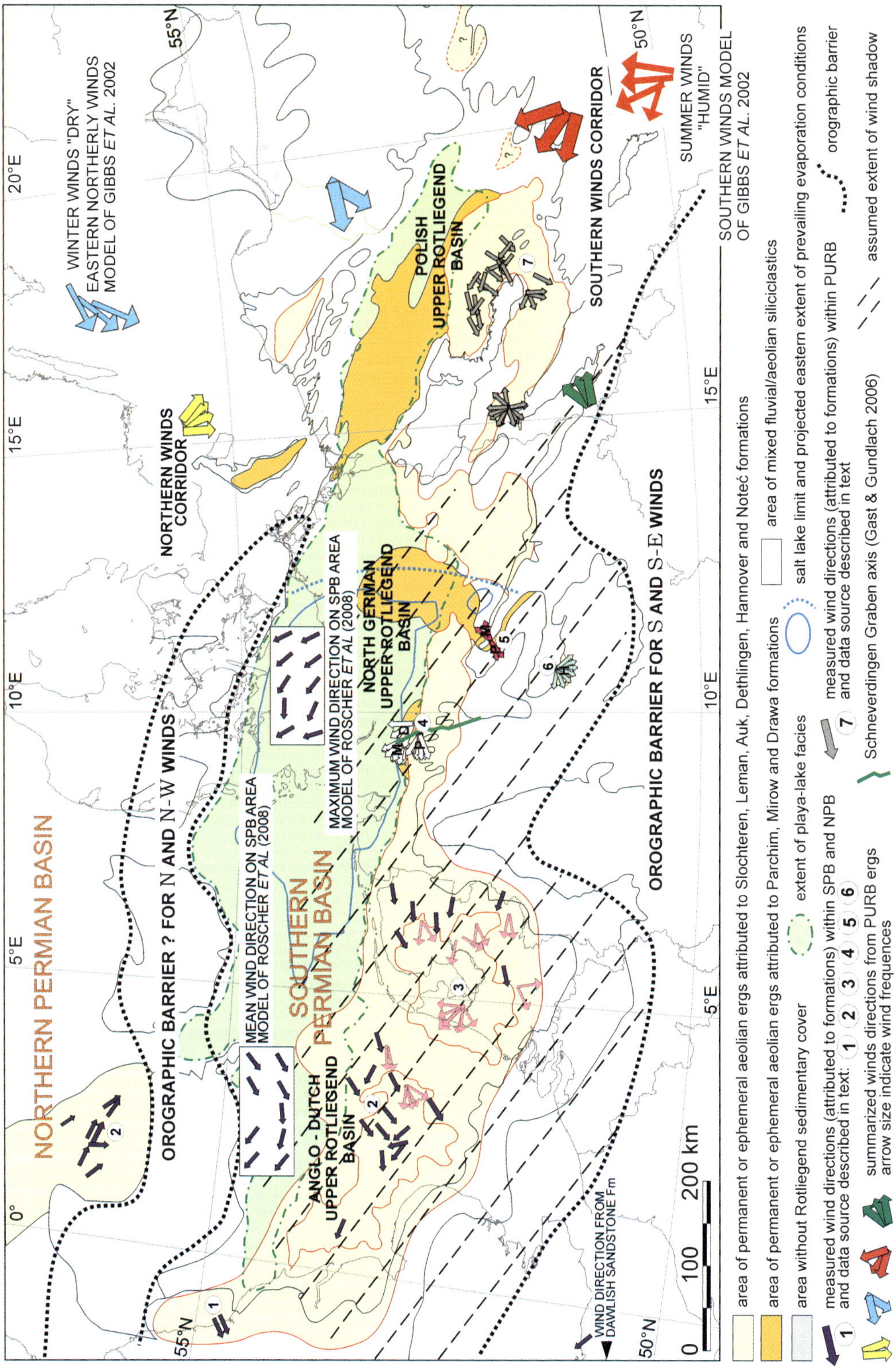
NORTHERN PERMIAN BASIN
SOUTHERN PERMIAN BASIN
ANGLO - DUTCH UPPER ROTLIEGEND BASIN
NORTH GERMAN UPPER ROTLIEGEND BASIN
POLISH UPPER ROTLIEGEND BASIN
OROGRAPHIC BARRIER ? FOR N AND N-W WINDS
OROGRAPHIC BARRIER FOR S AND S-E WINDS
NORTHERN WINDS CORRIDOR
SOUTHERN WINDS CORRIDOR
WINTER WINDS "DRY" EASTERN NORTHERLY WINDS MODEL OF GIBBS ET AL. 2002
SUMMER WINDS "HUMID"
SOUTHERN WINDS MODEL OF GIBBS ET AL. 2002
MEAN WIND DIRECTION ON SPB AREA MODEL OF ROSCHER ET AL (2008)
MAXIMUM WIND DIRECTION ON SPB AREA MODEL OF ROSCHER ET AL (2008)
WIND DIRECTION FROM DAWLISH SANDSTONE Fm
0°
5°E
10°E
15°E
20°E
50°N
55°N
0 100 200 km
area of permanent or ephemeral aeolian ergs attributed to Slochteren, Leman, Auk, Dethlingen, Hannover and Noteć formations
area of permanent or ephemeral aeolian ergs attributed to Parchim, Mirow and Drawa formations
area without Rotliegend sedimentary cover
area of mixed fluvial/aeolian siliciclastics
extent of playa-lake facies
salt lake limit and projected eastern extent of prevailing evaporation conditions
measured wind directions (attributed to formations) within SPB and NPB and data source described in text: 1 2 3 4 5 6
measured wind directions (attributed to formations) within PURB and data source described in text 7
summarized winds directions from PURB ergs arrow size indicate wind frequences
Schneverdingen Graben axis (Gast & Gundlach 2006)
orographic barrier
assumed extent of wind shadow

dune types varying between barchanoid ridges and elongated forms (possibly seif dunes and draas). In both cases different wind directions are postulated (unpublished data).

In both ECU 1 and 2, wind directions are recorded from four main sectors with prevailing winds from the east, NE and SE. These aeolian complexes developed simultaneously with playa-lake deposits (Fig. 2c, d). These playa-lake sediments are characterized by brown-red claystones, mudstones and very fine sandstones. The fine-grained sedimentary rocks contain anhydrite nodules and rare disturbed layers which may have been the result of evaporitic crystals (rock salt) growth. No aeolian deposits are preserved, indicating transport over the playa deposits. Both ACU 1 and 2 may record long time intervals, perhaps the longest of any of the PURB units. They reflect the last stage of Upper Rotliegend basin fill, when the basin reached maximum extent and lowest subsidence.

Late Permian palaeogeography using wind directions predicted from global wind circulation models

Late Permian wind directions: an overview

The methods of using wind directions determined from aeolian sandstones for palaeoclimatic models are successfully adopted and vice versa; wind directions predicted from global circulation models are tested in case studies (Parrish & Peterson 1988).

Glennie (1986) argued that the SPB entered the trade wind desert belt developed in the Permian of northern Pangea. Gibbs *et al.* (2002) reconstructed a palaeoclimatic model for Wordian time where the SPB latitudes displayed winter winds blowing from the east and NE and summer winds blowing generally from the south, with secondary winds from the north and NE. Gibbs *et al.* (2002) presented a model for average seasonal (December–January–February or DJF and June–July–August or JJA) surface wind velocity vectors, which shows distinct changes over the SPB area (Figs 6 & 8). Roscher *et al.* (2008) pointed out the importance of the equatorial Variscan mountain chain as a significant topographic barrier influencing palaeowind patterns.

Gibbs *et al.* (2002) pointed out the importance of seasonal winds without their general validation (excluding mean wind velocity vectors). In their model of palaeowind circulations and mean and maximum palaeowind directions, Roscher *et al.* (2008) attributed their origin to the lowermost atmospheric level. A general bimodal wind pattern therefore existed in the SPB area (Gibbs *et al.* 2002; Roscher *et al.* 2008) (Figs 6 & 8). Wind directions from the Wordian model (Gibbs *et al.* 2002) fit well with the observations from the SPB ergs, including these new data from the PURB (Fig. 2a, d). Roscher *et al.* (2008) calculated mean wind directions which also fit those measured and published from the SPB, including the NE palaeowind direction during the Middle Permian (270 Ma). However, Roscher *et al.* (2008) noticed that field observations based on data from Ellenberg *et al.* (1976) and Glennie (1983*a*, *b*) commonly show a bipolar distribution of palaeowind indicators, which can be attributed to changing wind directions. According to Roscher *et al.* (2008), observed bipolar wind directions for the SPB area are interpreted as being associated with long-lasting mean winds and short but intense maximum winds (Fig. 8). They stated that maximum winds may partially overprint mean wind indicators of a modelled and measured NE direction. The high-velocity southern winds postulated here are therefore partially supported (Ellenberg *et al.* 1976; Glennie 1983*a*, *b*). Moreover, Roscher *et al.* (2008) suggested that maximum winds may partially overprint mean NE wind indicators and are contrary to the southern seasonal winds of Gibbs *et al.* (2002). The key issue is whether these winds are long or short lasting, enabling the transport of sand from the source area to ergs that are hundreds of kilometres distant. The data with new evidence of southeasterly winds from the PURB are also supplemented by postulated secondary wind directions from the north and NW (Figs 2a–d, 3).

Wind patterns are summarized by George & Berry (1993, 1997) for the SPB part of the UK–Dutch aeolian ergs (Figs 4 & 6). Northeasterly prevailing trade winds are described as the main

Fig. 6. Southern Permian Basin (SPB) map after Gast *et al.* (2010) with summarized winds directions obtained from aeolian ergs attributed to the Upper Rotliegend. Data sources: (1) Clemmensen (1989) from Yellow Sands (western margin of SPB); (2) Glennie (1983*a*, *b*) and Glennie *et al.* (2003) from SPB (Anglo-Dutch Basin; Leman and Silverpit formations) and the Northern Permian Basin; (3) Van Wijhe *et al.* (1980); compilation of palaeowind directions interpreted from dipmeter measurements from nine boreholes from Upper Rotliegend aeolian deposits from onshore and offshore the Netherlands (southern part of the North Sea); (4) Gast (1993) from the SOZ well in Germany (P, Parchim Formation; M, Mirow Formation; D, Dethlingen Fromation); (5) Legler (2006); summarized older data from the Flechtingen Basin in Germany; (6) Paul (1997) from the Ilfeld Basin in Germany (H, Hannover Formation); (7) Kiersnowski *et al.* (2010*b* and unpublished) and Karnkowski (1999) from PURB.

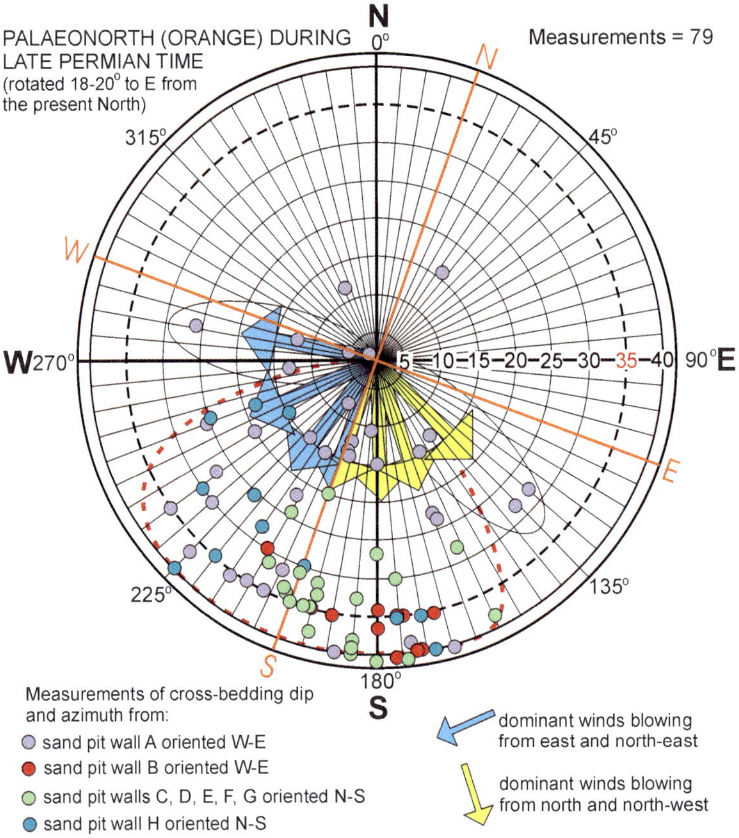

Fig. 7. Outcrop with Rotliegend aeolian sands from the sand-pit 'Elsa' near Ellrich, eastern Germany. Palaeowind directions are estimated from measurements of dune slip faces. Dips higher than 35° maximum sedimentary dip for dune slopes, resulted from later tilting of these area towards the present south. Completed measurements (Paul 1997) on the Schmidt Plot show complex dune system, composed of (possibly) two stacked ergs. Prevailing winds are from north and east.

component wind; secondary seasonal easterly winds transported aeolian sands from the lower area of the Netherlands. Southeasterly winds and rare but weak southwesterly winds (interpreted as katabatic) blew off the London–Brabant Massif.

The Glennie (1983*b*) wind model from the North Sea Rotliegend which 'inferred' a Mid North Sea High is based on predominantly easterly and northeasterly wind directions in the main SPB and southwesterly directions in the smaller Northern Permian and Moray Firth basins. Gibbs *et al.* (2002) claimed that 'we do not find any indication of such a relatively small-scale circulation pattern in the model (Wordian) results'. But wind directions from the north and NW, similar to Glennie (1983*a*, *b*), result from the Northern Permian and Moray Firth basins, and are relatively common in aeolian ergs from the PURB. However, Gibbs *et al.* (2002) stress the importance of seasonal wind directions, where 'model winds are strongest in winter (DJF), blowing uniformly from the east and NE trade wind direction'. In contrast, winds are weaker and more varied in summer (JJA). Gibbs *et al.* (2002) concluded that: 'one possibility for the measured southwesterly directions is that they were influenced by local fault-bounded mountain ranges, and the effect of local topography has been invoked in both the southern and northern North Sea areas (Glennie 1983*a*; George & Berry 1993)'. Roscher *et al.* (2008) pointed out the importance of the equatorial Variscan mountain chain as a significant topographic barrier influencing palaeowind pattern.

It should be noted that the summer and winter wind pattern simulations of Gibbs *et al.* (2002) are partly adjusted to the results of Glennie (1983*a*, *b*) and others. Similarly, Roscher *et al.* (2008) used results of Glennie (1983*a*, *b*) and Ellenberg *et al.*

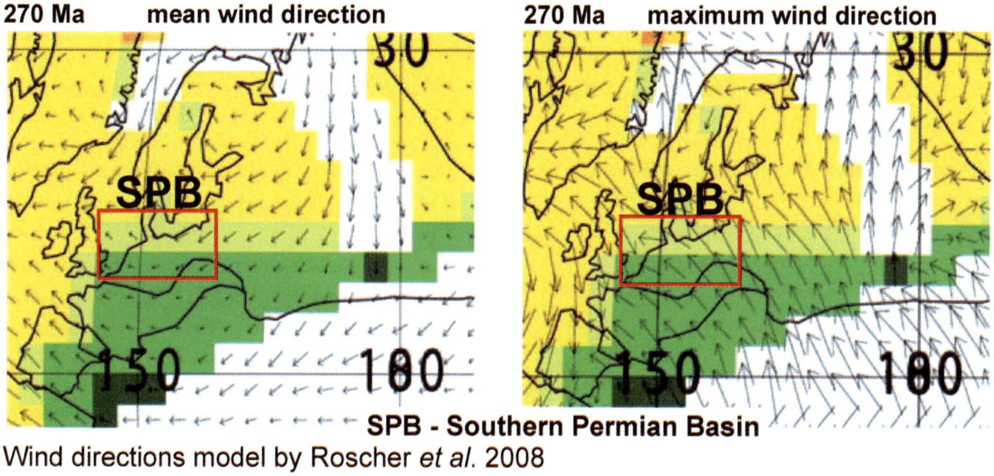

Fig. 8. Wind directions of the southern Permian Basin during Middle Permian (270 Ma) time (from Roscher *et al.* 2008). Left (right): arrows indicate direction and strength of mean (maximum) wind directions. The palaeogeography of the SPB area is rotated to present position.

(1976) for wind and climatic model adjustments. These results should be partly re-interpreted, incorporating interrelationships between dune forms and dominant wind directions (Steele 1983; Sneh 1988) as well as the differences between seasonal wind directions from small dune forms covering large forms versus principal wind directions attributed to long-lasting large aeolian forms such as draa complexes (Clemmensen 1989).

Gibbs *et al.* (2002) point out that: 'reliable directions may be obtained in the case of dunes that are transverse to the wind direction but not for oblique or longitudinal dunes'. However, the general wind directions from the Anglo-Dutch Basin (George & Berry 1993, 1997) are mostly from the NE and SE and the detection of dune forms may partly modify this picture. Gibbs *et al.* (2002) noted that dune fields in NE England were initially assumed to be dominated by northeasterly winds (Glennie 1983*b*). When interpreted by Sneh (1988) however, the description is partly different: 'a closer study re-interpreted the dune morphology as longitudinal and deposited under the influence of alternating northerly and southeasterly winds' (Chrintz & Clemmensen 1993). In this case the published wind pattern from dune forms interpretation should be re-examined in some areas. Sneh (1988) argued that present interpretation of dune patterns implies a change in wind direction of the order 45° anticlockwise (Fig. 4c).

Dunes types described from the UK–Dutch Permian basins by George & Berry (1997) (Figs 4 & 6) are, from bottom to top, Units 1–5 of the Rotliegend sequence: Units 1, 2 and 3 have mostly barchans; Unit 4 contains barchans and barchanoid ridges; and Unit 5 is composed of transverse dunes, barchanoid ridges and barchans (see George & Berry 1997; Figs 8 & 9). This change could be related to increasing dune (erg) area, probably best developed during the last dry phase (Unit 5). This dune description is consistent with dune types from the PURB. Howell & Mountney (1997) also refer to this number of units from the UK part of the SPB, but they concluded that each interval represents a number of climatic cycles (units; 12 are postulated). George & Berry (1997) divide each unit into wet (lower) and dry (upper) sub-units, which should be roughly coeval with 10 climatic cycles. The climatic cycles are connected with altered sand supply, wind velocity and prevailing erosion (deflation) or accumulation of dune sands. Recognition of climate-related depositional units (cycles) may therefore show changing dune pattern and different interpretations of wind directions.

According to George & Berry (1997), palaeogeographic models for the UK–Dutch part of the SPB developed 'wet' alluvial fans during humid periods and dry alluvial fans during dry periods. Both types of alluvial fans acted as a permanent source of 'fresh' sand to incorporate into aeolian depositional systems. Such a model of permanent alluvial-fluvial activity may also have been important in the PURB, necessary for a constant sand supply and the large extent and volume of aeolian sands. However, most of the accessible data from the PURB well cores contain only rare intercalations of alluvial (wadi) deposits in some of the basin fringe areas. Key to the existence of such permanent sand-feeding systems is a reliable source area, its relation to wind direction and its wind transporting potential. In the palaeogeographic models of George & Berry (1997) for the UK–Dutch part of the SPB, a permanent source area is the Anglo-Brabant Massif. Early source areas were covered by aeolian or alluvial deposits during basin expansion stages.

Late Permian wind directions: case studies

In the case of the PURB, the importance of wind directions should be also discussed from the point of view of sand source localization and supply potential. For the PURB, the permanent source of aeolian sands was the large alluvial plain developed in front of the Pre-Cambrian platform and Podlasie sub-basin area on the east and Małopolska Upland on the south (Fig. 2d). Part of this alluvial plain source area was later gradually overlapped by playa deposits as a result of basin subsidence and playa expansion related to hydrologic changes. Other secondary source areas were less important due to the small size of catchments feeding poorly developed alluvial fans and plains (i.e. Wolsztyn–Pogorzela High, Fig. 2d; Kiersnowski *et al.* 2010*b*).

Apart from deducing wind directions from dune forms from wells in the North Sea area (Glennie 1983*a*, *b*), well-described wind directions and recognition of dune forms can be examined from outcrops exposed at basin margins. A good example is Yellow Sands (Fig. 6) in NE England (Steele 1983; Clemmensen 1989; Chrintz & Clemmensen 1993; Turner & Smith 1997) forming linear draa as a result of the dominant wind direction of 235° (Figs 4c & 6). This correlates wells with the pattern of oblique draas, postulated by Sneh (1988) in the Northern Permian Basin (Fig. 4c) as a result of prevailing northerly winds.

At the central part of the SPB (North German Basin) there are very limited data showing wind directions from aeolian ergs, in particular from the Upper Rotliegend II deposits (Parchim, Mirow, Dethlingen and Hannover formation) (Ellenberg *et al.* 1976; Gast 1993; Legler 2006). The prevailing winds are from east and NE with minor evidence from other directions. There is only one piece of evidence of permanent southeasterly winds for the aeolian sand complex from the SOZ well from Schneverdingen Graben in the Lower Saxony Basin of eastern Germany (Fig. 5; Drong *et al.* 1982; Gast 1993; Gast & Gundlach 2006). In this complex the lower aeolian unit belonging to the Parchim and Mirow formations has an intermediary direction of eastern winds rotating to southern directions. The upper aeolian unit belonging to the Dethlingen Formation has intermediary, mostly southerly wind directions. In the case of the SOZ well this is evident from the clockwise rotation of wind directions during the time attributed to deposition of the Parchim, Mirow and Dethlingen formations. It must also be considered that the Hauptsandstein from Schneverdingen Graben (Drong *et al.* 1982), presently belonging to the Dethlingen Formation, could possibly instead represent the upper part of the Mirow Formation because the wind directions are more similar to wind directions from the PURB (Upper Drawa Formation). Schneverdingen Graben is relatively narrow and has an axis running roughly north–south (see Fig. 6 for location) with early eastern winds perpendicular to the graben axis. Perhaps the north–south-oriented graben walls forced dune form migration towards the south, not to the east. On other hand, it is also possible that the Schneverdingen Graben formed later and trapped and preserved aeolian sands that originated from dominantly easterly winds, forming its vast aeolian cover. It is also an option (as mentioned previously), that Schneverdingen Graben has longer depositional history and lower aeolian units with wind blowing from north to east (blue colour) should be attributed to pre-Parchim Formation.

At the southern margin of the North German Basin (palaeo-Beber Graben in the Flechtingen–Rosslau Uplift; see Fig. 6 for location) there are several outcrops with Upper Rotliegend II aeolian sandstones. According to Ellenberg *et al.* (1976), Kleditzsch & Kurze (1993) and Legler (2006), the oldest dunes (barchans to barchanoid ridges) incorporated within the Parchim Formation were described in the Altmark depression.

The main wind direction was from the NE (Fig. 6). The upper aeolian unit (Hünenküchen outcrop) in this basin is included in the Mirow Formation (Legler *et al.* 2004) and registers wind direction from WSW and SW. This wind direction 'anomaly', in comparison with a basin-wide direction from the NE, is interpreted by Glennie (1972) as a local direction forced by wadi topography. As investigated in the PURB this wind direction is rarely observed in the open ergs area, mostly in sections equivalents to the Dethlingen and Hannover Formation. However, the aeolian unit from the Hünenküchen outcrop is now included in the Mirow Formation (Legler *et al.* 2004); previously it was included in the Dethlingen Formation (Schneider *et al.* 1998) and even the Bahnsen Member (Ellenberg & Lützner 1995) or Wustrow Member, both belonging to the Hannover Formation (Fig. 9).

Rotliegend aeolian sandstones which crop out in the sand-pit known as 'Elsa' near Ellrich, eastern Germany consist of large-scale cross-bedding units taller than 45 m. These sands are attributed to Upper Rotliegend sandstones of the Walkenried Formation from the Ilfeld Basin (Fig. 6; Paul 1999). Palaeowind directions are estimated from measurements of dune slip faces. Dips higher than 35° maximum sedimentary dip for dune slopes resulted from later tilting of these areas to the present-day south. Completed dip azimuth measurements (Paul 1997) on the Schmidt Plot show a complex dune system, composed of possibly two stacked ergs, with dominant winds blowing from the east and NE as well as from the north and NW (Fig. 7).

SW of the SPB, the identification of a Middle–Late Permian (Artinskian–Wuchiapingian stages) palaeowind direction is from wave-dominated lacustrine sedimentary rocks (Viala-Rabejac and Salagou formations) of the Lodève Basin (France) where dominant winds were blowing from the NNE (N020° to N040°) towards the SSW (Pochat *et al.* 2005, fig. 7). After correction of 18° clockwise from the present-day north, 'Late Permian wind was blowing from the north (N002°E to N022°E) towards the south in this part of Europe' (Pochat *et al.* 2005). According to Roscher & Schneider (2006), the Viala–Rabejac Formation belongs to Upper Rotliegend I and the Salagou Formation belongs to Upper Rotliegend I and II. Thus, dominant wind direction during the Late Permian was stable for a very long time; this may be connected to the palaeogeomorphology of a 'wind corridor'.

Late Permian wind directions: significance of palaeogeomorphology

Such wind corridors may exist in areas of differentiated topography such as large tectonic grabens (see SPB margin topography and the Mid North Sea High) or rift valleys of vast mountains (Glennie 1986, 1990). According to Roscher & Schneider (2006), the elevation of the Variscides and other mountain ranges of the Trans-Pangea Mountain Chain (Keller & Hatcher 1999) is of major importance for European and global climate during Permo-Caboniferous time. Roscher & Schneider (2006) review different estimates for the altitudinal range of the Variscan Mountain chain. Existing numerical models include generalized topography varying from a maximum height of 1800 m (Gibbs *et al.* 2002) to an average elevation of *c.* 2000–2500 m (Keller & Hatcher 1999) to heights over 5000 m (Kutzbach & Ziegler 1994; Becq-Giraudon *et al.* 1996; Fluteau *et al.* 2001; Gibbs *et al.* 2002). The palaeotopography is key for ground winds transporting sand and forming aeolian dunes. On a basin scale, orographic barriers have a key role inducing rapid changes of wind direction (Glennie 1990) and controlling the distribution of moisture. It is assumed from Permian climate models (Roscher & Schneider 2006; Roscher *et al.* 2008) that winds from the palaeo-Tethys area brought moisture for precipitation over the SPB. The presumed scale of elevation of the Trans-Pangean Mountain Chain (see palaeotopography model of Roscher & Schneider 2006) was too small 'to induce heavy rainfall accompanied with a complete moisture loss at the southern flank of this supposed mountain chain and the topography was not enough to produce a continent-wide rain shadow'. This conclusion of Roscher & Schneider (2006) is in opposition to the model of Keller & Hatcher (1999), where the authors argued that 'winds rained out in front of mountain chain and moisture was retained and not precipitated' over the SPB area.

Here it is suggested that there is a more congenial intermediate solution to that of the Keller & Hatcher (1999) model. Roscher & Schneider (2006) stated that 'Trans-Pangean Mountain Chain did not exist as single orogen, and the height of individual mountain elements seems to be overestimated'. It is proposed that the high mountain group formed a border for surface winds and partly for moisture coming from the south and SE from the peri-Tethys area (Gibbs *et al.* 2002). Higher winds could bring some moisture across an orographic barrier. As a result, a rain shadow effect blocked most moisture from the western and central part of the SPB, where drier conditions (lower precipitation) with higher evaporation promoted the existence of perennial sabkhas and saline lakes (Fig. 6).

In the eastern part of the SPB (PURB), slightly different climatic conditions existed caused by the lack of significant barriers (framing the basin) for surface winds and moisture coming from the south and SE. The southern winds may lose significant part of their moisture due to orographic barriers located far south in the peri-Tethys area. Evidence supporting this model comes from the palinspastic reconstructions of Variscan and other orogenic basins preserved in the area of the present-day West Carpathians (Vozarova & Vozar 1988), as well as some assumptions that the Zechstein Sea had a link with the peri-Tethys domain (Kiersnowski 2001; Şengör & Atayman 2009). The evidence of palaeowind directions across the SPB gives support to the orographic barriers hypothesis. In the western part of the SPB (data from the UK–Dutch aeolian ergs), prevailing winds are from the east and NE with minor evidence of winds from the ESE, SE and south (Van Wijhe *et al.* 1980, fig. 7; George & Berry 1997, figs 8a–e and 9). Winds from the south and SE, as expected from the palaeoclimatic simulations of Gibbs *et al.* (2002) and Roscher *et al.* (2008), are lacking (Fig. 8). These data are only from the Upper Rotliegend II deposits from the Slochteren or Leman formations of the UK–Dutch basins. Perhaps the orographic barrier for southern winds existed only during Upper Rotliegend time (including roughly Capitanian and Wordian stages), formed from the Anglo-Brabant and Rhenish massifs along the Variscan orogenic belt.

Palaeowinds documented by Glennie (1983*a*, *b*) from the Rotliegend aeolian sandstones of the Northern Permian and Moray Firth basins (Scotland) and the Central and Viking Graben area (North Sea) (Glennie e*t al.* 2003) are from the north and NW directions. Gibbs *et al.* (2002) stated that they did not find any indication of such a relatively small-scale circulation pattern in their model results. They argued that one possibility for the measured southwesterly directions is that they were influenced by local fault-bounded mountain ranges and the effect of local topography has been invoked in both the southern and northern North Sea areas for these anomalies.

The northern winds observed in the PURB area are related to the postulated northern wind corridor. For the western and central parts of the SPB, the Ringkøbing-Fyn-Møns (Fig. 6) trend of highs was an orographic barrier for winds from the north and NW; in the eastern part (PURB area) such a direction is relatively common (Fig. 2a–d), probably due to a lack of significant topographic barriers in the present-day west Baltic Sea area.

The summer and winter wind pattern simulations of Gibbs *et al.* (2002) should be adjusted to take into account the results of Glennie (1983*a*, *b*) and of the PURB area. These adjustments should be influenced by the palaeotopography around the SPB, in particular remodelling the possible orographic barrier south of the western part of the SPB. Levelling of the postulated orographic barrier south of the SPB may have taken place during Upper Rotliegend time. Roscher & Schneider (2006) and Schneider *et al.* (2006) also postulated a shift through time, from east to west, of the maximum elevation coincident with the evolution of the Variscan Orogeny. However, Roscher & Schneider (2006) claim that the Hercynian system never exceeded an average of 2000 m, and therefore did not form a distinct orographic barrier. On the other hand, the origin of the postulated large depression in the SPB (Glennie 1983*a*) may increase altitude differences between the basin bottom and the top of the southern uplifts, thus enlarging the orographic effect and strengthening katabatic ground winds. In this case, the change of the wind direction during time is not evoked only by the clockwise rotation of the prevailing wind (in the case of SOZ well profile, Fig. 5), western Germany (Fig. 6), but rather by levelling or removing the orographic barrier to allow for southerly winds.

The main growth of the postulated orographic barrier south of the SPB may have taken place during Altmark tectonic movements (McCann 1998; Geissler *et al.* 2008) which, in the central part of the SPB, resulted from an erosional boundary, even disconformity, between Mirow and Dethlingen formations in Germany (Franke *et al.* 1996). As result of tectonic subsidence later, the orographic barrier may have formed from a postulated large depression (Glennie 1983*a*) that in turn increased the altitudes of adjoining mountains. A disconformity with erosional gap, perhaps representing a very long time period, is also recognized between the Drawa and Noteć formations in the PURB (Kiersnowski & Buniak 2006; Słowakiewicz *et al.* 2009).

In the eastern part of the SPB into the PURB, there is evidence of constant southern and southeastern winds bringing more moist air. The lower aeolian units (ERU, ERBU 1, EBYU and ERBU 2) attributed to the lower and upper Drawa Formation (equivalent to the Parchim and Mirow formations; Kiersnowski in Gast *et al.* 2010) mostly represent winds from the south and SE. The upper aeolian units (ECU 1 and ECU 2) attributed to the lower and upper Noteć Formation are equivalent to the Dethlingen and Hannover formations, representing prevailing winds from the east and NE with rare SE winds and less-common winds from the north, NW and SE (Figs 2a–d & 3).

The lower playa coeval with lower aeolian units is more 'wet' and clayey in comparison to the 'dry' playa which is sandier and contains evidence of former salt crusts and, in its upper portion, anhydrite streaks and nodules. The climate is interpreted as having the influence of southern 'monsoonal' winds prevailing first with the increasing addition of eastern winds. This generally fits the palaeowind simulations of Gibbs *et al.* (2002) and Roscher *et al.* (2008), taking into consideration such a long time span with possible changing topography and gradual movement northward of sands to cover the entire SPB area. This process of ongoing aridization in this area was deduced by Roscher & Schneider (2006) (Fig. 9).

For model simulations of Pangea A and B, Fluteau *et al.* (2001) deduced that Variscan elevation had to be *c.* 2000 m to account for Baltica (Southern Permian Basin) climate; this elevation is more realistic for Pangea B. These assumptions support the argumentations presented here.

The existence or lack of orographic barriers is probably one of the more important reasons for climatic- and facies-related differences between the western and eastern parts of the SPB. The PURB is partly located along the margins of the

NE Variscan belt and dominantly in the Variscan foredeep area. Southern winds from the beginning of the basin origin were not blocked by significant orographic barriers in the south. Additionally, the eastern part of the SPB was located closer to the palaeo-Tethys coastal zone (Gaetani *et al.* 2000) than the central and western parts of the SPB. Seasonal southern winds may have been strong and high enough (Roscher *et al.* 2008) to transport sand completely across the basin, even to the present-day Baltic Sea territory (Rønne Graben; Fig. 2a; Kiersnowski & Buniak 2006). This is attributed to a pronounced southern winds corridor; however, Gibbs *et al.* (2002) suggested in this area a simulated weak wind velocity vector for Sakmarian and Wordian times.

Climatic cyclicity and stratigraphic and depositional mechanisms

The evidence of climatic cyclicity within the Rotliegend sequence is clear. The cycle frequency and time attributed to cyclic sedimentary units depend on recorded cycle rank and time span assignment to particular lithostratigraphic formations or members. In the Dutch and UK portion of the SPB, the Upper Rotliegend is divided into units representing palaeoclimatic cycles: five units each contain a dry and wet phase (George & Berry 1994, 1997); five super-sequences are divided into twelve sequences (Yang & Baumfalk 1994, 1997; Yang & Nio 1994); and five intervals are divided into twelve units (Howell & Mountney 1997). For comparison,

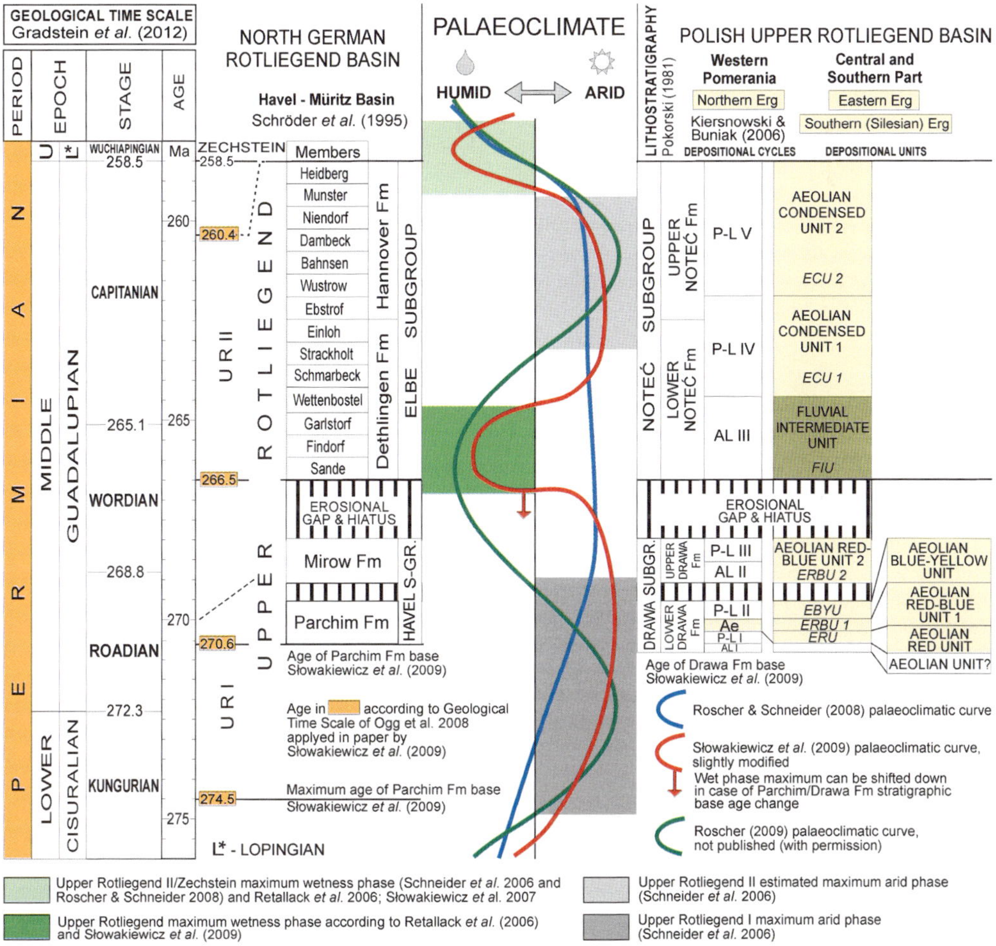

Fig. 9. Stratigraphic correlation of identified aeolian units within the PURB (Polish Upper Rotliegend Basin lithostratigraphy) with relation to Late Permian palaeoclimate fluctuations and NE North German Upper Rotliegend Basin lithostratigraphy and Geological Time Scale (Gradstein *et al.* 2012) for the Permian.

according to Van Ojik *et al.* (2011) the stratigraphic succession of the Upper Rotliegend in the Netherlands is divided into six major sequences. Six sequences attributed to climatic cycles, embracing the entire Upper Rotliegend (I and II) in the PURB, were distinguished by Karnkowski (1999); however, only the upper four are related to units with aeolian sedimentation.

In published climatic models, gradual climatic changes during Permian time are therefore presented (Roscher & Schneider 2006; Schneider *et al.* 2006); sediments indicative of rapid climatic changes are related to lower-rank climatic fluctuations or pseudo-cycles identified by frequent erosional events or stratigraphic condensation with the loss of sedimentary units. The interpretation of constant and stable climatic conditions in the SPB region depends on the scale of the width of climatic belts and the frequency of climatic belt migration to higher or lower latitudes (Perlmutter & Matthews 1989; Clemmensen *et al.* 1994; Roscher & Schneider 2006; Schneider *et al.* 2006). In palaeogeographic models of Pangea, the SPB area is located in lower latitudes (Livermore *et al.* 1986). Other palaeoclimatic models (Crowley *et al.* 1989) and models based on the compilation of palaeobotanical and sedimentary palaeoenvironments (Rees *et al.* 1999; Gibbs *et al.* 2002) support this lower latitude interpretation.

The main SPB axis during the Late Permian was located near 20°N in the Late Permian (255 Ma) (Crowley *et al.* 1989) or near 15–20°N (270 Ma) (Livermore *et al.* 1986), depending on palaeomagnetic data. The SPB area is postulated to have remained permanently in the range of lower latitudes and the distance from the peri-Tethys Ocean was constant. In consequence, a repeating or stable warm and dry climate would have stabilized the long-lasting existence of desert-dominated conditions during Rotliegend time. This issue was widely discussed regarding Pangean models of global circulation of air pressure highs and lows and changes of monsoon circulation (Glennie 1983*a*, *b*; Perlmutter & Matthews 1989, 1994; Parrish 1993).

The wind directions data of Glennie (1983*a*, *b*) mostly from the top Rotliegend aeolian deposits originated not later than during the latest Capitanian, *c.* 260 Ma (Słowakiewicz *et al.* 2009) and *c.* 258.5 Ma according to the latest geological time scale (Gradstein *et al.* 2012). This gives a 5–6.5 Ma interval (assuming 265 Ma given for the Wordian climatic model by Gibbs *et al.* 2002) for possible changes in wind circulation. However, palaeoclimatic simulations for the Wordian interval are fixed at 265 Ma; currently a 265 Ma age is correlated to early Capitanian (Fig. 9; Ogg *et al.* 2008). This age distinction is important in discussions concerning the change of wind directions over time and possible wind relationships to palaeotopography, also changeable during long time intervals due to tectonics.

Wind directions measured in PURB aeolian units agree with the simulations of Gibbs *et al.* (2002). It can therefore be assumed that each preserved aeolian unit with coherent wind direction may reflect an annual event. On the other hand, thick aeolian units with a constant wind direction are observed in the examined well cores and dipmeter data. This should be interpreted as examples of annual deposition and later preservation of thick aeolian units under conditions of abundant sand supply to areas with significant accommodation space (i.e. subsiding tectonic graben). Another interpretation is that these units represent a long-term stable prevailing wind direction without any 'admixture' of other wind directions. In this case, aeolian units represent stacked annual portions of sand with several large erosion gaps and it is rather unlikely that erosion removed all aeolian units showing other than preserved wind directions. In both interpretations the preserved aeolian deposits represent units of event stratigraphy (Kocurek & Havholm 1993; Howell & Mountney 1997). These units are separated by disconformities or long-lasting time gaps. In light of the simulations of Gibbs *et al.* (2002), the dilemma is whether aeolian profiles from the SPB Polish ergs represent a working model or 'anomalies' with constant long-lasting unified prevailing wind directions. Another important issue is the preservation potential of aeolian deposits and the interplay between sand source and supply and accommodation space in different parts of the basin.

The existence of significant disconformities argues against a long-lasting prevailing deposition of aeolian sand. This issue has not been previously discussed in published descriptions of wind-lain deposits and wind directions during the Upper Rotliegend (UR II) time in the Dutch–UK area. All summarized data most probably represent the topmost Rotliegend aeolian units and therefore a relatively short time span. These data can only be used for comparison with top Rotliegend sections within the Polish part of the SPB. Stratigraphically lower parts of Rotliegend sequence may exhibit wind direction patterns different from those recognized in the top units.

In general, Rotliegend stratigraphy is coherent across the entire SPB. In particular, the North German Basin and the PURB stratigraphy are in good compliance for lower units of the Upper Rotliegend (UR II) (Gebhardt *et al.* 1991; Hoffmann *et al.* 1997; Kiersnowski & Buniak 2006), but not for the SPB in the Dutch–UK area (Van Ojik *et al.* 2011). The key to a good palaeoclimatic model and simulations of changing wind patterns is the time span attributed to particular lithostratigraphic units and stratigraphic gaps and hiatuses. The time span of the Upper Rotliegend (UR II) is still under discussion and it currently ranges from 6 to over 14 Ma (Słowakiewicz *et al.* 2009). Such a large disparity is due to the variety of methods used for calibration and calculation of time for the deposition and erosion of Rotliegend deposits (Fig. 9). On the basis of Milankovitch cyclicity, Yang & Baumfalk (1994) argue that the upper part of the Upper Rotliegend (deposits from the Dutch part of the SPB) cover over 10 Ma. However, accepting the magnetostratigraphy scale of the whole Upper Rotliegend II of Menning *et al.* (2006), a span of 6–8 Ma is postulated, based on geological time scale (Ogg *et al.* 2008; Gradstein *et al.* 2012) IR (Illawarra Reversal) boundary and the Rotliegend–Zechstein boundary (Słowakiewicz *et al.* 2009). Therefore, the relative time span of the Upper Rotliegend (UR II) on the geological time scale (Gradstein *et al.* 2012) is still not fully established due to a lack of biostratigraphic levels and controversies regarding the Permian magnetostratigraphic scale (Menning *et al.* 2006; Słowakiewicz *et al.* 2009). The extension of UR II on the geological time scale has key implications for the palaeoclimatic model and simulations of changing wind pattern against models of Pangaea evolution and the palaeogeographic position of the SPB in the northern hemisphere.

The Upper Rotliegend (UR II) may indeed represent a longer time span (Słowakiewicz *et al.* 2009) than is generally assumed by German authors (Schröder *et al.* 1995). Consequently, early aeolian deposits could even have been deposited in the Kungurian (Fig. 9). The end of aeolian deposition within the SPB was terminated by Zechstein transgression, which began at 257.3 Ma (Menning *et al.* 2006), 260.4 Ma (Słowakiewicz *et al.* 2007, 2009) or the currently accepted 258.5 Ma (Gradstein *et al.* 2012). The presence of aeolian complexes (stacked ergs) in the Upper Rotliegend (UR II) certainly represents dry climate events over a long time period.

The lower aeolian units ERU, ERBU 1, EBYU and ERBU 2 seemingly belong to the lower and upper Drawa Formation and the fluvial unit FIU represents the lower Noteć Formation or possibly a new lithostratigraphic formation. The condensed units ECU 1 and ECU 2 represent the lower Noteć Formation (or its upper part) and the upper Noteć Formation. These were correlated with German lithostratigraphy (Schröder *et al.* 1995) and supplemented by curves showing palaeoclimatic oscillations (Schneider *et al.* 2006; Słowakiewicz *et al.* 2007, 2009) representing palaeoclimatic cycles by Roscher & Schneider (2006). The proposed extended time for the Upper Rotliegend from 258.5 to 270.6 Ma together with the proposed climatic curve and shifting down this time interval on the geologic time scale is presently in good agreement with the proposed climatic fluctuations of Roscher & Schneider (2006) (Fig. 9).

Conclusions

Deposition in the Southern Permian Basin (SPB) was complex due to different tectonic evolution of the western and eastern parts of the basin. Evolution of aeolian sand cover was determined by climate fluctuations and basin-wide tectonic movements. Early tectonic subsidence formed depressions for aeolian sand accumulation and preservation. Differential subsidence across the basin together with varied topography on basin margins resulted from the formation of orographic barriers for winds and topographic lows, shaping wind corridors. Lack of salt beds deposition within the playa deposits of the PURB, possibly indicate most fresh water conditions, caused by fluvial input. The playa deposits within the PURB appear wetter in comparison to playa/sabkha deposits from the central and western SPB.

Orographic barriers influenced local climatic conditions within the SPB area. A rain shadow desert with lower precipitation rates developed in the western and central part of the SPB. Consequently, the western part of the SPB (Anglo–Dutch Basin and partly North German Basin) was dryer in comparison to the PURB in the east. Aeolian deposits in the PURB formed the thickest and most complex system in the entire SPB. Lower aeolian units record climatic fluctuations (arid–humid cycles in Late Permian time). The upper aeolian units attributed to the whole SPB, including the Polish Rotliegend, formed in more constant and dry climatic conditions. The levelling of the SPB topography through incision and erosion together with basin expansion may have contributed to the change in wind patterns through time, from prevailing monsoon circulation winds (seasonal) to prevailing NE trade winds (more constant). This change could have occurred due to SPB migration northwards, and would have required a large time interval. This assumption forces a hypothesis regarding a longer time equivalent for Upper Rotliegend (UR I and II) deposition than is assumed on lithostratigraphic scales adjusted to the geologic time scale.

The correlation of sand covers within a particular erg area is more complex than was previously assumed. Excluding synsedimentary faulting, other factors such as local subsidence, large deeply deflated areas or large inter-draa corridors filled by aeolian deposits, changing base levels and groundwater levels and incision of valleys complicate correlation even between neighbouring sections. The simple model of stacked ergs separated by super-bounding surfaces across an erg area must be carefully reassessed.

The correlation across the Eastern Erg from the PURB area (roughly WE) is based on wind directions established from dip measurements from aeolian forms, adjusted to Permian north directions. Proposed well-to-well correlation shows sets of sedimentary units with roughly unified wind directions within an entire erg area. This correlation also shows the interpreted tectonic history of buried aeolian complexes manifested by changing thicknesses of recognized sedimentary units and the presence of erosional boundaries and discordances.

E. Gierlowski-Kordesch (Ohio State University, USA), B. Legler (University of Manchester) and an anonymous reviewer are acknowledged for their critical evaluations which led to significant improvements of the manuscript. The author is also grateful to T. Peryt (Polish Geological Institute-NRI, Warsaw) for his support and constructive comments, improving this paper. My special thanks go to N. Hoffmann, W. Lindert, B. Besley, K. Glennie and J. Paul for introducing me to the Rotliegend aeolian system during the field trips in Germany, England and Scotland. C. Breitkreuz, M. Geissler, E. Kallmeier and J. W. Schneider (TU Bergakademie Freiberg) are also acknowledged for fruitful discussions on the German Rotliegend sedimentary environments and palaeogeography. The Polish Oil and Gas Company, FX Energy Poland, ExxonMobil Deutschland, J. Kopp (LBGR Brandenburg) and P. Hoth (BGR Berlin) are acknowledged for kind access to cores and geophysical logs from Rotliegend wells. Finally, thanks to the editors for their valuable comments and scrupulous editing of successive versions of this work. This research was carried out during the Rotliegend project, part of *Petroleum Geological Atlas of the Southern Permian Basin Area* (2010) as a joint effort of National Geological Surveys related to the Southern Permian Basin area.

References

ALEKSANDROWSKI, P. & KIERSNOWSKI, H. 1998. Contemporary potential of detailed structural, sedimentological and geodynamic analysis of borehole profiles based on dipmeter data. *VII Krajowa Konferencja Naukowo-Techniczna, Koninki 1998*, 9–18 (In Polish with English summary).

ANDREOTTI, B., FOURRIÈRE, A., OULD-KADDOUR, F., MURRAY, B. & CLAUDIN, P. 2009. Giant eolian dune size determined by the average depth of the atmospheric boundary layer. *Nature*, **457**, 1120–1123.

BAILEY, R. J. & LLOYD, D. A. 2001. A log correlation of the Rotliegend of the northern Cleaver Bank High: the search for controls on reservoir sand distribution. *Petroleum Geoscience*, **7**, 351–358.

BECQ-GIRAUDON, J.-F., MONTENAT, C. & VAN DEN DRIESSCHE, J. 1996. Hercynian high-altitude phenomena in the French Massif Central: tectonic implications. *Palaeogeography, Palaeoclimatology, Palaeoecology*, **122**, 227–241.

BESSE, J., THÉVENIAUT, V. & COURTILLOT, V. 1996. Apparent polar wander paths for North America, Europe, Africa, Laurasia and west Gondwana since the Upper Carboniferous. *In*: NAIRN, A. E. M., RICOU, L. E., VRIELYNCK, B. & DERCOURT, J. (eds) *The Tethys Ocean, The Ocean Basins and Margins*. Plenum Press, New York, **8**, 71–97.

BLAKEY, R. C., PETERSON, F. & KOCUREK, G. 1988. Synthesis of late Paleozoic and Mesozoic eolian deposits of the Western Interior of the United States. *Sedimentary Geology*, **56**, 3–125.

CHRINTZ, T. & CLEMMENSEN, L. B. 1993. Draa reconstruction, the Permian Yellow Sands, northeast England. *In*: PYE, K. & LANCASTER, N. (eds) *Aeolian Sediments: Ancient and Modern*. International Association of Sedimentologists, Special Publications, 16, 151–161.

CLEMMENSEN, L. B. 1989. Preservation of interdraa and plinth deposits by the lateral migration of large linear draa (Lower Permian Yellow Sands, northeast England). *Sedimentary Geology*, **65**, 139–151.

CLEMMENSEN, L. B. & TIRSGAARD, H. 1990. Sand-drift surfaces: a neglected type of bounding surface. *Geology*, **18**, 1142–1145.

CLEMMENSEN, L. B., ØXNEVAD, I. E. I. & DE BOER, P. L. 1994. Climatic controls on ancient desert sedimentation: some late Paleozoic and Mesozoic examples from NW Europe and the Western Interior of the USA. *In*: DE BOER, P. L. & SMITH, D. G. (eds) *Orbital Forcing and Cyclic Sequences*. International Association of Sedimentologists, Special Publications, **19**, 439–457.

COGNE, J.-P., BRUN, J.-P. & VAN DEN DRIESSCHE, J. 1990. Paleomagnetic evidence for rotation during Stephano-Permian extension in southern Massif Central (France). *Earth and Planetary Science Letters*, **101**, 272–280.

CROWLEY, T. J., HYDE, W. T. & SHORT, D. A. 1989. Seasonal cycle variations on the supercontinent of Pangaea. *Geology*, **47**, 457–460.

DADLEZ, R., NARKIEWICZ, M., STEPHENSON, R. A., VISSER, M. T. M. & VAN WEES, J-D. 1995. Tectonic evolution of the Mid-Polish trough: modelling implications and significance for central European geology. *Tectonophysics*, **252**, 179–195.

DRONG, H-J., PLEIN, E., SANNEMANN, D., SCHUEPBACH, M. A. & ZIMDARS, J. 1982. Der Schneverdingen-Sandstein des Rotliegenden – eine äolische Sedimentfüllung alter Grabenstrukturen. *Zeitschrift der Deutschen Geologischen Gesellschaft*, **133**, 699–725.

ELLENBERG, J. & LÜTZNER, H. 1995. Flechtinger Scholle. *In*: PLEIN, E. (ed.) *Stratigraphie von Deutschland I; Norddeutsches Rotliegendbecken – Rotliegend-Monographie, Teil II*. Courier Forschungsinstitut Senckenberg, Frankfurt (Main), **183**, 84–96.

ELLENBERG, J., FALK, F., GRUMBT, E., LÜTZNER, H. & LUDWIG, A. O. 1976. Sedimentation des höheren Unterperms der Flechtinger Scholle. *Zeitschrift für Geologische Wissenschaften*, **5**, 705–737.

FLUTEAU, F., BESSE, J., BROUTIN, J. & RAMSTEIN, G. 2001. The Late Permian climate. What can be inferred from climate modeling concerning Pangea scenarios and Hercynian range altitude? *Palaeogeography, Palaeoclimatology, Palaeoecology*, **167**, 39–71.

FRANKE, D., HOFFMANN, R. & LINDERT, W. 1996. The Variscan Deformation Front in East Germany. Part 2: Tectonic Interpretation. *Zeitschrift für Angewandte Geologie*, **42**, 44–56.

FRYBERGER, S. G. 1986. Stratigraphic traps for petroleum in Wind-Laid Rocks. *AAPG Bulletin*, **70**, 1765–1776.

GAETANI, M., ARCHE, A. ET AL. 2000. Wordian paleogeographic map. *In*: DERCOURT, J., GAETANI, M., VIERYLNC, B., BARRIER, E., BIJU-DUVAL, B., BRUNET, M. F., CADET, J. P., CRASQUIN, S. & SANDULESCU, M. (eds) *Peri-Tethys Atlas and Explanatory Notes, Map 3*. CCGM/CGMW, Paris, 19–25.

GAST, R. E. 1991. The Perennial Rotliegend Saline Lake in Northwest Germany. *Geologisches Jahrbuch*, **A119**, 25–59.

GAST, R. E. 1993. Sequenzanalyse von äolischen Abfolgen im Rotliegenden und deren Verzahnung mit Küstensedimenten. *Geologisches Jahrbuch*, **A131**, 117–139.

GAST, R. & GUNDLACH, T. 2006. Permian strike slip and extensional tectonics in Lower Saxony, Germany. *Zeitschrift der Deutschen Gesellschaft für Geowissenschaften*, **157**, 41–56.

GAST, R., PASTERNAK, M., PISKE, J. & RASCH, H.-J. 1998. Das Rotliegend im nordostdeutschen Raum: Regionale Übersicht, Stratigraphie, Fazies und Diagenese. *In*: KARNIN, W.-D., MERKEL, D., PISKE, J. & SCHRETZENMAYER, S. (eds) *Geowissenschaftliche Ergebnisse der Kohlenwasserstoff-Exploration im Land Brandenburg und im Thüringer Becken in den Jahren 1991–1996 (Zechstein und Rotliegend)*. Geologisches Jahrbuch, **A149**, 59–79.

GAST, R., DUSAR, M. ET AL. 2010. Rotliegend. *In*: DOORNENBAL, H. & STEVENSON, A. G. (eds) *Petroleum Geological Atlas of the Southern Permian Basin Area*. EAGE Publications, Houten, The Netherlands, 101–122.

GAUPP, R., GAST, R. & FORSTER, C. 2000. Late Permian playa lake deposits of the Southern Permian Basin (Central Europe). *In*: GIERLOWSKI-KORDESCH, E. H. & KELTS, K. R. (eds) *Lake Basins Trough Space and Time*. AAPG, Tulsa, Studies in Geology, **46**, 75–86.

GEBHARDT, U., SCHNEIDER, J. & HOFFMANN, N. 1991. Modelle zur Stratigraphie und Beckenentwicklung im Rotliegenden der Norddeutschen Senke. Stratigraphic and basin development models for the Rotliegend of North German Basin. *Geologisches Jahrbuch*, **A127**, 405–427.

GEISSLER, M., BREITKREUZ, C. & KIERSNOWSKI, H. 2008. Late Paleozoic volcanism in the central part of the Southern Permian Basin (NE Germany, W Poland): facies distribution and volcano-topographic hiati. *In*: BAYER, U., GAJEWSKI, D., LITTKE, R. & , R. (Guest eds) *Sedimentary Basin Evolution: Subsidence, Salt Dynamics, Fluid Flow and Deformation. International Journal of Earth Sciences Special Issue*, 97, 973–989.

GEORGE, G. T. & BERRY, J. K. 1993. A new lithostratigraphy and depositional model for the Upper Rotliegend of the UK Sector of the Southern North Sea. *In*: NORTH, C. P. & PROSSER, D. J. (eds) *Characterization of Fluvial and Aeolian Reservoirs*. Geological Society, London, Special Publications, **73**, 291–319.

GEORGE, G. T. & BERRY, J. K. 1994. A new palaeogeographic and depositional model for the Upper Rotliegend, offshore Netherlands. *First Break*, **12**, 147–158.

GEORGE, G. T. & BERRY, J. K. 1997. Permian (Upper Rotliegend) synsedimentary tectonics, basin development and palaeogeography of the southern North Sea. *In*: ZIEGLER, K., TURNER, P. & DAINES, S. R. (eds) *Petroleum Geology of the Southern North Sea: Future Potential*. Geological Society, London, Special Publications, **123**, 31–61.

GIBBS, M. T., REES, P. M., KUTZBACH, J. E., ZIEGLER, A. M., BEHLING, P. J. & ROWLEY, D. B. 2002. Simulations of Permian climate and comparison with climate-sensitive sediments. *Journal of Geology*, **110**, 33–55.

GLENNIE, K. W. 1972. Permian Rotliegendes of Northwest Europe interpreted in light of modern desert sedimentation studies. *AAPG Bulletin*, **56**, 1048–1071.

GLENNIE, K. W. 1983*a*. Early Permian (Rotliegendes) Palaeowinds of the North Sea. *Sedimentary Geology*, **34**, 245–265.

GLENNIE, K. W. 1983*b*. Lower Permian Rotliegend desert sedimentation in the North Sea area. *In*: BROOKFIELD, M. E. & AHLBRANDT, T. S. (eds) *Eolian sediments and processes*. Developments in Sedimentology, Elsevier, Amsterdam, **38**, 521–541.

GLENNIE, K. W. 1985. Early Permian (Rotliegendes) palaeowinds of the North Sea – reply. *Sedimentary Geology*, **45**, 293–313.

GLENNIE, K. W. 1986. Early Permian – Rotliegend. *In*: GLENNIE, K. W. (ed.) *Introduction to the Petroleum Geology of the North Sea*. Blackwell Scientific Publications, London, 63–85.

GLENNIE, K. W. 1990. Outline of North Sea history and structural framework. *In*: GLENNIE, K. W. (ed.) *Introduction to the Petroleum Geology of the North Sea*. 3rd edn. Blackwell Science, London, 34–77.

GLENNIE, K. W. & BULLER, A. T. 1983. The Permian Weissliegend of N.W. Europe: the partial deformation of aeolian dune sands caused by the Zechstein transgression. *Sedimentary Geology*, **35**, 43–81.

GLENNIE, K. W., HIGHAM, J. & STEMMERIK, L. 2003. Permian. *In*: EVANS, D., GRAHAN, C., ARMOUR, A. & BATHURST, P. (eds & coordinators) *The Millenium Atlas: Petroleum Geology of the Central and Northern North Sea*. Geological Society, London, 91–103.

GRADSTEIN, F. M., OGG, J. G., SCHMITZ, M. & OGG, G. 2012. *The Geologic Time Scale 2012*. Elsevier, Amsterdam.

HELMUTH, H.-J. & SÜSSMUTH, S. 1993. Die lithostratigraphische Gliederung des jüngeren Oberrotliegenden (Oberrotliegendes II) in Nordostdeutschland. *In*: MÜLLER, E. P. & PORTH, H. (eds) *Perm im Ostteil der Norddeutschen Trough. Geologisches Jahrbuch*, **A131**, 31–55.

HOFFMANN, N., POKORSKI, J., LINDERT, W. & BACHMANN, H. 1997. Rotliegend stratigraphy, paleogeography and facies in eastern part of the central European Basin. *Proceedings of the XIII International Congress on Carboniferous-Permian, Kraków*. Prace Państwowego Instytutu Geologicznego, **157**, 75–86.

HOWELL, J. A. & MOUNTNEY, N. P. 1997. Climatic cyclicity and accommodation space in arid to semi-arid depositional systems: an example from the Rotliegend group of the Southern North Sea. *In*: ZIEGLER, K.,

TURNER, P. & DAINES, S. R. (eds) *Petroleum Geology of the Southern North Sea: Future Potential*. Geological Society, London, Special Publications, **123**, 63–86.

KARNKOWSKI, P. H. 1986. The nature of Zechstein transgression v. origin of the Weissliegendes in the Wielkopolska area (northern Fore-Sudetic Monocline, western Poland). (In Polish with English summary). *Geologia Sudetica*, **21**, 101–122.

KARNKOWSKI, P. H. 1999. Origin and evolution of the Polish Rotliegend Basin. *Polish Geological Institute Special Papers*, **3**, 1–93.

KARPETA, W. P. 1990. The morphology of Permian palaeodunes – a reinterpretation of the Bridgnorth Sandstone around Bridgnorth, England, in the light of modern dune studies. *Sedimentary Geology*, **69**, 59–75.

KELLER, G. R. & HATCHER, R. D. 1999. Some comparisons of the structure and evolution of the southern Appalachian-Ouachita orogen and problems of the trans-European suture zone region. *Tectonophysics*, **314**, 43–68.

KIERSNOWSKI, H. 1997. Depositional development of the Polish Upper Rotliegend Basin and evolution of its sediment source areas. *Geological Quarterly*, **41**, 433–456.

KIERSNOWSKI, H. 2001. Permian–Triassic deposits of the Liplas-Tarnawa Basin. *In*: MATYJA, H. (ed.) *Paleozoic Basement of the Central Segment of the Polish Outer Carpathians*. Prace Państowego Instytutu Geologicznego, Warsaw, **174**, 87–100. (In Polish with English summary).

KIERSNOWSKI, H. & BUNIAK, A. 2006. Evolution of the Rotliegend Basin of northwestern Poland. *Geological Quarterly*, **50**, 119–138.

KIERSNOWSKI, H. & TOMASZCZYK, M. 2010. Permian dune fields as geomorphic traps for gas accumulation: Upper Rotliegend – Zechstein Basin, SW Poland. *In*: SCHWARZ, E., GEORGIEFF, S., PIOVANO, E. & ARIZTEGUI, D. (eds) *Abstracts Volume, 18th International Sedimentological Congress*, Mendoza, Argentina, 944.

KIERSNOWSKI, H., PAUL, J., PERYT, T. M. & SMITH, D. B. 1995. Facies, paleogeography, and sedimentary history of the Southern Permian Basin in Europe. *In*: SCHOLLE, P. A., PERYT, T. M. & ULMER-SCHOLLE, D. S. (eds) *The Permian of the Northern Pangea, vol. 2: Sedimentary Basins and Economic Resources*. Springer-Verlag, Berlin, 119–136.

KIERSNOWSKI, H., BUNIAK, A., KUBERSKA, M. & SROKOWSKA-OKOŃSKA, A. 2010*a*. Tight gas accumulations in Rotliegend sandstones of Poland. *Przegląd Geologiczny*, **58**, 335–346 (In Polish with English summary).

KIERSNOWSKI, H., PERYT, T. M., BUNIAK, A. & MIKOŁAJEWSKI, Z. 2010*b*. From the intra-desert ridges to the marine carbonate island chain: middle to late Permian (Upper Rotliegend – Lower Zechstein) of the Wolsztyn – Pogorzela high, west Poland. *In*: SOMERVILLE, I. D., HENDERSON, Ch. M. & SHEN, S. (eds) *Special Issue: Lopingian (Late Permian) Stratigraphy of the World, Major Events and Environmental Change. Geological Journal*, **45**, 319–335.

KLEDITZSCH, O. & KURZE, M. 1993. Ergebnisse petrographischer Untersuchungen an Sandsteinen des tieferen Oberrotliegenden im Raum Altmark/Westmecklenburg. *Geologisches Jahrbuch*, **A131**, 141–178.

KOCUREK, G. A. 1988. First-order and super bounding surfaces in eolian sequences – Bounding surfaces revisited. *Sedimentary Geology*, **56**, 193–206.

KOCUREK, G. A. & HAVHOLM, K. 1993. Eolian event stratigraphy – a conceptual framework. *In*: WEIMER, P. & POSAMENTIER, H. (eds) *Siliclastic Sequence Stratigraphy: Recent Advances and Applications*. AAPG, Tulsa, Memoir, **58**, 393–409.

KRONER, U., MANSY, J-L. ET AL. 2008. Variscan tectonics. *In*: MCCANN, T. (ed.) *The Geology of Central Europe. Vol 1: Precambrian and Paleozoic*. Geological Society, London, 599–664.

KUTZBACH, J. E. & ZIEGLER, A. M. 1994. Simulation of Late Permian climate and biomes with an atmosphere-ocean model: comparisons and observations. *In*: ALLEN, J. R. L., HOSKINS, B. J., SELLWOOD, B. W., SPICER, R. A. & VALDES, P. J. (eds) *Palaeoclimates and their Modelling; with Special Reference to the Mesozoic era*. Chapman & Hall, London, 119–132.

LAMING, D. J. C. 1966. Imbrications, palaeocurrents and other sedimentary features in the Lower New Red Sandstone, Devonshire, England. *Journal of Sedimentary Petrology*, **36**, 940–959.

LANGFORD, R. P., PEARSON, K. M., DUNCAN, K. A., TATUM, D. M., ADAMS, L. & DEPRET, P-A. 2008. Aeolian topography as a control on deposition incorporating lessons from Modern Dune Seas: Permian Cedar Mesa Sandstone, SE Utah, U.S.A. *Journal of Sedimentary Research*, **78**, 410–422.

LEGLER, B. 2006. Faziesentwicklung im Südlichen Permbecken in Abhängigkeit von Tektonik, eustatischen Meeresspiegelschwankungen des Proto-Atlantiks und Klimavariabilität (Oberrotliegend, Nordwesteuropa). *Schriftenreihe der Deutschen Gesellschaft für Geowissenschaften*, **47**, 1–108.

LEGLER, B., ROSCHER, M. & SCHNEIDER, J. W. 2004. Das Bebergraben-Wadi – sedimenttransport in das Norddeutsche Permbecken. *In*: KUKLA, P., LITTKE, R., STOLLHOFEN, H. & SCHWARZER, D. (eds) *Sediment 2004, Kurzfassungen und Exkursionsführer*. Schriftenreihe DGG, Hannover, **33**, 91.

LEGLER, B., SCHNEIDER, J. W., GEBHARDT, U., MERTEN, D. & GAUPP, R. 2010. Lake deposits of moderate salinity as sensitive indicators of lake level fluctuations: example from the Upper Rotliegend saline lake (Middle–Late Permian, Northeast Germany). *Sedimentary Geology*, **234**, 56–69.

LINDERT, W., WEGNER, H.-U., ZAGORA, I. & ZAGORA, K. 1993. Ein neuer Perm-Aufschluß im Seegebiet östlich von Rügen. *Geologisches Jahrbuch*, **A131**, 351–360.

LIVERMORE, R. A., SMITH, A. G. & VINE, F. J. 1986. Late Paleozoic to early Mesozoic evolution of Pangea. *Nature*, **322**, 162–165.

LOOPE, D. B. 1985. Episodic deposition and preservation of eolian sands – a Late Paleozoic example from southeastern Utah. *Geology*, **13**, 73–76.

MAYNARD, J. M. & GIBSON, J. P. 2001. Potential for subtle traps in the Permian Rotliegend of the UK Southern North Sea. *Petroleum Geoscience*, **7**, 301–314.

MCCANN, T. 1998. Sandstone composition and provenance of the Rotliegend of the NE German Basin. *Sedimentary Geology*, **116**, 177–198.

MCCANN, T., KIERSNOWSKI, H. ET AL. 2008. Permian. *In*: MCCANN, T. (ed.) *The Geology of Central Europe*.

Vol 1: Precambrian and Paleozoic. Geological Society, London, 531–597.

McKie, T. 2011. A comparison of modern dryland depositional systems with the Rotliegend Group in the Netherlands. *In*: Gaupp, R. & Grötsch, J. (eds) *The Permian Rotliegend of the Netherlands*. SEPM, Tulsa, Special Publications, **98**, 89–103.

Menning, M., Alekseev, A. S. *et al.* 2006. Global time scale and regional stratigraphic reference scales of Central and West Europe, East Europe, Tethys, South China, and North America as used in the Devonian-Carboniferous-Permian Correlation Chart 2003 (DCP 2003). *Palaeogeography, Palaeoclimatology, Palaeoecology*, **240**, 318–372.

Mijnlieff, H. F. & Geluk, M. 2011. Palaeotopography-governed sediment distribution – a new predictive model for the Permian Upper Rotliegend in the Dutch sector of the Southern Permian Basin. *In*: Gaupp, R. & Grötsch, J. (eds) *The Permian Rotliegend of the Netherlands*. SEPM, Tulsa, Special Publications, **98**, 147–159.

Moscariello, A. 2011. Sedimentary facies, correlation, and architecture of Rotliegend reservoirs at the Southern Permian Basin margin: the P01-fa case study and the challenged myth of layer-cake stratigraphy. *In*: Gaupp, R. & Grötsch, J. (eds) *The Permian Rotliegend of the Netherlands*. SEPM, Tulsa, Special Publications, **98**, 177–190.

Newell, A. J. 2001. Bounding surfaces in a mixed aeolian-fluvial system (Rotliegend, Wessex Basin, SW UK). *Marine and Petroleum Geology*, **18**, 339–347.

Ogg, J. G., Ogg, G. & Gradstein, F. M. 2008. Permian Period. *In*: Ogg, J. G., Ogg, G. & Gradstein, F. M. (eds) *The Concise Geologic Time Scale*. Cambridge University Press, Cambridge, 85–93.

Parrish, J. T. 1993. Climate of the supercontinent Pangea. *The Journal of Geology*, **101**, 215–233.

Parrish, J. T. & Peterson, F. 1988. Wind directions predicted from global circulation models and wind directions determined from eolian sandstones of the western United States – a comparison. *Sedimentary Geology*, **56**, 261–282.

Paul, J. 1997. Rotliegend and Zechstein in Central Germany. Excursion Guidebook. University of Göttingen. Unpublished.

Paul, J. 1999. Evolution of a Permo-Carboniferous Basin: the Ilfeld Basin and its relationship to adjoining Permo-Carboniferous structures in Central Germany. *Neues Jahrbuch für Geologie und Paläontologie Abh., Stuttgart*, **214**, 211–236.

Perlmutter, M. A. & Matthews, M. D. 1989. Global cyclostratigraphy – a model. *In*: Cross, T. A. (ed.) *Quantitative Dynamic Stratigraphy*. Prentice Hall, Englewood Cliffs, New Jersey, 233–260.

Perlmutter, M. A. & Matthews, M. D. 1994. Global cyclostratigraphy: predicting stratigraphic response to Milankovitch Cycle. *In*: Cecil, C. B. & Edgar, N. T. (eds.) *Predictive stratigraphic analysis–concept and applications*. USGS Survey Bulletin, 2110, 6–7.

Plein, E. 1993. Bemerkungen zum Ablauf der paläogeographischen Entwicklung im Stefan und Rotliegend des Norddeutschen Beckens. *Geologisches Jahrbuch*, **A131**, 99–116.

Pochat, S., Van Den Driessche, J., Mouton, V. & Guillocheau, F. 2005. Identification of Permian palaeowind direction from wave-dominated lacustrine sediments (Lodève Basin, France). *Sedimentology*, **52**, 809–825.

Pokorski, J. 1989. Evolution of the Rotliegendes Basin in Poland. *Bulletin of the Polish Academy of Sciences, Earth Sciences*, **37**, 49–55.

Rees, P. M., Gibbs, M. T., Zigler, A. M., Kutzbach, J. E. & Behling, P. J. 1999. Permian climates: Evaulating model predictions using global paleobotanical data. *Geology*, **27**, 891–894.

Retallack, G. J., Metzger, C. A., Greaver, T., Jahren, A. H., Smith, R. M. H. & Sheldon, N. D. 2006. Middle-late Permian mass extinction on land. *GSA Bulletin*, **118**, 1398–1411.

Roscher, M. 2009. *Environmental reconstruction of the Late Palaeozoic*. Numeric modelling and geological evidences. PhD dissertation (unpublished), Freiberg.

Roscher, M. & Schneider, J. W. 2006. Permo-Carboniferous climate: Early Pennsylvanian to Late Permian climate development of central Europe in a regional and global context. *In*: Cassini, G. & Schneider, J. W. (eds) *Nonmarine Permian Biostratigraphy and Biochronology*. Geological Society, London, Special Publications, **265**, 95–136.

Roscher, M., Berner, U. & Schneider, J. W. 2008. A tool for the assessment of the paleo-distribution of source and reservoir rocks. *OIL GAS European Magazine*, **3**, 131–137.

Schneider, J. W. & Gebhardt, U. 1993. Litho- und Biofaziesmuster in intra- und extramontanen Senken des Rotliegend (Perm, Nord- und Ostdeutschland). *Geologisches Jahrbuch*, **A131**, 57–98.

Schneider, J. W., Schretzenmayr, S. T. & Gaitzsch, B. G. 1998. Rotliegend reservoirs at the margin of the Southern Permian Basin. Field trip F2; Field Trips to the EAGE Workshop Leipzig 1998. *Leipziger Geowisenschaften*, **7**, 15–44.

Schneider, J. W., Körner, F., Roscher, M. & Kroner, U. 2006. Permian climate development in the northern peri-Tethys area – the Lodève basin, French Massif Central, compared in a European and global context. *Palaeogeography, Palaeoclimatology, Palaeoecology*, **240**, 161–183.

Schröder, L., Plein, E. *et al.* 1995. Stratigraphische Neugliederung des Rotliegend im Norddeutschen Becken. *Geologisches Jahrbuch*, **148**, 3–21.

Şengör, A. M. C. & Atayman, S. (eds) 2009. *The Permian Extinction and the Tethys: An Exercise in Global Geology*. Geological Society of America, Boulder, Special Papers, **448**.

Sneh, A. 1988. Permian dune patterns in northwestern Europe challenged. *Journal of Sedimentary Petrology*, **58**, 44–51.

Słowakiewicz, M., Kiersnowski, H. & Wagner, R. 2007. Correlation of Permian marine and terrestrial sedimentary sequences within Polish and North American Basins with an attempt on paleoclimatic cyclicity and evidence in global 13C isotope change. Abstracts of the 16th International Congress on Carboniferous and Permian. *Journal of Stratigraphy*, **31** (Suppl.), 225–226.

Słowakiewicz, M., Kiersnowski, H. & Wagner, R. 2009. Correlation of the Middle and Upper Permian marine and terrestrial sedimentary sequences in Polish, German, and USA Western Interior Basins with reference to global time markers. *Palaeoworld*, **18**, 193–211.

Steele, R. P. 1983. Longitudinal draa in the Permian Yellow sands of North-East England. *In*: Brookfield, M. E. & Ahlbrandt, T. S. (eds) *Eolian Sediments and Processes. Developments in Sedimentology*, Elsevier, Amsterdam, **38**, 543–550.

Strömbäck, A. C. & Howell, J. A. 2002. Predicting distribution of remobilized aeolian facies using sub-surface data: the Weissliegend of the UK Southern North Sea. *Petroleum Geoscience*, **8**, 237–249.

Sweet, M. L. 1999. Interaction between aeolian, fluvial and playa environments in the Permian Upper Rotliegend Group, UK southern North Sea. *Sedimentology*, **46**, 171–187.

Turner, B. R. & Smith, D. B. 1997. A playa deposits of pre-Yellow Sands age (upper Rotliegend/Weissliegend) in the Permian of northeast England. *Sedimentary Geology*, **114**, 305–319.

Van Ojik, K., Böhm, A. R. *et al.* 2011. The rationale for an integrated stratigraphic framework of the Upper Rotliegend II depositional system in the Netherlands. *In*: Gaupp, R. & Grötsch, J. (eds) *The Permian Rotliegend of the Netherlands*. SEPM, Tulsa, Special Publications, **98**, 37–48.

Van Wijhe, D. H., Lutz, M. & Kaasschieter, J. P. H. 1980. The Rotliegend in the Netherlands and its gas accumulations. *Geologie en Mijnbouw*, **59**, 3–24.

Verdier, J. P. 1996. The Rotliegend sedimentation history of the Southern North Sea and adjacent countries. *In*: Rondeel, H. E., Batjes, D. A. J. & Nieuwenhuijs, W. H. (eds) *Geology of Gas and Oil under the Netherlands*. The Royal Geological and Mining Society of the Netherlands – KNGMG, Haarlem, 45–56.

Vozarova, A. & Vozar, J. 1988. *Late Paleozoic in West Carpathians*. Geologicky Ustav Dionyza Stura, Bratislava.

Yang, C. S. & Baumfalk, Y. A. 1994. Milankovitch cyclicity in the Upper Rotliegend Group of the Netherlands offshore. *In*: De Boer, P. L. & Smith, D. G. (eds) *Orbital Forcing and Cyclic Sequences*. IAS, Special Publications, Blackwell Science, Oxford, **19**, 47–61.

Yang, C. S. & Baumfalk, Y. A. 1997. Application of high-frequency cycle analysis in high-resolution sequence stratigraphy. *In*: Ziegler, K., Turner, P. & Daines, S. R. (eds) *Petroleum Geology of the Southern North Sea: Future Potential*. Geological Society, London, Special Publications, **123**, 181–203.

Yang, C. S. & Nio, S. D. 1994. Applications of high-resolution sequence stratigraphy to the Upper Rotliegend in the Netherlands offshore. *In*: Weimer, P. & Posamentier, H. (eds) *Siliclastic Sequence Stratigraphy: Recent Advances and Applications*. AAPG, Tulsa, Memoir, **58**, 285–316.

Ziegler, P. A. & Dèzes, P. 2006. Crustal Evolution of Western and Central Europe. *In*: Gee, D. G. & Stephenson, R. A. (eds) *European Lithosphere Dynamic*. Geological Society, London, Memoir, **32**, 43–56.

Palaeogeographic and palaeoclimate factors of salinity fluctuations in the eastern part of the Late Permian (Zechstein) European Basin: case study from the salt basin in Poland

GRZEGORZ CZAPOWSKI* & HANNA TOMASSI-MORAWIEC

Polish Geological Institute – National Research Institute, Rakowiecka 4, 00-975 Warsaw, Poland

**Corresponding author (e-mail: grzegorz.czapowski@pgi.gov.pl)*

Abstract: A study of the bromine content and facies in the salt successions (representing PZ1, PZ2 and PZ3 cycles) from the Late Permian (Zechstein) salt basin in Poland evidences fluctuations in mother brine concentrations. The study indicates that the local (better recorded in facies of the shallow basin and its margins) and regional changes were the result of Late Permian climate variations and eustatic sea-level rises and drops.

The observed trend for a lowering of brine concentration is attributed to more humid climate periods accompanied by frequent rainy seasons. Such a change from arid to more humid climatic conditions was observed in all three studied cycles, proving the variations of a dominant warm and dry climate, assumed for the inner part of the Pangea supercontinent in the Late Permian. The coastal zone of the Pangea supercontinent was more affected by the atmospheric perturbations, which were more frequent at the boundary between the continent and the ocean. The intracontinent areas were characterized with a more stable climate, thus interpreted by the authors, that the climate changes in the successions coming from the inner part of such intracontinental basin, far from the continent coasts (as the Polish part of the European Permian Basin) as the evidence for climate instability in the whole Zechstein period. The authors recorded the climate fluctuations in an area, which should be characterized with the most constant climate, particularly during the evaporite deposition. The prolonged more rainy periods that succeeded the warm and arid periods were recorded in the central part of the continent.

Eustatic sea-level fluctuations, responsible for large changes in brine concentrations, became independent from the climatic variations. Sea-level rises initiated huge marine water input and caused extreme refreshing of the bottom salt brine in deep-water basin parts. Two such events were registered in salts of the PZ1 cycle, and a further three in the evaporite succession of the PZ3 cycle. A single rapid drop in sea-level, resulting from a postulated tectonic close of the narrow North Atlantic seaway that conducted ocean water to the evaporite basin, was detected in the lower part of the PZ2 salt succession. This caused the evaporite basin cutoff, its successive desiccation and, consequently, the production of potash-bearing salts.

Evaporites, especially salts, are commonly considered to be an indicator of arid and warm climate conditions (e.g. Melvin 1991; Kendall 1992; Kendall & Harwood 1996; Warren 2006). Recently, they develop predominantly at equatorial to sub-equatorial altitudes, within coastal lagoons and pans delivered with marine waters (e.g. Shearman 1970; Orti Cabo *et al.* 1984; Logan 1987; Handford 1991) or within intracontinental depressions, such as salars in the intra-Andean plateau (Risacher *et al.* 2003) and inland salt lakes like the Great Salt Lake (e.g. Gwynn 1980; Handford 1982; Zhang 1987; Last 1989; Kilic & Kilic 2010), supplied directly with brines from dissolved older evaporites or Na–K–Mg enriched due to weathering of magmatic and volcanic rocks enriched in these cations. In coastal marine zones, such evaporite environments are subject to seasonal rainfalls during more humid seasons (e.g. Shearman 1970; Perthuisot 1975, 1976; Orti Cabo *et al.* 1984; Logan 1987), but in the distal inland basins wet periods are less frequent and are connected with specific atmospheric pressure perturbations (far inland migration of low-pressure fronts).

Salt series, generated in subsiding salt basins like the Central European Basin (Warren 2008), which have a constant supply of marine waters for evaporation, provided almost continuous salt precipitation until the total basin drawdown or a new marine transgression. Most salt rocks are devoid of any fossils, so their stratigraphic position is determined from data from the non-evaporitic deposits that lie below or above, or as interbeds. Salts precipitate as the last phase in the evaporation process, which begins with carbonates, then sulphates and, finally, salts. Repetition of this process creates several salt series that can be difficult to distinguish in halokinetic, folded salt structures. Fluctuations in basin brine concentration during chloride accumulation may be registered from the bromine content in salt

From: Gąsiewicz, A. & Słowakiewicz, M. (eds) 2013. *Palaeozoic Climate Cycles: Their Evolutionary and Sedimentological Impact*. Geological Society, London, Special Publications, **376**, 457–474.
First published online September 19, 2013, updated October 9, 2013, http://dx.doi.org/10.1144/SP376.16

units. Such variations are dependent on local (e.g. a subenvironment type, a storm surge, rainfall) and regional (e.g. global sea-level fluctuations, climate change, tectonic activity) factors, as is the case for siliciclastic and carbonate accumulation (e.g. Haq 1995). The character of fluctuations in basin brine concentrations may differ for each salt period, each of which may exhibit a characteristic pattern of bromine content distribution (e.g. Tomassi-Morawiec *et al.* 2009). This pattern and the limits of bromine content may also be used as an attribute for subdividing and correlating individual salt units, especially within highly tectonized (e.g. salt diapirs) and macroscopically similar evaporite complexes (Schramm *et al.* 2002; Tomassi-Morawiec & Czapowski 2006; Bornemann *et al.* 2008; Tomassi-Morawiec *et al.* 2009).

This paper interprets the observed fluctuations of Late Permian chloride brine concentrations in the Polish part of the evaporite European Permian Basin (EPB) and the effect of basin palaeogeography and possible regional to global seasonal climatic changes.

Upper Permian (Zechstein) evaporites

The Upper Permian (Zechstein) evaporites accumulated in the epicontinental sea where it advanced along the EPB onto the northern Pangea supercontinent (Fig. 1a, b). This sea contacted with the Panthalassa Superocean through the narrow North Atlantic seaway and the Boreal Sea (e.g. Ziegler 1990; Kiersnowski *et al.* 1995). In the eastern, distal and more isolated Polish part of the EPB (the Polish Permian Basin, PPB; Fig. 1b), both evaporitic salts and sulphates with a total thickness of over 1.8 km have been deposited during the evaporite PZ1 to PZ4 cycles of the Zechstein Group (Wagner 1987, 1994; Wagner & Peryt 1997, Peryt *et al.* 2010). The lower three cycles (PZ1 to PZ3) correspond with those (Table 1) distinguished within the classic Zechstein Group as defined in Germany (Richter-Bernburg 1955). The extreme distal location of the Polish part of the EPB allowed a more precise stratigraphic record for extra-regional phenomena such as climatic or eustatic sea-level fluctuations than the more proximal and deeper portions of the EPB.

Evaporites in the PZ1–PZ3 cycles (Table 1), with a total thickness of 1592 m, comprise 69.3% of the entire Zechstein stratigraphy in the Polish part of the EPB (Czapowski 2007). The evaporites were deposited during a relatively short time interval, estimated to range from >18 000 to >64 000 years depending on the assumed maximum or minimum precipitation rates (Czapowski 2007). These results are comparable with the calculated duration of 23 000 to 92 500 years for the Upper Jurassic evaporitic complex (785–1200 m thick) from the East Turkmenistan Basin (Baikov & Sedletskii 2001). The deposition time, estimated for the Zechstein evaporites, was determined to range from 0.28% (maximum precipitation rate) to 3.2% (minimum rate) of the calculated duration of Zechstein sedimentation (5–7 Ma, after Menning 1995; 6 Ma, after Słowakiewicz *et al.* 2009). In Poland, the Zechstein time period was dominated by deposition of non-evaporites (clastics, carbonates), non-deposition and erosion of just-accumulated sediments. Consequently, the observed salinity fluctuations in the short-lasting salt basins of successive Zechstein cycles seem to have occurred quite frequently, although their sedimentary control differed (e.g. climatic, eustatic), and were related to basin palaeobathymetry.

The last cycle in the PPB – PZ4 – which is the equivalent of cycles Z4 to Z8 in Germany (Wagner 1994), is characterized by frequent and voluminous accumulation of clayey salts (zuber units), limited sulphate deposition events, and transitions from desiccated marine to continental environments (Czapowski 1988, 1990; Wagner & Peryt 1997; Tomassi-Morawiec & Czapowski 2006). Because the PZ4 succession differed significantly from cycles Z4 to Z8 in Germany and represents singular development of the Polish part of the EPB at the end of the Permian, these strata were not considered in this study.

Evaporites of the first (PZ1) cycle (Table 1) include the Lower (A1d) and Upper (A1g) Anhydrite anhydrites (total thickness of *c.* 40 m); these strata bound the bottom and top of the Oldest Halite (Na1) salt unit, which is up to 300 m thick (Wagner 1994). Potash salts of this cycle (up to 40 m thick) occurred only locally on the northern margin of the basin (Czapowski *et al.* 1993) and became undistinguished as an individual lithostratigraphic unit (Wagner 1994). Salts of the second cycle (PZ2), with thickness up to 410 m (Wagner 1994), include the dominant Older Halite (Na2) unit and the Older Potash (K2) seam (12–30 m thick) (Czapowski & Bukowski 2010). Underlying (Basal Anhydrite; A2) and overlying (Screening Anhydrite; A2r) sulphate units are thin (up to 5 m). Salts (Younger Halite Na3) of the third cycle (PZ3) are up to 370 m thick, with the seam of Younger Potash (K3) unit being up to 35 m thick. The underlying Main Anhydrite (A3) unit is up to 40 m thick. The thickness and facies characteristics of each evaporite unit were controlled by the evaporite basin palaeogeography and the basin subsidence rate (e.g. Wagner *et al.* 1981; Czapowski *et al.* 1991, 1993; Wagner 1994; Czapowski 1995*a*, *b*; Kiersnowski *et al.* 1995; Słowakiewicz *et al.* 2009; Peryt *et al.* 2010).

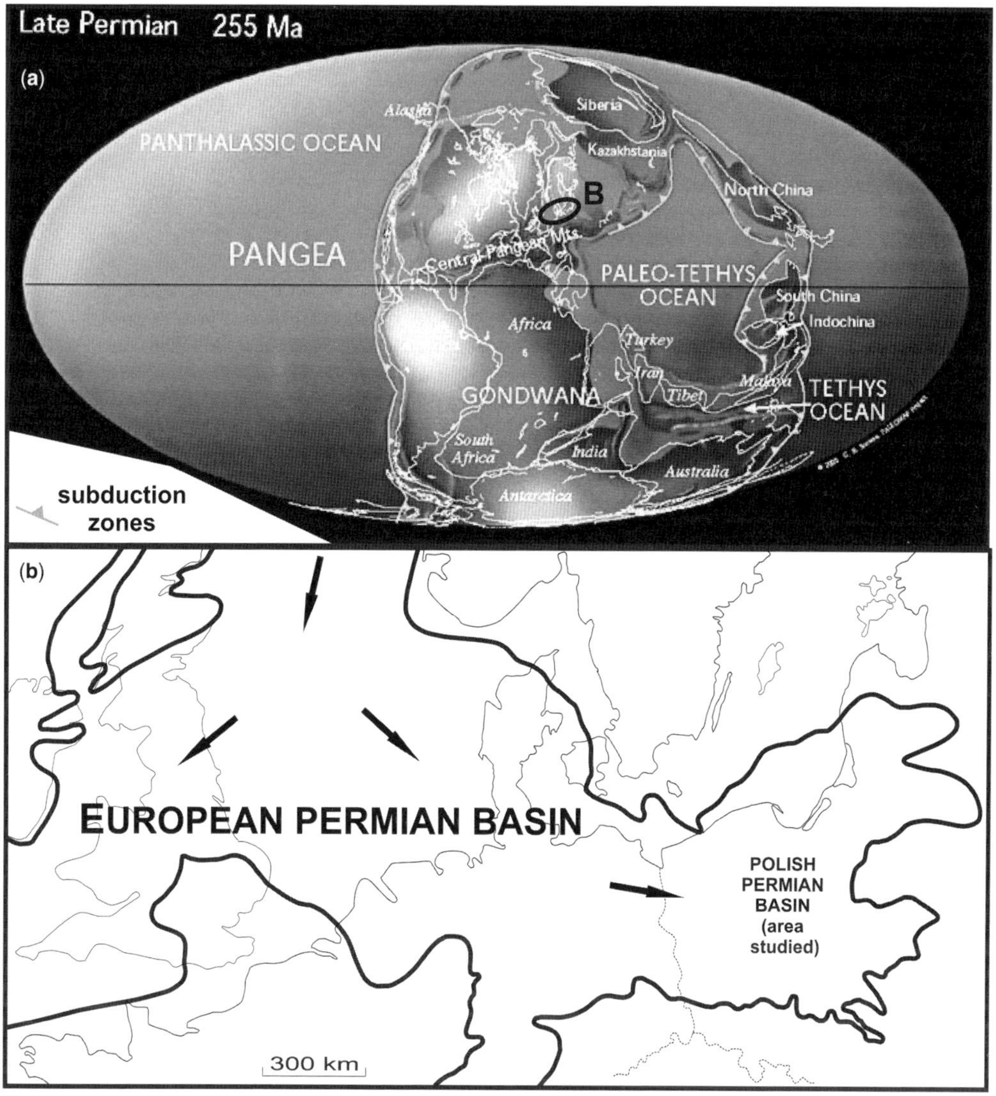

Fig. 1. Late Permian palaeogeography (**a**; after Scotese: www.scotese.com/newpage5.htm) with position of the European Permian Basins (**b**) and the studied Polish Permian Basin within (basin extent of the PZ2 cycle after Słowakiewicz & Mikołajewski 2011). Arrows indicate the fresh marine-water inflow from the ocean.

Salinity indicators in the salt succession

The unquestioned lithological indicators of major salinity variations within the EPB evaporites are sulphate-dominated interbeds within the halite succession. These sulphate interbeds register intensive salt brine dilution, which resulted from a change in precipitation regime from halite to sulphate phase. Thin sulphates within the dominant halite profiles may reflect the refreshing of halite brine by seawater floods of extreme storm surges. Changes referred to thicker sulphate interbeds could result from long-term temperature and humidity fluctuations (varied relation of rainy to dry periods – climatic factor) or from eustatic sea-level rise. Thus, they could also imply that the climate during salt deposition was not stable. However, in relation to thicker pure halite sequences this approach is groundless, and for salt profiles such fluctuations in brine concentration could essentially be reconstructed from geochemical data – mainly bromine content (e.g. Baar 1963; Siemann 2003;

Table 1. *Estimated thickness of evaporite units of PZ1 to PZ3 cycles in the central part of the Polish part of the European Permian Basin (after Wagner 1994; Menning et al. 2006; Czapowski 2007)*

Global stages (Menning *et al.* 2006)	Regional European stages (Menning *et al.* 2006)		Polish cycles (German cycle subdivision)	Lithostratigraphic unit (Wagner 1994)	Thickness (m) (Czapowski 2007)
	Central and West Europe	East Europe			
Changhsingian	Thuringian	Tatarian	PZ3 (Z3)	A3r (local on basin margin)	0.5–11
				Na3 + K3 + Na3t	370
				A3	40
Wuchiapingian			PZ2 (Z2)	A2r	1
				Na2 + K2 + Na2r	410
				A2	4
			PZ1 (Z1)	A1g	7
				Na1 + K1 (local in NW basin margin)	300
				A1d	30

Bornemann *et al.* 2008). This geochemical tool is adopted here to indicate climatic instability during Zechstein salt deposition.

During the evaporation of marine water, bromine is incorporated into the lattice of chloride minerals (e.g. halite), replacing a chlorine ion. Its content increases both in minerals and brines as evaporation progresses, but it mostly remains in liquid. Each salt mineral has its own distribution coefficient for bromine content within solid and liquid phases. For halite, the average value of this ratio is 0.14 ± 0.02 (Herrmann *et al.* 1973; Herrmann 1980; Herrmann & Schneider 1980). The calculated bromine content in the initially precipitated present-day primary halite is 68–75 ppm (Walaszko 1956; Holser 1966), but in initial first potash mineral precipitation the crystallizing halite contains *c.* 270 ppm bromine (Walaszko 1956).

As a result, the bromine content in the Upper Permian (Zechstein) salt sections of the PPB enables reconstruction of the salinity history of the evaporite basin (e.g. Tomassi-Morawiec & Czapowski 2007, 2010; Bornemann *et al.* 2008). The detection and correlation of salinity fluctuations in various parts of Zechstein evaporite basins allows us to define regional (related to the whole evaporite basin area) changes in mother brine concentration (Tomassi-Morawiec & Czapowski 2010). Furthermore, variations in local salinity, reflected by the environmental variations in evaporite subenvironments such as shallow salt lagoons, salinas and pans (more or less isolated from the open basin), have been noted (Tomassi-Morawiec & Czapowski 2010). Salt series, generated in conditions of a deep-water subsiding salt basin, with a constant supply of marine water, provided almost continuous salt precipitation until total basin drawdown or a new marine transgression. Such deep-water evaporite basins existed during the Late Permian in the EPB area, and the estimated initial depth of the halite-dominated basin of the PZ1 cycle in the PPB was 60–110 m (Czapowski 2007). The water depths of the Permian evaporite basins in West Texas and New Mexico reached hundreds of metres (Hovorka *et al.* 2007).

A very low bromine content ($<$20 ppm) is characteristic of secondary halite (Holser 1970, 1979; Wilgus & Holser 1984), the product of primary marine salt recycling or salts from continental evaporite pans with brines depleted in bromine (Handford 1982; Hardie *et al.* 1978, 1983; Hardie 1984; Pueyo 1991). Such salt recycling commonly results from the dissolution of accumulated chlorides by seasonal or accidental rainfalls and river floods or by fresh marine-water inflows to the evaporite lagoons during storm surges (Handford 1982; Hardie *et al.* 1978, 1983; Hardie 1984; Pueyo 1991).

Registered distinct fluctuations in bromine content in studied profiles of the same salt unit from the PPB were sometimes accompanied with the production of sulphates instead of chlorides (e.g. Czapowski 1987; Tomassi-Morawiec 1990; Czapowski *et al.* 1991, 1993; Czapowski 1993, 1995*b*; Tomassi-Morawiec 2003). This may reflect changes in basin salinity of (i) a regional character (if they can be correlated over most of the evaporite basin and are considered almost isochronous) or (ii) a local character (Tomassi-Morawiec 1990, 2003; Tomassi-Morawiec & Czapowski 2010). The changes, initiating prolonged tendencies towards a rise or drop in salinity, could be considered as regional events, such as (i) a changed fresh marine-water input from the ocean to the evaporated intracontinental sea through the closing seaway or (ii) a climatic change from arid to more humid conditions.

Sampling and analytical methods for bromine in salts

Two types of salt samples were taken to establish bromine content: a channel sample (a pile of salt 1–1.5 m long, cut from the core, crushed and averaged to the standard weight of an analytic sample) and a point sample (taken from a precise core depth). The analytical results were therefore attributed to the depth interval in the first case and to the defined depth of well or profile length in the second. The type of sampling depended on the analytical purpose: a channel sample was commonly taken to determine the resource characteristics of the salt, and a point sample to define the content of selected ions.

Bromine concentrations were determined in the Central Chemical Laboratory of the Polish Geological Institute. Furrow samples representing the Oldest Halite (Na1) unit from the Łeba Elevation area (northern Poland) were analysed (in the 1960s) using a modified Van Meulen jodometric method (Van der Meulen 1931; D'Ans & Höfer 1934). Analyses of the point samples (rock salts of the Na1, Na2 and Na3 units in other cores drilled in SW Poland) were performed using the X-Ray Fluorescence method (Rigaku Denki Co. Ltd, 1980s; PW 2400 Philips, 2000s). After grinding the sample, the halite crystals were selected for analyses from admixtures of sulphates and clay minerals.

In 2005–2007 the point samples of the rock salt units of the PZ2, PZ3 and PZ4 cycles were taken from mine galleries of the Kłodawa salt mine (central Poland). Bromine concentration was determined by XRF analysis (PW 2400 Philips). The accuracy of the method was *c.* 3%. The determination of major ions enabled the analysed bromine content to be related to the calculated halite content in the sample (Tomassi-Morawiec *et al.* 2009).

The most complete halite profiles (with quite abundant bromine data), located in various parts of the evaporite basin and representing other salt facies successions (Figs 2–4), were selected for salinity analysis. These profiles are devoid of tectonic influence, such as folding and faulting. Salt facies criteria were defined in previous papers (Czapowski 1987, 1995*a*; Czapowski *et al.* 1993; Vovnyuk & Czapowski 2007). The accessible core material in selected wells determined the sampling type (more core, channel samples; less core, point samples) and limited the amount of analytical data obtained. The interpreted bromine profiles were more precise with more abundant data, but general tendencies regarding bromine content were defined when there were fewer data.

Salinity fluctuations in Zechstein rock salts

The analysis of salinity fluctuations in the Zechstein evaporite (chloride phase) basins in Poland, based on selected salt profiles, had to be related to their location in the presumed image of basin palaeogeography (Czapowski 1987, 1993, 1995*a*, *b*; Czapowski *et al.* 1991, 1993). The complexity of the salt profiles (lack of erosional and non-depositional gaps) and the susceptibility of the deposition pattern to normal atmospheric (e.g. rainfalls, storm surges) and occasional regional (e.g. eustatic sea-level changes, tectonic closing of seaway) phenomena depended greatly on the depositional area. In shallow basin margins they were better and more frequently registered, as the thin brine body was more easily diluted by intense atmospheric and fresh marine water input. In the basinal part the change of quite stable density stratification with a chloride bottom brine, providing the almost constant and continued salt accumulation, required the spectacular and/or prolonged processes with regional effect. For the study, the salt sections were selected to represent both basinal and marginal environments.

Oldest Halite (Na1) unit

At the beginning of halite precipitation, which produced the Oldest Halite (Na1) salt unit, most of the Polish Permian Basin was relatively deep-water (>100 m; Czapowski 2007), with shallower margins (Fig. 2; Czapowski 1987, 1995*a*, *b*; Czapowski *et al.* 1993).

The representative halite profiles come from the following:

(1) The northern evaporite basin margin: sections of Czarny Młyn IG-1 and Głuszewo IG-1 wells, dominated by deep to shallow open salt basin facies and finalized with a salt lagoon. Bromine data from channel samples.
(2) The southwestern evaporite basin margin: sections of S-479 and Sieciejów P-5 wells with prevailing salt lagoon facies. Bromine data from point samples.

The lack of accessible sections from the basin centre limited the salinity analysis to the marginal areas.

Section Czarny Młyn IG-1 well (Fig. 2) represents the inner marginal sub-basin, semi-isolated from the central evaporite basin by a sulphate shoal (Czapowski 1987, 2007; Czapowski *et al.* 1993; Tomassi-Morawiec 2003). The large salt

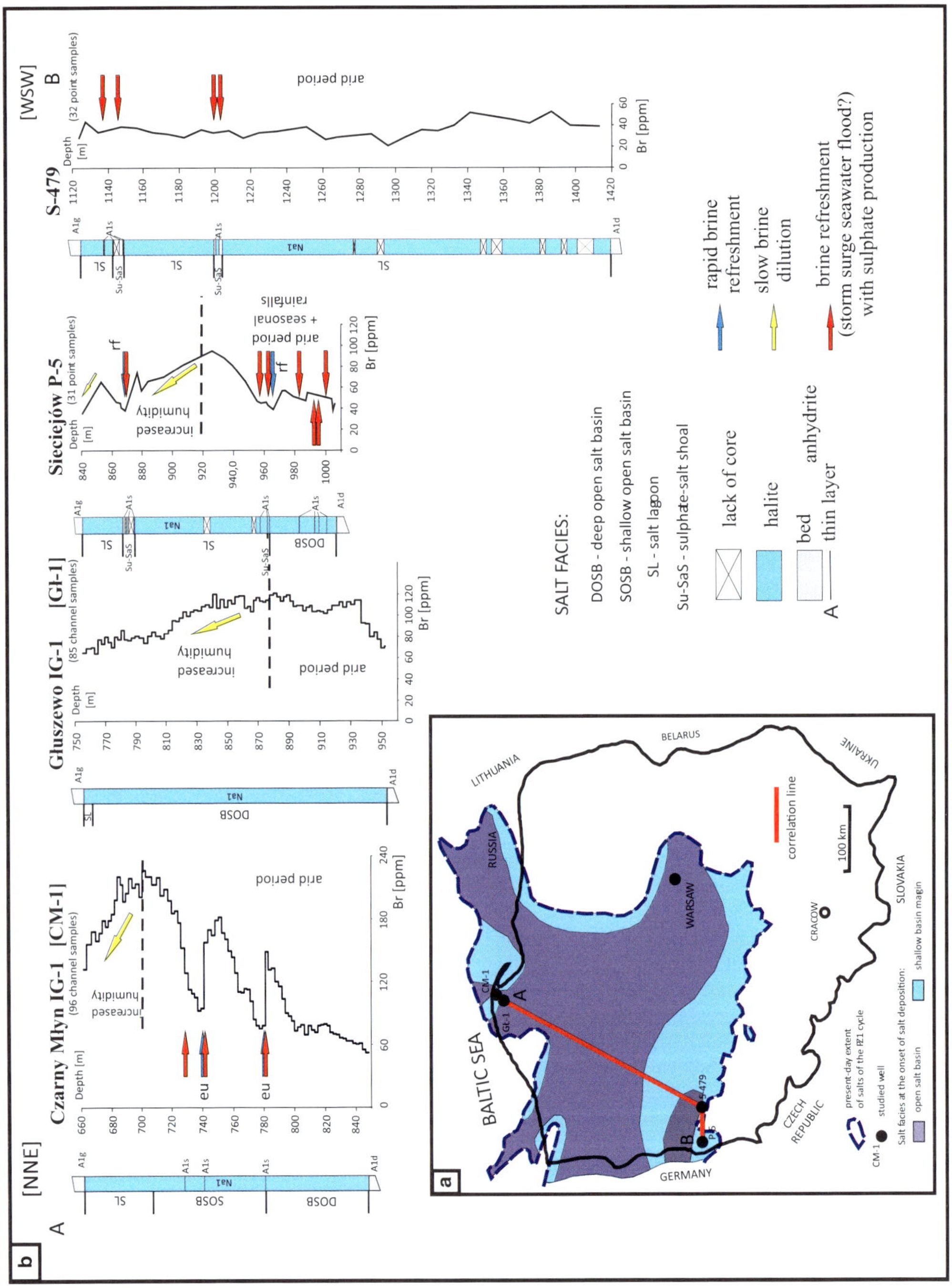
b
A [NNE]
Czarny Młyn IG-1 [CM-1]
(96 channel samples)
Depth [m]
Br [ppm]
increased humidity
arid period
eu
SL
SOSB
DOSB
Na1
A1g
A1s
A1d
Głuszewo IG-1 [Gł-1]
(85 channel samples)
Sieciejów P-5
(31 point samples)
arid period + seasonal rainfalls
rf
Su-SaS
S-479
(32 point samples)
[WSW] B
SALT FACIES:
DOSB - deep open salt basin
SOSB - shallow open salt basin
SL - salt lagoon
Su-SaS - sulphate-salt shoal
lack of core
halite
bed
anhydrite
thin layer
rapid brine refreshment
slow brine dilution
brine refreshment (storm surge seawater flood?) with sulphate production
a
BALTIC SEA
RUSSIA
LITHUANIA
BELARUS
UKRAINE
SLOVAKIA
CZECH REPUBLIC
GERMANY
WARSAW
CRACOW
correlation line
100 km
present-day extent of salts of the PZ1 cycle
studied well
Salt facies at the onset of salt deposition:
open salt basin
shallow basin margin

thicknesses (*c.* 187 m) and increased bromine content (up to 230 ppm) suggest a constant marine-water supply for evaporation, and limited outflow of dense brines. Three thin anhydrite interbeds register basin brine refreshments, and the lower two were accompanied by rapid bromine drops (of *c.* 70–90 ppm). Successive basin infill with salt resulted in basin shallowing (salt lagoon facies). Inputs of freshwater diluted the primary brine, resulting in a decrease in bromine in the upper section.

Section Głuszewo IG-1 well (Fig. 2), over 197 m thick and located in the outer marginal sub-basin (Czapowski 1987, 2007), represents deep open salt basin facies with a successive increase and decrease in brine concentration without evidence of rapid and high fluctuations in salinity. This salinity pattern resulted from the dominant conditions of a relatively deep basin environment, with the bottom dense brine almost untouched by any rapid, short episodes of fresh marine-water inflow expanding in the top part of the water column. These brines became more or less continuously supplied with salt brines, generated in shoals and inner sub-basins. The later basin shallowing due to sediment infill caused the bottom brine to be more subject to dilution by fresh seawater and meteoric water input (leading to the observed decrease in bromine in the upper part of the section).

In the southern basin margin one of the two selected sections (Sieciejów P-5 well; Fig. 2) is comparable with the Głuszewo IG-1 profile, although the bromine content is generally lower (below 100 ppm). Section Sieciejów P-5 well, over 166 m thick and dominated by salt lagoon facies (Czapowski 1995*b*), evidenced two episodes of significant brine refreshment, accompanied by sulphates precipitation and a bromine drop of *c.* 30–40 ppm. Another five fine brine dilutions (a registered single bromine drop of *c.* 10 ppm), producing sulphate interbeds, were noticed in the lower part of the profile. The next section of S-479 well, over 195 m thick, represented only the salt lagoon facies with halites and thin interbeds of sulphate generated from a low concentrated brine (bromine content between 20 and 50 ppm, average 40 ppm). Such constant, low salinity conditions resulted from the profile location in the internal salt lagoonal system, which was highly subject to fresh marine or meteorite water inflows. These inputs caused multiple episodes of primary salt recycling, also producing secondary low-bromine halite as well as four sulphate interbeds in the upper profile.

Older Halite (Na2) unit

Deposition of the Older Halite (Na2) unit across almost the entire Polish Permian Basin took place in the relatively deep evaporite basin, with narrow shallower margins (Fig. 3; Czapowski 1995*a*, *b*; Czapowski *et al.* 1991).

The representative halite profiles with bromine data from point samples come from (Fig. 3) the following:

(1) The basin axis: the profile from the Kłodawa salt mine *c.* 200 m thick, represented both by the halite (Older Halite) unit and the 'transitional beds' (Na2 + K2) to the potash (Older Potash [K2]) salts. This profile is dominated by a deep to shallow open salt basin and is terminated with salt lagoon facies.
(2) The SW basin slope: the section from Gorzów Wielkopolski IG-1 well, *c.* 120 m thick, started with a shallow open salt basin and terminated with a salt lagoon and salina facies.
(3) The SW shallow basin margin: a section from Lubiatów M-20 well, over 102 m thick, with salt lagoon and salina facies.

The bromine distribution in the profile from the Kłodawa salt mine divided the section into two parts:

(1) The lower (*c.* 90 m thick) part, with a slow gradual bromine content increase (up to 90 ppm), reflecting the successive increase in brine concentration due to permanent evaporation.
(2) The upper part, initiated with a rapid increase in salinity (up to 300 ppm), registering the moment of basin isolation from the outer marine-water input. The upper part of this profile represents the relatively high salinity conditions (bromine content varies from 210 to 330 ppm) with four distinct episodes of brine refreshments.

Two other sections from the wells were located on the basin slope and its shallow margin and included 10–14 point sample data. As a result, the bromine profiles (Fig. 3) illustrate only the general tendencies of basin salinity fluctuations. These fluctuations are characterized by a relatively sharp upward increase in bromine content (up to 260–280 ppm) with a single (Gorzów Wielkopolski IG-1 well) and three (Lubiatów M-20 well) episodes of freshwater inflows (for two of them the bromine drop ranges from 60 to 120 ppm), which sporadically (Lubiatów M-20 well) produced

Fig. 2. (**a**) Present-day extent and facies distribution of the Zechstein Oldest Halite (Na1) unit in Poland with (**b**) selected profiles of bromine content and salt facies. Zechstein lithostratigraphic units: A1d, Lower Anhydrite; A1g, Upper Anhydrite; A1s, intrasalt anhydrite bed. eu, brine dilution due to eustatic sea-level rise; rf, rainfall season.

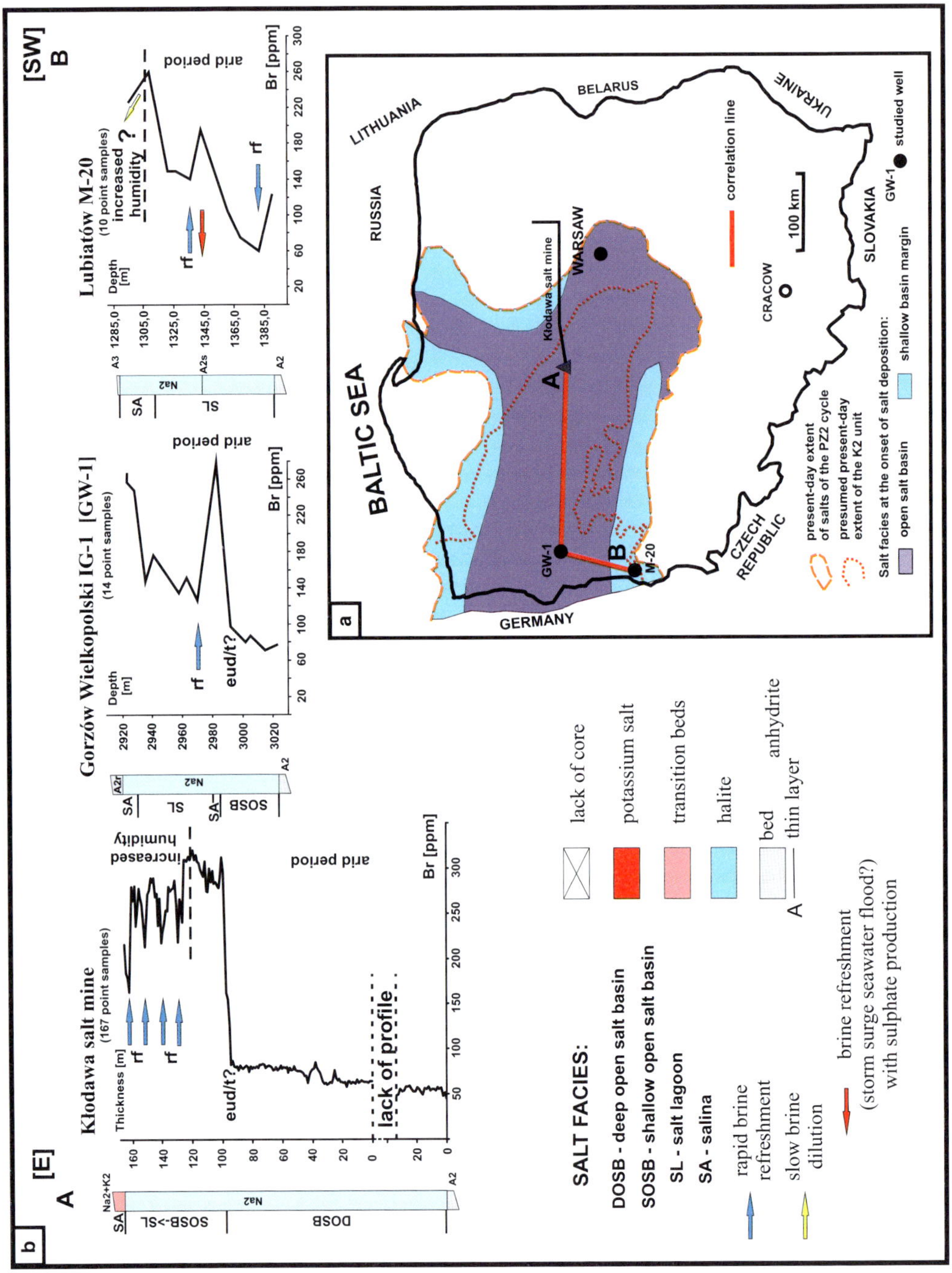
a
BALTIC SEA
LITHUANIA
RUSSIA
BELARUS
UKRAINE
SLOVAKIA
CZECH REPUBLIC
GERMANY
WARSAW
CRACOW
Kłodawa salt mine
GW-1
M-20
A
B
100 km
correlation line
studied well
present-day extent of salts of the PZ2 cycle
presumed present-day extent of the K2 unit
Salt facies at the onset of salt deposition:
open salt basin
shallow basin margin
b
A [E]
[SW] B
Kłodawa salt mine
(167 point samples)
Gorzów Wielkopolski IG-1 [GW-1]
(14 point samples)
Lubiatów M-20
(10 point samples)
Thickness [m]
Depth [m]
Br [ppm]
increased humidity
arid period
lack of profile
eud/t?
rf
SALT FACIES:
DOSB - deep open salt basin
SOSB - shallow open salt basin
SL - salt lagoon
SA - salina
lack of core
potassium salt
transition beds
halite
bed
anhydrite
thin layer
rapid brine refreshment
slow brine dilution
brine refreshment (storm surge seawater flood?) with sulphate production

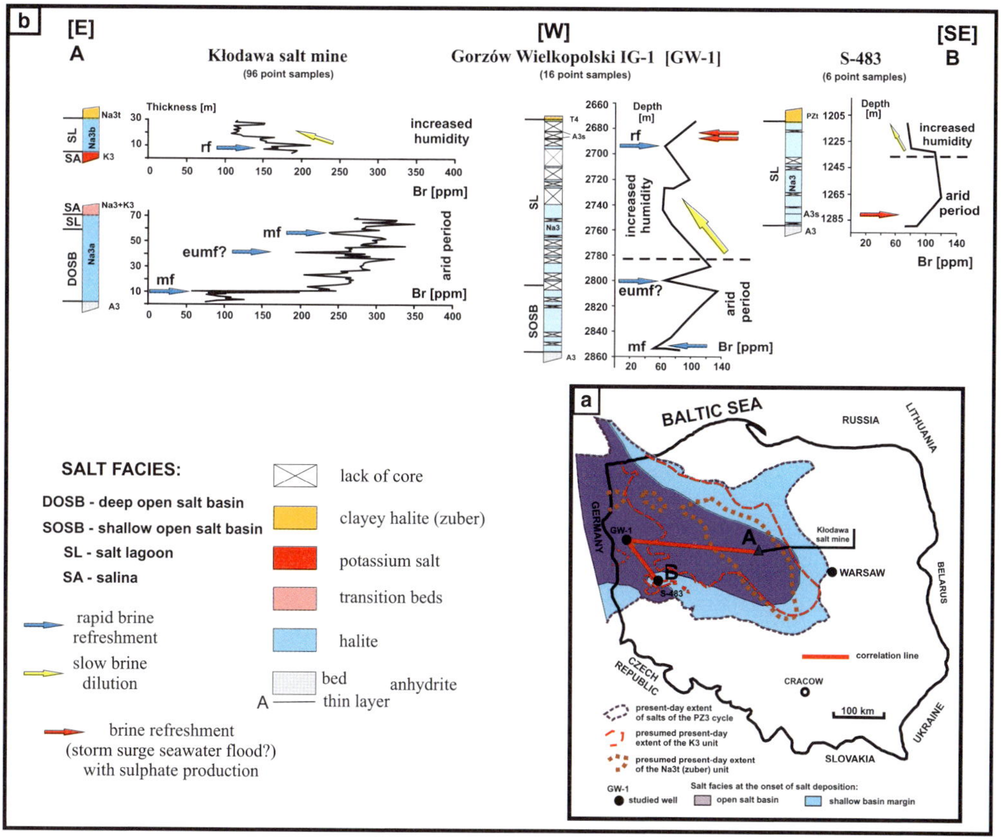

Fig. 4. (**a**) Present-day extent and facies distribution of the Zechstein Younger Halite (Na3: Na3a, lower part; Na3b, upper part), the Younger Potash (K3) and the Brownish Zuber (Na3t) in Poland with (**b**) selected profiles of bromine content and salt facies. Zechstein lithostratigraphic units: A3, Main Anhydrite; Na3 + K3, 'transitional beds'; T4, Red Pelite; A3s, intrasalt anhydrite bed; PZt, Top Terrigenous Series; mf, brine dilution by fresh marine-water inflow; eumf?, brine dilution by fresh marine-water inflow due to supposed eustatic sea-level rise; rf, rainfall season.

sulphates instead of halite as a result of high brine dilution. The general tendency of the bromine distribution in the Gorzów Wielkopolski IG-1 section is similar to that published by García-Veigas *et al.* (2011, their fig. 6).

In central Poland, the Older Halite is overlain successively by potash-bearing units: the 'transitional beds' (Na2 + K2) and Older Potash (K2), which have higher bromine content. In the Na2 + K2 unit the bromine concentration in the halite varied from 78 to 255 ppm, with a mean value of 178 ppm, and did not vary significantly (Bornemann *et al.* 2008; Tomassi-Morawiec *et al.* 2009). The Older Potash unit is characterized by a bromine content in the halite of 95 to 329 ppm, with a mean value of 180 ppm, and little variation as it decreases upwards (Czapowski *et al.* 2012). The PZ2 salt succession ended with a new thin halite unit – the Screening Older Halite (Na2r). Its bromine content, decreasing upward from 124 to 60 ppm with a mean value of 89 ppm (Tomassi-Morawiec *et al.* 2009), indicates the inflow of fresh marine water into the former potash basin and dilution of brine during the halite phase. This inflow probably resulted from a eustatic sea-level rise. Because of the significant halite content of the potash-bearing units and successive lowering of brine concentration in the overlying halite,

Fig. 3. (**a**) Present-day extent and facies distribution of the Zechstein Older Halite (Na2) and Older Potash (K2) units in Poland with (**b**) selected profiles of bromine content and salt facies. Zechstein lithostratigraphic units: A2, Basal Anhydrite; A2r, Screening Anhydrite; A3, Main Anhydrite; A2s, intrasalt anhydrite bed; Na2 + K2, 'transitional beds'. eud/t?, brine concentration increase due to eustatic drop or seaway tectonic closing; rf, rainfall season.

more humid climatic conditions were assumed for this depositional period.

Younger Halite (Na3) unit

Deposition of the Younger Halite (Na3) rock salts started at a similar palaeodepth (Fig. 4) as for the Older Halite salts (Fig. 3). The basin is characterized by a relatively deep-water centre and wide shallow margins (Czapowski 1993, 1995*a*, *b*).

Two representative halite profiles (Fig. 4) with bromine data from point samples are the same as for the Older Halite halites. The halite (Na3a and Na3b subunits) profile from the Kłodawa salt mine is *c.* 95 m thick and located in the basin axis. This halite is dominated by deep open salt basin facies in the lower part and ended with a salt lagoon facies. It is divided by the 'transitional beds' (Na3 + K3) and the thick potash interbed of the Younger Potash (K3) unit representing a salina facies. The section from the Gorzów Wielkopolski IG-1 well is located on the southwestern basin slope. The section is *c.* 108 m thick and began with shallow open salt basin facies and ended with a salt lagoon facies. The section S-483 well, *c.* 80 m thick, typical of a southwestern shallow basin margin, represents salt lagoon facies.

The Younger Potash (K3) seam and the 'transitional beds' (Na3 + K3) divide the salt succession in the Kłodawa salt mine profile into two subunits: the Lower (Na3a) and Upper (Na3b) Younger Halite salts. These represent different bromine distribution patterns (Fig. 4). The Lower Younger Halite (Na3a) halite section illustrated in its lower part a successive, sharp upward increase of bromine content from 70 to 300 ppm after its drop of *c.* 40 ppm. Upwards, a prolonged period of high salinity conditions (300–350 ppm) is then documented, reflecting the limited brine exchange, which was interrupted by a single episode of intense fresh marine-water input (bromine distinct drop of 160 ppm). Other small drops and salinity increases reflect minor salinity fluctuations, perhaps the result of surface inflows of fresh-marine or meteoric water in the shallowing basin. The bromine profile of the Upper Younger Halite (Na3b) subunit illustrates the different pattern of a gradual salinity decrease from 210 to 110 ppm. This may be due to brine dilution by fresh-marine and meteoric water input in the salt lagoon environment. Only a single episode of rapid major refreshment was observed there, as reflected by a bromine drop of 40 ppm.

In the central part of the PPB, the potash-bearing units – the 'transitional beds' (Na3 + K3) and the Younger Potash (K3) – are characterized by a higher bromine content than analogous series of the PZ2 cycle. The bromine concentration in halite from the Na3 + K3 unit varied from 255 to 329 ppm, with a mean value of 298 ppm and with little variation (Bornemann *et al.* 2008; Tomassi-Morawiec *et al.* 2009). The bromine content in halite within the Younger Potash unit changed from 125 to 502 ppm with a mean value of 262 ppm, and the bromine content decreased at the top of the unit (Czapowski *et al.* 2012). These data indicate that the climate was consistently warm and arid, which was favourable for the accumulation of potash salts. The first evidence for increased humidity came at the end of the period. The overlying Upper Younger Halite (Na3b) subunit corresponds with rapid brine being refreshed by a marine-water inflow which may have resulted from a eustatic sea-level rise.

The next two well sections of the Younger Halite unit include less core material and point sample data, so the presented bromine profiles (Fig. 4) illustrate only the general tendencies of basin salinity fluctuations. The section of Gorzów Wielkopolski IG-1 well exhibits a lower bromine content of 50–140 ppm, with a gradual increase in salinity in the lower part and a general decrease in the upper part of the section. Two episodes of distinct brine concentration decreases of 40–70 ppm were observed in the lower profile part, reflecting major brine refreshments. The general bromine profile is similar to that of García-Veigas *et al.* (2011; their fig. 6), but their single result of extremely low bromine content (26 ppm) was not confirmed in our study. In the top part of the Gorzów Wielkopolski IG-1 section a decrease in bromine (*c.* 30 ppm) precedes two events of significant brine dilution, as indicated by sulphate interbeds. Bromine data from a few point samples in the S-483 profile (Fig. 4) show a gradual increase and subsequent upwards decrease in brine salinity in a shallow lagoon environment. A single major brine refreshment produced a sulphate interbed and is observed in the lower part of the profile.

In the central part of the PPB, the PZ3 salt succession ended with a thick clayey halite series referred to as the Younger Clay Halite or the Brownish Zuber (Na3t) unit (Wagner 1994; Wagner & Peryt 1997; Tomassi-Morawiec & Czapowski 2006; Bornemann *et al.* 2008). The bromine content in the halite of this unit ranged from 46 to 176 ppm with a mean value of 109 ppm (Tomassi-Morawiec & Czapowski 2006; Tomassi-Morawiec *et al.* 2009), and the bromine content is generally lower in the upper part of the profile. Relatively low brine concentrations and frequent clay laminae and clay interbeds in the upper part of the profile indicate accumulation under more humid conditions. Longer rainy seasons favoured brine dilution, clay pelite transport and its deposition within the evaporite basin during the more wet season (Wagner &

Peryt 1997), and/or clay content increase due to salt recycling (Czapowski & Bukowski 2002; Czapowski *et al.* 2003).

Interpretation of salinity fluctuations

The end of the Kazanian and Tatarian stages (the East Europe stages corresponding with the Guadalupian and the Lopingian epochs of the Permian Global Stratigrapic Scale; Menning *et al.* 2006) marked the following 'Warm Mode' of Earth climate evolution (Frakes *et al.* 1992), resulting from extraordinary heating of the Earth at that time. Such warming partly resulted from the accelerated volcanogenic heat flow expressed by the development of *c.* 700 kyr Siberian lava traps (Khiel & Shields 2005). This period was characterized by limiting of the extent of former ice-rafting to high latitudes (65–80°) due to the warm surface temperatures (Khiel & Shields 2005) and the higher average temperatures and dominant arid conditions in the supercontinent centre (Frakes *et al.* 1992; Barron & Fawcett 1995; Fluteau *et al.* 2001). The minimum summer temperature in lands at average latitudes in the northern hemisphere became >15 °C higher than present-day temperatures, and the average precipitation rate was 1–2 mm a^{-1} (Khiel & Shields 2005). The climate warming accelerated at the Permian–Triassic boundary, and maximized the anoxic or euxinic conditions in the ocean in the latest Permian–Early Triassic period (Khiel & Shields 2005). This was responsible for mass extinction at that time. Extension of the Pangea supercontinent to almost both poles resulted in extreme seasonal conditions, with a dry continent interior (summer temperature >30–40 °C; Khiel & Shields 2005). Atmospheric water circulation between the more humid Pangea coasts and its interior was limited by uplifts of Variscan orogenes (*c.* 2 km high in the Europe area; Fluteau *et al.* 2001). This east–west mountain range, extending north of the equator with adjacent mountains and plateaux of subtropical and lower middle altitudes, determined the more wet tropical climate in the centre of the continent during the Kazanian stage, intensifying the summer monsoons in both hemispheres (Kutzbach & Ziegler 1993). The high temperature of the Panthalassa Superocean (*c.* 4 °C higher than present-day temperatures; Khiel & Shields 2005) produced an increase in atmospheric CO_2 and consequently a greenhouse effect, with a higher annual temperature at the Permian–Triassic transition (Barren & Fawcett 1995). This favoured longer dry seasons and rare but intense rainfall events. Such wetness fluctuations appear to be a common phenomenon during the Zechstein deposition (e.g. Gąsiewicz this volume, in press, and references therein). These conditions were best expressed by evaporite accumulation within the shallow Zechstein epicontinental sea, extending into the Central European Basin, located in the northern hot arid belt of low altitude (Chumakov & Zharkov 2003). The expansion of red beds and desert flora in the Pangea area (Parrish 1995) also documented a more arid and hot climate. Horizons of desiccation polygons found within the top part of the Younger Halite (Na3) in Poland and Germany (Tarka 1989; Garlicki 1991; Burliga 1995) confirmed occasional exposure of the basin bottom, very high air temperatures (>40 °C) and arid conditions that accompanied salt deposition. The generation of similar contraction polygons described in the Triassic salts in Cheshire (Tucker 1981) required a diurnal temperature variation of 60 °C. Other climatic modelling for the Late Permian (Roscher 2009; Roscher *et al.* 2011) questioned the global warming due to carbon greenhouse gases and suggested global cooling due to atmospheric pollution by sulphur aerosols and ash particles produced by the development of the Siberian lava traps. After this interpretation for most of the Late Permian, the global cold and dry climate was dominant, with more warm and wet periods at its beginning and end.

The Permian evaporites, comprising 27% of all the Phanerozoic evaporites, accumulated at the global lowstand period that resulted from the capture of water by glacier during the Early–Middle Permian and the following warm and dry climate of Late Permian–Middle Triassic times (Frakes *et al.* 1992). The discussed Zechstein evaporites of the EPB accumulated during the upper part of the Late Permian (Wuchiapingian to Changhsingian; Table 1). From the Late Kazanian to the Late Triassic were registered ten small eustatic sea-level fluctuations, with a total rise of *c.* 100 m and accumulation of widespread black marine shales (e.g. Copper Shale and Grey Shale units of the European Zechstein succession) corresponding to raised sea-level periods (Frakes *et al.* 1992).

Transgressions responsible for the Zechstein cyclic deposition (e.g. Legler *et al.* 2005) were considered to be the result of glacier-melting-related sea-level rise and/or subsidence changes along the Barents Sea due to rifting (García-Veigas *et al.* 2011; with references therein).

Geochemical (bromine and isotopes of sulphur and oxygen) data from the Polish Zechstein evaporates, supported by results from primary fluid inclusion studies (e.g. Kovalevych *et al.* 2000; García-Veigas *et al.* 2011), show that halite deposits in the PZ1 and most of the PZ3 and PZ4 cycles had precipitated from sulphate-rich brines similar in composition to modern evaporated seawater. However, the Older Halite (Na2) and the base of

Younger Halite (Na3) rock salts were produced by sulphate-depleted brines (calcium-rich). Such compositional changes in the brine, registered at the base of the Older Halite unit, was interpreted by García-Veigas *et al.* (2011) as the effect of global overturn of anoxic sulphidic deep waters from the Panthalassa stratified superocean. The euxinic waters were oxidized as they flowed through the North Atlantic seaway and became sulphate-depleted when expanding within the EPB (García-Veigas *et al.* 2011).

The brine salinity fluctuations in the Zechstein chloride basins of the EPB, registered by bromine content data and sulphate interbeds in the studied Zechstein salt successions of the PZ1 to PZ3 cycles in Poland, could be considered both as (i) the effect of climate seasonal variability and also (ii) a general change as a result of varied fresh marine-water input to the evaporite basin due to modified (tectonic uplift/lowering) passages with the ocean or eustatic sea-level rise/drop. These effects were recorded more or less clearly in the different parts of the evaporite basin.

In the deep-water part of the open salt basin, with near-stable density stratification and with the bottom halite-concentrated brines supplied constantly with ions from evaporite shallow margins, the produced salt commonly reflects an increase in salinity (e.g. Kłodawa profiles; Figs 3 & 4) and occasionally a later decrease (e.g. Głuszewo IG-1 profile; Fig. 2) due to brine dilution. In this deep-water environment, only a large fresh marine-water inflow, probably generated by eustatic rise or by extreme storm surges, could disturb the existing brine stratification and significantly modify the bottom brine salinity, as found in the lower part of the Na1 unit in Czarny Młyn IG-1 section (Fig. 2) and in the Na3a unit from the Kłodawa profile (Fig. 4). The minor inflow of fresh marine waters above the dense basin brines was better recorded in salts accumulated on the shallower salt basin slopes (the lower part of the Gorzów Wielkopolski IG-1 profile; Fig. 4). In the shallow margins the accumulated salt lagoon deposits as depicted in S-479 and Lubiatów M-20 profiles (Figs 2 & 3) recorded frequent rainfall (fresh meteoric waters) or storm surge (marine water inputs) events. Such inflows diluted the halite brines, dissolved the salt sediments, and could produce low-bromine salts or sulphate interbeds. A similar effect was observed in the shallow lagoonal and salina facies of salt deposits at the end of the initial deep basin successions (Sieciejów P-5 profile, Fig. 2; Gorzów Wielkopolski IG-1 profile, Fig. 4). Such successions better reflect the general climatic changes.

During deposition of PZ1 salts, three brine refreshments, shown by decreases in bromine content, were detected during the initial open basin facies, and that caused the salt brine dilution and precipitation of sulphates instead chlorides (Czarny Młyn IG-1 section, Fig. 2). These refreshments interrupted the observed continuous tendency towards an increase in brine concentration, attributed to the dry climatic period. The first two refreshments, registered by large bromine content drops, were interpreted as the effect of fresh marine-water input (overflow above the sulphate shoal) to the semi-isolated chloride sub-basin, and were generated by eustatic sea-level rise. The third minor brine refreshment was due to a seawater inflow as a result of an extreme storm surge. The observed upward brine concentration decrease in lagoonal facies resulted from more frequent rainfalls during a more humid climate. A similar bimodal pattern of brine concentration was observed in two other profiles (Głuszewo IG-1 and Sieciejów P-5 sections, Fig. 2), reflecting the climatic change from arid to more humid conditions. The deposits of open deep salt basin (Głuszewo IG-1 profile, Fig. 2), precipitated from stable bottom brines, did not register such eustatic fluctuations as in the Czarny Młyn IG-1 section. Basin and lagoonal successions of the semi-isolated southeastern sub-basin(Sieciejów P-5 profile, Fig. 2) and its shallow margin (S-479 section, Fig. 2) became more subject to the dissolution effect of intense rainfalls and storm seawater inflows. A lack of distinct bimodality in the bromine curve of the S-479 profile could not be precisely attributed to the arid or humid climatic periods recorded in other sections.

The rapid and high bromine content increase in the lower half of the PZ2 salt successions from the open deep to shallow-water basin (Kłodawa and Gorzów Wielkopolski IG-1 profiles, Fig. 3) was interpreted as the effect of limited fresh marine-water input to the evaporite basin. This input was caused by a eustatic sea-level drop or by partial closing (tectonic uplift?) of the North Atlantic seaway, this being the single way for delivery of ocean waters to the PEB. Minor bromine content fluctuations, observed upward in the shallow basin to the lagoonal deposits (Kłodawa profile, Fig. 3), were the effect of intense rainfall events. These corresponded with a general salinity decrease (Kłodawa profile), attributed to the more humid climate. The potash deposits of the transitional beds (Na2 + K2) and the Older Potash (K2) units occurred above the Older Halite (Na2) unit in Kłodawa section, and were the product of declining evaporite and a still subsiding restricted basin. The shallow marginal succession (Lubiatów M-20 profile, Fig. 3) recorded rainfall brine dilutions and brine refreshment with the formation of anhydrite (A2s) by pre-concentrated seawater. This profile evidenced a slight (*c.* 30 ppm) decrease in bromine content at the top, which could suggest a change from arid to humid climatic conditions. Above the Older Potash (K2) were found the thin halite deposits of the Screening Older Halite (Na2r) unit, which ended the PZ2 chloride succession in the basin centre (Wagner 1994; Tomassi-Morawiec *et al.* 2009; Tomassi-Morawiec & Czapowski 2010). These halites were an effect of post-potash brine refreshment by fresh marine-water input resulting from a eustatic sea-level rise.

In the PZ3 salt basin, three correlated refreshment events were defined in successions from the basin centre (Kłodawa profile) and SW slope (Gorzów Wielkopolski IG-1 profile; Fig. 4). The first, lowermost event, registered in deep-water facies, was interpreted as a fine brine dilution by fresh marine-water inflow caused by an extreme phenomenon (such as a prolonged storm surge). The second refreshment corresponded with a rapid and large salinity drop and is attributed to a eustatic sea-level rise. An analogous sea-level rise initiated halite deposition of the Upper Younger Halite (Na3b) unit after the potash-bearing series (Na3 + K3 & K3, Kłodawa profile; Fig. 4) accumulated under warm and arid climate conditions. Minor dilutions observed in the upper part of the salt succession in the basin centre (Kłodawa and Gorzów Wielkopolski IG-1 sections, Fig. 4) were the result of intense rainfalls. Brine refreshments by storm seawater inflows, indicated by sulphate precipitation, were registered in the profiles from the shallow basin margin (Gorzów Wielkopolski IG-1 and S-483 sections; Fig. 4). A climatic change from arid to more humid conditions probably started at the end of the Younger Potash unit deposition. The climate with more frequent humid seasons dominated during accumulation of the next halite (Na3b) unit and the characteristic clayey salts (the Younger Clay Halite [Na3t] unit) in the basin centre.

Registered fluctuations of brine concentration derived from the bromine content in salts of the PZ1 to PZ3 cycles of the Late Permian were interpreted as repetitive in each cycle of climatic change from arid to more humid conditions (Fig. 5). Halite deposition began under arid conditions and in the PZ1 and PZ2 cycles continued under periods of more frequent rainfall. During the PZ3 cycle, such a distinct climatic change occurred later, at the end of potash-bearing deposits. Assuming the common proposal of a dry and warm climate during the Late Permian in the northern hemisphere, especially in the centre of Pangea, the commented changes reflect minor climatic fluctuations within a generally dry regime. Variations in the degree of aridity within the generally warm and dry climate were suggested for the Zechstein succession in the PPB, with a tendency for a humidity increase from the end of the PZ3 cycle (Słowakiewicz *et al.* 2009). Similar wet seasons were responsible for expansion of the Late Permian (Rotliegend stage) lakes in the EPB in Germany, which, during the dry periods, evolved into saline basins with evaporites (Legler *et al.* 2010).

Limited constant ice cover at the poles and the snowline moving to the north and south during the Late Permian (Barren & Fawcett 1995) had less influence on the climate than in the Early Permian, with minimal fluctuations in ocean level as a result of ice melting. The climatic changes (from dry to short wet seasons) were correlated with changes in both solar luminosity and the Milankovitch Earth orbital geometries, which were most probably responsible for the Zechstein deposition cyclicity of fourth and fifth order (with periods of 20 000–400 000 years; Ross & Ross 1995). If we assume the calculated length of evaporite deposition at maximum and minimum precipitation rates

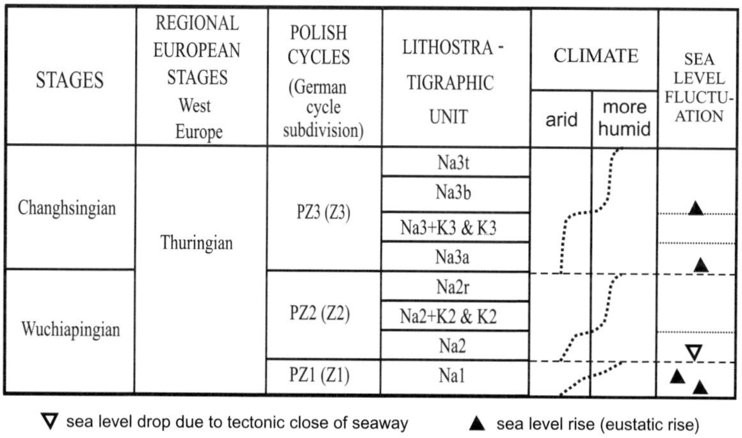

Fig. 5. Climate and sea-level changes during deposition of the Late Permian (Zechstein) salts in the PPB.

for evaporites during the three discussed cycles, then the PZ1 cycle lasted from 4500 to 36 000 years, the PZ2 cycle from 8400 to 85 000 years and the PZ3 cycle from 5300 to 43 000 years (Czapowski 2007). Thus, observed climatic changes in each cycle could correspond with the period of Milankovitch cycles (21 000 years; Ross & Ross 1995).

The cumulative effect of a long warm-period drawdown within the restricted Zechstein evaporite sea, especially in its most inland part (in Poland), and the later short episodes of intensive rainfalls and storm surges at the entrance of the North Atlantic seaway (created by more frequent atmospheric pressure perturbations at the coastal zone) could cause fresh marine-water inputs into an intracontinental salt basin. Such inflows diluted the basin brines to a greater or lesser extent depending on local water depth. The deep brines of the open deep-water basin and marginal sub-basins, with their limited output, became less diluted, for example, in the lower part of the PZ3 salt succession. Extreme changes in the brine salinity of bottom brines, often accompanied by a transition from halite to sulphate precipitation, were the result of abnormal brine refreshment by a huge fresh marine-water input, and reached the distal evaporite basin in Poland as a result of the proposed eustatic sea-level rise. Only a phenomenon such as this could move a sufficient volume of fresh oceanic waters via the relatively narrow North Atlantic seaway and into the basin, and two such events have been noted in the PZ1 cycle and three in the PZ3 cycle (Fig. 5). The last event was the flood that enabled the accumulation of halites (Na3b and Na3t units) in the renewed chloride basin after its partial drawdown during deposition of potash-bearing series (Na3 + K3 and K3 units). A single eustatic sea-level fall, probably due to tectonic closure of the North Atlantic seaway, was responsible for the successive basin desiccation, brine concentration and resulting accumulation of potash-bearing deposits (Na2 + K2 and K2 units) in the PZ2 cycle (Fig. 5). All these eustatic fluctuations were independent of the observed climatic changes.

Conclusions

The observed fluctuations in salt brine concentrations (after bromine content data and facies differentiation of salt successions belonging to the PZ1, PZ2 and PZ3 cycles) within the Late Permian (Zechstein) evaporite basin in Poland indicated local and regional changes, which are interpreted to result from climatic phenomena and eustatic sea-level variations. The distinct recording of these phenomena is highly dependent on their successive position in the evaporite basin palaeomorphology (deep or shallow parts).

The effects of minor climatic phenomena, such as intensive rainfalls and storm surges, diluted the chloride brines, which are better recorded in the shallow salt facies. This type of facies was dominant at the shallow evaporite basin margins and at its centre, where shallowing arose due to sediment infill at the end of the evaporite cycle. More frequent rainy seasons, attributed to a more humid climate, produced a general lowering of brine concentrations. Such a change from arid to more humid climatic conditions was observed in all three studied cycles. This indicated variations within a predominantly warm and dry climate, as commonly assumed for the Late Permian period, especially for the inner part of the Pangea supercontinent. Because of the rapid accumulation of evaporites over a period of perhaps several thousands to several tens of thousands of years, these climatic changes could be related to the Milankovitch cycles (periodicity of 21 kyr). Such variations in climate, registered in the salt successions, clearly document that the Late Permian climate in the interior of the northern supercontinent, far from the coastal zone, was also characterized by distinct humidity fluctuations, although the prevailing tendency was for a warm and arid climate, and the more rainy periods existed for a longer time.

Another factor, independent of climatic variations, was the change in brine concentration caused by fluctuations in eustatic sea-level. The bottom salt brine of deep-water basin parts was diluted only during extreme refreshing by prolonged marine-water inputs initiated by eustatic sea-level rises. Two such events were registered in the salts of the PZ1 cycle, and the next two sea-level rises in the evaporite succession of the PZ3 cycle. Only a single rapid sea-level drop, postulated to be of tectonic genesis (tectonic closing of the narrow North Atlantic seaway, conducting ocean waters to the evaporite basin), took place during deposition of the early PZ2 salts; this resulted in cutoff of the evaporite basin, its subsequent desiccation, and accumulation of potash-bearing deposits.

Anonymous reviewers and the editors are very much thanked for corrections, remarks and comments that greatly helped to improve and finally shape this paper.

References

Baar, C. A. 1963. Der Bromgehalt als stratigraphischer Indikator in Steinsalzlagerstatten. *Neues Jahrbuch für Mineralogie – Monatshefte*, **7**, 145–153.

Baikov, A. A. & Sedletskii, V. I. 2001. Same aspects of the genesis of halogenic sequences: evidence from the

Central Asian Halogenic Basin. *Lithology and Mineral Resources*, **36**, 582–593.

Barron, E. J. & Fawcett, P. J. 1995. The climate of Pangea: a review of climate model simulations of the Permian. *In*: Scholle, P. A., Peryt, T. M. & Ulmer-Scholle, D. S. (eds) *The Permian of Northern Pangea*. Springer-Verlag, Berlin, **1**, 37–52.

Bornemann, O., Schramm, M. *et al.* 2008. Wzorcowe profile bromowe cechsztyńskich solach kamiennych w Polsce i w Niemczech na przykładzie kopalni soli w Kłodawie i w Görleben (Standard bromine profiles of the Polish and German Zechstein salts [a case study from the Kłodawa and Gorleben salt mines]). *Geologos*, **14**, 73–90 (with English abstract).

Burliga, S. 1995. Desiccation polygons within the salt deposits of the Kłodawa Salt Mine. *Abstracts of XIII International Congress on Carboniferous-Permian, 28 Aug–2 Sept., Kraków, Poland*. Polish Geological Institute, Warszawa.

Chumakov, N. M. & Zharkov, M. A. 2003. Climate during the Permian–Triassic biosphere reorganizations. Article 2. Climate of the Late Permian and Early Triassic: general interferences. *Stratigraphy and Geological Correlations*, **II**, 361–375.

Czapowski, G. 1987. Sedimentary facies in the Oldest Rock Salt (Na1) of the Łeba elevation (northern Poland). *Lecture Notes of Earth Sciences*, **10**, 207–224.

Czapowski, G. 1988. Charakterystyka zubrów cyklu PZ4 w okolicach Piły [Characteristics of zuber rocks of PZ4 cycle near Piła]. *Przegląd Geologiczny*, **36**, 241–243 (with English abstract).

Czapowski, G. 1990. Kontynentalne osady chlorkowe w górnym cechsztynie Polski [Continental chloride deposits in the Upper Permian of Poland]. *Przegląd Geologiczny*, **38**, 370–374 (with English abstract).

Czapowski, G. 1993. Facies characteristics and distribution of the Zechstein (Upper Permian) salt deposits of PZ3 (Leine) Cycle in Poland. *Bulletin of Polish Academy of Sciences, Earth Sciences*, **41**, 229–237.

Czapowski, G. 1995*a*. Upper Permian (Zechstein) salt deposits on the Żary Pericline – characteristics, origin and economic value. *Prace Państwowego Instytutu Geologicznego*, **150**, 35–60.

Czapowski, G. 1995*b*. Salt facies of the Upper Permian. *XIII International Congress on Carboniferous-Permian, 28 Aug–2 Sept., Kraków, Poland*, Guide to excursions A3. Polish Geological Institute, Warszawa, 85–96.

Czapowski, G. 2007. Ocena głębokości zbiornika solnego i czasu depozycji chlorków sodu na przykładzie utworów najstarszej soli kamiennej (Na1) cyklu PZ1 cechsztynu w rejonie Zatoki Puckiej [Calculation of salt basin depth and deposition time of salts, exemplified by the Oldest Halite (Na1) unit from the Zechstein PZ1 cycle at the Puck Bay (northern Poland]. *Przegląd Geologiczny*, **55**, 573–581 (with English abstract).

Czapowski, G. & Bukowski, K. 2002. Genesis of clayey salt (zuber) facies (Upper Permian and Middle Miocene case studies from Poland). *16th International Sedimentological Congress*, Abstract Johannesburg, 8–13 July, 71–72.

Czapowski, G. & Bukowski, K. 2010. Geology and resources of salt deposits in Poland: the state of the art. *Geological Quarterly*, **54**, 509–518.

Czapowski, G., Antonowicz, L. & Peryt, T. 1991. Facies and paleogeography of the Zechstein (Upper Permian) Older Halite (Na2) in Poland. *Bulletin of Polish Academy of Sciences, Earth Sciences*, **38**, 45–55.

Czapowski, G., Peryt, T. M. & Antonowicz, L. 1993. Facies and paleogeography of the Zechstein (Upper Permian) Oldest Halite (Na1) in Poland. *Bulletin of Polish Academy of Sciences, Earth Sciences*, **41**, 217–227.

Czapowski, G., Bukowski, K., Tomassi-Morawiec, H., Langer-Kuźniar, A. & Peryt, T. M. 2003. Generation of mixed salt–siliciclastic rocks. *22nd IAS Meeting of Sedimentology, Opatija, Croatia, September 17–19 2003, Abstracts Book*, **38**. Institute of Geology, Zagreb.

Czapowski, G., Tomassi-Morawiec, H., Toboła, T. & Tadych, T. 2012. Geology, geochemistry and petrological characteristics of potash salt units from PZ2 and PZ3 Zechstein (Late Permian) cycles in Poland. *Geology, Geophysics & Environment*, **38**, 153–188.

D'Ans, J. & Höfer, P. 1934. Untersuchungen an Brom. *Zeitschrift für Angewandte Chemie*, **47**, 71–74.

Fluteau, F., Besse, J., Broutin, J. & Ramstein, G. 2001. The late Permian climate. What can be inferred from the climate modeling concerning Pangea scenarios and Hercynian range altitude? *Palaeogeography, Palaeoclimatology, Palaeoecology*, **167**, 39–71.

Frakes, L. A., Francis, J. E. & Syktus, J. I. 1992. *Climate Modes of the Phanerozoic*. Cambridge University Press, Glasgow, 1–274.

García-Veigas, J., Cendón, D. I., Pueyo, J. J. & Peryt, T. M. 2011. Zechstein saline brines in Poland, evidence of overturned anoxic ocean during the Late Permian mass extinction event. *Chemical Geology*, **290**, 189–201.

Garlicki, A. 1991. On some root structures of Zechstein salt deposits in Poland. *Zentralblatt für Geologie und Paläontologie*, **I**, 1211–1222.

Gąsiewicz, A. in press. Climatic control on the Late Permian Main Dolomite (Ca2) deposition in northern margin of the Southern Permian Basin and implications to its internal cyclicity. *In*: Gąsiewicz, A. & Słowakiewicz, M. (eds) *Palaeozoic Climate Cycles: Their Evolutionary and Sedimentological Impact*. Geological Society, London, Special Publications, **376**, http://dx.doi.org/10.1144/SP376.22.

Gwynn, J. W. (ed.) 1980. Great Salt Lake: a scientific, historical and economic overview. *Utah Department of Natural Resources Bulletin*, **116**, 1–400.

Handford, C. R. 1982. Sedimentology and evaporite genesis in a Holocene continental sabkha playa basin – Bristol Dry Lake, California. *Sedimentology*, **29**, 239–253.

Handford, C. R. 1991. Marginal marine halite. *In*: Melvin, J. L. (ed.) *Evaporites, Petroleum and Mineral Resources*. Developments in Sedimentology, Elsevier, Amsterdam, **50**, 1–68.

Haq, B. U. (ed.) 1995. *Sequence Stratigraphy and Depositional Response to Eustatic, Tectonic and Climatic*

Forcing. Kluwer Academic Publishers, Dordrecht/Boston/London.

HARDIE, L. A. 1984. Evaporites: marine or non-marine? *American Journal of Science*, **284**, 193–240.

HARDIE, L. A., SMOOT, J. P. & EUGSTER, H. P. 1978. Saline lakes and their deposits: a sedimentological approach. *IAS Special Publications*, **2**, 7–41.

HARDIE, L. A., LOWENSTEIN, T. K. & SPENCER, R. J. 1983. The problem of distinguishing between primary and secondary features in evaporites. *6th International Symposium on Salt, May 24–28, 1983, Toronto, Ontario, Canada*. Salt Institute Publisher, Alexandria, Virginia, USA.

HERRMANN, A. G. 1980. Bromide distribution between halite and NaCl-saturated seawater. *Chemical Geology*, **28**, 171–177.

HERRMANN, A. G. & SCHNEIDER, J. 1980. Saltworks – natural laboratories for microbiological and geochemical investigations during the evaporation of seawater. *5th Symposium on Salt, May 29–June 1, 1979, Hamburg, Germany*. Northern Ohio Geological Society, Kent State University, USA, 371–381.

HERRMANN, A. G., KNAKE, D., SCHNEIDER, J. & PETERS, H. 1973. Geochemistry of modern seawater and brines from salt pans: main components and bromine distribution. *Contributions to Mineralogy and Petrology*, **40**, 1–24.

HOLSER, W. T. 1966. Bromide geochemistry of salt rocks. *Second Symposium on Salt, May 3–5, 1965, Cleveland, USA*. Northern Ohio Geological Society, Kent State University, USA, 248–275.

HOLSER, W. T. 1970. *Bromide geochemistry of some non-marine salt deposits in the Southern Great Basin*. Mineralogical Society of America, Washington, D. C., Special Papers, **3**, 307–319.

HOLSER, W. T. 1979. Rotliegend evaporites, Lower Permian of Northwestern Europe. Geochemical confirmation of the non-marine origin. *Erdöl und Kohle-Erdgas-Petrochemie vereinigt mit Brennstoff-Chemie*, **32**, 159–162.

HOVORKA, S. D., HOLT, R. M. & POWERS, D. W. 2007. Depth indicators in Permian Basin evaporates. *In*: SCHREIBER, B. C., LUGLI, S. & MAÇBEL, M. (eds) *Evaporites Through Space and Time*. Geological Society, London, Special Publications, **285**, 335–364.

KENDALL, A. C. 1992. Evaporites. *In*: WALKER, R. G. & JAMES, N. P. (eds) *Facies Models*. Geological Assocation of Canada, Stittsville, 375–409.

KENDALL, A. C. & HARWOOD, G. M. 1996. Marine evaporites: arid shorelines and basins. *In*: READING, H. G. (ed.) *Sedimentary Environments. Processes, Facies and Stratigraphy*. Blackwell Science, Oxford, 281–324.

KIERSNOWSKI, H., PAUL, J., PERYT, T. M. & SMITH, D. B. 1995. Facies, paleogeography and sedimentary history of the Southern Permian Basin in Europe. *In*: SCHOLLE, P. A., PERYT, T. M. & ULMER-SCHOLLE, D. S. (eds) *The Permian of Northern Pangea*. Springer-Verlag, Berlin, **2**, 119–136.

KILIC, O. & KILIC, A. M. 2010. Salt crust mineralogy and geochemical evolution of the Salt Lake (Tuz Gölü), Turkey. *Scientific Research and Essays*, **5**, 1317–1324.

KHIEL, J. T. & SHIELDS, CH. A. 2005. Climate simulations of the latest Permian: implications for mass extinction. *Geology*, **33**, 757–760.

KOVALEVYCH, V. M., CZAPOWSKI, G., HAŁAS, S. & PERYT, T. M. 2000. Chemiczna ewolucja solanek cechsztyńskich basenów ewaporatowych Polski: badania inkluzji fluidalnych w halicie z poziomów soli Na1-Na4 [Chemical evolution of brines in the Zechstein (Upper Permian) evaporite basins of Poland: fluid inclusion study of halite from the rock salt units Na1 to Na4]. *Przegląd Geologiczny*, **48**, 448–454 (with English abstract).

KUTZBACH, J. E. & ZIEGLER, A. M. 1993. Simulation of Late Permian climate and biomes with an atmosphere–ocean model: comparison with observations. *Philosophical Transactions of the Royal Society of London B*, **341**, 327–340.

LAST, W. M. 1989. Sedimentology of saline playa in the northern Great Plains, Canada. *Sedimentology*, **36**, 109–123.

LEGLER, B., GEBHARDT, U. & SCHNEIDER, J. W. 2005. Late Permian non-marine transitional profiles in the central Southern Permian Basin, northern Germany. *International Journal of Earth Sciences*, **94**, 851–862.

LEGLER, B., SCHNEIDER, J. W., GEBHARDT, U., MERTEN, D. & GAUPP, R. 2010. Lake deposits of moderate salinity as sensitive indicators of lake level fluctuations: example from the Upper Rotliegend saline lake (Middle–Late Permian, Northeastern Germany). *Sedimentary Geology*, **234**, 56–69.

LOGAN, B. W. 1987. The MacLeod Evaporite Basin, Western Australia. *In*: *Holocene Environments, Sediments and Geological Evolution*. American Association of Petroleum Geologists, Tulsa, OK, Memoirs, **44**, 1–140.

MELVIN, J. L. (ed.) 1991. Evaporites, petroleum and mineral resources. *Developments in Sedimentology*, **50**, 1–556.

MENNING, M. 1995. A numerical time scale for the Permian and Triassic periods: An integrated time analysis. *In*: SCHOLLE, P. A., PERYT, T. M. & ULMER-SCHOLLE, D. S. (eds) *The Permian of Northern Pangea*. Springer-Verlag, Berlin, **1**, 77–97.

MENNING, M., ALEKSEEV, A. S. *ET AL*. 2006. Global time scale and regional stratigraphic reference scales of Central and West Europe, East Europe, Tethys, South China, and North America as used in the Devonian–Carboniferous–Permian Correlation Chart 2003 (DCP 2003). *Palaeogeography, Palaeoclimatology, Palaeoecology*, **240**, 318–372.

ORTI CABO, F., PUEYO MUR, J. J., GEISLER-CUSSEY, D. & DULAU, N. 1984. Evaporitic sedimentation in the coastal salinas of Santa Pola (Alicante, Spain). *Revista d'Investigacions Geologiques*, **38/39**, 169–220.

PARRISH, J. T. 1995. Geologic evidence of Permian climate. *In*: SCHOLLE, P. A., PERYT, T. M. & ULMER-SCHOLLE, D. S. (eds) *The Permian of Northern Pangea*. Springer-Verlag, Berlin, **1**, 53–61.

PERTHUISOT, J. P. 1975. *La Sebkha El Melah de Zarzis: genése et évolution d'un basin paralique. Travaux du Laboratoire de Géologie de l'Ecole Normale Superieur*, Paris, 1–252.

Perthuisot, J. P. 1976. Une Sebkha sulfatee sodique en pays sedimentaire. Le Sebkha Oum el Krialate (Sud Tunisien). *Geologie Mediterraneenne*, **III**, 265–274.

Peryt, T. M., Geluk, M., Mathiesen, A., Paul, J. & Smith, K. 2010. Zechstein. *In*: Doornenbal, J. C. & Stevenson, A. G. (eds) *Petroleum Geological Atlas of the Southern Permian Basin Area*. EAGE Publications, Houten, 123–147.

Pueyo, J. J. (ed.) 1991. *Génesis de formaciones evaporìticas: modelos andinos e ibèricos*. Publicacions de la Universitat de Barcelona, Barcelona.

Richter-Bernburg, G. 1955. Stratigraphische Gliederung des deutschen Zechsteins. *Zeitschrift der deutschen geologischen Gesellschaft*, **105**, 843–854 (in German).

Risacher, F., Alonso, H. & Salazar, C. 2003. The origin of brines and salts in Chilean salars: a hydrochemical review. *Earth-Science Reviews*, **63**, 249–293.

Roscher, M. 2009. *Environmental reconstruction of the Late Palaeozoic. Numeric modelling and geological evidences*. PhD dissertation, Freiberg.

Roscher, M., Stordal, F. & Svensen, H. 2011. The effect of global warming and global cooling on the distribution of the latest Permian climate zones. *Palaeogeography, Palaeoclimatology, Palaeoecology*, **309**, 186–200.

Ross, C. A. & Ross, J. R. P. 1995. Permian sequence stratigraphy. *In*: Scholle, P. A., Peryt, T. M. & Ulmer-Scholle, D. S. (eds) *The Permian of Northern Pangea*. Springer-Verlag, Berlin, **1**, 98–123.

Schramm, M., Bornemann, O., Wilke, F., Siemann, M. & Dijk, H. L. 2002. *Bromine analysis – a powerful tool to solve stratigraphical problems in exploration Boreholes for Salt Caverns. Technical Meeting Papers, SMRI, Fall 2002 Meeting*, Bad Ischl, Austria. Solution Mining Research Institute, Encinitas/USA.

Siemann, M. G. 2003. Extensive and rapid changes in seawater chemistry during the Phanerozoic: evidence from Br contents in basal halite. *Terra Nova*, **15**, 243–248.

Shearman, D. J. 1970. Recent halite rock, Baja California, Mexico. *Transactions of the Institute of Mining and Metallurgy*, **79**, 155–162.

Słowakiewicz, M. & Mikołajewski, Z. 2011. Upper Permian Main Dolomite microbial carbonates as potential source rocks for hydrocarbons (W Poland). *Marine and Petroleum Geology*, **28**, 1572–1591.

Słowakiewicz, M., Kiersnowski, H. & Wagner, R. 2009. Correlation of the Middle and Upper Permian marine and terrestrial sedimentary sequences in Polish, German, and USA Western Interior Basins with reference to global time markers. *Paleoworld*, **18**, 193–211.

Tarka, R. 1989. Thermal polygons in the Zechstein Salt. *Bulletin of the Polish Academy of Sciences, Earth Sciences*, **37**, 57–61.

Tomassi-Morawiec, H. 1990. Geochemia bromu w utworach najstarszej soli kamiennej w rejonie Zatoki Puckiej [Bromine geochemistry of the oldest rock salt in the Puck Bay region]. *Biuletyn Państwowego Instytutu Geologicznego*, **364**, 31–59 (with English abstract).

Tomassi-Morawiec, H. 2003. Charakterystyka geochemiczna najstarszej soli kamiennej (Na1) w rejonie Zatoki Puckiej [Geochemistry of the Zechstein oldest rock salt (Na1) in the Puck Bay region (northern Poland)]. *Przegląd Geologiczny*, **51**, 693–702 (with English abstract).

Tomassi-Morawiec, H. & Czapowski, G. 2006. Brom w skałach ilasto-solnych cechsztynu Polski [Bromine in Zechstein mixed clay–salt rocks of Poland]. *Przegląd Geologiczny*, **54**, 488–495 (with English abstract).

Tomassi-Morawiec, H. & Czapowski, G. 2007. Bromine stratigraphy for depositional history reconstruction of rock salts – a case study of Oldest Halite (Na1) formation (Upper Permian) in Poland. *Book of Abstracts, 25th Meeting of Sedimentology*. Conference & Cultural Centre of the University of Patras, Patras, Greece, 4–7 September 2007, 169.

Tomassi-Morawiec, H. & Czapowski, G. 2010. Salinity fluctuations and evaporate basin palaeogeography – bromine and facies record in the Upper Permian salt formation (Na1 unit) in Poland (Central Europe). *In*: Schwarz, E., Georgieff, S., Piovano, E. & Ariztegui, D. (eds) *Abstracts Volume, 18th International Sedimentological Congress, Mendoza, Argentina*, 872.

Tomassi-Morawiec, H., Czapowski, G., Bornemann, O., Schramm, M. & Misiek, G. 2009. Wzorcowe profile bromowe dla solnych utworów cechsztynu w Polsce [Standard bromine profiles of the Polish Zechstein Salts]. *Gospodarka Surowcami Mineralnymi*, **25**, 75–143 (with English abstract).

Tucker, J. B. 1981. Giant polygons in the Triassic salt of Cheshire England: a thermal contraction model for their origin. *Journal of Sedimentary Petrology*, **54**, 779–786.

van der Meulen, H. 1931. Bromo-iodometric investigations. *Chemisch Weckblad*, **28**, 239.

Vovnyuk, S. & Czapowski, G. 2007. *Generation of primary sylvite: the fluid inclusion data from the Upper Permian (Zechstein) evaporites, SW Poland. In*: Schreiber, B. C., Lugli, S. & Maçbel, M. (eds) *Evaporites Through Space and Time*. Geological Society, London, Special Publications, **285**, 275–284.

Wagner, R. 1987. Stratigraphy of the Uppermost Zechstein North-Western Poland. *Bulletin of the Polish Academy of Sciences, Earth Sciences*, **35**, 265–273.

Wagner, R. 1994. Stratygrafia i rozwój basenu cechsztyńskiego na Niżu Polskim [Stratigraphy and evolution of the Zechstein basin in the Polish Lowland]. *Prace Państwowego Instytutu Geologicznego*, **146**, 1–71 (with English abstract).

Wagner, R. & Peryt, T. M. 1997. Possibility of sequence stratigraphy subdivision of the Zechstein in the Polish Basin. *Geological Quarterly*, **41**, 457–474.

Wagner, R., Peryt, T. M. & Piątkowski, T. S. 1981. The evolution of the Zechstein sedimentary basin in Poland. *Proceedings of International Symposium on Central European Permian*, Wydawnictwa Geologiczne, Warszawa, Poland, 69–83.

Walaszko, M. G. 1956. Bromine geochemistry in salt generation and its content as the genetic-prospection criterion (Gieochimiya broma w prociesach galogieneza i ispolzovaniya soderzaniya broma w kathzestwie genetihzeskovo i poiskovo kryteria). *Geochimiya*, **6**, 33–48.

Warren, J. K. 2006. *Evaporites: Sediments, Resources and Hydrocarbons*. Springer-Verlag, Berlin, Heidelberg.

Warren, J. K. 2008. Salt as sediment in the Central European Basin system as seen from a deep time perspective. *In*: Littke, R., Bayer, U., Gajewski, D. & Nelskamp, S. (eds) *Dynamics of Complex Intracontinental Basins*, 249–276.

Wilgus, C. K. & Holser, W. T. 1984. Marine and nonmarine salts of Western Interior, United States. *American Association of Petroleum Geologists Bulletin*, **68**, 765–767.

Zhang, P. 1987. *Saline Lakes of the Qaidam Basin*. Science Publishers, Beijing.

Ziegler, M. A. 1990. *Geological Atlas of Western and Central Europe*, 2nd edn. Shell Internationale Petroleum Maatschappij B.V. Geological Society, London, 1–239.

Climatic control on the Late Permian Main Dolomite (Ca2) deposition in northern margin of the Southern Permian Basin and implications to its internal cyclicity

ANDRZEJ GĄSIEWICZ

Polish Geological Institute, ul. Rakowiecka 4, 00-975 Warszawa, Poland (e-mail: Andrzej.Gasiewicz@pgi.gov.pl)

Abstract: During the Late Permian (Zechstein), the Southern Permian Basin (SPB) of Europe was a shallow epicontinental and evaporitic sea surrounded by large continental masses. It has been assumed that both the cyclic Zechstein deposition and internal cyclicity of the lower Zechstein carbonate units resulted from glacio-eustatic fluxes of marine waters and evaporation in the stable arid climatic conditions.

The Main Dolomite (Ca2) is the carbonate unit of the second Zechstein cycle. The Main Dolomite (MD) deposits formed the Kamień Pomorski carbonate platform, which was attached to land at the northern margin of the SPB (NW Poland). Study of stable carbon and oxygen isotopic composition and major and minor elements distribution reveal palaeoclimatic instability induced by a relative rise in precipitation during MD deposition. A tendency towards increased climatic humidity towards the end of the unit reflected by an increased fluvial contribution argues for either a local- and/or regional-scale of temporal flow of wet humid air masses. Chemical weathering and erosion on the nearby Fennoscandian High caused dilution of carbonate accumulation by very fine siliciclastics in the proximal part of the platform.

A lack of internal cycles of a regional extent in the investigated platform implies that it was not global glacio-eustatic influxes but a combination of events, such as local relative sea-level fluctuations induced by climate impact and variation of the palaeoenvironmental parameters, that determined the development of the platform. Future research should investigate the effects of the climate and local tectonics to reveal the basinal dynamics and depositional history of the MD carbonate platforms.

The Late Permian Southern Permian Basin (SPB) of Europe extended from England to Latvia (Fig. 1a). This epicontinental sea resulted from extensive and differential subsidence of the central part of Europe (e.g. Ziegler 1990) due to thermal contraction of the lithosphere (Wilson *et al.* 2004). The SPB began to subside in early Zechstein times and was fed by open ocean waters coming through the relatively narrow Greenland–Norwegian strait in the north, forming the Zechstein Basin (Taylor 1998; Legler & Schneider 2008).

It has largely been assumed that the Zechstein depositional architecture and stratigraphic framework have essentially been controlled by the combined effect of tectonic processes (subsidence) and cyclic connections with an open ocean (Zechstein transgressions). Marine water inflows from the open ocean are linked to glacial–interglacial changes in the Late Permian sea-level (e.g. Smith 1980; Ziegler 1990; Taylor 1998; Peryt *et al.* 2010). Note, however, that extensive continental ice sheets were absent after the Early Permian (e.g. Frakes *et al.* 1994; Rees 2002 and references therein, Wopfner 2012; Isbell *et al.* 2012) and the Late Palaeozoic Ice Age, which ended at least 5–7 Ma before the mass extinction at the end of the Permian (Crowell 1995; Link 2009), that is, essentially before Zechstein deposition (Menning 1995). According to the presented view, the depositional series reflects cyclic open marine ingressions, subsequent basin isolations and seawater evaporation. In this model, relative low and high sea-levels controlled the sequence of the stratigraphic framework of the older Zechstein cycles (e.g. Tucker 1991; Strohmenger *et al.* 1996; Wagner & Peryt 1997; Kaiser *et al.* 2003; Słowakiewicz & Mikołajewski 2009). These main factors allowed the development of up to 1500–2000 m thick transgressive–regressive sedimentary cycles in the Polish part of the SPB (Peryt *et al.* 2010), even though it was located far from the connection with the open ocean to the north. The older three Zechstein carbonate–evaporite cycles are composed of terrigenous–carbonate–evaporitic series, while the younger (fourth cycle) is much thinner and composed of five terrigenous–evaporitic subcycles.

Widespread arid to semi-arid regions occupied the Pangaean supercontinent during the Late Permian (Parrish 1993). According to widely held views and modelling, the Zechstein deposition in Central–West Europe developed in low latitudes during times of reduced and ceased continental glaciations, and proceeded in generally hot summer temperatures and an arid climate, with a distinct aridization

From: Gąsiewicz, A. & Słowakiewicz, M. (eds) 2013. *Palaeozoic Climate Cycles: Their Evolutionary and Sedimentological Impact*. Geological Society, London, Special Publications, **376**, 475–521.
First published online October 17, 2013, http://dx.doi.org/10.1144/SP376.22

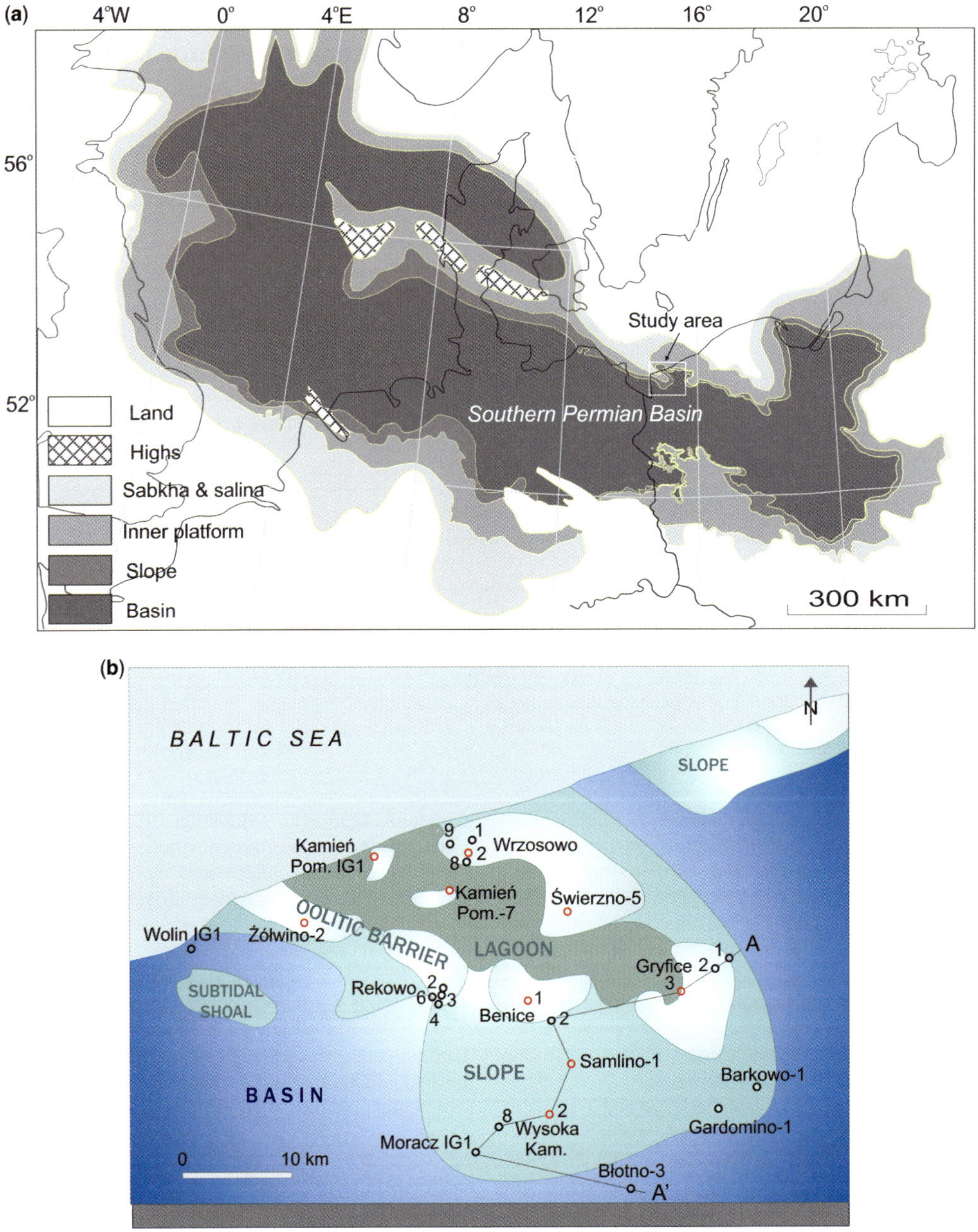

Fig. 1. (**a**) Palaeogeography of the SPB (after Słowakiewicz & Mikołajewski 2011, modified with respect to the area discussed in this paper). (**b**) Major palaeogeographic zones in the Main Dolomite Kamień Pomorski carbonate platform area: red circles, dolomitic cores analysed with respect to inorganic geochemistry; black circles, the Main Dolomite sections discussed in this paper. The location of Resko-1 well (discussed in the text) is not shown in this figure, as it is located far southeastward from the platform area. A–A′, cross-section (Fig. 3).

trend in the low- and mid-latitude continental interiors, particularly in northern Pangaea (e.g. Parrish 1995; Fluteau *et al.* 2001; Schneider *et al.* 2006; Roscher *et al.* 2008, 2011). The fourth Zechstein sedimentary cycle is linked to subcyclic arid–humid climatic changes (Wagner 1994; Geluk *et al.* 1997), but the distribution and development of the older Zechstein sedimentary cycles was largely dependent on local tectonic and palaeogeographic regimes (e.g. Wagner *et al.* 1978; Smith 1980; Taylor 1998; Peryt *et al.* 2010).

It is commonly assumed that the older Zechstein sedimentary cycles developed in generally stable arid climatic conditions. However, lithological variations within the Zechstein older depositional cycles, such as the presence of terrigenous material and both karstic and geochemical features within the sedimentary cycles (e.g. Turner & Magaritz 1986; Peryt & Scholle 1996; Sachse *et al.* 2004) clearly imply that the climatic conditions were rather unstable and influenced Zechstein sedimentary environments on various scales. In general, climatic fluctuations within the Zechstein depositional cycles have not yet been investigated any wider and thus remain a research challenge. The examination of stable carbon and oxygen isotopic data and selected major and trace elemental distributions may serve as tools to recognize the climatic impact on the Zechstein Main Dolomite (Ca2) sedimentary record. Material from weathered land, delivered to the sedimentary setting, as well as its mineral assemblage and geochemical composition, contains information regarding the climatic conditions and tectonic activity when the weathering process occurred (e.g. Magaritz 1987; Liu *et al.* 2012 and references therein). These data have traditionally been applied in palaeoclimatology as a proxy with which to reconstruct palaeoenvironmental changes. Indeed, sedimentary features influenced by climate can be recognized by using a combination of various geochemical proxies. This can then give information on climatic changes during the sedimentation process.

Hitherto, studies have proposed models for the development of the Zechstein Main Dolomite (MD) carbonate platforms that are related to regional sea-level changes in the SPB area. This study provides new data on the development of these platformal strata in the SPB region. It also reviews the interaction between palaeoenvironmental and palaeoclimatic conditions and their impact on the MD carbonate platform deposition at the northern SPB margin. The approach presented herein can be applied to this and other Zechstein carbonate and associated evaporite units throughout the basin. In particular, it would appear to be a helpful tool for understanding the factors controlling the development of the Zechstein carbonate platform and for drawing a climatic context, deepening our understanding of stratigraphic correlations and basin evolution, and completing the Late Permian climatic pattern.

Geological setting

During deposition of the second Zechstein cycle (Late Permian), peripheral parts of the SPB were occupied by numerous carbonate platforms (Kiersnowski *et al.* 1995). The MD of the Kamień Pomorski region (NW Poland) developed at the northern margin of the basin as a carbonate platform elongated in the SE direction, and most likely attached to the land (Fig. 1) (Gąsiewicz & Wichrowska in press). Palaeogeographic reconstruction of the Zechstein Pangaea indicates that the carbonate platform was located at a palaeolatitude of *c.* 13°N (Scotese & Langford 1995).

The lithofacies composition and arrangement, mineral and grain compositions, and sedimentary structures of the carbonate platform represent typical shallow-water regimes and allow three main palaeogeographical zones to be distinguished in the platform area: (1) an inner platform or lagoonal zone rimmed seaward by (2) a relatively thicker, strongly arched, narrow and sandy (essentially oolitic) barrier, and (3) a platform slope (Fig. 1b). The shallow inner platform zone was dominated by carbonate muddy lithofacies (mainly mudstones), locally laminated (or microbially laminated – biolaminated) with rare microbialite and grainy deposits, and contains relatively abundant sulphates. These deposits were protected from the open sea by a peritidal mobile carbonate sandy barrier system formed from uppermost subtidal to supratidal zones with local subaerial surfaces and tidal channels. Slope facies are dominated by subtidal laminated and non-laminated mudstones, locally slumped, with occasional debris flows and turbidites. The inner and slope platform laminites have variable thickness and usually consist of small successions of dark organic-rich wavy laminae alternating with lighter and much thicker organic-poor carbonate laminae or thin layers.

The MD in this area is underlain and overlain by sulphate (mostly anhydrite) units of Werra Anhydrite (A1) and Basal Anhydrite (A2), respectively (Fig. 2). The carbonate platform is relatively small and narrow and surrounded by open and deeper marine basin. The thickness of the platform carbonates varies substantially and increases basinwards. The unit is up to *c.* 90 m thick in the carbonate platform margin and slope zones, and then decreases, usually below 10–15 m thick, in the basinal area.

The MD platform consists mainly of dolomites and limestones with varying amounts of anhydrites. The MD basinal succession consists typically of laminated and clayey limestones, as in Błotno-3 and Wolin IG1. The most common carbonates of the platform sequence are fine to coarse crystalline fabric-replacive and fabric-preserved early diagenetic dolomites. In agreement with a common view, the dolomites have subsequently been affected by a local dedolomitization process (Gąsiewicz & Wichrowska in press). The common admixture of evaporites (anhydrites) is highly non-uniformly

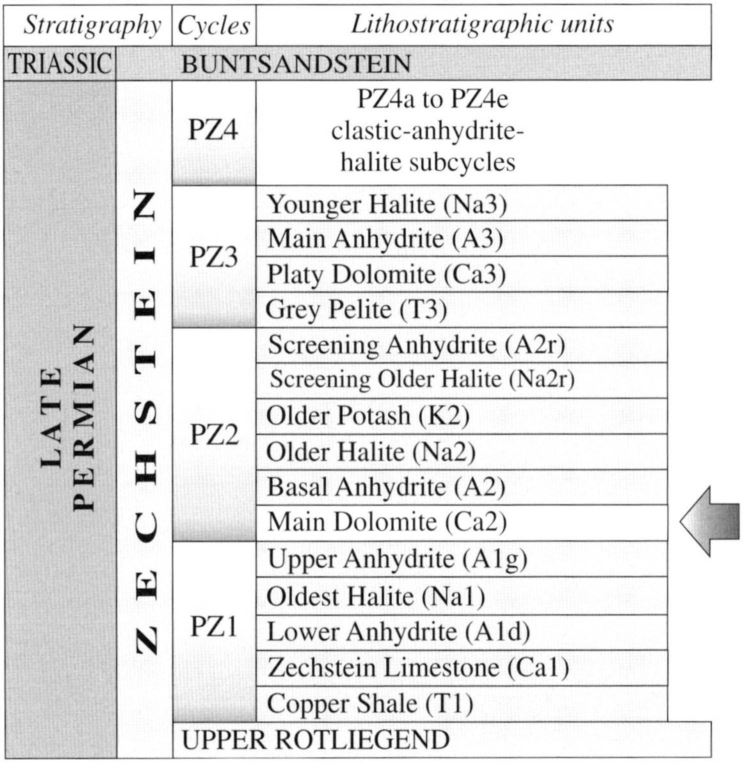

Fig. 2. Lithostratigraphic subdivision of the Zechstein of Poland. Arrowed is the Main Dolomite unit discussed in the text.

distributed throughout the platform. The inner platform dolomites are distinctly more enriched in anhydrites (up to *c.* 30% of the rock volume). Aragonite relics are observed occasionally. Both standard and scanning electron microscopy (SEM) petrography reveal that the MD carbonates contain common clay minerals (illite, kaolinite and chlorite), halite, locally abundant magnesite, minor amounts of pyrite and celestite, occasional secondary gypsum, trace amounts of fluorite, monocrystalline and polycrystalline quartz, plagioclase and feldspar grains, single mica flakes, rare zircons, opaque and heavy minerals as well as Fe and (possibly) Mn oxides. There is also common amorphous organic matter occurring in low concentrations (up to 0.7% total organic carbon, TOC). Clay minerals may infill fine pores or form diffuse incomplete coats around singular clastic and carbonate grains. The acid-resistant grains are dispersed in the carbonate matrix or form clastic streaks or laminae. These grains are composed of insoluble residue (IR). IR composition has not been studied either formerly or in this study.

The carbonate lithofacies contain various amounts of intraclasts, peloids and coated grains such as ooids, oncoids, pisoids and vadoids. Ooids are predominant in the platform margin zone and locally occur in the lagoon zone. Bioclasts are subordinate compared to ooids and consist of the remains of bivalves, foraminifera, ostracods, gastropods, brachiopods, bryozoans and rare algae. The bioclastic facies (wackestones) occur mainly in the inner part of the platform and rarely in the platform margin zone.

The lagoonal facies contain fenestrae, sulphate laminae and also granules and nodules, and are locally bioturbated. Graded and cross bedding, local truncation surfaces and intraclasts are associated with platform barrier facies. Thin coarser lags composed of reworked fragments of stromatolites and oolites are locally visible in the barrier facies. The platform slope deposits are commonly laminated and exhibit local slumping and folding.

The MD carbonate platform depositional sequence (Fig. 3) consists of laminated and non-laminated mudstones, pelolites, oolites, bioclastic wackestones to packstones (lumachelles), rudstones and microbialites (planar and domal stromatolites, and biolaminoids), whereas the basinal part adjacent to the platform is composed of dark-coloured,

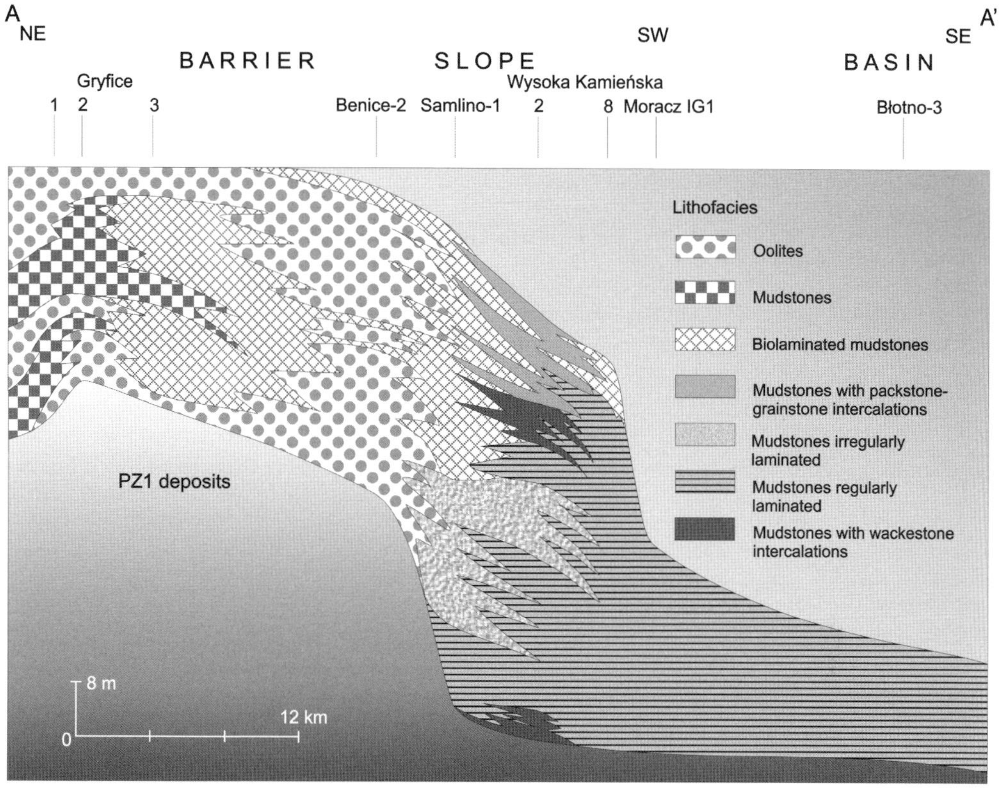

Fig. 3. Cross-section across the Kamień Pomorski carbonate platform margin illustrating the Main Dolomite major lithofacies distribution.

laminated muddy or clayey carbonates with numerous bioclasts (Gąsiewicz & Wichrowska in press).

The MD succession of the carbonate platform zones represents a continuous sedimentary record, with no regionally extended disconformities or noticeable hiatuses. Also, the transition of the MD carbonates into overlying A2 sulphates is continuous. The Ca2 lithofacies are arranged into a gradually shallowing-upward sequence, and most of this succession is represented by deposition during the sea-level highstand. In late MD times, carbonate deposition ceased and relatively rapidly passed into widespread deposition of the A2 sulphates. This coincides with the maximum PZ2 cycle regression of Słowakiewicz & Mikołajewski (2009) and Peryt *et al.* (2010).

Materials and methods

Almost fifty cores from the Kamień Pomorski Main Dolomite carbonate platform were studied with respect to sedimentology, petrography and organic geochemistry (Gąsiewicz & Wichrowska in press). Thirteen MD sections were analysed with respect to inorganic geochemistry. In order to minimize diagenetic disturbances this study is based exclusively on dolomite sections. Four wells (Wolin IG1, Moracz IG1, Błotno-3 and Resko-1) containing limestones or limestones and dolomites are excluded from this study. A total of 127 dolomitic samples from nine boreholes (Fig. 1b) were collected for bulk geochemistry for carbon and oxygen isotopes, as well as major and trace elements bulk carbonate composition, in order to characterize the carbonate platform dolomites. Besides relatively pure dolomites, the analytical base also included a few samples of dolomites enriched in sulphates (anhydrite), that is, locally containing up to 35% of $CaSO_4$ (>8 wt% of S). Only two samples contained 1.0–1.5 wt% of Sr, and most contained less than 0.1 wt% Sr, indicating that the admixture of Sr minerals in the analysed samples was generally insignificant.

To achieve the aim of this study, the three most complete MD cores located along the platform elongation were selected to provide representative sections for the MD inner platform, barrier and slope platform zones (Fig. 1b). These were the Kamień Pomorski IG-1, Benice-1 and Samlino-1 wells, respectively. The MD major lithofacies of these wells are presented in Figure 4, and chemical data are listed in Table 1.

The sampling avoided mineral veins and layers containing lithoclasts that clearly indicated rapid erosion and deposition. Dolomite samples for geochemical analyses were cleaned from the drilling mud and powdered using an agate mill. The elemental analyses were conducted using inductively coupled plasma mass spectrometry (ICP-MS; apparatus Philips PV8060) for Ca, Mg, S, Fe, Al, Ti, Na, K, P, Mn, Zn, Cd, Co, Cu, Mo, Ni, V, Cr, Pb, Sr and Ba concentrations. An atomic absorption spectrometer (AAS) provided quantitative analysis of Na and K concentrations. Major elements data are given in the oxide phase as wt% (reported as % hereafter) and trace elements in ppm. Each element has a different range of uncertainty. The range of uncertainty for the major elements ranges between 1 and 5%, and between 5 and 10% for trace elements. Because all or most of the analysed samples contained concentrations of Cd, Mo, Co and P below their detection limits (2 ppm, 10 ppm, 2 ppm and 0.001%, respectively), these elements are not included in the statistical analysis. IR was estimated as a measure of a product of dissolution of samples in diluted HCl and then roasting at 600 °C. TOC was determined by a coulometric method after removing carbonate phases from the samples and drying to a constant weight using a COULOMAT 702 Stroehlein apparatus. Chemical analyses were carried out by the Central Chemical Laboratory of the Polish Geological Institute.

The powdered and degassed whole-rock fraction analysed for bulk chemistry was further analysed for carbon and oxygen isotopic signatures. This was carried out using the conventional phosphoric acid method, slightly modified, according to the McCrea (1950) methodology. The condensed and cryogenically cleaned CO_2 gas was analysed using an MI 1305 mass spectrometer according to KH-2 and NBS-22 international standards, calibrated relative to the Peedee Belemnite (PDB) standard. The results are given as $\delta^{13}C$ and $\delta^{18}O$ values in units per mil (‰). The reproducibility of duplicate to triplicate samples was better than 0.05‰. Samples were analysed at the Institute of Geological Science, University of Wrocław (Poland).

For statistical purposes, the concentrations of some minor and trace elements (Ti, Ni and Pb) that were below their detection limits were conventionally taken as half of the detection boundaries. The concentrations of some elements were normalized to Al or Ti concentrations in order to avoid a dilution effect by the prevailing carbonate phases. Based on a very high positive correlation between Al and Ti in the analysed samples, it may be inferred that diagenesis did not influence the Al distribution. Thus, it can be assumed that these elements are immobile in the sedimentary settings, thereby providing a good base for comparisons.

Geochemical results

Carbon and oxygen isotopes and elemental concentrations vary vertically within each core sample and regionally across the platform. This is shown by the cumulative data for all analysed samples used in this paper (Table 2). Compared to shallow-water platform facies, the platform slope facies are enriched in most of the analysed elements. The carbonate barrier facies is characterized by a broader range of concentrations of the elements relative to the shallow-water and slope facies. The barrier zone facies are also marked by the comparatively lowest mean values of chemical elements. Most of the elements of the inner and barrier sections show increasing concentrations upwards, whereas most of the elements of the slope section show the opposite concentration trend. A vertical distribution of the concentrations of the major elemental components is presented in Figure 5.

Oxygen and carbon isotopic compositions

The MD platform dolomites are differentiated in terms of their carbon and oxygen isotopic compositions and commonly characterized by high $\delta^{13}C$ and $\delta^{18}O$ values (Tables 1 & 2). It is generally assumed that dolomitization during normal diagenesis decreases $\delta^{18}O$ values and negligibly alters $\delta^{13}C$ values (Tucker & Wright 1990). It is assumed that the carbonate platform dolomites represent the original isotopic record without alteration leading to significant changes in the $\delta^{18}O$ composition. It may be supported by insignificant correlation between the $\delta^{13}C$ and $\delta^{18}O$ values found in the lagoonal and slope zones and in the barrier system ($R^2 = 0.4$), (Table 3). This may also be confirmed by the low Mn/Sr ratios (commonly <1 in all investigated wells, except two samples, where they are as high as 2 and 2.2; Table 1), which indicate that the MD deposits retained primary $\delta^{13}C_{carb}$ (Kaufman & Knoll 1995).

Both the horizontal and vertical distribution patterns of the $\delta^{13}C$ and $\delta^{18}O$ values are non-uniform across the platform (Fig. 5). In general, there is no correlation between $\delta^{13}C$ and $\delta^{18}O$ values in the inner and slope platform settings, and only a moderate positive correlation ($R^2 = 0.49$) was found in the barrier zone deposits (Table 3). Particular MD sections from the inner and slope

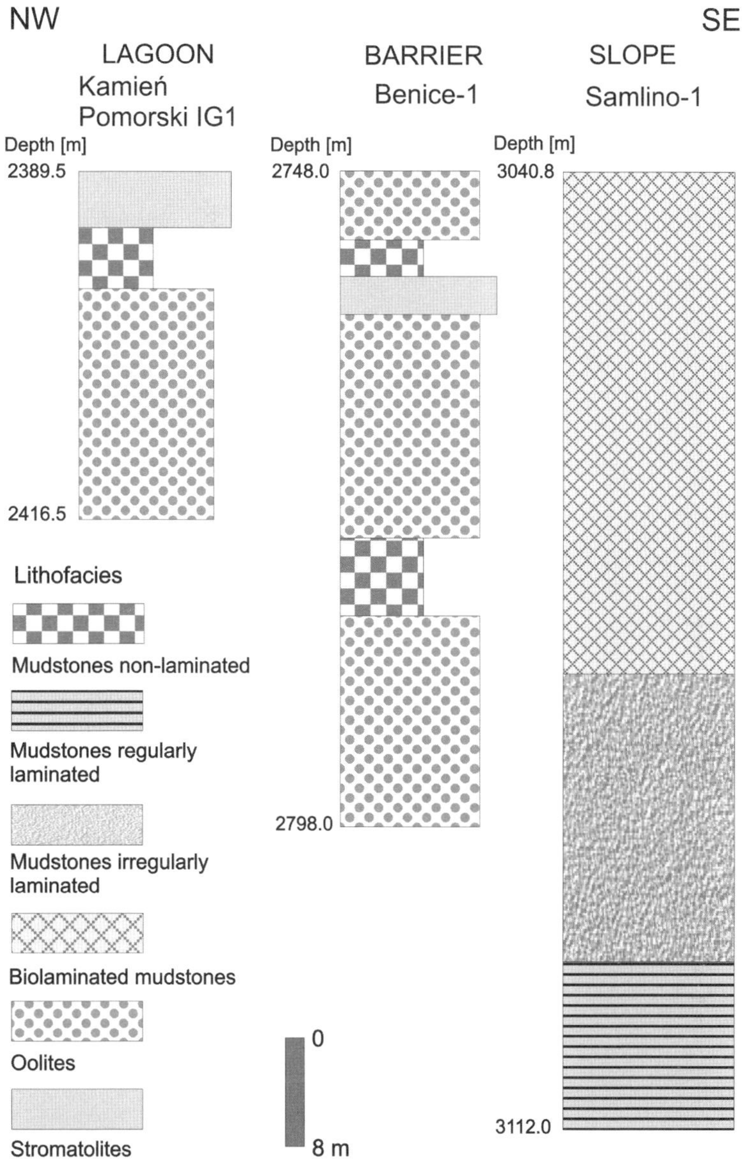

Fig. 4. General Main Dolomite lithofacies distribution in wells discussed in this paper.

facies do not show correlations between these isotope values. Similarly, the isotopic profiles from the barrier zone reveal no or only insignificant correlations between $\delta^{13}C$ values and $\delta^{18}O$ values.

The $\delta^{13}C$ values of the platform carbonates vary relatively widely, from +3.7‰ to +8.2‰ (mean of +6.9‰; Tables 1 & 2). The average $\delta^{13}C$ values are the same in the lagoonal and barrier systems (*c.* +7.2‰) and higher by *c.* 1‰ in the platform slope facies (mean of +6.1‰). This small enrichment (*c.* 1‰) in the average $\delta^{13}C$ value of the shallow platform (lagoonal and barrier) relative to slope dolomites may result from a slight increase in $\delta^{13}C$ in seawater composition during early stages of seawater evaporation (Lazar & Erez 1990). The relatively largest $\delta^{13}C$ value range occurs in barrier dolomites (4.2‰) and the smallest in the platform slope deposits (2.3‰). The inner platform dolomites exhibit a consistent trend towards lower $\delta^{13}C$ values in the upper 20 m. The platform barrier facies clearly shows variable tendencies in the vertical $\delta^{13}C$ value distribution, with a prevailing

Table 1. *Bulk isotopic data and major and selected trace elemental contents and their selected ratios for the Main Dolomite sections of the Kamień Pomorski carbonate platform discussed in detail in this paper.*

Palaeo-geographic zone	Borehole	Depth (m)	$\delta^{13}C$ (‰ PDB)	$\delta^{18}O$ (‰ PDB)	TOC (%)	Insoluble residue (IR) (%)	CaO (%)	MgO (%)	S (%)	Fe_2O_3 (%)	Al_2O_3 (%)	TiO_2 (%)	Na_2O (%)	K_2O (%)	MnO (%)	Zn (ppm)
Lagoon	Kamień	2390.0	4.68	1.82	0.20	5.9	17.80	25.42	6.58	0.23	0.41	0.003	0.11	0.05	0.025	7
	Pomorski-IG1	2396.4	6.68	2.97	0.13	10.6	26.72	20.22	2.03	0.30	0.93	0.007	0.15	0.14	0.014	9
		2400.1	6.97	3.21	0.11	2.4	29.96	16.31	5.35	0.05	0.13	0.001	0.05	0.02	0.012	6
		2404.3	8.14	2.77	0.15	7.4	30.29	12.50	8.61	0.17	0.37	0.003	0.05	0.07	0.007	38
		2406.2	7.63	2.77	0.08	12.7	26.80	15.68	4.18	0.06	0.19	0.002	0.07	0.06	0.009	7
		2409.1	7.64	2.92	0.07	2.9	29.55	18.04	4.16	0.05	0.20	0.002	0.08	0.07	0.012	87
		2409.5	7.02	1.43	0.05	5.3	28.64	16.11	5.00	0.04	0.14	0.001	0.07	0.06	0.011	6
		2413.1	7.07	1.34	0.07	3.1	27.53	20.90	0.83	0.06	0.21	0.002	0.07	0.10	0.018	6
		2415.0	7.22	1.23	0.03	2.6	23.87	6.12	8.98	0.03	0.05	0.001	0.03	0.01	0.006	5
		2416.2	7.38	2.26	0.05	2.4	28.00	19.21	1.76	0.07	0.18	0.002	0.04	0.05	0.020	5
Barrier	Benice-1	2738.2	7.25	2.03	0.06	2.0	28.87	17.10	3.21	0.07	0.26	0.002	0.04	0.03	0.008	8
		2742.1	7.83	3.15	0.08	2.6	27.30	19.46	0.12	0.06	0.12	<0.001	0.05	0.02	0.012	15
		2745.3	7.32	2.85	0.12	1.9	28.22	20.01	0.18	0.04	0.11	<0.001	0.05	0.02	0.014	13
		2748.8	7.74	2.52	0.10	1.9	28.55	20.07	0.53	0.04	0.15	0.001	0.05	0.02	0.011	12
		2750.8	7.81	3.06	0.16	4.6	26.65	19.46	0.28	0.07	0.31	0.002	0.07	0.05	0.017	25
		2756.5	7.52	1.94	0.08	2.5	28.34	18.40	2.03	0.05	0.17	0.001	0.07	0.04	0.017	11
		2762.0	7.53	1.86	0.06	2.4	28.30	19.18	1.30	0.02	0.07	<0.001	0.04	0.01	0.017	12
		2769.7	7.40	2.54	0.17	3.5	28.34	19.03	1.50	0.09	0.32	0.003	0.04	0.05	0.027	7
		2771.8	7.66	1.88	0.23	2.4	29.11	18.19	3.08	0.05	0.12	0.002	0.07	0.02	0.027	7
		2774.8	7.45	2.01	0.06	2.1	30.26	17.26	4.42	0.01	0.04	<0.001	0.04	0.01	0.019	5
		2776.1	7.49	1.82	0.08	2.8	28.78	18.61	1.80	0.01	0.04	<0.001	0.05	0.01	0.018	8
		2776.9	7.24	2.41	0.06	2.6	29.31	20.88	0.66	0.01	0.03	<0.001	0.04	0.01	0.021	7
		2777.3	7.14	2.01	0.07	2.1	30.51	16.84	4.81	0.01	0.04	<0.001	0.05	0.01	0.016	9
		2778.7	7.22	2.18	0.04	2.3	28.78	20.56	0.67	0.01	0.03	<0.001	0.09	0.01	0.019	6
		2781.3	6.90	1.87	0.12	2.1	29.42	19.69	1.64	0.01	0.05	<0.001	0.03	0.01	0.022	7
		2784.6	7.29	1.89	0.05	2.5	28.88	19.00	2.12	0.02	0.07	<0.001	0.05	0.01	0.018	9
		2787.2	7.30	2.02	0.05	2.6	28.42	21.10	0.40	0.02	0.08	<0.001	0.05	0.02	0.019	7
		2788.2	7.40	2.30	0.06	2.4	30.38	16.94	4.88	0.01	0.03	<0.001	0.03	0.01	0.014	6
		2790.0	7.04	1.92	0.14	6.0	27.51	20.04	1.36	0.11	0.48	0.003	0.07	0.06	0.024	6
		2792.0	7.28	2.06	0.10	2.2	29.65	20.00	2.11	0.02	0.10	<0.001	0.04	0.02	0.018	6
		2792.4	7.54	2.06	0.08	2.3	31.34	16.18	4.77	0.02	0.09	<0.001	0.04	0.01	0.019	6
		2793.1	7.17	2.24	0.13	3.8	30.90	14.66	6.86	0.04	0.08	0.001	0.04	0.02	0.005	4
Slope	Samlino-1	3039.8	6.47	0.49	0.31	7.9	27.00	19.65	0.57	0.21	0.47	0.003	0.04	0.13	0.017	9
		3040.0	5.62	0.60	0.50	17.3	24.10	16.40	1.53	0.31	0.59	0.004	0.07	0.17	0.017	9
		3041.0	5.50	0.85	0.25	7.5	25.98	18.10	2.05	0.08	0.30	0.002	0.04	0.08	0.015	7
		3041.4	6.54	1.22	0.13	9.2	26.43	19.74	0.08	0.11	0.41	0.003	0.04	0.11	0.018	9
		3041.5	6.00	0.85	0.26	4.9	26.65	20.18	0.26	0.04	0.18	0.001	0.04	0.05	0.017	5
		3042.0	5.97	0.20	0.24	9.5	25.62	19.15	0.11	0.15	0.58	0.004	0.04	0.14	0.015	5
		3043.0	6.48	0.30	0.13	5.4	27.12	20.50	0.25	0.08	0.33	0.002	0.03	0.10	0.017	6
		3044.0	6.19	0.69	0.16	4.9	27.85	20.85	0.46	0.08	0.26	0.002	0.04	0.07	0.016	8
		3045.0	6.11	1.17	0.14	5.0	27.90	20.58	0.29	0.10	0.26	0.002	0.03	0.07	0.016	30
		3045.9	6.23	0.60	0.11	7.5	26.45	20.01	0.43	0.14	0.42	0.003	0.04	0.12	0.015	8
		3046.9	6.43	0.94	0.12	3.3	27.76	21.37	0.22	0.06	0.25	0.002	0.03	0.06	0.015	7
		3047.5	6.56	0.87	0.53	7.3	26.58	19.81	0.60	0.13	0.42	0.003	0.03	0.11	0.015	14
		3047.8	6.54	0.44	0.14	4.8	26.62	19.58	0.69	0.06	0.27	0.002	0.03	0.07	0.015	5
		3048.8	5.48	0.54	0.11	5.1	26.94	20.34	0.48	0.10	0.38	0.003	0.03	0.11	0.014	4
		3049.7	6.51	0.24	0.20	5.1	26.95	20.29	0.47	0.07	0.18	0.002	0.03	0.05	0.014	5
		3050.7	6.47	0.15	0.21	8.1	26.49	19.85	0.45	0.12	0.39	0.003	0.04	0.11	0.014	10
		3051.6	6.54	0.84	0.18	5.9	27.48	19.64	0.98	0.09	0.34	0.002	0.03	0.10	0.014	7
		3052.2	6.60	1.12	0.21	5.7	27.20	20.17	0.44	0.10	0.41	0.003	0.03	0.11	0.015	10
		3052.6	6.54	0.52	0.20	7.7	26.61	19.64	0.76	0.10	0.40	0.003	0.04	0.11	0.014	5
		3054.5	6.44	0.23	0.15	7.1	26.25	19.90	0.24	0.15	0.55	0.003	0.04	0.14	0.014	6
		3055.4	6.30	1.18	0.21	4.8	27.90	20.12	0.77	0.11	0.29	0.002	0.03	0.08	0.015	6
		3056.4	4.47	0.62	0.27	6.1	26.20	19.90	0.30	0.10	0.34	0.002	0.04	0.10	0.015	4
		3057.3	6.30	0.76	0.21	7.6	25.77	19.48	0.17	0.11	0.48	0.003	0.05	0.13	0.015	8
		3058.0	6.46	0.70	0.16	3.6	27.23	20.69	0.15	0.08	0.33	0.002	0.04	0.08	0.015	5
		3058.2	6.24	1.02	0.18	5.2	27.24	20.30	0.50	0.08	0.24	0.002	0.03	0.06	0.015	9
		3060.0	6.38	0.48	0.13	3.9	28.28	21.52	0.18	0.11	0.41	0.002	0.04	0.11	0.016	6
		3060.4	6.62	1.14	0.13	9.9	26.00	19.32	0.15	0.24	0.80	0.003	0.04	0.22	0.016	7
		3061.0	6.24	0.32	0.40	4.5	27.88	20.02	0.79	0.11	0.30	0.002	0.03	0.08	0.013	7
		3061.9	6.37	0.00	0.20	15.1	24.17	18.00	0.20	0.25	0.74	0.003	0.05	0.20	0.015	12
		3062.8	6.30	0.77	0.28	7.6	26.59	20.12	0.29	0.14	0.63	0.004	0.04	0.17	0.013	8
		3063.7	6.54	0.99	0.17	8.9	25.89	19.10	0.30	0.21	0.73	0.004	0.05	0.19	0.014	7
		3065.5	6.07	1.19	0.21	7.3	26.29	19.61	0.14	0.20	0.65	0.005	0.04	0.16	0.016	11
		3068.4	6.16	0.30	0.24	8.3	27.47	20.30	0.16	0.16	0.57	0.004	0.04	0.14	0.015	8
		3069.3	6.54	0.71	0.31	5.5	27.01	20.24	0.10	0.10	0.40	0.002	0.03	0.10	0.016	5
		3078.2	6.37	1.17	0.21	22.3	22.16	16.56	0.14	0.33	0.98	0.005	0.05	0.22	0.019	14
		3081.5	6.60	1.04	0.46	11.2	27.26	15.84	3.72	0.22	0.91	0.006	0.05	0.23	0.014	9
		3100.1	6.72	1.25	0.18	6.0	27.05	20.19	0.21	0.11	0.54	0.004	0.04	0.13	0.017	40
		3105.4	6.36	0.86	0.15	3.7	28.12	20.87	0.22	0.04	0.17	0.002	0.03	0.05	0.013	27
		3107.4	6.48	0.94	0.19	3.6	27.94	20.69	0.33	0.06	0.26	0.002	0.03	0.07	0.013	195
		3111.2	6.31	1.65	0.23	4.5	27.11	20.11	0.29	0.17	0.32	0.004	0.04	0.06	0.014	17

<, below than detection limit

Cu (ppm)	Ni (ppm)	V (ppm)	Cr (ppm)	Pb (ppm)	Sr (ppm)	Ba (ppm)	Mn* 10000/Sr	Mn/Al	Ca/Al	V/Cr	Mn/Fe	K_2O/Al_2O_3	Ti/Al	Fe_2O_3/Al_2O_3	Ca/Mg	$CaO + MgO/Al_2O_3$
5	3	18	5	5	585	6	0.33	0.25	16.34	3.6	0.06	0.12	0.00233	0.56	0.83	105.4
6	5	11	10	<5	354	10	0.30	0.06	10.81	1.1	0.03	0.15	0.00239	0.32	1.56	50.5
7	<2	6	3	6	915	5	0.10	0.38	86.72	2.0	0.13	0.15	0.00245	0.38	2.17	355.9
2033	10	2	5	92	1076	6	0.05	0.08	30.81	0.4	0.02	0.19	0.00258	0.46	2.87	115.6
9	<2	8	4	<5	1021	4	0.07	0.19	53.08	2.0	0.08	0.32	0.00335	0.32	2.02	223.6
4	<2	8	2	<5	766	17	0.12	0.24	55.60	4.0	0.13	0.35	0.00318	0.25	1.94	238.0
5	<2	8	<2	<5	893	5	0.09	0.32	76.98	8.0	0.15	0.43	0.00227	0.29	2.10	319.6
3	<2	8	4	<5	596	26	0.23	0.35	49.33	2.0	0.16	0.48	0.00303	0.29	1.56	230.6
3	<2	4	<2	<5	927	6	0.05	0.49	179.65	4.0	0.11	0.20	0.00318	0.60	4.62	599.8
4	<2	<2	<2	<5	386	7	0.40	0.45	58.54	1.0	0.15	0.28	0.00353	0.39	1.72	262.3
10	<2	2	2	<5	374	19	0.16	0.13	41.78	1.0	0.06	0.12	0.00245	0.27	2.00	176.8
22	<2	3	6	28	97	14	0.95	0.41	85.61	0.5	0.11	0.17	0.00133	0.50	1.66	389.7
22	<2	2	3	15	112	13	0.96	0.52	96.54	0.7	0.19	0.18	0.00145	0.36	1.67	438.5
10	<2	2	2	20	135	19	0.63	0.30	71.62	1.0	0.15	0.13	0.00212	0.27	1.68	324.1
9	<2	54	4	14	123	25	1.06	0.22	32.35	13.5	0.13	0.16	0.00205	0.23	1.62	148.7
12	<2	5	<2	19	285	14	0.46	0.41	62.73	5.0	0.18	0.24	0.00187	0.29	1.82	274.9
16	<2	2	<2	12	182	6	0.72	0.99	152.13	2.0	0.46	0.14	0.00227	0.29	1.75	678.3
11	<2	13	4	<5	268	28	0.78	0.34	33.33	3.3	0.16	0.16	0.00298	0.28	1.76	148.0
10	<2	14	4	16	365	39	0.57	0.92	91.28	3.5	0.29	0.17	0.00530	0.42	1.89	394.2
11	<2	8	<2	15	475	8	0.31	1.94	284.67	8.0	1.02	0.25	0.00398	0.25	2.07	1188.0
8	<2	3	<2	21	231	34	0.60	1.84	270.75	3.0	0.97	0.25	0.00398	0.25	1.83	1184.8
10	<2	<2	<2	23	123	6	1.31	2.86	367.65	1.0	1.13	0.33	0.00530	0.33	1.66	1673.0
10	<2	2	<2	19	510	29	0.24	1.63	287.02	2.0	0.86	0.25	0.00398	0.25	2.14	1183.8
9	<2	<2	<2	<5	125	29	1.17	2.58	361.00	1.0	2.05	0.33	0.00530	0.17	1.66	1644.7
15	<2	2	<2	6	210	10	0.81	1.80	221.41	2.0	1.19	0.20	0.00318	0.20	1.77	982.2
11	<2	<2	<2	5	248	23	0.56	1.05	155.25	1.0	0.49	0.14	0.00227	0.29	1.80	684.0
12	<2	<2	<2	9	120	40	1.22	0.97	133.68	1.0	0.51	0.25	0.00199	0.25	1.59	619.0
9	<2	<2	<2	11	644	2	0.17	1.90	381.07	1.0	0.75	0.33	0.00530	0.33	2.12	1577.3
7	4	25	3	14	215	17	0.86	0.20	21.57	8.3	0.12	0.13	0.00199	0.23	1.62	99.1
8	3	11	<2	9	264	24	0.53	0.73	111.57	11.0	0.49	0.20	0.00159	0.20	1.75	496.5
10	<2	9	<2	5	715	20	0.20	0.86	131.04	9.0	0.51	0.11	0.00177	0.22	2.29	528.0
6	<2	4	2	6	712	5	0.05	0.26	145.35	2.0	0.07	0.25	0.00398	0.50	2.49	569.5
8	5	10	6	9	581	43	0.23	0.15	21.62	1.7	0.04	0.28	0.00203	0.45	1.63	99.3
7	7	11	8	13	814	52	0.16	0.12	15.37	1.4	0.03	0.29	0.00216	0.53	1.74	68.6
10	<2	10	4	12	14652	157	0.01	0.20	32.59	2.5	0.10	0.27	0.00212	0.27	1.70	146.9
19	4	15	6	10	244	14	0.57	0.18	24.26	2.5	0.09	0.27	0.00233	0.27	1.58	112.6
10	<2	6	3	<5	2882	123	0.05	0.39	55.71	2.0	0.23	0.28	0.00177	0.22	1.56	260.2
7	4	8	5	16	78	13	1.48	0.11	16.62	1.6	0.05	0.24	0.00219	0.26	1.58	77.2
7	3	7	4	12	1880	51	0.07	0.21	30.93	1.8	0.11	0.30	0.00193	0.24	1.57	144.3
9	3	5	3	13	547	30	0.23	0.25	40.31	1.7	0.11	0.27	0.00245	0.31	1.58	187.3
17	2	4	3	13	278	13	0.44	0.25	40.38	1.3	0.09	0.27	0.00245	0.38	1.60	186.5
6	2	6	5	12	2560	68	0.05	0.15	23.70	1.2	0.06	0.29	0.00227	0.33	1.56	110.6
93	4	4	3	9	1474	29	0.08	0.24	41.78	1.3	0.13	0.24	0.00254	0.24	1.54	196.5
11	3	6	4	16	10878	158	0.01	0.15	23.81	1.5	0.06	0.26	0.00227	0.31	1.59	110.5
9	<2	5	3	14	230	9	0.50	0.23	37.10	1.7	0.13	0.26	0.00236	0.22	1.61	171.1
7	2	8	4	9	2580	62	0.04	0.15	26.68	2.0	0.08	0.29	0.00251	0.26	1.57	124.4
7	<2	5	3	8	6208	98	0.02	0.32	56.34	1.7	0.11	0.28	0.00353	0.39	1.57	262.4
5	3	7	4	11	2582	54	0.04	0.15	25.56	1.8	0.06	0.28	0.00245	0.31	1.58	118.8
7	3	7	4	20	407	17	0.26	0.17	30.41	1.8	0.08	0.29	0.00187	0.26	1.66	138.6
16	2	8	4	9	2654	61	0.04	0.15	24.96	2.0	0.08	0.27	0.00233	0.24	1.60	115.5
5	<2	7	4	8	4941	100	0.02	0.14	25.03	1.8	0.08	0.28	0.00239	0.25	1.60	115.6
6	2	9	5	10	314	13	0.34	0.10	17.96	1.8	0.05	0.25	0.00173	0.27	1.56	83.9
14	3	7	3	8	4396	89	0.03	0.21	36.20	2.3	0.07	0.28	0.00219	0.38	1.64	165.6
4	<2	8	4	5	4105	49	0.03	0.18	29.00	2.0	0.08	0.29	0.00187	0.29	1.56	135.6
8	3	9	5	24	835	24	0.14	0.13	20.20	1.8	0.07	0.27	0.00199	0.23	1.57	94.3
5	<2	8	4	12	161	18	0.72	0.19	31.05	2.0	0.10	0.24	0.00193	0.24	1.56	145.2
37	<2	4	3	44	136	5	0.85	0.26	42.71	1.3	0.10	0.25	0.00265	0.33	1.59	198.1
6	<2	6	5	6	184	33	0.67	0.16	25.96	1.2	0.08	0.27	0.00155	0.27	1.56	121.5
8	4	11	7	7	113	17	1.09	0.08	12.23	1.6	0.04	0.28	0.00119	0.30	1.59	56.7
5	<2	5	4	<5	150	5	0.67	0.18	34.97	1.3	0.06	0.27	0.00212	0.37	1.65	159.7
2	5	11	8	20	88	14	1.31	0.08	12.29	1.4	0.03	0.27	0.00129	0.34	1.59	57.0
5	2	8	6	9	101	11	0.99	0.08	15.88	1.3	0.05	0.27	0.00202	0.22	1.56	74.1
3	4	9	7	10	103	11	1.05	0.08	13.35	1.3	0.04	0.26	0.00174	0.29	1.60	61.6
30	4	9	7	14	74	8	1.66	0.10	15.22	1.3	0.04	0.25	0.00245	0.31	1.59	70.6
12	3	8	5	9	102	40	1.13	0.11	18.14	1.6	0.05	0.25	0.00223	0.28	1.60	83.8
6	<2	6	4	<5	67	10	1.84	0.16	25.41	1.5	0.09	0.25	0.00159	0.25	1.58	118.1
35	8	18	9	27	81	9	1.81	0.08	8.51	2.0	0.03	0.22	0.00162	0.34	1.58	39.5
12	4	11	7	12	325	17	0.33	0.06	11.27	1.6	0.03	0.25	0.00210	0.24	2.04	47.4
44	4	11	5	36	91	10	1.44	0.13	18.85	2.2	0.08	0.24	0.00236	0.20	1.59	87.5
16	<2	6	2	18	94	7	1.06	0.31	62.24	3.0	0.18	0.29	0.00374	0.24	1.59	288.2
15	<2	6	3	8	81	4	1.24	0.20	40.44	2.0	0.12	0.27	0.00245	0.23	1.60	187.0
32	9	9	5	28	80	9	1.35	0.18	31.88	1.8	0.04	0.19	0.00398	0.53	1.60	147.6

Table 2. *Cumulative data base of the bulk-rock Main Dolomite samples form the studied carbonate platform palaeogeographic zones discussed in this paper*

Palaeo-geographic zone		$\delta^{13}C$ (‰ PDB)	$\delta^{18}O$ (‰ PDB)	TOC (%)	IR (%)	CaO (%)	MgO (%)	S (%)	Fe_2O_3 (%)	Al_2O_3 (%)	TiO_2 (%)	Na_2O (%)	K_2O (%)	MnO (%)	Zn (ppm)	Cu (ppm)	Ni (ppm)	V (ppm)	Cr (ppm)	Pb (ppm)	Sr (ppm)	Ba (ppm)
Lagoon	Mean	7.16	1.93	0.09	8.19	27.01	16.11	5.53	0.10	0.26	0.0022	0.07	0.06	0.02	14.44	147.19	2.50	5.69	3.19	10.38	791.00	8.97
	Standard deviation	0.80	0.98	0.05	8.22	3.63	4.63	2.63	0.08	0.22	0.002	0.03	0.03	0.007	20.95	506.65	2.94	4.48	2.32	22.14	252.82	6.39
	Range	3.47	3.84	0.17	26.5	12.49	19.3	8.34	0.27	0.88	0.007	0.12	0.13	0.021	83	2031	9	17	9	89	910	25.5
	Minimum	4.68	−0.63	0.03	2.2	17.8	6.12	0.83	0.03	0.05	0	0.03	0.01	0.01	4	2	1	1	1	3	354	0.5
	Maximum	8.15	3.21	0.2	28.7	30.29	25.42	9.17	0.3	0.93	0.01	0.15	0.14	0.03	87	2033	10	18	10	92	1264	26
	Number	16	16	16	16	16	16	16	16	16	16	16	16	16	16	16	16	16	16	16	16	16
	95% conf. interval	0.43	0.52	0.02	4.38	1.93	2.47	1.40	0.04	0.12	0.0009	0.02	0.02	0.003	11.16	269.97	1.57	2.39	1.23	11.80	134.72	3.40
Barrier	Mean	7.24	2.03	0.10	4.27	28.72	18.16	2.70	0.06	0.16	0.0012	0.05	0.03	0.018	10.47	7.54	1.60	4.63	2.46	10.10	321.44	17.51
	Standard deviation	0.52	0.75	0.05	3.93	1.22	3.41	3.00	0.080	0.20	0.001	0.018	0.04	0.007	8.11	7.79	1.54	7.50	2.20	14.26	231.12	22.46
	Range	4.18	4.19	0.28	27.2	6.49	20.42	15.11	0.54	1.23	0.007	0.09	0.24	0.028	49	46	10	53	12	112	1037	156.5
	Minimum	3.65	−0.33	0.03	1.4	24.85	0.68	0.12	0.01	0.01	0	0.03	0.01	0	2	1	1	1	1	3	97	0.5
	Maximum	7.83	3.86	0.31	28.6	31.34	21.1	15.23	0.54	1.24	0.01	0.12	0.24	0.03	51	47	11	54	13	115	1134	157
	Number	68	68	68	68	68	68	68	68	68	68	68	68	68	68	68	68	68	68	68	68	68
	95% conf. interval	0.12	0.18	0.01	0.95	0.30	0.82	0.73	0.02	0.05	0.0003	0.004	0.009	0.002	1.96	1.89	0.37	1.82	0.53	3.45	55.94	5.44
Slope	Mean	6.26	0.84	0.22	7.03	26.79	19.49	0.71	0.13	0.41	0.0027	0.04	0.11	0.015	14.35	13.58	2.77	7.91	4.58	12.63	1847.93	41.84
	Standard deviation	0.43	0.76	0.12	3.77	1.34	1.55	1.30	0.07	0.21	0.001	0.01	0.05	0.002	29.25	15.89	1.94	3.26	1.69	8.57	3203.78	46.31
	Range	2.31	5.09	0.61	19.60	8.33	7.74	7.89	0.30	0.93	0.006	0.08	0.22	0.01	192	92	8	17	8	41	14585	193
	Minimum	4.47	−0.98	0.04	2.70	22.16	13.78	0.08	0.03	0.05	0.0005	0.03	0.01	0.010	3	1	1	1	1	3	67	3
	Maximum	6.78	4.11	0.65	22.30	30.49	21.52	7.97	0.33	0.98	0.01	0.11	0.23	0.02	195	93	9	18	9	44	14652	196
	Number	43	43	43	43	43	43	43	43	43	43	43	43	43	43	43	43	43	43	43	43	43
	95% conf. interval	0.13	0.23	0.04	1.16	0.41	0.48	0.40	0.02	0.06	0.0004	0.004	0.02	0.0005	9.00	4.89	0.60	1.00	0.52	2.64	985.98	14.25

reversed trend. The platform slope facies reveal no or opposed $\delta^{13}C$ vertical trends, and most of the carbon isotope values are grouped around +6.3‰ to +6.5‰ (Fig. 5).

The $\delta^{18}O$ values of all the platform dolomites vary from −1‰ to +4.1‰, and the average value is +1.6‰ (Table 2). The average $\delta^{18}O$ values vary across palaeogeographic zones. The values are similar in the barrier and inner platform facies and are relatively higher (>1‰ at average) than the platform slope facies (Table 2). The lagoonal facies show relatively large-amplitude variations of the $\delta^{18}O$ values. A sharp increase by *c.* 1.5‰ in the lower part of the MD carbonates is followed by a distinct tendency to a lowering of the $\delta^{18}O$ values in the upper part. The same trend is visible in other wells from this palaeogeographic zone (e.g. Kamień Pomorski-7). In the barrier zone, the oxygen isotope values of the dolomites show a fluctuating but gradually increasing trend from the base to the top of the unit. In the platform slope facies, the $\delta^{18}O$ values reveal a slightly decreasing trend upwards, with a prominent fluctuation in the upper part of the unit. A general shift to higher $\delta^{18}O$ values from the base to the top of the unit is typical for the platform barrier sections, for example, in wells Benice-1 (Fig. 5), Żółwino-2, Świerzno-5, Wrzosowo-2 and Gryfice-3.

Major and minor elements distribution

Geochemical composition. Because of the dominant dolomitic lithology, CaO and MgO constitute the bulk geochemical composition of the platform deposits (Table 2). The platform dolomites also contain subordinate but common amounts of other major elements such as IR and S, occurring at relatively low abundance (a few percent or less on average). Average concentrations of other important components like TOC, Al_2O_3 and Fe_2O_3 are <0.5%, and TiO_2, Na_2O, K_2O and MnO occur in trace amounts (<0.1%). Except for TOC, these elements represent typical silicate components. Positive linear trends between Al_2O_3 and K_2O and TiO_2, as well as K_2O and Cr, expressed by high correlation coefficients (Table 3), indicate a primary influx of a clay minerals (aluminosilicate) fraction to the carbonate setting. The dolomites contain generally high Sr concentrations (*c.* 0.1% on average), much higher than other trace elements such as Zn, Cu, Pb and Ba (a few tens of ppm) or V, Cr and Ni (<10 ppm at average). The ranges of abundance of most elements are highly variable (Table 2).

Correlation matrices show two main elemental associations dominating the carbonate platform dolomites that relate to primary sedimentary conditions (Table 3). The first association comprises CaO, MgO and Sr and is related to carbonates, whereas the second is related to lithogenic elements. The other elements form subordinate to exotic elemental associations which, except for sulphur, are volumetrically insignificant and essentially related to various diagenetic processes.

Magnesium is inversely related to calcium in bulk dolomites (Table 3). A weak and positive correlation between these two elements occurs in the platform slope deposits (Table 3) where limestones (apparently of dedolomitic origin) also appear locally. It can be expected that a small admixture of limestone fraction in the bulk dolomitic lithology in this platform section has no substantial influence on the qualitative and quantitative chemical composition of elemental associations other than carbonate. Magnesium reveals a positive and weak to distinct correlation with some lithogenic elements found in inner platform dolomites (Table 3). Like Ca, MgO does not show correlation with Sr in the bulk dolomite analyses. Instead, it reveals complex relationships with other elements in particular palaeogeographic zones of the platform. A characteristic feature is a distinct and positive correlation between CaO and Sr, and a negative correlation between MgO and Sr, in the inner and barrier zones (Table 3). The association of Ca and Sr is typical for carbonates because Sr is usually incorporated into calcite (Morse & Mackenzie 1990). Generally high Sr concentrations in the platform dolomites are variable across the platform zones: the platform barrier zone is relatively Sr-poor on average, whereas the slope setting is Sr-rich (Table 2).

The second association is essentially related to IR concentrations, as the prominent component of the investigated deposits (Tables 1 & 2). Petrographic study of thin sections suggests that the IR is heterogeneous and dominated by various silicate minerals. In general, there is no correlation or only a weak to distinctly negative correlation between the elements of both elemental associations. There are variable but positive, from weak to very strong, correlations between silicate components (Fe_2O_3, Al_2O_3, TiO_2, Na_2O, K_2O) and IR (Table 3). Overall, significant to strongly positive correlations occur between elements of the silicate association in all the main platform settings (Table 3). The silicate components are generally variable, and weakly to strongly correlated with Ni, V, Cr and locally with Pb. Some alkaline elements (Na_2O, K_2O) and associated lithogenic elements like IR, Fe_2O_3, Al_2O_3, TiO_2 + Ni, V and Cr are weakly to highly positively correlated with TOC.

Secondary elemental associations (Table 3) are reflected by the following correlations: positive correlations between S and CaO and negative correlations between S and MgO; positive and weak to

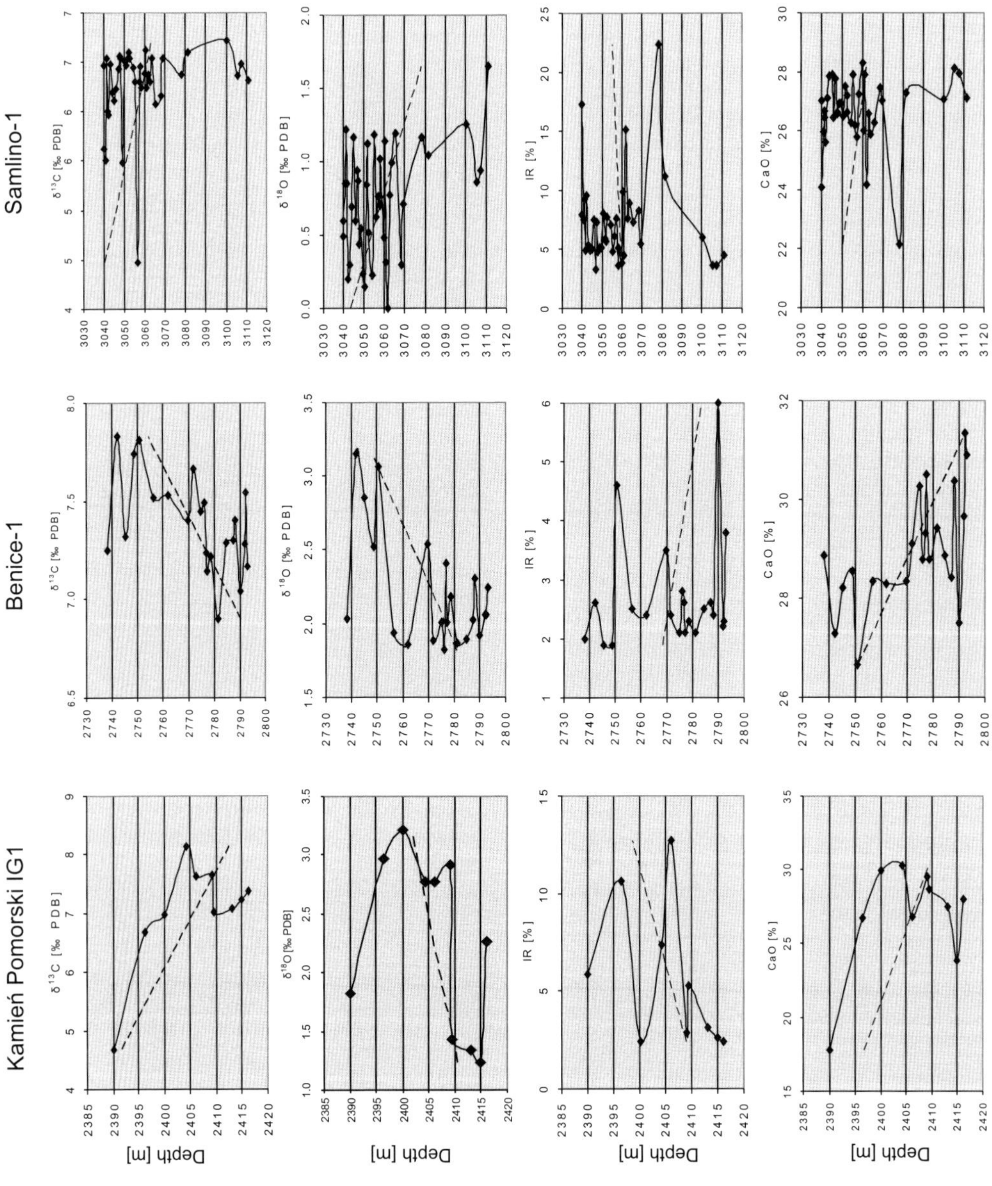
Samlino-1
Benice-1
Kamień Pomorski IG1
δ¹³C [‰ PDB]
δ¹⁸O [‰ PDB]
IR [%]
CaO [%]
Depth [m]

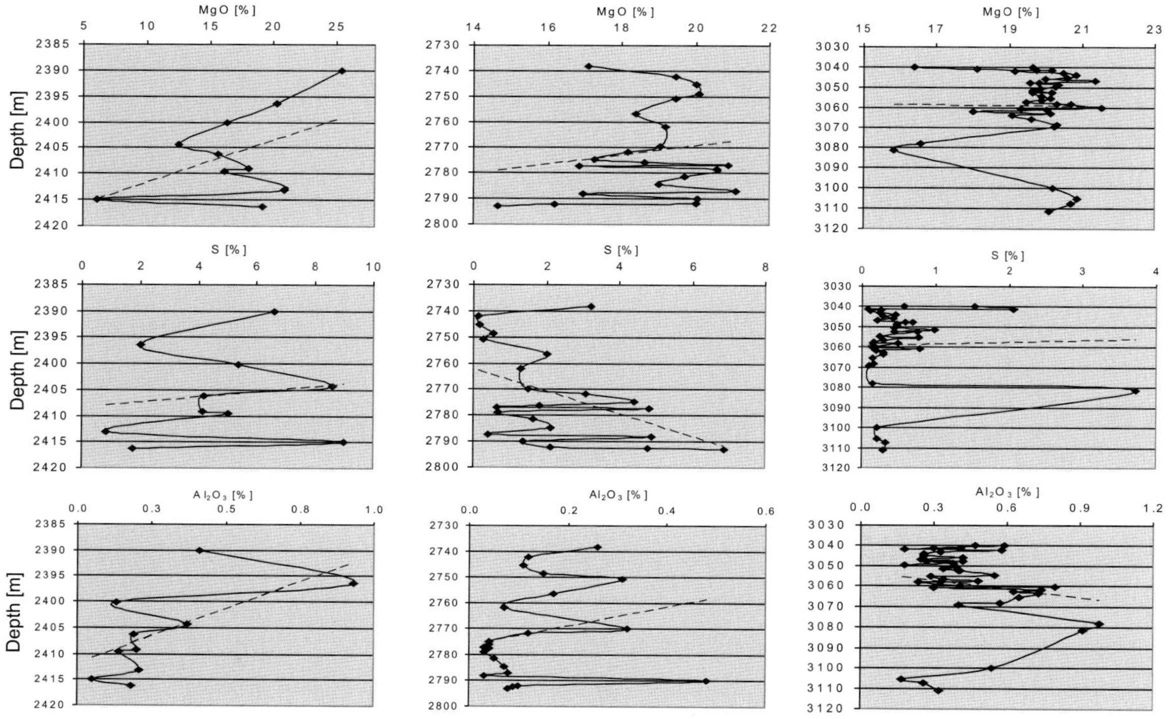

Fig. 5. Vertical distribution of bulk-rock $\delta^{13}C$ and $\delta^{18}O$ values and main elemental content in the Main Dolomite deposits section from three boreholes of the Kamień Pomorski carbonate platform representing each of the studied palaeogeographic zones, from the platform lagoonal to slope settings.

Table 3. *Correlation matrixes of bulk isotopic data and elemental concentrations for the main palaeogeographic zones in the investigated Main Dolomite carbonate platform area*

Palaeo-geographic zone		$\delta^{13}C$	$\delta^{18}O$	TOC	IR	CaO	MgO	S	Fe_2O_3	Al_2O_3	TiO_2	Na_2O	K_2O	MnO	Zn	Cu	Ni	V	Cr	Pb	Sr	Ba
Lagoon	$\delta^{13}C$	1																				
	$\delta^{18}O$	0.02	1																			
	TOC	−0.53	0.31	1																		
	IR	0.20	0.12	−0.12	1																	
	CaO	0.45	0.03	−0.28	−0.37	1																
	MgO	−0.50	0.12	0.52	−0.34	−0.20	1															
	S	−0.03	−0.12	0.16	0.26	−0.11	−0.66	1														
	Fe_2O_3	−0.49	0.23	0.72	0.19	−0.27	0.34	0.05	1													
	Al_2O_3	−0.40	0.30	0.56	0.08	−0.02	0.28	−0.07	0.90	1												
	TiO_2	−0.34	0.27	0.48	0.09	0.02	0.28	−0.14	0.87	0.98	1											
	Na_2O	−0.59	0.22	0.54	−0.10	−0.10	0.57	−0.30	0.77	0.85	0.83	1										
	K_2O	−0.21	0.27	0.26	−0.08	0.24	0.38	−0.43	0.59	0.80	0.83	0.79	1									
	MnO	−0.31	−0.45	0.15	−0.14	−0.18	0.61	−0.37	0.09	−0.11	−0.12	0.08	−0.12	1								
	Zn	0.25	0.33	0.03	−0.14	0.31	0.00	0.03	−0.04	0.03	0.07	0.11	0.19	−0.24	1							
	Cu	0.33	0.24	0.32	0.05	0.25	−0.26	0.35	0.25	0.13	0.13	−0.15	0.12	−0.33	0.30	1						
	Ni	0.04	0.32	0.31	0.40	0.12	−0.27	0.39	0.55	0.42	0.40	0.13	0.30	−0.22	0.18	0.75	1					
	V	−0.79	0.22	0.56	−0.19	−0.43	0.62	−0.20	0.49	0.43	0.39	0.77	0.41	0.09	0.05	−0.24	−0.05	1				
	Cr	−0.34	0.36	0.64	0.04	−0.03	0.35	−0.15	0.87	0.89	0.87	0.82	0.75	−0.08	−0.01	0.21	0.45	0.53	1			
	Pb	0.28	0.27	0.37	0.04	0.25	−0.26	0.41	0.27	0.18	0.17	−0.13	0.13	−0.37	0.29	0.99	0.75	−0.23	0.22	1		
	Sr	0.17	0.09	0.01	0.13	0.29	−0.63	0.75	−0.23	−0.14	−0.19	−0.30	−0.21	−0.64	0.13	0.33	0.24	−0.20	−0.15	0.40	1	
	Ba	−0.14	−0.05	−0.02	−0.35	0.31	0.27	−0.33	−0.02	0.24	0.26	0.29	0.58	0.04	0.30	−0.13	−0.12	0.15	0.18	−0.09	−0.04	1
Barrier	$\delta^{13}C$	1																				
	$\delta^{18}O$	0.42	1																			
	TOC	−0.11	−0.17	1																		
	IR	−0.05	0.14	0.01	1																	
	CaO	−0.16	0.05	−0.29	−0.14	1																
	MgO	0.14	−0.08	0.14	−0.81	−0.36	1															
	S	−0.13	0.13	−0.19	0.71	0.50	−0.97	1														
	Fe_2O_3	−0.05	0.03	0.37	0.21	−0.49	0.08	−0.11	1													
	Al_2O_3	−0.02	0.05	0.39	0.19	−0.54	0.11	−0.15	0.97	1												

	TiO_2	−0.02	0.05	0.39	0.18	−0.45	0.05	−0.07	0.90	0.93	1											
	Na_2O	0.14	0.37	−0.09	−0.15	−0.27	0.30	−0.32	0.03	0.07	0.06	1										
	K_2O	−0.06	0.06	0.37	0.22	−0.52	0.11	−0.14	0.98	0.97	0.90	0.11	1									
	MnO	0.03	−0.03	0.17	−0.29	−0.24	0.47	−0.44	0.05	0.02	0.02	0.37	0.09	1								
	Zn	−0.11	−0.21	0.16	−0.05	−0.11	0.04	−0.08	0.26	0.19	0.32	0.03	0.24	0.21	1							
	Cu	−0.13	−0.29	0.17	−0.23	0.07	0.05	−0.10	−0.02	−0.02	0.15	−0.17	−0.06	−0.11	0.54	1						
	Ni	0.00	0.13	0.32	0.25	−0.39	0.03	−0.04	0.88	0.85	0.82	0.10	0.87	0.13	0.27	0.01	1					
	V	0.11	0.16	0.35	0.07	−0.27	0.07	−0.10	0.35	0.46	0.46	0.17	0.40	0.09	0.23	0.09	0.32	1				
	Cr	0.00	0.13	0.40	0.20	−0.47	0.07	−0.10	0.92	0.89	0.88	0.05	0.90	0.07	0.33	0.11	0.82	0.41	1			
	Pb	0.04	0.14	0.01	−0.11	0.12	−0.02	0.05	0.12	0.07	0.30	0.08	0.08	0.04	0.57	0.68	0.23	0.15	0.26	1		
	Sr	−0.10	0.11	−0.20	0.55	0.57	−0.88	0.95	−0.09	−0.12	−0.06	−0.34	−0.13	−0.38	−0.08	−0.10	−0.05	−0.09	−0.11	0.02	1	
	Ba	0.10	0.01	0.02	−0.16	0.05	0.15	−0.14	−0.02	0.00	0.00	0.02	−0.03	−0.18	−0.11	0.00	−0.03	0.05	−0.05	−0.04	−0.16	1
Slope	$\delta^{13}C$	1																				
	$\delta^{18}O$	0.25	1																			
	TOC	0.06	0.30	1																		
	IR	0.02	−0.04	0.31	1																	
	CaO	−0.01	0.02	−0.35	−0.86	1																
	MgO	0.08	−0.46	−0.34	−0.51	0.26	1															
	S	−0.15	0.36	−0.01	−0.08	0.41	−0.73	1														
	Fe_2O_3	0.07	−0.04	0.32	0.88	−0.72	−0.40	−0.13	1													
	Al_2O_3	0.17	−0.17	0.14	0.78	−0.64	−0.24	−0.18	0.86	1												
	TiO_2	0.14	−0.16	0.16	0.62	−0.48	−0.14	−0.16	0.74	0.86	1											
	Na_2O	0.05	0.48	0.61	0.45	−0.52	−0.51	−0.02	0.41	0.20	0.08	1										
	K_2O	0.15	−0.18	0.17	0.77	−0.64	−0.24	−0.19	0.85	0.98	0.82	0.24	1									
	MnO	−0.13	−0.29	−0.31	0.35	−0.20	0.01	−0.07	0.29	0.28	0.21	−0.24	0.24	1								
	Zn	0.09	0.01	−0.06	−0.14	0.16	0.14	−0.08	−0.15	−0.12	−0.07	−0.11	−0.13	−0.16	1							
	Cu	0.18	0.23	−0.09	−0.03	−0.02	0.18	−0.18	−0.02	−0.01	0.13	−0.06	−0.09	0.14	0.08	1						
	Ni	0.13	0.10	0.17	0.65	−0.54	−0.27	−0.12	0.77	0.58	0.64	0.28	0.53	0.36	−0.09	0.32	1					
	V	−0.11	−0.28	0.02	0.64	−0.54	−0.21	−0.20	0.60	0.61	0.52	0.24	0.58	0.47	−0.01	−0.02	0.53	1				
	Cr	0.08	−0.13	0.24	0.84	−0.74	−0.30	−0.24	0.93	0.90	0.76	0.42	0.90	0.33	−0.14	−0.03	0.69	0.74	1			
	Pb	0.20	0.07	−0.06	0.24	−0.29	0.03	−0.19	0.21	0.23	0.34	−0.01	0.16	0.16	0.19	0.40	0.41	0.17	0.20	1		
	Sr	−0.14	0.25	0.43	−0.05	−0.13	−0.14	0.08	−0.20	−0.29	−0.28	0.24	−0.25	−0.27	−0.12	−0.07	−0.24	−0.14	−0.23	−0.16	1	
	Ba	−0.08	0.24	0.50	−0.03	−0.15	−0.11	−0.01	−0.17	−0.31	−0.32	0.38	−0.25	−0.23	−0.16	−0.12	−0.21	−0.13	−0.20	−0.26	0.91	1

distinct correlations between Mg and Mn; usually variable but distinct to strong correlations between Cu, Pb, Zn and Ni and occasionally with Fe and S (lagoon section); positive and strong correlations between S and Sr (shallow platform zones); and strongly positive correlations between Sr and Ba in the slope section.

Positive distinct to strong relationships occur between Cu, Pb, Zn, Ni and partly Fe_2O_3, especially in the inner and barrier settings (Table 3). They are positively but weakly correlated with S and TOC. Except for Ni, they are weakly correlated with the main lithogenic elements. The TOC concentration is commonly very low (commonly $<0.15\%$), in the shallow MD platform environments (Table 1).

Lateral and stratigraphic patterns. Both the horizontal and vertical tendencies of elemental distributions are meaningful in the context of this study. The horizontal distribution of averaged values of the elements is variable (Table 2). The mean concentrations of most of the analysed elements have comparable values in the three palaeogeographic zones, and relatively exceptional averaged amounts of S and Cu, as well as Sr and Ba, occur in the inner and platform slope dolomites, respectively. The lagoonal dolomites are enriched in averaged amounts of IR, S, Na_2O, Zn and Cu, whereas the platform barrier displays the highest concentrations of CaO only. The barrier zone deposits also exhibit relatively higher ranges of variations of most elements. Comparatively, the platform slope facies have higher averaged concentrations of most major and trace elements (TOC, MgO, Fe_2O_3, Al_2O_3, TiO_2, K_2O, Ni, V, Cr, Pb, Sr and Ba) than the two other platform zones.

The major and trace element concentrations also show a large variation along the MD sections, and their vertical distribution shows different trends in the three palaeogeographic zones of the carbonate platform (Table 1, Fig. 5). In general, the elemental trends mark the geochemical differentiation of the carbonate platform into two distinct areas, a shallow part that includes the inner and barrier zones, and a deeper one related to the platform slope. From the base to the top of the MD carbonates the elements may show increasing, decreasing and no or variable trends of concentration. Most of the analysed elements of the inner platform zone deposits, such as TOC, IR, Mg, Fe, Al, Ti, Na, K, Mn, Ni, V and Cr, reveal a general increasing upward trend of concentrations, and only a few (Ca, S, Sr and Ba) show a decreasing tendency. The distribution of still others (Zn, Cu, Pb) is variable. In contrast to the elemental patterns occurring in shallow platform environments, most elements from the slope sections show decreasing (TOC, IR, Fe, Al, Ti, P, Zn, Cu, Ni, V, Cr and Pb), increasing (K, Mn, Sr and Ba) or no (Ca, Mg, S and Na) trends upward the MD sections. The barrier deposits represent a transitional zone where comparable quantities of the elements exhibit increasing (Fe, Al, K, Zn, Cu, Cr and Pb) or decreasing (IR, Ca, S, Mn, Ni, Sr and Ba) abundance upward the MD section. Some elements of this platform zone (TOC, Mg, Ti, Na and V) also show a variable or no tendency of distribution toward the top of the unit.

Both the main and associated secondary elemental platform-frame components show differentiated relationships along the platform. In general, the correlation pattern between all elements is more differentiated in the lagoonal deposits (Table 3). The characteristic features of the inner platform dolomites are as follows: generally weak to distinct correlations of TOC and MgO with lithophile elements; relationships between MgO and MnO + Na_2O as well as between K_2O and Ba; and strong positive correlations between Cu, Pb and Ni, and between S and Sr (Table 3). There are also no correlations between IR and lithogenic elements and between CaO and S (apart from a relatively higher admixture of anhydrites in this platform setting). The barrier facies seem to show a similar pattern of correlation, but generally differ in having less frequent and often weaker co-variations between lithogenic elements, especially with respect to IR and TOC (Table 3). There is also a positive and distinct correlation between IR and both S and Sr; Ca, S and Sr; Zn, Cu and Pb; and Ni and Cr. There is a lack of relationships between Na_2O and typical lithogenic elements including IR and TOC. The slope setting is marked by weak correlations of Ca with S and MgO, as well as between MnO and Ni, V and Cr (Table 3). Typically, TOC concentrations in this platform zone display insignificant co-variation with silicate components and IR, and TOC significantly correlates only with Ba and Na_2O. There is also a very strong relationship between Ba and Sr.

Averaged concentrations of the main components (Table 2) are crucial for palaeogeographic features. The lagoonal zone is relatively depleted (43.1%) in carbonates (represented by a sum of CaO and MgO) compared with barrier and slope deposits (46.9% and 46.3%, respectively) (Fig. 6a). The horizontal distribution of averaged amounts of other main constituents like IR and S is also strongly variable. Generally, the carbonate platform deposits contain distinctly less IR compared to the surrounding basinal lithofacies (Gąsiewicz & Wichrowska in press). The inner, more coastward platform zone is enriched on average in IR concentration (8.2%) compared to the barrier and slope facies (4.3% and 7.0%, respectively) (Table 2). Typically, the lowest IR concentration is observed in carbonates of the platform barrier zone. The distribution of averaged concentrations of S exhibits a lowered tendency across the platform and changes from 5.5% in the lagoonal through 2.7% in the barrier to 0.7% in the slope facies. The distribution of averaged concentrations of composite concentrations of other subordinate silicates, mainly aluminosilicates ($Fe_2O_3 + Al_2O_3 + TiO_2 + K_2O$), has a similar pattern to the IR and varies across the platform from 0.5% through 0.3% to 0.7%. The total of the averaged concentrations of all of these main components of the platform indicates that the lagoonal zone is enriched in the major components (57.3%) compared to the platform barrier (54.2%) and slope (54.7%) zones (Fig. 6b).

Interpretation

Fossil carbonate platforms usually record regional trends of carbonate system development that are related to climate, oceanography and basin setting (Wilson 2012, and references therein). The use of various proxy geochemical signals enables us to detect possible palaeoclimatic influences recorded in ancient sedimentary settings. It also seems reasonable to expect that, in a generally shallow epicontinental and evaporitic sea like the SPB, surrounded by dry and large continental masses, climatic perturbations may be recorded distinctively.

Changes in the influx of meteoric water in ancient evaporite basins might be detectable by studying the mineralogical composition of detrital phases (clay, quartz and other clastic grains) and the chemical and stable-isotope composition of carbonate phases along a section from margin facies to the centre of the basin (Magaritz 1987). Consequently, terrigenous input to the carbonate-platform depositional environment is not an uncommon phenomenon in ancient and recent environments (e.g. Algeo *et al.* 2011 and Wilson 2012, and references therein) and has various implications for sedimentary basin evolution. Thus, relics of palaeoclimatic influences have to be differentially recorded across carbonate platforms, usually better in shallow and more proximal settings than in deeper and seaward settings. With respect to the Zechstein carbonate platforms, however, such an approach is rather unique to date. Examination of the regional geological background and bulk geochemical data reveals various relationships that reflect climatic impacts on the Kamień Pomorski carbonate platform deposition. Correlations and trends of geochemical proxies like carbon and oxygen values, major and trace element data, facilitate reconstruction of the general evolution of sedimentary environments and the associated climatic conditions during the MD deposition in the northern margin of the SPB.

General sedimentary conditions

The studied carbonate platform had a shallow open lagoon that was bordered to the SE by a narrow discontinuous sandy carbonate barrier. The oolitic barrier was formed by *in situ* precipitation of calcium carbonate in peritidal subenvironments forming as in modern environs, with abundant coated grains, mainly ooids, at a water depth of a few metres (e.g. Tucker & Wright 1990, and references therein). The platform margin was surrounded by a varying-width subtidal slope zone passing into the open Zechstein Sea.

The correlation pattern between Fe, Al, Ti, Na, K and IR (in the slope section) (Table 3) indicates they are related to the silicate fraction, implying a clastic derivation and lithophile provenance. The elemental co-variation between other non-lithogenic elements such as Ni, V, Cr and Pb suggests diagenetic enrichment or a hydrogenous origin and connection to the aluminosilicate fraction, of which Al and Ti are common components. The elemental co-variation suggests positive weak to strong correlations between the major silicate elements and some trace concentrations of Ni, V and Cr suggest their partial adsorption to clastic fraction components, especially clays, although bonding V to organic matter is also commonly noted (Scheffler *et al.* 2006).

The clastic contribution to the carbonate platform deposits reveals spatial and temporal variation reflected by changes in major and trace element compositions. A generally small admixture ($<10\%$ at average) of detrital fraction (a sum of IR, Fe_2O_3, Al_2O_3, TiO_2 and K_2O concentrations) in the three main platform zones (Table 3) does not allow us to classify the studied platform area as a carbonate-clastic shelf.

The MD platform is lacking the coarser soil-derived material that is usually transported by rivers, particularly during hot and humid episodes (cf. Potter 1978). It was located generally more seaward, and only fine- and very fine-grained clastic components could have been added to the growing platform sediments. In such a palaeogeographic situation, runoff was insufficient to prohibit carbonate development and delivered only very fine detrital material. It has therefore accumulated relatively more abundantly in lagoonal and slope settings. The observed palaeogeographic distribution of silicate elements in the investigated platform (Table 2) is in accord with the sediment distributions found in both recent and ancient carbonate platforms where coeval clastic influx is common in the coastal carbonates and typically low-relief features develop only in very shallow waters (e.g. Osleger 1998; Gischler *et al.* 2010; Algeo *et al.* 2011; Wilson 2012). The landward sections of

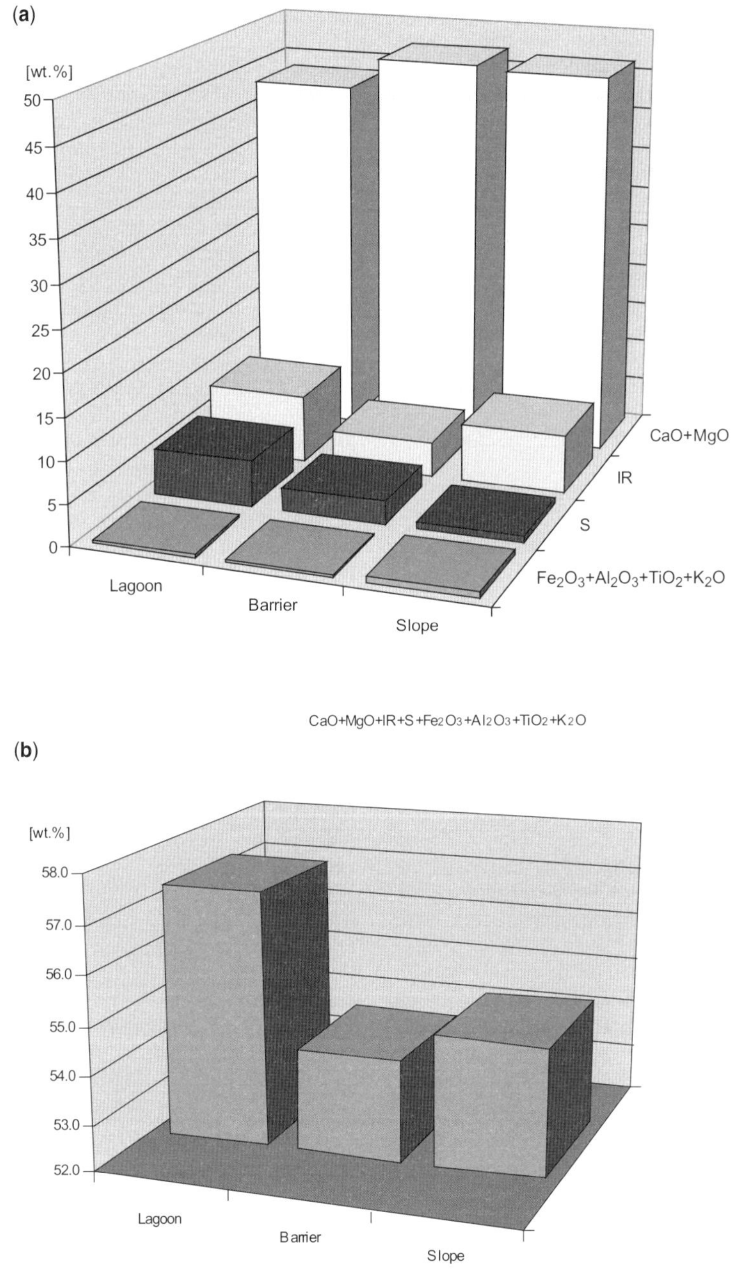

Fig. 6. Palaeogeographic distribution of bulk-rock main elemental concentrations in the Main Dolomite deposits. (**a**) Distribution of main elemental contents. (**b**) Distribution of the sum of bulk-rock main carbonate and silicate elements content in the carbonate platform palaeogeographic zones. The diagrams are based on cumulative elemental data collected in Table 2.

recent carbonate platforms include alluvial and marine siliciclastic and carbonate deposits, mid-shelf successions are dominated by carbonates, and outer shelf and basinward sections include carbonates and siliciclastics as well as lowstand alluvial siliciclastics indicative of shelf bypassing (e.g. Dunbar & Dickens 2003; Gischler *et al.* 2010). However, very little terrigeneous material may be supplied to carbonate deposition in arid climates, as in the recent Trucial Coast from the adjacent Arabian landmass (Purser & Evans 1973).

Complex relationships between elements like S, Mg, Mn, Cu, Pb, Zn, Ni and Fe are wholly or partly related to precipitation of a range of minerals including dolomite, celestite, gypsum/anhydrite and possibly sulphides in a diagenetic domain of the studied platform deposits. The diagenetic conditions of their formation are complex and may be reflected by elemental relationships. For example, Fe and Mn have very similar geochemical behaviour, but in the analysed deposits Fe shows a generally strongly positive correlation with lithogenic elements and Mn only local and very weak (and commonly no) correlation with lithogenic elements. The lack of correlation suggests that at least some Mn occurs in oxyhydroxide phases. Both elements may form (ferric and manganoan respectively) oxyhydroxides as a result of local dissolution of various mineral phases and subsequent release of elements to solution (Morse & Mackenzie 1990; Burdige 1993). Moreover, Mn is commonly a trace component of the Mg–Ca carbonate crystal lattice resulting from diagenetic processes (Morse & Mackenzie 1990). Correlations between Mn and MgO, visible in shallow platform dolomites, especially in the inner platform setting (Table 3), may be associated with Eh and pH conditions closer to neutral. The association of Mn and carbonates is common in carbonate-dominated sedimentary settings and is related to the chemistry of diagenetic conditions (redox cycling). Such conditions enable not only co-precipitation with Mg as a carbonate component or late mobilizations in more differentiated diagenetic stages, but also precipitation of Fe oxides and Mn oxides in marine pore waters (Morse & Mackenzie 1990; Burdige 1993). These processes can be facilitated by organic matter. The amounts of organic matter associated with carbonate deposition and the Fe and Mn minerals that are ultimately buried in the sediments are a function of the balance between organic production and destruction in sediments. This in turn depends on the 'carbon pump' and the effects of redox conditions leading to incorporation of Fe, Mn, organic carbon and other elements in sediments (e.g. Dean & Schwalb 2002). Similarly, a slightly higher P concentration in the platform slope deposits may result from either increased biological productivity (cf. Filippelli 1997) or reflect association with Fe oxyhydroxides (McManus *et al.* 1997).

Weak but positive correlation of Cu, Pb, Zn, Ni and partly Fe_2O_3 with S (probably a blurred effect of anhydrite admixtures) and locally high concentrations of Cu and Zn suggest the presence of diagenetic base metal sulphides. Enrichment of these elements may be related to microbial mediation during organic matter transformation in reducing diagenetic microenvironments (e.g. Schultz 2004; Brumsack 2006; Ehrlich & Newman 2009, and references therein) resulting from the common combination of weakly sulphidic pore-water chemistry (e.g. Brookins 1988) and metal supply via freshwater input. However, their weak relationships with TOC, even when normalized to Al, suggest that they may be partly associated with the lithogenic matter (and thus are of a hydrogenous nature) and with the oxyhydroxides.

As a consequence of the generally arid climatic and hot summer temperatures conditions during MD deposition, a salinity gradient was established across the restricted basin in the platform area. This is supported by various sedimentological and petrographic data from the platform. Locally abundant anhydrite accompanying lagoonal mudstones (also exemplified by an increase in S concentration; Table 1, Fig. 5) supports period(s) of hypersalinity. As may be inferred from the presented isotopic data, such conditions might have led to early diagenetic reflux of hypersaline brines through most of the shallow carbonate platform. The predominance of finely crystalline subhedral, non-zoned and same luminescence dolomites, common mimetic fabrics, and a lack of saddle dolomites and significant pre-dolomite compaction features point to early diagenetic dolomitization (Gąsiewicz & Wichrowska in press). However, in light of both the isotopic results and the petrosedimentological features, the platform dolomitization was more complex (Gąsiewicz & Wichrowska in press). Abundant evidence shows that the barrier oolitic facies were subjected to a variable degree of synsedimentary alteration induced by freshwater flow throughout more porous lithofacies (grainstones, packstones and wackestones). This suggests a mixing of saline and freshwaters as a cause of dolomitization in this platform setting. The dolomites are also present in the platform slope setting where, however, the marine isotopic composition of the dolomudstones may be explained by mixing of denser brines with connate seawater.

The Zechstein shallow carbonate platforms were well oxygenated and warm, as demonstrated by the relative diversity of preserved macro- and microfossils also common in other MD platforms (see extensive tabulation by Wagner 1994). Clearly, lower TOC concentrations in the MD inner and barrier zones compared to the slope setting (Table 2) resulted from deposition in a shallow, mostly peritidal setting where the produced organic matter was effectively degraded under oxic conditions.

Much more complex redox conditions occurred in the slope palaeoenvironment. The slope deposits are characterized by higher TOC averaged concentration (0.23%) and range (0.6%) than found in shallow platform settings (lagoonal, mean 0.09% and 0.10%; barrier, mean 0.2% and 0.3%, respectively). In this calmer palaeoenvironment, the upper part of the TOC profile shows a distinct fluctuation best expressed by a distinct tendency to lower concentrations of buried carbon in the uppermost 1–2 m interval (Table 1). This tendency apparently shows more oxygenated conditions with time. A similar conclusion may be drawn from the V/Cr ratio trends, which are commonly used as a proxy for redox conditions in ancient deposits (e.g. Jones & Manning 1994). The V/Cr ratios fluctuate highly in the MD sections of the inner and barrier platform zones and display only narrow changes in the slope setting (Table 1). Generally, the discussed trends are poorly outlined and show only a weak tendency to decrease upwards. Moreover, they are dissimilar to the trends of most of the silicate major elements. Thus, the V/Cr distribution could be partly explained by a lower concentration of clay minerals that adsorbed V, and partly by a diagenetic effect. The latter may be confirmed by fluctuations and locally higher values of the V/Cr ratio (>2) indicating locally anoxic conditions (Jones & Manning 1994). However, because of the close correlation of V and Cr with the detrital input expressed by their relation to Al_2O_3 concentration in the studied carbonates, the V/Cr ratio cannot be used as an indicator of redox conditions. Nevertheless, highly fluctuating Mn concentration and Mn/Fe ratio (Table 1) in all the platform zones seem to reflect a diagenetic effect on the local oxic/anoxic interface changes within the sediment column. This happens near the seafloor and below oxic bottom waters because Mn is a redox-sensitive element (Berner 1980; Morse & Mackenzie 1990).

The MD clastic material is associated with land-derived organic material. Fine-grained land-derived carbonaceous debris dispersed throughout the platform (discussed below) is commonly carried in suspension and typically concentrated in fluvial to turbidite overbank sediments (e.g. Wing 1984). Thus the fine organic debris suggests that the area studied was located more seaward and the platform relief was of low topography. Nevertheless, decreasing correlation coefficients between V and TOC across the platform (Table 3) from the lagoonal ($R^2 = +0.56$) to barrier ($R^2 = +0.35$) and slope ($R^2 = +0.02$) zones reflect changes in sedimentary regimes. However, mechanisms for cross-shelf transport of organic remains are still poorly constrained.

The generally higher TOC concentration in the slope deposits suggests increased palaeoproductivity in this platform zone. The deposits also seem to be relatively enriched in P concentrations compared to shallow platform deposits. The enrichment seem to be higher in the lower part of the MD section, as exemplified by Samlino-1 well, which contains more than 0.01% of P_2O_5 (max. 0.04%). In marine conditions, P is strongly coupled to the cycles of organic matter. This pattern of P distribution suggests that conditions for organic productivity and preservation were complex, but different and more favourable in the subtidal slope than in the other platform palaeoenvironments. This could be a result of a higher flux of primary organic matter produced in shallow marine areas and subsequent rapid burial or higher sedimentation rate (e.g. Filippelli 1997; Knies & Mann 2002, and references therein). A higher sedimentation rate in the platform area is in accord with sedimentological data.

Increased nutrient supply from the shallow platform and access to open marine conditions in the slope could easily stimulate high productivity in the MD surface waters and consequently would create locally increased oxygen demands. This would lead to dysoxic conditions at the sediment surface zone and to a general increase in preserved organic carbon volumes within the upper sediment column (e.g. Rullkötter 2000). Phosphorous burial efficiencies are generally lower during periods of increased palaeoproductivity and fluctuations in bottom water oxygen concentrations may have affect P burial and thus the dysoxic bottom water conditions could have enhanced or reduced autogenic P-mineral phase precipitation (Schenau *et al.* 2005). Consequently, the changing redox conditions would have explained the differences in P distribution in the MD slope deposits. Bulk sediment P data indicate P depletion over a large part of the MD slope section, implying a very effective regeneration of organic matter-bound P from the sediment under anoxic, probably even sulphidic bottom water conditions (März *et al.* 2009). The establishment of calm and locally reducing conditions is indicated by the occurrence of monotonous, darker laminated and non-laminated mudstones in the inner and especially in the platform slope deposits, which are generally enriched in TOC and contain pyrite grains. Distinct fluctuations and positive trends of Mn concentrations and Mn/Fe ratios (Table 1) along the slope MD section may support locally oxygen-depleted conditions. Such primary conditions could have facilitated the generation of intraformational hydrocarbons in the platform setting (Słowakiewicz & Gąsiewicz 2013).

Geochemical results

Across the MD succession of the platform, carbon and oxygen isotopic values typically fluctuate and show well-defined trends. They also vary between different palaeogeographic zones. Both the $\delta^{13}C$ and $\delta^{18}O$ vertical profiles demonstrate opposite trends along the section in the inner platform deposits. The isotope trends rise in the platform barrier section and decrease upwards in the slope section (Tables 1 & 2, Fig. 5). The stable oxygen and carbon isotope fluctuations and their variable trends from particular platform settings suggest that these features are generally primary and not subjected to late diagenetic alteration. This may be confirmed by no or insignificant correlations between $\delta^{13}C$ and $\delta^{18}O$ values in the main platform environments (Table 3). The isotope trends pattern across the platform suggests that evaporation and freshwater input were the factors influencing the isotopic features of shallow platform settings.

Carbon isotope data. The MD platform dolomites are generally characterized by high (6.9‰ on average) $\delta^{13}C$ values (Table 2). With respect to palaeogeographic environments the mean $\delta^{13}C$ values are essentially the same and slightly enriched in the inner and barrier zones (*c.* 7.2‰) compared to the platform slope facies (mean of 6.3‰). This increase may be a response to increased production and/or burial of organic material (Veizer & Hoefs 1976) or reflect sulphate removal (Magaritz & Schultze 1980, and references therein). The inverse correlation of average $\delta^{13}C$ values and TOC concentrations between the shallow platform zones and slope carbonates (Table 3) seems to argue for the latter mechanism.

The $\delta^{13}C$ trends recorded in the MD sections vary across the carbonate platform. They change from negative through positive to essentially none, from the more proximal (lagoonal) to distal (slope) part of the platform (Table 1). Moreover, the $\delta^{13}C$ values range is relatively wider in the lagoonal section (*c.* 3.5‰) and narrower (*c.* 1‰) in both the barrier and slope platform profiles (Fig. 5). It seems reasonable then to assume that the various $\delta^{13}C$ trends found in different core sections across the platform suggest they were controlled by a different combination of the same factors operating on open platform settings. In a normal open sedimentary system, carbon is fractionated between organic matter that is relatively concentrated in the lighter C isotope (^{12}C) and carbonate (inorganic) carbon that is heavier (enriched in ^{13}C). Consequently, variations in $\delta^{13}C$ carbonate usually result from a combined effect of participation of photosynthetic organisms preferentially taking up the light C isotope (^{12}C) and remineralization of organic matter by microbial consortia. Both processes lead to lower $\delta^{13}C$ values in the resultant carbonates (e.g. Irwin *et al.* 1977). The isotopic carbon composition is also influenced by dissolved inorganic CO_2, which is influenced by evaporation, temperature and freshwater inputs. It is interpreted thus that the fluctuating $\delta^{13}C$ values across the MD platform zones reflect a combined effect of changes in the relative decomposition of ^{13}C-depleted organic carbon and ^{13}C-enrichment of marine dissolved organic carbon by increased productivity and burial rate. In effect, shifts to more positive $\delta^{13}C$ values would reflect the increasing sequestration of ^{12}C-enriched organic material.

The fluctuations are clearly higher in shallow parts of the platform (the inner and barrier settings; Tables 1 & 2), which mirrors differentiated changes in the relative importance of the decomposition of ^{13}C-depleted organic carbon and ^{13}C-enrichment of the MD sea-dissolved inorganic carbon by increased productivity. In addition, these features imply that C-isotope trends along the lagoon to slope transect represent original (marine) composition values and were not fully affected by palaeogeography, facies changes or late diagenetic alteration. Such late alteration usually leads to diagenetic homogenization by overprinting previous values and consequently production of invariant $\delta^{13}C$ trends (cf. Carpenter & Lohmann 1997).

The lagoonal section reveals a distinct trend to more depleted $\delta^{13}C$ values upwards (Table 1). This may be explained by erosion of soils and fluxes of continent-derived water to the platform settings during MD times. Positive trends of aluminosilicate elements (Al, Fe), IR and TOC suggest an increase in land-derived material. Increased fluxes of nutrients from land would have created eutrophic conditions favouring surface productivity and microbialite growth at the bottom. This seems to be supported by increasing upwards TOC concentration (Table 1). Thus the relatively depleted $\delta^{13}C$ values in this zone may indicate increased oxidation of organic carbon and consequently a lower proportion of organic carbon burial (e.g. Veizer & Hoefs 1976). The increased weathering of land deposits that could have supplied isotopic light organic carbon from (photosynthetic) organic matter and its decomposition in a marine setting could have contributed to the relative depletion of the $\delta^{13}C$ values of carbonates marking the negative $\delta^{13}C$ values trend in this setting. The presence of land-derived ^{13}C-depleted kerogen is indicated by organic geochemistry data (Grotek 2002; Gąsiewicz & Wichrowska in press). However, a substantial flux of nutrients from the land area did not take place in the MD palaeoenvironment. First, shallow-marine stromatolitic or biolaminated lithofacies are secondary in this platform zone and the $\delta^{13}C$ data correlate weakly to distinctly negatively with most silicate elements and TOC. Second, land-derived organic material was of low abundance rather because the zone is characterized by relatively low TOC concentration (Table 2).

Nevertheless, the association of negative $\delta^{13}C$ values and positive silicate elemental trends of distribution in the studied succession (Table 1, Fig. 5) suggest that relatively more depleted $\delta^{13}C$ values may be partly related to an increased freshwater contribution (typically depleted in $\delta^{13}C$) during the MD deposition. This is in accord with the conclusions of Whittaker *et al.* (2011), who demonstrated that, in near-surface depositional systems, a wetter climate (an increase in annual rainfall) leads to lower $\delta^{13}C$ values.

The clear trend towards higher carbonate $\delta^{13}C$ values found in the MD platform barrier profile (Table 1, Fig. 5) may reflect a combined effect of land-derived nutrient delivery and consequent enhancement in marine productivity and organic carbon burial fluxes in this depositional system. Lower $\delta^{13}C$ values occur in thin biolaminated or microbialite layers where sulphate-reducing bacteria could remineralize organic matter and produce very light dissolved carbon, which could eventually be included in the precipitated carbonate (e.g. Ehrlich & Newman 2009, and references therein). However, microbialite facies in this setting are found occasionally and the barrier system was dominated by peritidal, sandy carbonate deposition, representing a relatively faster rate of sedimentation (unfavourable for microbialites) compared to other platform environments.

Carbonate sections from both shallow platform zones show distinct and opposite $\delta^{13}C$ trends. The platform slope section exhibits generally decreasing carbon isotope values (Table 1, Fig. 5). However, the $\delta^{13}C$ values fluctuate in the range of 1‰ within the upper 20 m, showing only a weak tendency to decrease. This suggests overprinting of primary isotopic values by early diagenetic fluids related to dolomitization. This platform zone is dominated by carbonate muddy lithofacies generated in a calm depositional environment favouring microbial activity and is reflected by biolaminated beds (Gąsiewicz & Wichrowska in press). The predominantly dark-coloured and millimetre-scale laminated slope mudstones ((bio)laminites), rare bioturbation and presence of microscopic pyrite through the section imply dysoxic conditions and that organic matter preservation potential was high throughout deposition (Słowakiewicz & Gąsiewicz 2013). A distinctly lower range of $\delta^{13}C$ values (2.3‰) of the dolomites from this deeper setting compared to the inner (3.5‰) and barrier (4.2‰) settings seems to be less affected by the environmental disturbances recorded in the shallow platform area. The lower average $\delta^{13}C$ value in the slope deposits (Table 2) may reflect either control of early diagenetic transformation of organic matter, and changes in the terrigenous and/or phytoplankton or autotrophic composition.

Oxygen isotope data. The MD platform dolomites have mostly positive $\delta^{18}O$ values reflecting the existence of saline waters, which is in accord with an arid climate and hot summer temperatures at that time. In such conditions, very shallow platform deposition, especially in the barrier zone, could have led to evaporation, which would have resulted in a high seawater $\delta^{18}O$ composition (Lloyd 1966) and consequently positive $\delta^{18}O$ values of the deposited carbonates (Allan & Matthews 1982). Shifts to more and less positive values can be best explained as a result of changes in freshwater runoff controlling the $\delta^{18}O$ composition of the seawater (Patterson & Walter 1994).

The $\delta^{18}O$ values vary across the platform. In the inner platform setting, $\delta^{18}O$ values increase in the lower 15 m and then decrease over 10 m. The barrier section shows a similar pattern: $\delta^{18}O$ values increase upward over a large portion of the section and then decrease in the upper 5 m. In the slope zone, $\delta^{18}O$ values decrease upward over the entire section (Table 1, Fig. 5). The dissimilar trends in $\delta^{18}O$ observed along the platform transect seem to be a function of isotopic composition of the platform water and reflect a combined effect of palaeogeography, evaporation and dilution by continental freshwater. In this context, a decrease in $\delta^{18}O$ values in the inner and barrier platform dolomites and the relatively large variations in Ca + Mg concentrations (Table 2) point to freshwater impact. Increased evaporation in this setting is revealed by a generally positive $\delta^{18}O$ values trend and an increased sulphate fraction and S content (Table 1, Fig. 5). As indicated by the major and minor geochemical data (see below), this proximal part of the platform, however, experienced an increasing supply of less saline continental waters, which would have generated a change in the $\delta^{18}O$ value trend in this setting. As a result, the $\delta^{18}O$ trend changes from increasing in the lower part to decreasing in the upper part of the MD succession (Table 1, Fig. 5). Overall, the $\delta^{18}O$ values, and the combination of decreasing $\delta^{18}O$ values with a lack of correlation between $\delta^{18}O$ and $\delta^{13}C$ values, weak correlation of $\delta^{18}O$ values with silicate content, as well as an increase in IR material and TOC amounts and a parallel decrease in carbonate content (exemplified by lowering upward CaO and MgO concentrations if normalized to Al_2O_3), argue for both evaporation and freshwater input as the main factors controlling the isotopic composition of the inner platform water. Towards the end of the MD period, the lagoon became very shallow and filled with carbonates mixed with silicate materials and increasing amounts of anhydrites, enabling finally widespread deposition of the A2 sulphates.

The distribution of $\delta^{18}O$ values in the barrier dolomites clearly indicates that they are not related to the relative abundances of calcite in the dolomites (as a common result of diagenesis), as reflected by changes in CaO and MgO concentrations (Fig. 5), but instead represent a consistent trend resulting from temporal changes in the chemistry of the dolomitizing fluids. Results of isotope analyses of the barrier dolomites suggest that the bulk carbonate oxygen isotope composition was variably affected by meteoric diagenesis. A moderate positive correlation between the $\delta^{13}C$ and $\delta^{18}O$ values ($R^2 = 0.55$) of the whole platform succession suggests that the platform depositional system underwent differentiated isotopic exchange with meteoric fluids. A relatively stronger early diagenetic meteoric impact took place in the barrier dolomites and is reflected by a lower correlation ($R^2 = 0.42$) between the isotope values (Table 3). Petrographic evidence indicates variable degrees of overprinting of the primary marine features by meteoric diagenesis. For example, accretion to sea-level fall and thus local subaerial exposures facilitated meteoric dissolution textures and vadose cements at the top of some oolitic bodies in the barrier zone (Gąsiewicz & Wichrowska in press). Such sedimentary structures are not found in the lagoonal and slope platform sections. This meteoric impact on shallow-water shoal facies was neither extensive nor thorough and probably occurred in a fluctuating manner as indicated by the $\delta^{18}O$ vertical distribution and correlation coefficients between the $\delta^{13}C$ and $\delta^{18}O$ values. This is supported by the limited dissolution features present in the platform.

The platform slope section is characterized by strong fluctuation of $\delta^{18}O$ values, and the average value (+0.8‰) is about two times lower than those occurring in the lagoon and barrier zones (+1.9‰ and +2.0‰, respectively). The platform slope dolomites are lacking primary meteoric diagenetic features. Both the lower oxygen isotope values and the trend toward decreasing $\delta^{18}O$ values upward through the slope section (Fig. 5) could have resulted from complex interactions between shallowing-up of the platform sedimentation (and thus increase of freshwater inputs) and open marine water controls.

Major and minor elements data. Different geochemical patterns delineate the three main palaeogeographic zones of the platform – that is, the inner platform, barrier and slope (Table 3, Figs 5 & 6) – and clearly delineate the shallow part of the platform (composed of the inner and barrier zones) with somewhat similar elemental features and the deeper, open marine setting (slope zone). In particular, the average admixture of IR and sulphur-related mineral phases (anhydrite) prevails volumetrically in the inner platform zone. The shallowest part of the platform (i.e. the barrier zone) is, on average, relatively impoverished in terms of most of the analysed elements that exhibit the largest distribution ranges. In contrast, the slope deposits representing deeper-water conditions where marine basin water buffered surficial conditions are relatively enriched in most of the elements. These elemental patterns indicate different sedimentary regimes operating in shallow and deeper parts of the plat-form area.

The MD carbonates building the platform are non-uniform in geochemical composition in both horizontal and vertical directions (Tables 1 & 2, Figs 5 & 6), suggesting physical mechanisms driving the deposition. The elemental distribution patterns suggest that the carbonate platform was sourced by siliciclastic material from adjacent land and point to its riverine derivation. Moreover, because of widespread aridity in the nearby continental interiors, the fluvial input could also be strengthened by dust transport to the coastal SPB area. Variable weak to very strong positive correlations between typically terrigenous components such as Fe, Al, Ti, K and locally (slope sediments) with IR (Table 3) implies a generally stable detrital source material area.

Sedimentological features of the MD deposits suggest that the combined influence of currents, wind and wave energies were not potent enough to transport coarse sediment from coastal areas to more distal settings of the platform, even under supposed storm events. These physical mechanisms explain the silicate material distribution in the shallow part of the platform. The intensities of wind and wave energies were able to transport and disperse fine, silt-sized material (clay fraction) throughout the platform area. This is best reflected by the horizontal distribution of IR concentrations (Fig. 6a). The highest average IR concentration occurs in the lagoonal zone, but a significant amount occurs also in the platform slope setting. Covariant but negative trends between the aluminosilicate elements in the platform slope deposits mark significantly different sedimentary and more open sea conditions. Enrichment of the slope zone deposits in common aluminosilicate elements and detrital particles (IR) indicates that these components could have crossed over the platform margin. The fine material from wash-over deposits associated with oolitic shoals could be distributed by channelling of tidal currents and waters loaded with much finer and slighter material could overcome or even distort the crest of the shallow bars belt, thereby flowing into more open marine environs (e.g. Ball 1967; Dravis 1979; Wanless & Tedesco 1993; Rankey & Reeder 2010). The different patterns of major element distributions in this platform zone (e.g. K and Mg) would result from restricted currents flowing around the platform and carrying various proportions of different clay minerals. Nevertheless, an elevated concentration of IR in the slope setting (Table 2) does not result from the distinctly lower sedimentation rates typical for peritidal settings. Instead, this deeper subtidal part of the platform is characterized by one of the highest Main Dolomite thicknesses in the area studied. This means that fine clastic materials could have accumulated in relatively higher amounts under low hydrodynamic regimes and higher sedimentation rates in this carbonate environment.

Both the stratigraphic and lateral distributions of both the isotopic major and minor elements across the platform indicate that the environmental effects are directly related to climatic perturbations, which are better recorded in the shallow platform setting than in the platform slope setting.

Palaeoclimatic proxies

Distribution, changes in concentration, and relationships between major and minor elements are the best proxies for the reconstruction of palaeoclimatic impact on the Kamień Pomorski carbonate platform deposition. Elements such as Mn, Sr, Ba, partly Fe, and base metals may partly result from transformations during diagenetic aging leading to the formation of complex but subordinate paragenetic mineral assemblage. This may be further supported by the regression lines of the Mn/Al ratios from different MD sections across the platform, which have the same decreasing trend upwards as the CaO/Al_2O_3 ratios (Table 1) and therefore reflect a dependence on carbonate concentrations. However, stratigraphic trends in Sr concentrations are not in concert with patterns of Fe distribution across the platform and with Mn concentrations in the lagoonal zone. This suggests that most elements of these accessory associations may be components of heavy minerals delivered by physical processes that were independent of diagenetic mineral transformations and consequently primary mineralogy. All these geochemical features point to changes in carbonate accumulation and both fluvial and aeolian delivery of detrital input that could have finally shaped the stratigraphic distribution of Fe, Mn and other metals abundances, as well as the organic matter contribution in different parts of the platform.

Carbonate accumulation. The development of the lithofacies on the platform depended on palaeogeographic zonation and was accompanied by a change in the proportions of major elements responsible for building up the carbonate framework. Ca and Mg, the main carbonate platform constituents, reveal differentiated relationships. These elements show a complex distribution pattern through space and time (Tables 1–3, Figs 5 & 6a). They show dissimilar trends of distribution – Ca concentration decreases upwards with the succession in the inner and barrier zones and remains generally undifferentiated in the slope setting, whereas Mg levels increase in the inner zone and it is undifferentiated in the rest of the platform area. This is partly a result of their geochemically antipathetic relation and partly the mixed carbonate mineralogical composition, which is non-uniform and includes dolomite, calcite and magnesite (although some Mg may also be concentrated in the aluminosilicate fraction, as indicated by a positive and weak to distinct correlation between Mg and some lithogenic elements found in inner platform dolomites). The Ca/Mg ratio exhibits variable trends of distribution upsection, with no trends in the barrier and slope settings, and a decreasing trend in the inner platform carbonates (Table 1, Fig. 7). The excess Mg concentration in this platform zone is simply linked to an increase in magnesite concentration in the upper part of the MD section and thus additionally suggests increase of evaporation in the lagoonal part of the platform. If, however, we normalize the CaO and MgO to Al_2O_3 or TiO_2 concentrations (both elements immobile at low temperatures), the trends of their stratigraphic distribution appear more clearly. The distribution of $CaO + MgO/Al_2O_3$ ratio, being a measure of carbonate concentration, shows decreasing upwards trends in the MD sections in the inner and barrier zones and is poorly differentiated in the platform slope environment (Table 1, Fig. 7). By excluding Mg as an essential diagenetic element related to dolomitization of the primary limestones framework, the visible decreasing trends in the shallow part of the platform seem to mark a decline of carbonate concentration up the MD sections.

A general tendency of a decrease in carbonate accumulation upwards the MD is accompanied by a parallel increase of some silicate components reflected by a change from relatively higher to low values of K/Al and Ti/Al ratios upwards (Table 1, Fig. 7). These features in the context of palaeogeographic conditions may have resulted from inputs of more aggressive meteoric waters leading to dilution of marine or saline waters and lowering carbonate precipitation. Such waters may also be responsible for an increase in carbonate dissolution and their local precipitation in the vadose zone, combined with concomitant precipitation in the phreatic zone (e.g. Morse & Mackenzie 1990). The studied carbonate platform deposits do not contain heavy dissolution structures leading to cave or karst formation, typical for the equatorial tropics and resulting in common leaching, cave and upstanding karst formation (Wilson 2012, and references therein). Instead, the meteoric relics of the studied platform are related to local and small-scale or poorly developed structures typical for more arid regions, such as macro- and microporosity (both fabric- and non-fabric-selective), vadose cementation, mineral stabilization (neomorphism), and occasional thin layers of vadose structures or possible initial palaeosols. These structures are found in the carbonate sandy shoal deposits and are absent from slope and lagoonal carbonates.

The meteoric influences are confirmed by the oxygen isotopic variations described above. Also, the variation in Sr concentrations in the platform dolomites may reflect a combination of both the palaeogeographic situation, that is salinity differentiation due to the impact of mixed marine–meteoric fluids (cf. Cander 1994), and diagenetic water/rock interaction of the shallow platform deposits during shallow burial diagenesis. Thus, various mineral transformations (especially dolomitization as the most noticeable diagenetic event in the area studied) might lead to Sr release from a limey framework (Machel 2004).

The small and local scale of these features suggests that the degree of meteoric incursion was not too extensive, and similar to that known from typically more arid regions (e.g. Morse & Mackenzie 1990; Sun & Esteban 1994). Therefore a trend to lower the carbonate content (and a relative deficit of carbonates during MD deposition) may reflect a tendency to an increase in more humid episodes during generally greater climatic aridity.

Generally, the shallow-water part of the platform dominated by dolomites contains locally a higher admixture of sulphates, S and various amounts of IR and silicate material. The abundance of silicate elements is non-uniform throughout the MD sections and changes palaeogeographically. Major silicate elements normalized to Al or Ti, which are usually considered to be stable during weathering and fluctuate relatively negligibly throughout the MD sections (mostly in the range up to *c.* 0.003% and 0.4%, respectively) (Table 2), may indicate changes in the detrital input. Very high covariation between TiO_2 and Al_2O_3 across the platform zones (Table 3) indicates that the detrital source has not changed substantially during the MD platform deposition. Both elements are very strongly correlated with Fe_2O_3, which displays distinctly positive enrichment trends in all platform zones. The strong and positive correlations of Fe with other silicate components (Table 3) combined with the Al-normalized Fe ratio, which is higher on average in the lagoon section (0.39) than in the barrier and slope zones (0.29 and 0.30, respectively) (Table 1), suggests that Fe was enriched in weathering products. Delivered to the platform area essentially by a riverine transport, it was widely distributed throughout the platform, especially in shallow settings.

Iron enrichment may be particularly common close to tropical river mouths from erosion of mature tropical soils in warm shallow-marine environments carrying Fe-rich colloids (Meunier 2005). Thus, covariation between Al_2O_3 and Fe_2O_3 and other clastic elements is related to changes in the patterns of chemical weathering products and possibly increased hydrolysis (mobility) of Fe-enriched aluminosilicates.

Fluvial input. The qualitative differences in silicate contents indicate different sedimentary conditions in both the shallow and slope platform settings with the relatively lowest range of concentration variability in the barrier setting. Typical silicate elements such as Fe, Al, Ti, K and Mn have been summed to assess the detrital input with time. The cumulative concentration of these elements increases upward in the inner and barrier sections and decreases upward in the platform slope section (Fig. 7). This pattern reflects a tendency to an increase of silicate material supply to the shallow MD platform area and generally to a decrease in the deeper-water part of the platform. Consequently, carbonate marine facies of the MD unit shift upwards towards a slightly more clay-rich composition that is also reflected by greater Al concentrations in the shallow-platform zones (Table 2, Fig. 5).

Among the aluminosilicate components, Al is the main constituent of the clay minerals that were essentially delivered to the platform area from the adjacent watershed. Although continental clay minerals remain seemingly inert when settled in marine waters, some major aluminosilicate elements (e.g. Na, K and Mg) are mobile during early diagenetic transformations (Meunier 2005). Standard and SEM petrographic observation showed that chlorite occurs in the platform lagoon and slope facies as chlorite grain coatings and pore fillings (Gąsiewicz & Wichrowska in press). However, the presence of Mg-rich clays is less indicative of this mobility because of common and extensive dolomitization. Similarly, Na content is not indicative because of a common halite.

A tendency of the K/Al ratios to decrease upsection visible in the inner and barrier deposits (Table 1, Fig. 7) suggests a change in land source and an increase in the smectite and/or kaolinite supply (cf. Sabatier *et al.* 2010, and references therein). The high K_2O/Al_2O_3 values, varying between 0.11 and 0.67 (0.25 on average) but mostly (67%) between 0.20 and 0.30 (Gąsiewicz & Wichrowska in press) typify the MD unit, reflecting a relatively high illite/kaolinite ratio in these deposits

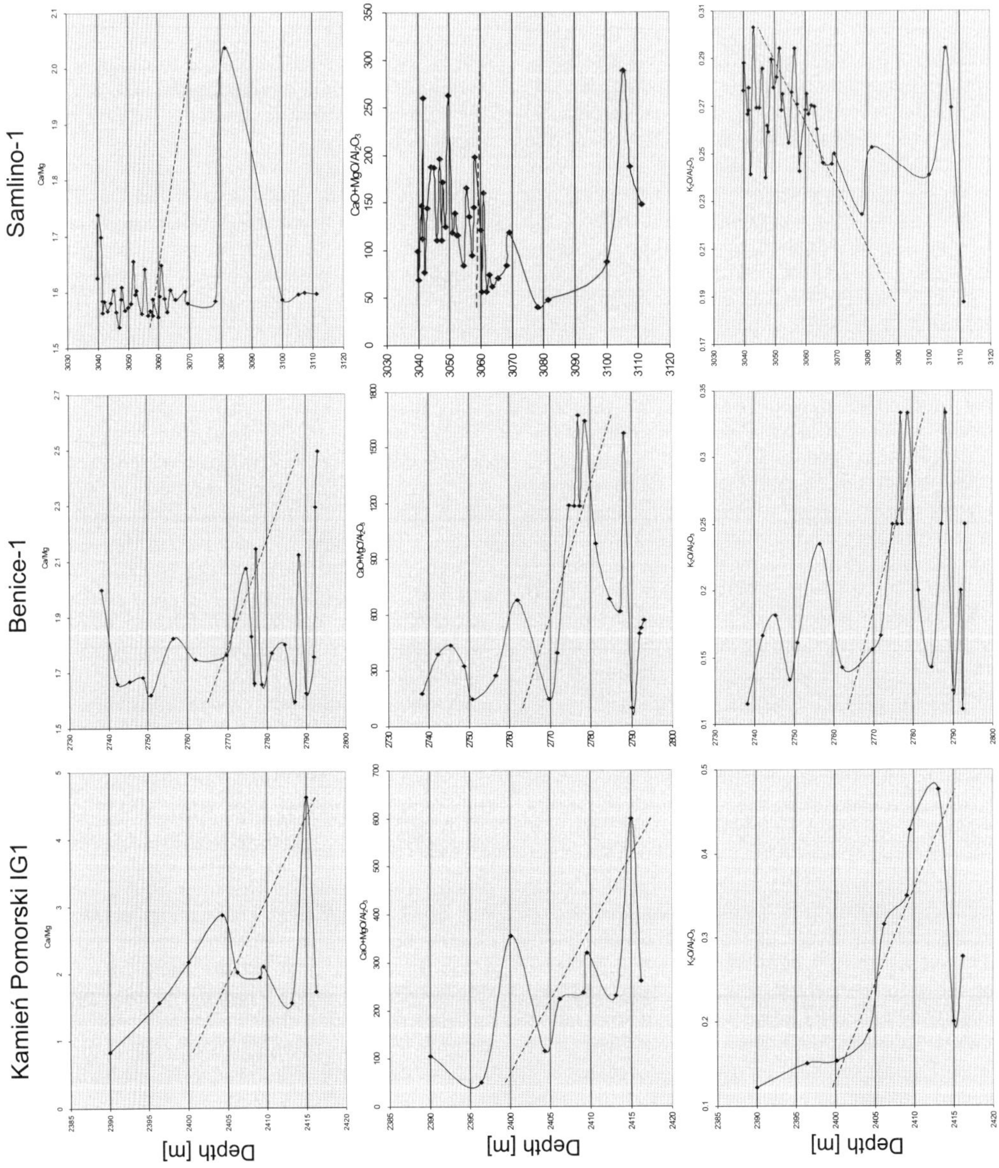
Samlino-1
Benice-1
Kamień Pomorski IG1
Ca/Mg
CaO+MgO/Al2O3
K2O/Al2O3
Depth [m]
Depth [m]
Depth [m]

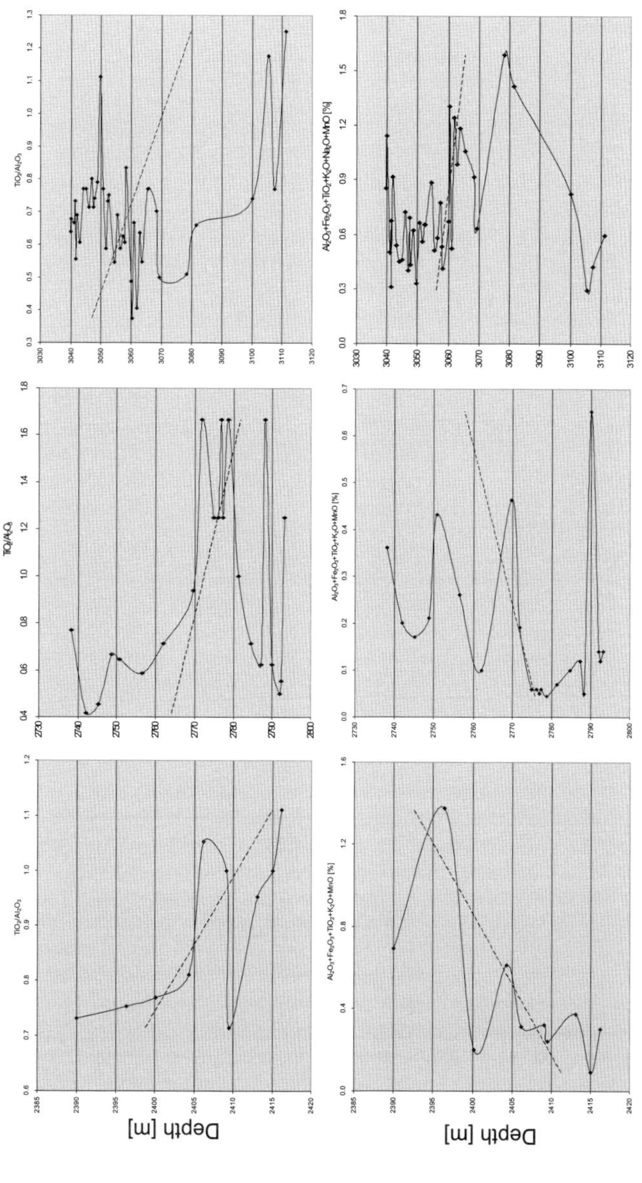

Fig. 7. Vertical trends of relationships between selected main elemental contents, main elements normalized to Al_2O_3 content and a sum of main silicate components along the Main Dolomite sections in the studied carbonate platform palaeogeographic zones: lagoon, Kamień Pomorski IG1; barrier, Banice-1; slope, Samlino-1.

(Weaver 1989; Ratcliffe *et al.* 2004). Thus, the K/Al ratios may be considered diagnostic for variations in clay minerals composition (Weaver 1989) during MD deposition. A positive and strong correlation between K and Al in the platform palaeogeographic zones (Table 3) suggests that lower concentrations of K in platform carbonates are not associated with higher concentrations of Al-bearing mineral (kaolinite). The major Al-bearing minerals are K-feldspar, illite and kaolinite. The K_2O/Al_2O_3 ratio is highly differentiated in lagoonal facies (0.27 on average), slightly lower in the barrier section (0.20 on average) and relatively equalized and higher in the slope setting (0.44 on average) (Table 1). This pattern suggests a common admixture of kaolinite throughout the platform, the presence of K-feldspars in the lagoonal zone, an admixture of illite in the barrier facies and a predominance of illite in the slope palaeoenvironment. The mineral distribution is confirmed by petrographic study of thin sections.

In the investigated deposits, K normalized to Al shows negative trends upward in both inner and barrier platform zones and a positive one in the platform slope zone. Almost excellent positive correlations of Al_2O_3 with K_2O (Table 3) and negative K_2O/Al_2O_3 trends in the lagoonal deposits (Fig. 7) imply that, together with a rise in Al-enriched clay minerals supply, the concentration of mobile alkaline elements (very leachable element K) decreased in parallel. This indicates that the deposited clayey fraction became progressively Al-enriched upsection (i.e. more kaolinitic; illite is K-rich) during MD sedimentation, suggesting a slight change in clay source material composition (Fig. 7).

The opposing trends of the K_2O/Al_2O_3 ratios in the shallow platform (inner + barrier) zones and slope settings (Fig. 7) suggest lateral changes in the clay mineralogy within the platform area resulting from sedimentary segregation of the material during transport. This seems to be reflected by different regression lines for the relationships of the two elements in the shallow and slope MD environments (Fig. 8). Such patterns have been explained by grain-size sorting, with kaolinite and illite accumulating closer to the coastline and the very fine-grained smectitic components are transported to more distal sites (Kühn *et al.* 2001, and references therein). However, the study of densely distributed surface sediments throughout the northeastern South China Sea and surrounding fluvial drainage basins showed that the clay mineral assemblage is mainly controlled by provenance supply and

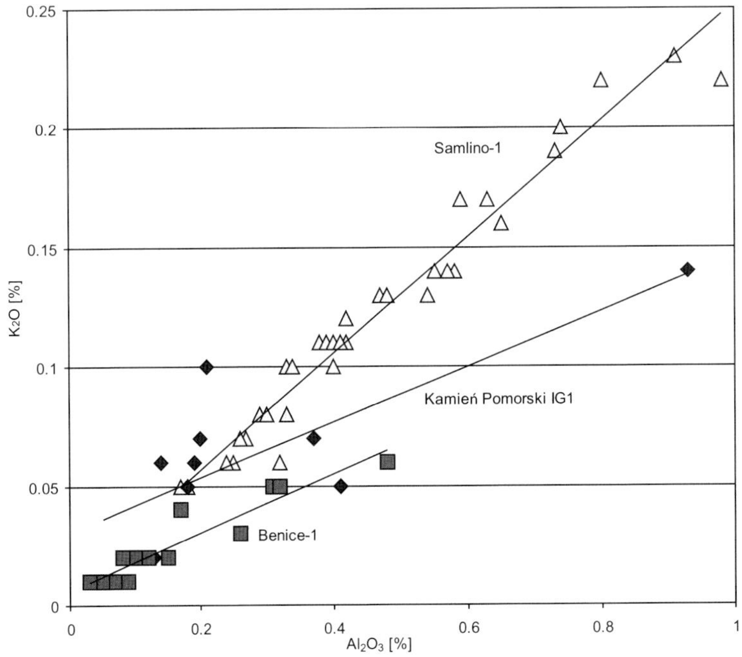

Fig. 8. Cross-plots representing relationships between K_2O and Al_2O_3 of bulk-rock analyses showing the differentiation of the Main Dolomite deposits across the carbonate platform. Note the different co-linear trends between shallow platform (lagoonal and barrier) and slope settings delineating different sedimentary environments reflected by different clay mineral compositions.

current transport, and it is not valid as an indication of direct palaeoclimatic conditions during the late Quaternary (Liu *et al.* 2010). It seems likely that marine current transport governed clay distribution in the MD platform area.

Strong and moderate correlations of K with typical lithogenic elements (Fe, Al, Ti, Na, Ni, V and Cr) across the platform (Table 3) and significant ones with IR only in the slope setting show close relationships with the siliciclastic contribution to the MD platform area. Specifically, the terrigenous supply is better articulated by the IR content variously dispersed across the platform setting (Table 2, Fig. 6a). The lagoonal zone is relatively enriched (mean of 8.2%) and the barrier depleted (4.3% on average), and the platform slope shows an intermediate (mean of 7%) IR content (Table 2). The amplified admixture of IR in the lagoonal and slope environments indicates palaeogeographic control. The similar distribution patterns of IR and typical aluminosilicate components suggests that the attached-to-land carbonate platform was influenced by river discharge and a change in clay source composition. This suggests weathering and leaching of the surrounding land surface and a differentiated supply to the coastal platform. The observed depletion of K/Al ratios up the MD sections (Table 1, Fig. 7) may be generally interpreted as an indicator of an intensification of river suspended matter delivery, as in the recent and ancient Mediterranean region (e.g. Wehausen & Brumsack 2000; Sabatier *et al.* 2010). Thus, the distinct vertical trends in relationships between typical clay components indirectly suggest that not only current transport (cf. Liu *et al.* 2010) but also palaeoclimatic changes influenced the clay mineral composition in and around the MD platform area with time. This points to the summer and winter monsoon winds, which could have dominated as in recent East Asia seasonal precipitation and runoff regimes, and finally determined the provenance supply (weathering distribution) and current transport processes (e.g. Liu *et al.* 2010). The weak correlation between Al_2O_3 concentration and $\delta^{18}O$ values (Table 3), high fluctuations of S concentration and its decreasing tendency upsection in the inner platform zone (Fig. 5) suggest a fluctuation of evaporation and a tendency to increased runoff, and may therefore support this interpretation.

Aeolian supply. It can be assumed that the continental runoff was not, as in recent environments, the only source of very fine clastic material delivered to the platform deposition, especially in the hyper-arid conditions that developed in the Pangaean interior where large deserts or desert-like areas developed to the west and in Northern Europe. Much of the vast inland Pangaea, particularly the northern part, experienced arid climates, and the dominant wind directions reconstructed for the Late Permian were blowing from the NE to SW in winter and conversely in the summer (e.g. Parrish 1993; Roscher *et al.* 2008, 2011). Thus, the generally limited riverine influx of siliciclastics (including hydroxides) from the northern continent to the platform setting was probably accompanied by some amounts of allochthonous material delivered by aeolian transport, as revealed by thin section analysis indicating the presence of fine grains of quartz, Ti and Fe oxides, zircon and other minerals, including some clay minerals (Gąsiewicz & Wichrowska in press). Some, such as secondary Fe oxides and hydroxides and clay minerals (kaolinite) are associated with oxisols (laterites) (Muenier 2005), suggesting wind-blown material from weathering of tropical soils.

The deflation process is characteristic of arid-climate sabkhas all over the world (e.g. Shinn 2011). High wind in hot continent areas may have caused the mobilization of coarser weather-resistant minerals (see e.g. Wehausen & Brumsack 2000) as well as clays (Meunier 2005) and calcium (e.g. Capo & Chadwick 1999). Wind is an important component of climate, and aeolian supply to marine deposition can be significant regionally. For example, in the recent Mediterranean Sea, aeolian dust supplied from North Africa could account for 20% of sediment deposition (Sabatier *et al.* 2010, and references therein). A similar situation occurs in Australia (Hesse & McTainsh 2003). The source of this input can be local or regional, and dust can travel great distances (e.g. Muhs *et al.* 1990; Wang *et al.* 2012). For such sediment contributions, variations in the Ti/Al ratios may be used as indicators to assign detrital input to local or regional conditions, as exemplified by the elevated Ti/Al ratio in aeolian-derived material (e.g. Wehausen & Brumsack 2000; Sabatier *et al.* 2010, and references therein).

However, with respect to the MD sections of the Kamień Pomorski platform, the variation in element/Al ratio could be partly a result of a differential contribution of various clay minerals, as suggested above. A weak linear relationship between Al_2O_3 and TiO_2 in the slope setting (Table 3) suggests multimineral control of TiO_2 (Ratcliffe *et al.* 2004) and varying amounts of Ti-bearing heavy minerals. Although, Ti may be generally associated with heavy mineral grains, its increasingly positive correlations with IR in more distal settings of the platform (Table 3) seem to suggest that, besides delivery by riverine inputs, Ti may also represent an aeolian input. The Ti/Al decreasing upwards trends occur in all three platform settings (Table 1, Fig. 7), indicating a consistent tendency that is interpreted as being related to an increase in humidity and thus a decline in the amount of dust during the MD interval. A distinct correlation between IR and S occurring in the platform barrier section and the decreasing trend of the elemental ratio probably reflect a closer relationship between evaporation and the input of aeolian dust, which could be more closely associated due to extremely shallow conditions and more frequent short-time subaerial exposures. Similarly, the very high correlation between Fe and Ti occurring in all platform zones may suggest that some of the Fe minerals were provided by wind. Analyses of recent and sub-recent aeolian depositional settings have shown that Fe-bearing minerals can be transported from land by wind-blown dust (e.g. Krystinik 1990; Gingele & Deckker 2005).

In conclusion, the IR data suggest aeolian dust input to the studied platform from Pangaean desert areas that developed in west and north (Scandinavia) Pangaea. Similar results can be inferred from Sachse *et al.* (2004), who, based on isotopic data, concluded that long-chain *n*-alkanes were transported through wind and/or rivers into the lagoon at the southern Zechstein Sea coast.

Organic matter contribution. The subaerial erosion adjacent to the MD platform Pangaean land and Fennoscandian High could only be associated with nutrient fluxes to coastal marine environments. It may be partly corroborated by a distinctly positive trend of the TOC values distribution in the lagoonal deposits (Table 1) and a positive correlation between TOC and components of the silicate fraction (e.g. Fe, Al, K, Ti, Ni, V and Cr; Table 3) throughout the platform (marked in the lagoonal section, weak in the barrier setting, and varied in the slope setting). The distinct correlation between TOC and typical silicate content implies that an increase in weathering intensity has occurred during the MD, and runoff could provide some land-derived detrital organic material. On the basis of these relationships it is suggested here that a progressive increase in chemical weathering and delivery of nutrient elements from land could lead to elevated nutrients and consequently to higher surface organic productivity (and/or preservation potential) in the platform slope and restricted sub-environments of the lagoonal zone (Słowakiewicz & Gąsiewicz 2013).

Although the continental interior of Pangaea may have been relatively inhospitable, the SPB coastal areas were probably more hospitable to plant life. The enhanced supply of clastic material to the platform area was accompanied by accumulation of plant-derived type III kerogen and fine organic material from higher plants. The input of plant material is indicated by the presence of humic compounds in the organic matter fraction of sediments like inertodetrinite and pyrofusinite. This material occurs, irregularly dispersed, throughout the platform, especially in the lagoonal (Kamień Pomorski-1 well) and slope (Samlino-1 well) platform settings (Gąsiewicz & Wichrowska in press, and references therein). At a basinal scale, these features signal a humid impact on MD deposition in the northern part of the SPB. The greater land runoff is simply related to warm or hot and wetter climatic conditions control when warm, humid air masses or storm tracks causing higher regional precipitation might cause stronger terrestrial runoff together with shifts in wind and current patterns (as in recent tropical platform areas; e.g. Wilson 2012). These influence soil generation, growth and destruction of terrestrial ecosystems.

Discussion

Geochemical results from the Kamień Pomorski MD carbonate platform provide a challenge to the assumed basin-wide internal cyclicity of the deposition and have various palaeoclimatic implications for the SPB area. These issues are discussed in the following.

The problem of Main Dolomite intra-cyclicity

General model. According to a commonly held view, the Zechstein Basin climate was predominantly arid, with hot summer temperatures, and the Zechstein cyclicity of sedimentation is attributed to periodic glaciations that controlled marine incursions (Peryt *et al.* 2010, and references therein). Accordingly, the sedimentation of older Zechstein cycles was controlled mainly by cyclic glacial eustasy and associated sea-level changes driven by repeated increases of sea-level and periodic opening/closing of the northern basin entry. This idea is core to recent sequence stratigraphic subdivisions within the Zechstein Formation and it is assumed to also explain the internal cyclicity of the older Zechstein carbonate units (e.g. Strohmenger *et al.* 1996; Wagner & Peryt 1997; Kaiser *et al.* 2003; Sachse *et al.* 2004; Peryt *et al.* 2010, and references therein). The model says that cyclic marine inflows of glacio-eustatic origin are responsible for regional (i.e. basin-scale extent) depositional intra-cycles and local(?) 'controlling factors include tectonics, climate and eustatic processes' (references cited above). In this understanding, the 'climatic' processes are simply related to arid conditions (responsible for evaporation). However, it remains unclear how vanishing glaciers in the Gondwana polar region could have caused

Zechstein cyclic sea-level rises of regional extent far to the north in the SPB epicontinental area, which was poorly interconnected with the world ocean. Note that, along the northwestern margin of the Pangaea (far to the west), sea-level was already at a near minimum towards the end of the Kungurian (Middle Permian; Zubin-Stathopoulos *et al.* 2012). Yet, more puzzling is how these dying out phases (expansion and contraction) of residual glaciers could have induced the inferred basin-scale high-frequency sea-level changes responsible for the invoked intra-Zechstein Limestone (Ca1) and intra-Main Dolomite cyclicity (e.g. Peryt & Scholle 1996; Kaiser *et al.* 2003; Peryt *et al.* 2010, all including references therein). Moreover, waned (very local) glaciations could not have affected the lower latitudes to any great extent, like during the Mesozoic when ice caps were either absent or negligible (cf. Price 1999, and references therein). Therefore, the glacio-interglacial mechanism invoked for MD internal cyclicity appears highly implausible.

Another mechanism for MD intra-cyclicity recently invoked was climatic forcing at Milankovitch frequency. According to Strohmenger *et al.* (1996) and Kaiser *et al.* (2003), the MD intraplatform high-frequency depositional cycles from the northern part of the SPB (NE Germany) are correlated with high amplitude and finer (fourth-order) sea-level fluctuations in the Milankovitch frequency band. Similarly, a climatic orbitally forced frequency is suggested for deposition of the Roker Formation (the MD equivalent in NE England; Mawson & Tucker 2009) and for each Zechstein cycle of salt successions in Poland (Czapowski & Tomassi 2013). The same cause and frequency, 'oscillatory in nature and resulting a cyclicity [of] environmental changes' within the Zechstein MD sedimentation in the SPB area, has recently been suggested by Peryt *et al.* (2010, and references therein). Based on the MD facies framework from the Kamień Pomorski carbonate platform, there is no way to contradict or confirm a short-lived Milankovitch cyclic sedimentation like that recognized in NE England (Mawson & Tucker 2009). However, such influences may not necessarily be reflected clearly in an environmentally more uniform carbonate-dominated platform deposition. This means that possible precessional climate signals forcing sea-level changes were too small. Nevertheless, the palaeoclimatic impact – gradual or fluctuating – may have exerted noteworthy control over MD deposition in the Kamień Pomorski carbonate platform. Therefore, taking into account the influences of Milankovitch frequencies on MD deposition, one may alternatively link them to variations in moisture related to changes in the transport of air masses around the SPB region.

According to the assumed general model of the MD subdivision, this unit should exhibit well-ordered cyclic variation and periodicity recorded by the sedimentary bundles correlatable throughout the basin. Various orders of relative sea-level variations are reported and discussed in the literature for the SPB region. Hitherto, interpretations of the MD assume that the unit is arranged into three depositional cycles that reflect high-amplitude (*c.* 40–120 m each) sea-level fluctuations (Peryt *et al.* 2010, and references therein). The ability to correlate the MD sections from different parts of the SPB is key to understanding sedimentary evolution changes in the carbonate platform belt surrounding the basin.

The MD depositional subcyclicity is based on lithofacies arrangements (especially the alternation of carbonate muddy/grainy lithofacies), their internal organization (shallowing-up or deepening-up trends) and the presence of characteristic sedimentary structures at the tops of cycles' boundaries. The latter are related to the erosional and exposure surfaces, brecciation, cementations and palaeosol horizons, solution features, intraclasts and others. Accordingly, interpretation of MD intra-cyclicity in the carbonate platform areas is based on identification of fourth-order (internally stacked from fifth-order) parasequences, which delineate cyclic sea-level changes (i.e. regionally extended falls/rises (transgressions/regressions) of sea-level) (e.g. Kaiser *et al.* 2003; Peryt *et al.* 2010). The larger-scale sequences (fourth-order) show regional and 'even wider correlation', whereas the small-scale (fifth-order) cycles are of local extent (Kaiser *et al.* 2003).

However, a clear cyclic organization of the MD deposits reflected either sedimentologically or geochemically (e.g. isotopic), correlatable not only throughout the basin but between adjacent regions (e.g. between the studied region and the Pomeranian carbonate platform), is practically unknown for the Polish Zechstein carbonate units. The invoked intra-MD high sea-levels should cyclically stimulate abundant amounts of surface platform sediments and an organic carbon contribution to the slope setting, but during sea-level lowstands, more sediments from the open sea (i.e. pelagic material). The effects of large sea-level changes such as those indicated by Peryt *et al.* (2010) should leave marked sedimentary features in both platform and slope platform settings. Similarly, cyclic sea-level falls and rises, especially when prolonged, should leave correlative isotopic shifts related to subaerial exposures or much shallower parts of the platform as a diagenetic effect of alteration by meteoric-water phreatic lenses (e.g. Matthews 1987). Moreover, carbonate muds accreting platformward and capping the whole sequence may serve as regionally important markers of sea-level changes. In the case of the Polish part of the Zechstein MD carbonate platforms, however, such a stratigraphic situation seems to be exceptional and has not been identified to date. Importantly, cyclic facies arrangements are generally not noticeable in most of the MD carbonate platforms in the Polish part of the SPB area. For example, Słowakiewicz & Mikołajewski (2009) recognized no cyclicty of the MD deposition in the Gorzów Block and in other parts of the regionally extensive Wielkopolska Platform of western Poland (M. Słowakiewicz, pers. comm. 2012).

There is no regionally extended subcyclicity expressed by the presence of correlatable distinct karstification and erosional surfaces as well as freshwater horizons, either in the barrier or in the shallow lagoonal deposits, that could suggest platform-wide internal cyclicity in the area discussed here. Moreover, the lagoonal MD sections, for example in the vicinity of Kamień Pomorski where numerous wells are concentrated within an area of several square kilometres, are built up of a diverse combination of grainy (mainly bioclastic and pelolitic), typically shallow subtidal and locally microbialite deposits, which neither alternate with tidal and/or supratidal caps nor form peritidal cycles. Instead, they often show one shallowing-upward trend. Accordingly, Gąsiewicz & Wichrowska (in press) could not identify either the two rapid rises in MD sea-level suggested by Peryt (1986) on the neighbouring Pomerania platform (Łeba Elevation) or two regressive larger-scale cycles in the MD sequence of nearby NE Germany (Kaiser *et al.* 2003). There is no evidence of two delineated climatic (desiccation) events reflecting a regional MD intracyclicity and 'correlated' with three eustatic sea-level cycles inferred by Sachse *et al.* (2004) from the Thuringian Basin (southern margin of the SPB). Moreover, the pattern of $\delta^{13}C$ changes across the Kamień Pomorski platform (Table 1) shows no sign of inferred maximum transgression and regression, which should be recorded as distinctive events resulting from the obvious different predominances of organic carbon composition for the two sea-level tracts. Furthermore, the MD $\delta^{13}C$ profiles, especially from the platform slope (permanently submarine) sections, do not reflect marked or cyclic changes in the carbon isotopic composition that have been postulated, for example, by Peryt & Scholle (1996) and Sachse *et al.* (2004).

Sedimentary features. The above discussion simply relates to interpretations of the lithofacies arrangement, various sedimentary structures like discontinuous surfaces and siliciclastic influxes, which are not always unequivocal when based on core material.

The interpretation of sedimentary features as shallowing- or deepening-upward subcycles (small-scale cycles), and consequently inferred aggradations or progradations of carbonate platform deposits with core material, remains a challenge. This is especially important when boreholes are distant from one another. In general, the MD platform successions display two main types of carbonate deposits: muddy lithofacies characteristic of a deeper-ramp system, and more grained (mainly oolitic grainstones and packstones) lithofacies are typical of a shoal system (Gąsiewicz & Wichrowska in press). The first group of lithofacies, represented by various carbonate mudstones, wackestones and local microbialites, and interpreted to reflect deposition in entirely subtidal shallow (permanently submerged) environments, could be related to a confined, near-shore water area located in the innermost part of the carbonate ramp (lagoon). The second, commonly comprising carbonate sands, is typical of shallow and extremely shallow, high-energy water conditions. Moreover, ooid shoals, including the tidal deltas, have relatively sharp boundaries with adjacent environments because of erosion or reworking of the depositional surface after initial deposition or replacement through time (Harris 2010, and references therein). It is commonly assumed that grainstones, being tidal deposits, are formed by increased accommodation space through eustatic sea-level rise, whereas low-energy, muddy carbonate facies tend to be more common during transgressive phases or in carbonate protected settings (e.g. Ball 1967; Wilson 2012, and references therein). Thus, mudstone–grainstone couplets, depending on their arrangement, may represent a regressive or transgressive (shallowing or deepening) path of the parasequence in the broadest sense of van Wagoner *et al.* (1990). Such rough reasoning seems to be generally applied to the interpretation of the spatial evolution of the Zechstein carbonate units.

Typically, platform carbonate sands contain variable, not infrequently significant, amounts of mud fraction. Fine material is winnowed by wave and tidal currents and transported in suspension during the passing of deep atmospheric lows or fronts and deposited in calm niches, in low-energy shadows in the lee of large barriers (e.g. Ball 1967; Wanless & Tedesco 1993; Rankey & Reeder 2010). However, grainy material is transported outside sand oolitic shoals, leading to interfingering of both oolitic and other carbonate muddy lithofacies. The 'muddy–grainy' facies alternation in the shallow platform interior may be related to trade winds that coupled with the open windward platform margin, and set up off-bank currents that locally winnowed the muds, leaving grainy layers (Rankey & Reeder 2010). Hence, intercalations of muddy lithofacies visible in oolitic bodies indicate changes of environmental conditions and thus a wandering facies framework and are not necessarily related to regional sea-level changes. In fact, there is no evidence that the lagoonal or platform slope mudstones studied were subaerially exposed at parasequence tops (thus marking discontinuous surfaces).

Similar conclusion results from the small-scale cyclicity in the MD oolitic sandy complexes. However, the peritidal parasequences display no systematic vertical trends in either the thickness or facies composition (Fig. 3). This heavily limits the lateral traceability of parasequences in the Kamień Pomorski platform. Detailed analysis showed that very closely located boreholes, such as those drilled within a few hundred metres of one another (in Rekowo (2, 3, 4, 6) and Wrzosowo (1, 2, 8, 9) localities, Fig. 1b), exhibit no correlatable subcyclicity within the platform barrier zone. An apparent facies cyclicity in oolitic shoal complexes observed locally seems to reflect secular changes observed in recent carbonate platforms. Ooid shoals, including the tidal deltas, undergo unidirectional and bidirectional sediment transport (Harris 2010, and references therein). The cumulative effects of the feedback among hydrodynamics, bathymetry and sedimentology add to systematic depositional trends in the sizes and shapes of sand bodies as well as the grain size and sorting of the bodies (Rankey & Reeder 2008) may form apparent cyclicity. In effect, the co-occurrence of all these different features, related to a convergence of highly supersaturated waters and tidally enhanced water circulation and leading together to the addition of oolitic sediment coupled with net on-platform transport, would encourage the expansion and aggradation of shoals as a whole (Rankey & Reeder 2011). In a fossil record, it may be easily interpreted as a rough environmental change related to sea-level variation, especially when based on drill core material. This makes it difficult to identify what system tract is recorded in the platform area.

Another problem is related to intraclastic lags, which are associated with discontinuous erosive surfaces found in the MD barrier zone deposits. They may simply be interpreted as transgressive deposits. However, the lags abundantly contain reworked stromatoclasts and oolite intraclasts and have been found only locally. Stromatolites and biolaminites are infrequent in the investigated deposits and are found most probably, as in the Bahamas, in high-energy carbonate sandy shore zones with continuous scour and sediment movement, grown in the least energetic upper portion of the tidal zone, on temporarily exposed margin channels (e.g. Gebelein 1976). Lateral migration of the levee facies over the channel floor could erode and rework such deposits and finally accumulate coarser material. Moreover, it is possible that surface patterns are commonly eroded or reworked after initial deposition (e.g. Harris 2010). Thus, both the occurrence and composition indicate that these coarser deposits represent either violent reworking or transport across considerable parts of the platform interior by the passage of fronts or tropical depressions (e.g. Ball 1967; Reeder & Rankey 2008), or indicate transport of reworked lithified material by the activity of tidal channel currents. Thus, such autochthonous storm lags are not traceable, even between nearby wells, and cannot be used for correlations as regional markers.

Discontinuity surfaces are crucial for stratigraphic subdivisions in carbonate platforms, but their nature in the MD unit remains poorly recognized. Subaerial surfaces exposed due to accretion, evaporative drawdown, or rainfall and aggressive waters commonly lead to dissolution, karstification and concomitant carbonate precipitation. In the case of the Kamień Pomorski carbonate platform, freshwater relics are sparse and volumetrically insignificant. They include rare and thin streaks of carbonate grains with singular vadoids (and occasional vadose pisoids), recrystallization of grains and matrix and dissolution, but lack karstification surfaces or true calcrete horizons. Consequently, distinctive subaerial exposure surfaces – marked by pedogenic features (sandy, oxidization crusts, root traces) or karstic structures (solution-enlarged joints and pits), associated with fenestrae, mud cracks, breccia or evaporite encrustations and so on. (see e.g. Tucker & Wright 1990; Algeo *et al.* 1992) – are absent from the studied platform. Instead, poor and occasionally pedogenic-like layers, locally found in the ooid shoal (barrier) deposits, form initial palaeosol structures from the precipitation of percolating calcite-saturated waters, resembling structures formed in seasonally arid conditions (Christ *et al.* 2012, and references therein) but absent from the fully arid regions (Wright 1994). Vadose structures and meteoric-originated diagenetic features, usually forming thin and poorly developed horizons within the platform deposits, are known to the author from oolitic barrier complexes of other Zechstein MD carbonate platforms (like the Pomerania and Wielkopolska platforms). These MD features rule out a high rainfall regime or signify short-lived exposures in a more distant platform setting, both insufficient to produce significant meteoric diagenetic signs and producing local and thin freshwater lenses only. Note, also, that low annual rainfall (a few hundred millimetres) occurring in subtropical semiarid shallow-water platform carbonate sands may produce weathering (dissolution) and freshwater vadose cementation (lithification), which are extremely short phenomena in geological terms (e.g. Halley & Harris 1979; Semeniuk & Searle 1985; Dravis 1996). In modern oolitic sediment in the marine realm, this rapid cementation is nearly syndepositional during nearly syndepositional subaerial exposure (Harris 2010, and references therein). Thus, the lack of true calcretes might result from the very short duration of exposures of these peritidal deposits coupled with weak climate variations. This highly complicates the possible correlation of such events in the carbonate platform setting, especially in core material. These more uniformly cemented layers and palaeosurfaces appear locally and seem to mark discontinuous and unconformable surfaces related to temporal exposures in the studied unit (Gąsiewicz & Wichrowska in press). However, they are not well correlated between particular wells and have no analogues in the lagoonal and slope settings and thus exclude the possibility of MD intra-cyclicity related to exposures or extreme shallowness of the wider platform areas. A scarcity of prolonged subaerial exposures in the examined MD platform suggests that the observed structures are no more than episodic (short-term) and local events occurring only within peritidal successions of the platform.

The unconformities may be created in variable ways. Sea-level changes influencing a rate of sediment production within the platform settings are not the only cause of stratigraphic discontinuities. The cemented layers occasionally found in the platform interior lithofacies are ambiguous evidence of sea-level changes. They may have resulted from submarine cementation during periods of relative stagnancy when sediments are not moved by currents (e.g. Ball 1967; Shinn 2011). Such marine cemented layers formed during highstand could easily be misidentified as flooding surfaces during sequence stratigraphy analysis (Shinn 2011). Thus, without a regional corroboration of such correlated discontinuities they remain episodic and local phenomena only.

On the other hand, it is widely accepted that relative sea-level changes provide the cycling stacking pattern in outer shelf settings (e.g. Osleger 1998; Tinker 1998). Such cycles are usually highlighted by a siliciclastic admixture. However, the MD geochemical profiles across the platform (Figs 5 & 7) do not exhibit any correlatable patterns that could link detrital input changes with larger regional sea-level fluctuations. This is in accord with a highstand sea-level comprising most of the MD interval. Moreover, the lower MD part is relatively carbonate-enriched, indicating higher carbonate accumulation during the transgressive and part of the highstand sea-level, whereas the carbonate-depleted upper part shows higher siliciclastic accumulation during the highstand and then, at the end of the MD, lowstand sea-level. Such a pattern of both carbonate and siliciclastic relationships is opposite the commonly observed distribution of sediments related to changes in sea-level: siliciclastic material supply is highest during marine transgressions and lowest during sea-level drops (e.g. Dunbar & Dickens 2003; Gischler *et al.* 2010, and references therein).

Relative sea-level rise usually leads to clastic sediment starvation on the shelves due to the trapping of land-derived material in newly formed nearshore accommodation space (e.g. Cattaneo & Steel 2003, and references therein). This is in accord with the geochemical data presented above, which suggests that the inner platform lagoon provided loci for the relative enrichment in silicates. This also suggests that land topography probably influenced the distribution of siliciclastic input to the carbonate platform.

The observed siliciclastic dilution of the platform deposits implies that accumulation in the marine slope setting was independent of terrigenous input. This deeper-water zone is relatively enriched in silicate material (IR + aluminosilicate elements). The relative abundance of the silicate fraction seaward achieved from increased trapping efficiency in the slope zone may have resulted from winnowing of shallow platform settings due to wave and current resuspension and transport outside the platform interior. Aluminosilicate fraction components such as silts and clays may be transported as aggregates and floccules into open marine environment over a large distance (Potter *et al.* 2005; Liu *et al.* 2010). These processes could have finally led to enrichment in clay-sized material, silt or sand fraction contents being additionally reinforced by dust supply. Variable distribution and increased siliciclastic sediment supply in the slope setting could finally be modified by the effects of syndepositional down-to-the-basin transport of carbonate material and/or governed by topography and possibly persistent NE–SW trade winds (cf. Rankey & Reeder 2010). The shallowing of the MD platform with time to tide levels suggests a relatively equalized palaeotopography. It also promoted hypersalinity, and a climatic instability affecting the sediment and water supply might have additionally influenced facies changes and their features across wider areas of the platform (cf. Bohacs *et al.* 2000).

Angular and rounded shallow-water-formed grains (like bioclasts, ooids, peloids and muddy grains) occur as grainy interlayers far down the platform slope facies in the middle part of the MD section in Gardomino-1, Barkowo-1 and Błotno-3 wells (Fig. 1b). This sediment shedding from the carbonate platforms into adjacent basins may be interpreted as carbonate turbidites and as 'highstand shedding'. However, the shedding is not controlled solely by sea-level, as carbonate overproduction on shallow platforms may lead to export of excess material to the slope (e.g. Schlager *et al.* 1994). Key environmental factors appear, such as physiographic setting and diagenetic potential, the latter influenced by the starting composition and mineralogy, the climate and the length of exposure to rainfall (e.g. Dravis 1996).

Stratigraphic modelling shows that although tectonic (tilting) rates were constant over time, the stratigraphic pattern in the tectonic stage autogenically produced cyclic stratal patterns comparable with those produced from allogenic effects and internally generated processes in the sediment transport system that resulted in cyclic strata without externally forced variations (Kim *et al.* 2012, and references therein). The prevailing highstand sea-level and shallowing-upward trend of the MD indicates that the carbonate platform area was relatively tectonically stable. It does not explain the observed geochemical features of the carbonate unit and implies that periodic, platform-wide sea-level changes invoked for the depositional framework of some other SPB carbonate platforms may have been induced by self-organized (autocyclic) processes. In shallow-water environments, a variety of complex feedback loops are known to operate, and they play a significant role in determining the resultant preserved architecture. Moreover, regional tectonic events may operate on a local scale, and jerky subsidence may produce differences in the number of sedimentary structures and lithofacies thickness, variation in their amounts and arrangements, and finally in their stacking patterns. In such a local intra-cyclicity model, episodic tectonic events punctuated longer periods of relative quiescence. Thus, it can be expected that the tectonics may also be responsible for a lack of correlation of the MD depositional subcycles between neighbouring MD carbonate platforms in the northern SPB.

The MD facies organization in the studied platform implies that amplitudes of sea-level changes induced by orbital insolation changes were too small to be found, or shallow-water dynamics might not have fully recorded high-frequency variations in the sedimentary succession. So far, the vertical distribution of the silicate fraction and carbonate dilution with time display high-amplitude variations that do not mark cyclic sea-level changes in the platform sequence.

The palaeoclimatic impact

The Late Permian Zechstein developed in low-latitude northern Pangaean areas in a generally arid climate with hot summer temperatures, characterized by marked aridiziation and global warming trends to the end of the Permian (e.g. Parrish 1995; Retallack 1995; Berner 2005; Schneider *et al.* 2006; Algeo *et al.* 2011; Roscher *et al.* 2008, 2011; Wopfner 2012). The area discussed here occurred between vast and hot areas of Pangaean deserts to the west and the open Palaeo-Tethys Ocean extending to the SE, not too distant from the SPB. The palaeogeographic situation and palaeoclimatic conditions heavily influenced carbonate deposition in the investigated MD platform.

Palaeogeographic controls. Widespread evaporitic and sabkha-like deposits associated with the Zechstein epicontinental basin are interpreted as an indicator of arid conditions. In such a dry climate, proximal parts of the SPB carbonate platforms were commonly occupied by hypersaline seawater due to intense evaporation. Palaeogeographic reconstruction of the Kamień Pomorski area shows (Gąsiewicz & Wichrowska in press) that the MD carbonate platform was not restricted to the marine inflows, and the associated sulphate precipitation in the shallow platform settings was common, as in the other SPB platforms. Open carbonate ramp deposition at the beginning of the MD interval in the Kamień Pomorski area evolved into a protected rimmed platform with a peritidal saline lagoonal environment. The inner part of the platform was rimmed by the high-energy tidal-flat oolitic shoal deposits dominated by wave action. The sheltered platform setting has been locally and/or episodically supersaturated with respect to calcium sulphate, leading locally to abundant gypsum/anhydrite and sometimes magnesite co-precipitation (lagoon area). Distinctly higher S concentrations in the lagoonal deposits (Tables 1 & 2) are in accord with the observed increase in anhydrite content, and suggest that the resultant shoreward salinity gradient was a consequence of the palaeogeographic pattern. Evaporation in this climate influenced the local water volume, but it is difficult to reconcile how it varied in space and time and thus led to local sea-level changes. Consequently, the Zechstein climate could have induced a negative water balance in the SPB area (evaporation > precipitation and runoff), leading progressively to increasingly saline water as it moved landwards. As the salinities increased with shallowing of the platform, especially in the shallow platform interior, it occasionally became diluted by humid episodes.

Palaeoclimatic effects. Climate is a first-order control on vegetation development and weathering intensity, with both factors regulating sediment flux from fluvial systems (e.g. Leeder *et al.* 1998).

It has been stated extensively that Late Permian conditions, except for the abundant forests in the central Transantarctic Mountains (e.g. Gulbranson *et al.* 2012), were unfavourable for terrestrial ecosystems, causing the Late Permian land surface to be largely devoid of plants (e.g. Berner 2005; Xie *et al.* 2005; Algeo *et al.* 2011). Palaeoclimatic reconstructions reveal that temperate and ever wet biomes were scattered throughout Pangaea, especially in tropical and coastal palaeoenvironments (e.g. Schneider *et al.* 2006; Scotese 2013).

The land surrounding the SPB was vegetated, as indicated by fossil plants and palynomorphs known from the Late Permian successions of Central and Western Europe (e.g. Fijałkowska 1995; Warrington 2008, and references therein). Terrestrially derived organic debris is dispersed throughout the MD platform. The distinct predominance of C_{29} over C_{27} steranes (68% to 20%) in Kamień Pomorski IG-1 confirms an admixture of land-derived organic matter composition (Grotek 2002). The existence of terrestrial higher plants and algae typical for coastal-marine deposits at that time is substantiated by a distinct increase in aromatic hydrocarbons relative to saturated hydrocarbons in biolaminites and mudstones in Samlino-1 well and the presence of long-chain *n*-alkanes and slight variations (from −25‰ to −27‰) in the $\delta^{13}C$ of kerogen in Moracz IG1 well (Gąsiewicz & Wichrowska in press). Both wells are located in the platform slope facies. The terrestrial organic material may have been enriched in lipids and lignin-related organic phases, which are strongly depleted in ^{13}C and thus could influence the observed variability in $\delta^{13}C$ composition of the MD kerogen. Terrestrial organic material (plant remains and/or soil organic matter) can be expected at the basin margin as a result of its proximity to the source area and short transport by river discharge. However, some organic constituents can easily be removed from the plant material by wind or rain or entrained as part of the soil and transported as dust over large distances into distal marine sediments (Naafs *et al.* 2012, and references therein). Wind mixing is likely to be responsible for the transport of macrophyte organic matter from shallow margins into deeper environments (McGlue *et al.* 2012). Thus, both the riverine and aeolian inputs would have contributed to the MD organic matter components in different places of the platform.

Enhanced land erosion and seaward transport is effective in warm and wetter climates (e.g. Wilson 2012; Liu *et al.* 2012). This could not be provided by the vast Pangaean deserts, but a differentiated situation is usually associated with widespread marine basins. As a result of evaporation in hot conditions, the SPB and its surroundings would be guaranteed enough moisture for precipitation. The high continentality of Pangaea on one side and large epicontinental sea on the other would easily generate different temperature regimes in the maritime northern SPB, which might in turn drive the warm, humid air masses. Thus, based on data from the studied platform, it may be argued that conditions favourable for plants could develop locally, and the terrestrial vegetation could have been aided by precipitation in the regional palaeoclimatic belt forming wet biomes around the SPB area.

The elemental relationships and their spatial distributions indicate enlarged supplies of terrigenous material (increase of aluminosilicate elements and IR) to the shallow-marine carbonate setting of the Kamień Pomorski platform. The covariant positive trends between these signals in the proximal part of the platform are best explained by a wetter MD climate. In this respect, increased weathering may be expected in areas close to the SPB, whereas an aeolian transport was possibly added from more distant barren sand areas of the Pangaean landmass interior. The clay mineral types and their regional distribution seem to be an effect of the palaeoclimatic impact. This is reflected by relatively more common and stronger correlations between silicate elements in the more proximal platform setting than in the more distal one, where they have been buffered by large marine basin. The presence of chlorite in the MD deposits suggests erosion of rocky land surface. Kaolinite could have been inherited from soils developed under the warm and humid climate (Muenier 2005). As has been shown by Grigsby (2001) and Liu *et al.* (2010), higher river discharge (wet seasons or flooding) and coastal current activity are essential factors for chlorite formation and distribution. It is therefore suggested here that a differential settlement pattern of chlorite content on the shallow platform and deeper slope settings was closely related to transport by local current systems operating temporally on the platform shallow and deeper slope in this northern SPB maritime zone.

Vertical changes in the chemical composition of the clastic components of the MD unit in the northern SPB may be interpreted in terms of shifts in climatic conditions prevailing in the continental source areas. The regressive trend of MD deposition in the Kamień Pomorski area was accompanied by enhanced weathering and erosion rates of palaeosols induced by a relative rise of precipitation in the closest land area. This is reflected by a gradual upward increase of IR and other aluminosilicate element concentrations, especially in the lagoonal and slope settings. The increasing fluvial transport of more resistant grains and clayey material was accompanied by a parallel decline in aeolian supply, as may be indicated by changes in the Ti/Al ratios (Table 1, Figs 5 & 7). Intensification of the humidity led to an increase in wetter climate conditions that could also suppress dust supply to the platform area, as into the subrecent Arabian Sea (Sirocko *et al.* 1993).

The highly fluctuating elemental distribution across the platform indicates that continental drying, erosion and transport of clastic materials from nearby land to the platform land area was an unstable process. As the distribution patterns of the MD silicate elements show (Figs 5 & 7, Tables 1 & 2), the detrital inputs and weathering were variable over time and probably reflect generally unstable climate weathering processes along the northern SPB margin. Enhanced weathering and supply of detrital material to the carbonate platform increased upwards throughout the MD unit, towards the A2 sulphate deposition.

The geochemical profiles along the MD sections from different platform zones do not show any cyclic pattern of distributions. This may imply that the source area(s) in the adjoining land fluctuated with respect to intensity and active erosion during the generally high MD sea-level in the area studied. The geochemical features may argue for both local and episodic humid impacts on MD deposition. Alternatively, they may reflect an increase of frequency of wetter episodes during MD sedimentation. A lack of cyclicity of the subaerially influenced layers and the presence of evaporites in associated lithofacies may argue for short humid periods or seasonally more humid conditions (occasionally recorded), rather than for a continuous increase in humidity of the platform area. This is recorded by an irregular pattern of vertical changes of both the platform carbonate facies as well as their geochemical features.

Both the palaeogeographic location and palaeoclimatic conditions in the northern SPB margin were favourable for an increase in seasonal rainfall. The southern Fennoscandian High catchments area could be seasonally dominated by precipitation linked with the Palaeo-Tethys monsoon. Therefore, it may be suggested that a parallel increase in precipitation during the MD might be related to increasing rainstorm flood pulses and intensity, which might be expected to occur as the Earth becomes warmer and is likely to amplify erosional effects (cf. Trenhaile 2010). This seems to be in accord with the general conclusions of climate modelling indicating that Late Permian climate seasonality increased as the Pangaean monsoon strengthened (Parrish 1995; Roscher *et al.* 2011). Moreover, the Tethyan margin of Gondwana was strongly influenced by the Permo-Triassic super monsoon system (Roscher *et al.* 2011, and references therein). Alternatively then, stronger seasonal precipitation fed by tropical air masses from the Tethys (rather than by a higher number of storms) could be transported northwards, bringing more precipitation to the area discussed here, as during the Pleistocene to the Great Basin when coastal California was still dry (Lyle *et al.* 2012).

The humid impact on MD deposition has also been suggested by other studies. For example, Peryt & Scholle (1996), based on geochemical features of the MD dolomites from the northern and southern parts of the SPB in Poland, suggested that the MD diagenetic alteration was probably associated with climatic changes (rainfall variations), and also that the northern margin of the Zechstein MD basin was more influenced by humidity than the southern one. In this case, they ascribed a humid impact to post-MD deposition control. Sachse *et al.* (2004) noted two intra-MD desiccation events of regional importance alternating with wetter periods in the Thuringian Basin (southern margin of the SPB). However, more eastwards, Peryt & Scholle (1996) did not identify either increased freshwater input or relevant drying out. Although the results of the present study are inconsistent with these interpretations, all these observations argue for local and temporal humid control on the MD deposition related to increases of transport or concentration of larger quantities of moisture into the MD platforms from maritime regions of the SPB. As may be inferred from the above-mentioned studies, the MD rainfall impact in the SPB area was regionally differentiated. Most probably it was controlled by palaeogeography, because the orographic pattern strongly influences precipitation distribution (e.g. Prömmel *et al.* 2013).

Dilution of carbonate deposition. An indirect result of an increase in humidity during MD deposition is an apparent decline of carbonate content upwards of the MD succession. This stratigraphic trend is best expressed in the shallow platform settings and is associated with a temporarily higher, strongly varying and locally elevated admixture of clastic material ($IR + Al_2O_3$ max. up to *c.* 30%). As discussed in the previous sections it can be inferred that variations in siliciclastic accumulation affected the bulk carbonate content in the platform area and were accompanied by freshwater input. Decreasing upwards, the content of carbonates, exemplified by a reduced concentration of total Ca and Mg (if normalized to Al) in the inner and barrier zones (Fig. 7), evidences a stratigraphic dilution of carbonate accumulation by silicates.

The carbonate content lowering landwards reflects a dependence on physiographic conditions of the marginal part of the platform area. This simply relates to a generally low gradient of the proximal part, which stored fluvial sediments behind the platform barrier belt. Carbonate dilution in the inner platform setting was accompanied by an influence of continental waters. The largest of CaO range, the lowest average MgO concentrations (Table 2), and high fluctuations in the Ca/Mg ratio in lagoonal facies are consistent with the model of humid impact on coastal deposition of the carbonate platform. In this platform zone the carbonate content normalized to Al is highest where $\delta^{18}O$ is lowest, and values of both parameters accordingly decrease in the upper part of the MD. These geochemical patterns can be linked to humid conditions exerting a negative influence on carbonate production and increased aluminosilicate concentration in this part of the platform.

In the barrier setting, a decrease of $CaO + MgO/Al_2O_3$ is accompanied by an increasing trend of $\delta^{18}O$ values upwards in the section (Table 1, Fig. 7). This pattern probably indicates that dilution (by silicates) of the barrier deposits was associated with an increased sedimentation rate and evaporation. Thus a progressive dilution of the Kamień Pomorski platform carbonates over time mirrors increased fluxes of weathered silicate material, suggesting a general increase in sedimentation rate towards the top of the MD unit. Consequently, the geochemical features of the MD platform do not seem to support the view that the final MD sea-level fall was directly caused by carbonate overproduction (e.g. Kaiser *et al.* 2003).

Summary: prospect for the development of a palaeoclimatic proxy

The Late Permian climate was dominated by an intracontinental desert comprising a huge arid to semiarid area covering large parts of Gondwana (Barron & Fawcett 1995; Fluteau *et al.* 2001; Korte *et al.* 2004; Roscher *et al.* 2011) and humid regions of low latitudes situated in the southeastern part of Laurussia (the Angara and Cathaysia provinces). For such a palaeogeographic situation, the wetter climatic impact on the deposition of the older Zechstein cycles and also on the MD succession is often ignored or even negated (e.g. Wagner 1994). However, the humid events certainly influenced the MD deposition, as revealed in this study, shedding light on the facies complexity of the MD deposition.

The MD geochemical features in the investigated area document that the MD climatic tendency changed from dry (and thus relatively low precipitation) in the lower part to relatively wetter conditions (through a gradual shift) during sedimentation of the upper part of the unit. It would be premature to generalize the results presented in this paper. However, it is worth considering that the Late Permian Zechstein climatic conditions, outside the distinctly humid areas in the low-latitude belt, were not always hot, nor were they permanently or progressively arid. Thus, the high aridity ascribed to the Zechstein units was differentiated and undoubtedly punctuated by palaeoclimatic oscillations, at least on the scale of carbonate units. In a comparative context, this may resemble the Pleistocene, when wetter climates resulted in a greater level of detrital material supply, and indeed many deserts were repeatedly affected during the Pleistocene by pluvial conditions (e.g. Capo & Chadwick 1999). However, although the MD sedimentary–diagenetic events developed generally in arid conditions and were locally and temporally punctuated by increased freshwater supply, caution should be advised when extrapolating local data to interpretations of basin-scale sea-level changes, especially if direct climatic effects are not considered in parallel. Therefore, it is important to note that the useful geochemical data reviewed here do not confirm the MD climate gradation from more ‘dry’ to more ‘wet’ end members, as could be inferred from lithological indicators such as evaporites, carbonates and siliciclastics.

The changes in the MD lithofacies framework and the geochemical features of the studied carbonate platform are interpreted as a combined result of local tectonics, evaporation, increased influx of silicates due to increased rainfall and thus freshwater, and nutrient inflows into the coastal waters of the MD Kamień Pomorski carbonate platform. Consequently, the $\delta^{13}C$ changes in the MD sections of different platform zones are a result of variations in the rate of organic carbon production and delivery relative to rates of organic carbon burial (Hayes *et al.* 1999) and can easily be viewed as more related to palaeogeography and changes in climatic conditions than cyclic sea-level fluctuations. Similarly, the bulk elemental analysis shows spatial–temporal changes in fluviatile discharge from elevated soil erosion as a result of increased humidity. Although, to some extent, palaeogeography influenced the allocation of continent-derived siliciclastics within the platform area, both the horizontal and vertical distributions of oxygen and carbon isotopes, as well as major and minor elements, point to changes in climatic conditions during MD deposition. The relative increase of surface humidity in later MD times inferred here could not, however, shut down the evaporation resulting from generally arid and hot surficial climatic conditions. Thus, a generally limited rainfall could not induce a substantial depletion of both carbon and oxygen isotope signatures.

This study documents a local intensification of wetter conditions in the northern part of the SPB. The nature of the relatively weaker and stronger climatic impact (rainfall regime) in the lower and upper MD, respectively, reflected by the changing trend in silicate fluxes (i.e. enhanced precipitation runoff from the adjacent land), is difficult to envisage. Climatic fluctuations are revealed that should be preserved on a similar scale in other MD coastal carbonate platforms, where apparently permanent arid conditions could succeed. However, because of a general lack of wider geochemical investigation of MD carbonate platforms throughout the SPB, it is difficult to infer if the climatic impact on the deposition of this Zechstein unit was only local or a symptom of wider, regional phenomena incuding the northern and southern margins of the basin. As indicated by Sachse *et al.* (2004) the intra-MD increase in evaporation in the Thuringian Basin (southern SPB margin) was interrupted by wetter events related to strong summer monsoonal rains originating from the Tethys Sea. These results suggest that the winds were able to pick up the moisture from the warm Palaeo–Tethys Ocean surface during their northward movement across the narrow Central Pangean Mountains belt, bringing the rains associated with the south or SW monsoon in the southern SPB margin (Fig. 9). This is in agreement with numeric modelling predicting that the European region (the northern Palaeo-Tethyan coast occurs in present southern and central Europe) was dominated by seasonal tropics, where most of the rain fell in summer time (Roscher *et al.* 2011). However, about 200–300 km eastward, at the same basin margin, palaeolatitude and similar palaeotopographic conditions, Peryt & Scholle (1996) did not find either 'strong rain events associated with the aforementioned monsoonal circulation patterns' or any heavy meteoric-water influences on MD carbonate deposition. In fact, they noted that 'extensive river input [was] largely lacking on the southern shelf', thus stressing impoverishment in freshwater influence and indirectly climate aridity of this SPB region. Thus, these prerequisites seem to suggest that the marginal belt of the SPB typically experienced locally wetter climatic impacts during MD deposition. The southern margin possibly received more limited moisture locally and temporarily in the form of monsoonal rains reduced by lee effects caused by the Central Pangaean Mountains.

Note also that more frequent rainy seasons, attributed to a more humid climate, were responsible for a general lowering of brine concentration, implying a change from arid to more humid climatic conditions in the older three Zechstein chlorides cycles (PZ1–PZ3) of Poland, proving subordinary variations of a dominant warm and dry climate

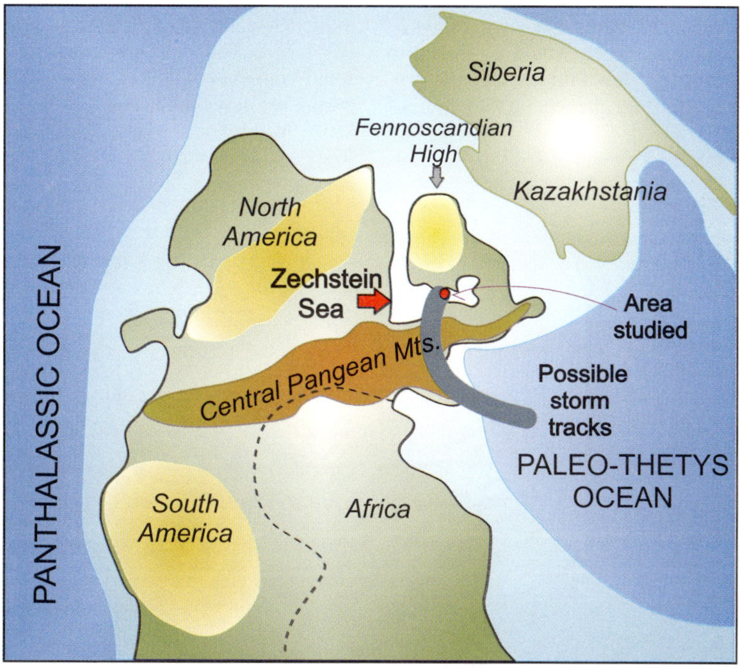

Fig. 9. Palaeogeographic reconstruction of the Pangaea during Late Permian (255 Ma) (after Scotese 2013 (PALEOMAP website), simplified and modified with respect to the Zechstein Sea extent). Light areas, desert and desert-like areas. Possible storm tracks were born in the Palaeo-Tethys Ocean and brought warm moist air wandering to the North and into the SPB area during the Main Dolomite period.

(Czapowski & Tomassi 2013). It is probable then that the warm, humid air masses brought by the prevailing south southwestern (SSW) trade winds or even monsoons tracks from the Tethyan Ocean were strong enough and moved northwards to the SPB area, meeting northern colder air (carried by the northeastern continental airflows) and resulting in increase in rainfall intensity in SPB maritime areas (Fig. 10). This factor had a significant impact on the sedimentary and diagenetic characteristics of the MD carbonate platform, as discussed above. Because of the quite different palaeogeographic conditions in the northern hemisphere, monsoons or heavy storms were probably not as strong as those of the southern hemisphere (Kutzbach & Gallimore 1989; Barron & Fawcett 1995; Horton *et al.* 2012). It could be hypothesized that, like in sub-recent tropical and subtropical regions, the change in climatic (rainfall) regime was possibly related to a southern displacement of the Intertropical Convergence Zone (sensitive enough to northern-hemisphere high-latitude climate cooling) to the warmer hemisphere (Muller *et al.* 2012). The Permian palaeolatitude of Poland was at or near the confluence of major global atmospheric circulation cells (Hadley and Ferrel cells) and small shifts in the positioning of these cells may have led to dramatic changes in regional rainfall patterns (Peryt & Scholle 1996).

In general, enhanced flood discharge followed enhanced precipitation, which would be linked to the northern hemisphere low-latitude climatic processes and is important for diagenetic alteration of the MD deposits. MD climate variation is crucial for interpretations of some diagenetic processes, especially for most extensive dolomitization, which often forms whole carbonate platforms. Peryt & Scholle (1996) proposed that the differences between the MD dolomites from the northern and southern margins of the SPB are best explained by regional variations in insignificant amounts of river water introduction during episodic exposure events. However, according to them, the alteration event was not synsedimentary but occurred later, 'shortly after sedimentation', during episodic regional sea-level lowstands. This suggestion is contradicted by the dataset presented in this work. Although tectonic uplift and related subaerial exposures of the limey platform could be important for the setup of hydrological circulation of dolomitizing fluids, this factor is not always operative and may not exert direct control on the dolomitization (Zhao & Jones 2012). Their interpretation also fails because diffusion of dolomitizing elements downwards through impermeable anhydrite beds is efficiently prevented or the flux is too small to allow for extensive platform dolomitization. Even more essential is that the associated lower A2 anhydrite sequences do not show relics of unconformities or discontinuities. Thus, the 'wetter climates' could not 'prevail during parts of the hiatus intervals reflecting episodic regional sea-level lowstands' inducing carbonate platforms dolomitization as Peryt & Scholle (1996) postulated. Instead, they acted contemporaneously with MD sedimentation, during sea-level highstand.

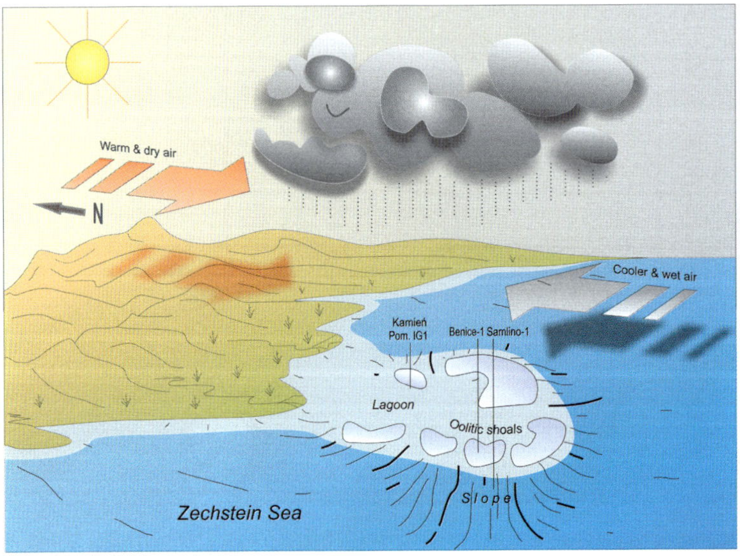

Fig. 10. Generalized model of the main palaeoclimatic factors operating during the Main Dolomite deposition in the Kamień Pomorski carbonate platform area.

The growth of the MD succession in the studied area was climatically controlled. The proposed model does not switch from an arid to a humid phase of climate. It assumes generally low freshwater input to the studied platform and a relative increase in wetter frequency within the generally arid tendency of the region (Fig. 11). Such interpretation is consistent with numeric climate modelling, which indicates wetness fluctuations in the lower part of the Zechstein (Roscher *et al.* 2008). To some extent this situation may resemble the Holocene sedimentary record of the Near East, where increasing aridity is associated with alternation between multiple wetter and dryer phases, leading to significant breaks in the cultural evolution (Migowski *et al.* 2006).

Major climatic proxy signals, when extracted and compared, may help to identify MD formation in other SPB regions, and possibly also its depositional cycles, and may attempt to correlate the sedimentary successions in distant locations. The above discussion would imply that new insights may be achieved, for example through examination of patterns of carbon isotopic co-variation with other geochemical proxies when extended to other MD carbonate platforms. The possibility of correlating local changes related to regional climatic trends may facilitate the reconstruction of the general evolution of sedimentary environments and the associated palaeoclimatic conditions during MD deposition. It would provide unparalleled stratigraphic control for other studies. Thus, future work should continue to drive specific questions related to the MD geological record, especially its stratigraphic subdivisions and correlations, which remain poorly constrained. Although further examination of these carbonates is required, earlier data seem to indicate the influence of both external and internal dynamics, ruling out MD intra-cyclicity events related to basin-scale sea-level changes of glacio-eustatic origin. Simultaneously, it cannot be excluded that climatic episodes operating on finer scales corresponding to the sixth order or less could efficiently obliterate more regional eustatic (tectonic) cycles. Specifically then, future research on SPB platforms should address the relative effects of climate on the Zechstein deposition in order to not only refine previous models resulting from stratigraphic sequence analyses, but also to recognize basinal dynamics and to support various modelling studies. However, the validity of this concept with respect to the Zechstein carbonate unit's stratigraphy has been queried very little so far. On the other hand, the idea of climatic impact on the Zechstein platform depositional architecture could not be realized as a driving model for solving obvious stratigraphic weaknesses regarding the subdivisions of the Zechstein carbonate units. Nevertheless, if palaeoclimatic impact commonly controlled, directly or indirectly, the Zechstein deposition, this would suggest a further confirmation of the hierarchy of processes that account for basin-scale

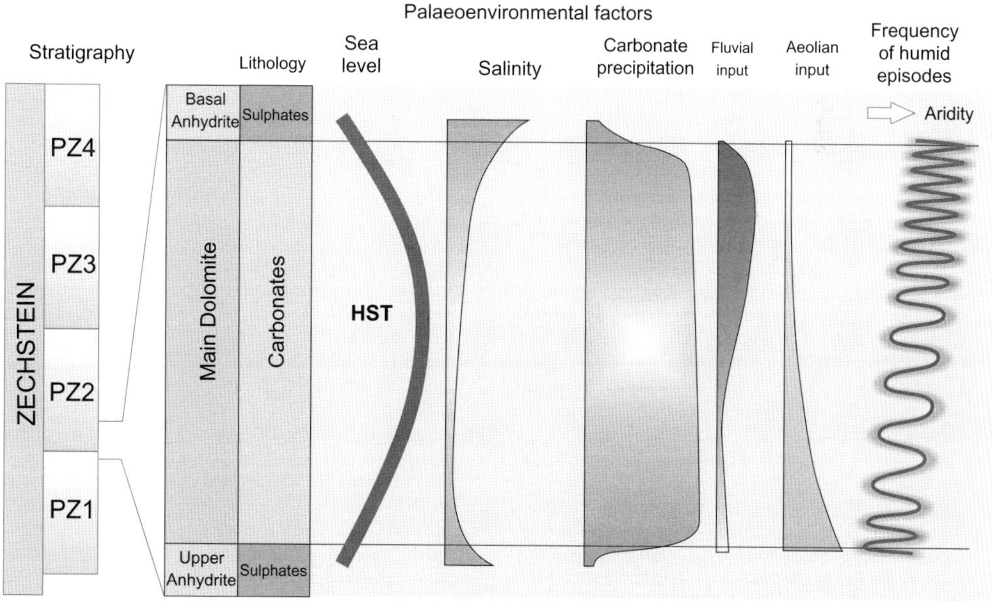

Fig. 11. Schematic record of main palaeoenvironmental factors operating during the prevailed highstand (HST) of the Main Dolomite carbonate platform development in the Kamień Pomorski area.

cyclical changes being an integral part of sequence stratigraphy. Hence, extrabasinal palaeoclimate-forcing processes may be a critical link to Zechstein stratigraphy.

Two GSL anonymous reviewers and M. Słowakiewicz are thanked for their valuable comments. They all improved the final version of the text. However, I would especially like to thank one anonymous reviewer who worked hard on this manuscript and made helpful comments. This research was partly supported by grants from the Ministry of Environmental Protection, Natural Resources and Forestry (2.13.1007.00.0) and the Polish Geological Institute (62.9711.1202.00.1).

References

ALGEO, T. J., WILKINSON, B. H. & LOHMANN, K. C. 1992. Meteoric-burial diagenesis of Middle Pennsylvanian limestones in the Orogrande Basin, New Mexico: water/rock interactions and basin geothermics. *Journal of Sedimentary Petrology*, **62**, 652–670.

ALGEO, T. J., CHEN, Z. Q., FRAISER, M. L. & TWITCHETT, R. J. 2011. Terrestrial–marine teleconnections in the collapse and rebuilding of Early Triassic marine ecosystems. *Palaeogeography, Palaeoclimatology, Palaeoecology*, **308**, 1–11.

ALLAN, J. R. & MATTHEWS, R. K. 1982. Isotopes signatures associated with early meteoric diagenesis. *Sedimentology*, **29**, 797–817.

BALL, M. M. 1967. Carbonate sand bodies of Florida and the Bahamas. *Journal of Sedimentary Petrology*, **3**, 556–591.

BARRON, E. J. & FAWCETT, P. J. 1995. The climate of Pangea: a review of climate model simulation of the Permian. *In*: SCHOLLE, P. A., PERYT, T. M. & ULMER-SCHOLLE, D. S. (eds) *The Permian of Northern Pangea. Paleogeography, Paleoclimates, Stratigraphy*. Springer-Verlag, Berlin, **1**, 37–52.

BERNER, R. A. 1980. *Early Diagenesis: A Theoretical Approach*. Princeton University Press, Princeton.

BERNER, R. A. 2005. The carbon and sulfur cycles and atmospheric oxygen from middle Permian to middle Triassic. *Geochimica et Cosmochimica Acta*, **69**, 3211–3217.

BOHACS, K. M., CARROLL, A. R., NEAL, J. E. & MANKIEWICZ, P. J. 2000. Lake-basin type, source potential, and hydrocarbon character: an integrated sequence-stratigraphic geochemical framework. *In*: Gierlowski-Kordesch, E. H. & Kelts, K. R. (eds) *Lake Basins Through Space and Time*. American Association of Petroleum Geologists, Studies in Geology, **46**, 3–34.

BROOKINS, D. G. 1988. *Eh-pH Diagrams for Geochemistry*. Springer-Verlag, Berlin.

BRUMSACK, H.-J. 2006. The trace metal content of recent organic carbon-rich sediments: implications for Cretaceous black shale formation. *Palaeogeography, Palaeoclimatology, Palaeoecology*, **232**, 344–361.

BURDIGE, D. J. 1993. The biogeochemistry of manganese and iron reduction in marine sediments. *Earth Science Reviews*, **35**, 249–284.

CANDER, H. S. 1994. An example of mixing-zone dolomite, Middle Eocene Avon Park Formation, Floridan aquifer system. *Journal of Sedimentary Research*, **64**, 615–629.

CAPO, R. C. & CHADWICK, O. A. 1999. Sources of strontium and calcium in desert soil and calcrete. *Earth Planetary Scientific Letters*, **170**, 61–72.

CARPENTER, S. J. & LOHMANN, K. C. 1997. Carbon isotope ratios of Phanerozoic marine cements: re-evaluating the global carbon and sulfur systems. *Geochimica et Cosmochimica Acta*, **61**, 4831–4846.

CATTANEO, A. & STEEL, R. J. 2003. Transgressive deposits: a review of their variability. *Earth-Science Reviews*, **62**, 187–228.

CHRIST, N., IMMENHAUSER, A. *ET AL*. 2012. Triassic Latemar cycle tops – subaerial exposure of platform carbonates under tropical arid climate. *Sedimentary Geology*, **265–266**, 1–29.

CROWELL, J. C. 1995. The ending of the Late Paleozoic Ice Age during the Permian period. *In*: SCHOLLE, P. A., PERYT, T. M. & ULMER-SCHOLLE, D. S. (eds) *The Permian of Northern Pangea. Vol. 1: Paleogeography, Paleoclimates, Stratigraphy*. Springer-Verlag, Berlin, 62–74.

CZAPOWSKI, G. & TOMASSI-MORAWIEC, H. 2013. Palaeogeographic and palaeoclimate factors of salinity fluctuations in the eastern part of the Late Permian (Zechstein) European Basin: the case study from the salt basin in Poland. *In*: GĄSIEWICZ, A. & SŁOWAKIEWICZ, M. (eds) *Palaeozoic Climate Cycles: Their Evolutionary and Sedimentological Impact*. Geological Society, London, Special Publications, **376**. First published online September 19, 2013, updated October 9, 2013, http://dx.doi.org/10.1144/SP376.16.

DEAN, W. E. & SCHWALB, A. 2002. The lacustrine carbon cycle as illuminated by the waters and sediments of two hydrologically distinct headwater lakes in North-Central Minnesota, U.S.A. *Journal of Sedimentary Research*, **72**, 416–431.

DRAVIS, J. 1979. Rapid and widespread generation of Recent oolitic hardgrounds on a high energy Bahamian platform, Eleuthera Bank, Bahamas. *Journal of Sedimentary Petrology*, **49**, 195–208.

DRAVIS, J. J. 1996. Rapidity of freshwater calcite cementation – implications for carbonate diagenesis and sequence stratigraphy. *Sedimentary Geology*, **107**, 1–10.

DUNBAR, G. B. & DICKENS, G. R. 2003. Late Quaternary shedding of shallow-marine carbonate along a tropical mixed siliciclastic–carbonate shelf: Great Barrier Reef, Australia. *Sedimentology*, **50**, 1061–1077.

EHRLICH, H. L. & NEWMAN, D. K. 2009. *Geomicrobiology*, 5th edn. CRC Press/Taylor & Francis, Boca Raton.

FIJAŁKOWSKA, A. 1995. Palynostratigraphy and palynofacies of the Permian–Triassic transitional sequence in the Żary Pericline (SW Poland). *Geological Quarterly*, **39**, 307–332.

FILIPPELLI, G. M. 1997. Controls on phosphorus concentration and accumulation in oceanic sediments. *Marine Geology*, **139**, 231–240.

FLUTEAU, F., BESSE, J., BROUTIN, J. & RAMSTEIN, G. 2001. The Late Permian climate. What can be inferred from climate modeling concerning Pangea scenarios and Hercynian range altitude? *Palaeogeography, Palaeoclimatology, Palaeoecology*, **167**, 39–71.

FRAKES, L. A., FRANCIS, J. E. & SYKTUS, J. I. 1994. *Climate Modes of the Phanerozoic*. Cambridge University Press, Cambridge.

GĄSIEWICZ, A. & WICHROWSKA, M. in press. Sedymentologia, diageneza i geochemia utworów cechsztyńskiego dolomitu głównego (Ca2) platformy węglanowej Kamienia Pomorskiego. *Prace Państwowego Instytutu Geologicznego* (in Polish).

GEBELEIN, C. D. 1976. Open marine subtidal and intertidal stromatolites (Florida, the Bahamas and Bermuda). *In*: WALTER, M. R. (ed.) *Stromatolites*. Developments in Sedimentology, Elsevier, Amsterdam, **20**, 381–388.

GELUK, M. C., VAN WEES, J. D., GRÖNLOH, H. & VAN ADRICHEM BOOGAERT, H. A. 1997. Palaeogeography and palaeotectonics of the Zechstein (Upper Permian) in the Netherlands. *Prace Państwowego Instytutu Geologicznego*, **157**(Part 2), 63–75.

GINGELE, F. X. & DECKKER, P. D. 2005. Late Quaternary fluctuations of paleoproductivity in the Murray Canyons Area, south Australian continental margin. *Palaeogeography, Palaeoclimatology, Palaeoecology*, **220**, 361–373.

GISCHLER, E., GINSBURG, R. N., HERRLE, J. O. & PRASAD, S. 2010. Mixed carbonates and siliciclastics in the Quaternary of southern Belize: Pleistocene turning points in reef development controlled by sea-level change. *Sedimentology*, **57**, 1049–1068.

GRIGSBY, J. D. 2001. Origin and growth mechanism of authigenic chlorite in sandstones of the lower Vicksburg Formation, South Texas. *Journal of Sedimentary Research*, **71**, 27–36.

GROTEK, I. 2002. Biomarkery jako wskaźniki facji organicznych w utworach dolomitu głównego. *Biuletyn Państwowego Instytutu Geologicznego*, **402**, 51–74.

GULBRANSON, E. L., ISBELL, J. L., TAYLOR, E. L., RYBERG, P. E., TAYLOR, T. N. & FLAIG, P. P. 2012. Permian polar forests: deciduousness and environmental variation. *Geobiology*, **10**, 479–495.

HALLEY, R. B. & HARRIS, P. M. 1979. Freshwater cementation of a 1000-year-old oolite. *Journal of Sedimentary Petrology*, **49**, 969–988.

HARRIS, P. M. 2010. Delineating and quantifying depositional facies patterns in carbonate reservoirs: insight from modern analogs. *American Association of Petroleum Geologists Bulletin*, **94**, 61–86.

HAYES, J. M., STRAUSS, H. & KAUFMAN, A. J. 1999. The abundance of ^{13}C in marine organic matter and isotopic fractionation in the global biogeochemical cycle of carbon during the past 800 Ma. *Chemical Geology*, **161**, 103–125.

HESSE, P. P. & MCTAINSH, G. H. 2003. Australian dust deposits: modern processes and the Quaternary record. *Quaternary Sciences Reviews*, **22**, 2007–2035.

HORTON, D. E., POULSEN, C. J., MONTAÑEZ, I. P. & DIMICHELE, W. A. 2012. Eccentricity-paced late Paleozoic climate change. *Palaeogeography, Palaeoclimatology, Palaeoecology*, **331–332**, 150–161.

IRWIN, H., CURTIS, C. & COLEMAN, M. L. 1977. Isotopic evidence for source of diagenetic carbonates formed during burial of organic-rich sediments. *Nature*, **269**, 209–213.

ISBELL, J. L., HENRY, L. C., REID, C. M. & FRAISER, M. L. 2012. Sedimentology and palaeoecology of lonestone-bearing mixed clastic rocks and cold-water carbonates of the Lower Permian Basal Beds at Fossil Cliffs, Maria Island, Tasmania (Australia): insight into the initial decline of the late Palaeozoic ice age. *In*: GĄSIEWICZ, A. & SŁOWAKIEWICZ, M. (eds) *Palaeozoic Climate Cycles: Their Evolutionary and Sedimentological Impact*. Geological Society, London, Special Publications, **376**. First published online November 26, 2012, http://dx.doi.org/10.1144/SP376.2

JONES, B. & MANNING, D. A. C. 1994. Comparison of geochemical indices used for the interpretation of palaeoredox conditions in ancient mudstones. *Chemical Geology*, **111**, 111–129.

KAISER, R., NÖTH, S. & RICKEN, W. 2003. Sequence stratigraphy with emphasis on platform-related parasequences of the Zechstein 2 carbonate (Ca2) – the Northern Platform margin of the Southern Permian Basin (NE Germany). *International Journal of Earth-Sciences (Geologische Rundschau)*, **92**, 54–67.

KAUFMAN, A. J. & KNOLL, A. H. 1995. Neoproterozoic variations in the C-isotopic composition of seawater: stratigraphic and biogeochemical implications. *Precambrian Research*, **73**, 27–49.

KIERSNOWSKI, H., PAUL, J., PERYT, T. M. & SMITH, D. B. 1995. Facies, paleogeography, and sedimentary history of the Southern Permian Basin in Europe. *In*: SCHOLLE, P. A., PERYT, T. M. & ULMER-SCHOLLE, D. S. (eds) *The Permian of Northern Pangea*. Springer-Verlag, Berlin, Sedimentary Basins and Economic Resources, **2**, 119–136.

KIM, W., FOUKE, B. W., PETTER, A. L., QUINN, T. M., KERANS, C. & TAYLOR, F. 2012. Sea-level rise, depth-dependent carbonate sedimentation and the paradox of drowned platforms. *Sedimentology*, **59**, 1677–1694.

KNIES, J. & MANN, U. 2002. Depositional envirnment and source rock potential of Miocene strata from the central Fram Strait: introduction of a new computing tool for simulating organic facies variations. *Marine and Petroleum Geology*, **19**, 811–828.

KORTE, C., KOZUR, H. W., JOACHIMSKI, M. M., STRAUSS, H., VEIZER, J. & SCHWARK, L. 2004. Carbon, sulfur, oxygen and strontium isotope records, organic geochemistry and biostratigraphy across the Permian/Triassic boundary in Abadeh, Iran. *International Journal of Earth-Sciences (Geologische Rundschau)*, **93**, 565–581.

KRYSTINIK, L. F. 1990. Early diagenesis in continental eolian deposits. *In*: FRYBERGER, S. G., KRYSTINIK, L. F. & SCHENK, C. J. (eds) *Modern and Ancient Eolian Deposits: Petroleum Exploration and Production, Rocky Mountain Section*. Rocky Mountain Section, Society of Economic Paleontologists and Mineralogists, Denver, 8.1–8.11.

KUTZBACH, J. E. & GALLIMORE, R. G. 1989. Pangaean climates – megamonsoons of the megacontinent. *Journal of Geophysical Research*, **94**, 3341–3357.

KÜHN, V., HEYDEMANN, A. & USDOWSKI, E. 2001. Mineralogy of the Upper Albian sediments from the Kirchrode I borehole with special emphasis on thr clay mineralogy. *Palaeogeography, Palaeoclimatology, Palaeoecology*, **174**, 145–179.

LAZAR, B. & EREZ, J. 1990. Extreme ^{13}C depletions in seawater-derived brines and their implicalions for the past geochemical carbon cycle. *Geology*, **18**, 1191–1194.

LEEDER, M. R., HARRIS, T. & KIRKBY, M. J. 1998. Sediment supply and climate change: implications for basin stratigraphy. *Basin Research*, **10**, 7–18.

LEGLER, B. & SCHNEIDER, J. W. 2008. Marine ingressions into the Middle/Late Permian saline lake of the Southern Permian Basin (Rotliegend, Northern Germany) possibly linked to sea-level highstands in the Arctic rift. *Palaeogeography, Palaeoclimatology, Palaeoecology*, **267**, 102–114.

LINK, P. K. 2009. 'Icehouse' (cold) climates. *In*: GORNITZ, V. (ed.) *Encyclopedia of Paleoclimatology and Ancient Environments*. Springer-Verlag, Berlin, 463–471.

LIU, Z., COLIN, C. ET AL. 2010. Clay mineral distribution in surface sediments of the northeastern South China Sea and surrounding fluvial drainage basins: source and transport. *Marine Geology*, **277**, 48–60.

LIU, Z., WANG, H., HANTORO, W. S., SATHIAMURTHY, E., COLIN, C., ZHAO, Y. & LI, J. 2012. Climatic and tectonic controls on chemical weathering in tropical Southeast Asia (Malay Peninsula, Borneo, and Sumatra). *Chemical Geology*, **291**, 1–12.

LLOYD, R. M. 1966. Oxygen isotope enrichment of seawater by evaporation. *Geochimica et Cosmochimica Acta*, **30**, 801–814.

LYLE, M., HEUSSER, L. ET AL. 2012. Out of the Tropics: the Pacific, Great Basin Lakes, and Late Pleistocene water cycle in the Western United States. *Science*, **337**, 1629–1633.

MACHEL, H. G. 2004. Concepts and models of dolomitization: a critical reappraisal. *In*: BRAITHWAITE, C. J. R., RIZZI, G. & DARKE, G. (eds) *The Geometry and Petrogenesis of Dolomite Hydrocarbon Reservoirs*. Geological Society, London, Special Publications, **235**, 7–63.

MAGARITZ, M. 1987. A new explanation for cyclic deposition in marine evaporite basins: meteoric water input. *Chemical Geology*, **62**, 239–250.

MAGARITZ, M. & SCHULTZE, K.-H. 1980. Carbon isotope anomaly of the Permian Period. *In*: FÜCHTBAUER, H. & PERYT, T. (eds) *The Zechstein Basin with Emphasis on Carbonate Sequences. Contributions to Sedimentology*, E. Schweitzerbart'sche Verlagsbuchhandlung (Naegele u. Obermiller), Stuttgart, **9**, 269–277.

MÄRZ, C., BECKMANN, B., FRANKE, C., VOGT, C., WAGNER, T. & KASTEN, S. 2009. Geochemical environment of the Coniacian–Santonian western tropical Atlantic at Demerara Rise. *Palaeogeography, Palaeoclimatology, Palaeoecology*, **273**, 286–301.

MATTHEWS, R. K. 1987. Eustatic controls on near-surface carbonate diagenesis. *In*: HALLEY, R. B. & MATTHEWS, R. K. (eds) *Carbonate Depositional Environments: Modern and Ancient*. Part 6: Diagenesis 2. *Colorado School of Mines Quarterly*, **82**, 17–40.

MAWSON, M. & TUCKER, T. 2009. High-frequency cyclicity (Milankovitch and millennial-scale) in slope-apron carbonates: Zechstein (Upper Permian), north-east England. *Sedimentology*, **56**, 1905–1936.

MCCREA, J. M. 1950. The isotopic chemistry of carbonates and a paleotemperature scale. *Journal of Chemical Physics*, **18**, 849–857.

MCGLUE, M. M., ELLIS, G. S., COHEN, A. S. & SWARZENSKI, P. W. 2012. Playa-lake sedimentation and organic matter accumulation in an Andean piggyback basin: the recent record from the Cuenca de Pozuelos, north-west Argentina. *Sedimentology*, **59**, 1237–1256.

MCMANUS, J., BERELSON, W. M., COALE, K. H., JOHNSON, K. S. & KILGORE, T. E. 1997. Phosphorus regeneration in continental margin sediments. *Geochimica et Cosmochimica Acta*, **61**, 2891–2907.

MENNING, M. 1995. A numerical timescale for the Permian and Triassic periods: an integrated time analysis. *In*: SCHOLLE, P. A., PERYT, T. M. & ULMER-SCHOLLE, D. S. (eds) *The Permian of Northern Pangea. Paleogeography, Paleoclimates, Stratigraphy*. Springer-Verlag, Berlin, **1**, 77–97.

MEUNIER, A. 2005. *Clays*. Springer-Verlag, Berlin.

MIGOWSKI, C., STEIN, M., PRASAD, S., NEGENDANK, J. F. W. & AGNON, A. 2006. Holocene climate variability and cultural evolution in the Near East from the Dead Sea sedimentary record. *Quaternary Research*, **66**, 421–431.

MORSE, J. W. & MACKENZIE, F. T. 1990. *Geochemistry of Sedimentary Carbonates*. Elsevier, Amsterdam, Developments in Sedimentology, **48**.

MUHS, D. R., BUSH, C. A., ROWLAND, T. R. & STEWART, C. C. 1990. Geochemical evidence of Saharan dust parent material for soils developed on Quaternary limestones of Caribbean and western Atlantic islands. *Quaternary Research*, **33**, 157–177.

MULLER, J., MCMANUS, J. F., OPPO, D. W. & FRANCOIS, R. 2012. Strengthening of the Northeast Monsoon over the Flores Sea, Indonesia, at the time of Heinrich event 1. *Geology*, **40**, 635–638.

NAAFS, B. D. A., HEFTER, J., ACTON, G., HAUG, G. H., MARTÍNEZ-GARCIA, A., PANCOST, R. & STEIN, R. 2012. Strengthening of North American dust sources during the late Pliocene (2.7 Ma). *Earth and Planetary Science Letters*, **317–318**, 8–19.

OSLEGER, D. A. 1998. Sequence architecture and sea level dynamics of late Permian shelfal facies, Guadalupe Mountains, southern New Mexico. *Journal of Sedimentary Research*, **68**, 327–346.

PARRISH, J. T. 1993. Climate of the supercontinent Pangea. *Journal of Geology*, **101**, 215–233.

PARRISH, J. T. 1995. Geological evidence of Permian climate. *In*: SCHOLLE, P. A., PERYT, T. M. & ULMER-SCHOLLE, D. S. (eds) *The Permian of Northern Pangea. Paleogeography, Paleoclimates, Stratigraphy*. Springer-Verlag, Berlin, **1**, 53–61.

PATTERSON, W. P. & WALTER, L. M. 1994. Depletion of ^{13}C in seawater CO_2 on modern carbonate platforms: significance for the carbon isotopic record of carbonates. *Geology*, **22**, 885–888.

PERYT, T. M. 1986. The Zechstein (Upper Permian) Main Dolomite deposits of the Leba Elevation, Northern Poland; facies and depositional history. *Facies*, **14**, 151–200.

PERYT, T. M. & SCHOLLE, P. A. 1996. Regional setting and role of meteoric water in dolomite formation and diagenesis in an evaporite basin: studies in the Zechstein

(Permian) deposits of Poland. *Sedimentology*, **43**, 1005–1023.

Peryt, T. M., Geluk, M. C., Mathiesen, A., Paul, J. & Smith, K. 2010. Zechstein. *In*: Doornenbal, J. C. & Stevenson, A. G. (eds) *Petroleum Geological Atlas of the Southern Permian Basin Area*. EAGE Publications, Houten, 123–147.

Potter, P. E. 1978. Petrology and chemistry of modern big river sands. *Journal of Geology*, **86**, 423–449.

Potter, P., Maynard, J. & Depetris, P. 2005. *Mud and Mudstones*. Springer-Verlag, Berlin.

Price, G. D. 1999. The evidence and implications of polar ice during the Mesozoic. *Earth-Science Reviews*, **48**, 183–210.

Prömmel, K., Cubasch, U. & Kaspar, F. 2013. A regional climate model study of the impact of tectonic and orbital forcing on African precipitation and vegetation. *Palaeogeography, Palaeoclimatology, Palaeoecology*, **369**, 154–162.

Purser, B. H. & Evans, G. 1973. Regional sedimentation along the Trucial Coast, SE Persian Gulf. *In*: Purser, B. H. (ed.) *The Persian Gulf. Holocene Carbonate Sedimentation and Diagenesis in a Shallow Epicontinental Sea*. Springer-Verlag, Berlin, 211–231.

Rankey, E. C. & Reeder, S. L. 2010. Controls on platform-scale patterns of surface sediments, shallow Holocene platforms, Bahamas. *Sedimentology*, **57**, 1545–1565.

Rankey, E. C. & Reeder, S. L. 2011. Holocene oolitic marine sand complexes of the Bahamas. *Journal of Sedimentary Research*, **81**, 97–117.

Ratcliffe, K. T., Wright, A. M., Hallsworth, C., Morton, A., Zaitlin, B. A., Potocki, D. & Wray, D. S. 2004. Alternative correlation techniques in the petroleum industry: an example from the (Lower Cretaceous) Basal Quartz, Southern Alberta. *American Association of Petroleum Geologists Bulletin*, **88**, 1419–1432.

Reeder, S. L. & Rankey, E. C. 2008. Interactions between tidal flows and ooid shoals, northern Bahamas. *Journal of Sedimentary Research*, **78**, 175–186.

Rees, P. M. 2002. Land-plant diversity and the end-Permian mass extinction. *Geology*, **30**, 827–830.

Retallack, G. J. 1995. Permian–Triassic extinction on land. *Science*, **267**, 77–80.

Roscher, M., Berner, U. & Schneider, J. W. 2008. A tool for the assessment of the paleo-distribution of source and reservoir rocks. *OIL GAS European Magazine*, **3**, 131–137.

Roscher, M., Stordal, F. & Svensen, H. 2011. The effect of global warming and global cooling on the distribution of the latest Permian climate zones. *Palaeogeography, Palaeoclimatology, Palaeoecology*, **309**, 186–200.

Rullkötter, J. 2000. Organic matter, the driving force for early diagenesis. *In*: Schulz, H. D. & Zabel, M. (eds) *Marine Geochemistry*. Springer-Verlag, Berlin, 129–172.

Sabatier, P., Dezileau, L., Briqueu, L., Colin, C. & Sian, G. 2010. Clay minerals and geochemistry record from northwest Mediterranean coastal lagoon sequence: implications for paleostorm reconstruction. *Sedimentary Geology*, **228**, 205–217.

Sachse, D., Radke, J., Gaupp, R., Schwark, L., Lüniger, G. & Gleixner, G. 2004. Reconstruction of palaeohydrological conditions in a lagoon during the 2nd Zechstein cycle through simultaneous use δD values of individual n-alkanes and $\delta^{18}O$ and $\delta^{13}C$ values of carbonates. *International Journal of Earth Sciences (Geologische Rundschau)*, **93**, 554–564.

Scheffler, K., Buehmann, D. & Schwark, L. 2006. Analysis of late Palaeozoic glacial to postglacial sedimentary successions in South Africa by geochemical proxies – response to climate evolution and sedimentary environment. *Palaeogeography, Palaeoclimatology, Palaeoecology*, **240**, 184–203.

Schenau, S. J., Reichart, G. J. & De Lange, G. J. 2005. Phosphorus burial as a function of paleoproductivity and redox conditions in Arabian Sea sediments. *Geochimica et Cosmochimica Acta*, **69**, 919–931.

Schlager, W., Reijmer, J. J. G. & Droxler, A. 1994. Highstand shedding of carbonate platforms. *Journal of Sedimentary Research*, **64**, 270–281.

Schneider, J. W., Körner, F., Roscher, M. & Kroner, U. 2006. Permian climate development in the northern peri-Tethys area – the Lodève basin, French Massif Central, compared in a European and global context. *Palaeogeography, Palaeoclimatology, Palaeoecology*, **240**, 161–183.

Schultz, R. B. 2004. Geochemical relationships of Late Paleozoic carbon-rich shales of the Midcontinent, USA: a compendium of results advocating changeable geochemical conditions. *Chemical Geology*, **206**, 347–372.

Scotese, C. R. 2013. PALEOMAP website. http://www.scotese.com.

Scotese, C. R. & Langford, R. P. 1995. Pangea and the paleogeography of the Permian. *In*: Scholle, P. A., Peryt, T. M. & Ulmer-Scholle, D. S. (eds) *The Permian of Northern Pangea. Paleogeography, Paleoclimates, Stratigraphy*. Springer-Verlag, Berlin, **1**, 3–19.

Semeniuk, V. & Searle, D. J. 1985. Distribution of calcrete in Holocene coastal sands in relationship to climate, southwestern Australia. *Journal of Sedimentary Petrology*, **55**, 86–95.

Shinn, E. A. 2011. Interplay between Holocene sedimentation and diagenesis, and implications for hydrocarbon exploitation: return to the sabkha of Ras Umm Said, Qatar. *In*: Kendall, C. G. St. C. & Alsharhan, A. (eds) *Quaternary Carbonate and Evaporite Sedimentary Facies and Their Ancient Analogues: A Tribute to Douglas James Shearman*. International Association of Sedimentologists, Special Publications, **43**, 133–148.

Sirocko, F., Sarnthein, M., Erlenkeuser, H., Lange, H., Arnold, M. & Duplessy, J. C. 1993. Century-scale events in monsoonal climate over the past 24,000 years. *Nature*, **364**, 322–324.

Słowakiewicz, M. & Gąsiewicz, A. 2013. Palaeoclimatic imprint, distribution and genesis of Zechstein Main Dolomite (Upper Permian) petroleum source rocks in Poland: sedimentological and geochemical rationales. *In*: Gąsiewicz, A. & Słowakiewicz, M. (eds) *Palaeozoic Climate Cycles: Their Evolutionary and Sedimentological Impact*. Geological Society,

London, Special Publications, **376**. First published online September 25, 2013, http://dx.doi.org/10.1144/SP376.21.

Słowakiewicz, M. & Mikołajewski, Z. 2009. Sequence stratigraphy of the Upper Permian Zechstein Main Dolomite carbonates in western Poland: a new approach. *Journal of Petroleum Geology*, **32**, 215–234.

Słowakiewicz, M. & Mikołajewski, Z. 2011. Upper Permian Main Dolomite microbial carbonates as potential source rocks for hydrocarbons (W Poland). *Marine and Petroleum Geology*, **28**, 1572–1591.

Smith, D. B. 1980. The evolution of the English Zechstein Basin. *In*: Füchtbauer, H. & Peryt, T. (eds) *The Zechstein Basin with Emphasis on Carbonate Sequences*. Contributions to Sedimentology, E. Schweitzerbart'sche Verlagsbuchhandlung (Naegele u. Obermiller), Stuttgart, **9**, 7–34.

Strohmenger, C., Voigt, E. & Zimdars, J. 1996. Sequence stratigraphy and cyclic development of Basal Zechstein carbonate-evaporite deposits with emphasis on Zechstein 2 off-platform carbonates (Upper Permian, northeast Germany). *Sedimentary Geology*, **102**, 33–54.

Sun, S. Q. & Esteban, M. 1994. Paleoclimatic controls on sedimentation, diagenesis and reservoir quality: lessons from Miocene carbonates. *American Association of Petroleum Geologists Bulletin*, **78**, 519–543.

Taylor, J. C. M. 1998. Upper Permian–Zechstein. *In*: Glennie, K. W. (ed.) *Introduction to the Petroleum Geology of the North Sea*. Blackwell Scientific Publications, Oxford, 153–189.

Tinker, S. W. 1998. Shelf-to-basin facies distributions and sequence stratigraphy of a steep-rimmed carbonate margin: Capitan depositional system, McKittrick Canyon, New Mexico and Texas. *Journal of Sedimentary Research*, **68**, 1146–1174.

Trenhaile, A. S. 2010. Modeling cohesive clay coast evolution and response to climate change. *Marine Geology*, **27**, 711–720.

Tucker, M. E. 1991. Sequence stratigraphy of carbonate-evaporite basins: models and application to the Upper Permian (Zechstein) of northeast England and adjoining North Sea. *Journal of the Geological Society of London*, **148**, 1019–1036.

Tucker, M. E. & Wright, V. P. 1990. *Carbonate Sedimentology*. Blackwell, Oxford.

Turner, P. & Magaritz, M. 1986. Chemical and isotopic studies of a core of marl slate from NE England: influence of freshwater influx into the Zechstein sea. *In*: Harwood, G. M. & Smith, D. B. (eds) *The English Zechstein and Related Topics*. Geological Society, London, Special Publications, **22**, 19–29.

van Wagoner, J. C., Mitchum, R. M., Campion, K. M. & Rahmanian, V. D. 1990. *Siliciclastic Sequence Stratigraphy in Well Logs, Cores, and Outcrops; Concepts for High-Resolution Correlation of Time and Facies*. American Association of Petroleum Geologists, Tulsa, Methods in Exploration Series, 7.

Veizer, J. & Hoefs, J. 1976. The nature of $^{18}O/^{16}O$ and $^{13}C/^{12}C$ secular trends in sedimentary carbonate rocks. *Geochimica et Cosmochimica Acta*, **40**, 1387–1395.

Wagner, R. 1994. Stratygrafia i rozwój basenu cechsztyńskiego na Niżu Polskim. *Prace Państwowego Instytutu Geologicznego*, **146**, 1–71.

Wagner, R. & Peryt, T. M. 1997. Possibility of sequence stratigraphic subdivison of the Zechstein in the Polish Basin. *Geological Quarterly*, **41**, 457–474.

Wagner, R., Piątkowski, T. S. & Peryt, T. M. 1978. Polski basen cechsztyński. *Przegląd Geologiczny*, **26**, 673–686 (in Polish).

Wang, X., Xia, D., Zhang, C., Lang, L., Hua, T. & Zhao, S. 2012. Geochemical and magnetic characteristics of fine-grained surface sediments in potential dust source areas: implications for tracing the provenance of Aeolian deposits and associated palaeoclimatic change in East Asia. *Palaeogeography, Palaeoclimatology, Palaeoecology*, **323–325**, 123–132.

Wanless, H. R. & Tedesco, I. P. 1993. Comparison of oolitic sand bodies generated by tidal v. wind-wave agitation. *In*: Keith, B. D. & Zuppann, C. W. (eds) *Mississippian Oolites and Modern Analogs*. American Association of Petroleum Geologists, Studies in Geology, **35**, 199–225.

Warrington, G. 2008. Palynology of the Permian succession in the Hilton Borehole, Vale of Eden, Cumbria, UK. *Proceedings of the Yorkshire Geological Society*, **57**, 123–130.

Weaver, C. E. 1989. *Clays, Muds, and Shales*. Elsevier, Amsterdam, Developments in Sedimentology, **44**.

Wehausen, R. & Brumsack, H.-J. 2000. Chemical cycles in Pliocene sapropel-bearing and sapropelbarren eastern Mediterranean sediments. *Palaeogeography, Palaeoclimatology, Palaeoecology*, **158**, 325–352.

Whittaker, T. E., Hendy, C. H. & Hellstrom, J. C. 2011. Abrupt millennial-scale changes in intensity of Southern Hemisphere westerly winds during marine isotope stages 2–4. *Geology*, **39**, 455–458.

Wilson, M. E. J. 2012. Equatorial carbonates: an earth systems approach. *Sedimentology*, **59**, 1–31.

Wilson, M., Neumann, E.-R., Davies, G. R., Timmerman, M. J., Heeremans, M. & Larsen, B. T. 2004. Permo-Carboniferous magmatism and rifting in Europe: introduction. *In*: Wilson, M., Neumann, E.-R, Davies, G. R., Timmerman, M. J., Heeremans, M. & Larsen, B. T. (eds) *Permo-Carboniferous Magmatism and Rifting in Europe*. Geological Society, London, Special Publications, **223**, 1–10.

Wing, S. L. 1984. Relation of paleovegetation to geometry and cyclicity of some fluvial carbonaceous deposits. *Journal of Sedimentary Petrology*, **54**, 52–66.

Wopfner, H. 2012 Late Palaeozoic–Early Triassic deposition and climates between Samfrau and Tethys: a review. *In*: Gąsiewicz, A. & Słowakiewicz, M. (eds) *Palaeozoic Climate Cycles: Their Evolutionary and Sedimentological Impact*. Geological Society, London, Special Publications, **376**, First published online November 26, 2012, http://dx.doi.org/10.1144/SP376.4

Wright, V. P. 1994. Paleosols in shallow marine carbonate sequences. *Earth-Sciences Reviews*, **35**, 367–395.

Xie, S., Pancost, R. D., Yin, H., Wang, H. & Evershed, R. P. 2005. Two episodes of microbial change coupled with Permo/Triassic faunal mass extinction. *Nature*, **434**, 494–497.

Zhao, H. & Jones, B. 2012. Genesis of fabric-destructive dolostones: a case study of the Brac Formation (Oligocene), Cayman Brac, British West Indies. *Sedimentary Geology*, **267–268**, 36–54.

Ziegler, P. A. 1990. *Geological Atlas of Western and Central Europe*. Shell International Petroleum Maatschappij, The Netherlands.

Zubin-Stathopoulos, K. D., Beauchamp, B., Davydov, V. I. & Henderson, C. M. 2012. Variability of Pennsylvanian–Permian carbonate associations and implications for northwest Pangea palaeogeography, East-Central British Columbia, Canada. *In*: Gąsiewicz, A. & Słowakiewicz, M. (eds) *Palaeozoic Climate Cycles: Their Evolutionary and Sedimentological Impact*. Geological Society, London, Special Publications, **376**. First published online November 26, 2012, http://dx.doi.org/10.1144/SP376.1

Palaeoclimatic imprint, distribution and genesis of Zechstein Main Dolomite (Upper Permian) petroleum source rocks in Poland: Sedimentological and geochemical rationales

MIROSŁAW SŁOWAKIEWICZ[1,2]* & ANDRZEJ GĄSIEWICZ[2]

[1]*School of Chemistry, University of Bristol, Organic Geochemistry Unit, Cantock's Close, BS8 1TS, UK*

[2]*Polish Geological Institute, Rakowiecka 4, 00-975 Warsaw, Poland*

**Corresponding author (e-mail: m.slowakiewicz@gmail.com)*

Abstract: Arid palaeoclimatic conditions and marine current circulation influenced organic matter productivity and its distribution and preservation in the Upper Permian (Zechstein) Main Dolomite (Ca2) source rocks in Poland. The gaseous and liquid hydrocarbons occurring within the Main Dolomite show distinctive geochemical characteristics which distinguish them from those derived from Carboniferous or other Permian strata and indicate that the main type of organic matter from which they were derived was of microbial origin. It is suggested here that the majority of the Ca2 hydrocarbons come from *in situ* organic matter (intraformational deposits). The lagoonal and slope zones are enriched in total organic carbon (TOC) compared to both the oolite shoal and basinal facies. These sedimentary settings were dominated by various biolaminated microbialite facies. The basinal areas, previously believed to be the location of the main source rocks for hydrocarbons found in the Main Dolomite, appear to have little source rock potential. Geochemical and geological features such as high organic productivity in lagoonal and slope zones of the Main Dolomite indicate that at least some of the hydrocarbons may have been generated from *in situ* deposited and degraded organic matter. This conclusion may improve prediction of the occurrence of hydrocarbon deposits in the Ca2.

The Late Permian was a warmhouse time as the Late Palaeozoic Ice Age finished in the Early Permian with some small ice areas persisting until the end of the Middle Permian (e.g. Chumakov & Zharkov 2002, 2003; Jones *et al.* 2006). Minimum summer temperatures on land at the mid-latitudes of the northern hemisphere were >15 °C higher than the present day (22.4 °C) and average precipitation rate was 1–2 mm a^{-1} (Khiel & Shields 2005). The extension of the Pangea supercontinent to almost both poles resulted in a dry continental interior (summer temperatures <45 °C in the arid subtropical regions at about 20–25°N; Roscher *et al.* 2011). Atmospheric water circulation between the more humid Pangea coasts and its interior was limited by the position of the continents and the Variscan Mountains (elevation of *c.* 2 km; Fluteau *et al.* 2001; Roscher *et al.* 2011). However, water circulation and the position of land masses became crucial, not only for global temperature changes but also for regional and local climate reconstructions and the assessment of the palaeodistribution of reservoir and source rocks (Parrish 1993; Roscher & Schneider 2006; Roscher *et al.* 2008).

Subtropical climate, temperature, salinity and evaporation affected the Polish part of the intercontinental evaporitic Southern Permian Basin (SPB). The SPB was formed in an East–West-trending depocentre located in northern Pangea at palaeolatitudes of 10–15°N (present central and eastern Europe) in Late Permian time (Fig. 1). The formation of the epicontinental Zechstein Sea in the subtropical belt of the northern hemisphere under warm and arid conditions facilitated formation of sabkha-type systems around the basin. Sabkha deposits resembling the modern Abu Dhabi sabkha environment resulted in the deposition of petroleum reservoir and source rocks in the Main Dolomite (Ca2), which were particularly important in the Polish part of the SPB. Although the Zechstein Limestone (Ca1) has the highest gas-bearing potential in general it does not contain source rocks; it is the Main Dolomite (Ca2) which has numerous oil and gas deposits, mostly within platform carbonates in western Poland (Figs 2 & 3).

In general, it has been assumed that hydrocarbons occurring within the Zechstein carbonates in the Polish Lowlands are both of Carboniferous and of intraformational Permian origin (e.g. Kotarba *et al.* 2000*a*, *b*; Kosakowski *et al.* 2003; Kotarba *et al.* 2006; Kotarba & Wagner 2007). An 'intraformational origin' is considered by us to be hydrocarbons formed *in situ* from source rocks occurring within the same stratigraphic interval. The most

From: GĄSIEWICZ, A. & SŁOWAKIEWICZ, M. (eds) 2013. *Palaeozoic Climate Cycles: Their Evolutionary and Sedimentological Impact*. Geological Society, London, Special Publications, **376**, 523–538.
First published online September 25, 2013, http://dx.doi.org/10.1144/SP376.21

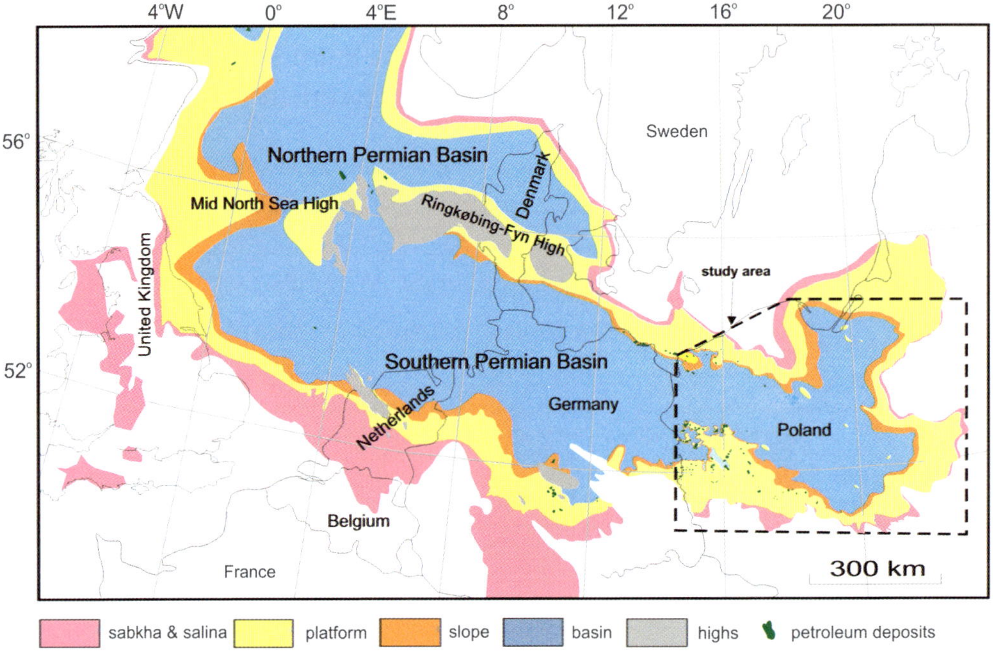

Fig. 1. Present-day extent and palaeogeographical zones of the Main Dolomite basin in Europe with the study area (after Słowakiewicz & Mikołajewski 2011; Słowakiewicz *et al.* 2013).

intriguing and as yet poorly understood problem in the petroleum geology of Poland is the source of Main Dolomite hydrocarbons. While the origin of Carboniferous hydrocarbons present in the Zechstein is quite well understood and confined to structural features of the basement, the origin of intraformational hydrocarbons is highly puzzling with regard to their source rocks; consequently, they are much less predictable.

It has commonly been postulated that the Main Dolomite hydrocarbons were generated in deeper basin and then subsequently migrated into shallow coastal or isolated but porous platform deposits. However, various detailed sedimentological and geochemical studies clearly show that this source has a low organic matter content. Instead, various other facies within the Main Dolomite platform carbonates are rich in organic matter, suggesting that the origin of the intraformational hydrocarbons is essentially confined to these palaeogeographic units. This then introduces the question of primary productivity and the impact of palaeoclimate on the source rock formation. It is well known that climate and nutrients control primary productivity and the rate of organic matter generation. The areas that are warm and arid, where Main Dolomite carbonates were deposited, are generally more productive. Additionally, fluctuating sea levels driven by evaporation and recorded in Main Dolomite (Ca2) carbonates from the Fore-Sudetic Monocline (e.g. Słowakiewicz & Mikołajewski 2009) and palaeoclimatic perturbations (Gąsiewicz, this volume, in press) might have controlled the seawater–porewater redox conditions which would have played a key role in nutrient cycling and availability, pH, sediment and organic matter accumulation and decomposition.

The focus of this paper is the palaeoclimatic influence on organic matter productivity, the origin of intraformationally generated Zechstein Main Dolomite hydrocarbons and the distribution of source rocks in the Polish part of the SPB. An understanding of these issues is of great commercial interest and could lead to improved petroleum reservoir prediction.

Geological background

The Zechstein of Poland (Fig. 2) occupies the eastern part of the intracratonic SPB, one of two major sub-basins that developed in Europe during Permian times. In the centre of the Polish part of the SPB only thin carbonate mudstones were deposited during the time of Ca2 sedimentation, whereas relatively thick platform carbonates developed around the margins of the basin as well as on intrabasinal highs and on top of thick platforms

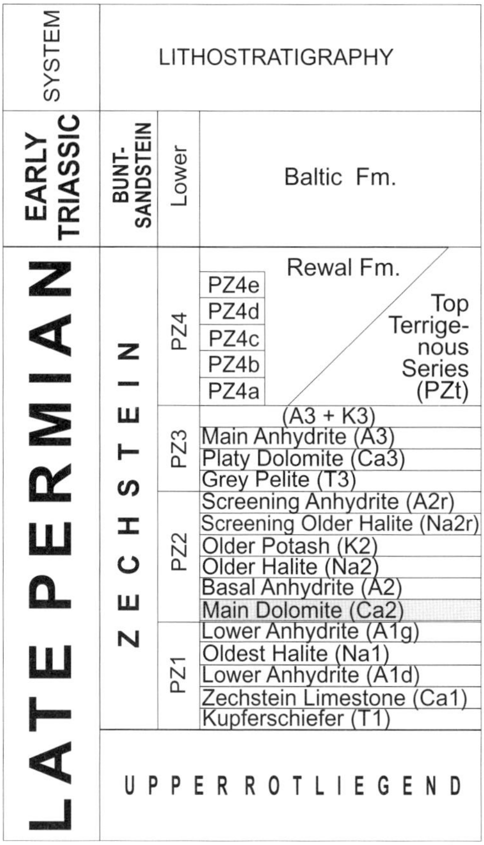

Fig. 2. Lithostratigraphy of the Zechstein in Poland (after Wagner 1994). A3 + K3: Younger Halite + Younger Potash.

formed by the first cycle Werra Anhydrite (A1). It is commonly suggested that the A1 platforms determined the morphology of the overlying Ca2 platforms (e.g. Wagner 1994), but the palaeorelief of the Ca2 carbonate platforms was also controlled by evaporation (evaporative drawdown) and gradual sea-level fall (shallowing-up depositional system and accretion of carbonate platforms), factors determined by local climate impact.

During early Zechstein time the SPB began to subside as a result of thermal contraction of the lithosphere (Wilson *et al.* 2004) and increased local tectonic activity and salt mobilization. It has been suggested earlier that these processes could subsequently have been responsible for potential migration pathways for hydrocarbons in the Ca2 during strong subsidence of the central part of the basin. This view seems to correspond to the well-rooted hypothesis that the central basin was the place from which the Main Dolomite hydrocarbons were derived.

The Main Dolomite carbonates are overlain by the Basal Anhydrite (A2) and underlain by the Werra Anhydrite (A1) evaporites, and by thick (100–400 m) Older Halite (Na2) and thinner (200–300 m) Oldest Halite (Na1) rock salts, respectively. These evaporites acted as seals for hydrocarbons (Fig. 2).

The thickness of the Main Dolomite carbonates is usually within the range 30–120 m on the platforms, whereas slope and basin-floor successions are typically 20–60 m and <10 m thick, respectively. Locally in western Poland, the Ca2 succession can be up to 180–190 m thick (e.g. Babimost-1, Kargowa-1 wells), but generally it is only 30–40 m in thickness. Eight principal depositional systems can be recognized within the northern and southern margins of the Ca2 succession and these are summarized in Table 1. They include: sabkha, tidal flat with tidal channels, restricted lagoon, outer and inner barrier/shoal, foreshore-backshore, slope, toe-of-slope apron and basin. These settings are characterized by a range of mostly abiogenic carbonate lithofacies changing from evaporites (nodular to bedded anhydrite) to relatively thick sequences of structureless, laminated or bedded mudstone, wackestone, packstone and grainstone, as well as secondary rudstone, floatstone and others. Within these, various microbially mediated or influenced lithofacies have been recognized such as biolaminated mudstone and wackestone, various stromatolites, biolaminoids and clotted thrombolites (Fig. 4).

The Ca2 reservoir in Poland occurs at a present-day subsurface depth of 1200–4400 m, such that palaeotemperatures have been high enough to generate hydrocarbons at some localities (Polish Geological Institute, archival data).

Features and distribution of intraformational petroleum deposits

The Ca2 hydrocarbons do not all have the same source, that is, they are of both Carboniferous and intraformational origin. Although most Permian petroleum deposits are of mixed composition, tabulation of the hydrocarbon deposits found in the Zechstein according to the isotopic characteristics of their kerogen clearly shows that the intraformational hydrocarbons have different and distinct isotopic values compared to those of Carboniferous or Rotliegend origin. As can be seen in Figure 5, gases in the Main Dolomite framework occupy a discrete area of the nitrogen cross-plot compared with those from the Carboniferous, Rotliegend and Zechstein Limestone. The tabulation also reveals that the Carboniferous hydrocarbons occur mainly in the SE whereas the intraformational

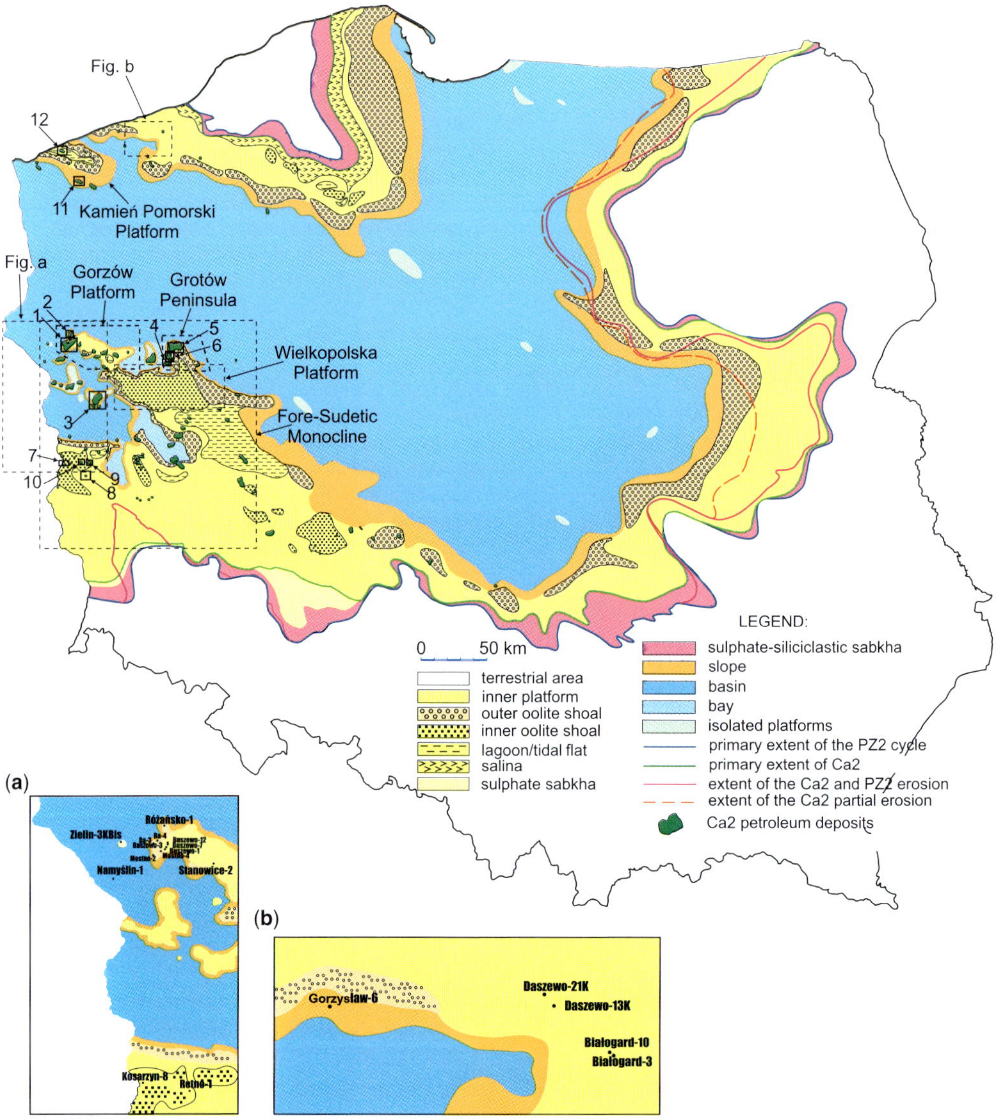

Fig. 3. Palaeogeographic map of the Main Dolomite in Poland with oil and gas fields (modified after Dyjaczyński *et al.* 2000). (**a**) and (**b**) show localities of wells discussed in the text. 1–12: hydrocarbon deposits discussed in the text within the area of the Fore-Sudetic Monocline: 1, Barnówko-Mostno-Buszewo oil and gas field; 2, Różańsko oil field; 3, Sulęcin gas field; 4, Lubiatów oil field; 5, Grotów oil field; 6, Międzychód gas field; 7, Kosarzyn oil field; 8, Czeklin oil field; 9, Retno oil field; 10, Struga oil field; and in West Pomerania: 11, Wysoka Kamieńska oil field; and 12, Kamień Pomorski oil field.

hydrocarbons occur mostly in northwestern and western Poland.

The data presented in Figure 6 and Table 2 show that Ca2 gases are almost solely generated via methyl-type fermentation, and some of them might even have originated from the conversion of oil to methane during anaerobic biodegradation (secondary microbial methane). Furthermore, Figure 7 shows that Kosarzyn, Różańsko, Retno and Struga Main Dolomite oil deposits in western Poland (Fore-Sudetic Monocline) have different $\delta^{13}C$ isotopic data from other crude oils, with values between −27‰ and −25.5‰ PDB for saturated hydrocarbons and −27‰ to −25‰ PDB for

Table 1. *Summary of depositional systems associated with the northern and southern margins of the Main Dolomite basin in Poland*

Depositional systems	Facies and sedimentary structures
Sabkha	Sandy carbonate + carbonate + clayey nodular anhydrite
Tidal flat	Heavily dolomitized microbialites, microbially influenced carbonate (stromatolite, thrombolite, oncolite, biolaminite) and muddy + bioclastic wackestone-packstone; tidal channel and tempestite with cycles of coarse- to fine-grained carbonate; desiccation cracks, fenestrae, erosion surfaces, microtepees, ripple marks, lenticular and flaser-like bedding
Foreshore-backshore	Dolomitized oolitic-vadolitic packstone and grainstone, compaction of coated grains, vadose cements, cross-bedded and -stratified up to 10 m thick, graded-bedding
Restricted and open lagoon (subtidal)	Structureless and flat laminated dolomitized mudstones and peloidal wackestone-packstone, locally microbial boundstone; relatively abundant fauna (bivalves, ostracods, gastropods, foraminifers); flaser-like and lenticular bedding
Outer and inner shoal	Dolomitized oolitic grainstone and packstone, local thin muddy biolaminite and tidal channel-fill; variously developed ooids + local oncolites, thromboids and pisoids; cross-stratification and flaser-like bedding, vadose to phreatic structures, local thin calcretes
Slope	Calcareous, dolomitized and dedolomitized, flat and wavy laminated mudstones, breccia, debrite, anhydrite breccia and conglomerate interbeds with thin medium- to coarse-grain-dominated beds (turbidites); thicker units of floatstone, rudstone, microbial boundstone (biolaminite, stromatolite and thrombolite), mudstone and bioclastic wackestone-packstone; grain-dominated beds consist of ooids, peloids and bioclasts; soft-sediment deformation, slump structures, inverse grading (debrites) and normal grading (turbidites); slope angle ranged from 2° to 3°; accretionary slope sequence and locally a rimmed-type or a scarp-type of carbonate platform margin
Toe-of-slope apron (lower slope)	Calcareous, dolomitized and dedolomitized interbedded units of oolitic packstone, carbonate conglomerates and debris-flow (debrites?); fragments of redeposited microbialite and carbonate-platform-margin grains; flaser-like and lenticular bedding, local cross-stratification and graded bedding; deposition from turbidity currents, cohesive flows and suspension
Basin	Very thin laminated clayey calcareous mudstone and coquina; local hardgrounds; hemipelagic deposition from suspension and condensation

aromatic hydrocarbons from the western part of the Fore-Sudetic Monocline. This carbon isotopic difference was interpreted as indicating that these four crude oils may have been generated from organic matter deposited under different sedimentary conditions, that is, derived from a Carboniferous source of organic matter (Kotarba *et al.* 2000*a*), as a result of thermal maturation or a combination of the two. From a plot of carbon isotope values of saturated to aromatic hydrocarbon fractions (Fig. 7), it can also be deduced that the source of the oil deposits was marine (algal and microbial) and terrestrial organic matter. In general, densities of Ca2 crude oils vary from 0.746 to 0.896 g cm^{-3} (°API = 25.8–57.2) and contain 0.09–1.67 wt% sulphur (Kotarba *et al.* 2000*a*). These are light crude oils sourced with relatively low sulphur content, typical of carbonate-evaporite environments (e.g. Ali *et al.* 2002) in which production of microbial organic matter is high.

TOC distribution and organic matter maturity

The total organic carbon (TOC) is highly variable throughout the Ca2 basin in Poland; all TOC values summarized in Table 3 are therefore given as averages. In general, the lowest TOC values are in the basin-centre zone and they increase towards the platform-interior, excluding the oolite shoal zones.

The TOC of platform-interior lithofacies exhibits <0.6 wt% but sporadically can demonstrate values as high as 6 wt%. In the oolite shoal zone values of up to *c.* 0.7 wt% are known but values of up to 0.2 wt% are more typical. Within slope and basin-floor lithofacies, values for TOC are typically 0.5 and <0.3 wt%, respectively (Table 3). The TOC values of stromatolites are <0.4 and <1.5 wt% in laminated and biolaminated mudstone of slopes and tidal pools, respectively.

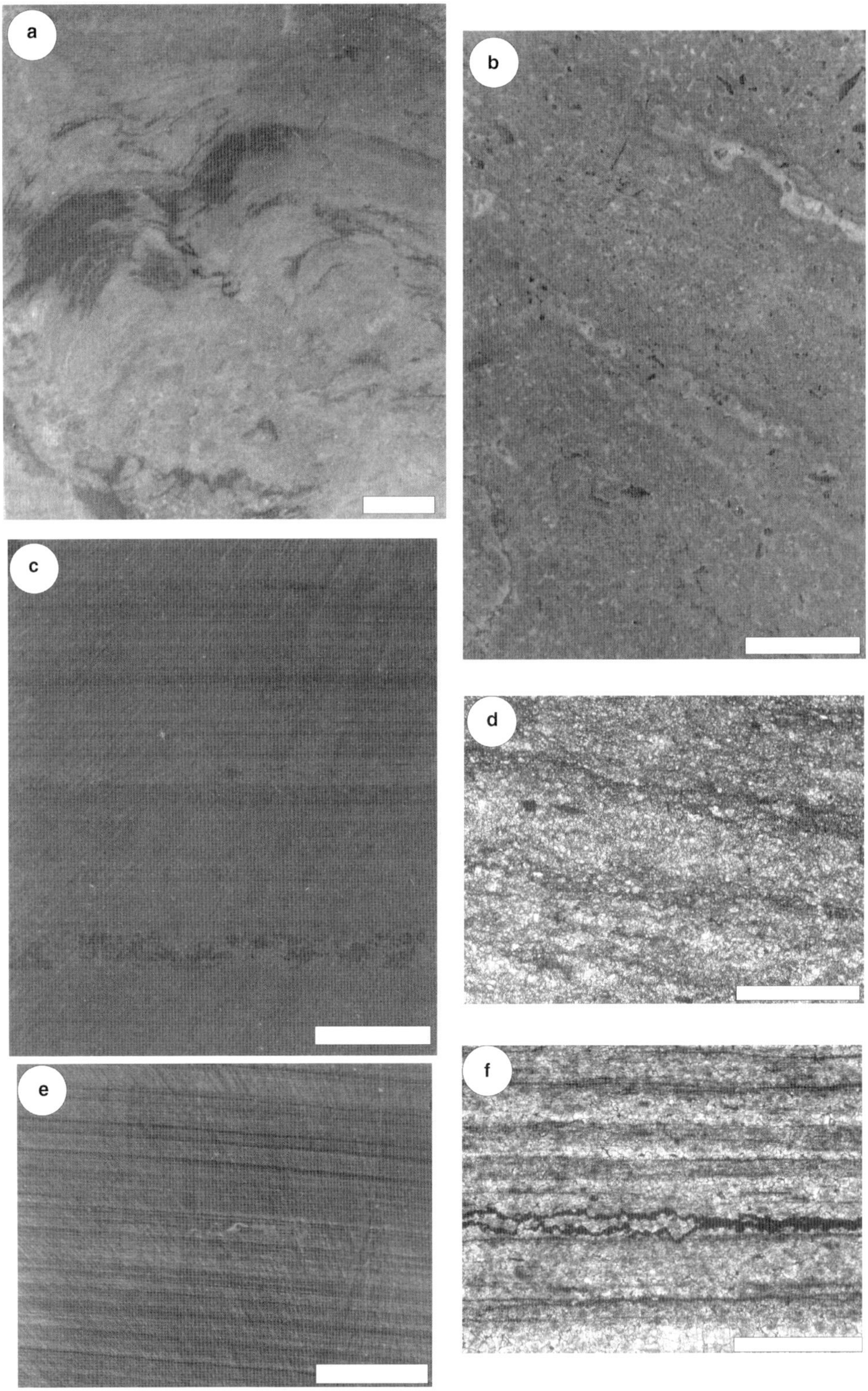
a
b
c
d
e
f

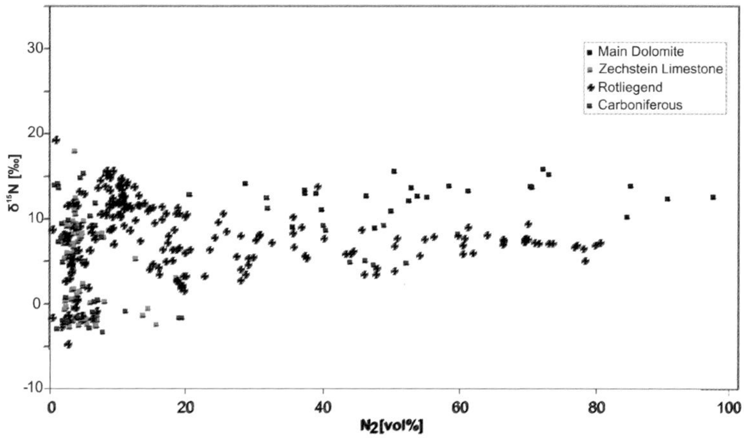

Fig. 5. Correlation of stable nitrogen isotope composition of gaseous nitrogen and its concentration in natural gases from the Carboniferous, Rotliegend, Zechstein Limestone and Main Dolomite deposits (data after Kotarba 1998; Lokhorst 1998; Kotarba *et al.* 2000*b*).

The original calculated TOC (TOC_o) mean values also vary according to palaeogeographic zone: from *c.* 0.77 wt% in the platform interior to *c.* 1.0 wt% in the oolite shoal zone; *c.* 0.65 wt% within the slope deposits; and *c.* 0.46 wt% in the basin zone. According to Kotarba & Wagner (2007) the highest original TOC values were confined to oolite shoal areas and reach up to 5 wt%. Such a relatively high TOC_o value is highly unrepresentative, since these values were obtained from thin microbialite intercalations present within the oolites. Note also that modern oolite shoals are commonly submarine deserts, extremely unfavourable for the production, accumulation and preservation of organic matter (e.g. Purdy 1964; Kenig *et al.* 1990; Rankey & Reeder 2011). Moreover, deposited organic matter is commonly rapidly and completely oxygenated and destroyed in such environments. A more realistic average TOC value for the Ca2 oolite shoal lithofacies is <0.15 wt% (e.g. Kotarba *et al.* 2003; Słowakiewicz *et al.* 2013; Gąsiewicz, this volume, in press); it therefore appears unlikely that hydrocarbons were generated from organic matter originally occurring in the carbonate platform oolite shoal zones. The preserved TOC values therefore suggest that the relatively more productive palaeogeographic zones were confined to platform interiors and slopes. These correlate with the abundance of microbialites deposited within these platform zones where the TOC_o mean values were distinctly higher; for example, in stromatolites and thrombolites TOC is typically *c.* 0.4 wt% and in the biolaminated mudstone may even reach 2 wt% (Słowakiewicz *et al.* 2013).

In general, oil-prone type II kerogen is predominant with admixtures of type I and gas-prone type III in Main Dolomite carbonates. This suggests different origins for the organic matter, that is, microbial and/or algal and/or terrestrial. This may also suggest that at least some kerogen values might have been overprinted during thermal maturation. Thermal maturity of organic matter from the entire Ca2 basin in Poland at T_{max} varies from 264 °C to 530 °C but is mostly within the range 410–450 °C (Grotek 1998; Kotarba *et al.* 1998; Kotarba & Wagner 2007). The T_{max} values in particular palaeogeographic zones vary over the range 409–440 °C in the inner part of the carbonate platform, 385–455 °C in the oolite shoal zone, 264–496 °C (sporadically to 530 °C) within the slope deposits and 412–447 °C in the basinal facies (Table 3). Such a vast range of T_{max} values from immature-mature-to-overmature source rocks does not correlate well with either the TOC distribution (described above) or with the microbialite lithofacies distribution. This suggests that the maturation process was controlled by several factors such as different regional/local geothermal gradients, overprinting by diagenetic processes, subsidence rate and/or relatively early diagenetic processes in particular

Fig. 4. Examples of various microbialites (a–d) and basinal (e–f) facies from the Main Dolomite in Poland: (**a**) columnar stromatolites, platform-interior facies; (**b**) clotted texture of a thrombolite, platform-interior facies; (**c**) biolaminated mudstone, platform slope facies; (**d**) photomicrograph of (c) of biolaminated mudstone, platform slope facies, scale 0.25 mm; (**e**) thinly laminated calcareous mudstone, basinal facies; and (**f**) photomicrograph of (e) of thinly laminated basinal calcareous mudstone with sporadic clayey laminae.

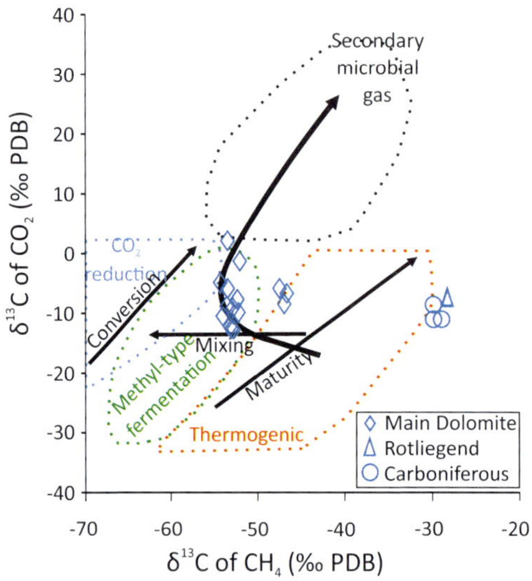

Fig. 6. $\delta^{13}C$ of methane and $\delta^{13}C$ of CO_2 isotopic composition of gases from the Carboniferous, Rotliegend and Main Dolomite deposits with genetic fields after Whiticar (1999) and Milkov (2011). Data after Kotarba (1998) and Kotarba *et al.* (2000*b*, 2005).

parts of the Ca2 basin. Based on this parameter, it is therefore difficult to infer the source and time of the Main Dolomite organic matter transformation and hydrocarbon generation.

A widely used maturity parameter is vitrinite reflectance (VR_r); reliable values of vitrinite reflectance lie within the range 0.5–1.30%, which is equivalent to the early, main to late stage of oil generation ('oil window'). The VR_r values increase towards the basin centre, reaching a maximum of 3.80% (Grotek 1998). The values also change according to palaeogeographic zones: 0.94–2.51% in the inner part of the carbonate platform; 0.58–1.65% in the oolite shoal zone; 0.85–2.10% within the slope deposits; and 1.40–3.80% in the basinal zone (Table 2). This distribution suggests that most of the organic matter was transformed into liquid hydrocarbon phases as inferred from the VR_r results for the platform-interior and slope zones, where the primary organic matter productivity was supposedly relatively high. In the context considered here, it must be stressed that very high VR_r values found in the basinal zone (dark bituminous carbonate laminites) could generate gas. However, no gas deposits have been found within them and long migration pathways (tens to hundreds of kilometres), through irregularly fractured basinal lithofacies to potential toe-of-slope or oolite shoal traps, are rather unrealistic. This is based on the irregular distribution of faults or their absence, suggesting they do not correspond to potential hydrocarbon routes.

From a consideration of the above data it must be concluded that the TOC distribution in the Ca2 basin in Poland is highly variable. Low TOC values in the basin-centre correspond well with the absence of petroleum accumulations or even their

Table 2. *$\delta^{13}C$ of methane and $\delta^{13}C$ of CO_2 isotopic composition of gases from the Carboniferous, Rotliegend and Main Dolomite deposits (data after Kotarba* et al. *2000b, 2005).*

Well name	Stratigraphy	$\delta^{13}C$ (CH_4) PDB	$\delta^{13}C$ (CO_2) PDB
Barnówko-3	Ca2	−52.3	−12.3
Barnówko-4	Ca2	−52.1	−9.2
Buszewo-1	Ca2	−52.5	−6
Buszewo-3	Ca2	−52.6	−9
Buszewo-7	Ca2	−52	−12.8
Buszewo-12	Ca2	−51.6	−10
Cychry-4	Ca2	−47.8	−8.6
Kosarzyn-8	Ca2	−48.2	−6
Mostno-2	Ca2	−51.5	−1.4
Mostno-4	Ca2	−52.2	−12.7
Namyślin-1	Ca2	−53.1	−5
Retno-1	Ca2	−47.6	−6.8
Różańsko-1	Ca2	−51.7	−7.8
Stanowice-2	Ca2	−52.9	−10.6
Zielin-3Kbis	Ca2	−52.5	2
Białogard-3	Rotliegend	−28.3	−7.2
Białogard-10	Rotliegend	−28.1	−7.2
Gorzysław-6	Carboniferous	−30.1	−12.2
Daszewo-13K	Carboniferous	−30.4	−8.9
Daszewo-21K	Carboniferous	−29.2	−11.8

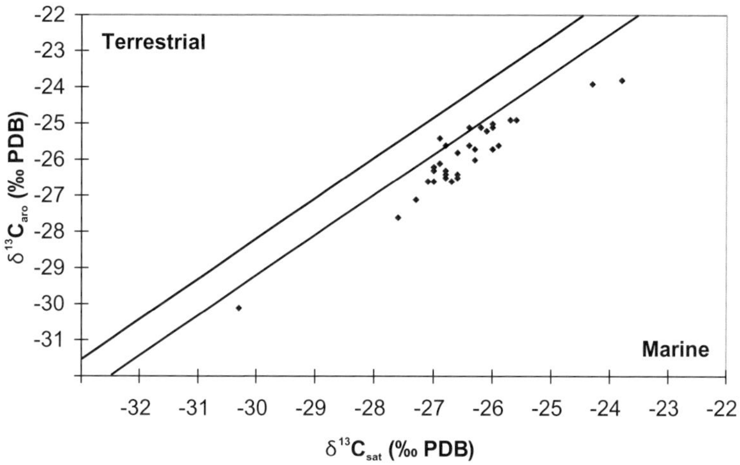

Fig. 7. Saturated hydrocarbons ($\delta^{13}C_{sat}$) v. aromatic hydrocarbons ($\delta^{13}C_{aro}$) of Main Dolomite crude oils after categories of Sofer (1984): 1, Petrykozy-7; 2, Struga-1; 3, Retno-1; 4, Kosarzyn-1; 5, Różańsko-1. Data after Kotarba *et al.* (2000*a*).

traces there. This also correlates well with the highest VR_r and T_{max} values being in the basin-centre, which suggests overmature basinal source rocks. Moreover, stromatolites and thrombolites, which have very low TOC values, cannot be regarded as potential source rocks for hydrocarbons; instead the biolaminated (stromatolitic) mudstone and wackestone localized in the platform-interior lagoons (pools) and upper slope zones should be considered as having better source rock potential. Localization of biolaminated lithofacies is also closely associated with the distribution of petroleum deposits, which in general are confined to slope or platform-interior zones. They also occur in the close vicinity of oolite shoals where hydrocarbons are often trapped.

Discussion

A theoretical model including the main factors influencing the distribution of organic matter, the hydrocarbon factory and formation of source rocks in intertidal and subtidal Main Dolomite deposits may be proposed on the basis of observations made of the Ca2 carbonates and TOC distribution and other published information on comparable organic matter accumulation in modern subtropical environments (e.g. Risk & Rhodes 1985; Kenig 2011). This model comprises climatic imprint, organic productivity and sediment accumulation (including carbonate platform accretion) and preservation of hydrocarbons.

Climatic control

The trend of increasing aridization which began during the Westphalian in the northern part of the Pangea, marked by the spreading of red-bed facies in the European basins, continued until the end of the Permian (e.g. Roscher *et al.* 2011). The formation of thick evaporites (anhydrites and salts) with relatively thin carbonates in the SPB indicates that they formed in semi-arid to arid conditions

Table 3. *Summary of average total organic carbon (TOC), thermal maturity (T_{max}) and vitrinite reflectance (VR_r) values confined to particular depositional systems associated with the northern and southern margins of the Main Dolomite sea in Poland (Grotek 1998; Kotarba* et al. *1998; Kotarba & Wagner 2007)*

Depositional systems	Average TOC (wt%)	T_{max} (°C)	VR_r
Sabkha	–	<420	<0.60
Tidal flat	<0.5	409–440	0.94–2.51
Restricted lagoon	<0.7		
Outer barrier and inner shoal, foreshore-backshore	<0.2	264–496	0.58–1.65
Slope	<0.6	264–496	0.85–2.10
Toe-of-slope apron (lower slope)	<0.7		
Basin	<0.3	412–447	1.40–3.80

similar to those of modern saltern (subaqueous) environments (e.g. the Arabian Gulf and Red Sea area). It is accepted that the arid climate controlled the evaporation rate of the Zechstein Sea which, together with sea-level changes and evaporitic drawdown, resulted in a gradual salinity increase of Zechstein seawater (e.g. Wagner & Peryt 1997). While the Zechstein Limestone carbonates of the first Zechstein cycle were deposited under normal-salinity seawater (evidenced by abundant fauna), the Main Dolomite (second carbonate cycle) and Platy Dolomite (third carbonate cycle) were formed under higher-salinity conditions (evidenced by restricted fauna). However, it is only the Ca2 which produced significant hydrocarbon source rocks. Additionally, subtropical arid climatic conditions accompanied by palaeoclimatic instability induced by a relative rise of precipitation (Gąsiewicz, this volume, in press) enhanced the accumulation of organic carbon through increased production of phytoplankton and bacterial-type organic matter on slopes and in the platform interior, resulting in the formation of the microbially mediated source rocks of the Main Dolomite. On the other hand, a humid climate does not favour the development of microbial mats (or microbialites) because mat development is greatly reduced and the organic layers are thin in such climatic conditions, commonly as a result of an increase in burrowing and grazing organisms and macrophytes. The basinal part had the features of a starved basin (subsidence rate higher than sediment production rate) and so was filled with thin hemipelagic and essentially lime mudstone.

Platform organic productivity

The predominantly dark-coloured basinal lime mudstone of the Main Dolomite implies dysoxic or anoxic conditions in the deeper basin. Casual mechanisms may involve a nutrient-rich layer in shallow waters, increased productivity and resulting O_2 consumption at the bottom. This might have resulted in an O_2-depleted water layer below and the burial of increased volumes of organic matter. However, the TOC distribution across the Main Dolomite carbonate platform zones indicates its higher concentrations in the inner and slope environments. This suggests that higher organic productivity was generally associated with relatively calm and shallower environment of the accreting carbonate platform settings. Such settings, like recent coastal marine environments, could be characterized by high organic matter productivity. The recent environments contain up to 83% of total organic matter (Canfield *et al.* 1998) and up to *c.* 87% of total buried organic matter (Berner 1982). Coastal sediments typically contain 1–2% of organic carbon (Canfield *et al.* 1998). Although phytoplankton is the main donor of organic matter (>90%) in modern environments, in specific sedimentary niches annual organic matter supply may be significant (Post 1981; Head *et al.* 1995) and even higher than phytoplankton delivery (e.g. the Black Sea; Romankevich 1984). Some sediments provide a bacterial biomass up to 1% of dry sediment weight and constitute up to 50% of sedimentary TOC (Gong & Hollander 1997).

The shape of marine basins and the distribution of continental masses have an important effect on currents induced by wind action and the location of high-productivity zones (Killops & Killops 2005). Surface waters of hypersaline lagoons are extremely rich in plankton and nekton, enabling a 20–30 times higher rate of organic matter production in evaporitic basins than in an open ocean (Peterson & Hite 1969; Evans & Kirkland 1988; Hite & Anders 1991). Interactions between temperature, salinity, nutrient supply and oxygen levels lead to productivity enrichment in partly isolated aquifers, and sedimentary accretion isolates burial sub-environments and hence leads to anoxic conditions. Moreover, marine water evaporation results in additional concentration of nutrients and water exchange by tides and currents provided with additional nutrients. As a result, hypersaline lagoons are characterized by extremely high organic productivity (e.g. Phleger & Ewing 1962; Sonnefeld 1984; Logan 1987; Javor 1989).

The northeasterly trade winds which blew in low latitudes of the northern hemisphere from the subtropical high-pressure belts towards the equatorial low-pressure system in winter (in summer SSW direction) (Roscher *et al.* 2011) might have had a significant imprint on marine surface currents in Zechstein time. These currents might have developed on Main Dolomite platforms (shelves) where shallowing and associated increase in current velocity could have resulted in upwelling and enhanced primary productivity in platform slope settings. As noted above, the TOC content in the Main Dolomite platform slope settings is often relatively higher. This suggests efficient water exchange between the shallow platform and deeper basin, and thus current activity facilitating production and accumulation of organic matter.

The high organic productivity in partly isolated basins usually leads to water column stratification and anoxia. This redox stratification or oxycline is observed in present-day isolated basins. In photic zones of aquifers and sediments photosynthesizing algae and bacteria occur and where there is an O_2 deficit (in bottom waters and deeper-water sediments), anaerobic bacteria effectively decompose organic matter (e.g. Jørgensen 1983; Postgate 1984; Ehrlich & Newman 2009). Shallow areas of carbonate platform aquifers are associated with common microbial activity, usually forming various microbialite lithofacies (well documented in the Zechstein Formation). Interactions between their physical setting and biota in general established complex microbiotic ecosystems, which are also stratified. Both detrital organic matter and biomass form by *in situ* growth of prokaryotes and eukaryote-enabled biochemical reactions. Microbes are not limited by substrate availability in such environments, as is typical in shallow open-marine environments, that is, with normal salinity.

Microbial activity produces various gases in interstitial pore water including H_2S, H_2, CO, CO_2 and CH_4 (e.g. Ehrlich & Newman 2009). Production of methane is known from abundant laminated sediments (organic mats) associated with hypersaline lakes such as Solar Lake (e.g. Giani *et al.* 1984; Kenig *et al.* 1990), sabkhas such as those in Sinai (Friedman & Krumbein 1985), Abu Dhabi sabkha, salinas in south Australia (Kendall & Skipwith 1969; Warren & Kendall 1985) and in other evaporite environments (Schreiber 1988; Warren 2006). Methane is the result of the decay of organic matter in already very shallowly buried sediments and accumulates in interstitial pores and beneath the mat layers. $\delta^{13}C-CO_2$ values of the tabulated hydrocarbon deposits (Table 2, Fig. 6) suggest that at least part of the carbon dioxide in the Main Dolomite gases primarily originates from the degradation of soluble organic matter via methyl-type fermentation processes (e.g. Irwin *et al.* 1977; Wycherley *et al.* 1999; Milkov 2011).

It is also important to stress that liquid hydrocarbons are generated in evaporitic environments and have also been found in bottom sediments of many recent hypersaline lakes (e.g. Sonnenfeld 1984; Javor 1989). Moreover, differential amounts of dispersed hydrocarbons are commonly identified in fossil evaporites of different ages. Also, liquid and gaseous hydrocarbons are common in relatively undistributed salt and potash sequences, and are evidently formed *in situ* (Peterson & Hite 1969; Sonnefeld 1984). Additionally, Connan *et al.* (1986) concluded that microbial biomass (halophilic bacteria and methanogenic archaea, cyanobacteria, etc.) contributes significantly to the kerogens from sabkha sediments, and heavy, sulphur-rich oils in carbonates from evaporites are of early generation. These early diagenetic hydrocarbons could also be added to the final late diagenetic Main Dolomite hydrocarbons.

Basinal source rocks and carbonate platform oolite shoal reservoirs

The calcareous thinly laminated claystone and mudstone (commonly named the Stinkschiefer), the basin-floor equivalent of the Main Dolomite, occur in the centre of the Ca2 basin and are commonly thought to be crucial in understanding the origin of Ca2 hydrocarbons. These strata are considered as source rocks for the Ca2 hydrocarbons (e.g. Füchtbauer 1962; Depowski *et al.* 1978; Kotarba & Wagner 2007; Pletsch *et al.* 2010), as well as for a shale gas system (Hartwig & Schulz 2010). According to this view, the central basin area is the hydrocarbon kitchen with migration through thin and relatively low-porosity rocks into high-porosity platform-margin carbonates. As a consequence, the potential of hydrocarbons being discovered in Ca2 deposits was for a long time focused on the platform-margin grainstone shoal facies. This strategy has met with much success; however, hydrocarbon accumulations have also unexpectedly been discovered within other palaeogeographic zones of Ca2 platforms. For instance, in the Kamień Pomorski platform (Fig. 3) petroleum accumulations were found within inner platform and slope facies, within strata which generally have modest reservoir properties. The deposits do not occur in grainstone but in mudstone reservoirs. Another example includes the Sulęcin oilfield (discovered in 1972) at the toe of an isolated platform and the Lubiatów oilfield (the second-largest oil deposit in Poland discovered in 2002) also at the toe of a carbonate platform. These discoveries have changed exploration strategies so that more recent exploration activity has moved to the slopes around the Ca2 platforms. However, these new exploration targets require further development since several dry holes have been drilled, for example in the Lubiatów area. Not only were no hydrocarbon accumulations found in these wells, but in some cases seismic interpretation did not confirm the exploration concept which predicted thick and porous dolomites in these locations (Czekański *et al.* 2010). This again suggests that hydrocarbon exploration in Poland based on structural-palaeogeographic criteria, especially with regard to the Main Dolomite carbonates, needs a reappraisal.

The petroleum accumulations found on the carbonate platform slopes and lagoons (i.e. out of oolite shoal reservoir systems) are intriguing because there is no satisfactory explanation for their presence, even if it is assumed that in some cases there are fractured reservoirs. Excluding a Carboniferous origin for the hydrocarbons, there is still no reliable information on from where, how or why they could have migrated intraformationally.

Main Dolomite microbialites as source rocks

As stated above, basinal facies regarded as the only or essential hydrocarbon source rocks are highly inadequate for many reasons. To address this discrepancy and look for intraformational source rocks associated with the carbonate platform facies, Kotarba & Wagner (2007) attempted to link the generation of hydrocarbons with the occurrence of microbialites. Consequently, according to these authors, routes of hydrocarbon migration from source rocks to reservoir rocks within the Main Dolomite strata were a few kilometres at most.

Kotarba & Wagner (2007) deduced that most of the primary TOC content measured in Ca2 carbonates in relation to that measured today was transformed into productive hydrocarbon deposits (e.g. the Lubiatów oil deposit, which has an estimated total in-place generated reserve of *c.* 42 million barrels of oil, and the Barnówko-Mostno-Buszewo oil pool with reserves of oil of *c.* 77 million barrels). However, the calculated original $TOC_o = 0.8-2.0$ wt% (similar values obtained by Kotarba & Wagner 2007) confirms the hypothesis that at such low TOC concentrations, it is improbable that such large accumulations will be obtained.

Moreover, by counting biolaminated sections of profiles at least 1 m thick, Kotarba & Wagner (2007) distinguished four types of biolaminated beds defined by summarizing their total thickness per Ca2 individual section. They distinguished between sections with rare biolaminae comprising up to 10% of a fixed section thickness, frequent as 10–20%, common as 20–30% and very common as 30–40%. This classification has nothing to do with the primary content of organic matter, however. Note that their calculation is not based upon TOC content but distribution of biolaminae in the particular Ca2 profiles. However, this approach assumes that all the deposited organic carbon was transformed into hydrocarbons but ignores several important factors responsible for hydrocarbon generation such as complex diagenetic processes of organic matter degradation, biochemical interactions and mineral precipitation. Moreover, this also ignores a complex set of diagenetic rock properties affecting hydrocarbon preservation such as recrystallization, cementation and occlusion of pores, vugs and fractures.

Finally, Kotarba & Wagner (2007) emphasized the inaccuracy and subjectivity of their methodology and made an estimate of the generation potential of Ca2 microbial deposits. Their calculations included algal and microbial source rocks and they found that in the Grotów Peninsula area (Lubiatów-Międzychód-Grotów or LMG oil-gas deposit area, Fig. 3) that with a source rock containing a calculated TOC_o content of 1.0–5.0 wt%, the generation potential varies from 12.4 to 80.5 kg HC m^{-3}. Such increased TOC_o and generation potential values are simply too high and still do not explain the large hydrocarbon volumes such as those in the LMG or Barnówko-Mostno-Buszewo deposits.

As can be deduced from high TOC values (Table 3), production of organic matter was highest in the depositional environments which are (1) commonly not bioturbated due to the likely hypersalinity of the platform water developed mainly in platform-interior zones (i.e. restricted lagoons and tidal pools) or (2) relatively low-energy where *in situ* organic matter could have accumulated and been buried relatively quickly and so subject to relatively little reworking. Slopes were supplied by partly redeposited organic matter from the platform-interior (storm activity) and/or accumulated *in situ* (only on low inclined slopes) below fair-weather wave base (see depositional model of Słowakiewicz & Mikołajewski 2011). Restricted lagoons and slopes which contain organic-rich laminated and biolaminated mudstone and wackestone should therefore be considered as the main hydrocarbon factories.

The preservation of hydrocarbons in sedimentary sequences

In situ generated gaseous and liquid hydrocarbons can only be preserved for any length of time in anaerobic environments because the increasing salinity of brines and formation of waters decreases the solubility of hydrocarbons and their relative resistance (especially saturated fractions) to microbial degradation (Sonnenfeld 1984). Retention of hydrocarbons in a sedimentary record is also facilitated by the activity of microbial ecosystems (microbialites). Microbial mat that covers substrates acts like a sealing plastic tissue, and prohibits the escape of gas from the underlying sediment leading to biostabilization (Noffke 2010). Moreover, in photic zone photoautotrophs proliferate, especially cyanobacteria, whereas the deeper settings (of aquifers and sediments) are settled by chemoautotrophic and heterotrophic-in-origin microbiota. They are a likely primary source of induced carbonate precipitation (e.g. Castanier *et al.* 1999; Dupraz *et al.* 2004), and the quantitative contribution of large populations of heterotrophic bacteria to carbonate mineral deposition may be much greater than often presumed. In aqueous environments (apart from the deep ocean) their potential contribution to calcium carbonate sedimentation is much higher than that of autotrophic or abiotic processes (Wright & Oren 2005).

Lithification of microbialite lithofacies is relatively fast in geological terms, almost syndepositional, and results from the growth and degradation of organic matter, common concomitant catalysis and mineral precipitation (e.g. Kendall &

Skipwith 1969; Monty 1976; Gerdes & Krumbein 1987; Papineau *et al.* 2005; Gischler *et al.* 2008). It cannot be excluded that at least part of the high concentrations of H_2S (0–50%; related plausibly to bacterial sulphate reduction or late diagenesis) and nitrogen (3–97%) in Main Dolomite petroleum deposits and inferred average stable sulphur isotopes $\delta^{34}S$ of *c.* 11‰ for Ca2 carbonates suggest efficient organic matter transformation already at early diagenesis. The main controls are sedimentation rate, temperature, nutrient levels and composition of both the ambient sea and pore waters, reinforced by marine currents, flushing of pore spaces by carbonate saturated waters, and so on. A range of diagenetic processes, along with evaporite precipitation, therefore lead to further sealing of the sediment framework and enhanced preservation of organic matter, their compounds and products of biochemical reactions in place. The hydrocarbon preservation potential also increases in evaporitic settings which, like the Southern Permian Basin, are usually characterized by high subsidence rates and consequently high rates of sedimentation, especially of evaporites.

Summary

The conditions of accumulation and especially the generation of hydrocarbons found within the Main Dolomite of Poland are not well constrained and this has resulted in reduced exploration success. Moreover, all the available sedimentological, geochemical and structural data comprising the current model of emplacement for hydrocarbon generation are inadequate to explain the diverse production reserves of hydrocarbons in the Ca2 strata. The basinal lithofacies commonly considered as potential source rocks for hydrocarbons are ruled out because of low thicknesses (<10 m), low TOC values (<0.3 wt%), high burial depths (usually >4000 m), high values of vitrinite reflectance ($VR_r > 1.40$) and long migration pathways through relatively tight lithofacies (from tens to hundreds of kilometres) to traps localized within carbonate sandy shoal facies or at the toe-of-slope aprons of the Ca2 carbonate platforms. As a consequence, the influence of the whole range of factors in petroleum generation should be reconsidered and other closely related but poorly recognized (e.g. trade-wind activity, sources of organic matter, facies changes) factors should be developed more extensively. One promising line of research is to determine the conditions of organic matter preservation in relation to their depositional environments, in terms of both their facies development and redox conditions.

From the data and considerations presented above, it may be concluded that:

(1) climate favoured the distribution and production of organic matter on slopes and in the platform interior;
(2) the main type of organic matter in the source rocks was microbial in origin although microbialites occurring in well-oxygenated zones (TOC values $<0.4\%$) are not able to produce the discovered hydrocarbon volumes; instead biolaminated stromatolitic mudstone and wackestone deposited in restricted lagoons and on tidal flats (tidal pools) and in upper parts of slopes may have significant source rock potential;
(3) the Main Dolomite basinal areas were most probably not the main and/or the only source rocks for the hydrocarbons;
(4) the Ca2 hydrocarbons are different in origin but are mostly intraformational, that is, they were derived from *in situ* deposited and degraded organic matter; and
(5) the Main Dolomite hydrocarbon deposits occur in platformal lithofacies.

Conclusions

The source rocks proposed here for the hydrocarbons within the Zechstein Main Dolomite in Poland are different from those currently accepted and this deduction moves the main location of hydrocarbon generation from deeper basinal areas to carbonate platforms. This is suggested by the existence of microbialites and volumetrically important microbially biolaminated lithofacies in the platformal carbonates. The close association of these organic-rich shallow-water facies with very good reservoir facies confined to carbonate platform margins (thick carbonate sandy shoal systems) created an excellent petroleum system. The hydrocarbons could have migrated from the platform interior or upper slope sources to the very porous oolitic complexes, thus forming the intraformational deposits which are often documented in the Main Dolomite platform carbonates.

The accumulation of organic carbon under a warm and arid tropical climate in Late Permian time in the Polish part of the Southern Permian Basin favoured increased production of organic matter in the Ca2 sea and hence the formation of source rocks.

Two possible sources of hydrocarbons occur: carbonate platform-interior facies and upper slope facies. In light of the data presented above, the potential of the deeper basin for source rocks appears less important. There is still a lack of convincing information regarding the geochemical differences between the hydrocarbons from the slope and platform-interior areas, as well as a definitive migration pathway.

The TOC distribution indicates that the slope and platform-interior environments are enriched in organic carbon compared to the basinal and oolite shoal facies. The former appear to be the most promising potential exploration target since the thickest intervals of potential microbial source rock occur there. Porous or fractured laminated mudstone, turbiditic and redeposited oolitic packstone and grainstone are very good reservoir rocks sealed by evaporites, which make them excellent traps for hydrocarbons which could have been generated in neighbouring carbonate platform zones and migrated to these traps, or generated and left *in situ*.

Both the hydrocarbons themselves and the lithofacies framework, as well as their relationships in the Main Dolomite carbonate platform areas, need much more extensive study than has previously been undertaken. The recognition of this situation may improve prediction of hydrocarbon deposits in the Polish Zechstein.

We would like to thank M. Tucker and A. Milkov for helpful suggestions regarding this paper. This research was supported by the 'Mobility Plus' Programme postdoctoral fellowship (M.S.) of the Ministry of Science and Higher Education of Poland and Royal Dutch Shell.

References

ALI, M. F., AL-KHADHRAWI, M. R., PERZANOWSKI, H. & HLPEREN, H. I. 2002. Central Saudi Arabian crude oils: a geochemical investigation. *Petroleum Science and Technology*, **20**, 633–654.

BERNER, R. A. 1982. Biogeochemical cycles of carbon and sulfur and their effect on atmospheric oxygen over Phanerozoic time. *Palaeogeography, Palaeoclimatology, Palaeoecology*, **75**, 97–122.

CANFIELD, D. E., BOUDREAU, B. P., MUCCI, A. & GUNDERSEN, J. K. 1998. The early diagenetic formation of organic sulfur in the sediments of Mangrove Lake, Bermuda. *Geochimica et Cosmochimica Acta*, **62**, 767–781.

CASTANIER, S., LE MÉTAYER-LEVREL, G. & PERTHUISOT, J.-P. 1999. Ca-carbonates precipitation and limestone genesis – the microbiogeologist point of view. *Sedimentary Geology*, **126**, 9–23.

CHUMAKOV, N. M. & ZHARKOV, M. A. 2002. Climate during Permian–Triassic Biosphere Reorganizations, Article 1: Climate of the Early Permian. *Stratigraphy and Geological Correlation*, **10**, 586–602.

CHUMAKOV, N. M. & ZHARKOV, M. A. 2003. Climate during the Permian–Triassic Biosphere Reorganizations. Article 2. Climate of the Late Permian and Early Triassic: General Inferences. *Stratigraphy and Geological Correlation*, **11**, 361–375.

CONNAN, J., BOUROULLEC, D., DESSORT, D. & ALBRECHT, P. 1986. The microbial input in carbonate-anhydrite facies of a sabkha palaeoenvironment from Guatemala: A molecular approach. *Organic Geochemistry*, **10**, 29–50.

CZEKAŃSKI, E., KWOLEK, K. & MIKOŁAJEWSKI, Z. 2010. Hydrocarbon fields in the Zechstein Main Dolomite (Ca2) on the Gorzów Block (NW Poland). *Przegląd Geologiczny*, **58**, 695–703 (in Polish with English abstract).

DEPOWSKI, S., PERYT, T. M., PIĄTKOWSKI, T. S. & WAGNER, R. 1978. Sedimentation and paleogeography of the Main Dolomite and its oil- and gas-bearing. *Przegląd Geologiczny*, **26**, 141–146 (in Polish with English summary).

DUPRAZ, C., VISSCHER, P. T., BAUMGARTNER, L. K. & REID, R. P. 2004. Microbe-mineral interactions: early carbonate precipitation in a hypersaline lake (Eleuthera Island, Bahamas). *Sedimentology*, **51**, 745–765.

DYJACZYŃSKI, K., PAPIERNIK, B., PERYT, T. M., PROTAS, S. & WAGNER, R. 2000. Mapa paleogeograficzna dolomitu głównego (Ca2) 1: 500 000. *In*: KOTARBA, M. (ed.) *Bilans i potencjał węglowodorowy dolomitu głównego basenu permskiego Polski*. Archiwum PIG, Warszawa (in Polish).

EHRLICH, H. L. & NEWMAN, D. K. 2009. *Geomicrobiology*, 5th edn. CRC Press/Taylor & Francis, Boca Raton.

EVANS, R. & KIRKLAND, D. W. 1988. Evaporitic environments as a source of petroleum. *In*: SCHREIBER, B. C. (ed.) *Evaporites and Hydrocarbons*. Columbia University Press, New York, 256–299.

FLUTEAU, F., BESSE, J., BROUTIN, J. & RAMSTEIN, G. 2001. The late Permian climate. What can be inferred from the climate modelling concerning Pangea scenarios and Hercynian range altitude? *Palaeogeography, Palaeoclimatology and Palaeoecology*, **167**, 39–71.

FRIEDMAN, G. M. & KRUMBEIN, W. E. (eds) 1985. Hypersaline ecosystems. The Gavish Sabkha. *Ecological Studies*. Springer, Heidelberg, **53**, 484.

FÜCHTBAUER, H. 1962. Fazies, Porosität und Gasinhalt der Karbonatgesteine des norddeutschen Zechsteins. *Zeitschrift der Deutschen Gesellschaft für Geowissenschaften*, **114**, 484–531.

GĄSIEWICZ, A. in press. Climatic control on the Late Permian Main Dolomite (Ca2) deposition in northern margin of the Southern Permian Basin and implications to its internal cyclicity. *In*: GĄSIEWICZ, A. & SŁOWAKIEWICZ, M. (eds) *Palaeozoic Climate Cycles: Their Evolutionary and Sedimentological Impact*. Geological Society, London, Special Publications, **376**, http://dx.doi.org/10.1144/SP376.22

GERDES, G. & KRUMBEIN, W. E. 1987. Biolaminated deposits. *Lecture Notes in Earth Sciences*. Springer, Heidelberg, **9**, 188.

GIANI, D., GIANI, L., COHEN, Y. & KRUMBEIN, W. E. 1984. Methanogenesis in the hypersaline Solar Lake (Sinai). *FEMS Microbiology Letters*, **25**, 219–224.

GISCHLER, E., GIBSON, M. A. & OSCHMANN, W. 2008. Giant Holocene freshwater microbialites, Laguna Bacalar, Quintana Roo, Mexico. *Sedimentology*, **55**, 1293–1309.

GONG, C. & HOLLANDER, D. 1997. Differential contribution of bacteria to sedimentary organic matter in oxic and anoxic environments, Santa Monica Basin, California. *Organic Geochemistry*, **26**, 545–563.

GROTEK, I. 1998. Thermal maturity of organic matter in the Zechstein deposits of the Polish Lowlands area. *Prace Państwowego Instytutu Geologicznego*, **165**, 255–260 (in Polish with English summary).

HARTWIG, A. & SCHULZ, H.-M. 2010. Applying classical shale gas evaluation concepts to Germany – Part I: The basin and slope deposits of the Stassfurt Carbonate (Ca2, Zechstein, Upper Permian) in Brandenburg. *Chemie der Erde*, **70**, 77–91.

HEAD, I., FARRIMOND, P. & LARTER, S. 1995. Organic carbon in sediments. *Nature*, **373**, 293.

HITE, R. J. & ANDERS, D. E. 1991. Petroleum and evaporites. *In*: MELVIN, J. L. (ed.) *Evaporites, Petroleum and Mineral Resources*. Developments in Sedimentology, Elsevier, Amsterdam, **50**, 349–411.

IRWIN, H., CURTIS, C. & COLEMAN, M. L. 1977. Isotopic evidence for source of diagenetic carbonates formed during burial of organic-rich sediments. *Nature*, **269**, 209–213.

JAVOR, B. 1989. *Hypersaline Environments*. Springer-Verlag, Berlin.

JONES, A., FRANK, T. D. & FIELDING, C. R. 2006. Clod climate in the eastern Australian mid to late Permian may reflect cold upwelling waters. *Palaeogeography, Palaeoclimatology, Palaeoecology*, **237**, 370–377.

JØRGENSEN, B. B. 1983. The microbial sulphur cycle. *In*: KRUMBEIN, W. E. (ed.) *Microbial Geochemistry*. Blackwell Science Ltd, Oxford, 91–124.

KENDALL, C. G. ST. C. & SKIPWITH, P. A. D'E. 1969. Holocene shallow-water carbonate and evaporite sediments of Khor al Bazam, Abu Dhabi, Southwest Persian Gulf. *AAPG Bulletin*, **53**, 841–869.

KENIG, F. 2011. Distribution of organic matter in the transgressive and regressive Holocene sabkha sediments of Abu Dhabi, United Arab Emirates. *IAS Special Publication*, **43**, 277–298.

KENIG, F., HUC, A. Y., PURSER, B. H. & OUDIN, J.-L. 1990. Sedimentation, distribution and diagenesis of organic matter in a recent carbonate environment, Abu Dhabi, U.A.E. *Organic Geochemistry*, **16**, 735–747.

KHIEL, J. T. & SHIELDS, C. A. 2005. Climate simulation of the latest Permian: Implications for mass extinction. *Geology*, **33**, 757–760.

KILLOPS, S. & KILLOPS, V. 2005. *Introduction to Organic Geochemistry*. Blackwell Publishing, Oxford.

KOSAKOWSKI, P., BURZEWSKI, W. & KOTARBA, M. J. 2003. Petroleum potential of the Main Dolomite strata of the Kamień Pomorski area (northern Poland). Part 2 – Petroleum generation and expulsion processes. *Przegląd Geologiczny*, **51**, 663–672 (in Polish with English abstract).

KOTARBA, M. J. 1998. Geochemistry, Poland. *In*: LOKHORST, A. (ed.) *Northwest European Gas Atlas – Composition and Isotope Ratios of Natural Gases*. Netherlands Institute of Applied Geoscience TNO report, Haarlem, CD-ROM.

KOTARBA, M. & WAGNER, R. 2007. Generation potential of the Zechstein Main Dolomite (Ca2) carbonates in the Gorzów Wielkopolski-Międzychód-Lubiatów area, geological and geochemical approach to microbial-algal source rock. *Przegląd Geologiczny*, **55**, 1025–1036.

KOTARBA, M. J., KOSAKOWSKI, P., KOWALSKI, A. & WIĘCŁAW, D. 1998. Preliminary geochemical characteristics of organic matter and hydrocarbon potential of the Zechstein Main Dolomite deposits in the Polish Lowlands. *Prace Państwowego Instytutu Geologicznego*, **165**, 227–234.

KOTARBA, M. J., WIĘCŁAW, D. & KOWALSKI, A. 2000*a*. Composition, origin and habitat of oils in the Zechstein Main Dolomite strata of the western part of the Fore-Sudetic area (SW Poland). *Przegląd Geologiczny*, **48**, 436–442 (in Polish with English summary).

KOTARBA, M. J., WIĘCŁAW, W. & STECKO, Z. 2000*b*. Composition, origin and habitat of natural gases in the Zechstein Main Dolomite strata of the western part of the Fore-Sudetic area (SW Poland). *Przegląd Geologiczny*, **48**, 429–435 (in Polish with English summary).

KOTARBA, M. J., KOSAKOWSKI, P., WIĘCŁAW, D. & KOWALSKI, A. 2003. Petroleum potential of Main Dolomite strata of the Kamień Pomorski area (northern Poland). Part 1 – source rock. *Przegląd Geologiczny*, **51**, 587–594 (in Polish with English summary).

KOTARBA, M. J., POKORSKI, J., GRELOWSKI, C. & KOSAKOWSKI, P. 2005. Origin of natural gases accumulated in Carboniferous and Rotliegend strata on the Baltic part of the Western Pomerania. *Przegląd Geologiczny*, **53**, 425–433 (in Polish with English summary).

KOTARBA, M. J., PERYT, T. M., KOSAKOWSKI, P. & WIĘCŁAW, D. 2006. Organic geochemistry, depositional history and hydrocarbon generation modelling of the Upper Permian Kupferschiefer and Zechstein Limestone strata in south–west Poland. *Marine and Petroleum Geology*, **23**, 371–386.

LOGAN, B. W. 1987. The MacLeod evaporite basin, Western Australia: Holocene environments, sediments and geological evolution. *AAPG Memoir*, **44**, 140.

LOKHORST, A. (ed.) 1998. *NW European Gas Atlas – Composition and Isotope Ratios of Natural Gases*. Published by BGR, BGS, GEUS, NITG-TNO and PGI, Delft. ISBN: 90-72869-60-5 (CD-ROM).

MILKOV, A. V. 2011. Worldwide distribution and significance of secondary microbial methane formed during petroleum biodegradation in conventional reservoirs. *Organic Geochemistry*, **42**, 184–207.

MONTY, C. V. L. 1976. The origin and development of cryptalgal fabrics. *In*: WATER, M. R. (ed.) *Stromatolites*. Developments in Sedimentology, Elsevier, Amsterdam, **20**, 193–251.

NOFFKE, N. 2010. *Geobiology: Microbial Mats in Sandy Deposits from Archean Era to Today*. Springer-Verlag, Heidelberg.

PAPINEAU, D., WALKER, J. J., MOJZSIS, S. J. & PACE, N. R. 2005. Composition and structure of microbial communities from stromatolites of Hamelin Pool in Shark Bay, Western Australia. *Applied and Environmental Microbiology*, **71**, 4822–4832.

PARRISH, J. T. 1993. Climate of the Supercontinent Pangea. *Journal of Geology*, **101**, 215–233.

PETERSON, J. A. & HITE, R. J. 1969. Pennsylvanian evaporite-carbonate cycles and their relation to petroleum occurrence, Southern Rocky Mountains. *AAPG Bulletin*, **53**, 884–908.

PHLEGER, F. B. & EWING, G. C. 1962. Sedimentology and oceanography of coastal lagoons in Baja, California, Mexico. *GSA Bulletin*, **73**, 145–182.

PLETSCH, T., APPEL, J. *ET AL.* 2010. Petroleum generation and migration. *In*: DOORNEBAL, J. C. & STEVENSON, A. G. (eds) *Petroleum Geological Atlas of the Southern Permian Basin Area.* EAGE Publications b.v., Houten, 225–253.

POST, F. J. 1981. Microbiology of the Great Salt Lake north arm. *Hydrobiologia*, **81**, 59–69.

POSTGATE, J. R. 1984. *The Sulfate-Reducing Bacteria.* 2nd edn. Cambridge University Press, Cambridge.

PURDY, E. G. 1964. Recent calcium carbonate facies of the Great Bahama-Bank. 2. Sedimentary facies. *Journal of Geology*, **71**, 427–497.

RANKEY, E. C. & REEDER, S. L. 2011. Holocene oolitic marine sand complexes of the Bahamas. *Journal of Sedimentary Research*, **81**, 97–117.

RISK, M. J. & RHODES, E. J. 1985. From mangrove to petroleum precursors: An example from tropical northeast Australia. *AAPG Bulletin*, **69**, 1230–1240.

ROMANKEVICH, E. A. 1984. *Geochemistry of Organic Matter in the Ocean.* Springer-Verlag, Berlin.

ROSCHER, M. & SCHNEIDER, J. W. 2006. Permo-Carboniferous climate: Early Pennsylvanian to Late Permian climate development of central Europe in a regional and global context. *In*: LUCAS, S. G., CASSINIS, G. & SCHNEIDER, J. W. (eds.) *Non-Marine Permian Biostratigraphy and Biochronology.* Geological Society, London, Special Publication, **265**, 95–136.

ROSCHER, M., BERNER, U. & SCHNEIDER, J. W. 2008. A tool for the assessment of the paleodistribution of source and reservoir rocks. *Oil Gas – European Magazine*, **34**, 131–137.

ROSCHER, M., STORDAL, F. & SVENSEN, H. 2011. The effect of global warming and global cooling on the distribution of the latest Permian climate zones. *Palaeogeography, Palaeoclimatology, Palaeoecology*, **309**, 186–200.

SCHREIBER, B. 1988. *Evaporites and Hydrocarbons.* Columbia University Press, New York.

SŁOWAKIEWICZ, M. A. & MIKOŁAJEWSKI, Z. 2009. Sequence stratigraphy of the Upper Permian Zechstein Main Dolomite carbonates in western Poland: a new approach. *Journal of Petroleum Geology*, **32**, 215–234.

SŁOWAKIEWICZ, M. & MIKOŁAJEWSKI, Z. 2011. Upper Permian Main Dolomite microbial carbonates as potential source rocks for hydrocarbons (W Poland). *Marine and Petroleum Geology*, **28**, 1572–1591.

SŁOWAKIEWICZ, M., TUCKER, M. E., PANCOST, R. D., PERRI, E. & MAWSON, M. 2013. Upper Permian (Zechstein) microbialites: supratidal through deep subtidal deposition, source rock, and reservoir potential. *AAPG Bulletin*, **97**, 1921–1936.

SOFER, Z. 1984. Stable carbon isotope compositions of crude oils; application to source depositional environments and petroleum alteration. *AAPG Bulletin*, **68**, 31–49.

SONNENFELD, P. 1984. *Brines and Evaporites.* Academic Press Inc., Orlando.

WAGNER, R. 1994. Stratygrafia i rozwój basenu cechsztyńskiego na Niżu Polskim [Stratigraphy and evolution of the Zechstein basin in the Polish Lowland]. *Prace Państwowego Instytutu Geologicznego*, **146**, 1–71 (in Polish with English summary).

WAGNER, R. & PERYT, T. M. 1997. Possibility of sequence stratigraphy subdivision of the Zechstein in the Polish Basin. *Geological Quarterly*, **41**, 457–474.

WARREN, J. K. 2006. *Evaporites. Sediments, Resources and Hydrocarbons.* Springer-Verlag, Berlin.

WARREN, J. K. & KENDALL, C. G. ST. C. 1985. Comparison of sequences formed in marine sabkha (subaerial) and salina (subaqueous) settings – modern and ancient. *AAPG Bulletin*, **69**, 1013–1023.

WHITICAR, M. J. 1999. Carbon and hydrogen isotope systematics of bacterial formation and oxidation of methane. *Chemical Geology*, **161**, 291–314.

WILSON, M., NEUMANN, E.-R., DAVIES, G. R., TIMMERMAN, M. J., HEEREMANS, M. & LARSEN, B. T. 2004. Permo-Carboniferous magmatism and rifting in Europe: introduction. *In*: WILSON, M., NEUMANN, E.-R., DAVIES, G. R., TIMMERMAN, M. J., HEEREMANS, M. & LARSEN, B. T. (eds) *Permo-Carboniferous Magmatism and Rifting in Europe.* Geological Society, London, Special Publications, **223**, 1–10.

WRIGHT, D. T. & OREN, A. 2005. Nonphotosynthetic bacteria and the formation of carbonates and evaporites through time. *Geomicrobiology Journal*, **22**, 27–53.

WYCHERLEY, H., FLEET, A. & SHAW, H. 1999. Some observations on the origins of large volumes of carbon dioxide accumulations in sedimentary basins. *Marine and Petroleum Geology*, **16**, 489–494.

Permian deposits and the Permian–Triassic boundary in Croatia: palaeoclimatic implications based on palaeontological and geochemical data

KARMEN FIO[1]*, JASENKA SREMAC[1], IGOR VLAHOVIĆ[2], IVO VELIĆ[3] & JORGE E. SPANGENBERG[4]

[1]*University of Zagreb, Faculty of Science, Horvatovac 102a, HR-10000 Zagreb, Croatia*

[2]*University of Zagreb, Faculty of Mining, Geology and Petroleum Engineering, Pierottijeva 6, HR-10000 Zagreb, Croatia*

[3]*Geolog d.o.o. and Croatian Geological Summer School, Pančićeva, 5 HR-10000 Zagreb, Croatia*

[4]*Institute of Earth Sciences, University of Lausanne, Building Geopolis, CH-1015 Lausanne, Switzerland*

**Corresponding author (e-mail: karmen.fio@geol.pmf.hr)*

Abstract: The Late Palaeozoic deposits in Croatia were derived from the Palaeotethyan shelf areas along the northern Gondwana passive continental margin and consist of mixed carbonate–clastic facies. These sedimentary rocks accumulated in the equatorial area along northern Gondwana under a dominantly hot climate and laid the foundations for the future Adriatic Carbonate Platform (AdCP).

Detailed classification and geochemical studies of the Permian and Permian–Triassic deposits in most parts of Croatia are still in progress, especially concerning palaeoclimate and regional palaeogeography. The best preserved Permian outcrops can be found in the Velebit mountain area, including continuous Permian–Triassic sections. Stable isotopes of carbon and oxygen from carbonates, with characteristic positive $\delta^{13}C_{carb}$ excursion during the Late Permian and negative $\delta^{13}C_{carb}$ excursion at the Permian–Triassic transition combined with palaeontological and sedimentological data, helped in palaeoclimate estimations and the positioning of the chemostratigraphical Permian–Triassic Boundary (PTB). Sedimentological, palaeontological and geochemical data indicate that deposition in the Permian took place mainly in shallow-water tropical areas under oligotrophic conditions. High carbon isotope ratios in Capitanian carbonate rocks (+4 to +6‰ VPDB) mark the 'Kamura event' characterized by temporary cooling, high productivity and burial rate. Evaporites in the central part of the Karst Dinarides and red beds of the 'Gröden type' indicate a warm and mostly arid climate. Calculations based on the oxygen isotopic compositions of whole-rock carbonate and fossil samples estimate latest Permian temperatures up to 26 °C.

Palaeozoic deposits in Croatia are quite diverse and outcrop in 11 tectonically limited areas within the Internal and External (Karst) Dinarides, as well as occurring as clasts in younger sediments, or in drilling-core samples (Fig. 1) (Sremac 2005). These sediments represent an ancestor of the Adriatic Carbonate Platform (AdCP), and were derived from the Carboniferous–Lopingian epeiric platform in an equatorial area along northern Gondwana (e.g. Ramovš *et al.* 1990; Kiessling *et al.* 2003). The deposition took place on a large platform with occasionally emerged areas and intraplatform basins or, more probably, on several smaller platforms, separated by continental or basinal barriers, where mixed carbonate–clastic facies had been deposited (Sremac 2005; Vlahović *et al.* 2005; Vozárová *et al.* 2009).

This paper is a short overview of the Permian and Permian–Triassic deposits in Croatia, and their implication on the palaeoclimate due to the known palaeontological and geochemical data. Special emphasis will be given to the best investigated Upper Permian successions of the Mt Velebit and Lika region in SW Croatia, with their alternation of dolomites and occurrence of organically rich dark limestones (Figs 1 & 2). The fossil content of the carbonate deposits point to their origin in mostly shallow subtidal and intertidal environments (Sremac 2005). These data, supported by the isotopes of carbon and oxygen from carbonates (Polšak & Pezdič 1978; Fio *et al.* 2010), enabled a comparison with existing data from studied and surrounding areas, and palaeoclimate interpretations that pointed to dominant subtropical–tropical climate.

From: GĄSIEWICZ, A. & SŁOWAKIEWICZ, M. (eds) 2013. *Palaeozoic Climate Cycles: Their Evolutionary and Sedimentological Impact*. Geological Society, London, Special Publications, **376**, 539–548.
First published online February 27, 2013, http://dx.doi.org/10.1144/SP376.8

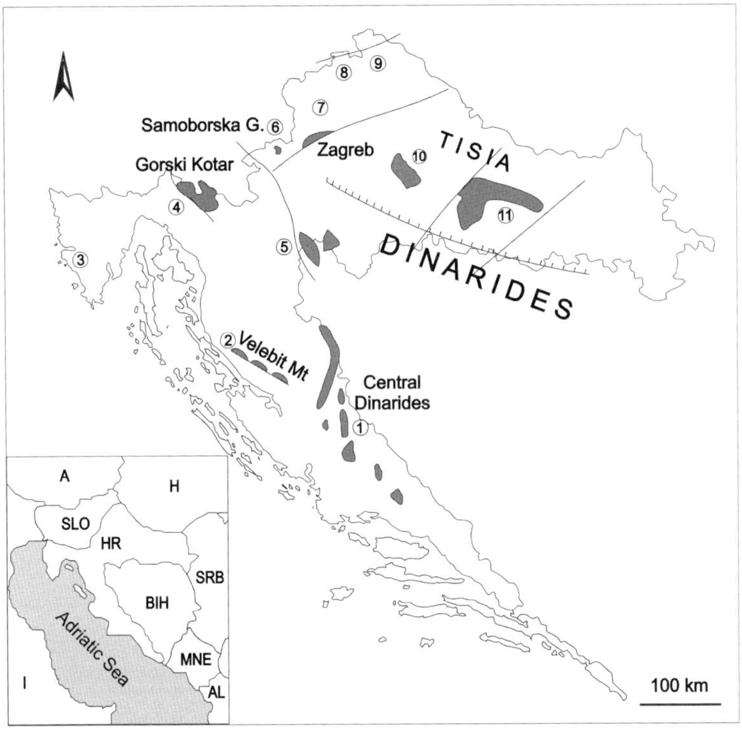

Fig. 1. Areas with Palaeozoic deposits in Croatia (modified after Sremac 2005): (1) evaporite complex in the central part of the Karst Dinarides; (2) the area of Mt Velebit; (3) borehole in the Rovinj area; (4) the Gorski Kotar region; (5) the Banovina and Kordun region; (6) Mt Samoborska Gora; (7) Mt Medvednica; (8) Mt Ravna Gora; (9) borehole Zebanec; (10) Mt Moslavačka Gora; and (11) the Slavonian Mountains.

Carbonates and evaporites from the area of Central Dinarides and red beds of the Gröden type also point to a hot and arid climate.

Methods

This paper is an overview of the published articles on the Late Palaeozoic deposits in Croatia. These deposits, however, still require more detailed studies and a proper classification as only informal lithostratigraphical units are commonly used. Previously published articles include mostly palaeontological and sedimentological data (Salopek 1942, 1948; Kochansky-Devidé 1973; Sokač *et al.* 1974, 1976; Palinkaš & Sremac 1989; Palinkaš *et al.* 1993; Aljinović & Sremac 1997; Sremac & Aljinović 1997; Aljinović *et al.* 2003, 2008; Sremac 2005).

A geochemical approach (Polšak & Pezdič 1978; Fio *et al.* 2010) helped in understanding Late Palaeozoic sedimentation and its palaeoclimatic implications. Polšak & Pezdič (1978) included stable isotope composition of oxygen, determined from limestones and fossil skeletons, to determine Late Palaeozoic climate and an approximate palaeogeographical position of the Dinarides. A study carried out by Fio *et al.* (2010) included stable isotopes of carbonates ($\delta^{13}C$, $\delta^{18}O$), organic matter ($\delta^{13}C$, $\delta^{15}N$), and major, trace and rare earth element compositions of marine carbonate rocks of Late Permian–Early Triassic age, with the main concern being the determination of the exact Permian–Triassic boundary (PTB) in the Mt Velebit area. Oxygen isotopes have shown a gradual decrease in values from Upper Permian to Lower Triassic samples, and a probable increase in temperature. However, owing to possible diagenetic alterations, especially in dolomitic rocks, these results should be taken with caution.

Late Palaeozoic successions in Croatia

In the Gorski Kotar area (see (4) on Fig. 1) during the Late Carboniferous–Middle Permian, deep- to shallow-marine conditions are proposed, with a shallowing-upwards forearc sequence that was deformed during the Late Permian (Aljinović *et al.* 2003 and references therein). The terrigenous clastics of the intraplatform basins found in the Gorski

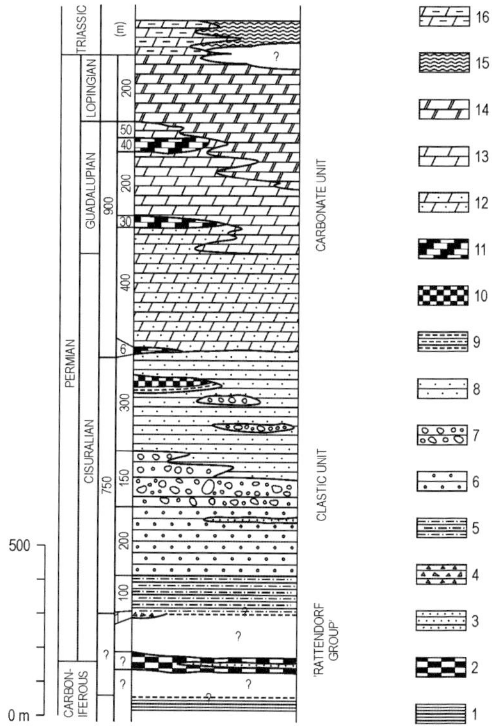

Fig. 2. Shematic column of the Upper Palaeozoic and Lower Triassic deposits in the Mt Velebit area (modified after Kochansky-Devidé 1973). Legend: (1) Carboniferous shales; (2) Lower Permian Rattendorf limestones (thickness and exact position of the Rattendorf limestones is unknown and therefore marked by question marks); (3) sandstones in Auernig beds; (4)–(10) Trogkofel equivalents locally called 'Košna beds': (4) basal breccia; (5) Permian green pyritic sandstones; (6) Permian quartz conglomerates; (7) conglomerates with limestone pebbles; (8) Permian sandstones (grey, red, yellow); (9) Middle Permian shales; (10) Trogkofel limestones; (11)–(14) Permian carbonate unit with limestones and dolomites: (11) Middle–Upper Permian black limestones; (12) dark-grey spotty dolomites; (13) light-coloured saccharoidal dolomites; (14) dense, bedded dolomite of the Upper Permian–Transitional Dolomite (TD); (15) Lower Triassic shales; and (16) Lower Triassic dolomites–Sandy Dolomite (SD). Note: the continuous transition from the Permian to the Triassic strata is currently located at two exposures in the Velebit mountain area: Rizvanuša and Paklenica (Fio *et al.* 2010).

Kotar are mostly of Pennsylvanian, Early and Middle Permian age (Milanović 1981; Aljinović & Sremac 1997), with the Trogkofel beds occurring in the Middle Permian (Kochansky-Devidé 1973). An absence of fossils and intensive tectonics complicates the chronostratigraphical determination of these deposits. A depositional basin was probably a part of a fan-delta depositional system, marked by the presence of marine black shales, coarse-grained deposits from alluvial fans and plant detritus, indicating the proximity of the hinterland. Fossil taxa in the Gorski Kotar region indicate Lower Carboniferous–Lower Permian age (Sremac & Aljinović 1997). Even though foraminifers *Frondina* and *Crypotoseptida* of the lagenid taxa have been determined only from Upper Permian sediments, their presence does not provide sufficient evidence for a Late Permian age (Sremac & Aljinović 1997). For the Permian–Triassic transition some authors have proposed a stratigraphical break caused by a pause in sedimentation (Aljinović & Sremac 1997; Sremac & Aljinović 1997; Aljinović *et al.* 2003, 2008), while others have suggested continuous clastic shallow-marine deposition from the Late Permian into the Early Triassic (Herak & Tomić 1995).

Late-diagenetic barite was precipitated within the porous Permian clastic deposits and Lower Triassic ooid grainstones in the Gorski Kotar area (Aljinović *et al.* 2000). In parts of this region, an early-diagenetic barite and pyrite stratabound ore deposit, where sulphides were formed by bacteriogenic sulphate reduction in a muddy peritidal environment, mark the Permian–Triassic Boundary (PTB) (Palinkaš & Sremac 1989; Palinkaš *et al.* 1993).

Predominantly clastic Late Palaeozoic deposits crop out also in the Banovina and Kordun regions, and the Medvednica, Marijagorička Brda and Samoborska Gora mountains. Late Palaeozoic rocks of the Slavonian Mountains in northern Croatia belong to the Tisia Megaterrane, which forms the basement of the Pannonian Basin, and consists mostly of clastic and volcanic deposits (Vozárová *et al.* 2009). Permian carbonate clasts occur also within Middle Triassic conglomerates and breccias of the Medvednica and Ravna Gora mountains (Sremac & Kochansky-Devidé 1982; Sremac 2005) (Fig. 1).

The Central Dinarides area (central and northern Dalmatia, SE Lika and western Bosnia) includes Late Permian and Early Triassic evaporites and volcanic rocks (Ivanović *et al.* 1977; Šušnjara *et al.* 1992; Tišljar 1992). Evaporites and associated carbonate and clastic deposits in this area were determined as being Upper Permian on the basis of palynological and sulphur isotope data, as well as Lower Triassic age of the deposits directly overlying them at some localities (Šušnjara *et al.* 1992). Evaporites were deposited in arid shallow-marine environments or, according to Tišljar (1992), along the edges of the epeiric marine basin, in a large belt on desert carbonate–shoreline coastal sabkhas, which existed during the period of general regression and permanent coastal progradation caused by continuous relative sea-level fall. These evaporative settings evolved into sabkha and playa environments, and, finally, into lacustrine, alluvial and continental environments, with the re-establishment of shallow-marine deposition occurring at the beginning of the Triassic (Tišljar 1992).

Continuous deposition from the Permian to the Triassic

In Croatia, a continuous transition from Permian to Triassic is only found in the carbonate successions (Aljinović & Sremac 1997). Best-known Upper Palaeozoic areas with dominant shallow-marine carbonates crop out in the Lika region and Velebit mountain areas (Figs 1 & 2) (Salopek 1942, 1948; Kochansky-Devidé 1973; Sokač *et al.* 1974, 1976; Kochansky-Devidé 1982; Ramovš *et al.* 1990; Fio *et al.* 2010).

In the Mt Velebit area, the Upper Carboniferous–Lower Cisuralian (Aselian and Sakmarian) is marked by the fossiliferous Rattendorf beds (containing *Daixina alpina*, *Occidentoschwagerina extensa*, *Zellia heritschi*) and sandstones within the Auernig beds. The Upper Cisuralian (Artinskian) is composed of sandstones and conglomerates deposited in marine depressions (Lower Permian clastic unit or 'Košna Formation'). During the Guadalupian and Lopingian, carbonate sedimentation prevailed (local equivalent of the Gröden and Bellerophone beds in the Alps), during which a more than 900 m- thick succession of dolomites with three zones of organic-rich limestones and clastic rocks were deposited (Middle and Upper Permian carbonate unit or 'Velebit Formation', *sensu* Flügel 1977). This succession is clearly visible in the Brušane area (Fig. 2).

The Upper Permian is represented by a well-defined platform carbonate succession, characterized predominantly by dolomites with three thin zones of black limestone. Most common fossil remains from the Permian First or Lower Limestone Zone are fusulinid foraminifera (*Eoverbeekina salopeki*) and recrystallized dasycladales (*Mizzia*), with sporadic occurrence of nautilids, gastropods, brachiopods and crinoids. The following early-diagenetic dolomites contain abundant recrystallized fragments of corals *Waagenophyllum*, and are rich in dasycladales, fusulinids, brachiopods and molluscs (*Bellerophon*, *Pleurotomaria*). The Second or Middle Limestone Zone is most fossiliferous, commonly represented by platform algal 'meadows' with sporadic mounds and patch-reefs. Limestones are rich in calcareous algae (*Mizzia velebitana*, *Mizzia yabei*, *Gymnocodium bellerophontis*, *Permocalculus tenellus*), foraminifera (*Neoschwagerina craticulifera*) and brachiopods (productids, spirifers). The Guadalupian–Lopingian boundary or 'Kamura event' was recently recognized in this area, with unusually high carbon isotope ratios for carbonate rocks ($\delta^{13}C_{carb}$ from +4 to +6‰ VPDB), suggesting a high productivity and burial rate during the Capitanian (Isozaki *et al.* 2011). Above the Middle Limestone Zone, light-coloured 'saccharoid' or *Neoschwagerina* late-diagenetic dolomite can be found. The Third or Upper Limestone Zone comprises a fossil assemblage similar to the second zone, with a more common appearance of gymnocodiacean *Permocalculus fragilis* and fusulinid *Yabeina syrtalis*, as well as new brachiopod taxa. It is overlain by the cryptocrystalline Transitional Dolomite (TD) with rare fossils, followed by the Sandy Dolomite (SD) unit; their boundary was considered by Salopek (1942, 1948) and Sokač *et al.* (1974, 1976) to represent the PTB. Stable isotope and trace element analyses, combined with palaeontological and sedimentological data, enabled the chemostratigraphical PTB to be positioned within the lowermost part of the Sandy Dolomite unit; that is, 11 m above the lithological boundary (Fio *et al.* 2010) in the Rizvanuša section near Brušane.

The Late Palaeozoic climate along the eastern coast of the Pangaea supercontinent

During the Late Carboniferous and Permian, the continental mass of the Pangaea supercontinent extended from pole to pole. It was surrounded by the Panthalassa Ocean in the west and the Palaeo-Tethys in the east, with formation of the Tethys Ocean commencing within the southern part of the Palaeo-Tethys at the end of the Early Permian (Stampfli *et al.* 2001). Remnants of the Upper Permian–Middle Triassic mélanges in the area of the Dinarides are evidence of the Palaeo-Tethys final closure (Stampfli & Borel 2002).

The Late Carboniferous (Pennsylvanian)–Early Permian period was generally characterized by the Gondwana glaciation and a prevailing continental arid climate in lower latitudes, which resulted in lower accumulation rates of organic matter and a lower intensity of chemical weathering. Consequently, higher CO_2 concentrations in the atmosphere enhanced the greenhouse effect (e.g. Erwin 1994; Retallack *et al.* 2011). The late Middle Permian to early Middle Triassic interval therefore represents the transition from the Late Palaeozoic ice-house to the Mesozoic greenhouse climate mode (Palaeozoic–Mesozoic transition interval), which started at the end of the Guadalupian (Isozaki *et al.* 2007 and references therein). Decreasing $\delta^{13}C_{carb}$ values suggest lower productivity through the latest Permian, with a considerable negative

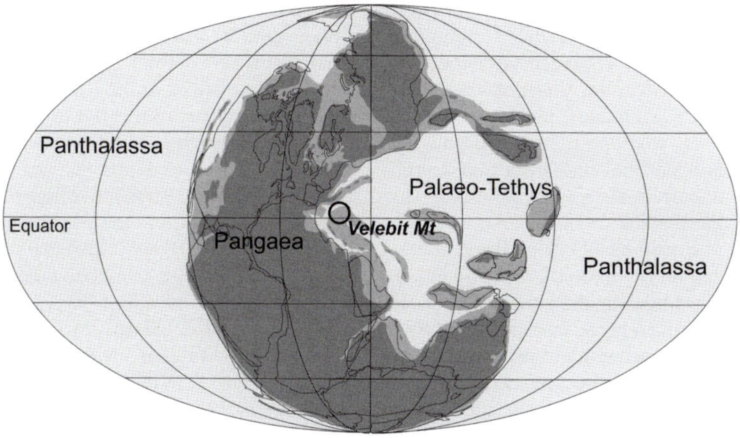

Fig. 3. Late Permian (Guadalupian) palaeogeographical map with the marked position of the Velebit Mt in the western part of the Palaeo-Tethys (modified after Kiessling *et al.* 2003). Middle grey colour marks assumed shelf areas.

shift at the PTB interval. This is in line with a global scenario, where the Earth was becoming significantly warmer than present, especially in the higher latitudes, and that oceans during the latest Permian were globally approximately 4 °C warmer than today (e.g. Kidder & Worsley 2004; Kiehl & Shields 2005).

Croatian Upper Palaeozoic sedimentary rocks were deposited in the western Palaeo-Tethys area, up to 20° from the Equator (Stampfli & Borel 2002; Kiessling *et al.* 2003), with a dominant subtropical–tropical climate and high sea temperatures (Fig. 3). During the Carboniferous, there was an important influence from the uplift of the Variscides and their subsequent erosion. Pennsylvanian tropical vegetation existed on the islands.

A hot arid zone in the Late Permian is supported by the existence of red clastic beds in the Lika and Gorski Kotar regions, and evaporite deposits in the Central Dinarides (Polšak & Pezdič 1978 and references therein). The presence of patch-reefs and mud-mounds in the Velebit Mt area, with rugose corals and abundant bryozoa, calcispongia and brachiopods, together with large fusulines, bivalves and abundant algae in protected areas (Fig. 4), especially within the Second or Middle Limestone Zone, indicates a shallow-water area with oligotrophic conditions (Sremac 2005; Aljinović *et al.* 2008). High productivity within the Third or Upper Limestone Zone (Brušane area in the Velebit Mt) is supported by the relatively high total organic carbon concentrations (TOCs *c.* 0.9 wt.%: Tišljar *et al.* 1991; Isozaki *et al.* 2011), with values of up to 5.3 wt% (Velić 2007), and high $\delta^{13}C_{carb}$ values (+4 to +6‰ VPDB: Isozaki *et al.* 2011), marking the ‘Kamura event’ (Isozaki *et al.* 2011).

Late Palaeozoic climate in the Karst Dinarides: palaeontological evidence and estimations based on oxygen isotopes

Upper Permian dolomites in the Velebit mountain area are mostly early diagenetic (Tišljar & Vlahović 2001), deposited in subtidal, intertidal and supratidal environments, as a result of global trends in sea-level oscillations (Sremac 2005). Transgressive–regressive phases and stressed environment events were particularly common in the latest Permian, causing the decrease in biodiversity and the appearance of the opportunistic taxa.

The uppermost Permian TD unit includes dolobiomicrites and dolobiopelmicrites, with complete

Fig. 4. Fossiliferous black limestones of the Upper Permian Second or Middle Limestone Zone (Baške Oštarije, Velebit Mt) with abundant algae, and common bryozoa, calcispongia, brachiopods, molluscs and large fusulines.

fossils and bioclasts of algae (dasycladales, gymnocodiaceans, solenoporaceans), foraminifera (*Earlandia*, *Ammodiscus*, *Glomospira*, *Tuberitina*, *Pachyphloia*, *Globivalvulina*, *Permodiscus*, *Staffela*, *Hemigordius*, Lagenidae), plus fragments of sponges, ostracods, gastropods, echinoderms and brachiopods. Conodonts have not been found, probably because depositional environment was too shallow. The lithological boundary between the TD (Fig. 5a) and overlying SD (Fig. 5b) is marked by a sudden occurrence of ooids and siliciclastic grains (Fig. 6). This lithological change was previously considered as the chronostratigraphical PTB. However, detailed geochemical, palaeontological and sedimentological analyses of these deposits (Fio *et al.* 2010) have shown that this lithological change represents a latest Permian regressive phase, marked by an enrichment in concentrations of most major, trace and rare earth elements, which was followed by the last occurrence of the Permian fossils in the lower part of the SD unit (Fig. 7). Dolobiopelmicrites in the lower part of the SD unit contain very rare fossil fragments. The upper part of the studied SD unit contains dolobiopelsparite with thin bivalve and/or ostracod shells, and an exchange of dolopelsparite, dolointrasparite and dolomicrite. The $\delta^{13}C_{carb}$ values in the Rizvanuša section of the Velebit Mt range from -0.8 to $+2.0$‰ VPDB for the Upper Permian, and from -1.3 to 0.9‰ for the Lower Triassic, with a distinct negative excursion from $+1.4$ to -0.6‰ marking the chemostratigraphical PTB within the lowermost part of the SD unit (Fig. 7). This event is also marked by the parallel excursion in organic carbon content and changes in element concentrations. Oxygen isotope ratios show a gradual decrease from the Upper Permian to the Lower Triassic, with values ranging from -3.2 to -1.3‰ VPDB (Fig. 7) (for details see Fio *et al.* 2010).

Oxygen isotopic data from marine limestones and their calcite or aragonite fossil components and cements can provide evidence for changes in palaeotemperature and ocean composition (e.g. Anderson & Arthur 1983; Marshall 1992; Bemis *et al.* 1998 and references therein; Korte *et al.* 2005). These palaeotemperature estimations – based on foraminifera tests, mollusc and brachiopod shells, and carbonate rocks – are dependent on the estimations of the $\delta^{18}O$ value of the contemporaneous sea water (e.g. $\delta^{18}O$ values close to 0‰ VSMOW: Korte *et al.* 2005) and on the preservation of the primary isotopic signature of the marine carbonates. Diagenetic processes and the interaction with hydrothermal and metamorphic fluids may change the primary environmental isotopic signal. Rapid lithification of carbonate sediments during early diagenesis can help in the preservation of the original isotopic composition, and $\delta^{18}O$ data gained from whole-rock and well-preserved

Fig. 5. Outcrops on the Rizvanuša section (Velebit Mt) with (a) the Upper Permian Transitional Dolomite unit (TD) and (b) the uppermost Permian–Lower Triassic Sandy Dolomite unit (SD).

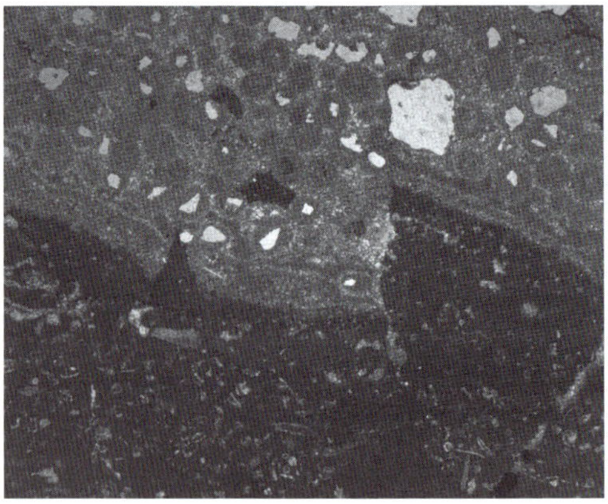

Fig. 6. Photomicrograph of the erosional contact between the Transitional Dolomite (lower) and Sandy Dolomite (upper part) units in the Rizvanuša section (Velebit Mt), previously considered as the PTB (from Fio *et al.* 2010). The width of the photomicrograph is 5 mm.

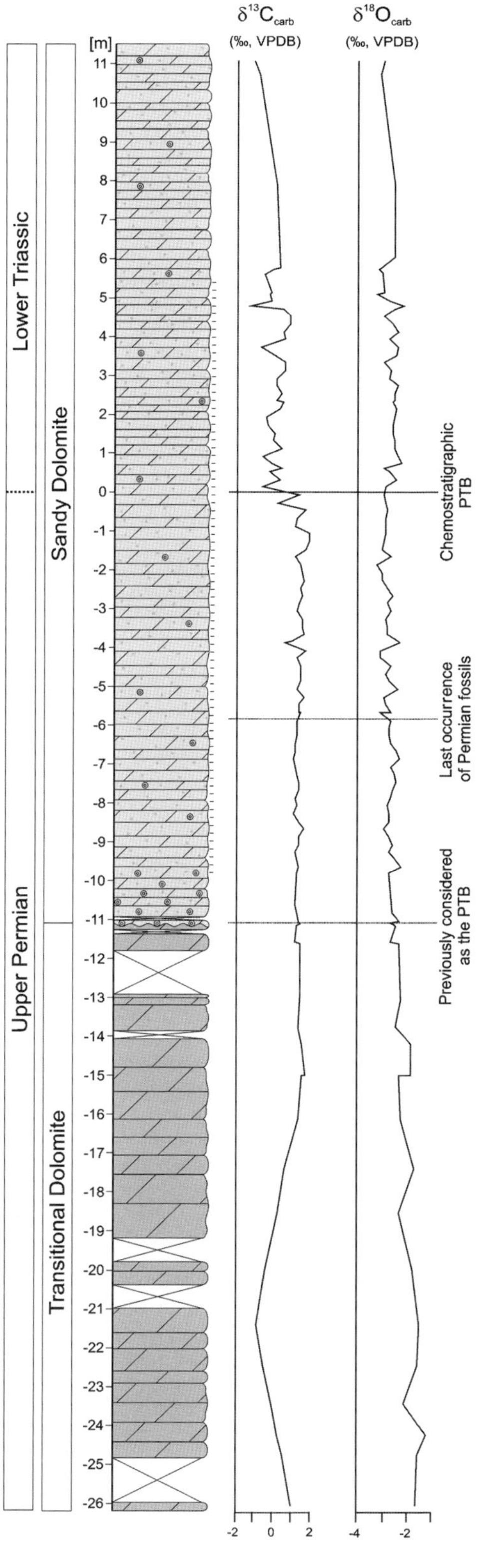

Fig. 7. Shematic geological column of the Rizvanuša section (Velebit Mt), showing the Upper Permian Transitional Dolomite (TD) and uppermost Permian and Lower Triassic Sandy Dolomite unit (SD) (modified after Fio *et al.* 2010). The lithostratigraphical boundary between the TD and SD was previously considered as the PTB. The position of the chemostratigraphical PTB is marked by the negative shift in $\delta^{13}C_{carb}$ and is set within the SD unit. Oxygen isotopes ($\delta^{18}O_{carb}$) display more negative values towards the Triassic.

brachiopods may often show comparable records (cf. Korte *et al.* 2005; Kearsey *et al.* 2009). Palaeotemperature estimations based on oxygen isotopes from Carboniferous and Permian whole-rock carbonate samples and skeletons of brachiopods, gastropods and corals from the Dinarides and Slovenian Alps were initially carried out by Polšak & Pezdič (1978). Using the calcium carbonate–water palaeotemperature scale of Epstein *et al.* (1953), they estimated average temperatures for the Early Permian of 24 °C, the Middle Permian of 21 °C and the Late Permian of 26 °C.

The Rizvanuša section of the Velebit Mt comprises mostly microfossil remains, and only whole-rock isotopic analyses were possible. Late Permian whole-rock $\delta^{18}O_{carb}$ varies from −3.2 to −1.3‰ VPDB (average −2.6 ± 0.4‰), and for the Early Triassic from −3.2 to −2.2‰ (−2.7 ± 0.3‰) (Fio *et al.* 2010). In view of the very low and negative correlation of carbon and oxygen isotopes ($r^2 = -0.04$, $n = 111$), diagenetic modifications of the $\delta^{13}C_{carb}$ values were excluded (Fio *et al.* 2010). Also, as a result of the narrow range of low negative $\delta^{18}O_{carb}$ values obtained from throughout the section, it could be assumed that the measured values are close to the original depositional signatures. Conodonts have not been found at the studied locality, but Middle Triassic conodonts from the neighbouring Donje Pazarište locality have shown that temperatures ranging from 60 to 150 °C can be estimated (Colour Alteration Index (CAI) from 1.5 to 2: H.J. Gawlick pers. comm. 2009). Reflecting the average geothermal gradient, these results indicate that the TD and SD units have most probably not been affected by temperatures higher than 170 °C.

Assuming a sea water $\delta^{18}O$ value of 0‰ VSMOW (Korte *et al.* 2005), the whole-rock $\delta^{18}O_{carb}$ values obtained show an increase in temperature from the Permian to the Triassic, and compare well with the estimations made by Polšak & Pezdič (1978), suggesting that the Late Permian mean sea temperatures in the Dinarides and neighbouring Alps were up to 26 °C, which are, today, typical temperatures for the tropical areas close to the Equator (up to 20°N and 20°S of the Equator). The palaeotemperature curve made by Scotese

(2008) displays end-Permian temperatures of up to 27 °C, Kiehl & Shields (2005) palaeotemperature models indicated western Palaeo-Tethys temperatures ranging from 28 to 30 °C, and Kearsey *et al.* (2009) estimated latest Permian temperatures between 26 and 29 °C. A similar trend with more negative $\delta^{18}O_{carb}$ values from the Permian–Triassic samples have been noted in the PTB sections of the Austrian Alps, where changes in oxygen isotopes were interpreted as an increase of between 5 and 6 °C in global temperature or the influence of meteoric water (Magaritz & Holser 1991; Kidder & Worsley 2004). The $\delta^{18}O_{carb}$ v. $\delta^{13}C_{carb}$ values for carbonates from the Velebit Mt area can be compared with Upper Permian and Lower Triassic carbonates from neighbouring Slovenia (Idrijca section: Dolenec *et al.* 2001; Schwab & Spangenberg 2004) and Hungary (Bálvány: Haas *et al.* 2006, 2007). Higher $\delta^{18}O$ values in samples from the Velebit Mt mark the higher dolomite content (cf. Horacek *et al.* 2007).

Conclusions

Permian and Permian–Triassic deposits in Croatia have mostly been investigated for their sedimentological and palaeontological features, with only few geochemical studies, especially with regard to the palaeotemperature estimations.

This brief overview has been concerned with the main areas of Upper Palaeozoic strata in Croatia, and the main characteristics of the Permian and Permian–Triassic deposits and their implications for the palaeoclimate. The presence of reefal structures with common sponges, foraminifera, corals, algae and brachiopods indicate a subtropical–tropical climate and high sea temperatures, while evaporites and red beds of the 'Gröden type' point to a warm arid climate in the Permian of the western Palaeo-Tethys. Stable isotope data gained from fossil shells and whole-rock agree with the suptropical–tropical climate, with Late Permian temperatures of up to 26 °C. A probable increase in temperatures towards the Early Triassic can be assumed from the constant more negative $\delta^{18}O_{carb}$ values. The palaeontological, sedimentological and stable isotope data presented here, combined with the suggested temperatures, correspond well to the global palaeotemperature models for that time.

For financing and providing necessary facilities for this research we would like to acknowledge the Swiss National Foundation, SCOPES project no. IB7320-110885, and the Croatian Ministry of Science, Education and Sports, project nos. 119-1951293-1162, 195-1953068-0242 and 195-1953068-2704. The authors would also like to thank the anonymous reviewers who helped to improve the manuscript.

References

Aljinović, D. & Sremac, J. 1997. The Middle and Upper Permian deposits in Gorski Kotar: facts and misconceptions. *Geologia Croatica*, **50**, 231–242.

Aljinović, D., Slovenec, D., Durn, G., Slavković, R., Miko, S. & Kolarić, D. 2000. Mineral occurences in Palaeozoic clastic and Lower Triassic carbonate rocks in Gorski Kotar (Croatia). *In*: Vlahović, I. & Biondić, R. (eds) *Excursion Guide-book, Second Croatian Geological Congress, Cavtat–Dubrovnik, Proceedings*. Institute of Geology, Zagreb, 59–64.

Aljinović, D., Sremac, J. & Ibrahimpašić, H. 2003. Permian and Lower Triassic facies of Velebit Mt. and Gorski Kotar region (NW part of External Dinarides). *In*: Vlahović, I. & Tišljar, J. (eds) *Evolution of Depositional Environments from the Palaeozoic to Quaternary in the Karst Dinarides and the Pannonian Basin. Field Trip Guidebook. 22nd IAS Meeting of Sedimentology, Opatija*. Institute of Geology, Zagreb, 185–206.

Aljinović, D., Isozaki, Y. & Sremac, J. 2008. The occurrence of giant bivalve Alatoconchidae from the Yabeina zone (Upper Guadalupian, Permian) in European Tethys. *Gondwana Research*, **13**, 275–287.

Anderson, T. F. & Arthur, M. A. 1983. Stable isotopes of oxygen and carbon and their applications to sedimentological and paleoenvironmental problems. *In*: Arthur, M. A., Anderson, T. F., Kaplan, I. R., Veizer, J. & Land, L. S. (eds) *Stable Isotopes in Sedimentary Geology*. SEPM, Short course, 1–151.

Bemis, B. E., Spero, H. J., Bijma, J. & Lea, D. W. 1998. Reevaluation of the oxygen isotopic composition of planktonic foraminifera: experimental results and revised paleotemperature equations. *Paleoceanography*, **13**, 150–160.

Dolenec, T., Lojen, S. & Ramovš, A. 2001. The Permian–Triassic boundary in Western Slovenia (Idrijca Valley Section): magnetostratigraphy, stable isotopes, and elemental variations. *Chemical Geology*, **175**, 175–190.

Epstein, S., Buchsbaum, R., Lowenstam, H. A. & Urey, B. C. 1953. Revised carbonate-water isotopic temperature scale. *Bulletin of the Geological Society of America*, **64**, 1315–1326.

Erwin, D. H. 1994. The Permo-Triassic extinction. *Nature*, **367**, 231–235.

Fio, K., Spangenberg, J. E., Vlahović, I., Sremac, J., Velić, I. & Mrinjek, E. 2010. Stable isotope and trace element stratigraphy across the Permian–Triassic transition: a redefinition of the boundary in the Velebit Mountain, Croatia. *Chemical Geology*, **278**, 38–57.

Flügel, E. 1977. Environmental models for Upper Paleozoic benthic calcareous algal communities. *In*: Flügel, E. (ed.) *Fossil Algae*. Springer, Berlin, 314–343.

Haas, J., Demény, A., Hips, K. & Vennemann, T. W. 2006. Carbon isotope excursions and microfacies changes in marine Permian–Triassic boundary sections in Hungary. *Palaeogeography, Palaeoclimatology, Palaeoecology*, **237**, 160–181.

Haas, J., Demény, A., Hips, K., Zajzon, N., Weiszburg, T. G., Sudar, M. & Pálfy, J. 2007. Biotic and environmental changes in the Permian–Triassic boundary interval recorded on a western Tethyan ramp in the Bükk Mountains, Hungary. *Global and Planetary Change*, **55**, 136–154.

HERAK, M. & TOMIĆ, V. 1995. Continental subduction tectonics in the high karst Dinarides of Western Croatia. *Geologia Croatica*, **48**, 161–166.

HORACEK, M., BRANDNER, R. & ABART, R. 2007. Carbon isotope record of the P/T boundary and the Lower Triassic in the Southern Alps: evidence for rapid changes in storage of organic carbon. *Palaeogeography, Palaeoclimatology, Palaeoecology*, **252**, 347–354.

ISOZAKI, Y., KAWAHATA, H. & MINOSHIMA, K. 2007. The Capitanian (Permian) Kamura cooling event: The beginning of the Paleozoic–Mesozoic transition. *Palaeoworld*, **16**, 16–30.

ISOZAKI, Y., ALJINOVIĆ, D. & KAWAHATA, H. 2011. The Guadalupian (Permian) Kamura event in European Tethys. *Palaeogeography, Palaeoclimatology, Palaeoecology*, **308**, 12–21.

IVANOVIĆ, A., SIKIRICA, V., MARKOVIĆ, S. & SAKAČ, K. 1977. *Basic Geological Map of SFRY 1:100 000. The Drniš Sheet.* State Geological Institute, Beograd.

KEARSEY, T., TWITCHETT, R. J., PRICE, G. D. & GRIMES, S. T. 2009. Isotope excursions and palaeotemperature estimates from the Permian/Triassic boundary in the Southern Alps (Italy). *Palaeogeography, Palaeoclimatology, Palaeoecology*, **279**, 29–40.

KIDDER, D. L. & WORSLEY, T. R. 2004. Causes and consequences of extreme Permo-Triassic warming to globally equable climate and relation to the Permo–Triassic extinction and recovery. *Palaeogeography, Palaeoclimatology, Palaeoecology*, **203**, 207–237.

KIEHL, J. T. & SHIELDS, C. A. 2005. Climate simulation of the latest Permian: Implications for mass extinction. *Geology*, **33**, 757–760.

KIESSLING, W., FLÜGEL, E. & GOLONKA, J. 2003. Patterns of Phanerozoic carbonate platform sedimentation. *Lethaia*, **36**, 195–226.

KOCHANSKY-DEVIDÉ, V. 1973. Trogkofel deposits in Croatia. *Geološki vjesnik*, **26**, 41–51.

KOCHANSKY-DEVIDÉ, V. 1982. Upper Palaeozoic of the Velebit Mt. (In Croatian.) *In*: SREMAC, J. (ed.) *IGCP Project No. 5, Yugoslav Working Group Field Meeting, Field guide to Velebit Mt.*, IGCP, Paris, 8–15.

KORTE, C., KOZUR, H. W. & VEIZER, J. 2005. $\delta^{13}C$ and $\delta^{18}O$ values of Triassic brachiopods and carbonate rocks as proxies for coeval seawater and palaeotemperature. *Palaeogeography, Palaeoclimatology, Palaeoecology*, **226**, 287–306.

MAGARITZ, M. & HOLSER, W. T. 1991. The Permian–Triassic of the Gartnerkofel-1 Core (Carnic Alps, Austria): carbon and oxygen isotope variation. *Abhandlungen der Geologischen Bundesanstalt*, **A45**, 149–163.

MARSHALL, J. D. 1992. Climatic and oceanographic isotopic signals from the carbonate rock record and their preservation. *Geological Magazine*, **129**, 143–160.

MILANOVIĆ, M. 1981. Carboniferous microfossil associations from Gorski Kotar, Hrvatsko Zagorje and Banija. *Palaeontologia Jugoslavica*, **28**.

PALINKAŠ, L. & SREMAC, J. 1989. Barite bearing stromatolites at the Permian–Triassic boundary in Gorski Kotar (Croatia, Yugoslavia). *Memorie della Societa Geologica Italiana*, **40**, 259–264.

PALINKAŠ, L., PEZDIČ, J. & ŠINKOVEC, B. 1993. The Lokve Barite deposit, Croatia: an example of early diagenetic sedimentary ore deposit. *Geologia Croatica*, **46**, 97–106.

POLŠAK, A. & PEZDIČ, J. 1978. Paleotemperature relations of the Carboniferous and Permian periods in the Dinarids and Alps based on the oxygen isotope method and their paleogeographic applications. *Geološki Vjesnik*, **30/1**, 167–187.

RAMOVŠ, A., HINTERLECHNER-RAVNIK, A. ET AL. 1990. Stratigraphic correlation forms of Yugoslav Paleozoic. *Rendiconti Società Geologica Italiana*, **12**, 359–383.

RETALLACK, G. J., SHELDON, N. D. ET AL. 2011. Multiple Early Triassic greenhouse crisis impeded recovery from Late Permian mass extinction. *Palaeogeography, Palaeoclimatology, Palaeoecology*, **308**, 233–251.

SALOPEK, M. 1942. O gornjem paleozoiku Velebita u okolici Brušana i Baških Oštarija. [On the Upper Palaeozoic of the Velebit Mt. in the vicinity of Brušane and Baške Oštarije.] *Rad – Yugoslavian Academy of Science and Art*, **274**, 218–272.

SALOPEK, M. 1948. O gornjem paleozoiku sjeveroistočnog podnožja Velebita i Like. [On the Upper Palaeozoic of the northeastern slope of the Velebit Mt. and Lika region.] *Natural Sciences – Yugoslavian Academy of Science and Art*, **24**, 101–143.

SCHWAB, V. & SPANGENBERG, J. E. 2004. Organic geochemistry across the Permian–Triassic transition at the Idrijca Valley, Western Slovenia. *Applied Geochemistry*, **19**, 55–72.

SCOTESE, C. R. 2008. *PALEOMAP Project.* www.scotese.com (accessed 29 December 2012).

SOKAČ, B., NIKLER, L., VELIĆ, I. & MAMUŽIĆ, P. 1974. *Basic Geological Map of SFRY 1:100 000. The Gospić Sheet.* State Geological Institute, Beograd.

SOKAČ, B., ŠĆAVNIČAR, B. & VELIĆ, I. 1976. *Basic Geological Map of SFRY 1:100 000. Geology of Gospić Sheet.* State Geological Institute, Beograd.

SREMAC, J. 2005. Equatorial shelf of the Palaeozoic supercontinent – cradle of the Adriatic Carbonate Platform. *Geologia Croatica*, **58/1**, 1–19.

SREMAC, J. & KOCHANSKY-DEVIDÉ, V. 1982. Palaeozoic in Croatia (short review). (In Croatian.) *In*: SREMAC, J. (ed.) *IGCP Project No. 5, Yugoslav Working Group Field Meeting, Field guide to Velebit Mt.*, IGCP, Paris, 41–42.

SREMAC, J. & ALJINOVIĆ, D. 1997. Upper Palaeozoic fossils from clastic sedimentary rocks in the Gorski Kotar region. *Geologia Croatica*, **50**, 187–199.

STAMPFLI, G. M. & BOREL, G. 2002. A plate tectonic model for the Paleozoic and Mesozoic constrained by dynamic plate boundaries and restored synthetic oceanic isochrons. *Earth and Planetary Science Letters*, **196**, 17–33.

STAMPFLI, G. M., BOREL, G., CAVAZZA, W., MOSAR, J. & ZIEGER, P. A. (eds) 2001. *The Paleotectonic Atlas of the Peritethyan Domain.* (CD–ROM.) European Geophysical Society, Munich.

ŠUŠNJARA, A., SAKAČ, K., JELEN, B. & GABRIĆ, A. 1992. Upper Permian evaporites and associated rocks of Dalmatia and borderline area of Lika and Bosnia. *Geologia Croatica*, **45**, 95–114.

TIŠLJAR, J. 1992. Origin and depositional environments of the evaporite and carbonate complex (Upper Permian) from the Central Part of the Dinarides (Southern Croatia and Western Bosnia). *Geologia Croatica*, **45**, 115–127.

TIŠLJAR, J. & VLAHOVIĆ, I. 2001. Dolomitization and dolomites within the carbonate platform deposits of the Dinaric Karst region. *In*: DRAGIČEVIĆ, I. & VELIĆ, I. (eds) *First Scientific Meeting Carbonate Platform or Carbonate Platforms of Dinarids, Zagreb, 1–2 October*. The Mining-Geology-Petroleum Engineering Bulletin, Zagreb.

TIŠLJAR, J., VLAHOVIĆ, I., SREMAC, J., VELIĆ, J., VESELI, V. & STANKOVIĆ, D. 1991. Excursion A – Velebit Mt., Permian–Jurassic. *In*: VLAHOVIĆ, I. & VELIĆ, I. (eds) *Some Aspects of the shallow Water Sedimentation on the Adriatic Carbonate Platform (Permian to Eocene). Second International Symposium on the Adriatic Carbonate Platform, Excursion Guidebook.* Institute of Geology, Zagreb, 3–52.

VELIĆ, J. 2007. *Geologija ležišta nafte i plina.* (In Croatian) Faculty of Mining, Geology and Petroleum Engineering, University of Zagreb, Zagreb.

VLAHOVIĆ, I., TIŠLJAR, J., VELIĆ, I. & MATIČEC, D. 2005. Evolution of the Adriatic Carbonate Platform: Palaeogeography, main events and depositional dynamics. *Palaeogeography, Palaeoclimatology, Palaeoecology*, **220**, 333–360.

VOZÁROVÁ, A., EBNER, F. ET AL. 2009. Late Variscan (Carboniferous to Permian) environments in the Circum Pannonian Region. *Geologica Carpathica*, **60**, 71–104.

Facies and climate changes across the Permian–Triassic boundary in the North German Basin: insights from a high-resolution organic carbon isotope record

MICHAEL HIETE[1]*, HEINZ-GERD RÖHLING[2], CARMEN HEUNISCH[2] & ULRICH BERNER[3]

[1]*Center for Environmental Systems Research (CESR), University of Kassel, 34109 Kassel, Germany*

[2]*Landesamt für Bergbau, Energie und Geologie, Stilleweg 2, 30655 Hannover, Germany*

[3]*Bundesanstalt für Geowissenschaften und Rohstoffe (BGR), Stilleweg 2, 30655 Hannover, Germany*

**Corresponding author (e-mail: hiete@uni-kassel.de)*

Abstract: The transition from the uppermost Permian Zechstein Group to the lowermost Triassic Lower Buntsandstein Subgroup in the North German Basin is characterized by a shift from sabkha to playa and to fluviatile facies indicative of a shift to a more humid climate and associated with a major reorganization of the sedimentary system. These changes are recorded in a high-resolution carbon isotope profile of bulk organic matter spanning from the upper Zechstein Group of Changhsingian age to the Calvörde Formation of the Lower Buntsandstein Subgroup of Changhsingian–Indusian age. The organic carbon isotopes reflect a varying composition of the organic matter. Dependent on the facies type, which is mainly controlled by humidity, the proportions of cyanobacterial, phytoplankton and land-plant-derived organic matter with differing organic carbon isotope values change in the samples. A negative organic carbon isotope excursion shortly below the supposed Permian–Triassic Boundary (PTB) is interpreted as a horizon rich in isotopically light phytoplankton, representing a perennial playa lake stage. Although certain effects of global carbon cycle perturbations around the PTB on the analysed organic carbon isotope profile cannot be excluded, the facies and climatic changes associated with the PTB events are thought to be more important.

Numerous profiles worldwide of both marine and terrestrial facies indicate that the Permian–Triassic boundary (PTB) is associated with major environmental changes leading to the most severe mass extinction in Earth's history (e.g. Erwin 1993, 2006; Retallack 1995; Sepkoski 1995; Walliser 1995). These environmental changes are reflected in geochemical proxies. Carbon isotopes of both carbonate and organic matter in marine and terrestrial environments show several excursions and a shift towards lighter isotope values around the PTB, which are mostly interpreted as the feedback of an addition of isotopically light carbon to the atmosphere in the course of perturbations of the global carbon cycle (Holser 1997; cf. also Korte & Kozur 2010 for a recent review). As possible mechanisms for these excursions, *inter alia* a destabilization and oxidation of methane clathrates, enhanced volcanic degassing, a stagnant anoxic ocean, and the dying-off and consequent oxidation of marine and terrestrial biota, as well as of soils, are discussed (e.g. Erwin 1995; Bowring *et al.* 1999; Korte & Kozur 2010). Some of these hypotheses have been analysed in modelling studies for their plausibility (cf. Berner 2002). Some authors interpret the carbon isotope excursions of carbonates as at least partially the result of meteoric diagenesis (for discussions see Scholle 1995; Heydari *et al.* 2001; Korte & Kozur 2010) as in many profiles traces of subaerial alteration/diagenesis are visible. However, meteoric diagenesis cannot explain the negative carbon isotope excursions found in organic matter that is far less affected by diagenesis.

Here, we present a high-resolution profile of carbon isotopes for bulk organic matter of continental sediments of a core from the North German Basin. The drill core spans and covers the transition from uppermost Permian Zechstein Group to lowermost Triassic Lower Buntsandstein Subgroup, including the PTB. The transition is associated with a change from an evaporitic sabkha facies to a playa lake and fluvial facies showing cycles of a different order. In addition, performed kerogen analyses, as well as previously published

From: GĄSIEWICZ, A. & SŁOWAKIEWICZ, M. (eds) 2013. *Palaeozoic Climate Cycles: Their Evolutionary and Sedimentological Impact*. Geological Society, London, Special Publications, **376**, 549–574.
First published online February 27, 2013, http://dx.doi.org/10.1144/SP376.9

inorganic geochemical and palynostratigraphical data of the same core (Hiete *et al.* 2006), allow us to discuss the organic carbon isotope profile in the light of facies and climatic changes.

Geological setting and stratigraphical background

The North German Basin in the Southern Permo-Triassic Basin

During Upper Permian and Lower Triassic times, two elongated ESE–WNW-striking basins existed in Northern and Central Europe: the Northern Permo-Triassic Basin; and the Southern Permo-Triassic Basin (SPTB, cf. Bachmann *et al.* 2010). The SPTB stretched over more than 1.500 km from eastern Britain in the west to central Poland in the east. The northern boundary of the SPTB was characterized by a chain of highs or swells, the Mid North Sea High, the Ringkøbing–Fyn High, the Stevens High and the Rügen Swell (Fig. 1). The Fennoscandian shield limited the basin in the east. Connections to the NPTB, and thus to the Greenland–Norway Seaway, existed via the Central Graben, the Horn Graben and the Slagelse Trough, and the Mid-Polish Trough in connection with the Storebælt. To the west, the SPTB was enclosed by the London–Brabant and the Ardennian–Gallian massifs, whereas the southern limit of the SPTB was less pronounced. The Variscan mountain belt, with the Vindelician and Bohemian massifs as its remnants, separated the basin from the Tethys realm, although several inlets and through-lets developed in the Roet Formation representing the Upper Buntsandstein Subgroup (cf. Fig. 1) and connected the basin with the open ocean.

Internally, the SPTB was subdivided into three larger sub-basins: the Anglo-Dutch Basin in the west; the North German Basin (NGB) in the centre; and the Mid-Polish Trough in the east. Within the area of the NGB, synsedimentary fault movement occurred in the Glückstadt Graben, possibly in the Horn Graben (Röhling 1991; Baldschuhn *et al.* 2001); outside of these grabens broad, undulating, swells and lows caused a pattern of minor thinning and thickening (Röhling 1991, 2013; Geluk & Röhling 1997, 1999).

Facies development in the North German Basin

In the upper part of the Zechstein Group a sabkha and then a playa lake facies developed in the central part of the NGB, in the Anglo-Dutch Basin and in western Poland. A marine influence of boreal origin decreased until continental conditions prevailed. Minor argillic and detritic sediment influx generally came from the south (the London–Brabant Massif, and the Rhenish and Bohemian massifs) and from the NE (the East European Platform) but not from the Ringkøbing–Fyn High in the NW (Beutler & Schüler 1987; Röhling 2013).

In the Lower Buntsandstein Subgroup more clastic sediments were deposited in the NGB and the other sub-basins in a fluvio-lacustrine environment, intermittently influenced by marine ingressions (Lepper & Röhling 1998; Beutler & Szulc 1999; Röhling 2013). The major portion of the fluvial sediments has its origin from the Rhenish and the Bohemian massifs, and had been transported from the south by a SSW–NNE-trending ephemeral river system. The main distribution system followed the Hessian Depression, flanked by the Rhenish Massif in the west and the Eichsfeld–Altmark Swell in the east. Minor influx reached the basin via the Trier Embayment west of the Rhenish Massif and via the Thuringian–West-Brandenburg Trough east of the Eichsfeld–Altmark Swell. The proximal facies of the southern margin is characterized by wedges comprising coarse clastic material and conglomeratic components. They were deposited by a braided river system. Towards the north the facies grades into finer-grained mudflat and playa sediments, and, finally, in the centre of the basin into lacustrine and hypersaline to shallow-marine clastics with oolites, stromatolitic limestones and carbonate cements. Sediment influx from the north, from the Fennoscandian Platform, was captured in the NPTB and did not overstep the Ringkøbing–Fyn barrier high. Only in the Mid-Polish Trough is a Scandinavian influence noticeable (Beutler & Schüler 1987; Röhling 2013).

Basin subsidence was driven essentially by thermal relaxation of the lithosphere (Ziegler 1990; Geluk 2005; Bachmann *et al.* 2010) resulting in a continuous deposition of 1000 m of Buntsandstein Group sediments in the NGB and up to 4000 m in major rift structures (Röhling 1991, 2013; Bachmann *et al.* 2010). Pulse-like rifting processes led to unconformities that define the boundaries between the main basic lithostratigraphical units in the Buntsandstein of the NGB, as well as in the adjacent sub-basins. The palaeostructural differentiation into mainly NNE-trending swells and troughs is reflected by local transgressions and thickness variation patterns where the swells are characterized by slightly reduced thicknesses of the subcycles. In addition, sedimentation was probably influenced by short-lived climatic events, expressed in a cyclicity of different order (base-level cycles with a fining-upwards trend) in the Lower Buntsandstein Subgroup.

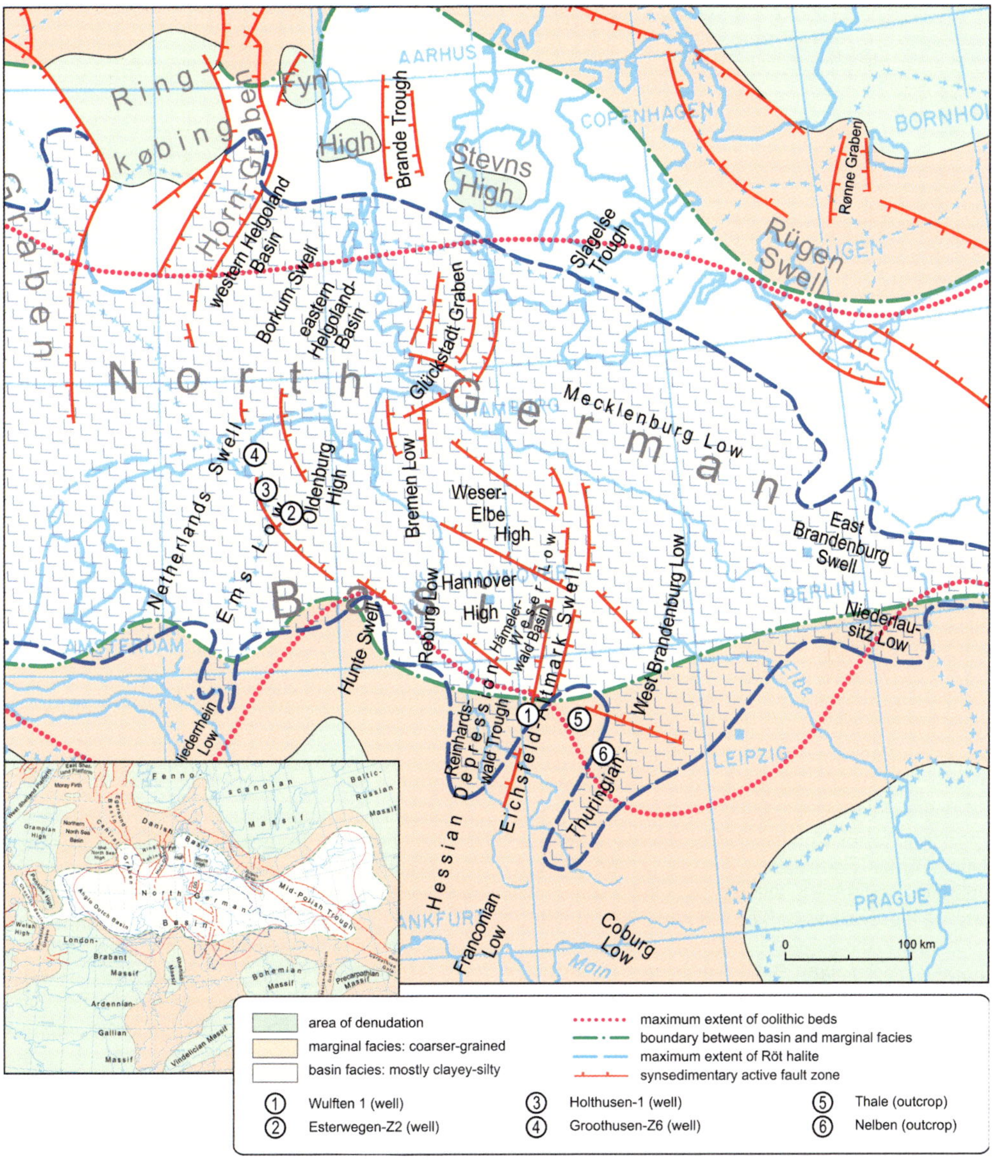

Fig. 1. Palaeogeographical map of the Northern and Southern Permo-Triassic Basins, including the positions of the Esterwegen Z2, Groothusen Z6, Holthusen Z1 and Wulften-1 drillholes, as well as the Nelben and Thale outcrops (based on Bertelsen 1980; Day *et al.* 1981; IGCP 1986; Ziegler 1990; Dadlez *et al.* 1998).

Lithostratigraphy

Zechstein Group. In the NGB the Zechstein Group, whose base (the Kupferschiefer) is date-marked to 257.3 ± 1.6 Ma (Re–Os) (Brauns *et al.* 2003), is lithostratigraphically subdivided into seven cycles or formations: z1–z7 (cf. Best 1989; Käding 2000) (Fig. 2), with the upper cycles of Changhsingium age (Deutsche Stratigraphische Kommission 2002). Of the seven cycles, only z1–z3 show the complete succession of marine transgressive clays, carbonates, anhydrites and rock salts, whereas z4–z6 lack marine limestones. The seventh cycle, the Fulda Formation, takes for several reasons an exceptional position. An unconformity that is visible on swells subdivides z7 into a lower and an upper part. Whereas the lower part of z7 (Lower Fulda Formation, z7low) shows both a basinal facies with

Series	Stage	Duration (Ma)
Middle Triassic	Anisian	6.0
Lower Triassic	Olenekian	5.0
	Indusian	2.0
Upper Permian	Changhsingian	4.0
	Wuchiapingian	5.5

Group	Subgroup	Formation	Duration (Ma)
Muschelkalk			
Buntsandstein Group	Upper (so)	Röt Fm.	1.5
	Middle (sm)	smS Solling Fm. smH Hardegsen Fm. smD Detfurth Fm. smV Volpriehausen Fm. smQ Quickborn Fm.	4.5
	Lower (su)	suB Bernburg Fm.	1.0
		suC Calvörde Fm.	1.0
Zechstein Group		z7 Fulda Fm. (Upper Fulda Fm. / Lower Fulda Fm.) z6 Friesland Fm.	1.0
		z5 Ohre Fm. z4 Aller Fm.	1.0
		z3 Leine Fm.	1.5
		z2 Staßfurt Fm.	1.5
		z1 Werra Fm.	2.0
Rotliegend			

Fig. 2. Lithostratigraphical subdivision of the Zechstein Group and Buntsandstein Group in the North German Basin (based on Deutsche Stratigraphische Kommission 2002).

red clay, anhydrite and around 20 m of rock salt and a proximal facies with red clastics for the upper part (Upper Fulda Formation, z7up) no evaporites are reported (Best 1989). The unconformity marks the base of the Triassic in The Netherlands, whereas, in Germany, the base of the Triassic Buntsandstein Group is placed at the first oolitic sandstone interval above z7up; that is, about 25 m higher (Best 1988, 1989; Röhling 1993, 2013; SKPT 1993, 2011; Roman 2004; Geluk 2005; Hiete *et al.* 2006).

Lower Buntsandstein Subgroup. Traditionally, the Buntsandstein Group, as the lower group of the tripartite German Triassic, is lithostratigraphically subdivided into the Lower, Middle and Upper Buntsandstein Subgroups of Changhsingian–Anisian age (cf. Fig. 2) (Boigk 1959, 1961; Deutsche Stratigraphische Kommission 2002; SKPT 2011). The Lower Buntsandstein Subgroup, with a thickness of 300–350 m, was deposited in a large intracratonic basin in a playa to fluvio-lacustrine environment with deposition of predominantly fine-grained sandstones, siltstones, claystones and interbedded oolitic limestones for which intermittent sedimentation is typical (Röhling 1991, 1993, 2013; Geluk & Röhling 1997, 1999). It is subdivided into the Calvörde Formation and the Bernburg Formation (Fig. 2), which is well known for its oolitic sandstones (stromatolites) (Paul & Peryt 2000). The Calvörde and Bernburg formations differ from each other both by colour and by the texture of the clay- and siltstones (cf. Roman 2004). A basin-wide unconformity (the Volpriehausen unconformity) marks the top of the Bernburg Formation (Röhling 1991, 1993, 2013; Geluk & Röhling 1997, 1999).

Both the Calvörde Formation and the Bernburg Formation show cycles of different order, ranging from a few millimetres to several tens of metres (Hoppe 1968; Kunert 1964; Geluk & Röhling 1997; Hiete 2004). The most prominent cycles are base-level cycles of about 10–25 m in thickness in the NGB, which contain four–five subordinate base-level cycles (Szurlies 2001). In the Calvörde Formation a total of 10 (denoted as suC1–suC10), and in the Bernburg Formation another 14, of these cycles can be distinguished and traced over several hundreds of kilometres in wireline logs, cores and outcrops from the UK to western Poland (Brüning 1984, 1986; Röhling 1986, 1991, 1993, 2013; Best 1988, 1989; Geluk & Röhling 1997, 1999; Roman *et al.* 2001; Szurlies 2001; Szurlies *et al.* 2012).

These correlations are further supported by supraregional lithological marker horizons such as the Crumbled Shale Bed Interval in z7low ('Bröckelschieferbank'), the Banded Shale Facies in z7up ('Bänderschieferfazies'), the Grey–Green

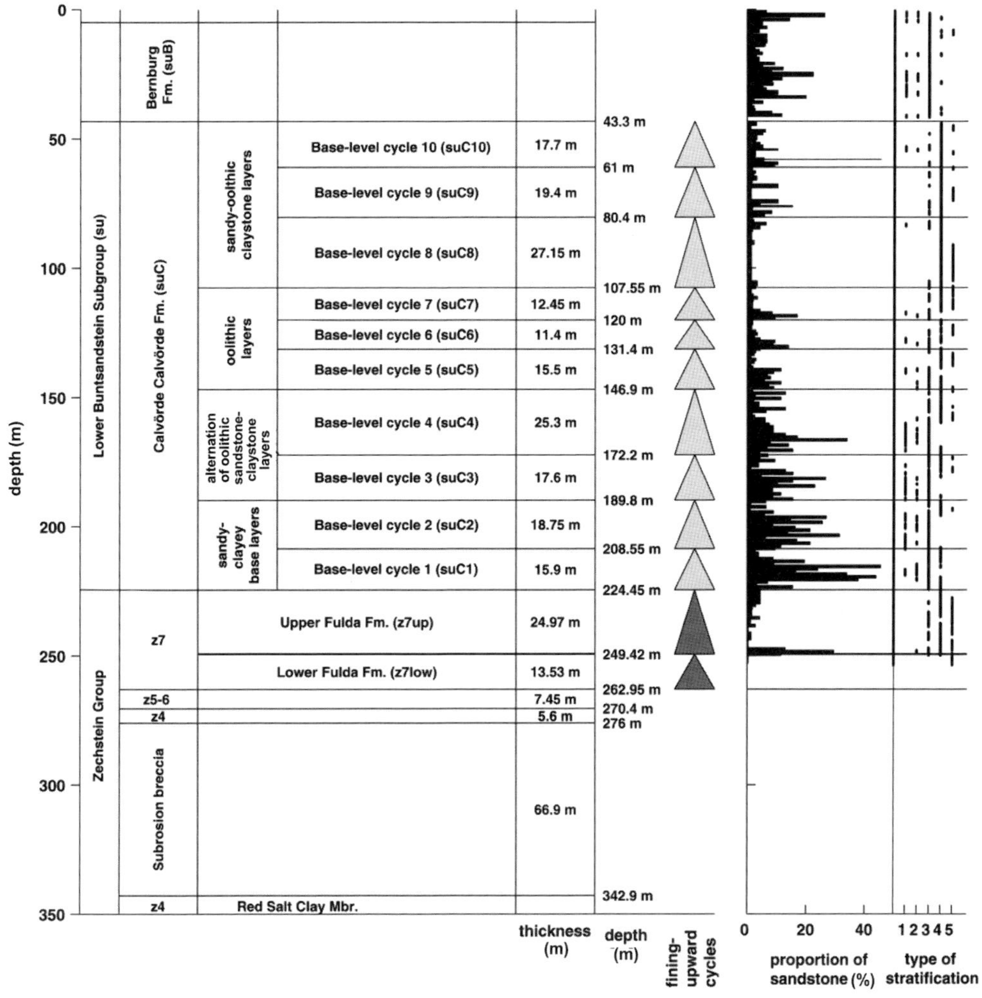

Fig. 3. Lithostratigraphical subdivision of Upper Zechstein and Calvörde Formation of the Lower Buntsandstein Subgroup of the Wulften-1 drill core. Within the Calvörde Formation, 10 fining-upwards cycles can be distinguished (suC1–suC10); sandstone proportion per metre; type of stratification: (1) for cross-bedding, (2) for horizontal bedding, (3) for ripple bedding, (4) for flaser bedding, (5) for no bedding (after Brüning *et al.* 1981; Brüning 1986).

Boundary Bed Interval directly below z7low ('grau-grüne Grenzbank'), the Grey Bed Interval in suC2 ('Graubankbereich'), the Thale Horizon in suC3 ('Thale-Horizont') and a Gamma Horizon in suC5–suC7 (Seidel 1965; Schulze 1969; Brüning 1986, 1988; Szurlies 1999, 2001; Szurlies *et al.* 2012). Owing to their large lateral extent, these cycles are interpreted as climatically driven, possibly representing short eccentricity cycles of about 100 kyr in duration (Clemmensen *et al.* 1994; Geluk & Röhling 1999; Szurlies 2001), resulting in an estimated duration of 1.0 Myr for the Calvörde Formation and 1.4 Myr for the Bernburg Formation (Deutsche Stratigraphische Kommission 2002; Szurlies *et al.* 2012). The base-level cycles suC1–suC10, with their intercalated sand-, silt- and claystones, show a fining-upwards environment with the ideal successions of a playa system (Hofmann *et al.* 2000; Szurlies 2001):

- alluvial fan → sand plain → clay plain (playa) (if present) → ephemeral playa lake with oolite facies near the shore and clayey–siltstone facies in the basin in case of a low erosional base;
- sand plain → oolite → ephemeral playa lake facies in case of a high erosional base.

As the flooding reached its maximum in suC3 (Thale horizon with a thick oolitic sandstone: cf.

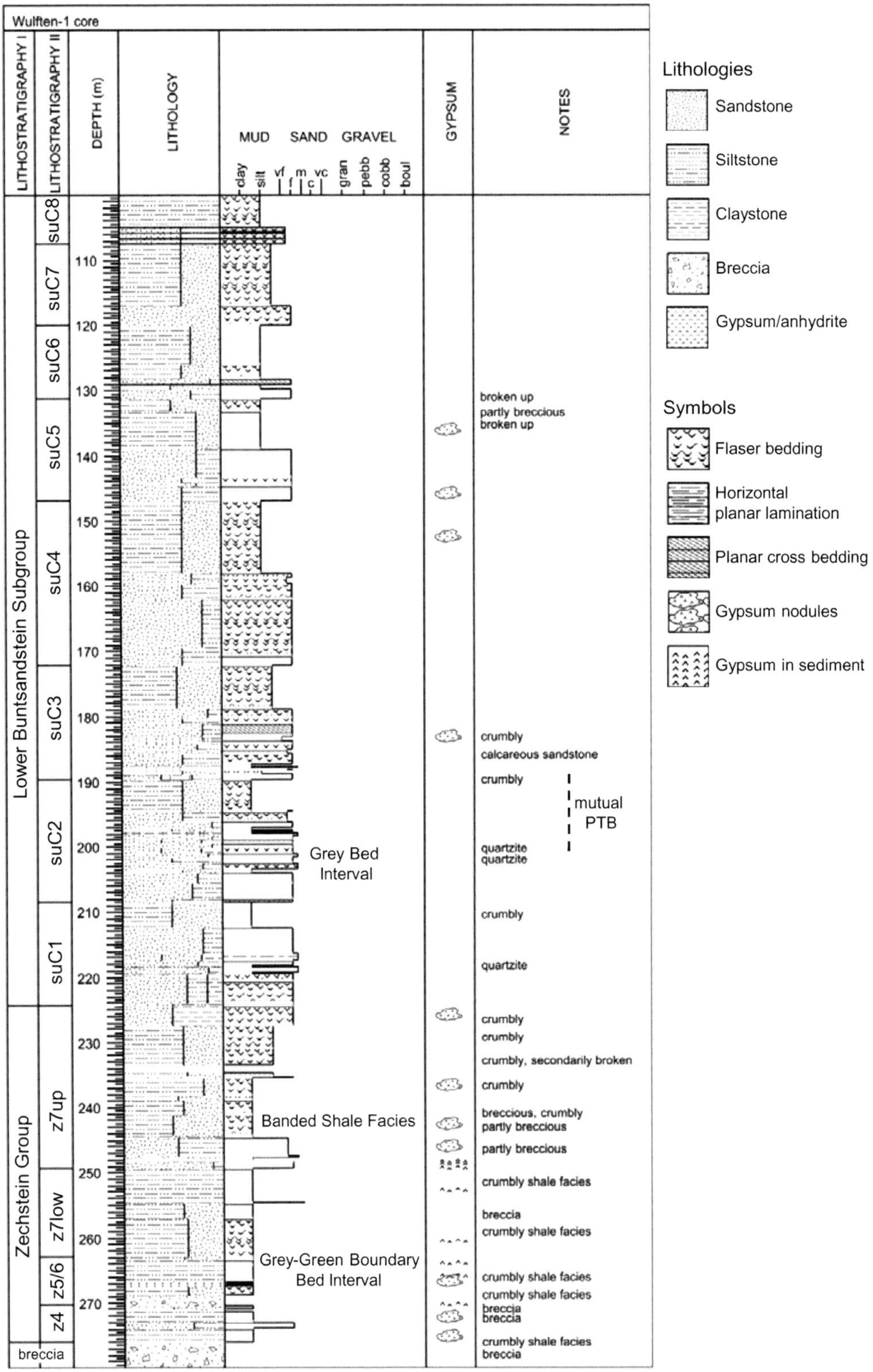
Wulften-1 core
LITHOSTRATIGRAPHY I
LITHOSTRATIGRAPHY II
DEPTH (m)
LITHOLOGY
MUD SAND GRAVEL
clay silt vf f m c vc gran pebb cobb boul
GYPSUM
NOTES
Lower Buntsandstein Subgroup
Zechstein Group
breccia
suC8
suC7
suC6
suC5
suC4
suC3
suC2
suC1
z7up
z7low
z5/6
z4
110
120
130
140
150
160
170
180
190
200
210
220
230
240
250
260
270
broken up
partly breccious
broken up
crumbly
calcareous sandstone
crumbly
mutual PTB
quartzite
quartzite
Grey Bed Interval
crumbly
quartzite
crumbly
crumbly
crumbly, secondarily broken
crumbly
Banded Shale Facies
breccious, crumbly
partly breccious
partly breccious
crumbly shale facies
breccia
crumbly shale facies
Grey-Green Boundary Bed Interval
crumbly shale facies
crumbly shale facies
breccia
breccia
crumbly shale facies
breccia
Lithologies
Sandstone
Siltstone
Claystone
Breccia
Gypsum/anhydrite
Symbols
Flaser bedding
Horizontal planar lamination
Planar cross bedding
Gypsum nodules
Gypsum in sediment

Szurlies 1999), the maximum of oolite formation moved towards the basin centre from the older to the younger base-level cycles (Szurlies 2001).

Sedimentation in a playa environment occurs after precipitation events and is therefore in time-scales of years intermittent but continuous. More humid conditions are probably associated with more frequent sedimentation events such that the lower part of the Calvörde Formation, which contains the PTB, is supposedly characterized by rather short intervals between sedimentation events.

Bio- and magnetostratigraphy

Both the Zechstein Group and the Lower Buntsandstein Subgroup were deposited under arid and saline conditions resulting in an environment hostile to (most) life and unfavourable preservation conditions (Ecke 1986; van der Zwan & Spaak 1992; Heunisch 1999). Accordingly, biostratigraphical correlation is mainly restricted to horizons bearing miospores and conchostracans (Ecke 1986; Kozur 1989, 1993*a*, *b*, 1999*a*, *b*; Schulz 1994; Backhaus 1996; Hiete *et al.* 2006; Kozur & Weems 2010).

An overview of palynostratigraphy in the German Triassic can be found in works of Heunisch (1999) and Kürschner & Herngreen (2010). Ecke (1986) placed the boundary between the Permian *virkkiae–microcorpus*-phase and the lowermost Triassic *obsoleta–playfordii*-phase in Lower Saxony between the oolite horizons of suC3 and suC4. Based on the sporomorph zone *Lundbladispora obsoleta–Lunatisporites noviaulensis* and the *Otynisporites eotriassicus* megaspores association, Kozur (1999*a*) placed the Grey Bed Interval of suC2 in the uppermost Permian. Hiete *et al.* (2006) identified the lowermost Triassic sporomorph association *papillatus–playfordii*-phase with *Endosporites papillatus* and *Densoisporites playfordii* 2 m above the boundary between suC2 and suC3. These findings are in agreement with biostratigraphical results based on conchostracans. The transition from the Permian conchostracan zone *Falsisca postera* to the Triassic *Falsisca verchojanica* zone occurs at the top of the Grey Bed Interval in suC2 (Kozur 1999*b*; Kozur & Weems 2010).

Many studies analysed the magnetic polarity in the Permian–Triassic boundary interval worldwide (cf. Hounslow & Muttoni 2010). In the NGB, there is a complete magnetostratigraphical profile covering the upper Zechstein Group and the Calvörde Formation (see the Discussion later) (Szurlies 2001, 2004, 2007; Szurlies *et al.* 2003, 2012). According to Szurlies (2007) and Szurlies *et al.* (2012), the long zone of the normal polarity zone (LT1n) containing the PTB extends from z7up to the basal part of suC7, whereas Hounslow & Muttoni (2010) argued that it extends as far as suC9. Irrespective of these discussions, based on these bio- and magnetostratigraphical findings, the mutual PTB in the NGB can be placed in suC2 above the Grey Bed Interval.

Chemostratigraphy

Compared to the large number of geochemical studies, including studies on stable isotopes of carbon, oxygen and sulphur covering the PTB interval worldwide, such studies are scarce for the NGB and most of them cover inorganic geochemical analyses of a few samples only (e.g. Füchtbauer & Goldschmidt 1959; Ludwig 1962; Hartmann 1963; Kästner *et al.* 1966; Kühn & Radzinski 1972; Hug 2004). A high-resolution inorganic geochemical profile (major and minor elements using X-ray fluorescence (XRF) analyses) spanning z4–suC7 using an analysis of the Wulften-1 drill core is provided in Hiete (2004) and Hiete *et al.* (2006). Bachmann *et al.* (2003) reported a minimum of organic carbon isotopes in the oolite horizon of suC2 and a decrease in the carbonate carbon isotopes from the oolite horizon of suC1 to a minimum in the oolite horizon of suC2. Korte & Kozur (2005) reported 25 analyses of carbonate carbon and oxygen isotopes covering suC1–suC5 from the Thale section (about 60 km east of the Wulften-1 core) and noted a $\delta^{13}C_{carb}$ minimum of −4.21‰ at the base of suC2 followed by an increase to −1‰ near the suC2–suC3 boundary, and a second minimum of −3.3‰ followed by values of −2.5‰ above (also presented in Korte *et al.* 2007). Bachmann & Kozur (2004) and Hansen (2006) presented organic carbon isotope values of a 4 m-thick interval spanning from the upper suC1 to the lower suC2 of the Nelben section (about 100 km east of the Wulften-1 core) with a negative excursion of $\delta^{13}C_{org}$ down to −30‰ VPDB (see the Discussion later).

Materials and methods

Zechstein and Buntsandstein in Wulften-1 borehole

The drill cored section of the Wulften-1 borehole (about 10 km SW of the town of Osterode,

Fig. 4. Lithological profile of the geochemically analysed interval of the Wulften-1 core showing marker horizons and the mutual position of the PTB, which is placed within suC2 above the Grey Bed Interval based on bio- and magnetostratigraphical findings (cf. the text for details).

Germany; Gauss–Krüger co-ordinates: 35 82 120, 57 27 480) contains the Zechstein Formations z3–z7, as well as the complete Calvörde Formation (Fig. 3). Within z4, a subrosion breccia is intercalated. The lithostratigraphical subdivision in z4–z7 and suC1–suC10, adapted from Brüning *et al.* (1981) and Brüning (1984, 1986), is supported by prominent lithomarker horizons such as the Grey–Green Boundary Interval, the Crumbled Shale Bed Interval, the Banded Shale Facies and the Grey Bed Interval (see above, Fig. 4). Palaeographically, Wulften-1 at the margin of the Eichsfeld–Altmark Swell took an intermediate position between the playa lake facies of the central NGB and the sandflat and fluvial facies in Hesse and further south (cf. Hug 2004; Hug & Gaupp 2006 for the more marginal facies of the upper Zechstein in Hesse).

Zechstein in the Holthusen Z1 borehole

The cored section of the Holthusen Z1 borehole (about 10 km SW of the town of Leer, Germany; Gauss–Krüger co-ordinates: 25 91 769, 58 89 604) comprises the complete Zechstein. Lithostratigraphical division of the core was supported by lithological markers such as z4 anhydrite, and correlation with nearby boreholes Esterwegen Z2 and Groothusen Z6, which can be correlated with the Wulften-1 borehole (Röhling 1991; Geluk & Röhling 1999; Szurlies *et al.* 2012). Palaeogeographically, Holthusen Z1 took a basinal position in the Ems Trough.

Analyses

Organic petrography. Polished sections of 14 samples of selected greyish-green lithologies and 27 samples of red and brown lithologies representing claystones, siltstones and sandstones (all Wulften-1 drill core) were prepared. Reflectance of the vitrinite macerals was measured according to DIN 22020, part 5 (DIN 1986).

Kerogen analyses. Selected greyish-green lithologies of the Wulften-1 drill core were sampled for kerogen analyses. In total, 31 samples were taken and prepared according to internal standards (preparation with HCl and HF, ultrasonic sieving >10 μm, strew slides with glycerine jelly as mounting media).

Organic carbon isotopes. Sediments of Wulften-1 were sampled from the drill core at about 10 cm intervals between 280 and 103 m depth below surface (z5–z7 and suC1–suC8). From the subrosion breccia and the underlying Grey Salt Clay Member of z3 and Red Salt Clay Member of z4, 12 additional samples were taken. Of the Holthusen Z1 drill core, 66 samples were taken. For all samples, inorganic geochemical measurements were performed using XRF analysis (cf. Hiete 2004; Hiete *et al.* 2006). Carbon isotopes of bulk organic matter ($\delta^{13}C_{org}$-(bulk)) were measured in every second sample of the Wulften-1 drill core (in total 929 samples) and all samples of the Holthusen Z1 drill core using Finnigan MAT Delta^{Plus} isotope ratio mass spectrometer coupled to a Carlo Erba NC 2500 (Thermo Quest) element analyser (cf. Fry *et al.* 1992). To avoid corrosion, samples with $(SO_3) > 1\%$ as determined by XRF were excluded. Prior to analysis, samples were crushed and ground to a particle size of smaller than 40 μm, then carbonates were removed with HCl (20%) and samples were dried. Some samples contain calcite and dolomite, respectively (cf. Hiete *et al.* 2006), but there is no clear relationship between $\delta^{13}C_{org}$ value and dolomite content. Additional $\delta^{13}C_{org}$ measurements were performed on selected samples originally prepared for kerogen analyses; that is, on acid-insoluble organic particulate matter (POM). Very low contents of organic matter made it necessary to increase the sample weight to a maximum and to add V_2O_5 to enhance combustion (for details see Hiete 2004). $\delta^{13}C_{org}$ values are reported relative to the international VPDB standard and calibrated against NBS-22 with a $\delta^{13}C_{org}$ value of -30.03‰ (IAEA 2007). The standard deviation of $\delta^{13}C_{org}$ for the calibrated and internal laboratory standard was 0.15‰.

Results

Particulate organic matter (POM) is sparse and largely dominated by black wood and detrital graphite (Fig. 5, Table 1). For example, in the analysed samples of the Wulften-1 drill core, black wood (highly coalified, opaque vitrinite or inertinite) dominates with, on average, 75% of the counts (42–99.5%), of which 92% (80–100%) is equally dimensioned. The proportion of brown wood (vitrinite) is, on average, another 8.6% (0.5–19.5%). Vitrinite reflectance was around 1.1–1.4% for the higher coalified and 0.3–0.5% for the lower coalified vitrinite. Coalification and the degree of roundness suggests that black wood is allochthonous and brown wood is autochthonous. Other components are tissue with 7.4% (0–26.5%), amorphous organic matter (AOM) with 4.9% (0–40%) and resins (resinite) with 0.1% (0–1%). In addition, small amounts of detritic grains of schists are present. The sample of the Grey Salt Clay Member differs from the samples from 278 to 105 m by a higher proportion of POM, mostly weakly coalified vitrinite and fluorescent bitumen, and by inertinite. Structureless tissue dominates over structured tissue from the

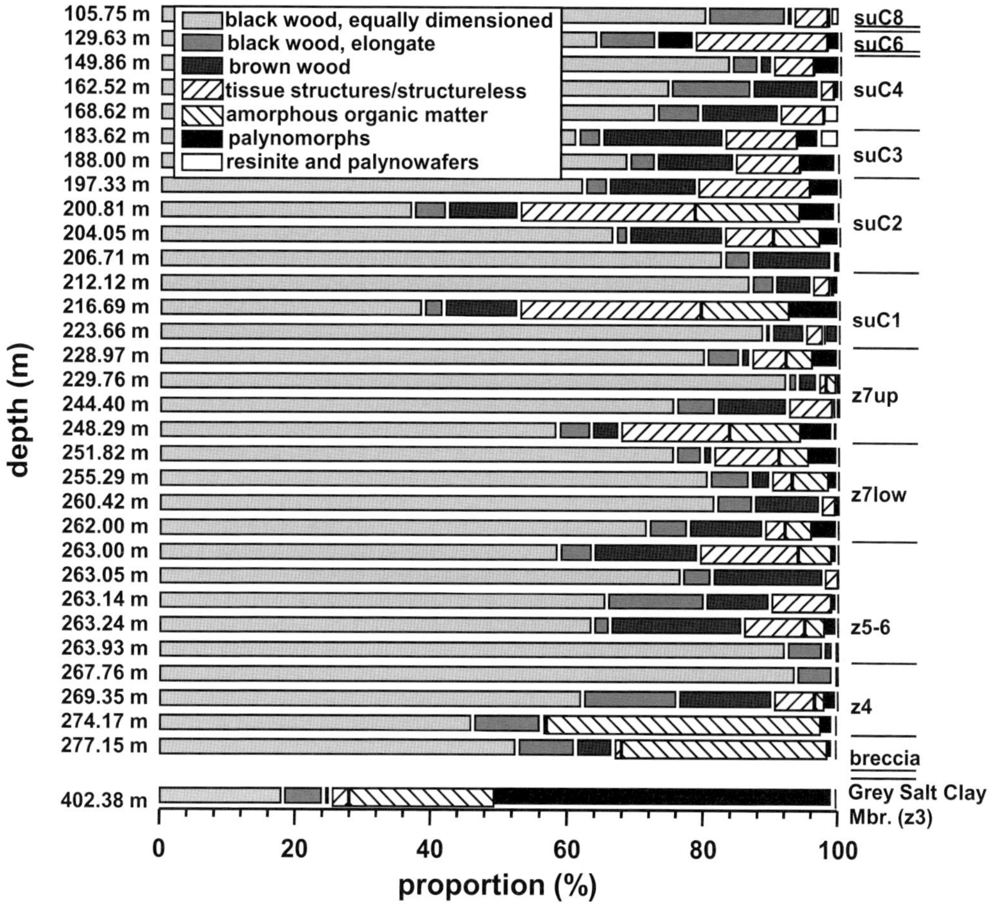

Fig. 5. Kerogen composition of the Wulften-1 drill core showing the dominance of equally dimensioned black wood in POM in most of the samples. The sample at 402.38 m (Grey Salt Clay Member (z3)) is different by its high share of palynomorphs, and the samples at 216.69 m (suC1) and 200.81 m (suC2) by their relatively high shares of tissue.

upper part of z7low onwards. Amorphous organic matter is very rare above suC2.

Only nine samples of the Grey Salt Clay Member, suC1–suC4, and of suC6 contained sufficient amounts of palynomorphs for a more detailed analysis (Fig. 6, Table 2). Around a quarter (19–31%) of the palynomorphs were indeterminable due to poor preserving conditions. The sample of the Grey Salt Clay Member differs from the other samples by its high proportion of palynomorphs, reaching 50% compared to an average of 2.3% (0–7%) in the other samples. In addition, the spectrum of the sample of the Grey Salt Clay Member is dominated by (bisaccate) pollen (77.5%), whereas spores (2.5%) and other plankton (2%) play a minor role only. In the samples of the Calvörde Formation, however, phytoplankton such as prasinophyceae and/or zygnemataceae (single-celled green algae with a brackish to (acidified) freshwater habitat) dominate in the spectrum with 40% (27–49%), followed by pollen with 20% (0–32%) and spores with 7% (0–18%). Resedimented palynomorphs, especially of Permian age, were common. Fungal remains and cell clusters were frequent, especially in suC2 and suC3, but not in an amount comparable with, for example, the fungal spike in the Tesero Oolite Horizon (TOH) in the Alps (Massari *et al.* 1994). Perhaps some of the fungal spores belong to zygnemataceae (e.g. Afonin *et al.* 2001).

The samples prepared for kerogen analyses show $\delta^{13}C_{org}$-(POM) values between −24.9 to −23.4‰ for acid-insoluble POM, with some values reaching −26.4 to −25.9‰. The $\delta^{13}C_{org}$-(bulk) values of the Wulften-1 sediments varied in z5 to suC8 to between −28.4 and −21.4‰ (Table 3).

Table 1. *Data of kerogen analyses of Wulften-1 core*

Depth (m)	Black wood, equally dimensioned	Black wood, elongate	Brown wood	Tissue structured/ structureless	Amorphous organic matter	Paly-nomorphs	Resinite and paly-nowafers	Total
149.86	168	8	4	12	0	8	0	200
162.52	150	24	20	4	0	2	0	200
168.62	146	13	23	13	0	0	5	200
183.62	123	7	36	21	0	7	6	200
188.00	138	8	23	19	0	11	1	200
197.33	125	7	26	33	0	9	0	200
200.81	75	10	21	51	31	11	1	200
204.05	134	4	28	14	14	6	0	200
206.71	166	8	24	1	0	0	1	200
212.12	174	7	11	5	1	0	2	200
216.69	78	6	22	53	26	15	0	200
223.66	178	2	10	5	1	4	0	200
228.97	161	10	3	10	8	8	0	200
229.76	185	3	6	2	3	1	0	200
244.40	152	12	21	13	1	1	0	200
248.29	144	12	10	39	26	12	2	245
251.82	152	8	3	19	9	9	0	200
255.29	162	12	6	6	11	3	0	200
260.42	164	11	20	4	0	1	0	200
262.00	144	12	22	6	8	8	0	200
263.00	118	10	31	29	10	2	0	200
263.05	154	9	33	4	0	0	0	200
263.14	132	29	19	18	0	2	0	200
263.24	128	5	39	18	6	4	0	200
263.93	185	11	3	1	0	0	0	200
267.76	188	11	1	0	0	0	0	200
269.35	125	28	28	12	3	4	0	200
274.17	93	20	1	0	81	4	1	200
277.15	106	17	11	2	61	2	1	200
105.75	161	23	2	10	0	1	3	200
129.63	129	17	11	39	0	4	0	200

The Grey Salt Clay (z3 clay) is extremely heavy, with $\delta^{13}C_{org}$-(bulk) values of up to −7.7‰. Within the $\delta^{13}C_{org}$ (bulk) curve, several zones can be distinguished (Table 3, Fig. 7):

In samples of the Holthusen Z1, organic carbon isotope values range from −3.8 to −27.0‰ from z4–z6 to the base of suC (Fig. 8). To definitely exclude that the extremely heavy values are the result of not entirely removed carbonates, in some samples carbonates were removed twice. The extreme values of up to −3.8‰ were confirmed, although the three samples measured twice (after one and two decarbonation steps) showed still very heavy but considerably lighter values (Fig. 8) (e.g. −14.8‰ instead of −8.6‰), which is explained by inhomogeneities and loss of organic matter during the decarbonation process. Overall, the carbon isotope values increase from around −25‰ from the base of z4–z6 to maximum values of up to −3.8‰ in the central part of z7low, to again lighter values of about −22‰ near the base of 7up, and around −24‰ in the remainder of z7up and the base of suC. Single samples show strong variations to heavier values, for example, in z4/z6 and in z7up.

Discussion

Factors affecting bulk sediment organic carbon isotope values in the analysed samples

The organic carbon isotope values of the analysed bulk sediment samples are the result of several factors affecting: (i) the composition of deposited organic matter (the different fractions making up bulk organic matter such as phytoplankton or land-plant-derived organic matter show different carbon isotopic values); (ii) the preservation of

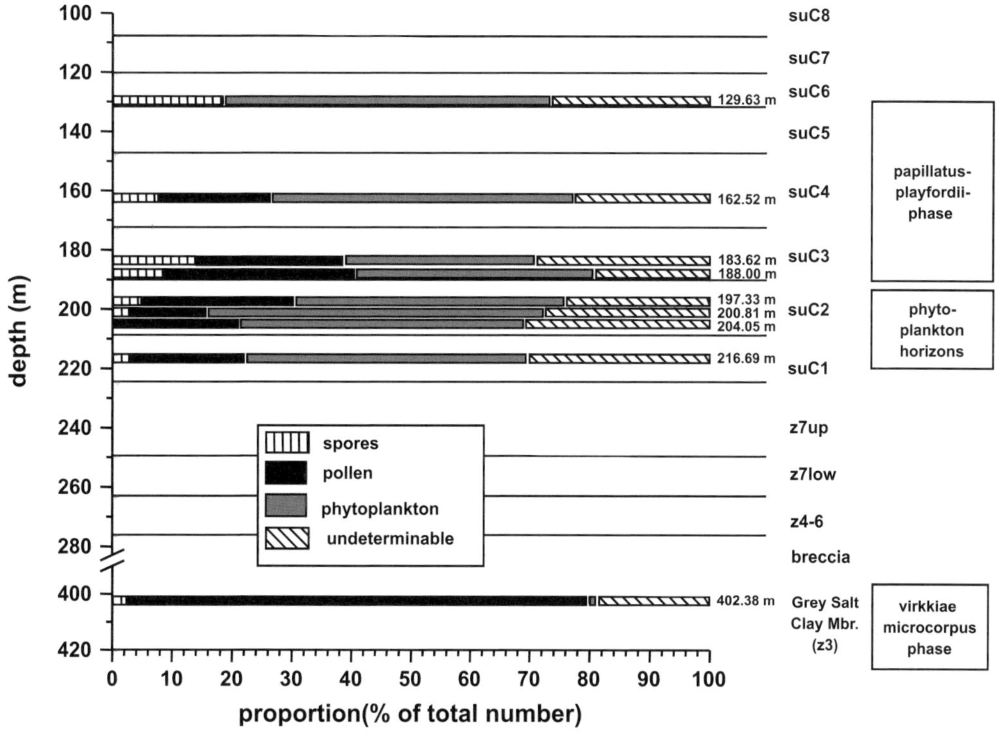

Fig. 6. Palynospectra and palynostratigraphical subdivision of the Wulften-1 drill core. Whereas pollen largely dominate in the sample at 402.38 m (Grey Salt Clay Member (z3)), the other samples contain more than 30% phytoplankton. Note the high shares of undeterminable palynomorphs, ranging between 19 and 31%. The sample of the Grey Salt Clay Member can be assigned to the *virkkiae–microcorpus*-phase *sensu* Ecke (1986) of uppermost Permian age and the analysed samples from suC3 to suC6 to the lowermost Triassic *papillatus–playfordii*-phase (cf. Hiete *et al.* 2006). The analysed samples of suC1 and suC2 contain large amounts of phytoplankton.

these fractions after deposition; (iii) the isotopic values of the different fractions, which are controlled by a number of environmental factors such as water stress or salinity; and (iv) the organic carbon isotope value of atmospheric CO_2 and thus also of dissolved CO_2, which is believed to have changed around the PTB due to perturbations of the global carbon cycle. In the following, the effects of different factors, including climate, on organic carbon isotope values of bulk samples are discussed for sediments deposited in a sabkha and playa environment, respectively.

Composition of organic matter. Both the kerogen and organic petrographical analyses show that the organic matter in the bulk samples of the

Table 2. *Data of palynomorph analyses of Wulften-1 core*

Depth (m)	Spores	Pollen	Phytoplankton	Undeterminable	Total
129.63	7	0	21	10	38
162.52	11	27	72	32	142
183.62	29	52	67	60	208
188.00	19	69	86	41	215
197.33	8	42	74	39	163
200.81	6	28	119	58	211
204.05		14	31	20	65
216.69	6	41	99	63	209

Table 3. *Identified zones in organic carbon isotope profile of Wulften-1 core*

Zone	Depth (m)	Description of organic carbon isotope variations
A	402.36, 402.93	Both samples of the Grey Salt Clay Member show extremely heavy values of −13.4 and −7.3‰
B	343.69–282	Red Salt Clay Member with −23.1‰ and subrosion breccia with 25.8 to −22.4‰
C	281.0–277.75	Breccia material with relatively light values of −27 to −26‰
D	277.65–276.65	At the base of mostly greyish breccia sediment jumps to heavier values of −24 to −23‰
E	276.45–272.55	With −25.5 to −24‰, more strongly varying values within breccia and possibly z4 material
F	272.55–271.55	Jump to heavier values of around −21.5‰ directly above the z4 cover anhydrite
G	271.55–243.15	This zone with heavier values of more than −25‰ extends up into the Banded Shale Facies of z7up and is characterized by an almost steady trend towards lighter values. High (SO_3)-content inhibited analyses of the Grey–Green Boundary Bed Interval
G1	271.55–257.25	Interval z5–z6 with values of −22.5 to −21‰; lower part of z7low with −23.0 to −22.0‰ and in a single sample down to −24.5‰
G2	257.2–254.95	Subrosion breccia material with −24.8 to −21.8‰
G3	254.45–243.15	Almost steady decrease from about −23.3 to −22.3‰ in the interval 254.45–251.85 m (with two outliers up to −24.4‰) down to −25.1‰ at 249.65 m. This is followed by a jump back to heavier values of around −23‰ 20 cm below the z7low–z7up boundary. Whereas this boundary is marked by a change from siltstones to a poorly sorted base sandstone, the interval 243.0–244.5 m with the jump to lighter carbon isotope values is lithologically inconspicuous
H	242.85–233.95	With values of −26 to −25‰ the central interval of z7up is considerably lighter than the underlying sediments
I	233.75–177.24	This interval includes the highest part of z7up, suC1, suC2 and about two-thirds of suC3 with values of −28 to −25‰ with very light, mostly grey, intervals; in particular, the Grey Bed Interval of suC2 with strong negative excursions between 203.5 and 201.0 m and from 198 to 197 m. The lower boundaries of suC1, suC2 and suC3 are isotopically inconspicuous
J	177.14–161.33	Interval with heavier values of around −25‰ and with some grey intervals showing lighter values included
K	161.05–131.35	Interval with light values of around −26 to −25‰ and negative excursions down to −26.5‰; for example, from 147.8 to 148.4 m at the top of suC4. These peaks are not bound to grey lithologies
L	131.25–120.95	With −25‰ isotopically relatively heavy interval that is almost congruent with suC6
M	120.65–103.05	Interval with lighter values of around −26 to −25‰, in particular in the lower part of suC7, between 115 and 109 m; that is, upper part of suC7, slightly heavier

Wulften-1 drill core is a mixture of allochthonous and autochthonous material with different maturities and of different origins, and that wooden plant material dominates in (acid-insoluble) POM (Fig. 9).

The palynological analyses reveal further that the spectrum of palynomorphs, which constitute only a small portion of acid-insoluble POM, consists mainly of pollen, spores and phytoplankton, as well as fungal remains. Therefore, the finely dispersed organic matter, which supposedly represents the major part of the organic matter in the samples, is probably composed of very small particles and amorphous material of land-plant and phytoplankton origin, as well as fungal and probably also bacterial organic matter. Consequently, the organic carbon isotopic values of the bulk samples represent a mix signal reflecting the changing composition of organic matter, as well as the changing organic carbon isotopic values of the components. Similar isotopic mix signals in the PTB profiles are reported, for example, by Foster *et al.* (1997, 1999) and Hermann *et al.* (2010).

In the analysed samples of the Wulften-1 core, acid-insoluble POM is dominated by wooden material and shows values of −24.9 to −23.4‰ (Fig. 7). Such values are typical for terrestrial organic matter of Permian and Triassic age (Peters-Kottig 2003; Hiete & Berner cited in Hiete 2004), whereas marine phytoplankton is isotopically

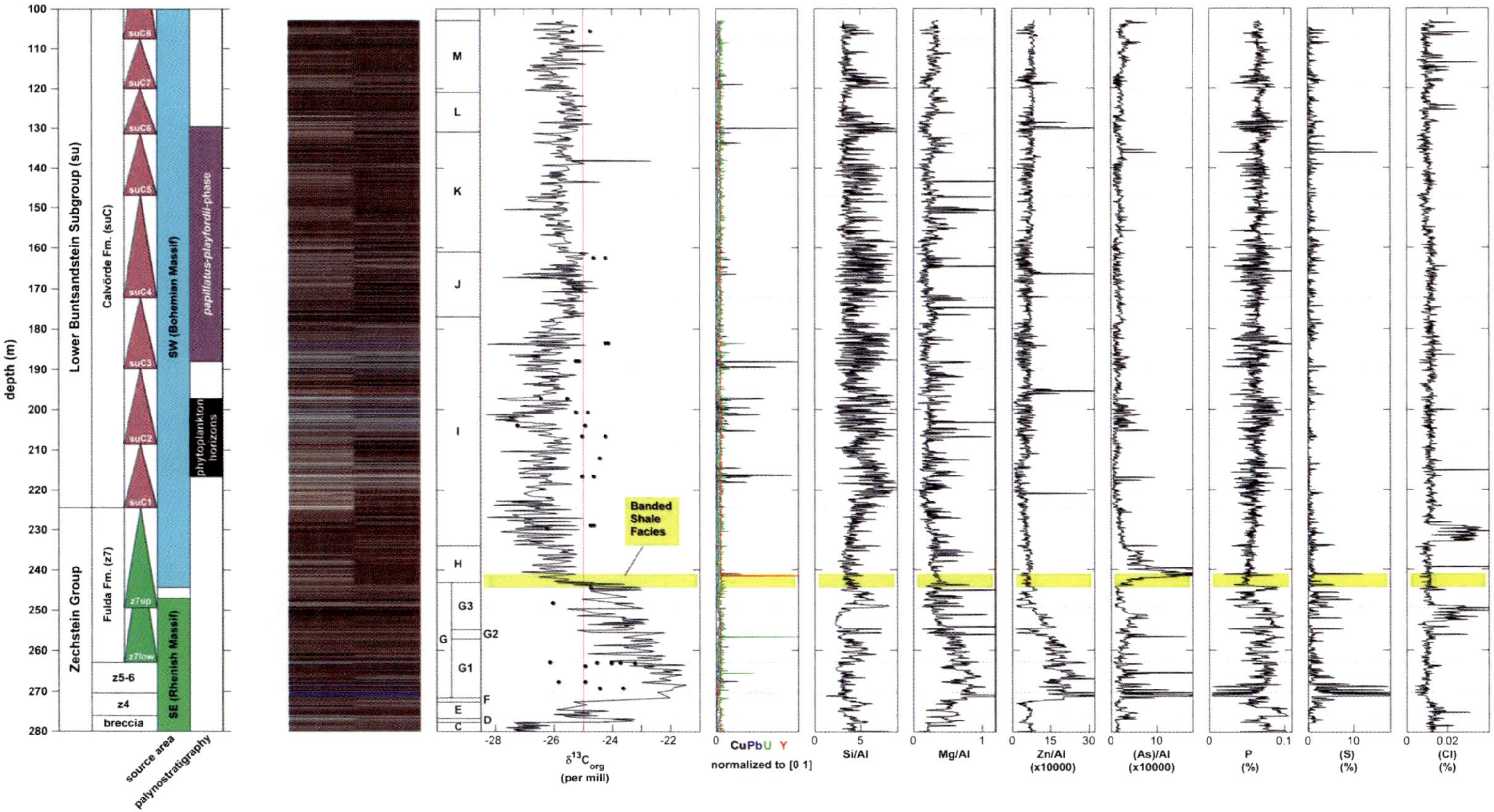

Fig. 7. Profiles of $\delta^{13}C_{org}$ (bulk) and selected inorganic geochemical indicators of the upper Zechstein Group and the Calvörde Formation of the Wulften-1 drill core. In the basal part of z7up, the source area changes from SE to SW. The colour chart shows the Grey Bed Interval in suC2 and dark red-brown colours in the uppermost part of the profile. To facilitate interpretation, several zones (C–M) are identified in the organic carbon isotope profile (cf. Table 3). The yellow band marks the position of the Banded Shale Facies, in which organic carbon isotope values drop by about 3‰. The black circles show organic carbon isotope values of analysed acid-insoluble POM from kerogen analyses. Selected inorganic geochemical analyses are shown to facilitate interpretation.

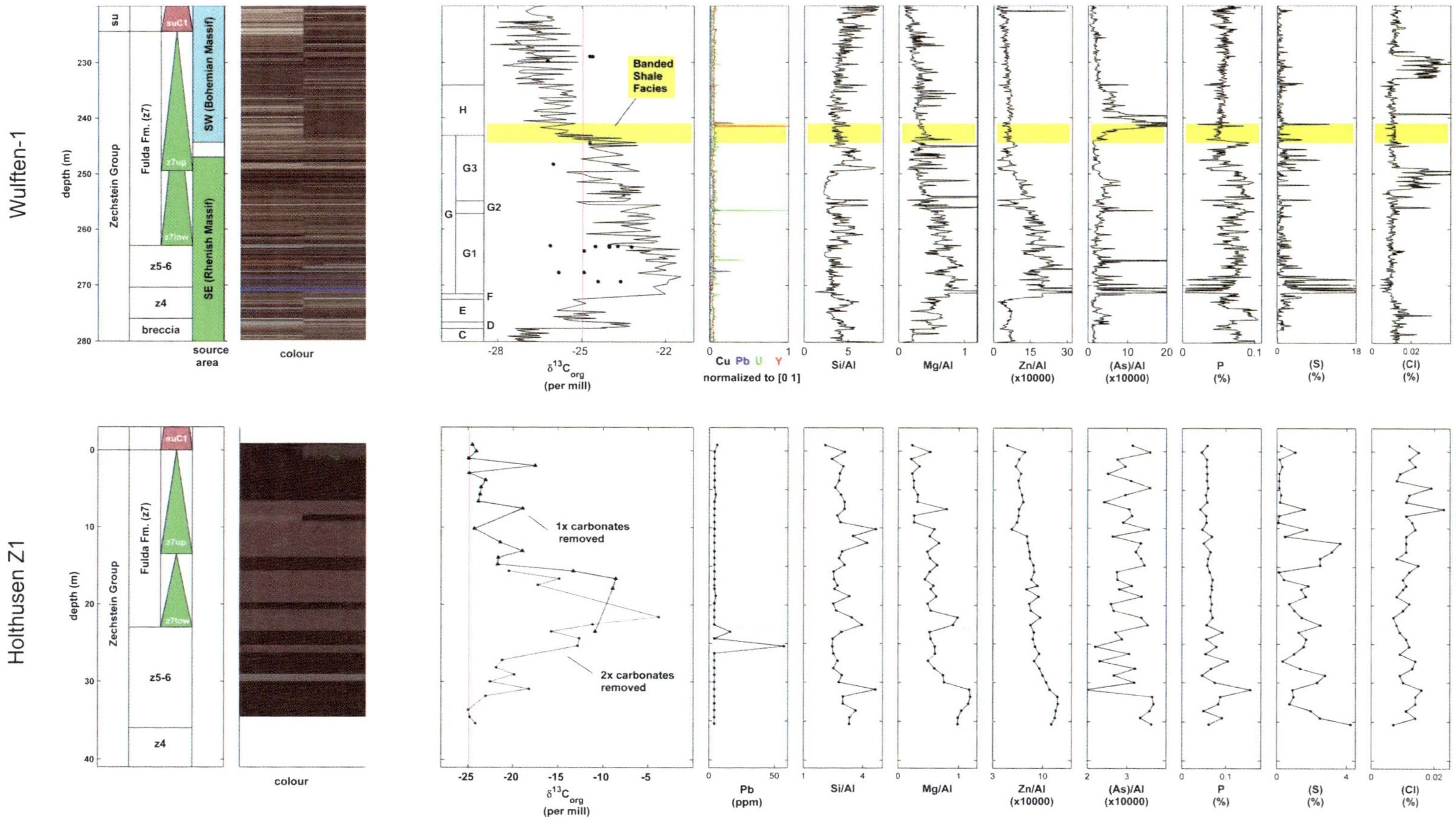

Fig. 8. Profiles of $\delta^{13}C_{org}$ and selected inorganic geochemical indicators of the upper Zechstein of the Wulften-1 and Holthusen Z1 drill cores (cf. Fig. 7 for explanations).

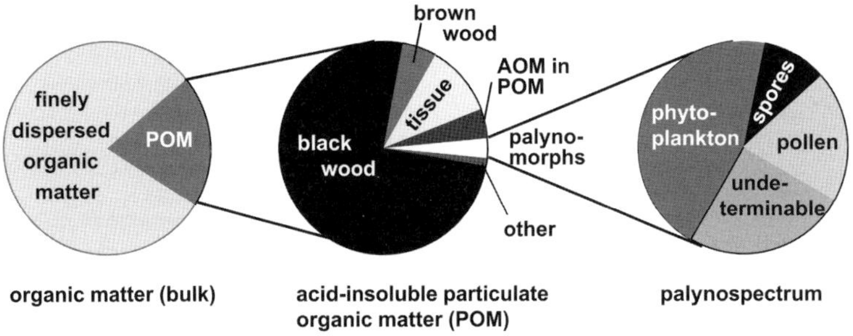

Fig. 9. Composition of organic matter in the Wulften-1 drill core samples (proportions are exemplary only): (left) organic matter composition in bulk samples (organic carbon isotope value measured in the $\delta^{13}C_{org}$ (bulk) profile); (middle) composition of acid-insoluble POM as determined in the kerogen analyses (for some samples organic carbon isotope values were determined, cf. black circles in Figs 7 & 8); (right) composition of the palynospectrum.

lighter with values of −29 to −27‰ (Hayes *et al.* 1999). For PTB sections in Australia, Foster *et al.* (1997, 1999) assumed values of −24‰ for terrestrial and −30‰ for marine organic matter; Hermann *et al.* (2010) reported values of −22 and −32‰, respectively, for sections in Greenland.

Preservation of organic matter after deposition. Although respiration of organic matter is not associated with a significant fractionation (Herczeg & Fairbanks 1987), decomposition and diagenesis may lead to a fractionation of up to 2‰ due to different isotopic values of the products and differences in the decomposability (cf. Jackson *et al.* 1978; Dean *et al.* 1986; Galimov 1990; Hayes 1993; Meyers 1994; Laws *et al.* 1995; Lehmann *et al.* 2002). More important for the analysed samples are, however, differences in the decomposition rates of the different fractions of organic matter. As decomposition rates of organic matter are high during dry phases in playa and sabkha environments (Schütt & Baumhauer 1996; Schütt 2000), and as phytoplankton and bacterial organic matter is more easily decomposed than terrestrial land-plant material, organic carbon isotope values of sediments deposited in phases with frequent drying episodes tend to show values similar to those of land-plant material (i.e. *ca.* −24‰), irrespective of the organic carbon isotope value of the freshly deposited organic matter.

Factors altering carbon isotope values of land plant, phytoplankton and microbial organic matter. The typical organic isotope values for land-plant material or phytoplankton presented above are altered by several factors, most of them are linked to climate.

For land plants, deviations of up to 10‰ have been observed depending on water and osmotic stress, temperature and photosynthesis rate (e.g. O'Leary 1981; Tieszen 1991; Nguyen Tu *et al.* 1999, 2002; Arens *et al.* 2000). Therefore, the herbaceous pioneer vegetation of the well-drained floodplains (lowlands) in the Lower Buntsandstein Subgroup marked by lycopsid spores (van der Zwan & Spaak 1992) probably shows a different $\delta^{13}C_{org}$ value to the Late Permian coniferous flora of the arid hinterland marked by bisaccate pollen, even if the $\delta^{13}C$ value of atmospheric CO_2 had not changed around the PTB.

The $\delta^{13}C_{org}$ value of phytoplankton and photoautotroph microbenthos is influenced *inter alia* by the dissolved CO_2 concentration in water [$CO_{2(aq)}$], which in turn mainly depends on growth rate, salinity and pH, as well as other factors such as nitrogen availability (e.g. Calder & Parker 1973; Fogel & Cifuentes 1993; Hayes 1993; Laws *et al.* 1995; Schidlowski *et al.* 1994; Ramesh & Krishnamurthy 1998). Superheavy organic carbon isotope values of up to −3‰ are reported for microbial mats in hypersaline environments (Schidlowski *et al.* 1984; Javor 1989; Krause-Mirus 1991; Talbot & Johannessen 1992; Grice *et al.* 1998*a*; Valero-Garcés 2000*a*, *b*; Trichet *et al.* 2001). However, interpreting superheavy organic carbon isotope values as caused by microbial mats or hypersaline conditions alone may fall short as not all microbial mats show such superheavy values and as isotopically superheavy phytoplankton is also reported for freshwater lakes (Deuser *et al.* 1968; Talbot & Johannessen 1992; Schidlowski *et al.* 1994). Therefore, very high bioproductivity leading to $CO_{2(aq)}$ limitation and the subsequent use of HCO_3^- as an additional carbon source is considered as the main reason for superheavy organic carbon isotope values in green algae and cyanobacteria, with salinity, temperature and pH altering their growth conditions (Schidlowski *et al.* 1994). The presence of microbial mats is, in general, bound to hypersaline conditions – they are missing, for example, in deeper brackish

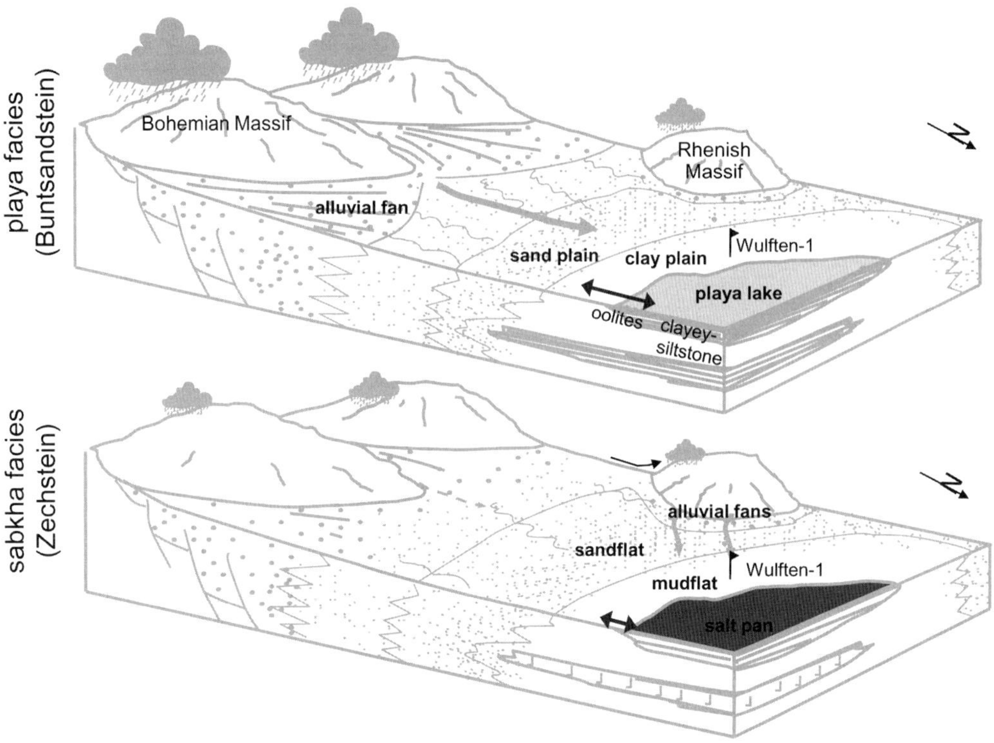

Fig. 10. Idealized facies succession in the upper Zechstein Group sabkha and the Lower Buntsandstein Subgroup playa facies in the North German Basin, including the change in source area (cf. the text for the significance of the organic carbon isotope values).

playa lake stages (Valero-Garcés *et al.* 2000*b*) – and might be inhibited by certain trace elements in the water (Hines *et al.* 1992). Superheavy carbon isotope values may be also indicative of closed systems such as evaporating brines, in which isotopically light organic matter is exported and air–water CO_2 exchange is reduced (Schidlowski *et al.* 1985; Stiller *et al.* 1985).

Implications for interpreting organic carbon isotopes in a sabkha and playa facies. For the sabkha/playa sediments of the upper Zechstein Group and the Lower Buntsandstein Subgroup, $\delta^{13}C_{org}$-(bulk) values depend first of all on the environmental conditions, in particular humidity. These conditions control the proportions of the components phytoplankton with about −32 to −27‰, land-plant-derived organic matter with about −25 to −22‰ and cyanobacterial organic matter of microbial mats with more than about −13‰ in the sediments. Variations of $\delta^{13}C_{org}$ values of these components are thought to play a subordinate role. For a typical sabkha facies succession (from a proximal to a distal position) (Warren 2006) with

alluvial fan ⟶ sandflat ⟶ dry mudflat
⟶ saline (wet) mudflat ⟶ ephemeral lake
(*saltpan*) ⟶ perennial lake (saltpan)

the following organic matter composition and resulting $\delta^{13}C_{org}$ values can be expected (Fig. 10). For the alluvial fan, sandflat and dry but also highly saline wet mudflat facies, it can be assumed that mainly allochthonous terrestrial organic matter with −25 to −23‰ dominates. Supposedly, in the ephemeral lake facies, the addition of isotopically heavier cyanobacterial organic matter of microbial mats was possible and was dependent on growth conditions such as salinity and nutrient supply, as well as preservation conditions. In the perennial lake facies with somewhat greater water depths either isotopically heavier phytoplankton material from green algae blooms (cf. Grice *et al.* 1998*b* for Dead Sea sediments) or, in the case of a low productive environment, allochthonous terrestrial organic matter supplied by aeolian transport might dominate.

A dominance of allochthonous terrestrial organic matter from the highlands with $\delta^{13}C_{org}$ values from

-25 to $-23‰$ can be assumed for the alluvial fan and dry clay plain (Fig. 10). For the sandflat, wet playa facies and fluviatile sediments, similar isotopic values can be assumed, although autochthonous organic matter of lowland vegetation (sandflat and fluviatile sediments) and of halophytic vegetation (wet clay playa) might have been added. Lighter $\delta^{13}C_{org}$ values can be assumed for the ephemeral playa lake facies and ephemeral ponds in the fluviatile facies due to organic matter of phytoplankton origin. Better preservation conditions for algae material and a lower input of sediments, and thus organic matter from land, probably resulted in more negative isotopic values.

Upper Zechstein Group

It is suggested here that superheavy organic carbon isotope values of up to $-3.8‰$, as measured in samples of the Grey Salt Clay Member of the Wulften-1 drill core (Zone A in Table 3) and of z4–z6, z7low of the Holthusen Z1 drill core, can be ascribed to highly productive microbial mats ecosystems, probably in a saline pan environment. For the Wulften-1, this finding is in agreement with other findings such as the typical sabkha facies, the abundance of magnesite, dolomite and gypsum/anhydrite (Brüning *et al.* 1981; Hiete *et al.* 2006; Warren 2000), the absence of phytoplankton (Ecke 1986; Schulz 1998) and a high bituminite to vitrinite ratio in the sample of the Grey Salt Clay Member, indicating a high productivity and low organic matter supply from land. Inertinite present in these samples is indicative of vegetation fires (Bojesen-Koefoed *et al.* 1997; Uhl & Kerp 2003) in the xerophytic conifer flora in the hinterland characterized by bisaccate taeniate pollen (Ecke 1986; Visscher & Brugman 1986; van der Zwan & Spaak 1992). Note that inertinite, probably originating from wild fires, is absent above 290 m in the Wulften-1 drill core (i.e. from z4 to z6 onwards), pointing to the onset of more humid conditions in the hinterland in the upper Zechstein Group. Superheavy $\delta^{13}C_{org}$ values within the z7low of the Holthusen Z1 drill core can likewise be interpreted with an ephemeral lake (saltpan) facies replacing the perennial lake (saltpan) facies in the upper part of the unstructured z4–z6 interval and being replaced itself by a perennial playa lake facies in the lower part of z7up, with at least two recurrences of the saltpan facies still in z7up. The saltpan environment is clearly reflected in the high sulphur content, which is considerably lower from the lower third of z7up onwards (Fig. 8).

The $-24.4‰$ $\delta^{13}C_{org}$ value of the Red Salt Clay Member sample (Zone B) and the other samples of z4 (Zone E) in the range -26 to $-24‰$ of the Wulften-1 drill core, as well as the samples near the base of the interval comprising z5–z6 close to the $-25‰$ value of the Holtusen Z1 drill core, are typical of land-plant material. Obviously, environmental conditions were rather unfavourable for microbial mats and phytoplankton there due, for example, to extremely high salinity, a higher water depth or other factors such as inhibiting trace elements.

The breccia (Zone C) intercalated in z4 in the Wulften-1 drill core contains *inter alia* sediments of the Lower Buntsandstein Subgroup (Brüning *et al.* 1981) with values of around $-27‰$, which corresponds well to (lighter) values measured in the Calvörde Formation.

The almost steady trend towards lighter $\delta^{13}C_{org}$ values in Zone G of the Wulften-1 drill core from about $-22‰$ in the upper z4 to about $-25‰$ in the lower z7up is remarkable. It is paralleled in principle but not in all details (e.g. unlike $\delta^{13}C_{org}$ the Zn/Al ratio does not return to higher values in the lower z7up) by decreasing ratios of Mg/Al and Zn/Al, and higher concentrations of phosphorous (Fig. 8). Note that Mg/Al and Zn/Al ratios decrease similarly in the Holthusen Z1 drill core, whereas the $\delta^{13}C_{org}$ values do not show a similar steady decrease. Hiete *et al.* (2006) discussed possible reasons for the almost parallel decreasing trends of Mg/Al and Zn/Al ratios, as well as the decrease in phosphorous concentration. If an enrichment of Mg, Zn and phosphorous in microbial mats (Lyons & Gaudette 1985) were the reason, the curves of Mg/Al and Zn/Al in the Holthusen Z1 core would be expected to be lower at the base of the z5–z6 interval, where $\delta^{13}C_{org}$ values point to a mainly terrestrial origin of organic matter, and to show a maximum in the z7low similar to $\delta^{13}C_{org}$. Therefore, other controlling factors, probably comprising pH, redox conditions and in particular salinity, interacting in a complex manner are likely (Hiete *et al.* 2006).

More distinct than in the Holthusen Z1 drill core, the lower part of the z7up in the Wulften-1 drill core marks a transition from the heavier $\delta^{13}C_{org}$ values measured in the Zechstein samples to values of mostly $< -25‰$ above. This transition, as well as an extreme enrichment of arsenic (Fig. 8), falls into a characteristic alternation of 0.5–1.5 cm-thick bands of sand-, silt- and claystone that resembles 'Palaeozoic banded shale' (Banded Shale Facies, see earlier) (Kulick 1966).

These well-sorted 'heterolithic' layers with ripple lamination and desiccation cracks were deposited when vast sheet floods were discharging into extremely shallow ephemeral lakes with a large extension to the south of the basin (Hug 2004; Hug & Gaupp 2006). The good lamination indicates the absence of bioturbation and a good drainage in an environment with a low sedimentation rate

(Koslowski 1986). Therefore, some parallels exist with the thinly bedded laminites or heterolithic layers reported from the PTB profiles in the Karoo as basin-wide event beds representing 'dead zones' (e.g. Botha & Smith 2006; Ward *et al.* 2012). However, as Gastaldo *et al.* (2009) and Gastaldo & Neveling (2012) have questioned the isochronism of the laminites in the Karoo, we expect deposition in a similar palaeoenvironment but not an underlying global event. In the NGB, the Banded Shale Facies interfingers at proximal positions even with conglomeratic sandstones and marks a fundamental reconfiguration of the basin with an unconformity on swells and a change in the source area from the Rhenish Massif (SW) to the Bohemian Massif (SE) for the Wulften-1 core, along with the change from the Zechstein Group typical sabkha to the Buntsandstein Group typical playa and fluvial facies (cf. Brüning 1988; Hiete 2004; Hiete *et al.* 2006). This unconformity is not present in the more marginal facies of the Hessian Depression, however, where an unconformity and, in the southern part, also a soil formation is noticeable at the top of the z7up (Hug 2004). Hug (2004) and Hug & Gaupp (2006) pointed out that this reorganization in the z7up is likely to be both the effect of a more humid climate and of tectonic activity affecting sediment sources and accommodation space, thus leading to an increase in the estimated sedimentation rate from 0.02 mm/a in the upper Zechstein Group (z4–z7) to 0.15 mm/a in the Calvörde Formation We interpret this increase as probably related to the Permian–Triassic crisis as enhanced soil erosion resulting in a higher sediment load reflected in a change in river morphology as well as an in increase in sedimentation rates are also

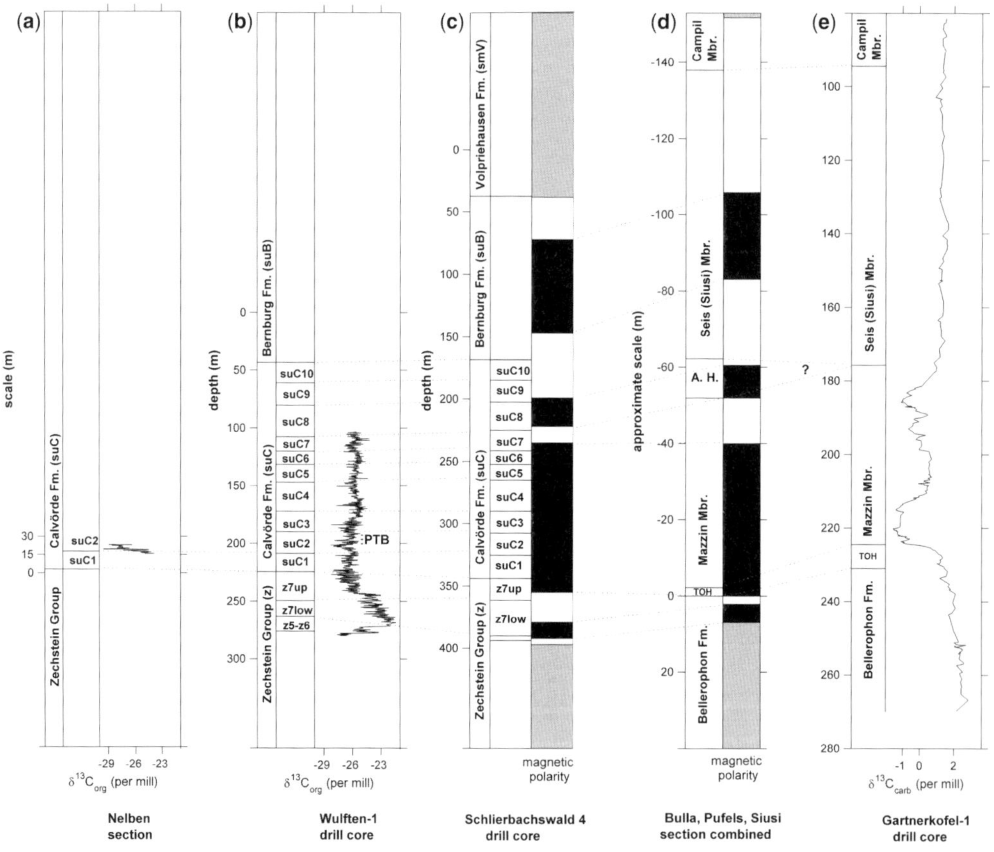

Fig. 11. Correlation of $\delta^{13}C_{org}$ (bulk) profile of the Wulften-1 drill core, with a $\delta^{13}C_{org}$ profile of the Nelben section and a $\delta^{13}C_{carb}$ profile of the Gartnerkofel-1 drill core. (**a**) $\delta^{13}C_{org}$ of the Nelben section in the North German Basin (Bachmann & Kozur 2004). (**b**) $\delta^{13}C_{org}$ (bulk) profile of the Wulften-1 drill core with a mutual position of the PTB. (**c**) Magnetostratigraphy of the Schlierbachswald 4 drill core in the North German Basin (Szurlies *et al.* 2003, 2012). (**d**) Magnetostratigraphy of the combined Bulla, Pufels and Siusi sections in the Southern Alps (Scholger *et al.* 2000; Brandner *et al.* 2009). (**e**) $\delta^{13}C_{carb}$ profile of the Gartnerkofel-1 core in the Southern Alps (Magaritz & Holser 1991).

reported from PTB profiles in Australia, Italy, South Africa and the UK (Olson 1989; Ward *et al.* 2000, 2005; Smith & Ward 2001; Sephton *et al.* 2001, 2003; Michaelsen 2002; Steiner *et al.* 2003; Retallack 2005; Botha & Smith 2006; Algeo & Twitchett 2010).

The Calvörde Formation of the Lower Buntsandstein Subgroup

For the Calvörde Formation, only samples of the Wulften-1 drill core were analysed for $\delta^{13}C_{org}$, making a distinction between local and regional effects difficult as no comparable $\delta^{13}C_{org}$ profiles in the NGB exist. All acid-insoluble POM samples show heavier $\delta^{13}C_{org}$ values than the bulk samples, such that the finely dispersed organic matter probably contains isotopically lighter organic matter of phytoplankton origin, which is present in all analysed palynospectra (Fig. 6).

Within the samples of the Calvörde Formation, $\delta^{13}C_{org}$ shows a positive rank correlation with *inter alia* Fe and hematite, and a negative correlation with Na, Si and albite (correlation with elements are significant at the 0.01 level). The negative correlation with Si can be attributed to grain size as silt- and sandstones were deposited during more humid phases with more favourable growth conditions for phytoplankton and the preservation of organic matter. The negative correlation with albite supposedly results from a lower albite content in suC4–suC7. The positive correlation with Fe and hematite is attributed to the high content of organic matter (mainly isotopically light phytoplankton) deposited during playa lake stages, which resulted in reducing conditions in the sediment such that hematite is dissolved and Fe^{2+} is removed. Accordingly, samples of greyish colour are among the isotopically lightest $\delta^{13}C_{org}$ values in the profile (Fig. 7) and are supposedly depleted in Fe. Dark-brown clayey samples from the upper parts of the base-level cycles representing dry–wet clay plains contain more hematite and are isotopically heavier.

Zone I, ranging from the upper part of the z7up to the uppermost part of suC3, shows the lightest $\delta^{13}C_{org}$ values pointing to an ephemeral to perennial playa lake facies. This interpretation is supported by phytoplankton horizons in suC1 and suC2 (Ecke 1986; Schulz 1994; Heunisch 2001), and maximum flooding at the base of suC3 (Szurlies 2001). The negative organic carbon isotopic excursion for a few metres down to −28.3‰ in the Grey Bed Interval of suC2 falls within these phytoplankton horizons. The greyish colour and abundant pyrite point to synsedimentary anoxic conditions in the sediment as a result of plankton blooms with a high supply of organic matter. Similar negative excursions but thinner can be found in the uppermost z7up and lowermost suC1. Therefore, we attribute this excursion first of all to a facies favourable for the supply and preservation of isotopically light phytoplankton organic matter in the sediment (Hiete *et al.* 2003; Hiete 2004). Similarly, Foster *et al.* (1997, 1999) interpreted organic carbon isotope variations in PTB profiles in Australia solely by a varying ratio of land-plant- and phytoplankton-derived organic matter. Also, Hermann *et al.* (2010) attributed some of the variations in a marine PTB section in Greenland to such changes. A negative $\delta^{13}C_{org}$ excursion in suC2 (there down to −30‰) is also visible in the Nelben section of the same region (Fig. 11) (Bachmann & Kozur 2004; Hansen 2006). These authors interpreted the excursion as a carbon cycle perturbation at the PTB, although the presented interval covers only the uppermost suC1 and the lower suC2 so making it impossible to judge whether the preceding negative $\delta^{13}C_{org}$ excursions in the z7up and suC1 are also present in this section. Correlation of the Wulften-1 drill core with the Gartnerkofel-1 drill core of the Southern Alps shows, nevertheless, that the isotopically light interval in the Wulften-1 spanning approximately from suC1 to suC3 (approximately equivalent with Zone I) with several phytoplankton horizons (Fig. 6) can be correlated with the negative $\delta^{13}C_{carb}$ excursion in the lowermost Mazzin Member of the Werfen Formation (Fig. 11). Therefore, it is likely that the strong influx of freshwater resulting in the phytoplankton horizons and the lighter $\delta^{13}C_{org}$ values in the lower quarter of the Calvörde Formation in the North German Basin is linked to the events around the PTB, which have resulted in the negative excursion in $\delta^{13}C_{carb}$ in the Southern Alps. In fact, Brandner (1987) and Kozur (1989) reported a change from slightly hypersaline conditions in the Bellerophon Formation to partially brachyhaline conditions in some profiles of the Southern Alps. A change to a braided river system resulting from a sudden increase in humidity is also known from the lowermost Triassic Lower Alpine Buntsandstein Formation (Krainer 1993).

Within Zone I, the base-level cycles suC1 and suC3 show the expected succession of lighter $\delta^{13}C_{org}$ values at the base and heavier ones at the top, with less favourable conditions for phytoplankton productivity and its preservation as a more arid climate prevails. This succession is also visible for suC7 but hardly recognizable for suC4–suC6. Zone J covering the upper suC3 and the lower suC4 is clearly set apart from the sediments below and above as a block of slightly heavier $\delta^{13}C_{org}$ values with only a few lighter values at the base part of suC4. The $\delta^{13}C_{org}$ values suggest the splitting up of the suC4 near the upper boundary of Zone J; a splitting-up of the suC4 was also supposed by Korte & Kozur (2005) and by Hiete *et al.* (2006) based on findings from inorganic geochemical data. But even then, only the lower cycle shows the expected $\delta^{13}C_{org}$ succession of a drying-up base-level cycle, whereas the upper cycle and the remaining part of Zone K and Zone L do not. This is likely to be related to a higher aridity in suC4–suC7 marked by desiccation cracks and, possibly, primary anhydrite/gypsum enrichments in their upper parts (cf. Brüning *et al.* 1981), which resulted in a lower phytoplankton bioproductivity and a higher decomposition rate of organic matter. Consequently, the ideal succession of a drying-upwards base-level cycle is no longer recorded in the $\delta^{13}C_{org}$ profile.

Conclusions

In the North German Basin, the Permian–Triassic boundary interval is characterized by a transition from a sabkha facies in the uppermost Permian Zechstein Group to a playa and fluviatile facies in the lowermost Triassic Lower Buntsandstein Subgroup. The Lower Triassic Buntsandstein Subgroup shows cycles of different order, with base-level cycles from 10 to 25 m as the most prominent ones. Owing to their large extent in the basin, these cycles are used for further lithostratigraphical subdivision and interpreted as being climatically driven, possibly short eccentricity cycles. In the Lower Triassic Buntsandstein Subgroup, 24 of these cycles were identified of which 10 were in the Calvörde Formation (suC1–suC10). Based on palynomorphs and conchostracans, as well as magnetostratigraphy, the PTB is placed in the upper part of suC2.

A high-resolution carbon isotope profile of bulk organic matter ($\delta^{13}C_{org}$-bulk) of the Wulften-1 drill core taking an intermediate position in the basin and additional samples from the Holthusen Z1 drill core from the basin centre served to analyse the complex interplay between the facies transition, the climatically driven cycles of different order and the events around the PTB.

Facies and, in particular, humidity affect both the supply and the preservation of the different organic materials, and thus controls the $\delta^{13}C_{org}$ values. Therefore, for a typical playa succession, the lightest $\delta^{13}C_{org}$ values can be expected for the perennial playa lake facies where the supply of isotopically light phytoplankton organic matter is high and the preservation conditions are favourable. In ephemeral playa lakes, frequent drying results in a reduced supply and an increase in the decomposition of phytoplankton material compared to organic matter of terrestrial origin. Therefore, intermediate $\delta^{13}C_{org}$ values are likely for the alluvial fan, sandflat and dry but also the highly saline wet mudflat facies in which allochthonous and terrestrial organic matter from the highlands and partly autochthonous from the lowlands dominate. Heavy and super-heavy $\delta^{13}C_{org}$ values may arise in the perennial and ephemeral playa lake as well as in the (wet) saline mudflat facies if dissolved CO_2 becomes limited due to a high primary productivity or if CO_2 becomes enriched in ^{13}C in a closed saline system. The former may be the result of a high nutrient supply, unfavourable conditions for predators due to high salinity or a reduced CO_2 diffusion through biofilms in microbial mats.

The $\delta^{13}C_{org}$-profile of the Wulften-1 drill core clearly reflects the facies transition from a sabkha to a playa and fluvial facies. The main shift in $\delta^{13}C_{org}$ values falls within the Banded Shale Facies near the base of the Upper Fulda Formation (z7up) and is associated with a major reorganization of the sedimentary system. This shift takes place approximately 20 m below the Zechstein–Buntsandstein boundary as defined in Germany and about 50 m below the supposed position of the PTB. The $\delta^{13}C_{org}$ profile of the more basinal Holthusen Z1 does not show a similar shift. Furthermore, the complex relationship between facies type and $\delta^{13}C_{org}$ make it difficult to identify the lithostratigraphically defined base-level cycles with their typical facies successions in the $\delta^{13}C_{org}$ profile of the Wulften-1 drill core. Only suC1–suC3 clearly show the expected $\delta^{13}C_{org}$ variations.

Heavy $\delta^{13}C_{org}$ values in z4 of the Lower Fulda Formation (z7low) of the Wulften-1 drill core and super-heavy values in z5–z7 of the Holthusen Z1drill core are interpreted as indicative of hypersaline conditions of a sabkha (very dry conditions). Light $\delta^{13}C_{org}$ values down to −28‰ in the upper part of the Upper Fulda Formation and the lower third of the Calvörde Formation, and in particular suC2 of the Wulften-1 drill core, are interpreted as a perennial playa lake facies pointing to more humid conditions. This isotopically light interval attributed to isotopically light phytoplankton organic matter and a more humid climate can be correlated with the negative $\delta^{13}C_{carb}$ excursion in the lowermost Mazzin Member of the Werfen Formation of the Gartnerkofel-1 drill core of the Southern Alps where there are also signs for an increased freshwater influx.

In summary, the high-resolution $\delta^{13}C_{org}$ profile of bulk organic matter recorded severe changes in the facies and climate, as well as in the deposition system from the upper Zechstein to the Lower Buntsandstein in the North German Basin. The measured $\delta^{13}C_{org}$ variations and, in particular, a negative excursion slightly below the PTB placed in the second base-level cycle suC2 are explained by facies and climate changes. However, the events around the PTB such as volcanic degassing with acid rain and the dying-off of vegetation are thought to have had a strong influence on these facies and on climate changes.

This paper is dedicated to the memory of our colleagues D. Panten and T. Wippermann. The research of M. Hiete was supported by a collaboration of the Bundesanstalt für Geowissenschaften und Rohstoffe (Federal Institute for Geosciences and Natural Resources, BGR) and the Technical University of Braunschweig. We are particularly grateful to J. Koch and D. Panten for their help with the analyses. The authors like to thank O. Richter for encouraging this work. We also thank BGR and NLfB for the permission to publish the results.

References

AFONIN, S. A., BARINOVA, S. S. & KRASSILOV, V. A. 2001. A bloom of *Tympanicysta* Balme (green algae of zygnematalean affinities) at the Permian–Triassic boundary. *Geodiversitas*, **23**, 481–487.

ALGEO, T. J. & TWITCHETT, R. J. 2010. Anomalous early Triassic sediment fluxes due to elevated weathering rates and their biological consequences. *Geology*, **38**, 1023–1026.

ARENS, N. C., JAHREN, A. H. & AMUNDSON, R. 2000. Can C3 plants faithfully record the carbon isotopic composition of atmospheric carbon dioxide? *Paleobiology*, **26**, 137–164.

BACHMANN, G. H. & KOZUR, H. W. 2004. The Germanic Triassic: correlations with the international chronostratigraphic scale, numerical ages and Milankovitch cyclicity. *Hallesches Jahrbuch für Geowissenschaften*, **B26**, 17–62.

BACHMANN, G. H., KOZUR, H. W. & SZURLIES, M. 2003. Cyclostratigraphy, magnetostratigraphy and microspherules of the continental: Permian–Triassic boundary interval, Germany. *In*: WONG, T. E. (ed.) *XVth International Congress on Carboniferous and Permian Stratigraphy, Utrecht, 10–16 August*. Royal Netherlands Academy of Arts and Sciences, Amsterdam, 36–38.

BACHMANN, G.-H., GELUK, M. C. *ET AL.* 2010. Triassic. *In*: DOORNENBAL, J. C. & STEVENSON, A. G. (eds) *Petroleum Geological Atlas of the Southern Permian Basin Area*. EAGE Publications, Houten, 149–173.

BACKHAUS, E. 1996. Die biostratigraphische Bewertung der Faunen im Buntsandstein des germanischen Beckens. *Jahresbericht und Mitteilungen des Oberrheinischen Geologischen Vereins*, **78**, 257–279.

BALDSCHUHN, R., BINOT, F., FLEIG, S. & KOCKEL, F. 2001. *Geotektonischer Atlas von Nordwest-Deutschland und dem deutschen Nordsee-Sektor: Strukturen, Strukturentwicklung, Paläogeographie. Deutsch-Englisch*. Schweizerbart, Stuttgart.

BERNER, R. A. 2002. Examination of hypotheses for the Permo-Triassic boundary extinction by carbon cycle modeling. *Proceedings of the National Academy of Sciences of the United States of America*, **99**, 4172–4177.

BERTELSEN, F. 1980. *Lithostratigraphy and depositional history of the Danish Triassic*. Reitzel, Copenhagen.

BEST, G. 1988. Die Grenze Zechstein/Buntsandstein in Nordwestdeutschland und in der südlichen Nordsee nach Bohrlochmessungen (Gamma-Ray- und Sonic-Log). *Geologisches Jahrbuch Hessen*, **116**, 19–22.

BEST, G. 1989. Die Grenze Zechstein–Buntsandstein in Nordwest-Deuschland nach Bohrlochmessungen. *Zeitschrift der Deutschen Geologischen Gesellschaft*, **140**, 73–85.

BEUTLER, G. & SCHÜLER, F. 1987. Probleme und Ergebnisse der lithostratigraphischen Korrelation der Trias am Nordrand der Mitteleuropäischen Senke. *Zeitschrift für Geologische Wissenschaften*, **15**, 421–436.

BEUTLER, G. & SZULC, J. 1999. Die paläogeographische Entwicklung des Germanischen Beckens in der Trias und die Verbindung zur Tethys. *In*: HAUSCHKE, N. & WILDE, V. (eds) *Trias Eine ganze andere Welt Mitteleuropa im frühen Erdmittelalter*. Dr. Friedrich Pfeil, Munich, 71–80.

BOIGK, H. 1959. Zur Gliederung und Fazies des Buntsandsteins zwischen Harz und Emsland. *Geologisches Jahrbuch*, **76**, 597–636.

BOIGK, H. 1961. Ergebnisse und Probleme stratigraphisch-paläogeographischer Untersuchungen im Buntsandstein Nordwestdeutschlands. *Geologisches Jahrbuch*, **78**, 123–134.

BOJESEN–KOEFOED, J. A., PETERSEN, H. I., SURLYK, F. & VOSGERAU, H. 1997. Organic petrography and geochemistry of inertinite-rich mudstones, Jakobsstigen Formation, Upper Jurassic, northeast Greenland: indications of forest fires and variations in relative sea-level. *International Journal of Coal Geology*, **34**, 345–370.

BOTHA, J. & SMITH, R. M. H. 2006. Rapid vertebrate recuperation in the Karoo Basin of South Africa following the End-Permian extinction. *Journal of African Earth Sciences*, **45**, 502–514.

BOWRING, S. A., ERWIN, D. H. & ISOZAKI, Y. 1999. The tempo of mass extinction and recovery: the end-Permian example. *Proceedings of the National Academy of Sciences of the United States of America*, **96**, 8827–8828.

BRANDNER, R. 1987. Ozeanographische und klimatische Veränderungen an der Perm-Trias-Grenze in der westlichen Tethys. *In*: KOCH, R., MÜLLER, G. & SCHMITZ, W. (eds) *Tagungsband 2. Treffen deutschsprachiger Sedimentologen 24–27 Mai. Heidelberger Geowissenschaftliche Abhandlungen*, **8**, 44–45.

BRANDNER, R., HORACEK, M., KEIM, L. & SCHOLGER, R. 2009. The Pufels/Bulla road section: deciphering environmental changes across the Permian–Triassic Boundary to the Olenekian by integrated litho-, magneto- and isotope stratigraphy. A field trip guide. *Jahreszeitschrift zur Alpengeologie*, **6**, 116–132.

BRAUNS, C. M., PÄTZOLD, T. & HAACK, U. 2003. A Re–Os study bearing the age of Kupferschiefer at Sangerhausen (Germany). *In*: WONG, T. E. (ed.) *XVth International Congress on Carboniferous and Permian Stratigraphy, Utrecht, 10–16 August*. Royal Netherlands Academy of Arts and Sciences, Amsterdam, 66.

BRÜNING, U. 1984. *Zur Stratigraphie und Litofazies des Unteren Buntsandsteins in Südniedersachsen und Nordhessen*. PhD thesis, University of Würzburg.

BRÜNING, U. 1986. Stratigraphie und Lithofazies des Unteren Buntsandsteins in Südniedersachsen und Nordhessen. *Geologisches Jahrbuch*, **A90**, 3–125.

BRÜNING, U. 1988. Die Zechstein/Buntsandstein-Grenze in Niedersachsen und Hessen – Schwermineraluntersuchungen als Beitrag zur Geologie des Bröckelschiefers. *Geologisches Jahrbuch Hessen*, **116**, 23–44.

BRÜNING, U., RICHTER-BERNBURG, G. & SIMON, P. 1981. *Bundesbohrprogramm, Teilprojekt Kupferschiefer, Bohrakte*. Niedersächsisches Landesamt für Bodenforschung (NLfB), Hannover.

CALDER, J. A. & PARKER, P. L. 1973. Geochemical implications of induced changes in $\delta^{13}C$ fractionation by blue-green algae. *Geochimica et Cosmochimica Acta*, **37**, 133–140.

CLEMMENSEN, L. B., OXNEVAD, I. E. I. & DE BOER, P. L. 1994. Climatic controls on ancient desert sedimentation, some late Paleozoic and Mesozoic examples form NW Europe and the western interior of the USA. *In*: DE BOER, P. L. & SMITH, D. G. (eds) *Orbital Forcing and Cyclic Sequences*. Blackwell, Oxford, 439–457.

DADLEZ, R., MAREK, S. & POKORSKI, J. 1998. *Atlas paleogeograficzny epikontynentalnego permu i mezozoiku w Polsce, 1:2 500 000. [Paleogeographical Atlas of the Epicontinental Permian and Mesozoic in Poland, 1:2 500 000.]* Państwowy Instytut Geologiczny, Warszawa.

DAY, G. A., COOPER, B. A., ANDERSON, C., BURGERS, W. F. J., RONNEVIK, H. C. & SCHÖNEICH, H. 1981. Regional seismic structure maps of the North Sea. *In*: ILLING, L. V. & HOBSON, G. D. (eds) *Petroleum Geology of the Continental Shelf of Northwest Europe*. Heyden, London, 76–84.

DEAN, W. E., ARTHUR, M. A. & CLAYPOOL, G. E. 1986. Depletion of $\delta^{13}C$ in Cretaceous marine organic matter, source, diagenetic, or environmental change? *Marine Geology*, **70**, 119–159.

DEUSER, W. G., DEGENS, E. T. & GUILLARD, R. R. L. 1968. Carbon isotope relationships between plankton and sea water. *Geochimica et Cosmochimica Acta*, **32**, 657–660.

DEUTSCHE STRATIGRAPHISCHE KOMMISSION 2002. Stratigraphische Tabelle von Deutschland 2002, Geo-Forschungszentrum (Potsdam) and Senckenberg Forschungsinstitut (Frankfurt a.M.). World Wide Web address: http://www.stratigraphie.de/std/index.html (accessed 7 June 2012).

DIN. 1986. *Investigations of the Raw Material in Hard-Coal-Mining – Microscopical Examinations of Hard Coal, Coke and Briquettes – Part 5: Reflectance Measurements on Vitrinites*. Deutsches Institut Fur Normung, **DIN 22020-5:1986-09**.

ECKE, H.-H. 1986. *Palynologie des Zechsteins und Unteren Buntsandsteins im Germanischen Becken*. PhD thesis, University of Göttingen.

ERWIN, D. H. 1993. *The Great Paleozoic Crisis, Life and Death in the Permian*. Columbia University Press, New York.

ERWIN, D. H. 1995. The end-Permian mass extinction. *In*: SCHOLLE, P. A., PERYT, T. M. & ULMER-SCHOLLE, D. S. (eds) *The Permian of Northern Pangea Volume 1, Paleogeography, Paleoclimates, Stratigraphy*. Springer, Berlin, 20–34.

ERWIN, D. H. 2006. *Extinction: How Life on Earth Nearly Ended 250 Million Years Ago*. Princeton University Press, Princeton, NJ.

FOGEL, M. L. & CIFUENTES, L. A. 1993. Isotope fractionation during primary production. *In*: ENGEL, M. & MACKO, S. A. (eds) *Organic Geochemistry: Principles and Applications*. Plenum Press, New York, 73–98.

FOSTER, C. B., LOGAN, G. A., SUMMONS, R. E., GORTER, J. D. & EDWARDS, D. S. 1997. Carbon isotopes, kerogen types and the Permian–Triassic boundary in Australia: implications for exploration. *APPEA Journal*, **37**, 472–488.

FOSTER, C. B., LOGAN, G. A. & SUMMONS, R. E. 1999. The Permian–Triassic boundary in Australia – organic carbon isotopic anomalies relate to organofacies, not a 'biogeochemical' event. *In*: *Ninth Annual V. M. Goldschmidt Conference, 22–27 August, 1999, Cambridge, Massachusetts*. Lunar and Planetary Institute, Houston, TX, abstract no. 7301.

FRY, B., BRAND, W., MERSCH, F. J., THOLKE, K. & GARRITT, R. 1992. Automated analysis system for coupled $\delta^{13}C$ and $\delta^{15}N$ measurements. *Analytical Chemistry*, **64**, 288–291.

FÜCHTBAUER, H. & GOLDSCHMIDT, H. 1959. Die Tonminerale der Zechsteinformation. *Beiträge zur Mineralogie und Petrographie*, **6**, 320–345.

GALIMOV, E. M. 1990. Biogeochemistry of stable carbon isotopes. *Mitteilungen aus dem Geologisch-Paläontologischen Institut der Universität Hamburg*, **69**, 9–20.

GASTALDO, R. A. & NEVELING, J. 2012. The terrestrial Permian–Triassic boundary event is a nonevent: REPLY. *Geology*, **40**, e257, doi: 10.1130/G32975Y.1

GASTALDO, R. A., NEVELING, J., CLARK, C. K. & NEWBURY, S. S. 2009. The terrestrial Permian–Triassic boundary event bed is a nonevent. *Geology*, **37**, 199–202.

GELUK, M. C. 2005. *Stratigraphy and tectonics of Permo-Triassic basins in the Netherlands and surrounding areas*. PhD thesis, Utrecht University.

GELUK, M. C. & RÖHLING, H.-G. 1997. High-resolution sequence stratigraphy of the Lower Triassic 'Buntsandstein' in The Netherlands and northwestern Germany. *Geologie en Mijnbouw*, **76**, 227–246.

GELUK, M. C. & RÖHLING, H.-G. 1999. High resolution sequence stratigraphy of the Lower Triassic Buntsandstein: a new tool for basin analysis. *In*: BACHMANN, G. H. & LERCHE, I. (eds) *The Epicontinental Triassic. Zentralblatt für Geologie und Paläontologie, Teil I*, (1998), **7–8**, 727–745.

GRICE, K., SCHOUTEN, S., NISSENBAUM, A., CHARRACH, J. & DAMSTE, J. S. S. 1998*a*. A remarkable paradox: sulfurised freshwater algal Botryococcus braunii lipids in an ancient hypersaline euxinic ecosystem. *Organic Geochemistry*, **28**, 195–216.

GRICE, K., SCHOUTEN, S., NISSENBAUM, A., CHARRACH, J. & DAMSTE, J. S. S. 1998*b*. Isotopically heavy carbon in the C21 to C25 regular isoprenoids in halite-rich deposits from the Sdom Formation, Dead Sea Basin, Israel. *Organic Geochemistry*, **28**, 349–359, doi: 10.1016/S0146-6380(98)00006-0

HANSEN, H. J. 2006. Stable isotopes of carbon from basaltic rocks and their possible relation to atmospheric isotope excursions. *Lithos*, **92**, 105–116.

HARTMANN, M. 1963. Einige geochemische Untersuchungen an Sandsteinen aus Perm und Trias. *Geochimica et Cosmochimica Acta*, **27**, 459–499.

HAYES, J. M. 1993. Factors controlling $\delta^{13}C$ contents of sedimentary organic compounds, principles and evidence. *Marine Geology*, **113**, 111–125.

HAYES, J. M., STRAUSS, H. & KAUFMAN, A. J. 1999. The abundance of $\delta^{13}C$ in marine organic matter and isotopic fractionation in the global biogeochemical cycle of carbon during the past 800 Ma. *Chemical Geology*, **161**, 103–125.

HERCZEG, A. L. & FAIRBANKS, R. G. 1987. Anomalous carbon isotope fractionation between atmospheric CO_2 and dissolved inorganic carbon induced by intense photosynthesis. *Geochimica et Cosmochimica Acta*, **51**, 895–899.

HERMANN, E., HOCHULI, P. A., BUCHER, H., VIGRAN, J. O., WEISSERT, H. & BERNASCONI, S. M. 2010. A close-up view of the Permian–Triassic boundary based on expanded organic carbon isotope records from Norway (Trøndelag and Finnmark Platform). *Global and Planetary Change*, **74**, 156–167.

HEUNISCH, C. 1999. Die Bedeutung der Palynologie für Biostratigraphie und Fazies in der Germanischen Trias. *In*: HAUSCHKE, N. & WILDE, V. (eds) *Trias Eine ganze andere Welt Mitteleuropa im frühen Erdmittelalter*. Dr. Friedrich Pfeil, Munich, 207–262.

HEUNISCH, C. 2001. *Mikropaläontologischer Untersuchungsbericht: Palynologische Untersuchung des Unteren Buntsandsteins und höheren Zechsteins der Bohrung Wulften-1*. Unpublished report, 881_2192. Niedersächsisches Landesamt für Bodenforschung (NLfB), Hannover.

HEYDARI, E., WADE, W. J. & HASSANZADEH, J. 2001. Diagenetic origin of carbon and oxygen isotope compositions of Permian–Triassic boundary strata. *Sedimentary Geology*, **143**, 191–197.

HIETE, M. 2004. *Umweltveränderungen in der Permo-Trias, Geochemische Charakterisierung – Zeitreihenanalyse – Modellierung*. PhD thesis, Technische Universität Braunschweig.

HIETE, M., BERNER, U., HEUNISCH, C. & RÖHLING, H. G. 2003. Organic carbon isotopic excursions near the Permian–Triassic boundary in NW-Germany: global v. regional effects. *In*: WONG, T. E. (ed.) *XVth International Congress on Carboniferous and Permian Stratigraphy, Utrecht, 10–16 August*. Royal Netherlands Academy of Arts and Sciences, Amsterdam, 224–227.

HIETE, M., BERNER, U., HEUNISCH, C. & RÖHLING, H.-G. 2006. A high-resolution inorganic geochemical profile across the Zechstein–Buntsandstein boundary in the North German Basin. *Zeitschrift der Deutschen Gesellschaft für Geowissenschaften*, **157**, 77–105, doi: 10.1127/1860-1804/2006/0157-0077

HINES, M. E., LYONES, W. B., LENT, R. M. & LONG, D. T. 1992. Sedimentary biogeochemistry of an acidic, saline groundwater discharge zone in Lake Tyrrell, Victoria, Australia. *Chemical Geology*, **96**, 53–65.

HOFMANN, A., TOURANI, A. & GAUPP, R. 2000. Cyclicity of Triassic to Lower Jurassic continental red beds of the Argana valley, Morocco: implications for palaeoclimate and basin evolution. *Palaeogeography, Palaeoclimatology, Palaeoecology*, **161**, 229–266.

HOLSER, W. T. 1997. Geochemical events documented in inorganic carbon isotopes. *Palaeogeography, Palaeoclimatology, Palaeoecology*, **132**, 173–182.

HOPPE, W. 1968. Zyklen und Rhythmen, Zeit- und Faziesgrenzen im Buntsandstein. *Geologie*, **17**, 1130–1137.

HOUNSLOW, M. W. & MUTTONI, G. 2010. The geomagnetic polarity timescale for the Triassic: linkage to stage boundary definitions. *In*: LUCAS, S. G. (ed.) *The Triassic Timescale*. Geological Society, London, Special Publications, **334**, 61–102.

HUG, N. 2004. Sedimentgenese und Paläogeographie des höheren Zechstein bis zur Basis des Buntsandstein in der Hessischen Senke. *Geologische Abhandlungen Hessen*, **113**, 1–238.

HUG, N. & GAUPP, R. 2006. Palaeogeographic reconstruction in red beds by means of genetically related correlation: results from the upper Zechstein (Late Permian). *Zeitschrift der Deutschen Gesellschaft für Geowissenschaften*, **157**, 107–120, doi: 10.1127/1860-1804/2006/0157-0107

IAEA 2007. *NBS 22, IAEA-CH-3, IAEA-CH-6, IAEA-CH-7, USGS24 Reference Sheet*. International Atomic Energy Agency, Vienna. World Wide Web Address: http://nucleus.iaea.org/rpst/Documents/RS_NBS22_USGS24_IAEA-CH-3-6-7.pdf (accessed 3 August 2007).

IGCP 1986. Lithologic–palaeogeographic map Middle Bunter (Triassic), inset map 1:10 000 000 (first version). *In*: *IGCP Project No. 86, Southwest Border of the East European Platform*. IGCP, Paris.

JACKSON, T. A., FRITZ, P. & DRIMME, R. 1978. Stable carbon isotope ratios and chemical properties of kerogen and extractable organic matter in Pre-Phanerozoic and Phanerozoic sediments, their interrelations and possible paleobiological significance. *Chemical Geology*, **21**, 335–350.

JAVOR, B. 1989. *Hypersaline Environments: Microbiology and Biogeochemistry*. Springer, Berlin.

KÄDING, K.-C. 2000. Die Aller-, Ohre-, Friesland- und Fulda-Folge (vormals Bröckelschiefer-Folge) – Stratigraphie und Verbreitung des z4 bis z7 im Zechstein–Becken. *Kali und Steinsalz*, **13**, 86–96.

KÄSTNER, H., KÜHN, W. & LANGBEIN, R. 1966. Stratigraphische, petrographische und geochemische Untersuchungen im Unteren und Mittleren Buntsandstein des thüringischen Werra-Kaligebietes. *Geologie*, **15**, 461–483.

KORTE, C. & KOZUR, H. W. 2005. Carbon isotope trends in continental lake deposits of uppermost Permian to Lower Olenekian: germanic Lower Buntsandstein (Calvörde and Bernburg Formations). *Hallesches Jahrbuch für Geowissenschaften*, **B19**, 87–94.

KORTE, C. & KOZUR, H. W. 2010. Carbon-isotope stratigraphy across the Permian–Triassic boundary: a review. *Journal of Asian Earth Sciences*, **39**, 215–235.

KORTE, C., KOZUR, H. W. & BACHMANN, G. H. 2007. Carbon isotope values of Triassic lacustrine and hypersaline playa-lake carbonates: lower Buntsandstein and middle Keuper (Germany). *Hallesches Jahrbuch für Geowissenschaften*, **29**, 1–10.

KOSLOWSKI, W. 1986. *Zur sedimentären Entwicklung des Buntsandsteins: Ein Vergleich zwischen dem Ostrand des Rheinischen Schiefergebirges und dem Raum um*

Bad Hersfeld/Hessen. PhD thesis, University of Marburg.

Kozur, H. 1989. The Permian–Triassic boundary in marine and continental sediments. *Zentralblatt für Geologie und Paläontologie, Teil I*, (1998), **11–12**, 1245–1277.

Kozur, H. W. 1993*a*. Annoted correlation of the Germanic Buntsandstein and Keuper. *In*: Lucas, S. G. & Morales, M. (eds) *The Nonmarine Triassic. Museum of Natural History and Science Bulletin*, **3**, 243–248.

Kozur, H. W. 1993*b*. Range charts of conchostracans in the Germanic Buntsandstein. *In*: Lucas, S. G. & Morales, M. (eds) *The Nonmarine Triassic. Museum of Natural History and Science Bulletin*, **3**, 249–253.

Kozur, H. W. 1999*a*. Aspekte der Wende Perm/Trias. *In*: Hauschke, N. & Wilde, V. (eds) *Trias Eine ganze andere Welt Mitteleuropa im frühen Erdmittelalter*. Dr. Friedrich Pfeil, Munich, 23–36.

Kozur, H. W. 1999*b*. The correlation of the Germanic Buntsandstein and Muschelkalk with the Tethyan scale. *Zentralblatt für Geologie und Paläontologie, Teil I*, (1998), **7–8**, 701–725.

Kozur, H. W. & Weems, R. E. 2010. The biostratigraphic importance of conchostracans in the continental Triassic of the northern hemisphere. *In*: Lucas, S. G. (ed.) *The Triassic Timescale*. Geological Society, London, Special Publications, **334**, 315–417, doi: 10.1144/SP334.13

Krainer, K. 1993. The Alpine Buntsandstein formation of the Drau Range (Eastern Alps, Austria): transition form fluvial to shallow marine facies. *In*: Lucas, S. G. & Morales, M. (eds) *The Nonmarine Triassic. Museum of Natural History & Science Bulletin*, **3**, 267–275.

Krause-Mirus, A. 1991. *Initial establishment and development of a coastal hypersaline ecosystem in association with a microbial mat community: microbial composition, water chemistry and sedimentation*. PhD thesis, University of Mainz.

Kühn, W. & Radzinski, K.-H. 1972. Bemerkungen zu den geochemischen Verhältnissen und den Bildungsbedingungen des Unteren und Mittleren Buntsandsteins im südöstlichen Harzvorland. *Geologie*, **21**, 786–801.

Kulick, J. 1966. Der Untere Buntsandstein am Südrande des Kellerwaldes (Nordhessen). *Notizblatt des Hessischen Landesamtes für Bodenforschung*, **94**, 182–200.

Kunert, R. 1964. Kleinzyklen in der Oberen Schiefertonzone su5 des nordöstlichen Harzvorlandes. *Berichte der Geologischen Gesellschaft in der DDR, Sonderheft*, **2**, 35–40.

Kürschner, W. M. & Herngreen, G. F. W. 2010. Triassic palynology of central and northwestern Europe: a review of palynofloral diversity patterns and biostratigraphic subdivisions. *In*: Lucas, S. G. (ed.) *The Triassic Timescale*. Geological Society, London, Special Publications, **334**, 263–283, doi: 10.1144/SP334.11.

Laws, E. A., Popp, N. N., Bidigare, R. R., Kennicutt, M. C. & Macko, S. A. 1995. Dependence of phytoplankton carbon isotopic composition on growth rate and $[CO_2]_{aq}$: Theoretical considerations and experimental results. *Geochimica et Cosmochimica Acta*, **59**, 1131–1138.

Lehmann, M. F., Bernasconi, S. M., Barbieri, A. & McKenzie, J. A. 2002. Preservation of organic matter and alteration of its carbon and nitrogen isotope composition during simulated and in situ early sedimentary diagenesis. *Geochimica et Cosmochimica Acta*, **66**, 3573–3584.

Lepper, J. & Röhling, H.-G. 1998. Buntsandstein. *In*: Bachmann, G. H. (ed.) *Excursions of the International Symposium on the Epicontinental Triassic. Hallesches Jahrbuch für Geowissenschaften*, **B6**, 27–34.

Ludwig, G. 1962. Eine Methode zum Nachweis hydrodynamisch-paläogeographisch bedingter Metallverteilungen (texturell-metallfazielle Integralmethode). *Erzmetall*, **15**, 201–206.

Lyons, W. B. & Gaudette, H. E. 1985. Trace metals concentrations in sediments from the Gavish Sabkha. *In*: Friedman, G. M. & Krumbein, W. E. (eds) *Hypersaline Ecosystems, the Gavish Sabkha*. Ecological Studies, **53**, Springer, Berlin, 346–349.

Magaritz, M. & Holser, W. T. 1991. The Permian–Triassic of the Gartnerkofel-1 core (Carnic Alps, Austria), carbon and oxygen isotope variation. *In*: Holser, W. T. & Schönlaub, H. P. (eds) *The Permian–Triassic boundary in the Carnic Alps of Austria (Gartnerkofel Region). Abhandlungen der Geologischen Bundesanstalt*, **45**, 149–164.

Massari, F., Neri, C., Pittau, P., Fontana, D. & Stefani, C. 1994. Sedimentology, palynostratigraphy and sequence stratigraphy of a continental to shallow marine rift-related succession: Upper Permian of the eastern Southern Alps (Italy). *Memorie di Scienze Geologiche, Padova*, **46**, 119–243.

Meyers, P. A. 1994. Preservation of elemental and isotopic source identification of sedimentary organic matter. *Chemical Geology*, **114**, 289–302.

Michaelsen, P. 2002. Mass extinction of peat-forming plants and the effect on fluvial styles across the Permian–Triassic boundary, northern Bowen basin, Australia. *Palaeogeography, Palaeoclimatology, Palaeoecology*, **179**, 173–188.

Nguyen Tu, T. T., Bocherens, H. *et al.* 1999. Ecological distribution of Cenomanian terrestrial plants based on 13C/12C ratios. *Palaeogeography, Palaeoclimatology, Palaeoecology*, **145**, 79–93.

Nguyen Tu, T. T., Kvacek, J., Ulicny, D., Bocherens, H., Mariotti, A. & Broutin, J. 2002. Isotope reconstruction of plant palaeoecology, case study of Cenomanian floras from Bohemia. *Palaeogeography, Palaeoclimatology, Palaeoecology*, **183**, 43–70.

O'Leary, M. 1981. Carbon isotope fractionation in plants. *Phytochemistry*, **20**, 553–567.

Olson, E. C. 1989. Problems of Permo-Triassic terrestrial vertebrate extinctions. *Historical Biology*, **2**, 17–35.

Paul, J. & Peryt, T. M. 2000. Kalkowsky's stromatolites revisited (Lower Triassic Buntsandstein, Harz mountains, Germany). *Palaeogeography, Palaeoclimatology, Palaeoecology*, **161**, 435–458.

Peters-Kottig, W. H. 2003. *Untersuchungen zur Kohlenstoffisotopenzusammensetzung von terrestrischem organischen Material des jüngeren Paläozoikums*. PhD thesis, Westfälische Wilhelms-Universität Münster.

Ramesh, R. & Krishnamurthy, R. V. 1998. $\delta^{13}C$ of marine organic matter and ocean pH. *Geochemical Journal*, **32**, 65–69.

RETALLACK, G. J. 1995. Permian–Triassic life crisis on land. *Science*, **267**, 77–81.

RETALLACK, G. J. 2005. Earliest Triassic claystone breccias and soil-erosion crisis. *Sedimentology*, **75**, 679–695.

RÖHLING, H.-G. 1986. *Die Gliederung des Unteren und Mittleren Buntsandsteins nach Bohrlochmessungen (Gamma-Ray und Sonic-Log) im Nordwestdeutschen Becken*. Unpublished report. Bundesanstalt für Geowissenschaften und Rohstoffe (BGR), Hannover.

RÖHLING, H.-G. 1991. A lithostratigraphic subdivision of the Lower Triassic in the northwest German lowlands and the German sector of the North Sea based on gamma-ray and sonic logs. *Geologisches Jahrbuch*, **A119**, 3–24.

RÖHLING, H.-G. 1993. Der Untere Buntsandstein in Nordwest- und Nordostdeutschland – Ein Beitrag zur Vereinheitlichung der stratigraphischen Nomenklatur. *Geologisches Jahrbuch*, **A142**, 149–183.

RÖHLING, H.-G. 2013. Der Buntsandstein im Norddeutschen Becken. *In*: LEPPER, J. & RÖHLING, H.-G. (eds) *Stratigraphie von Deutschland X. Buntsandstein*. Schriftenreihe der Deutschen Gesellschaft für Geowissenschaften, Hannover.

ROMAN, A. 2004. *Sequenzstratigraphie und Fazies des Unteren und Mittleren Buntsandsteins im östlichen Teil des Germanischen Beckens (Deutschland, Polen)*. PhD thesis, Martin-Luther-Universität Halle-Wittenberg.

ROMAN, A., SZEWCZYK, J. & KRZYWIEC, P. 2001. On possibilities of application of natural gamma ray core logging to stratigraphical and sedimentological studies and to basin analysis (results of Lower Triassic pilot studies of wells Wrzesnia IG-1 and Gorzow Wielkopolski IG-1, NW Poland). *Przeglad Geologiczny*, **49**, 867–872.

SCHIDLOWSKI, M., MATZIGKEIT, U. & KRUMBEIN, W. E. 1984. Superheavy organic carbon from hypersaline microbial mats, assimilatory pathways and geochemical implications. *Naturwissenschaften*, **71**, 303–308.

SCHIDLOWSKI, M., MATZIGKEIT, U., MOOK, W. G. & KRUMBEIN, W. E. 1985. Carbon isotope geochemistry and 14C ages of microbial mats from the Gavish Sabkha and the Solar Lake. *In*: FRIEDMAN, G. M. & KRUMBEIN, W. E. (eds) *Hypersaline Ecosystems, the Gavish Sabkha*. Ecological Studies, **53**, Springer, Berlin, 381–401.

SCHIDLOWSKI, M., GORZAWSKI, H. & DOR, I. 1994. Carbon isotope variations in a solar pond microbial mat: Role of environmental gradients as steering variables. *Geochimica et Cosmochimica Acta*, **58**, 2289–2298, doi: 10.1016/0016-7037(94)90011-6

SCHOLGER, R., MAURITSCH, H. J. & BRANDNER, R. 2000. Permian–Triassic boundary magnetostratigraphy from the Southern Alps (Italy). *Earth and Planetary Science Letters*, **176**, 495–508.

SCHOLLE, P. A. 1995. Carbon and sulfur isotope stratigraphy of the Permian and adjacent intervals. *In*: SCHOLLE, P. A. & PERYT, T. M. (eds) *The Permian of Northern Pangea Volume 1, Paleogeography, Paleoclimates, Stratigraphy*. Springer, Berlin, 133–149.

SCHULZ, E. 1994. *Sporenstratigraphische und fazielle Gliederung des Unteren und Mittleren Buntsandsteins im Nordostdeutschen Raum*. Interim report, Deutsche Forschungsgemeinschaft (DFG), Bonn.

SCHULZ, E. 1998. Concerning palynostratigraphy and facies divisions of the Buntsandstein in the Northeast German region. *In*: BACHMANN, G.-H. (ed.) *Excursions of the International Symposium on the Epicontinental Triassic. Hallesches Jahrbuch für Geowissenschaften*, **B6**, 160–161.

SCHULZE, G. 1969. Der Untere Buntsandstein der Scholle von Calvörde und benachbarter Gebiete. *Geologie*, **18**, 5–20.

SCHÜTT, B. 2000. Holocene paleohydrology of playa-lake systems in central Spain – a reconstruction based on mineral composition of the lacustrine sediments. *Quaternary International*, **73/74**, 7–27.

SCHÜTT, B. & BAUMHAUER, R. 1996. Playasedimente aus dem zentralen Ebrobecken/Spanien als Indikatoren für holozäne Klimaschwankungen – ein vorläufiger Bericht. *Petermanns Geographische Mitteilungen*, **140**, 33–42.

SEIDEL, G. 1965. Zur geologischen Entwicklungsgeschichte des Thüringer Beckens. *Geologie*, **14**, 1–115.

SEPHTON, M. A., VEEFKIND, R. J., LOOY, C. V., VISSCHER, H., BRINKHUIS, H. & DE LEEUW, J. W. 2001. Lateral variations in End-Permian organic matter in Northern Italy. *In*: BUFFETAUT, E. & KOEBERL, C. (eds) *Geological and Biological Effects of Impact Events*. Springer, Berlin, 11–24.

SEPHTON, M., LOOY, C., BRINKHUIS, H. & VISSCHER, H. 2003. From land to sea – Haemorrhage of soil deposits during the end-Permian crisis. *In*: WONG, T. E. (ed.) *XVth International Congress on Carboniferous and Permian Stratigraphy, Utrecht, 10–16 August*. Royal Netherlands Academy of Arts and Sciences, Amsterdam, 489–490.

SEPKOSKI, J. J., JR. 1995. Patterns of Phanerozoic extinction, a perspective from global data bases. *In*: WALLISER, O. H. (ed.) *Global Events and Event Stratigraphy in the Phanerozoic, Results of International Interdisciplinary Cooperation in the IGCP Project 216 'Global Biological Events in Earth History'*. Springer, Berlin, 35–51.

SKPT 1993. Beschlüsse zur Festlegung der lithostratigraphischen Grenzen Zechstein/Buntsandstein/Muschelkalk und zu Neubenennungen im Unteren Buntsandstein in der Bundesrepublik Deutschland. *Nachrichten der Deutschen Geologischen Gesellschaft*, **49**, 76–81.

SKPT 2011. Beschlüsse der Deutschen Stratigraphischen Kommission 1991–2010. *Zeitschrift der deutschen Gesellschaft für Geowissenscahften*, **162**, 1–18. DOI: 10.1127/1860-1804/2011/0162-0001

SMITH, R. M. H. & WARD, P. D. 2001. Pattern of vertebrate extinctions across an event bed at the Permian–Triassic boundary in the Karoo basin of South Africa. *Geology*, **29**, 1147–1150.

STEINER, M. B., ESHET, Y., RAMPINO, M. R. & SCHWINDT, D. M. 2003. Fungal abundance spike and the Permian–Triassic boundary in the Karoo Supergroup, (South Africa). *Palaeogeography, Palaeoclimatology, Palaeoecology*, **194**, 405–414.

STILLER, M., ROUNICK, J. S. & SASHA, S. 1985. Extreme carbon-isotope enrichments in evaporating brines. *Nature*, **316**, 434–435.

SZURLIES, M. 1999. Zyklenstratigraphie und Gamma-Ray-Log-Korrelation im Unteren Buntsandstein (Untere Trias) des nördlichen Harzvorlandes. *In*: BACHMANN, G.-H. (ed.) *Excursions of the International Symposium on the Epicontinental Triassic. Hallesches Jahrbuch für Geowissenschaften*, **B21**, 35–53.

SZURLIES, M. 2001. *Zyklische Stratigraphie und Magnetostratigraphie des Unteren Buntsandsteins in Mitteldeutschland*. PhD thesis, University Halle-Wittenberg.

SZURLIES, M. 2004. Magnetostratigraphy: the key to a global correlation of the classic Germanic Trias – case study Volpriehausen Formation Middle Buntsandstein), Central Germany. *Earth and Planetary Science Letters*, **227**, 395–410.

SZURLIES, M. 2007. Latest Permian to Middle Triassic cyclo-magnetostratigraphy from the Central European Basin, Germany: Implications for the geomagnetic polarity timescale. *Earth and Planetary Science Letters*, **261**, 602–619.

SZURLIES, M., BACHMANN, G. H., MENNING, M., NOWACZYK, N. R. & KÄDING, K.-C. 2003. Magnetostratigraphy and high-resolution lithostratigraphy of the Permian–Triassic boundary interval in Central Germany. *Earth Planetary Science Letters*, **212**, 263–278.

SZURLIES, M., GELUK, M. C., KRIJGSMAN, W. & KÜRSCHNER, W. M. 2012. The continental Permian–Triassic boundary in the Netherlands: implications for the geomagnetic polarity time scale. *Earth and Planetary Science Letters*, **317–318**, 165–176.

TALBOT, M. R. & JOHANNESSEN, T. 1992. A high resolution paleoclimatic record for the last 27,500 years in tropical West Africa form the carbon and nitrogen isotopic composition of lacustrine organic matter. *Earth Planetary Science Letters*, **110**, 23–37.

TIESZEN, L. L. 1991. Natural variations in the carbon isotope values of plants: implications for archaeology, ecology, and paleoecology. *Journal of Archaeological Science*, **18**, 227–248.

TRICHET, J., DEFARGE, C., TRIBBLE, J., TRIBBLE, G. & SANSONE, F. 2001. Christmas Island lagoonal lakes, models for the deposition of carbonate–evaporite–organic laminated sediments. *Sedimentary Geology*, **140**, 177–189.

UHL, D. & KERP, H. 2003. Wildfires in the Late Palaeozoic of Central Europe – the Zechstein Upper Permian of NW–Hesse (Germany. *Palaeogeography, Palaeoclimatology, Palaeoecology*, **199**, 1–15.

VALERO-GARCÉS, B. L., DELGADO-HUERTAS, A., NAVAS, A., MACHÍN, J., GONZÁLEZ-SAMPÉRIZ, P. & KELTS, K. 2000*a*. Quaternary palaeohydrological evolution of a playa lake: Salada Mediana, Central Ebro Basin, Spain. *Sedimentology*, **47**, 1135–1156.

VALERO-GARCÉS, B. L., GONZÁLEZ-SAMPÉRITZ, P., DELGADO-HUERTAS, A., NAVAS, A., MACHÍN, J. & KELTS, K. 2000*b*. Lateglacial and late Holocene environmental and vegetational change in Salada Mediana, Central Ebro Basin, Spain. *Quaternary International*, **7374**, 29–46.

VAN DER ZWAN, C. J. & SPAAK, P. 1992. Lower to Middle Triassic sequence stratigraphy and climatology of The Netherlands, a model. *Palaeogeography, Palaeoclimatology, Palaeoecology*, **91**, 277–290.

VISSCHER, H. & BRUGMAN, W. A. 1986. The Permian–Triassic boundary in the Southern Alps: a palynological approach. *Memorie della Societa Geologica Italiana*, **34**, 121–128.

WALLISER, O. H. 1995. Patterns and causes of global events. *In*: WALLISER, O. H. (ed.) *Global Events and Event Stratigraphy in the Phanerozoic, Results of International Interdisciplinary Cooperation in the IGCP Project 216 'Global Biological Events in Earth History'*. Springer, Berlin, 7–19.

WARD, P. D., MONTGOMERY, D. R. & SMITH, R. 2000. Altered river morphology in South Africa related to the Permian–Triassic extinction. *Science*, **289**, 1740–1743.

WARD, P. D., BOTHA, J. *ET AL*. 2005. Abrupt and gradual extinction among late Permian land vertebrates in the Karoo Basin, South Africa. *Science*, **307**, 709–714.

WARD, P. D., RETALLACK, G. J. & SMITH, R. M. H. 2012. The terrestrial Permian–Triassic boundary event bed is a nonevent: Comment. *Geology*, **40**, e256, doi: 10.1130/G31473C.1.

WARREN, J. 2000. Dolomite: occurrence, evolution and economically important associations. *Earth Science Reviews*, **52**, 1–81.

WARREN, J. K. 2006. *Evaporites: Sediments, Resources and Hydrocarbons*. Springer, Berlin.

ZIEGLER, P. A. 1990. *Geological Altas of Western and Central Europe*, 2nd edn. Geological Society, London.

Index

Page numbers in *italic* denote Figures. Page numbers in **bold** denote Tables.